Algebra

PRODUCTS AND FACTORS:

$(ax+b)(cx+d) = acx^2 + (ad+bc)x + bd$

$a^2 - b^2 = (a+b)(a-b)$

$a^3 - b^3 = (a-b)(a^2+ab+b^2)$

$a^3 + b^3 = (a+b)(a^2-ab+b^2)$

RULES OF EXPONENTS:

$x^n = x \cdot x \cdot \cdots \cdot x$ (n factors), $n \in \mathbb{N}$

$x^{-n} = \dfrac{1}{x^n}$

$x^m \cdot x^n = x^{m+n}$

$(x^m)^n = x^{mn}$

$x^0 = 1$

$x^{1/n} = \sqrt[n]{x}$

$\left(\dfrac{x}{y}\right)^n = \dfrac{x^n}{y^n}$

$(xy)^n = x^n y^n$

QUADRATIC FORMULA: If $ax^2 + bx + c = 0$, then $x = \dfrac{-b \pm \sqrt{b^2 - 4ac}}{2a}$.

Trigonometry

DEFINITIONS:

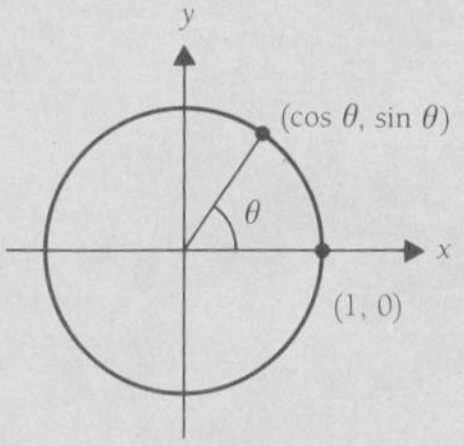

$x^2 + y^2 = 1$

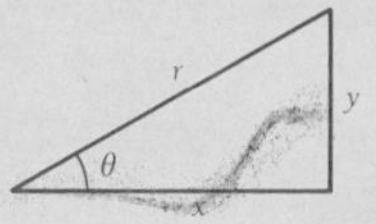

$\cos\theta = \dfrac{x}{r}$ $\sin\theta = \dfrac{y}{r}$

FUNDAMENTAL IDENTITIES:

$\sin^2\theta + \cos^2\theta = 1$

$1 + \tan^2\theta = \sec^2\theta$

$1 + \cot^2\theta = \csc^2\theta$

$\tan\theta = \dfrac{\sin\theta}{\cos\theta}$ $\cot\theta = \dfrac{\cos\theta}{\sin\theta}$

$\sec\theta = \dfrac{1}{\cos\theta}$ $\csc\theta = \dfrac{1}{\sin\theta}$

ADDITION AND SUBTRACTION FORMULAS:

$\sin(\theta \pm \phi) = \sin\theta\cos\phi \pm \cos\theta\sin\phi$

$\cos(\theta \pm \phi) = \cos\theta\cos\phi \mp \sin\theta\sin\phi$

$\tan(\theta \pm \phi) = \dfrac{\tan\theta \pm \tan\phi}{1 \mp \tan\theta\tan\phi}$

DOUBLE-ANGLE FORMULAS:

$\sin 2\theta = 2\sin\theta\cos\theta$

$\tan 2\theta = \dfrac{2\tan\theta}{1-\tan^2\theta}$

$$\begin{aligned}\cos 2\theta &= \cos^2\theta - \sin^2\theta \\ &= 2\cos^2\theta - 1 \\ &= 1 - 2\sin^2\theta\end{aligned}$$

HALF-ANGLE FORMULAS:

$\sin^2\theta = \dfrac{1-\cos 2\theta}{2}$

$\cos^2\theta = \dfrac{1+\cos 2\theta}{2}$

$$\begin{aligned}\tan\theta &= \frac{\sin 2\theta}{1+\cos 2\theta} \\ &= \frac{1-\cos 2\theta}{\sin 2\theta}\end{aligned}$$

CALCULUS

AND ANALYTIC GEOMETRY

CALCULUS

AND ANALYTIC GEOMETRY

Lynn E. Garner
Brigham Young University

Dellen Publishing Company
San Francisco, California

Collier Macmillan Publishers
London

Divisions of Macmillan, Inc.

On the cover: "Untitled" was executed by Los Angeles artist Charles Arnoldi in 1986. The work, measuring $16 \times 12 \times 2\frac{1}{4}$ inches, is acrylic on laminated plywood and has been sculpted using a chain saw. Arnoldi's work may be seen at the Fuller Gross Gallery in San Francisco, the James Corcoran Gallery in Los Angeles, and the Charles Cowles Gallery in New York City. His work is included in the permanent collections of the Albright–Knox Museum in Buffalo, New York, the Chicago Art Institute, the Museum of Modern Art in New York City, the Metropolitan Museum, the Los Angeles County Museum, and the San Francisco Museum of Modern Art.

Printed in the United States of America

Permissions: Dellen Publishing Company
400 Pacific Avenue
San Francisco, California 94133
Orders: Dellen Publishing Company
c/o Macmillan Publishing Company
Front and Brown Streets
Riverside, New Jersey 08075

Collier Macmillan Canada, Inc.

Library of Congress Cataloging-in-Publication Data

Garner, Lynn E.
Calculus and analytic geometry.

Includes index.
1. Calculus. 2. Geometry, Analytic. I. Title.
QA303.G34 1988 515'.15 87-13088
ISBN 0-02-340590-2

Printing: 2 3 4 5 6 7 8 9 Year: 8 9 0 1
ISBN 0-02-340590-2

This book is dedicated to my wife
Kaye
who has remained a steadfast and supportive companion
through
trials and triumphs,
discouragement and hope,
and many late hours.

CONTENTS

PREFACE

Calculus is one of the greatest intellectual achievements of the human race, a discipline with internal order and form of surprising beauty, as well as a tool of unsurpassed usefulness. Probably no other discipline of comparable scope yields larger dividends for the intellectual investment, in terms of logical reasoning, complex problem-solving skills, and useful applications.

Philosophy

It is my belief that calculus can be taught interestingly, logically, and thoroughly, and need not appear charmless, arcane, or cumbersome to the student. The aim of this book is to improve the performance of both the teacher and the student in their respective roles in the transmission of the calculus to the rising generation.

Two tenets have guided the formation of this book, one philosophical and one pedagogical. The philosophical tenet is that the calculus should be presented in an order that is as logical as possible, but not divorced from its historical and cultural setting. The internal order and logical beauty of the subject are then evident, and the inclusion of historical notes and biographical sketches help create a context in which the new knowledge can be placed. For example, the importance of the fundamental theorem of calculus is emphasized, both logically and historically. A logical organization and interesting sidelights enhance the learning process.

The pedagogical tenet is based on the fact that the vast majority of students who study calculus use mathematics primarily as a tool. Therefore, the development of each topic in this text culminates in a discussion of all the standard applications (plus a few) of that topic. Typically, a new topic is motivated by a physical or geometric idea or problem. This gives the student an interpretation or application of the topic, an example of something that it is "good for." Then the theory and techniques of the topic are developed, followed by illustrations of its applications.

This order of development allows for realistic applications to be discussed, unrestricted by lack of techniques, and it also leaves the student with a strong impression of the usefulness of the subject.

Level of Rigor

It is a fact of life that clarity and precision must be traded off at times. The most clear statement is probably a poem, but it has little precision. The most precise statement is undoubtedly a legal document, but it is so far from clear that we hire someone to explain it to us. This "uncertainty principle" involving clarity and precision prevails in the development of the calculus, where some sort of compromise must be made. Things must be clear enough for the student to understand, but precise enough that the student will not misunderstand.

Basic concepts and technical terms of the calculus have been developed using as much precision as necessary. They are too crucial to be discussed on an intuitive basis alone, because too many misconceptions persist unless the appropriate amount of rigor is used. On the other hand, I have tried to avoid appearing to be rigorous. Technical terms are used precisely and without apology, but matter-of-factly. An informal conversational style, appropriate motivation, careful logical development, and illustrative examples all contribute to clarity without loss of precision.

In the statements of theorems, I prefer to err on the side of clarity. Hopefully, a context is established in which a given theorem may be stated simply, without explicitly detailing all its conditions, limitations, and hypotheses. If the student knows the context and understands what the theorem means, he or she can supply some of the missing details.

The purpose of proof is manifold. Some proofs are given primarily to teach the anatomy of a proof. Other proofs are given to clarify the basic definitions, or previous theorems, or their interrelationships. Some proofs are given because the theorem is unbelievable without supporting argument. When none of these purposes is served, it is best to omit the proof, or in some cases relegate it to the exercises. Proofs given in this text always appear directly following the theorems.

Organization

The guiding forces of the organization of topics have been logical development and theory before applications. The needs of other disciplines also dictate some of the organization (such as both differentiation and integration in the first semester), since many students of calculus are majors in departments other than mathematics.

The first chapter of the book is a review of precalculus topics, done so as to focus on the specific prerequisites for calculus. This chapter can be covered lightly if the student is well-prepared. Notice that Chapter 1 includes a review of the trigonometric functions, preparing for their use throughout the text.

Chapter 2 introduces the notion of limit, upon which the rest of calculus is based. Here, we use careful motivation, precise definitions, and logical development to arrive quickly at results that the student will use as rules. Limits at infinity and infinite limits are also dealt with here, making for their smooth entry later in graphical problems.

The applications of limits give us derivatives (Chapter 3), definite integrals (Chapter 5), and infinite series (Chapter 9). Each of these topics is introduced by showing how the notion of limit solves some kind of problem. These topics are then developed thoroughly, generally focusing on techniques first and then many applications.

In many calculus books, the transcendental functions have been postponed until after the development of both differentiation and integration, usually in an effort to get the required topics into one semester without overloading it. Here, the trigonometric functions are used from the beginning, but the logarithmic, exponential, and inverse trigonometric functions are postponed until Chapter 6. Having the trigonometric functions enables a nontrivial illustration of the chain rule, and the formulas for the derivatives of the trigonometric functions are learned long before the antidifferentiation formulas appear, thereby avoiding much confusion for the student.

Placing the logarithmic functions immediately after the development of the definite integral illustrates the use of the fundamental theorem. L'Hôpital's rules also appear in Chapter 6, where they are needed for discussing the limits of transcendental functions. This also makes them available for use in Chapter 9.

Improper integrals are covered at the end of Chapter 8, preparing for the integral test in Chapter 9. Chapter 9 completes the development of the calculus of a single variable.

The transition from one variable to several is foreshadowed in Chapters 10–13, where two trends predominate: analytic geometry moves from the plane into space, and parametric equations lead to vector functions. The calculus of several variables is then presented in a manner paralleling the development in the first eight chapters of the book. A special effort has been made to maintain the strength of the earlier development through Chapters 14–17. A final chapter on differential equations emphasizes applications.

Examples and Exercises

The role of examples in the text is twofold: First, they illustrate the theory, techniques, or applications under discussion with concrete instances; and then they become models for the student to follow in working exercises.

Exercise sets are perhaps the most important part of the book. The role of exercises, of course, is to help the student develop basic skills in manipulations and techniques, extend the theory in various directions, and pose realistic problems to which the calculus can be applied. Generally speaking, large sets of paired, graded problems are presented in each section. Answers to the odd-numbered problems appear in Appendix E.

Generally, the exercises are not marked in any way. Other than noting that the exercise sets proceed from the easy problems to the difficult, no explicit identification of challenging problems is made. To mark a problem as a challenge predisposes students to think it is difficult, when perhaps they might be able to see an easy way to work the problem if they were not afraid to tackle it.

Exercises that can be solved appropriately using a calculator or computer are sprinkled throughout the text. Generally, they are not marked in any special way. Part of the learning process for the student involves recognizing such problems, wherever they are met. From previous experience, it is assumed that the student will automatically reach for the calculator when any computations are needed. Computing machinery is now so much an established part of the student's life that it should be treated matter-of-factly as the great tool that it is.

However, given the classroom situation, it is unwise to leave exercise sets totally unannotated. Realistic problems often lead to algebraic complications and computational complexities that are unreasonable, at least in quantity, for most course settings. Such problems have been included deliberately, but have been annotated to warn instructors that large doses of such problems work hardships on students. For some problems, caution is given in the statement of the problem. For problems involving an unusual level of algebraic manipulation or an exceptionally complicated answer, the symbol ^{A} is placed next to the number of the problem. For problems that involve long numerical calculations requiring a calculator or a computer, the symbol ^{C} is placed next to the number of the problem.

The last section in each chapter is a chapter review, consisting mainly of exercises. The Review Exercises are a set of routine and intermediate-level problems to which the instructor can point and say, "If you can work all these problems, then you can pass the test on this chapter." The Additional Exercises are of a miscellaneous character, and include combinations of ideas, extensions of the theory, exotic applications, and intersecting sidelights. Most of the challenging exercises in the text are in these sets, but an effort has been made to keep them possible for the student to solve. How often I have found such sets to be entirely unusable!

Calculus and Computers

The revolution in mathematics and science caused by the personal computer has been taken into account in the preparation of this textbook. More numerical methods than usual are included: the bisection method, Newton's method, and several methods of numerical integration (with error estimates) are presented, in addition to the usual development and analysis of error that accompany Taylor's polynomials. The use of computers in calculus is easily introduced using any of these topics.

As was mentioned before, however, computers should not be introduced for the sake of computers, but computational devices can well be used to more fully develop the calculus when that is appropriate. Here, it is done without much ado, so that for the student, the emphasis remains on the theory and applications of calculus.

In the preparation of this book, the computer has played a major role. All of the graphs and surfaces pictured in the text have been prepared from computer-generated graphs, thus ensuring their accuracy; many three-dimensional computer-generated graphics enhance the development of functions of two or three variables.

Supplementary Materials

A *Student Solutions Manual* by Gerald L. Alexanderson, Alice J. Kelly, and Leonard F. Klosinski is available to students at a nominal cost. Complete solutions to all the odd-numbered exercises are given.

An *Instructor's Solutions Manual,* in which complete solutions to all the even-numbered exercises are given, is available to instructors free of charge.

DellenTest, a computer-generated random test system, is also available to instructors free of charge. This program runs on an IBM PC equipped with any of several commonly used dot-matrix printers. The program generates almost an unlimited number of chapter tests and final examinations, each different from the others, quickly and easily. At the same time, the program generates a student worksheet with answer column and an answer key exactly matching the answer column. Graphing grids are included on the student worksheet, and the answer keys show the completed graphs.

Acknowledgments

While the responsibility for any shortcomings of this book is entirely mine, it is definitely the case that it would not exist but for the efforts of many others.

I am endebted to Don Dellen and his very able staff for the final form of this book. I wish to thank Phyllis Niklas for the editing and her attention to a thousand production details; Janet Bollow for the design and layout of the text; and George Morris of Scientific Illustrators for superb graphics and artwork.

No project like this is possible without the able assistance of many reviewers. To the following go thanks for reviewing the manuscript:

Gerald L. Alexanderson	Santa Clara University
Phillip M. Anselone	Oregon State University
Stephen L. Campbell	North Carolina State University
Barbara Cortzen	DePaul University
William M. Cunnea	Washington State University
Platon C. Deliyannis	Illinois Institute of Technology
Gary L. Ebert	University of Delaware
Bruce Edwards	University of Florida
Mark Finkelstein	University of California, Irvine
Fletcher I. Gross	University of Utah
Vern L. Heeren	American River College
Kendell Hyde	Weber State College
Harvey B. Keynes	University of Minnesota
Leonard F. Klosinski	Santa Clara University
Joseph Krebs	Boston College

H. E. Lacey	Texas A&M University
John W. Lee	Oregon State University
John L. Leonard	University of Arizona
Jeff E. Lewis	University of Illinois at Chicago
Joe L. Mott	Florida State University
Thomas B. Muenzenberger	Kansas State University
Jerome L. Paul	University of Cincinnati
David C. Royster	University of North Carolina, Charlotte
William C. Schulz	Northern Arizona University
Claudia J. Taylor	Brevard Community College
Peter Ungar	New York University
Vladeta Vuckovic	University of Notre Dame
Jerry M. Wolfe	University of Oregon

Special thanks go to William C. Schulz for help with the historical notes and biographical sketches, and to Karl J. Smith for his generous loan of the photographs of the sunflower and chambered nautilus.

For proofreading galleys, I wish to thank Garret J. Etgen of the University of Houston and Leslee Francis and Ronald D. Jamison of Brigham Young University.

For working the exercises, I am greatly endebted to Gerald L. Alexanderson, Alice J. Kelly, Leonard F. Klosinski, and Nedra Shunk of Santa Clara University, Michael Thibodeaux of Menlo College, and Kevin Black.

I am especially grateful for the facilities and creative atmosphere in the Mathematics Department of Brigham Young University, and for the encouragement I received from department chairmen Kenneth Hillam and Peter Crawley. I also wish to thank a generation of students at Brigham Young University for the great amount of calculus they taught me.

Lynn E. Garner

CALCULUS

AND ANALYTIC GEOMETRY

CHAPTER 1

Preliminaries

INTRODUCTION

What Is Calculus?

Calculus is a part of mathematics identified by a particular notation or method of reasoning. In Latin, *calculus* originally meant "pebble," and since pebbles were used in tallying and reckoning, *calculus* came to mean a system of reckoning or calculating.

There are several subjects called *calculus* in modern mathematics — the propositional calculus and predicate calculus of logic, the vector calculus and calculus of variations in applied mathematics, and the differential calculus and integral calculus of Newton and Leibniz. It is to these last two that we refer when we speak of *the* calculus.

The modern course in introductory calculus usually contains much more than just the calculus. We begin with a brief review of topics from college algebra and trigonometry, and include a healthy dose of analytic geometry and such extensions of the calculus as infinite series, vector field theory, and differential equations. Throughout, there are applications to the sciences and other subjects.

As we have inherited it, calculus is a landmark in intellectual history. It is a prime example of a logical discipline, and in the depth of its results we find a beauty of form and coherence.

A Little History

The prehistory of calculus reaches back into ancient Greece, where areas or volumes were approximated by dissecting a figure into a number of figures whose areas or volumes were already known. This technique was further developed by Eudoxus and Theaetetos about 400 BC. It became the basis of the *method of exhaustion,* which

Isaac Newton was born on Christmas day, 1642. He showed no particular talent as a young student, but managed to attend Cambridge University. He began his career by helping to edit the geometrical lectures of his teacher, Isaac Barrow, for publication. During his year at home in the country avoiding the plague (1665–1666), Newton's genius suddenly blossomed; he discovered the fundamental theorem of calculus, the law of universal gravitation, and the composition of white light. By 1680 he had worked out these discoveries to his satisfaction, and turned his attention to chemistry and theology. His aim was to include these subjects with his earlier work on gravitation and the calculus in a grand synthesis of knowledge. The theological work was concerned mainly with revising the dates of Biblical events, and went reasonably well; his chemistry was largely alchemy, and produced no results. In 1696, Newton left Cambridge to accept an appointment as Chancellor of the Mint. While there, he invented the milled edge used on silver coins to prevent shaving or clipping of coins, and he was involved in successfully capturing and prosecuting counterfeiters. He became president of the Royal Society (the British scientific association) in 1703, and remained its president until his death in 1727. By the time of his death, it was generally recognized that Newton had revolutionized the study of science, and he was buried in Westminster Abbey with a ceremony befitting a king.

we will consider further when we study integral calculus. This method was used by Archimedes to solve what were very difficult problems at the time, the most famous of which was his determination of the surface area of a sphere.

The mathematicians of early modern times were aware of the method and results of Archimedes, but because of the difficulty of the method, results were few. John Wallis (1616–1703) was able to find the area under the curve $y = x^k$. William Neil (in 1687) succeeded in modifying the method to find the length of a curve. All such results depended on the summation of infinite series, and this topic was also vigorously pursued.

From our present point of view, the key idea necessary for the development of a systematic theory was the derivative, which grew out of the problem of tangents. This problem, the construction of a tangent line to a curve at a given point, was touched on only very briefly by the ancients, and was first considered in a systematic way by Fermat and Descartes about 1637. Fermat's approach to the problem is essentially our own, but only in the context of Descartes's coordinate geometry could it be applied effectively. The methods of Fermat and Descartes became part of Isaac Barrow's lectures at Cambridge, which were attended by Isaac Newton in the early 1660's. Barrow and others were aware of a connection between the method of exhaustion and the problem of tangents, but they failed to see that there lay the key to a systematic development. Calculus as a unified discipline dates from the discovery of the significance of the connection between the problem of tangents and the determinations of areas. Newton and Leibniz are both credited with this discovery.

Sir Isaac Newton worked out the details of the connection during 1665 and 1666, while on an enforced leave of absence from Cambridge because of the plague. He also formulated the law of universal gravitation about the same time, and combining it with the calculus, was able to give a description of the mechanics of the solar system. This work was finally published in 1687 in Newton's *Principia,* and while not using calculus in any very explicit way, the work did demonstrate that the new methods could be used to solve physical problems. It established Newton as the leading scientist of his time. His work on the calculus proper was not published until 1704, but a preliminary version was circulated among his friends, and his methods became fairly well known long before then.

Gottfried Wilhelm Leibniz became interested in the complex of problems underlying the calculus during conversations with Christian Huygens (a Dutch mathematician, physicist, and astronomer) about 1673. By 1675 he had realized the significance of the connection between tangents and areas, and turned his attention to finding systematic methods of calculation to exploit these ideas. He worked out notations for differential and integral calculus, and published the differential part in 1684. He taught his methods to the Bernoulli brothers, who in turn taught Euler. All

Gottfried Wilhelm Leibniz was born in 1646. He graduated at age 19 from the University of Leipzig, studied law, and in 1669 became librarian for the royal family in Hanover, in which position he remained for the rest of his life. At the age of 20, he conceived the idea of a universal symbolic system that would reduce logical thought to a sort of algebra. He did not have time to develop this scheme beyond the most rudimentary steps, but his idea was sound, as evidenced by the symbolic logic that now exists. The influence of this scheme on his development of the calculus was profound, for his method clearly shows his desire to have the notation do much of the work. By shifting much of the burden of the calculus to the notation, Leibniz made it accessible to people of quite ordinary intelligence, who could now solve problems that had challenged Archimedes and Newton.

Most of Leibniz's mathematical work was done at odd moments in his legal and diplomatic career. One unsuccessful diplomatic venture was arranging a reunification conference between the Catholic and Evangelical churches. As an expert genealogist, he was able to find (or manufacture) evidence to support his employer's claim to half the thrones of Europe. He was of great service to science by establishing the journal *Acta Eriditorum,* in which the first articles on the calculus were published; by founding the Berlin Academy; and by designing the St. Petersburg Academy, which was implemented after his death. He shared his work on the calculus with the Bernoulli brothers, who mastered and extended his work.

During the dispute with Newton's followers over the invention of the calculus, Leibniz had few public defenders. He suffered such neglect toward the end of his life that at his funeral in 1716, only his secretary attended.

modern textbooks trace their notation back to Euler's textbooks, and our present notations differ hardly at all from those of Leibniz.

An unnecessary dispute arose between English scientists and those on the Continent over just who should be credited with the invention of the calculus. Largely due to the influence of the Bernoullis and the popularity of Euler, continental mathematicians generally followed the work of Leibniz, while English mathematicians became rather isolated and followed Newton's lead in concerning themselves primarily with mathematical physics for about a century and a half.

What Can Calculus Do?

Calculus is unparalleled as a tool for the study of physical phenomena, and is used increasingly in the biological, social, and management sciences. Without calculus a student can understand science only superficially. Almost all of higher mathematics assumes the background and mental maturity gained by a thorough study of the calculus.

Calculus is usually the course in mathematics that introduces *the continuous* and *the infinite*. Most of precalculus mathematics is finite in nature, and applies to isolated, discrete processes that proceed one step at a time. Calculus goes beyond the finite, and applies to continuous processes, as illustrated on the facing page.

Concepts of Calculus

The new idea that gives calculus this added power is the concept of *limit*. This idea first arose in ancient Greek times in the "paradoxes" of Zeno of Elea (*c*. 450 BC) and in the method of exhaustion of Eudoxus (*c*. 360 BC) and Archimedes (*c*. 220 BC). Newton and Leibniz founded the calculus on the notion of "infinitesimals," or infinitely small quantities, but they gave no logically rigorous justification. Eventually, in about 1765, d'Alembert (a French mathematician and philosopher) proposed the theory of limits as a replacement for infinitesimals. During the 1820's, Augustin Louis Cauchy developed the theory of limits, placing the calculus on a logical foundation and giving the calculus essentially its present form. Limits will be considered in detail beginning in Chapter 2. We will not consider infinitesimals, even though Abraham Robinson (in 1966) gave a logical foundation for their development.

One aspect of limit concerns the infinite divisibility of time and space and leads to the *differential calculus*. Such topics as continuity and rates of change are included under this heading, and will be studied beginning in Chapters 2 and 3. Another aspect of limit is concerned with the existence of the sum of infinitely many quantities and leads to the *integral calculus*. Here, topics such as area, arc length, and infinite series are included; these are discussed in Chapters 5–9.

These two aspects of the limit idea appear to be quite different but are actually very closely related. The link that combines differential calculus and integral calculus into a single discipline is known as the *fundamental theorem of calculus*. This important result is studied in Chapter 5.

WITHOUT CALCULUS, WE CAN:

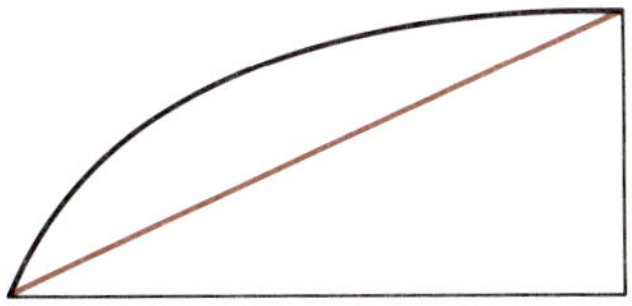

Find an average rate of change

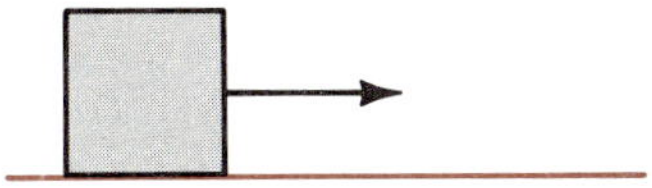

Describe motion at a constant rate

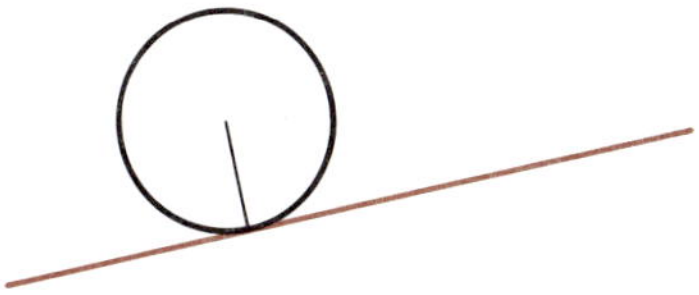

Find a tangent line to a circle

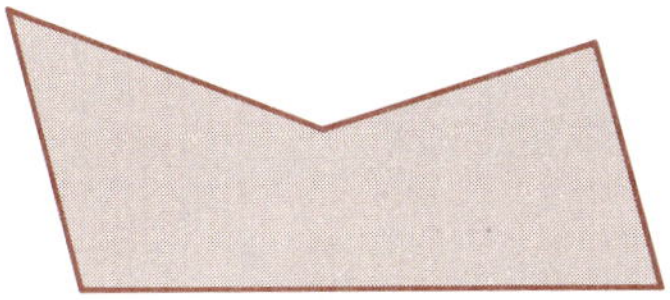

Find the area of a polygon

Find the length of a straight line

Find the work done by a constant force

WITH CALCULUS, WE CAN:

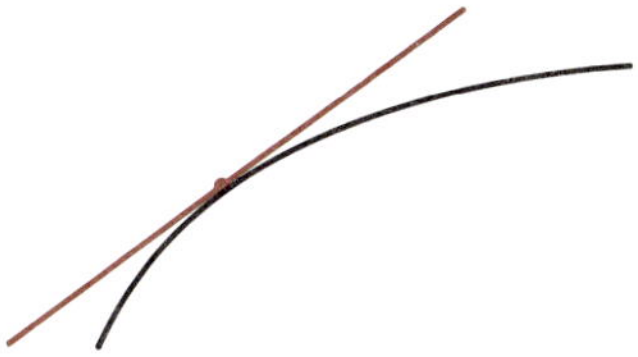

Find an instantaneous rate of change

Describe motion at a variable rate

Find a tangent line to any curve

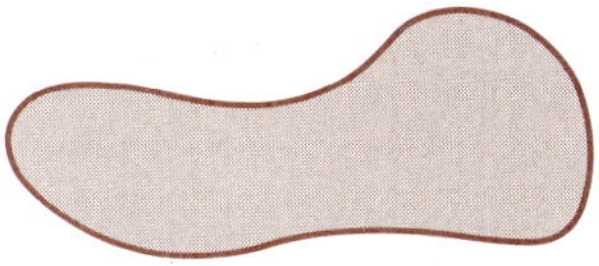

Find the area of a figure bounded by curves

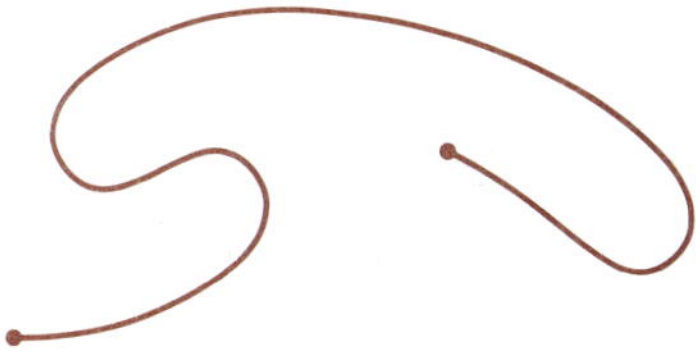

Find the length of a curve

Find the work done by a variable force

It is, of course, the usefulness of calculus that has made it a popular course of study, for it occupies a central role in mathematics and science. The calculus has been the language of science ever since the time of Newton; modern science and technology would have been impossible without it. Calculus is becoming increasingly important as it is applied to new areas of study, such as economics, medicine, and management.

Prerequisites

Calculus is typically formulated in terms of real numbers, applied to functions, and interpreted geometrically. Therefore, the student of calculus needs to be familiar with the algebraic properties of the real number system, the use and notation of functions, and the methods of elementary analytic geometry. These topics are usually covered in the standard precalculus or college algebra and trigonometry course.

The specific properties of real numbers and functions upon which the calculus is based are reviewed in this chapter. Some new ideas are introduced among the old ones, so that the calculus can be developed efficiently once the review is completed. Geometric ideas will play a constant role in our work and will be reviewed, expanded, or introduced as needed.

The most exacting requirement of the calculus is the ability to carry out complicated algebraic manipulations rapidly and without error. The average calculus student is guilty far more often of simple errors in arithmetic and algebra than of higher-level conceptual errors. The only way to overcome these simple errors is by constant attention to detail and lots of practice.

It is not easy to study calculus. But few subjects offer as much in return: You will learn an amazing amount about logic, order, problem-solving, and discipline, in addition to the calculus itself.

SECTION 1.1

Real Numbers and the Real Line

It is convenient to use the language of set theory in discussing the real number system, so here is a brief review of set terminology and notation.

Sets

A **set** is a collection of objects that are called **elements** of the set. The notation $x \in S$ means that the object x is an element of the set S. The notation $S = \{a, b, c, \ldots\}$ means that S is the set whose elements are a, b, c, and so forth. If P is a particular property, the notation $S = \{x | x \text{ satisfies } P\}$ means that S is the set of all objects x that satisfy the property P.

The **empty set** is the set with no elements, and is often denoted by $\emptyset$.

A set S is a **subset** of set T, denoted $S \subseteq T$, if every element of S is also an element of T. Sets S and T are **equal**, denoted $S = T$, if they have exactly the same elements —that is, if each is a subset of the other.

The **union** of sets S and T is the set of objects belonging to either of the sets S or T, and is written

$$S \cup T = \{x | x \in S \text{ or } x \in T\}.$$

The **intersection** of sets S and T is the set of objects belonging to both the sets S and T, and is written

$$S \cap T = \{x | x \in S \text{ and } x \in T\}.$$

Real Numbers

The set of real numbers is often denoted by $\mathbb{R}$. Occasionally, we shall refer to the following subsets of $\mathbb{R}$:

The set of **counting numbers**: $\mathbb{N} = \{1, 2, 3, \ldots\}$

The set of **integers**: $\mathbb{Z} = \{\ldots, -2, -1, 0, 1, 2, 3, \ldots\}$

The set of **rational numbers**: $\mathbb{Q} = \left\{ x \,\middle|\, x = \frac{a}{b}, \quad a, b \in \mathbb{Z}, b \neq 0 \right\}$

Any number that is not rational is called **irrational**.

The **system of real numbers** consists of the set $\mathbb{R}$ together with the usual operations of addition and multiplication and the usual notions of equality and order. For reference, a list of the axioms of the real number system is given at the end of this section.

The Real Line

The real number system can be studied geometrically by identifying each real number with a point on a line. This is done by selecting a point on the line, called the **origin**, as a reference point. The number 0 is identified with the origin. Then a second point is selected, usually to the right of the origin, and is identified with the number 1. This second point is called the **unit point**, and the distance between the origin and the unit point is called the **unit distance**.

Next, each positive real number x is identified with the point that is distance x from the origin (using the unit distance as a scale) and on the same side of the origin as the unit point. The negative of a positive real number x is denoted by $-x$, and is identified with the point that is distance x from the origin and on the opposite side of the origin from the unit point. This is illustrated in Figure 1.1. Such a line, with its points identified with real numbers, is called a **real number line**. Since all real

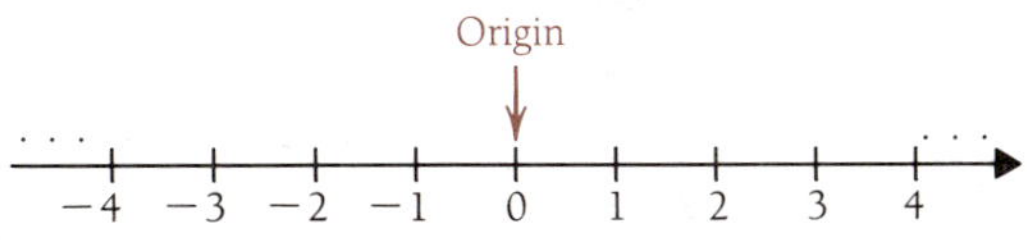

FIGURE 1.1 The real line

number lines are essentially the same in reflecting the properties of the real number system, we refer to any one of them simply as **the real line.**

If a particular point on the real line corresponds to the real number x, it is customary to refer to the point as "the point x" rather than as "the point corresponding to x." This slight abuse of the language makes for simpler terminology and greater clarity.

Order and Betweenness

The familiar relations **less than** ($<$) and **greater than** ($>$) can be interpreted on the real line as "to the left of" and "to the right of," respectively. The properties of $<$ are listed at the end of this section among the axioms. They also hold for $>$, and for the relations **less than or equal to** ($\leq$) and **greater than or equal to** ($\geq$).

If a, b, and x are points on the real line such that $a < x$ and $x < b$, we write $a < x < b$, and say that x is **between** a and b.

Intervals

An interval is a certain kind of subset of the real line. The several types of intervals are described in Table 1.1, in which a and b are points such that $a < b$.

TABLE 1.1 INTERVALS

Type	Interval notation	Set notation	Graph
Open	(a, b)	$\{x \mid a < x < b\}$	
Closed	$[a, b]$	$\{x \mid a \leq x \leq b\}$	
Half-open	$[a, b)$	$\{x \mid a \leq x < b\}$	
	$(a, b]$	$\{x \mid a < x \leq b\}$	
Infinite	(a, ∞)	$\{x \mid x > a\}$	
	$[a, \infty)$	$\{x \mid x \geq a\}$	
	$(-\infty, b)$	$\{x \mid x < b\}$	
	$(-\infty, b]$	$\{x \mid x \leq b\}$	
	$(-\infty, \infty)$	$\mathbb{R}$	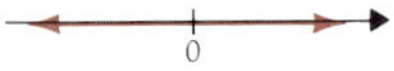

Thus, an **open interval** from a to b is the line segment with endpoints a and b, but the endpoints do not belong to the interval. The **closed interval** from a to b is the same segment, including the endpoints a and b. A **half-open interval** could also be called "half-closed," and contains only one endpoint of the segment. The interval $(a, b]$ is "open at a" and "closed at b," while the interval $[a, b)$ is "closed at a" and "open at b."

The **infinite intervals** represent rays, or half-lines, except for $(-\infty, \infty)$, which is the entire line. The symbol ∞ represents "infinity." It does not represent a number, but rather the idea of "unboundedness." Consequently, an "endpoint" at $-\infty$ or ∞ is never included in an interval; for example, we write $[2, \infty)$, but never $[2, \infty]$.

Example 1.1 Rewrite the following intervals using inequalities:

(a) $(1, 5)$ (b) $[-1, 4]$ (c) $(4, 20]$ (d) $(-\infty, 1]$

Solution We simply apply the definition of each interval notation, as in Table 1.1, to get the following:

(a) $(1, 5) = \{x|1 < x < 5\}$ (b) $[-1, 4] = \{x|-1 \le x \le 4\}$
(c) $(4, 20] = \{x|4 < x \le 20\}$ (d) $(-\infty, 1] = \{x|x \le 1\}$ ■

Example 1.2 Rewrite the following sets in interval notation:

(a) $\{x|10 \le x < 50\}$ (b) $\{x|-4 < x \le -2\}$
(c) $\{x|x < -20\}$ (d) $\{x|b < x < b + 2\}$

Solution Using the inequalities as definitions, we apply the corresponding interval notation from Table 1.1 to get the following answers:

(a) $[10, 50)$ (b) $(-4, -2]$ (c) $(-\infty, -20)$ (d) $(b, b + 2)$ ■

Example 1.3 Rewrite the set $\{x|x^2 + x - 6 < 0\}$ in interval notation.

Solution The stated condition $x^2 + x - 6 < 0$ can be written in factored form as $(x + 3)(x - 2) < 0$. That is, the product of the numbers $x + 3$ and $x - 2$ must be negative. Therefore, one of the numbers must be positive and the other, negative. Since $x + 3$ is greater than $x - 2$, it must be that

$$x + 3 > 0 \quad \text{and} \quad x - 2 < 0,$$
$$x > -3 \quad \text{and} \quad x < 2,$$
$$-3 < x < 2.$$

Hence, the given set is $\{x|-3 < x < 2\} = (-3, 2)$. ■

Example 1.4 Rewrite the set $\{x|x^2 + 2x - 8 \ge 0\}$ in interval notation.

Solution The condition $x^2 + 2x - 8 \ge 0$ can be rewritten as $(x + 4)(x - 2) \ge 0$. Thus, it must be that both of the numbers $x + 4$ and $x - 2$ are nonnegative, or both are nonpositive. Both will be nonnegative if the smaller, $x - 2$, is nonnegative, and both will be nonpositive if the larger, $x + 4$, is nonpositive. Thus, we must have

$$\begin{aligned} x - 2 \geq 0 \quad &\text{or} \quad x + 4 \leq 0, \\ x \geq 2 \quad &\text{or} \quad x \leq -4. \end{aligned}$$

The original set is therefore the same as

$$\begin{aligned} \{x | x \geq 2 \text{ or } x \leq -4\} &= \{x | x \geq 2\} \cup \{x | x \leq -4\} \\ &= [2, \infty) \cup (-\infty, -4]. \end{aligned}$$

■

Absolute Value and Distance

The **absolute value** of a real number x is denoted by $|x|$ and is defined by

$$|x| = \begin{cases} x & \text{if } x \geq 0 \\ -x & \text{if } x < 0. \end{cases}$$

Thus, $|x|$ is the distance between point x and the origin on the real line. Note that the absolute value of a real number is always nonnegative.

■ **Example 1.5** Find the absolute values of the numbers.:

(a) 3.25 (b) -2

Solution By the definition, we have:

(a) $|3.25| = 3.25$ (b) $|-2| = -(-2) = 2$

■

Recall that if x is a nonnegative real number, then the square root of x is the *unique nonnegative* real number $\sqrt{x}$ with the property that $(\sqrt{x})^2 = x$. It follows that $\sqrt{x^2} = |x|$. For example, $\sqrt{(-5)^2} = \sqrt{25} = 5 = |-5|$.

The distance between points on the real line also can be expressed using absolute value. If x_1 and x_2 are any two points on the real line, then the distance between them is their difference, the larger minus the smaller. If we do not know which is larger, we can write the distance as $|x_2 - x_1|$ (see Figure 1.2). Because the distance is the same either way, it is apparent that

$$|x_1 - x_2| = |x_2 - x_1|.$$

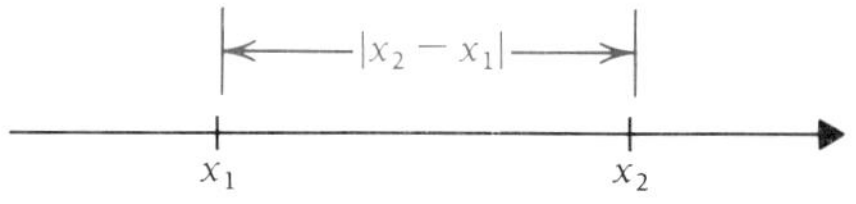

FIGURE 1.2 Distance as absolute value

■ **Example 1.6** Find the distance between the points:

(a) 2 and 5 (b) 3 and -4

Solution

(a) The distance between 2 and 5 is $|5-2|=|3|=3$. Notice that $|2-5|=|-3|=3$ also.

(b) The distance between 3 and -4 is $|-4-3|=|-7|=7$. ■

Absolute Value and Intervals

Let p be a positive real number. If the real number x has the property that $|x|<p$, then, since $|x-0|<p$, the distance between x and the origin must be less than p. Therefore, x must lie between $-p$ and p, and $\{x \mid |x|<p\}=\{x \mid -p<x<p\}=(-p, p)$ (see Figure 1.3). On the other hand, if $|x|>p$, then x cannot lie in the interval $[-p, p]$, and $\{x \mid |x|>p\}=\{x \mid x<-p \text{ or } x>p\}=(-\infty, -p) \cup (p, \infty)$ (see Figure 1.4).

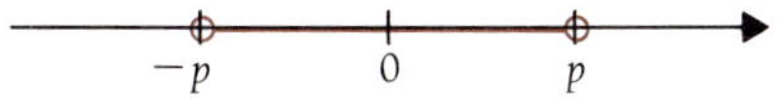

FIGURE 1.3 $|x|<p$

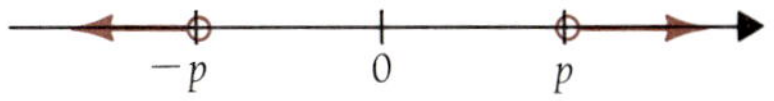

FIGURE 1.4 $|x|>p$

Now let p and r be real numbers with p positive. If x is a real number such that $|x-r|<p$, then the distance between r and x is less than p. Therefore, x must lie between $r-p$ and $r+p$, or $\{x \mid |x-r|<p\}=\{x \mid r-p<x<r+p\}=(r-p, r+p)$ (see Figure 1.5). On the other hand, if $|x-r|>p$, then x cannot lie in the interval $[r-p, r+p]$, or $\{x \mid |x-r|>p\}=\{x \mid x<r-p \text{ or } x>r+p\}=(-\infty, r-p) \cup (r+p, \infty)$ (see Figure 1.6).

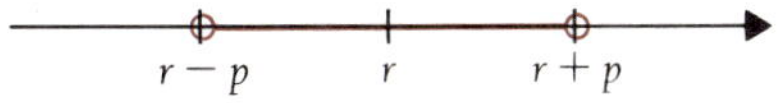

FIGURE 1.5 $|x-r|<p$

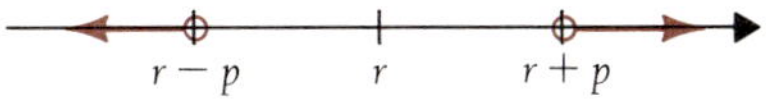

FIGURE 1.6 $|x-r|>p$

Similar statements to the above can be made involving $\leq$ and $\geq$. These facts are summarized in Table 1.2.

TABLE 1.2 INEQUALITIES

With absolute value	Without absolute value	Interval notation
1. $\lvert x\rvert<p$	$-p<x<p$	$(-p, p)$
2. $\lvert x\rvert\leq p$	$-p\leq x\leq p$	$[-p, p]$
3. $\lvert x\rvert>p$	$x>p$ or $x<-p$	$(-\infty, -p) \cup (p, \infty)$
4. $\lvert x\rvert\geq p$	$x\geq p$ or $x\leq -p$	$(-\infty, -p] \cup [p, \infty)$
5. $\lvert x-r\rvert<p$	$r-p<x<r+p$	$(r-p, r+p)$
6. $\lvert x-r\rvert\leq p$	$r-p\leq x\leq r+p$	$[r-p, r+p]$
7. $\lvert x-r\rvert>p$	$x>r+p$ or $x<r-p$	$(-\infty, r-p) \cup (r+p, \infty)$
8. $\lvert x-r\rvert\geq p$	$x\geq r+p$ or $x\leq r-p$	$(-\infty, r-p] \cup [r+p, \infty)$

Table 1.2 is a convenient reference in solving inequalities. Consider the following examples.

Example 1.7 Solve the following inequalities:

(a) $|x - 4| < 3$ (b) $|2x - 5| < 2$ (c) $|3x + 2| > \frac{1}{2}$ (d) $|6 - 2x| \leq 4$

Solution

(a) By line 5 of Table 1.2, the solution set is the interval $(4 - 3, 4 + 3) = (1, 7)$.

(b) By line 5 of Table 1.2 again, we can rewrite the inequality as

$$\begin{aligned} 5 - 2 < 2x &< 5 + 2, \\ 3 < 2x &< 7, \\ \tfrac{3}{2} < x &< \tfrac{7}{2}. \end{aligned}$$

Hence, the solution set is $(\frac{3}{2}, \frac{7}{2})$.

(c) We rewrite the inequality as $|3x - (-2)| > \frac{1}{2}$ and use Table 1.2 with x replaced by $3x$. By line 7, the inequality can be rewritten as

$$\begin{aligned} 3x &> -2 + \tfrac{1}{2} \quad \text{or} \quad 3x < -2 - \tfrac{1}{2}, \\ x &> -\tfrac{1}{2} \quad \text{or} \quad x < -\tfrac{5}{6}. \end{aligned}$$

Therefore, the solution set is $(-\infty, -\frac{5}{6}) \cup (-\frac{1}{2}, \infty)$.

(d) We first rewrite the inequality as $|2x - 6| \leq 4$, and then use line 6 of Table 1.2 to get

$$\begin{aligned} 6 - 4 \leq 2x &\leq 6 + 4, \\ 2 \leq 2x &\leq 10, \\ 1 \leq x &\leq 5. \end{aligned}$$

The solution set is the closed interval $[1, 5]$. ■

Absolute Value with Operations

It is clear that if p is a positive real number and $|x| = p$, then either $x = p$ or $x = -p$. It is also the case that the absolute value of a product or quotient is the product or quotient, respectively, of the absolute values—that is,

$$|ab| = |a| \cdot |b| \quad \text{and} \quad \left|\frac{a}{b}\right| = \frac{|a|}{|b|}.$$

However, with sums and differences, similar rules do *not* apply, as discussed below.

The Triangle Inequality

An important property of the absolute values of real numbers is that the absolute value of a sum is less than or equal to the sum of the absolute values. That is, for real numbers a and b,

$$|a + b| \leq |a| + |b|.$$

This statement is known as the **triangle inequality**. It gets its name from the fact that the length of a side of a triangle is always less than the sum of the lengths of the other two sides.

To derive the triangle inequality, we first note that either $|a| = a$ or $|a| = -a$, and therefore either $a = |a|$ or $a = -|a|$. Hence,

$$-|a| \leq a \leq |a|.$$

Similarly,

$$-|b| \leq b \leq |b|.$$

Adding the corresponding parts of these two inequalities (see Axiom 16 below), we get

$$-(|a| + |b|) \leq a + b \leq |a| + |b|,$$

from which the triangle inequality follows by line 2 of Table 1.2.

A difference rule for absolute values is found in the exercises (see Problem 71).

Supplement: Axioms of the Real Number System

The following axiomatic description of the real number system is given as a reference. The **system of real numbers** consists of the set $\mathbb{R}$, together with the operations of **addition** and **multiplication** and the relations **equals**, **less than**, and their combinations **(less than or equal to, greater than, greater than or equal to)**, satisfying the following axioms:

EQUALITY AXIOMS

1. **Reflexivity** If $x \in \mathbb{R}$, then $x = x$.
2. **Symmetry** If $x, y \in \mathbb{R}$ and $x = y$, then $y = x$.
3. **Transitivity** If $x, y, z \in \mathbb{R}$, $x = y$, and $y = z$, then $x = z$.
4. **Additivity** If $x, y, z \in \mathbb{R}$ and $x = y$, then $x + z = y + z$.
5. **Multiplicativity** If $x, y, z \in \mathbb{R}$ and $x = y$, then $xz = yz$.

FIELD AXIOMS

6. **Closure** If $x, y \in \mathbb{R}$, then $x + y \in \mathbb{R}$ and $xy \in \mathbb{R}$.
7. **Commutativity** If $x, y \in \mathbb{R}$, then $x + y = y + x$ and $xy = yx$.
8. **Associativity** If $x, y, z \in \mathbb{R}$, then $x + (y + z) = (x + y) + z$ and $x(yz) = (xy)z$.
9. **Distributivity** If $x, y, z \in \mathbb{R}$, then $x(y + z) = xy + xz$.
10. **Zero** There exists $0 \in \mathbb{R}$ such that if $x \in \mathbb{R}$, then $x + 0 = x$.
11. **One** There exists $1 \in \mathbb{R}$ such that if $x \in \mathbb{R}$, then $1x = x$.
12. **Additive inverses** If $x \in \mathbb{R}$, then there exists $-x \in \mathbb{R}$ such that $x + (-x) = 0$.
13. **Multiplicative inverses** If $x \in \mathbb{R}$ and $x \neq 0$, then there exists $1/x \in \mathbb{R}$ such that $x(1/x) = 1$.

ORDER AXIOMS

14. **Transitivity** If $x, y, z \in \mathbb{R}$, $x < y$, and $y < z$, then $x < z$.
15. **Trichotomy** If $x \in \mathbb{R}$, then exactly one of the following is true: $x < 0$, $x = 0$, $x > 0$.

16. **Additivity** If $x, y, z \in \mathbb{R}$ and $x < y$, then $x + z < y + z$.
17. **Multiplicativity** If $x, y, z \in \mathbb{R}$, $x < y$, and $z > 0$, then $xz < yz$.

THE COMPLETENESS AXIOM

Let $S \subseteq \mathbb{R}$. A number $u \in \mathbb{R}$ is called an **upper bound** of S if and only if $x \leq u$ for every $x \in S$. If S has an upper bound, we say S is **bounded above**. A number $b \in \mathbb{R}$ is a **least upper bound** of S if and only if (1) b is an upper bound of S and (2) if u is any upper bound of S, then $b \leq u$.

18. **Completeness** Every nonempty subset of $\mathbb{R}$ that is bounded above has a least upper bound.

Exercises 1.1

Complete the following table, using the first line as a sample:

Open	*(1, 2)*	$\{x \mid 1 < x < 2\}$	
1.	2.	$\{x \mid 1 \leq x \leq 2\}$	3.
Infinite	4.	5.	
6.	7.	$\{x \mid x \leq 4\}$	8.
9.	$(-1, 1]$	10.	11.
12.	13.	14.	
15.	16.	$\{x \mid \lvert x \rvert < 2\}$	17.
18.	19.	$\{x \mid \lvert x \rvert \leq 4\}$	20.

Rewrite each set in interval notation.

21. $\{x \mid -1 < x \leq 3\}$
22. $\{x \mid -p \leq x < p\}$
23. $\{x \mid c - d < x < c + d\}$
24. $\{x \mid a \leq x \leq a + 1\}$
25. $\{x \mid x > 0\}$
26. $\{x \mid x \geq 1\}$
27. $\{x \mid x < -1\}$
28. $\{x \mid x \leq 5\}$
29. $\{x \mid x^2 + 5x + 6 \leq 0\}$
30. $\{x \mid x^2 - 4 > 0\}$
31. $\{x \mid 3x^2 - x - 2 \leq 0\}$
32. $\left\{x \,\middle|\, \dfrac{x+2}{3x+1} < 0\right\}$

Rewrite each interval using inequalities.

33. $[2, 6]$
34. $[-1, 1)$
35. $(-\infty, 0]$
36. $(-2, \infty)$
37. $(c - d, c + d)$
38. $[-1, 1]$
39. $[0, \infty)$
40. $(-\infty, 0)$

Rewrite each interval using absolute values and inequalities.

41. $(-1, 1)$
42. $[-2, 2]$
43. $[0, 2]$
44. $(1, 2)$
45. $(c - d, c + d)$
46. $[p - 1, p + 1]$
47. $(a, a + 1)$
48. (a, b)

Give the solution set of each inequality.

49. $|x + 2| \geq 3$
50. $|x + 4| \leq 2$
51. $|x - 3| > 1$
52. $|x - 2| < \frac{1}{2}$
53. $|x| < -1$
54. $|x| > 2$
55. $|x| > 1$
56. $|x - 5| < 6$

57. $|3x+7|\geq 1$ 58. $|4-x|>3$ 59. $|3-2x|\geq 5$ 60. $|3(x-2)-4|\leq 1$

61. $|x+2|\geq|x-3|$ [*Hint:* Consider the cases $x-3\geq 0$ and $x-3<0$ separately.]

62. $|2x-1|\leq|x+4|$ 63. $\dfrac{|x-2|}{|2x-3|}<1$ 64. $\dfrac{|2x+3|}{|x-2|}\leq 4$

Rewrite each set as a single interval.

65. $(0, 2)\cap(1, 3)$ 66. $(0, 2)\cup(1, 3)$ 67. $[-1, 0]\cup(0, 1)$ 68. $[0, 5]\cap(4, 5)$

69. Does any point lie in all the intervals of the form $(0, 1/n)$, $n\in\mathbb{N}$?

70. Show that $|a|^2=a^2$ for any $a\in\mathbb{R}$.

71. Show that $|a|-|b|\leq|a-b|$. [*Hint:* Notice that $|a|=|(a-b)+b|$.]

72. Let $S\subseteq\mathbb{R}$. Describe what you think is meant by a **lower bound** of S.

73. What is meant by a **greatest lower bound?**

74. Prove that if a nonempty set of real numbers has a lower bound, then it has a greatest lower bound. (See Problems 72 and 73.)

Give the least upper bound (lub) and the greatest lower bound (glb) of each of the following sets; if a bound does not exist, so state:

75. $(3, 5)$ 76. $(-3, 4]$ 77. $(-\infty, 0)$ 78. $[1, \infty)$

79. $S=\left\{\dfrac{1}{n}\,\middle|\, n\in\mathbb{N}\right\}$ 80. $\mathbb{N}$

81. Explain why a least upper bound, when it exists, is unique.

82. Explain why division by 0 is excluded, by trying to assign a value to the expression $x/0$, $x\in\mathbb{R}$.

83. Find the flaw in the following "proof" that $1=2$:

$$\begin{aligned} a&=b,\\ a^2&=ab,\\ a^2-b^2&=ab-b^2,\\ (a+b)(a-b)&=b(a-b),\\ a+b&=b,\\ 2b&=b,\\ 2&=1. \end{aligned}$$

84. Find the flaw in the following "proof" that $4=5$:

$$\begin{aligned} -20&=-20,\\ 16-36&=25-45,\\ 16-36+\frac{81}{4}&=25-45+\frac{81}{4},\\ \left(4-\frac{9}{2}\right)^2&=\left(5-\frac{9}{2}\right)^2,\\ 4-\frac{9}{2}&=5-\frac{9}{2},\\ 4&=5. \end{aligned}$$

SECTION 1.2

Lines in the Plane

In the previous section, algebra was used to relate sets of real numbers to subsets of a line. In this section, algebraic methods are used to denote subsets of the plane. The description of geometric objects by algebraic methods is called **analytic geometry**. The idea of analytic geometry was first published in 1637 by Descartes; analytic geometry is also called **cartesian geometry** in his honor. Descartes's fellow Frenchman and contemporary Pierre de Fermat also worked on the development of analytic geometry.

The great advantage of cartesian geometry is its blending of geometric and algebraic methods. A geometric set can be described in terms of a set of numbers that satisfy certain conditions. Then, by manipulating the conditions algebraically, we can discover properties of the geometric set. Conversely, geometric pictures help us to understand the algebraic conditions. Geometry and algebra thus enrich each other.

Rectangular Coordinates

We set up a **rectangular** (or **cartesian**) **coordinate system** in the plane by choosing two copies of the real line, one horizontal and one vertical, so that the origin on each line is their point of intersection (see Figure 1.7). The two lines are called **coordinate axes**; the horizontal axis is called the ***x*-axis**, and the vertical one is called the ***y*-axis**. The axes divide the plane into four parts, called **quadrants**, numbered counterclockwise.

René Descartes (1596–1654) was born into an old noble French family and educated in a manner befitting a gentleman and soldier. His health was delicate, and he was allowed to remain in bed until midday. He continued this habit almost all his life, and claimed that his most productive thinking occurred during those mornings.

While in his twenties, he participated in several military campaigns, but at about age thirty he became interested in philosophy. He spent the next 20 years in Holland, elaborating his philosophical system and relating it to the mathematics and science of the time. He carried on an extensive correspondence on scientific and philosophical topics through an intermediary in Paris, but kept his address secret.

Descartes's development of analytic geometry was his most important mathematical contribution and was published in an appendix to his philosophical work *Discourse on Method,* in 1637. He also discovered several facts about the roots of polynomial equations, the most famous being Descartes's rules of signs.

Toward the end of his life, he became a tutor to Queen Kristina of Sweden. She demanded her philosophy lessons at 5 AM, and the combination of early rising and Swedish winter undermined Descartes's health and caused his premature death.

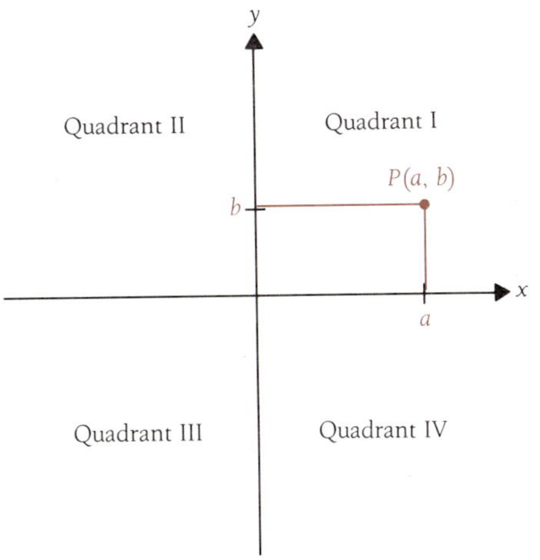

FIGURE 1.7 The cartesian coordinate system

Now each point P in the plane is labeled with a pair of real numbers, called **coordinates** of P, as follows. If the vertical line through P meets the x-axis at the point a, and the horizontal line through P meets the y-axis at b, then the coordinates of P are (a, b). The first coordinate a is also called the **x-coordinate** or **abscissa** of P, and the second coordinate b is also called the **y-coordinate** or **ordinate** of P.

Every point in the plane has a pair of coordinates, and every pair (x, y) of real numbers is the coordinate pair of some point in the plane. This is easily seen by reversing the above procedure. Points on the x-axis have coordinates of the form $(x, 0)$, and those on the y-axis have coordinates of the form $(0, y)$. The common point of the two axes has coordinates $(0, 0)$ and is called the **origin** of the coordinate system. Hereafter, we refer to the point with coordinates (x, y) as "the point (x, y)."

The Distance Formula

A simple application of analytic geometry yields a formula for the distance between two points. If $P_1(x_1, y_1)$ and $P_2(x_2, y_2)$ are points, we let $|P_1P_2|$ denote the distance between them. Let P_3 be the point (x_1, y_2) (see Figure 1.8). Then the right triangle

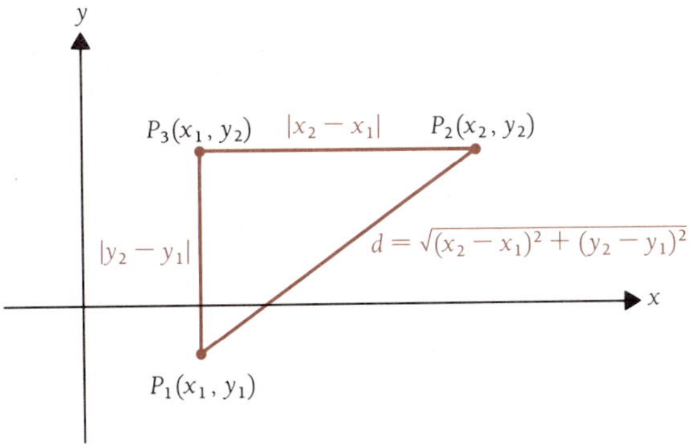

FIGURE 1.8 The distance formula

$P_1P_2P_3$ has legs of lengths $|P_1P_3| = |y_2 - y_1|$ and $|P_2P_3| = |x_2 - x_1|$. Therefore, by the Pythagorean theorem, the distance d between P_1 and P_2 is

$$d = |P_1P_2| = \sqrt{(x_2 - x_1)^2 + (y_2 - y_1)^2}.$$

This is the **distance formula**. (Note that the absolute value signs are not needed under the radical, by Problem 70 of Section 1.1.)

Example 1.8 Find the distance between points $P(3, -1)$ and $Q(5, 8)$.

Solution By the distance formula,

$$|PQ| = \sqrt{(5-3)^2 + (8+1)^2} = \sqrt{85}.$$

The Midpoint Formula

Another useful formula enables us to find the midpoint of a segment. Using the same points as in Figure 1.8, let $M(x, y)$ be the midpoint of the segment P_1P_2. If $S = (x_1, y)$ and $T = (x, y_2)$ (see Figure 1.9), then triangles P_1SM and MTP_2 are congruent. Therefore, $x - x_1 = x_2 - x$ and $y - y_1 = y_2 - y$, from which it follows that

$$x = \frac{x_1 + x_2}{2} \quad \text{and} \quad y = \frac{y_1 + y_2}{2}.$$

This is the **midpoint formula**.

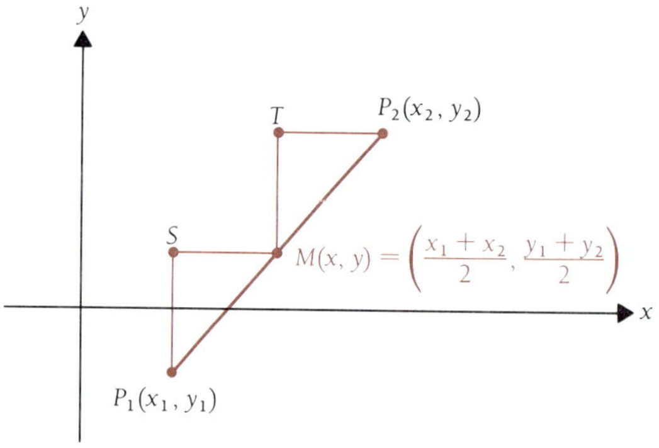

FIGURE 1.9 The midpoint formula

Example 1.9 Find the point midway between (4, 1) and (−2, 5).

Solution By the midpoint formula, the midpoint is

$$\left(\frac{4-2}{2}, \frac{1+5}{2}\right) = (1, 3).$$

Lines and Slope

Let l be a line in the plane, and let $P_1(x_1, y_1)$ and $P_2(x_2, y_2)$ be any two points on l. The (horizontal) difference $x_2 - x_1$ is called the **run** from P_1 to P_2, and the (vertical) difference $y_2 - y_1$ is called the **rise** from P_1 to P_2. This is illustrated in Figure 1.10. The **slope** of line l is defined to be the ratio of rise to run,

$$\text{Slope} = \frac{\text{Rise}}{\text{Run}}.$$

(If the run is 0, the slope is not defined, because division by 0 is not defined; more will be said of this case in a moment.) Thus, in terms of the coordinates of P_1 and P_2, the slope m of line l is

$$m = \frac{y_2 - y_1}{x_2 - x_1}, \qquad x_2 - x_1 \neq 0.$$

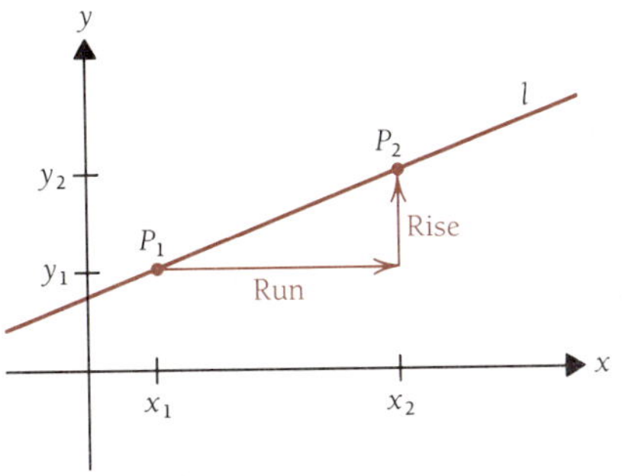

FIGURE 1.10 Slope $= \dfrac{\text{Rise}}{\text{Run}}$

Although the slope of l has been defined in terms of points P_1 and P_2, the slope of l is actually independent of P_1 and P_2. For example, if points P_1 and P_2 are interchanged, then both rise and run will change sign, but their ratio is unchanged (see Figure 1.11). If two other points Q_1 and Q_2 are chosen on l (see Figure 1.12), the properties of similar triangles tell us that the same slope will be obtained from them.

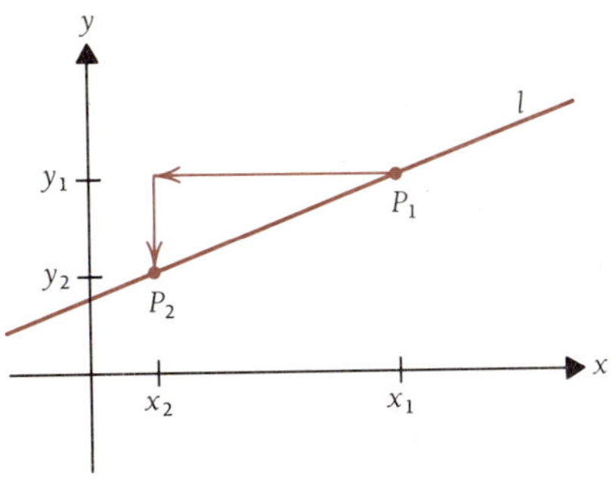

FIGURE 1.11 Slope is independent of order

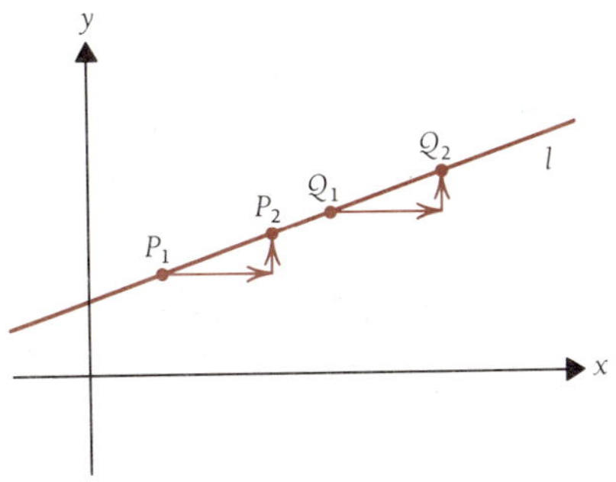

FIGURE 1.12 Slope is independent of points

If two lines are parallel, the properties of similar triangles again tell us that they have the same slope (see Figure 1.13). Thus, the slope of a line determines its direction. Specifically, there can be only one line of a given slope through a given point.

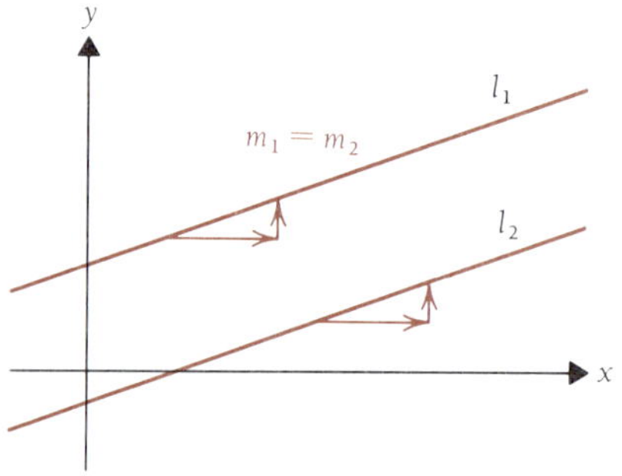

FIGURE 1.13 Parallel lines have the same slope

Inspection of a line enables us to say something about its slope. If the line rises from left to right, as in Figure 1.14, then the rise and run will have the same sign and the slope is positive. If the line falls from left to right, as in Figure 1.15, then the rise and run will have opposite signs and the slope is negative.

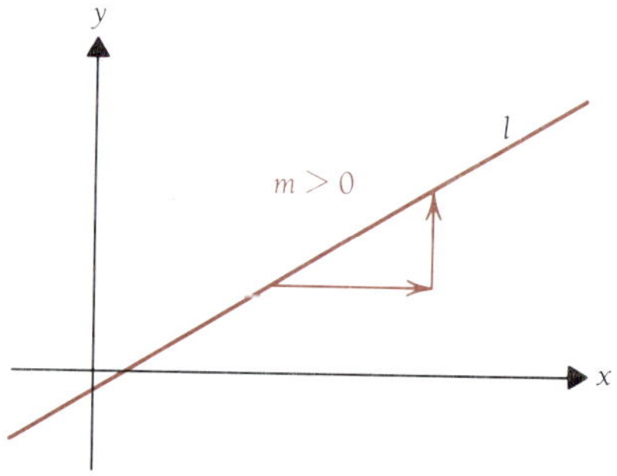

FIGURE 1.14 A line with positive slope rises from left to right

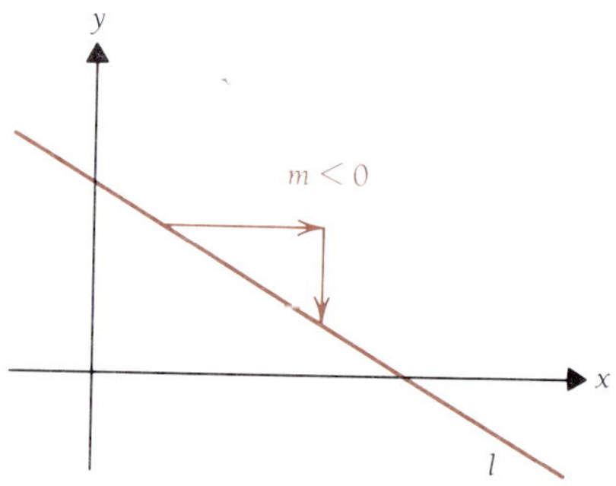

FIGURE 1.15 A line with negative slope falls from left to right

If a line is horizontal, then the rise is 0 and the slope is 0 (see Figure 1.16). If a line is vertical, as in Figure 1.17, then the run is 0 and the slope is undefined. Thus, a vertical line has no slope. It is important to note the difference: *A horizontal line has 0 slope, and a vertical line has no slope at all.* Figure 1.18 summarizes these ideas about slopes.

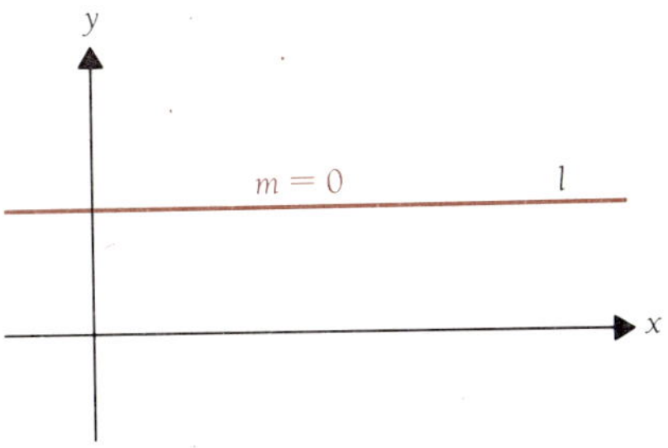

FIGURE 1.16 A horizontal line has 0 slope

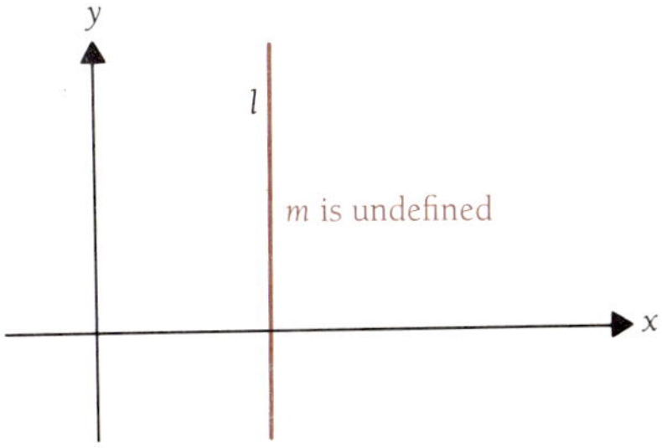

FIGURE 1.17 A vertical line has no slope

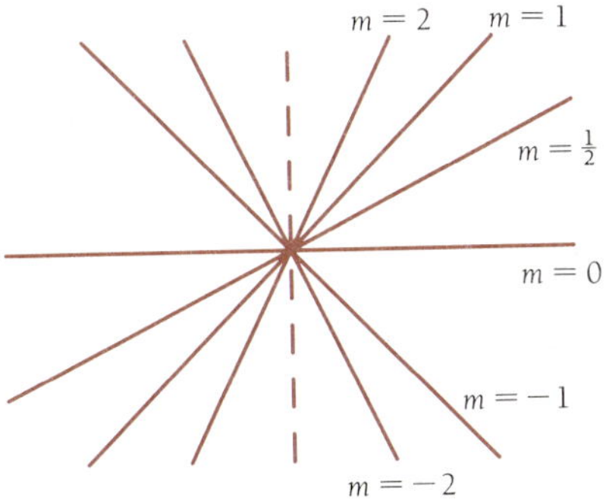

FIGURE 1.18 Slopes of various lines

Equation of a Line

Suppose we are given a line l that passes through point $P_1(x_1, y_1)$ and has slope m. We can find an equation in the two variables x and y such that line l is exactly the set of points (x, y) whose coordinates satisfy the equation. To find such an equation, we note that a point $P(x, y)$ will lie on l if and only if the line determined by points P and P_1 is l. In particular, their slopes must be the same, or

$$m = \frac{y - y_1}{x - x_1}.$$

That is,

$$y - y_1 = m(x - x_1). \tag{1}$$

Equation (1) is called a **linear equation**, since it is the equation of a line. Note that the variables x and y appear only to the first power. Because it was determined by a point and the slope, equation (1) is called the **point–slope form** of a linear equation.

Example 1.10 Find an equation of the line through point $(-2, 1)$ and having slope 3.

Solution Using the point–slope form (1) with $x_1 = -2$, $y_1 = 1$, and $m = 3$, we get

$$y - 1 = 3[x - (-2)],$$
$$y - 1 = 3(x + 2).$$

If a line l meets the y-axis, the point of intersection has the form $(0, b)$. The number b is called the **y-intercept**, or simply the **intercept**, of line l. A line with slope m and intercept b has an equation of the form

$$y = mx + b, \tag{2}$$

obtained by letting $x_1 = 0$ and $y_1 = b$ in equation (1). Equation (2) is called the **slope–intercept form** of a linear equation. The slope and intercept can be read directly from the equation as the coefficient of x and the constant term, once y has been isolated.

Example 1.11 Find an equation for the line with slope $m = -2$ and intercept $b = 4$.

Solution By equation (2), the line must have the equation

$$y = -2x + 4.$$

Example 1.12 Find the slope and intercept of the line whose equation is $y - 1 = 3(x + 2)$.

Solution We write the equation in slope–intercept form by solving for y:

$$y - 1 = 3(x + 2),$$
$$y - 1 = 3x + 6,$$
$$y = 3x + 7.$$

Thus, the line has slope 3 and intercept 7.

If a line is specified by two points, we can compute the slope of the line using the coordinates of the two points. Then using the point–slope form (1), we obtain an equation of the form

$$y - y_1 = \frac{y_2 - y_1}{x_2 - x_1}(x - x_1). \tag{3}$$

This is called the **two-point form** of a linear equation.

Example 1.13 Find an equation for the line through the two points $(-1, 3)$ and $(2, 1)$.

Solution Using the two-point form (3), we get

$$y - 3 = \frac{1-3}{2+1}(x+1),$$

$$y - 3 = \frac{-2}{3}(x+1).$$

This can be further simplified to the slope–intercept form

$$y = -\frac{2}{3}x + \frac{7}{3}.$$ ■

All the lines we have considered so far have had slope; none of them have been vertical lines. Now, let's find the equation of a vertical line. Let line l be vertical and meet the x-axis at point $(a, 0)$. We call a the **x-intercept** of l. Then every point on l will have its first coordinate equal to a, and the equation

$$x = a$$

exactly specifies the points on l.

■ **Example 1.14** Find an equation for the vertical line through the point $(-5, 8)$.

Solution Since the x-intercept of the line is -5, the equation of the line is seen to be $x = -5$. ■

As another example of a vertical line, note that the y-axis has equation $x = 0$.
All linear equations can be written in the form

$$Ax + By + C = 0, \tag{4}$$

which is called the **standard form** of a linear equation. If $B = 0$, the equation represents a vertical line, since the equation becomes $x = -C/A$. If $B \neq 0$, the equation may be rewritten as

$$y = -\frac{A}{B}x - \frac{C}{B},$$

and thus represents a line with slope $-A/B$ and intercept $-C/B$. If $A = 0$ in equation (4), then the equation represents a horizontal line.

■ **Example 1.15** Write the equation below in standard form:

$$y = -\frac{2}{3}x + \frac{7}{3}.$$

Solution Rearranging and clearing the denominators, we get the equation

$$2x + 3y - 7 = 0,$$

which we note by comparing with equation (4) is in standard form. ■

The equation forms are summarized in Table 1.3.

TABLE 1.3 EQUATIONS OF A LINE

Form	Equation	Remarks
Point–slope	$y - y_1 = m(x - x_1)$	Through (x_1, y_1), with slope m
Slope–intercept	$y = mx + b$	Slope m, y-intercept b
Two-point	$y - y_1 = \dfrac{y_2 - y_1}{x_2 - x_1}(x - x_1)$	Through (x_1, y_1) and (x_2, y_2)
Standard	$Ax + By + C = 0$	Vertical if $B = 0$, horizontal if $A = 0$

Perpendicular Lines

We have remarked already that parallel lines have the same slope. Now we investigate the relationship between the slopes of perpendicular lines.

Suppose lines l_1 and l_2 are perpendicular and that neither of them is vertical (see Figure 1.19). Let the lines have slopes m_1 and m_2, respectively. Further, let l_1 and l_2 meet at point P, and let l_3 be a horizontal line not through P. Let l_1 and l_2 meet l_3 at Q_1 and Q_2, respectively, and let the vertical line through P meet l_3 at R, as shown in Figure 1.19. The slope of l_1 is

$$m_1 = \frac{|PR|}{|Q_1R|},$$

and the slope of l_2 is

$$m_2 = \frac{-|PR|}{|RQ_2|}.$$

We know that triangles PQ_1R and Q_2PR are similar, because PR is the altitude on the hypotenuse of a right triangle (or, you may check the angles). Hence,

$$\frac{|PR|}{|Q_1R|} = \frac{|Q_2R|}{|PR|},$$

and therefore,

$$m_1 = \frac{-1}{m_2}.$$

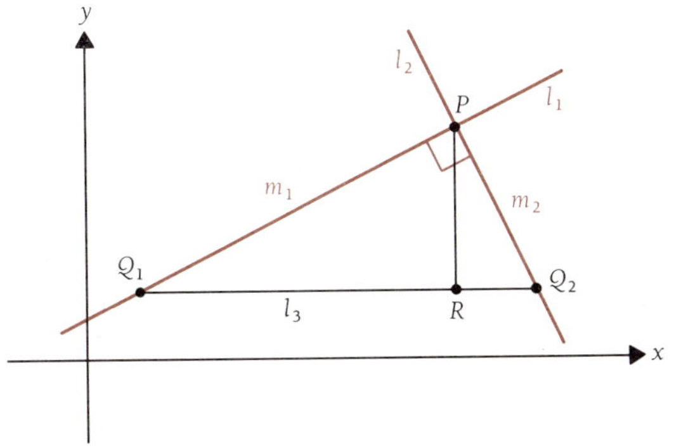

FIGURE 1.19 Slopes of perpendicular lines: $m_1 m_2 = -1$

That is, the slopes are *negative reciprocals* of each other. We also can write

$$m_1 m_2 = -1,$$

and conclude that perpendicular lines have slopes whose product is -1.

Example 1.16 Find an equation of the line through the point $(5, -2)$ that is perpendicular to the line $y = 3x - 2$.

Solution The slope of the given line is 3, so the slope of the line we seek is $-\frac{1}{3}$. From the point–slope form, we get

$$y + 2 = -\tfrac{1}{3}(x - 5),$$
$$y = -\tfrac{1}{3}x - \tfrac{1}{3}$$

in slope–intercept form, or

$$x + 3y + 1 = 0$$

in standard form. ■

Exercises 1.2

Find the distance between each pair of points.

1. $(1, 3)$ and $(2, -1)$
2. $(1, -3)$ and $(2, -1)$
3. $(\frac{1}{2}, -5)$ and $(-3, 0)$
4. $(1, \sqrt{3})$ and $(-4, 8)$

Find the point midway between each pair of points.

5. $(1, 3)$ and $(2, -1)$
6. $(1, -3)$ and $(2, -1)$
7. $(\frac{1}{2}, -5)$ and $(-3, 0)$
8. $(1, \sqrt{3})$ and $(-4, 8)$

Find the equation in slope–intercept form for each line.

9. Through $(-2, 1)$ and with slope 3
10. Through $(2, 9)$ and with slope $-\frac{2}{3}$
11. Through $(-1, 1)$ and with slope -1
12. Through $(a, 1)$ and with slope 1
13. With slope 2 and intercept 4
14. With slope $\frac{1}{2}$ and intercept -1

15. With slope -2 and intercept 6
16. With slope m and intercept 1
17. Through (1, 3) and (2, -6)
18. Through (-4, 1) and (-9, 3)
19. Through (2.3, 5.4) and (-4.5, 7.2)
20. Through (5, 2570) and (6, 2980)
21. Horizontal and through (-2, 1)
22. Horizontal and through (8, 9)
23. Horizontal and through (0, b)
24. Horizontal and through (x_0, y_0)
25. Vertical and through (1, 4)
26. Vertical and through (-2, -5)
27. Vertical and through (0, b)
28. Vertical and through (x_0, y_0)
29. Through (5, 3) and parallel to $3x - y - 2 = 0$
30. Through (-4, -1) and parallel to $3x + y - 5 = 0$
31. Through (5, $\sqrt{3}$) and perpendicular to $3x - y - 5 = 0$
32. Through ($\sqrt{2}$, -4) and perpendicular to $3x + y - 2 = 0$
33. Through (1.43, 2.79) and with slope -0.04
34. With slope 1.9 and intercept 1.06
35. Through (1.05, 0.216) and (1.06, 0.229)
36. With intercept -4.3 and parallel to the line through (2.4, 1.25) and (2.5, 1.13)

Find the slope and intercept of the line, and sketch its graph.

37. $x - 3y = 5$
38. $2x + 5y = 0$
39. $x = \frac{2}{3}y - \frac{1}{3}$
40. $\frac{x}{2} + \frac{y}{6} = 1$
41. $3x = 2y$
42. $5x - 8y = -19$
43. $0.25x + 0.35y = 2.48$
44. $1.05x - 3.66y = 9.32$
45. Show that the line with x-intercept a and y-intercept b has equation

$$\frac{x}{a} + \frac{y}{b} = 1.$$

This equation is called the **intercept form** of a linear equation.

46. Find an equation for the line with x-intercept -2 and y-intercept 3. (See Problem 45.)
47. Find the intercepts of the line $3x - 4y - 12 = 0$.
48. Find the intercepts of the line perpendicular to $3x - 4y - 12 = 0$ at the point $(6, \frac{3}{2})$.
49. Find the equation of the perpendicular bisector of the segment with endpoints (1, 0) and (0, 1).
50. Repeat Problem 49 for the points (x_1, y_1) and (x_2, y_2).
51. The line $5x + 4y - 20 = 0$ cuts a triangle off the first quadrant. Find the area of the triangle.
52. Use the distance formula to determine whether the points $A(0, 8)$, $B(4, 16)$, and $C(13, 0)$ are **collinear** (that is, on the same line).
53. Repeat Problem 52, using a linear equation.
54. Repeat Problems 52 and 53 for the points $A(0, 8)$, $B(8, 3)$, and $C(-2, 16)$.
55. Find an equation for the line through the origin and perpendicular to the line $5x + 4y - 20 = 0$.

56. Find the distance from the origin to the line $5x + 4y - 20 = 0$, as measured along the perpendicular.

57. Repeat Problem 56 for the line $Ax + By + C = 0$.

58. Find the point on the segment joining points (4, 1) and (2, 3) that divides the segment into two segments with the ratio 1 to 2.

59. Repeat Problem 58 for the ratio m to n.

60. Repeat Problem 59 for the points (x_1, y_1) and (x_2, y_2).

61. Recall that the value of the 2×2 determinant

$$\begin{vmatrix} a & b \\ c & d \end{vmatrix}$$

is $ad - bc$ (see Appendix A). Show that the line through points (x_1, y_1) and (x_2, y_2) is given by

$$\begin{vmatrix} y_1 & 1 \\ y_2 & 1 \end{vmatrix} x + \begin{vmatrix} 1 & x_1 \\ 1 & x_2 \end{vmatrix} y + \begin{vmatrix} x_1 & y_1 \\ x_2 & y_2 \end{vmatrix} = 0.$$

62. Show that the point of intersection of the lines $A_1x + B_1y + C_1 = 0$ and $A_2x + B_2y + C_2 = 0$ has coordinates

$$x = \frac{\begin{vmatrix} B_1 & C_1 \\ B_2 & C_2 \end{vmatrix}}{\begin{vmatrix} A_1 & B_1 \\ A_2 & B_2 \end{vmatrix}} \quad \text{and} \quad y = \frac{\begin{vmatrix} C_1 & A_1 \\ C_2 & A_2 \end{vmatrix}}{\begin{vmatrix} A_1 & B_1 \\ A_2 & B_2 \end{vmatrix}},$$

provided that

$$\begin{vmatrix} A_1 & B_1 \\ A_2 & B_2 \end{vmatrix} \neq 0.$$

63. Show that if

$$\begin{vmatrix} A_1 & B_1 \\ A_2 & B_2 \end{vmatrix} = 0,$$

then the lines $A_1x + B_1y + C_1 = 0$ and $A_2x + B_2y + C_2 = 0$ are the same or parallel.

64. Show that the three points (x_1, y_1), (x_2, y_2), and (x_3, y_3) are collinear if and only if

$$\begin{vmatrix} x_1 & y_1 & 1 \\ x_2 & y_2 & 1 \\ x_3 & y_3 & 1 \end{vmatrix} = 0.$$

[*Hint:* Use Problem 61. See Appendix A, Theorem A.3, for the value of a 3×3 determinant.]

65. Show that the lines $A_1x + B_1y + C_1 = 0$, $A_2x + B_2y + C_2 = 0$, and $A_3x + B_3y + C_3 = 0$ are **concurrent** (that is, pass through the same point) if and only if

$$\begin{vmatrix} A_1 & B_1 & C_1 \\ A_2 & B_2 & C_2 \\ A_3 & B_3 & C_3 \end{vmatrix} = 0.$$

66. A **median** of a triangle is the segment joining a vertex to the midpoint of the opposite side. The three medians of a triangle are concurrent, meeting at the **centroid**. Find the coordinates of the centroid of the triangle with vertices $A(0, 0)$, $B(b, 0)$, and $C(c, h)$.

[A]67. Find the coordinates of the centroid of the triangle with vertices (x_1, y_1), (x_2, y_2), and (x_3, y_3). (See Problem 66.)

SECTION 1.3

Circles

In this section, we discuss the representation of circles in the rectangular coordinate system. First, recall that a **circle** is the set of points in the plane that are a given distance from a given point. The given distance is called the **radius** of the circle, and the given point is called the **center**.

Equation of a Circle

In the rectangular coordinate system, let (h, k) be the center of a circle of radius r. Then a point (x, y) will be on the circle (see Figure 1.20) if and only if the distance between (x, y) and (h, k) is r, or

$$r = \sqrt{(x-h)^2 + (y-k)^2},$$

which is equivalent to

$$(x-h)^2 + (y-k)^2 = r^2. \tag{1}$$

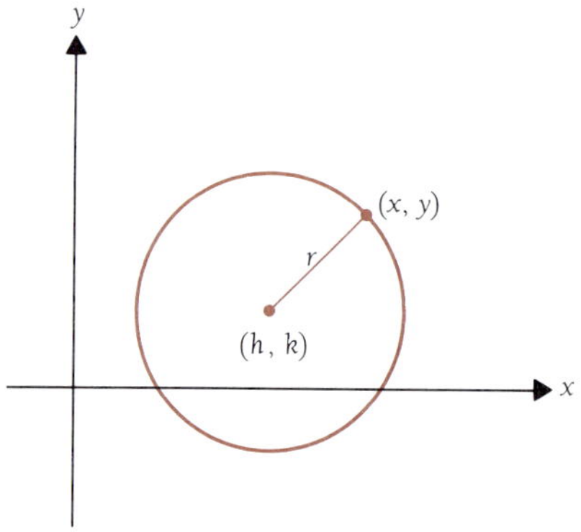

FIGURE 1.20 Point (x, y) is on the circle when $(x-h)^2 + (y-k)^2 = r^2$

■ **Example 1.17** Find an equation of the circle with:

(a) Center $(2, -3)$ and radius 4 (see Figure 1.21)
(b) Center at the origin and radius r (see Figure 1.22)

Solution **(a)** By equation (1), the equation is

$$(x-2)^2 + (y+3)^2 = 16.$$

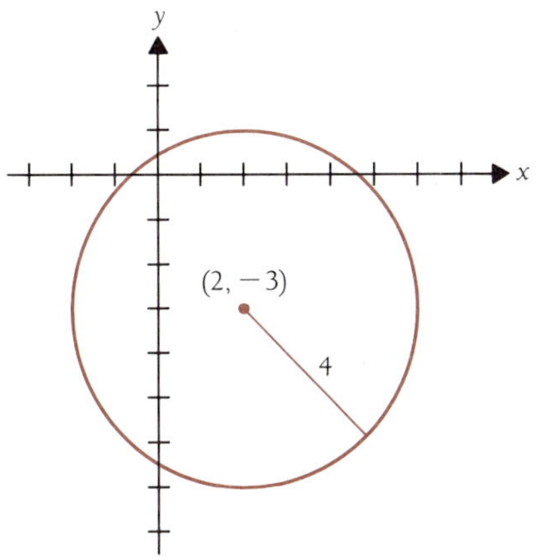

FIGURE 1.21 The circle $(x-2)^2+(y+3)^2=16$

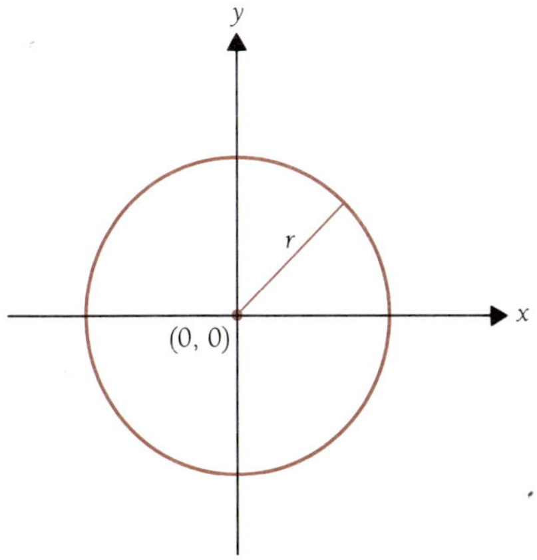

FIGURE 1.22 The circle $x^2+y^2=r^2$

(b) Again by equation (1), the equation reduces to

$$x^2+y^2=r^2.$$

The **unit circle** is the circle with center at the origin and radius 1 (see Figure 1.23). By Example 1.17(b), it has equation

$$x^2+y^2=1.$$

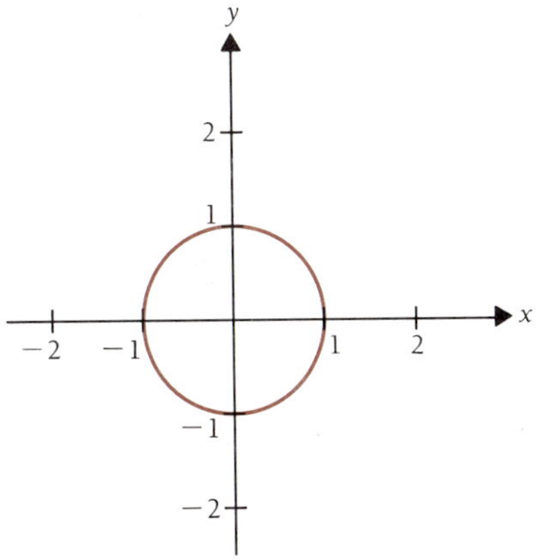

FIGURE 1.23 The unit circle, $x^2+y^2=1$

We can expand and rearrange equation (1) as follows:

$$(x-h)^2+(y-k)^2=r^2,$$
$$x^2-2hx+h^2+y^2-2ky+k^2=r^2,$$
$$x^2+y^2-2hx-2ky+(h^2+k^2-r^2)=0.$$

This last equation has the form

$$x^2 + y^2 + ax + by + c = 0, \tag{2}$$

with a, b, and c being constants. Any circle has an equation of this form.

Example 1.18 Write the equation of the circle in Example 1.17(a) in the form (2).

Solution We expand and simplify:

$$\begin{aligned}(x-2)^2 + (y+3)^2 &= 16,\\ x^2 - 4x + 4 + y^2 + 6y + 9 - 16 &= 0,\\ x^2 + y^2 - 4x + 6y - 3 &= 0.\end{aligned}$$

■

Completing the Square

The question now arises whether any equation of the form (2) is the equation of a circle. We can answer this question by using the process called **completing the square**, which goes as follows: We first regroup the terms in equation (2):

$$\begin{aligned}x^2 + y^2 + ax + by + c &= 0,\\ x^2 + ax \quad + y^2 + by \quad &= -c. \end{aligned} \tag{3}$$

Then we add numbers to both sides of equation (3) so as to form x and y trinomials that are perfect squares. To make $x^2 + ax$ into a perfect square, we add the square of half the coefficient of x, which is $a^2/4$, and similarly for the y trinomial:

$$x^2 + ax + \frac{a^2}{4} + y^2 + by + \frac{b^2}{4} = -c + \frac{a^2}{4} + \frac{b^2}{4}.$$

Finally, we factor:

$$\left(x + \frac{a}{2}\right)^2 + \left(y + \frac{b}{2}\right)^2 = \frac{a^2 + b^2 - 4c}{4}. \tag{4}$$

If the constant term of equation (4) is positive, the equation can be rewritten as

$$\left(x + \frac{a}{2}\right)^2 + \left(y + \frac{b}{2}\right)^2 = \left(\frac{\sqrt{a^2 + b^2 - 4c}}{2}\right)^2,$$

which is the equation of a circle with center $(-a/2, -b/2)$ and radius $\sqrt{a^2 + b^2 - 4c}/2$.

If the constant term of equation (4) is 0, the equation is

$$\left(x + \frac{a}{2}\right)^2 + \left(y + \frac{b}{2}\right)^2 = 0,$$

which is satisfied only by the point $(-a/2, -b/2)$. We sometimes call equation (2) or (4) a **point circle** when $a^2 + b^2 - 4c = 0$.

Finally, if the constant term in equation (4) is negative, then no point with real coordinates satisfies equation (2), and hence equation (2) does not represent any real curve. We sometimes call equation (2) the equation of an **imaginary circle** when $a^2 + b^2 - 4c < 0$.

Example 1.19 Describe the curve with equation $x^2 + y^2 - 3x + 4y = 6$.

Solution First, we complete the square:

$$x^2 - 3x \qquad + y^2 + 4y \qquad = 6,$$

$$x^2 - 3x + \frac{9}{4} + y^2 + 4y + 4 = 6 + \frac{9}{4} + 4,$$

$$\left(x - \frac{3}{2}\right)^2 + (y + 2)^2 = \frac{49}{4}.$$

Thus, the curve is a circle with center $(\frac{3}{2}, -2)$ and radius $\frac{7}{2}$ (see Figure 1.24).

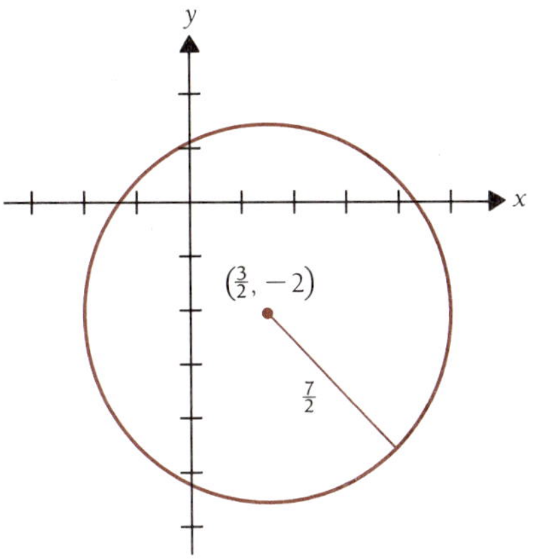

FIGURE 1.24 The circle $x^2 + y^2 - 3x + 4y = 6$ ■

Lines and Circles

Let a circle have center C and radius r, and let l be a line. Line l meets the circle in two, one, or no points, according as the distance from C to l is less than r, equal to r, or greater than r (see Figure 1.25). If l meets the circle in two points, we call l a

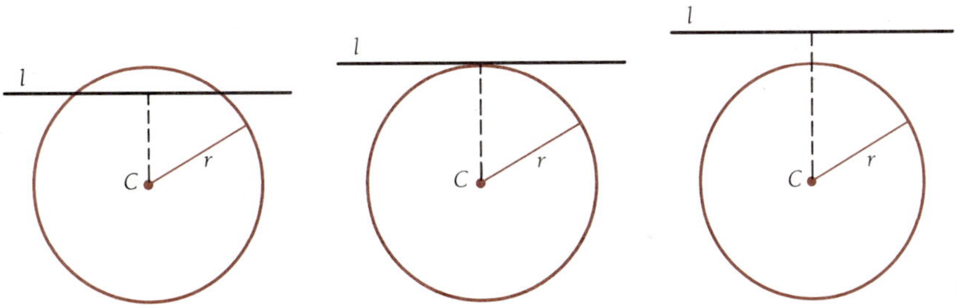

(a) Secant line (b) Tangent line (c) Nonintersecting line

FIGURE 1.25 Relationships of lines and circles

secant line. If l meets the circle in exactly one point, we call l a **tangent line** of the circle.

Tangent Lines

Suppose line l is a tangent line to the circle $(x - h)^2 + (y - k)^2 = r^2$ at the point (x_1, y_1) (see Figure 1.26). Then the radius joining (h, k) to (x_1, y_1) is perpendicular to l. Since the slope of the line through (h, k) and (x_1, y_1) is

$$m = \frac{y_1 - k}{x_1 - h},$$

the slope of line l must be

$$\frac{-1}{m} = \frac{x_1 - h}{k - y_1}. \tag{5}$$

The equation of l can now be found.

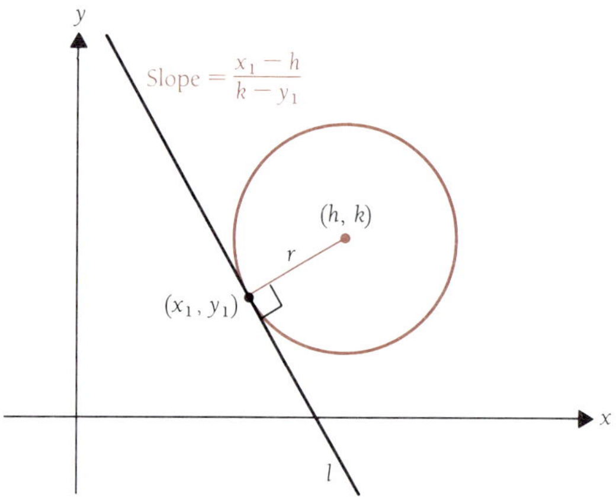

FIGURE 1.26 Slope of a tangent line to the circle $(x - h)^2 + (y - k)^2 = r^2$

■ **Example 1.20** The circle $(x - 2)^2 + (y + 3)^2 = 16$ passes through the three points $(2, 1)$, $(6, -3)$, and $(2 + \sqrt{7}, 0)$. Find an equation of the tangent line to the circle at each of these points.

Solution The circle is shown in Figure 1.27. Using the formula (5) for the slope of the tangent line, we find that the slope of the tangent line at $(2, 1)$ is

$$\frac{2 - 2}{-3 - 1} = 0.$$

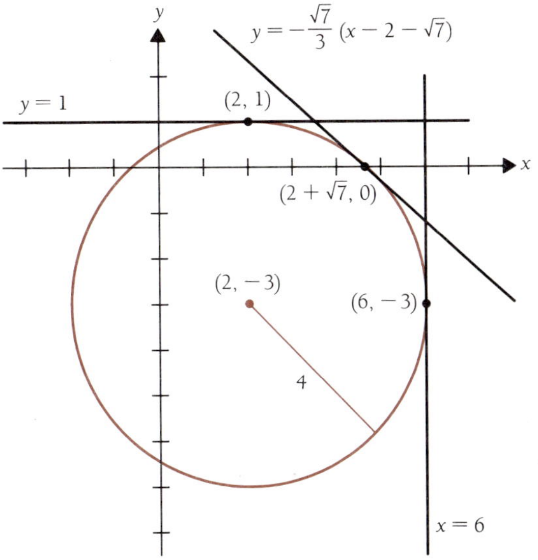

FIGURE 1.27 Tangent lines to the circle $(x-2)^2 + (y+3)^2 = 16$ through various points

Thus, the tangent line at (2, 1) is horizontal and has equation

$$y = 1.$$

The tangent line at (6, −3) has slope

$$\frac{6-2}{-3+3},$$

which is undefined. Hence, the tangent line is vertical and has equation

$$x = 6.$$

The tangent line at $(2 + \sqrt{7}, 0)$ has slope

$$\frac{2+\sqrt{7}-2}{-3-0} = -\frac{\sqrt{7}}{3},$$

so has equation

$$y = -\frac{\sqrt{7}}{3}(x - 2 - \sqrt{7}).$$

■

Exercises 1.3

Find an equation of the circle with the given point as center and the given number as radius.

1. (0, 0); 3
2. (0, 0); $\frac{1}{2}$
3. (−7, 1); 2
4. (−4, −2); 5
5. $(3, -\frac{2}{3})$; $\frac{2}{3}$
6. $(-1, \sqrt{2})$; $\sqrt{5}$
7. $(a, 3)$; $1 + \sqrt{2}$
8. $(2 - \sqrt{3}, -\sqrt{3})$; $\sqrt{3}/2$

Describe the following curves, giving the center and radius of each circle:

9. $x^2 + y^2 = 4$
10. $x^2 + y^2 = 7$
11. $x^2 + y^2 = 0$
12. $x^2 + y^2 = -1$
13. $x^2 + y^2 + 2x + y = 4$
14. $x^2 + y^2 - 6y = -2$
15. $x^2 + y^2 - 6x = 8$
16. $x^2 + y^2 + 2x - 4y = 0$
17. $x^2 + y^2 - 2x + 3y + 9 = 0$
18. $x^2 + y^2 - 4x = -7$
19. $x^2 + y^2 - 4x + 6y + 13 = 0$
20. $x^2 + y^2 - 8x = -16$
21. $x^2 + y^2 = 2ax$
22. $x^2 + y^2 = ay$
23. $x^2 + y^2 = a(x + y)$
24. $x^2 + y^2 + ax + by = 0$

Find an equation of the circle fitting the given description.

25. Center (0, 0) and through point (3, −4)
26. Center (4, 7) and through point (−1, −1)
27. Center (3, −4) and through the origin
28. Center (h, k) and through the origin
29. Center (1, 2) and tangent to the line $y = 5$
30. Center (−2, 3) and tangent to the line $x = 4$
31. Center (4, 1) and tangent to the y-axis
32. Center the origin and tangent to the line $Ax + By + C = 0$

A33. Radius 4 and through points (2, 0) and (0, 4)

A34. Radius 6 and through points (2, −4) and (2, −2)

35. Radius 2, center in the first quadrant, and tangent to both axes
36. Radius 2 and tangent to both lines $y = x$ and $y = 0$
37. Center (1, 5) and tangent to the circle $x^2 + y^2 = 4$
38. Center (−1, 3) and tangent to the circle $x^2 + y^2 = 2x$
39. Tangent to both axes and the line $x = 2$

A40. Tangent to both axes and the line $x + y = 1$

41. Through the three points (2, −1), (3, 0), and (4, −1)
42. Through the three points (0, 0), (1, 5), and (−5, 5)
43. Through the three points (2, −1), (3, 0), and (0, 0)
44. Through the three points (0, 0), (a, b), and (c, d)

Find an equation of the line tangent to the given circle at the given point.

45. $x^2 + y^2 = 4$; (2, 0)
46. $x^2 + y^2 = 4$; $(1, \sqrt{3})$
47. $x^2 + y^2 - 2x - 6y - 6 = 0$; (5, 3)
48. $x^2 + y^2 + 5y - 9 = 0$; $(-\sqrt{3}, 1)$
49. Find the equations of the lines through the point (4, 0) that are tangent to the circle $x^2 + y^2 = 4$.
50. Find the equations of the upper and lower semicircles of the unit circle by solving the equation of the circle for y.
51. Find the equations of the upper and lower semicircles of the circle with radius r and center (h, k).
52. Show that if $x^2 + ax + b = (x + c)^2$, then $c = a/2$ and $b = a^2/4$.

SECTION 1.4

Functions and Graphs

In nature, we observe quantities that change. Some quantities change with the passage of time, while others change in response to changes in yet other quantities. In order to describe the relationship between varying quantities, the notion of *function* was invented. If quantities are satisfactorily related by a function and properties of the function are understood, then predictions about the quantities can be made.

A major purpose of calculus is to study the properties of functions. We will review in this section the definition, terminology, and notation of functions and their graphs. Some of the ways functions apply to the real world will also be mentioned.

The Definition of Function

We begin with the definition of a function as a mathematical object.

DEFINITION 1.1

Let X be a set of real numbers. A **function** f with **domain** X is a rule that assigns to each number $x \in X$ exactly one real number $f(x)$. The number $f(x)$ assigned to x by f is called the **image** of x under f, or the **value** of f at x. We read $f(x)$ as "f of x." The set of all images of elements of X (that is, the set of all values of f) is called the **range** of f.

Notice that we have restricted the domain of a function to being a set of real numbers. While this is an artificial restriction from a mathematical point of view, it is the only kind of function with which we will be concerned for the time being. The domain of a function is the set of real numbers for which the function is defined, or for which values of the function are given.

Notice also that according to Definition 1.1, the values of a function must be real numbers. That is, Definition 1.1 defines what is technically called a *real-valued function of a single real variable.*

Historical Note on Functions

The above definition of function is the result of a long process of generalization. Descartes's concept was that a function was a polynomial. This idea was generalized to include power series, which may be regarded as polynomials of infinite degree (see Section 9.5), about 1650. The next concept was that a function was a formula, an expression involving the variable x and sums, products, limits, and previously defined functions. The same formula had to be used for all values of x, however. This sufficed until the work of d'Alembert and Euler on the vibrating string (about 1748), when it became necessary to divide the domain of the function into different parts and use a different formula for each part. In 1755, Euler gave our definition, but no use was made of this general concept until the 1820's. Lobachevski (1834) and Dirichlet (1837) added the requirement of continuity, but Hankel returned to our definition in 1870.

Functions Described by Formulas

The simplest way to describe a function is by use of a formula. A formula tells explicitly how to find the image of a number by carrying out arithmetic operations. Consider the following examples.

Example 1.21 State in words the rule given by each of the following formulas:

(a) $f(x) = 2x + 1$ (b) $g(x) = \sqrt{2 - x}$

Solution

(a) The function f assigns to x the number 1 more than twice x. That is, to find the image of x, multiply x by 2 and add 1.
(b) To find the image of x under g, subtract x from 2 and find the square root of the difference. ■

Example 1.22 Write a formula for each of the following rules:

(a) The function f assigns to x the number 8 more than 3 times x.
(b) The function g assigns to x the square of x, diminished by 4.

Solution

(a) Since 8 more than 3 times x is $3x + 8$, a formula is $f(x) = 3x + 8$.
(b) The square of x, diminished by 4, is $x^2 - 4$, so a formula is $g(x) = x^2 - 4$. ■

Example 1.23 For each of the following functions, find $f(0)$, $f(-2)$, $f(2)$, $f(p)$, and $f(p^2)$:

(a) $f(x) = \dfrac{1}{x}$ (b) $f(x) = \sqrt{x}$

Solution

(a) For $f(x) = 1/x$, we see that $f(0)$ is not defined; that is, 0 is not in the domain of f. We have $f(-2) = -\frac{1}{2}$, $f(2) = \frac{1}{2}$, $f(p) = 1/p$, and $f(p^2) = 1/p^2$.
(b) For $f(x) = \sqrt{x}$, we have $f(0) = 0$, $f(2) = \sqrt{2}$, $f(p) = \sqrt{p}$, and $f(p^2) = |p|$. However, $f(-2) = \sqrt{-2}$ is not a real number, so -2 is not in the domain of f. In fact, p is in the domain of f only if p is nonnegative. ■

In these examples, the domains of the functions have not been specified. Whenever the domain of a function is not specified, we take as its domain the largest set of real numbers for which the function assigns images.

Example 1.24 Find the domains of the following functions:

(a) $f(x) = \dfrac{x + 1}{x - 2}$ (b) $g(x) = x^2 - 4$ (c) $F(x) = \sqrt{4 - x^2}$

Solution

(a) The only restriction on the domain of f is that x cannot equal 2, since division by 0 is undefined. The domain of f is therefore $(-\infty, 2) \cup (2, \infty)$.
(b) For g, there is no restriction on the values of x, since $x^2 - 4$ is a real number no matter what real number x is. Thus, the domain of g is all of $\mathbb{R}$.

(c) In the formula for F, the radicand must be nonnegative. That is, we must have

$$\begin{aligned} 4 - x^2 &\geq 0, \\ x^2 &\leq 4, \\ |x| &\leq 2, \\ -2 \leq x &\leq 2. \end{aligned}$$

Thus, the domain of F is $[-2, 2]$. ■

Finding the range of a function is usually less important and often more difficult than finding the domain. We will consider finding ranges in Section 1.5.

Variation

If f is a function with domain X, the equation $y = f(x)$ defines y as a function of x. We call x an **independent variable**, because x takes on values from X independently. We call y a **dependent variable**, because the values of y depend on the values of x. The equation $y = f(x)$ is said to define a **functional variation** between x and y. Several kinds of functional variation come up in applications, so we will mention them here.

Direct variation is defined by an equation of the form

$$y = kx,$$

where k is a constant. In this equation, y is said to **vary directly** as x, or to be **directly proportional** to x.

Inverse variation is defined by an equation of the form

$$y = \frac{k}{x},$$

where k is a constant. Here, y is said to **vary inversely** as x, or to be **inversely proportional** to x.

Example 1.25 The distance traveled at a constant speed is proportional to the time spent traveling. If a distance of 5 miles is traveled in $\frac{1}{2}$ hour, express the distance traveled at that speed as a function of time.

Solution Since distance d varies directly as time t, they are related by an equation of the form

$$d = kt,$$

where k is a constant. If $d = 5$ when $t = \frac{1}{2}$, then $5 = k(\frac{1}{2})$, or $k = 10$, so that

$$d = 10t.$$

This formula expresses d as a function of t. ■

Functions as Models

The last example indicates one way in which functions can be applied. Once a functional relationship between d and t is known, we can find either one when a value for the other is given. Thus, for instance, we can determine how long it would take to travel 1000 miles at the rate of 5 miles per $\frac{1}{2}$ hour, without actually doing it. This type of application is so commonplace that we take it for granted.

The same principles apply in more complicated cases. Consider the following experiment: A heavy object is dropped from various heights, and the time it takes to reach the ground is recorded. For example, if the object is dropped from a height of $s = 100$ feet, it takes 2.50 seconds to reach the ground. The results of the experiment are given in Table 1.4.

The objective now is to determine the nature of the functional relationship between s and t. That there is a functional relationship seems obvious; the distance the object falls naturally depends on the time it spends falling. It is easy to see from Table 1.4 that s increases as t increases, so the variation is not inverse. It is not direct, though, for s increases regularly while t increases by smaller and smaller amounts. After some further guessing and figuring, perhaps, we decide to determine whether s is proportional to the square of t. We extend Table 1.4 to get Table 1.5.

TABLE 1.4

Distance (ft) s	Time (sec) t
25	1.25
50	1.77
75	2.16
100	2.50
125	2.80
150	3.06
175	3.31
200	3.54

TABLE 1.5

s	t	t^2	s/t^2
25	1.25	1.56	16.03
50	1.77	3.13	15.97
75	2.16	4.67	16.06
100	2.50	6.25	16.00
125	2.80	7.84	15.94
150	3.06	9.36	16.03
175	3.31	10.96	15.97
200	3.54	12.53	15.96

From the last column of Table 1.5, we have approximately

$$\frac{s}{t^2} = 16,$$

or

$$s = 16t^2.$$

This formula gives s as a function of t: s is proportional to the square of t.

This function is said to **model** the situation of a heavy object falling through relatively short distances near the earth's surface. If this model, or formula, is verified by all our observations, we say the model **fits** well. A model that fits well may become known as a **law**.

Galileo Galilei (1564–1642) was the son of a Florentine mathematician who kept him in ignorance of mathematics for fear that the boy would abandon the study of medicine if he were exposed to mathematics. His fears were justified, for when Galileo accidentally overheard a lesson in geometry, his scientific interests immediately awakened.

Galileo became a scientist and mathematician, and went on to contribute to virtually the entire range of natural science. His most important theoretical work consisted of expressing natural relationships as relationships between numbers; we would now call such relationships *functions* or *mathematical models*. For example, he was able to describe the position of a falling body at any time. Such relationships suggested that nature obeys definite mathematical laws, and paved the way for Newton's work.

Galileo was convinced on mathematical grounds that the new heliocentric theory of Copernicus was correct, but not until he looked through a telescope that he himself had built and saw the moons of Jupiter, did he become enthusiastic about supporting it. His enthusiasm eventually put him in trouble with religious authorities, and he spent the last 8 years of his life under a loose form of house arrest.

A model such as $s = 16t^2$ is not intended to fit in a wide variety of circumstances. We would not expect this formula to describe what happens if air resistance has a sizeable effect, or if we were on the moon instead of the earth. Thus, scientific laws apply only to the kinds of situations for which they were formulated.

The formulation of models to describe natural phenomena is the basic activity of science. We are indebted to Galileo for establishing the scientific method. By distrusting qualitative descriptions of physical phenomena, and requiring quantitative descriptions before searching for causes, he cast the mold for modern scientific research.

Many functions and the formulas that describe them are commonly used as models. The use of models to describe or predict physical phenomena is called the *application of mathematics*. We will point out a number of applications as we proceed.

The Graph of a Function

It is usually possible to represent a function graphically. This is a great help in studying the properties of a function, because a "picture" of that function makes it much easier to understand. The geometric representation of a function is called its *graph*.

DEFINITION 1.2 The **graph** of a function f is the set of all points $(x, f(x))$ in the coordinate plane, with x in the domain of f.

Thus, if the function f has domain X, then

$$\text{Graph of } f = \{(x, y) | y = f(x) \text{ and } x \in X\}.$$

Normally, the graph of a function is some kind of curve in the plane. Since it contains infinitely many points in most cases, we cannot plot them all. We select a representative sample of points and plot them so we can see what the graph must look like. Actually, there may be no simple way to tell what the graph is like, based on only a few points. One purpose of the calculus is to enable us to describe the graph of a function without plotting a multitude of points. Lacking that knowledge at first, we tend to plot lots of points in sketching a graph. (A calculator may be of considerable help.)

Consider a simple example.

Example 1.26 Sketch the graph of $f(x) = x + 2$.

Solution We start out by selecting several values for x and computing the corresponding values $f(x)$. Then we plot the points $(x, f(x))$ obtained. Table 1.6 shows some possible choices. The points listed in the table are plotted in Figure 1.28. These points appear to lie along a straight line, and this guess is reinforced if we plot more points. The line is shown in Figure 1.29.

TABLE 1.6

x	$f(x)$	Point
-3	-1	$(-3, -1)$
-2	0	$(-2, 0)$
-1	1	$(-1, 1)$
0	2	$(0, 2)$
1	3	$(1, 3)$
2	4	$(2, 4)$

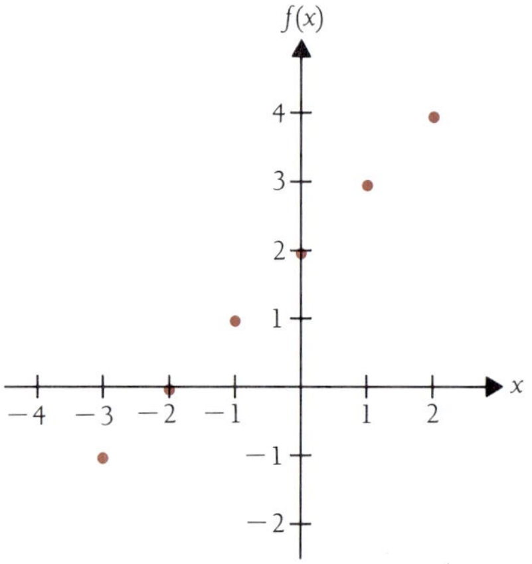

FIGURE 1.28 Plotting points on a graph

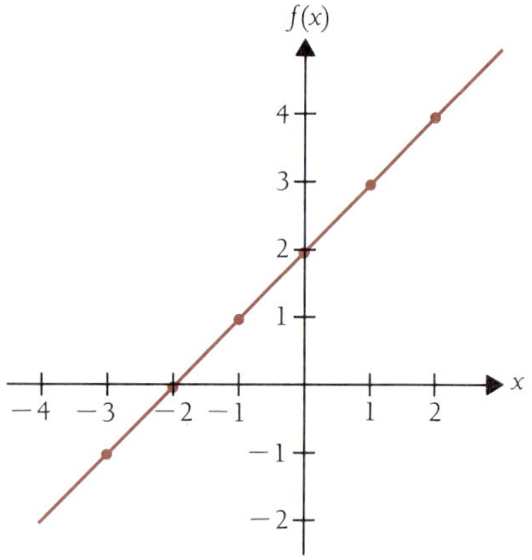

FIGURE 1.29 Joining the points with a curve ■

■ **Example 1.27** Show that the graph of $f(x) = x + 2$ really is the line shown in Figure 1.29.

Solution The graph of f is the set of points

$$\{(x, y) | y = x + 2\}.$$

Since $y = x + 2$ is the equation of a line with slope 1 and intercept 2, the points of the graph do indeed lie on the line in Figure 1.29. Having noticed in advance that the graph of f would be a line, two points would have been a sufficient sample to plot. ■

Some Types of Functions

Many functions of certain types come up repeatedly in our work. Here we note some of the more common ones and give some specific examples and their graphs.

DEFINITION 1.3 A **constant function** is defined by a formula of the form

$$f(x) = c, \qquad c \text{ a constant.}$$

■ **Example 1.28** Sketch the graph of the function $f(x) = 3$.

Solution Every point on the graph of this constant function is of the form $(x, 3)$. The graph is therefore the line $y = 3$ shown in Figure 1.30.

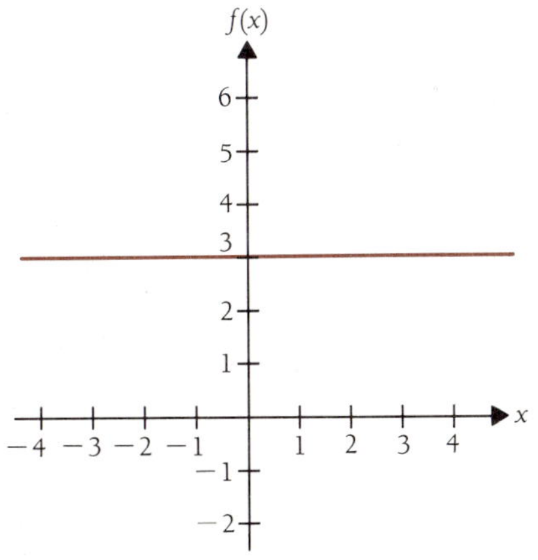

FIGURE 1.30 The graph of the constant function $f(x) = 3$ ■

The graph of a constant function is always a horizontal line.

DEFINITION 1.4

A **linear function** is defined by a formula of the form

$$f(x) = ax + b,$$

with a and b constants, and $a \neq 0$.

Example 1.29 Sketch the graph of the function $f(x) = 2x - 3$.

Solution The graph of this linear function is the line $y = 2x - 3$ shown in Figure 1.31.

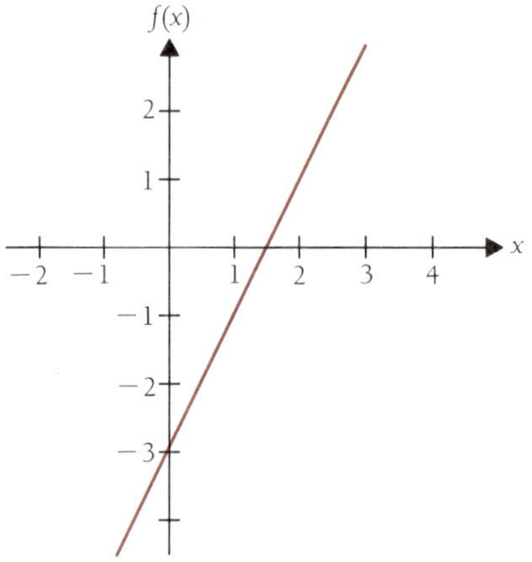

FIGURE 1.31 Graph of the linear function $f(x) = 2x - 3$

The graph of a linear function is always an oblique line.

DEFINITION 1.5

A **quadratic function** is defined by a formula of the form

$$f(x) = ax^2 + bx + c,$$

with a, b, and c constants, and $a \neq 0$.

Example 1.30 Sketch the graph of the function $f(x) = x^2 + 2x - 3$.

Solution We construct points on the graph of this quadratic function as shown in Table 1.7,

TABLE 1.7

x	$f(x)$	Point
−4	5	(−4, 5)
−3	0	(−3, 0)
−2	−3	(−2, −3)
−1	−4	(−1, −4)
0	−3	(0, −3)
1	0	(1, 0)
2	5	(2, 5)

and plot them in the coordinate plane. Joining the points with a smooth curve, we obtain the graph in Figure 1.32.

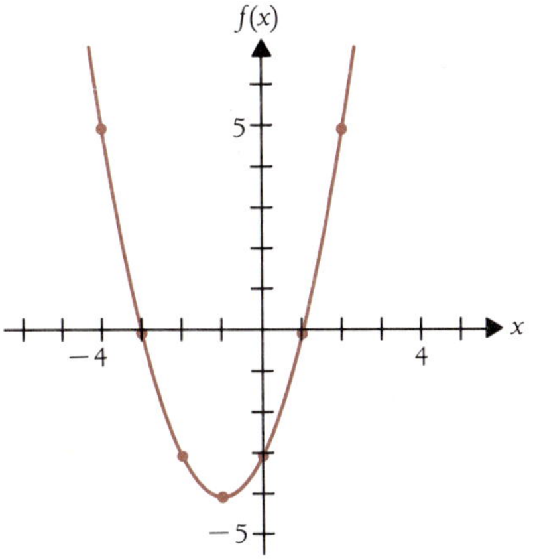

FIGURE 1.32 Graph of the quadratic function $f(x) = x^2 + 2x - 3$

The graph of a quadratic function is always a parabola.

DEFINITION 1.6

A **polynomial function** is defined by a formula of the form

$$f(x) = a_n x^n + a_{n-1} x^{n-1} + \cdots + a_1 x + a_0,$$

with n a nonnegative integer and $a_0, a_1, \ldots, a_n$ constants.

Constant, linear, and quadratic functions are examples of polynomial functions. So is a **power function**, defined by a formula of the form

$$f(x) = x^n,$$

with n a positive integer. The **degree** of a polynomial is the largest exponent that occurs in its formula.

Example 1.31 Sketch the graph of $f(x) = x^3$.

Solution A few points on the graph of this power function of degree 3 are (0, 0), (1, 1), (2, 8), $(\frac{1}{2}, \frac{1}{8})$, (−1, −1), (−2, −8), and $(-\frac{1}{2}, -\frac{1}{8})$. Plotting these points, and joining them with a curve, we get the graph shown in Figure 1.33 (at the top of the next page).

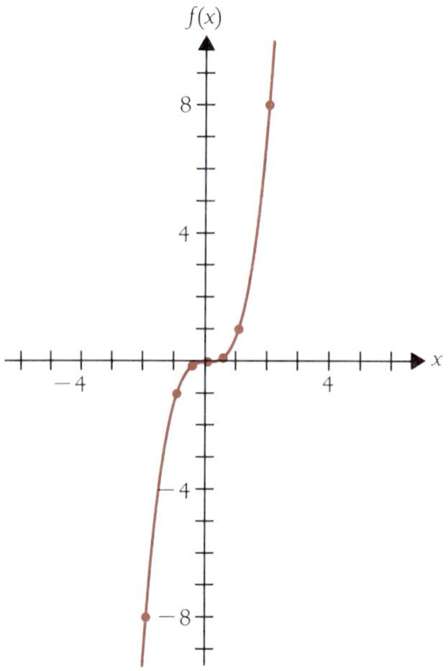

FIGURE 1.33 Graph of the power function $f(x) = x^3$ ■

DEFINITION 1.7

A **rational function** is the ratio of two polynomial functions. That is, if p and q are polynomial functions, then the function

$$f(x) = \frac{p(x)}{q(x)}$$

is a rational function.

■ **Example 1.32** Sketch the graph of $f(x) = 1/x$.

Solution First, note that the domain of this rational function does not include 0; this means that its graph cannot touch the y-axis. Plotting a few points, such as $(1, 1)$, $(2, \frac{1}{2})$, $(\frac{1}{2}, 2)$, $(-1, -1)$, $(-2, -\frac{1}{2})$, $(-\frac{1}{2}, -2)$, and filling in, we get the graph in Figure 1.34. Because $|1/x|$ gets smaller as $|x|$ gets larger, the graph gets close to the x-axis but does not touch it. The same is true of the y-axis. ■

The graphs of rational functions often have vertical asymptotes, like the y-axis in Figure 1.34, and horizontal asymptotes, like the x-axis in Figure 1.34. Loosely speaking, an *asymptote* is a line that the graph gets arbitrarily close to as the graph recedes from the origin.

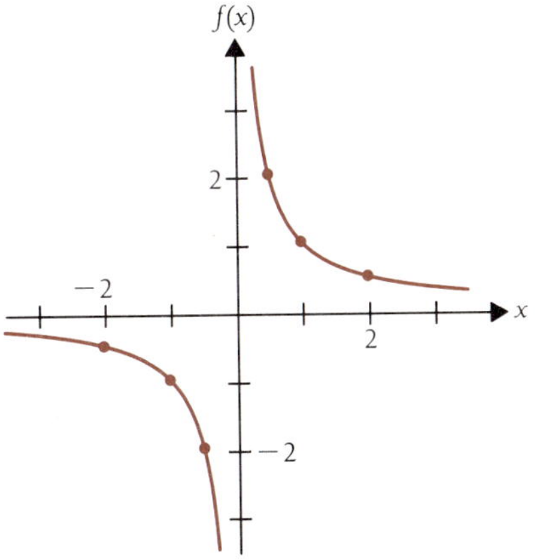

FIGURE 1.34 The graph of $f(x) = \frac{1}{x}$

DEFINITION 1.8

A **root function** is defined by a formula of the form

$$f(x) = x^{1/n} = \sqrt[n]{x},$$

with n a positive integer. The domain of a root function is the set of all real numbers if n is odd and the set of nonnegative real numbers if n is even.

■ **Example 1.33** Sketch the graph of the function $f(x) = x^{1/2}$.

Solution This is, of course, the square root function, $f(x) = \sqrt{x}$. Since 2 is even, the domain of this function is $[0, \infty)$. Plotting a few points, such as $(0, 0)$, $(1, 1)$, $(2, \sqrt{2})$, $(4, 2)$, etc., we get the graph in Figure 1.35.

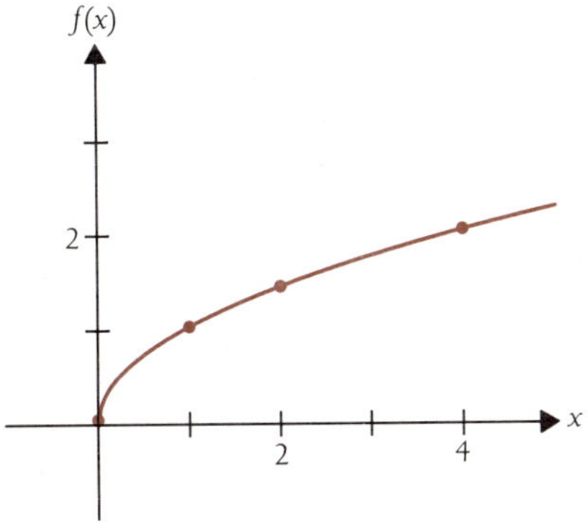

FIGURE 1.35 The graph of $f(x) = \sqrt{x}$ ■

The next two examples show two other types of functions.

■ **Example 1.34** Sketch the graph of the **absolute value function**, $f(x) = |x|$.

Solution Recalling the definition of absolute value, we can rewrite the formula for f as

$$f(x) = \begin{cases} x & \text{if } x \geq 0 \\ -x & \text{if } x < 0. \end{cases}$$

The graph of f is therefore the union of the two rays,

$$y = x, \quad x \geq 0, \qquad \text{and} \qquad y = -x, \quad x < 0,$$

shown in Figure 1.36.

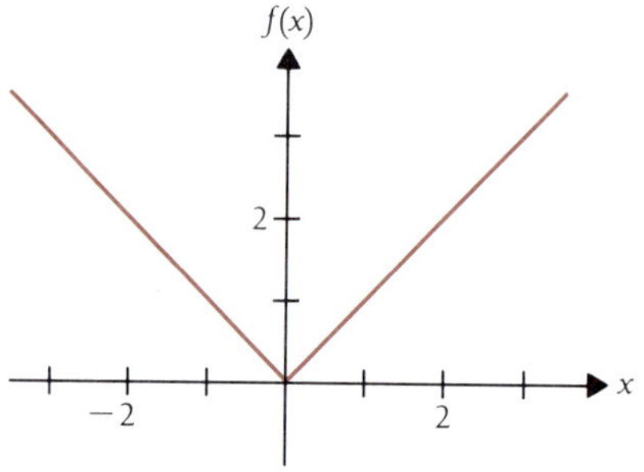

FIGURE 1.36 The graph of $f(x) = |x|$ ■

■ **Example 1.35** The **greatest integer in** $\mathbf{x}$ is defined to be the greatest integer not exceeding x, and is denoted by $\lfloor x \rfloor$. Sketch the graph of the **greatest integer function,** $f(x) = \lfloor x \rfloor$.

Solution First, note that if x is an integer, then $\lfloor x \rfloor = x$. Then, observe that if n is an integer and $n < x < n + 1$, then $\lfloor x \rfloor = n$. These considerations lead to the graph shown in Figure 1.37.

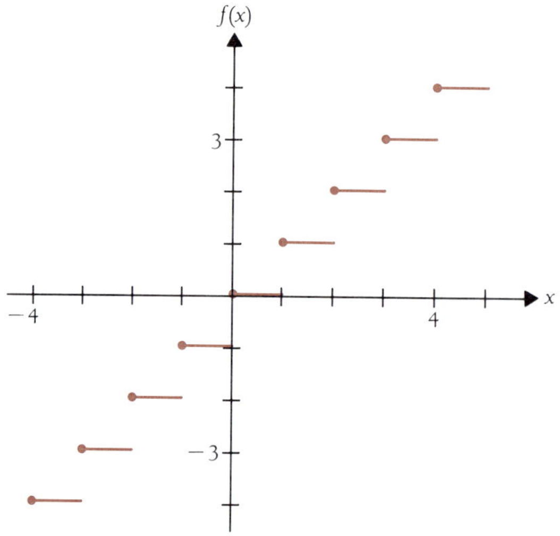

FIGURE 1.37 The graph of $f(x) = \lfloor x \rfloor$ ■

The greatest integer function is an example of a **step function**, whose graph consists of horizontal rays and/or line segments.

We will meet many other types of functions as we proceed.

The Vertical Line Test

Not just any curve is the graph of a function. Since each number x in the domain of a function f has just one image, no two points on the graph of f can have the same first coordinate. This means that no vertical line can meet the graph of f more than once. This fact is called the **vertical line test.**

By the vertical line test, we see, for example, that Figure 1.38 shows the graph of a function, but Figure 1.39 shows a curve that is not the graph of a function.

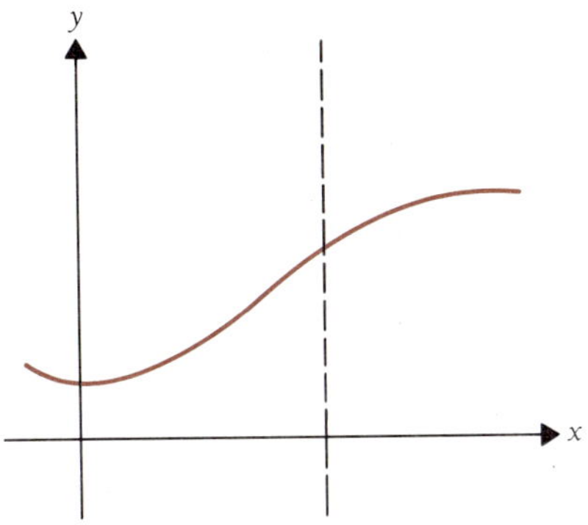

FIGURE 1.38 The graph of a function

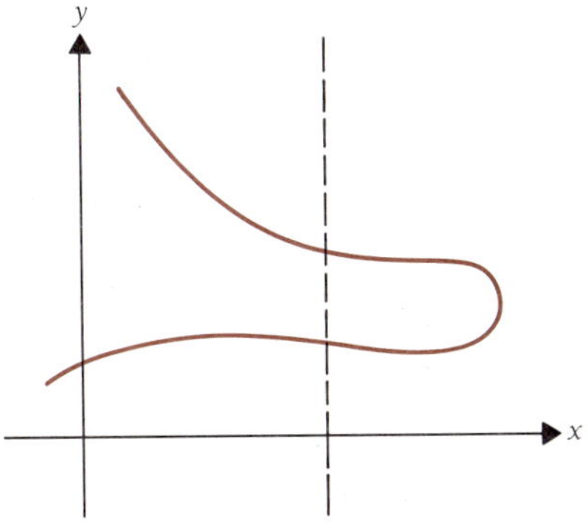

FIGURE 1.39 Not the graph of a function

The Graph of an Equation

Let $F(x, y)$ be an expression involving the variables x and y. Then

$$F(x, y) = 0$$

is an equation in x and y. The **graph** of this equation is the set of points (x, y) in the plane satisfying the equation. The graph of an equation is usually a curve of some kind.

For example, we saw in Section 1.3 that the equation

$$x^2 + y^2 - 1 = 0$$

has as its graph the unit circle (Figure 1.40, page 48).

The equation

$$xy = 0$$

is satisfied by the points (x, y) such that either $x = 0$ (the y-axis) or $y = 0$ (the x-axis). Thus, the graph of $xy = 0$ is the pair of coordinate axes (Figure 1.41).

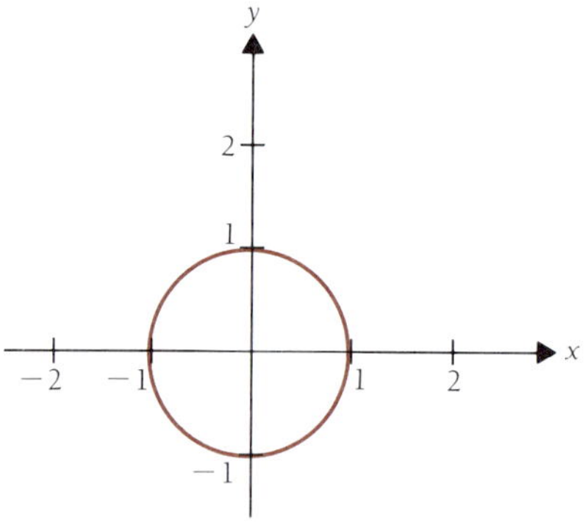

FIGURE 1.40 The unit circle, $x^2 + y^2 = 1$

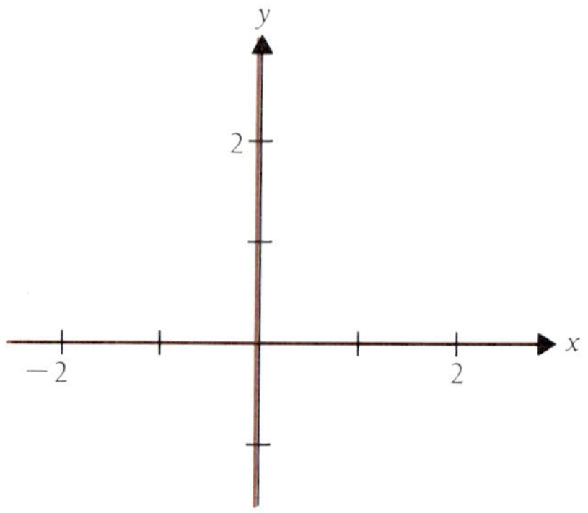

FIGURE 1.41 The coordinate axes, $xy = 0$

A curve is said to be **symmetric about the *y*-axis** in case the point (x, y) is on the curve if and only if the point $(-x, y)$ is on the curve (see Figure 1.42). This means the y-axis is an axis of symmetry of the curve. A curve is **symmetric about the *x*-axis** in case (x, y) is on the curve if and only if $(x, -y)$ is on the curve (see Figure 1.43). This means the x-axis is an axis of symmetry. A curve is **symmetric about the origin** in case (x, y) is on the curve if and only if $(-x, -y)$ is on the curve (see Figure 1.44).

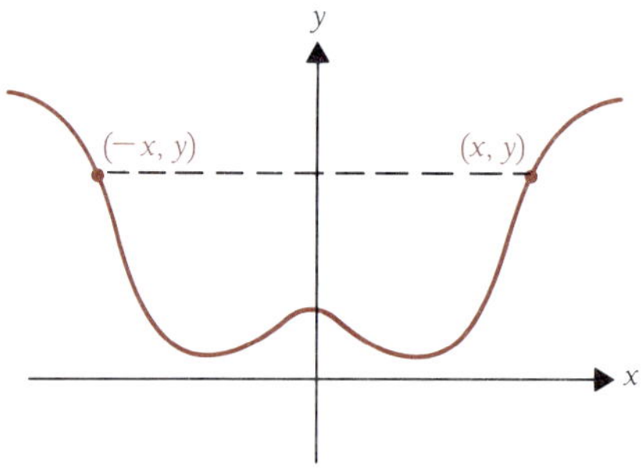

FIGURE 1.42 Symmetry about the y-axis

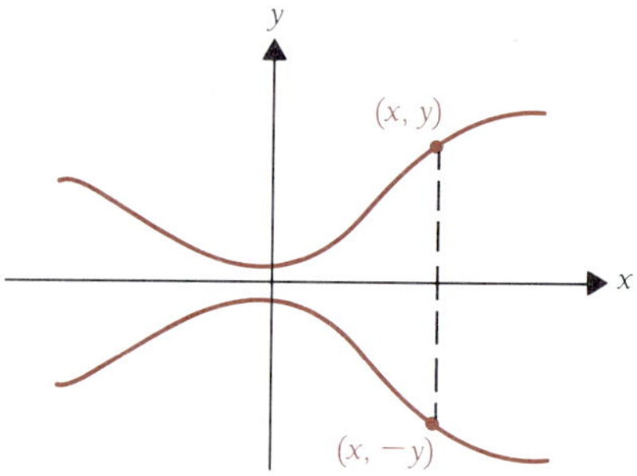

FIGURE 1.43 Symmetry about the x-axis

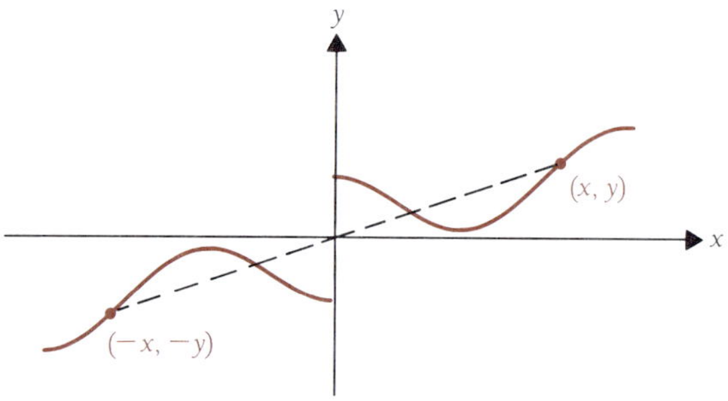

FIGURE 1.44 Symmetry about the origin

Example 1.36 Determine the symmetry of the graph of the equation $x^2 - y^3 - 1 = 0$.

Solution Let point (x, y) be on the curve. The point $(-x, y)$ is also on the curve, because $(-x)^2 - y^3 - 1 = x^2 - y^3 - 1 = 0$, so there is symmetry about the y-axis. However, neither of the points $(x, -y)$ and $(-x, -y)$ is on the curve, as you may check, so we have no other symmetry. ■

Finally, we wish to say a word about the effect on the graph of a function of additional constants. If a constant is added to the functional value, it moves the graph of the function up or down (see Figure 1.45). If a constant is added to the independent variable, it moves the graph of the function to the left or right (see Figure 1.46). If the functional value is multiplied by a constant, the shape or orientation of the graph is changed (see Figure 1.47). These ideas will be used repeatedly in our work.

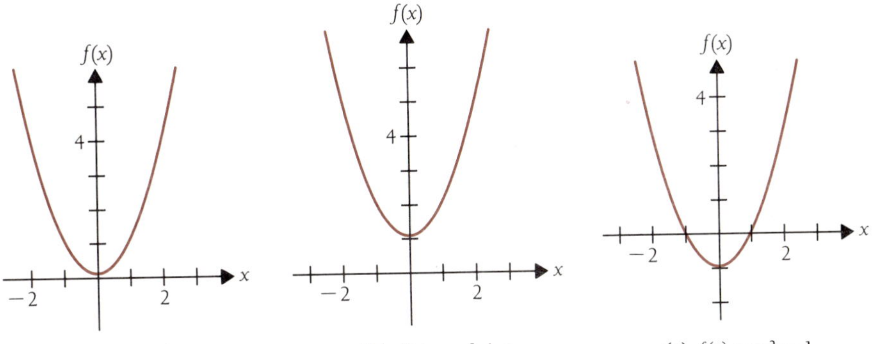

(a) $f(x) = x^2$ (b) $f(x) = x^2 + 1$ (c) $f(x) = x^2 - 1$

FIGURE 1.45 Moving a graph up or down

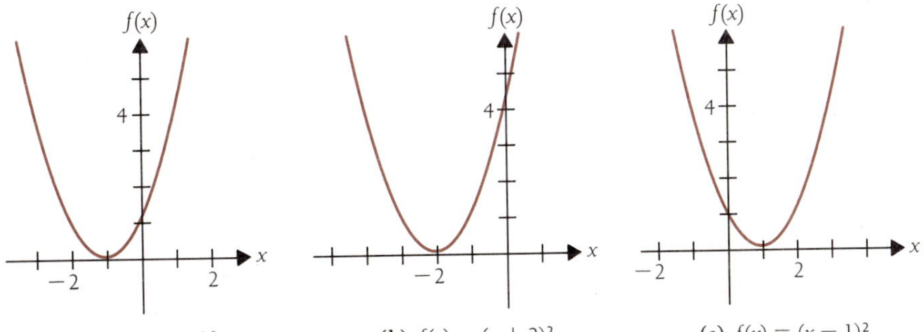

(a) $f(x) = (x + 1)^2$ (b) $f(x) = (x + 2)^2$ (c) $f(x) = (x - 1)^2$

FIGURE 1.46 Moving a graph left or right

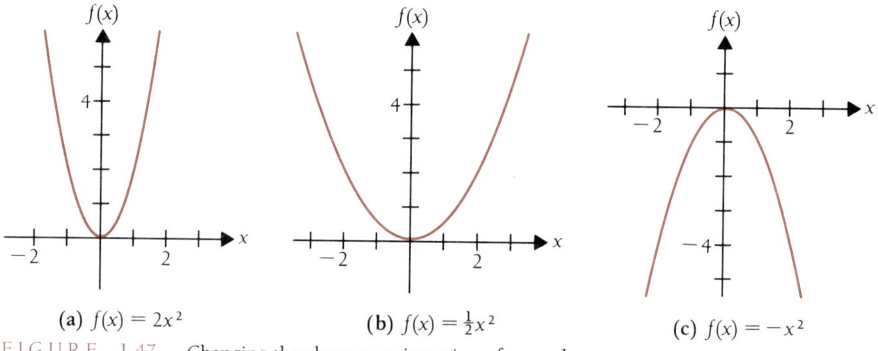

FIGURE 1.47 Changing the shape or orientation of a graph

Exercises 1.4

State in words the rule defined by each of the following functions:

1. $f(x) = x + 10$
2. $F(x) = \sqrt{x} - 4$
3. $g(x) = \dfrac{1}{x-1}$
4. $f(x) = x^2 - \dfrac{1}{x}$

Give a formula for each of the following rules:

5. The image of x under f is 1 more than twice x.
6. The image of x under g is 1 more than x, divided by 1 less than x.
7. $f(x)$ is the absolute value of 1 less than 3 times x.
8. $g(x)$ is the reciprocal of the greatest integer in x.

For $f(x) = 4x - 1$, find the following:

9. $f(1)$
10. $f(0)$
11. $f(-\frac{1}{3})$
12. $f(b)$
13. $f(2b + 1)$
14. $f(f(2))$

For $g(x) = 2x^2 - x + 8$, find the following:

15. $g(-1)$
16. $g(0)$
17. $g(\frac{1}{2})$
18. $g(h)$
19. $g(x + h)$
20. $g(x + h) - g(x)$

Find the domain of each function.

21. $f(x) = 5x + 2$
22. $f(x) = x$
23. $g(x) = 3$
24. $f(x) = \dfrac{1}{x^2}$
25. $g(x) = \dfrac{4}{x-3}$
26. $h(x) = \dfrac{2}{x(x+1)(x-7)}$
27. $f(x) = \sqrt{x-3}$
28. $h(x) = \sqrt{4x+7}$
29. $f(x) = \sqrt{\dfrac{4}{x-3}}$
30. $f(x) = \dfrac{2}{3 - \sqrt{x}}$
31. $f(x) = \sqrt{x^2 - 4}$
32. $H(x) = \sqrt{2 + 3x^2}$

Give formulas expressing the following functional relationships:

33. The force F applied to a spring is proportional to the compression x of the spring.
34. The height h of a spruce tree varies directly as its age a.

35. The rate r at which a battery charges is inversely proportional to the time t it has been charging.

36. The gravitational force g of the earth on an object varies inversely as the square of the distance r of the object from the center of the earth.

37. The drag D on an aircraft is proportional to the square of the velocity v.

38. The temperature T of a gas varies directly as the product of the pressure P and volume V. (We say T varies **jointly** as P and V.)

39. If an object weighs 100 pounds at the earth's surface (3950 miles from the center of the earth), use the formula from Problem 36 to determine how much it would weigh 100 miles above the earth's surface.

40. If a force of 12 pounds compresses a spring 2 inches, use the formula from Problem 33 to find out how much force is required to compress it 5 inches.

41. The period of a pendulum is the time required for it to swing from one extreme to the other and back again. For a pendulum of length l feet, the period p in seconds is given in the table. Find a relationship between l and p.

Length (ft) l	Period (sec) p
1	0.90
2	1.27
3	1.56
4	1.80
5	2.01
6	2.21

42. The illuminance E (in lumens per square meter) of a surface a distance d meters from a certain light source is given in the table. Find a relationship between E and d.

Distance (m) d	Illuminance (lumen/m²) E
0.25	2400
0.50	600
0.75	267
1.00	150
1.25	96
1.50	67

Sketch the graph of each function.

43. $f(x) = x + 1$

44. $f(x) = 3x - 4$

45. $f(x) = x^2, \quad x \in [-1, 3]$

46. $f(x) = 2x - 1, \quad x \in [0, 2]$

47. $f(x) = \dfrac{2}{x}$

48. $f(x) = \dfrac{2}{x+3}$

49. $f(x) = \begin{cases} 5 - x & \text{if } x < 5 \\ (5 - x)^2 & \text{if } x > 5 \end{cases}$

50. $f(x) = \begin{cases} 1 - x & \text{if } x < 0 \\ x^2 & \text{if } x \geq 0 \end{cases}$

51. $f(x) = \dfrac{x}{|x|}$ 52. $f(x) = x + |x|$ 53. $f(x) = |x - 1|$

54. $f(x) = |x| - 1$ 55. $f(x) = 2 + |x|$ 56. $f(x) = |2 + x|$

57. $f(x) = |2x|$ 58. $f(x) = x - |x|$ 59. $f(x) = x^{1/5}$

60. $f(x) = 1 - x^{1/3}$

Plot the graphs of the following sets of functions on the same set of axes:

61. $f(x) = x^2, \quad f(x) = -x^2, \quad f(x) = x^2 + 1, \quad f(x) = 2x^2$

62. $f(x) = x^2 - 4x, \quad f(x) = -x^2 + 4x$

63. $f(x) = x^3, \quad f(x) = -x^3, \quad f(x) = 1 - x^3, \quad f(x) = (1 - x)^3$

64. $f(x) = \sqrt{x + 1}, \quad f(x) = \sqrt{x} + 1$

The curve $y = f(x)$ is shown in the figure.

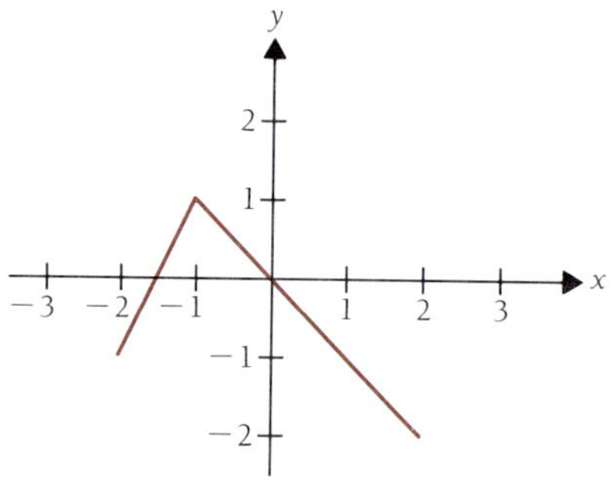

Sketch the following graphs:

65. $y = -f(x)$ 66. $y = |f(x)|$ 67. $y = 2 + f(x)$ 68. $y = \lfloor f(x) \rfloor$

The curve $y = f(x)$ is shown in the figure.

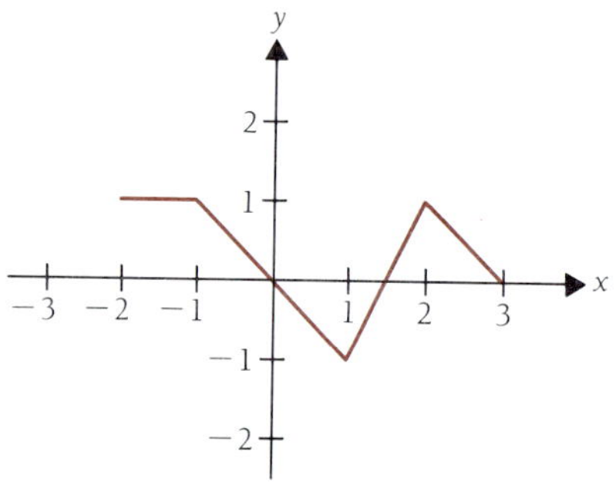

Sketch the following graphs:

69. $y = -f(x)$ 70. $y = |f(x)|$ 71. $y = 2 + f(x)$ 72. $y = \lfloor f(x) \rfloor$

Determine the symmetry of the graphs of the following equations:

73. $xy - 4 = 0$
74. $x^2 - y^2 = 0$
75. $xy^2 - 1 = 0$
76. $xy^3 - 8 = 0$
77. $x^2 + x^2y^4 - y^6 = 0$
78. $y - |x| = 0$
79. $y^2 + |x| = 0$
80. $x - \lfloor y^2 \rfloor = 0$
81. If the graph of the equation $F(x, y) = 0$ has symmetry about both axes, must it have symmetry about the origin?
82. If the graph of $F(x, y) = 0$ has symmetry about the origin and about one axis, must it have symmetry about the other axis also?
83. The greatest integer in x, $\lfloor x \rfloor$, is sometimes called the **floor** of x. The **ceiling** of x is the function $\lceil x \rceil$ defined as the least integer not less than x. Show that if n is an integer, then $\lceil n \rceil = n$, and if $n < x \leq n + 1$, then $\lceil x \rceil = n + 1$. Sketch the graph of $f(x) = \lceil x \rceil$.
84. The **truncation function** is defined as the function that truncates (cuts off) the decimal part of any real number and is denoted by trunc(x). For example, trunc(3.45) = 3 and trunc(−2.3) = −2. Show that

$$\operatorname{trunc}(x) = \begin{cases} \lfloor x \rfloor & \text{if } x \geq 0 \\ \lceil x \rceil & \text{if } x < 0. \end{cases}$$

SECTION 1.5

Combinations of Functions; Inverses

Functions can be combined in various ways to get other functions. The simplest of these operations use the arithmetic of the real number system.

Algebraic Combinations

DEFINITION 1.9

If f and g are functions, the **sum** $f + g$, the **difference** $f - g$, the **product** fg, and the **quotient** f/g are the functions defined by

$$(f + g)(x) = f(x) + g(x),$$
$$(f - g)(x) = f(x) - g(x),$$
$$(fg)(x) = f(x)g(x),$$
$$\left(\frac{f}{g}\right)(x) = \frac{f(x)}{g(x)}.$$

■ **Example 1.37**

Let $f(x) = 2x$ and $g(x) = x - 1$. Find formulas for $f + g$, $f - g$, fg, and f/g.

Solution

$$(f + g)(x) = 2x + (x - 1) = 3x - 1,$$
$$(f - g)(x) = 2x - (x - 1) = x + 1,$$
$$(fg)(x) = 2x(x - 1) = 2x^2 - 2x,$$
$$\left(\frac{f}{g}\right)(x) = \frac{2x}{x - 1}.$$

■

DEFINITION 1.10

If f is a function and c is a real number, the **scalar product** of c and f is the function defined by

$$(cf)(x) = c \cdot f(x).$$

Example 1.38 If $f(x) = 2x$ and $c = 3.5$, find a formula for cf.

Solution

$$(cf)(x) = (3.5)(2x) = 7x.$$

■

DEFINITION 1.11

If f and g are functions, a **linear combination** of f and g is any function of the form $af + bg$, with a and b real numbers.

It is easy to generalize Definition 1.11 to a linear combination of any number of functions.

Example 1.39 Express the function $h(x) = x + 4$ as a linear combination of the functions $f(x) = 2x$ and $g(x) = x - 1$.

Solution We seek real numbers a and b such that

$$h(x) = af(x) + bg(x),$$

or

$$\begin{aligned} x + 4 &= a(2x) + b(x - 1) \\ &= 2ax + bx - b \\ &= (2a + b)x - b. \end{aligned}$$

Thus, we must have

$$\begin{aligned} -b &= 4, \qquad \text{and} \qquad & 2a + b &= 1, \\ b &= -4, & a &= \tfrac{5}{2}. \end{aligned}$$

Thus,

$$h(x) = \tfrac{5}{2}f(x) - 4g(x).$$

■

Many functions are built up out of simpler functions by forming combinations. For example, a polynomial function is a linear combination of power functions, and a rational function is the quotient of two polynomial functions.

The sum, difference, and product of two functions are defined wherever both functions are defined. Thus, if functions f and g have domains X_f and X_g, respectively, then the functions $f + g$, $f - g$, and fg have domains $X_f \cap X_g$.

Almost the same thing is true for the quotient of two functions, but the quotient f/g will not be defined at any point x for which $g(x) = 0$. If g has domain X_g, the set

$$S_g = \{x | x \in X_g \text{ and } g(x) \neq 0\}$$

is called the **support** of g. Therefore, the domain of f/g is

$$X_f \cap S_g.$$

Function Composition

Functions also can be combined by first applying one function and then another.

DEFINITION 1.12 If f and g are functions, the **composition** $g \circ f$ of f and g is the function defined by

$$(g \circ f)(x) = g(f(x)).$$

(Note that g is written first, but applied last.)

■ **Example 1.40** Let $f(x) = 2x$ and $g(x) = x - 1$. Find formulas for $g \circ f$ and $f \circ g$.

Solution

$$(g \circ f)(x) = g(f(x)) = g(2x) = (2x) - 1 = 2x - 1,$$
$$(f \circ g)(x) = f(g(x)) = f(x - 1) = 2(x - 1) = 2x - 2.$$ ■

■ **Example 1.41** Repeat Example 1.40 for the functions $f(x) = \sin x$ and $g(x) = x^2$.

Solution

$$(g \circ f)(x) = g(f(x)) = g(\sin x) = (\sin x)^2 = \sin^2 x,$$
$$(f \circ g)(x) = f(g(x)) = f(x^2) = \sin(x^2).$$ ■

Note that $g \circ f$ is not the same as $f \circ g$ in Examples 1.40 and 1.41. This is the way it usually works out. That is, function composition is not commutative.

The composition $g \circ f$ is defined for all x in the domain of f such that $f(x)$ is in the domain of g. If S is any set, the set

$$f^{-1}(S) = \{x | x \in X_f \text{ and } f(x) \in S\}$$

is called the **inverse image** of S under f. Thus, the domain of $g \circ f$ is the inverse image of X_g under f, or $f^{-1}(X_g)$. (Note here that the "-1" in f^{-1} is not to be thought of as an exponent, but as part of the functional notation.)

In certain settings, it is necessary to recognize a function as the composition of other functions, and to find functions of which it is the composition. It is usually possible to do this in many ways. Here is a simple example.

■ **Example 1.42** Find functions f and g such that $(g \circ f)(x) = \sqrt{x - 3}$.

Solution We can describe $g \circ f$ in words as "subtract 3 and find the square root." Hence, $f(x) = x - 3$ ("subtract 3") and $g(x) = \sqrt{x}$ ("find the square root") are two functions having the desired composition. Another (though less natural) pair might be $f(x) = x + 1$ and $g(x) = \sqrt{x - 4}$. ■

It is easy to see that the sum, product, or composition of polynomial functions is again a polynomial function. In a similar way, any algebraic combination of rational functions is again a rational function. These, and any combinations and compositions of power and root functions, are examples of **algebraic functions**. Any combination or composition of algebraic functions is again an algebraic function.

A function that is not algebraic is called **transcendental**. Examples of transcendental functions are trigonometric and logarithmic functions.

One-to-One Functions

A function is a rule for mapping each element of the domain to a unique element of the range. In general, different elements of the domain may have the same image. Functions for which this is not the case have some special properties.

DEFINITION 1.13 A function f is **one-to-one** in case no two elements of the domain have the same image under f.

It follows from the definition that if f is one-to-one and $f(x_1) = f(x_2)$, then $x_1 = x_2$. Equivalently, for each y in the range of f, there is exactly one x in the domain of f such that $f(x) = y$.

■ **Example 1.43** Determine whether each function is one-to-one.

(a) $f(x) = x^5$ (b) $f(x) = x^2 - 1$

Solution

(a) Since distinct numbers have distinct fifth powers, this function is one-to-one. To argue it symbolically, suppose $x_1^5 = x_2^5$. Then $(x_1/x_2)^5 = 1$. Now, if $x_1/x_2 > 1$, then $(x_1/x_2)^5 > 1$; similarly, if $x_1/x_2 < 1$, then $(x_1/x_2)^5 < 1$. Thus, we must have $x_1/x_2 = 1$, so that $x_1 = x_2$.

(b) This function is not one-to-one, since, for example, $f(-3) = 8 = f(3)$, and two different numbers have the same image. ■

A simple graphic test may be used to determine whether a function is one-to-one. If a horizontal line cannot meet the graph of f more than once, then f is one-to-one. Conversely, if some horizontal line meets the graph of f twice, say at points (x_1, y) and (x_2, y), then the two numbers x_1 and x_2 in the domain have the same image y, and f is not one-to-one. This test is called the **horizontal line test**. The horizontal line test shows that the graph in Figure 1.48 is that of a one-to-one function, but the graph in Figure 1.49 represents a function that is not one-to-one.

The Inverse of a Function

If f is a one-to-one function with domain X and range Y, then there exists a function g with domain Y and range X such that for all $x \in X$, $g \circ f(x) = x$, and for all $y \in Y$,

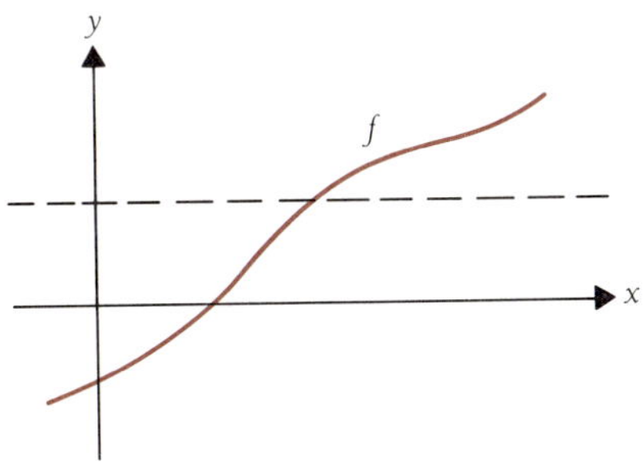

FIGURE 1.48 A one-to-one function

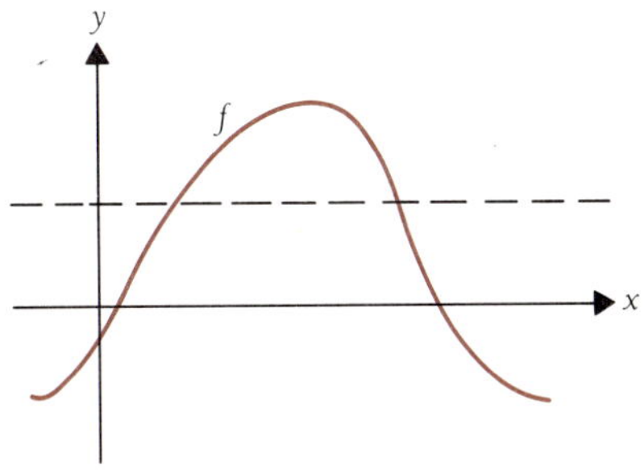

FIGURE 1.49 A function that is not one-to-one

$f \circ g(y) = y$. To construct g, we note that if $x \in X$, then $f(x) = y \in Y$. Moreover, no other element of X has image y under f, because f is one-to-one. Hence, if we define $g(y) = x$, then g will be a function. It then follows that for $x \in X$,

$$g \circ f(x) = g(f(x)) = g(y) = x,$$

and for $y \in Y$,

$$f \circ g(y) = f(g(y)) = f(x) = y.$$

The function g we have constructed is called an **inverse** of f, and it can be shown to be unique. The inverse of f is usually denoted by f^{-1}, in which the "-1" is not to be regarded as an exponent, but as a symbol for the inverse. We have the following formal definition.

DEFINITION 1.14 If f is a function, the **inverse** of f is the function f^{-1} such that for each x in the domain of f,

$$f^{-1} \circ f(x) = x, \tag{1}$$

and for each x in the domain of f^{-1},

$$f \circ f^{-1}(x) = x. \tag{2}$$

Our construction of g above is the proof of the following theorem.

THEOREM 1.1 A function f has an inverse function f^{-1} if and only if f is one-to-one.

The identity (2) gives us a technique for finding the inverse of a function if it exists. If the inverse does not exist, that fact will also become apparent by the same technique. Consider some examples.

Example 1.44 Find the inverse of each function, if it exists.

(a) $f(x) = 3x + 2$ **(b)** $f(x) = x^3 - 4$
(c) $f(x) = x^2$ **(d)** $f(x) = x^2, \quad x \in [0, \infty)$

Solution **(a)** Using equation (2) and the definition of f, we have

$$\begin{aligned} x &= f \circ f^{-1}(x) \\ &= 3f^{-1}(x) + 2. \end{aligned}$$

Now solving for $f^{-1}(x)$, we get

$$f^{-1}(x) = \frac{x - 2}{3}.$$

(b) Proceeding as in part (a), we get

$$\begin{aligned} x &= f \circ f^{-1}(x) \\ &= [f^{-1}(x)]^3 - 4, \\ [f^{-1}(x)]^3 &= x + 4, \\ f^{-1}(x) &= (x + 4)^{1/3}. \end{aligned}$$

This function is the inverse of f.

(c) We note first of all that this function is not one-to-one, so it will have no inverse function. However, if we proceed as in parts (a) and (b), we get

$$\begin{aligned} x &= f \circ f^{-1}(x) \\ &= [f^{-1}(x)]^2. \end{aligned}$$

We now see that this equation does not define f^{-1} as a function of x, since both $\sqrt{x}$ and $-\sqrt{x}$ satisfy the equation if substituted for $f^{-1}(x)$. Thus, f^{-1} is not a function, and f has no inverse function.

(d) This function *is* one-to-one, because distinct numbers have distinct squares in the specified domain. Using equation (2), we get

$$x = [f^{-1}(x)]^2$$

as in part (c), but this time we can solve the equation uniquely for $f^{-1}(x)$, because the range of f^{-1}, which is the domain of f, has no negative numbers in it. We thus get the function

$$f^{-1}(x) = \sqrt{x}, \qquad x \in [0, \infty).$$

This is the inverse of f. ■

Parts (c) and (d) of Example 1.44 illustrate a principle that will be of importance later on. If a function is not one-to-one, we usually can find some portion of the domain over which the function *is* one-to-one. By restricting the function to that part of the domain, an inverse function can be found. This technique is used to define the inverse trigonometric functions, for example.

It is helpful to see how the graph of a function f and the graph of its inverse f^{-1} are related. Since the domain and range of f are the range and domain, respectively, of f^{-1}, we have essentially interchanged the roles of x and y in the function f to obtain f^{-1}. In other words, if f is defined by the equation

$$y = f(x),$$

then f^{-1} is defined by the equation

$$x = f(y),$$

which is equivalent to

$$y = f^{-1}(x).$$

Thus, the graph of f^{-1} is obtained from the graph of f by interchanging the coordinates of each point. Since interchanging the coordinates of a point reflects the point across the line $y = x$, the graph of f^{-1} is just the reflection of the graph of f across the line $y = x$. This is illustrated in Figures 1.50 and 1.51.

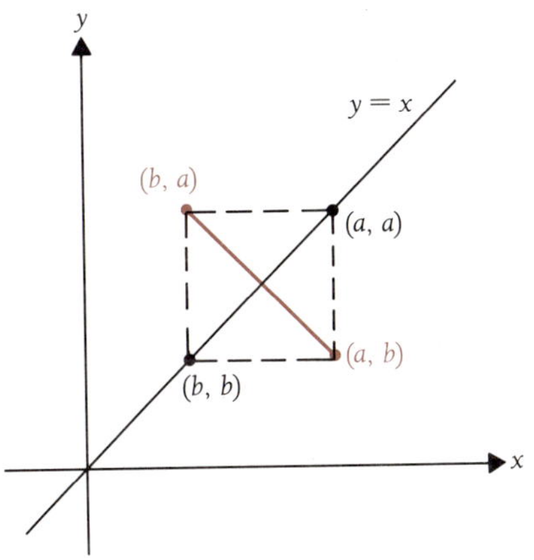

FIGURE 1.50 Interchanging coordinates

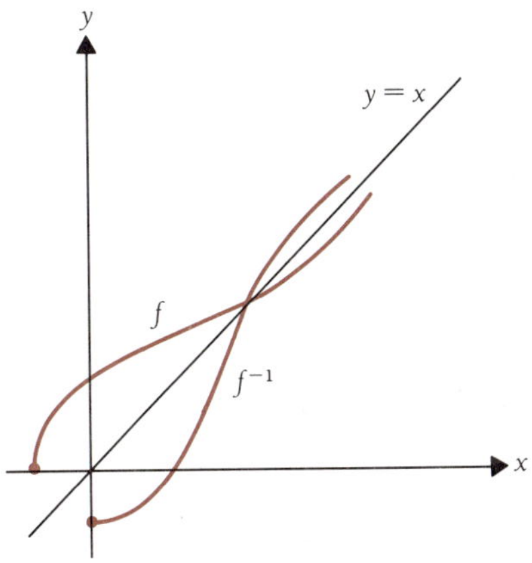

FIGURE 1.51 A function and its inverse

Finding the Range of a Function

The notion of inverse of a function also gives us a technique for finding the range of a function, based on the fact that the range of a function f is the domain of its inverse f^{-1}. Even when the "inverse" is not a function, the same technique will work, though care must be taken in some cases. Consider the following examples.

Example 1.45 Find the domains and ranges of the following functions:

(a) $f(x) = \dfrac{x+4}{x-3}$ (b) $f(x) = \dfrac{1}{\sqrt{x-3}}$ (c) $f(x) = \dfrac{1}{\sqrt{4-x^2}}$

Solution (a) The domain of this rational function is easily seen to be the set of all real numbers except 3. To find the range, we find the inverse of f. In order to simplify the process, we will write $y = f(x)$, interchange x and y [thus, expressing y as $f^{-1}(x)$], and solve for y:

$$y = \frac{x+4}{x-3},$$

$$x = \frac{y+4}{y-3},$$

$$xy - 3x = y + 4,$$

$$xy - y = 3x + 4,$$

$$(x-1)y = 3x + 4,$$

$$y = \frac{3x+4}{x-1} = f^{-1}(x).$$

It is readily seen that f^{-1} is defined for all real numbers except $x = 1$. Thus, the range of f, which is the domain of f^{-1}, is the set of all real numbers except 1.

(b) Since $x - 3$ must be positive for this function to be defined, we see that the domain of f is $(3, \infty)$. The range clearly consists of positive numbers. To be more specific, we find f^{-1}:

$$y = \frac{1}{\sqrt{x-3}},$$

$$x = \frac{1}{\sqrt{y-3}},$$

$$x\sqrt{y-3} = 1,$$

$$x^2(y-3) = 1,$$

$$y - 3 = \frac{1}{x^2},$$

$$y = 3 + \frac{1}{x^2} = f^{-1}(x).$$

Thus, we see that only $x = 0$ is not in the domain of f^{-1}, but since the range of f has only positive elements, the domain of f^{-1} is actually the set of positive real numbers. That is, the range of f is $(0, \infty)$.

(c) Since $4 - x^2$ must be positive, we find that the domain of f is $(-2, 2)$. We again note that the range consists only of positive numbers. Next, we find f^{-1}:

$$y = \frac{1}{\sqrt{4-x^2}},$$

$$x = \frac{1}{\sqrt{4-y^2}},$$

$$x^2(4 - y^2) = 1,$$
$$y^2 = 4 - \frac{1}{x^2},$$
$$y = \pm\sqrt{4 - \frac{1}{x^2}}.$$

Now we see that f^{-1} does not turn out to be a function, which is what we expected. However, we can still ask for what values of x the expression is defined. The radicand $4 - (1/x^2)$ must be nonnegative, or

$$4 - \frac{1}{x^2} \geq 0,$$
$$\frac{1}{x^2} \leq 4,$$
$$x^2 \geq \frac{1}{4},$$
$$|x| \geq \frac{1}{2},$$

so that

$$x \geq \frac{1}{2} \quad \text{or} \quad x \leq -\frac{1}{2}.$$

Since the range of f has only positive members, the range of f is therefore $[\frac{1}{2}, \infty)$. ■

Exercises 1.5

For each pair of functions, find $f + g$, $f - g$, fg, f/g, $g \circ f$, and $f \circ g$.

1. $f(x) = x + 3, \quad g(x) = x - 3$
2. $f(x) = x^2, \quad g(x) = x + 2$
3. $f(x) = x, \quad g(x) = \dfrac{1}{x}$
4. $f(x) = x - 4, \quad g(x) = |x|$
5. $f(x) = |x|, \quad g(x) = |x - 2|$
6. $f(x) = 1 + x^2, \quad g(x) = \sqrt{x}$
7. $f(x) = \lfloor x \rfloor, \quad g(x) = x^2$
8. $f(x) = 1 - x, \quad g(x) = \dfrac{1}{x^2 + 1}$

Find a formula for $h \circ g \circ f$.

9. $f(x) = 2x, \quad g(x) = x - 1, \quad h(x) = x + 3$
10. $f(x) = x - 5, \quad g(x) = \sqrt{x}, \quad h(x) = x^3$

Given the graphs of functions f and g as shown in the figures, sketch the graphs of the following:

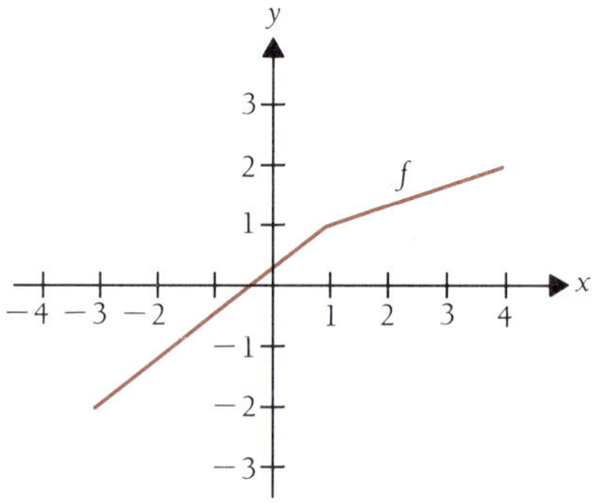

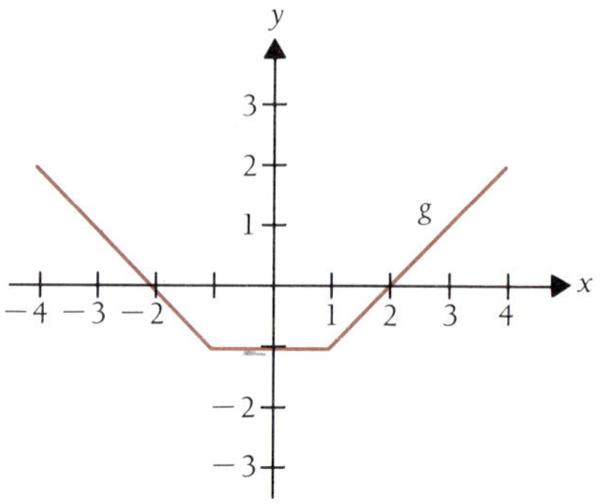

11. $-f$ **12.** $2f$ **13.** $f+g$ **14.** $f-g$

15. fg **16.** f/g **17.** $2f-g$ **18.** $1/f$

Find functions f and g such that $g \circ f = h$.

19. $h(x) = x^2 + 9$ **20.** $h(x) = \dfrac{1}{\sqrt{x}}$ **21.** $h(x) = \sqrt{\dfrac{1}{x-3}}$ **22.** $h(x) = \dfrac{1}{\sqrt{x}} - 3$

Find functions f, g, and h such that $h \circ g \circ f = F$.

23. $F(x) = x^3 - 2$ **24.** $F(x) = \sqrt{x^2 - 4}$ **25.** $F(x) = (x+1)^{3/2}$ **26.** $F(x) = |x-2|^5$

Give the domain of each function.

27. $f(x) = \dfrac{x+1}{x-1}$ **28.** $f(x) = \dfrac{3x}{x^2-7}$ **29.** $f(x) = \dfrac{x-1}{x^3-x^2}$

30. $f(x) = (x-4)^{2/3}$ **31.** $f(x) = (1-x)^{1/4}$ **32.** $f(x) = \dfrac{\sqrt{x-1}}{3x}$

For each pair of functions, give the domains of $f+g$, f/g, and $g \circ f$.

33. $f(x) = \sqrt{x-3}$, $g(x) = x^2 - 1$ **34.** $f(x) = \sqrt{3-x}$, $g(x) = x^2 - 4$

35. $f(x) = \sqrt{4-x^2}$, $g(x) = x + 1$ **36.** $f(x) = \sqrt{x^2-16}$, $g(x) = |x|$

Give the domain and range of each function.

37. $f(x) = 3 - \dfrac{1}{2x}$ **38.** $f(x) = x^3 - 4$

39. $f(x) = \dfrac{x-12}{2x+1}$ **40.** $f(x) = \dfrac{x+2}{x-2}$

41. $f(x) = (x-1)^{1.5}$ **42.** $f(x) = \sqrt{2x-3}$

43. $f(x) = (x^2+4)^{1/3}$ **44.** $f(x) = (x^2-1)^{-1/2}$

Tell whether the function is one-to-one.

45. $f(x) = 2x - 5$
46. $f(x) = x^4$
47. $f(x) = -x^3$
48. $f(x) = \dfrac{6}{x}$
49. $f(x) = \dfrac{6}{x^2}$
50. $f(x) = \dfrac{3}{x(x-1)}$
51. $f(x) = \sqrt{x}$
52. $f(x) = x\sqrt{x}$

Verify that functions f and g are inverses of each other.

53. $f(x) = 2x, \quad g(x) = \dfrac{x}{2}$
54. $f(x) = 4x + 6, \quad g(x) = \dfrac{x-6}{4}$
55. $f(x) = \dfrac{1}{x}, \quad g(x) = \dfrac{1}{x}$
56. $f(x) = (x-9)^{1/3}, \quad g(x) = x^3 + 9$

Find the inverse function if it exists; if not, so state.

57. $f(x) = 2x + 1$
58. $f(x) = 3x - 2$
59. $f(x) = x^3 - 1$
60. $f(x) = 4 - x^2$
61. $f(x) = \dfrac{2}{x+1}$
62. $f(x) = \dfrac{3x}{x-2}$
63. $f(x) = \dfrac{x}{x-5}$
64. $f(x) = x^{1/3}$

Sketch the graph of f^{-1}, given the graph of f in the figure.

65.

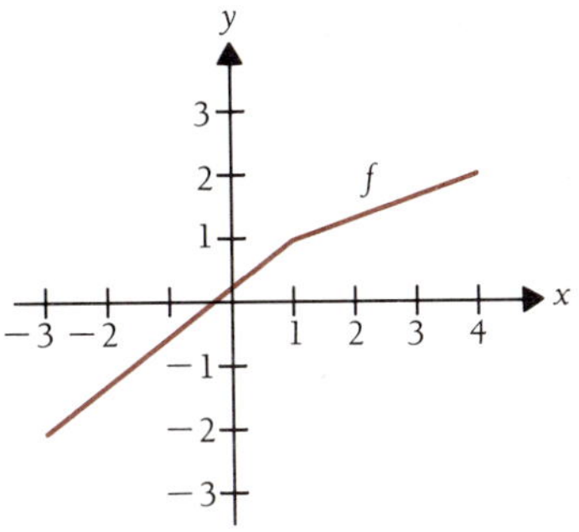

66.

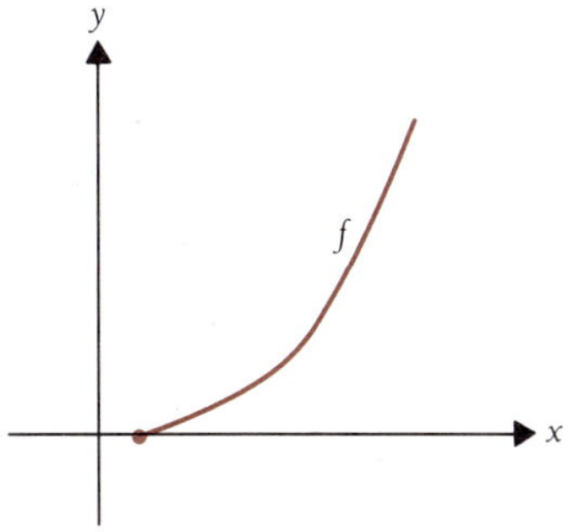

67.

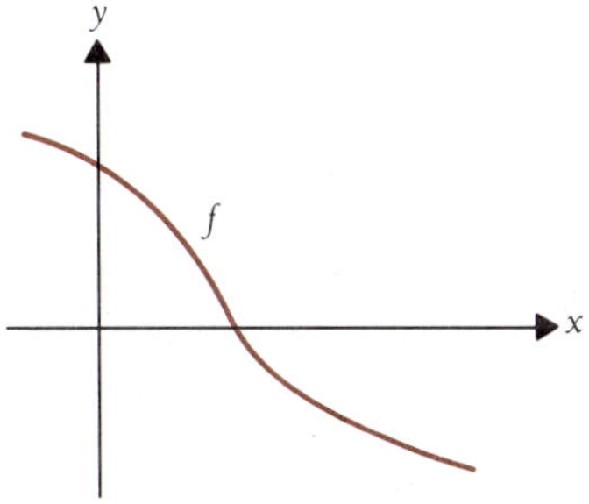

68.

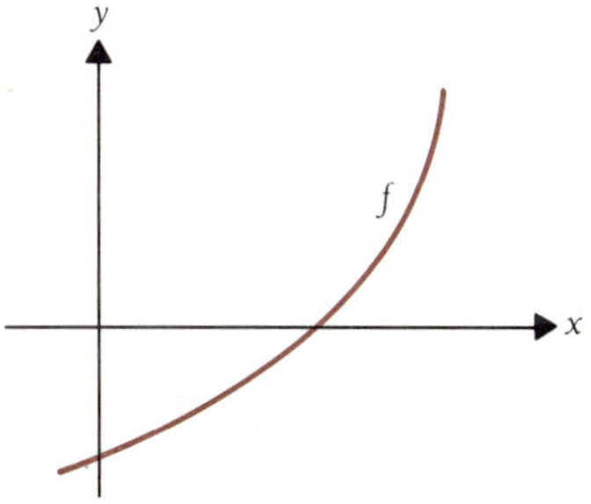

69. Let $f(x) = (x-2)^2$. By suitably restricting the domain of f, obtain a one-to-one function and find its inverse.

Repeat Problem 69 for the following functions:

70. $f(x) = (x-1)^4$
71. $f(x) = \dfrac{1}{x^2}$
72. $f(x) = \sqrt{4 - x^2}$

73. Prove that if a function f has an inverse function, then f is one-to-one.
74. If g is the inverse of f, explain why f is the inverse of g.
75. If f is a function such that f^{-1} is also a function, explain why f^{-1} is one-to-one.
76. Show that the graph of at most one linear function can pass through two given points.
77. Show that the graph of at most one polynomial function of degree n can pass through $n+1$ given points.

Find the inverse image of the given set S under the given function f.

78. $S=[0, 1]$; $f(x)=2x-3$
79. $S=(0, 1)$; $f(x)=4x+7$
80. $S=(0, \infty)$; $f(x)=4-x^2$
81. $S=(-\infty, 1]$; $f(x)=\dfrac{1}{x}$

Find the support of each of the following functions:

82. $f(x)=\dfrac{1}{x}$
83. $f(x)=\lfloor x \rfloor$
84. $f(x)=\sqrt{4-x^2}$
85. $f(x)=\begin{cases} 0 & \text{if } x \leq 2 \\ x-3 & \text{if } x>2 \end{cases}$
86. Show that the line $y=x$ is the perpendicular bisector of the segment with endpoints (a, b) and (b, a). (This shows that interchanging the coordinates of a point reflects it across the line $y=x$.)

SECTION 1.6

Review of the Trigonometric Functions

The trigonometric functions are perhaps the most familiar of the transcendental functions. They are also called **circular functions**, because they arise naturally in the study of circular motion. In this section, we will review briefly the definitions and basic properties of the trigonometric functions, in terms of the unit circle. It is assumed that you have already studied trigonometry thoroughly in a previous course. This section can serve as a reference.

Definitions

Let O be the origin of a rectangular coordinate system, and let U be the point $(1, 0)$. The unit circle Γ, whose equation is $x^2+y^2=1$, has center O and passes through U. Now, if t is any real number, let $P(t)$ be the point on Γ constructed as follows: If t is nonnegative, then $P(t)$ is located a distance t around Γ from U in the counterclockwise direction, as shown in Figure 1.52. If t is negative, $P(t)$ is located a distance $|t|$ around Γ from U in the clockwise direction, as shown in Figure 1.53. Let $P(t)$ have coordinates (x, y).

The **trigonometric functions** are the **sine**, **cosine**, **tangent**, **cotangent**, **secant**, and **cosecant** functions. They are denoted by $\sin t$, $\cos t$, $\tan t$, $\cot t$, $\sec t$, and $\csc t$, and defined in terms of the coordinates of $P(t) = (x, y)$ by

$$\sin t = y, \qquad \cos t = x,$$

$$\tan t = \frac{y}{x}, \qquad \cot t = \frac{x}{y},$$

$$\sec t = \frac{1}{x}, \qquad \csc t = \frac{1}{y}.$$

The trigonometric functions are thus functions of the real number t.

We define the **radian measure** of an angle by declaring that there are π radians in two right angles. Thus, there are π radians in 180°, and therefore, $\pi/2$ radians in 90° and 2π radians in 360°. We will consistently use radian measure for angles in this book. If you use a calculator to find the values of the trigonometric functions, be sure the calculator is in radian mode.

In the unit circle, the radian measure of a central angle is numerically equal to the length of the arc it subtends. Thus, in Figure 1.52, the number t is the radian measure of the angle UOP. In Figure 1.53, we also let t be the measure of $\angle UOP$, and adopt the convention of using negative numbers as measures of angles in the clockwise direction. Thus, the trigonometric functions are also functions of angles.

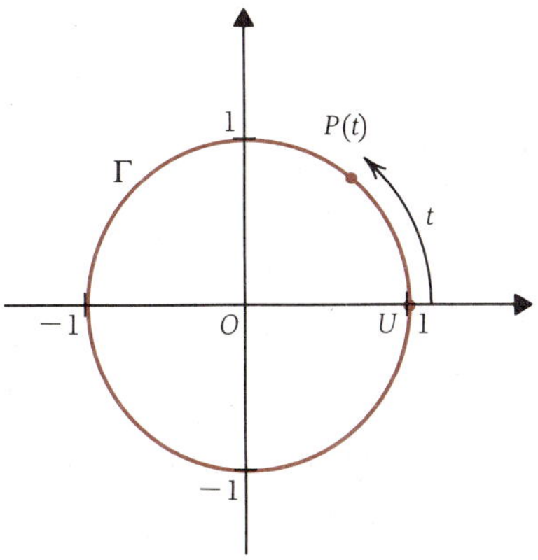

FIGURE 1.52 $P(t)$, t positive

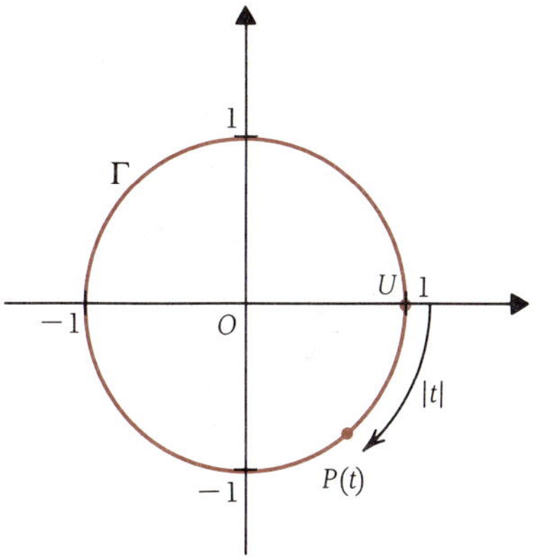

FIGURE 1.53 $P(t)$, t negative

In terms of an acute angle of a right triangle, the trigonometric functions can be described as ratios of the sides of the triangle. Referring to Figure 1.54 (page 66), the trigonometric ratios are

$$\sin t = \frac{\text{Opposite side}}{\text{Hypotenuse}}, \qquad \cos t = \frac{\text{Adjacent side}}{\text{Hypotenuse}},$$

$$\tan t = \frac{\text{Opposite side}}{\text{Adjacent side}}, \qquad \cot t = \frac{\text{Adjacent side}}{\text{Opposite side}},$$

$$\sec t = \frac{\text{Hypotenuse}}{\text{Adjacent side}}, \qquad \csc t = \frac{\text{Hypotenuse}}{\text{Opposite side}}.$$

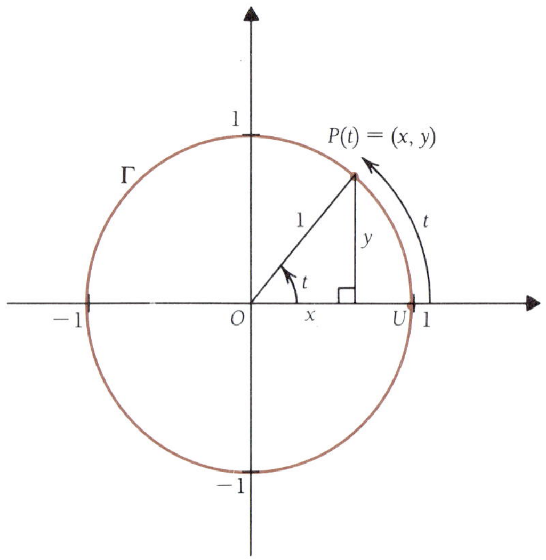

FIGURE 1.54 Right triangle definitions of the circular functions

Graphs of the Trigonometric Functions

A function f is **periodic** if there exists a positive real number p such that $f(x + p) = f(x)$ for all x in the domain of f. The least such p, if there is one, is called the **period** of f.

Since the unit circle has circumference 2π, the points $P(t)$ and $P(t + 2\pi)$ are the same for all t. Therefore, the trigonometric functions are periodic; the tangent and cotangent functions have period π, and the others have period 2π. It is also clear from the definitions that not all the trigonometric functions are going to be defined for all values of t. These facts are evident in the graphs of the trigonometric functions shown in Figures 1.55–1.60. It is best to think of the properties of the circular functions in terms of their graphs. Note that we use x as the independent variable in the graphs.

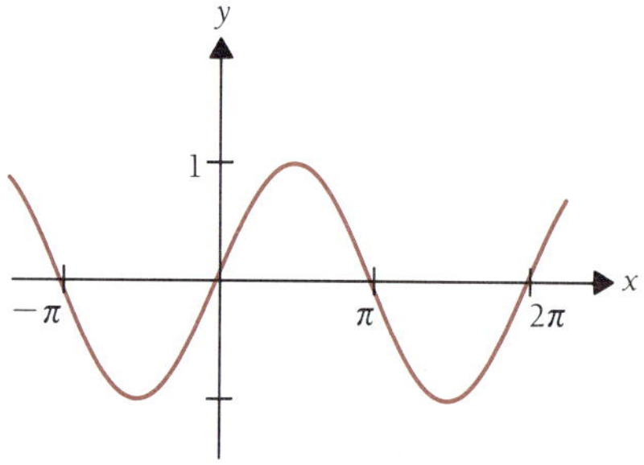

FIGURE 1.55 $y = \sin x$

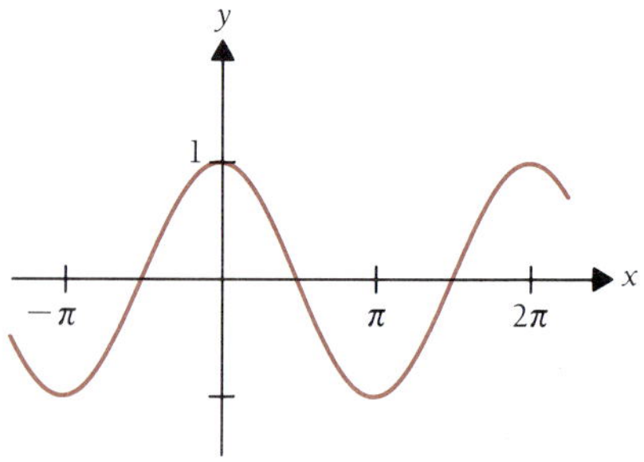

FIGURE 1.56 $y = \cos x$

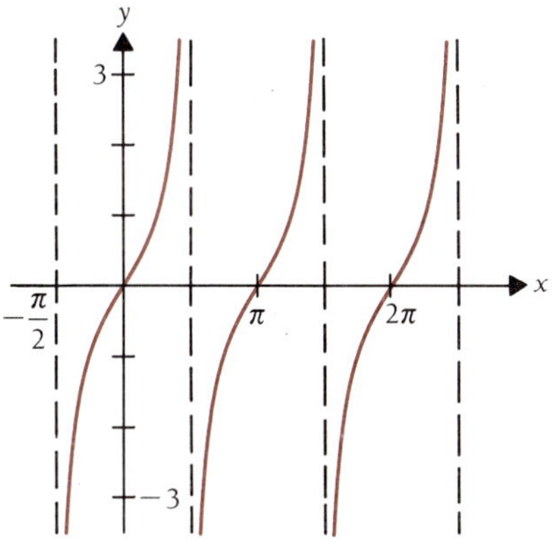

FIGURE 1.57 $y = \tan x$

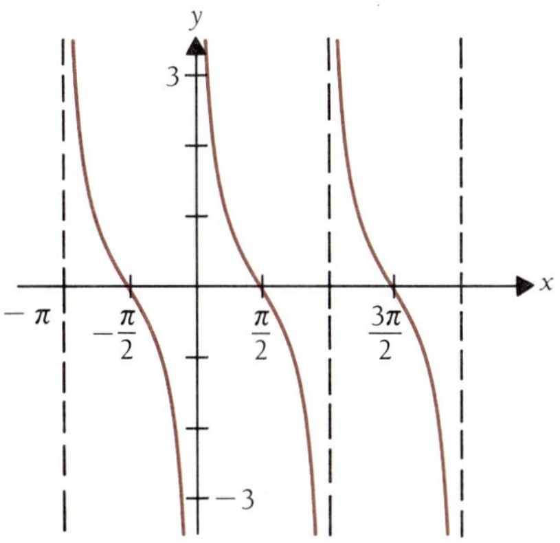

FIGURE 1.58 $y = \cot x$

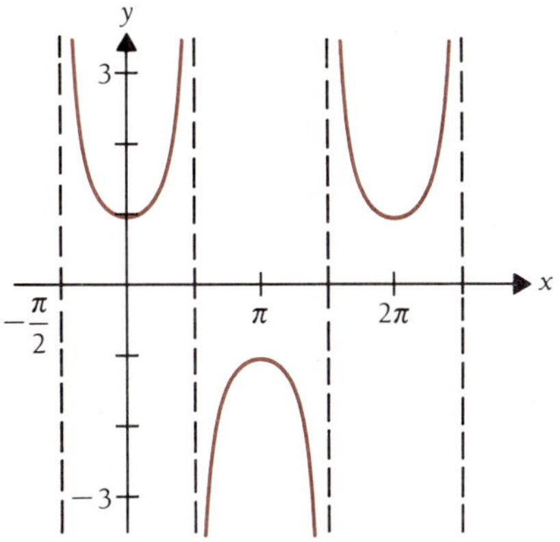

FIGURE 1.59 $y = \sec x$

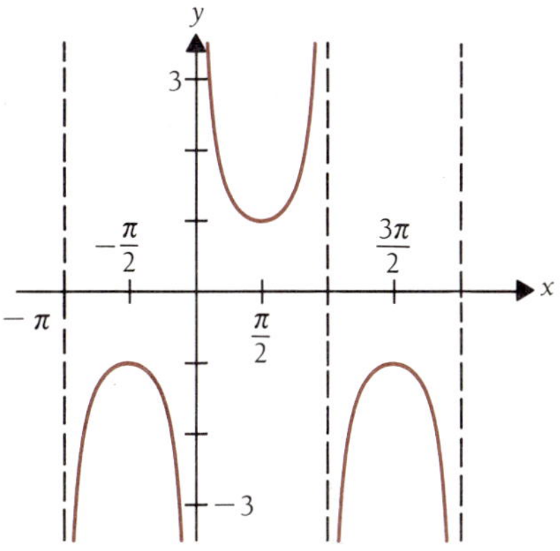

FIGURE 1.60 $y = \csc x$

Certain values of the independent variable occur often enough that the corresponding values of the trigonometric functions should be memorized. The values in Table 1.8 are sufficient to commit to memory. Values of the other functions at these points can be found using the identities listed on pages 68–69. Values of the functions for other points x can be found in tables or by using a calculator.

TABLE 1.8

$x =$	0	$\pi/6$	$\pi/4$	$\pi/3$	$\pi/2$
$\sin x =$	0	$\frac{1}{2}$	$\sqrt{2}/2$	$\sqrt{3}/2$	1
$\cos x =$	1	$\sqrt{3}/2$	$\sqrt{2}/2$	$\frac{1}{2}$	0
$\tan x =$	0	$\sqrt{3}/3$	1	$\sqrt{3}$	—

Trigonometric Identities

From the definitions of the trigonometric functions and the equation of the unit circle come the following **fundamental identities** relating the functions:

FUNDAMENTAL IDENTITIES

$$\tan x = \frac{\sin x}{\cos x} \tag{1}$$

$$\cot x = \frac{\cos x}{\sin x} \tag{2}$$

$$\sec x = \frac{1}{\cos x} \tag{3}$$

$$\csc x = \frac{1}{\sin x} \tag{4}$$

$$\cos^2 x + \sin^2 x = 1 \tag{5}$$

$$1 + \tan^2 x = \sec^2 x \tag{6}$$

$$\cot^2 x + 1 = \csc^2 x \tag{7}$$

These formulas, and also all those that follow, hold for all values of x for which the functions are defined.

Many other trigonometric identities will be useful. They include the following:

REDUCTION FORMULAS

$$\sin(-x) = -\sin x \tag{8}$$

$$\cos(-x) = \cos x \tag{9}$$

$$\sin(x + \tfrac{\pi}{2}) = \cos x \tag{10}$$

$$\cos(x + \tfrac{\pi}{2}) = -\sin x \tag{11}$$

$$\sin(\tfrac{\pi}{2} - x) = \cos x \tag{12}$$

$$\cos(\tfrac{\pi}{2} - x) = \sin x \tag{13}$$

ADDITION AND SUBTRACTION FORMULAS

$$\sin(x \pm y) = \sin x \cos y \pm \cos x \sin y \tag{14}$$

$$\cos(x \pm y) = \cos x \cos y \mp \sin x \sin y \tag{15}$$

$$\tan(x \pm y) = \frac{\tan x \pm \tan y}{1 \mp \tan x \tan y} \tag{16}$$

DOUBLE-ANGLE FORMULAS

$$\sin 2x = 2 \sin x \cos x \tag{17}$$

$$\cos 2x = \cos^2 x - \sin^2 x \tag{18}$$

$$= 1 - 2\sin^2 x \tag{19}$$

$$= 2\cos^2 x - 1 \tag{20}$$

$$\tan 2x = \frac{2 \tan x}{1 - \tan^2 x} \tag{21}$$

HALF-ANGLE FORMULAS

$$\sin^2 x = \tfrac{1}{2}(1 - \cos 2x) \tag{22}$$
$$\cos^2 x = \tfrac{1}{2}(1 + \cos 2x) \tag{23}$$
$$\tan x = \frac{1 - \cos 2x}{\sin 2x} \tag{24}$$
$$= \frac{\sin 2x}{1 + \cos 2x} \tag{25}$$

PRODUCT-TO-SUM FORMULAS

$$\sin x \sin y = \tfrac{1}{2}[\cos(x - y) - \cos(x + y)] \tag{26}$$
$$\sin x \cos y = \tfrac{1}{2}[\sin(x + y) + \sin(x - y)] \tag{27}$$
$$\cos x \cos y = \tfrac{1}{2}[\cos(x + y) + \cos(x - y)] \tag{28}$$

We shall have occasion to refer to these identities from time to time. Their derivations are reviewed in the exercises.

Example 1.46 The first four fundamental identities express each of the other functions in terms of the sine and cosine functions. Express each of the functions in terms of the tangent function alone.

Solution The easy ones are

$$\tan x = \tan x \qquad \text{and} \qquad \cot x = \frac{1}{\tan x}.$$

Since $\sec^2 x = 1 + \tan^2 x$, we have

$$\sec x = \pm\sqrt{1 + \tan^2 x},$$

with the appropriate sign chosen according to the quadrant in which x lies. Similarly,

$$\csc x = \pm\sqrt{1 + \cot^2 x} = \pm\sqrt{1 + \frac{1}{\tan^2 x}}$$
$$= \pm\frac{\sqrt{\tan^2 x + 1}}{\tan x}.$$

Finally,

$$\sin x = \tan x \cos x = \frac{\tan x}{\sec x}$$
$$= \pm\frac{\tan x}{\sqrt{1 + \tan^2 x}}$$

and

$$\cos x = \frac{1}{\sec x} = \pm\frac{1}{\sqrt{1+\tan^2 x}}.$$

■

■ Example 1.47 If x is in Quadrant III and $\tan x = \frac{5}{12}$, find $\sin x$, $\cos x$, $\sec x$, $\sin 2x$, and $\tan(x/2)$.

Solution If x is in Quadrant III, then $\sin x$ is negative. Using the results of Example 1.46, we get

$$\begin{aligned}\sin x &= -\frac{\tan x}{\sqrt{1+\tan^2 x}}\\ &= \frac{-\frac{5}{12}}{\sqrt{1+\frac{25}{144}}} = -\frac{5}{13}.\end{aligned}$$

Similarly,

$$\begin{aligned}\cos x &= -\frac{1}{\sqrt{1+\tan^2 x}} \quad &\text{and} \quad \sec x &= -\sqrt{1+\tan^2 x}\\ &= -\frac{1}{\sqrt{1+\frac{25}{144}}} = -\frac{12}{13} & &= -\sqrt{1+\tfrac{25}{144}} = -\frac{13}{12}.\end{aligned}$$

Using the sine double-angle formula, we now get

$$\sin 2x = 2\sin x\cos x = 2\left(-\frac{5}{13}\right)\left(-\frac{12}{13}\right) = \frac{120}{169}.$$

Finally, we can find $\tan(x/2)$ from the half-angle formula (24), with $x/2$ replacing x:

$$\tan\frac{x}{2} = \frac{1-\cos x}{\sin x} = \frac{1+\frac{12}{13}}{-\frac{5}{13}} = -5.$$

■

Exercises 1.6

Using Table 1.8 and the identities, find exact radical values for the following:

1. $\sin\left(-\frac{\pi}{4}\right)$
2. $\cos\frac{2\pi}{3}$
3. $\sin\frac{\pi}{8}$
4. $\tan\frac{\pi}{12}$
5. $\sec\frac{\pi}{3}$
6. $\cot\frac{5\pi}{4}$
7. $\cos\left(-\frac{7\pi}{12}\right)$
8. $\cos\frac{7\pi}{12}$

Using a calculator, give four-decimal approximations to the following:

9. $\sin 1$
10. $\cos 2$
11. $\tan 0.55$
12. $\sec 1.74$
13. $\sin\sqrt{2}$
14. $\csc\frac{\pi}{10}$
15. $\tan 0.03\pi$
16. $\cos\frac{\pi\sqrt{2}}{19}$

Given that $x \in (0, \pi/2)$ and $\cos x = \frac{2}{3}$, find the values of the following:

17. $\sin x$
18. $\tan x$
19. $\sec x$
20. $\cot x$
21. $\cos 2x$
22. $\sin 2x$
23. $\sin\frac{x}{2}$
24. $\tan\frac{x}{2}$

25. Explain why the coordinates of the point $P(x)$ are $(\cos x, \sin x)$.

Problems 26–39 are organized to give a logical derivation of the identities of this section.

26. Let x and y be real numbers. Using the coordinates of $P(x)$, $P(y)$, and $P(x-y)$ (see the figure), the distance formula, and the fact that the distance between $P(x)$ and $P(y)$ is the same as the distance between $P(x-y)$ and $U = P(0)$, derive the cosine subtraction formula (15),

$$\cos(x-y) = \cos x \cos y + \sin x \sin y.$$

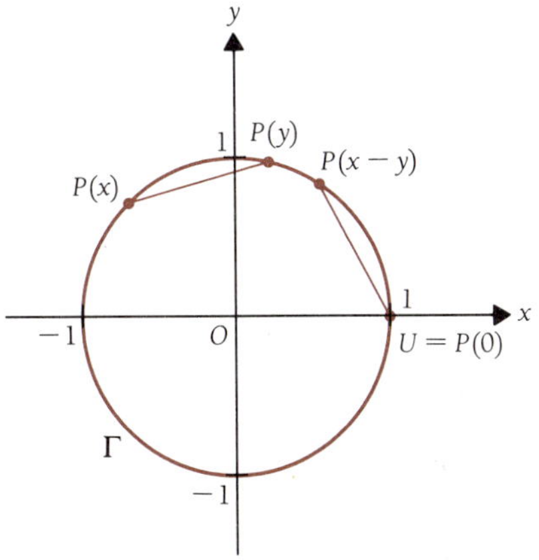

Using the cosine subtraction formula (15) and the fundamental identities, derive the following:

27. Formula (9) **28.** Formula (13) **29.** Formula (12) **30.** $\tan(\frac{\pi}{2} - x) = \cot x$

31. From the definition of $\sin x$, deduce that $\sin(-x) = -\sin x$.

32. Using formula (8) and the cosine subtraction formula (15), derive the cosine addition formula—the other half of formula (15).

33. Derive both parts of formula (14), using the results of the above problems.

34. Derive both parts of formula (16) from formulas (14) and (15).

35. Derive the double-angle formulas from the addition formulas.

36. Derive formulas (22) and (23) from formulas (19) and (20).

37. Derive formula (24) from the cosine subtraction formula (15), using $x = 2x - x$.

38. Derive formula (25) from formula (24).

39. Derive the product-to-sum formulas by expanding the right-hand side of each one.

40. Solving the two equations

$$x + y = A,$$
$$x - y = B,$$

use formula (27) to derive the **sum-to-product** formula

$$\sin A + \sin B = 2 \sin\left(\frac{A+B}{2}\right) \cos\left(\frac{A-B}{2}\right).$$

Sketch the graphs of the following functions:

41. $f(x) = \sin 2x$
42. $f(x) = 3 \cos \frac{x}{2}$
43. $f(x) = \tan \pi x$
44. $f(x) = 2 \sin 3x$
45. $f(x) = 2 \sin \frac{x}{3}$
46. $f(x) = 4 \cos 4x$
47. $f(x) = \frac{1}{2} \tan 2x$
48. $f(x) = 2 \sec 3x$
49. $f(x) = |\sin x|$
50. $f(x) = \sin(\pi - x)$
51. $f(x) = \frac{|\sin x|}{\sin x}$
52. $f(x) = |\cos x|$
53. $f(x) = 1 + \sin x$
54. $f(x) = 2 - \cos x$
55. $f(x) = x + \sin x$
56. $f(x) = \cos x - \sin x$
57. If l is a line with nonzero slope, the **inclination** ϕ of l is defined to be the angle between the positive x-axis and the upper half of l (see the figure). The inclination of a horizontal line is defined to be 0, and the inclination of a vertical line is defined to be $\pi/2$. Show that the slope of a line is the tangent of its inclination—that is, $m = \tan \phi$.

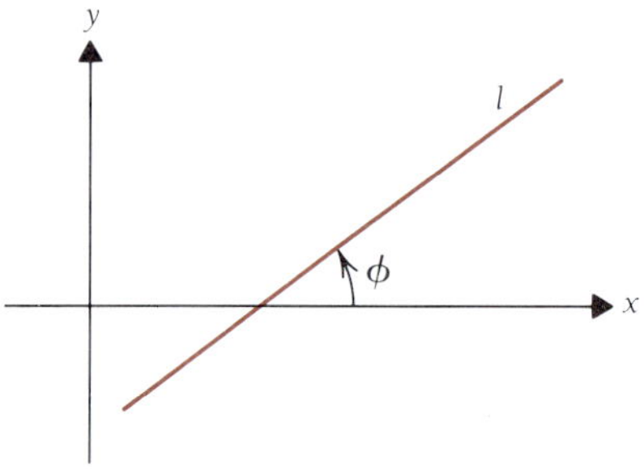

58. Given lines l_1 and l_2 with slopes m_1 and m_2, show that the angle θ at which the two lines meet satisfies

$$\tan \theta = \frac{m_2 - m_1}{1 + m_1 m_2}.$$

[*Hint:* Refer to Problem 57. Use inclinations and the tangent subtraction formula (16).]

SECTION 1.7

Chapter Review

Important Concepts

History of calculus
Impact of calculus
Set
Element of a set
Empty set
Subset
Equal sets
Union
Intersection
Real numbers
Counting numbers
Integers
Rational numbers
Irrational numbers
The real line
Intervals
Absolute value
Triangle inequality
Axioms of the real number system
Rectangular coordinate system
Distance formula
Midpoint formula
Slope
Equations of a line

Point–slope form
Slope–intercept form
Two-point form
Standard form
Equation of a circle
Unit circle
Completing the square
Secant line
Tangent line
Function
Domain
Range
Variation
Independent variable
Dependent variable
Models
Graph
Constant function
Linear function
Quadratic function
Polynomial function
Power function
Rational function
Root function
Absolute value function
Greatest integer function
Vertical line test
Symmetry
Linear combination
Function composition
One-to-one function
Horizontal line test
Inverse function
Trigonometric functions
Radians
Periodic function
Trigonometric identities

Review Exercises

Graph each set on the real line.

1. $[0, 1]$
2. $[-1, 1]$
3. $(0, 1)$
4. $(-2, 2)$
5. $(3, \infty)$
6. $(-\infty, 0)$
7. $(-\infty, -1] \cup [1, \infty)$
8. $(-\infty, 1] \cap [-1, \infty)$
9. $\{x \mid |x| < 3\}$
10. $\{x \mid |x| \geq 2\}$
11. $\{x \mid |x - 5| \leq 4\}$
12. $\{x \mid |x - \frac{1}{2}| > \frac{3}{8}\}$

Find the distance between the given points.

13. $(5, 1)$ and $(3, 2)$
14. $(-1, -1)$ and $(2, -3)$

Find the point midway between the given points.

15. $(0, 0)$ and $(2, 4)$
16. $(-1, 6)$ and $(3, 2)$

Find an equation for the line.

17. Through $(3, 1)$ and with slope 7
18. Through $(3, 1)$ and with slope -0.5
19. Through $(3, 1)$ and $(3, 5)$
20. Through $(3, 1)$ and $(-4, 3)$
21. Through $(2, 4)$ and perpendicular to $x - y + 4 = 0$
22. Through $(6, 3)$ and parallel to $x - y + 4 = 0$
23. The perpendicular bisector of the segment joining $(4, -1)$ and $(3, 3)$
24. With x-intercept 2 and y-intercept -1

Find the equation for the circle.

25. With center $(1, -1)$ and radius $\sqrt{2}$
26. With center $(1, -2)$ and through the point $(1, 0)$
27. With center $(3, 7)$ and tangent to the line $y = -1$
28. Through points $(1, 0)$, $(0, 1)$, and $(4, 4)$

Describe the curve.

29. $x^2 + y^2 - 2x + 4y - 1 = 0$
30. $x^2 + y^2 - 8y + 12 = 0$
31. $x^2 + y^2 - 2x + 4y - 4 = 0$
32. $x^2 + y^2 = 4x$

Give the domain and range of each function.

33. $f(x) = 2x + 5$
34. $f(x) = \dfrac{7}{x^2}$
35. $f(x) = \dfrac{2}{x - 3}$
36. $f(x) = \dfrac{x}{|x|}$
37. $f(x) = \dfrac{x^2 + 1}{x^2 - 1}$
38. $f(x) = \sqrt{1 - \sin x}$
39. $f(x) = \dfrac{1 + \sec x}{\sin x}$
40. $f(x) = \tan \dfrac{x}{3}$

Sketch the graph of each function.

41. $f(x) = \dfrac{3}{x}$
42. $f(x) = |x - 2|$
43. $f(x) = |x| - 2$
44. $f(x) = |x - 2| + 3$
45. $f(x) = \sqrt{x - 1}$
46. $f(x) = x + \sin x$
47. $f(x) = \dfrac{\sin x}{x}$
48. $f(x) = 1 + \dfrac{x}{x^2 + 1}$

Let $f(x) = x^2$, $g(x) = \sin x$, and $h(x) = 1/x$. Find expressions for each of the following:

49. $f(4)$
50. $h(\sqrt{3})$
51. $f(x + 2)$
52. $g(x - a)$
53. $g(h(-1))$
54. $g(x) + f(\sin x)$
55. $h(f(2x))$
56. $f(h(2x))$

Find the inverse function if it exists.

57. $f(x) = -1 - x^3$
58. $f(x) = \dfrac{3}{x + 4}$
59. $f(x) = 1 + \cos x, \quad x \in [0, \pi/2]$
60. $f(x) = 4 - \tan^2 x, \quad x \in (0, \pi/2)$

Additional Exercises

1. Is the intersection of two closed intervals always a closed interval? Ever a closed interval? Is the union of two closed intervals a closed interval?
2. Describe the union and intersection of two open intervals.
3. What points (if any) belong to all the open intervals $(0, 1)$, $(0, \frac{1}{2})$, $(0, \frac{1}{3})$, . . . , $(0, 1/n)$, . . . ?
4. What points (if any) belong to all the closed intervals $[0, 1]$, $[0, \frac{1}{2}]$, $[0, \frac{1}{3}]$, . . . , $[0, 1/n]$, . . . ?
5. What points (if any) belong to all the open intervals $(-1, 1)$, $(-\frac{1}{2}, \frac{1}{2})$, $(-\frac{1}{3}, \frac{1}{3})$, . . . , $(-1/n, 1/n)$, . . . ?
6. What points (if any) belong to all the intervals $(1, \infty)$, $(2, \infty)$, $(3, \infty)$, . . . , (n, ∞), . . . ?

Solve for x.

7. $x + 1 = |2x - 3|$
8. $|4x| = |2x - 3|$
9. $|2x - 3| = 4|x + 1|$
10. $|x^2 - a^2| = 1$
11. $|x^2 - x - 2| < 0$
12. $|x^2 - 4| > 0$
13. $|\sin x| = 0.5$
14. $|\sin^2 x - 1| > 0$
15. Show that if x and y are of the same sign, then $|x + y| = |x| + |y|$.
16. Show that if x and y are of opposite signs, then $|x + y| < |x| + |y|$.
17. Give an alternate proof of the triangle inequality.

18. Sketch the graph of $|x| + |y| = 1$.

19. Shade the region corresponding to $|x| + |y| < 1$ in the graph drawn in Problem 18.

20. Three vertices of a parallelogram are (0, 0), (2, 4), and (1, 3). Find the fourth vertex. (There are three solutions.)

21. Is the point (5, 4) on, inside, or outside the circle $x^2 + y^2 - 4x - 2y - 11 = 0$?

22. Where is the center of a circle that is tangent to both the x-axis and the y-axis?

23. For what values of c will the line $x + y = c$ meet the circle $x^2 + y^2 = c$?

24. Find the equations of the lines through the point (4, 1) that are tangent to the unit circle.

25. Show that

$$\begin{vmatrix} x^2 + y^2 & x & y & 1 \\ x_1^2 + y_1^2 & x_1 & y_1 & 1 \\ x_2^2 + y_2^2 & x_2 & y_2 & 1 \\ x_3^2 + y_3^2 & x_3 & y_3 & 1 \end{vmatrix} = 0$$

is the equation of the circle through the three points (x_1, y_1), (x_2, y_2), and (x_3, y_3).

26. Find the equation of the line bisecting the angle between the lines $y = m_1 x + b$ and $y = m_2 x + b$.

27. Repeat Problem 26 for the lines $A_1 x + B_1 y + C_1 = 0$ and $A_2 x + B_2 y + C_2 = 0$. [*Caution:* This problem is not for the faint-hearted!]

28. Show analytically that the bisectors of the angles of a triangle are concurrent. (The point of intersection is called the **incenter** of the triangle.) [*Caution:* This problem illustrates that an analytic proof of a simple geometric theorem can be *extremely* complicated.]

29. Let l be a line, and let the line through the origin O that is perpendicular to l meet l at F. Let r be the distance from O to F, and let α be the inclination of the line OF. Show that l has equation

$$x \cos \alpha + y \sin \alpha = r.$$

(This equation is called the **polar form** of a linear equation.)

Write a formula for the function defined by the given rule.

30. The number that is 2 less than x

31. The area of a triangle with base b and height 3

32. The area of a regular hexagon with side s

33. The earnings on \$100 invested at 5% simple interest after t years

34. Fahrenheit temperature T is a linear function of the Celsius temperature t. Find $T(t)$, given that $0°C = 32°F$ and $100°C = 212°F$.

35. Suppose that the value to a collector of a postage stamp is inversely proportional to the number of copies of that stamp that are known to exist. If 100,000 copies of a certain stamp exist and the value of that stamp is \$1, what is the value of a stamp of which 500 copies exist?

36. Suppose that the diameter of the trunk of a tree is proportional to the $\frac{3}{2}$ power of its height. If a tree 16 meters high has a trunk 1.1 meters in diameter, what is the diameter of the trunk of a tree that is 36 meters high?

Sketch the graphs of the following functions:

37. $f(x) = 4 - x^2$
38. $f(x) = 5 - x^2$
39. $f(x) = -(4 - x^2)$
40. $f(x) = 2(4 - x^2)$
41. $f(x) = 4 - (x - 1)^2$
42. $f(x) = 4 - (x + 2)^2$
43. $f(x) = |4 - x^2|$
44. $f(x) = -|4 - x^2|$
45. $f(x) = (4 - x^2)^2$
46. $f(x) = \sqrt{4 - x^2}$
47. Explain why the function $r(x) = \lfloor x + \frac{1}{2} \rfloor$ is called the **rounding function**.

Sketch the graphs of the following functions:

48. $f(x) = 3\lfloor x \rfloor$
49. $f(x) = \lfloor 3x \rfloor$
50. $f(x) = \left\lfloor \dfrac{x}{2} \right\rfloor$
51. $f(x) = \dfrac{\lfloor x \rfloor}{2}$
52. $f(x) = \lfloor 2x - 1 \rfloor$
53. $f(x) = 2\lfloor x \rfloor - 1$
54. $f(x) = \lfloor \sin x \rfloor$
55. $f(x) = \sin\lfloor x \rfloor$
56. $f(x) = \lfloor |x| \rfloor$
57. $f(x) = |\lfloor x \rfloor|$

Give the domain and range of each function and sketch its graph.

58. $f(x) = \dfrac{3}{x(x - 1)}$
59. $f(x) = \dfrac{x}{x + 3}$
60. $f(x) = \dfrac{1}{\sqrt{x^2 - 1}}$
61. $f(x) = (x^2 - 1)^{1/3}$
62. $f(x) = |x - 5|$
63. $f(x) = \dfrac{1}{|x - 5|}$
64. $f(x) = \sin(4 - x^2)$
65. $f(x) = 1 - \lfloor \cos x \rfloor$
66. Give the domains and ranges of the functions

$$f(x) = \frac{x^2 - 4}{x - 2} \quad \text{and} \quad g(x) = x + 2.$$

Are f and g the same function?

For $f(x) = (2/x) - x$, find the following:

67. $f\left(\dfrac{1}{x}\right)$
68. $f(x + h)$
69. $f(x + h) - f(x)$
70. $\dfrac{f(x + h) - f(x)}{h}$
71. If

$$f(x) = \frac{x - 2}{x + 2} \quad \text{and} \quad g \circ f(x) = \frac{x + 2}{x - 2},$$

find $g(x)$.

72. Print each capital letter of the alphabet on a rectangular coordinate system. Which are the graphs of functions? Which are the inverses of functions?
73. Give an example of a function f such that $f(a + b) = f(a) + f(b)$ for all a and b in the domain of f.
74. Give an example of a function g such that $g(a + b)$ is never equal to $g(a) + g(b)$.
75. Give an example of a function F such that $F(ab) = F(a)F(b)$ for all a and b in the domain of F.
76. Give an example of a function G such that $G(ab)$ is never equal to $G(a)G(b)$.
77. Find a quadratic function whose graph passes through the points (0, 0), (1, 4), and (2, 3).
78. For what values of a and b will the function $f(x) = ax + b$ be its own inverse?

Find the inverse of the given function, and sketch the graph of each function and its inverse in the same coordinate system.

79. $f(x) = (x^3 - 2)^5$

80. $f(x) = \dfrac{1 + x}{1 - x}$

For which of the following domains does the function $f(x) = x^2$ have an inverse function?

81. $(-\infty, \infty)$

82. $[0, \infty)$

83. $(-\infty, 0]$

84. $[-2, 4]$

85. $[1, 5]$

86. $[-5, -2]$

Using the table, calculate the following, if possible:

$x =$	-1	0	1	2
$f(x) =$	3	2	0	4
$g(x) =$	-2	2	-2	-1
$h(x) =$	0	-1	2	1

87. $f \circ g \circ h(-1)$

88. $f \circ g \circ h(0)$

89. $f \circ g \circ h(1)$

90. $f \circ g \circ h(2)$

91. $h \circ g \circ f(1)$

92. $h \circ g \circ f(0)$

93. Find a rational function whose domain is all real numbers except 1, 2, and $\sqrt{3}$.

Recall that a function f is periodic if there exists a positive number p such that $f(x + p) = f(x)$ for all x in the domain of f, and that the least such p, if there is one, is called the period of f.

94. Show that a constant function is periodic but has no period.

95. Prove that the sine function is periodic and has period 2π.

96. Prove that the tangent function is periodic and has period π.

97. Prove that $f(x) = \sin ax$ has period $2\pi/a$.

98. Find the period of $f(x) = A \sin 3x + B \cos 5x$.

A function f is **even** *if and only if $f(-x) = f(x)$ for all x. A function f is* **odd** *if and only if $f(-x) = -f(x)$ for all x.*

99. Show that the graph of an even function is symmetric about the y-axis and that the graph of an odd function is symmetric about the origin (see Figures 1.42 and 1.44).

Determine whether the following functions are even, odd, or neither:

100. $f(x) = x^2$

101. $f(x) = x^3$

102. $f(x) = x|x|$

103. $f(x) = x^2 + x^3$

104. $f(x) = \sin x$

105. $f(x) = \cos x$

106. $f(x) = x \sin x$

107. $f(x) = \tan x$

108. Show that a linear combination of even functions is even and that a linear combination of odd functions is odd.

109. What can you say about the product of two even functions?

110. What is true of the product or quotient of two odd functions?

111. What can be said of the product of an odd function and an even function?

112. Show that the formula

$$f(x) = \frac{f(x) + f(-x)}{2} + \frac{f(x) - f(-x)}{2}$$

expresses f as the sum of an even function and an odd function.

CHAPTER 2

Limits and Continuity

INTRODUCTION

Calculus is usually the first course in which one studies limits. In fact, calculus may be described as algebra and trigonometry with limits. The major concern of calculus is the study of various aspects of the limit concept.

Limits first arose in connection with the study of motion. Zeno of Elea (*c.* 450 BC) introduced the notion in his famous paradoxes, in which he proved that certain intuitively acceptable properties of the infinite were incompatible with common sense. Galileo and Newton used the idea of a limit, in a certain sense, to describe instantaneous speed.

The idea of limit also arose in connection with geometric problems. Archimedes (*c.* 230 BC) used limits to calculate the area of a circle and the volume and surface area of a sphere, and to find an approximate value of π. The same idea arose again in the 17th century in connection with the problem of finding the tangent line to a curve.

The modern idea of limit was first proposed by d'Alembert (*c.* 1765) and systematically developed by Cauchy during the 1820's as part of the study of the real number system. The present-day student of calculus studies limits essentially as Cauchy developed them. His treatment of limits brings together the earlier ideas of limit and unifies them into a single theme. The ε,δ definition of limit which we will use was formulated by the German mathematician Karl Weierstrass (1816–1897).

There are three ways that limits are used in the modern calculus course: limits of functions, limits of sums, and limits of sequences. The study of limits of functions leads to the derivative and the calculation of rates of change. The study of limits of sums leads to the definite integral and such applications as areas and volumes. The study of limits of sequences leads to infinite series. In this chapter, we concentrate on the limit of a function. Our approach will be graphic and computational, leading to a formal definition of limit and to rules for evaluating limits of functions. We will

Augustin Louis Cauchy (1789–1857) was probably the most noted French mathematician of his time. His talent for mathematics was recognized by Laplace and Lagrange while he was still a boy, and his original research began while he was still in his teens. His 789 papers range over the whole of mathematics. The rigorous treatment of calculus, based on the notion of limit, became widely accepted because of Cauchy's lectures and textbooks, which were models of clear exposition. His development of the calculus of complex numbers through the line integral established complex analysis as an independent discipline within mathematics. His work on permutation groups helped pave the way for the introduction of modern abstract algebra, and he also made fundamental contributions to the theory of wave propagation and elasticity.

Cauchy belonged to the minority of Frenchmen who remained loyal to Charles X when that king was invited to vacate the throne in 1830. He spent 8 years in exile with the king, and even when he returned, he felt that he could not take an oath of loyalty to any other French government. In 1852, the government gave in and specifically exempted Cauchy from the required loyalty oath.

include a study of one-sided limits, limits at infinity, and infinite limits. Closely related to the limit concept is the notion of continuity. We will close the chapter with a discussion of continuity and the properties of continuous functions.

SECTION 2.1

Introduction to Limits

Having studied the basic properties of the real number system and the elementary properties of functions, we now turn to the first main task of calculus, the study of limits.

Limits and Real Numbers

The idea of limit is built into the real number system. It comes from the fact that numbers are very closely packed on the real line—not spaced like beads on a string. Between any two real numbers, no matter how close together, there are infinitely many other real numbers. Thus, there are infinitely many real numbers within a specified distance from any given real number, no matter how small the distance.

Let x be a variable that takes on real number values, and let c be a fixed real number. To say that **x approaches c** means that x takes on values arbitrarily close to c. We use the notation $x \to c$ to stand for "x approaches c." If $x \to c$, then x differs from c by arbitrarily small amounts, so that for any positive number δ, the expression $|x - c|$ has values less than δ.

Limits of Functions

Now we consider what happens to the values $f(x)$ of a function f as x approaches some real number c. If $f(x)$ approaches some real number L as x approaches c, we call L the **limit** of f as x approaches c. Let's consider an example.

Example 2.1 Find the limit of the function

$$f(x) = \frac{x^2 - 9}{x - 3}$$

as x approaches 3 (see Figure 2.1).

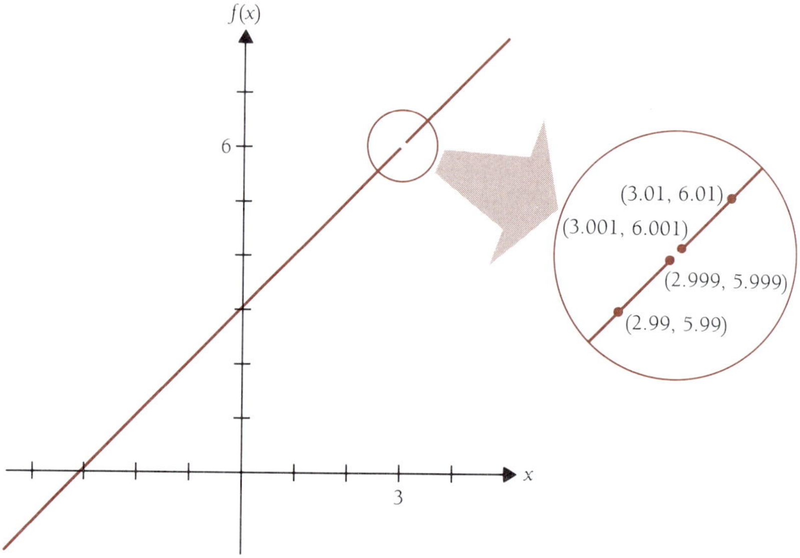

FIGURE 2.1 The graph of $f(x) = \frac{x^2 - 9}{x - 3}$ near (3, 6)

Solution This function is defined for all real numbers except 3. As long as x is different from 3, we can write

$$f(x) = \frac{x^2 - 9}{x - 3} = \frac{(x + 3)(x - 3)}{x - 3} = x + 3.$$

Now we choose some real numbers close to 3, and look at the values of f at each of these numbers. We could have the following table, for example:

$x =$	2	2.5	2.9	2.99	2.999	2.9999
$f(x) =$	5	5.5	5.9	5.99	5.999	5.9999

Here is another possibility:

$x =$	4	3.5	3.1	3.01	3.001	3.0001
$f(x) =$	7	6.5	6.1	6.01	6.001	6.0001

There are many other possibilities, also. But from these, it appears that as x approaches 3, $f(x)$ gets close to 6. That is, if we choose x close enough to 3, we can force $f(x)$ to be as close to 6 as we wish. We conclude that 6 is the limit of f as x approaches 3. ■

Using the arrow symbolism, we can write the result of Example 2.1 as

$$f(x) \to 6 \qquad \text{as} \qquad x \to 3.$$

We also write

$$\lim_{x \to 3} f(x) = 6,$$

which is read, "the limit of $f(x)$ as x approaches 3 is 6." In general, if $f(x) \to L$ as $x \to c$, we write

$$\lim_{x \to c} f(x) = L.$$

The following examples illustrate that the limit of a function may or may not be the value of the function at the point in question and that a limit may not exist (that is, there may be no real number that is the limit).

■ **Example 2.2** Find $\lim_{x \to 1}(x^2 + 2)$.

Solution Selecting some numbers progressively closer to 1, we have the following table:

$x =$	0	0.5	0.9	0.99	0.999	. . .	1.001	1.01	1.1	1.5	2
$f(x) =$	2	2.25	2.81	2.98	2.998	. . .	3.002	3.02	3.21	4.25	6

Thus, it appears that $(x^2 + 2) \to 3$ as $x \to 1$, and the limit is 3. Note that $f(1) = 3$. This function is shown in Figure 2.2 at the top of the next page.

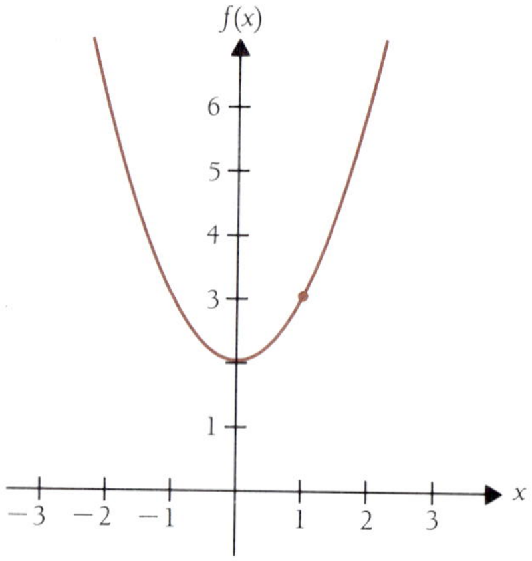

FIGURE 2.2 The graph of $f(x) = x^2 + 2$ near (1, 3)

Example 2.3 Given

$$g(x) = \begin{cases} (x-2)^2 & \text{if } x \neq 2 \\ 4 & \text{if } x = 2, \end{cases}$$

find $\lim_{x\to 2} g(x)$.

Solution Using the same approach as in Example 2.2, we pick some numbers close to 2. The rule for g tells us that as long as x is not equal to 2, the image of x is $(x-2)^2$. We have, for example, the following table:

$x =$	0	1	1.5	1.9	1.99	. . .	2.01	2.1	2.5	3	4
$f(x) =$	4	1	0.25	0.01	0.0001	. . .	0.0001	0.01	0.25	1	4

We therefore conclude that as x approaches 2 and is not equal to 2, $g(x) \to 0$. That is, we can force $g(x)$ to be as close to 0 as we like by choosing x close enough to 2, but

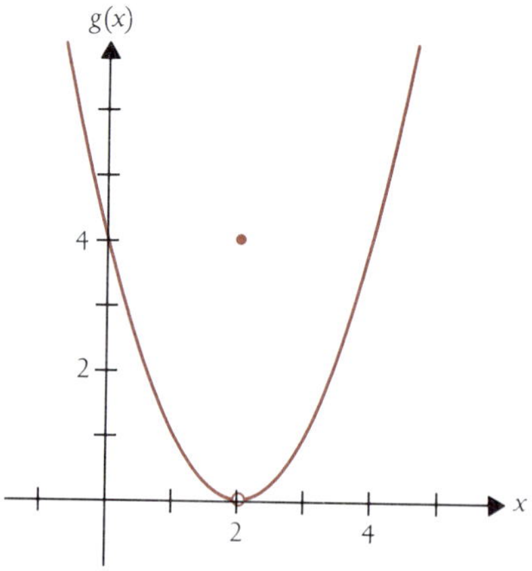

FIGURE 2.3 The graph of $g(x) = \begin{cases} (x-2)^2 & \text{if } x \neq 2 \\ 4 & \text{if } x = 2 \end{cases}$ near $x = 2$

not equal to 2. We therefore see that the limit is 0. The key is what happens *near* 2, and g(2) does not affect the limit. The graph of g is shown in Figure 2.3. ■

■ **Example 2.4** Find $\lim_{x \to 0}(1/x)$.

Solution The graph of $f(x) = 1/x$ is shown in Figure 2.4, from which it is evident that the value of $1/x$ does not approach any particular number as $x \to 0$. A table of values bears out this guess:

$x =$	1	0.5	0.1	0.01	0.001	...	−0.001	−0.01	−0.1	−0.5	−1
$f(x) =$	1	2	10	100	1000	...	−1000	−100	−10	−2	−1

When the function does not approach a number, we say the function has **no limit**, or the limit **does not exist**. Thus, the limit as $x \to 0$ of $1/x$ does not exist.

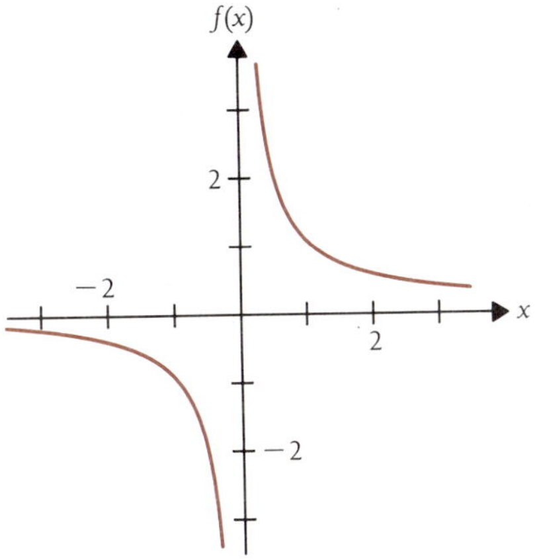

FIGURE 2.4 The graph of $f(x) = \dfrac{1}{x}$ ■

■ **Example 2.5** Find

$$\lim_{x \to 0} \frac{\sin x}{x}.$$

Solution We construct a table of values for $f(x) = (\sin x)/x$ with x close to 0. Using a calculator (set in radian mode), we find the following values:

$x =$	1	0.5	0.1	0.01	0.001	...
$f(x) =$	0.8415	0.9589	0.9983	0.99998	0.9999998	...

Negative values of x give us the same values for f as these. We therefore guess that the limit is 1. We will prove this result in Section 2.3. The graph of f is shown in Figure 2.5.

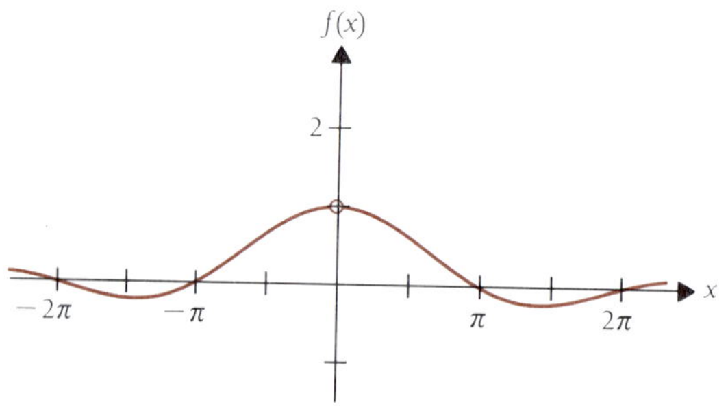

FIGURE 2.5 The graph of $f(x) = \dfrac{\sin x}{x}$

■ **Example 2.6** If f is the function whose graph is shown in Figure 2.6, then find $\lim_{x\to 1} f(x)$.

Solution It is evident that if x is close to 1 and larger than 1, then $f(x)$ will be close to 2. But if x is close to 1 and less than 1, then $f(x)$ will be close to 3. Thus, we cannot force $f(x)$ to be arbitrarily close to any single number by choosing x close to 1. Therefore, f has no limit as $x \to 1$, or, in other words, the limit does not exist.

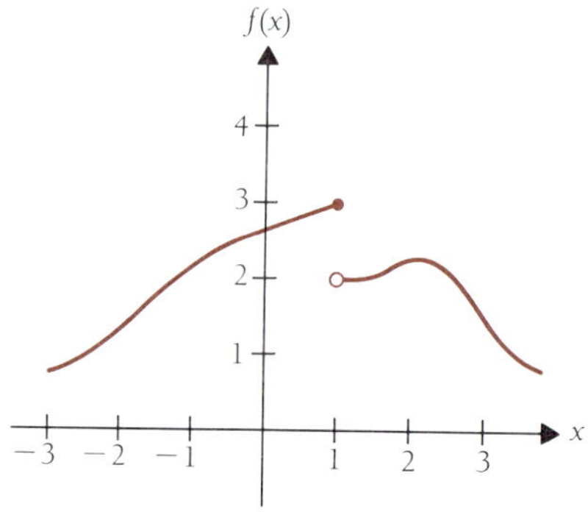

FIGURE 2.6

The numerical and graphic arguments we have been using are not to be mistaken for proofs. Before we can prove anything about limits, we must be much more precise about what a limit is.

A Closer Look at Limits

Let's return to Example 2.1 and look more closely at what is going on. As we chose numbers closer and closer to 3, the values of

$$f(x) = \frac{x^2 - 9}{x - 3}$$

got closer and closer to 6 (at least for the numbers we chose). We could not choose $x = 3$, because f was not defined at 3.

Suppose we want $f(x)$ to be within 0.01 of 6; how close to 3 must x be chosen? To answer this question, we will require the difference between $f(x)$ and 6 to be less than 0.01, and see what we can learn about x. We will use the fact that $f(x) = x + 3$ as long as $x \neq 3$. We have

$$|f(x) - 6| = |(x + 3) - 6| = |x - 3|,$$

so that requiring

$$|f(x) - 6| < 0.01$$

is the same as saying

$$|x - 3| < 0.01.$$

In order for this condition to be true, we must have

$$-0.01 < x - 3 < 0.01,$$

or

$$2.99 < \quad x \quad < 3.01.$$

That is, by choosing x within 0.01 of 3, we can make $f(x)$ be within 0.01 of 6 (see Figure 2.7).

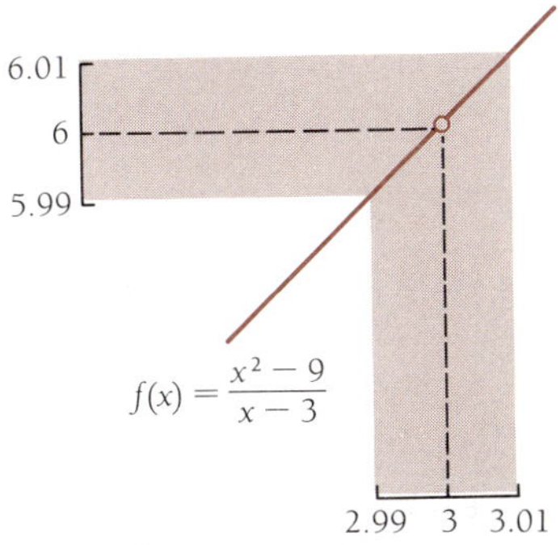

FIGURE 2.7 If x is close to 3, then $f(x)$ is close to 6

From what we have done so far, it is clear that in this example, if we choose x within a distance of δ of 3, then we force $f(x)$ to be within the same distance δ of 6. It is now evident that we can force $f(x)$ to be arbitrarily close to 6 simply by choosing x that close to 3.

Let's look at another example. Suppose we are asked to find the limit of $f(x) = 2x + 3$ as $x \rightarrow 2$. We evaluate f for various numbers close to 2, finding each time that the result is close to 7. Consider the following table, for example:

$x =$	1.9	1.99	1.999	. . .	2.001	2.01	2.1
$f(x) =$	6.8	6.98	6.998	. . .	7.002	7.02	7.2

Now let's be more precise about how $f(x)$ approaches 7 as x approaches 2. Suppose we choose x within 0.003 of 2, so that

$$|x - 2| < 0.003.$$

Then we wish to find the difference between $f(x)$ and 7, which is

$$\begin{aligned}|f(x) - 7| &= |(2x + 3) - 7| \\ &= |2x - 4| \\ &= 2|x - 2|. \end{aligned} \tag{1}$$

Hence, when $|x - 2| < 0.003$, we find that (see Figure 2.8)

$$|f(x) - 7| < 2(0.003) = 0.006.$$

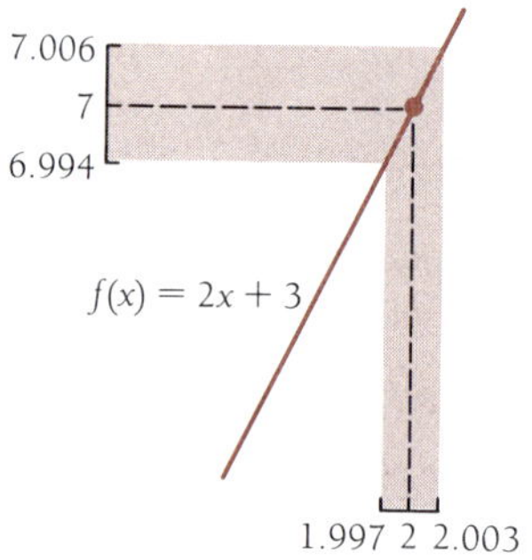

FIGURE 2.8 If x is close to 2, then $f(x)$ is close to 7

Suppose we now choose x within 0.0001 of 2. Then, from equation (1), we can see that

$$|f(x) - 7| < 0.0002,$$

or that $f(x)$ is within 0.0002 of 7. It should be clear that if we choose x within δ of 2, then $f(x)$ will be within 2δ of 7, for any positive number δ.

Now let's turn the problem around. Suppose we want to force $f(x)$ to be within some distance ε of 7. How close to 2 must x be chosen? Well, since choosing x within δ of 2 makes $f(x)$ within 2δ of 7, we can just require that $2\delta < \varepsilon$. That is, $\delta < \varepsilon/2$. Thus, if we choose x within $\varepsilon/2$ of 2, then we force $f(x)$ to be within ε of 7.

Another way to say this is as follows: For any positive number ε, no matter how small, we can find a number δ (in this case, $\varepsilon/2$) such that if x is within δ of 2, then $f(x)$ is within ε of 7. In other words,

$$|f(x) - 7| < \varepsilon \qquad \text{whenever} \qquad |x - 2| < \delta.$$

This relationship is shown graphically in Figure 2.9.

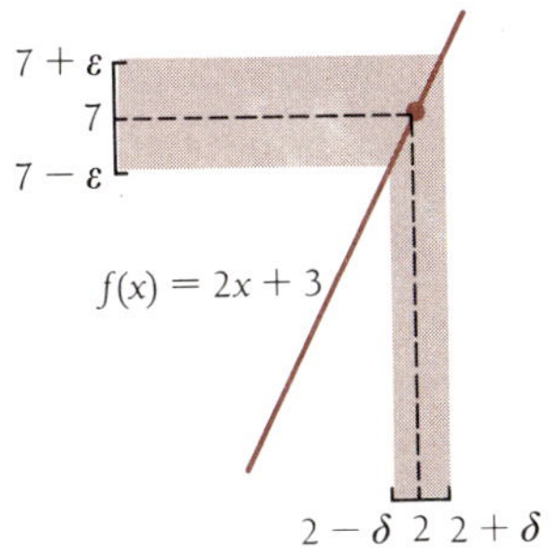

FIGURE 2.9 If x is within δ of 2, then $f(x)$ is within ε of 7

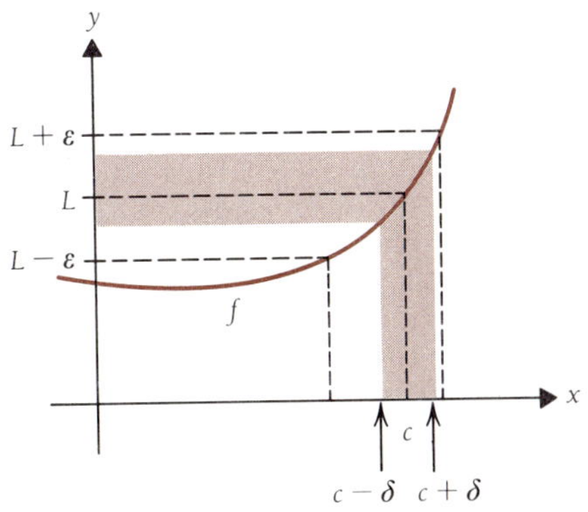

FIGURE 2.10 Given ε, there is a δ such that if x is within δ of c, then $f(x)$ is within ε of L

Generalizing from this example, suppose for a function f defined near the real number c there is a number L such that for any positive number ε there exists a number δ such that $|f(x) - L| < \varepsilon$ whenever $|x - c| < \delta$. Then we have the situation pictured in Figure 2.10. Since we can force $f(x)$ to be arbitrarily close to L by choosing x close enough to c, we want to say that L is the limit of $f(x)$ as x approaches c.

Definition of Limit

We are now ready to give the definition of limit.

DEFINITION 2.1 Let a function f be defined in an open interval containing c, except possibly at c. Then **L is the limit of $f(x)$ as x approaches c**, written

$$\lim_{x \to c} f(x) = L,$$

if and only if for any $\varepsilon > 0$ there exists $\delta > 0$ such that

$$|f(x) - L| < \varepsilon \quad \text{when} \quad 0 < |x - c| < \delta.$$

The definition specifies that f must be defined near c, but not necessarily at c itself. Thus, we can talk about the limit of f at c without f being defined at c, as in Example 2.1. Note that the specification $0 < |x - c| < \delta$ prevents x from actually being equal to c, just in case f is not defined at c.

We will illustrate the definition of limit with some examples.

Example 2.7 Given that $\lim_{x \to 3}(3x - 1) = 8$, find δ such that

$$|(3x - 1) - 8| < 0.0015 \qquad \text{when} \qquad 0 < |x - 3| < \delta.$$

Solution We first discover a relationship between $|(3x - 1) - 8|$ and $|x - 3|$. Simplifying the former, we get

$$|(3x - 1) - 8| = |3x - 9| = 3|x - 3|.$$

Thus, if we want $|(3x - 1) - 8| < 0.0015$, then

$$3|x - 3| < 0.0015,$$
$$|x - 3| < 0.0005.$$

We may therefore choose $\delta = 0.0005$, and the problem is solved. ■

Example 2.8 Prove that $\lim_{x \to c}(mx + b) = mc + b$.

Solution In order to prove that a limit has a certain value, we must show that the definition of limit is satisfied. In this case, we must show that, given any $\varepsilon > 0$, we can find $\delta > 0$ such that if $0 < |x - c| < \delta$, then $|(mx + b) - (mc + b)| < \varepsilon$. First, note that

$$|(mx + b) - (mc + b)| = |m|\,|x - c|.$$

Thus, if we choose $\delta = \varepsilon/|m|$, then $|x - c| < \delta$ implies that $|x - c| < \varepsilon/|m|$, so that

$$\begin{aligned} |(mx + b) - (mc + b)| &= |m|\,|x - c| \\ &< |m| \cdot \frac{\varepsilon}{|m|} \\ &= \varepsilon. \end{aligned}$$

That is, the definition of limit is satisfied and the proof is complete. ■

Example 2.9 Explain why the function $f(x) = \sqrt{x}$ does not have a limit as $x \to 0$.

Solution The function $f(x) = \sqrt{x}$, shown in Figure 2.11, is not defined for $x < 0$. That is, it is not defined in an open interval containing 0, as Definition 2.1 requires. Therefore, the limit does not exist.

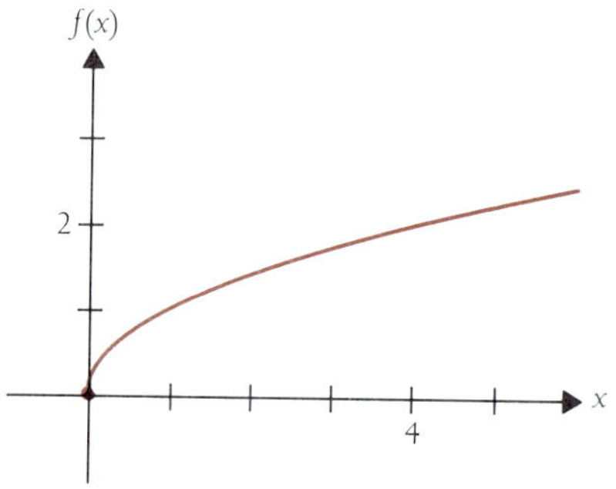

FIGURE 2.11 The graph of $f(x) = \sqrt{x}$ ■

Somehow, the situation in Example 2.9 seems artificial. Intuitively, it seems that the limit really ought to exist and equal 0. All we have to do is to keep x on the positive side of 0. This is a perfectly good approach, which we will see in Section 2.4.

Historical Note

Definition 2.1 was formulated about 1860 by Karl Weierstrass. The essential idea behind the definition had long been recognized, and definitions easily translated into this ε,δ form go back as far as Leibniz. This final step in precision was made necessary by the increasingly subtle nature of the functions under investigation. Our current form of calculus is in all essentials identical to the presentations of Weierstrass and Cauchy.

Exercises 2.1

Using graphic and computational methods, evaluate the following limits. If the limit does not exist, so state.

1. $\lim_{x\to\pi}(2x-3)$
2. $\lim_{x\to-1}(4-2x)$
3. $\lim_{x\to-2}(2x+|x|)$
4. $\lim_{x\to5}\dfrac{1}{x-4}$
5. $\lim_{x\to-3}\dfrac{x^2+x-6}{x+3}$
6. $\lim_{x\to0}x^{-2}$
7. $\lim_{x\to2}(x-2)^{-1}$
8. $\lim_{x\to2}\lfloor x\rfloor$
9. $\lim_{x\to3/2}\lfloor x\rfloor$
10. $\lim_{x\to0}\dfrac{x}{\sin x}$
11. $\lim_{x\to\pi/2}\tan x$
12. $\lim_{x\to\pi/2}\sin x$
13. $\lim_{x\to0}\dfrac{\sin x}{1+\tan x}$
14. $\lim_{x\to2}(\sin^2 x+\cos^2 x)$
15. $\lim_{x\to3}\dfrac{x}{|x|}$
16. $\lim_{x\to0}(x^2+1)^{-1}$
17. Find $\lim_{x\to1}f(x)$, given
$$f(x)=\begin{cases}0 & \text{if } x=1\\ x^2+4 & \text{if } x\neq 1.\end{cases}$$
18. Find $\lim_{x\to1}f(x)$, given
$$f(x)=\begin{cases}x & \text{if } x<1\\ -1 & \text{if } x\geq 1.\end{cases}$$
19. Let
$$g(x)=\frac{x^2+2}{2x}.$$
Let x_1 be any convenient number, and find $x_2=g(x_1)$, $x_3=g(x_2)$, and so on. Do the numbers produced seem to approach a limit? Now solve the equation $x=g(x)$. What do you observe?

Repeat Problem 19 for each of the following functions:

20. $g(x)=\dfrac{x^2+4}{2x}$
21. $g(x)=\sqrt{x}+4$
22. $g(x)=1+\dfrac{3}{x}$

In Problems 23–26, $f(x)\to L$ as $x\to c$. Specify how close x must be to c in order for $f(x)$ to be within 0.001 of L.

23. $f(x)=2x-1,\quad L=5,\quad c=3$
24. $f(x)=\dfrac{1}{x},\quad L=\dfrac{1}{2},\quad c=2$
25. $f(x)=x^2+1,\quad L=5,\quad c=2$
26. $f(x)=\sqrt{x},\quad L=4,\quad c=16$

Given the limit and the value of ε, find a number δ satisfying the definition of limit.

27. $\lim_{x\to 3}(2x-1)=5, \quad \varepsilon=0.001$

28. $\lim_{x\to -1}(x+4)=3, \quad \varepsilon=0.0024$

29. $\lim_{x\to 0}(2x-5)=-5, \quad \varepsilon=0.0004$

30. $\lim_{x\to 2}(4-3x)=-2, \quad \varepsilon=0.0009$

31. $\lim_{x\to 4}(\sqrt{x}-1)=1, \quad \varepsilon=0.002$

32. $\lim_{x\to 3}\sqrt{2x+3}=3, \quad \varepsilon=0.036$

33. $\lim_{x\to 3}\frac{1}{x}=\frac{1}{3}, \quad \varepsilon=0.0001$

34. $\lim_{x\to 1}\frac{x}{x-4}=-\frac{1}{3}, \quad \varepsilon=0.09$

35. Prove that $\lim_{x\to 9}\sqrt{x}=3$.

36. Prove that $\lim_{x\to 1}(x^2+4)=5$.

Given that $\lim_{x\to c}f(x)=3$ and $\lim_{x\to c}g(x)=-5$, guess the values of the following:

37. $\lim_{x\to c}[f(x)+g(x)]$

38. $\lim_{x\to c}5f(x)$

39. $\lim_{x\to c}f(x)g(x)$

40. $\lim_{x\to c}\frac{f(x)}{g(x)}$

41. Suppose that f is a function such that $\lim_{x\to -2}f(x)=L$, and for all x, $f(x)>5$. Which of the following statements are true?
 (a) $L>5$ (b) $L>4.5$ (c) $L>4.99$ (d) $L\geq 5$

Given the function f whose graph is shown in the figure, give the value of the limit of f for each of the following cases. If the limit does not exist, so state.

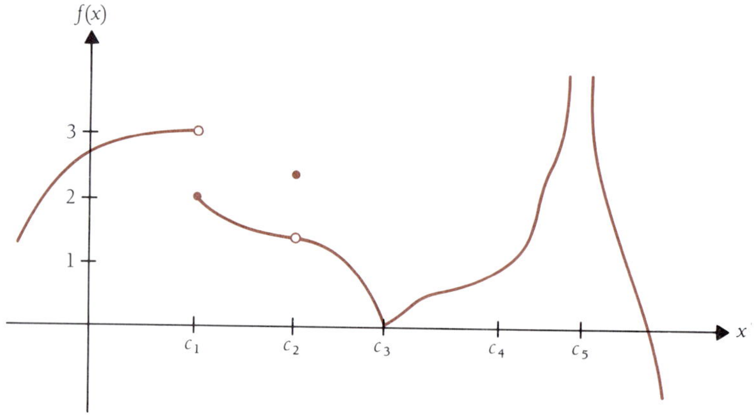

42. As $x\to c_1$

43. As $x\to c_2$

44. As $x\to c_3$

45. As $x\to c_4$

46. As $x\to c_5$

SECTION 2.2

Limit Theorems and Evaluation of Limits

In this section, we study the evaluation of limits. We have seen how to guess what the limit is and how to prove that the limit actually is what we guessed it to be, at least for some simple cases. But it is too tedious to use the definition of limit every time we want to find a limit, and it is unsafe, and sometimes impossible, just to guess the limit and let it go at that. Therefore, we develop in this section some simple results and rules, called **limit theorems**, that greatly simplify the finding of limits.

After deriving the limit theorems, we will show how to apply them to get the limits of a wide variety of functions.

Some Limit Theorems

We present in this section the simplest of the limit theorems; others will appear in the next section. While they are stated as theorems, their proofs are omitted for the most part.

The first theorem tells us that if a limit exists, it is unique.

THEOREM 2.1 If $f(x) \to L$ as $x \to c$ and $f(x) \to M$ as $x \to c$, then $L = M$.

The proof of this result is based on the fact that $f(x)$ cannot be arbitrarily close to two distinct numbers for the same value of x.

The next four theorems deal with the arithmetic of limits. We will outline the proof of only one of them; hints for the proofs of the others can be found in the exercises.

THEOREM 2.2 If $\lim_{x\to c} f(x) = L$ and $\lim_{x\to c} g(x) = M$, then $\lim_{x\to c}[f(x) + g(x)]$ exists and equals $L + M$. In short, **the limit of a sum is the sum of the limits.**

Proof Let $\varepsilon > 0$ be given. Since $f(x) \to L$ as $x \to c$, there exists δ_1 such that $|f(x) - L| < \varepsilon/2$ when $0 < |x - c| < \delta_1$. Since $g(x) \to M$ as $x \to c$, there exists δ_2 such that $|g(x) - M| < \varepsilon/2$ when $0 < |x - c| < \delta_2$. If δ is the smaller of δ_1 and δ_2, then when $0 < |x - c| < \delta$, we have

$$\begin{aligned} |[f(x) + g(x)] - (L + M)| &= |(f(x) - L) + (g(x) - M)| \\ &\le |f(x) - L| + |g(x) - M| \\ &< \frac{\varepsilon}{2} + \frac{\varepsilon}{2} \\ &= \varepsilon. \end{aligned}$$

Thus, the definition of limit is satisfied, and the theorem is proved. Note how we have used the triangle inequality in the second line of the derivation above.

The short statement, "the limit of a sum is the sum of the limits" is an easily remembered description of the theorem, and does not pretend to state all the particulars of the theorem. The hypotheses of Theorem 2.2 are that the limits of two functions exist as x approaches c, and the conclusion is that the limit of their sum as

x approaches c also exists and is equal to the sum of the limits of the two functions. You can easily see why our shortened statement of the theorem is necessarily incomplete.

THEOREM 2.3 If $\lim_{x\to c} f(x) = L$ and $\lim_{x\to c} g(x) = M$, then $\lim_{x\to c}[f(x) - g(x)]$ exists and equals $L - M$. In short, **the limit of a difference is the difference of the limits.**

THEOREM 2.4 If $\lim_{x\to c} f(x) = L$ and $\lim_{x\to c} g(x) = M$, then $\lim_{x\to c} f(x)g(x)$ exists and equals LM. In short, **the limit of a product is the product of the limits.**

THEOREM 2.5 If $\lim_{x\to c} f(x) = L$ and $\lim_{x\to c} g(x) = M \neq 0$, then $\lim_{x\to c}[f(x)/g(x)]$ exists and equals L/M. In short, **the limit of a quotient is the quotient of the limits**, provided the quotient of the limits is defined.

Some Simple Applications

The next three results enable us to apply the preceding limit theorems; they get us started, so to speak.

THEOREM 2.6 For any constant k, $\lim\limits_{x\to c} k = k$.

Proof For $f(x) = k$ and $L = k$, we have $|f(x) - L| = 0$, so the definition of limit can be satisfied by choosing any δ whatever. Stated another way, it is clear that k stays close to k, no matter what x does.

THEOREM 2.7 $\lim_{x\to c} x = c$.

It is obvious that by choosing x close enough to c, we can force x to be as close to c as we like.

THEOREM 2.8 $\lim_{x\to c}[1/(x - c)]$ does not exist.

The graph of $f(x) = 1/(x - c)$ is shown in Figure 2.12. It is evident that the closer x is to c, the larger in magnitude (or absolute value) $f(x)$ is. Since $f(x) > 0$ for $x > c$ and $f(x) < 0$ for $x < c$, there is no number L such that $f(x)$ is forced close to L by choosing x close to c. That is, the limit does not exist.

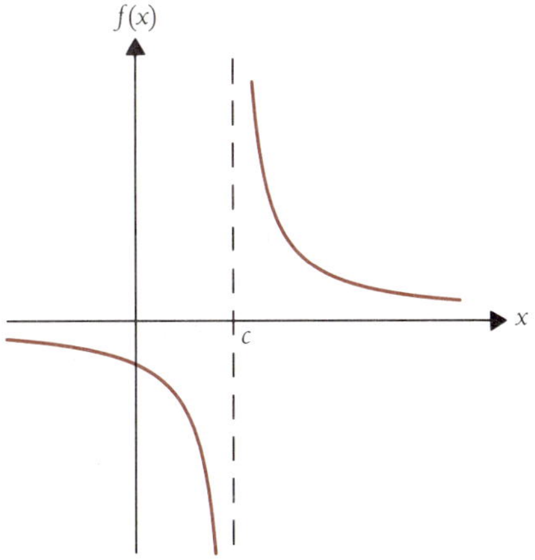

FIGURE 2.12 $f(x) = \dfrac{1}{x-c}$ has no limit at c

Evaluation of Limits

We can put these theorems together now to get values for a wide variety of limits. Consider the following example.

■ **Example 2.10** Give reasons, from the limit theorems and simple applications, for each of the following limits:

(a) $\lim_{x \to 2} (x + 1) = 3$ (b) $\lim_{x \to 2} 3x = 6$ (c) $\lim_{x \to 2} x^2 = 4$

(d) $\lim_{x \to 2} x^3 = 8$ (e) $\lim_{x \to 2} \dfrac{1}{x} = \dfrac{1}{2}$ (f) $\lim_{x \to 2} (x^2 - 3x + 5) = 3$

(g) $\lim_{x \to 2} \dfrac{3x}{x + 1} = 2$ (h) $\lim_{x \to 2} \dfrac{3x}{x - 2}$ does not exist

Solution

(a) By Theorem 2.7, $x \to 2$ as $x \to 2$, and by Theorem 2.6, $1 \to 1$ as $x \to 2$. Hence, by Theorem 2.2, $(x + 1) \to 3$ as $x \to 2$.
(b) By Theorems 2.6 and 2.7 again, $3 \to 3$ and $x \to 2$ as $x \to 2$, so by Theorem 2.4, $3x \to 6$ as $x \to 2$.
(c) Since x^2 is the product of x with itself, and $x \to 2$ as $x \to 2$ by Theorem 2.7, Theorem 2.4 again implies that $x^2 \to 4$ as $x \to 2$.
(d) Again we apply Theorem 2.4, using the facts that $x \to 2$ and $x^2 \to 4$ to get $x^3 \to 8$.
(e) Since $x \to 2 \neq 0$ and $1 \to 1$, Theorem 2.5 says that $1/x \to \frac{1}{2}$ as $x \to 2$.
(f) This result follows from parts (c) and (b) above, together with Theorems 2.6, 2.2, and 2.3.
(g) This result comes from parts (a) and (b), together with Theorem 2.5.

(h) Just suppose that the limit does exist, and that it equals L. Since $\lim_{x\to 2} 3x = 6$, we would then have

$$\begin{aligned}\frac{L}{6} &= \left(\lim_{x\to 2}\frac{1}{3x}\right)\left(\lim_{x\to 2}\frac{3x}{x-2}\right)\\ &= \lim_{x\to 2}\left(\frac{1}{3x}\right)\left(\frac{3x}{x-2}\right)\\ &= \lim_{x\to 2}\frac{1}{x-2}.\end{aligned}$$

But this is impossible, by Theorem 2.8. Therefore, the original limit cannot exist. ■

This rather extended example illustrates how functions that are put together with arithmetic operations have limits that can be evaluated by using the limit theorems. We can make some specific generalizations, as follows. (Keep in mind that these statements are in short form, and are therefore incomplete.)

RULE 2.1 **The limit of a constant times a function is the constant times the limit of the function.** That is,

$$\lim_{x\to c} k\cdot f(x) = k\lim_{x\to c} f(x).$$

This statement is a combination of Theorems 2.4 and 2.6, and is illustrated by Example 2.10(b).

RULE 2.2 **The limit of a polynomial function as $x\to c$ is found by evaluating the polynomial function at c.** That is, if $p(x)$ is a polynomial function, then

$$\lim_{x\to c} p(x) = p(c).$$

This fact comes by repeated application of Theorems 2.2, 2.3, 2.4, 2.6, and 2.7, since a polynomial function is just a linear combination of powers of x. It is illustrated by Example 2.10(f).

RULE 2.3 **If a rational function is defined at c, then the limit of the function as $x\to c$ is found by evaluating the function at c.** That is, if $f(x) = p(x)/q(x)$ is the ratio of polynomials and the denominator $q(c)\neq 0$, then

$$\lim_{x\to c} f(x) = f(c).$$

This comes from applying Theorem 2.5 to Rule 2.2. It is illustrated by Example 2.10(g).

RULE 2.4

If a rational function is not defined at c, but its numerator is nonzero at c, then the limit of the function as $x \to c$ does not exist. That is, if $f(x) = p(x)/q(x)$ is the ratio of polynomials, $q(c) = 0$, and $p(c) \neq 0$, then $\lim_{x \to c} f(x)$ does not exist. If the limit did exist, say $\lim_{x \to c} f(x) = L$, then since $p(x) = f(x)q(x)$, we would have

$$p(c) = \lim_{x \to c} p(x) = \lim_{x \to c} f(x)q(x) = L \cdot q(c) = 0,$$

contrary to hypothesis. Thus, the limit cannot exist. This is illustrated by Example 2.10(h). Note that this result is also a generalization of Theorem 2.8.

There is one major case that is not included in the above rules, namely the case that a rational function is not defined at c and both the numerator and denominator vanish at c. Let's go back to an example of that sort.

■ **Example 2.11** Prove that

$$\lim_{x \to 3} \frac{x^2 - 9}{x - 3} = 6.$$

(See Example 2.1.)

Solution Let $f(x) = (x^2 - 9)/(x - 3)$. Given $\varepsilon > 0$, we must find a number $\delta > 0$ such that $|f(x) - 6| < \varepsilon$ when $0 < |x - 3| < \delta$. Note that the proviso $0 < |x - 3|$ prevents x from equaling 3. But as long as $x \neq 3$, the function $f(x)$ reduces to $x + 3$. Therefore, we see that $|f(x) - 6| = |(x + 3) - 6| = |x - 3|$ will be less than ε as long as δ is less than ε and $0 < |x - 3| < \delta$. That is, the definition of limit allows us to simplify f by factoring out $x - 3$. Hence,

$$\lim_{x \to 3} \frac{x^2 - 9}{x - 3} = \lim_{x \to 3} (x + 3) = 6.$$ ■

The technique of Example 2.11 applies in general, and leads to the following rule.

RULE 2.5

If both the numerator and denominator of a rational function are 0 at c, then the limit of the function as $x \to c$ is the same as the limit of the simplified function obtained by factoring out $x - c$.

The following example illustrates the use of this rule.

■ **Example 2.12** Find the following limits:

(a) $\lim_{x \to 0} \dfrac{x(x - 1)}{x(x + 1)}$ (b) $\lim_{x \to 3} \dfrac{\frac{1}{3} - 1/x}{x - 3}$ (c) $\lim_{x \to 5} \dfrac{x^2 - 25}{x^2 - 10x + 25}$

Solution

(a) Since both the numerator and denominator vanish at $x = 0$, we simplify by factoring x out of both numerator and denominator, getting

$$\lim_{x\to 0} \frac{x(x-1)}{x(x+1)} = \lim_{x\to 0} \frac{x-1}{x+1} = \frac{-1}{1} = -1.$$

(b) Again we have both numerator and denominator approaching 0, so we simplify:

$$\lim_{x\to 3} \frac{\frac{1}{3} - 1/x}{x-3} = \lim_{x\to 3} \frac{x-3}{3x(x-3)} = \lim_{x\to 3} \frac{1}{3x} = \frac{1}{9}.$$

(c) We proceed here just as we have been doing:

$$\lim_{x\to 5} \frac{x^2-25}{x^2-10x+25} = \lim_{x\to 5} \frac{(x+5)(x-5)}{(x-5)^2} = \lim_{x\to 5} \frac{x+5}{x-5}.$$

But by Rule 2.4, this last limit does not exist. Hence, the original limit does not exist either. ■

We can now find the limit of any rational function by following the rules stated above. But many functions are not rational functions. To find their limits conveniently, we need some more limit theorems.

More Limit Theorems

First, we need a limit theorem to tell us about inverse functions. The following theorem is a little technical, but it does the trick for us.

THEOREM 2.9 If the function f has an inverse function f^{-1} on an open interval containing point a, and $\lim_{x\to a} f(x) = b$, then $\lim_{x\to b} f^{-1}(x)$ exists and is a.

The proof of this theorem depends on the fact that the domain and range of f are the range and domain, in that order, of f^{-1}. It is illustrated in Figure 2.13. You can

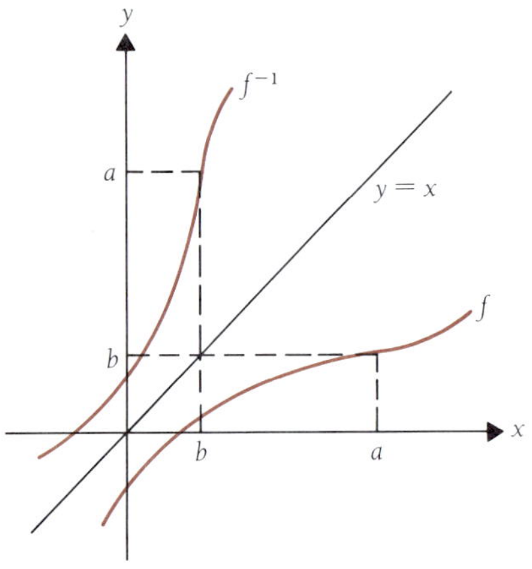

FIGURE 2.13 If $f(x) \to b$ as $x \to a$, then $f^{-1}(x) \to a$ as $x \to b$

find some hints for constructing the proof in the exercises. Instead of proving Theorem 2.9 here, we will use it on some examples.

Example 2.13 Prove that if $c > 0$, then $\lim_{x\to c}\sqrt{x} = \sqrt{c}$.

Solution As long as $c > 0$, the difficulty mentioned in Example 2.9 (Section 2.1) does not occur. To prove that $\sqrt{x} \to \sqrt{c}$ as $x \to c$, we note that the square root function is the inverse of the power function $f(x) = x^2$. Then as $x \to \sqrt{c}$, $f(x) \to (\sqrt{c})^2 = c$ (we simply evaluate a polynomial to find its limit). Therefore, by Theorem 2.9, with $a = \sqrt{c}$ and $b = c$, we get $\sqrt{x} \to \sqrt{c}$ as $x \to c$. ■

Example 2.13 can be generalized for any root function by using $x^{1/n}$ in place of $\sqrt{x}$. If n is even, then 0 is an endpoint of the domain of the root function $x^{1/n}$, and the same restriction as in Example 2.13 applies. If n is odd, the root function is defined on all of $\mathbb{R}$, and there is no restriction. We restate these comments as follows.

RULE 2.6 **The limit of the root function $f(x) = x^{1/n}$ as $x \to c$ is found by evaluating the function at c, provided $c > 0$ if n is even.**

Example 2.14 Find the limits:

(a) $\lim_{x\to 4}\sqrt{x}$ (b) $\lim_{x\to 9} x\sqrt{x}$

Solution (a) Applying Rule 2.6 directly, we have

$$\lim_{x\to 4}\sqrt{x} = \sqrt{4} = 2.$$

(b) Since $x\sqrt{x} = x^{3/2} = (\sqrt{x})^3$, we first use Rule 2.6 and then Theorem 2.4 to get

$$\lim_{x\to 9} x\sqrt{x} = (\sqrt{9})^3 = 3^3 = 27.$$ ■

This example illustrates that the other limit theorems can be combined with Theorem 2.9 to find limits of combinations of root functions with other functions. We can also use a form of Rule 2.5 when root functions are involved, but the simplification of the function may involve more than just factoring out $x - c$. A rationalization of the numerator or denominator may be helpful, as in the following examples.

Example 2.15 Find the limits:

(a) $\lim_{x\to 1}\dfrac{\sqrt{x}-1}{x-1}$ (b) $\lim_{x\to 4}\dfrac{1/\sqrt{x}-\frac{1}{2}}{x-4}$

Solution **(a)** Here, we can use a factorization technique:

$$\lim_{x\to 1}\frac{\sqrt{x}-1}{x-1}=\lim_{x\to 1}\frac{\sqrt{x}-1}{(\sqrt{x}+1)(\sqrt{x}-1)}$$
$$=\lim_{x\to 1}\frac{1}{\sqrt{x}+1}$$
$$=\frac{1}{2}.$$

(b) Here, we find that some algebraic simplification, followed by rationalizing the numerator, leads to the limit:

$$\lim_{x\to 4}\frac{1/\sqrt{x}-\frac{1}{2}}{x-4}=\lim_{x\to 4}\frac{2-\sqrt{x}}{2\sqrt{x}(x-4)}\cdot\frac{2+\sqrt{x}}{2+\sqrt{x}}$$
$$=\lim_{x\to 4}\frac{4-x}{2\sqrt{x}(x-4)(2+\sqrt{x})}$$
$$=\lim_{x\to 4}\frac{-1}{2\sqrt{x}(2+\sqrt{x})}$$
$$=\frac{-1}{16}.$$

■

Another limit theorem that is extremely useful tells us how to handle composite functions when finding limits. After we have discussed continuity, this theorem will be stated much more simply, but for now the following version will do.

THEOREM 2.10 If $\lim_{x\to c} f(x) = L$ and g is a function such that $g(x) \to g(L)$ as $x \to L$, then $\lim_{x\to c} g \circ f(x) = g(L)$.

Another way to say the same thing is that if g is a function whose limit we can find by evaluating, then we can find the limit of $g(f(x))$ by evaluating g at the limit of f. That is,

$$\lim_{x\to c} g(f(x)) = g(\lim_{x\to c} f(x)).$$

Consider an example or two.

■ **Example 2.16** Find the limits:

(a) $\lim_{x\to 5}(3x-7)^{2/3}$ **(b)** $\lim_{x\to 3}\frac{x-3}{\sqrt{x+1}-2}$

Solution

(a) Since $3x - 7 \to 8$ as $x \to 5$, we use Theorem 2.10 to get

$$\lim_{x\to 5} (3x - 7)^{2/3} = 8^{2/3} = 4.$$

(b) Since $x + 1 \to 4$ as $x \to 3$, it follows that $\sqrt{x+1} \to 2$ by Theorem 2.10. Therefore, in this limit, both numerator and denominator approach 0. In looking for a way to simplify, we find that rationalizing the denominator helps:

$$\begin{aligned}\lim_{x\to 3} \frac{x-3}{\sqrt{x+1}-2} &= \lim_{x\to 3} \frac{(x-3)(\sqrt{x+1}+2)}{(\sqrt{x+1}-2)(\sqrt{x+1}+2)} \\ &= \lim_{x\to 3} \frac{(x-3)(\sqrt{x+1}+2)}{(x+1)-4} \\ &= \lim_{x\to 3} (\sqrt{x+1}+2) \\ &= 4.\end{aligned}$$

■

We will continue with some more limit theorems in the next section.

Exercises 2.2

Find the limit, if it exists. If not, so state.

1. $\lim_{x\to -2} 4x$
2. $\lim_{x\to 0.5} (-3x)$
3. $\lim_{x\to 0} \frac{2}{x}$
4. $\lim_{x\to 3} \frac{-1}{3-x}$
5. $\lim_{x\to 1} (4x - 3)$
6. $\lim_{x\to 0} (x^2 - 6x + 5)$
7. $\lim_{x\to 0} \frac{x-1}{x-2}$
8. $\lim_{x\to -2} \frac{x-1}{x-2}$
9. $\lim_{x\to 4} \frac{x^2+4}{x^2-4}$
10. $\lim_{x\to 2} \frac{x^2-4}{x^2+4}$
11. $\lim_{x\to 2} \frac{x-1}{x-2}$
12. $\lim_{x\to 2} \frac{x^2-4}{x+2}$
13. $\lim_{x\to 1} \frac{x^3+1}{x^3-1}$
14. $\lim_{x\to -2} \frac{x^2-4}{x+2}$
15. $\lim_{x\to 0} \frac{x^2-x}{3x}$
16. $\lim_{x\to 0} \frac{x^2-x}{x^3+2x}$
17. $\lim_{x\to 3} \frac{9-x^2}{x-3}$
18. $\lim_{x\to 1} \frac{x-1}{1-x^2}$
19. $\lim_{x\to -1} \frac{x-1}{1-x^2}$
20. $\lim_{x\to 2} \frac{\frac{1}{2}-1/x}{x-2}$
21. $\lim_{x\to 1} \frac{1/x-1}{x-1}$
22. $\lim_{x\to 3} \frac{2-6/x}{x^2-9}$
23. $\lim_{x\to -1} \frac{3/x-3x}{x^4-1}$
24. $\lim_{x\to \sqrt{2}} \frac{x-\sqrt{2}}{x^2-2}$
25. $\lim_{x\to 3} \sqrt{x+1}$
26. $\lim_{x\to 9} \left(1 - \frac{3}{\sqrt{x}}\right)$
27. $\lim_{x\to 1} \frac{3x-3}{\sqrt{x}-1}$
28. $\lim_{x\to 4} \frac{\sqrt{x}-2}{\sqrt{x}-4}$
29. $\lim_{x\to 2} \frac{3-\sqrt{x+7}}{x-2}$
30. $\lim_{x\to 7} \frac{\sqrt{x+2}}{x+2}$
31. $\lim_{x\to 1} \frac{\sqrt{x}+1}{\sqrt{x}-1}$
32. $\lim_{x\to 0} \frac{x^2+1}{\sqrt{x}}$
33. $\lim_{x\to 2} \frac{\sqrt{2x+1}-\sqrt{5}}{x-2}$
34. $\lim_{x\to 3} \frac{\sqrt{3x}-3}{x^2-9}$
35. $\lim_{x\to -1} \frac{\sqrt{3-x}-2}{x+1}$
36. $\lim_{x\to 1} \frac{x^4-x}{\sqrt{7x+9}-4}$
37. $\lim_{x\to 0} x^{3/2}$
38. $\lim_{x\to -1} x^{2/3}$
39. $\lim_{x\to 0} \frac{x^{3/5}-x}{2x}$
40. $\lim_{x\to 1} \frac{x^{2/3}-1}{x-1}$

Give reasons from the results of this section for the following statements:

41. $\lim_{x \to 4} (x + 3) = 7$

42. $\lim_{x \to 2} (x^3 + 4x^2 - x + 6) = 28$

43. $\lim_{x \to 1} \frac{x^2}{x - 1}$ does not exist

44. $\lim_{x \to 2} \frac{x^3 - 8}{x^2 - 4} = \lim_{x \to 2} \frac{x^2 + 2x + 4}{x + 2} = 3$

45. $\lim_{x \to 5} 14g(x) = 14 \lim_{x \to 5} g(x)$

46. $\lim_{x \to 2} \sqrt{g(x)} = \sqrt{\lim_{x \to 2} g(x)}$

47. If $\lim_{x \to 2} f(x) = 3$, then $\lim_{x \to 3} f^{-1}(x) = 2$.

48. $\lim_{x \to c} \frac{f(x)}{g(x)} = \frac{\lim_{x \to c} f(x)}{\lim_{x \to c} g(x)}$

49. Use Theorem 2.2 and Rule 2.1 to prove Theorem 2.3.

50. Prove Theorem 2.1.

51. Prove Theorem 2.4 for the case $L = 0$.

52. Prove Theorem 2.4 for the case $L \neq 0$. [*Hint:* Write $f(x)g(x) - LM$ as $f(x)g(x) - f(x)M + f(x)M - LM$ and proceed as in the proof of Theorem 2.2, letting $\delta_1 = \varepsilon/2|M|$ and $\delta_2 = \varepsilon|M|/(2|LM| + \varepsilon)$.]

53. If $\lim_{x \to c} g(x) = M$ and $M \neq 0$, prove that

$$\lim_{x \to c} \frac{1}{g(x)} = \frac{1}{M}.$$

54. Use Problem 53 and Theorem 2.4 to prove Theorem 2.5.

55. Prove that $\lim_{x \to c} f(x) = 0$ if and only if $\lim_{x \to c} |f(x)| = 0$.

56. Prove that the following are equivalent:

$$\lim_{x \to c} f(x) = L, \qquad \lim_{x \to c} [f(x) - L] = 0, \qquad \lim_{x \to c} |f(x) - L| = 0.$$

57. Find a function f such that for some particular number c, $\lim_{x \to c} f(x)$ does not exist, but $\lim_{x \to c} |f(x)|$ does exist.

58. Prove Theorem 2.10.

59. Prove Theorem 2.9 for the case that f is increasing (that is, if $p < q$, then $f(p) < f(q)$ for all p and q). [*Hint:* Use Figure 2.13.]

SECTION 2.3

More Limit Theorems; Trigonometric Limits

A few more limit theorems are introduced here and then are applied to the trigonometric functions.

More Limit Theorems

Our first result concerns the relationship between the sizes of two functions and their limits.

THEOREM 2.11

If $\lim_{x\to c} f(x) = L$, $\lim_{x\to c} g(x) = M$, and $f(x) \le g(x)$ for all x in some open interval containing c, then $L \le M$.

The proof of this theorem is left as an exercise. It is illustrated in Figure 2.14. It also has some immediate consequences that are very useful.

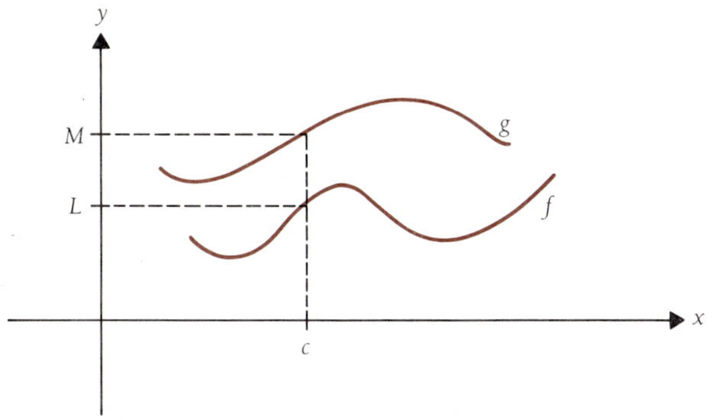

FIGURE 2.14 If $f(x) \le g(x)$, then $\lim_{x\to c} f(x) \le \lim_{x\to c} g(x)$

THEOREM 2.12

If $\lim_{x\to c} f(x) = \lim_{x\to c} g(x) = L$ and h is a function such that $f(x) \le h(x) \le g(x)$ for all x in an open interval containing c, then $\lim_{x\to c} h(x)$ exists and equals L.

It is easy to see how Theorem 2.12 follows from Theorem 2.11 if $\lim_{x\to c} h(x)$ exists; see also Figure 2.15. We will usually refer to Theorem 2.12 as the **squeeze play**.

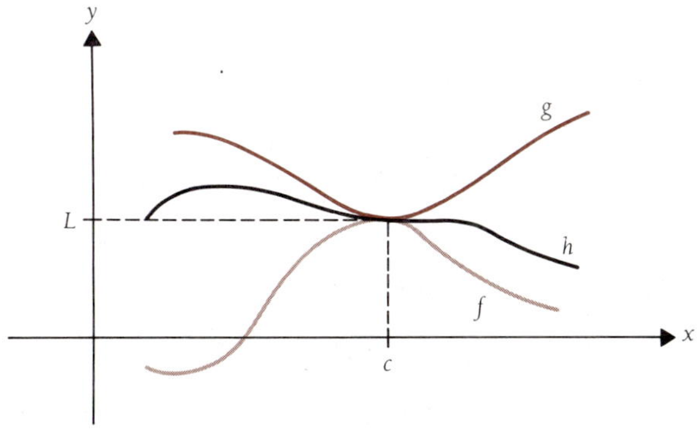

FIGURE 2.15 If $f(x) \le h(x) \le g(x)$ and $\lim_{x\to c} f(x) = \lim_{x\to c} g(x) = L$, then $\lim_{x\to c} h(x) = L$

■ **Example 2.17** If $0 \leq |f(x)| \leq |x|$, find $\lim_{x\to 0} f(x)$.

Solution Since $\lim_{x\to 0} 0 = 0$ and $\lim_{x\to 0}|x| = 0$, the squeeze play asserts that

$$\lim_{x\to 0}|f(x)| = 0,$$

as illustrated in Figure 2.16. Hence, by Problem 55 of Section 2.2,

$$\lim_{x\to 0} f(x) = 0.$$

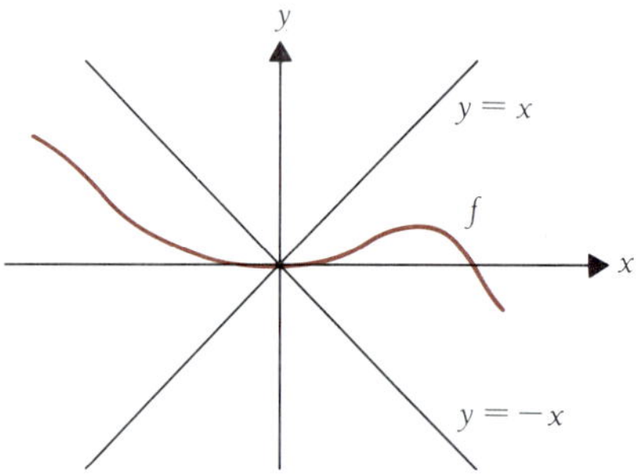

FIGURE 2.16 If $|f(x)| \leq |x|$, then $\lim_{x\to 0} f(x) = 0$ ■

Before stating the next theorem, we need a definition.

DEFINITION 2.2 A function f is **bounded on set S** if and only if there exists a positive real number B such that $|f(x)| \leq B$ for all $x \in S$. We say f is **bounded** when f is bounded on its domain.

Thus, for example, the cosine function is bounded, because $|\cos x| \leq 1$ for all real x. On the other hand, the function $f(x) = 1/x$ is not bounded, since $|1/x|$ can be made larger than any given real number B by choosing x close enough to 0.

THEOREM 2.13 If $\lim_{x\to c} f(x) = 0$ and the function g is bounded on an open interval containing c, then $\lim_{x\to c} f(x)g(x) = 0$.

Proof Let $|g(x)| \leq B$ for some real number B. Then

$$0 \leq |f(x)g(x)| \leq |f(x)| \cdot B$$

for all x in an interval containing c. Since

$$\lim_{x\to c}[|f(x)| \cdot B] = B \cdot \lim_{x\to c}|f(x)| = 0,$$

it follows from the squeeze play that

$$\lim_{x \to c} |f(x)g(x)| = 0.$$

The result then follows by Problem 55 of Section 2.2.

Example 2.18 Find $\lim_{x \to 0} x^2 \cos x$.

Solution Since $x^2 \to 0$ as $x \to 0$ and $\cos x$ is bounded, the limit is 0 by Theorem 2.13. ■

Some Trigonometric Limits

Now we discuss the limits of trigonometric functions. We will establish several claims that will allow us to find the limits of a variety of combinations of the trigonometric functions. It is instructive to see how the claims follow from previous results, so we will outline their proofs. Finally, we will summarize these results as general rules, and apply them to a variety of situations.

We will start simple and build our way up. We start with the sine function at 0.

CLAIM 1 $\lim_{x \to 0} \sin x = 0$.

Proof Since $x \to 0$, we may assume that $-\pi/2 \leq x \leq \pi/2$. For $0 \leq x \leq \pi/2$, we see from Figure 2.17 that

$$0 \leq \sin x \leq x.$$

For $-\pi/2 \leq x < 0$, we have

$$x \leq \sin x < 0.$$

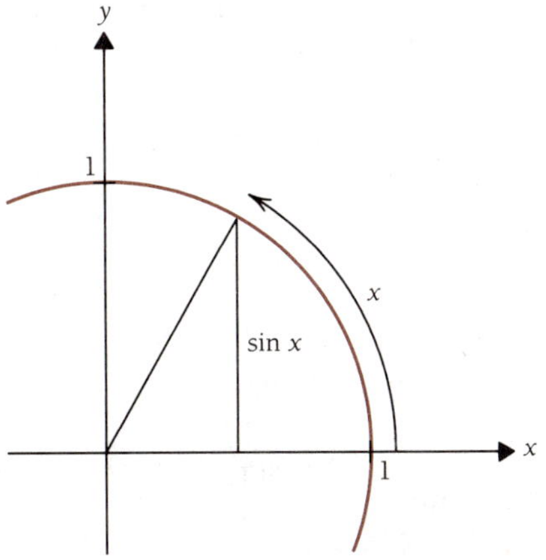

FIGURE 2.17 $\sin x < x$

(You should verify this.) Hence, for $|x| \le \pi/2$, we have

$$0 \le |\sin x| \le |x|.$$

Since $0 \to 0$ and $|x| \to 0$ as $x \to 0$, it follows from the squeeze play that

$$\lim_{x \to 0} |\sin x| = 0.$$

The claim is then proved by applying Problem 55 of Section 2.2.

CLAIM 2 $\lim_{x \to c} \sin x = \sin c$ for any real number c.

Proof First, note that the sine function is defined for all real numbers c, and by Claim 1, the limit of the sine function at 0 can be found by evaluating the sine at 0. Now there is a trigonometric identity that states

$$\sin x - \sin c = 2 \sin \frac{(x-c)}{2} \cos \frac{(x+c)}{2}.$$

As $x \to c$, $(x - c) \to 0$, so by Claim 1 and Theorem 2.10,

$$\sin \frac{(x-c)}{2} \to 0.$$

Since the cosine function is bounded, Theorem 2.13 tells us that

$$\lim_{x \to c} (\sin x - \sin c) = 0.$$

The claim now follows from Problem 56 of Section 2.2.

Another way to state Claim 2 is that **the limit of the sine function as $x \to c$ is found by evaluating the sine function at c.**

■ **Example 2.19** Find the limits:

(a) $\lim_{x \to \pi} \sin x$ (b) $\lim_{x \to \pi/2} \sin\left(x - \frac{\pi}{3}\right)$

Solution (a) By Claim 2, $\lim_{x \to \pi} \sin x = \sin \pi = 0$.

(b) By Theorem 2.10 and Claim 2, we have

$$\lim_{x \to \pi/2} \sin\left(x - \frac{\pi}{3}\right) = \sin\left(\frac{\pi}{2} - \frac{\pi}{3}\right) = \sin \frac{\pi}{6} = \frac{1}{2}.$$ ■

CLAIM 3 $\lim_{x\to c} \cos x = \cos c$ for all real numbers c.

Proof Using the identity $\cos x = \sin(\pi/2 - x)$, we have

$$\begin{aligned}\lim_{x\to c} \cos x &= \lim_{x\to c} \sin\left(\frac{\pi}{2} - x\right)\\ &= \sin\left[\lim_{x\to c}\left(\frac{\pi}{2} - x\right)\right] \qquad \text{(By Theorem 2.10)}\\ &= \sin\left(\frac{\pi}{2} - c\right)\\ &= \cos c.\end{aligned}$$

■ **Example 2.20** Find the limits:

(a) $\lim_{x\to 0} \cos x$ (b) $\lim_{x\to \pi/4} \cos x$ (c) $\lim_{x\to \pi/4} \tan x$

Solution (a) By Claim 3, we have $\lim_{x\to 0} \cos x = \cos 0 = 1$.
(b) In the same way, we have $\lim_{x\to \pi/4} \cos x = \cos(\pi/4) = \sqrt{2}/2$.
(c) Here, we also use Theorem 2.5:

$$\lim_{x\to \pi/4} \tan x = \lim_{x\to \pi/4} \frac{\sin x}{\cos x} = \frac{\sqrt{2}/2}{\sqrt{2}/2} = 1.$$ ■

These claims lead to the following general statement.

THEOREM 2.14 If f is any arithmetic combination or composition of trigonometric and algebraic functions and is *defined at c*, then $\lim_{x\to c} f(x) = f(c)$.

Proof Since any trigonometric function can be expressed in terms of the sine function and/or cosine function, this theorem follows from Claims 2 and 3 and the results of Section 2.2.

■ **Example 2.21** Find

$$\lim_{x\to \pi/2} \frac{\csc x + \tan 2x}{1 + \sin x}.$$

Solution Since the function

$$\frac{\csc x + \tan 2x}{1 + \sin x}$$

is defined at $\pi/2$, we simply evaluate to find the limit:

$$\lim_{x\to\pi/2} \frac{\csc x + \tan 2x}{1 + \sin x} = \frac{\csc(\pi/2) + \tan \pi}{1 + \sin(\pi/2)} = \frac{1+0}{1+1} = \frac{1}{2}.$$

For a combination of trigonometric and algebraic functions *not defined at c*, there are two cases we will consider. If the function can be written as a fraction whose denominator approaches 0 and whose numerator does not approach 0, then the limit does not exist. This is analogous to Rule 2.4.

If the function has both its numerator and denominator approaching 0, then further results are needed. The first of these is the following theorem.

THEOREM 2.15

$$\lim_{x\to 0} \frac{\sin x}{x} = 1.$$

Proof Since $x \to 0$, we will assume $-\pi/2 < x < \pi/2$. For $0 < x < \pi/2$, we display x as a central angle in the unit circle, as in Figure 2.18. We have

$$\text{Area of } \Delta OBC < \text{Area of sector } OUB < \text{Area of } \Delta OUD. \tag{1}$$

The area of ΔOBC is $\frac{1}{2}\cos x \sin x$. The area of a sector of radius r and central angle θ is $A = \frac{1}{2}r^2\theta$, so the area of sector OUB is $x/2$. The area of ΔOUD is $\frac{1}{2}\tan x$. (By the way, the fact that UD has length $\tan x$ suggests where the tangent function got its name.) The inequality (1) therefore becomes

$$\frac{1}{2}\cos x \sin x < \frac{1}{2}x < \frac{1}{2}\tan x.$$

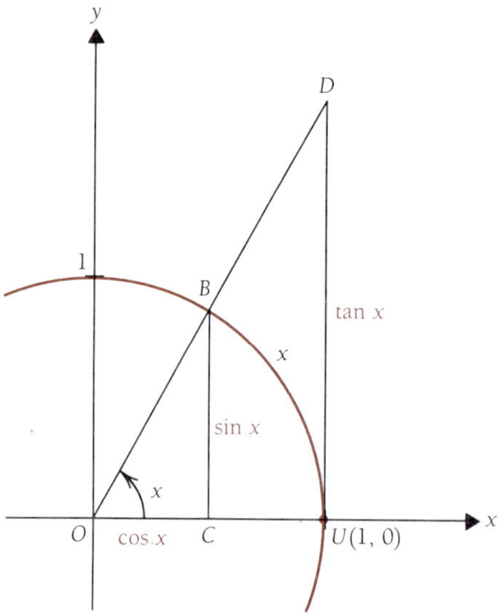

FIGURE 2.18 Area $\Delta OBC <$ Area sector $OUB <$ Area ΔOUD

Multiplying by 2 and dividing by sin x (which is positive), we get

$$\cos x < \frac{x}{\sin x} < \frac{1}{\cos x}.$$

Since each term here is positive, we can invert all terms, getting

$$\frac{1}{\cos x} > \frac{\sin x}{x} > \cos x. \qquad (2)$$

In a similar manner, we obtain the same relationship (2) when we take $-\pi/2 < x < 0$ (which you should verify). This means that inequality (2) holds for all $x \in (-\pi/2, \pi/2)$ except for $x = 0$.

Now $\cos x \to 1$ as $x \to 0$, so also $1/(\cos x) \to 1$. The theorem therefore follows from inequality (2) by the squeeze play.

The following examples illustrate how Theorem 2.15 can be applied.

Example 2.22 Find the limits:

(a) $\lim_{x\to 0} \frac{x}{\sin x}$ **(b)** $\lim_{x\to 0} \frac{\tan x}{x}$ **(c)** $\lim_{x\to 0} \frac{\sin 2x}{x}$

(d) $\lim_{x\to 0} \frac{1-\cos x}{x}$ **(e)** $\lim_{x\to 0} \frac{\tan 3x}{\tan 5x}$ **(f)** $\lim_{x\to \pi} \frac{\sin x}{x-\pi}$

Solution **(a)** Since $x/(\sin x)$ is just the reciprocal of $(\sin x)/x$, the limit is the reciprocal of 1. That is,

$$\lim_{x\to 0} \frac{x}{\sin x} = 1.$$

(b) Using a simple identity, we get

$$\lim_{x\to 0} \frac{\tan x}{x} = \lim_{x\to 0} \left(\frac{\sin x}{x} \cdot \frac{1}{\cos x}\right) = 1 \cdot 1 = 1.$$

(c) Here, we could use an identity, but it is easier to use the following trick:

$$\lim_{x\to 0} \frac{\sin 2x}{x} = \lim_{x\to 0} \left(\frac{2}{2} \cdot \frac{\sin 2x}{x}\right) = 2 \lim_{x\to 0} \frac{\sin 2x}{2x} = 2 \cdot 1 = 2.$$

That $(\sin 2x)/2x \to 1$ follows from Theorem 2.15, since $2x \to 0$ as $x \to 0$.

(d) With this limit, we do some algebraic manipulation before taking the limit:

$$\lim_{x\to 0} \frac{1-\cos x}{x} = \lim_{x\to 0} \left(\frac{1-\cos x}{x} \cdot \frac{1+\cos x}{1+\cos x}\right)$$
$$= \lim_{x\to 0} \frac{1-\cos^2 x}{x(1+\cos x)}$$

$$\lim_{x\to 0} \frac{1-\cos^2 x}{x(1+\cos x)} = \lim_{x\to 0} \frac{\sin^2 x}{x(1+\cos x)}$$
$$= \lim_{x\to 0}\left(\frac{\sin x}{x}\cdot\frac{\sin x}{1+\cos x}\right)$$
$$= 1\cdot\frac{0}{2} = 0.$$

(e) Here, we use some simple identities again:

$$\lim_{x\to 0}\frac{\tan 3x}{\tan 5x} = \lim_{x\to 0}\left(\frac{3\sin 3x}{3x}\cdot\frac{5x}{5\sin 5x}\cdot\frac{\cos 5x}{\cos 3x}\right)$$
$$= 3\cdot\frac{1}{5}\cdot\frac{1}{1} = \frac{3}{5}.$$

(f) Here, we use the identity $\sin x = \sin(\pi - x)$:

$$\lim_{x\to\pi}\frac{\sin x}{x-\pi} = \lim_{x\to\pi}\frac{-\sin(\pi-x)}{\pi-x} = -1.$$

Theorem 2.15 applies because $(\pi - x) \to 0$ as $x \to \pi$. ■

Summarizing these examples, we conclude that if both the numerator and denominator of a function approach 0, then we try to rewrite the function and use the limit theorems and Theorem 2.15.

Exercises 2.3

Find the limit, if it exists.

1. $\lim_{x\to\pi/4} \sec x$
2. $\lim_{x\to-\pi/6} \tan x$
3. $\lim_{x\to\pi/3} \cot x$
4. $\lim_{x\to 2\pi/3} \csc x$
5. $\lim_{x\to 0} \frac{\sin 2x}{3x}$
6. $\lim_{x\to 0} \frac{2x}{\tan x}$
7. $\lim_{x\to 0} \frac{\tan 2x}{\sin 5x}$
8. $\lim_{x\to\pi/2} \frac{\cos x}{\pi/2 - x}$
9. $\lim_{x\to\pi/4} \frac{\tan(\pi - 4x)}{x - \pi/4}$
10. $\lim_{x\to\pi/2} \frac{2x-\pi}{\cos x}$
11. $\lim_{x\to 0} \frac{1-\cos 2x}{x}$
12. $\lim_{x\to 0} \frac{2x}{1-\cos x}$
13. $\lim_{x\to 0} x\csc x$
14. $\lim_{x\to\pi} \frac{\sin x}{x}$
15. $\lim_{x\to 0} \frac{\cos x}{x}$
16. $\lim_{x\to 0} \frac{\sin x}{x^2 - x}$
17. $\lim_{x\to 0} |x| \cos x$
18. $\lim_{x\to 1} (x^2 - 1)\tan x$
19. $\lim_{x\to\pi} (x^3 - 5x^2 + 3)\sin x$
20. $\lim_{x\to\pi} \sin x \cos x$
21. $\lim_{x\to 4} \sin\sqrt{x}$
22. $\lim_{x\to 4} \sqrt{\sin x}$
23. $\lim_{x\to 0} \cos\sqrt{x}$
24. $\lim_{x\to\pi} \sqrt{\sin x}$

Let f be a function such that $f(x) \geq x^3$ for all x, and let g be a function such that $x^3 \leq g(x) \leq f(x)$ for all x. Given that $\lim_{x\to 0} f(x) = 0$, $\lim_{x\to 1} f(x) = 1$, $\lim_{x\to 2} f(x) = 8$, and $\lim_{x\to -1} f(x) = 0$, find the following, if possible:

25. $\lim_{x\to 0} g(x)$ **26.** $\lim_{x\to 1} g(x)$ **27.** $\lim_{x\to 2} g(x)$ **28.** $\lim_{x\to -1} g(x)$

Tell whether the given function is bounded on the given interval. If so, give a number B satisfying Definition 2.2.

29. $f(x) = \dfrac{1}{x}$; $(1, 2)$ **30.** $f(x) = \dfrac{1}{x}$; $(0, 1)$

31. $f(x) = \tan x$; $(-1, 1)$ **32.** $f(x) = \tan x$; $(1, 2)$

Tell whether the given function is bounded. If so, give a bound B satisfying Definition 2.2.

33. $f(x) = \sin x$ **34.** $f(x) = \dfrac{\sin x}{x}$ **35.** $f(x) = \dfrac{1}{x^2+1}$ **36.** $f(x) = \sqrt{x}$

37. Prove Theorem 2.11. [*Hint:* Suppose $L > M$, and let $\varepsilon = (L - M)/2$.]

38. Assuming that $\lim_{x\to c} h(x)$ exists in Theorem 2.12, prove that $\lim_{x\to c} h(x) = L$.

39. In Theorem 2.12, prove that $\lim_{x\to c} h(x)$ exists.

40. Prove the identity

$$\sin x - \sin c = 2 \sin \frac{x-c}{2} \cos \frac{x+c}{2}.$$

$$\left[\textit{Hint: } x = \frac{x-c}{2} + \frac{x+c}{2}.\right]$$

SECTION 2.4

One-Sided and Infinite Limits

We now look at some special kinds of limits. In this section, we introduce the concepts of one-sided limits, limits at infinity, and infinite limits.

One-Sided Limits

The notation $x \to c$ is used to mean that x approaches c. It requires x to be on either "side" of c; that is, x is either less than c or greater than c. It may be that we wish to have x approach c but always stay larger than c. If that is the case, we say that x approaches c **from the right** or **from above**, and we use the notation

$$x \to c^+.$$

If x approaches c but always stays less than c, we say that x approaches c **from the left** or **from below**, and we use the notation

$$x \to c^-.$$

If f is a function defined near c, the limits of $f(x)$ as $x \to c^+$ or $x \to c^-$ are called **one-sided limits** of f. The **right-hand limit** of f at c is denoted by

$$\lim_{x \to c^+} f(x),$$

and the **left-hand limit** of f at c is denoted by

$$\lim_{x \to c^-} f(x).$$

The formal definition is as follows.

DEFINITION 2.3

Let f be a function defined in the interval (c, d). Then **L is the limit of $f(x)$ as x approaches c from the right**, written as

$$\lim_{x \to c^+} f(x) = L,$$

if and only if for any $\varepsilon > 0$ there exists $\delta > 0$ such that $|f(x) - L| < \varepsilon$ when $0 < x - c < \delta$. If the function f is defined in the interval (b, c), then **L is the limit of $f(x)$ as x approaches c from the left**, written as

$$\lim_{x \to c^-} f(x) = L,$$

if and only if for any $\varepsilon > 0$ there exists $\delta > 0$ such that $|f(x) - L| < \varepsilon$ when $0 < c - x < \delta$.

You should compare these definitions with the definition of limit given in Section 2.1.

It may be the case that a one-sided limit of f exists at c when the ordinary "two-sided" limit does not. Consider the following examples.

■ **Example 2.23**

Explain why $\lim_{x \to 0^+} \sqrt{x}$ exists when $\lim_{x \to 0} \sqrt{x}$ does not.

Solution

Recall from Example 2.9 (Section 2.1) that $\lim_{x \to 0} \sqrt{x}$ does not exist because $\sqrt{x}$ is not defined in an open interval containing 0, but only for $x \geq 0$ (see Figure 2.11). In $\lim_{x \to 0^+} \sqrt{x}$, it is only required that the function be defined for x just to the right of 0. To show that $\lim_{x \to 0^+} \sqrt{x}$ exists, let $\varepsilon > 0$ be given. Then $|\sqrt{x} - 0| < \varepsilon$ when $0 < x < \varepsilon^2$. Thus, the right-hand limit exists and equals 0. ■

■ **Example 2.24**

Show that if $\lim_{x \to c} f(x) = L$, then both $\lim_{x \to c^+} f(x) = L$ and $\lim_{x \to c^-} f(x) = L$.

Solution

If by choosing x close enough to c, we can force $f(x)$ to be arbitrarily close to L, with x on either side of c, then we can force $f(x)$ to be arbitrarily close to L by choosing x close enough to c and on a given side of c. That is, if $0 < |x - c| < \delta$ implies that $|f(x) - L| < \varepsilon$, then $0 < x - c < \delta$ implies that $|f(x) - L| < \varepsilon$, and $0 < c - x < \delta$ implies that $|f(x) - L| < \varepsilon$. Thus, both one-sided limits exist and also have value L. ■

Example 2.25 Find the one-sided limits of $f(x) = |x|/x$ at 0.

Solution The graph of $f(x) = |x|/x$ is shown in Figure 2.19. For the right-hand limit, we note that if $x > 0$, then $f(x) = 1$. Hence,

$$\lim_{x \to 0+} \frac{|x|}{x} = \lim_{x \to 0+} 1 = 1.$$

(For the last equal sign, we may appeal to Example 2.24.) For the left-hand limit, if $x < 0$, then $f(x) = -1$, so

$$\lim_{x \to 0-} \frac{|x|}{x} = \lim_{x \to 0-} (-1) = -1.$$

Note that because these two one-sided limits are different, the two-sided limit cannot exist (by Example 2.24).

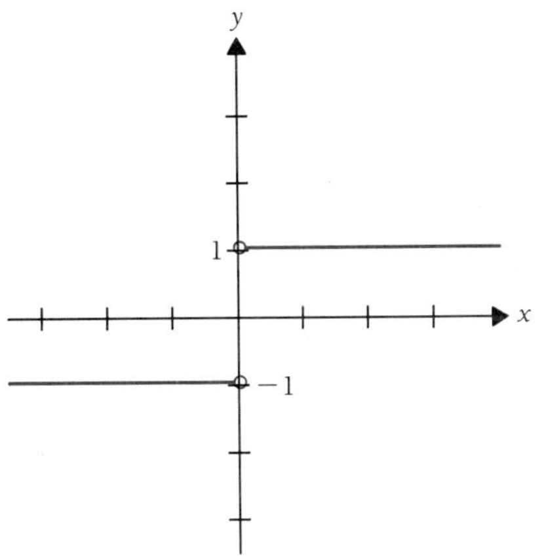

FIGURE 2.19 The graph of $f(x) = \dfrac{|x|}{x}$ ■

Example 2.26 Find the one-sided limits of the function below at 1:

$$f(x) = \frac{x^2 - 1}{x - 1}.$$

Solution Since $f(x) = x + 1$ for $x \neq 1$, we can find the two-sided limit,

$$\lim_{x \to 1} \frac{x^2 - 1}{x - 1} = \lim_{x \to 1} (x + 1) = 2.$$

Therefore, by Example 2.24, both one-sided limits exist and have value 2. ■

The relationship between one- and two-sided limits illustrated by these examples can be summarized in the following theorem.

THEOREM 2.16 The two-sided limit of f at c exists and equals L if and only if both one-sided limits of f at c exist and equal L.

Half of this theorem was proved in Example 2.24. You will be invited to prove the other half in the exercises.

Example 2.27 Discuss the one- and two-sided limits of the following functions f at the given value of c:

(a) $f(x) = \lfloor x \rfloor; \quad c = 3$ (b) $f(x) = \begin{cases} x^2 - 4 & \text{if } x < 2 \\ x + 1 & \text{if } x \geq 2; \end{cases} \quad c = 2$

Solution (a) The graph of $f(x) = \lfloor x \rfloor$ is shown in Figure 2.20. For x near 3 and $x > 3$, $\lfloor x \rfloor = 3$, so

$$\lim_{x \to 3^+} \lfloor x \rfloor = 3.$$

For x near 3 and $x < 3$, $\lfloor x \rfloor = 2$, so

$$\lim_{x \to 3^-} \lfloor x \rfloor = 2.$$

Since the one-sided limits are unequal, the two-sided limit cannot exist.

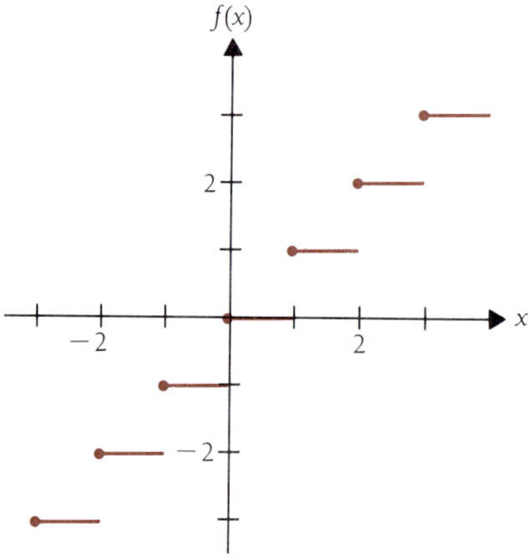

FIGURE 2.20 The graph of $f(x) = \lfloor x \rfloor$

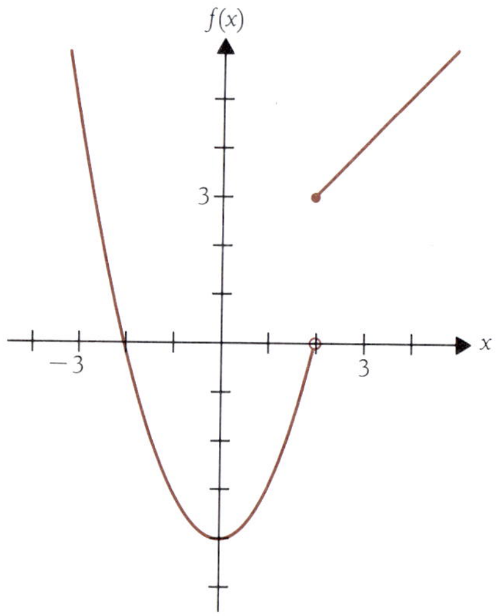

FIGURE 2.21 The graph of $f(x) = \begin{cases} x^2 - 4 & \text{if } x < 2 \\ x + 1 & \text{if } x \geq 2 \end{cases}$

(b) The graph of this function is shown in Figure 2.21. For x near 2 and $x < 2$, $f(x) = x^2 - 4$, so

$$\lim_{x \to 2^-} f(x) = \lim_{x \to 2^-} (x^2 - 4) = \lim_{x \to 2} (x^2 - 4) = 0.$$

For x near 2 and $x > 2$, $f(x) = x + 1$, so

$$\lim_{x \to 2^+} f(x) = \lim_{x \to 2^+} (x + 1) = \lim_{x \to 2} (x + 1) = 3.$$

Once again the two-sided limit does not exist, because the one-sided limits are unequal. ■

■ **Example 2.28** Discuss the one- and two-sided limits of the function below at $c = 0$:

$$f(x) = x \sin \frac{1}{x}.$$

Solution The graph of $f(x) = x \sin(1/x)$ is shown in Figure 2.22. If the part of the graph near 0 were enlarged, you could see the same pattern of oscillations continue. (This function is an example of a *damped oscillation*.) Since $\sin(1/x)$ is bounded and $x \to 0$, the two-sided limit must be 0. Thus, both one-sided limits will have value 0 also.

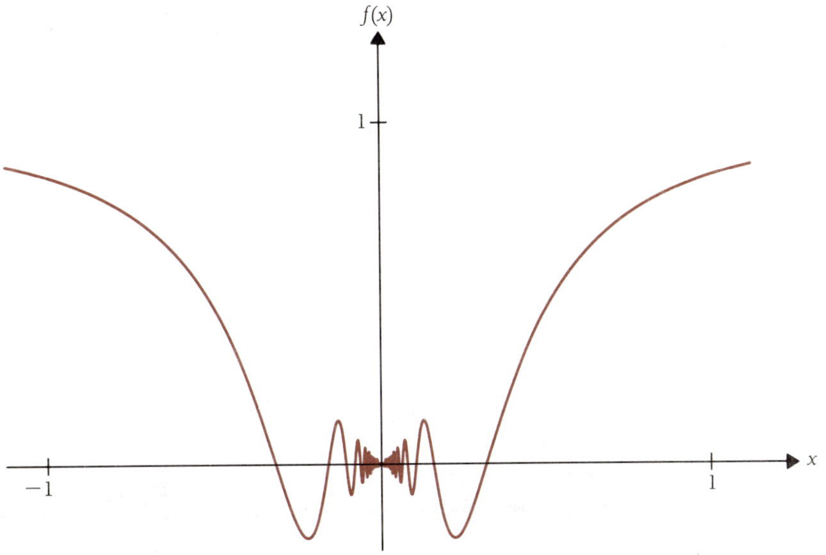

FIGURE 2.22 The graph of $f(x) = x \sin \frac{1}{x}$ ■

Because of the similarity in the definitions of one- and two-sided limits, the same limit theorems apply to one-sided limits. All the limit theorems of the previous sections can be restated and proved for limits as $x \to c^+$ or as $x \to c^-$. It follows, then, that the techniques of evaluation that we have already developed also apply to one-sided limits.

Limits at Infinity

We say x **gets large** or **approaches infinity**, denoted

$$x \longrightarrow \infty,$$

in case x becomes larger than any positive number. We say x **gets large negatively** or **approaches minus infinity**, denoted

$$x \longrightarrow -\infty,$$

if x becomes less than any negative number. An open ray of the form (a, ∞) is called a **neighborhood of infinity**, and one of the form $(-\infty, b)$ is called a **neighborhood of minus infinity**. The limit of a function as $x \longrightarrow \infty$ or as $x \longrightarrow -\infty$ is called a **limit at infinity**. Following is a formal definition.

DEFINITION 2.4

Let a function f be defined in a neighborhood of infinity. Then **L is the limit of $f(x)$ as x approaches infinity**, written

$$\lim_{x \to \infty} f(x) = L,$$

if and only if for any $\varepsilon > 0$ there exists $B > 0$ such that $|f(x) - L| < \varepsilon$ when $x > B$. If f is defined in a neighborhood of minus infinity, then **L is the limit of $f(x)$ as x approaches minus infinity**, written

$$\lim_{x \to -\infty} f(x) = L,$$

if and only if for any $\varepsilon > 0$ there exists $B > 0$ such that $|f(x) - L| < \varepsilon$ when $x < -B$.

The following example points out the basic properties of limits at infinity.

■ **Example 2.29** Show that $\lim_{x \to \infty}(1/x) = 0$.

Solution The function $f(x) = 1/x$ is graphed in Figure 2.23. According to Definition 2.4, we must show that by choosing x large enough, we can force $1/x$ to be as close to 0 as we like. Given $\varepsilon > 0$, we note that $1/x < \varepsilon$ when $x > 1/\varepsilon$, so the definition is satisfied by choosing $B = 1/\varepsilon$. Graphically, the line $y = 0$ (the x-axis) is a horizontal asymptote of $f(x) = 1/x$. ■

There are two things that we want to notice particularly in Example 2.29. The first is that a limit at infinity is always associated with a horizontal asymptote of the graph. That is, if $\lim_{x \to \infty} f(x) = L$, then the line $y = L$ is a horizontal asymptote of the graph of f. The same is true for limits as $x \longrightarrow -\infty$. We will see several illustrations of this fact in the following examples.

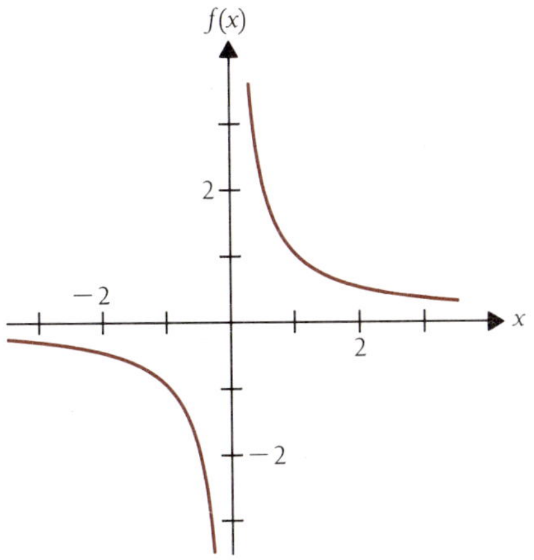

FIGURE 2.23 The graph of $f(x) = \dfrac{1}{x}$

The second thing to notice in Example 2.29 is that $1/x \to 0^+$ as $x \to \infty$. In fact, if $u = 1/x$, then $u \to 0^+$ if and only if $x \to \infty$. This leads to the following useful theorem (the proof is left as an exercise).

THEOREM 2.17 The limits

$$\lim_{x\to\infty} f(x) \qquad \text{and} \qquad \lim_{u\to 0^+} f\left(\frac{1}{u}\right)$$

either both exist or both fail to exist. If they both exist, they have the same value. The same is true of the limits

$$\lim_{x\to -\infty} f(x) \qquad \text{and} \qquad \lim_{u\to 0^-} f\left(\frac{1}{u}\right).$$

Thus, we can use Theorem 2.17 to transform a limit at infinity into a one-sided limit. This may make the evaluation of the limit easier, as will be illustrated in the examples below. Also, Theorem 2.17 points out that the theory of limits at infinity is the same as the theory of one-sided limits. Hence, the limit theorems and techniques for evaluating limits from the previous sections also apply to limits at infinity.

Example 2.30 Find the following limits, if they exist:

(a) $\displaystyle\lim_{x\to\infty} \frac{x+2}{x-3}$ **(b)** $\displaystyle\lim_{x\to\infty} \frac{3x^2+2x+1}{x^2+3x-4}$

(c) $\displaystyle\lim_{x\to -\infty} \frac{1}{x+3}$ **(d)** $\displaystyle\lim_{x\to\infty} (\sqrt{x^2+4x} - x)$

Solution

(a) We will show two ways to evaluate this limit. One way is to divide both the numerator and denominator by x, and use Example 2.29:

$$\lim_{x\to\infty}\frac{x+2}{x-3}=\lim_{x\to\infty}\frac{1+(2/x)}{1-(3/x)}=\frac{1+0}{1-0}=1.$$

The other way is to use Theorem 2.17:

$$\lim_{x\to\infty}\frac{x+2}{x-3}=\lim_{u\to0^+}\frac{(1/u)+2}{(1/u)-3}=\lim_{u\to0^+}\frac{1+2u}{1-3u}=\frac{1+0}{1-0}=1.$$

The graph of $f(x)$ is shown in Figure 2.24. Note that the line $y=1$ is a horizontal asymptote; note also that the placement of the graph is evident from the identity

$$\frac{x+2}{x-3}=1+\frac{5}{x-3}.$$

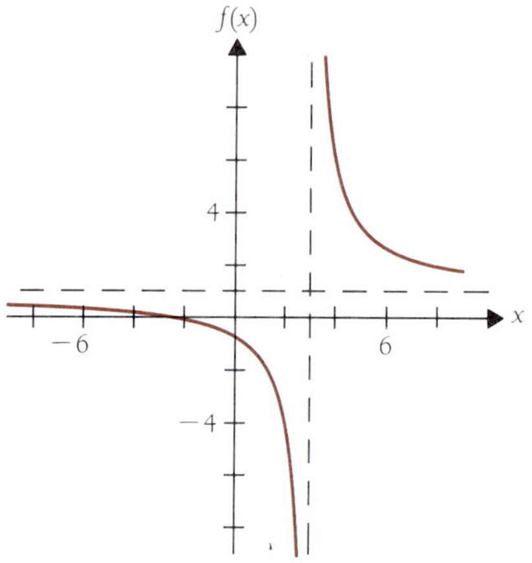

FIGURE 2.24 The graph of $f(x)=\dfrac{x+2}{x-3}$

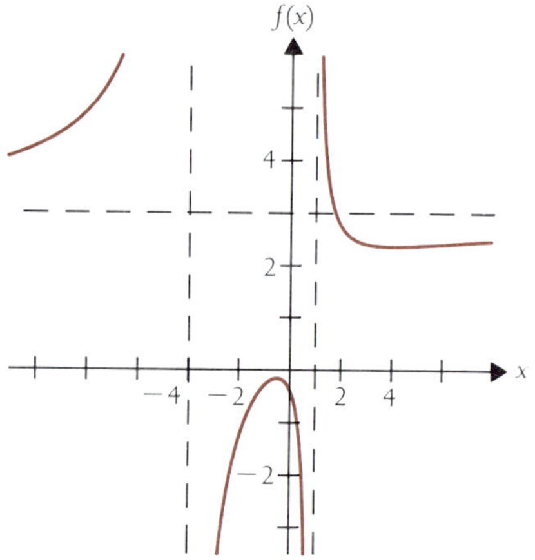

FIGURE 2.25 The graph of $f(x)=\dfrac{3x^2+2x+1}{x^2+3x-4}$

(b) In this limit, dividing the numerator and denominator through by x^2 (the highest power of x) is the simplest approach:

$$\lim_{x\to\infty}\frac{3x^2+2x+1}{x^2+3x-4}=\lim_{x\to\infty}\frac{3+(2/x)+(1/x^2)}{1+(3/x)-(4/x^2)}=\frac{3+0+0}{1+0-0}=3.$$

The graph of $f(x)$ is shown in Figure 2.25. The line $y=3$ is a horizontal asymptote.

(c) Here we use Theorem 2.17:

$$\lim_{x\to-\infty}\frac{1}{x+3}=\lim_{u\to 0^-}\frac{1}{(1/u)+3}=\lim_{u\to 0^-}\frac{u}{1+3u}=\frac{0}{1+0}=0.$$

The line $y=0$ is a horizontal asymptote of the graph of f, as shown in Figure 2.26.

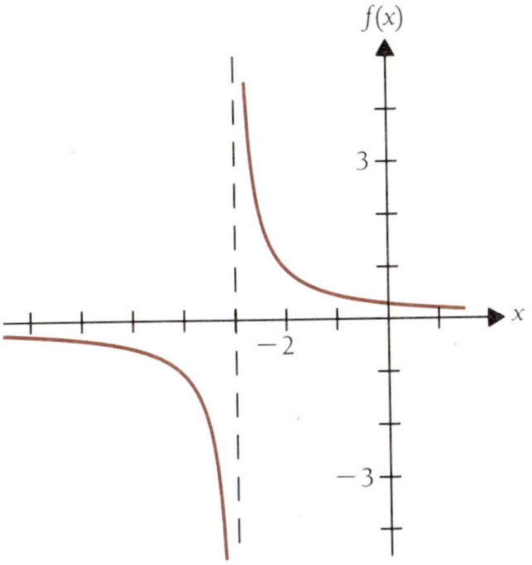

FIGURE 2.26 The graph of $f(x)=\dfrac{1}{x+3}$

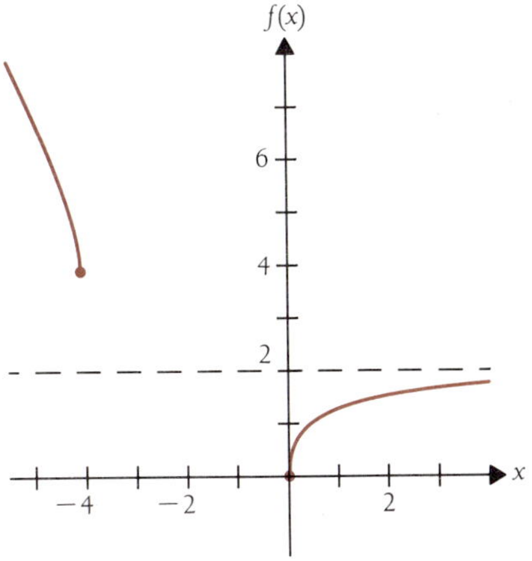

FIGURE 2.27 The graph of $f(x)=\sqrt{x^2+4x}-x$

(d) With this limit, we use Theorem 2.17, followed by some algebra, including rationalizing the numerator:

$$\begin{aligned}\lim_{x\to\infty}(\sqrt{x^2+4x}-x) &= \lim_{u\to 0^+}\left(\sqrt{\frac{1}{u^2}+\frac{4}{u}}-\frac{1}{u}\right)\\ &= \lim_{u\to 0^+}\frac{\sqrt{1+4u}-1}{u}\cdot\frac{\sqrt{1+4u}+1}{\sqrt{1+4u}+1}\\ &= \lim_{u\to 0^+}\frac{(1+4u)-1}{u(\sqrt{1+4u}+1)}\\ &= \lim_{u\to 0^+}\frac{4}{\sqrt{1+4u}+1}\\ &= \frac{4}{2}=2.\end{aligned}$$

The line $y=2$ is a horizontal asymptote of the graph of f, as shown in Figure 2.27. ■

Infinite Limits

We have seen examples of functions f for which $\lim_{x\to c} f(x)$ exists and functions for which $\lim_{x\to c} f(x)$ does not exist. Now we consider functions whose limits fail to exist for a particular reason—namely, that the functional values approach ∞ or $-\infty$.

If the values of a function f grow larger and larger without bound as $x \to c$, we say $f(x)$ **gets large** or **approaches infinity**, written

$$f(x) \to \infty \qquad \text{or} \qquad \lim_{x\to c} f(x) = \infty.$$

If the values of f become less than any negative number, we say $f(x)$ **gets large negatively** or **approaches minus infinity**, written

$$f(x) \to -\infty \qquad \text{or} \qquad \lim_{x\to c} f(x) = -\infty.$$

If $\lim_{x\to c} f(x) = \pm\infty$, we say the limit of f at c is **infinite**. We apply the same terminology to the behavior of f in one-sided limits or limits at infinity. Here is a formal definition.

DEFINITION 2.5

Let a function f be defined in an open interval containing c, except possibly at c itself. Then **the limit of $f(x)$ as x approaches c is infinity**, written

$$\lim_{x\to c} f(x) = \infty,$$

if and only if for any $B > 0$ there exists $\delta > 0$ such that $f(x) > B$ whenever $0 < |x - c| < \delta$. Similarly, **the limit of $f(x)$ as x approaches c is minus infinity**, written

$$\lim_{x\to c} f(x) = -\infty,$$

if and only if for any $B > 0$ there exists $\delta > 0$ such that $f(x) < -B$ whenever $0 < |x - c| < \delta$.

This definition tells us in each case not only that the limits fail to exist, but also the manner in which they fail to exist. Similar definitions can be made for such statements as $\lim_{x\to c^+} f(x) = \infty$, $\lim_{x\to\infty} f(x) = -\infty$, etc.

■ **Example 2.31**

Show that $\lim_{x\to 0^+}(1/x) = \infty$.

Solution

This fact is obvious from the graph of $f(x) = 1/x$, which was shown in Figure 2.23. To show that the definition is satisfied, we must show that we can choose x close enough to 0, but positive, to make $f(x)$ as large as we like. Given $B > 0$, we see that $1/x > B$ when $0 < x < 1/B$. Thus, the definition is satisfied by choosing $\delta = 1/B$. Graphically, the line $x = 0$ (the y-axis) is a vertical asymptote of the graph. ■

Example 2.31 points out the general notion that if the one- or two-sided limit of f at c is infinite, then the line $x = c$ is a vertical asymptote of the graph of f. Looking back at Example 2.30(a), we can see that

$$\lim_{x \to 3^+} \frac{x+2}{x-3} = \infty \qquad \text{and} \qquad \lim_{x \to 3^-} \frac{x+2}{x-3} = -\infty$$

(see Figure 2.24). Also, in Example 2.30(c),

$$\lim_{x \to -3^+} \frac{1}{x+3} = \infty \qquad \text{and} \qquad \lim_{x \to -3^-} \frac{1}{x+3} = -\infty$$

(see Figure 2.26). For the function in Example 2.30(b),

$$f(x) = \frac{3x^2 + 2x + 1}{x^2 + 3x - 4},$$

we see from Figure 2.25 that

$$\lim_{x \to 1^+} f(x) = \infty, \qquad \lim_{x \to -4^-} f(x) = \infty, \qquad \lim_{x \to 1^-} f(x) = -\infty, \qquad \lim_{x \to -4^+} f(x) = -\infty.$$

The limit theorems previously developed do not, in general, apply to infinite limits without modification. For example, the statement, "the limit of a sum is the sum of the limits" has no meaning for infinite limits, since ∞ cannot be treated like a real number. However, we can say that if $f(x) \to L$ and $g(x) \to \infty$ as $x \to c$, then $[f(x) + g(x)] \to \infty$ as $x \to c$. Some more results similar to this can be found in the exercises.

When evaluating limits that turn out to be infinite, it is best to keep in mind the definition and simple results like those in Example 2.31. Consider the following examples.

Example 2.32 Discuss the following limits:

(a) $\lim_{x \to 0^+} \left(\frac{1}{x} + 2\right)$ (b) $\lim_{x \to -\infty} (x^2 + 3x)$ (c) $\lim_{x \to \infty} (3x - x^3)$

(d) $\lim_{x \to \infty} \frac{x^2}{x+1}$ (e) $\lim_{x \to \infty} \frac{x^2}{x^3+1}$

Solution

(a) Since $1/x \to \infty$ as $x \to 0^+$, it is clear that the limit is ∞. That is, we can make $(1/x) + 2$ large by choosing x close enough to 0. The graph is the same as Figure 2.23, but raised 2 units so that the horizontal asymptote is $y = 2$.

(b) The graph of the function $f(x) = x^2 + 3x$ is a parabola opening upward, so we expect the limit as $x \to -\infty$ to be ∞. Writing the function as $f(x) = x(x + 3)$, and noting that both $x \to -\infty$ and $(x + 3) \to -\infty$ as $x \to -\infty$, the product must be large positively, and the limit is indeed ∞.

(c) The function $f(x) = 3x - x^3$ has as its graph the curve shown in Figure 2.28 (page 120), so we expect the limit as $x \to \infty$ to be $-\infty$. Since $f(x) = x(3 - x^2) = -x(x^2 - 3)$, $f(x)$ does get large negatively as $x \to \infty$.

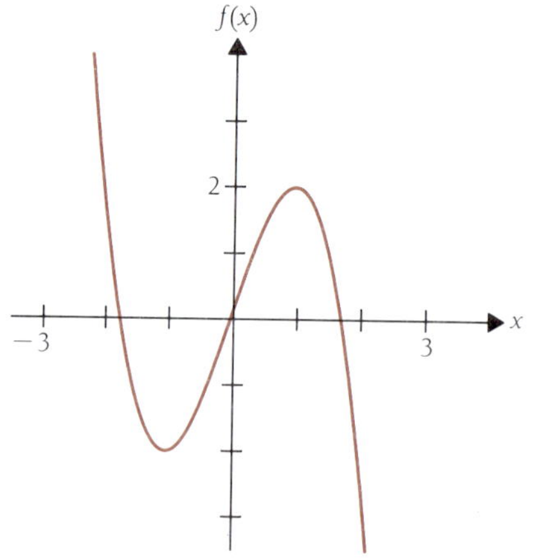

FIGURE 2.28 The graph of $f(x) = 3x - x^3$

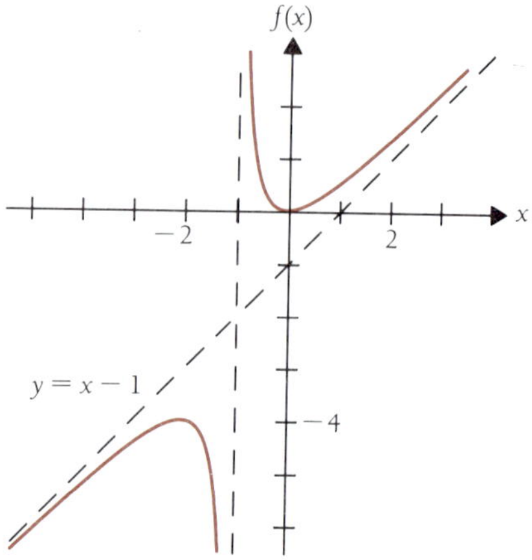

FIGURE 2.29 The graph of $f(x) = \dfrac{x^2}{x+1}$

(d) This function $f(x)$ can be written as

$$f(x) = \frac{x^2}{x+1} = x - 1 + \frac{1}{x+1},$$

and since $1/(x+1) \to 0$ as $x \to \infty$, the line $y = x - 1$ is an asymptote of the graph of f, as shown in Figure 2.29. Thus, we see that $f(x) \to \infty$ as $x \to \infty$. (We also could have used Theorem 2.17 and Example 2.31 here.)

(e) This limit calls for a simple application of Theorem 2.17:

$$\lim_{x\to\infty} \frac{x^2}{x^3+1} = \lim_{u\to 0^+} \frac{1/u^2}{(1/u^3)+1} = \lim_{u\to 0^+} \frac{u}{1+u^3} = 0.$$

Here, we can think of the denominator as growing large faster than the numerator. In contrast, the limit in part (d) had a numerator that grew faster than the denominator. The graph of $f(x)$ is shown in Figure 2.30.

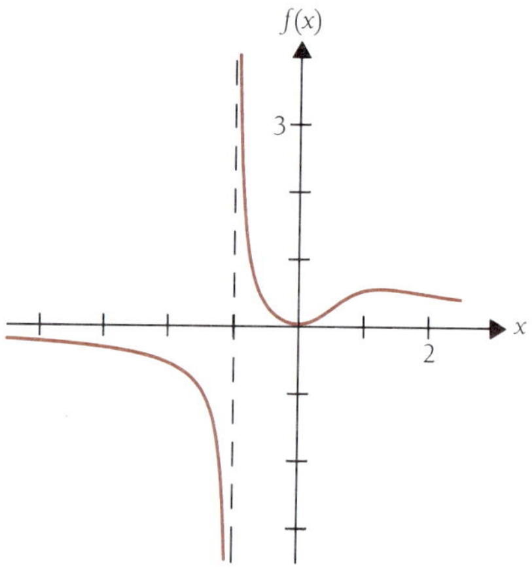

FIGURE 2.30 The graph of $f(x) = \dfrac{x^2}{x^3+1}$ ■

By now we have dealt with several kinds of limits. Their definitions are summarized in Table 2.1.

TABLE 2.1

$\lim_{x\to c} f(x) = L$	if and only if for every $\varepsilon > 0$ there exists $\delta > 0$ such that if $0 < \lvert x - c \rvert < \delta$, then $\lvert f(x) - L \rvert < \varepsilon$.
$\lim_{x\to c^+} f(x) = L$	if and only if for every $\varepsilon > 0$ there exists $\delta > 0$ such that if $0 < x - c < \delta$, then $\lvert f(x) - L \rvert < \varepsilon$.
$\lim_{x\to c^-} f(x) = L$	if and only if for every $\varepsilon > 0$ there exists $\delta > 0$ such that if $0 < c - x < \delta$, then $\lvert f(x) - L \rvert < \varepsilon$.
$\lim_{x\to \infty} f(x) = L$	if and only if for every $\varepsilon > 0$ there exists $B > 0$ such that if $x > B$, then $\lvert f(x) - L \rvert < \varepsilon$.
$\lim_{x\to -\infty} f(x) = L$	if and only if for every $\varepsilon > 0$ there exists $B > 0$ such that if $x < -B$, then $\lvert f(x) - L \rvert < \varepsilon$.
$\lim_{x\to c} f(x) = \infty$	if and only if for every $B > 0$ there exists $\delta > 0$ such that if $0 < \lvert x - c \rvert < \delta$, then $f(x) > B$.
$\lim_{x\to c} f(x) = -\infty$	if and only if for every $B > 0$ there exists $\delta > 0$ such that if $0 < \lvert x - c \rvert < \delta$, then $f(x) < -B$.

Exercises 2.4

Find the limit. If the limit is infinite or does not exist, so state.

1. $\lim_{x\to 3^+} \dfrac{x^2-9}{x-3}$
2. $\lim_{x\to 3^-} \dfrac{x^2-9}{x-3}$
3. $\lim_{x\to 1^+} \dfrac{x-1}{x^2-1}$
4. $\lim_{x\to 1^-} \dfrac{x-1}{x^2-1}$
5. $\lim_{x\to 4^+} \dfrac{|x-4|}{x-4}$
6. $\lim_{x\to 4^-} \dfrac{|x-4|}{x-4}$
7. $\lim_{x\to 2^+} \dfrac{1}{x-2}$
8. $\lim_{x\to 2^-} \dfrac{1}{x-2}$
9. $\lim_{x\to 2} \dfrac{1}{x-2}$
10. $\lim_{x\to -2} \dfrac{1}{x+2}$
11. $\lim_{x\to 1^+} \dfrac{x}{x^2-1}$
12. $\lim_{x\to 1^-} \dfrac{x}{x^2-1}$
13. $\lim_{x\to \infty} \dfrac{x+2}{x-1}$
14. $\lim_{x\to \infty} \dfrac{x^2-4}{x^2-5}$
15. $\lim_{x\to -\infty} \dfrac{2x-1}{x+2}$
16. $\lim_{x\to \infty} \dfrac{4-x}{3x-1}$
17. $\lim_{x\to -\infty} \dfrac{x^2+2x}{x^2-1}$
18. $\lim_{x\to -\infty} \dfrac{1-x-x^2}{1-x^2}$
19. $\lim_{x\to -\infty} \dfrac{x^2}{x^3+4}$
20. $\lim_{x\to -\infty} \dfrac{x^2+4}{4-x}$
21. $\lim_{x\to \infty} x \sin \dfrac{1}{x}$
22. $\lim_{x\to \infty} \dfrac{\sin x}{x}$
23. $\lim_{x\to -\infty} \dfrac{\cos x}{x^2+4}$
24. $\lim_{x\to -\infty} \dfrac{1-\cos x}{x}$

25. $\lim_{x\to\infty}(\sqrt{x+1}-\sqrt{x})$

26. $\lim_{x\to\infty}(\sqrt{x^2+ax}-\sqrt{x})$

27. $\lim_{x\to\infty}\dfrac{\sqrt{x^2+1}}{x}$

28. $\lim_{x\to-\infty}\dfrac{\sqrt{x^2+1}}{x}$

29. $\lim_{x\to\infty}\dfrac{\sqrt{x^2+4}}{x+4}$

30. $\lim_{x\to-\infty}\dfrac{x+4}{\sqrt{x^2+4}}$

31. $\lim_{x\to\infty}\dfrac{\sqrt{x^2+1}-x}{x}$

32. $\lim_{x\to\infty}\dfrac{\sqrt{x^4+1}-x}{x^2+1}$

33. $\lim_{x\to 3^+}\lfloor x\rfloor$

34. $\lim_{x\to 3^-}\lfloor x\rfloor$

35. $\lim_{x\to n^+}\lfloor x\rfloor,\quad n\in\mathbb{Z}$

36. $\lim_{x\to n^-}\lfloor x\rfloor,\quad n\in\mathbb{Z}$

37. $\lim_{x\to 1/2}\lfloor x\rfloor$

38. $\lim_{x\to 1/2}\lfloor 2x\rfloor$

39. $\lim_{x\to \pi/2^+}\lfloor \sin x\rfloor$

40. $\lim_{x\to \pi/2^-}\lfloor \sin x\rfloor$

Give definitions of the following limits similar to the definitions given in Table 2.1:

41. $\lim_{x\to c^+} f(x)=\infty$

42. $\lim_{x\to c^-} f(x)=\infty$

43. $\lim_{x\to c^+} f(x)=-\infty$

44. $\lim_{x\to c^-} f(x)=-\infty$

45. $\lim_{x\to \infty} f(x)=\infty$

46. $\lim_{x\to -\infty} f(x)=\infty$

47. $\lim_{x\to \infty} f(x)=-\infty$

48. $\lim_{x\to -\infty} f(x)=-\infty$

49. Prove Theorem 2.16.

50. Prove Theorem 2.17.

Suppose $\lim_{x\to c} f(x)=\infty$ *and* $\lim_{x\to c} g(x)=L$. *Prove the following:*

51. $\lim_{x\to c}[f(x)+g(x)]=\infty$

52. $\lim_{x\to c}\dfrac{g(x)}{f(x)}=0$

53. If $L>0$, then $\lim_{x\to c} f(x)g(x)=\infty$.

54. If $L<0$, then $\lim_{x\to c} f(x)g(x)=-\infty$.

55. **(a)** Find $\lim_{x\to\infty}(\sqrt{x^2+2ax}-x)$.

(b) Explain why $\lim_{x\to\infty}[f(x)-g(x)]$ is not necessarily 0 when $\lim_{x\to\infty}f(x)=\infty$ and $\lim_{x\to\infty}g(x)=\infty$. [Use part (a).]

SECTION 2.5

Continuity

We have already remarked that points on the real line are not spaced like beads on a string, but are so densely packed that between any two real numbers there are infinitely many real numbers. The packing is so dense, in fact, that if we cut the line, the cut cannot occur between numbers, but must occur at a number. This property, which is a consequence of the completeness property of the real numbers, is named the **Dedekind cut property**, after the German mathematician Richard Dedekind (1831–1916) who first stated it. Such properties as the completeness property and the Dedekind cut property are referred to as **continuity properties** of the real number system. Intuitively speaking, the real line is continuous, meaning that it has no breaks in it.

Similarly, we speak of continuity properties of functions. For example, the mercury level in a thermometer is a continuous function of the temperature, because very small changes in the heat content of the air give rise to very small differences in the thermometer readings. But the cost of first class postage to mail a letter is not a continuous function of the weight of the letter, since a very small change in the weight of a letter can make a large difference in the cost of postage.

To be precise, we use the limit concept to define continuity.

Continuity at a Point

Our first definition below tells what it means for a function to be continuous at a particular point. In a sense, it is saying that there is no break in the graph of the function at that point. Before we state this definition, in order to avoid confusion, we must make a distinction between some kinds of points in the domain of a function. We say point c is an **interior point** of the domain of f if f is defined in an open interval containing c. We say c is an **endpoint** of the domain of f if c is not an interior point but f is defined in an interval of the form $(a, c]$ or the form $[c, b)$. (It is possible for a point in the domain of a function to be neither an interior point nor an endpoint, but we will not need to worry about that.) For instance, a function defined on $(-\infty, \infty)$, such as a polynomial function, has only interior points in its domain. A function defined everywhere except at an isolated point c, such as a rational function, also has only interior points in its domain.

DEFINITION 2.6 A function f is **continuous at point c** if and only if c is in the domain of f, and:

1. If c is an interior point of the domain of f, then $\lim_{x\to c} f(x)$ exists and equals $f(c)$.
2. If c is an endpoint of the domain of f and f is defined in an interval of the form $(a, c]$, then $\lim_{x\to c^-} f(x)$ exists and equals $f(c)$.
3. If c is an endpoint of the domain of f and f is defined in an interval of the form $[c, b)$, then $\lim_{x\to c^+} f(x)$ exists and equals $f(c)$.

The following example illustrates this definition.

■ **Example 2.33** Explain why a rational function is continuous at any point of its domain.

Solution If f is a rational function and c is a point of its domain, then c must be an interior point of the domain of f. Also, the limit of f at c is found by evaluating, by Rule 2.3 of Section 2.2. That is, $\lim_{x\to c} f(x) = f(c)$, and by Definition 2.6, f is continuous at c. ■

The same argument can be applied to most of the functions we have seen. We have, for example, the following theorem.

THEOREM 2.18 Any function that is an arithmetic combination of algebraic and trigonometric functions is continuous at any point of its domain.

Example 2.34 Explain why the greatest integer function is not continuous at an integer.

Solution First, the greatest integer function $f(x) = \lfloor x \rfloor$ is defined for all real numbers, so every integer is an interior point of its domain. If n is an integer, then the two-sided limit of $\lfloor x \rfloor$ does not exist, by the results of Problems 35 and 36 of Section 2.4. Therefore, $\lfloor x \rfloor$ is not continuous at n. ■

We also can turn this idea around, and use continuity to find limits. If the function f is continuous at point c, then the limit of f at c is simply $f(c)$. For example, it is now possible to state in compact form the limit theorem, Theorem 2.10.

THEOREM 2.19 If $\lim_{x\to c} f(x) = L$ and g is continuous at L, then $\lim_{x\to c} g \circ f(x) = g(L)$.

Points of Discontinuity

A function f may fail to be continuous at a point c for a variety of reasons. For example, c may not be in the domain of f, or the limit of f at c may not exist (for one of several reasons), or the limit may exist but not equal $f(c)$. For certain points of this sort, we have a special name.

DEFINITION 2.7 For a given function f, a **point of discontinuity** is a point c such that either

1. c is in the domain of f, but f is not continuous at c; or
2. c is not in the domain of f, but f is defined in intervals of the form (a, c) and (c, b).

If c is a point of discontinuity of f, we say f is **discontinuous at c**, or **has a discontinuity at c.** Discontinuities are classified according to the reason that the function fails to be continuous at that point.

If $\lim_{x\to c} f(x)$ exists but does not equal $f(c)$, either because f is not defined at c or because $f(c)$ is not the value of the limit, then f is said to have a **removable discontinuity** at c. It is called "removable" because f can be defined or redefined at c so as to be continuous at c.

Example 2.35 Show that the following functions have removable discontinuities at the points indicated. Tell how to define or redefine f so that the point of discontinuity is removed.

(a) $f(x) = \dfrac{x^2 - 9}{x - 3}$; $c = 3$ (b) $f(x) = \begin{cases} x^2 & \text{if } x < 0 \\ x & \text{if } x > 0; \end{cases}$ $c = 0$

(c) $f(x) = \lfloor \sin x \rfloor$; $c = \pi/2$

Solution

(a) We recall from Example 2.11 (Section 2.2) that the limit of this function as $x \to 3$ exists and equals 6, but since the function is not defined at 3, it is discontinuous there. If f were defined to be 6 at $x = 3$, then f would be continuous at 3. This function is illustrated in Figure 2.31.

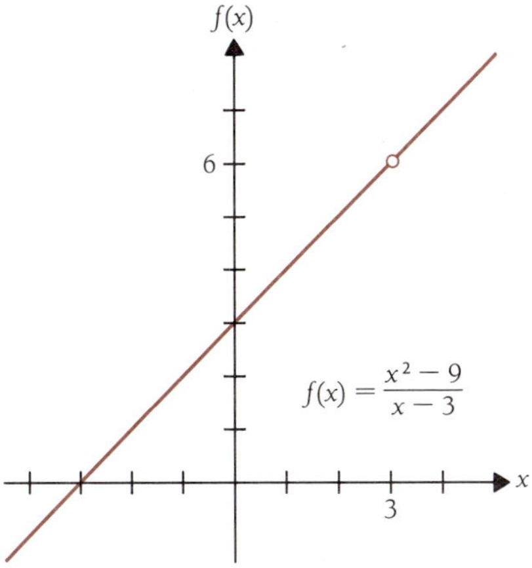

FIGURE 2.31 A removable discontinuity

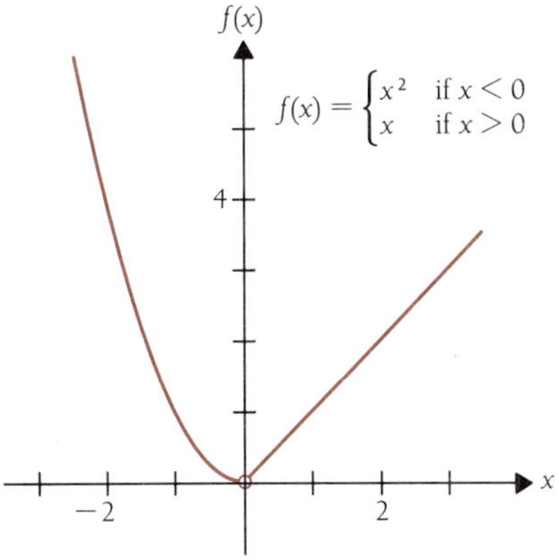

FIGURE 2.32 A removable discontinuity

(b) Note that f is not defined at 0, so f is discontinuous there. However, as $x \to 0^-$, $f(x) = x^2 \to 0$, and as $x \to 0^+, f(x) = x \to 0$. Thus, the two-sided limit of f at 0 exists and is 0. If f were defined to be 0 at 0, then f would be continuous there. This function is illustrated in Figure 2.32.

(c) We first sketch the graph of this function, as shown in Figure 2.33. For $0 < x < \pi/2$ and for $\pi/2 < x < \pi$, we have $0 < \sin x < 1$, so that $\lfloor \sin x \rfloor = 0$, and hence, $\lfloor \sin x \rfloor \to 0$ as $x \to \pi/2$. However, $f(\pi/2) = 1$, so f has a removable discontinuity at $\pi/2$. We need to redefine f to be 0 at $\pi/2$ to remove the discontinuity.

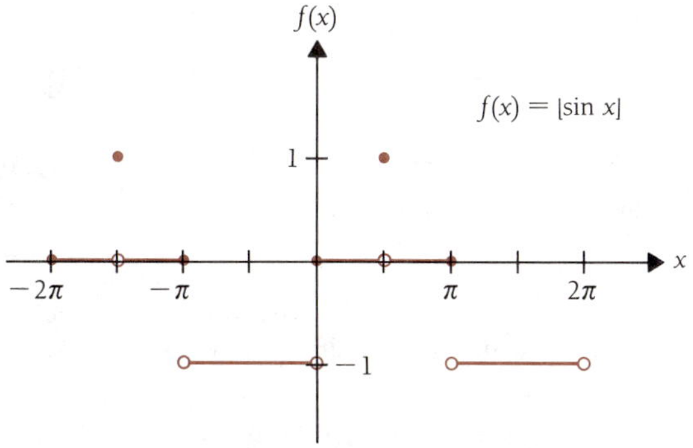

FIGURE 2.33 A removable discontinuity at $\pi/2$ ■

A point of discontinuity that is not removable is called **essential.** There are several types of essential discontinuities, of which we will name a couple. First, if c is a point of discontinuity of f for which the one-sided limits $\lim_{x\to c^-} f(x)$ and $\lim_{x\to c^+} f(x)$ exist but have different values, f is said to have a **jump discontinuity** at c. The following example provides illustration.

Example 2.36 Show that the following functions have jump discontinuities at the indicated points:

(a) $f(x) = \dfrac{|x|}{x}$; $c = 0$ (b) $f(x) = \lfloor x \rfloor$; $c = 2$

Solution (a) The graph of this function is sketched in Figure 2.34. For $x < 0, |x|/x = -1$, so that $f(x) \to -1$ as $x \to 0^-$. For $x > 0$, $|x|/x = 1$, so that $f(x) \to 1$ as $x \to 0^+$. Therefore, f has a jump discontinuity at 0. Note that f is not defined at 0.

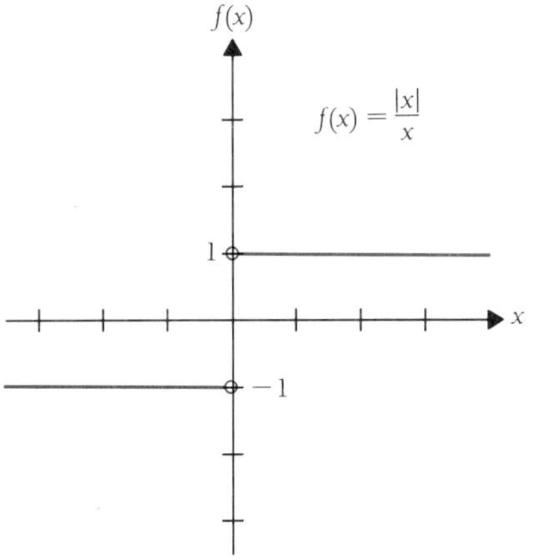

FIGURE 2.34 A jump discontinuity

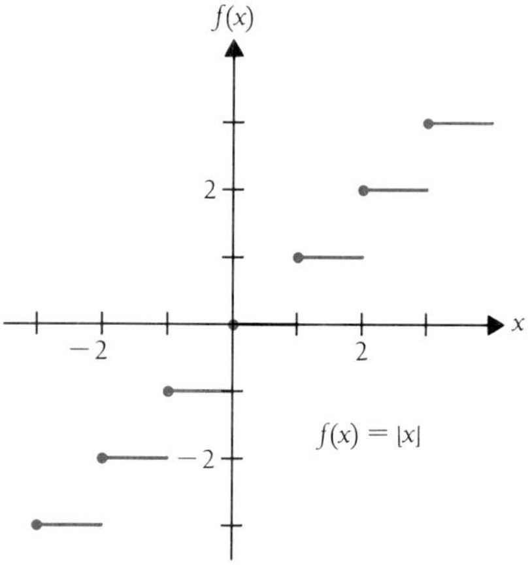

FIGURE 2.35 Jump discontinuities

(b) As we have seen previously,

$$\lim_{x\to 2^-} \lfloor x \rfloor = 1, \qquad \text{but} \qquad \lim_{x\to 2^+} \lfloor x \rfloor = 2.$$

Thus, f has a jump discontinuity at 2. In fact, f has a jump discontinuity at every integer. Note also that f is defined at every integer. The graph is shown in Figure 2.35. ■

Another type of essential discontinuity occurs if c is a point of discontinuity of f such that one of the one-sided limits of f at c is infinite. Then f is said to have an **infinite discontinuity** at c. Again, f may or may not be defined at c.

■ **Example 2.37** Show that the following functions have infinite discontinuities at the given points:

(a) $f(x) = \dfrac{1}{x^2}$; $c = 0$ (b) $f(x) = \begin{cases} 1/x & \text{if } x \neq 0 \\ 0 & \text{if } x = 0; \end{cases}$ $c = 0$

Solution (a) It is clear from the graph of this function shown in Figure 2.36 that there is an infinite discontinuity at 0, since both one-sided limits are infinite. Note that f is not defined at 0.

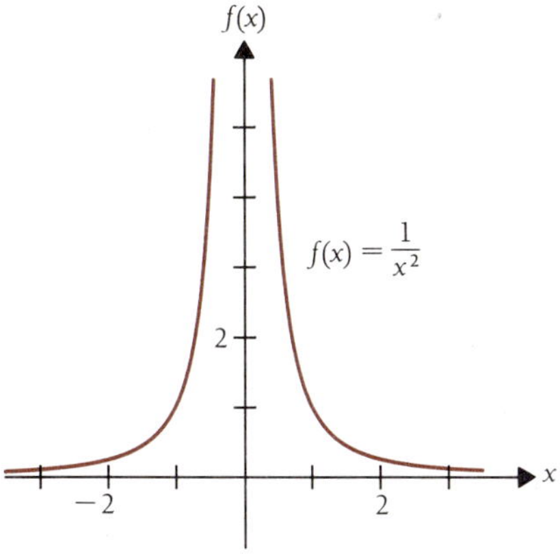

FIGURE 2.36 An infinite discontinuity

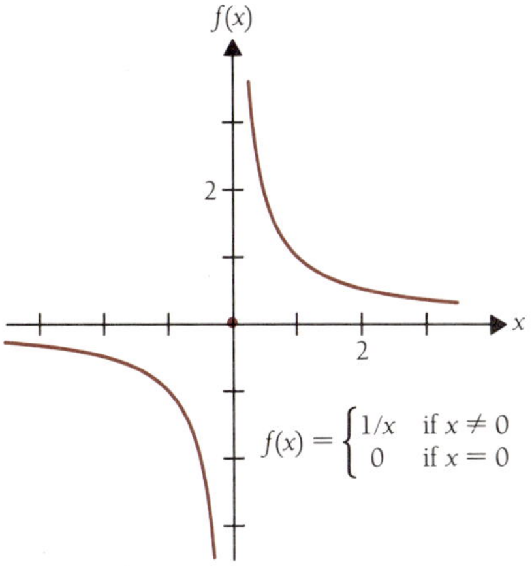

FIGURE 2.37 An infinite discontinuity

(b) Again, both one-sided limits are infinite, so the function has an infinite discontinuity at 0. This time f is defined at 0. The graph is shown in Figure 2.37. ■

There are other kinds of essential discontinuities, but we will not meet any of them, except possibly in the exercises. The following example will summarize our discussion of points of discontinuity.

■ **Example 2.38** Find and classify the discontinuities of the function f defined by

$$f(x) = \begin{cases} 0 & \text{if } x \leq 0 \text{ and } x \neq -2 \\ 2 & \text{if } 0 < x \leq 2 \\ 1/(x-2) & \text{if } x > 2. \end{cases}$$

Solution First, we sketch the graph of f, as shown in Figure 2.38 (page 128). Since f is not defined at -2, there is a point of discontinuity there. It is removable, because the limit of f at -2 exists (it is 0). At 0, f has a jump discontinuity, since the left-hand limit at 0 is 0 and the right-hand limit at 0 is 2. At 2, f has an infinite discontinuity, since the right-hand limit at 2 is infinite.

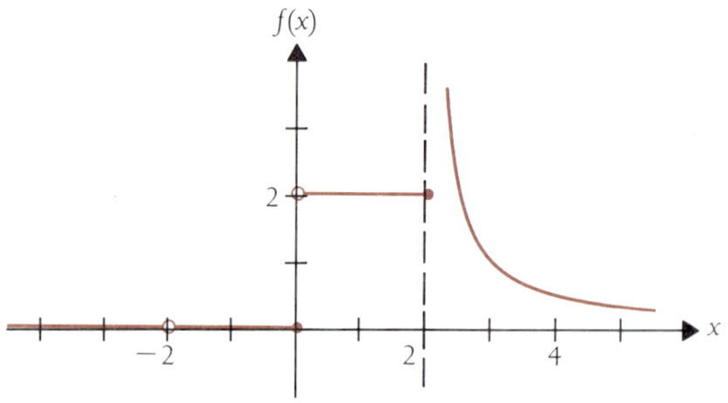

FIGURE 2.38 The graph of $f(x) = \begin{cases} 0 & \text{if } x \leq 0 \text{ and } x \neq -2 \\ 2 & \text{if } 0 < x \leq 2 \\ 1/(x-2) & \text{if } x > 2 \end{cases}$ ■

Continuity on an Interval

In addition to being interested in continuity at a point, we are interested in more global properties of continuity. We usually want to know whether a function is continuous on some set of points. We first consider continuity on intervals.

DEFINITION 2.8

The function f is said to be **continuous on the open interval (a, b)** if and only if f is continuous at every point of (a, b). We say f is **continuous on the closed interval $[a, b]$** if and only if

1. f is continuous on (a, b),
2. $\lim_{x \to a^+} f(x) = f(a)$, and
3. $\lim_{x \to b^-} f(x) = f(b)$.

Graphically, a function that is continuous on $[a, b]$ is represented by an unbroken curve joining the point $(a, f(a))$ to the point $(b, f(b))$, as illustrated in Figure 2.39.

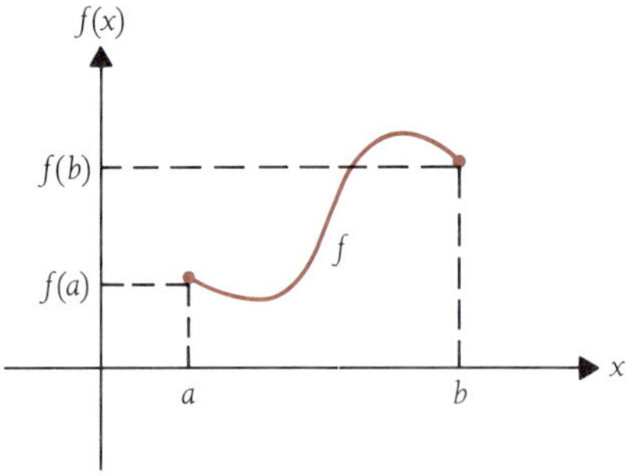

FIGURE 2.39 A function that is continuous on a closed interval

■ **Example 2.39** Discuss the continuity of the greatest integer function on each of the intervals $(1, 2)$, $[1, 2]$, $[1, 1.5]$, and $(0.5, 1.5)$.

Solution Since $\lfloor x \rfloor$ is constant on $(1, 2)$ (see Figure 2.40), it is clearly continuous there. For the interval $[1, 2]$, we note that

$$\lim_{x \to 2^-} \lfloor x \rfloor = 1,$$

while $\lfloor 2 \rfloor = 2$. Thus, condition 3 in Definition 2.8 fails, and $\lfloor x \rfloor$ is not continuous on $[1, 2]$. It is easy to see from Figure 2.41 that $\lfloor x \rfloor$ is continuous on $[1, 1.5]$, but not on $(0.5, 1.5)$, because of the jump discontinuity at 1.

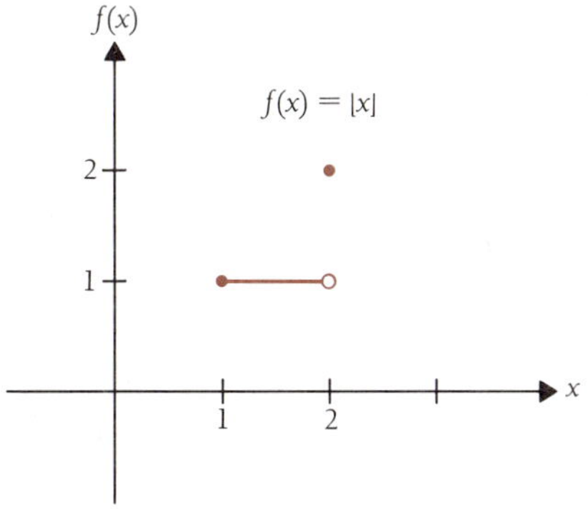

FIGURE 2.40 The greatest integer function over [1, 2]

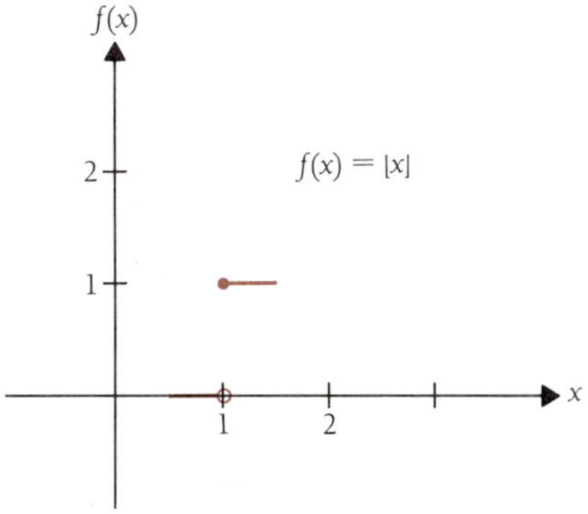

FIGURE 2.41 The greatest integer function over [0.5, 1.5]

■

More generally, if set S is a union of intervals, we define a function f to be **continuous on S** in case f is continuous on every interval contained in S. (The discussion of continuity on more general sets than this is left to more advanced courses.) We also say that a function is **continuous** when it has no points of discontinuity.

■ **Example 2.40** Determine the largest possible set S on which each of the following functions is continuous:

(a) $f(x) = x^2 - x + 4$ **(b)** $f(x) = \sqrt{x}$ **(c)** $f(x) = 1/x$

Solution

(a) Since this polynomial function is defined and continuous at every real number, f is continuous on $\mathbb{R}$. The same is true of every polynomial.

(b) The square root function is continuous on its domain, which is $[0, \infty)$. The only thing to check is that the right-hand limit at 0 is 0, which it is.

(c) This rational function has an infinite discontinuity at 0, so it is not continuous there. It is continuous at every other point, however, so this function is continuous on the set of nonzero real numbers. ■

By using the appropriate limit theorems, we can prove the following theorem. (This is left as an exercise.)

THEOREM 2.20

If functions f and g are continuous at point c, then the combinations $af + bg$ and fg are continuous at c. If $g(c) \neq 0$, then f/g is also continuous at c. If f is continuous at c, g is continuous at $f(c)$, and c is an interior point or endpoint of the domain of $g \circ f$, then $g \circ f$ is also continuous at c.

■ **Example 2.41**

Explain why the function $f(x) = \sqrt{x^2 + 9}$ is continuous on $\mathbb{R}$.

Solution

We can express f as the composition $g \circ h$ of the functions $g(x) = \sqrt{x}$ and $h(x) = x^2 + 9$. Since h is continuous on $\mathbb{R}$, the range of h is in the domain of g, and g is continuous on its domain, it follows from Theorem 2.20 that f is continuous on $\mathbb{R}$. ■

Continuity on a Closed Interval

A function that is continuous on a closed interval has some special properties that will be of importance to us. We will state these properties as theorems without proof, and then look at some of their consequences.

THEOREM 2.21

Extreme Value Theorem If the function f is continuous on the closed interval $[a, b]$, then f has a maximum value and a minimum value on the interval $[a, b]$.

That is, there is some point $c \in [a, b]$ such that $f(c) = M$, and for all $x \in [a, b]$, $f(x) \leq M$. Also, there is some point $d \in [a, b]$ such that $f(d) = m$, and for all $x \in [a, b]$, $f(x) \geq m$. The numbers M and m are the maximum and minimum values, respectively, of f on the interval $[a, b]$.

THEOREM 2.22

Intermediate Value Theorem If the function f is continuous on the closed interval $[a, b]$ and v is any value between the minimum and the maximum of f on $[a, b]$, then f takes on the value v.

That is, there exists at least one point $c \in [a, b]$ such that $f(c) = v$ if v is between m and M.

In other words, these theorems say that the graph of a function that is continuous on a closed interval has a high point and a low point, and takes on every value in between (see Figure 2.42).

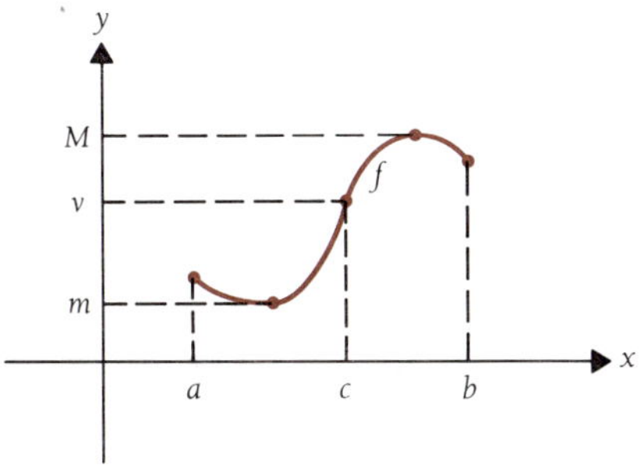

FIGURE 2.42 The extreme value and intermediate value theorems

Here are some applications of the intermediate value theorem.

Example 2.42 Prove that the polynomial

$$p(x) = x^3 - x^2 - 2x + 2$$

has a zero between -2 and 2. [A **zero** of $p(x)$ is a solution of $p(x) = 0$.]

Solution This polynomial is continuous on the interval $[-2, 2]$, because all polynomials are continuous everywhere. Since $p(-2) = -6$ and $p(2) = 2$, it must be that 0 is a value between the minimum and maximum of p on $[-2, 2]$. By the intermediate value theorem, there must be a point $z \in [-2, 2]$ such that $p(z) = 0$. Thus, z is a zero of p. We see from the graph of p in Figure 2.43 that there are actually three zeros of p in the interval $[-2, 2]$.

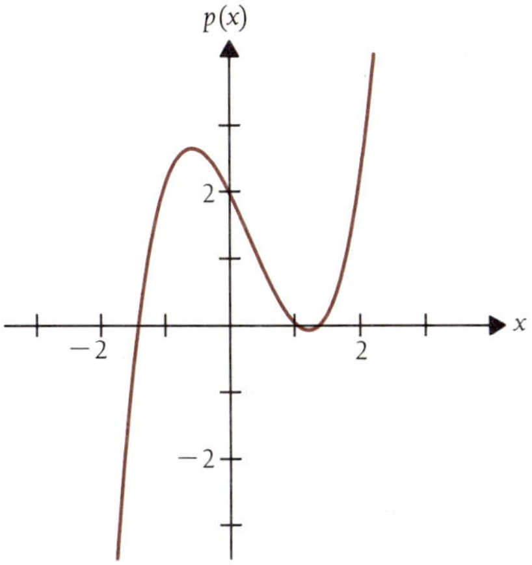

FIGURE 2.43 The graph of $p(x) = x^3 - x^2 - 2x + 2$ ■

Example 2.43 Prove that at some time in your life, your weight in pounds was exactly equal to your age in weeks.

Solution It is reasonable to make the assumption that age and weight change continuously in time, so that your age $a(t)$ and your weight $w(t)$ are continuous functions of t.

Therefore, the function $f(t) = a(t) - w(t)$ is also continuous. If $t = 0$ is the time of your birth, then $a(0) = 0$ and $w(0) > 0$, so that $f(0) < 0$. If $t = N$ is the present time, then undoubtedly $a(N) > w(N)$, so that $f(N) > 0$. Thus, 0 is a number between the minimum and maximum of f, and since f is continuous on $[0, N]$, there was a time T such that $f(T) = 0 = a(T) - w(T)$. Hence, $a(T) = w(T)$. ■

An Application: The Bisection Method

The intermediate value theorem affords us a means of approximating a zero of a continuous function. If f is continuous on $[a, b]$ and $f(a)f(b) < 0$, then f has a zero in $[a, b]$, as was illustrated in Example 2.42. We may use this fact systematically to shorten the interval containing the zero.

The method simply cuts the interval $[a, b]$ in half, or bisects it. If c is the midpoint of $[a, b]$ and $f(c) = 0$, then c is a zero of f. If $f(c) \neq 0$, then either $f(a)f(c) < 0$ or $f(c)f(b) < 0$. In the first case, $[a, c]$ contains a zero of f; and in the second case, $[c, b]$ contains a zero. We may repeat the procedure with the new interval, continuing as many times as we like. This process is called the **bisection method.**

As an algorithm, the bisection method can be stated as follows:

1. Given $[a, b]$ such that $f(a)f(b) < 0$, let $c = (a + b)/2$.
2. If $f(c) = 0$, quit; c is a zero of f.
3. If $f(c) \neq 0$, then:
   (a) If $f(a)f(c) < 0$, replace b by c.
   (b) If $f(a)f(c) > 0$, replace a by c.
4. If an appropriate stopping criterion is met, quit; otherwise, go to Step 1. An "appropriate stopping criterion" is met when one of the following is true:
   (a) The procedure has been applied as many times as we wanted.
   (b) The interval $[a, b]$ is short enough to give a zero of satisfactory precision.
   (c) Available resources (such as study time, computer time, paper, or patience) are exhausted.

The following example illustrates the use of the bisection method.

■ **Example 2.44** Find an interval of length less than 0.01 containing a zero of the function $f(x) = x^3 + x - 3$.

Solution From Descartes's rule of signs, we see that this polynomial function has only one real zero, and it is positive. By evaluating f at a few points, we quickly discover that $f(1) = -1$ and $f(2) = 7$, so the continuous function f has its only zero in the interval $[1, 2]$ (see Figure 2.44). Now we use the bisection method, noting that we must keep an interval with f negative at the left endpoint and f positive at the right endpoint. When the length of the interval is less than 0.01, we can quit. If we start with an interval of length 1, we expect it to take 7 bisections to get an interval of length less than 0.01 (since $2^7 = 128$). We summarize the procedure in Table 2.2 and Figures 2.45–2.47.

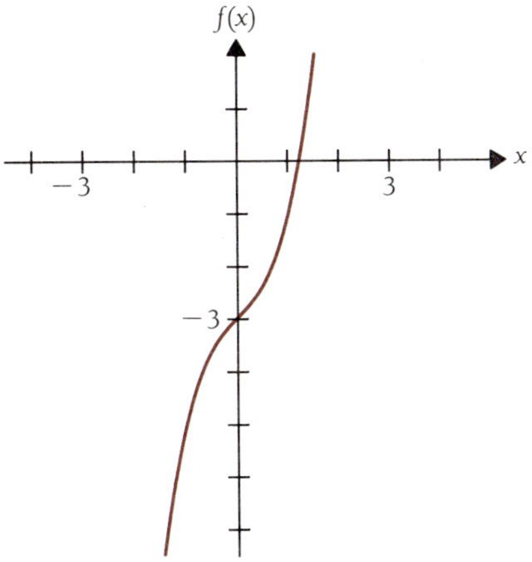

FIGURE 2.44 The graph of $f(x) = x^3 + x - 3$

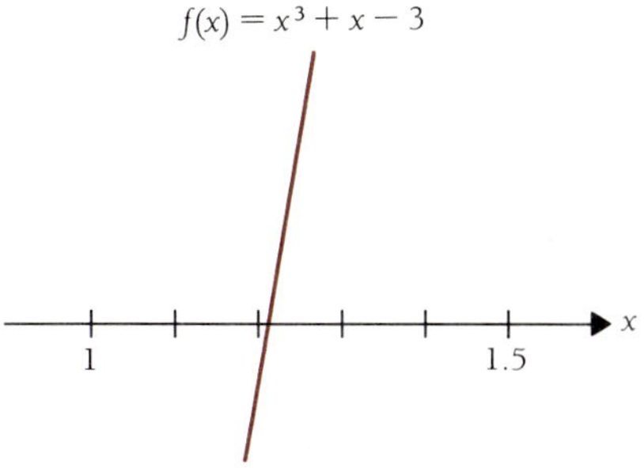

FIGURE 2.45 The graph of $f(x)$ over a smaller interval

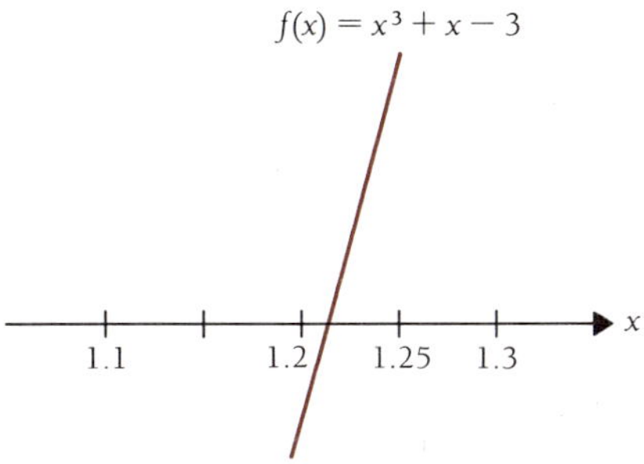

FIGURE 2.46 The graph of $f(x)$ over a smaller interval yet

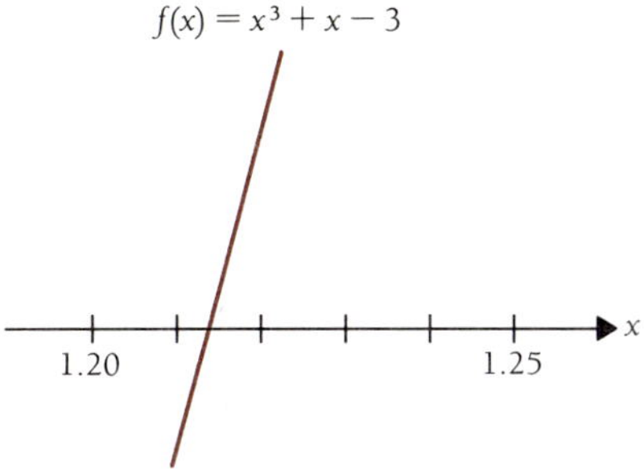

FIGURE 2.47 The graph of $f(x)$ over an even smaller interval

TABLE 2.2

New $[a, b]$	Is $a - b < 0.01$?	$c = \dfrac{a+b}{2}$	$f(c)$	Replace a or b?
[1, 2]	No	1.5	1.875	b
[1, 1.5]	No	1.25	0.203	b
[1, 1.25]	No	1.125	−0.451	a
[1.125, 1.25]	No	1.1875	−0.138	a
[1.1875, 1.25]	No	1.21875	0.029	b
[1.1875, 1.21875]	No	1.203125	−0.055	a
[1.203125, 1.21875]	No	1.2109375	−0.013	a
[1.2109375, 1.21875]	Yes			

Thus, correct to two decimal places, the real zero of $f(x) = x^3 + x - 3$ is 1.21. ■

The bisection method always works (within human or machine limitations), though it can be a little slow. A couple of variations of the bisection method are the *method of false position* and the *secant method*, which are mentioned in the Additional Exercises at the end of this chapter. These variations are faster, but they may not always work. The complete study of numerical methods such as these is done in a course on numerical analysis.

Exercises 2.5

Determine whether the given function is continuous at the given point c, and sketch the graph of the function. If the function is discontinuous, classify the type of discontinuity.

1. $f(x) = \dfrac{x^2 - 4}{x - 2}$; $c = 2$
2. $f(x) = \dfrac{x - 3}{x^2 - 9}$; $c = 3$
3. $f(x) = \begin{cases} -0.25 & \text{if } x = -2 \\ \dfrac{x+2}{x^2-4} & \text{if } x \neq -2; \end{cases}$ $c = -2$
4. $f(x) = \begin{cases} 0 & \text{if } x = 1 \\ \dfrac{x-1}{x^2-1} & \text{if } x \neq 1; \end{cases}$ $c = 1$
5. $f(x) = 2$ if $x \neq 0$; $c = 0$
6. $f(x) = x^{-2}$; $c = 0$
7. $f(x) = \dfrac{x + 3}{x - 3}$; $c = -3$
8. $f(x) = \dfrac{x + 3}{x - 3}$; $c = 3$
9. $f(x) = \begin{cases} \dfrac{1}{x-1} & \text{if } x < 1 \\ 0 & \text{if } x \geq 1; \end{cases}$ $c = 1$
10. $f(x) = \begin{cases} x^{-2} & \text{if } x \neq 0 \\ 0 & \text{if } x = 0; \end{cases}$ $c = 0$
11. $f(x) = \begin{cases} x^{-2} & \text{if } x < 0 \\ x & \text{if } x \geq 0; \end{cases}$ $c = 0$
12. $f(x) = \begin{cases} x + 1 & \text{if } x \leq 0 \\ x - 1 & \text{if } x > 0; \end{cases}$ $c = 0$
13. $f(x) = \tan x$; $c = \pi/2$
14. $f(x) = \tan x$; $c = \pi/4$
15. $f(x) = \sec 2x$; $c = \pi/4$
16. $f(x) = \dfrac{\cos x}{1 + \sin x}$; $c = 3\pi/2$
17. $f(x) = |x - 2|$; $c = 2$
18. $f(x) = \dfrac{x - 2}{|x - 2|}$; $c = 2$
19. $f(x) = \lfloor \cos x \rfloor$; $c = 0$
20. $f(x) = \lfloor \cos x \rfloor$; $c = \pi/2$

Determine the largest set of real numbers on which each of the following functions is continuous, and classify any points of discontinuity:

21. $f(x) = \dfrac{1}{x^2 + 1}$
22. $f(x) = \dfrac{1}{x^2 + 2x + 1}$
23. $f(x) = x + \dfrac{1}{x}$
24. $f(x) = \begin{cases} 4 - x^2 & \text{if } x < 0 \\ x + 4 & \text{if } x > 0 \end{cases}$
25. $f(x) = \tan 2x$
26. $f(x) = \sec^2 x - 1$
27. $f(x) = \sqrt{4 - x^2}$
28. $f(x) = \dfrac{1}{\sqrt{4 - x^2}}$

Determine whether the following functions are continuous on the given intervals:

29. $f(x) = \lfloor x \rfloor$ on $[0, 0.75]$
30. $f(x) = \lfloor x \rfloor$ on $[0, 1]$
31. $f(x) = \dfrac{1}{x}$ on $[1, 3]$
32. $f(x) = \dfrac{1}{x}$ on $(0, 1)$

33. $f(x) = \dfrac{x+1}{x-1}$ on $(-2, 0)$

34. $f(x) = \dfrac{x+1}{x-1}$ on $(0, 2)$

35. $f(x) = \dfrac{x+1}{x-1}$ on $(0, 1)$

36. $f(x) = \dfrac{x+1}{x-1}$ on $[0, 1]$

37. $f(x) = \sin 2x$ on $[-2\pi, 2\pi]$

38. $f(x) = \sec 2x$ on $[0, \pi]$

39. $f(x) = \sec 3x$ on $(0, \pi/6)$

40. $f(x) = \sec 3x$ on $[0, \pi/6]$

Let $p(x) = x^4 - 2x^3 - 4x^2 + 4x + 4$.

41. Show that p has a zero between -1 and 0.

42. Show that p has a zero between -2 and -1.

43. Does p have a zero between 1 and 2?

44. Does p have a zero between 0 and 1?

Apply the bisection method four times, starting with the given interval, to obtain a smaller interval containing a zero of the given function.

^{C}45. $f(x) = x^2 - 2$; $[1.4, 1.5]$

^{C}46. $f(x) = x^3 - 2$; $[1.2, 1.3]$

^{C}47. $f(x) = x^4 - 2x^3 - 4x^2 + 4x + 4$; $[-1, 0]$

^{C}48. $f(x) = x^4 - 2x^3 - 4x^2 + 4x + 4$; $[-2, -1]$

^{C}49. $f(x) = x - \cos x$; $[0, 1]$ (Be sure x is in radians!)

^{C}50. $f(x) = \dfrac{1}{x^2+1} - \sin x$; $[0, 1]$

Find a zero of the given function, correct to two decimal places.

^{C}51. $f(x) = x^3 + x + 3$

52. $f(x) = 4x^3 - 3x^2 + 4x - 3$

^{C}53. $f(x) = x^2 - \cos x$

^{C}54. $f(x) = x \sin x + 1$

55. Starting with an interval of length L, the bisection method is applied n times. How long is the resulting interval?

56. Starting with an interval of length 1, how many applications of the bisection method are needed to approximate a zero correct to three decimal places?

57. An automobile travels without refueling until it runs out of gasoline. Prove that at some point on the journey, the number of liters of fuel in the tank is equal to the number of kilometers traveled.

58. A rocket is launched into the air and eventually burns all its fuel. Prove that at some point in the rocket's flight the amount of fuel left is equal to the distance traveled to that point.

59. Prove Theorem 2.20.

60. Find functions f and g, neither of which is continuous at 0, such that $f + g$ is continuous at 0.

61. Show that the functions

$$f(x) = x + 2 \qquad \text{and} \qquad g(x) = \frac{1}{x-2}$$

are both continuous at 0, but that their composition $g \circ f$ is not continuous at 0.

62. Explain why the example in Problem 61 does not contradict Theorem 2.20.

63. Prove that if the function f is continuous on (a, b) and $a < c < d < b$, then f is continuous on $[c, d]$.

64. Show that the function $f(x) = \sin(1/x)$ has an essential discontinuity at 0 that is neither a jump discontinuity nor an infinite discontinuity.

SECTION 2.6

Chapter Review

Important Concepts

Definition of limit	Limit at infinity	Jump discontinuity
Limit theorems	Infinite limit	Infinite discontinuity
Evaluating limits	Continuity at a point	Continuity on an interval
Bounded function	Point of discontinuity	Extreme value theorem
Trigonometric limits	Removable discontinuity	Intermediate value theorem
One-sided limit	Essential discontinuity	Bisection method

Review Exercises

Find the limit if it exists. If the limit does not exist, so state.

1. $\lim_{x\to -2} 8$
2. $\lim_{x\to 0} 5$
3. $\lim_{x\to -1} x$
4. $\lim_{x\to 4} x$
5. $\lim_{x\to 1} (x^2 + 3x + 5)$
6. $\lim_{x\to 4} (2x - 5)$
7. $\lim_{x\to 2} \dfrac{1}{2x - 1}$
8. $\lim_{x\to -3} \dfrac{x}{x - 3}$
9. $\lim_{x\to -2} \dfrac{1}{x + 2}$
10. $\lim_{x\to 3} \dfrac{x}{x - 3}$
11. $\lim_{x\to 5} \dfrac{x - 3}{x - 5}$
12. $\lim_{x\to 3} \dfrac{x - 3}{x - 5}$
13. $\lim_{x\to 5^+} \dfrac{x - 3}{x - 5}$
14. $\lim_{x\to 5^-} \dfrac{x - 3}{x - 5}$
15. $\lim_{x\to -5} \dfrac{x - 5}{x^2 - 25}$
16. $\lim_{x\to 5} \dfrac{x - 5}{x^2 - 25}$
17. $\lim_{x\to 5^-} \dfrac{x - 5}{x^2 - 25}$
18. $\lim_{x\to 5^+} \dfrac{x - 5}{x^2 - 25}$
19. $\lim_{x\to -2^+} \dfrac{|x + 2|}{x + 2}$
20. $\lim_{x\to -2^-} \dfrac{|x + 2|}{x + 2}$
21. $\lim_{x\to 1} \dfrac{1 - \sqrt{2 - x}}{1 - x}$
22. $\lim_{x\to 2} \dfrac{\sqrt{x + 2} - 2}{2 - x}$
23. $\lim_{x\to 1^+} [x]$
24. $\lim_{x\to 1^-} [x]$
25. $\lim_{x\to 1} [x]$
26. $\lim_{x\to 2} [x]$
27. $\lim_{x\to -\infty} \dfrac{3x + 5}{x - 4}$
28. $\lim_{x\to \infty} (x - x^2)$
29. $\lim_{x\to -\infty} (x - x^3)$
30. $\lim_{x\to \infty} \dfrac{1 - x^2}{1 - 4x^2}$
31. $\lim_{x\to \infty} (x - \sqrt{x^2 + 1})$
32. $\lim_{x\to -\infty} (x^2 - \sqrt{x^4 - x})$
33. $\lim_{x\to 0} \dfrac{2x}{\sin x}$
34. $\lim_{x\to 0} \dfrac{1 - \cos 3x}{x}$
35. $\lim_{x\to -\infty} \dfrac{\sin x}{x}$
36. $\lim_{x\to -\infty} \dfrac{\cos x}{x}$
37. $\lim_{x\to 0} \dfrac{\sin 2x}{\sin 3x}$
38. $\lim_{x\to 0} \dfrac{\sin 3x}{\tan \pi x}$
39. $\lim_{x\to \pi/2} \dfrac{\sin x}{\tan x}$
40. $\lim_{x\to \pi/4} \dfrac{1 + \cos x}{\tan 2x}$

Determine the largest set of real numbers on which the function is continuous. Classify any discontinuities.

41. $f(x) = x^3 - x$

42. $f(x) = \dfrac{2x}{x^2 + 4}$

43. $f(x) = \dfrac{1}{x^3 - x}$

44. $f(x) = \dfrac{2x}{x^2 - 4}$

45. $f(x) = \dfrac{1}{1 - \tan x}$

46. $f(x) = \dfrac{1}{1 - \sin x}$

47. $f(x) = \begin{cases} x^2 + 1 & \text{if } x < 0 \\ x & \text{if } x \geq 0 \end{cases}$

48. $f(x) = \begin{cases} 1/x & \text{if } x < 0 \\ x^2 & \text{if } x \geq 0 \end{cases}$

49. $\sigma(x) = \begin{cases} -1 & \text{if } x < 0 \\ 0 & \text{if } x = 0 \\ 1 & \text{if } x > 0 \end{cases}$ (This function is called the **signum function**.)

50. $f(x) = \sigma(\lfloor x \rfloor)$

51. $f(x) = \lfloor \sigma(x) \rfloor$

52. $f(x) = |\sigma(x)|$

53. $f(x) = \sigma(|x|)$

54. $f(x) = \sigma(\sin x)$

55. $f(x) = \dfrac{|x|}{x} - \sigma(x)$

56. $f(x) = \sigma\left(\dfrac{1}{x^2 + 1}\right)$

57. $f(x) = \left\lfloor \dfrac{1}{x^2 + 1} \right\rfloor$

58. $f(x) = x - \lfloor x \rfloor$

59. $f(x) = \lfloor \sin x \rfloor$

60. $f(x) = \begin{cases} x \sin(1/x) & \text{if } x \neq 0 \\ 0 & \text{if } x = 0 \end{cases}$

61. Using the results of this chapter, explain why a polynomial of odd degree always has a real zero.

62. Show that at some point in the life of a person, his or her height in inches is equal to his or her weight in pounds. (Make reasonable assumptions about the height and weight of the average person.) Does the same result hold if we measure height in centimeters and weight in kilograms?

63. If one were to travel from a point on the equator, where the temperature was 100°F, to the North Pole, where the temperature was −20°F, explain why at some point in the trip the temperature would be equal to the position in degrees north latitude.

Additional Exercises

Find the limit, if it exists.

1. $\lim\limits_{t\to\infty} [(t^2 + 1)^{-1/2} - (t^2 - 1)^{-1/2}]$

2. $\lim\limits_{x\to\infty} (\sqrt{2x^2 + 1} - \sqrt{x^2 - x})$

3. $\lim\limits_{x\to 1^-} (1 - x + \lfloor x \rfloor - \lfloor 1 - x \rfloor)$

4. $\lim\limits_{x\to 1^-} \dfrac{x - 1/x}{(x - 1)^2}$

5. $\lim\limits_{x\to 2^+} \dfrac{x - 1}{|x - 2|}$

6. $\lim\limits_{x\to 0^+} \dfrac{x|x|}{x^3}$

7. $\lim\limits_{x\to 0^-} \dfrac{x|x|}{x^3}$

8. $\lim\limits_{t\to\infty} \dfrac{|t^3|}{t^3}$

9. $\lim_{x \to 3^+} \dfrac{\lfloor x \rfloor^2 - 9}{x - 3}$ 10. $\lim_{t \to 0} \dfrac{(2 + t)^{1/3} - 2^{1/3}}{t}$ 11. $\lim_{x \to 1} \dfrac{x^{2/3} - 1}{x^{1/3} - 1}$ 12. $\lim_{x \to \infty} \dfrac{x^{2/3} - 1}{x^{1/3} - 1}$

Given the functions

$$f(x) = \begin{cases} -x & \text{if } x \le 1 \\ 3 + x & \text{if } x > 1 \end{cases} \quad \text{and} \quad g(x) = \begin{cases} x^2 & \text{if } x \le 1 \\ 2 - x & \text{if } x > 1, \end{cases}$$

find the following limits, if they exist:

13. $\lim_{x \to 1^-} f(x)$ 14. $\lim_{x \to 1^+} f(x)$ 15. $\lim_{x \to 1} f(x)$ 16. $\lim_{x \to 1^-} g(x)$

17. $\lim_{x \to 1^+} g(x)$ 18. $\lim_{x \to 1} g(x)$ 19. $\lim_{x \to 1^-} f(g(x))$ 20. $\lim_{x \to 1^+} f(g(x))$

21. Given that $\lim_{x \to 2} f(x) = 6$ and $\lim_{x \to 2} g(x) = \frac{1}{2}$, find

$$\lim_{x \to 2} \frac{1 + f(x)}{g(x)}.$$

22. Given a positive number B, find a number δ such that if $0 < x < \delta$, then $1/x > B$.
23. Given a positive number ε, find a number B such that if $x > B$, then $1/\sqrt{x} < \varepsilon$.
24. Prove that $\lim_{x \to c} |x| = |c|$.
25. If

$$f(x) = \frac{x^2 + 3x - 10}{x - 2}$$

for $x \ne 2$, define $f(2)$ so that f is continuous.

26. Find an example of a function f such that the limit of $f(x)$ as $x \to 3$ exists but $f(3)$ is not defined.
27. Find an example of a function f such that the limit of $f(x)$ as $x \to 3$ does not exist but $f(3)$ is defined.
28. Find an example of a function f such that the limit of $f(x)$ as $x \to 3$ exists and $f(3)$ is defined, but f is not continuous at 3.
29. Let f be a function defined on all of $\mathbb{R}$ such that f is continuous at 0 and has the property that $f(a + b) = f(a) + f(b)$ for all $a, b \in \mathbb{R}$. Prove that f is continuous.
30. Is the function $f(x) = \lfloor |x| \rfloor$ continuous at 0?
31. For what values of x is the function $f(x) = \lfloor x \rfloor - x$ continuous?
32. Is the function $f(x) = |\lfloor x \rfloor|$ continuous at 0?
33. The **Dirichlet function** D is defined by

$$D(x) = \begin{cases} 1 & \text{if } x \text{ is rational} \\ 0 & \text{if } x \text{ is irrational.} \end{cases}$$

Show that D is discontinuous at every point.

34. The **ruler function** R is defined as follows: if x is irrational, then $R(x) = 0$. If x is rational and $x = p/q$, reduced to lowest terms, then $R(x) = 1/q$. Show that R is continuous at c if c is irrational and discontinuous at c if c is rational. (See if you can tell where this function got its name, by considering its graph.)
35. Show that the Dirichlet function is the composition of the ruler function with the signum function—that is, $D(x) = \sigma(R(x))$. (See Problem 49 in the Review Exercises and Problems 33 and 34 above.)

36. Explain why $D(R(x))$ is continuous.

37. Is $R(\sigma(x))$ continuous at 0?

38. For what values of x is $R(\lfloor x \rfloor)$ continuous?

39. Discuss the continuity of $\sin(R(x))$.

40. Discuss the continuity of $R(\sin x)$.

41. Is $\sin\lfloor x \rfloor$ continuous at 0?

42. Given $f(x) = x^2 + 3$ on the interval $[0, 3]$ and a number $\varepsilon > 0$, find a number δ such that if $x_1, x_2 \in [0, 3]$ and $|x_1 - x_2| < \delta$, then $|f(x_1) - f(x_2)| < \varepsilon$. (This property is called **uniform continuity** and holds for any function that is continuous on a closed interval.)

43. Let a function f be periodic with period p and continuous on $[0, p]$. Prove that f is continuous on all of $\mathbb{R}$.

44. If p is a polynomial function of odd degree, prove that the function $f(x) = 1/p(x)$ has a point of discontinuity.

45. Is $R(x)$ periodic?

46. Is $\tan(R(x))$ periodic?

Find the limit, if it exists.

47. $\lim_{x \to 0^+} \dfrac{\sec x}{x}$

48. $\lim_{x \to \infty} \dfrac{\sin 3x}{x}$

49. $\lim_{x \to 0} x \sin\left(\dfrac{1}{x}\right)$

50. $\lim_{x \to \infty} \dfrac{\cos x}{x + 1}$

51. $\lim_{x \to \infty} \dfrac{x \sin x}{x + 1}$

52. $\lim_{x \to \infty} \sin\lfloor x \rfloor$

Let a regular n-sided polygon be inscribed in a circle of radius r. Let O be the center of the circle, and let A and B be adjacent vertices of the polygon (see the figure).

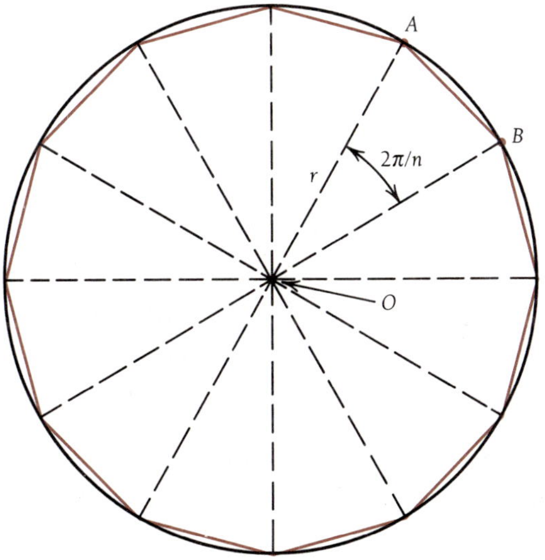

53. Express the length of side AB in terms of r and n, using the fact that $\angle AOB = 2\pi/n$ radians.

54. Express the perimeter P of the polygon in terms of r and n.

55. Find the circumference C of the circle by finding the limit of the perimeter as $n \to \infty$.

56. Find the area of the circle by taking a limit.

57. Prove that the ratio of the circumference to the diameter of a circle is the same for all circles. [*Hint:* Use an inscribed n-gon and take a limit. *Caution:* You may not use any formula for the circumference of the circle in terms of the radius, for that presumes the result you are trying to prove.]

A function is said to be **right-continuous** *at c if and only if $\lim_{x\to c^+} f(x) = f(c)$ and* **left-continuous** *at c if and only if $\lim_{x\to c^-} f(x) = f(c)$. Tell whether the given function is left-continuous, right-continuous, both, or neither at the given point.*

58. $f(x) = \lfloor x \rfloor;\quad c = 2$

59. $f(x) = \lfloor x \rfloor;\quad c = 2.5$

60. $f(x) = |x|;\quad c = 0$

61. $f(x) = \dfrac{x^2 - 9}{x - 3};\quad c = 3$

62. $f(x) = \lfloor \sin x \rfloor;\quad c = \pi/2$

63. $f(x) = \lfloor \sin x \rfloor;\quad c = 0$

64. $f(x) = \lfloor \sin x \rfloor;\quad c = \pi$

65. $f(x) = \sigma(x);\quad c = 0$

66. Prove that the function f is continuous at c if and only if f is both right-continuous at c and left-continuous at c.

67. State the definition of continuity on a closed interval in terms of right and left continuity.

68. The **method of false position** is a variation on the bisection method for approximating the zero of a function. Instead of letting c be the midpoint of $[a, b]$, we let c be the point at which the line joining $(a, f(a))$ to $(b, f(b))$ crosses the x-axis (see the figure). Show that

$$c = \frac{af(b) - bf(a)}{f(b) - f(a)} = a + \frac{(a - b)f(a)}{f(b) - f(a)}.$$

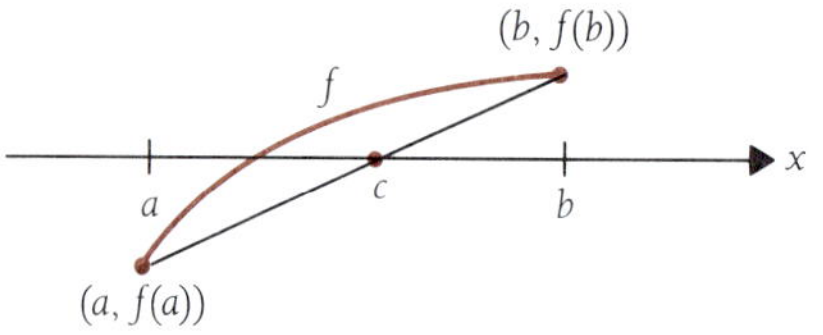

Use the method of false position four times to approximate a zero of the given function, starting with the given interval.

69. $f(x) = x^3 + x - 3;\quad [1, 2]$

70. $f(x) = x - \cos x;\quad [0, 1]$

71. $f(x) = 1 + x - x^4;\quad [1, 2]$

72. $f(x) = 3 - \tan x;\quad [1, 1.5]$

73. Discuss some advantages and disadvantages of the method of false position, as compared with the bisection method.

74. Look up the *secant method* in a numerical analysis book, and redo Problems 69–72 by that method. Discuss the advantages and disadvantages of the secant method as compared to the bisection method and the method of false position.

CHAPTER 3

The Derivative

INTRODUCTION

During the 16th and 17th centuries, two problems arose that occupied the minds of natural philosophers. One of these was a discussion of motion, particularly planetary motion, and the quantitative description of position and velocity. Such people as Kepler and Galileo worked on this problem. The other problem arose out of the newly developed study of analytic geometry, and concerned the tangent line to a curve. Such people as Descartes, Roberval (1602–1675), Torricelli (1609–1647), and Fermat addressed this problem.

Newton and Leibniz solved both of these problems, using the idea of the *derivative*. We will show in this chapter how both the tangent problem and the velocity problem lead to the use of limits, which in turn lead to the derivative. We will discuss the slope of a curve and the instantaneous rate of change of a quantity, the two primary interpretations of the derivative. We will also consider various techniques for computing the derivative, along with some applications.

It is with this chapter that we begin in earnest our study of the differential calculus.

SECTION 3.1

Tangent Lines and Slope

In this section, we apply the limit concept to the geometric problem of describing the line that is tangent to a curve at a given point. After discussing what is meant by a tangent line, we show how to use limits to find the equation of a line tangent to the graph of a function at a given point.

Johannes Kepler (1571–1630), German astronomer and mathematician, inherited the astronomical observations of the Danish astronomer Tycho Brahe. After 20 years of grueling arithmetic, Kepler was able to announce his three laws of planetary motion, which he had guessed and then verified from Brahe's data. The mathematical importance of these laws was that Newton was able to derive them from his theory of gravitation, using calculus. This led to immediate acceptance of Newton's new mathematical view of the universe.

One of Kepler's laws states that the orbit of a planet is an ellipse; this brought conic sections back into prominence. Kepler made other contributions to analytic geometry, studied semiregular polyhedra, used something like integration to approximate areas and volumes, and was first to consider points and lines "at infinity," as studied in projective geometry.

The professions of astronomy and astrology were not distinct in those days as they are now, so during lean times, Kepler earned a living casting horoscopes.

Pierre de Fermat (1601?–1665) worked as a legal officer for the French government in the vicinity of Toulouse. During his spare time, he played with mathematics, becoming the most celebrated amateur mathematician of all time. He invented analytic geometry at the same time Descartes did, and used it to find tangent lines to curves much as we do today. He then applied this material in finding maxima and minima of curves. One by-product of this work was Fermat's *principle of least time* in optics, which states that a ray of light follows the path that takes the least time. His crowning achievement was in number theory, which he inherited from the Greeks but changed profoundly with his own methods and discoveries. He also helped found the theory of probability.

Fermat published little, but carried on a voluminous correspondence with many leading scientists of his day. He announced many of his results in these letters, without proof, as a sort of challenge. All but one of Fermat's theorems have subsequently been proved—some after many years. The one exception, his "last theorem," was found in the margin of one of his books after he died. It states that the equation $x^n + y^n = z^n$ has no solution in integers if $n > 2$. In the 300 years since that time, the effort spent on trying to prove this result has created entire new disciplines within mathematics, but no proof or counterexample has yet been found.

What Is a Tangent Line?

Recall from elementary plane geometry that a line is tangent to a circle at a given point P on the circle if the line meets the circle at P and nowhere else, as in Figure 3.1. For a curve other than a circle, the condition that a line meets the curve at exactly one point does not necessarily define a tangent line. A line may meet a curve at only one point P, as in Figure 3.2, and yet fail to be a tangent line, or it may be a tangent line and yet meet the curve more than once, as in Figure 3.3.

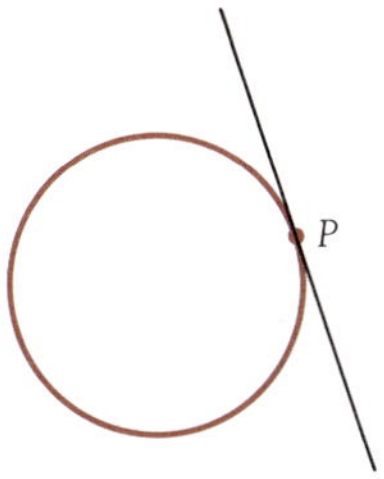

FIGURE 3.1 Tangent line to a circle

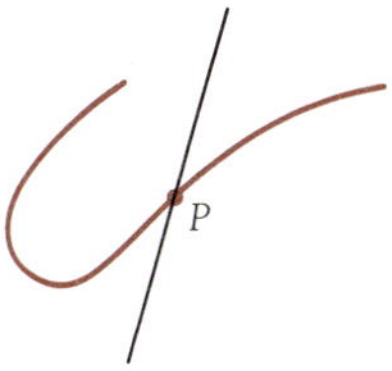

FIGURE 3.2 Not a tangent line

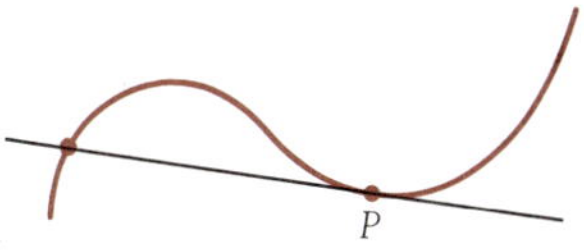

FIGURE 3.3 This tangent line crosses the curve

To decide what is meant by tangency, let's look more closely at the case of the circle. Consider a circle with line t tangent to the circle at P, as in Figure 3.4, and with a secant line s cutting the circle at P and at Q. As Q is moved around the circle toward P, the line s rotates about P, moving closer to the position of the tangent line t. In fact, as Q approaches P, s approaches t. We can thus think of the tangent line t at a point P as the limiting position of a variable secant line $s = PQ$ as Q approaches P.

This same point of view yields tangent lines for other curves as well as for circles.

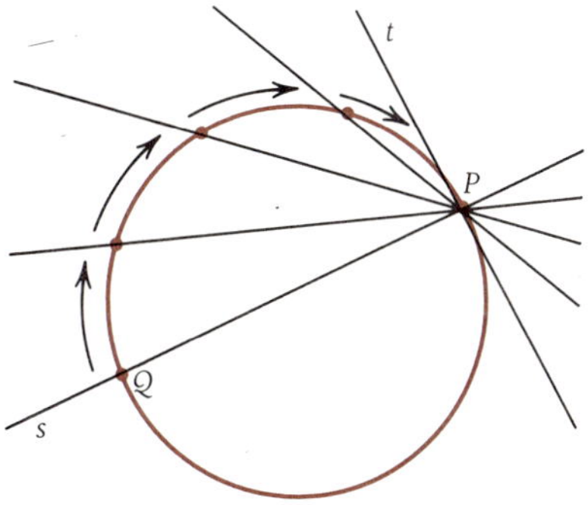

FIGURE 3.4 A secant line approaches the tangent line

The Tangent Line to a Graph

We start with a curve that is the graph of a function, $y = f(x)$. Let P be the point $(c, f(c))$, as in Figure 3.5 on page 144. If Q is another point $(c + h, f(c + h))$ on the curve, then Q will approach P as h approaches 0, provided that the curve is continuous at P. To allow Q to approach P from either side, we must allow h to be either positive or negative.

Now, as $h \to 0$, $Q \to P$, and the secant line $s = PQ$ approaches the position of the tangent line t at P, provided that the curve is smooth at P. But if the curve has a sharp corner at P, as in Figure 3.6, then the secant lines approach different positions as Q approaches P from opposite sides, and the curve can have no (unique) tangent line at P.

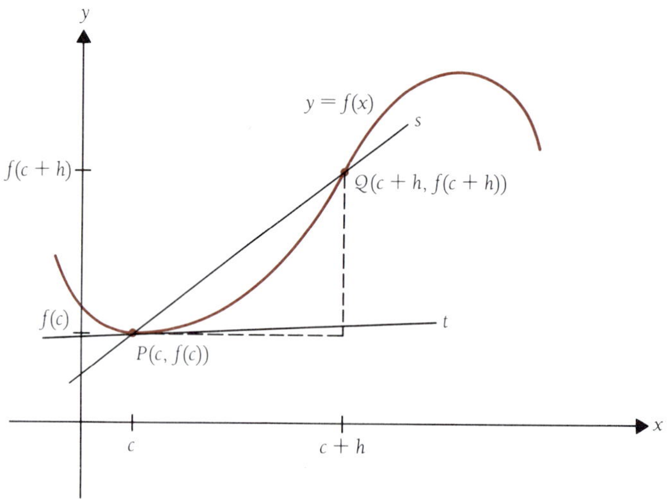

FIGURE 3.5 Secant line to the graph of a function

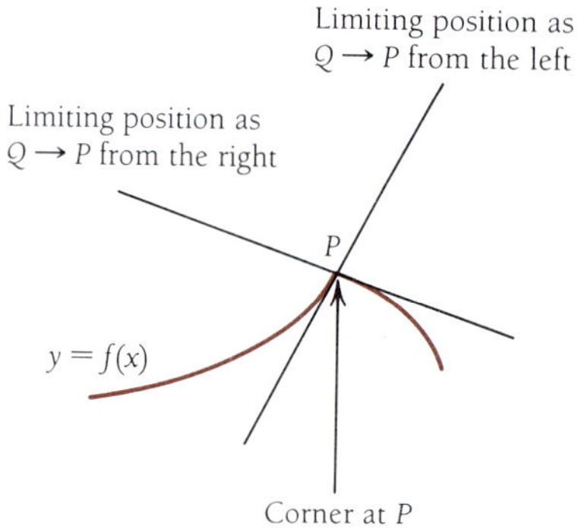

FIGURE 3.6 A curve with a corner has no tangent line there

Slope

From Figure 3.5, we see that the slope of the secant line PQ is the rise, $f(c+h) - f(c)$, divided by the run, $(c+h) - c = h$. That is, the slope of the secant line is

$$m_{\text{sec}} = \frac{f(c+h) - f(c)}{h}. \tag{1}$$

Therefore, the slope of the tangent line is the limit of m_{sec} as $h \to 0$, if the limit exists. We are thus led to make the following definition.

DEFINITION 3.1 The **tangent line** to the curve $y = f(x)$ at the point $(c, f(c))$ is the line through that point with slope

$$m_{\tan} = \lim_{h \to 0} \frac{f(c+h) - f(c)}{h},$$

provided the limit exists. If f is continuous and the limit is infinite, the tangent line is vertical. If the limit does not exist for any other reason, we say the curve *has no tangent line* at the point $(c, f(c))$.

The slope of the tangent line to a curve at point P is also called the **slope of the curve** at P.

If $m_{\tan}$, given in Definition 3.1, is the slope of the tangent line to the curve at the point $(c, f(c))$, then it follows from the point–slope form of the equation of a line that the tangent line has the equation

$$y - f(c) = m_{\tan}(x - c). \tag{2}$$

Example 3.1 Find the equation of the tangent line to the curve $y = x^2 + 1$ at the point (2, 5).

Solution We first compute the slope of the tangent line:

$$\begin{aligned} m_{\tan} &= \lim_{h \to 0} \frac{f(2+h) - f(2)}{h} \\ &= \lim_{h \to 0} \frac{[(2+h)^2 + 1] - 5}{h} \\ &= \lim_{h \to 0} \frac{(4 + 4h + h^2 + 1) - 5}{h} \\ &= \lim_{h \to 0} \frac{4h + h^2}{h} \\ &= \lim_{h \to 0} (4 + h) \\ &= 4. \end{aligned}$$

Then, from equation (2), we get the equation of the tangent line:

$$\begin{aligned} y - 5 &= 4(x - 2), \\ y &= 4x - 3. \end{aligned}$$

This example is illustrated in Figure 3.7.

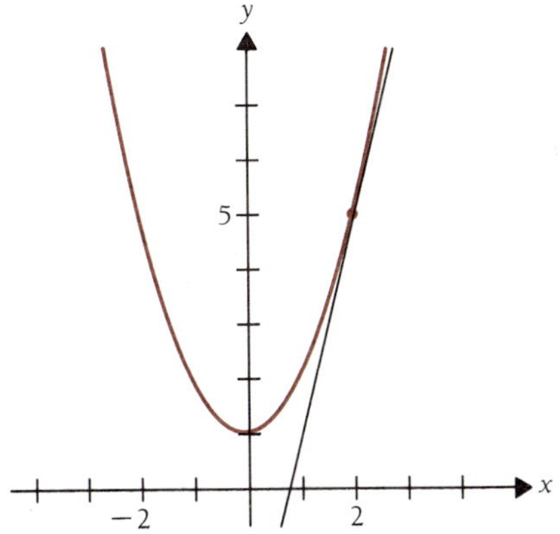

FIGURE 3.7 Tangent line to $y = x^2 + 1$ at (2, 5)

Example 3.2 Find the equation of the tangent line to the curve $y = \sqrt{x}$ at the point (4, 2).

Solution Again, we first find the slope:

$$\begin{aligned} m_{\tan} &= \lim_{h\to 0} \frac{f(4+h) - f(4)}{h} \\ &= \lim_{h\to 0} \frac{\sqrt{4+h} - 2}{h} \cdot \frac{\sqrt{4+h} + 2}{\sqrt{4+h} + 2} \\ &= \lim_{h\to 0} \frac{(4+h) - 4}{h(\sqrt{4+h} + 2)} \\ &= \lim_{h\to 0} \frac{1}{\sqrt{4+h} + 2} \\ &= \frac{1}{4}. \end{aligned}$$

The tangent line is therefore (see Figure 3.8)

$$y - 2 = \frac{1}{4}(x - 4),$$

$$x - 4y + 4 = 0.$$

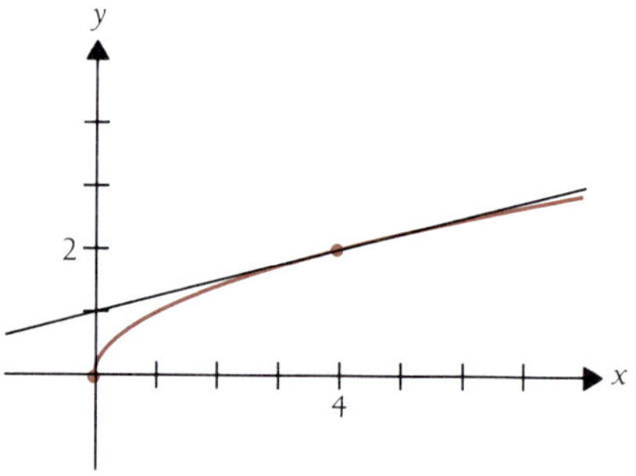

FIGURE 3.8 Tangent line to $y = \sqrt{x}$ at (4, 2)

Example 3.3 Explain why the curve $y = |x^2 - 4|$ has no tangent line at $x = 2$.

Solution For this curve, the slope at (2, 0) is

$$m_{\tan} = \lim_{h\to 0} \frac{f(2+h) - f(2)}{h}$$
$$= \lim_{h\to 0} \frac{|(2+h)^2 - 4| - 0}{h}$$
$$= \lim_{h\to 0} \frac{|4h + h^2|}{h}$$
$$= \lim_{h\to 0} \frac{|h| \cdot |4+h|}{h}.$$

But this limit does not exist, because the two one-sided limits are unequal (as you should verify). Thus, the curve $y = |x^2 - 4|$ has no tangent line at $x = 2$. The graph in Figure 3.9 shows that the curve has a corner at $x = 2$. The graph of $y = x^2 - 4$ is shown in Figure 3.10 for comparison.

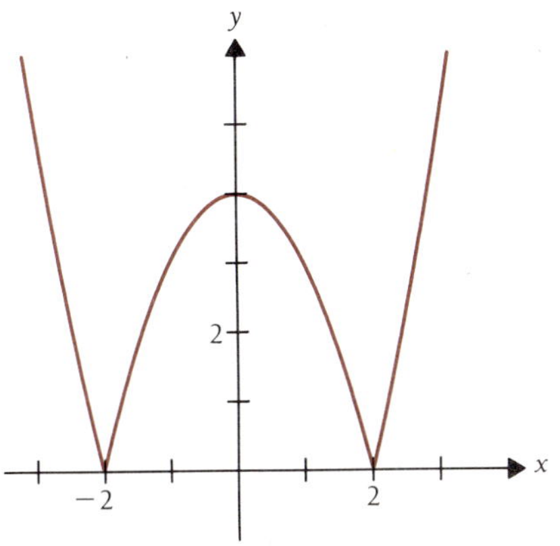

FIGURE 3.9 The graph of $f(x) = |x^2 - 4|$

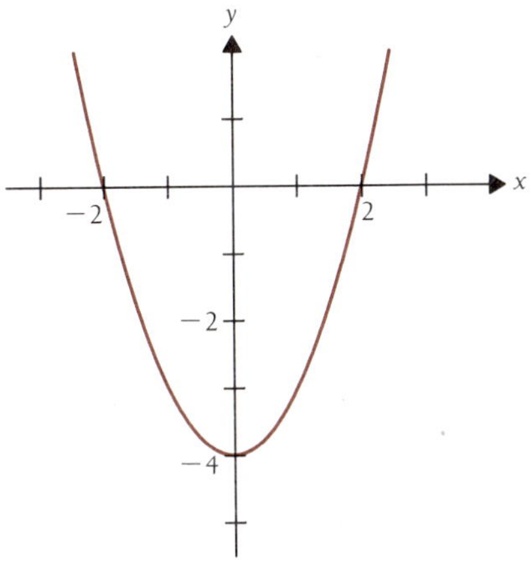

FIGURE 3.10 The graph of $f(x) = x^2 - 4$ ■

Vertical Tangent Lines

The slope of a vertical line is undefined. It is not difficult to convince yourself that the limit of the slope as a line becomes vertical is infinite. Thus, to say that the tangent line to a curve at point P is vertical is to say that the slope $m_{\tan}$ is ∞ or $-\infty$.

■ **Example 3.4** Find the line tangent to the curve $y = x^{1/3}$ at $c = 0$.

Solution When we compute $m_{\tan}$, we get

$$\begin{aligned} m_{\tan} &= \lim_{h\to 0} \frac{f(0+h)-f(0)}{h} \\ &= \lim_{h\to 0} \frac{h^{1/3}}{h} \\ &= \lim_{h\to 0} h^{-2/3} \\ &= \infty. \end{aligned}$$

Since $m_{\tan}$ is infinite, the tangent line is vertical, and is the line $x = 0$, the y-axis (see Figure 3.11).

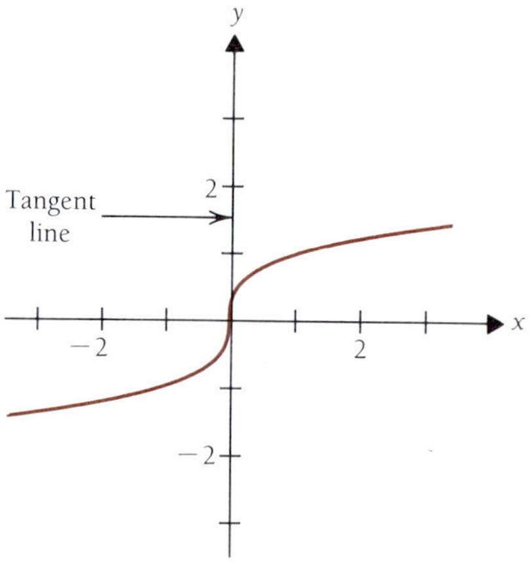

FIGURE 3.11 The curve $y = x^{1/3}$ has a vertical tangent line at 0

■

Normal Lines

A line is said to be **normal** to a curve at point P if it is perpendicular to the tangent line at P (see Figure 3.12). Thus, if the tangent line has slope $m_{\tan}$, then the normal line will have slope $-1/m_{\tan}$. It is clear that if the tangent line is horizontal, then the normal line will be vertical, and vice versa.

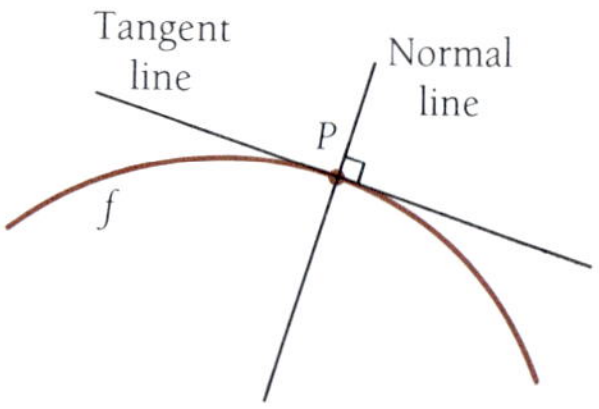

FIGURE 3.12 The normal line to a graph is perpendicular to the tangent line

■ **Example 3.5** Find the equation of the normal line to the curve $y = \sqrt{x}$ at (4, 2).

Solution From Example 3.2, the tangent line has slope $\frac{1}{4}$, so the normal line has slope -4. The equation is therefore (see Figure 3.13)

$$y - 2 = -4(x - 4),$$
$$y = -4x + 18.$$

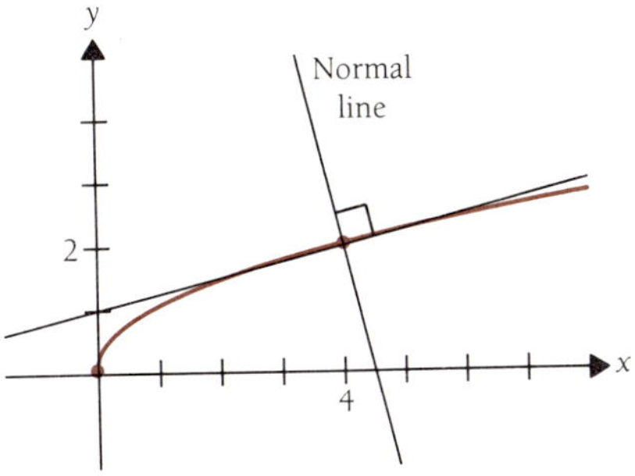

FIGURE 3.13 Normal line to the curve $y = \sqrt{x}$ at (4, 2) ■

Exercises 3.1

Compute m_{tan}, if it exists, for each curve at the given value of c.

1. $y = 3x;\quad c = 2$
2. $y = 3x;\quad c = -3$
3. $y = 2 - x;\quad c = 0$
4. $y = 4x - 3;\quad c = 2$
5. $y = mx + b;$ any c
6. $y = x^2;\quad c = 0$
7. $y = x^2;\quad c = 3$
8. $y = x^2 - 4;\quad c = 1$
9. $y = x - x^2;\quad c = 1$
10. $y = 3x^2 + 2x;\quad c = 2$
11. $y = 2x^2 + 3x - 4;\quad c = -1$
12. $y = ax^2 + bx + k;$ any c
13. $y = -\dfrac{2}{x};\quad c = 3$
14. $y = x^{-2};\quad c = 3$
15. $y = \dfrac{1}{x+1};\quad c = 2$
16. $y = \dfrac{x}{x+3};\quad c = -2$
17. $y = x^{2/3};\quad c = 0$
18. $y = x^{1/5};\quad c = 1$
    [*Hint:* $x - 1 = (x^{1/5} - 1)(x^{4/5} + x^{3/5} + x^{2/5} + x^{1/5} + 1)$.]
19. $y = \sin x;\quad c = 0$
20. $y = \cos x;\quad c = 0$

Find an equation of the tangent line, if it exists, to the given curve at the given point.

21. $y = x^2;\quad (2, 4)$
22. $y = x^2 - x;\quad (1, 0)$
23. $y = \dfrac{1}{x};\quad (1, 1)$
24. $y = \dfrac{1}{1-x};\quad (2, -1)$
25. $y = x^{2/3};\quad (8, 4)$
    [*Hint:* $x^{2/3} - 4 = (x^{1/3} + 2)(x^{1/3} - 2)$
    and $x - 8 = (x^{1/3} - 2)(x^{2/3} + 2x^{1/3} + 4)$.]
26. $y = 1 + (x - 2)^{1/3};\quad (10, 3)$
    [*Hint:* $x - 10 = (x - 2) - 8 = [(x - 2)^{1/3} - 2][(x - 2)^{2/3} + 2(x - 2)^{1/3} + 4]$.]

27. $y=\sqrt{x+1}$; $(3, 2)$

28. $y=\dfrac{1}{\sqrt{x}}$; $(1, 1)$

29. $y=|x|$; $(1, 1)$

30. $y=|x|$; $(0, 0)$

31. $y=\sin x$; $(0, 0)$

32. $y=\cos x$; $(\pi/4, \sqrt{2}/2)$

Explain why there is no tangent line to the given curve at the given point.

33. $y=\dfrac{1}{x}$; $c=0$

34. $y=\dfrac{x}{x+2}$; $c=-2$

35. $y=|x-1|$; $c=1$

36. $y=\dfrac{|x-3|}{x-3}$; $c=3$

37. $y=|1-x^2|$; $c=-1$

38. $y=\sqrt{x}$; $c=0$

39. $y=x^{2/3}$; $c=0$

40. $y=|\sin x|$; $c=\pi$

Find an equation for the normal line to the given curve at the given point.

41. $y=x^2$; $(2, 4)$

42. $y=x^2-x$; $(\frac{1}{2}, -\frac{1}{4})$

43. $y=x^{2/3}$; $(8, 4)$ [See Problem 25.]

44. $y=1+(x-2)^{1/3}$; $(2, 1)$

45. $y=\sin x$; $(0, 0)$

46. $y=\cos x$; $(0, 1)$

47. $y=\sqrt{4-x^2}$; $(0, 2)$

48. $y=\sqrt{4-x^2}$; $(\sqrt{2}, \sqrt{2})$

49. Find an equation like equation (2) in the text for the normal line to the curve $y=f(x)$ at the point $(c, f(c))$.

SECTION 3.2

Rates of Change

One of the most prominent features of nature is change. In order to describe natural processes mathematically, an idea of rate of change is necessary. To illustrate, consider a falling object. The ancient Greeks thought that an object had a quality called "heaviness" that made it fall if dropped. They believed that the heavier an object was, the faster it would fall. Galileo decided to perform an experiment to see how much faster a heavier object would fall. Once he began measuring and comparing speeds of falling objects, he concluded that the speed with which an object falls is independent of its heaviness. The Greek thinkers had not discovered their mistake (at least, in part) because they did not quantify their ideas.

The question, "How fast?" asks for a rate of change. In this section, we see how limits can be used to describe rates of change.

Motion

Let's consider motion as a typical example of change. The **position** of an object is its location relative to some reference, usually a coordinate system. **Motion** is simply change in position in some direction. The **rate** of motion is the rate of change of position—that is, the motion per unit time—and is called **velocity**. The magnitude of velocity, disregarding its direction, is called **speed**. The rate of change of velocity is called **acceleration**.

For now, we will consider only motion that takes place along a straight line, which is called **rectilinear motion**. An object in rectilinear motion can be thought of as a point moving along the real line. We will generally denote position by s, velocity by v, and acceleration by a. The speed is then $|v|$.

Average Velocity

Suppose that at an initial time t, an object in rectilinear motion is at position $s(t)$ on the line. Suppose that at a later time $t + h$, the object is at position $s(t + h)$. The **displacement** of the object is the difference $s(t + h) - s(t)$. The **time elapsed** is $(t + h) - t = h$. The **average velocity** of the object during the time interval $[t, t + h]$ is the quotient

$$v_{\text{ave}} = \frac{\text{Displacement}}{\text{Time elapsed}} = \frac{s(t+h) - s(t)}{h}. \tag{1}$$

If the object travels in one direction only, then the displacement of the object is the same as the distance it travels. Taking the distance to be positive, we then call the average velocity the **average speed** or **rate**, denoted by r. Formula (1) is then the familiar rule, "rate equals distance divided by time," or

$$r = \frac{d}{t}. \tag{2}$$

■ **Example 3.6** Find the average speed:

(a) Of a train that travels 320 miles in 8 hours
(b) Of an object that falls 400 feet in 5 seconds

Solution **(a)** Using equation (2), we have

$$r = \frac{320}{8} = 40 \text{ miles per hour.}$$

(b) Similarly,

$$r = \frac{400}{5} = 80 \text{ feet per second.}$$

■

If the position of a moving object is given as a function of time, the equation defining s as a function of t is called the **equation of motion** of the object. Given the equation of motion, the average velocity during any time interval $[t, t + h]$ can be computed from equation (1).

Example 3.7 An object falling in a vacuum falls a distance $s(t) = 16t^2$ feet in the first t seconds. Find the average velocity:

(a) During the first 5 seconds **(b)** During the interval [2, 5]

Solution **(a)** The average velocity during [0, 5] is found from equation (1) with $s(t) = 16t^2$, $t = 0$, and $h = 5$:

$$v_{\text{ave}} = \frac{s(0+5) - s(0)}{5} = \frac{16 \cdot 5^2 - 0}{5} = 80 \text{ feet per second.}$$

(b) The average velocity during [2, 5] is found the same way, with $t = 2$ and $h = 3$:

$$v_{\text{ave}} = \frac{s(2+3) - s(2)}{3} = \frac{16 \cdot 5^2 - 16 \cdot 2^2}{3} = 112 \text{ feet per second.}$$

■

Instantaneous Velocity

While it may be useful to know the average velocity of an object, it may not provide a very useful description of the motion of the object throughout the time interval. For example, an automobile that travels 30 miles in 1 hour has an average velocity of 30 miles per hour. The car's speedometer, however, probably was not reading 30 miles per hour for the entire hour. The speedometer reading could have varied quite a bit during that hour; the car could even have been at rest for part of the time. The speedometer reading is an *instantaneous velocity,* the velocity of the car at the instant that the speedometer is read.

Over a long time interval, the average and instantaneous velocities may differ markedly. Over a short time interval, however, it is not likely that the average and instantaneous velocities differ greatly. For example, if our automobile travels 44 feet in 1 second, then its average velocity is still 44 feet per second = 30 miles per hour. But during that second, it is very likely that the speedometer reading was close to 30 miles per hour, because it is difficult to vary the speed of a car in a very short time interval.

Because the average velocity approaches the instantaneous velocity as the time interval gets shorter, we are led to define the **instantaneous velocity** as the limit of the average velocity as the time lapsed approaches 0. That is,

$$v(t) = \lim_{h \to 0} \frac{s(t+h) - s(t)}{h}. \tag{3}$$

From now on, *velocity* will mean *instantaneous velocity* unless otherwise specified.

Example 3.8 For the object in Example 3.7, find the velocity at the end of the fifth second.

Solution We find the velocity using equation (3) with $s(t) = 16t^2$ and $t = 5$:

$$\begin{aligned} v(5) &= \lim_{h\to 0} \frac{s(5+h) - s(5)}{h} \\ &= \lim_{h\to 0} \frac{16(5+h)^2 - 16 \cdot 5^2}{h} \\ &= \lim_{h\to 0} \frac{160h + 16h^2}{h} \\ &= \lim_{h\to 0} (160 + 16h) \\ &= 160 \text{ feet per second.} \end{aligned}$$

■

Example 3.9 A certain ball thrown into the air has height $s(t) = 48t - 16t^2$ feet above the ground after t seconds. Find $v(0)$, $v(1)$, and $v(3)$.

Solution Using equation (3), we have

$$\begin{aligned} v(0) &= \lim_{h\to 0} \frac{s(h) - s(0)}{h} \\ &= \lim_{h\to 0} \frac{48h - 16h^2}{h} \\ &= \lim_{h\to 0} (48 - 16h) \\ &= 48 \text{ feet per second,} \end{aligned}$$

$$\begin{aligned} v(1) &= \lim_{h\to 0} \frac{s(1+h) - s(1)}{h} \\ &= \lim_{h\to 0} \frac{48(1+h) - 16(1+h)^2 - 32}{h} \\ &= \lim_{h\to 0} \frac{16h - 16h^2}{h} \\ &= \lim_{h\to 0} (16 - 16h) \\ &= 16 \text{ feet per second,} \end{aligned}$$

and

$$\begin{aligned} v(3) &= \lim_{h\to 0} \frac{s(3+h) - s(3)}{h} \\ &= \lim_{h\to 0} \frac{-48h - 16h^2}{h} \\ &= \lim_{h\to 0} (-48 - 16h) \\ &= -48 \text{ feet per second.} \end{aligned}$$

The positive velocities indicate that the ball is going up, and the negative velocity indicates that it is coming down. You might find it interesting to compute the height of the ball at each of the given times. ■

We will discuss motion again in Sections 4.7 and 13.2.

Other Rates of Change

Other rates of change are defined in the same way as velocity. If $Q(t)$ is a quantity that changes with time, the **average rate of change** of Q during the time interval $[t, t+h]$ is

$$Q_{\text{ave}} = \frac{Q(t+h) - Q(t)}{h}. \tag{4}$$

The **instantaneous rate of change** of Q at time t is the limit of the average rate of change during $[t, t+h]$ as $h \to 0$, and is denoted by Q', so that

$$Q'(t) = \lim_{h \to 0} \frac{Q(t+h) - Q(t)}{h}. \tag{5}$$

Example 3.10 While a tank is draining, the amount of oil in the tank after t minutes is $Q(t) = 50t^2 - 200t + 200$ gallons. Find the rate at which the oil begins to drain.

Solution We seek the rate of change of the quantity of oil in the tank at time $t = 0$. From equation (5), we find

$$\begin{aligned} Q'(0) &= \lim_{h \to 0} \frac{Q(h) - Q(0)}{h} \\ &= \lim_{h \to 0} \frac{(50h^2 - 200h + 200) - 200}{h} \\ &= \lim_{h \to 0} (50h - 200) \\ &= -200 \text{ gallons per minute.} \end{aligned}$$

The minus sign indicates that the quantity of oil in the tank is decreasing. Thus, the tank begins to drain at the rate of 200 gallons per minute. ■

Quantities that are not functions of time, but of other variables, still change with respect to their independent variables. Thus, if F is a function of x, then the **rate of change of F with respect to x** is the limit

$$F'(x) = \lim_{h \to 0} \frac{F(x+h) - F(x)}{h}.$$

Following is an example of an application of such a situation.

Example 3.11 The cost of producing a quantity x of a product is a function of the number x, and is denoted by $C(x)$. The revenue, or income, from the sale of quantity x of a product is a function of the amount sold, and is denoted by $R(x)$. The rate of change of C with respect to x is called the **marginal cost**, and is measured in dollars per unit. The marginal cost is approximately the cost of producing the last unit of the product.

The rate of change of R with respect to x is called the **marginal revenue**, and is measured in dollars per unit. The marginal revenue is approximately the income from the sale of the last unit of the product.

A certain manufacturer finds that the cost of producing x cases of a soft drink per month is $C(x) = 5x + 0.0002x^2 + 2000$ dollars, and the revenue from the sale of x cases per month is $R(x) = 12x - 0.001x^2$ dollars. Find the marginal cost and the marginal revenue at the production level of 1000 cases per month.

Solution The marginal cost is

$$\begin{aligned} C'(1000) &= \lim_{h\to 0} \frac{C(1000+h) - C(1000)}{h} \\ &= \lim_{h\to 0} \frac{5.4h - 0.0002h^2}{h} \\ &= \lim_{h\to 0} (5.4 - 0.0002h) \\ &= \$5.40 \text{ per case,} \end{aligned}$$

and the marginal revenue is

$$\begin{aligned} R'(1000) &= \lim_{h\to 0} \frac{R(1000+h) - R(1000)}{h} \\ &= \lim_{h\to 0} \frac{10h - 0.001h^2}{h} \\ &= \lim_{h\to 0} (10 - 0.001h) \\ &= \$10 \text{ per case.} \end{aligned}$$

This means that at the production level of 1000 cases per month, it is costing about \$5.40 to produce each case, and each case sold brings in about \$10. ■

Exercises 3.2

An object falls $s(t) = 16t^2$ feet in the first t seconds. Find the average velocity of the object during each of the following time intervals:

1. [3, 6] **2.** [3, 5] **3.** [3, 4] **4.** [3, 3.5]

5. [4, 5] **6.** [4, 4.5] **7.** [4, 4.1] **8.** [4, 4.01]

9. Find the velocity of the falling object described for Problems 1–8 at time $t = 3$. Compare with the results of Problems 1–4.

10. Find the velocity of the falling object described for Problems 1–8 at time $t = 4$. Compare with the results of Problems 5–8.

The velocity of a ball thrown into the air is $v(t) = 64 - 32t$ feet per second at time t. Find the average acceleration of the ball during each of the following time intervals:

11. [2, 4] **12.** [2, 3] **13.** [2, 2.5] **14.** [2, 2.1]

15. [0, 1] **16.** [0, 0.5] **17.** [0, 0.1] **18.** [0, 0.01]

19. Find the acceleration of the ball described for Problems 11–18 at $t = 2$. Compare with the results of Problems 11–14.

20. Find the acceleration of the ball described for Problems 11–18 at $t = 0$. Compare with the results of Problems 15–18.

As a swimming pool drains, the number of gallons of water in the pool after t hours is given by $Q(t) = 5000t^2 - 20{,}000t + 20{,}000$. *Find the following:*

21. The average rate at which the pool drains during the first hour
22. The average rate at which the pool drains during the first half-hour
23. The rate of drainage at the end of the first hour
24. The rate of drainage at the end of the first half-hour
25. The rate of drainage at the beginning of the first hour
26. The length of time it will take for the pool to empty

A fiber processor produces x tons of fiber every day. The cost of processing x tons of fiber per day is

$$C(x) = 5000 + 10x + (0.001)x^2$$

dollars. Find the following:

27. The average rate of change of cost as the processing level changes from $x = 600$ to $x = 700$

28. The marginal cost at the processing level of $x = 600$

The above fiber process results in revenue (in dollars) from sales of x tons of fiber according to the formula

$$R(x) = 250x - 0.1x^2.$$

Find the following:

29. The average rate of change of revenue as the sales volume changes from $x = 600$ to $x = 700$

30. The marginal revenue at the sales volume of $x = 600$

The profit P(x) realized from the fiber process described above is the difference between revenue and cost—that is,

$$P(x) = R(x) - C(x).$$

31. Using the cost and revenue functions given above, find a formula for profit.

32. Find the average rate of change of profit as the production and sales level change from $x = 600$ to $x = 700$.

33. Find the **marginal profit**, the rate of change of P with respect to x, at the production and sales level of $x = 600$.

34. The equation of motion of an object is $s(t) = At^2 + Bt + C$. Show that the velocity of the object is $v(t) = 2At + B$ and the acceleration of the object is $a(t) = 2A$.

35. Show that if y is directly proportional to x, then the rate of change of y with respect to x is the constant of proportionality.

36. Show that if y is inversely proportional to x, then the rate of change of y with respect to x is inversely proportional to the square of x.

37. The figure shows the position of an object as a function of time. Identify the slope of one of the lines shown as the average velocity of the object over the time interval $[a, b]$, and identify the slope of the other line as the instantaneous velocity at time a.

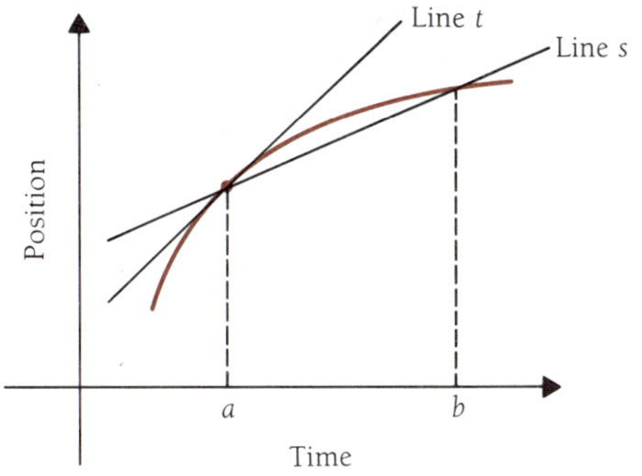

SECTION 3.3

The Derivative

In Section 3.1, we saw that the slope of the curve $y = f(x)$ at the point $(c, f(c))$ is

$$m_{\tan} = \lim_{h \to 0} \frac{f(c+h) - f(c)}{h}.$$

In Section 3.2, we saw that the velocity of an object with equation of motion $s = f(t)$ is

$$v(t) = \lim_{h \to 0} \frac{f(t+h) - f(t)}{h}.$$

It is no accident that these results are similar; they are both examples of a special kind of limit called the *derivative*.

In this section, we give the formal definition of the derivative of a function, along with the usual notations for the derivative. The interpretations and simplest applications of the derivative will be mentioned. We will also discuss the concept of differentiability and its relationship to continuity.

Definition of Derivative

DEFINITION 3.2

Let a function f be defined in an open interval (a, b). For $x \in (a, b)$, the **derivative of f at x** is the limit

$$f'(x) = \lim_{h \to 0} \frac{f(x+h) - f(x)}{h}, \qquad (1)$$

if the limit exists. The function f' defined by equation (1) is also called the **derived function** of f.

Because of our work in the previous two sections, we have two standard interpretations of the derivative. One of them is the slope of the curve $y = f(x)$ at $(c, f(c))$, since $f'(c) = m_{\tan}$. The other interpretation is the rate of change of f with respect to x; that is, if $Q(x) = f(x)$, then $Q'(x) = f'(x)$. In particular, if the position of an object is $s(t)$, then the velocity is $v(t) = s'(t)$ and the acceleration is $a(t) = v'(t)$.

To summarize the results of the two previous sections, we can say that if $y = f(x)$ is the equation of a curve, then the slope of the curve at point $x = c$ is $f'(c)$, and the equation of the tangent line is

$$y - f(c) = f'(c)(x - c).$$

If $y = f(x)$ gives y as a function of x, then the rate of change of y with respect to x, evaluated at $x = c$, is $f'(c)$. In both these cases, $f'(c)$ is the derived function f' evaluated at c,

$$f'(c) = \lim_{h \to 0} \frac{f(c + h) - f(c)}{h}.$$

Differentiable Functions

The derivative of a function may or may not exist at a particular point x, according as the limit (1) exists. If the derivative of a function exists, we use that fact as a special property of the function.

DEFINITION 3.3

Let a function f be defined in an open interval containing the point c. The function f is **differentiable at c** if and only if the derivative $f'(c)$ exists. If f is differentiable at every point of a set S, we say f is **differentiable on S**. If f is differentiable at every point of its domain, we say simply that f is **differentiable**.

Example 3.12 Determine the set of values x for which the given function is differentiable.

(a) $f(x) = x^3 + 4x$ (b) $f(x) = \sqrt{x}$

Solution (a) We first try to compute the derivative:

$$\begin{aligned} f'(x) &= \lim_{h \to 0} \frac{f(x + h) - f(x)}{h} \\ &= \lim_{h \to 0} \frac{(x + h)^3 + 4(x + h) - (x^3 + 4x)}{h} \\ &= \lim_{h \to 0} \frac{x^3 + 3x^2h + 3xh^2 + h^3 + 4x + 4h - x^3 - 4x}{h} \\ &= \lim_{h \to 0} (3x^2 + 3xh + h^2 + 4) \\ &= 3x^2 + 4. \end{aligned}$$

Since this expression exists for all real x, this function f is differentiable on all of $\mathbb{R}$, which is the domain of f.

(b) Again, we start by computing the derivative:

$$\begin{aligned} f'(x) &= \lim_{h\to 0} \frac{f(x+h)-f(x)}{h} \\ &= \lim_{h\to 0} \frac{\sqrt{x+h}-\sqrt{x}}{h} \cdot \frac{\sqrt{x+h}+\sqrt{x}}{\sqrt{x+h}+\sqrt{x}} \\ &= \lim_{h\to 0} \frac{(x+h)-x}{h(\sqrt{x+h}+\sqrt{x})} \\ &= \lim_{h\to 0} \frac{1}{\sqrt{x+h}+\sqrt{x}} \\ &= \frac{1}{2\sqrt{x}}. \end{aligned}$$

Since this derived function is defined for $x > 0$, and the domain of f includes all those values, f is differentiable on $(0, \infty)$. ■

An Alternative Form for $f'(c)$

An alternative way to write $f'(c)$ can be obtained from the formula

$$f'(c) = \lim_{h\to 0} \frac{f(c+h)-f(c)}{h}.$$

We let $x = c + h$, so that $h = x - c$, and $x \to c$ if and only if $h \to 0$. Substituting these into the formula above, we get

$$f'(c) = \lim_{x\to c} \frac{f(x)-f(c)}{x-c}. \tag{2}$$

This formula can be used to compute $f'(c)$.

■ **Example 3.13** For $f(x) = x^3 + 4x$, compute $f'(1)$ in two ways.

Solution From Example 3.12, we know that $f'(x) = 3x^2 + 4$. Therefore, we have $f'(1) = 3(1)^2 + 4 = 7$. This is one way to find the derivative at 1. Another way is to use formula (2):

$$\begin{aligned} f'(1) &= \lim_{x\to 1} \frac{f(x)-f(1)}{x-1} \\ &= \lim_{x\to 1} \frac{x^3+4x-5}{x-1} \\ &= \lim_{x\to 1} \frac{(x-1)(x^2+x+5)}{x-1} \\ &= \lim_{x\to 1} (x^2+x+5) \\ &= 7. \end{aligned}$$

■

Differentiability and Continuity

As was mentioned in Section 3.1, a curve can have a tangent line at a point P only if the curve is continuous at P. Therefore, if a function is differentiable at a certain point, it must be continuous there. We state this fact formally as a theorem, and provide a proof below.

THEOREM 3.1 If the function f is differentiable at c, then f is continuous at c.

Proof If f is differentiable at c, then f is defined in an interval containing c, and $f'(c)$ exists. Hence,

$$\begin{aligned}\lim_{x\to c}[f(x)-f(c)] &= \lim_{x\to c}\frac{f(x)-f(c)}{x-c}\cdot(x-c)\\ &= \lim_{x\to c}\frac{f(x)-f(c)}{x-c}\cdot\lim_{x\to c}(x-c)\\ &= f'(c)\ \lim_{x\to c}(x-c)\\ &= f'(c)\cdot 0\\ &= 0.\end{aligned}$$

Therefore, $\lim_{x\to c}f(x)=f(c)$, and f is continuous at c.

The converse of Theorem 3.1 is not true. That is, a function can be continuous at a point without being differentiable there. As was remarked in Section 3.1, a curve that has a corner at point P has no tangent line at P, so the corresponding function is not differentiable there. Here is a concrete example.

■ **Example 3.14** Show that the function $f(x)=|x-c|$ is continuous at c, but is not differentiable at c.

Solution The function $f(x)=|x-c|$ is continuous at c, since

$$\lim_{x\to c}f(x)=\lim_{x\to c}|x-c|=0=f(c).$$

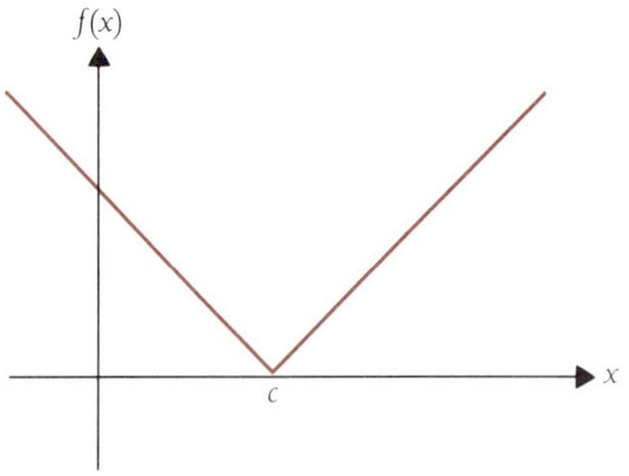

FIGURE 3.14 The graph of $f(x)=|x-c|$

If we try to compute the derivative, we get

$$f'(c) = \lim_{x \to c} \frac{f(x) - f(c)}{x - c}$$

$$= \lim_{x \to c} \frac{|x - c|}{x - c}.$$

But this limit does not exist, because the one-sided limits are unequal. Therefore, f is not differentiable at c. This function is shown in Figure 3.14. ■

Notations for the Derivative

The use of f' to represent the derivative of f was introduced by Lagrange. We extend this *prime notation* a little and write y' or $y'(x)$ to mean f' or $f'(x)$ when $y = f(x)$. This is similar to the notation used by Newton, who wrote a dot over the letter instead of a prime after it.

Another notation that is just as widely used as the prime notation is the *double d notation* introduced by Leibniz. In Leibniz's notation, if $y = f(x)$, then we use dy/dx for $y'(x)$. The symbol dy/dx is read, "dee wye by dee eks" or, simply, "dee wye dee eks," and means the derivative of y with respect to x. For the number $f'(c)$, we write

$$\left.\frac{dy}{dx}\right|_{x=c},$$

which is read, "dy/dx evaluated at $x = c$."

The Leibniz notation has the advantage of naming the independent variable. For example, the volume of a cylinder of height h and radius r is given by $V = \pi r^2 h$. The rate of change of volume with respect to radius is then dV/dr, while the rate of change of volume with respect to height is dV/dh. In such a context as this, it would be confusing to use the notation V'; the Leibniz notation is a must here.

A third notation is the use of D or D_x as a **differential operator**. In this notation, an expression in x is preceded by D_x to denote the derivative of the expression. That is, $D_x f(x)$ means $f'(x)$. You will often see part of the Leibniz notation, d/dx, used in place of the operator D_x.

Exercises 3.3

For each function f, find $f'(x)$ and tell for what values of x the function f is differentiable.

1. $f(x) = 4x + 3$
2. $f(x) = 5 - x$
3. $f(x) = \pi x + 1$
4. $f(x) = ax + b$
5. $f(x) = x^2$
6. $f(x) = 7x^2 - x$
7. $f(x) = 1 + 3x - 5x^2$
8. $f(x) = ax^2 + bx + c$
9. $f(x) = x^3$
10. $f(x) = 2x^3 + 4x$
11. $f(x) = 3x^3 + 2x^2$
12. $f(x) = 1 - x - x^3$
13. $f(x) = \dfrac{1}{x}$
14. $f(x) = \dfrac{1}{x - 1}$
15. $f(x) = \dfrac{x}{x + 1}$
16. $f(x) = \dfrac{2x + 1}{3x - 2}$
17. $f(x) = \sqrt{x - 3}$
18. $f(x) = \sqrt{x^2 + 1}$
19. $f(x) = \dfrac{1}{\sqrt{x + 2}}$
20. $f(x) = \dfrac{1 + \sqrt{x}}{x}$
21. $f(x) = \sin x$

22. $f(x) = \cos x$

23. $f(x) = \dfrac{1}{x^2 + 1}$

24. $f(x) = \dfrac{\sin x}{x}$

For the given function s, find $v = ds/dt$ and $a = dv/dt$.

25. $s = t - t^2$

26. $s = 3t + t^2$

27. $s = 64 - 16t^2$

28. $s = 64t - 16t^2$

29. $s = t - \dfrac{1}{t}$

30. $s = t\sqrt{t}$

31. $s = \sin 2t$

32. $s = 2 \sin t$

33. The circumference of a circle in terms of its radius is $C = 2\pi r$. Find dC/dr.

34. The area of a circle in terms of its radius is $A = \pi r^2$. Find dA/dr.

35. The volume of a cone of height h and base radius r is $V = \pi r^2 h/3$. Find dV/dr and dV/dh.

36. The volume of a box of length x, width y, and depth z is $V = xyz$. Find dV/dx, dV/dy, and dV/dz.

37. The cost of producing x units of a product is

$$C(x) = 10{,}000 + 250x - 0.4x^2.$$

Find the marginal cost dC/dx and its value at $x = 300$, $(dC/dx)|_{x=300}$.

38. The revenue from the sale of x units of a product is

$$R(x) = 460x - 0.02x^2.$$

Find the marginal revenue dR/dx and its value at $x = 300$, $(dR/dx)|_{x=300}$.

39. For $s = 80t - 16t^2$, find $(ds/dt)|_{t=0}$ and $(ds/dt)|_{t=4}$.

40. For $s = 3 \sin t$, find $v(\pi) = (ds/dt)|_{t=\pi}$, and then find $(dv/dt)|_{t=\pi}$.

For each function, find $f'(2)$, if it exists.

41. $f(x) = x^4$

42. $f(x) = x^2 + 5x$

43. $f(x) = \dfrac{x^2 - 4}{x - 2}$

44. $f(x) = \dfrac{1}{x - 2}$

45. $f(x) = \dfrac{3}{x}$

46. $f(x) = \sqrt{x} + 1$

47. $f(x) = \sin \pi x$

48. $f(x) = \cos \pi x$

49. Show that if f is a constant function, then $f'(x) = 0$.

50. Show that if $y = ax + b$, then $dy/dx = a$.

51. The graph of a function f is shown in the accompanying figure (a), and its derivative is shown in figure (b), so that for each x, the slope of f at x is the value $f'(x)$. The graph of the function g is given in figure (c). Sketch the graph of g'.

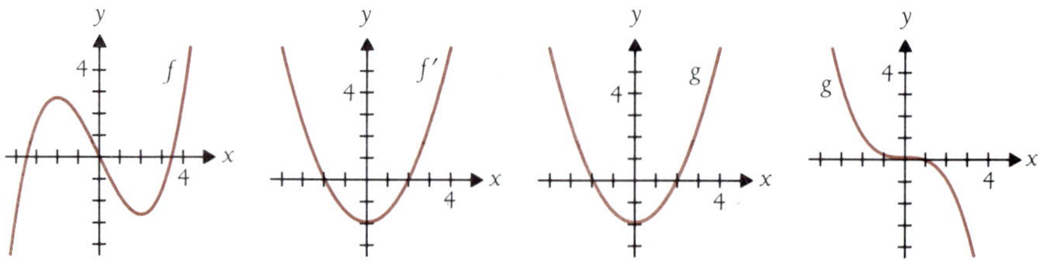

(a) (b) (c) (d)

52. Repeat Problem 51 for the function g graphed in figure (d).

SECTION 3.4

Differentiation Formulas

The process of finding derivatives is called **differentiation.** The only way we have, at present, to find derivatives is by using the definition of derivative, Definition 3.2. As you have seen in the last set of exercises, this can lead to some very tedious computations. Fortunately, there are a number of rules, called **differentiation formulas**, that reduce the process of differentiation to a relatively simple mechanical procedure. In this section, we develop the basic formulas for differentiating algebraic and trigonometric functions.

The Power Rule

The most useful differentiation formula is called the **power rule**; it is used to differentiate a power function. At this point, we can prove it only for positive integral powers.

THEOREM 3.2

The Power Rule If n is a nonnegative integer and $f(x) = x^n$, then f is differentiable on all of $\mathbb{R}$, and

$$f'(x) = nx^{n-1}. \tag{1}$$

Proof First, for the case $n = 0$, $f(x) = x^0 = 1$, and f is a constant function. Therefore, its graph is a horizontal line, and its slope or derivative is 0. But the given formula yields $f'(x) = 0x^{-1} = 0$, so the theorem is true when $n = 0$.

Now, for n a positive integer, we have from the definition of derivative,

$$\begin{aligned} f'(x) &= \lim_{h\to 0} \frac{f(x+h) - f(x)}{h} \\ &= \lim_{h\to 0} \frac{(x+h)^n - x^n}{h}. \end{aligned}$$

By the binomial theorem,

$$(x+h)^n = x^n + nx^{n-1}h + \frac{n(n-1)}{2}x^{n-2}h^2 + \cdots + h^n,$$

so that

$$(x+h)^n - x^n = h\left[nx^{n-1} + \frac{n(n-1)}{2}x^{n-2}h + \cdots + h^{n-1}\right],$$

and therefore,

$$\frac{(x+h)^n - x^n}{h} = nx^{n-1} + \frac{n(n-1)}{2}x^{n-2}h + \cdots + h^{n-1}.$$

As a consequence,

$$\frac{(x+h)^n - x^n}{h} \to nx^{n-1} \qquad \text{as} \qquad h \to 0,$$

and

$$f'(x) = nx^{n-1}.$$

In Leibniz notation, the power rule is stated as

$$\frac{d(x^n)}{dx} = nx^{n-1}. \tag{2}$$

The following example illustrates the use of the power rule.

Example 3.15 Find dy/dx:

(a) $y = x$ **(b)** $y = x^3$ **(c)** $y = x^{45}$

Solution **(a)** Since $x = x^1$, the power rule tells us that

$$\frac{dy}{dx} = 1x^0 = 1.$$

That is, if $f(x) = x$, then $f'(x) = 1$. (The line $y = x$ has slope 1.)

(b) By the power rule directly, if $y = x^3$, then

$$\frac{dy}{dx} = 3x^2.$$

(c) Again, by direct application of the power rule, if $y = x^{45}$, then

$$\frac{dy}{dx} = 45x^{44}.$$

■

Rules for Combinations of Functions

Now that we can differentiate some of the power functions, we turn to rules for differentiating arithmetic combinations of functions. We will list the rules as theorems, indicating the proofs of some of them below. The proofs primarily use appropriate limit theorems.

THEOREM 3.3 **The Scalar Product Rule** The derivative of a constant times a function is the constant times the derivative of the function. That is, if f is a differentiable function and k is a constant, then kf is differentiable and

$$D_x[kf(x)] = k\,D_x f(x) \qquad \text{or} \qquad (kf)' = kf'.$$

Proof Let $g(x) = kf(x)$. Then

$$\begin{aligned} g'(x) &= \lim_{h\to 0} \frac{g(x+h)-g(x)}{h} \\ &= \lim_{h\to 0} \frac{kf(x+h)-kf(x)}{h} \\ &= \lim_{h\to 0} k \cdot \frac{f(x+h)-f(x)}{h} \\ &= k \lim_{h\to 0} \frac{f(x+h)-f(x)}{h} \\ &= kf'(x). \end{aligned}$$

Example 3.16 If $f(x) = 5x^2$, find $f'(x)$ and $f'(3)$.

Solution By the scalar product rule,

$$\begin{aligned} D_x(5x^2) &= 5\, D_x(x^2) \\ &= 5(2x^1) \\ &= 10x. \end{aligned}$$

Note the use of the power rule in the second line above.

Now, since $f'(x) = 10x$, we find by evaluating that $f'(3) = 10 \cdot 3 = 30$. ■

Two special applications of Theorems 3.2 and 3.3 are worthy of mention here. They are the simple but often-used formulas

$$D_x k = 0 \qquad \text{and} \qquad D_x kx = k.$$

We will leave it to you to show exactly how these follow from the differentiation formulas, and instead, consider them geometrically. The graph of $y = k$ is a horizontal line (see Figure 3.15), so the slope y' is 0. Similarly, the graph of $y = kx$ (Figure 3.16) is a line with slope $y' = k$.

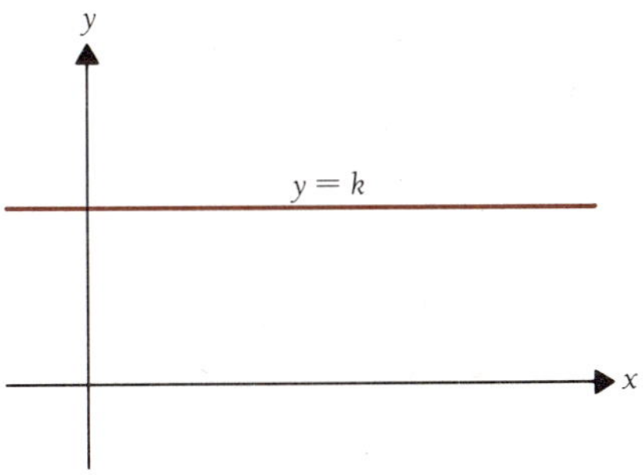

FIGURE 3.15 A constant function has 0 slope

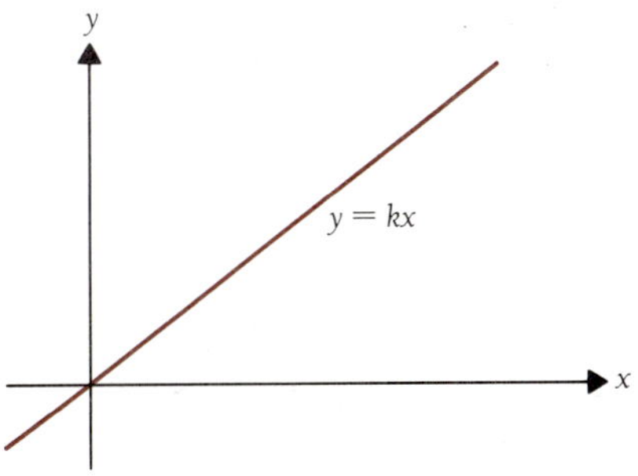

FIGURE 3.16 The line $y = kx$ has slope k

THEOREM 3.4

The Sum Rule The derivative of a sum is the sum of the derivatives. That is, if f and g are differentiable functions, then $f+g$ is differentiable and

$$D_x[f(x)+g(x)] = D_x f(x) + D_x g(x) \qquad \text{or} \qquad (f+g)' = f' + g'.$$

The proof of this theorem is the straightforward application of the fact that the limit of a sum is the sum of the limits. You will be invited to give the proof in the exercises. This rule also can be extended to the sum of any finite number of functions:

$$(f_1 + f_2 + \cdots + f_n)' = f_1' + f_2' + \cdots + f_n'.$$

Example 3.17 If $y = 5x^2 + 2x^3$, find dy/dx and $(dy/dx)|_{x=0.5}$.

Solution By the sum rule and the scalar product rule, we have

$$\begin{aligned}\frac{dy}{dx} &= D_x(5x^2 + 2x^3)\\ &= D_x(5x^2) + D_x(2x^3)\\ &= 5\,D_x x^2 + 2\,D_x x^3\\ &= 5(2x) + 2(3x^2)\\ &= 10x + 6x^2.\end{aligned}$$

Therefore,

$$\left.\frac{dy}{dx}\right|_{x=0.5} = 10(0.5) + 6(0.5)^2 = 6.5.$$

■

THEOREM 3.5

The Difference Rule The derivative of a difference is the difference of the derivatives. That is, if f and g are differentiable functions, then $f-g$ is differentiable and

$$D_x[f(x)-g(x)] = D_x f(x) - D_x g(x) \qquad \text{or} \qquad (f-g)' = f' - g'.$$

Example 3.18 If $y = x^2 + 3x - 5$, find y' and $y'(1)$.

Solution Using all the rules we have so far, we get

$$\begin{aligned}y' &= (x^2 + 3x - 5)'\\ &= (x^2)' + (3x)' - (5)'\\ &= 2x^1 + 3 - 0\\ &= 2x + 3.\end{aligned}$$

Then we find that

$$y'(1) = 2(1) + 3 = 5.$$

■

THEOREM 3.6

The Product Rule The derivative of the product of two functions is the derivative of the first times the second plus the first times the derivative of the second. That is, if f and g are differentiable functions, then fg is differentiable and

$$D_x[f(x)g(x)] = [D_x f(x)]g(x) + f(x)[D_x g(x)] \qquad \text{or} \qquad (fg)' = f'g + fg'.$$

Proof Let $F(x) = f(x)g(x)$. We will show that F is differentiable by finding its derivative, and in the process establish the formula in the theorem.

$$\begin{aligned} F'(x) &= \lim_{h\to 0} \frac{F(x+h) - F(x)}{h} \\ &= \lim_{h\to 0} \frac{f(x+h)g(x+h) - f(x)g(x)}{h} \\ &= \lim_{h\to 0} \frac{f(x+h)g(x+h) - f(x)g(x+h) + f(x)g(x+h) - f(x)g(x)}{h} \\ &= \lim_{h\to 0} \left[\frac{f(x+h) - f(x)}{h} \cdot g(x+h) + f(x) \cdot \frac{g(x+h) - g(x)}{h}\right]. \end{aligned}$$

Now,

$$\lim_{h\to 0} \frac{f(x+h) - f(x)}{h} = f'(x) \qquad \text{and} \qquad \lim_{h\to 0} \frac{g(x+h) - g(x)}{h} = g'(x).$$

Also, since g is differentiable, g is continuous, so $g(x+h) \to g(x)$ as $h \to 0$. Hence, by the limit theorems,

$$F'(x) = f'(x)g(x) + f(x)g'(x),$$

and the theorem is proved.

■ **Example 3.19** If $f(x) = (x^2+1)(x^3-1)$, find $f'(x)$ and $f'(-1)$.

Solution Using the product rule, we have

$$\begin{aligned} f'(x) &= (x^2+1)'(x^3-1) + (x^2+1)(x^3-1)' \\ &= (2x+0)(x^3-1) + (x^2+1)(3x^2-0) \\ &= 2x(x^3-1) + 3x^2(x^2+1). \end{aligned}$$

Evaluating at -1, we have

$$f'(-1) = 2(-1)[(-1)^3 - 1] + 3(-1)^2[(-1)^2 + 1] = 10.$$ ■

THEOREM 3.7

The Reciprocal Rule The derivative of the reciprocal of a function is the negative of the derivative of the function, divided by the square of the function. That is, if g is a differentiable function, then $1/g$ is differentiable wherever it is defined, and

$$\frac{d}{dx}\left[\frac{1}{g(x)}\right] = -\frac{D_x g(x)}{[g(x)]^2} \qquad \text{or} \qquad \left(\frac{1}{g}\right)' = -\frac{g'}{g^2}.$$

A proof of the reciprocal rule can be obtained by differentiating the product $g(x) \cdot 1/g(x)$. Since the product is 1, the derivative is 0. The proof will be left as an exercise.

Example 3.20 Given

$$y = \frac{1}{x^2 + 1},$$

find dy/dx and $(dy/dx)|_{x=0}$.

Solution By direct application of the reciprocal rule, we have

$$\frac{dy}{dx} = \frac{-2x}{(x^2 + 1)^2}.$$

Therefore,

$$\left.\frac{dy}{dx}\right|_{x=0} = 0.$$

With the reciprocal rule, we can prove that the power rule holds for any integer, not just nonnegative integers. Let m be a negative integer, and let $n = -m$. Then n is a positive integer. If $f(x) = x^m$, then also

$$f(x) = \frac{1}{x^n}$$

and

$$\begin{aligned} f'(x) &= -\frac{nx^{n-1}}{(x^n)^2} \\ &= -nx^{n-1-2n} \\ &= (-n)x^{(-n)-1} \\ &= mx^{m-1}. \end{aligned}$$

The power rule thus holds for any integer exponent.

Example 3.21 If $y = x^{-4}$, find y' and $y'(3)$.

Solution By the power rule extended to negative exponents,

$$y' = (-4)x^{-4-1} = -4x^{-5}.$$

Evaluating at $x = 3$, we get

$$y'(3) = -4(3)^{-5} = -\frac{4}{243}.$$

THEOREM 3.8

The Quotient Rule The derivative of the quotient of two functions is the derivative of the numerator times the denominator, minus the numerator times the derivative of the denominator, all divided by the square of the denominator. That is, if f and g are differentiable functions, then f/g is differentiable wherever it is defined, and

$$\frac{d}{dx}\left[\frac{f(x)}{g(x)}\right] = \frac{[D_x f(x)]g(x) - f(x)[D_x g(x)]}{[g(x)]^2} \quad \text{or} \quad \left(\frac{f}{g}\right)' = \frac{f'g - fg'}{g^2}.$$

It is easy to see how this rule is the combination of the product rule and the reciprocal rule, applied to $f/g = f \cdot (1/g)$. The proof is left as an exercise.

■ **Example 3.22** If

$$f(x) = \frac{x+1}{x-1},$$

find $f'(x)$ and $f'(0.2)$.

Solution Using the quotient rule, we have

$$\begin{aligned} f'(x) &= \frac{(x+1)'(x-1) - (x+1)(x-1)'}{(x-1)^2} \\ &= \frac{(1)(x-1) - (1)(x+1)}{(x-1)^2} \\ &= \frac{-2}{(x-1)^2}. \end{aligned}$$

Hence,

$$f'(0.2) = \frac{-2}{(0.2-1)^2} = -3.125.$$

■

At this point, we have rules allowing us to differentiate any rational function. We simply need to recognize the function as a combination of power functions, and apply the appropriate differentiation formulas.

Now we look at the trigonometric functions.

Derivatives of the Trigonometric Functions

The sine and cosine functions have appeared in a few exercises on differentiation, but now we turn to finding the derivatives of all the trigonometric functions.

There are six important formulas for differentiating the trigonometric functions. We will state the formulas as a theorem, indicating how some of the formulas are derived.

THEOREM 3.9 The trigonometric functions are differentiable wherever they are defined, and

$$D_x \sin x = \cos x, \tag{3}$$
$$D_x \cos x = -\sin x, \tag{4}$$
$$D_x \tan x = \sec^2 x, \tag{5}$$
$$D_x \cot x = -\csc^2 x, \tag{6}$$
$$D_x \sec x = \sec x \tan x, \tag{7}$$
$$D_x \csc x = -\csc x \cot x. \tag{8}$$

Proof To derive rule (3), we use the definition of derivative:

$$\begin{aligned}\frac{d}{dx}\sin x &= \lim_{h\to 0}\frac{\sin(x+h)-\sin x}{h}\\ &= \lim_{h\to 0}\frac{\sin x\cos h+\cos x\sin h-\sin x}{h}\\ &= \lim_{h\to 0}\left[(\sin x)\cdot\frac{\cos h-1}{h}+(\cos x)\cdot\frac{\sin h}{h}\right].\end{aligned}$$

Now by Theorem 2.15,

$$\frac{\sin h}{h}\to 1 \quad \text{as} \quad h\to 0,$$

and it follows from Example 2.22(d), in Section 2.3, that

$$\frac{\cos h-1}{h}\to 0 \quad \text{as} \quad h\to 0.$$

Hence,

$$\begin{aligned}D_x \sin x &= (\sin x)\cdot 0+(\cos x)\cdot 1\\ &= \cos x,\end{aligned}$$

and rule (3) is proved. The derivation of rule (4) is similar, and will be left as an exercise.

To get rule (5), we use the identity $\tan x = (\sin x)/(\cos x)$, along with rules (3) and (4) and the quotient rule:

$$\begin{aligned}\frac{d}{dx}\tan x = \frac{d}{dx}\frac{\sin x}{\cos x} &= \frac{(D_x \sin x)(\cos x)-(\sin x)(D_x \cos x)}{(\cos x)^2}\\ &= \frac{(\cos x)(\cos x)-(\sin x)(-\sin x)}{\cos^2 x}\\ &= \frac{1}{\cos^2 x}\\ &= \sec^2 x.\end{aligned}$$

Rule (6) is derived in a similar manner.

Finally, to get rule (7), we use the fact that $\sec x = 1/\cos x$, along with the reciprocal rule:

$$\begin{aligned}\frac{d}{dx}\sec x &= \frac{d}{dx}\frac{1}{\cos x} = \frac{-D_x \cos x}{(\cos x)^2}\\ &= \frac{-(-\sin x)}{\cos^2 x}\\ &= \frac{1}{\cos x}\cdot\frac{\sin x}{\cos x}\\ &= \sec x \tan x.\end{aligned}$$

Rule (8) arises in a similar way.

With the formulas in Theorem 3.9 and the other rules already developed, we can now differentiate any arithmetic combination of rational and trigonometric functions. We will close this section with one last example.

Example 3.23 Differentiate:

(a) $f(x) = x \tan x$ **(b)** $y = \sin^2 x$ **(c)** $y = \dfrac{\sin x}{1 + \cos x}$

Solution **(a)** Using the product rule, we get

$$\begin{aligned}f'(x) &= 1 \tan x + x \sec^2 x\\ &= \tan x + x \sec^2 x.\end{aligned}$$

(b) Since $\sin^2 x = \sin x \sin x$, we can use the product rule:

$$\begin{aligned}\frac{dy}{dx} &= \cos x \sin x + \sin x \cos x\\ &= 2 \sin x \cos x.\end{aligned}$$

(c) Using the quotient rule,

$$\begin{aligned}y' &= \frac{\cos x\,(1 + \cos x) - \sin x\,(-\sin x)}{(1 + \cos x)^2}\\ &= \frac{\cos x + \cos^2 x + \sin^2 x}{(1 + \cos x)^2}\\ &= \frac{1}{1 + \cos x}.\end{aligned}$$

Exercises 3.4

Differentiate.

1. $f(x) = x^3$
2. $f(x) = x^{-7}$
3. $f(x) = 2x^4$
4. $f(x) = -5x^{-3}$
5. $y = 6$
6. $y = 8\pi$

7. $y = 2x^2 + 3x - 6$
8. $y = 3x^5 - 5x - 20$
9. $y = x^2 + x^{-2}$
10. $y = x^3 + 5 - x^{-2}$
11. $y = 2x(x - 3)$
12. $y = x^3(4x + 3)$
13. $f(x) = (x^2 + 3x + 1) \sin x$
14. $f(x) = (2x^5 - 4x^2)(3x^{10} + x^2)$
15. $f(x) = (x^2 + 3x + 1)^2$
16. $f(x) = (2x + 1)^3$
17. $y = \dfrac{x}{x + 1}$
18. $y = \dfrac{x^2 - 1}{2x - 3}$
19. $y = \dfrac{1}{2x - 1}$
20. $y = (x^4 + x^2 - 1)^{-1}$
21. $y = \dfrac{2x}{x^2 - 1}$
22. $y = \dfrac{2x^3 - x - 1}{x^2 + 4x}$
23. $y = \dfrac{\sin x}{x}$
24. $y = \dfrac{x^2 - 1}{\cos x}$
25. $f(x) = x(x + 1)(x^2 + 1)$
26. $f(x) = ax(bx + c)(px + q)$
27. $f(x) = \dfrac{x \tan x}{x^2 + 1}$
28. $f(x) = \dfrac{4x + 3}{2x \sin x}$
29. $y = 2x \sec x$
30. $y = \sec^2 x$
31. $y = \sec x \tan x$
32. $y = 2 \sin x \cos x$
33. $y = \dfrac{1}{x} - \dfrac{1}{\sin x}$
34. $y = \cos^2 x - 1$
35. $y = x^2 \tan x$
36. $y = \dfrac{x \sin x}{1 - \cos x}$

Find $v = ds/dt$ and $a = dv/dt$. Then find $v(t_0)$ and $a(t_0)$.

37. $s = 50 + 80t - 16t^2$; $t_0 = 0$
38. $s = t + \dfrac{1}{t}$; $t_0 = 5$
39. $s = 400 - 16t^2$; $t_0 = 4$
40. $s = \dfrac{1}{t + 1}$; $t_0 = 4$
41. $s = 4 \sin t$; $t_0 = \pi$
42. $s = 3 \cos t$; $t_0 = \pi$
43. $s = 1 - 4 \sin t$; $t_0 = 0$
44. $s = t - 5 \cos t$; $t_0 = \pi/2$

Find an equation of the line tangent to the given curve at the given point.

45. $y = x^3 - x$; $(1, 0)$
46. $y = 3x^4 - \dfrac{1}{x}$; $(1, 2)$
47. $y = x^{-2}$; $(-1, 1)$
48. $y = \dfrac{x^2}{x - 1}$; $(2, 4)$
49. $y = \sin x$; $(0, 0)$
50. $y = \sin x$; $(\pi/2, 1)$
51. $y = \sin x$; $(\pi, 0)$
52. $y = \sin x$; $(\pi/4, \sqrt{2}/2)$
53. $y = \cos x$; $(\pi/2, 0)$
54. $y = 2 \sin x$; $(0, 0)$
55. $y = x - \cos x$; $(0, -1)$
56. $y = x^2 + 2 \sin x$; (π, π^2)
57. Find the rate of change of the area of a square with respect to its side.
58. Find the rate of change of the volume of a sphere with respect to its radius.
59. Express the volume of a cube as a function of its edge, and find the rate of change of the volume with respect to the edge.
60. Express the volume of a cube as a function of half its edge, and find the rate of change of the volume with respect to half the edge.

61. Find the point(s) on the curve $y = x^3$ at which the tangent line has slope 1.

62. Find the point(s) on the curve $y = x^2$ at which the tangent line has y-intercept -2.

63. Prove that if k is a constant, then $D_x k = 0$, using Theorems 3.2 and 3.3.

64. Prove that if k is a constant, then $D_x kx = k$, using Theorems 3.2 and 3.3.

65. Prove the sum rule, using the definition of derivative.

66. Prove the difference rule, using the sum and scalar product rules and writing $f - g$ as $f + (-1)g$.

67. Prove the difference rule, using the definition of derivative.

68. Prove the reciprocal rule.

69. Prove the quotient rule, using the product and reciprocal rules.

70. Prove that $D_x \cos x = -\sin x$.

71. Prove that $D_x \cot x = -\csc^2 x$.

72. Prove that $D_x \csc x = -\csc x \cot x$.

SECTION 3.5

The Chain Rule

The differentiation formulas of the previous section apply to arithmetic combinations of functions. In this section, we develop a formula called the *chain rule*, which is used to differentiate the composition of functions. Knowing how to differentiate the composition of functions will enable us to find the derivatives of many complicated functions.

Recall that the composition of the functions f and g is the function $g \circ f(x) = g(f(x))$. We will think of f as the "inside" function and g as the "outside" function.

The Chain Rule

The composition of two functions is not usually a function to which the differentiation formulas we have developed thus far will apply. The chain rule is a means for differentiating the composition of two functions.

THEOREM 3.10

The Chain Rule If the function f is differentiable at c and the function g is differentiable at $f(c)$, then the composition $g \circ f$ is differentiable at c, and

$$(g \circ f)'(c) = g'(f(c)) \cdot f'(c).$$

In Leibniz notation, if $y = g(u)$ and $u = f(x)$, then

$$\frac{dy}{dx} = \frac{dy}{du}\frac{du}{dx}.$$

Proof By the alternative definition of derivative, formula (2) in Section 3.3,

$$\begin{aligned}(g \circ f)'(c) &= \lim_{x\to c} \frac{g \circ f(x) - g \circ f(c)}{x - c} \\ &= \lim_{x\to c} \frac{g(f(x)) - g(f(c))}{x - c}. \qquad (1)\end{aligned}$$

Now, we let a be any point in the domain of g, and we define a new function

$$h(x) = \begin{cases} \dfrac{g(x) - g(a)}{x - a} & \text{if } x \neq a \\ g'(a) & \text{if } x = a. \end{cases}$$

Then h is continuous at a, since

$$\begin{aligned}\lim_{x\to a} h(x) &= \lim_{x\to a} \frac{g(x) - g(a)}{x - a} \\ &= g'(a) \\ &= h(a).\end{aligned}$$

Also, by definition of h, we have

$$g(x) - g(a) = h(x)(x - a). \qquad (2)$$

If we replace x by $f(x)$ and let $a = f(c)$, then equation (2) becomes

$$g(f(x)) - g(f(c)) = h(f(x))[f(x) - f(c)].$$

Putting this into equation (1), we get

$$\begin{aligned}(g \circ f)'(c) &= \lim_{x\to c} \frac{h(f(x))[f(x) - f(c)]}{x - c} \\ &= \lim_{x\to c} h(f(x)) \lim_{x\to c} \frac{f(x) - f(c)}{x - c} \\ &= h(f(c)) \cdot f'(c) \\ &= h(a) \cdot f'(c) \\ &= g'(a) \cdot f'(c) \\ &= g'(f(c)) \cdot f'(c).\end{aligned}$$

This establishes the chain rule.

The chain rule tells us that to find the derivative of a composition of functions, we differentiate the outside function, evaluate it at the inside function, and then multiply by the derivative of the inside function. That is, for any x at which the functions $g \circ f$, f, and g are differentiable,

$$D_x g(f(x)) = g'(f(x)) \cdot f'(x).$$

Example 3.24 Use the chain rule to differentiate $F(x) = \sin 2x$. Then use trigonometric identities to rewrite F in another form (without composition), and differentiate without using the chain rule.

Solution The function $F(x) = \sin 2x$ is the composition of the outside function $g(x) = \sin x$ and the inside function $f(x) = 2x$. By the chain rule, then, the derivative is

$$\begin{aligned} D_x \sin 2x &= (\cos 2x)\, D_x(2x) \\ &= (\cos 2x)(2) \\ &= 2 \cos 2x. \end{aligned}$$

Alternatively, we can use the sine double-angle formula to rewrite F, and without using the chain rule, we get

$$\begin{aligned} D_x \sin 2x &= D_x(2 \sin x \cos x) \\ &= 2 \cos x \cos x + 2 \sin x\, (-\sin x) \\ &= 2(\cos^2 x - \sin^2 x) \\ &= 2 \cos 2x. \end{aligned}$$

(This last equality is the result of applying the cosine double-angle formula.) ■

Example 3.25 Differentiate:

(a) $y = (x^2 + 1)^{-3}$ **(b)** $f(x) = \cos 3x$ **(c)** $f(x) = \cos^3 x$

Solution **(a)** Here, the inside function is, "square and add 1," and the outside function is, "raise to the -3 power." If we let $y = u^{-3}$ and $u = x^2 + 1$, then the Leibniz form of the chain rule gives

$$\begin{aligned} \frac{dy}{dx} = \frac{dy}{du}\frac{du}{dx} &= (-3u^{-4})(2x) \\ &= -3(x^2 + 1)^{-4}(2x) \\ &= -6x(x^2 + 1)^{-4}. \end{aligned}$$

(b) Here, the outside function is the cosine, and the inside function is $3x$. The chain rule gives us

$$\begin{aligned} f'(x) &= (-\sin 3x)(3) \\ &= -3 \sin 3x. \end{aligned}$$

(c) If we rewrite f as $f(x) = (\cos x)^3$, we recognize f as the composition of the inside function $\cos x$ with the outside function x^3. The chain rule therefore tells us that

$$\begin{aligned} f'(x) &= 3(\cos x)^2(-\sin x) \\ &= -3 \cos^2 x \sin x. \end{aligned}$$

■

■ Example 3.26 If $y = u^3 - 2u$ and $u = x - 1/x$, find $(dy/dx)|_{x=2}$.

Solution By the chain rule,

$$\frac{dy}{dx} = \frac{dy}{du}\frac{du}{dx} = (3u^2 - 2)\left(1 + \frac{1}{x^2}\right).$$

Now, when $x = 2$, $u = 2 - \frac{1}{2} = \frac{3}{2}$, so

$$\left.\frac{dy}{dx}\right|_{x=2} = [3(\tfrac{3}{2})^2 - 2]\left(1 + \frac{1}{2^2}\right) = \frac{95}{16}.$$

■

■ Example 3.27 Given $f'(3) = -1$ and $g(x) = 4 - x^2$, find the slope of the curve $y = f(g(x))$ at $x = 1$.

Solution The slope is

$$\begin{aligned} y'(1) &= f'(g(1)) \cdot g'(1) \\ &= f'(3) \cdot g'(1) \\ &= -g'(1). \end{aligned}$$

Since $g'(x) = -2x$, we have $g'(1) = -2$, and the slope is

$$y'(1) = -(-2) = 2.$$

■

■ Example 3.28 If $y = f(x^2 + 1)$ and you know that $f'(x) = 1/(x + 1)$, what is dy/dx?

Solution

$$\begin{aligned} \frac{dy}{dx} &= f'(x^2 + 1)(2x) \\ &= \frac{2x}{(x^2 + 1) + 1} \\ &= \frac{2x}{x^2 + 2}. \end{aligned}$$

■

The Generalized Power Rule

One application of the chain rule gives us a formula for the derivative of the power of a function. This formula is a generalization of the power rule.

THEOREM 3.11 **The Generalized Power Rule** If f is a differentiable function and n is an integer, then the power of the function

$$y = [f(x)]^n$$

is differentiable, and

$$\frac{dy}{dx} = n[f(x)]^{n-1}f'(x).$$

The proof of this theorem is a simple application of the chain rule, and will be left as an exercise. The following examples illustrate its use.

■ **Example 3.29** Find y'.

(a) $y = (5x - 3)^4$ (b) $y = (x^2 - x^{-2})^{-4}$ (c) $y = \sin^2 x \cos^3 x$

Solution

(a) $y' = 4(5x - 3)^3(5) = 20(5x - 3)^3$.

(b) $y' = (-4)(x^2 - x^{-2})^{-5}[2x - (-2)x^{-3}] = -4(x^2 - x^{-2})^{-5}(2x + 2x^{-3})$.

(c) Here, we use the generalized power rule in the process of using the product rule:

$$\begin{aligned} y' &= 2 \sin x \cos x \cos^3 x + \sin^2 x\ (3) \cos^2 x\ (-\sin x) \\ &= 2 \sin x \cos^4 x - 3 \sin^3 x \cos^2 x. \end{aligned}$$ ■

■ **Example 3.30** Differentiate $f(x) = \sec^3 4x = (\sec 4x)^3$.

Solution Here, we have a threefold composition, the cube of the secant of 4 times x. We therefore apply the generalized power rule first, followed by an application of the chain rule:

$$\begin{aligned} f'(x) &= 3 \sec^2 4x \sec 4x \tan 4x\ (4) \\ &= 12 \sec^3 4x \tan 4x. \end{aligned}$$ ■

This last example illustrates that the chain rule can be applied repeatedly. In Leibniz notation, if y is a function of u, u is a function of v, and v is a function of x, then

$$\frac{dy}{dx} = \frac{dy}{du}\frac{du}{dv}\frac{dv}{dx}.$$

This formula can be extended to the composition of any number of functions.

Differentiation Formulas in Chain Rule Form

We close this section with a list of the differentiation formulas developed up to this point, most with the chain rule built into them. In the following list, we assume k is a constant, n is an integer, and u and v are functions of x. The differentiation formulas are then as follows:

Constant rule	$\dfrac{dk}{dx} = 0$
Power rule	$\dfrac{d(u^n)}{dx} = nu^{n-1}\dfrac{du}{dx}$
Scalar product rule	$\dfrac{d(ku)}{dx} = k\dfrac{du}{dx}$

Sum rule	$\dfrac{d(u+v)}{dx}=\dfrac{du}{dx}+\dfrac{dv}{dx}$
Difference rule	$\dfrac{d(u-v)}{dx}=\dfrac{du}{dx}-\dfrac{dv}{dx}$
Product rule	$\dfrac{d(uv)}{dx}=\dfrac{du}{dx}v+u\dfrac{dv}{dx}$
Reciprocal rule	$\dfrac{d}{dx}\left(\dfrac{1}{u}\right)=-\dfrac{1}{u^2}\dfrac{du}{dx}$
Quotient rule	$\dfrac{d}{dx}\left(\dfrac{u}{v}\right)=\dfrac{\dfrac{du}{dx}v-u\dfrac{dv}{dx}}{v^2}$
Trigonometric rules	$\dfrac{d(\sin u)}{dx}=\cos u\dfrac{du}{dx}$
	$\dfrac{d(\cos u)}{dx}=-\sin u\dfrac{du}{dx}$
	$\dfrac{d(\tan u)}{dx}=\sec^2 u\dfrac{du}{dx}$
	$\dfrac{d(\cot u)}{dx}=-\csc^2 u\dfrac{du}{dx}$
	$\dfrac{d(\sec u)}{dx}=\sec u\tan u\dfrac{du}{dx}$
	$\dfrac{d(\csc u)}{dx}=-\csc u\cot u\dfrac{du}{dx}$

Exercises 3.5

Find dy/dx.

1. $y=(3x+5)^2$
2. $y=(2x-1)^5$
3. $y=(1-x)^7$
4. $y=(4-3x)^5$
5. $y=\left(\dfrac{2}{x}\right)^4$
6. $y=\left(x-\dfrac{1}{x}\right)^2$
7. $y=\sin^3 x$
8. $y=(1-\cos x)^3$
9. $y=(x^2+4)^3$
10. $y=(x^2-8)^7$
11. $y=(2x+x^4)^{-5}$
12. $y=(x^6-3x^{-2})^9$
13. $y=\sin 3x$
14. $y=\tan 5x$
15. $y=\sec\dfrac{x}{2}$
16. $y=\cos 2x$
17. $y=3\sin x+\cos 3x$
18. $y=\tan x-\sec 2x$
19. $y=\cos 4x-\cos x$
20. $y=3\sin 5x-5\sin 3x$
21. $y=\sin x\cos 4x$
22. $y=\cos^2 3x$
23. $y=\sin^4 2x$
24. $y=\tan x\sec^2 x$
25. $y=\sin(2x+3)$
26. $y=x\sin x$
27. $y=x^2\cos x$
28. $y=\tan(x^2)$
29. $y=\sin(\tan x)$
30. $y=\cos(x\sin x)$
31. $y=x^3\sin(x^2)$
32. $y=x\cos(x^2)+x^2\cos^2 x$
33. $y=\dfrac{2x-1}{(4x^3-2x)^3}$

34. $y = \dfrac{3x^5}{x^4 - 6}$

35. $y = \dfrac{3x^5}{(x^4 - 6)^2}$

36. $y = \dfrac{\sin^3 x}{x^2}$

37. $y = \dfrac{(x^2 + 3)^2}{2x - 3}$

38. $y = \dfrac{x^2 + 4}{(x^2 + 8)^3}$

39. $y = \left(\dfrac{5}{3x^4}\right)^7$

40. $y = \left(\dfrac{x}{3x + 1}\right)^8$

A41. $y = \dfrac{(x + 1)(x + 2)^2}{(x + 3)^3(x + 4)^4}$

42. $y = \dfrac{(2x + 3)^2 \sin x}{x^2 + 4}$

43. $y = (4x - 9)^5(3x + 2)^{-7}$

44. $y = x^7(x^2 + 1)^{-2} \tan 3x$

45. $y = 4 \sin^2 5x$

46. $y = \sin^2 3x \cos x$

47. $y = \cos^3 2x$

48. $y = \sec^3 2x$

49. $y = x \sin 2x$

50. $y = x^2 \cos 3x$

51. $y = 2x \cos^2 x$

52. $y = \tan(x^2)$

53. $y = \sin(2x - 1)$

54. $y = \cos^2(3x - 2)$

55. $y = (x + 1) \csc(x + 1)$

56. $y = x^2 \sin(x^3)$

57. $y = \dfrac{x}{2 + \sin x}$

58. $y = \dfrac{\cos x}{1 - \cos x}$

59. $y = \dfrac{\sin 2x}{x^2 + 1}$

60. $y = \dfrac{\tan x}{(1 + \sin^2 x)^2}$

61. $y = \cos[x + \sin(x + 1)]$

62. $y = \tan[x^2 + \cos(x^2 + 1)]$

63. $y = [2x + (3x + 1)^2 + 4x^3]^4$

64. $y = \{x^2 + [x^3 + (x^4 + 1)^2]^3\}^4$

65. $y = 3u + 7; \quad u = x^6 - 1$

66. $y = u^2 - 2; \quad u = 4x + 3$

67. $y = \cos u; \quad u = x^2 + 4$

68. $y = u^2 + 4; \quad u = \sin 3x$

69. $y = 4t + 3; \quad t = 8 \tan x$

70. $y = 8r^2; \quad r = \sin x \cos x$

71. $y = u^3 + 1; \quad u = v - 2; \quad v = x^3 + 1$

72. $y = 1 + u^2; \quad u = \sin v; \quad v = x^2$

Find $v = ds/dt$ and $a = dv/dt$, and then find $v(t_0)$ and $a(t_0)$.

73. $s = (2 - t)^{-3}; \quad t_0 = 1$

74. $s = (8 - 4t)^2; \quad t_0 = 3$

75. $s = 4 \sin 3t; \quad t_0 = \pi/3$

76. $s = -5 \cos \pi t; \quad t_0 = 0.5$

77. If $y = g(f(x))$, $g'(1) = -7$, and $f(x) = x^2 - 8$, find $(dy/dx)|_{x=3}$.

78. If $y = g(f(x))$, $f(2) = 7$, $f'(2) = -2$, and $g(x) = x^2$, find $(dy/dx)|_{x=2}$.

79. If $f'(x) = \sqrt{x}$ and $y = f(x^2 + 4)$, find y' and $y'(0)$.

80. If $g'(x) = x^{-1}$ and $f(x) = (1 - x^3)^{-2}$, find $D_x g(f(x))$.

81. Try to differentiate $y = \cos 3x$ without using the chain rule.

82. Prove the generalized power rule.

83. Recalling that x degrees $= \pi x/180$ radians, find $D_x \sin x^\circ$.

84. Explain why degrees are not often used in connection with differentiation of the trigonometric functions.

SECTION 3.6

Higher-Order Derivatives

If f is a differentiable function, its derivative f' is just another function. If f' is also differentiable, we can differentiate it to get what is called a *higher-order derivative* of f. In this section, we examine this process of repeated differentiation.

The Second Derivative

If f is a differentiable function whose derivative f' is also differentiable, we say f is **twice differentiable**. The derivative of f' is called the **second derivative** of f, and is denoted by f'' (read "f prime prime" or "f double prime" or "f second"). We call f' the **first derivative** of f by contrast. Other notations for the second derivative of $y = f(x)$ are

$$y'', \qquad \frac{d^2y}{dx^2} = \frac{d}{dx}\frac{dy}{dx}, \qquad \frac{d^2f(x)}{dx^2}, \qquad D_x^2 f.$$

An interpretation of the second derivative as a rate of change is already familiar to us. The velocity of a moving object is the first derivative of the position function, and the acceleration

$$a = \frac{dv}{dt} = \frac{d^2s}{dt^2}$$

is the second derivative of position.

A graphic interpretation of the second derivative as the rate of change of slope of a curve will be considered in detail later.

The process of finding the second derivative of a function is straightforward: Just differentiate twice. There is no shortcut that skips the first derivative. Consider an example or two.

Example 3.31 Find the second derivative:

(a) $f(x) = 5x^2 - 6x + 1$ **(b)** $f(x) = (3x - 5)^4$ **(c)** $f(x) = \sec 2x$

Solution **(a)** The first derivative is

$$f'(x) = 10x - 6,$$

and the second derivative is

$$f''(x) = 10.$$

(b) The first derivative is

$$\begin{aligned} f'(x) &= 4(3x-5)^3(3) \\ &= 12(3x-5)^3, \end{aligned}$$

and the second derivative is

$$\begin{aligned} f''(x) &= 12 \cdot 3(3x-5)^2(3) \\ &= 108(3x-5)^2. \end{aligned}$$

(c) The first derivative is

$$\begin{aligned} f'(x) &= \sec 2x \tan 2x\ (2) \\ &= 2 \sec 2x \tan 2x, \end{aligned}$$

and the second derivative is (found using the product rule and chain rule)

$$f''(x) = 2[(\sec 2x \tan 2x)(2)] \tan 2x + 2 \sec 2x\, [(\sec^2 2x)(2)]$$
$$= 4 \sec 2x \tan^2 2x + 4 \sec^3 2x.$$ ■

Higher-Order Derivatives

Derivatives of derivatives of a function are called **higher-order derivatives** of the function. The derivative of the second derivative is called the **third derivative**, and its derivative is called the **fourth derivative**, and so on.

Higher-order derivatives are useful, even though there seem to be few physical interpretations of derivatives beyond the second. Derivatives of higher order are used extensively in applications such as Taylor polynomials, which we will consider later.

The most common notations for higher-order derivatives are

$$f'',\quad f''',\quad f^{(4)},\quad f^{(5)},\quad \ldots,\quad f^{(n)},\quad \ldots,$$
$$D_x^2,\quad D_x^3,\quad D_x^4,\quad D_x^5,\quad \ldots,\quad D_x^n,\quad \ldots,$$
$$\frac{d^2y}{dx^2},\quad \frac{d^3y}{dx^3},\quad \frac{d^4y}{dx^4},\quad \frac{d^5y}{dx^5},\quad \ldots,\quad \frac{d^ny}{dx^n},\quad \ldots.$$

The process of finding higher-order derivatives is no more complicated than finding the first derivative; we just differentiate the required number of times.

■ **Example 3.32** Find the third derivative of $y = \tan 3x$.

Solution Differentiating repeatedly, we have

$$y' = 3 \sec^2 3x,$$
$$y'' = 3(2 \sec 3x)(3 \sec 3x \tan 3x)$$
$$= 18 \sec^2 3x \tan 3x,$$
$$y''' = 18(2 \sec 3x)(3 \sec 3x \tan 3x) \tan 3x + 18 \sec^2 3x\, (3 \sec^2 3x)$$
$$= 108 \sec^2 3x \tan^2 3x + 54 \sec^4 3x.$$ ■

■ **Example 3.33** Find the nth derivative of $f(x) = 1/x$.

Solution We will write $f(x) = x^{-1}$, and start differentiating:

$$f'(x) = (-1)x^{-2},$$
$$f''(x) = (-1)(-2)x^{-3},$$
$$f'''(x) = (-1)(-2)(-3)x^{-4},$$
$$\ldots$$
$$f^{(n)}(x) = (-1)(-2)(-3) \cdot \cdots \cdot (-n)x^{-(n+1)}$$
$$= (-1)^n n!\, x^{-(n+1)}.$$

(Here, $n!$ is "n factorial," and is the product $n! = 1 \cdot 2 \cdot 3 \cdot \cdots \cdot n$.) It is not usually the case that a simple formula for the nth derivative of a function can be found, as is the case for this function. ■

Exercises 3.6

Find the first and second derivatives.

1. $f(x) = x^2 + 3x - 1$
2. $f(x) = 5x^4 - x^3 + 2x + 5$
3. $f(x) = \dfrac{1}{1-x}$
4. $f(x) = \dfrac{x}{x^2+1}$
5. $y = (2x+7)^9$
6. $y = (x^3+1)^2$
7. $y = \sin^2 x$
8. $y = \cos^2 x$
9. $y = \sin ax$
10. $y = \cos ax$
11. $y = \tan x$
12. $y = \sec x$
13. $y = \tan ax$
14. $y = \sec ax$
15. $y = \sin^n x$
16. $y = \cos^n x$
17. $f(x) = \sin^2 ax$
18. $f(x) = \sec^2 ax$
19. $f(x) = x \sin x$
20. $f(x) = x^2 \cot x$
21. $y = \dfrac{x}{\tan x}$
22. $y = \dfrac{x}{1 + \cos x}$
23. $y = \dfrac{\sin^2 x}{1 + \cos^2 x}$
24. $y = \dfrac{1 - \cos 2x}{x^2 + 2x}$

Find the third derivative.

25. $y = \dfrac{1}{1-x}$
26. $y = \dfrac{1}{x^2+1}$
27. $y = \sin ax$
28. $y = \tan ax$

Find the fourth derivative.

29. $f(x) = 3x^5 - 2x^4 + 7x^3 + 18x^2 - 45x - 36$
30. [A] $f(x) = \sec x$
31. Find the sixth derivative of $y = (x+3)^{-2}$.
32. Find the eighth derivative of $y = (x+4)^{10}$.

Find a formula for the nth derivative.

33. $y = \dfrac{1}{x+1}$
34. $y = \dfrac{1}{1-x}$
35. $y = \sin ax$
36. $y = \cos ax$

Find the acceleration a at the given value of t.

37. $s = 32 + 80t - 16t^2;\quad t = 3$
38. $s = t^2 - \dfrac{1}{t};\quad t = 1.5$
39. $s = \dfrac{1}{t + t^2};\quad t = 3$
40. $s = 6t + \dfrac{1}{2}at^2;\quad t = 0$
41. $s = 3 \sin 5t;\quad t = \pi$
42. $s = 3 \sin 5t;\quad t = 2\pi$
43. $s = 2 \sin \pi t;\quad t = 0.25$
44. $s = 4 \sin 2\pi t;\quad t = 0.125$
45. If p is a polynomial function of degree n, show that $p^{(n+1)}(x) = 0$.
46. If an object has position s, velocity v, and acceleration a, and v can be expressed as a function of s, use the chain rule to show that

$$a = v\,\frac{dv}{ds}.$$

SECTION 3.7

Increments and Differentials

In this section, we discuss increments, introduce differentials, and explain their relationship to the derivative. We also demonstrate the application of differentials to approximation and error estimation.

Increments

Suppose $y = f(x)$ is a differentiable function of x. If the independent variable x changes in value from x_1 to x_2, the amount of change (which can be either positive or negative) is called an **increment** of x, and is denoted by Δx. That is,

$$\Delta x = x_2 - x_1. \tag{1}$$

Since the dependent variable y depends on x, a change in x produces a change in y (see Figure 3.17). As x changes from x_1 to x_2, y changes from $y_1 = f(x_1)$ to $y_2 = f(x_2)$. The amount of change in y is called the **increment** of y, denoted by Δy, and

$$\Delta y = y_2 - y_1 = f(x_2) - f(x_1). \tag{2}$$

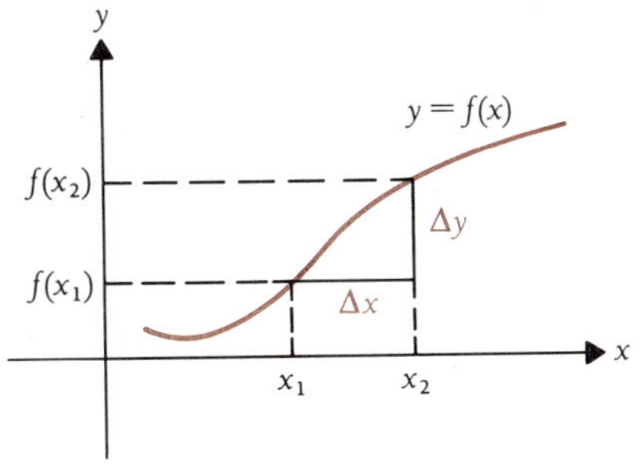

FIGURE 3.17 Increments in x and y

From equation (1) we can write

$$x_2 = x_1 + \Delta x,$$

so that equation (2) can be rewritten as

$$\Delta y = f(x_1 + \Delta x) - f(x_1).$$

We now drop the subscript, and say that as x changes from x to $x + \Delta x$, y changes from y to $y + \Delta y$, with

$$\Delta y = f(x + \Delta x) - f(x). \tag{3}$$

Thus, the increment Δy produced by the increment Δx depends on both x and Δx, according to formula (3).

Example 3.34 The side of a square changes from 6 to 6.1 feet. Find the change in the area of the square.

Solution The area of a square in terms of the side s is

$$A = s^2.$$

By equation (3), an increment Δs in s produces an increment

$$\begin{aligned} \Delta A &= (s + \Delta s)^2 - s^2 \\ &= s^2 + 2s\Delta s + (\Delta s)^2 - s^2 \\ &= 2s\Delta s + (\Delta s)^2 \end{aligned}$$

in A. For this problem, we let $s = 6$ feet and $\Delta s = 0.1$ foot. Then

$$\Delta A = 2(6)(0.1) + (0.1)^2 = 1.21 \text{ square feet.}$$

A geometric interpretation is shown in Figure 3.18.

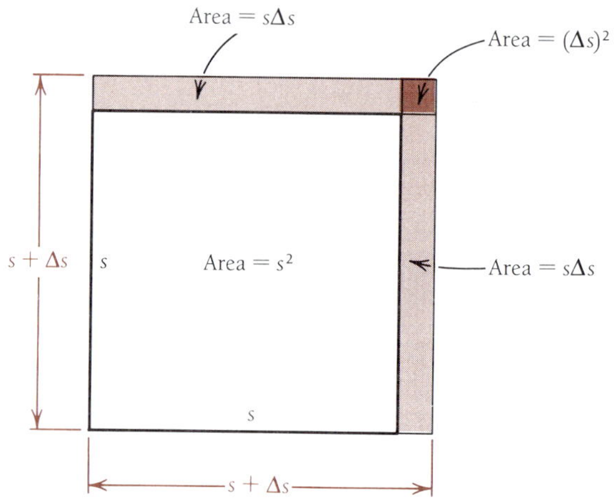

FIGURE 3.18 If the side of a square increases by Δs, then the area increases by $2s\Delta s + (\Delta s)^2$

If the increment Δx is not 0, then from equation (3) we can obtain

$$\frac{\Delta y}{\Delta x} = \frac{f(x + \Delta x) - f(x)}{\Delta x}.$$

Then

$$\begin{aligned} \lim_{\Delta x \to 0} \frac{\Delta y}{\Delta x} &= \lim_{\Delta x \to 0} \frac{f(x + \Delta x) - f(x)}{\Delta x} \\ &= f'(x) \end{aligned}$$

by the definition of derivative. That is, the ratio of the increments approaches the derivative as $\Delta x \to 0$. For Δx small, then, the ratio $\Delta y/\Delta x$ is close to $f'(x)$. Let ε be the difference,

$$\varepsilon = \frac{\Delta y}{\Delta x} - f'(x). \tag{4}$$

Then $\varepsilon \to 0$ as $\Delta x \to 0$. Multiplying both sides of equation (4) by Δx, we get

$$\Delta y = f'(x)\Delta x + \varepsilon \Delta x.$$

When Δx is small, so is ε, so that the product $\varepsilon\Delta x$ is smaller yet. Thus, Δy is approximately equal to $f'(x)\Delta x$—that is,

$$\Delta y \approx f'(x)\Delta x. \tag{5}$$

Differentials

We now introduce some new variables. If $y = f(x)$, the **differential of x** is a new independent variable, denoted by dx. The **differential of y** is a new dependent variable, denoted by dy, and defined by the equation

$$dy = f'(x)\, dx. \tag{6}$$

That is, dy depends on *both* of the independent variables x and dx.

■ **Example 3.35** If $y = x^3$, find dy for $x = 2$ and $dx = 0.05$.

Solution By equation (6),

$$\begin{aligned} dy &= 3x^2\, dx \\ &= 3(2)^2(0.05) = 0.6. \end{aligned}$$

■

Differentials in Linear Approximations

Comparing equations (5) and (6), we see that if $dx = \Delta x$ and Δx is small, then $\Delta y \approx dy$. That is, dy is an approximation to Δy when dx is the increment Δx. The accuracy of the approximation depends on the size of Δx and on the nature of the function f.

The nature of this approximation can be seen by studying Figure 3.19 on page 186. Let t be the line tangent to the graph of f at the point $P(x, f(x))$, and let $Q(x + dx, f(x + dx))$ be a second point on the graph. Now let R be the point $(x + dx, f(x))$, and let line t meet line QR at S. Since $f(x + dx) = y + \Delta y$, the distance from R to Q is Δy, while the distance from R to S is dy. That is, the increment Δy is the change along the graph of f, while the differential dy is the change along the tangent line to the graph. The use of dy to approximate Δy is called a **linear approximation.** The tangent line is used to approximate the graph of f.

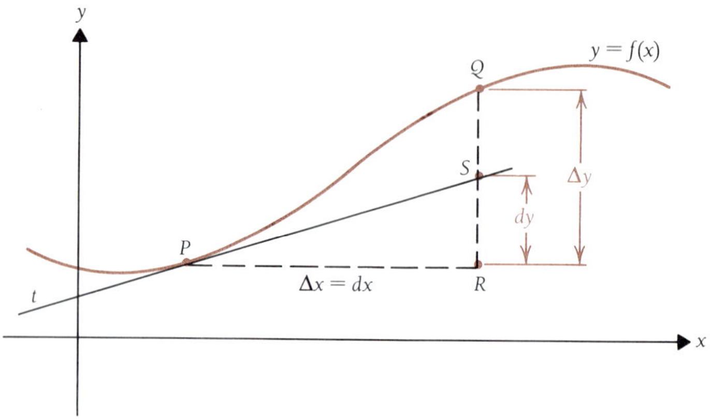

FIGURE 3.19 The differential approximation is a linear approximation

Example 3.36 Approximate $(1.05)^4$ using differentials, and compare the linear estimate with the calculated value.

Solution Let $y = x^4$; then $(1.05)^4$ is $y + \Delta y$ when $x = 1$ and $\Delta x = 0.05$. Letting $dx = \Delta x = 0.05$, from equation (6) we have

$$\begin{aligned} \Delta y \approx dy &= 4x^3\,dx \\ &= 4(1)^3(0.05) = 0.2. \end{aligned}$$

When $x = 1$, $y = 1$, so

$$(1.05)^4 = y + \Delta y \approx 1 + dy = 1.2.$$

Using a calculator, we find the calculated value to be 1.21550625. (Clearly, the calculator result is better and easier.) ■

Example 3.37 If the radius of a circle increases by 3%, use differentials to approximate the increase in the area of the circle.

Solution The area of a circle of radius r is $A = \pi r^2$. The increase in area is ΔA, which we will approximate by the differential. If the radius increases by 3%, then the change in radius is $0.03r$, and we have

$$\begin{aligned} \Delta A \approx dA &= 2\pi r\,dr \\ &= 2\pi r(0.03r) \\ &= 0.06\pi r^2 = 0.06A. \end{aligned}$$

Thus, we see that the increase in area is approximately 6%. (Note that it is difficult to use a calculator in this setting!) ■

Example 3.38 The circumference of a certain tree trunk is 156 inches, with an error in measurement of not more than 0.5 inch. Use differentials to estimate the maximum possible error in computing the diameter of the tree trunk.

Solution The diameter D is given in terms of the circumference C by $D = C/\pi$. If we let the possible error in C be the differential dC, then $dC = \pm 0.5$. The possible error in D is approximately

$$dD = \frac{1}{\pi}\, dC = \frac{\pm 0.5}{\pi} \approx \pm 0.16.$$

That is, the error will not be more than about 0.16 inch. ■

Finding Differentials

The differential $dy = f'(x)\, dx$ of the function $y = f(x)$ is found by differentiating f. The symbol df is also used to denote dy. Thus, $df = f'(x)\, dx$.

The formulas for finding differentials are just restatements of the differentiation formulas in differential form. We restate here the formulas from Section 3.5 in differential notation:

Constant rule	$dk = 0$
Power rule	$d(u^n) = nu^{n-1}\, du$
Scalar product rule	$d(ku) = k\, du$
Sum rule	$d(u + v) = du + dv$
Difference rule	$d(u - v) = du - dv$
Product rule	$d(uv) = v\, du + u\, dv$
Reciprocal rule	$d\left(\frac{1}{u}\right) = -\frac{1}{u^2}\, du$
Quotient rule	$d\left(\frac{u}{v}\right) = \frac{v\, du - u\, dv}{v^2}$
Trigonometric rules	$d(\sin u) = \cos u\, du$
	$d(\cos u) = -\sin u\, du$
	$d(\tan u) = \sec^2 u\, du$
	$d(\cot u) = -\csc^2 u\, du$
	$d(\sec u) = \sec u \tan u\, du$
	$d(\csc u) = -\csc u \cot u\, du$

Note that the chain rule is built into these formulas. The differential notation also shows off the chain rule especially well.

Example 3.39 If $y = \sin^2 3x$, find dy.

Solution Using the formulas on page 187, we have

$$
\begin{aligned}
dy &= d(\sin^2 3x) \\
&= 2 \sin 3x \, d(\sin 3x) \\
&= 2 \sin 3x \cos 3x \, d(3x) \\
&= 2 \sin 3x \cos 3x \; 3 \, dx \\
&= 6 \sin 3x \cos 3x \, dx.
\end{aligned}
$$

■

From equation (6), we can write

$$\frac{dy}{dx} = f'(x)$$

whenever $dx \neq 0$. This means that the differential dy varies in such a way that the ratio of dy to dx is always the derivative of the function. Thus, the Leibniz notation for the derivative may be thought of as the ratio of the differentials.

Exercises 3.7

1. The edge of a cube changes from 6 feet to 6.1 feet. Find the change in the volume of the cube, and then use differentials to estimate the change in volume.
2. Sketch an illustration similar to Figure 3.18 for Problem 1. [Do not worry about drawing it to scale.]
3. The radius of a circle changes from 6 feet to 6.1 feet. Find the change in the area of the circle, and then use differentials to estimate the change in area.
4. Sketch an illustration similar to Figure 3.18 for Problem 3. [Do not worry about the scale.]

Use differentials to approximate the following, and compare the results with the values you get on your calculator:

5. $(1.01)^5$
6. $(1.001)^{10}$
7. $\dfrac{1}{0.99}$
8. $(1.03)^{-4}$
9. $\sqrt{63}$
10. $(63)^{1/3}$
11. $(1729)^{2/3}$
12. $(30)^{1/5}$

Find dy.

13. $y = x^3$
14. $y = x^{-6}$
15. $y = x^2 + 2x$
16. $y = 8x - \dfrac{8}{x}$
17. $y = (x^2 + 3)^4$
18. $y = \dfrac{x - 1}{x^2}$
19. $y = \dfrac{x^2 - 4}{x^2 + 4}$
20. $y = \dfrac{x}{x - 1}$
21. $y = \sin 4x$
22. $y = \tan 5x$
23. $y = \cos^3 2x$
24. $y = \sec^3 2x$
25. $y = \dfrac{1 - \tan^2 x}{1 - \cot^2 x}$
26. $y = (1 - \sin x - \cos x)^2$
27. $y = \cot(x^3)$
28. $y = \sin(\tan x)$
29. Show that if the side of a square increases by 1%, the area increases by approximately 2%.

30. Show that if the radius of a circle decreases by 5%, then the area of the circle decreases by about 10%.

31. The radius of a sphere increases by 1%. Approximately how much does the volume change?

32. The edge of a cube decreases by 2%. Approximately how much does the volume change?

33. The radius of a sphere is 1.54 meters, with an error in measurement of at most 0.01 meter. Use differentials to estimate the maximum possible error in computing the volume of the sphere.

34. Answer Problem 33 for the surface area of the sphere.

35. The edge of a cube is 18.6 inches, with an error of not more than 0.1 inch. Use differentials to estimate the possible error in computing the volume of the cube.

36. Answer Problem 35 for the total surface area of the cube.

37. At a point 100 feet from the base of a cliff, the angle of elevation to the top of the cliff is $\pi/3 \pm 0.002$ radians. Using differentials, estimate the height of the cliff, including the possible error.

C38. A truck on a hillside subtends an angle of 0.05 ± 0.001 radian at the eye of the observer. The truck is known to be 30 feet long. Use differentials to estimate the distance from the observer to the truck, including the possible error.

39. The number of bacteria in a certain culture is known to be proportional to the time of incubation. If after 5 hours of incubation, the number of bacteria is $12{,}000 \pm 100$, what are the minimum and maximum numbers that can be expected after 8 hours?

C40. The force on an electron exerted by a charged plate varies inversely as the square of the distance from the electron to the plate. If at a distance of 0.005 meter, the force is 0.0003 ± 0.00001 newton, approximately what are the minimum and maximum values for the force at a distance of 0.004 meter?

41. Show that the approximation $\Delta y \approx dy$ is the "best" linear approximation by showing that if any line other than the tangent line is used to approximate the graph of the function, then greater error occurs for small dx.

SECTION 3.8

Implicit Differentiation

Not every function can be defined conveniently by an equation of the form $y = f(x)$. Sometimes, the most natural way to describe a relationship between two variables is with an equation in the two variables, such as $x^2 + y^2 = 1$. In this section, we discuss functions that are defined in this way.

Implicit Functions

If f is a function of x, then the equation $y = f(x)$ is said to define y as a function of x **explicitly**. All the functions we have seen thus far have been defined explicitly.

Now, let $g(x, y)$ be an expression in the two variables x and y. If there is a function f such that $g(x, f(x)) = 0$ for each x in the domain of f, then the equation $g(x, y) = 0$ is said to define y as a function of x **implicitly**.

Example 3.40 Show that the equation $x^2y - 2x + y = 0$ implicitly defines the function that can be written explicitly as

$$f(x) = \frac{2x}{x^2+1}.$$

Solution In the given equation, we replace y by the function:

$$\begin{aligned} x^2y - 2x + y &= x^2f(x) - 2x + f(x) \\ &= x^2\left(\frac{2x}{x^2+1}\right) - 2x + \frac{2x}{x^2+1} \\ &= (x^2+1)\left(\frac{2x}{x^2+1}\right) - 2x \\ &= 0. \end{aligned}$$

Since the equation is satisfied, it does indeed implicitly define y as $f(x)$. In this case, the function $f(x)$ can be found explicitly by solving the original equation for y. ■

Example 3.41 Find a function defined implicitly by the equation $x - y^3 = 1$.

Solution If we solve the given equation for y, we get the function

$$y = f(x) = (x-1)^{1/3}.$$

It is easily verified that this function satisfies the original equation when substituted for y. ■

Example 3.42 Show that the equation $x - y^2 = 0$ defines many functions implicitly.

Solution One solution for y in the equation $x - y^2 = 0$ is $y = \sqrt{x}$ (see Figure 3.20), which is clearly a function defined implicitly by the equation. But $y = -\sqrt{x}$ (see Figure 3.21) is also a solution, so it is another function so defined.

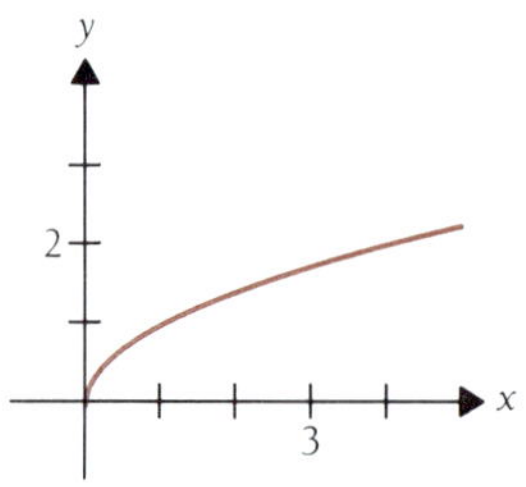

FIGURE 3.20
The graph of $f(x) = \sqrt{x}$

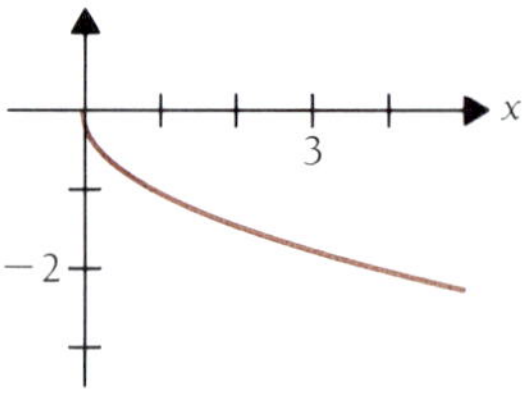

FIGURE 3.21
The graph of $f(x) = -\sqrt{x}$

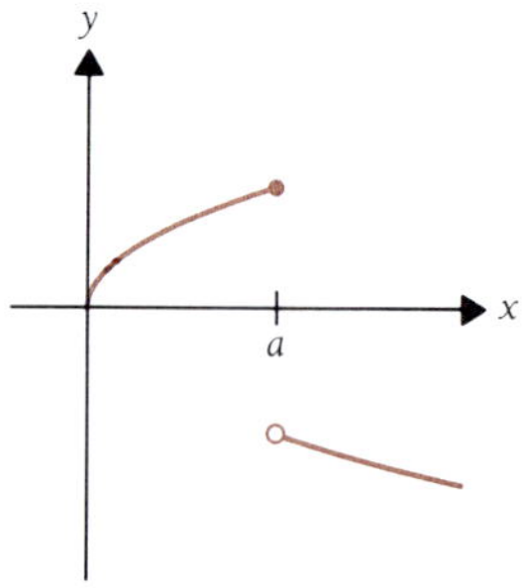

FIGURE 3.22
The graph of

$$f_a(x) = \begin{cases} \sqrt{x} & \text{if } 0 \le x \le a \\ -\sqrt{x} & \text{if } x > a \end{cases}$$

If a is any positive real number, let the function f_a be defined by

$$f_a(x) = \begin{cases} \sqrt{x} & \text{if } 0 \le x \le a \\ -\sqrt{x} & \text{if } x > a \end{cases}$$

(see Figure 3.22). Then $y = f_a(x)$ is also a solution of the equation, and therefore is a function defined implicitly by it. With a little thought, you can see how to define many more such functions. Thus, the equation $x - y^2 = 0$ implicitly defines infinitely many functions. ■

As the above examples show, it is often possible to find explicit formulas for functions defined implicitly. Sometimes, this can be done by solving the given equation for y in terms of x. However, it often cannot be done uniquely, and in many cases it cannot be done at all. For example, the equation

$$x^6 - x^2y^3 - y^5 = 0$$

defines y implicitly as a function of x (see Figure 3.23), but it is impossible to solve algebraically this fifth-degree equation for y. Thus, we must learn to deal with implicit functions.

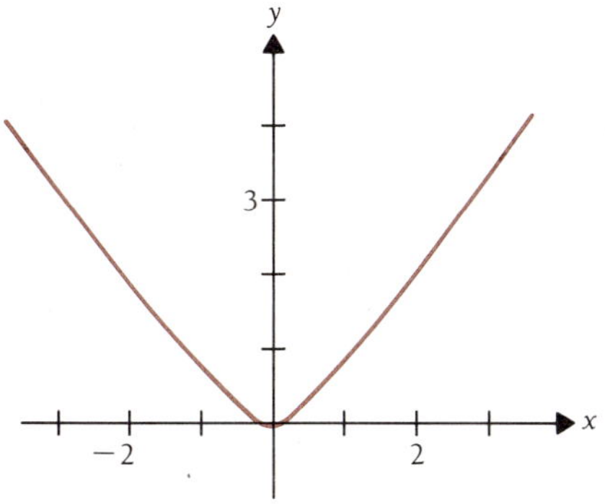

FIGURE 3.23 The graph of $x^6 - x^2y^3 - y^5 = 0$

It should be noted here that not every equation in x and y will define a function implicitly. For example, the equation

$$x^2 + y^2 = -1$$

does not define y as a function of x, since no pair (x, y) of real numbers satisfies the equation.

We have also seen, from Example 3.42, that a function defined implicitly may or may not be differentiable at a given point. The function $y = \sqrt{x}$ is differentiable for any positive number x, but the function $y = f_a(x)$ is not differentiable at a.

The general questions of whether a given equation in x and y defines y as a function of x and whether such a function is differentiable are treated in advanced calculus. Unless it is obviously not the case, we will always assume that a given equation in x and y defines y implicitly as a differentiable function of x.

Now we turn to the problem of finding the derivative.

Implicit Differentiation

When a function is defined implicitly, its derivative may be found by a process called **implicit differentiation**. We will illustrate with an example.

Let's assume that the equation

$$3x^2 + y^2 - 1 = 0 \tag{1}$$

defines y implicitly as a differentiable function of x. Since y is a function of x, the expression $3x^2 + y^2 - 1$ is also a function of x. Since that expression is equal to 0, it is constant, and its differential must be 0, or

$$d(3x^2 + y^2 - 1) = 0. \tag{2}$$

Simplifying the left-hand side of equation (2) by using the differentiation formulas of Section 3.7, we get

$$6x\,dx + 2y\,dy - 0 = 0. \tag{3}$$

Now we can solve equation (3) for dy, getting

$$dy = -\frac{3x}{y}\,dx,$$

and the derivative is

$$\frac{dy}{dx} = -\frac{3x}{y}.$$

Instead of solving equation (3) for dy, we could have rewritten (3) as

$$6x + 2y\frac{dy}{dx} = 0 \tag{4}$$

or as

$$6x + 2yy' = 0, \tag{5}$$

and then solved for dy/dx or y'. But note that we also could have gone directly from equation (1) to (4) or (5), simply by differentiating both sides of (1) using the chain rule on the function y of x.

We can verify that for some of the functions defined by equation (1), the derivative $y' = -3x/y$ obtained above is correct. This can be done by writing the functions explicitly and differentiating them in the usual way. For instance, $f(x) = \sqrt{1 - 3x^2}$ and $g(x) = -\sqrt{1 - 3x^2}$ are two of the functions defined by equation (1) implicitly (these are the continuous ones). Using the definition of derivative, we can get (you should verify this)

$$f'(x) = \frac{-3x}{\sqrt{1 - 3x^2}} = \frac{-3x}{f(x)}.$$

Also, since $g(x) = -f(x)$,

$$g'(x) = -f'(x) = \frac{3x}{\sqrt{1 - 3x^2}} = \frac{-3x}{g(x)}.$$

Thus, $y' = -3x/y$ is the derivative for either $y = f(x)$ or $y = g(x)$. In fact, whatever function is defined implicitly by equation (1), if that function is differentiable at x, then its derivative is $y' = -3x/y$.

In every instance, implicit differentiation proceeds in the same way. Given the defining equation, we differentiate both sides, remembering to use the chain rule when differentiating any expression involving y, and set the derivatives of the two sides equal to each other. Then we solve for y'.

Example 3.43 Find y':

(a) $xy + y^2 = 3$ (b) $x \cos y = xy^2$

Solution (a) Differentiating both sides (don't forget the product rule!) and setting them equal, yields

$$\begin{aligned} 1y + xy' + 2yy' &= 0, \\ (x + 2y)y' &= -y, \\ y' &= \frac{-y}{x + 2y}. \end{aligned}$$

(b) Implicit differentiation gives us

$$\begin{aligned} 1 \cos y + x(-\sin y)y' &= 1y^2 + x(2yy'), \\ -(x \sin y + 2xy)y' &= y^2 - \cos y, \\ y' &= \frac{\cos y - y^2}{x \sin y + 2xy}. \end{aligned}$$

■

Example 3.44 Given $x^2 - y^2 = 5$, find y' and y''.

Solution To find y', we differentiate implicitly:

$$\begin{aligned} 2x - 2yy' &= 0, \\ 2yy' &= 2x, \\ y' &= \frac{x}{y}. \end{aligned}$$

To find y'', we differentiate this last equation implicitly, and substitute for y' from the above equation:

$$\begin{aligned} y'' &= \frac{1y - xy'}{y^2} \\ &= \frac{y - x(x/y)}{y^2} \\ &= \frac{y^2 - x^2}{y^3}. \end{aligned}$$

We may leave y'' in this form, or we may use the given equation to simplify it to

$$y'' = \frac{-5}{y^3}.$$

■

■ **Example 3.45** Find the equation of the line tangent to the curve

$$x^3 + y^3 = 3xy$$

at the point $(\frac{2}{3}, \frac{4}{3})$.

Solution This curve, called the *folium of Descartes*, is sketched in Figure 3.24. Before we can find the equation of the tangent line, we must find the slope at the given point. To find the slope, we differentiate implicitly:

$$\begin{aligned} 3x^2 + 3y^2y' &= 3y + 3xy', \\ 3(y^2 - x)y' &= 3(y - x^2), \\ y' &= \frac{y - x^2}{y^2 - x}. \end{aligned}$$

Now, substituting $x = \frac{2}{3}$, $y = \frac{4}{3}$, we get

$$y'(\tfrac{2}{3}, \tfrac{4}{3}) = \frac{\frac{4}{3} - \frac{4}{9}}{\frac{16}{9} - \frac{2}{3}} = \frac{4}{5}.$$

The equation of the tangent line is therefore

$$\begin{aligned} y - \tfrac{4}{3} &= \tfrac{4}{5}(x - \tfrac{2}{3}), \\ y &= \tfrac{4}{5}x + \tfrac{4}{5}. \end{aligned}$$

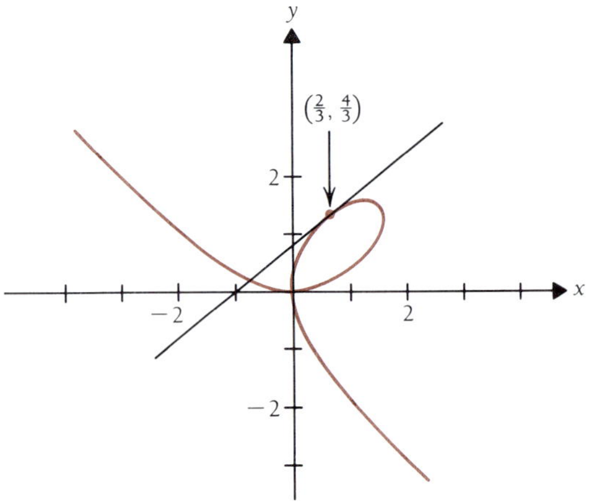

FIGURE 3.24 Tangent line to the folium of Descartes, $x^3 + y^3 = 3xy$

■

Derivatives of Inverse Functions

If the equation $y = f(x)$ defines y explicitly as a function of x, then the equation $x = f(y)$ defines y implicitly as the inverse function of f, provided f has an inverse function. We can therefore use implicit differentiation to find the derivative of the inverse of a function.

Suppose f is a differentiable function having inverse function f^{-1}. If

$$y = f^{-1}(x),$$

then

$$f(y) = x.$$

Differentiating implicitly, we get

$$f'(y)y' = 1,$$

so that

$$y' = \frac{1}{f'(y)}.$$

Since $y = f^{-1}(x)$, we therefore have

$$D_x[f^{-1}(x)] = \frac{1}{f'(f^{-1}(x))}.$$

This is the outline of the proof of the following theorem.

THEOREM 3.12 **The Inverse Function Rule** If the function f is differentiable and has an inverse function f^{-1}, then f^{-1} is differentiable for all x such that $f'(f^{-1}(x)) \neq 0$, and

$$D_x[f^{-1}(x)] = \frac{1}{f'(f^{-1}(x))}.$$

A graphic verification of this theorem is helpful. Consider the curves $y = f(x)$ and $y = f^{-1}(x)$ shown in Figure 3.25 (page 196). Let t be the tangent line to the graph of f^{-1} at point (a, b); let τ be the tangent line to the graph of f at the point (b, a). Just as the graphs of f and f^{-1} are reflections of each other in the line $y = x$, the lines t and τ are reflections of each other.

Let t have slope $m = D_x[f^{-1}(x)]|_{x=a}$. Let τ have slope $\mu = f'(b) = f'(f^{-1}(a))$. Since t and τ are reflections of each other, their slopes are reciprocals of each other. That is, the rise and run of t become the run and rise of τ. Therefore,

$$D_x[f^{-1}(x)]|_{x=a} = m = \frac{1}{\mu} = \frac{1}{f'(b)} = \frac{1}{f'(f^{-1}(a))}.$$

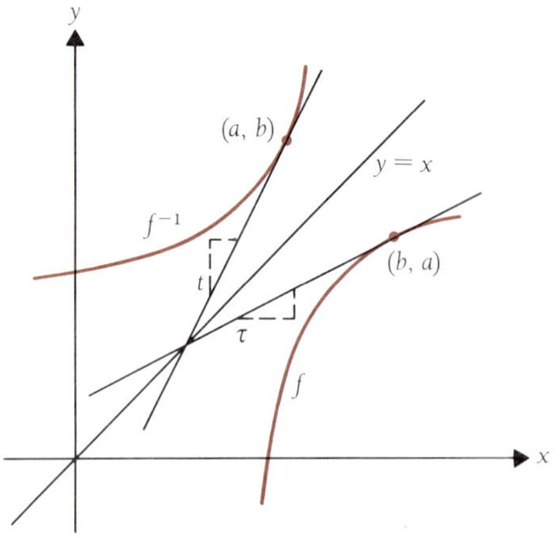

FIGURE 3.25 Tangent lines to a function and its inverse

Rather than use the inverse function rule directly, it is usually better to use implicit differentiation, as in the derivation of the theorem. We will illustrate this in showing that the power rule holds for any rational exponent.

First, we prove the power rule for root functions. If n is a positive integer, then the root function

$$f(x) = x^{1/n}$$

is defined on $[0, \infty)$; it is also defined on $(-\infty, 0)$ if n is odd. Since the power function

$$f^{-1}(x) = x^n$$

is the inverse of the root function f, we can use implicit differentiation(or the inverse function rule) to find the derivative of the root function. Letting

$$y = x^{1/n},$$

we have

$$y^n = x.$$

Differentiating implicitly, we get

$$\begin{aligned} ny^{n-1}y' &= 1, \\ y' &= \frac{1}{n} y^{-n+1} \\ &= \frac{1}{n} (x^{1/n})^{-n+1} \\ &= \frac{1}{n} x^{1/n-1}. \end{aligned}$$

Thus, the power rule holds for root functions.

Example 3.46 Find y':

(a) $y = \sqrt{x}$ (b) $y = \sqrt{x^2 + 4}$

Solution (a) Expressing the square root as a fractional power, we have

$$y = x^{1/2}.$$

Hence, by the power rule for root functions,

$$y' = \tfrac{1}{2}x^{-1/2} = \frac{1}{2\sqrt{x}}.$$

(b) Here, we can use the generalized power rule with a root function (that is, we use the chain rule). First writing

$$y = (x^2 + 4)^{1/2},$$

we get

$$y' = \tfrac{1}{2}(x^2 + 4)^{-1/2}(2x) = x(x^2 + 4)^{-1/2} = \frac{x}{\sqrt{x^2 + 4}}.$$

■

Now we can prove the power rule for any rational exponent. Let r and s be integers, and let

$$f(x) = x^{r/s}.$$

If r/s is negative, let r be negative and s be positive. Then we can write

$$f(x) = (x^{1/s})^r.$$

By the generalized power rule and the power rule for root functions,

$$f'(x) = r(x^{1/s})^{r-1}\frac{1}{s}x^{1/s-1} = \frac{r}{s}x^{r/s-1/s+1/s-1} = \frac{r}{s}x^{r/s-1}.$$

This proves that the power rule holds for any rational exponent.

Example 3.47 Find y', and tell for what values of x the derivative exists.

(a) $y = x^{2/3}$ (b) $y = x^{5/3}$ (c) $y = x^{3/2}$

Solution

(a) By the power rule, if $y = x^{2/3}$, then

$$y' = \tfrac{2}{3}x^{-1/3} = \frac{2}{3x^{1/3}}.$$

Since this function is defined for all nonzero values of x, the function $f(x) = x^{2/3}$ is differentiable except at the origin (see Figure 3.26).

(b) For $y = x^{5/3}$, the derivative is

$$y' = \tfrac{5}{3}x^{2/3},$$

which is defined for all x. Thus, the function $f(x) = x^{5/3}$ is differentiable on all of $\mathbb{R}$ (see Figure 3.27).

(c) For $y = x^{3/2}$, the derivative is

$$y' = \tfrac{3}{2}x^{1/2} = \tfrac{3}{2}\sqrt{x}.$$

Since the domain of this function is $[0, \infty)$, the function $f(x) = x^{3/2}$ is differentiable on $(0, \infty)$ (see Figure 3.28).

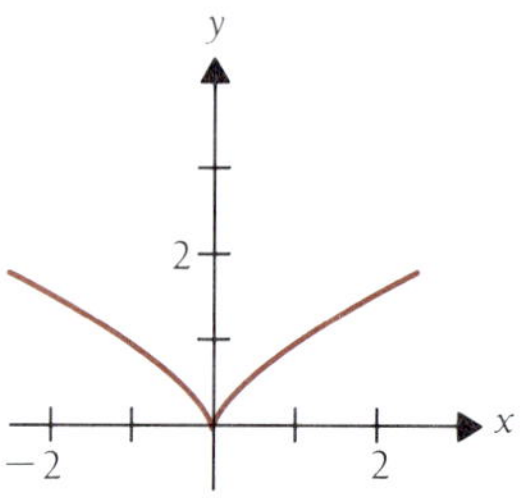

FIGURE 3.26
The graph of $f(x) = x^{2/3}$

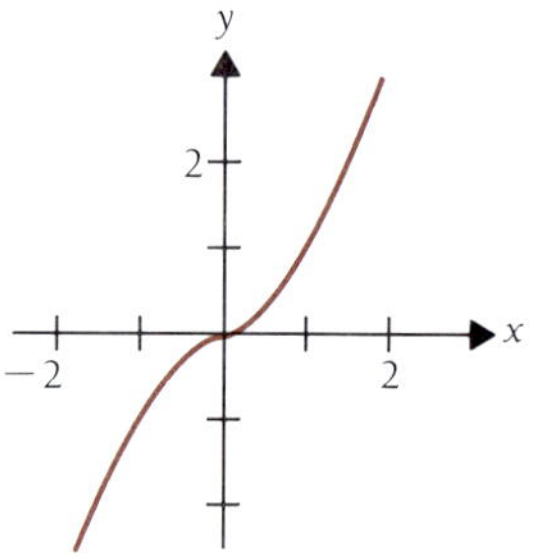

FIGURE 3.27
The graph of $f(x) = x^{5/3}$

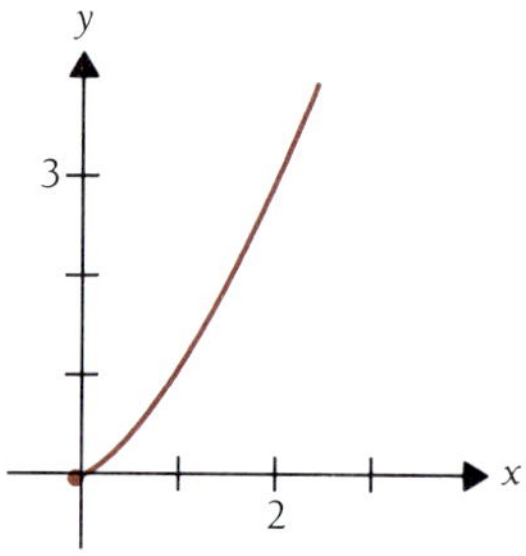

FIGURE 3.28
The graph of $f(x) = x^{3/2}$ ■

■ **Example 3.48** Find y':

(a) $y = |x|$ (b) $y = |f(x)|$

Solution

(a) Using the fact that

$$y = |x| = \sqrt{x^2} = (x^2)^{1/2},$$

we get

$$y' = \tfrac{1}{2}(x^2)^{-1/2}(2x) = \frac{x}{(x^2)^{1/2}} = \frac{x}{|x|}.$$

This derivative exists at any nonzero x.

(b) Using part (a) and the chain rule, we get, for $y = |f(x)|$,

$$y' = \frac{f(x)f'(x)}{|f(x)|}.$$

■

Exercises 3.8

Find y'.

1. $x^2 + y^2 = 1$
2. $x^2 - 2y^2 = 1$
3. $x^2 + 2xy + y^2 = 5$
4. $x^2 - xy - y^2 = 0$
5. $x^2 = y^3$
6. $x^6 - x^2y^3 - y^5 = 0$
7. $x^2 + y^{-2} = 4$
8. $xy^{-3} = x + y$
9. $y = x^{-1/2}$
10. $y = x$
11. $y = \sqrt{x^2 + 1}$
12. $y = \sqrt{4 - x^2}$
13. $\sqrt{x} + \sqrt{y} = 1$
14. $x^{2/3} + y^{2/3} = 1$
15. $\dfrac{\sqrt{x}}{\sqrt{y}} = 1 - x$
16. $\sqrt{x} - \sqrt{y} = xy$
17. $x = \sin y$
18. $x = \tan y$
19. $x \sin y = y$
20. $x \cos y = x + y$
21. $y^2 = \sin x$
22. $y^2 = \tan x$
23. $xy = \sin(x + y)$
24. $\cos xy = 0$
25. $\sin xy = x + y$
26. $\tan\sqrt{x + y} = 2x$
27. $\tan(\sin y) = x - y$
28. $x + 2y = \sec(x + y)$
29. $y = |2x|$
30. $y = |x^3|$
31. $y = |3x - 1|$
32. $y = |9 - x^2|$
33. $y = |\sin x|$
34. $y = \sin|x|$
35. $|x| + |y| = 1$
36. $x + \tan|y| = 1$

Find y''.

37. $x^2 + y^2 = 1$
38. $x^2 - 2y^2 = 1$
39. $\sqrt{x} + \sqrt{y} = 1$
40. $x^{2/3} + y^{2/3} = 1$
41. $y = |x|$
42. $y = |3x + 2|$
43. $x^2 + xy = 6$
44. [A] $\sin x = -\cos y$
45. $y = \sqrt{1 - x^2}$
46. $y = \sqrt{x - 3}$
47. $y = \dfrac{\sqrt{x}}{\sqrt{x - 1}}$
48. $y = \dfrac{2x}{1 + \sqrt{x}}$

Tell the values of x for which the function is differentiable.

49. $f(x) = x^{3/4}$
50. $f(x) = x^{2/5}$
51. $f(x) = \sqrt{\sin x}$
52. $f(x) = \tan\dfrac{1}{\sqrt{x}}$

Find the equation of the tangent line to the given curve at the given point.

53. $y = x^{4/3}$; $(8, 16)$
54. $y = \sqrt{4 - x^2}$; $(1, \sqrt{3})$
55. $x^{2/3} + y^{2/3} = 1$; $(\frac{1}{8}, 3\sqrt{3}/8)$
56. $y = |\sin x|$; $(5\pi/4, \sqrt{2}/2)$
57. $x^3 + y^3 = 3xy$; $(\frac{4}{3}, \frac{2}{3})$
58. $x^3 + y^3 = 3xy$; $(\frac{3}{2}, \frac{3}{2})$
59. $x^3 + y^3 = 3xy$; $(2^{1/3}, 4^{1/3})$
60. $x^3 + y^3 = 3xy$; $(4^{1/3}, 2^{1/3})$
61. $(x^2 + y^2)^2 = 2xy$; $(1/\sqrt{2}, 1/\sqrt{2})$
62. $(x^2 + y^2)^2 = 2xy$; $(-1/\sqrt{2}, -1/\sqrt{2})$
63. [A] $(x^2 + y^2)^2 = 2xy$; $(3^{1/4}/2^{3/2}, 3^{3/4}/2^{3/2})$
64. [A] $(x^2 + y^2)^2 = 2xy$; $(-3^{1/4}/2^{3/2}, -3^{3/4}/2^{3/2})$

Find explicit expressions for two different functions defined implicitly by each of the following equations, and sketch their graphs:

65. $x^2 + y^2 = 4$
66. $x^2 + y^2 = 4x$
67. $|x| + |y| = 1$
68. $x^4 + y^4 = 1$
69. Show that the derivative of an even function is an odd function.
70. Show that the derivative of an odd function is an even function.
71. Draw the version of Figure 3.25 in which $a = b$.
72. Draw the version of Figure 3.25 in which t and τ are parallel, and derive the result of the inverse function rule in that case.

SECTION 3.9

Chapter Review

Important Concepts

Tangent line	Equation of motion	Reciprocal rule
Secant line	Marginal cost	Quotient rule
Slope of a curve	Marginal revenue	Derivatives of trigonometric functions
Normal line	Marginal profit	Chain rule
Rate of change	Derivative	Generalized power rule
Motion	Differentiable function	Second derivative
Position	Differential operator	Higher-order derivatives
Velocity	Leibniz notation	Increment
Acceleration	Differentiation formulas	Differential
Speed	Power rule	Linear approximation
Rectilinear motion	Scalar product rule	Differential notation
Displacement	Sum rule	Implicit function
Average rate	Difference rule	Implicit differentiation
Instantaneous rate	Product rule	Inverse function rule

Review Exercises

Find the derivative.

1. $y = 4x$
2. $y = -3x^2$
3. $y = 3x^{-2}$
4. $y = 2\pi$
5. $y = \dfrac{2}{x}$
6. $y = \dfrac{x}{2}$
7. $y = x^{3/5}$
8. $y = \frac{4}{5}x^{5/4}$
9. $y = 2x + 5$
10. $y = 1 - x$
11. $y = x^2 + 4x - 6$
12. $y = x^2 + x^{-3}$
13. $y = x\sqrt{x}$
14. $y = x\sqrt{x^2 + 1}$
15. $y = (x + 5)^8$
16. $y = (x^3 + 3x^2 + 4)^9$
17. $y = (1 - x)^{-3}$
18. $y = (1 - x - x^2)^{3/5}$
19. $y = \left(x + \dfrac{1}{x}\right)^{-1}$
20. $y = (x^3 + x^{-3})^{-6}$
21. $y = (x^2 + 4)^5 - (x^2 + 5)^4$
22. $y = (3x + 5)^4(7x + 9)^{-11}$
23. $y = \dfrac{x + 3}{x^2}$
24. $y = \dfrac{x^2 + 4x + 3}{2x}$
25. $y = \dfrac{x^2 + 1}{(x + 3)^4}$
26. $y = \dfrac{(x^2 + 4)^8}{(x + 2)^3}$
27. $y = (x + 4)^5(2x - 1)^{-4}$

28. $y = x^{-4} \sin x$
29. $y = 2 \sin x$
30. $y = \sin 2x$
31. $y = \sin^2 x$
32. $y = \sin x^2$
33. $y = \cot 2x$
34. $y = \cos^4 x$
35. $y = \dfrac{x}{\sin 2x}$
36. $y = \dfrac{\sin 3x}{\tan 2x}$
37. $y = 3|x|$
38. $y = 2\sqrt{x}$
39. $y = |3 + 2x|$
40. $y = |x + \sqrt{x}|$
41. $y = x \sin^2 x$
42. $y = \tan|x|$
43. $y = |\cos x|$
44. $y = \sec x \sec 4x$
45. $x^2 + xy = 9$
46. $x^2y^2 + y^2 = 2x$
47. $x^2 + 2y = x\sqrt{y}$
48. $\sqrt{x^2 + y^2} = x$
49. $\sqrt{x} - 7 = 2y + y^3$
50. $x^{1/3} + y^{2/3} = 2$
51. $\sqrt{x} - \sqrt{y} = 1$
52. $|x| + |y| = 5$
53. $\sin xy = 0.5$
54. $y^3 + y = \sin x$
55. $\sin(x + y) = y^2$
56. $\sin^2 x = y - \cos^2 x$

Find the second derivative.

57. $y = (x^2 + 9)^3$
58. $y = \dfrac{x + 2}{x - 2}$
59. $y = \sqrt{x}$
60. $y = x^{1/3}$
61. $y = \sec 3x$
62. $y = \tan^2 x$
63. $y = \sin^2 x + \sin x + 1$
64. $y = x \sin x$
65. $x^2 + xy = 9$
66. $x^{1/3} + y^{2/3} = 2$
67. $\cos xy = 0$
68. $x = \tan y$

Find a formula for the nth derivative.

69. $y = \dfrac{1}{x + 1}$
70. $y = \dfrac{2}{x - 4}$
71. $y = x^m$
72. $y = \cos 2x$

Find equations for the tangent and normal lines to the given curve at the given point.

73. $y = x^2 + 4$; $(1, 5)$
74. $y = \sin x$; $(\pi/6, \frac{1}{2})$
75. $x^2 - y^2 = 1$; $(2, \sqrt{3})$
76. $y = \dfrac{1}{x^2 + 1}$; $(1, \frac{1}{2})$
77. $y = \sqrt{x + 1}$; $(3, 2)$
78. $y = \tan x$; $(\pi/4, 1)$
79. $y = \sin x - 2 \cos x$; $(\pi/4, -\sqrt{2}/2)$
80. $y = \dfrac{\sin x}{x}$; $(\pi, 0)$
81. $9x^2 + 4y^2 = 36$; $(\sqrt{3}, -\frac{3}{2})$
82. $9x^2 + 4y^2 = 36$; $(0, 3)$
83. $9x^2 - 4y^2 = 36$; $(\sqrt{5}, \frac{3}{2})$
84. $9x^2 - 4y^2 = 36$; $(2, 0)$

Find the velocity and acceleration at time t, and also at the given time.

85. $s = 4t^2$; $t = 2$
86. $s = 500 - 16t^2$; $t = 5$
87. $s = 10 - \dfrac{5}{t}$; $t = 5$
88. $s = t - \sin t$; $t = \pi/2$
89. $s = t \sin t$; $t = \pi$
90. $s = 1 - \cos \pi t$; $t = 2$
91. $s = 4 \cos(3t - 1)$; $t = 0$
92. $s = A \sin(bt - c)$; $t = 0$
93. The side of an equilateral triangle increases by 2%. Use differentials to estimate the change in the perimeter and the change in the area of the triangle.

94. A candy company makes a chocolate bar in the shape of a cylinder. If the cost of ingredients rises 8% and the company decides to make the bar smaller rather than raise its price, use differentials to estimate how the diameter of the candy bar will change if its length remains the same.

95. The area of a certain triangle is given by $A = \frac{1}{2}ac \sin B$. If B increases by 2%, how does the area change?

96. In Problem 95, if B increases by 2% and c decreases by 2% at the same time, what is the effect on A?

Additional Exercises

1. A number c is a **k-fold zero** of a polynomial function p if and only if there exists a polynomial function q such that $q(c) \neq 0$ and

$$p(x) = (x - c)^k q(x)$$

for all x. Verify that $c = 2$ is a 2-fold **(double)** zero of

$$p(x) = x^3 - 3x^2 + 4.$$

2. Verify that $c = 2$ is a 3-fold **(triple)** zero of

$$p(x) = x^4 + x^3 - 2x^2 + 4x - 24.$$

3. Let f and g be polynomial functions. The curves $y = f(x)$ and $y = g(x)$ have a **k-fold intersection** at $x = c$ if and only if c is a k-fold zero of the function $p = f - g$. Verify that the curve $y = x^2 - 4x + 4$ has a double intersection with the x-axis.

4. Verify that the curve $y = x^3$ has a triple intersection with the x-axis.

5. Prove that if the line $y = mx + b$ is tangent to the polynomial curve $y = p(x)$ at $x = c$, then the line and the curve have a double intersection at $x = c$.

6. Prove that if the line $y = mx + b$ and the polynomial curve $y = p(x)$ have a double intersection at $x = c$, then $m = p'(c)$.

7. Show that the scalar product rule for differentiation is a special case of the product rule.

8. An alternate proof of the power rule for positive integer exponents uses the alternative definition of derivative,

$$f'(c) = \lim_{x \to c} \frac{f(x) - f(c)}{x - c},$$

and the factorization

$$x^n - c^n = (x - c)(x^{n-1} + cx^{n-2} + c^2x^{n-3} + \cdots + c^{n-1}).$$

Verify this factorization, and construct the proof of the power rule.

9. Derive the following formula for the second derivative of a composition of functions $y = f(u)$, $u = g(x)$:

$$\frac{d^2y}{dx^2} = \frac{d^2y}{du^2}\left(\frac{du}{dx}\right)^2 + \left(\frac{dy}{du}\right)\frac{d^2u}{dx^2}.$$

10. Derive the formula for the second derivative of a product:

$$(fg)'' = f''g + 2f'g' + fg''.$$

11. Derive the formula for the nth derivative of a sum:

$$\frac{d^n(u+v)}{dx^n} = \frac{d^n u}{dx^n} + \frac{d^n v}{dx^n}.$$

12. Derive the "binomial" formula for the nth derivative of a product:

$$(fg)^{(n)} = f^{(n)}g + nf^{(n-1)}g' + \frac{n(n-1)}{2}f^{(n-2)}g''$$
$$+ \frac{n(n-1)(n-2)}{3!}f^{(n-3)}g^{(3)} + \cdots + nf'g^{(n-1)} + fg^{(n)}.$$

13. Derive the alternative form of the inverse function rule:

$$\frac{dx}{dy} = \frac{1}{dy/dx}.$$

Find the points at which the following curves have vertical tangent lines:

14. $x = y^3$ 15. $x = y^2 - 2y + 1$ 16. $x^2 + y^2 = 2x$ 17. $2x^2 + 3y^2 = 6$

The one-sided limits

$$f'(c^+) = \lim_{x\to c^+} \frac{f(x) - f(c)}{x - c} \quad \text{and} \quad f'(c^-) = \lim_{x\to c^-} \frac{f(x) - f(c)}{x - c}$$

are called ***one-sided derivatives*** *of f at c; $f'(c^+)$ is called the* ***right-hand derivative,*** *and $f'(c^-)$ is called the* ***left-hand derivative.*** *Find the one-sided derivatives $f'(2^+)$ and $f'(2^-)$ for each of the following functions:*

18. $f(x) = 4 - x^2$ 19. $f(x) = \sqrt{4 - x^2}$ 20. $f(x) = |4 - x^2|$

21. $f(x) = (x - 2)^{2/5}$ 22. $f(x) = |\sin \pi x|$ 23. $f(x) = |\sin x|$

24. $f(x) = \begin{cases} x & \text{if } x < 2 \\ 4 - x & \text{if } x \geq 2 \end{cases}$

25. Prove that the function f is differentiable at c if and only if both of the one-sided derivatives of f at c exist and have the same value.

26. Prove that if $f'(c^+)$ exists, then f is right-continuous at c.

A curve $y = f(x)$ has a ***corner*** *at $x = c$ if and only if f is continuous at c and either (1) both $f'(c^+)$ and $f'(c^-)$ exist and are finite, but $f'(c^+)$ and $f'(c^-)$ have different values, or (2) one of $f'(c^+)$ and $f'(c^-)$ is finite and the other is infinite. Find the corners on each of the following curves:*

27. $y = |x - c|$ 28. $y = |x^2 - a^2|$ 29. $y = |x^3|$ 30. $y = |x^{1/3}|$

31. $y = 1 - |1 - x^2|$ 32. $y = \left|1 - \frac{1}{x}\right|$ 33. $y = |\sin x|$ 34. $y = |\cos \pi x|$

A curve $y = f(x)$ has a ***cusp*** *at $x = c$ if and only if f is continuous at c and either $f'(c^+) = \infty$ and $f'(c^-) = -\infty$, or $f'(c^+) = -\infty$ and $f'(c^-) = \infty$. Find the cusps of the following curves:*

35. $y = x^{2/3}$ 36. $y = (1 - x^{2/3})^{3/2}$ 37. $y = |x^{1/3}|$ 38. $y = (x - 3)^{2/5}$

Find an equation of the tangent line to the given curve at the given point.

39. $x^2 + y^2 = 4$; $(-1, \sqrt{3})$

40. $3(x^2 + y^2)^2 = 25(x^2 - y^2)$; $(2, 1)$

41. $(x^2 + 8y - 16)^2 = y^2(16 - x^2)$; $(0, 4)$

42. $x^{2/3} + y^{2/3} = 1$; $(-\frac{1}{8}, \frac{3}{8}\sqrt{3})$

43. Let f be a polynomial function. Prove that if both $f(c) = 0$ and $f'(c) = 0$, then c is a double zero of f.

44. State and prove a similar result to Problem 43 for a triple zero of f.

45. Prove that if c is such that both $f'(c^+) = \infty$ and $f'(c^-) = \infty$ [or both $f'(c^+) = -\infty$ and $f'(c^-) = -\infty$], then f has a vertical tangent line at $x = c$.

CHAPTER 4

Applications of Differentiation

INTRODUCTION

Having developed methods for finding derivatives, we are now ready to use derivatives in solving a wide variety of "real-world" problems, as well as some theoretical problems. In this chapter, we shall do such things as compute the relative velocity of two ships sailing on different courses; prove that a moving object has a velocity at some point equal to its average velocity; apply the first and second derivatives to the graphing of functions; determine the relative dimensions of the most economical tin can; find out how many trees to plant in an orchard to obtain maximum yield; determine the position of a projectile at any time from its initial position and velocity; and find out how long it takes to charge a battery. These are only a small sample of the types of problems that can be solved using differentiation. Most of these cannot be solved without using differentiation.

Perhaps it is appropriate here to discuss the philosophy of applying mathematics to the real world. A physical situation is usually much more complicated than we are able to consider. Fortunately, it is usually possible to ignore many features of the physical situation, however, because the things we are interested in depend on only a few facts about the situation. Therefore, rather than apply mathematical ideas directly to a real situation, we construct a **mathematical model** of the physical situation. This is done by abstracting and idealizing the actual situation, hopefully retaining only the features of reality that are pertinent to the problem at hand. Then mathematical methods are applied to the mathematical model, yielding a mathematical solution. This solution is then translated back into the language of the real situation, to obtain a real-world prediction. Finally, the "fit" of the model is gauged by comparing the prediction with what is actually observed. If the fit is good enough, we are satisfied with the modeling process. If the fit is not good enough, we try again with another model. This process is summarized schematically in the chart on the next page.

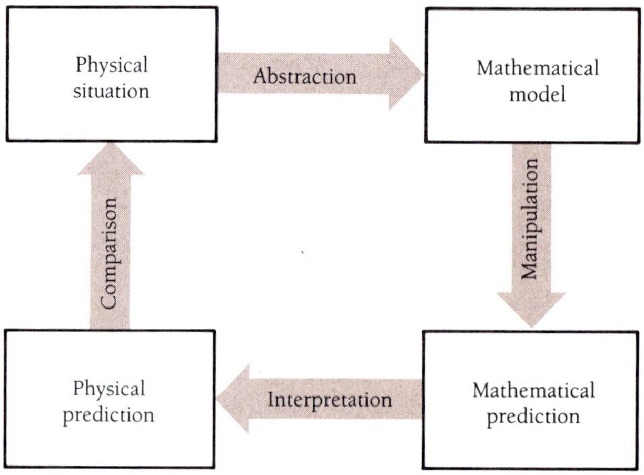

THE PROCESS OF MATHEMATICAL MODELING

With the applications found in this chapter, we begin to see the power of the calculus in solving problems.

SECTION 4.1

Related Rates

In many real-life settings, quantities change in time, or are functions of time. Often, two or more such functions are related by an equation. If the rates of change of all but one of them are known, it is usually possible to determine the rate of change of the remaining function. Problems asking for the rate of change of one function, when the rates of change of related functions are given, are called **related rates** problems.

Consider the following example.

■ **Example 4.1** A man 6 feet tall walks directly away from a lamp post whose light is situated 20 feet above the ground. If the man walks at the rate of 7 feet per second, at what rate does his shadow grow?

Solution We have represented in Figure 4.1 the situation described; the light at L casts a shadow FS of the man FH. Let x be the distance from the lamp post to the man, and let s be the length of the shadow. We need to find the rate at which the shadow grows, which is ds/dt, and we are given that $PL = 20$, $FH = 6$, and $dx/dt = 7$.

We first find out how s and x are related. Since the triangles SPL and SFH are similar, we have

$$\frac{SF}{SP} = \frac{FH}{PL},$$

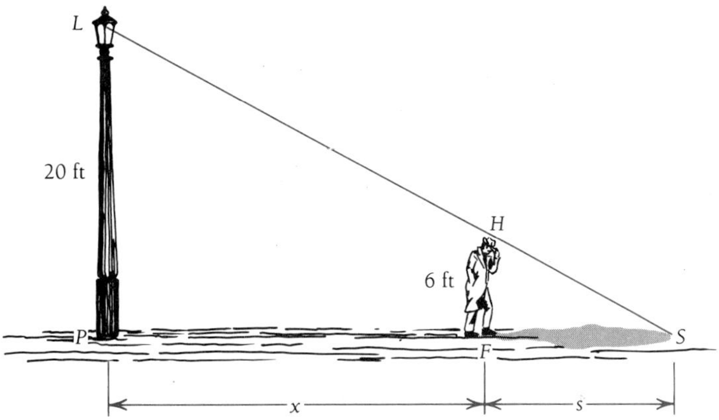

FIGURE 4.1 A lamp post casts a shadow

or

$$\frac{s}{x+s} = \frac{6}{20}.$$

Solving for s, we get

$$s = \frac{3x}{7}.$$

Now we differentiate this equation implicitly with respect to time t, getting

$$\frac{ds}{dt} = \frac{3}{7}\frac{dx}{dt}.$$

Substituting 7 feet per second for dx/dt, we get

$$\frac{ds}{dt} = 3 \text{ feet per second.}$$

This is the rate at which the shadow grows. ■

To solve related rates problems such as this, it is necessary to identify the varying quantities and then find an equation (model) that relates them. Strategies for finding an equation include considering similar triangles, the Pythagorean theorem, and formulas for area, volume, etc. Drawing a picture often helps. Then differentiating both sides of the equation found (usually implicitly) will yield an equation relating the rates of change of the varying quantities. Finally, the known values of variables and rates are substituted into the equation to obtain the unknown rate. Some of these known values may be given, and others may have to be computed.

A caution or two: Make sure that varying quantities are represented by variables, not constants. And make sure that differentiation takes place *before* particular values of the variables are substituted in the equation.

To summarize:

1. Draw a sketch of the general situation, if possible.
2. Represent each changing quantity by a variable.
3. Find an equation relating the variables.
4. Differentiate the equation with respect to time t.
5. Substitute values for the variables and rates given.
6. Solve for the remaining unknown rate. Check to see if your answer fits the problem.

The following examples illustrate these steps.

Example 4.2 A certain water tank is an inverted cone with height 12 feet and top diameter 8 feet (see Figure 4.2). Water is being pumped into the tank at the rate of 50 cubic feet per minute. How fast is the water level rising when the water in the tank is 6 feet deep?

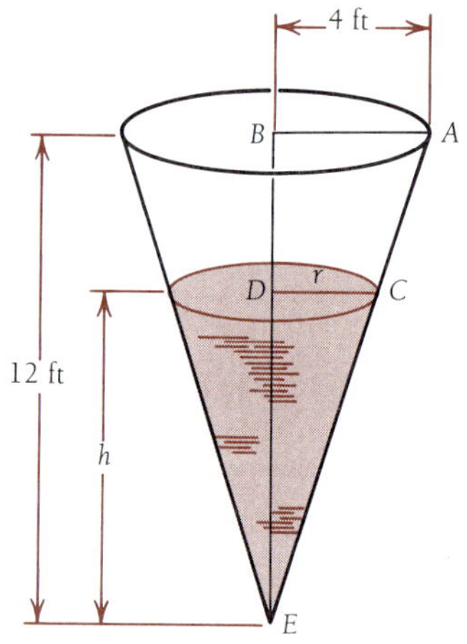

FIGURE 4.2 A conical water tank

Solution Since the depth of the water changes, we let the variable h represent its depth. Since the radius of the "cone of water" in the tank changes, we will denote the radius by r. The problem asks us to find $(dh/dt)|_{h=6}$. We are given, in addition to the dimensions of the tank, the rate at which water enters the tank. That is, we are given the rate of change of the volume V of water in the tank. Thus, we seek an equation that relates the depth h to the volume V. One obvious equation is the formula for the volume of water in the tank, namely

$$V = \frac{1}{3}\pi r^2 h. \tag{1}$$

Another equation, which relates r and h to the dimensions of the tank, is obtained from the similar triangles ABE and CDE, namely

$$\frac{r}{h} = \frac{4}{12},$$

or

$$r = \frac{h}{3}.$$

Putting this expression for r into equation (1), we get

$$V = \frac{\pi h^3}{27}.$$

Now we differentiate this equation with respect to t, getting

$$\frac{dV}{dt} = \left(\frac{\pi}{27}\right) 3h^2 \frac{dh}{dt}.$$

Finally, we substitute the values given corresponding to $h = 6$, and solve for $(dh/dt)|_{h=6}$. We get

$$50 = \left(\frac{\pi}{9}\right) (6)^2 \left.\frac{dh}{dt}\right|_{h=6},$$

so that

$$\left.\frac{dh}{dt}\right|_{h=6} = \frac{25}{2\pi} \text{ feet per minute.}$$

This is the rate at which the water level is rising at that moment. ■

■ **Example 4.3** Air is being pumped into a spherical balloon at the rate of 2 cubic feet per minute. How is the radius changing at the instant that the circumference of the balloon is 8 feet?

Solution Let r be the radius of the balloon and let C be the circumference. We must find $(dr/dt)|_{C=8}$. We are given the rate at which air enters the balloon, so we know how the volume changes. The formula

$$V = \frac{4}{3}\pi r^3$$

relates the radius and volume. Differentiating, we get

$$\frac{dV}{dt} = 4\pi r^2 \frac{dr}{dt},$$

which relates the rates of change. Solving for dr/dt, we get

$$\frac{dr}{dt} = \frac{1}{4\pi r^2}\frac{dV}{dt}.$$

Finally, we evaluate this expression at the instant that $C = 8$. Since $C = 2\pi r$, we have $r = C/2\pi$, or $r = 4/\pi$ when $C = 8$. Hence,

$$\left.\frac{dr}{dt}\right|_{C=8} = \frac{2}{4\pi(4/\pi)^2} = \frac{\pi}{32} \approx 0.1 \text{ foot per minute.}$$ ■

■ **Example 4.4** The beacon in a lighthouse revolves 6 times per minute. A straight road passes the lighthouse, with its nearest point P located $\frac{1}{2}$ mile from the beacon light (see Figure 4.3). At a point 1 mile down the road from P, at what rate does the ray from the beacon move along the road?

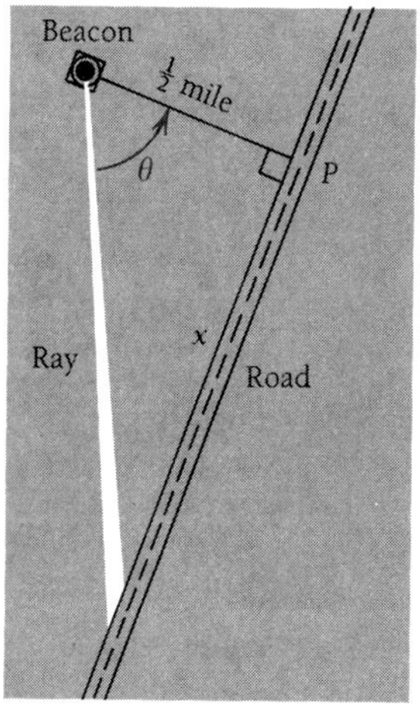

FIGURE 4.3 A beacon ray moves along a road

Solution This problem requires us to relate the motion of the light beam along the road to the rate of rotation of the beacon. Let x be the distance between the point P and the point at which the ray meets the road. Let θ be the angle between the ray and the line from the lighthouse to P. Then by right-triangle properties,

$$x = \frac{1}{2}\tan\theta.$$

Differentiating with respect to t, we get

$$\frac{dx}{dt} = \frac{1}{2}\sec^2\theta\,\frac{d\theta}{dt}.$$

Now we put in the given values. When $x = 1$, then $\tan\theta = 2$, so $\sec^2\theta = 1 + \tan^2\theta = 5$. Also,

$$\frac{d\theta}{dt} = 6 \text{ revolutions per minute} = 12\pi \text{ radians per minute.}$$

(We must use radians, since trigonometric functions are involved.) Hence,

$$\begin{aligned}\left.\frac{dx}{dt}\right|_{x=1} &= \frac{1}{2}(5)(12\pi) \\ &= 30\pi \text{ miles per minute} \\ &= 1800\pi \text{ miles per hour.}\end{aligned}$$

■

Exercises 4.1

1. How is the area of the square changing when the side s is 4 feet in length and growing at the rate of 2 inches per minute?

2. How is the area of a circle changing when the radius is 10 centimeters and decreasing at the rate of 3 millimeters per second?
3. The base b of the triangle is growing at the rate of 2 meters per hour, and the altitude h is decreasing at the rate of $\frac{1}{2}$ meter per hour. At one instant, the base and altitude are both 5 meters. How is the area changing at that instant?

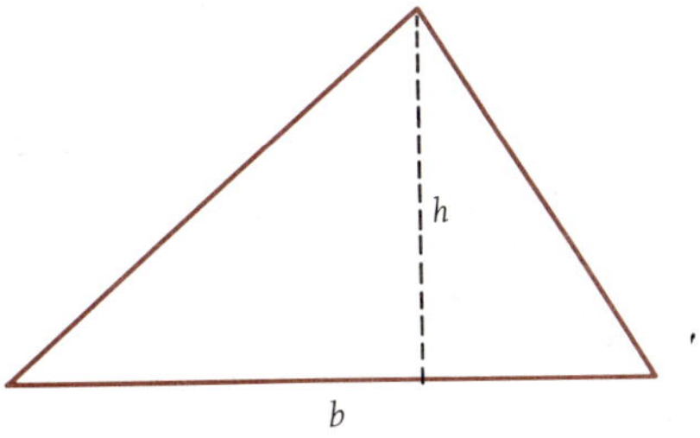

4. The side of an equilateral triangle decreases at the rate of 1 inch per second. When the area is 100 square inches, how is the area changing?
5. A cylindrical barrel, standing vertically, is being filled with water. It is noted that the water level rises at the rate of 3 inches per minute. If the barrel is 2 feet in diameter, at what rate is water flowing into the barrel?

6. Sand is dumped off a conveyor belt into a pile at the rate of 15 cubic feet per minute. The sand pile is shaped like a cone whose height and base diameter are always equal (see the figure). At what rate is the height of the pile growing when the pile is 10 feet high?

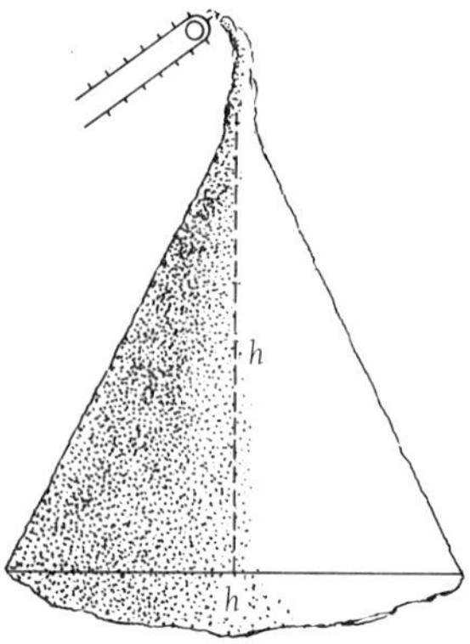

7. A kite is flying at an angle of elevation of 30° (see the figure). The kite string is being taken in at the rate of 1 foot per second. If the angle of elevation does not change, how fast is the kite losing altitude?

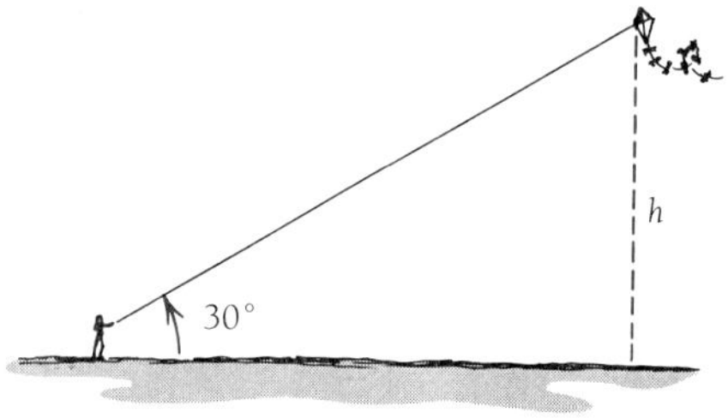

8. A plane is flying along the flight path shown in the figure at the rate of 300 miles per hour. How fast is the plane losing altitude?

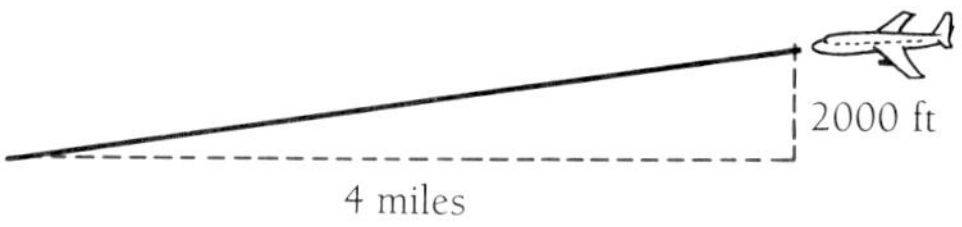

9. A skier goes over the ramp shown in the figure at the rate of 15 meters per second. As he leaves the ramp, how fast is he rising?

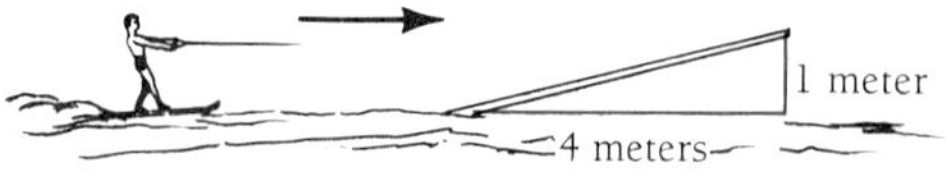

10. A ladder 30 feet long leans against a vertical wall. The foot of the ladder is sliding away from the wall at the rate of 2 feet per second. How fast is the top of the ladder sliding down the wall when the foot of the ladder is 10 feet from the wall?

11. A 12 foot ladder leans against a vertical wall, and a person stands on the rung at the middle of the ladder. The foot of the ladder is being pulled away from the wall at the rate of 6 inches per second. How fast is the person being lowered when the foot of the ladder is 4 feet from the wall?

12. A parachutist descends at the rate of 3 meters per second, and a jeep is driving toward the landing spot at the rate of 9 meters per second. When the parachutist is 60 meters from the ground, the jeep is 80 meters from the landing spot. At what rate is the distance between them changing at that instant?

13. A truck crosses an intersection and continues northward at 20 miles per hour. Ten seconds later, a car crosses the intersection and continues eastward at 25 miles per hour. At what rate is the distance between the car and the truck changing after another 60 seconds?

14. A glider is being towed by a cable attached to a power winch. The glider is flying at a constant altitude of 50 feet, and the cable is being taken in at the rate of 80 feet per second. How fast is the glider moving when there are 300 feet of cable out?

15. Two ships sail from the same port at the same time. Ship A travels east at 12 knots, and ship B travels northeast at 8 knots. One hour after sailing, how is the distance between the ships changing? [*Hint:* Use the law of cosines; 1 knot = 1 nautical mile per hour; 1 nautical mile = 6080.2 feet.]

16. Two ships leave port at the same time, ship A sailing southeast at 10 knots and ship B sailing north at 12 knots. After 2 hours, how is the distance between them changing?

17. In Problem 14, at what rate is the angle of elevation of the glider from the winch changing?

18. In Quito on March 21, the sun is visible for exactly 12 hours and at noon is directly overhead. A certain building is 100 meters high. At what rate is the length of its shadow changing at 4:00 PM?

19. An object is being pulled by a rope through a pulley as shown in the figure. If the rope is being pulled at the rate of 10 feet per minute, how fast is the object moving when $x = 12$ feet?

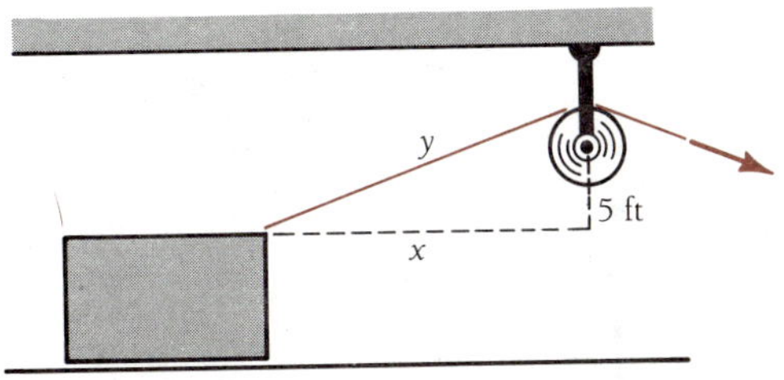

20. Rework Problem 19 for the pulley system shown at the top of the next page. Assume that the pulleys are negligible in size. [*Hint:* If y is the length of rope between pulleys, then the distance between pulleys is $y/2$.]

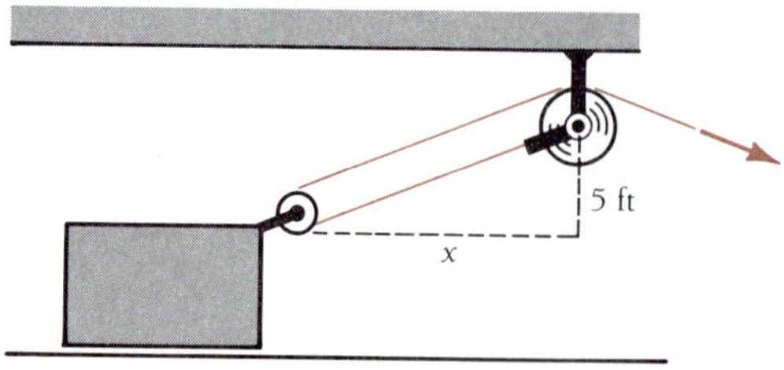

21. If an object is being lifted by the pulley system in this figure, and the rope is pulled at the rate of 12 feet per minute, how fast is the object rising?

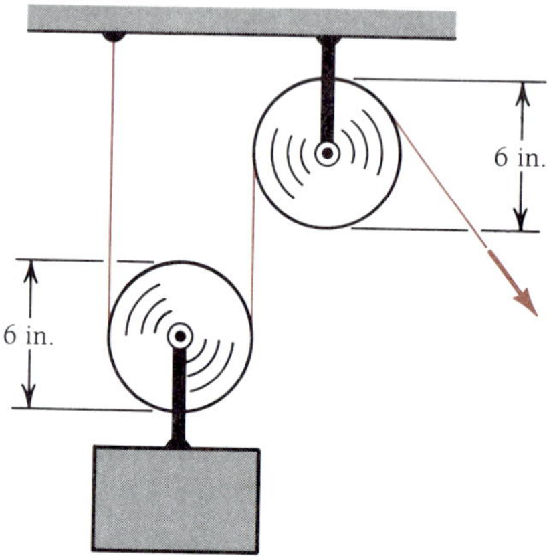

22. Rework Problem 21 for the pulley system shown in this figure.

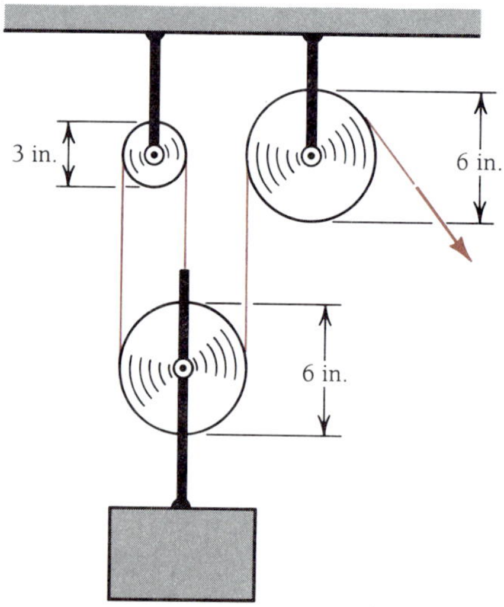

23. A gas has temperature T, pressure P, and volume V, related by the law $PV = kT$, where k is a constant. If V is decreasing at the rate of 12 cubic centimeters per second, and T is increasing at the rate of 3° per second, how is the pressure changing when $T = 120°\text{C}$ and $V = 200$ cubic centimeters?

24. According to the theory of relativity, the mass M of a moving object is

$$M = \frac{mc}{\sqrt{c^2 - v^2}},$$

where m is the rest mass, c is the speed of light, and v is the speed of the object. What is the rate of change of the mass M if the velocity of an object is $0.4c$ and the acceleration is $-0.02c$?

25. For a certain manufacturer, the cost of producing x refrigerators per month is

$$C(x) = 200x - 0.006x^2 + 14{,}000$$

dollars. The revenue from the sale of x refrigerators per month is

$$R(x) = 350x - 0.02x^2 - 10{,}000$$

dollars. If sales keep pace with production and production is increasing at the rate of 20 refrigerators per month, how are profits changing when the production level reaches 5000 per month?

26. Three competing species of antelope in the same grazing area have populations x, y, and z that satisfy the relationship

$$3x + 5y + 30z = 10{,}000.$$

If x is increasing at the rate of 20 per year and y is decreasing at the rate of 15 per year, how is z changing?

27. In a certain habitat, coyotes and rabbits are the principal animals. Because of their predator–prey relationship, the numbers c of coyotes and r of rabbits satisfy the condition

$$500(c - 200)^2 + (r - 10{,}000)^2 = 300{,}000.$$

It is estimated that there are about 10,500 rabbits and 210 coyotes, and the number of rabbits is decreasing at the rate of 50 per month. How is the number of coyotes changing?

28. A certain bacterial culture creates a circular colony in a Petri dish, with a density of about 8000 per square centimeter. A colony that was 0.8 centimeter in diameter at 8:00 AM is found to be 1.4 centimeters in diameter at 11:00 AM. Assuming that the number of bacteria increases at a constant rate, how big will the colony be at 4:00 PM?

29. The circumference of a spherical balloon expands at the rate of 6 inches per second. When the balloon is 2 feet in diameter, at what rate is air being pumped into the balloon?

30. A spherical snowball loses volume in melting at a rate proportional to the surface area. Show that the radius decreases at a constant rate.

31. Tax revenues are proportional to the product of the number of workers and the average salary. If the average salary is increasing at the rate of 1.2% per month and the number of workers is decreasing at the rate of 0.3% per month, how are tax revenues changing?

32. An airliner flying due east at 650 miles per hour at an altitude of 29,000 feet passes directly over a cargo plane flying southwest at 480 miles per hour at an altitude of 27,500 feet. At the instant the planes are vertically aligned, how is the distance between them changing?

33. For the planes in Problem 32, how is the distance between them changing 1 second after they are vertically aligned?

SECTION 4.2

The Mean Value Theorem

The mean value theorem, or "theorem of the mean," is an important theoretical result. In addition to being extremely useful in the development of the theory of the calculus, it has some interesting physical applications.

Before we can state the mean value theorem, we need to discuss some preliminary ideas.

Extreme Values

Let f be a function defined on the closed interval $[a, b]$. We say M is the (**global** or **absolute**) **maximum** of f on $[a, b]$ if and only if for some $c \in [a, b]$, $f(c) = M$, and for all $x \in [a, b]$, $f(x) \leq M$. If $f(c) = M$, we say f has its maximum **at** c. Similarly, a number m is the (**global** or **absolute**) **minimum** of f on $[a, b]$ if and only if for some $c \in [a, b]$, $f(c) = m$, and for all $x \in [a, b]$, $f(x) \geq m$. An **extremum** of f is the maximum or the minimum. These ideas are illustrated in Figure 4.4. (See also the extreme value theorem, Theorem 2.21.)

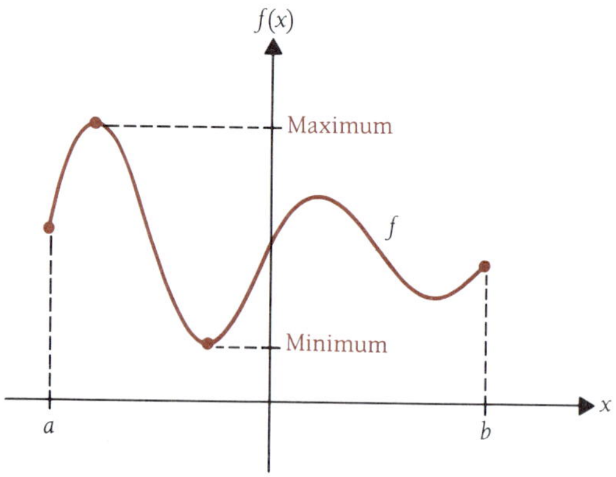

FIGURE 4.4 The extreme value theorem

Critical Points

If f is a function defined in an open interval containing the point c, we call c a **critical point** (or **critical number**) of f if and only if either $f'(c) = 0$ or $f'(c)$ does not exist. Thus, for example, a point of discontinuity in the domain of f is a critical point, since the derivative cannot exist there.

Example 4.5 Find the critical points of the following functions:

(a) $f(x) = \dfrac{1}{x^2 + 1}$ (b) $g(x) = x^3 - 3x$

(c) $F(x) = \begin{cases} 2 + (x-1)^{1/3} & \text{if } x \geq 0 \\ 1 - x & \text{if } x < 0 \end{cases}$

Solution (a) Noting that f is defined and continuous on all of $\mathbb{R}$, we compute the derivative:

$$f'(x) = \frac{-2x}{(x^2+1)^2}.$$

Since f' is defined for all x and is 0 only when $x = 0$, we see that 0 is the only critical point of f (see Figure 4.5).

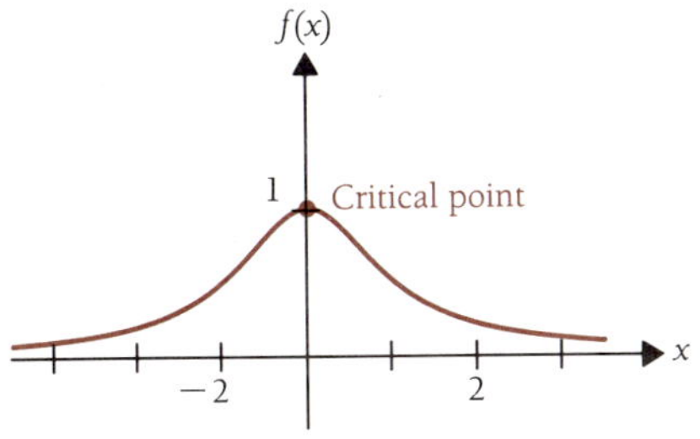

FIGURE 4.5 The graph of $f(x) = \dfrac{1}{x^2+1}$

(b) Since g is a polynomial, it is defined, continuous, and differentiable for all x. That means that if g has a critical point, if must be where the derivative is 0. The derivative is

$$g'(x) = 3x^2 - 3 = 3(x+1)(x-1),$$

which is 0 at $x = 1$ and at $x = -1$. The critical points of g are therefore 1 and -1 (see Figure 4.6, page 218).

(c) We note that F is continuous for all x (see Figure 4.7), and its derivative is

$$F'(x) = \begin{cases} \frac{1}{3}(x-1)^{-2/3} & \text{if } x > 0 \\ -1 & \text{if } x < 0. \end{cases}$$

The derivative is never 0, but it is undefined at 0 and at 1. Thus, 0 and 1 are the critical points of F.

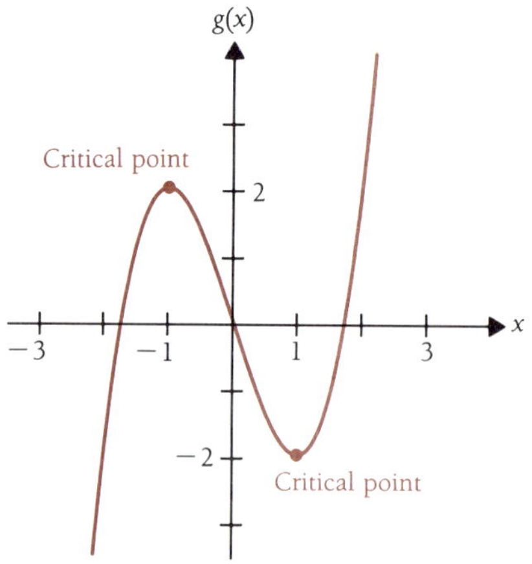

FIGURE 4.6 The graph of $g(x) = x^3 - 3x$

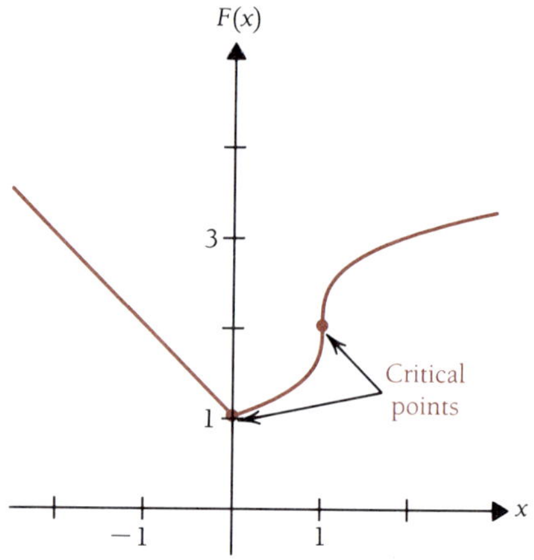

FIGURE 4.7 The graph of $F(x) = \begin{cases} 2 + (x-1)^{1/3} & \text{if } x \geq 0 \\ 1 - x & \text{if } x < 0 \end{cases}$ ■

The next theorem explains why critical points are useful.

THEOREM 4.1 If the function f has an extremum at c, then either c is an endpoint of the domain of f or c is a critical point of f.

Proof Let f have an extremum at c, and suppose c is not an endpoint of the domain of f. Either f has a derivative at c or it does not. If not, then c is a critical point of f. If $f'(c)$ does exist, suppose that f has its maximum at c (the case of the minimum is similar). Recall that

$$f'(c) = \lim_{x \to c} \frac{f(x) - f(c)}{x - c}.$$

If $f'(c) > 0$, as in Figure 4.8, then we can choose x close enough to c so that

$$\frac{f(x) - f(c)}{x - c} > 0.$$

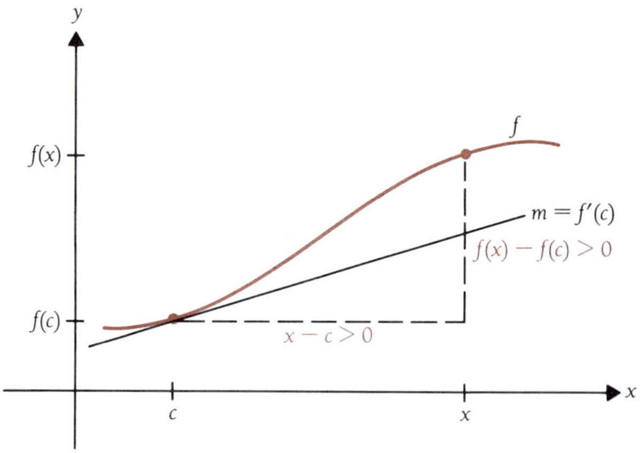

FIGURE 4.8 Theorem 4.1

We can also choose $x > c$ at the same time. This means $x - c > 0$, which forces $f(x) - f(c) > 0$. But then $f(x) > f(c)$, contradicting the fact that f has a maximum at c. Hence, it cannot be that $f'(c) > 0$. In a similar manner, we can show that $f'(c) < 0$ is not possible either. Therefore, $f'(c) = 0$, and c is a critical point of f.

A little later, we will use this theorem directly for finding extrema, but for now we will use it to prove another theorem.

Rolle's Theorem

The next theorem is a special case of the mean value theorem, due to the French mathematician Michel Rolle.

THEOREM 4.2 **Rolle's Theorem** If the function f is continuous on the closed interval $[a, b]$ and differentiable on the open interval (a, b), and $f(a) = f(b) = 0$, then there is at least one point $c \in (a, b)$ such that $f'(c) = 0$.

Proof Either f is constant on $[a, b]$ or it is not. If so, then $f'(x) = 0$ for all $x \in (a, b)$, and the theorem holds for any choice of $c \in (a, b)$. If f is not constant on $[a, b]$, then either f is positive at some point or it is not.

If f is positive at some point, then the maximum of f on $[a, b]$ (which exists by the extreme value theorem) must be positive. The maximum cannot occur at a or at b, so occurs at some point $c \in (a, b)$. Then by Theorem 4.1, c is a critical point of f. Since f is differentiable in (a, b), $f'(c)$ exists, so $f'(c) = 0$.

If f is not positive at any point, but is not constant on $[a, b]$, then f must be negative at some point. It follows that the minimum of f must occur at some point $c \in (a, b)$, and $f'(c) = 0$ as before.

Rolle's theorem says that if a differentiable function has value 0 at points a and b, then somewhere between a and b the graph must have a horizontal tangent line (see Figure 4.9). There could, of course, be several such points between a and b. If the function fails to be continuous on the closed interval or is not differentiable in the entire open interval, then no horizontal tangent line need exist (see Figures 4.10 and 4.11).

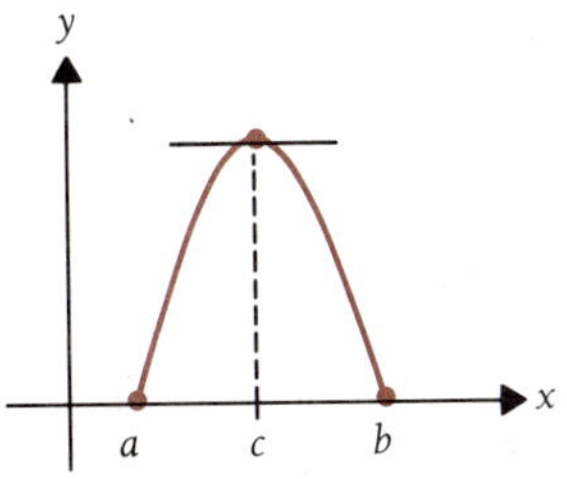

FIGURE 4.9
Rolle's theorem

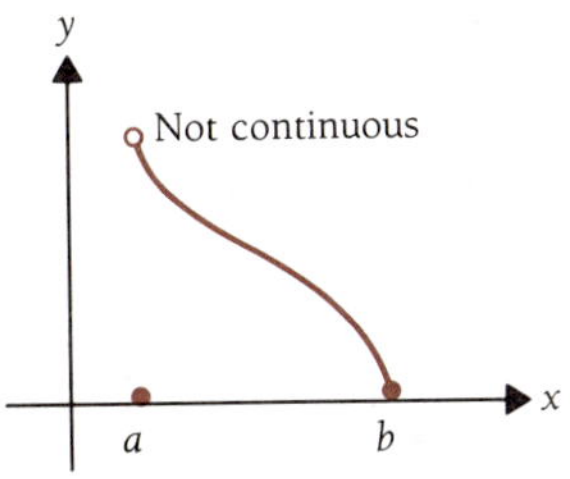

FIGURE 4.10
Rolle's theorem does not hold

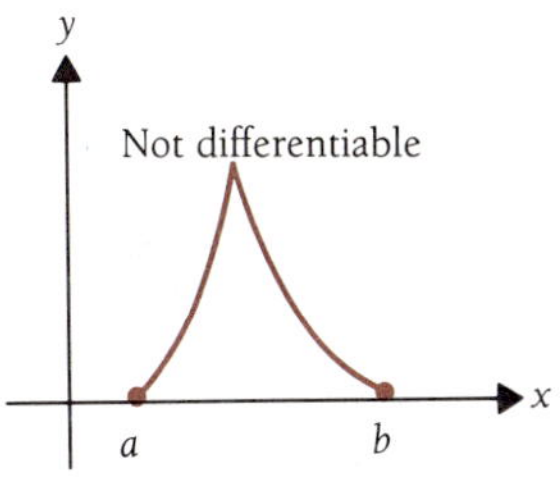

FIGURE 4.11
Rolle's theorem does not hold

Michel Rolle (1652–1719) was a mathematician of the French Académie des Sciences. He raised serious questions about the validity of calculus, describing it as a collection of ingenious fallacies. His famous theorem was stated incidentally in a little-known book on algebra and geometry, published in 1691. In a more famous book, *Traité d'algèbre,* that appeared in 1690, Rolle was the first to state that a nonzero number has n distinct nth roots.

Example 4.6 Verify that the function

$$f(x) = x^2 - 5x + 4$$

satisfies Rolle's theorem on [1, 4].

Solution Since $f(1) = f(4) = 0$ and f is continuous and differentiable, f satisfies the hypotheses of Rolle's theorem. The derivative, $f'(x) = 2x - 5$, is 0 at $x = 2.5$, which is between 1 and 4, so the conclusion of Rolle's theorem also holds (see Figure 4.12).

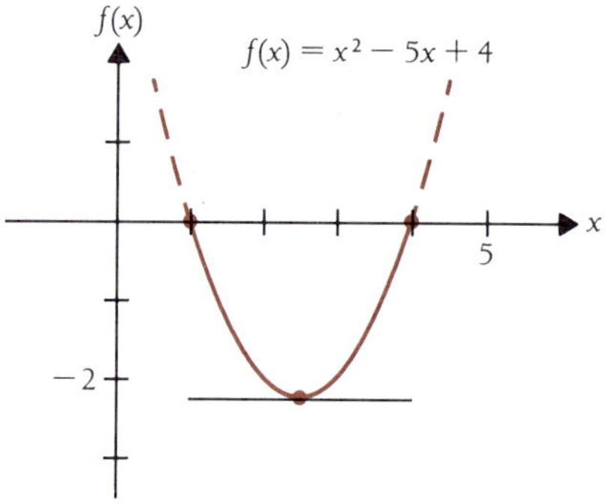

FIGURE 4.12 Satisfies Rolle's theorem ■

Example 4.7 Explain why the function

$$f(x) = 3 - |x - 2|$$

(Figure 4.13) does not satisfy Rolle's theorem on [−1, 5].

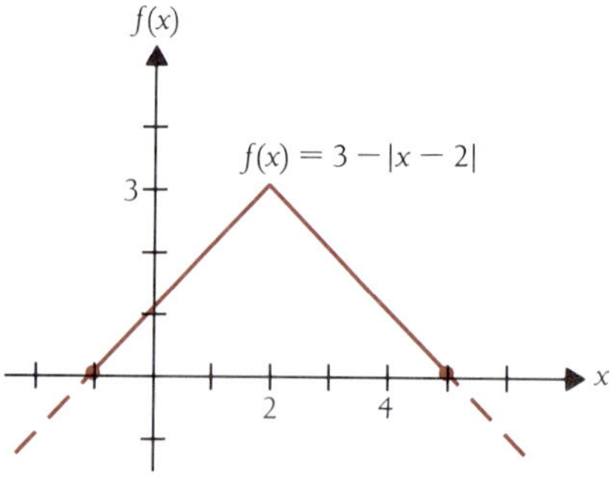

FIGURE 4.13 Does not satisfy Rolle's theorem

Solution While $f(-1) = f(5) = 0$ and f is continuous on the interval $[-1, 5]$, f fails to be differentiable at $x = 2$. Thus, all the hypotheses of Rolle's theorem are not satisfied. The conclusion of Rolle's theorem also fails, since the graph has no horizontal tangent line. ■

■ **Example 4.8** An automobile trip begins and ends with the car at rest. Assuming the velocity is a differentiable function, explain why the acceleration of the car is 0 at some point.

Solution If the velocity is differentiable, then the velocity function satisfies the hypotheses of Rolle's theorem. Hence, there is a time during the trip when the derivative of the velocity, namely the acceleration, is 0. ■

The Mean Value Theorem

We are at last ready to state and prove the main theorem of this section.

THEOREM 4.3 **The Mean Value Theorem** If the function f is continuous on the closed interval $[a, b]$ and differentiable on the open interval (a, b), then there is at least one point $c \in (a, b)$ such that

$$f'(c) = \frac{f(b) - f(a)}{b - a}.$$

Proof The expression

$$\frac{f(b) - f(a)}{b - a}$$

is the slope of the line through the points $(a, f(a))$ and $(b, f(b))$ (see Figure 4.14). We call this line the **secant line** of f on $[a, b]$. Hence, we seek a point c such that the tangent line to the graph of f at c is parallel to the secant line of f on $[a, b]$.

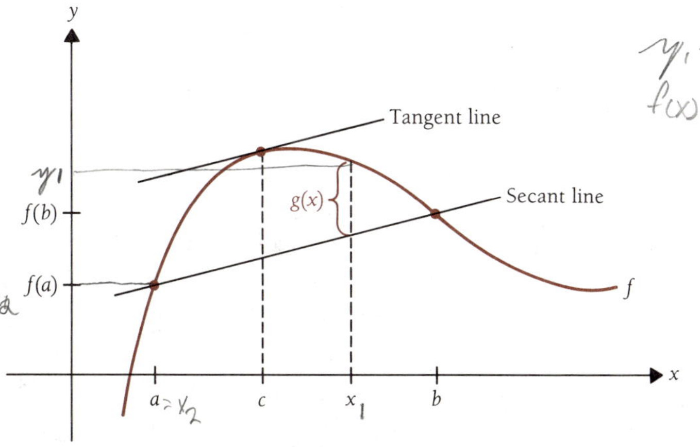

FIGURE 4.14 The mean value theorem

Let x be any point in $[a, b]$. The vertical distance from the secant line to the graph of f at x is

$$g(x) = f(x) - f(a) - \frac{f(b) - f(a)}{b - a}(x - a).$$

This can be verified by finding the equation of the secant line (you will be asked to do so in the exercises). Now, since f is continuous on $[a, b]$, and g is just a combination of f with a polynomial, g is also continuous on $[a, b]$. Since f is differentiable in (a, b), so is g; in fact,

$$g'(x) = f'(x) - \frac{f(b) - f(a)}{b - a}.$$

Moreover, as you can verify, $g(a) = g(b) = 0$. Thus, g satisfies the hypotheses of Rolle's theorem. Hence, there is at least one point $c \in (a, b)$ such that $g'(c) = 0$. This means that

$$f'(c) - \frac{f(b) - f(a)}{b - a} = 0,$$

and the conclusion of the theorem follows.

Thus, the mean value theorem says that a differentiable function will have a tangent line parallel to the secant line on any closed interval. In addition to this graphic interpretation, there is also a simple physical interpretation for the mean value theorem in terms of velocity.

■ **Example 4.9**

Explain why at some point during an automobile trip, the actual speed of the car is the same as the average speed for the entire trip.

Solution

Let $f(t)$ be the distance traveled by the car up to time t. Let the trip last through the time interval $[a, b]$. The difference $f(b) - f(a)$ is the total distance traveled by the car, and the difference $b - a$ is the time elapsed. Hence, the quotient

$$\frac{f(b) - f(a)}{b - a}$$

is the average speed of the car. Assuming that f is continuous and differentiable, which are safe assumptions in this case, the mean value theorem says that there must be a point $c \in (a, b)$ at which the derivative $f'(c)$ is equal to the above quotient. But $f'(c)$ is the speed of the car, and the quotient is the average speed; thus, at $t = c$, the speed of the car is the average speed for the trip. ■

■ **Example 4.10**

The position of a ball thrown into the air is $s(t) = 88t - 16t^2$ during the interval $[0, 4]$. Find the instant at which the velocity of the ball is equal to its average velocity over $[0, 4]$.

Solution The average velocity of the ball is

$$\frac{s(4) - s(0)}{4 - 0} = 24,$$

and the velocity at time t is

$$s'(t) = 88 - 32t.$$

Hence, we set

$$88 - 32t = 24$$

and solve for t. The solution is $t = 2$, which lies in the interval $[0, 4]$, and the problem is solved. ■

An Application: Newton's Method

Another application of the mean value theorem gives another method of estimating a root of an equation (or equivalently, the zero of a function). It is called **Newton's method**, and uses the tangent line to a curve to estimate where the curve crosses the x-axis.

If z is a zero of the function f, then z is a root of the equation $f(x) = 0$, and the curve $y = f(x)$ crosses the x-axis at z, as in Figure 4.15. If x_1 is an approximation to z, let t be the tangent line to the curve at $(x_1, f(x_1))$. If t crosses the x-axis at x_2, then x_2 will be closer to z than x_1 was (if all goes well). By iterating (repeating) this procedure, we can (hopefully) get arbitrarily close to z.

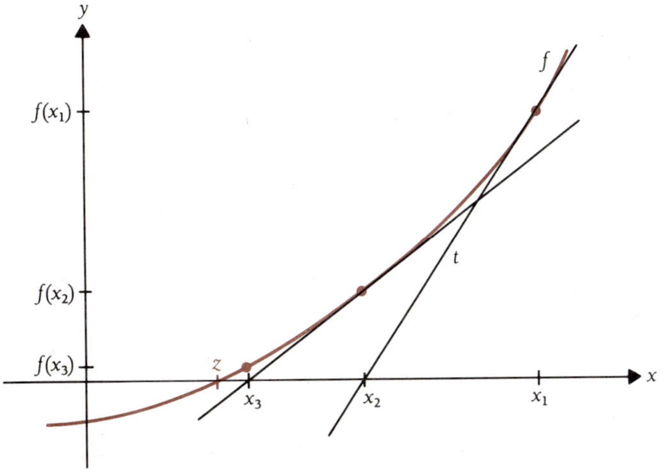

FIGURE 4.15 Newton's method is tangent approximation

The line t has slope $f'(x_1)$ and is on the point $(x_1, f(x_1))$, so has equation

$$y - f(x_1) = f'(x_1)(x - x_1).$$

If t crosses the x-axis at x_2, then t passes through the point $(x_2, 0)$. This means

$$0 - f(x_1) = f'(x_1)(x_2 - x_1),$$

or, solving for x_2,

$$x_2 = x_1 - \frac{f(x_1)}{f'(x_1)},$$

provided that $f'(x_1) \neq 0$, of course.

In general, if x_k is the kth approximation to the zero z, then

$$x_{k+1} = x_k - \frac{f(x_k)}{f'(x_k)} \tag{1}$$

is the next approximation. Formula (1) is the **iterating formula** for Newton's method.

Of course, x_{k+1} will be closer to z than x_k was only if the function f has the right properties. The following theorem gives sufficient conditions for the success of Newton's method.

THEOREM 4.4

If z is a zero of the function f, and in some interval containing z, f is twice differentiable, f' is nonzero, and f'' is bounded, then there is a smaller interval (a, b) containing z such that if $x_1 \in (a, b)$ and for any integer k,

$$x_{k+1} = x_k - \frac{f(x_k)}{f'(x_k)},$$

then for any $\varepsilon > 0$ there exists $n \in \mathbb{N}$ such that $|z - x_n| < \varepsilon$.

Rather than try to prove this theorem, we will leave an outline of its proof as a series of exercises.

Newton's method is applied by selecting an initial estimate x_1 of the zero and using the iterating formula (1) until the x_k's are sufficiently close to each other. For example, if $x_{k+1} = x_k$ to three decimal places, then that value is a three-decimal approximation to the zero.

■ **Example 4.11**

Solve the equation $x^2 - \cos x = 0$.

Solution

Let $f(x) = x^2 - \cos x$. Then we solve the equation by finding the zeros of f. By sketching a rough graph of f (Figure 4.16), we see that there are two zeros of f, one near 0.8 and the other near -0.8, by symmetry. Since f satisfies all the requirements of Theorem 4.4 near 0.8, we will use Newton's method to estimate the zero there.

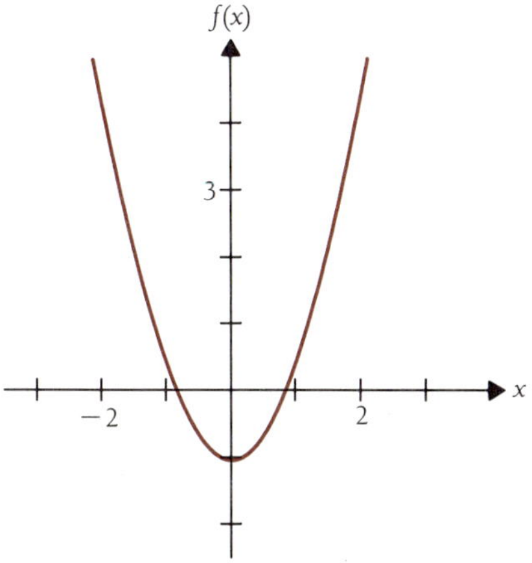

FIGURE 4.16 The graph of $f(x) = x^2 - \cos x$

Since $f'(x) = 2x + \sin x$, the iterating formula (1) becomes

$$x_{k+1} = x_k - \frac{x_k^2 - \cos x_k}{2x_k + \sin x_k}.$$

Using a programmable calculator (in radian mode!) with $x_1 = 0.8$, we get successively

$$x_2 = 0.82447,$$
$$x_3 = 0.824132377,$$
$$x_4 = 0.824132312,$$
$$x_5 = 0.824132312.$$

Thus, the zero is 0.824132312, to nine decimal places. By symmetry, the other zero is the negative of this one. ■

By starting fairly close to the zero in the above example, we got good results with Newton's method. If x_1 is not close enough to the zero, the method may not work. In some cases, it is desirable to use something like the bisection method (Section 2.5) to get a good first estimate for x_1.

Exercises 4.2

Find the critical points of each function.

1. $f(x) = 5 - x, \quad x \in [0, 3]$
2. $f(x) = 4 - x^2, \quad x \in [-1, 3]$
3. $f(x) = x^3 - x^4$
4. $f(x) = x^3 - 12x$
5. $f(x) = 2x + \dfrac{8}{x}$
6. $f(x) = x^2 - \dfrac{1}{x}$
7. $f(x) = |x^2 - 9|$
8. $f(x) = \sqrt{x^2 + 1}$
9. $f(x) = (x - 1)^{2/3}$
10. $f(x) = (x + 2)^{1/5}$

11. $f(x) = \dfrac{x}{x^2 + 4}$

12. $f(x) = \dfrac{x^2}{x^2 + 4}$

13. $f(x) = \sin x + \cos x$

14. $f(x) = \sin x - 2 \cos x$

15. $f(x) = 2 \cos x + \sin 2x$

16. $f(x) = 3 \sin \dfrac{x}{2} + \dfrac{1}{2} \cos 2x$

17. $f(x) = x + \sin x$

18. $f(x) = x + 2 \sin x$

19. $f(x) = \sin^2 x$

20. $f(x) = |\sin x|$

Find some c in the given interval (a, b) such that

$$f'(c) = \frac{f(b) - f(a)}{b - a}.$$

21. $f(x) = x^2$; $(0, 2)$

22. $f(x) = x^3$; $(-1, 1)$

23. $f(x) = \dfrac{1}{x}$; $(1, 4)$

24. $f(x) = \dfrac{x}{x + 1}$; $(0, 1)$

25. $f(x) = \sin x$; $(0, \pi/2)$

26. $f(x) = \cos x$; $(0, 3\pi/2)$

27. $f(x) = \sin^2 x$; $(0, \pi/2)$

28. $f(x) = \tan x$; $(-\pi/4, \pi/4)$

Explain why no point exists in the given interval that satisfies the conclusion of the mean value theorem.

29. $f(x) = |x|$; $[-1, 1]$

30. $f(x) = \sigma(x)$; $[-1, 1]$
(See Problem 49 in the Review Exercises for Chapter 2.)

31. $f(x) = \sigma(x)$; $[0, 1]$

32. $f(x) = \tan x$; $[0, \pi]$

33. $f(x) = \lfloor x \rfloor$; $[1, 2]$

34. $f(x) = \sec x$; $[0, \pi]$

35. The position of a moving point is $s(t) = 3t - t^3$. Find the instant at which the velocity of the point is equal to the average velocity during the time interval $[0, 3]$.

36. The number Q of gallons of water in a draining tank after t hours is $Q(t) = 10{,}000 - 10{,}000t + 2500t^2$. Find the instant at which the rate of drainage equals the average rate at which the tank drains. [*Hint:* First find the time it takes for the tank to drain.]

37. Verify that the vertical distance from the secant line to the graph of f in Figure 4.14 is

$$g(x) = f(x) - f(a) - \frac{f(b) - f(a)}{b - a}(x - a).$$

38. Verify that $g(a) = g(b) = 0$ in Problem 37.

39. Prove that if f satisfies the hypotheses of the mean value theorem, then there is some $c \in (a, x)$ such that

$$f(x) = f(a) + f'(c)(x - a).$$

Let $f(x) = x^3 - 3x + 3$. Apply Newton's method, starting with the given value for x_1, and explain the results you obtain.

40. $x_1 = 1$

41. $x_1 = 0$

42. $x_1 = -2$

C43. $x_1 = 0.1$

C44. $x_1 = 2$

ᶜ*Use Newton's method to approximate the zero in the indicated interval of the given function to four decimal places.*

45. $f(x) = x^3 + x + 1; \quad z \in (-1, 0)$
46. $f(x) = 2x^3 - x^2 + 2; \quad z \in (-1, 0)$
47. $f(x) = x^5 - 5x^4 + 3x^3 + 11x^2 - 24x + 14; \quad z \in (3, 4)$
48. $f(x) = x^5 - 5x^4 + 3x^3 + 11x^2 - 24x + 14; \quad z \in (-2, -1)$
49. $f(x) = x^6 - 22x^4 + 9x^2 + 102; \quad z \in (4, 5)$
50. $f(x) = x^6 - 22x^4 + 9x^2 + 102; \quad z \in (1, 2)$
51. $f(x) = x^6 - 22x^4 + 9x^2 + 102; \quad z \in (-2, -1.5)$
52. $f(x) = x^6 - 22x^4 + 9x^2 + 102; \quad z \in (-5, -4)$
53. $f(x) = x - \cos x; \quad z \in (0, \pi)$
54. $f(x) = x - 1 + \sin 2x; \quad z \in (0, 1)$
55. $f(x) = x^2 - \cos x; \quad z \in (0.5, 1)$
56. $f(x) = x^2 - \sin x; \quad z \in (0.5, 0.6)$

ᶜ*Find all the zeros of f. [Some of these problems are very time-consuming!]*

57. $f(x) = x^3 - 7x^2 + x + 2$
58. $f(x) = x^3 + x^2 - 9x + 1$
59. $f(x) = x^4 + x^3 - 8x^2 - 7x + 7$
60. $f(x) = x^4 + 2x^3 - 13x^2 - 22x + 22$
61. $f(x) = 3x^5 + 4x^4 - 70x^3 - 106x^2 + 139x - 26$
62. $f(x) = 11x^5 - 4x^4 - 209x^3 + 76x^2 + 858x - 312$
63. $f(x) = \cos 3x + 0.2x$
64. $f(x) = \sin x - 0.5x$

Problems 65–71 outline a proof of Theorem 4.4.

65. A point z is a **fixed point** of a function g if and only if $g(z) = z$. Prove that a zero of the differentiable function f that is not a critical point of f is a fixed point of the function
$$g(x) = x - \frac{f(x)}{f'(x)}.$$

66. Let f be a differentiable function whose derivative is nonzero in an interval (a, b), and let
$$g(x) = x - \frac{f(x)}{f'(x)}.$$
Prove that if $z \in (a, b)$ is a fixed point of g, then z is a zero of f.

67. Prove that if z is a fixed point of the differentiable function g, $|g'(x)| < 1$ for all x in some interval (a, b) containing z, and $x_1 \in (a, b)$, then $x_2 = g(x_1)$ is closer to z than x_1 is. [*Hint:* Apply the mean value theorem to g on the closed interval with endpoints x_1 and z.]

68. Prove that if z is a zero of a function f that is twice differentiable in some interval containing z, f' is nonzero, and f'' is bounded in that interval, then there is another

(smaller) interval (a, b) containing z such that if

$$g(x) = x - \frac{f(x)}{f'(x)},$$

then $|g'(x)| < 1$ for all $x \in (a, b)$.

69. Prove that in Problem 68, we can actually find an interval (a, b) such that if $x \in (a, b)$, then $|g'(x)| < 0.1$.

70. Prove that if z is a zero of a function f that is twice differentiable in an interval containing z, and f' is nonzero and f'' is bounded in that interval, then there is an interval (a, b) containing z such that if $x_1 \in (a, b)$ and

$$x_2 = x_1 - \frac{f(x_1)}{f'(x_1)},$$

then

$$|z - x_2| < (0.1)|z - x_1|.$$

71. Prove Theorem 4.4.

72. Sketch the graph of a function defined on a closed interval but having no maximum or minimum there. [*Hint:* The function must not be continuous.]

73. Sketch the graph of a function that is continuous on an open interval but has no maximum or minimum there.

74. Sketch the graph of a function that is continuous on a closed interval $[a, b]$ and has its extrema at the points a and b.

SECTION 4.3

Applications to Graphing Functions

The derivatives of a function give us some valuable information about the graph of the function. We have already seen that the critical points are related to extrema, and we will work with this relationship a great deal more in the next section. In this section, we will see that critical points are related to the graphic property of *monotonicity* of the function.

Figure 4.17 shows the graphs of a function f and its derivatives f' and f''. Learning how the properties of the graphs of f' and f'' are related to the properties of the graph of f is the objective of this section.

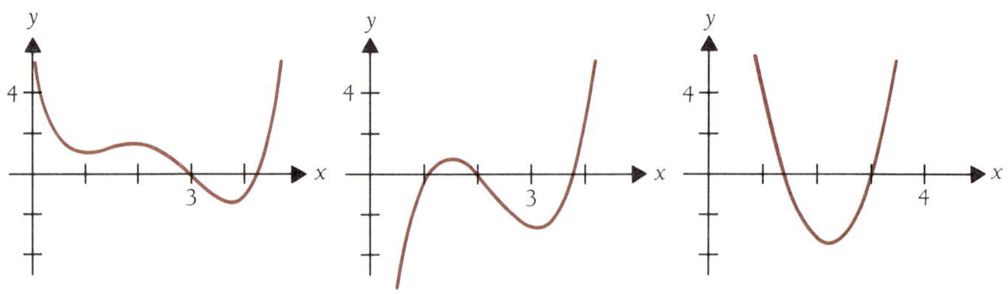

(a) Graph of f (b) Graph of f' (c) Graph of f''

FIGURE 4.17 A function and its first two derivatives

Increasing and Decreasing Functions

The graphic interpretation of the derivative of a function is the slope of the graph of the function. Thus, the first thing that the derivative tells us about a graph is its direction.

For example, if the derivative of a function f is positive, then the slope of the graph is positive, or the graph goes upward as followed from left to right (see Figure 4.18). If the derivative is negative, the graph goes downward (see Figure 4.19).

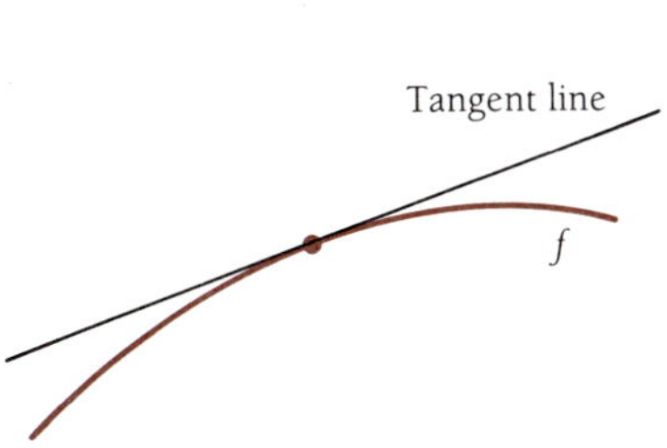

FIGURE 4.18 A graph with positive slope

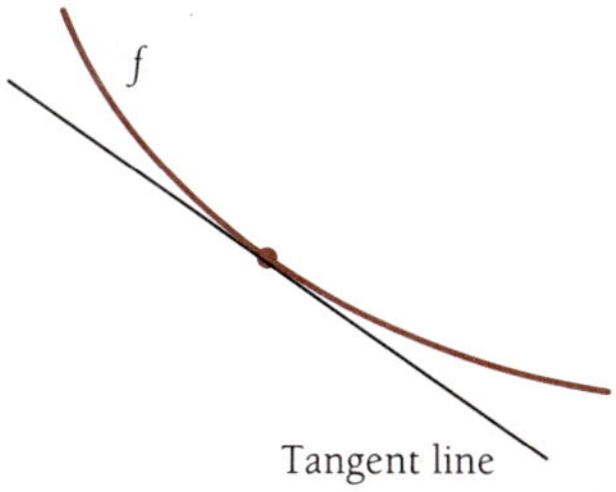

FIGURE 4.19 A graph with negative slope

The following definition gives us more precise terminology.

DEFINITION 4.1

A function f defined in the interval (a, b) is:

(a) **Increasing** in (a, b) if and only if $x_1 < x_2$ implies $f(x_1) < f(x_2)$ for all $x_1, x_2 \in (a, b)$
(b) **Decreasing** in (a, b) if and only if $x_1 < x_2$ implies $f(x_1) > f(x_2)$ for all $x_1, x_2 \in (a, b)$

The function f is called **monotonic** in (a, b) if f is increasing or decreasing in (a, b). The monotonicity of a function is linked to the derivative by the following theorem, which is a consequence of the mean value theorem.

THEOREM 4.5

Let the function f be differentiable in (a, b).

(a) If $f'(x) > 0$ for all $x \in (a, b)$, then f is increasing in (a, b).
(b) If $f'(x) < 0$ for all $x \in (a, b)$, then f is decreasing in (a, b).

Proof

We will prove only part (a); the proof of part (b) is similar. Let $x_1, x_2 \in (a, b)$ with $x_1 < x_2$. We must show that if $f'(x) > 0$ for all $x \in (a, b)$, then $f(x_1) < f(x_2)$. By the mean value theorem applied to f on $[x_1, x_2]$, there is some point $c \in (x_1, x_2)$ such that

$$f(x_2) - f(x_1) = f'(c)(x_2 - x_1) > 0.$$

That is, $f(x_1) < f(x_2)$, and f is increasing.

■ **Example 4.12** Discuss the monotonicity of the following functions:

(a) $f(x) = x^2$ (b) $f(x) = x^5 + x$
(c) $f(x) = x^{-2}$ (d) $f(x) = 3x^4 - 8x^3 + 2$
(e) $f(x) = 4x^3 - 3x^2 - 18x + 12, \quad x \in [-2, 2]$ (f) $f(x) = \sin 2x$

Solution

(a) The derivative of $f(x) = x^2$ is $f'(x) = 2x$, which is positive when $x > 0$ and negative when $x < 0$. Therefore, f is increasing in $(0, \infty)$ and decreasing in $(-\infty, 0)$, as can be seen in Figure 4.20.
(b) The derivative of $f(x) = x^5 + x$ is $f'(x) = 5x^4 + 1$, which is positive for all $x \in \mathbb{R}$. This function is therefore increasing on all of $\mathbb{R}$ (see Figure 4.21).

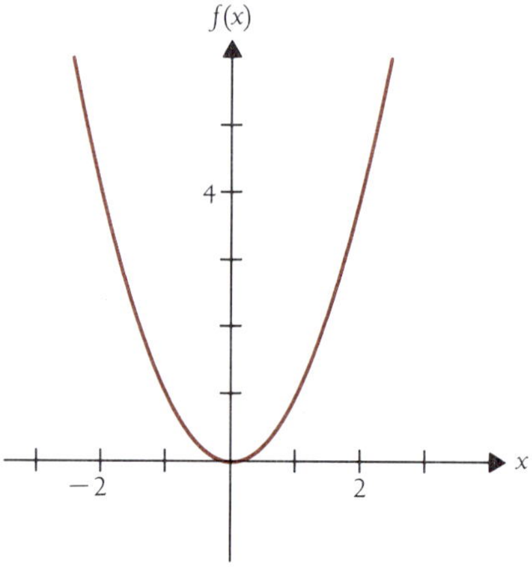

FIGURE 4.20 The graph of $f(x) = x^2$

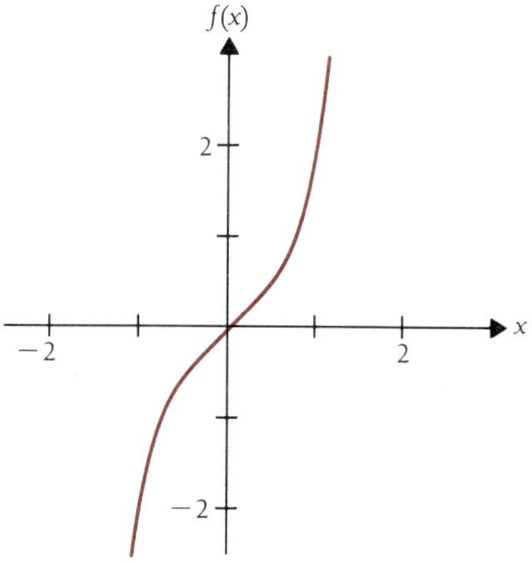

FIGURE 4.21 The graph of $f(x) = x^5 + x$

(c) The derivative of $f(x) = x^{-2}$ is $f'(x) = -2x^{-3}$, which is positive if $x < 0$ and negative if $x > 0$. Thus, f is increasing on $(-\infty, 0)$ and decreasing on $(0, \infty)$ (see Figure 4.22).
(d) The derivative of $f(x) = 3x^4 - 8x^3 + 2$ is $f'(x) = 12x^3 - 24x^2 = 12x^2(x - 2)$. The critical points of f are then 0 and 2. We find by noting the signs of x^2 and $(x - 2)$ that

$f'(x)$ is negative if $x < 0$,
$f'(x)$ is negative if $0 < x < 2$, and
$f'(x)$ is positive if $x > 2$.

Thus, f is decreasing in $(-\infty, 0)$ and $(0, 2)$ [so is actually decreasing in $(-\infty, 2)$] and increasing in $(2, \infty)$, as seen in Figure 4.23.
(e) The derivative of the function $f(x) = 4x^3 - 3x^2 - 18x + 12$ is $f'(x) = 12x^2 - 6x - 18 = 6(2x - 3)(x + 1)$, so the critical points of f are -1 and $\frac{3}{2}$. Both of these points are in the given domain. We find that if $-2 < x < -1$, then both $2x - 3$ and $x + 1$ are negative, so $f'(x) > 0$. If $-1 < x < \frac{3}{2}$, then

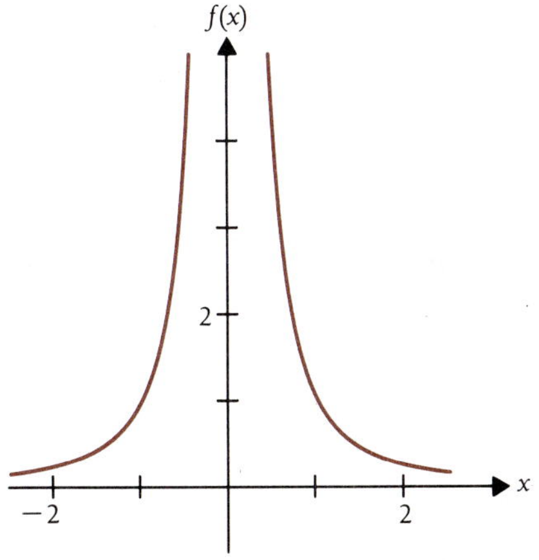

FIGURE 4.22 The graph of $f(x) = x^{-2}$

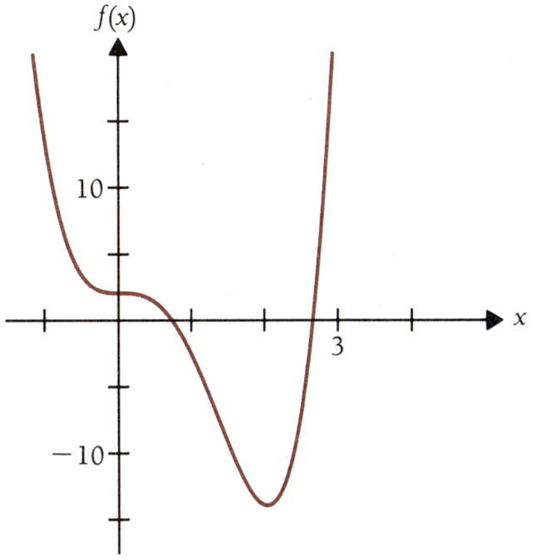

FIGURE 4.23 The graph of $f(x) = 3x^4 - 8x^3 + 2$

$2x - 3 < 0$ and $x + 1 > 0$, so $f'(x) < 0$. If $\frac{3}{2} < x < 2$, then $f'(x) > 0$. Hence, f is increasing in $(-2, -1)$ and in $(\frac{3}{2}, 2)$ and decreasing in $(-1, \frac{3}{2})$ (see Figure 4.24).

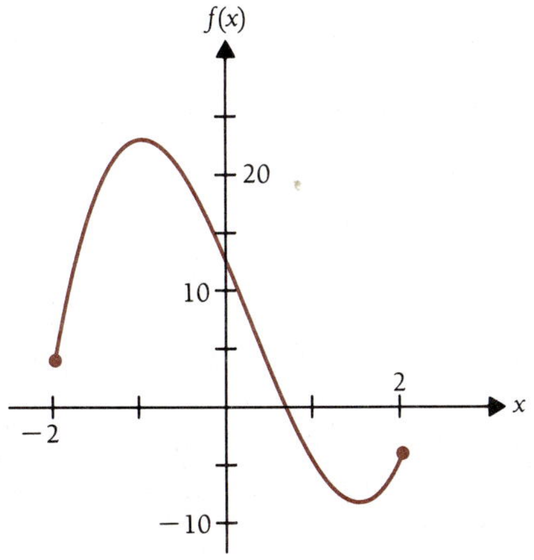

FIGURE 4.24 The graph of $f(x) = 4x^3 - 3x^2 - 18x + 12$

(f) The derivative of $f(x) = \sin 2x$ is $f'(x) = 2 \cos 2x$. To find out where f is increasing, we find out where $\cos 2x$ is positive. Since the cosine is positive in Quadrants I and IV, we must have

$$\frac{-\pi}{2} + 2n\pi < 2x < \frac{\pi}{2} + 2n\pi,$$

$$\frac{-\pi}{4} + n\pi < x < \frac{\pi}{4} + n\pi.$$

Thus, f is increasing in any interval of the form

$$\left(n\pi - \frac{\pi}{4}, n\pi + \frac{\pi}{4}\right), \qquad n \in \mathbb{Z}.$$

In an analogous fashion, we find that f is decreasing in any interval of the form

$$\left(n\pi + \frac{\pi}{4}, n\pi + \frac{3\pi}{4}\right), \qquad n \in \mathbb{Z}$$

(see Figure 4.25).

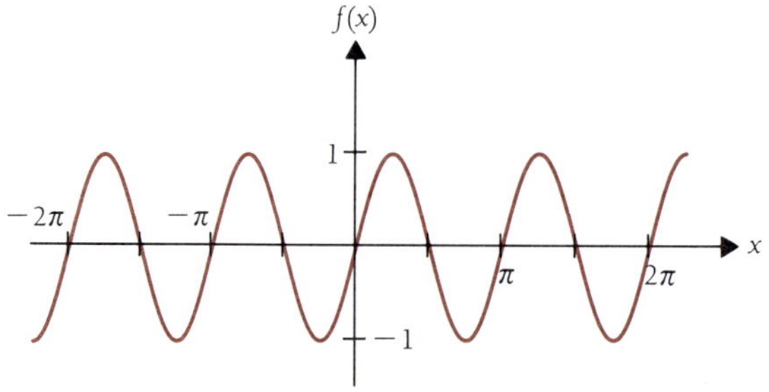

FIGURE 4.25 The graph of $f(x) = \sin 2x$ ■

We note from these examples that the points of the domain at which a function changes from increasing to decreasing or vice versa are always critical points of the function. However, the function need not change monotonicity at every critical point.

Concavity

Now we look at the second derivative, and see what it tells us about the graph of a function. If we interpret f'' as the rate of change of f', then f'' tells us how the slope of the graph is changing. That is, if f'' is positive, then f' is increasing, and the graph is getting steeper, or curving upward (see Figure 4.26). If f'' is negative, then the graph of f is curving downward (see Figure 4.27).

FIGURE 4.26
A curve that is concave upward

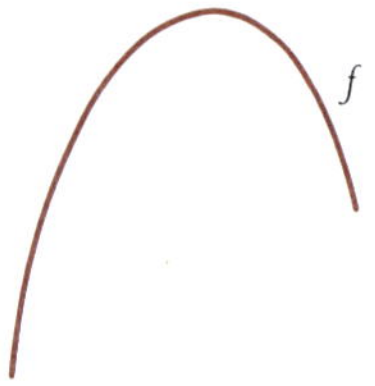

FIGURE 4.27
A curve that is concave downward

DEFINITION 4.2

A function f that is differentiable in (a, b) is said to be **concave upward** in (a, b) if and only if f' is increasing in (a, b). The function f is **concave downward** in (a, b) if and only if f' is decreasing in (a, b).

The following theorem is now obvious.

THEOREM 4.6

Let function f be twice differentiable in (a, b).

(a) If $f''(x) > 0$ for all $x \in (a, b)$, then f is concave upward in (a, b).
(b) If $f''(x) < 0$ for all $x \in (a, b)$, then f is concave downward in (a, b).

■ **Example 4.13**

Discuss the concavity of the following functions:

(a) $f(x) = x^5 + x$ **(b)** $f(x) = x^{-2}$
(c) $f(x) = 3x^4 - 8x^3 + 2$ **(d)** $f(x) = \sin 2x$

Solution

(a) The first derivative of $f(x) = x^5 + x$ is $f'(x) = 5x^4 + 1$, so the second derivative is $f''(x) = 20x^3$. We see that $f''(x) > 0$ when $x > 0$, and $f''(x) < 0$ when $x < 0$. Thus, f is concave upward in $(0, \infty)$ and concave downward in $(-\infty, 0)$ (see Figure 4.21).

(b) The first derivative of $f(x) = x^{-2}$ is $f'(x) = -2x^{-3}$, so the second derivative is $f''(x) = 6x^{-4}$. Since $f''(x) > 0$ for all nonzero x, we see that f is concave upward in $(-\infty, 0)$ and in $(0, \infty)$ (see Figure 4.22).

(c) The first derivative of $f(x) = 3x^4 - 8x^3 + 2$ is $f'(x) = 12x^3 - 24x^2$, so the second derivative is $f''(x) = 36x^2 - 48x = 12x(3x - 4)$. We can see that

$f''(x)$ is positive if $x < 0$,
$f''(x)$ is negative if $0 < x < \frac{4}{3}$,
$f''(x)$ is positive if $x > \frac{4}{3}$.

Therefore, f is concave upward in $(-\infty, 0)$ and in $(\frac{4}{3}, \infty)$, and concave downward in $(0, \frac{4}{3})$ (see Figure 4.23).

(d) The first derivative of $f(x) = \sin 2x$ is $f'(x) = 2\cos 2x$, so the second derivative is $f''(x) = -4\sin 2x$. Thus, $f''(x) > 0$ wherever $\sin 2x < 0$. A glance at Figure 4.25 tells us that $f''(x) > 0$ in any interval of the form

$$\left(\frac{\pi}{2} + n\pi,\ \pi + n\pi\right), \qquad n \in \mathbb{Z},$$

so f is concave upward on those intervals. Similarly, $f''(x) < 0$ and f is concave downward on any interval of the form

$$\left(n\pi,\ \frac{\pi}{2} + n\pi\right), \qquad n \in \mathbb{Z}.$$

■

A point in the domain of f at which concavity changes is called an **inflection point** of f. To find the inflection points of a function f, we simply determine the concavity of f, and note the points at which the concavity changes; those that are in the domain of f are inflection points. We see from the above examples that the points of inflection of f are critical points of f'. Not all critical points of f' are inflection points of f, though, as the next example will show.

Example 4.14 Find the points of inflection of the following functions:

(a) $f(x) = 3x^5 - 5x^4$ (b) $f(x) = x + x^{7/5}$ (c) $f(x) = \sin 2x$

Solution

(a) The derivatives of $f(x) = 3x^5 - 5x^4$ are $f'(x) = 15x^4 - 20x^3$ and $f''(x) = 60x^3 - 60x^2 = 60x^2(x-1)$. The second derivative is 0 when $x = 0$ and $x = 1$. For $x < 0$ and for $0 < x < 1$, $f''(x) < 0$, so f is concave downward in $(-\infty, 0)$ and in $(0, 1)$. Thus, 0 is not an inflection point of f, since the concavity of f does not change at 0. For $x > 1$, $f''(x) > 0$, and f is concave upward in $(1, \infty)$. Thus, 1 is an inflection point of f (see Figure 4.28).

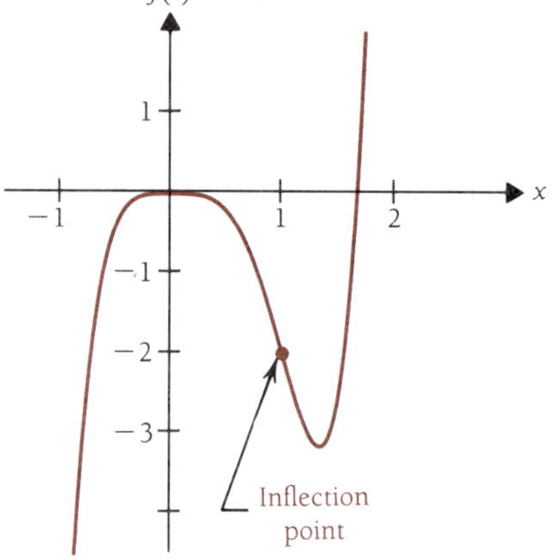

FIGURE 4.28 The graph of $f(x) = 3x^5 - 5x^4$

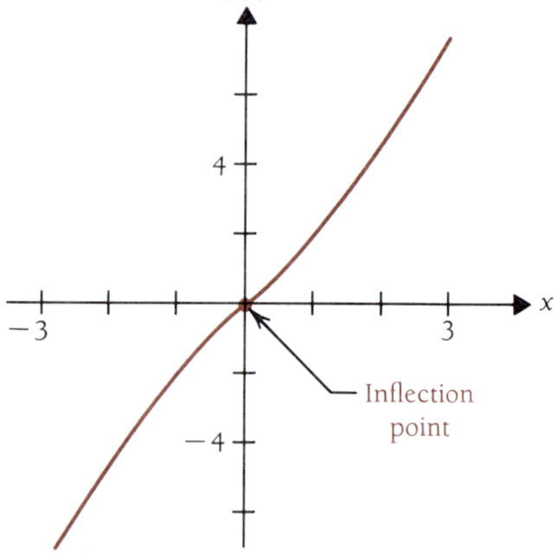

FIGURE 4.29 The graph of $f(x) = x + x^{7/5}$

(b) The derivatives of $f(x) = x + x^{7/5}$ are $f'(x) = 1 + \frac{7}{5}x^{2/5}$ and $f''(x) = \frac{14}{25}x^{-3/5}$. The second derivative is negative if $x < 0$, undefined at $x = 0$, and positive if $x > 0$. The concavity changes at 0, and 0 is an inflection point of f (see Figure 4.29).

(c) By Example 4.13(d) and Figure 4.25, we can see that the inflection points of $f(x) = \sin 2x$ are the points

$$x = \frac{n\pi}{2}, \qquad n \in \mathbb{Z}.$$ ■

We have now developed the tools needed to sketch the graph of a function without having to plot a host of points. Consider the following cases.

Sketching Graphs of Polynomial Functions

Sketching the graph of a polynomial function is fairly straightforward at this point. The following algorithm outlines the steps required. The proofs of the claims made will be left as exercises, since they follow directly from the theory we have already developed.

Polynomial Graph Algorithm To sketch the graph of the polynomial function p:

1. Find the critical points of p, which are the zeros of p'. For each critical point c, plot the point $(c, p(c))$ and draw a horizontal tangent line through it.
2. Determine the monotonicity of p. The monotonicity of p cannot change except at critical points. Thus:
   (a) If c_1 and c_2 are consecutive critical points of p and $p(c_1) < p(c_2)$, then p is increasing in (c_1, c_2); if $p(c_1) > p(c_2)$, then p is decreasing in (c_1, c_2).
   (b) If c_1 is the least (that is, farthest left) critical point of p, a is any number less than c_1, and $p'(a) > 0$, then p is increasing in $(-\infty, c_1)$; if $p'(a) < 0$, then p is decreasing in $(-\infty, c_1)$.
   (c) If c_n is the greatest (that is, farthest right) critical point of p, b is any number greater than c_n, and $p'(b) > 0$, then p is increasing in (c_n, ∞); if $p'(b) < 0$, then p is decreasing in (c_n, ∞).
3. Find the critical points of p', which are the zeros of p''. For each such point c, plot the point $(c, p(c))$. These values are the possible inflection points of p.
4. Determine the concavity of p, which is the same as determining the monotonicity of p'. That is, apply Step 2 to p'.
5. Perhaps plot a few more well-chosen points, particularly if the points already plotted are far apart, and then join the points with a smooth curve, using the tangent lines, monotonicity, and concavity properties discovered above.

The only hard work in this algorithm may be finding the zeros of p' and p''. The techniques of college algebra (synthetic division, deflation, the quadratic formula, rational zeros, Descartes's rules of signs) can be used to find zeros of many polynomials. Additional techniques include the bisection method (discussed in Section 2.5) and Newton's method (discussed in Section 4.2).

Here are a couple of examples.

Example 4.15 Sketch the graph of the function $p(x) = x^3 + 6x^2 + 9x - 2$.

Solution First, we find the derivative to be

$$\begin{aligned} p'(x) &= 3x^2 + 12x + 9 \\ &= 3(x^2 + 4x + 3) \\ &= 3(x + 3)(x + 1). \end{aligned}$$

The zeros of the derivative are -3 and -1, so these are the critical points of p. Since $p(-3) = -2$ and $p(-1) = -6$, we plot the points $(-3, -2)$ and $(-1, -6)$ and draw horizontal lines through them to serve as tangent lines (see Figure 4.30).

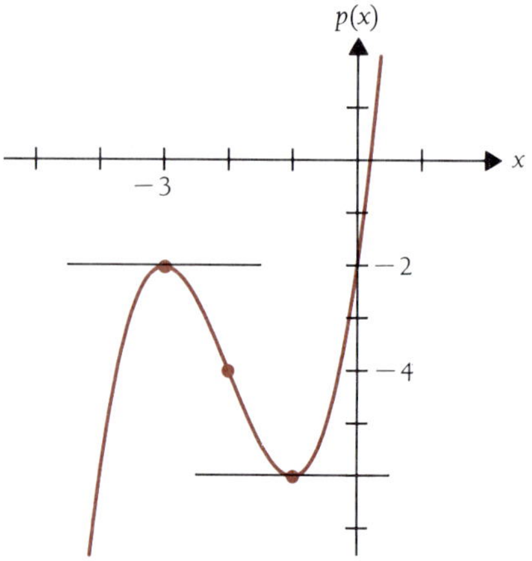

FIGURE 4.30 The graph of $p(x) = x^3 + 6x^2 + 9x - 2$

Next, we find the monotonicity of p. Since $p(-3) > p(-1)$, we know p is decreasing in $(-3, -1)$. Since $p'(-4) = 6 > 0$, p is increasing in $(-\infty, -3)$. Since $p'(0) = 9 > 0$, p is increasing in $(-1, \infty)$.

The second derivative of p is

$$\begin{aligned} p''(x) &= 6x + 12 \\ &= 6(x + 2). \end{aligned}$$

The only zero is -2, so that is the only critical point of p'—and the only possible inflection point. Since $p(-2) = -4$, we plot the point $(-2, -4)$.

The concavity of p is found by noting that for $x < -2$, $p''(x) < 0$, and for $x > -2$, $p''(x) > 0$. Thus, p is concave downward on $(-\infty, -2)$ and concave upward on $(-2, \infty)$, and -2 is an inflection point.

Perhaps plotting one or two more points, we get the graph in Figure 4.30. ■

■ **Example 4.16** Sketch the graph of the function

$$p(x) = 3x^5 - 25x^4 + 80x^3 - 120x^2 + 80x - 24.$$

Solution Following the algorithm, we first compute the derivative:

$$p'(x) = 15x^4 - 100x^3 + 240x^2 - 240x + 80.$$

We find by college algebra methods that the zeros of p' are 2 (of multiplicity 3) and $\frac{2}{3}$, so these are the critical points of p. Since $p(2) = -8$ and $p(\frac{2}{3}) = -\frac{392}{81}$, we plot the points $(2, -8)$ and $(\frac{2}{3}, -\frac{392}{81})$ and draw horizontal tangent lines through them, as in Figure 4.31.

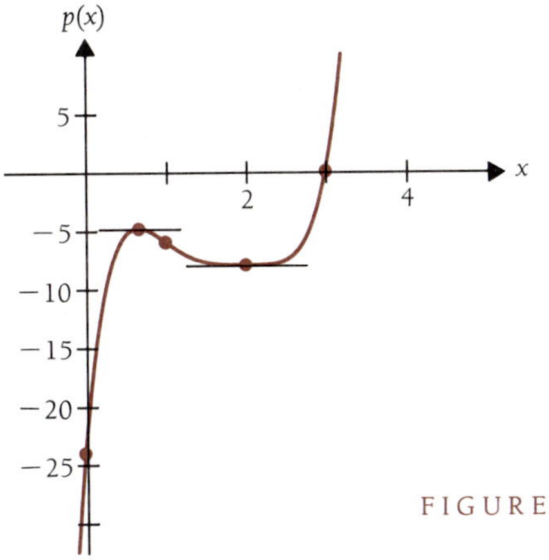

FIGURE 4.31 The graph of $p(x) = 3x^5 - 25x^4 + 80x^3 - 120x^2 + 80x - 24$

Next, we determine the monotonicity of p. Since $p'(0) = 80 > 0$, p is increasing in $(-\infty, \frac{2}{3})$. Since $p(\frac{2}{3}) > p(2)$, p is decreasing in $(\frac{2}{3}, 2)$. Finally, $p'(3) = 35 > 0$, so p is increasing again in $(2, \infty)$.

To find the critical points of p', we compute the second derivative:

$$p''(x) = 60x^3 - 300x^2 + 480x - 240.$$

The zeros of p'' are 2 (of multiplicity 2) and 1. The point $(2, -8)$ is already plotted, and since $p(1) = -6$, we plot the point $(1, -6)$. The values 1 and 2 are possible inflection points.

Now we determine the concavity of p. Since $p''(0) = -240 < 0$, p is concave downward in $(-\infty, 1)$. Since $p'(1) = -5$ and $p'(2) = 0$, we have $p'(1) < p'(2)$, so p' is increasing in $(1, 2)$, or p is concave upward in $(1, 2)$. This makes 1 an inflection point. Since $p''(3) = 120 > 0$, p is also concave upward in $(2, \infty)$. Thus, 2 is not an inflection point.

To assist in accurately placing the curve, we plot more points, such as $(0, -24)$ and $(3, 0)$, noting the slopes $p'(0) = 80$ and $p'(3) = 35$. Finally, we sketch in the curve, using our knowledge of monotonicity and concavity, as in Figure 4.31. More points may be plotted if greater accuracy is desired. ■

Sketching Graphs of Rational Functions

A rational function is of the form

$$f(x) = \frac{p(x)}{q(x)}$$

with p and q polynomial functions. We will assume that p and q have no common polynomial factors; that is, f is expressed "in lowest terms."

A rational function may have asymptotes—something a polynomial does not have. A vertical asymptote is associated with each zero of the denominator of the function. For f as above, if $q(c) = 0$, then the line $x = c$ is a vertical asymptote. Horizontal asymptotes of f are associated with limits of f at infinity. If $\lim_{x\to\infty} f(x) = L$ or $\lim_{x\to-\infty} f(x) = L$, then the line $y = L$ is a horizontal asymptote.

The process of sketching the graph of a rational function can be outlined as follows.

Rational Graph Algorithm To sketch the graph of the rational function $f(x) = p(x)/q(x)$ (in lowest terms):

1. Find and sketch the vertical asymptotes by finding the zeros of q. Also investigate the graph of f near any vertical asymptote $x = c$ by finding the one-sided limits of f at c.
2. Find and sketch any horizontal asymptotes by finding the limits of f as $x \to \infty$ and as $x \to -\infty$.
3. Find the critical points of f, which are the zeros of f'. For each critical point c, plot the point $(c, f(c))$ and draw a horizontal tangent line through it.
4. Determine the monotonicity of f, noting that monotonicity can change at a critical point or at a vertical asymptote.
5. Find the zeros of f''. These are possible inflection points of f. For each such point c, plot the point $(c, f(c))$.
6. Determine the concavity of f, noting that concavity can change at any of the points found in Step 5 or at a vertical asymptote.
7. Perhaps plot a few more well-chosen points, and then sketch the graph using the asymptotic behavior, monotonicity, and concavity discovered above.

Consider an example.

■ **Example 4.17** Sketch the graph of the rational function

$$f(x) = \frac{4x - 5}{x^2 - 1}.$$

Solution Following the algorithm, we first note that the denominator $x^2 - 1$ has zeros at 1 and -1, so the lines $x = 1$ and $x = -1$ are vertical asymptotes. We sketch these in Figure 4.32. Next, we look at the one-sided limits of f at 1 and -1 and find that $f \to \infty$ as $x \to 1^-$ and as $x \to -1^+$, and $f \to -\infty$ as $x \to 1^+$ and as $x \to -1^-$. This tells us how the graph behaves near the vertical asymptotes.

We note next that $f \to 0$ as $x \to \infty$ and as $x \to -\infty$, so the x-axis (the line $y = 0$) is a horizontal asymptote.

To find the critical points of f, we compute the derivative:

$$\begin{aligned} f'(x) &= \frac{-4x^2 + 10x - 4}{(x^2 - 1)^2} \\ &= \frac{-2(2x - 1)(x - 2)}{(x^2 - 1)^2}. \end{aligned}$$

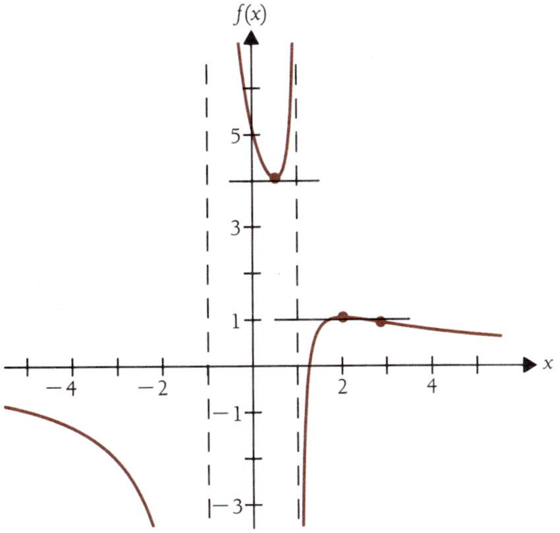

FIGURE 4.32 The graph of $f(x) = \dfrac{4x-5}{x^2-1}$

The critical points are therefore the zeros of f', namely $\frac{1}{2}$ and 2. (The points ± 1 are not critical points of f, because they are not in the domain of f.) We find $f(\frac{1}{2}) = 4$ and $f(2) = 1$, so we plot the points $(\frac{1}{2}, 4)$ and $(2, 1)$ and draw horizontal tangent lines through them.

Now we determine the monotonicity of f. We see that f' is negative if $x < -1$ or if $-1 < x < \frac{1}{2}$ or if $x > 2$, and f' is positive if $\frac{1}{2} < x < 1$ or if $1 < x < 2$. This means that f is decreasing in $(-\infty, -1)$, in $(-1, \frac{1}{2})$, and in $(2, \infty)$, and increasing in $(\frac{1}{2}, 1)$ and in $(1, 2)$.

Next, we look for possible inflection points of f by finding the second derivative:

$$f''(x) = \frac{8x^3 - 30x^2 + 24x - 10}{(x^2-1)^3}.$$

The only zeros of f'' are the zeros of the numerator. The numerator turns out to have only one real zero, which we find by Newton's method to be approximately 2.85. Since $f(2.85) \approx 0.90$, we plot the point $(2.85, 0.90)$.

We find the concavity as follows: For $x < -1$, $f'' < 0$; for $-1 < x < 1$, $f'' > 0$; for $1 < x < 2.85$, $f'' < 0$; and for $x > 2.85$, $f'' > 0$. Thus, f is concave downward on $(-\infty, -1)$ and on $(1, 2.85)$, and f is concave upward on $(-1, 1)$ and on $(2.85, \infty)$.

Finally, we sketch in the curve as shown in Figure 4.32. Plotting a few more points helps in locating the graph a little more exactly. ■

Other Graphing Problems

The graphs of other functions can be found in essentially the same way as those of rational functions. A few other features may arise, however, such as restricted domains, vertical tangent lines, and cusps. The following algorithm outlines a procedure.

General Graph Algorithm To sketch the graph of a function f:

1. Determine the domain of f. Check for vertical asymptotes, and determine the one-sided limits at endpoints of the domain and at points of discontinuity.
2. Check for horizontal asymptotes by finding limits of f at ∞ and $-\infty$.
3. Find the critical points of f, which are those numbers c such that either $f'(c) = 0$ or $f'(c)$ is undefined. Plot each point $(c, f(c))$.
   (a) For each c such that $f'(c) = 0$, draw a horizontal tangent line through $(c, f(c))$.
   (b) For each c in the domain of f such that $f'(c)$ is undefined, find the one-sided derivatives of f at c, and check for vertical tangent lines and cusps (see Problems 35–38 in the Additional Exercises at the end of Chapter 3).
4. Determine the monotonicity of f.
5. Find the possible inflection points of f, which are the numbers c in the domain of f' such that either $f''(c) = 0$ or $f''(c)$ is undefined. Plot the points $(c, f(c))$.
6. Determine the concavity of f.
7. Plot some additional points to improve accuracy if desired, and then draw the curve, using all the information obtained above.

We will illustrate with an example.

■ **Example 4.18** Sketch the graph of the function

$$f(x) = \frac{x^{2/3}}{x^2 - 1}.$$

Solution We first note that $x = 1$ and $x = -1$ are vertical asymptotes, and $y = 0$ is a horizontal asymptote. The behavior of f near the vertical asymptotes is given by noting that $f \to \infty$ as $x \to 1^+$ and as $x \to -1^-$, and $f \to -\infty$ as $x \to 1^-$ and as $x \to -1^+$.

We next seek the critical points by differentiating. The derivative simplifies to

$$f'(x) = \frac{-2(2x^2 + 1)}{3x^{1/3}(x^2 - 1)^2},$$

which is defined (and nonzero) for all x in the domain of f except $x = 0$. Since $f'(0^-) = \infty$ and $f'(0^+) = -\infty$, we have a cusp at 0. Checking the sign of f', we find f is increasing in $(-\infty, -1)$ and $(-1, 0)$, and decreasing in $(0, 1)$ and $(1, \infty)$.

We seek possible inflection points by finding the second derivative, which reduces to

$$f''(x) = \frac{2(14x^4 + 23x^2 - 1)}{9x^{4/3}(x^2 - 1)^3}.$$

We find the zeros of f'' by finding the zeros of $14x^4 + 23x^2 - 1$; since this polynomial is in quadratic form, it is not difficult to find that its real zeros are at

$$x = \pm\sqrt{\frac{3\sqrt{65} - 23}{28}} \approx \pm 0.21.$$

These are possible inflection points. Since $f(\pm 0.21) \approx -0.36$, we plot the points $(0.21, -0.36)$ and $(-0.21, -0.36)$, as shown in Figure 4.33. Now we check the sign of f'' and find that f is concave upward in $(-\infty, -1)$, $(-0.21, 0)$, $(0, 0.21)$, and $(1, \infty)$; and f is concave downward in $(-1, -0.21)$ and $(0.21, 1)$. Perhaps plotting a few more points, and using symmetry, we complete the graph as in Figure 4.33.

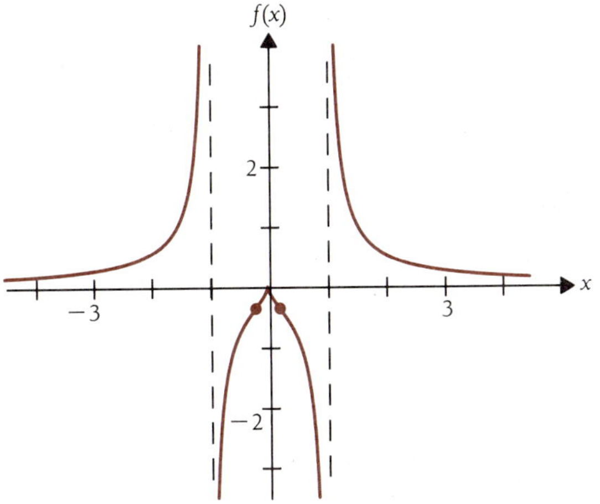

FIGURE 4.33 The graph of $f(x) = \dfrac{x^{2/3}}{x^2 - 1}$ ■

Exercises 4.3

Discuss the monotonicity and concavity of the following functions, giving all the critical points and points of inflection. Sketch the graph in each case.

1. $f(x) = 3x + 1$
2. $f(x) = x^2 - 2x + 1$
3. $f(x) = x^3 - 3x + 2$
4. $f(x) = x - x^4$
5. $f(x) = 4x^3 - 21x^2 - 24x + 5$
6. $f(x) = x^3 - x^6$
7. $f(x) = 1 - \dfrac{1}{x}$
8. $f(x) = x - \dfrac{1}{x}$

A9. $f(x) = |16 - x^2|$

A10. $f(x) = |x^2 - 5x + 6|$

11. $f(x) = \sqrt{2x + 1}$
12. $f(x) = 1 - x^{2/3}$
13. $f(x) = x + x^{7/3}$

A14. $f(x) = |2 - x^{1/3}|$

15. $f(x) = \sin x + \cos x, \quad x \in [0, 2\pi]$
16. $f(x) = \sin x - 2\cos x, \quad x \in [0, 2\pi]$
17. $f(x) = x + 2\sin x$
18. $f(x) = x + \cos 4\pi x$

19. $f(x) = \sin^2 x$

20. $f(x) = \tan^2 x$

c21. $f(x) = \dfrac{x-5}{x^2-9}$

c22. $f(x) = \dfrac{x-5}{x^2-16}$ [Newton's method]

c23. $f(x) = \dfrac{x-2}{x^2-9}$

24. $f(x) = \dfrac{x^2-4}{x-1}$

A tangent line to the graph of f at an inflection point is called an **inflectional tangent**. *Find equations of all the inflectional tangents of the following curves:*

25. $y = x^3 - 3x^2$

26. $y = 4x^3 - 2x^2 + 3$

27. $y = 3x^5 - 10x^3$

28. $y = 0.3(x^6 + 3x^5)$

29. $y = 2x^{1/3}$

30. $y = 2 + x^{1/5}$

31. $y = \sin x, \quad x \in [0, 2\pi]$

32. $y = \tan x, \quad x \in (-\pi/2, \pi/2)$

33. If c_1 and c_2 are consecutive critical points of a polynomial function p, then $p(c_1) \neq p(c_2)$. Explain why. [*Hint:* Rolle's theorem will help.]

34. With c_1 and c_2 as in Problem 33, prove that if $p(c_1) < p(c_2)$, then p is increasing in (c_1, c_2).

35. Prove that if c_0 is the least critical point of the polynomial function p and $a < c_0$, then $p'(a) < 0$ implies p is decreasing in $(-\infty, c_0)$.

36. Prove that the critical points of f' for a rational function f are always numbers c such that $f'(c) = 0$, and never numbers c such that f' is undefined at c.

37. Give an example of a function f for which the domains of f and f' include all real numbers, but for which f'' is not defined for some real number.

SECTION 4.4

Maxima and Minima

We are now ready to resume our study of extrema. In this section, we will discuss types of extrema, and show how the derivatives of a function can be used to find and test for extrema.

In Section 4.2, we defined the global or absolute maximum and minimum of a function on a closed interval. Here, we modify the definition and introduce *local* or *relative extrema.*

Relative Extrema

Consider the function f whose graph is shown in Figure 4.34. Over the interval $[a, b_1]$, the absolute maximum of f is at c_1, but over the interval $[a, b_2]$, f has its absolute maximum at c_2. To indicate that f has some sort of maximum at c_1, without regard for what interval is being considered, we introduce the idea of a *relative maximum.*

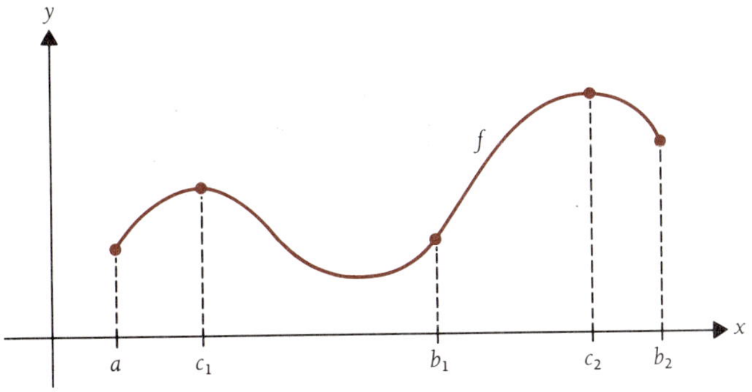

FIGURE 4.34 Relative maxima at c_1 and at c_2

DEFINITION 4.3 Let a function f be defined in an interval containing c. We say f has a **relative maximum** at c if and only if there exists an interval (a, b) containing c such that $f(x) \leq f(c)$ for all $x \in (a, b)$. Similarly, we say f has a **relative minimum** at c if and only if there exists an interval (a, b) containing c such that $f(x) \geq f(c)$ for all $x \in (a, b)$. A **relative extremum** is a relative maximum or a relative minimum. A relative extremum is also called a **local extremum.**

Thus, a relative maximum of f occurs at c when the value of f at c is at least as great as the value of f at nearby points. Looking at Figure 4.34 again, we see that f has a relative maximum at c_1 and at c_2, and a local minimum just to the left of b_1.

Endpoint Extrema

An extremum of a function that occurs at an endpoint of the domain is given a separate name.

DEFINITION 4.4 Let a function f be defined on the closed interval $[a, b]$. We say f has a **relative endpoint maximum at *a*** if and only if there exists a point $c \in (a, b)$ such that $f(a) \geq f(x)$ for all $x \in [a, c]$, and f has a **relative endpoint minimum at *a*** if and only if there is a point $c \in (a, b)$ such that $f(a) \leq f(x)$ for all $x \in [a, c]$. We say f has a **relative endpoint maximum at *b*** if and only if there is a point $c \in (a, b)$ such that $f(b) \geq f(x)$ for all $x \in [c, b]$, and f has a **relative endpoint minimum at *b*** if and only if there is a point $c \in (a, b)$ such that $f(b) \leq f(x)$ for all $x \in [c, b]$.

For example, the function in Figure 4.35 (at the top of the next page) has a relative endpoint maximum at a and a relative endpoint minimum at b.

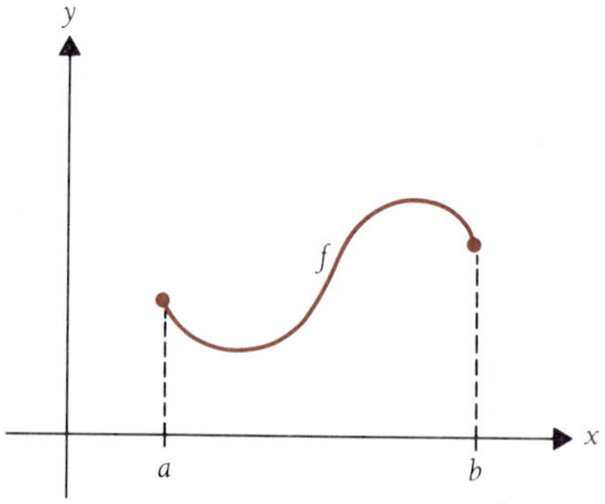

FIGURE 4.35 Relative endpoint extrema

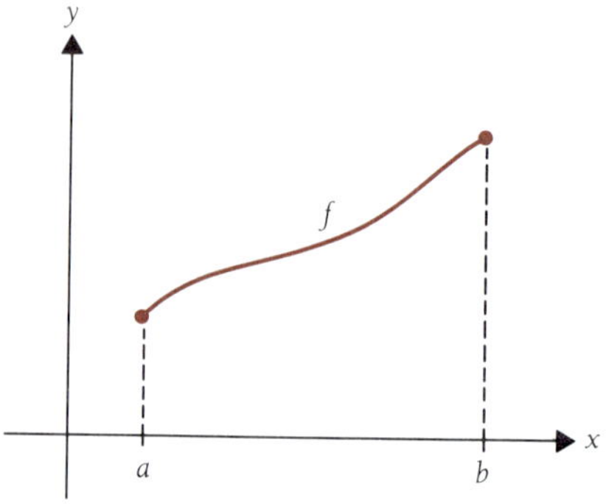

FIGURE 4.36 Absolute endpoint extrema

An **absolute endpoint extremum** of a function is simply an absolute extremum that occurs at an endpoint of the domain. For example, the function in Figure 4.36 has an absolute endpoint minimum at a and an absolute endpoint maximum at b.

We can also extend the definition of absolute extremum given in Section 4.2 to functions other than those defined on a closed interval.

DEFINITION 4.5

Let f be a function defined at c. We say f has an **absolute maximum** at c if and only if $f(c) \geq f(x)$ for all x in the domain of f. We say f has an **absolute minimum** at c if and only if $f(c) \leq f(x)$ for all x in the domain of f. An absolute extremum is also called a **global extremum**.

Finding Extrema

The first step in finding the extrema of a function is to find out where they can occur. We proved in Theorem 4.1 that absolute extrema can occur only at critical points of a function or at endpoints of its domain. The same is true for relative extrema, as you can see if you examine the proof of Theorem 4.1. Thus, the first step in finding the extrema of a function is to find the critical points of the function and the endpoints of its domain, if any.

We will illustrate with some examples.

■ **Example 4.19**

Find the extrema of the following functions:

(a) $f(x) = \dfrac{1}{x^2+1}$ **(b)** $f(x) = \sqrt{4-x}$ **(c)** $f(x) = x^2 - x, \quad x \in [-1, 1]$

Solution

(a) The derivative of

$$f(x) = \frac{1}{x^2+1}$$

is

$$f'(x) = \frac{-2x}{(x^2+1)^2},$$

which is defined for all x, and is 0 only for $x = 0$. Thus, 0 is the only critical point of f. Also, there are no endpoints of the domain of f, so the only place an extremum can occur is at $x = 0$. Since $f(0) = 1$ and $f(x) < 1$ if $x \neq 0$ (as you should verify), f has an absolute maximum at 0. The graph of f is shown in Figure 4.37.

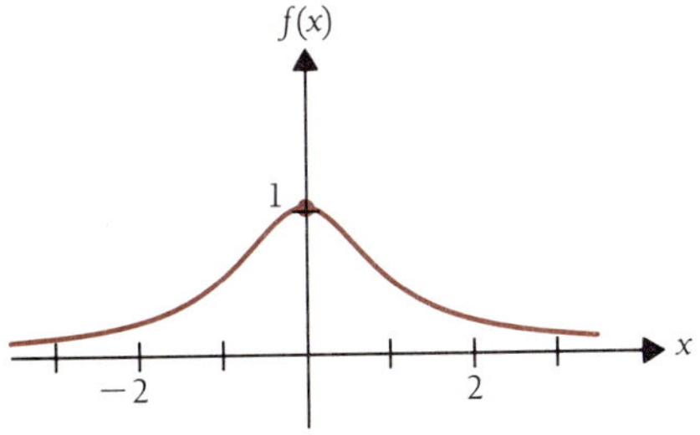

FIGURE 4.37 The graph of $f(x) = \frac{1}{x^2+1}$

(b) The domain of $f(x) = \sqrt{4-x}$ is $(-\infty, 4]$, so an extremum may occur at the endpoint 4. The derivative of f is

$$f'(x) = \frac{-1}{2\sqrt{4-x}},$$

which is defined for $x < 4$, and is always negative. Thus, f has no critical points. Since $f(4) = 0$ and $f(x) > 0$ for $x < 4$, f has an absolute endpoint minimum at 4 (see Figure 4.38).

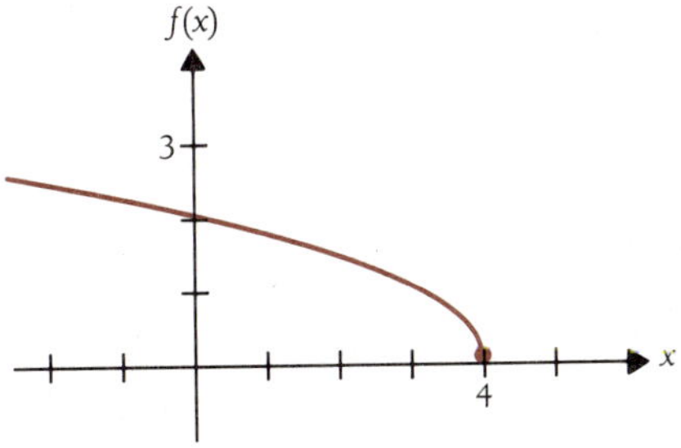

FIGURE 4.38 The graph of $f(x) = \sqrt{4-x}$

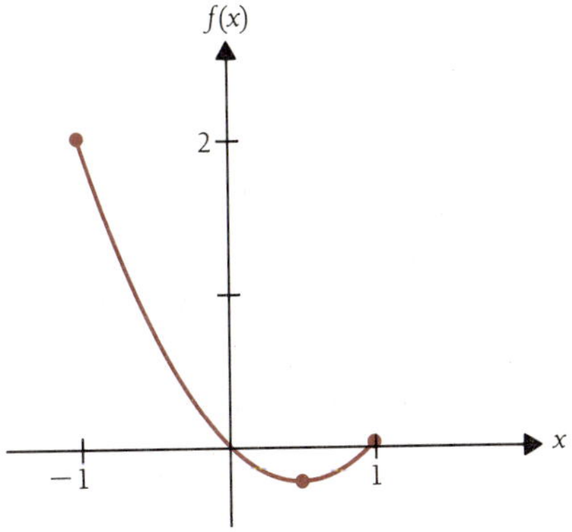

FIGURE 4.39 The graph of $f(x) = x^2 - x$ over $[-1, 1]$

(c) The endpoints -1 and 1 of the domain of this function are immediate candidates for the location of extrema. The derivative of

$$f(x) = x^2 - x$$

is

$$f'(x) = 2x - 1,$$

which is 0 at $x = \frac{1}{2}$. Since $\frac{1}{2} \in [-1, 1]$, we can also have an extremum at $\frac{1}{2}$. Since $f(-1) = 2$, $f(\frac{1}{2}) = -\frac{1}{4}$, and $f(1) = 0$, we can see (Figure 4.39) that f has an absolute endpoint maximum at -1 and an absolute minimum at $\frac{1}{2}$. Also, f has a relative endpoint maximum at 1. ■

The First Derivative Test

It is evident from these examples that the graphic properties of a function can help us determine its extrema. First, we consider what monotonicity tells us.

Suppose a continuous function f is defined in (a, b) and has a critical point c in (a, b). If f is increasing in (a, c) and decreasing in (c, b), as in Figure 4.40(a), then f must have a relative maximum at c. If f is decreasing in (a, c) and increasing in (c, b), as in Figure 4.40(b), then f has a relative minimum at c. If the monotonicity of f does not change at c, as in Figure 4.40(c), then f has no extremum at c. Similar considerations lead to endpoint extrema, as illustrated in Figure 4.41.

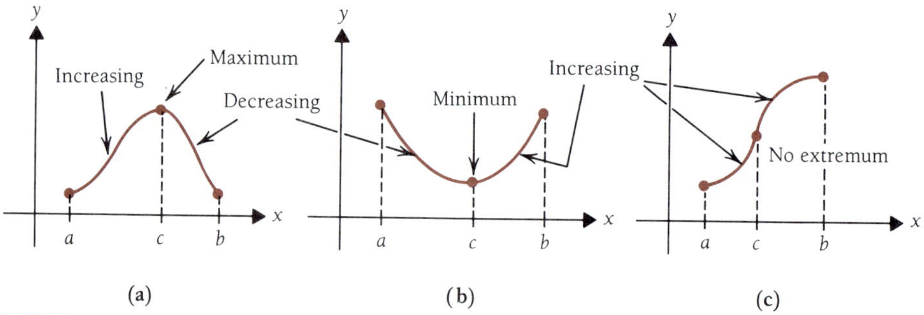

FIGURE 4.40 The first derivative test

Because of the relationship between monotonicity and the derivative, we therefore have a test for extrema, as follows.

THEOREM 4.7

The First Derivative Test Suppose that c is a critical point of a function f, and that f is continuous at c and differentiable in intervals of the form (a, c) and (c, b).

1. If $f'(x) > 0$ for $x \in (a, c)$ and $f'(x) < 0$ for $x \in (c, b)$, then f has a relative maximum at c.
2. If $f'(x) < 0$ for $x \in (a, c)$ and $f'(x) > 0$ for $x \in (c, b)$, then f has a relative minimum at c.
3. If $f'(x) > 0$ for $x \in (a, c) \cup (c, b)$, or if $f'(x) < 0$ for $x \in (a, c) \cup (c, b)$, then f has no extremum at c.

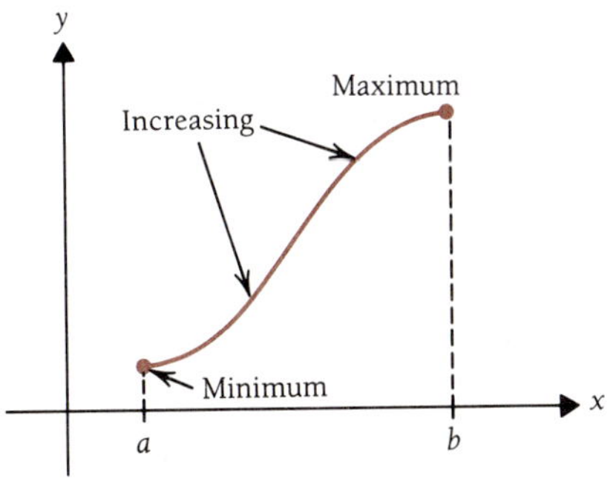

(a)

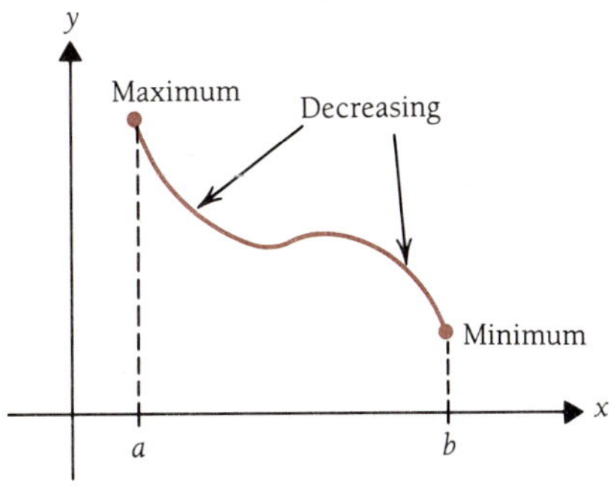

(b)

FIGURE 4.41 Endpoint extrema in the first derivative test

This theorem has an obvious modification for the determination of endpoint extrema. The proof of the theorem requires continuity at c; Figure 4.42 shows why f must be continuous at c for the first derivative test to apply.

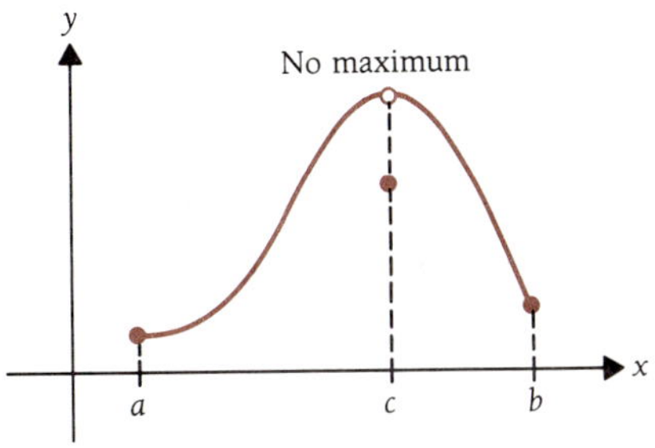

FIGURE 4.42 Continuity is necessary in the first derivative test

The following examples illustrate the use of the first derivative test.

■ **Example 4.20** Find the extrema of the following functions:

(a) $f(x) = x^4$ **(b)** $f(x) = 1 - x^4$ **(c)** $f(x) = x^3, \quad x \in [-1, 2]$

Solution **(a)** The derivative of

$$f(x) = x^4$$

is

$$f'(x) = 4x^3,$$

which is 0 only at $x = 0$. Since $f'(x) < 0$ for $x < 0$ and $f'(x) > 0$ for $x > 0$, f has a relative minimum at 0, by part 2 of the first derivative test. Since $x^4 > 0$

for $x \neq 0$, the relative minimum turns out to be an absolute minimum (see Figure 4.43).

(b) The derivative of

$$f(x) = 1 - x^4$$

is

$$f'(x) = -4x^3,$$

which is 0 only at $x = 0$. Since $f'(x) > 0$ for $x < 0$ and $f'(x) < 0$ for $x > 0$, f has a maximum at 0. This maximum is absolute, as can be seen in Figure 4.44.

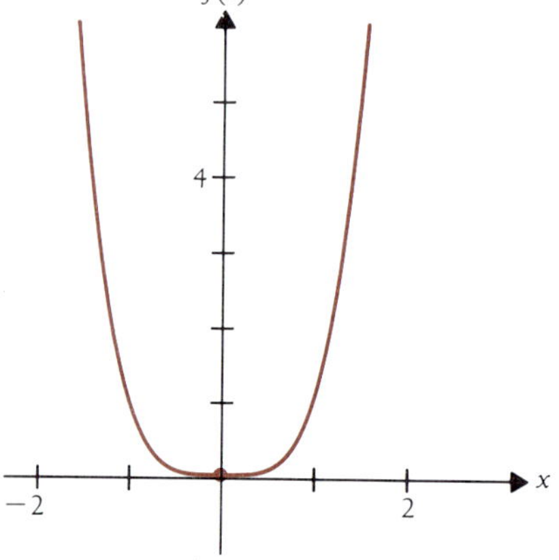

FIGURE 4.43 The graph of $f(x) = x^4$

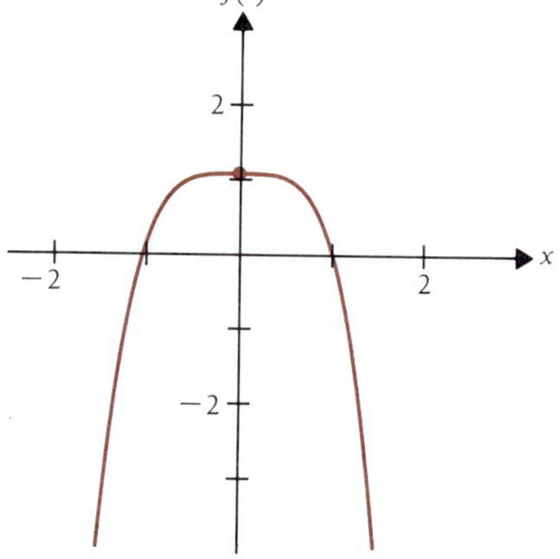

FIGURE 4.44 The graph of $f(x) = 1 - x^4$

(c) The derivative of

$$f(x) = x^3$$

is

$$f'(x) = 3x^2,$$

which is 0 at $x = 0$ and positive otherwise. Thus, by part 3 of the first derivative test, f has no extremum at 0. Because f is increasing in $[-1, 2]$, f has an endpoint minimum at -1 and an endpoint maximum at 2. Both of these extrema are absolute (see Figure 4.45). ■

The Second Derivative Test

The relationship of the second derivative to concavity provides another test for extrema under certain conditions. The idea of the test is as follows. Suppose a function f has 0 derivative at a point c, so that the graph of f has a horizontal tangent

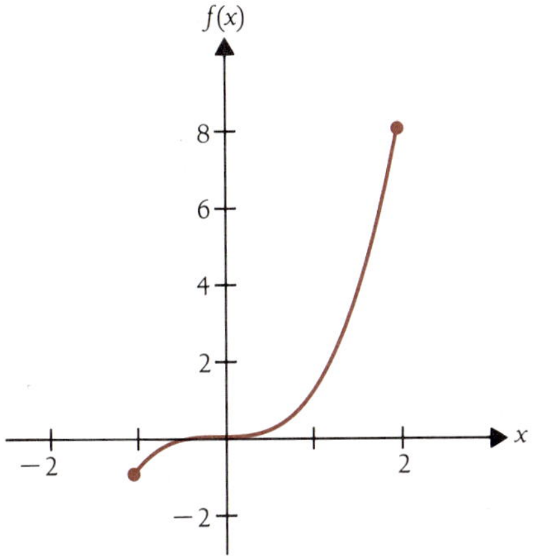

FIGURE 4.45 The graph of $f(x) = x^3$ over $[-1, 2]$

line at c. Suppose also that the graph is concave upward in an interval (a, b) containing c, as shown in Figure 4.46(a). Then f' is increasing in (a, b), and is 0 at c, so must be negative in (a, c) and positive in (c, b). Therefore, f has a relative minimum at c by the first derivative test. Similarly, if the graph is concave downward in (a, b), as shown in Figure 4.46(b), then f has a relative maximum at c.

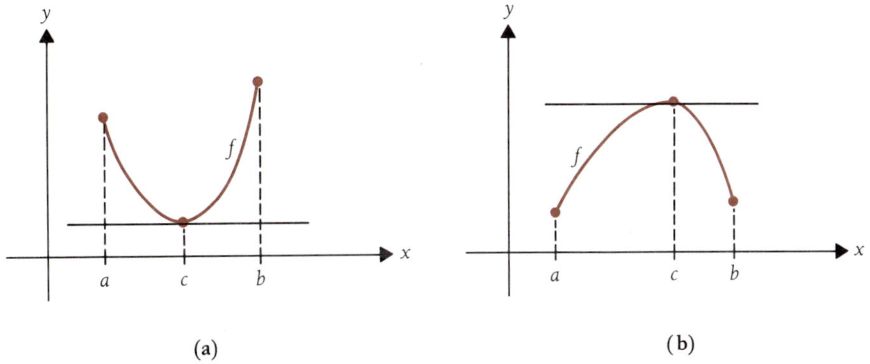

FIGURE 4.46 The second derivative test

These ideas constitute the proof of the next theorem.

THEOREM 4.8

The Second Derivative Test Let a function f be twice differentiable in (a, b), and suppose c is a point such that $f'(c) = 0$.

1. If $f''(c) < 0$, then f has a relative maximum at c.
2. If $f''(c) > 0$, then f has a relative minimum at c.
3. If $f''(c) = 0$, then the test fails.

To say that the test fails in case $f''(c) = 0$ means that anything may happen, and therefore we may draw no conclusion. Each function in Example 4.20 satisfies the hypotheses of the second derivative test near 0, and in each case, $f''(c) = 0$. But one of the functions has a maximum, one has a minimum, and the third has no extremum at 0 (see Figures 4.43, 4.44, and 4.45). When the second derivative test fails like this, we must use the first derivative test.

Here are some more examples.

Example 4.21 Locate the extrema of the following functions:

(a) $f(x) = 3x^5 - 5x^4$ (b) $f(x) = \dfrac{x}{x^2 + 1}$ (c) $f(x) = \cos x - 2 \sin x$

Solution (a) The derivative of

$$f(x) = 3x^5 - 5x^4$$

is

$$f'(x) = 15x^4 - 20x^3 = 5x^3(3x - 4).$$

Thus, the only critical points of f are 0 and $\frac{4}{3}$. To apply the second derivative test, we find the second derivative, which is

$$f''(x) = 60x^3 - 60x^2 = 60x^2(x - 1).$$

Since $f''(0) = 0$, the second derivative test fails there. But since $3x - 4$ is negative for x near 0, the first derivative f' is positive for x just less than 0 and negative for x just greater than 0. Hence, by the first derivative test, f has a relative maximum at 0. Since $f''(\frac{4}{3}) > 0$, f has a relative minimum at $\frac{4}{3}$, by the second derivative test (see Figure 4.47).

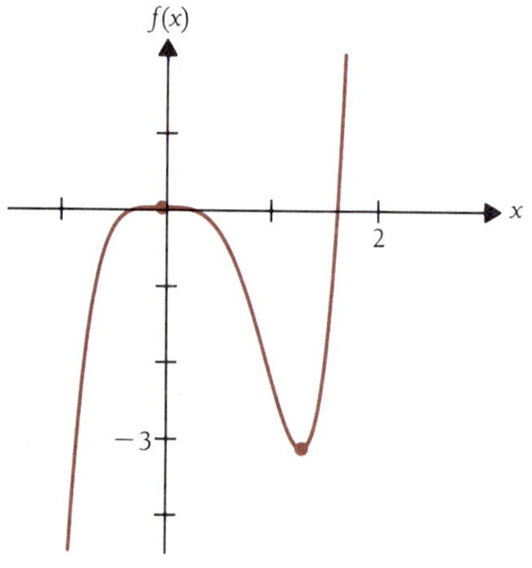

FIGURE 4.47 The graph of $f(x) = 3x^5 - 5x^4$

(b) The first derivative of

$$f(x) = \frac{x}{x^2 + 1}$$

is

$$f'(x) = \frac{1 - x^2}{(x^2 + 1)^2},$$

so that the critical points of f are -1 and 1. The second derivative is

$$f''(x) = \frac{2x^3 - 6x}{(x^2 + 1)^3},$$

and we find that $f''(-1) = \frac{1}{2} > 0$ and $f''(1) = -\frac{1}{2} < 0$. Thus, f has a relative minimum at -1 and a relative maximum at 1. We see from the graph (Figure 4.48) that these extrema turn out to be absolute.

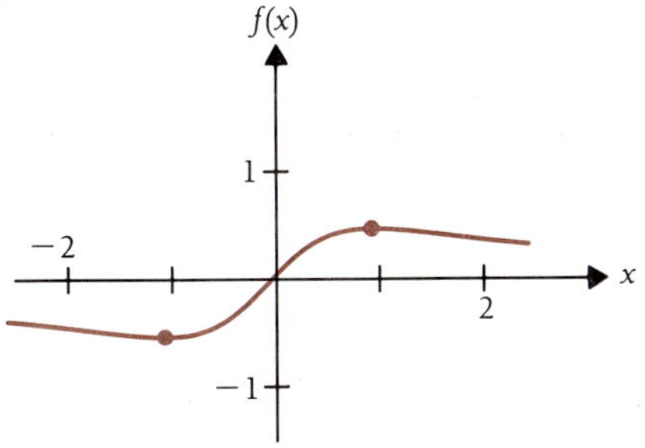

FIGURE 4.48 The graph of $f(x) = \frac{x}{x^2 + 1}$

(c) The derivative of

$$f(x) = \cos x - 2 \sin x$$

is

$$f'(x) = -\sin x - 2 \cos x.$$

Setting $f'(x) = 0$ and solving, we find

$$\begin{aligned} -\sin x - 2 \cos x &= 0, \\ \sin x &= -2 \cos x, \\ \tan x &= -2, \\ x &\approx -1.1071 + n\pi, \qquad n \in \mathbb{Z}. \end{aligned}$$

If we let $\theta = -1.1071$, then the critical points of f are the numbers $\theta_n = \theta + n\pi$. The second derivative of f is

$$f''(x) = -\cos x + 2 \sin x.$$

Evaluating f'' at θ_n, we get

$$\begin{aligned} f''(\theta_n) &= -\cos\theta_n + 2\sin\theta_n \\ &= -\cos(\theta + n\pi) + 2\sin(\theta + n\pi) \\ &= -\cos\theta\cos n\pi + \sin\theta\sin n\pi + 2(\sin\theta\cos n\pi + \cos\theta\sin n\pi). \end{aligned}$$

Since $\cos n\pi = (-1)^n$ and $\sin n\pi = 0$, we have

$$\begin{aligned} f''(\theta_n) &= (-1)^n(-\cos\theta + 2\sin\theta) \\ &= (-1)^n[-\cos(-1.1071) + 2\sin(-1.1071)] \\ &= (-1)^n[-\cos(1.1071) - 2\sin(1.1071)] \\ &= (-1)^{n+1}[\cos(1.1071) + 2\sin(1.1071)]. \end{aligned}$$

Since 1.1071 is in the first quadrant, we see that f'' will be positive if n is odd and negative if n is even. Therefore, f has a relative maximum at each point $-1.1071 + 2n\pi$, and a relative minimum at each point $-1.1071 + (2n+1)\pi \approx 2.0345 + 2n\pi$, $n \in \mathbb{Z}$ (see Figure 4.49).

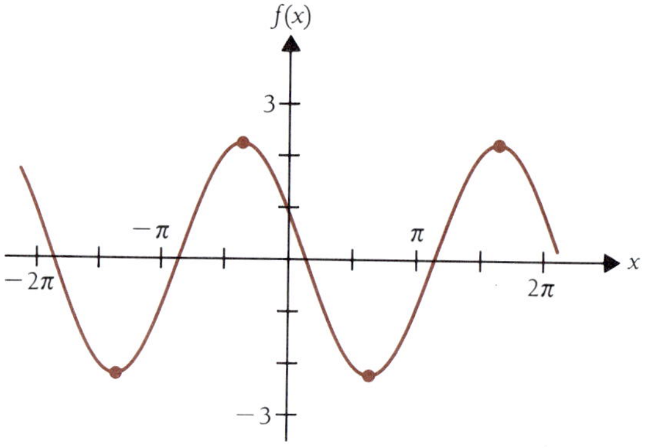

FIGURE 4.49 The graph of $f(x) = \cos x - 2\sin x$ ■

Thus far we have concentrated on finding the *location* of the relative and absolute extrema of a function. To find the extreme values themselves, we simply evaluate the function at the points at which the extrema occur.

Some Special Cases

The extreme value theorem stated that a function that is continuous on a closed interval has an absolute maximum and an absolute minimum there. To find the absolute extrema, we can evaluate the function at each endpoint of its domain and at each critical point. The greatest value obtained is the absolute maximum, and the least is the absolute minimum.

The first and second derivative tests yield only relative extrema, in general. To tell whether a given relative extremum is also an absolute extremum, we must know the global properties of the function. One special case will arise frequently in which it is easy to tell. If a continuous function has only one point at which an extremum can exist, and we determine by one of the tests that there *is* a relative extremum there, then it is also an absolute extremum (see Problem 36 below).

Exercises 4.4

For each function, locate the extrema; find the extreme values; and classify each extremum as endpoint, relative, or absolute.

1. $f(x) = 3x + 1, \quad x \in [-3, 5]$
2. $f(x) = x^2 - 2x + 1$
3. $f(x) = x^3 - 3x + 2$
4. $f(x) = 4x^3 - 21x^2 - 24x + 5$
5. $f(x) = x - x^4$
6. $f(x) = x^3 - x^6$
7. $f(x) = 1 + \dfrac{1}{x}, \quad x \in [-2, 0) \cup (0, 2]$
8. $f(x) = x - \dfrac{1}{x}$
9. $f(x) = 16 - x^2$
10. $f(x) = x^2 - 5x + 6$
11. $f(x) = (x - 1)^{2/3}, \quad x \in [0, 3]$
12. $f(x) = (x + 2)^{1/5}, \quad x > -4$
13. $f(x) = \dfrac{x}{x^2 + 4}$
14. $f(x) = \dfrac{x^2}{x^2 + 4}$
15. $f(x) = \sqrt{x - 1}$
16. $f(x) = x^{4/5}$
17. $f(x) = \begin{cases} x^2 - 1 & \text{if } x \le 1 \\ x - 1 & \text{if } x > 1 \end{cases}$
18. $f(x) = \begin{cases} -x & \text{if } x < 0 \\ (x - 1)^{1/3} & \text{if } x \ge 0 \end{cases}$
19. $f(x) = \sqrt{x^2 - 4}$
20. $f(x) = \sqrt{4 - x^2}$
21. $f(x) = x\sqrt{x^2 - 1}$
22. $f(x) = x + \sqrt{1 - x^2}$
23. $f(x) = \sin^2 x$
24. $f(x) = \tan^2 x$
25. $f(x) = \begin{cases} |x - 2| & \text{if } x \ne 2 \\ 3 & \text{if } x = 2 \end{cases}$
26. $f(x) = \begin{cases} x + 3 & \text{if } x < 2 \\ 4 & \text{if } x = 2 \\ 2x & \text{if } x > 2 \end{cases}$
27. $f(x) = \sin x + \cos x$
28. ^{c} $f(x) = \sin 3x + \cos 2x$
29. $f(x) = x + \sin x$
30. $f(x) = x - 2 \sin x$
31. $f(x) = \sin x - \cos 2x$
32. ^{c} $f(x) = 3 \sin x + \cos 3x$
33. Show that if f' is continuous and c is an inflection point of f, then f' has an extremum at c.
34. Devise a third derivative test for inflection points.
35. Is it true that if a function is continuous on a closed interval and has exactly one relative or endpoint maximum there, then that maximum is an absolute maximum?
36. Prove that if a continuous function defined over an open interval has only one critical point and has a relative extremum there, then it has an absolute extremum there.

SECTION 4.5

Extremum Problems

The theory of extrema has many applications to real-life problems that arise outside of mathematics. The techniques that we have been developing can be applied to mathematical models of such problems, enabling us to find relative or absolute maxima or minima of a physical quantity once it is modeled as a function.

The pattern for solving most extremum problems is as follows:

1. First, express the quantity to be minimized or maximized as a function of a single variable. The function may be given, or it may take some manipulation to get it. A sketch is often extremely helpful.
2. Differentiate the function, and find the critical points.
3. Test for extrema at the critical points and at endpoints of the domain, if any, and find the solution to the problem.

We will illustrate with several examples.

Examples of Extremum Problems

Example 4.22 An open rectangular box is to be made from a 10 inch by 16 inch piece of cardboard by cutting a square out of each corner, folding up the sides, and taping the sides together at the corners. What size square should be cut out of each corner so that the resulting box has maximum volume?

Solution Let x be the side of the square to be cut from each corner (see Figure 4.50). Then the dimensions of the box will be $16 - 2x$ by $10 - 2x$ by x. Hence, the volume will be

$$\begin{aligned} V &= (16 - 2x)(10 - 2x)x \\ &= 160x - 52x^2 + 4x^3. \end{aligned}$$

This function V is our model, and we want to find its absolute maximum. Actually, we do not need to find the maximum value of V, but only the value of x that yields a maximum.

Note that the domain of V is $(0, 5)$, because x must be positive and at the same time less than half the width of the cardboard. Since V is a polynomial, the only

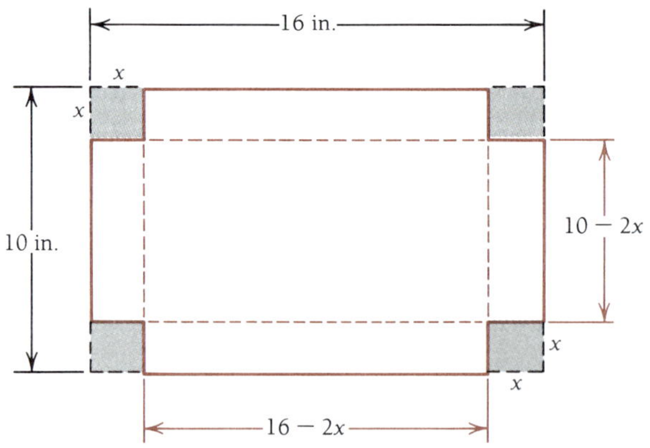

FIGURE 4.50

critical points are those for which the derivative is 0. The derivative is

$$\begin{aligned} V'(x) &= 160 - 104x + 12x^2 \\ &= 4(20 - 3x)(2 - x), \end{aligned}$$

which is 0 when $x = \frac{20}{3}$ or $x = 2$. Since $\frac{20}{3}$ is not in the domain (0, 5), we consider only the critical point $x = 2$. Using the second derivative test, we find

$$V''(x) = -104 + 24x,$$

and

$$V''(2) = -56 < 0,$$

so that V has a relative maximum at $x = 2$. It follows from Problem 36 of Section 4.4 that V has an absolute maximum at $x = 2$. Thus, a 2 inch square should be cut from each corner of the cardboard to give the box maximum volume. ■

Example 4.23 A farmer has enough fencing to build 40 feet of fence. He wishes to build a rectangular pen next to his barn, with the barn wall forming one side of the pen. What dimensions should he make the pen so as to enclose the greatest possible area?

Solution Let the pen have length l and width w, as in Figure 4.51. Then $l + 2w$ feet of fence are required, so $l + 2w = 40$. This tells us that $l = 40 - 2w$, and thus, the area enclosed is

$$\begin{aligned} A &= lw \\ &= (40 - 2w)w \\ &= 40w - 2w^2. \end{aligned}$$

FIGURE 4.51

The domain of A is $0 < w < 20$. To maximize A, we first find the critical points. The derivative is

$$A'(w) = 40 - 4w,$$

which is 0 when $w = 10$. Using the second derivative test,

$$A''(w) = -4 < 0,$$

so A is maximum when $w = 10$ (see again Problem 36 of Section 4.4). Hence, $l = 20$, and the dimensions of the pen should be 20 feet by 10 feet. ■

Example 4.24 A wire of length L is to be cut into two pieces. One piece is bent to form a circle and the other piece is bent to form a square. How should the wire be cut so that the total area of the two figures is a maximum? (Allow for the possibility of just one figure or the other.)

Solution Let x be the length of the piece bent into a circle. Then $L - x$ is the length of the piece bent into a square (see Figure 4.52). A circle with circumference x has radius $x/2\pi$, so the area of the circle is

$$\pi \left(\frac{x}{2\pi}\right)^2 = \frac{x^2}{4\pi}.$$

A square with perimeter $L - x$ has side $(L - x)/4$, so its area is

$$\frac{(L-x)^2}{16}.$$

Thus, the total area of the two figures is

$$A = \frac{x^2}{4\pi} + \frac{(L-x)^2}{16}.$$

The domain of A is $[0, L]$. If $x = 0$, we use all the wire for the square; and if $x = L$, we use all the wire for the circle. If $x = 0$, then

$$A = \frac{L^2}{16};$$

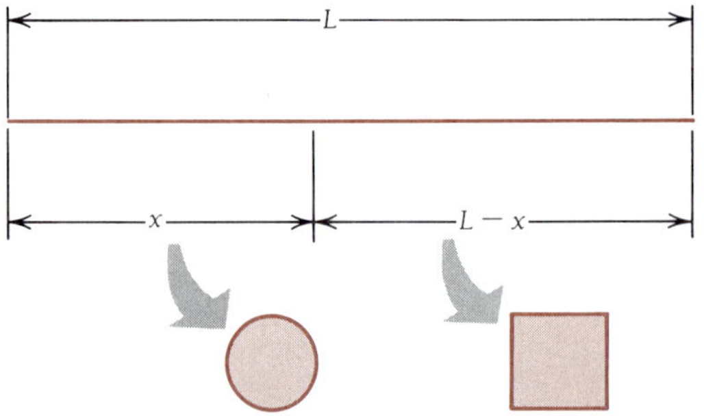

FIGURE 4.52

if $x = L$, then

$$A = \frac{L^2}{4\pi}.$$

We see, then, that for the two extreme cases, $x = L$ yields the greater area; so if only one figure is formed, it should be the circle.

We find the critical points of A by differentiation:

$$\frac{dA}{dx} = \frac{x}{2\pi} - \frac{(L-x)}{8}$$
$$= \frac{(4+\pi)x - \pi L}{8\pi},$$

which is 0 when

$$x = \frac{\pi L}{4+\pi}.$$

Now, the second derivative is

$$\frac{d^2A}{dx^2} = \frac{1}{2\pi} + \frac{1}{8},$$

which is always positive. Hence, by the second derivative test, A has a minimum when $x = \pi L/(4+\pi)$. Thus, A is maximum when $x = L$, and all the wire should be used for a circle. ■

■ **Example 4.25** A cylindrical tin can is to be manufactured that will hold a certain volume V. If materials for the ends of the can are twice as expensive as materials for the sides, what shape of can is most economical to manufacture?

Solution We assume that the cost is proportional to the surface area, since the materials for a tin can have uniform thickness. (We ignore the costs of forming the can, which are about the same for cans of any size.) Let r be the radius of the base of the can, and let h be its height. The volume is then

$$V = \pi r^2 h,$$

so that

$$h = \frac{V}{\pi r^2}.$$

Now, the surface area of the sides is $2\pi rh$, and the two ends have combined area $2\pi r^2$. Thus, the cost is proportional to

$$C = 2\pi rh + 2(2\pi r^2)$$
$$= 2\pi r\left(\frac{V}{\pi r^2}\right) + 4\pi r^2$$
$$= \frac{2V}{r} + 4\pi r^2.$$

The domain of C is $(0, \infty)$, at least theoretically. Now we seek the critical points. The derivative is

$$\frac{dC}{dr} = \frac{-2V}{r^2} + 8\pi r.$$

The derivative is 0 when

$$r = \left(\frac{V}{4\pi}\right)^{1/3}.$$

By the second derivative test, C has its absolute minimum at this value of r, as you should verify. [*Hint:* Problem 36 of Section 4.4.] For this value of r,

$$h = \frac{V}{\pi r^2} = \frac{V}{\pi (V/4\pi)^{2/3}},$$

so that

$$h = 4\left(\frac{V}{4\pi}\right)^{1/3}.$$

Thus, the most economical can will have height equal to four times the radius, or twice the diameter. ■

■ **Example 4.26** Find the point on the curve $y = \sqrt{x}$ that is nearest to the point $(1, 0)$.

Solution If (x, y) is a point on the curve $y = \sqrt{x}$, then its distance from $(1, 0)$ is

$$D = \sqrt{(x-1)^2 + y^2}.$$

But $y = \sqrt{x}$, so

$$\begin{aligned} D &= \sqrt{(x-1)^2 + x} \\ &= \sqrt{x^2 - x + 1}. \end{aligned}$$

The domain of D is $[0, \infty)$, since that is the domain of the curve. The derivative is

$$\frac{dD}{dx} = \frac{2x-1}{2\sqrt{x^2 - x + 1}},$$

which is 0 when $x = \frac{1}{2}$. Thus, $x = \frac{1}{2}$ is a critical point. The use of the second derivative test is a bit tedious, so instead we write

$$D = \sqrt{(x - \tfrac{1}{2})^2 + \tfrac{3}{4}}$$

by completing the square. This shows that $D = \frac{1}{2}\sqrt{3}$ when $x = \frac{1}{2}$, and D is always greater than that when $x \neq \frac{1}{2}$. Thus, the point $(\frac{1}{2}, 1/\sqrt{2})$ is the point on the curve nearest to $(1, 0)$. ■

■ **Example 4.27** A swimmer in a lake is 70 meters from shore when he discovers he can swim no more. The lifeguard who hears his cry for help is 100 meters down the shore from the point nearest the swimmer. The lifeguard can run at a rate of 6 meters per second along the shore, and can average 1 meter per second in the water. How far along the shore should the lifeguard run before jumping into the water so as to reach the swimmer as quickly as possible?

Solution Let x be the distance along the shore the lifeguard should run (see Figure 4.53). Then the distance the lifeguard must swim is $\sqrt{(100-x)^2+70^2}$. Since the time spent is the distance traveled divided by the rate, the total time the lifeguard takes to reach the swimmer is

$$T=\frac{x}{6}+\sqrt{(100-x)^2+70^2}.$$

The domain of T is $x \in [0, 100]$. If $x=0$ (the lifeguard immediately jumps into the lake), then

$$T=\sqrt{100^2+70^2}\approx 122 \text{ seconds.}$$

If $x=100$ (the lifeguard swims the shortest distance possible), then

$$T=\frac{100}{6}+\sqrt{70^2}\approx 86.7 \text{ seconds.}$$

Seeking critical points, we differentiate:

$$\frac{dT}{dx}=\frac{1}{6}-\frac{100-x}{\sqrt{(100-x)^2+70^2}}.$$

Setting the derivative equal to 0, we get

$$\begin{aligned}\sqrt{(100-x)^2+70^2}&=6(100-x),\\(100-x)^2+70^2&=36(100-x)^2,\\35(100-x)^2&=70^2,\\(100-x)^2&=140,\\100-x&=\pm 2\sqrt{35},\\x&=100\pm 2\sqrt{35}.\end{aligned}$$

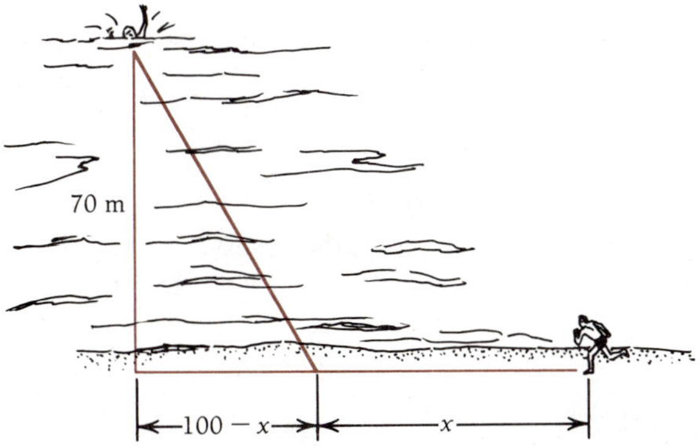

FIGURE 4.53

Since the point $100 + 2\sqrt{35}$ is not in the domain of T, we ignore it. At $x = 100 - 2\sqrt{35}$, we find that

$$T \approx 85.7 \text{ seconds.}$$

Since this is a minimum, the lifeguard should run $100 - 2\sqrt{35} \approx 88.2$ meters along the shore before jumping into the lake. ■

■ **Example 4.28** A bus line offers group tours for a fare of \$2 per person when the number of persons in the group does not exceed 12. If the size of the group exceeds 12 persons, the fare for each person is reduced by 10¢ for each person in excess of 12. That is, for 13 persons, the fare is \$1.90 per person; for 14 persons, the fare is \$1.80 per person; and so forth. What size group will result in the largest revenue for the company?

Solution Let n be the number of persons in the group. Then n exceeds 12 by $n - 12$, so each person's fare is $2.00 - 0.10(n - 12)$ dollars. Hence, the total revenue, or fare paid by the whole group, is

$$\begin{aligned} R &= n[2.00 - 0.10(n - 12)] \\ &= n(2.00 - 0.10n + 1.20) \\ &= n(3.20 - 0.10n) \\ &= 3.20n - 0.10n^2. \end{aligned}$$

The domain of R is the set of positive integers, but we will proceed as though the domain of R were $(0, \infty)$ and see what happens. We seek a critical point by setting the derivative equal to 0:

$$\begin{aligned} R'(n) = 3.20 - 0.20n &= 0, \\ n &= 16. \end{aligned}$$

By the second derivative test, R has a maximum at $n = 16$. This maximum is absolute by Problem 36 of Section 4.4. Luckily, this result is in the actual domain of R. ■

■ **Example 4.29** A sign hangs over the sidewalk with its lower edge 4 feet above eye level. If the sign is 3 feet tall, at what point on the sidewalk will the sign subtend the largest angle at an observer's eye?

Solution Let the observer be a distance x from the point directly below the sign, and let α be the angle subtended by the sign at the observer's eye (see Figure 4.54). Let β be the angle of elevation to the bottom of the sign. From the figure, it is easily seen that

$$\tan(\alpha + \beta) = \frac{7}{x} \qquad \text{and} \qquad \tan \beta = \frac{4}{x}.$$

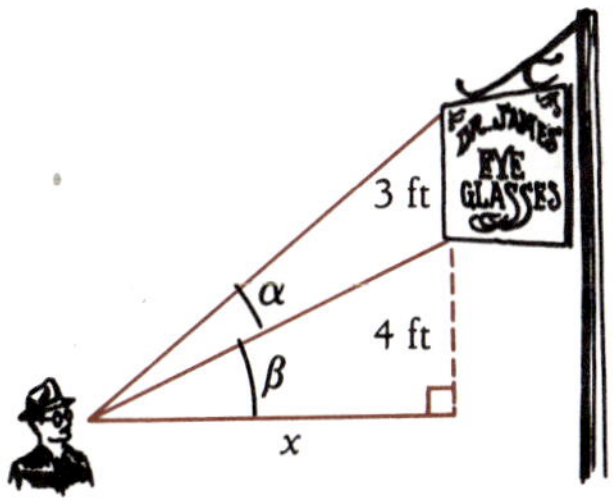

FIGURE 4.54

We use the tangent addition formula and substitute for $\tan \beta$:

$$\tan \alpha + \tan \beta = \left(\frac{7}{x}\right)(1 - \tan \alpha \tan \beta),$$

$$\tan \alpha + \frac{4}{x} = \left(\frac{7}{x}\right)\left[1 - \left(\frac{4}{x}\right) \tan \alpha\right],$$

$$\left(1 + \frac{28}{x^2}\right) \tan \alpha = \frac{3}{x},$$

$$\tan \alpha = \frac{3x}{x^2 + 28}.$$

This equation gives α implicitly as a function of x. We seek the value of x that makes α a maximum. Differentiating implicitly, we have

$$\sec^2 \alpha \frac{d\alpha}{dx} = \frac{3(x^2 + 28) - 3x(2x)}{(x^2 + 28)^2},$$

$$\frac{d\alpha}{dx} = \frac{84 - 3x^2}{(x^2 + 28)^2 \sec^2 \alpha}.$$

This is 0 when $x^2 = 28$, or $x = 2\sqrt{7}$. Thus, at a distance (along the sidewalk) of $2\sqrt{7} \approx 5.3$ feet, the sign will subtend the maximum angle. That it is indeed an absolute maximum is clear from the geometry of the situation. ■

A variety of applications following this pattern will be found in the exercises. The steps are always the same as those outlined at the beginning of this section: Draw a figure, express the quantity to be maximized or minimized as a function of a single variable, differentiate, find the critical points, and test the critical points and endpoints, if any, for extrema.

Exercises 4.5

1. In Example 4.22, suppose the dimensions of the sheet of cardboard are 8 inches by 15 inches. What are the dimensions of the resulting box of maximum volume?
2. Repeat Problem 1 for a square sheet of cardboard measuring s inches on a side.
3. A rectangular pasture is to be fenced off beside a river, with no need for fencing along the river. If there is material available for 400 yards of fence, what is the area of the largest possible pasture that can be enclosed?

4. A dog owner wishes to build several kennels by making a long rectangular pen and putting three fences across it to make four small pens (see the figure). If there is 100 feet of fencing available, what dimensions should the long rectangle be in order to enclose the maximum area?

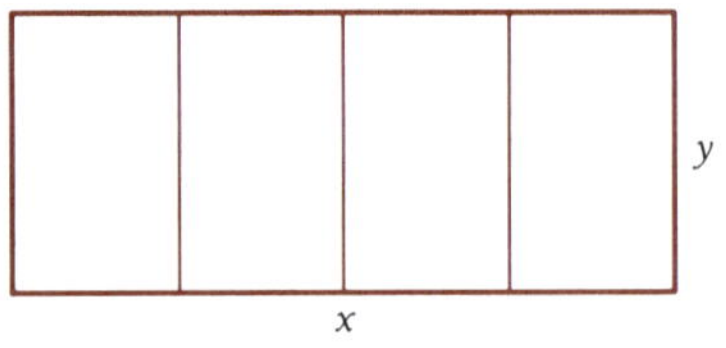

5. A piece of commercial property is to be enclosed by fencing on the front and two sides. Fencing for the sides costs \$2.50 per foot, and fencing for the front costs \$4.00 per foot. What are the dimensions of the largest rectangular lot that can be so enclosed for \$800?

6. The strength of a beam is proportional to the product of its width W and the square of the depth D (see the figure). What are the dimensions of the strongest beam that can be cut from a circular log of radius r?

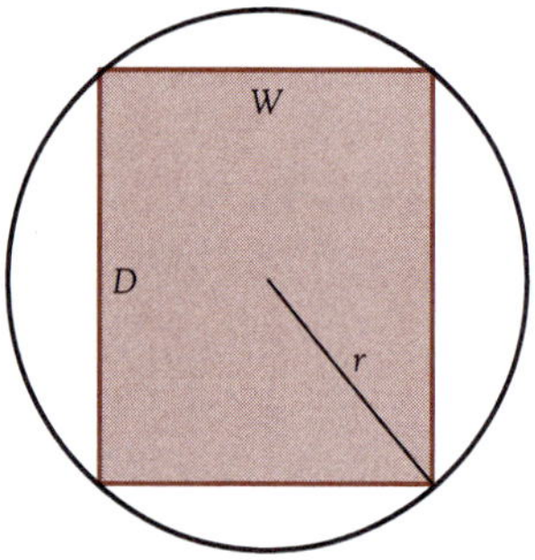

7. Show that among rectangles with a given perimeter, the square has the greatest area.

8. Find two numbers whose sum is 50 and whose product is a maximum.

9. A rectangle has its base on the x-axis and its upper vertices on the curve $y = 4 - x^2$ (see the figure). Find the largest possible area of such a rectangle.

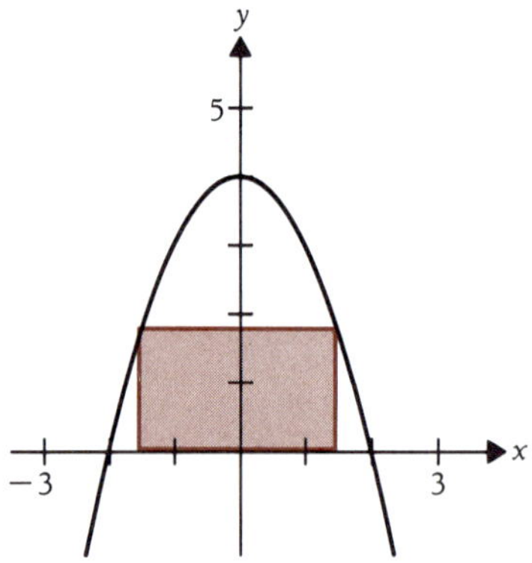

C10. A rectangle has its base on the x-axis and its upper vertices on the curve $y = \sin x$, $0 < x < \pi$ (see the figure). What are the dimensions of the rectangle having maximum area?

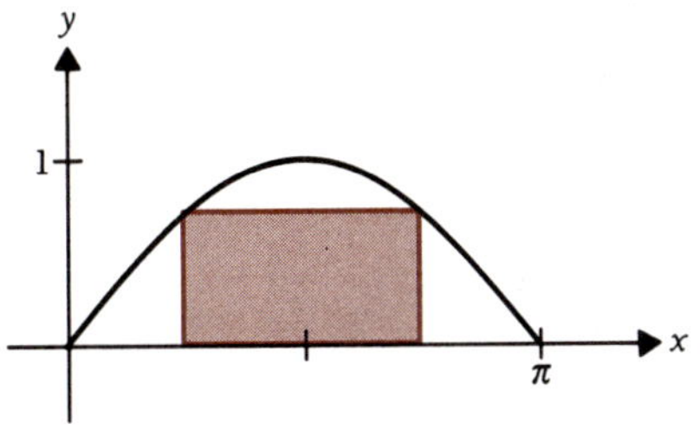

11. An isosceles triangle has base b and height h. A rectangle is inscribed in the triangle, with one side lying on the base (see the figure). What are the dimensions of the largest such rectangle?

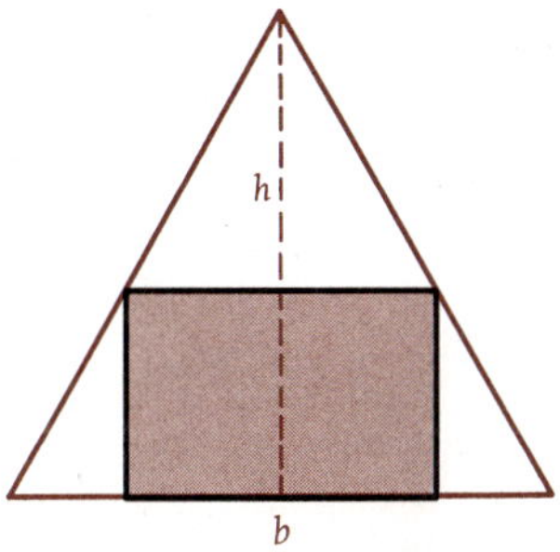

12. A trapezoid has three sides of length L (see the figure). What is the maximum area the trapezoid can have?

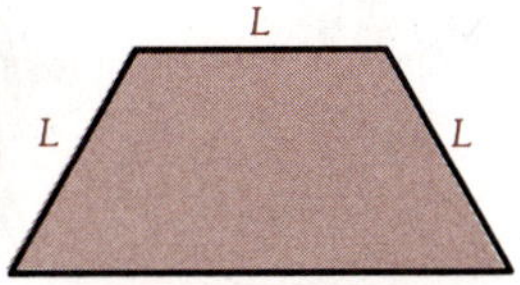

13. Find the volume of the largest cylinder that can be inscribed in a hemisphere of radius r (see the figure).

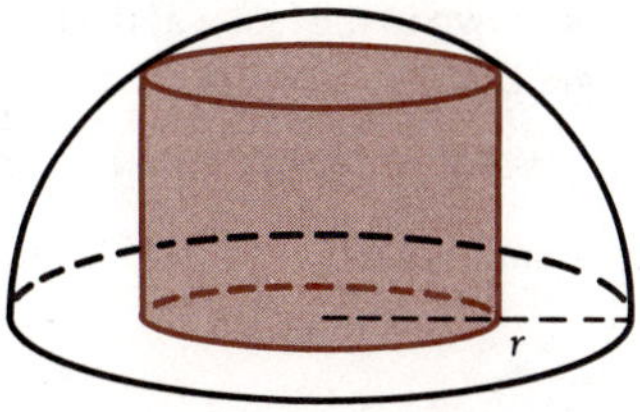

14. Find the volume of the largest right circular cylinder that can be inscribed in a cone of height h and base radius r (see the figure).

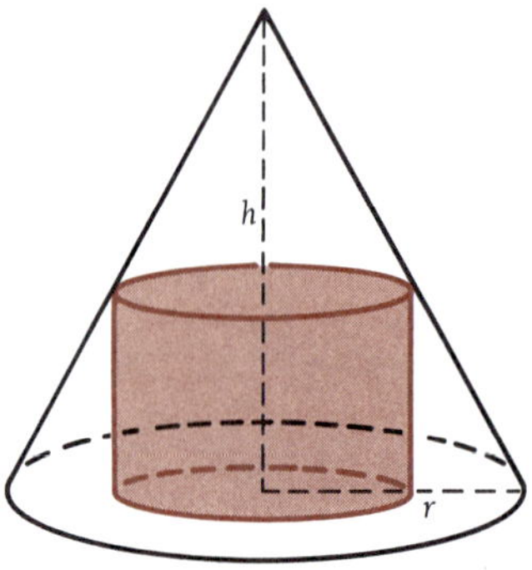

15. A church window is a rectangle surmounted by a semicircle (see the figure). The semicircle is to be stained glass, which admits half as much light per unit area as the clear glass of the rectangle. If the perimeter of the window is 204 inches, what should the dimensions of the rectangle be in order that the maximum amount of light is admitted?

16. Show that among regular convex polygons with fixed perimeter, the greater the number of sides, the greater the area.

17. A cell in a honeycomb is a regular hexagonal prism, open at the front and having a trihedral apex at the back, as shown in the figure at the top of the next page. The surface area of a cell with apex angle t, $0 < t < \pi/2$, is

$$S(t) = 6ab + (\tfrac{3}{2}b^2)(\sqrt{3}\csc t - \cot t).$$

Show that the surface area is a minimum if $t \approx 0.955$. (This is the angle actually found in beehives.)

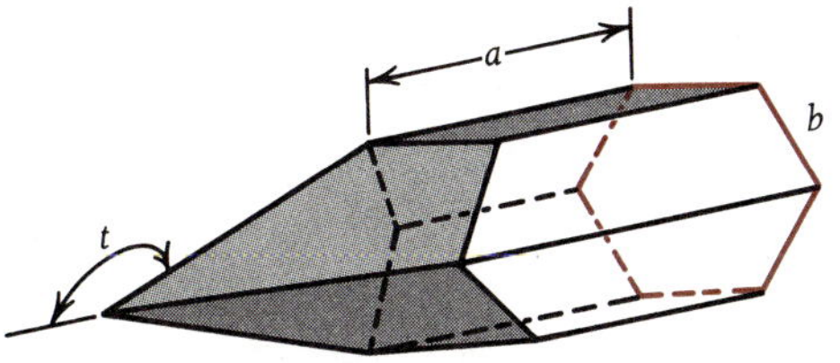

18. The owner of a professional football team finds that if t tickets are sold for a home game, then the profit will be

$$P(t) = 12t - (0.0001)t^2 - 200{,}000.$$

How many tickets should be sold for a maximum profit? What does this say (if anything) about how large the team's stadium should be?

19. In Example 4.24, how should the wire be cut so that the total area is a minimum?

20. A wire of length L is cut into two pieces; one piece is bent into a square and the other is bent into an equilateral triangle. How should the wire be cut so that the total area of the figures is:

(a) A maximum? **(b)** A minimum?

(Allow for just one of the two figures in each case.)

21. A box with a square base is to hold 500 cubic inches. Material for the top and bottom of the box costs 72¢ per square foot, and material for the sides costs 48¢ per square foot. Find the dimensions and cost of the most economical box.

22. A box with a square base is to contain 1000 cubic centimeters. Material for the bottom of the box costs 6¢ per square centimeter; for the sides, 10¢ per square centimeter; and for the top, 12¢ per square centimeter. What are the dimensions of the most economical box?

23. A tin can is a cylinder with two ends. Show that the can with the least amount of material and containing a given volume has height equal to its diameter.

24. A certain company makes a tin can from a rectangle and two squares of sheet metal. The rectangle is formed into a cylinder, and ends for the cylinder are stamped from the two squares. The "corners" of the squares are wasted (see the figure). Find the relative dimensions of the rectangle and squares that yield the most economical can containing a given volume.

25. In a certain orchard, it is found that if 40 trees are planted per acre, the yield per tree is 6 bushels of fruit. For each additional tree per acre, the yield per tree drops by $\frac{1}{8}$ bushel. How many trees per acre give the greatest yield per acre?

26. In a particular apartment complex of 80 units, it is found that all units remain occupied when the rent is \$400 per month. For each \$20 increase in the rent, one unit becomes vacant, on the average. Occupied units require \$40 per month for maintenance, while vacant units require none. Fixed costs for the buildings are \$24,000 per month. What rent should be charged for maximum profit?

27. It costs a certain manufacturer $4000 + 150n$ dollars to make n television sets per week. If n sets per week are produced, they will sell for $250 - 0.5n$ dollars each. How many television sets per week should be produced for maximum profit?

28. Explain why the economist says that if profit is maximized, then marginal cost equals marginal revenue. (See Example 3.11.)

29. A piano dealer sells 160 pianos per year. It costs \$250 per year to store one piano. When a shipment of pianos is ordered, it costs \$50 to process the order and \$120 per piano to have the pianos delivered. How many times per year, and in what lot size, should the dealer order pianos so as to minimize inventory costs? [*Hint:* If the dealer orders n pianos at a time and does not reorder until they are sold, then average stock on hand will be $n/2$, and orders will have to be placed $160/n$ times per year.]

30. A sporting goods store sells 360 tennis racquets per year. It costs \$1 per year to store a tennis racquet. To order racquets, it costs \$5 plus \$0.50 per racquet. How many times per year, and in what lot size, should tennis racquets be ordered to minimize inventory costs?

31. A **reproduction curve** for an animal species is a function $f(P)$ that gives next year's population of the animal as a function of this year's population P. Given an initial population P, the increase $f(P) - P$ during the next year may be removed, by a natural predator, for example, without depleting the population. Such a removal is called **sustainable.** The number of animals removed each year in a sustainable removal is thus

$$r(P) = f(P) - P.$$

If f is differentiable, show that the maximum sustainable removal occurs at that value of P for which $f'(P) = 1$.

32. A certain animal species has the reproduction curve $f(P) = P(11 - P)$, with P measured in thousands. Find the maximum sustainable removal and the population at which it occurs.

33. A tapestry 14 feet high hangs on a wall with its lower edge 2 feet above eye level (see the figure at the top of the next page). How far from the wall should a viewer stand so as to get the best view of the tapestry? (The "best" view is with θ as large as possible.)

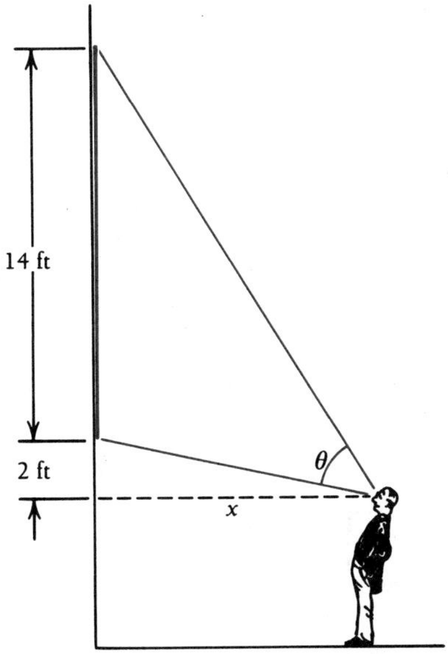

34. A corridor 3 feet wide has a right-angled turn in it (see the figure). What is the length of the longest thin rod that will pass horizontally around the corner without bending?

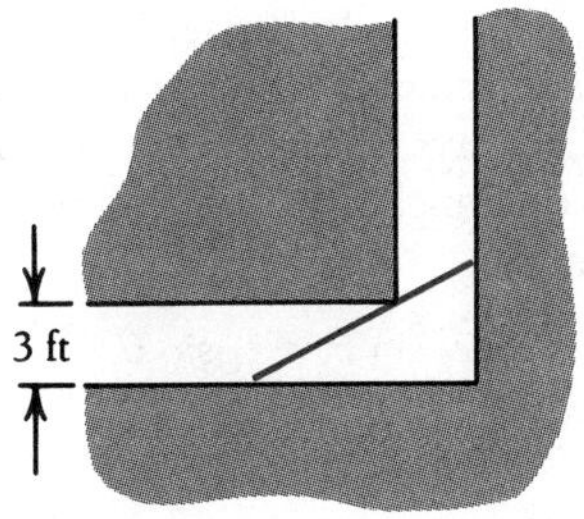

C35. Find the point on the curve $xy = x + 1$ that is nearest the origin.

36. Find the point on the curve $y = \sqrt{4 - x}$ that is nearest the origin.

37. A crook decides to hold up a pedestrian along a certain block. At each end of the block there is a street light, but there are no other lights on that block. Maintenance problems being as they are, one of the street lights is six times as bright as the other. The intensity of light at a point is jointly proportional to the brightness of the source and the reciprocal of the square of the distance from the source. If the crook wants to accost his victim at the darkest point along the street, where should he station himself?

38. A student estimates that her learning efficiency E depends on the number t of hours since she awoke by the formula

$$E(t) = 62.7t - 16t^2 + 1.47t^3 - 0.044t^4.$$

If she awakens at 8:00 AM, when does she reach her maximum learning efficiency?

39. A fence 6 feet high is 2 feet from a building (see the figure). How long is the shortest ladder that will rest on the ground and against the building, clearing the fence?

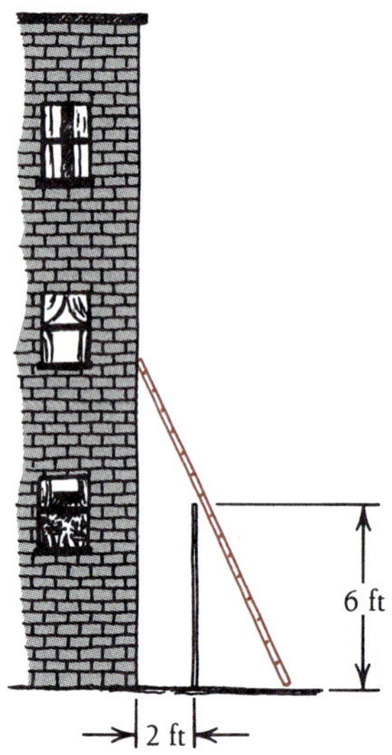

40. In optics, **Fermat's principle** states that a light ray going from A to B, located as in the figure, will follow the path that takes less time than any other path with the same endpoints. If light passes from a medium in which it has speed c_1 into another medium in which it has speed c_2, the light rays will be bent, or **refracted** (see the figure). Using Fermat's principle, show that the angle θ_i of the **incident ray** is related to the angle θ_r of the **refracted ray** by the formula

$$c_2 \sin \theta_i = c_1 \sin \theta_r$$

(known as **Snell's law**).

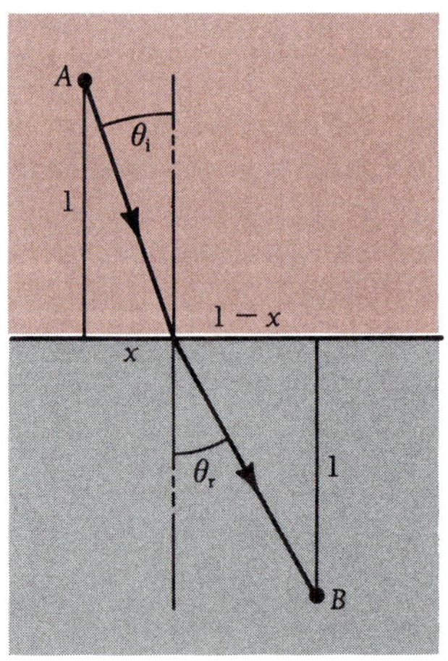

41. In Example 4.28, how would we have interpreted the result if the maximum of R had occurred at $n = 16.2$, a value not actually in the domain of R?

SECTION 4.6

Antiderivatives

The rate of change of one variable with respect to another can be found by differentiation if an equation relating the two variables is known. For example, the velocity and acceleration of an object can be found by differentiating if the equation of motion is known.

However, in many instances it is much easier to measure the rate of change than it is to tell how the two variables are related. For example, an object released in the earth's gravitational field is subject to a known acceleration, but its subsequent velocity and position may not be known. Thus, it often occurs, especially in physical applications, that the derivative of a function is known when the function itself is not yet known. In this section, we will consider the problem of finding a function whose derivative is given.

The Antiderivative

The operation of differentiation is used to find the derivative of a function. The inverse operation of finding a function when its derivative is given is called **antidifferentiation.** Just as f' is called the derivative of a function f, we will call f an **antiderivative** of f'. Following is a formal definition.

DEFINITION 4.6

An **antiderivative** of a function f is any function g such that $g'(x) = f(x)$ for each x in the domain of f.

■ **Example 4.30**

Find an antiderivative of:

(a) $2x + 3$ **(b)** $12x^2$

Solution

(a) We find, by trial and error, that the function $x^2 + 3x$ has $2x + 3$ as its derivative, so $x^2 + 3x$ is an antiderivative of $2x + 3$.

(b) The antiderivative must involve x^3 if x^2 is to be part of the derivative. Also, the coefficient 12 can only arise if 4 is already a coefficient. Thus, we see that $4x^3$ is an antiderivative of $12x^2$. ■

The question as to what kinds of functions have antiderivatives will be considered more fully in the next chapter. We will see later that if a function is continuous on a closed interval, then it has an antiderivative on that interval.

It is not difficult to see that a function can have more than one antiderivative. In fact, if g is an antiderivative of f and C is a constant, then $g(x) + C$ also has derivative $f(x)$, because the derivative of a constant is 0.

Example 4.31 Describe the antiderivatives of the function $f(x) = 2x$.

Solution We quickly see that the antiderivative must involve x^2, so any function of the form $g(x) = x^2 + C$ is an antiderivative of f, where C is any constant. ∎

Thus, two functions that differ by a constant have the same derivative. Conversely, two functions that have the same derivative differ only by a constant, as the next two theorems show. The first theorem tells us what kind of function can have 0 as its derivative.

THEOREM 4.9 If the function f is differentiable in (a, b) and $f'(x) = 0$ for every $x \in (a, b)$, then f is constant on (a, b).

Proof Suppose f is not constant on (a, b). Then there exist points x_1 and x_2 in (a, b) with $x_1 < x_2$ and such that $f(x_1) \neq f(x_2)$. By the mean value theorem applied to f on the interval $[x_1, x_2]$, there is a point $c \in (x_1, x_2)$ such that

$$f'(c) = \frac{f(x_2) - f(x_1)}{x_2 - x_1}.$$

But then $f'(c) \neq 0$, a contradiction; therefore, f must be constant on (a, b).

This theorem assures us that only a constant function can have its derivative identically 0.

THEOREM 4.10 If functions F and G are differentiable in (a, b) and $F'(x) = G'(x)$ for each $x \in (a, b)$, then there is a constant C such that $G(x) = F(x) + C$ for all $x \in (a, b)$.

Proof Let $h(x) = G(x) - F(x)$. Then $h'(x) = G'(x) - F'(x) = 0$ in (a, b). Hence, by Theorem 4.9, h is constant on (a, b). That is, there is a constant C such that $G(x) - F(x) = h(x) = C$, or $G(x) = F(x) + C$ for all $x \in (a, b)$.

Thus, we may conclude that if $F(x)$ is one antiderivative of $f(x)$, then every antiderivative of $f(x)$ will be of the form $F(x) + C$, C a constant. We might think of $F(x) + C$ as the **general antiderivative** of f. In Example 4.31, for instance, all the possible antiderivatives are of the form $x^2 + C$.

The Indefinite Integral

The general antiderivative of a function has a special name.

DEFINITION 4.7 The **indefinite integral** of a function f is the general antiderivative of f, and is denoted by the symbol

$$\int f(x)\,dx. \tag{1}$$

The symbol $\int$ in the above definition is called the **integral sign**. The function $f(x)$ in expression (1) is called the **integrand**. The variable x is referred to as a **dummy variable**, because if x is replaced by any other symbol throughout expression (1), the meaning is unchanged. That is, all the expressions

$$\int f(x)\,dx, \qquad \int f(u)\,du, \qquad \int f(t)\,dt, \qquad \text{and} \qquad \int f(\theta)\,d\theta$$

represent the indefinite integral of f.

Just as the symbol D_x is a differential operator, we may think of the symbol $\int \ldots dx$ as an "antidifferential" operator. While D_x preceding a function yields the derivative of the function, the symbol $\int \ldots dx$ surrounding a function yields the general antiderivative, or indefinite integral, of the function.

If F is an antiderivative of f, then we may write

$$\int f(x)\,dx = F(x) + C. \tag{2}$$

Since it is understood that C is an arbitrary real number, the expression $F(x) + C$ represents all the antiderivatives of f.

If F is an antiderivative of f, then we also have $F'(x) = f(x)$. Hence,

$$\int d(F(x)) = \int F'(x)\,dx = \int f(x)\,dx = F(x) + C. \tag{3}$$

Finding Antiderivatives

In general, the process of antidifferentiation is more difficult than differentiation. Here, we will consider only some simple cases, and postpone a broader approach to later chapters. The following antidifferentiation formulas follow from (2) and (3), and can be verified by differentiation:

$$\int du = u + C, \tag{4}$$

$$\int x^n\,dx = \frac{x^{n+1}}{n+1} + C, \qquad n \neq -1, \tag{5}$$

$$\int [f(x) + g(x)]\,dx = \int f(x)\,dx + \int g(x)\,dx, \tag{6}$$

$$\int k \cdot f(x)\,dx = k \int f(x)\,dx, \tag{7}$$

$$\int \cos u \, du = \sin u + C, \tag{8}$$

$$\int \sin u \, du = -\cos u + C, \tag{9}$$

$$\int \sec^2 u \, du = \tan u + C, \tag{10}$$

$$\int \csc^2 u \, du = -\cot u + C, \tag{11}$$

$$\int \sec u \tan u \, du = \sec u + C, \tag{12}$$

$$\int \csc u \cot u \, du = -\csc u + C. \tag{13}$$

The following examples illustrate the use of these formulas.

Example 4.32 Find the antiderivatives:

(a) $\int dx$ (b) $\int x^4 \, dx$ (c) $\int x^{-3} \, dx$

(d) $\int x^{2/3} \, dx$ (e) $\int 4x^2 \, dx$ (f) $\int k \, dx$

(g) $\int (x^4 + x^{-4}) \, dx$ (h) $\int (x^2 + 3x + 2) \, dx$ (i) $\int (2 \cos x - 3 \sin x) \, dx$

(j) $\int \frac{1 - \sin x}{\cos^2 x} \, dx$

Solution

(a) $\int dx = x + C$ by formula (4).

(b) $\int x^4 \, dx = \frac{x^5}{5} + C$ by formula (5).

(c) $\int x^{-3} \, dx = \frac{x^{-2}}{-2} + C = \frac{-1}{2x^2} + C$ by formula (5).

(d) $\int x^{2/3} \, dx = \frac{x^{5/3}}{\frac{5}{3}} + C = \frac{3x^{5/3}}{5} + C$ by formula (5).

(e) $\int 4x^2 \, dx = 4 \int x^2 \, dx = 4\left(\frac{x^3}{3} + C\right) = \frac{4x^3}{3} + 4C.$

Here, $4C$ is still an arbitrary constant, so we customarily write simply C for $4C$. Thus,

$$\int 4x^2 \, dx = \frac{4x^3}{3} + C.$$

(f) $\int k\,dx = k\int dx = k(x + C) = kx + C.$

(g)
$$\int (x^4 + x^{-4})\,dx = \int x^4\,dx + \int x^{-4}\,dx$$
$$= \frac{x^5}{5} + C_1 + \frac{x^{-3}}{-3} + C_2$$
$$= \frac{x^5}{5} - \frac{1}{3x^3} + C_1 + C_2.$$

Since $C_1 + C_2$ is still an arbitrary constant, we write

$$\int (x^4 + x^{-4})\,dx = \frac{x^5}{5} - \frac{1}{3x^3} + C.$$

(h)
$$\int (x^2 + 3x + 2)\,dx = \int x^2\,dx + 3\int x\,dx + 2\int dx$$
$$= \frac{x^3}{3} + C_1 + 3\left(\frac{x^2}{2} + C_2\right) + 2(x + C_3)$$
$$= \frac{x^3}{3} + \frac{3x^2}{2} + 2x + C.$$

(i)
$$\int (2\cos x - 3\sin x)\,dx = 2\int \cos x\,dx - 3\int \sin x\,dx$$
$$= 2\sin x - 3(-\cos x) + C$$
$$= 2\sin x + 3\cos x + C.$$

(j) Using trigonometric identities, we find that

$$\int \frac{1 - \sin x}{\cos^2 x}\,dx = \int (\sec^2 x - \sec x\tan x)\,dx$$
$$= \tan x - \sec x + C. \qquad \blacksquare$$

We want to generalize some of the antidifferentiation formulas given above. It is not difficult to verify the following by differentiation:

$$\int \cos au\,du = \frac{1}{a}\sin au + C \tag{14}$$

$$\int \sin au\,du = -\frac{1}{a}\cos au + C \tag{15}$$

These formulas will be quite useful to us. Consider this simple example.

Example 4.33 Find $\int \cos 3x\,dx$.

Solution By formula (14), we have

$$\int \cos 3x\,dx = \frac{1}{3}\sin 3x + C. \qquad \blacksquare$$

Differential Equations

An equation that involves the derivative or differential of a function $y = f(x)$ is called a **differential equation**. Such equations arise very frequently in the applications of mathematics.

To **solve** a differential equation is to find a function y, usually by antidifferentiation, that satisfies the equation. In general, a differential equation does not have just one solution, since a function has many antiderivatives. Often, particular values of the function y are given; these values are called **boundary conditions**. From among the many solutions of a differential equation we may choose the **particular solution** satisfying the boundary conditions.

■ **Example 4.34** Solve the differential equation

$$\frac{dy}{dx} = x^2 - 1,$$

subject to the boundary condition $y(3) = 2$.

Solution Noting that the equation gives the derivative of y explicitly as a function of x, it follows that y is one of the antiderivatives of the function of x,

$$y = \frac{x^3}{3} - x + C \tag{16}$$

for some constant C. Equation (16) is the **general solution** of the differential equation. In Figure 4.55, each curve is a solution of the equation. Now using the boundary condition $y(3) = 2$, equation (16) gives us

$$2 = \frac{3^3}{3} - 3 + C,$$

$$C = -4.$$

Hence, the particular solution of the differential equation satisfying the boundary condition is

$$y = \frac{x^3}{3} - x - 4.$$

Only the curve with $C = -4$ in Figure 4.55 satisfies the condition that $y = 2$ when $x = 3$. ■

A **separable** differential equation is one that can be written in the form (using differential notation)

$$f(y)\,dy = g(x)\,dx. \tag{17}$$

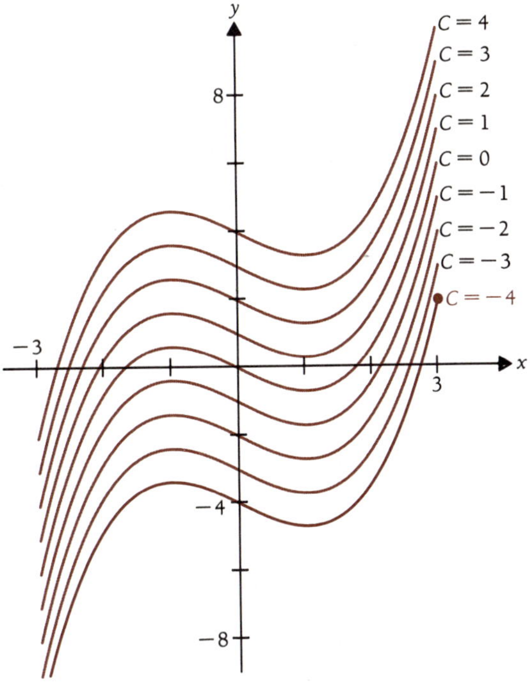

FIGURE 4.55 Graphs of curves $y = \frac{x^3}{3} - x + C$

When a differential equation is written in this form, we say that the variables x and y are **separated**, because they are on opposite sides of the equation. Now, since the differentials $f(y)\, dy$ and $g(x)\, dx$ are equal, their antiderivatives are the same, or

$$\int f(y)\, dy = \int g(x)\, dx.$$

Finally, if F is an antiderivative of f and G is an antiderivative of g, we can write

$$F(y) = G(x) + C,$$

and we have the general solution of the original equation. We can leave it in this implicit form, or perhaps we can solve for y explicitly.

Here are some examples.

Example 4.35 Solve the differential equation

$$\frac{dy}{dx} = xy^2,$$

subject to the boundary condition $y = 1$ when $x = 0$. (A boundary condition with $x = 0$ is also called an **initial condition**.)

Solution Separating the variables, we have

$$y^{-2}\,dy = x\,dx,$$
$$\int y^{-2}\,dy = \int x\,dx,$$
$$\frac{-1}{y} = \frac{x^2}{2} + C.$$

Setting $x = 0$ and $y = 1$, we find that $C = -1$. Therefore,

$$\frac{-1}{y} = \frac{x^2}{2} - 1,$$

or, solving for y,

$$y = \frac{-2}{x^2 - 2}.$$

■

■ **Example 4.36** Solve the differential equation

$$y'y^2 + y' = x,$$

subject to the condition $y(1) = 0$.

Solution Setting $y' = dy/dx$ and separating the variables, we get

$$(y^2 + 1)\,dy = x\,dx,$$
$$\int (y^2 + 1)\,dy = \int x\,dx,$$
$$\frac{y^3}{3} + y = \frac{x^2}{2} + C,$$
$$2y^3 + 6y - 3x^2 = 6C.$$

This equation defines the general solution implicitly. Now, letting $x = 1$ and $y = 0$, we find $6C = -3$. Hence, the particular solution is

$$2y^3 + 6y - 3x^2 = -3.$$

■

■ **Example 4.37** Solve the differential equation

$$\frac{dy}{dx} = x \sec y.$$

Solution We solve by separating the variables:

$$\cos y\,dy = x\,dx,$$
$$\int \cos y\,dy = \int x\,dx,$$
$$\sin y = \frac{x^2}{2} + C.$$

This equation defines the general solution implicitly. If a boundary condition were given, the constant C could be evaluated. ■

■ **Example 4.38** Suppose water evaporates at a rate proportional to the area of the surface of the water. A conical water tank of depth 1 meter and top diameter 1 meter is originally full of water. After 1 hour of exposure to a warm stream of dry air, the water level has fallen 1 centimeter due to evaporation. Under these conditions, how long will it take for all of the water to evaporate?

Solution Let V be the volume of water in the tank at any time t; let h be the depth of the water and r, the radius of the corresponding surface of the water (see Figure 4.56). That is, V, h, and r are all functions of t. From properties of similar triangles, we see that $r = h/2$. We can therefore express V in terms of h:

$$V = \frac{\pi r^2 h}{3} = \frac{\pi}{3}\left(\frac{h}{2}\right)^2 h = \frac{\pi h^3}{12}.$$

Differentiating implicitly with respect to t, we get

$$\begin{aligned} \frac{dV}{dt} &= \frac{\pi(3h^2)}{12}\frac{dh}{dt} \\ &= \frac{\pi h^2}{4}\frac{dh}{dt}. \end{aligned}$$

This gives us the rate of change of V in terms of the rate of change of h. But we are told that the evaporation rate is proportional to the surface area of the water, or

$$\frac{dV}{dt} = k(\pi r^2) = \frac{k\pi h^2}{4},$$

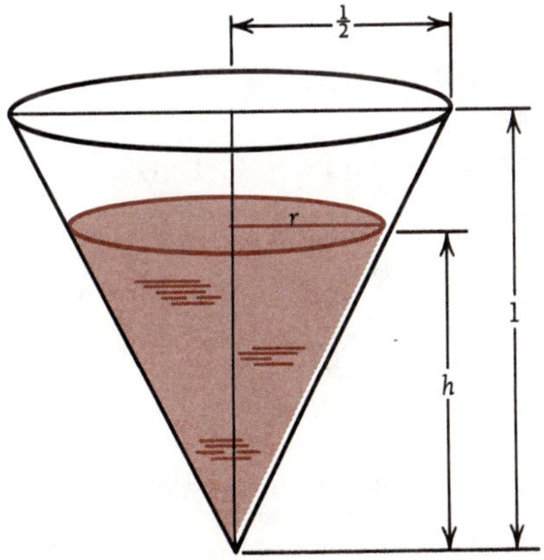

FIGURE 4.56 A conical water tank

where k is the proportionality constant. Combining the last two equations, we get

$$\frac{dh}{dt} = k.$$

This means that the water level falls at a constant rate. Using the fact that it took 1 hour to fall 1 centimeter, we can guess that it will take 100 hours to fall 1 meter. This is verified by solving the above differential equation:

$$dh = k\,dt,$$
$$h = kt + C.$$

When $t = 0$, we have $h = 1$ meter, because the tank was initially full. Hence, $C = 1$, and

$$h = kt + 1.$$

When $t = 1$ hour, $h = 0.99$ meter, and we find that $k = -0.01$. Thus,

$$h = 1 - 0.01t.$$

It follows that when $h = 0$, $t = 100$. ■

■ **Example 4.39** A certain tree grows in height at a rate inversely proportional to its height. If it grows 3 feet in the first 3 years, how tall will it be at the end of 4 years?

Solution Let h be the height of the tree at any time t. Then dh/dt is the rate at which the tree grows, so that for some proportionality constant k,

$$\frac{dh}{dt} = \frac{k}{h}.$$

We solve this equation by separating the variables:

$$h\,dh = k\,dt,$$
$$\int h\,dh = \int k\,dt,$$
$$\frac{h^2}{2} = kt + C,$$
$$h^2 = 2kt + C.$$

Assuming $h = 0$ when $t = 0$, we find $C = 0$. Thus,

$$h^2 = 2kt.$$

Now, $h = 3$ when $t = 3$, so $k = \frac{3}{2}$, and

$$h^2 = 3t,$$
$$h = \sqrt{3t}.$$

Having solved the equation, we answer the question by letting $t = 4$. The height will be $h = \sqrt{12} = 2\sqrt{3} \approx 3.5$ feet. ■

Exercises 4.6

Carry out the antidifferentiation.

1. $\int x\,dx$
2. $\int x^5\,dx$
3. $\int x^{-1/2}\,dx$
4. $\int \sqrt[3]{x}\,dx$
5. $\int 3x\,dx$
6. $\int \frac{2}{\sqrt{x}}\,dx$
7. $\int \frac{1}{2}\,dx$
8. $\int 4x^5\,dx$
9. $\int (x - x^2)\,dx$
10. $\int (x^{1/2} + x^{-1/2})\,dx$
11. $\int (x^2 + 4x + 3 - x^{-2})\,dx$
12. $\int (4x^9 - 3x\sqrt{x})\,dx$
13. $\int 3\sin x\,dx$
14. $\int (\cos x - \sin x)\,dx$
15. $\int \sin 2x\,dx$
16. $\int \cos 5x\,dx$
17. $\int \sec 2x \tan 2x\,dx$
18. $\int \cot \pi x \csc \pi x\,dx$
19. $\int \csc^2 4x\,dx$
20. $\int \cos 5x\,dx$
21. $\int (x^2 + \sin x)\,dx$
22. $\int (\cos x - x^3)\,dx$
23. $\int (\sqrt{x} - \sin 2x)\,dx$
24. $\int \frac{x+1}{x^3}\,dx$

Solve the differential equation subject to the given boundary condition.

25. $\frac{dy}{dx} = \frac{3x}{2y}$; $y = 0$ when $x = 0$
26. $\frac{dy}{dx} = \frac{\sqrt{x}}{y^3}$; $y = 0$ when $x = 0$
27. $y^2y' = 2x$; $y = 1$ when $x = 0$
28. $yy' = 3(1 + x^2)$; $y = 2$ when $x = 1$
29. $y^2y' = \sin x$; $y = 1$ when $x = 0$
30. $y' = x\cos^2 y$; $y = \pi/4$ when $x = 0$
31. $1 + y^2y' = x^3$; $y = 2$ when $x = 0$
32. $1 - y^2y' = x^3$; $y = 2$ when $x = 0$
33. Suppose the population of a certain state increases at a constant rate. If the population was 650,000 in 1960 and 780,000 in 1970, when is the population expected to reach 1 million?
34. In a certain bacterial culture, the rate of growth is proportional to the $\frac{2}{3}$ power of the number of bacteria present. After 10 hours of incubation, there were 8000 bacteria; and after 15 hours, there were 15,625 bacteria. With how many bacteria did the culture start?
35. A certain battery charges at a rate inversely proportional to the charge it contains. If it takes 2 hours to fully charge the battery when it has no initial charge, how long will it take to bring the battery from no charge up to one-half its capacity?
36. In Problem 35, how long will it take to fully charge the battery if it is already one-third charged?
37. A hailstone, in forming, gains volume at a rate proportional to its surface area. If it takes 30 seconds to form a hailstone 1 centimeter in diameter, how long will it take to form a hailstone 3 centimeters in diameter?
38. The slope of a certain curve at point (x, y) is $\frac{2}{3}x$. If the curve passes through the origin, find its equation.

39. Before it reaches maturity, a certain species of tree grows in height at a rate inversely proportional to the square root of its height. If it takes 4 years for a tree to reach 10 feet in height, how tall will it be after 8 years?

40. A lake is stocked with 10,000 fish. Thereafter, the fish population grows at a rate proportional to the square root of the population. After 4 years, the fish population is estimated to be 12,100. How many fish were there in the lake after the first year?

SECTION 4.7

Rectilinear Motion

One of the earliest applications of calculus was to problems of motion. In fact, calculus was invented largely to describe motion more completely. In Section 3.2, we discussed rectilinear motion briefly. In this section, we apply differentiation and antidifferentiation to the study of motion along a straight line.

Rectilinear Motion

As mentioned earlier, by **rectilinear motion** is meant motion along a straight line, or in a one-dimensional coordinate system. Examples of rectilinear motion might be a stone thrown vertically into the air or a car traveling along a turnpike. The **position** of an object in rectilinear motion is given by a single number x, which is the coordinate of the object in the coordinate system. Thus, x might be the height of the stone above the ground or the distance of the car from a certain tollbooth. In any rectilinear motion problem, some reference point must be selected as the origin of the coordinate system, and one direction must be selected as the positive direction.

If the position of an object in rectilinear motion is given as a function of time t, then the equation $x = f(t)$ is called the **equation of motion** of the object. We can assume that the motion is completely described if we have the equation of motion, since by differentiating (as we saw in Section 3.2), we can find the velocity

$$v(t) = \frac{dx}{dt} = f'(t)$$

and acceleration

$$a(t) = \frac{dv}{dt} = \frac{d^2x}{dt^2} = f''(t).$$

If the velocity v is constant, the motion is said to be **uniform**. If v is positive, the motion is in the direction of increasing x, or the positive direction; if v is negative, then x is decreasing. Thus, if v changes sign, the object reverses its direction.

Similarly, if the acceleration a is positive, then the velocity v is increasing; when v is also positive, we say the object is "speeding up." If a is negative and v is positive, the object is "slowing down" or "decelerating." The motion is uniform if and only if the acceleration is 0.

Finding the Equation of Motion

Since velocity is the derivative of position, the equation of motion of an object whose velocity is known can be found by antidifferentiation. Similarly, if the acceleration is known, the velocity (and hence the position) can be found in the same way. Consider the following examples.

Example 4.40 **Uniform Motion** If an object in uniform motion, with velocity v, has initial position x_0 at time $t = 0$, show that the equation of motion of the object is the linear function

$$x = x_0 + vt.$$

Solution In uniform motion, the velocity v is constant. Thus, the motion is described by the differential equation

$$\frac{dx}{dt} = v.$$

Solving, we get

$$dx = v\,dt,$$

$$\int dx = \int v\,dt,$$

$$x = vt + C.$$

Setting $x = x_0$ when $t = 0$, we find $C = x_0$, and the equation of motion is

$$x = vt + x_0.$$

■

Example 4.41 **Constant Acceleration** If an object with constant acceleration a has initial position x_0 and initial velocity v_0 at time $t = 0$, show that the equation of motion of the object is the quadratic function

$$x = x_0 + v_0 t + \tfrac{1}{2} at^2.$$

Solution With the acceleration a being constant, the velocity of the object is found from the differential equation

$$\frac{dv}{dt} = a.$$

Solving, we get

$$dv = a\,dt,$$

$$\int dv = \int a\,dt,$$

$$v = at + C.$$

Setting $v = v_0$ when $t = 0$, we find $C = v_0$, so that

$$v = v_0 + at.$$

This is actually the differential equation

$$\frac{dx}{dt} = v_0 + at,$$

which we solve in the usual way:

$$dx = (v_0 + at)\,dt,$$

$$\int dx = \int (v_0 + at)\,dt,$$

$$x = v_0 t + \tfrac{1}{2}at^2 + C.$$

Setting $x = x_0$ when $t = 0$, we find $C = x_0$, and the equation of motion is

$$x = x_0 + v_0 t + \tfrac{1}{2}at^2.$$

■

■ **Example 4.42** A car accelerates from 0 to 60 miles per hour in 30 seconds. If acceleration is constant, how far does the car travel in that time?

Solution If we choose the car's starting point as the origin of our coordinate system, then the distance traveled will be the position at $t = 30$ seconds. We therefore have the initial conditions $x_0 = 0$ and $v_0 = 0$, so that the equation of motion is

$$x = \tfrac{1}{2}at^2,$$

from Example 4.41. Differentiating, we have

$$v = at.$$

Then we can set $v = 60$ miles per hour when $t = 30$ seconds $= \frac{1}{120}$ hour, and find $a = 7200$ miles per hour/hour. Therefore, the equation of motion is

$$x = 3600t^2.$$

Thus, when $t = \frac{1}{120}$,

$$x = 3600\left(\tfrac{1}{120}\right)^2 = \tfrac{1}{4} \text{ mile}.$$

That is, the car travels $\frac{1}{4}$ mile in the 30 seconds.

■

■ **Example 4.43** A stone is thrown upward with an initial velocity of 96 feet per second. How high does it rise? (Use 32 feet per second/second downward as the acceleration of gravity.)

Solution Choosing the starting point as the origin and upward as the positive direction, we have $x_0 = 0$, $v_0 = 96$, and $a = -32$. By Example 4.41, the stone's equation of motion is

$$x = 96t - 16t^2.$$

To find out how high the stone rises, we simply find the maximum value of x. Differentiating, we get

$$v = 96 - 32t,$$

which is 0 when $t = 3$. At $t = 3$,

$$x = 96(3) - 16(3)^2 = 144 \text{ feet}.$$

Thus, the stone rises 144 feet. ■

■ **Example 4.44** An object weighing 8 pounds slides across a rough surface, experiencing a friction force equal in magnitude to the square root of the velocity. If the object is released at a velocity of 16 feet per second, describe its subsequent motion.

Solution The object obviously slows down until it stops, after which it remains at rest. To be more precise, we will choose the point of release as the origin and the direction of motion as the positive direction. The friction force is opposite to the direction of motion, so we may write $F = -\sqrt{v}$. By **Newton's second law**, $F = ma$, where m is the mass of the object. The weight of the object is the product of its mass and the acceleration of gravity, so that $32m = 8$, or $m = \frac{1}{4}$ slug.* We therefore have

$$-\sqrt{v} = \tfrac{1}{4}a,$$
$$a = -4\sqrt{v}.$$

This is the differential equation

$$\frac{dv}{dt} = -4\sqrt{v},$$

which we solve in the usual way:

$$v^{-1/2}\, dv = -4\; dt,$$
$$\int v^{-1/2}\, dv = -4 \int dt,$$
$$2v^{1/2} = -4t + C.$$

Putting in $v = 16$ when $t = 0$, we find $C = 8$. Thus,

$$2v^{1/2} = -4t + 8,$$
$$v = (4 - 2t)^2.$$

* In the British system, an acceleration is measured in feet per second per second, a mass is measured in slugs, and a force is measured in pounds.

From this equation, we can tell that the object comes to rest when $t = 2$ seconds, after which it remains at rest. To find the position of the object, we rewrite the last equation as a differential equation, and solve:

$$\frac{dx}{dt} = 16 - 16t + 4t^2,$$
$$dx = (16 - 16t + 4t^2)\,dt,$$
$$x = 16t - 8t^2 + \tfrac{4}{3}t^3 + C.$$

Since $x = 0$ when $t = 0$, we get $C = 0$, or

$$x = 16t - 8t^2 + \tfrac{4}{3}t^3.$$

When $t = 2$, $x = \frac{32}{3}$; therefore, the object slides $\frac{32}{3}$ feet before coming to rest. ■

Simple Harmonic Motion

Trigonometric functions can be used to describe a certain kind of rectilinear motion known as **simple harmonic motion**. In simple harmonic motion, there is a point, called the **equilibrium point**, at which the acceleration is 0. At every other point, the acceleration is directed toward the equilibrium point and is proportional in magnitude to the distance from the equilibrium point.

If we let the origin be the equilibrium point, then the acceleration at point x can be expressed as

$$a = -\omega^2 x$$

for some proportionality constant ω^2. This is the same as the differential equation

$$\frac{d^2x}{dt^2} = -\omega^2 x. \qquad (1)$$

■ **Example 4.45** Show that any function of the form

$$x = C \sin \omega t + D \cos \omega t \qquad (2)$$

is a solution of equation (1).

Solution Differentiating equation (2), we get

$$x'(t) = C\omega \cos \omega t - D\omega \sin \omega t,$$
$$x''(t) = -C\omega^2 \sin \omega t - D\omega^2 \cos \omega t$$
$$= -\omega^2 x.$$

Thus, equation (1) is satisfied. ■

It can be shown that all the solutions of equation (1) are of the form (2), but we shall not do so here. The arbitrary constants C and D are determined by boundary conditions, if any are given. A more convenient form than equation (2) is possible, as the next example shows.

Example 4.46 Show that equation (2) can be written in the form

$$x = A\sin(\omega t + \phi). \tag{3}$$

Solution If C and D are both 0, let $A = 0$; then equations (2) and (3) are obviously the same. If C and D are not both 0, we let

$$A = \sqrt{C^2 + D^2},$$

and then we can write equation (2) as

$$x = A\left(\frac{C}{A}\sin\omega t + \frac{D}{A}\cos\omega t\right).$$

Since the point $(C/A, D/A)$ is on the unit circle, there is a number ϕ such that $\cos\phi = C/A$ and $\sin\phi = D/A$. Hence,

$$\begin{aligned} x &= A\,(\cos\phi\sin\omega t + \sin\phi\cos\omega t) \\ &= A\sin(\omega t + \phi). \end{aligned}$$

■

When the solution to equation (1) is written in the form

$$x = A\sin(\omega t + \phi),$$

the number A is called the **amplitude** of the motion, and is the maximum displacement from the equilibrium point (see Figure 4.57). The number $2\pi/\omega$ is called the **period** of the motion, because the function $\sin(\omega t + \phi)$ is periodic with period $2\pi/\omega$. The number $-\phi/\omega$ is called the **phase shift**, and is a time at which the object is at the equilibrium point.

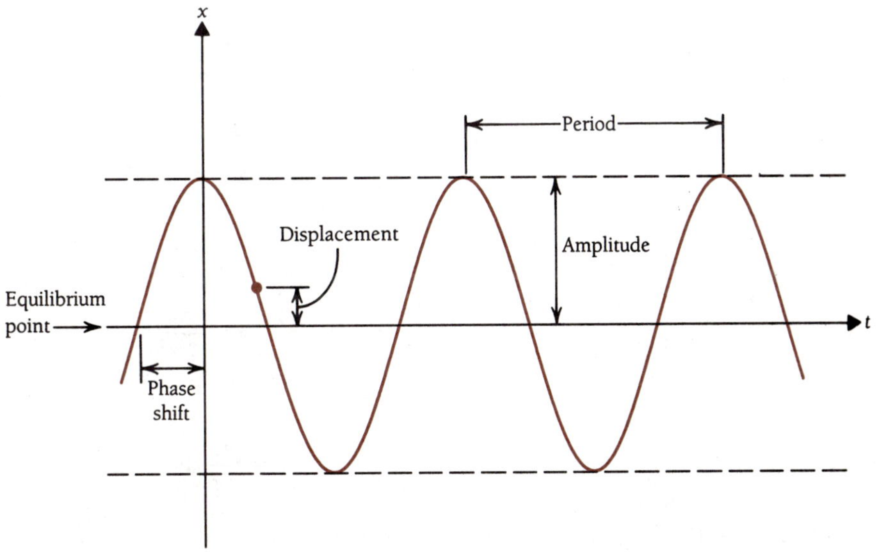

FIGURE 4.57 Simple harmonic motion

Example 4.47 Show that when $v = 0$, $x = \pm A$, and when $x = 0$, $|v|$ is a maximum.

Solution Given $x = A\sin(\omega t + \phi)$, we find

$$v = A\omega\cos(\omega t + \phi).$$

If $v = 0$, then $\cos(\omega t + \phi) = 0$, or $\omega t + \phi = \pm\pi/2$. Hence, $\sin(\omega t + \phi) = \pm 1$, and

$$x = A\sin(\omega t + \phi) = \pm A.$$

Also,

$$|v| = A\omega|\cos(\omega t + \phi)| \leq A\omega.$$

But when $x = 0$, $\sin(\omega t + \phi) = 0$, so that $\cos(\omega t + \phi) = \pm 1$. Therefore, $v = \pm A\omega$ when $x = 0$, and $|v|$ is a maximum. ■

Example 4.48 An object in simple harmonic motion has period p and satisfies the initial conditions $x = 0$ and $v = v_0$ when $t = 0$. Find its equation of motion.

Solution We know that the equation of motion is of the form $x = A\sin(\omega t + \phi)$, with A, ω, and ϕ yet to be determined. Since the period is p, we have $2\pi/\omega = p$, or $\omega = 2\pi/p$. Thus,

$$x = A\sin\left(\frac{2\pi t}{p} + \phi\right).$$

Setting $t = 0$ and $x = 0$, we next find that $A\sin\phi = 0$. Assuming that $A \neq 0$, so that the object does move, we get $\sin\phi = 0$, and we take $\phi = 0$. Differentiating the expression for x above, we get

$$v = \frac{2\pi A}{p}\cos\left(\frac{2\pi t}{p} + \phi\right).$$

Now, putting in $\phi = 0$, $t = 0$, and $v = v_0$, we get

$$v_0 = \frac{2\pi A}{p}\cos\phi = \frac{2\pi A}{p}.$$

Therefore,

$$A = \frac{pv_0}{2\pi},$$

and the equation of motion is

$$x = \frac{pv_0}{2\pi}\sin\frac{2\pi t}{p}.$$ ■

Example 4.49 An object in simple harmonic motion has period 2 and amplitude 3, and is at equilibrium when $t = 0$. Find its maximum speed.

Solution

By Example 4.47, the object has maximum speed when $x = 0$. But $x = 0$ when $t = 0$, so $|v_0|$ is the maximum speed. From Example 4.48, the amplitude is

$$A = \frac{pv_0}{2\pi} = \frac{2v_0}{2\pi} = \frac{v_0}{\pi} = 3,$$

so that $v_0 = 3\pi$. Thus, 3π is the maximum speed. ■

■ **Example 4.50**

An object in simple harmonic motion has maximum speed 3 and amplitude 3. Find its equation of motion.

Solution

The equation of motion is of the form $x = A\sin(\omega t + \phi)$. We are told that $A = 3$, so

$$x = 3\sin(\omega t + \phi).$$

When $x = 0$, $\omega t + \phi$ is a multiple of π, so

$$v = 3\omega\cos(\omega t + \phi) = \pm 3\omega.$$

But this velocity is one of maximum speed, so $3\omega = 3$ and $\omega = 1$. Thus,

$$x = 3\sin(t + \phi).$$

We cannot determine ϕ from the given data. ■

■ **Example 4.51**

The force required to stretch or compress a certain spring by an amount x from its natural length is proportional to x. The spring is hung vertically, and a weight is attached to it. When the weight is at rest, the spring is 1 inch longer than its natural length. Then the weight is pulled down 1 inch farther, and released. Describe the subsequent motion of the weight.

Solution

Let us select the point where the weight rests motionless as the origin of our coordinate system, and downward as the positive direction. If the weight is at position x, then the spring is stretched to a length $x + 1$ beyond its natural length (see Figure 4.58), so the force exerted by the spring is $-k(x + 1)$. The minus sign indicates that the force is opposed to the position, or toward the origin. At rest, or $x = 0$, the spring exerts force $-k$. If m is the mass of the weight and g is the acceleration of gravity, then gravity exerts force mg. But the net force is 0 when the weight is at rest, so $-k + mg = 0$, or $k = mg$. Thus, at any point x, the spring exerts force $-mg(x + 1)$, and gravity exerts force mg. The net force on the weight at position x is therefore $-mgx$. But by Newton's second law, the force is also ma, so

$$a = -gx.$$

This equation tells us that the weight will be in simple harmonic motion about its rest point. By comparison with equation (1), we see that the equation of motion of the object will be of the form

$$x = A\sin(t\sqrt{g} + \phi).$$

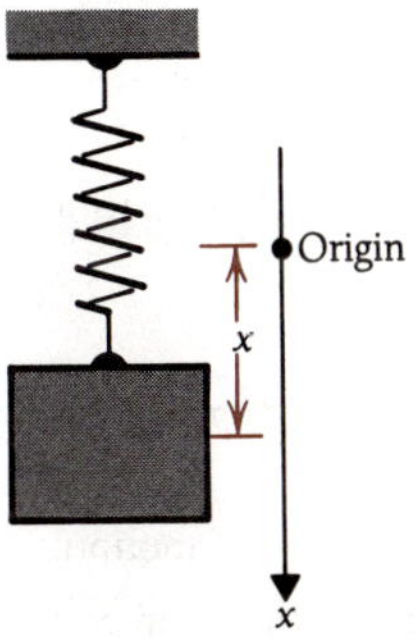

FIGURE 4.58

Using $x = 1$ and $v = 0$ when $t = 0$, we find that $A = 1$ and $\phi = \pi/2$. Thus, the equation of motion is

$$x = \sin\left(t\sqrt{g} + \frac{\pi}{2}\right). \qquad \blacksquare$$

Exercises 4.7

Find the equation of motion.

1. $v = -4$; $x_0 = 0$
2. $v = 2t$; $x_0 = 0$
3. $a = 4$; $v_0 = 24$; $x_0 = 0$
4. $a = 2t$; $v_0 = 60$; $x_0 = 0$
5. $a = 1 - t$; $v_0 = -24$; $x_0 = 55$
6. $a = 3v^{2/3}$; $v_0 = 2$; $x_0 = 0$
7. An automobile traveling 88 feet per second begins decelerating at the constant rate of 8 feet per second/second. How far does it travel before stopping?
8. A drag racer can travel $\frac{1}{4}$ mile from a standing start in 12 seconds. Assuming constant acceleration, what is the racer's top speed?
9. An arrow is shot into the air with an initial velocity of 60 feet per second upward. How high does it rise?
10. A bullet is fired vertically into the air with a muzzle velocity of 1260 feet per second. Assuming only the force of gravity acts on the bullet, how high does it rise? Does this assumption seem reasonable?
11. An object is dropped from a balloon at a height of 2500 feet. Neglecting air resistance, what is the velocity of the object on impact?
12. Two objects move along the same line. The first has $a = -8$, $v_0 = 12$, and $x_0 = -9$, and the second has $a = 2$, $v_0 = -8$, and $x_0 = 16$. What is the minimum distance between the two objects?
13. Suppose that the acceleration of a certain object falling through a viscous medium is inversely proportional to the square root of its velocity. If the object falls 16 millimeters in the first second, how far does it fall in 8 seconds? (Use $v_0 = 0$.)
14. Two cars approach the same intersection from perpendicular directions, one at 30 miles per hour and the other at 45 miles per hour. At one instant, they are both 132 feet from the center of the intersection. If both cars continue uniformly on their way, what will be the minimum distance between them?
15. A certain boat moving across the water experiences a drag force proportional to the square of the velocity. The boat is traveling 30 feet per second when its power ceases. If its speed after 10 seconds is 15 feet per second, what is its speed after 30 seconds?
16. An object in simple harmonic motion satisfies $x = 0$ and $v = 4$ when $t = 0$. If the period is 4π, find the amplitude.
17. In Problem 16, find the equation of motion.
18. An object in simple harmonic motion has amplitude 6 and period 3. Find its maximum speed.
19. In Problem 18, if $x = 2$ when $t = 0$, find the phase shift and the equation of motion.
20. Show that the equation $x = C \sin \omega t + D \cos \omega t$ can be written in the form $x = A \cos(\omega t + \phi)$.
21. An object in simple harmonic motion has period p and is subject to the initial conditions $x = x_0 \neq 0$ and $v = v_0$ when $t = 0$. Find its equation of motion.

22. Show that an object in simple harmonic motion experiences its maximum acceleration when $x = -A$.

23. Find the maximum acceleration of the object in Problem 18.

24. A block hangs on a spring, stretching the spring 2 inches beyond its natural length. If the block is lifted 3 inches and released, describe the subsequent motion of the block.

25. What is the maximum speed of the block in Problem 24?

SECTION 4.8

Chapter Review

Important Concepts

Mathematical model	Concavity	Particular solution
Related rates	Inflection point	Boundary condition
Absolute maximum	Graphing algorithm	Separable differential equation
Absolute minimum	Relative maximum	Initial condition
Extremum	Relative minimum	Rectilinear motion
Critical point	Endpoint extremum	Uniform motion
Rolle's theorem	First derivative test	Constant acceleration
Mean value theorem	Second derivative test	Newton's second law
Newton's method	Extremum problems	Simple harmonic motion
Iteration formula	Antiderivative	Equilibrium point
Increasing function	Indefinite integral	Amplitude
Decreasing function	Differential equation	Period
Monotonic function	General solution	Phase shift

Review Exercises

Given $f(x)$, find $c \in (a, b)$ such that $f(b) - f(a) = f'(c)(b - a)$.

1. $f(x) = x^3$; $[a, b] = [-1, 1]$

2. $f(x) = \dfrac{x}{x+1}$; $[a, b] = [0, 1]$

3. $f(x) = \sin x$; $[a, b] = [0, 2\pi]$

4. $f(x) = \tan x$; $[a, b] = [0, 1]$

For each function f:

(a) Find the critical points.
(b) Describe the monotonicity.
(c) Describe the concavity.
(d) Give the points of inflection.
(e) Give the relative and absolute extreme values and their locations.
(f) Sketch the graph of the function.

5. $f(x) = x^4 - 5x^2 + 4$

6. $f(x) = -3x^4 + 24x^3 - 66x^2 + 72x - 23$

7. $f(x) = 5x^3 - x^5$

8. $f(x) = x^6 - 6x^2$

9. $f(x) = \dfrac{x^2}{x+1}$

10. $f(x) = \dfrac{x^3}{x^2 - 9}$

11. $f(x) = \cos 2x$

12. $f(x) = 2 \sin x + 3 \cos x$, $x \in [0, 2\pi]$

13. A rectangle is inscribed in an equilateral triangle, with one side on a side of the triangle. If the triangle has side s, what is the maximum area of the rectangle?

14. A rectangle is inscribed in a right triangle, with two sides on the legs of the triangle. If the legs have lengths a and b, what is the maximum area of the rectangle?

15. A manufacturer can sell x items per week at the selling price of $S(x) = 100 - 0.01x$ cents each, and it costs $C(x) = 10x + 5000$ cents to produce the x items. What is the production level for which profit is a maximum?

16. It costs a candle maker $C(x) = 1200 + 2x$ cents to produce x fancy candles per day, and the candles can be sold for $S(x) = 500 - 12x$ cents each. How many candles should be made per day to maximize profit, and what is the maximum daily profit?

17. An open-topped cylindrical bucket is to hold a volume V. Find the dimensions of the bucket that requires the least sheet metal to construct.

18. A window is in the shape of a rectangle surmounted by a semicircle with diameter equal to the width of the rectangle. If the perimeter of the window is P, what dimensions will give the window the greatest area?

19. The pressure P, volume V, and temperature T of a certain gas are related by the formula $PV = kT$, where k is a constant. At a certain instant, $P = 10$ atmospheres, $V = 100$ cubic centimeters, P is increasing at the rate of 2 atmospheres per minute, and T is decreasing at the rate of 10°C per minute. How is V changing?

20. An isosceles triangle has two sides of length 5 centimeters. If the base is increasing at the rate of 1 centimeter per second, how is the height changing when the base is 8 centimeters long? How is the vertex angle changing at that instant?

21. What is the maximum area of the triangle in Problem 20?

22. A cone with base radius r and height h has lateral surface area $S = \pi r\sqrt{r^2 + h^2}$. If the volume $V = \pi r^2 h/3$ of the cone is fixed, what ratio of r to h makes S a minimum?

23. A sign 3 feet high is placed over a sidewalk with its lower edge 7 feet above eye level. At what point on the sidewalk does the sign subtend the greatest angle at the pedestrian's eye?

24. A conical cup is to be constructed from a circular sheet of paper by cutting out a sector and then joining the straight edges of the remaining piece with tape. Find the central angle of the sector to be removed that will maximize the volume of the cup.

25. Show that the function $f(x) = x + \sin x$ has no relative extrema.

26. The least force that will pull an object of weight W along the floor when acting at an angle α with the horizontal is given by

$$F = \frac{kW}{\cos\alpha + k\sin\alpha},$$

with k a positive constant. What angle α minimizes F?

Solve each differential equation, subject to the given boundary condition.

27. $\dfrac{dy}{dx} = \dfrac{2x}{y}$; $y(0) = 3$

28. $\dfrac{dy}{dx} = \dfrac{2x}{1 - y^2}$; $y(1) = 2$

29. $y' = x^{2/3}y^2$; $y(0) = -1$

30. $x^2yy' = 1 + x^3$; $y(1) = 0$

31. $y' = x\sec y$; $y(0) = 0$

32. $x^2y' = \sqrt{1 + y}$; $y(1) = 3$

Antidifferentiate.

33. $\int (4x - x^2 - x^3)\, dx$

34. $\int (x^2 - 16x + 5)\, dx$

35. $\int (ax + b)\, dx$

36. $\int (ax^2 + bx + c)\, dx$

37. $\int a\sqrt{x}\, dx$

38. $\int \frac{a}{\sqrt{x}}\, dx$

39. $\int (3 + x^{-1/3})\, dx$

40. $\int 2x^{-2/5}\, dx$

41. $\int \cos 4x\, dx$

42. $\int \sec 3x \tan 3x\, dx$

Find the equation of motion of an object in uniform motion satisfying the given conditions.

43. $x_0 = 2;\quad v = 6$

44. $x_0 = -2;\quad v = -3$

Find the equation of motion of an object with constant acceleration satisfying the given conditions.

45. $x_0 = 0;\quad v_0 = 9;\quad a = -6$

46. $x_0 = v_0 = -40;\quad a = 40$

Find the equation of motion of an object satisfying the given conditions.

47. $x_0 = 1;\quad v = 3x^2$

48. $x_0 = 4;\quad v_0 = 0;\quad a = t^2$

49. A certain major league pitcher can throw a baseball 132 feet per second. If he can throw it vertically with the same speed, how high can he throw it?

50. What initial velocity is required to throw an object to a height of h feet?

51. A stunt diver dives from a cliff 144 feet above the sea. With what velocity does he hit the water?

52. The acceleration of a particle is proportional to the square of the velocity. If $v_0 = 0.5$, find the velocity as a function of time.

53. At 10 PM on May 31, a comet is sighted, and its speed is estimated at 12,500 kilometers per hour. At 10 PM on June 10, its speed is 15,000 kilometers per hour; and at 10 PM on June 25, its speed is 14,000 kilometers per hour. Assuming that the acceleration of the comet is a linear function of time, on what day does the comet reach its maximum speed? What is its maximum speed?

54. An automobile starts from rest and accelerates at a steadily decreasing rate, so that after 30 seconds the acceleration is 0 and the speed is 60 miles per hour. How far does the car travel in those 30 seconds?

Find the equation of motion of an object in simple harmonic motion satisfying the given conditions.

55. $x_0 = 0;\quad v_0 = 10;\quad p = 1$

56. $A = 5;\quad x_0 = 2;\quad v_0 = 3$

Additional Exercises

1. Show by example that the converse of Rolle's theorem is not true.
2. Use Rolle's theorem and induction on n to prove that a real polynomial function of degree n has at most n real zeros.

3. Show that the equation $6x^5 + 13x + 1 = 0$ has exactly one real solution.
4. Show that the equation $x^3 + 9x^2 + 33x - 8 = 0$ has exactly one real solution.
5. Let f be a polynomial function. Show that between two consecutive zeros of f' there is at most one zero of f.
6. Show that the equation $x^3 + ax + b = 0$ has exactly one real solution if $a \geq 0$, and at most one real solution between $-a/\sqrt{3}$ and $a/\sqrt{3}$ if $a < 0$.
7. Let functions f and g be continuous on $[a, b]$ and differentiable in (a, b), and suppose $f'(x) = g'(x)$ for each $x \in [a, b]$. Show that $f(x) - g(x) = f(a) - g(a)$ for each $x \in [a, b]$.
8. Let functions f and g be continuous on $[a, b]$ and differentiable in (a, b), and suppose $f(a) = g(a)$ and $f(b) = g(b)$. Show that $f'(c) = g'(c)$ for some $c \in (a, b)$.
9. Prove that a quadratic function has no inflection points.
10. Prove that a polynomial function of degree n has at most $n - 2$ points of inflection.
11. Show that the function

$$f(x) = \frac{x+1}{x^2+1}$$

has three inflection points that are collinear.

12. Let f be a differentiable function. Prove that if $|f'(x)| \leq 1$ for all x, then $|f(x_1) - f(x_2)| \leq |x_1 - x_2|$ for all x_1 and x_2.
13. Find the extrema of $f(x) = x(x^2 + r^2)^{-3/2}$ in $[0, \infty)$.
14. Find the extrema of $f(x) = x|x|$.
15. If f is a polynomial function and z is a double zero of f, show that z is also a zero of f'.
16. If f is a polynomial function and z is a k-fold zero of f, show that z is also a zero of $f^{(k-1)}$.
17. Find A and B so that the curve $y = x^2 + Ax + B$ will be tangent to the line $y = 2x + 1$ at the point $(1, 3)$.
18. Let f be a differentiable function. Show that if PQ is the longest or shortest line segment that joins P to a point Q on the curve $y = f(x)$, then line PQ is normal to the curve.
19. A square field is surrounded by a sidewalk. A cyclist can ride twice as fast on the sidewalk as she can on the field. She wishes to ride from one corner of the field to the midpoint of a nonadjacent side. How much of her path should be across the field in order to minimize travel time?
20. The interior of a 400 meter track is a rectangle with a semicircle at each end. Find the dimensions of the rectangle that has maximum area.
21. A tourist rents a motorboat to make a 30 mile trip across a lake. The boat costs \$10 per hour, and gasoline costs \$2 per gallon. The boat uses

$$\frac{v}{20} + \frac{2v^2}{45}$$

gallons of gasoline per hour at speed v. What is the most economical speed at which the tourist can cross the lake?

22. A truck driver charges \$25 per hour plus $3 + 0.01v$ dollars per mile to drive at v miles per hour. What speed will minimize the cost per mile?

23. Find the point on the curve $x^2 - y^2 = 4$ that is closest to the point (4, 1). [*Hint:* Distance d is a minimum when d^2 is a minimum.]

24. Find the points on the curve $x^2 + 4y^2 = 4$ that are nearest and farthest from the point (3, 4).

25. A paper cup is an open cone. To hold 8 ounces of water, it must have a volume of 4.43 cubic inches. Find the dimensions of such a cup that requires the least amount of material.

26. Find the volume of the largest rectangular solid that can be cut from a solid hemisphere of radius r.

27. A hatter keeps a small shop that costs \$370 per month for rent and utilities. Materials for a single hat cost \$2, and if x hats are made per day, they can be sold for $10 - 0.25x$ dollars each. Miscellaneous and advertising costs total \$110 per month. What is the hatter's maximum possible daily profit?

28. A bicycle manufacturer agrees to sell 200 bicycles to a dealer at \$90 each, and to reduce the cost per bicycle on the entire order by 50¢ for each bicycle over 200. What is the maximum possible dollar value of an order under this arrangement?

29. Two ranches are located on the same side of a straight river, one 16 miles from the river and the other 10 miles from the river. The ranches are 10 miles apart. It is desired to place a pumping station on the river and run pipelines from it to each ranch. Where should the pumping station be located so as to minimize the lengths of the pipelines?

30. Solve Problem 29 if it is possible to run a single pipeline to one ranch and then the other.

31. Solve Problem 29 without calculus by imagining one of the ranches to be on the other side of the river.

32. In optics, Fermat's principle (Problem 40, Section 4.5) states that a light ray travels along a path that minimizes time. If a light ray travels from point A to a mirror M and then to a point B, show that the principle of least time requires that the angles of incidence and reflection at M are congruent.

33. **The Problem of Viviani** A transversal of parallel lines l and m meets l at A and m at B. A second point C on l is given, and a point D is chosen on m, on the opposite side of AB from C. Let E be the intersection of AB and CD. How should D be chosen so that the sum of the areas of triangles ACE and BDE is a minimum?

34. Show that the rate of change of the volume of a sphere with respect to its radius is numerically equal to the surface area.

35. Find the rate of change of the volume of a cube with respect to its edge.

36. Find the rate of change of the volume of a cube with respect to half its edge.

37. A point moves along the curve $y = 2x^2$ so that the y-coordinate increases uniformly at the rate of 2 units per second. At what point is the x-coordinate increasing at the rate of 2 units per second?

38. The roof of a silo is a hemisphere of radius 10 feet. A uniform layer of ice covers the roof. If the thickness is decreasing at the rate of $\frac{1}{4}$ inch per hour, how is the volume of the ice changing?

39. The marginal cost in a certain business is $C'(x) = 3 - 0.1x$. Find the cost if fixed costs are \$200.

40. The marginal revenue in a certain business is $R'(x) = 12 - 0.2x$. Find the revenue, assuming that $R(0) = 0$.

41. A stone is dropped from a bridge and strikes the water below 4 seconds later. How high is the bridge?

42. A stone is dropped down a well, and a splash is heard 3 seconds later. If sound travels at 1080 feet per second, how far down is the water level?

43. A train traveling 60 kilometers per hour decelerates at the rate of 2 meters per second/second. How far does it travel before stopping?

44. A car traveling 45 miles per hour runs a stop sign. Immediately, a patrol car starts from rest at the stop sign, accelerating at the constant rate of 4 feet per second/second. How long will it take for the patrol car to overtake the car, and what will be the patrol car's top speed?

45. Two sides of a triangle are 8 feet and 12 feet, and the included angle is $\pi/6$. If the 8 foot side is increasing at the rate of 1 foot per minute and the other two sides do not change, how is the included angle changing?

46. A flagpole stands 100 feet away. The angle of elevation to the top of the pole is 60°, with a possible error of 0.25°. Estimate the possible error in computing the height of the flagpole.

47. Find the maximum value of $y = 5 \cos x - 12 \sin x$.

48. A corridor 4 feet wide has a right-angled turn in it. How long can a thin rod be and still pass horizontally around the turn without bending?

49. A steel beam 20 feet long is being carried horizontally down a passageway 8 feet wide and into a corridor that meets the passageway at right angles. How wide must the corridor be so that the beam can go around the corner? (Neglect the width of the beam.)

50. **An Extension of the Mean Value Theorem** Let f and f' be continuous on $[a, b]$, and let f'' exist in (a, b). Prove that there exists $c \in (a, b)$ such that

$$f(b) - f(a) = f'(a)(b - a) + \tfrac{1}{2}f''(c)(b - a)^2.$$

[*Hint:* Define K by $f(b) - f(a) = f'(a)(b - a) + K(b - a)^2$, and apply Rolle's theorem to the function $F(x) = f(x) - f(a) - f'(a)(x - a) - K(x - a)^2$.]

51. Prove that if the rational function $f(x) = p(x)/q(x)$ has denominator of degree greater than the degree of the numerator, then the x-axis is a horizontal asymptote of its graph.

52. Prove that if the numerator and denominator of a rational function have the same degree, then the graph of the rational function has a horizontal asymptote other than the x-axis.

53. Prove that if the numerator of a rational function has degree 1 greater than the degree of the denominator, then the graph of the function has an oblique line as an asymptote.

Sketch the graphs of the following functions:

54. $f(x) = \dfrac{x^2 - 4}{x^3 - x}$ 55. $f(x) = \dfrac{x^2 - 4}{x^4 - x}$ 56. $f(x) = \dfrac{x^3 - 8}{x^3 - x}$ 57. $f(x) = \dfrac{x^3 - 8}{x^2 - x}$

CHAPTER 5

The Definite Integral

INTRODUCTION

In Chapters 2–4, we developed most of the differential calculus. Now we turn to a development of the integral calculus.

In this chapter, we discuss the computation of area, using the *method of exhaustion*. This leads to the notion of *Riemann sum*, which in turn leads to the *definite integral*. After discussing properties of the definite integral, we prove the *fundamental theorem of calculus*. Then we introduce the substitution technique of integration, and close with some simple applications of the definite integral.

We will initially motivate and interpret the definite integral in terms of area. However, just as the derivative has several uses, the definite integral can also be used in many different settings. A few applications appear at the end of this chapter, and many more in Chapter 8.

SECTION 5.1

Area as the Limit of a Sum

The problem of computing areas has been of interest ever since governments began taxing farmers. If a region has straight-line boundaries, its area can be found using elementary plane geometry. (The region can be cut up into triangles, and the areas of the triangles can be added.) But if a region has some curved boundaries, we must either be content with approximations or use calculus to aid in finding the area. In this section, we will see how the limit concept is related to the problem of finding areas.

The earliest recorded efforts to measure areas were made by the ancient Egyptians and Babylonians. They devised rules for finding areas of simple figures, but for even so simple a figure as a circle, the rules gave rather crude approximations. The ancient Greeks were the first to systematically collect measurement formulas. They even used the idea of limit, or at least of successive approximation, to obtain exact formulas for some figures and fairly accurate formulas for others. Their achievements are even more impressive when we reflect that the Greeks had no algebraic notation whatever and a system of numeration just a little more awkward than Roman numerals.

The Method of Exhaustion

The Greeks introduced a method called the **method of exhaustion** to obtain some of their formulas. The method consisted of inscribing in the figure to be measured a sequence of polygons with successively greater numbers of sides. The areas of these polygons eventually "exhausted" the area of the given figure, and thus, very close approximations could be obtained. The only real limitation was the ability to calculate.

Archimedes was a leading proponent of this method. For example, he approximated a circle by an inscribed regular polygon of 512 sides, and thus obtained

Archimedes (*c.* 287 – 212 BC), the greatest scientist of antiquity, and possibly of all time, spent most of his life in the then Greek city of Syracuse in Sicily. He was of aristocratic birth, and may have been related to Hieron, the ruler of the city. He was remarkably clever with mechanical devices. When Hieron requested a demonstration of mechanical principles, Archimedes single-handedly dragged a large ship up on the beach with a system of compound pulleys. To emphasize the power of the lever, he is supposed to have said, "Give me a place to stand, and I will move the earth." His talents were put to practical use in the defense of Syracuse against attack by the Romans. He built some devices that lifted Roman ships out of the water and other devices that used mirrors to concentrate sunlight and set Roman ships on fire. (This latter device was doubted for centuries, but proved feasible in 1971 by Professor Stamatis and the Greek army.) Archimedes was killed by a Roman soldier when Syracuse finally fell (through treachery).

Archimedes also discovered principles of buoyancy and hydrostatics, and invented machines used in irrigation. He extended the Greek numeration system to include very large numbers, and contributed significantly to euclidean geometry. Although he was not the first to use the ideas of calculus, he fashioned them into a coherent and flexible tool. He used the method of exhaustion to solve much more difficult problems than his predecessors. His own favorite was the determination of the surface area of a sphere. His work survived the Dark Ages in Constantinople, and was the groundwork upon which Barrow, Newton, and Leibniz built the calculus.

approximate formulas for the circumference and area of a circle. (See Problems 53–57 in the Additional Exercises at the end of Chapter 2.) He concluded that the ratio of the circumference to the diameter (which we now denote by π) exceeded 3 by less than $\frac{1}{7}$ but more than $\frac{10}{71}$. He undoubtedly did not actually draw the figure, but carried out the corresponding calculations. (You might be able to think of another reason for the name of the method!) If Archimedes had been in possession of a system of algebraic notation, he might well have invented the calculus 1800 years before Newton and Leibniz did.

Area under the Graph of a Function

In the spirit of Archimedes's work, we will apply the method of exhaustion to a region bounded on one side by the graph of a function. Let f be any function that is continuous and nonnegative over the interval $[a, b]$. Let R be the region bounded on the left by the line $x = a$, on the right by the line $x = b$, below by the x-axis, and above by the curve $y = f(x)$. In Figure 5.1, R is the shaded region. We will call R the **region below f over $[a, b]$**. We call the area of R the **area under f over $[a, b]$**.

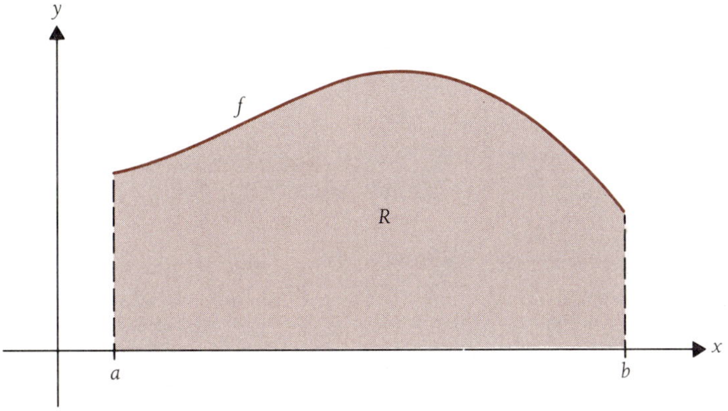

FIGURE 5.1 The region below the graph of a function

Because R has a curve as one boundary, we may not have any elementary formula for the area of R. To find the area of R, we place inside the region R figures whose areas we can compute, and then "exhaust" R by a limiting procedure. In this case, we will inscribe rectangles in R.

We begin by dividing up the interval $[a, b]$ into n pieces, or **subintervals**, all of the same length (see Figure 5.2, page 298). We do so by selecting points

$$x_0 = a,$$

$$x_1 = a + \frac{b-a}{n},$$

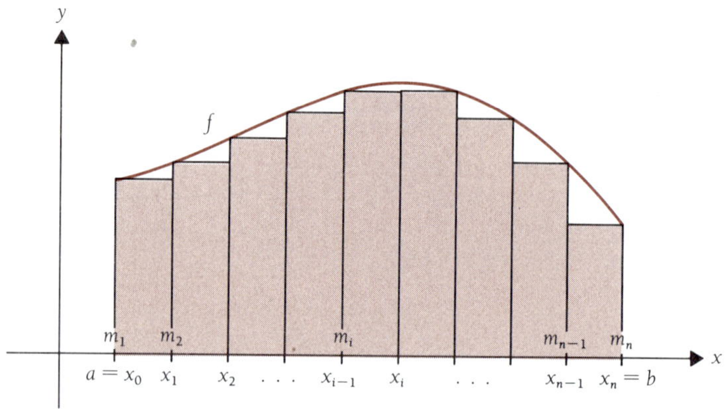

FIGURE 5.2 Inscribed rectangles under the graph of a function

$$x_2 = x_1 + \frac{b-a}{n} = a + \frac{2(b-a)}{n},$$

$$\cdots$$

$$x_i = x_{i-1} + \frac{b-a}{n} = a + \frac{i(b-a)}{n},$$

$$\cdots$$

$$x_n = a + \frac{n(b-a)}{n} = b.$$

We call $[x_{i-1}, x_i]$ the **ith subinterval**; the length of each subinterval is $\Delta x = (b-a)/n$.

Now, for each subinterval $[x_{i-1}, x_i]$, let m_i be a point at which f has its absolute minimum over that subinterval. That is, let $f(m_1)$ be the minimum of f over $[x_0, x_1]$, let $f(m_2)$ be the minimum of f over $[x_1, x_2]$, and so forth.

We now construct the rectangles as shown in Figure 5.2. On the ith subinterval, we construct a rectangle of height $f(m_i)$. Since its width is $\Delta x = (b-a)/n$, the area of the ith rectangle is

$$f(m_i)\frac{b-a}{n}.$$

The total area of these rectangles is the sum of their individual areas, namely,

$$\begin{aligned} A_n &= f(m_1)\Delta x + f(m_2)\Delta x + \cdots + f(m_n)\Delta x \\ &= [f(m_1) + f(m_2) + \cdots + f(m_n)]\Delta x \\ &= [f(m_1) + f(m_2) + \cdots + f(m_n)]\frac{b-a}{n}. \end{aligned}$$

Let A be the area under f over $[a, b]$. The area of the rectangles A_n is less than A, but if n is increased, the difference between A_n and A becomes smaller. This is evident

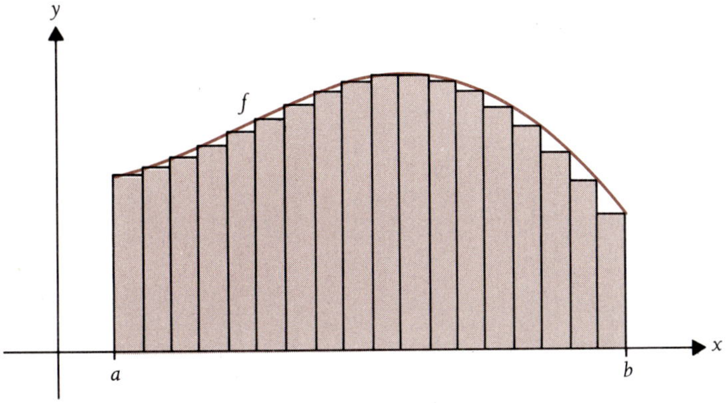

FIGURE 5.3 More rectangles exhaust more of the area

from Figure 5.3, which shows the same region as Figure 5.2, but with a larger number of rectangles.

Thus, $A_n \to A$ as $n \to \infty$. That is, $A = \lim_{n\to\infty} A_n$, or

$$A = \lim_{n\to\infty} [f(m_1) + f(m_2) + \cdots + f(m_n)] \frac{b-a}{n}. \tag{1}$$

That this limit exists and equals A can be proved rigorously using properties of the continuous function f, as we will show later.

The method of exhaustion just demonstrated can be modified without changing the end result. For instance, if we let M_i be a point at which f has its absolute maximum over the ith subinterval, and we construct a rectangle of height $f(M_i)$ there, then we get a set of rectangles whose union contains the region R (see Figure 5.4). Since the area of the ith rectangle is $f(M_i)\,(b-a)/n$, the combined area of all the rectangles is

$$B_n = [f(M_1) + f(M_2) + \cdots + f(M_n)] \frac{b-a}{n}.$$

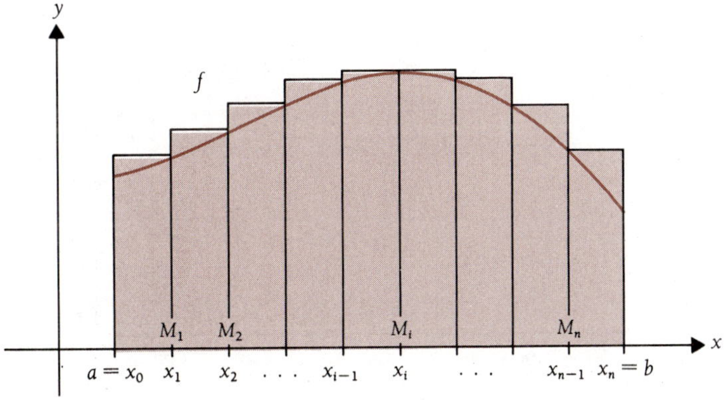

FIGURE 5.4 Outer rectangles covering the region below f

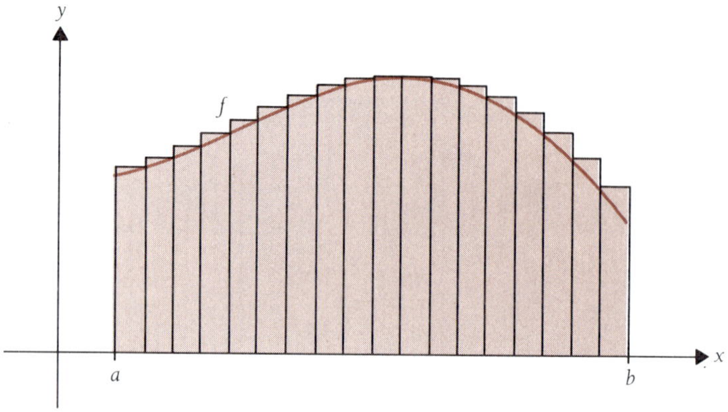

FIGURE 5.5 More rectangles are closer to the area

This area is greater than A. As n increases, the difference $B_n - A$ decreases (see Figure 5.5), so that $B_n \to A$ as $n \to \infty$. That is, $A = \lim_{n\to\infty} B_n$, or

$$A = \lim_{n\to\infty} [f(M_1) + f(M_2) + \cdots + f(M_n)] \frac{b-a}{n}. \tag{2}$$

It follows that if we let t_i be any point in the ith subinterval, and construct a rectangle of height $f(t_i)$ over $[x_{i-1}, x_i]$, as in Figure 5.6, then the combined area of these rectangles will be an approximation to A, and will approach A as $n \to \infty$. That is,

$$A = \lim_{n\to\infty} [f(t_1) + f(t_2) + \cdots + f(t_n)] \frac{b-a}{n}. \tag{3}$$

(The idea of the squeeze play for limits can be used here.)

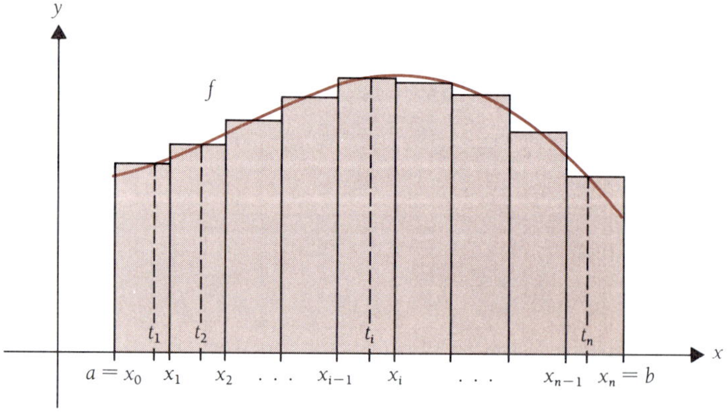

FIGURE 5.6 Rectangles of height $f(t_i)$

The limit (1) is sometimes called the **inner area** of the region R, because it is a limit of areas of rectangles lying inside R. The limit (2) is sometimes called the **outer area** of R, by contrast. For a continuous function, the inner and outer areas are the same, and their common value is defined to be the **area under the graph**.

For more general functions, the inner and outer areas need not be the same. Thus, we are led to a criterion for deciding whether there exists area under an arbitrary nonnegative function: We say the region under f,

$$R = \{(x, y) | a \leq x \leq b \text{ and } 0 \leq y \leq f(x)\},$$

has area if and only if the inner and outer areas are the same.

Example 5.1 Show that the region under the Dirichlet function

$$D(x) = \begin{cases} 1 & \text{if } x \text{ is rational} \\ 0 & \text{if } x \text{ is irrational} \end{cases}$$

over the interval $[0, 1]$ does not have area.

Solution Since any rectangle lying entirely under the function (see Figure 5.7) must have height 0, the inner area must be 0. On the other hand, any outer rectangle must have height 1, so that the outer area is 1. Thus, the region under this function does not have area.

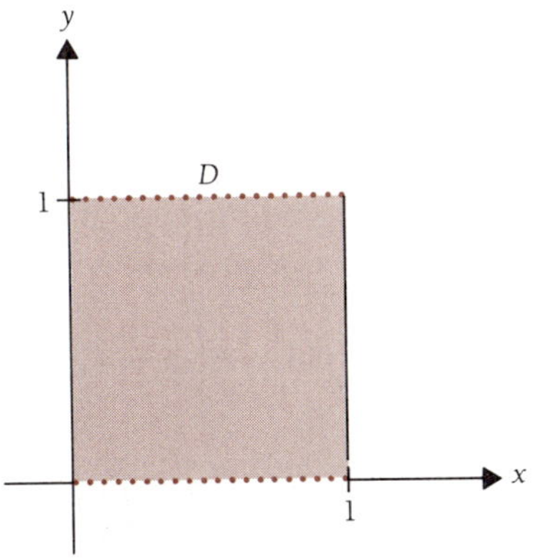

FIGURE 5.7 The Dirichlet function over [0, 1]

Summation Notation

Summation notation is a shorthand device for expressing the sum of many terms. It is also called **sigma notation**, because it is symbolized using the Greek letter capital sigma, Σ.

If f is a function and m and n are integers such that $m \leq n$, then the notation

$$\sum_{i=m}^{n} f(i) \tag{4}$$

stands for the sum

$$f(m) + f(m + 1) + f(m + 2) + \cdots + f(n).$$

The symbol Σ in expression (4) is called a **summation sign**; the letter i in (4) is called the **index of summation**. We call m and n the **lower** and **upper limits** of the summation; the expression $f(i)$ is called the **summand**. The index i is called a "dummy" index; if some other symbol were used instead of i, the notation would still mean the same thing.

The following examples illustrate the use of summation notation.

Example 5.2 Expand the following expressions (that is, write them without summation notation):

(a) $\sum_{i=2}^{5} i^2$ (b) $\sum_{i=0}^{6} \frac{i}{i+1}$

Solution (a) Here, $f(i) = i^2$, and

$$\sum_{i=2}^{5} i^2 = 2^2 + 3^2 + 4^2 + 5^2 = 54.$$

(b) Here, $f(i) = i/(i + 1)$, and

$$\sum_{i=0}^{6} \frac{i}{i+1} = 0 + \frac{1}{2} + \frac{2}{3} + \frac{3}{4} + \frac{4}{5} + \frac{5}{6} + \frac{6}{7}.$$

■

Example 5.3 Write the following using summation notation:

(a) $1 + 3 + 5 + \cdots + 49$

(b) The sum of the first 2000 positive integers.

Solution (a) Since $2i - 1$ represents a typical odd number, we may let $f(i) = 2i - 1$, with i ranging from 1 to 25, so that

$$1 + 3 + 5 + \cdots + 49 = \sum_{i=1}^{25} (2i - 1).$$

(b) The sum of the first 2000 positive integers can be written as

$$\sum_{i=1}^{2000} i.$$

■

Certain properties of summation notation follow easily from the commutative, associative, and distributive laws of arithmetic. For example, we have

$$\sum_{i=m}^{n} [f(i) + g(i)] = \sum_{i=m}^{n} f(i) + \sum_{i=m}^{n} g(i) \tag{5}$$

and

$$\sum_{i=m}^{n} cf(i) = c\sum_{i=m}^{n} f(i). \tag{6}$$

Thus, for instance, we may use properties (5) and (6) to rewrite

$$\sum_{i=1}^{n} (i^2 + 4i + 3)$$

as

$$\sum_{i=1}^{n} i^2 + 4\sum_{i=1}^{n} i + 3\sum_{i=1}^{n} 1.$$

Certain other identities involving summation notation that will be useful to us are given below. They can be proved in a variety of ways, perhaps the most straightforward of which is mathematical induction (see Appendix C).

$$\sum_{i=1}^{n} 1 = \underbrace{1 + 1 + 1 + \cdots + 1}_{n \text{ times}} = n \tag{7}$$

$$\sum_{i=1}^{n} i = 1 + 2 + 3 + \cdots + n = \frac{n(n+1)}{2} \tag{8}$$

$$\sum_{i=1}^{n} i^2 = 1^2 + 2^2 + 3^2 + \cdots + n^2 = \frac{n(n+1)(2n+1)}{6} \tag{9}$$

$$\sum_{i=1}^{n} i^3 = 1^3 + 2^3 + 3^3 + \cdots + n^3 = \frac{n^2(n+1)^2}{4} \tag{10}$$

You may find it useful to examine these formulas for a few specific values of n. You will be invited to prove some of them in the exercises.

Computing Area

Now we will use the limit idea to compute some areas by the method of exhaustion.

Let f be a continuous function, positive over $[a, b]$, and let $[a, b]$ be divided into n subintervals of equal length $\Delta x = (b - a)/n$. For each $i = 1, 2, \ldots, n$, let t_i be any point in the ith subinterval $[x_{i-1}, x_i]$. Then the area under f over $[a, b]$ is

$$A = \lim_{n\to\infty} [f(t_1) + f(t_2) + \cdots + f(t_n)]\Delta x. \tag{11}$$

This is the same as formula (3). In summation notation, equation (11) is written

$$A = \lim_{n\to\infty} \sum_{i=1}^{n} f(t_i)\Delta x. \tag{12}$$

Now let's compute some areas.

Example 5.4 Find the area under $f(x) = 2x$ over the interval $[0, 1]$.

Solution In order to use formula (12), we divide the interval $[0, 1]$ into n subintervals, each of length $\Delta x = 1/n$. Thus, we let

$$x_0 = 0, \quad x_1 = \frac{1}{n}, \quad x_2 = \frac{2}{n}, \quad \ldots, \quad x_n = 1.$$

Now we choose t_i to be any point in the ith subinterval $[(i-1)/n, i/n]$. For convenience, we will choose t_i to be the right-hand endpoint in each case, $t_i = i/n$, as shown in Figure 5.8.

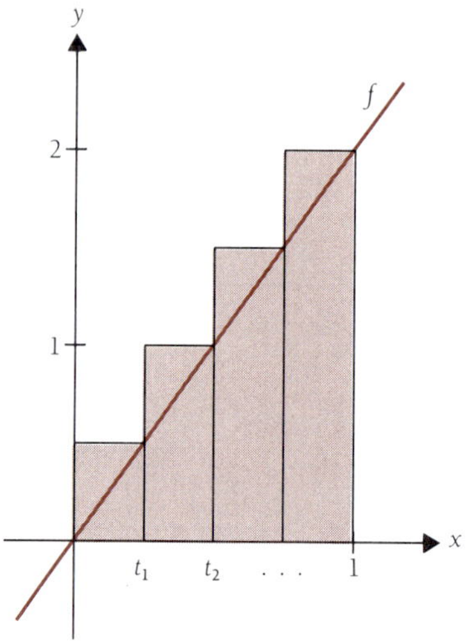

FIGURE 5.8 Finding the area under $f(x) = 2x$ over $[0, 1]$

Now $f(t_i) = 2t_i = 2i/n$, and formula (12) becomes

$$A = \lim_{n\to\infty} \sum_{i=1}^{n} \left(\frac{2i}{n}\right)\left(\frac{1}{n}\right).$$

Since n is constant relative to i, we can use formula (6) to rewrite this as

$$A = \lim_{n\to\infty} \left(\frac{2}{n^2}\right) \sum_{i=1}^{n} i.$$

Then using formula (8), we get

$$\begin{aligned} A &= \lim_{n\to\infty} \left(\frac{2}{n^2}\right) \frac{n(n+1)}{2} \\ &= \lim_{n\to\infty} \frac{n+1}{n} \\ &= 1. \end{aligned}$$

Since the region below f over $[0, 1]$ is a triangle of base 1 and height 2, you can see that the method of exhaustion gives the correct answer. ■

■ **Example 5.5** Find the area under $f(x) = x^2$ over $[0, 1]$.

Solution We approach this problem in the same way as we did Example 5.4, choosing $t_i = i/n$. This is shown in Figure 5.9. With $f(t_i) = t_i^2 = i^2/n^2$, formula (12) becomes

$$\begin{aligned} A &= \lim_{n\to\infty} \sum_{i=1}^{n} \left(\frac{i^2}{n^2}\right)\left(\frac{1}{n}\right) \\ &= \lim_{n\to\infty} \left(\frac{1}{n^3}\right) \sum_{i=1}^{n} i^2 \\ &= \lim_{n\to\infty} \left(\frac{1}{n^3}\right) \frac{n(n+1)(2n+1)}{6} \\ &= \lim_{n\to\infty} \frac{2n^2+3n+1}{6n^2} \\ &= \frac{1}{3}. \end{aligned}$$

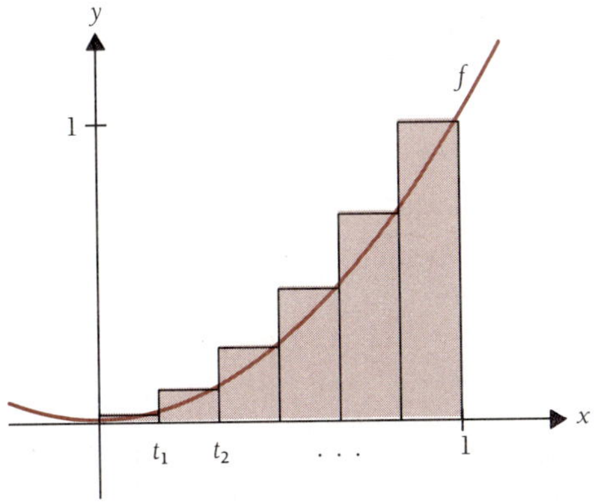

FIGURE 5.9 Finding the area under $f(x) = x^2$ over $[0, 1]$ ■

■ **Example 5.6** Find the area under $f(x) = 8 - x^3$ over $[1, 2]$.

Solution We subdivide $[1, 2]$ into n subintervals, each of length $\Delta x = (2-1)/n = 1/n$. We therefore have

$$x_0 = 1, \quad x_1 = 1 + \frac{1}{n}, \quad x_2 = 1 + \frac{2}{n}, \quad \ldots, \quad x_i = 1 + \frac{i}{n}, \quad \ldots, \quad x_n = 2.$$

We may choose any t_i in $[1 + (i - 1)/n, 1 + i/n]$. For the sake of simplicity, we will choose $t_i = 1 + i/n$. Then

$$\begin{aligned} f(t_i) &= 8 - t_i^3 \\ &= 8 - \left(1 + \frac{i}{n}\right)^3 \\ &= 7 - \frac{3i}{n} - \frac{3i^2}{n^2} - \frac{i^3}{n^3}. \end{aligned}$$

Now formula (12) becomes

$$\begin{aligned} A &= \lim_{n\to\infty} \sum_{i=1}^{n} \left(7 - \frac{3i}{n} - \frac{3i^2}{n^2} - \frac{i^3}{n^3}\right)\left(\frac{1}{n}\right) \\ &= \lim_{n\to\infty} \left[\left(\frac{7}{n}\right)\sum_{i=1}^{n} 1 - \left(\frac{3}{n^2}\right)\sum_{i=1}^{n} i - \left(\frac{3}{n^3}\right)\sum_{i=1}^{n} i^2 - \left(\frac{1}{n^4}\right)\sum_{i=1}^{n} i^3\right] \\ &= \lim_{n\to\infty} \left[\left(\frac{7}{n}\right) n - \left(\frac{3}{n^2}\right)\frac{n(n+1)}{2} \right. \\ &\qquad \left. - \left(\frac{3}{n^3}\right)\frac{n(n+1)(2n+1)}{6} - \left(\frac{1}{n^4}\right)\frac{n^2(n+1)^2}{4}\right] \\ &= \lim_{n\to\infty} \left[7 - \frac{3}{2}\left(1 + \frac{1}{n}\right) - \frac{1}{2}\left(1 + \frac{1}{n}\right)\left(2 + \frac{1}{n}\right) - \frac{1}{4}\left(1 + \frac{1}{n}\right)^2\right] \\ &= 7 - \frac{3}{2} - 1 - \frac{1}{4} \\ &= \frac{17}{4}. \end{aligned}$$

■

In these examples, we have been able to evaluate the limits, because we were able to rewrite the summations as simple formulas. In other situations, we may not be able to replace the summations, and hence may not be able to evaluate the limits. In the general case, we may have to be content with expressing the area as a limit of a sum.

■ **Example 5.7** Express the area under $f(x) = 1/x$ over $[1, 3]$ as the limit of a sum. (We will evaluate the limit of this sum later.)

Solution We subdivide the interval $[1, 3]$ into n subintervals, each of length $2/n$. Thus,

$$x_0 = 1, \quad x_1 = 1 + \frac{2}{n}, \quad x_2 = 1 + \frac{4}{n}, \quad \ldots, \quad x_i = 1 + \frac{2i}{n}, \quad \ldots, \quad x_n = 3.$$

If we let

$$t_i = x_i = 1 + \frac{2i}{n},$$

then

$$f(t_i) = \frac{1}{t_i} = \frac{n}{n + 2i}.$$

Hence, formula (12) gives us

$$A = \lim_{n\to\infty} \sum_{i=1}^{n} \left(\frac{n}{n+2i}\right)\left(\frac{2}{n}\right)$$
$$= \lim_{n\to\infty} \sum_{i=1}^{n} \frac{2}{n+2i}.$$

■

Exercises 5.1

1. Let $f(x) = 2x$ and let $[a, b] = [0, 1]$.
   (a) Give the numbers x_0, x_1, x_2, x_3, and x_4 that divide the interval $[a, b]$ into four subintervals of equal length.
   (b) Give the numbers m_i and $M_i (i = 1, 2, 3, 4)$ at which f has its minimum and maximum, respectively, over the ith subinterval.
   (c) Evaluate the inner area A_4 and the outer area B_4 (see Figure 5.8).
   (d) Similarly, find A_8 and B_8.

Repeat Problem 1 for each of the following functions and intervals:

2. $f(x) = 2x + 1$; $[1, 2]$
3. $f(x) = 3 - x$; $[0, 1]$
4. $f(x) = x^2$; $[0, 1]$
5. $f(x) = 4 - x^2$; $[-2, 2]$
6. $f(x) = \sqrt{x}$; $[0, 4]$
7. $f(x) = \dfrac{1}{x}$; $[1, 3]$
8. $f(x) = \dfrac{x}{x^2 + 1}$; $[0, 1]$

Rewrite without using summation notation.

9. $\sum_{i=1}^{3} 2i$
10. $\sum_{i=2}^{8} \frac{i}{2}$
11. $\sum_{i=-2}^{3} \frac{1}{i+4}$
12. $\sum_{i=5}^{9} \frac{i+1}{i^2+1}$
13. $\sum_{i=1}^{8} i^2$
14. $\sum_{i=0}^{9} (i^2 + 1)$
15. $\sum_{i=2}^{7} \frac{1}{i}$
16. $\sum_{i=1}^{6} \frac{i+1}{i^2}$

Rewrite using summation notation.

17. $1 + 3 + 5 + 7 + 9$
18. $\frac{2}{3} + \frac{3}{4} + \frac{4}{5} + \frac{5}{6} + \frac{6}{7}$
19. $\frac{1}{2} + \frac{1}{4} + \frac{1}{8} + \cdots + \frac{1}{128}$
20. $1 + \frac{1}{2} + \frac{1}{3} + \frac{1}{4} + \cdots + \frac{1}{25}$
21. $3 + 6 + 9 + \cdots + 90$
22. $1 + 4 + 7 + \cdots + 100$
23. $2 + \frac{3}{2} + \frac{4}{3} + \cdots + \frac{50}{49}$
24. $0.3 + 0.03 + 0.003 + \cdots + 0.000000000003$
25. $f(x_1) + f(x_2) + \cdots + f(x_n)$
26. $a_0 + a_1 + a_2 + \cdots + a_k$
27. $x_1^2 + x_2^2 + x_3^2 + \cdots + x_n^2$
28. $2x_1 + 3x_2 + 4x_3 + \cdots + (j + 1)x_j$

Prove by induction (see Appendix C):

29. $1 + 2 + 3 + \cdots + n = \dfrac{n(n+1)}{2}$

30. $1^2 + 2^2 + 3^2 + \cdots + n^2 = \dfrac{n(n+1)(2n+1)}{6}$

31. $1^3 + 2^3 + 3^3 + \cdots + n^3 = \dfrac{n^2(n+1)^2}{4}$

32. $1^4 + 2^4 + 3^4 + \cdots + n^4 = \dfrac{n(n+1)(2n+1)(3n^2+3n-1)}{30}$

Find the area under f over the given interval.

33. $f(x) = x + 4$; $[0, 1]$
34. $f(x) = 3 - x$; $[0, 3]$
35. $f(x) = 2x + 1$; $[0, 3]$
36. $f(x) = 2x - 1$; $[1, 3]$
37. $f(x) = x^2$; $[0, 2]$
38. $f(x) = x^3$; $[0, 4]$
39. $f(x) = x^2 - x$; $[1, 2]$
40. $f(x) = x^4$; $[0, 1]$

Express the area under f over the given interval as the limit of a sum.

41. $f(x) = \dfrac{1}{x}$; $[1, 2]$
42. $f(x) = x^{-2}$; $[0, 1]$
43. $f(x) = \sin x$; $[0, \pi]$
44. $f(x) = \dfrac{1}{x^2+1}$; $[0, 1]$

45. Derive the formula for the area of a circle of radius r by carrying out the following steps:
   (a) Inscribe a regular n-sided polygon in the circle. Divide the polygon into triangles by drawing radii from the center of the circle to the vertices of the polygon. Using the fact that there are 2π radians in a circle, and elementary facts from trigonometry, find an expression for the area of one of the triangles.
   (b) Show that the area of the polygon is

$$A_n = \pi r^2\left(\frac{n}{\pi}\sin\frac{\pi}{n}\cos\frac{\pi}{n}\right).$$

   (c) Take the limit of the area A_n as $n \to \infty$.

46. Using the procedure of Problem 45, find the formula for the circumference of a circle of radius r.

47. Find the inner and outer areas for the function

$$f(x) = \begin{cases} 2 & \text{if } x \text{ is rational} \\ 1 & \text{if } x \text{ is irrational} \end{cases}$$

over the interval $[0, 1]$.

48. Repeat Problem 47 for the ruler function (see Problem 34 in the Additional Exercises at the end of Chapter 2).

SECTION 5.2

Riemann Sums and the Definite Integral

In this section, we generalize the procedure for expressing area as the limit of a sum. We introduce *Riemann sums* and the *definite integral*, and consider the problem of evaluating definite integrals. The approach given here is similar to that originally given by G. F. B. Riemann.

Georg Friedrich Bernhard Riemann (1826–1866), German mathematician and professor at Göttingen, was fortunate in having a succession of teachers who recognized his extraordinary talent and did their best to develop it. His personal shyness contrasted oddly with the boldness and originality of his mathematical thought. His mathematical gift included the ability to view broad areas of mathematics and their essential properties, unhampered by minor details. While other men worked on a single problem, he was able to see what was common to a whole class of such problems and solve them all at once. In the theory of elliptic functions, he invented new organizing principles from which the hard work of his predecessors could be easily derived. In the process, he invented the Riemann surface, and stimulated the new field of topology. He invented a family of noneuclidean geometries upon which Einstein based his theory of relativity. A conjecture put forth by Riemann in the theory of prime numbers has far-reaching consequences, but has defied the efforts of the ablest mathematicians.

Riemann visited Italy in his last few years in an attempt to improve his health. His work there with Italian mathematicians produced several important new ideas. He died in Italy at the age of 39.

Riemann Sums

In the previous section, we expressed the area under a function as the limit of a sum. To do so, we assumed first of all that the function f was continuous and nonnegative over the interval $[a, b]$. Then we divided $[a, b]$ into subintervals $[x_0, x_1]$, $[x_1, x_2]$, . . . , $[x_{n-1}, x_n]$ of equal length $\Delta x = (b - a)/n$. After choosing point t_i in the ith subinterval, we formed the product $f(t_i)\Delta x$ for each i, and then the sum

$$\sum_{i=1}^{n} f(t_i)\Delta x. \tag{1}$$

Now we generalize this procedure in order to form a sum of products whose limit is more useful than just for finding area. First of all, we let f be any function that is defined on $[a, b]$. Next, we form a subdivision of $[a, b]$ whose subintervals need not all be the same length. We will let $\Delta x_i = x_i - x_{i-1}$ denote the length of the ith subinterval. [If, for each i, $\Delta x_i = (b - a)/n$, we will call the subdivision **regular**.] The sum (1) therefore becomes

$$\sum_{i=1}^{n} f(t_i)\Delta x_i, \tag{2}$$

which is called a **Riemann sum for f over $[a, b]$, relative to the subdivision $s = \{x_0, x_1, \ldots, x_n\}$.**

Note that there are many Riemann sums for f relative to a given subdivision s, because of the many possible choices for $t_1, t_2, \ldots, t_n$. Note also that the sum (1), which is an approximation to the area under a positive function f, is a Riemann sum for f relative to a regular subdivision.

For each summand $f(t_i)\Delta x_i$ in (2), if $f(t_i)$ is positive, then the summand is the area of the rectangle of height $f(t_i)$ and width Δx_i; this rectangle is above the x-axis (see Figure 5.10). If $f(t_j)$ is negative, then $f(t_j)\Delta x_j$ is the negative of the area of a rectangle of height $|f(t_j)|$ and width Δx_j; this rectangle is below the x-axis. Thus, a Riemann sum in general is a sum of areas of rectangles above the x-axis, minus a sum of areas of rectangles below the x-axis.

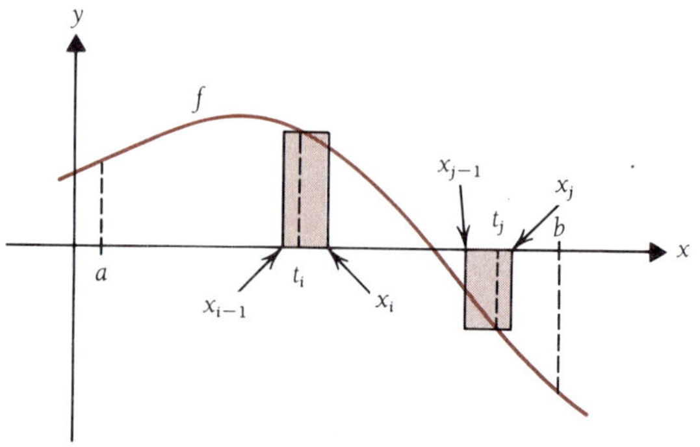

FIGURE 5.10 Interpreting a Riemann sum as area

Limits of Riemann Sums

Let s be a subdivision of interval $[a, b]$. The **norm** of s, denoted by $\|s\|$, is the length of the longest subinterval of s. That is, $\|s\| =$ maximum $\{\Delta x_i\}$. The limit of a Riemann sum is defined relative to the norm of s.

DEFINITION 5.1 Let R_s be a Riemann sum for the function f over the interval $[a, b]$, relative to a subdivision s. We say R_s has **limit** L **as** $\|s\| \to 0$, denoted $\lim_{\|s\|\to 0} R_s = L$, if and only if for any $\varepsilon > 0$ there exists $\delta > 0$ such that if $\|s\| < \delta$, then $|R_s - L| < \varepsilon$.

Letting $R_s = \sum_{i=1}^{n} f(t_i)\Delta x_i$, Definition 5.1 says that

$$\lim_{\|s\|\to 0} \sum_{i=1}^{n} f(t_i)\Delta x_i = L$$

in case, for any $\varepsilon > 0$ there is a $\delta > 0$ such that if $\|s\| < \delta$, then

$$\left|\sum_{i=1}^{n} f(t_i)\Delta x_i - L\right| < \varepsilon.$$

This holds for any choices whatever of the t_i's if the limit exists. You may wish to compare this definition of limit with those in Chapter 2.

Note that for a regular subdivision, $\Delta x_i = \|s\| = (b - a)/n$, so that $\|s\| \to 0$ is equivalent to $n \to \infty$. This shows that finding area in Section 5.1 by evaluating limits is a special case of finding the limit of a Riemann sum.

By analogy with the problem of finding areas, we may conclude that the limit of a Riemann sum, if it exists, can be interpreted as the difference between the area under the function above the x-axis and the area above the function under the x-axis (see Figure 5.11).

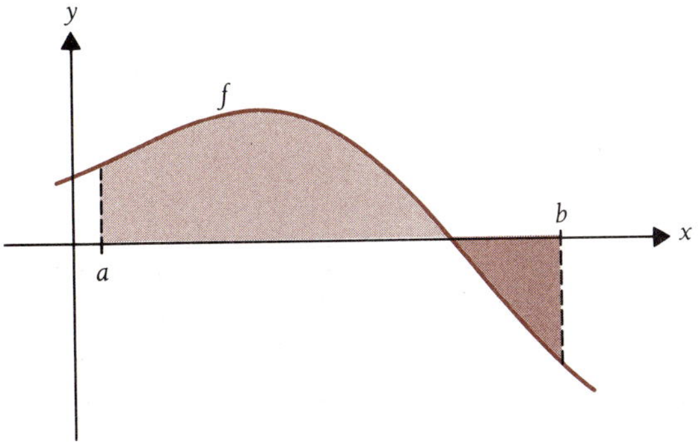

FIGURE 5.11 Interpreting the limit of a Riemann sum as area

The Definite Integral

The limit of a Riemann sum for f does not always exist, depending on the properties of f. Those functions for which Riemann sums do have limits are given a special name, and the limits are also denoted in a special way.

DEFINITION 5.2

Let the function f be defined on the interval $[a, b]$. If the limit of the Riemann sum for f over $[a, b]$ exists, we say the function f is **integrable on $[a, b]$**, and we call the limit of the Riemann sum the **definite integral of f on $[a, b]$**, denoted by

$$\int_a^b f(x)\,dx.$$

That is,

$$\lim_{\|s\| \to 0} \sum_{i=1}^{n} f(t_i)\Delta x_i = \int_a^b f(x)\,dx$$

when the limit exists.

Notation for the Definite Integral

The notation $\int_a^b f(x)\,dx$ for the definite integral has several parts that deserve comment. The **integral sign**, $\int$, is a symbol for "sum." The number a is called the

lower limit, and the number b is called the **upper limit**; together they are called **limits of integration.** (Here, "limit" is used in the sense of "bound," as it is in summation notation.) The function $f(x)$ is called the **integrand**. The variable x is a "dummy" variable; if it is replaced by another symbol, the meaning does not change. That is, such expressions as $\int_a^b f(t)\,dt$, $\int_a^b f(z)\,dz$, $\int_a^b f(\theta)\,d\theta$, etc., all mean the same as $\int_a^b f(x)\,dx$.

The definite integral looks a lot like the indefinite integral, which we saw in Section 4.6. However, it is important to remember that they stand for different things. The indefinite integral $\int f(x)\,dx$ is the general antiderivative, a function; while the definite integral $\int_a^b f(x)\,dx$ is a number.

Properties of Integrable Functions

It is proved at the end of this section that *a function that is continuous on $[a, b]$ is also integrable on $[a, b]$*. Since most of the functions we consider are continuous, we will rarely encounter a function that fails to be integrable.

However, not every integrable function is continuous. Many functions with points of discontinuity are also integrable. In general, if a function has only a finite number of points of discontinuity in $[a, b]$ and is bounded in $[a, b]$ (that is, there is a number B such that $|f(x)| \leq B$ for all $x \in [a, b]$), then f is integrable in $[a, b]$.

Another large class of integrable functions is the class of monotonic functions (see Definition 4.1). A proof that a monotonic function is integrable appears at the end of this section. Thus, for example, the greatest integer function $f(x) = \lfloor x \rfloor$ is integrable.

It is the case that *any function that is integrable on $[a, b]$ is bounded on $[a, b]$*. For, if a function f has no upper bound, say, and s is any subdivision, then f is unbounded over some subinterval $[x_{k-1}, x_k]$; hence, t_k can be chosen so that $f(t_k)\Delta x_k$ is as large as we please. Hence, $R_s = \sum_{i=1}^{n} f(t_i)\Delta x_i$ can be made as large as we please, so $\lim_{\|s\|\to 0} R_s$ will not exist.

Evaluating the Definite Integral

If a function f is integrable on $[a, b]$, then the definite integral $\int_a^b f(x)\,dx$ is the limit of a Riemann sum. In particular, if regular subdivisions are used, then

$$\int_a^b f(x)\,dx = \lim_{n\to\infty} \sum_{i=1}^{n} f(t_i)\,\frac{b-a}{n}. \tag{3}$$

If the sum in (3) can be rewritten without the summation sign for some choice of the t_i's, then the definite integral can be evaluated as the limit. This was done in Section 5.1 for nonnegative functions, but now the restriction that f be nonnegative is unnecessary.

■ **Example 5.8** Evaluate $\int_{-1}^{2} x^3\,dx$.

Solution Subdividing the interval $[-1, 2]$ into n equal parts, we find

$$x_0 = -1, \quad x_1 = -1 + \frac{3}{n}, \quad x_2 = -1 + \frac{6}{n},$$

$$\ldots, \quad x_i = -1 + \frac{3i}{n}, \quad \ldots, \quad x_n = 2.$$

Letting $t_i = x_i$, equation (3) becomes

$$\begin{aligned}
\int_{-1}^{2} x^3\,dx &= \lim_{n\to\infty} \sum_{i=1}^{n} x_i^3\left(\frac{3}{n}\right) \\
&= \lim_{n\to\infty} \sum_{i=1}^{n} \left(-1 + \frac{3i}{n}\right)^3\left(\frac{3}{n}\right) \\
&= \lim_{n\to\infty} \frac{3}{n}\left[-\sum_{i=1}^{n} 1 + \left(\frac{9}{n}\right)\sum_{i=1}^{n} i - \left(\frac{27}{n^2}\right)\sum_{i=1}^{n} i^2 + \left(\frac{27}{n^3}\right)\sum_{i=1}^{n} i^3\right] \\
&= \lim_{n\to\infty} \frac{3}{n}\left[-n + \frac{9}{n}\cdot\frac{n(n+1)}{2} - \frac{27}{n^2}\cdot\frac{n(n+1)(2n+1)}{6} + \frac{27}{n^3}\cdot\frac{n^2(n+1)^2}{4}\right] \\
&= \lim_{n\to\infty}\left[-3 + \frac{27}{2}\left(1 + \frac{1}{n}\right) - \frac{27}{2}\left(1 + \frac{1}{n}\right)\left(2 + \frac{1}{n}\right) + \frac{81}{4}\left(1 + \frac{1}{n}\right)^2\right] \\
&= -3 + \frac{27}{2} - 27 + \frac{81}{4} \\
&= \frac{15}{4}.
\end{aligned}$$

■

A definite integral may also be evaluated by interpreting it in terms of known areas.

■ **Example 5.9** Evaluate $\int_{1}^{3} (x + 1)\,dx$.

Solution This integral can be interpreted as the area of the trapezoid with sides $y = x + 1$, $x = 1$, $y = 0$, and $x = 3$ (see Figure 5.12, page 314). Thus,

$$\int_{1}^{3} (x + 1)\,dx = 6.$$

■

■ **Example 5.10** Evaluate $\int_{0}^{1} \sqrt{1 - x^2}\,dx$.

Solution Since the curve $y = \sqrt{1 - x^2}$ is the upper half of the unit circle (see Figure 5.13), this integral is the area of one quarter of the circle, the part lying over $[0, 1]$. Hence,

$$\int_{0}^{1} \sqrt{1 - x^2}\,dx = \frac{\pi}{4}.$$

■

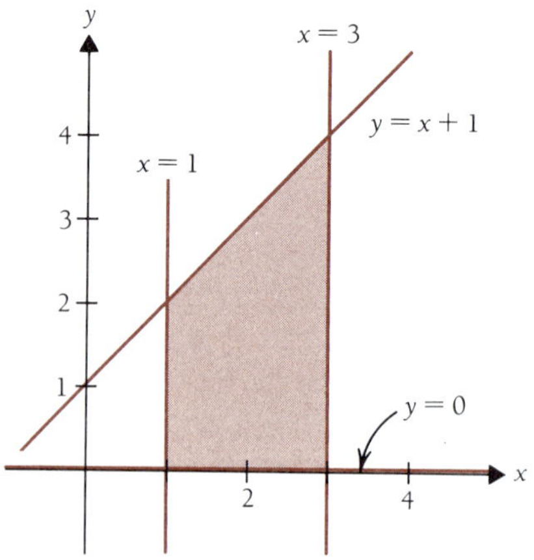

FIGURE 5.12 The region under $y = x + 1$ over $[1, 3]$

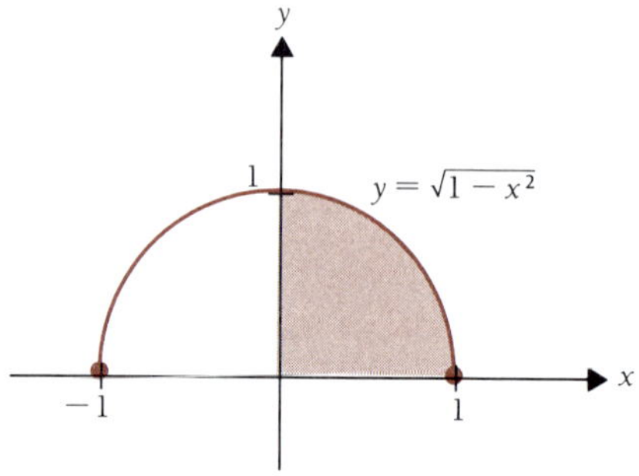

FIGURE 5.13 The region under $y = \sqrt{1 - x^2}$ over $[0, 1]$

■ **Example 5.11** Evaluate $\int_{-2}^{4} \lfloor x \rfloor \, dx$.

Solution From the graph of $y = \lfloor x \rfloor$ (see Figure 5.14), we can interpret this integral as a difference of areas. By simply counting the square units, we find that

$$\int_{-2}^{4} \lfloor x \rfloor \, dx = 6 - 3 = 3.$$

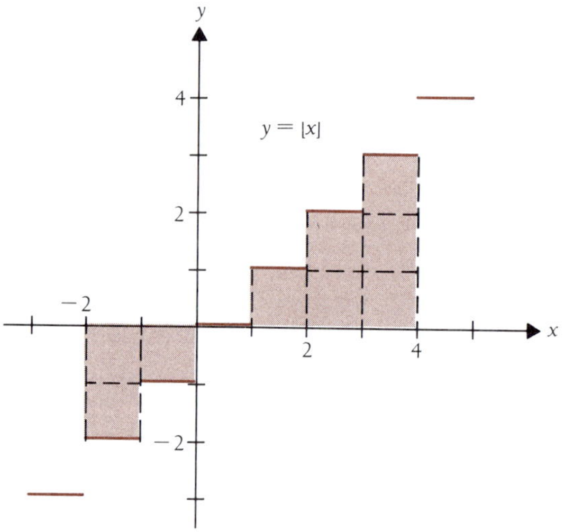

FIGURE 5.14 The region between $y = \lfloor x \rfloor$ and the x-axis in the interval $[-2, 4]$ ■

The methods illustrated here for evaluating definite integrals are a little cumbersome and do not apply to very many kinds of functions. Much more general (and easier) methods will be developed in the sections that follow.

Equation (3) also gives us a method for approximating a definite integral. By selecting values for n and each t_i, we can compute the Riemann sum $\sum_{i=1}^{n} f(t_i)(b-a)/n$, and thereby obtain an approximation to the integral $\int_a^b f(x)\,dx$. We will have a great deal more to say about approximations later.

Proofs of Integrability (Optional)

We will conclude this section by proving that continuous functions and monotonic functions are integrable.

THEOREM 5.1 If function f is continuous on $[a, b]$, then f is integrable on $[a, b]$.

Before beginning the proof, we must add to our knowledge of continuity. Recall that if f is continuous on $[a, b]$, then for any $c \in [a, b]$, $\lim_{x\to c} f(x) = f(c)$. That is, for any $\varepsilon > 0$, there is some $\delta > 0$ such that if $0 < |x - c| < \delta$, then $|f(x) - f(c)| < \varepsilon$. Note that here, the value of δ chosen may depend not only on ε, but also on the point c. However, because f is continuous on a closed interval, δ can be chosen independently of c. That is, if f is continuous on $[a, b]$, then it can be shown that for any $\varepsilon > 0$, there is a number $\delta > 0$, which depends only on ε and the interval $[a, b]$, such that if $x_1, x_2 \in [a, b]$ and $|x_1 - x_2| < \delta$, then $|f(x_1) - f(x_2)| < \varepsilon$. This property of f is called **uniform continuity** (see Problem 42 in the Additional Exercises at the end of Chapter 2). Stated another way, any function that is continuous on a closed interval is uniformly continuous there. Uniform continuity is used to prove integrability.

Proof of Theorem 5.1 Consider any subdivision $s = \{x_0, x_1, \ldots, x_n\}$ of $[a, b]$. The maximum Riemann sum for f relative to s is obtained by choosing t_i so that $f(t_i)$ is a maximum over $[x_{i-1}, x_i]$. If M_i is the point in the ith subinterval at which f has its maximum, then the maximum Riemann sum is

$$R_s = \sum_{i=1}^{n} f(M_i)\Delta x_i.$$

Similarly, if m_i is the point in the ith subinterval at which f has its minimum, then the minimum Riemann sum for f over $[a, b]$ relative to s is

$$r_s = \sum_{i=1}^{n} f(m_i)\Delta x_i.$$

Now we will show that the Riemann sums have a limit by showing that the difference between the maximum and minimum Riemann sums approaches 0. That is, the maximum and minimum Riemann sums have the same limit, so every Riemann sum has that same limit (think of the squeeze play). This will prove that f is integrable on $[a, b]$.

Let $\varepsilon > 0$ be given. Select δ such that if $x_1, x_2 \in [a, b]$ and $|x_1 - x_2| < \delta$, then $|f(x_1) - f(x_2)| < \varepsilon/(b - a)$—that is, f is uniformly continuous on $[a, b]$. Now let s be a subdivision such that $\|s\| < \delta$. Then the difference between the maximum and minimum Riemann sums is

$$R_s - r_s = \sum_{i=1}^{n} f(M_i)\Delta x_i - \sum_{i=1}^{n} f(m_i)\Delta x_i$$
$$= \sum_{i=1}^{n} [f(M_i) - f(m_i)]\Delta x_i.$$

Now, M_i and m_i are both in the ith subinterval, so $|M_i - m_i| < \delta$, and consequently, $|f(M_i) - f(m_i)| < \varepsilon/(b-a)$. Hence (see Figure 5.15),

$$R_s - r_s < \sum_{i=1}^{n} \left(\frac{\varepsilon}{b-a}\right)\Delta x_i$$
$$= \left(\frac{\varepsilon}{b-a}\right)\sum_{i=1}^{n} \Delta x_i$$
$$= \left(\frac{\varepsilon}{b-a}\right)(b-a) = \varepsilon.$$

Thus, $\lim_{\|s\|\to 0}(R_s - r_s) = 0$.

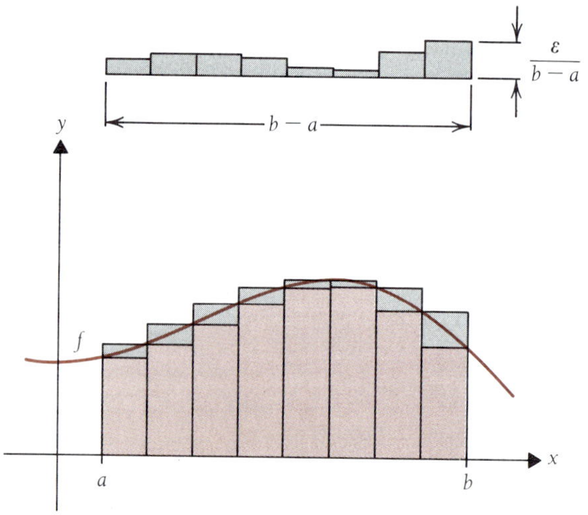

FIGURE 5.15 Making the difference between maximum and minimum Riemann sums small in the case of a continuous function

THEOREM 5.2 If the function f is monotonic on $[a, b]$, then f is integrable in $[a, b]$.

Proof We will prove the theorem for f increasing; the other case is similar. Again, we will show that the maximum and minimum Riemann sums for f differ by an arbitrarily small amount.

If $s = \{x_0, x_1, \ldots, x_n\}$ is a subdivision of $[a, b]$, then since f is increasing, the maximum of f over $[x_{i-1}, x_i]$ occurs at x_i, and the minimum occurs at x_{i-1}. Hence, the maximum Riemann sum is

$$R_s = \sum_{i=1}^{n} f(x_i)\Delta x_i,$$

and the minimum Riemann sum is

$$r_s = \sum_{i=1}^{n} f(x_{i-1})\Delta x_i.$$

Now, $\Delta x_i \leq \|s\|$, so that

$$\begin{aligned} R_s - r_s &= \sum_{i=1}^{n} f(x_i)\Delta x_i - \sum_{i=1}^{n} f(x_{i-1})\Delta x_i \\ &= \sum_{i=1}^{n}[f(x_i) - f(x_{i-1})]\Delta x_i \\ &\leq \sum_{i=1}^{n} [f(x_i) - f(x_{i-1})]\|s\| \\ &= [f(x_n) - f(x_0)]\|s\| \\ &= [f(b) - f(a)]\|s\|. \end{aligned}$$

Since $f(b) \neq f(a)$, we can force $R_s - r_s < \varepsilon$ by letting $\|s\| < \varepsilon/[f(b) - f(a)]$ (see Figure 5.16). This completes the proof.

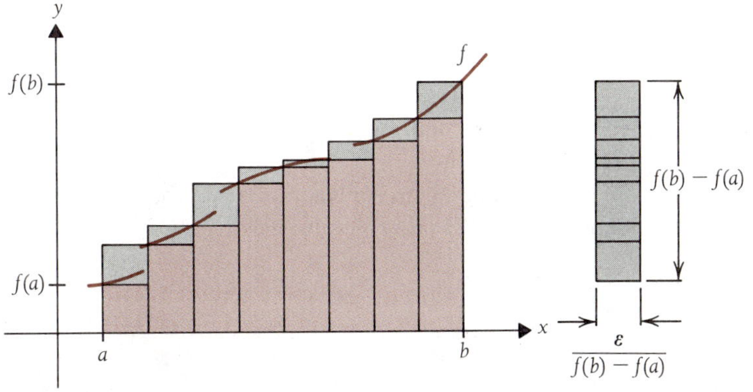

FIGURE 5.16 Making the difference between the maximum and minimum Riemann sums small in the case of a monotonic function

Exercises 5.2

Evaluate each definite integral as the limit of a sum.

1. $\int_0^1 2x\,dx$
2. $\int_{-1}^1 x^2\,dx$
3. $\int_0^1 (x - 2)\,dx$
4. $\int_{-2}^0 x^3\,dx$

5. $\int_1^2 (3x - 1)\, dx$ 6. $\int_0^2 (2 - 3x)\, dx$ 7. $\int_0^2 (x^2 - 2x)\, dx$ 8. $\int_1^4 (x - x^3)\, dx$

Evaluate each definite integral by interpreting it as an area.

9. $\int_0^3 x\, dx$ 10. $\int_1^4 (x + 2)\, dx$ 11. $\int_{-2}^4 |x|\, dx$ 12. $\int_{-1}^2 |x - 1|\, dx$

13. $\int_1^5 \lfloor x \rfloor\, dx$ 14. $\int_4^7 \lfloor x \rfloor\, dx$ 15. $\int_{-1}^1 \sqrt{1 - x^2}\, dx$ 16. $\int_{-2}^2 \sqrt{4 - x^2}\, dx$

Rewrite each limit as a definite integral.

17. $\lim_{n\to\infty} \sum_{i=1}^{n} \frac{2i}{n^2}$ 18. $\lim_{n\to\infty} \sum_{i=1}^{n} \frac{i^3}{n^4}$

19. $\lim_{n\to\infty} \left(\frac{2}{n+2} + \frac{2}{n+4} + \cdots + \frac{2}{n+2n} \right)$

20. $\lim_{n\to\infty} \left(\sin\frac{\pi}{n} + \sin\frac{2\pi}{n} + \sin\frac{3\pi}{n} + \cdots + \sin\frac{n\pi}{n} \right)\left(\frac{\pi}{n}\right)$

21. Show that $\int_n^{n+1} \lfloor x \rfloor\, dx = n$ by interpreting the integral as an area.

22. Evaluate $\int_1^n \lfloor x \rfloor\, dx$.

23. Approximate $\int_1^2 (1/x)\, dx$ by letting $n = 8$ and $t_i = x_i$ in equation (3).

24. Approximate $\int_0^\pi \sin x\, dx$ by letting $n = 4$ and $t_i = x_i$ in equation (3).

25. Approximate $\int_0^1 x^2\, dx$ by letting $n = 5$ and $t_i = (x_i + x_{i-1})/2$ in equation (3).

26. Approximate $\int_0^1 x^2\, dx$ by letting $n = 10$ and $t_i = (x_i + x_{i-1})/2$ in equation (3).

27. Show that $\int_a^b dx = b - a$. [*Hint:* The integrand is $f(x) = 1$.]

28. Let c be a constant. Explain why $\int_a^b c\, dx = c(b - a)$.

SECTION 5.3

Properties of the Definite Integral

The definite integral has several properties that are useful for both theoretical and practical reasons. In this section, we discuss these properties.

Interval Properties

According to Definition 5.2, the definite integral $\int_a^b f(x)\, dx$ exists if f is integrable on interval $[a, b]$. Thus, the notation for the definite integral tacitly assumes that $a < b$. We wish first of all to define the definite integrals $\int_a^a f(x)\, dx$ and $\int_b^a f(x)\, dx$.

DEFINITION 5.3

If f is defined at a, then

$$\int_a^a f(x)\, dx = 0. \tag{1}$$

If $a < b$ and $\int_a^b f(x)\,dx$ exists, then

$$\int_b^a f(x)\,dx = -\int_a^b f(x)\,dx. \tag{2}$$

Statement (1) makes sense if the integral is interpreted in terms of area, as illustrated in Figure 5.17. Statement (2) also may be stated as, "Switching the limits (or reversing the direction of integration) changes the sign" (see Figure 5.18). To see why this changes the sign, note that the Riemann sums defining $\int_a^b f(x)\,dx$ involve $\Delta x_i = x_i - x_{i-1} > 0$, while the Riemann sums for $\int_b^a f(x)\,dx$ involve $\Delta x_i = x_{i-1} - x_i < 0$.

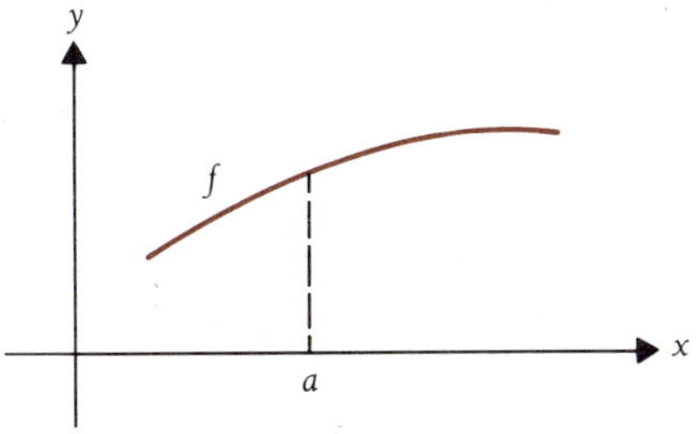

FIGURE 5.17 $\int_a^a f(x)\,dx = 0$

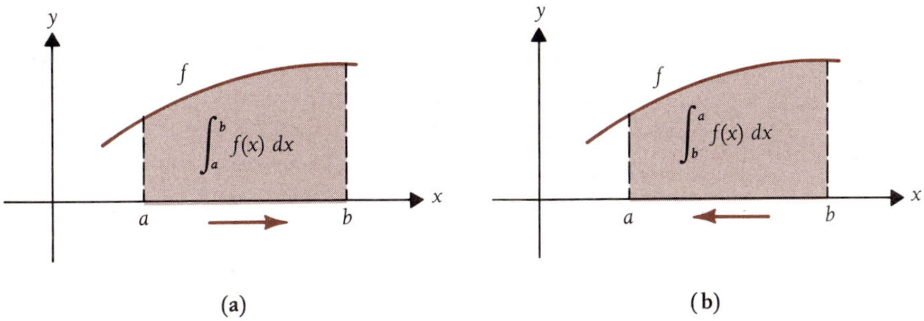

FIGURE 5.18 $\int_b^a f(x)\,dx = -\int_a^b f(x)\,dx$

With this expanded notation, we can now discuss the properties of the definite integral. We will state the properties as theorems; the proofs given in several cases are just outlines of proofs, with some of the more technical details omitted.

The first property concerns integrability on intervals with a common endpoint.

THEOREM 5.3 If $a < b < c$ and the function of f is integrable on $[a, b]$ and on $[b, c]$, then f is integrable on $[a, c]$ and

$$\int_a^c f(x)\,dx = \int_a^b f(x)\,dx + \int_b^c f(x)\,dx.$$

This result certainly makes sense when interpreted in terms of area, as illustrated in Figure 5.19. The proof is omitted.

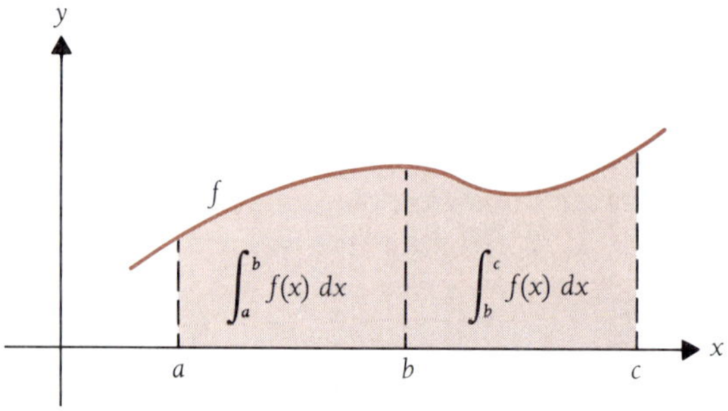

FIGURE 5.19 $\int_a^b f(x)\,dx + \int_b^c f(x)\,dx = \int_a^c f(x)\,dx$

THEOREM 5.4 If the function f is integrable in an interval I, then f is integrable in any subinterval of I. If a, b, and c are points in I, then regardless of the order of a, b, and c,

$$\int_a^c f(x)\,dx = \int_a^b f(x)\,dx + \int_b^c f(x)\,dx.$$

Rather than prove this theorem, we simply remark that the first assertion can be established by considering Riemann sums. The second assertion then can be shown by applying Definition 5.3 and Theorem 5.3 to any ordering of the points a, b, and c.

Example 5.12 Show that

$$\int_0^1 x\,dx + \int_1^2 x\,dx = \int_0^2 x\,dx$$

by interpreting the integrals as areas.

Solution From Figure 5.20, we see that

$$\int_0^1 x\,dx = \frac{1}{2}, \qquad \int_1^2 x\,dx = \frac{3}{2},$$

and

$$\int_0^2 x\,dx = 2,$$

so the result holds.

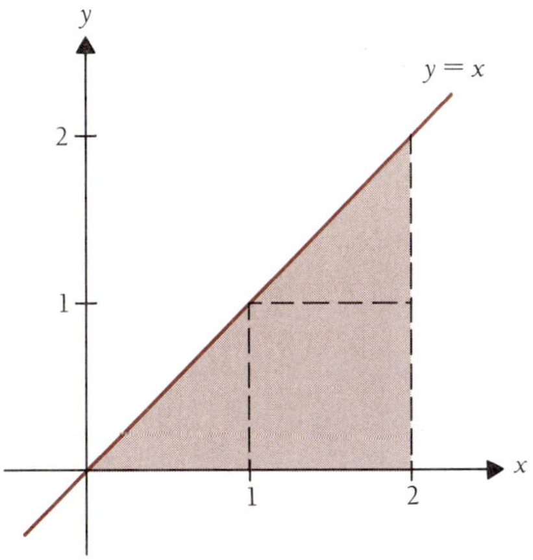

FIGURE 5.20 $\int_0^1 x\,dx + \int_1^2 x\,dx = \int_0^2 x\,dx$

Linearity Properties

The next theorem concerns scalar multiples of integrable functions.

THEOREM 5.5 If the function f is integrable on $[a, b]$ and k is a constant, then the function kf is integrable on $[a, b]$, and

$$\int_a^b kf(x)\,dx = k\int_a^b f(x)\,dx.$$

Proof

$$\begin{aligned}\int_a^b kf(x)\,dx &= \lim_{\|s\|\to 0} \sum_{i=1}^{n} kf(t_i)\Delta x_i \\ &= \lim_{\|s\|\to 0} k\sum_{i=1}^{n} f(t_i)\Delta x_i \\ &= k\lim_{\|s\|\to 0} \sum_{i=1}^{n} f(t_i)\Delta x_i \\ &= k\int_a^b f(x)\,dx.\end{aligned}$$

Theorem 5.5 states that a constant factor of the integrand is also a factor of the definite integral, and may be "factored out" or "moved across the integral sign." We may interpret this theorem geometrically by viewing k as a vertical scale factor, each value of $kf(x)$ being k times the value of $f(x)$ (see Figure 5.21, page 322). Hence, the area under kf is k times the area under f.

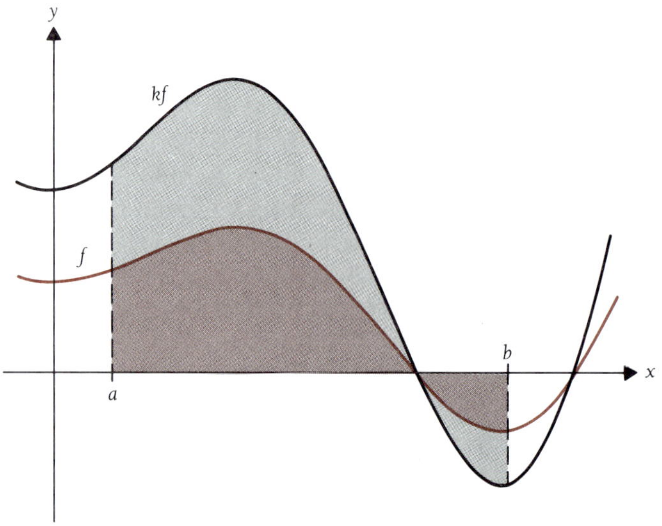

FIGURE 5.21 $\int_a^b kf(x)\,dx = k\int_a^b f(x)\,dx$

Example 5.13 Given that $\int_1^3 (x+1)\,dx = 6$, find $\int_1^3 (5x+5)\,dx$.

Solution By Theorem 5.5,

$$\begin{aligned}\int_1^3 (5x+5)\,dx &= \int_1^3 5(x+1)\,dx \\ &= 5\int_1^3 (x+1)\,dx \\ &= 5(6) = 30.\end{aligned}$$

The next theorem deals with a sum of integrable functions.

THEOREM 5.6 If functions f and g are both integrable on $[a, b]$, then $f+g$ is integrable on $[a, b]$, and

$$\int_a^b [f(x)+g(x)]\,dx = \int_a^b f(x)\,dx + \int_a^b g(x)\,dx.$$

The proof of this theorem is based on the fact that the sum of Riemann sums for f and g is a Riemann sum for $f+g$, so that the sum of the definite integrals of f and g is the definite integral of the sum of f and g. Interpreted geometrically, the area under f plus the area under g is the area under $f+g$ (see Figure 5.22).

Example 5.14 Given that $\int_0^1 x^2\,dx = \frac{1}{3}$ and $\int_0^1 \sqrt{1-x^2}\,dx = \pi/4$, find $\int_0^1 (x^2+\sqrt{1-x^2})\,dx$.

Solution Using Theorem 5.6, we can easily see that

$$\int_0^1 (x^2+\sqrt{1-x^2})\,dx = \frac{1}{3}+\frac{\pi}{4}.$$

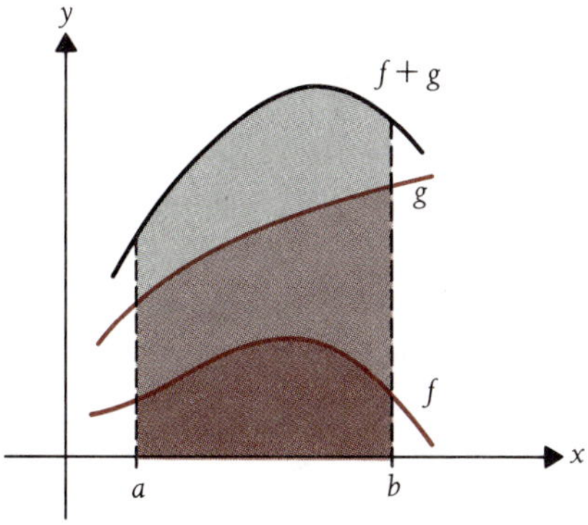

FIGURE 5.22 $\int_a^b [f(x) + g(x)]\, dx = \int_a^b f(x)\, dx + \int_a^b g(x)\, dx$

Theorems 5.5 and 5.6 can be combined and generalized for any linear combination of integrable functions. That is, if $f_1, f_2, \ldots, f_n$ are any integrable functions over $[a, b]$ and $k_1, k_2, \ldots, k_n$ are any constants, then the function $k_1 f_1 + k_2 f_2 + \cdots + k_n f_n$ is integrable, and

$$\int_a^b \left[\sum_{i=1}^{n} k_i f_i(x) \right] dx = \sum_{i=1}^{n} k_i \int_a^b f_i(x)\, dx. \tag{3}$$

Property (3) is called the **linearity property** of the definite integral, and is perhaps the most commonly used property of all.

Other Properties

If a function f is continuous and positive over $[a, b]$, then the definite integral $\int_a^b f(x)\, dx$ is the area under f, and is also positive. This proves the following theorem.

THEOREM 5.7 If $f(x) \geq 0$ for all $x \in [a, b]$ and f is integrable on $[a, b]$, then $\int_a^b f(x)\, dx \geq 0$.

Theorem 5.7 allows us to compare the integrals of two functions.

THEOREM 5.8 If functions f and g are integrable on $[a, b]$ and $f(x) \leq g(x)$ for all $x \in [a, b]$, then

$$\int_a^b f(x)\, dx \leq \int_a^b g(x)\, dx.$$

Proof If $f(x) \leq g(x)$, then $g(x) - f(x) \geq 0$ for all $x \in [a, b]$. Hence, by property (3) and Theorem 5.7, we have

$$\int_a^b g(x)\, dx - \int_a^b f(x)\, dx = \int_a^b [g(x) - f(x)]\, dx \geq 0,$$

and the conclusion of the theorem follows.

Theorem 5.8 is intuitively evident when given a geometric interpretation, as illustrated in Figure 5.23. If $f \leq g$ over $[a, b]$, then the area under f cannot exceed the area under g.

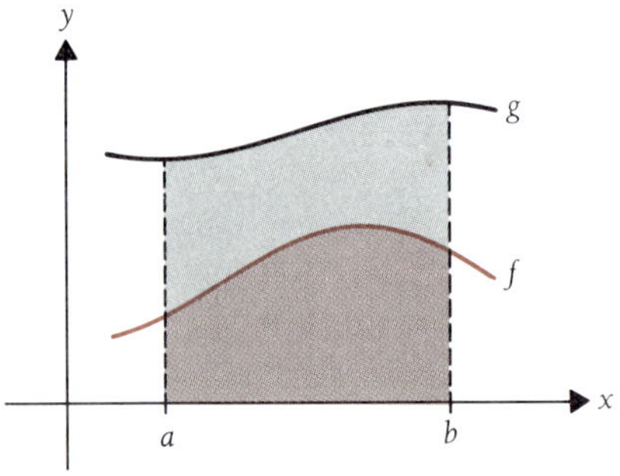

FIGURE 5.23 If $f(x) \leq g(x)$, then $\int_a^b f(x)\,dx \leq \int_a^b g(x)\,dx$

These considerations lead up to an important theoretical result, which follows.

THEOREM 5.9 **The Mean Value Theorem for Definite Integrals** If the function f is continuous on $[a, b]$, then there exists a point $z \in [a, b]$ such that

$$\int_a^b f(x)\,dx = f(z)(b-a).$$

Proof If f is continuous on $[a, b]$, then f has a maximum M and a minimum m in $[a, b]$. This means that for all $x \in [a, b]$,

$$m \leq f(x) \leq M.$$

Hence, by Theorem 5.8,

$$\int_a^b m\,dx \leq \int_a^b f(x)\,dx \leq \int_a^b M\,dx.$$

Now, by Problem 28 of Section 5.2, which says that $\int_a^b c\,dx = c(b-a)$, we can write

$$m(b-a) \leq \int_a^b f(x)\,dx \leq M(b-a), \tag{4}$$

so that

$$m \leq \frac{1}{b-a}\int_a^b f(x)\,dx \leq M.$$

Finally, since f is continuous, the intermediate value theorem says that there is some $z \in [a, b]$ such that

$$f(z) = \frac{1}{b-a} \int_a^b f(x)\, dx,$$

and the conclusion of the theorem follows.

It is instructive to look at the geometric interpretation of Theorem 5.9. If the function f is nonnegative, then $\int_a^b f(x)\, dx$ represents the area under f. The quantity $f(z)(b-a)$ is the area of a rectangle with height $f(z)$ and base $b-a$ (see Figure 5.24). Thus, Theorem 5.9 tells us that at the point z, the functional value $f(z)$ is an "average" of the functional values $f(x)$ in the sense of being the height of a rectangle whose area is the same as the area under f.

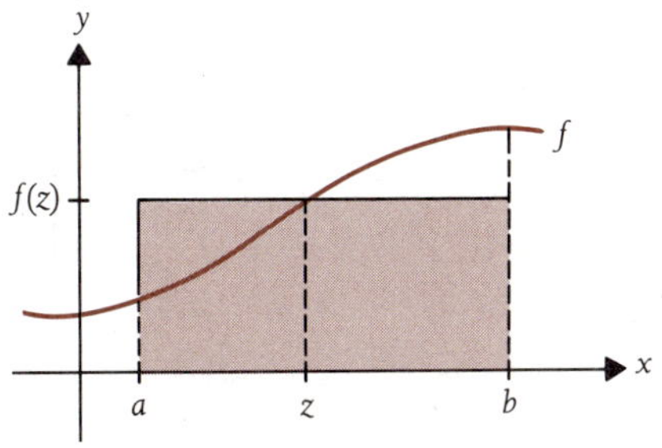

FIGURE 5.24 $\int_a^b f(x)\, dx = f(z)(b-a)$

Example 5.15 Find $z \in [0, 1]$ such that for $f(x) = x^2$,

$$\int_0^1 f(x)\, dx = f(z)(1-0).$$

Solution Recall from previous work (Example 5.5, Section 5.1) that $\int_0^1 x^2\, dx = \frac{1}{3}$. Hence, we require $f(z) = z^2 = \frac{1}{3}$, so $z = 1/\sqrt{3}$. ■

Equation (4) in the proof of Theorem 5.9 is another form of the mean value theorem that is useful for finding bounds.

Example 5.16 Find bounds on the value of $\int_1^2 (1/x)\, dx$.

Solution On [1, 2], the function $f(x) = 1/x$ has minimum $\frac{1}{2}$ and maximum 1. The length of [1, 2] is 1. Hence, equation (4) tells us that

$$\frac{1}{2} \le \int_1^2 \frac{1}{x}\, dx \le 1.$$ ■

Finally, we state without proof an integral version of the triangle inequality.

THEOREM 5.10

If the function f is integrable on $[a, b]$, then its absolute value $|f|$ is integrable on $[a, b]$, and

$$\left|\int_a^b f(x)\,dx\right| \leq \int_a^b |f(x)|\,dx.$$

Exercises 5.3

Let

$$I = \int_0^3 \sqrt{x^2+9}\,dx, \qquad J = \int_1^3 \sqrt{x^2+9}\,dx, \qquad \text{and} \qquad K = \int_1^4 \sqrt{x^2+9}\,dx.$$

Express the following definite integrals in terms of I, J, and K:

1. $\int_0^1 \sqrt{x^2+9}\,dx$
2. $\int_3^4 \sqrt{x^2+9}\,dx$
3. $\int_1^4 5\sqrt{x^2+9}\,dx$
4. $\int_3^4 3\sqrt{x^2+9}\,dx$
5. $\int_4^1 \sqrt{x^2+9}\,dx$
6. $\int_3^1 \sqrt{x^2+9}\,dx$
7. $\int_4^0 2\sqrt{x^2+9}\,dx$
8. $\int_4^3 8\sqrt{x^2+9}\,dx$
9. $\int_3^3 \sqrt{x^2+9}\,dx$
10. $\int_0^3 7\sqrt{x^2+9}\,dx$
11. $\int_0^1 \frac{1}{2}\sqrt{x^2+9}\,dx$
12. $\int_0^4 (-5)\sqrt{x^2+9}\,dx$

Let $P = \int_a^b p(x)\,dx$, $Q = \int_a^b q(x)\,dx$, *and* $R = \int_a^b r(x)\,dx$. *Express the following definite integrals in terms of P, Q, and R:*

13. $\int_a^b 4p(x)\,dx$
14. $\int_a^b 6q(x)\,dx$
15. $\int_a^b [p(x) + r(x)]\,dx$
16. $\int_a^b [q(x) - r(x)]\,dx$
17. $\int_a^b [3p(x) - 5q(x) + 8r(x)]\,dx$
18. $\int_a^b [r(x) - 4p(x) - 4q(x)]\,dx$
19. $\int_a^b (-\frac{2}{3})p(x)\,dx$
20. $\int_a^b [5r(x) + 2q(x)]\,dx$
21. $\int_b^a 3p(x)\,dx$
22. $\int_b^a [4q(x) - r(x)]\,dx$
23. $\int_b^b [3p(x) + r(x)]\,dx$
24. $\int_a^a [4r(x) + q(x) - p(x)]\,dx$

Find a value for z satisfying Theorem 5.9 by finding an average value for the integrand over the interval of integration. If no such z exists, explain why.

25. $\int_1^3 (x+1)\,dx$
26. $\int_1^5 (x+1)\,dx$
27. $\int_0^1 \sqrt{1-x^2}\,dx$
28. $\int_{-1}^1 \sqrt{1-x^2}\,dx$
29. $\int_0^3 |x|\,dx$
30. $\int_0^{2.5} |x|\,dx$
31. $\int_{-1}^2 |x|\,dx$
32. $\int_0^3 |x|\,dx$

Find upper and lower bounds for the values of the definite integrals in Problems 33–36.

33. $\int_0^4 \sqrt{x}\, dx$ 34. $\int_{-1}^2 x^5\, dx$ 35. $\int_0^3 \frac{1}{x+1}\, dx$ 36. $\int_0^3 \sqrt{x^2+9}\, dx$

37. Show that

$$\int_0^n \lfloor x \rfloor\, dx = \frac{n(n-1)}{2}$$

if n is a positive integer.

38. Show that

$$\int_m^n \lfloor x \rfloor\, dx = \frac{n(n-1) - m(m-1)}{2}$$

for positive integers m and n.

39. Use an area interpretation to explain why Theorem 5.10 holds.

40. Use an area interpretation to show that

$$\int_{-b}^b f(x)\, dx = 2\int_0^b f(x)\, dx$$

if f is an even function, and

$$\int_{-b}^b f(x)\, dx = 0$$

if f is an odd function.

SECTION 5.4

The Fundamental Theorem of Calculus

There are some difficulties in evaluating a definite integral as a limit of Riemann sums, as we saw in Section 5.2. With the aid of a computer, we can easily obtain approximate results; but we can get exact results for only a few functions by that method, and the work is extremely tedious.

Fortunately, there is a much simpler method that yields exact results for many more functions and requires relatively little effort. It is based on the relationship between the definite integral and the antiderivative, a relationship known as the *fundamental theorem of calculus*. The fundamental theorem was first recognized—at least for a few special cases—by Isaac Barrow, the teacher of Isaac Newton. But it was Newton who first stated it in generality and exploited its use.

The Function $\int_a^x f(t)\, dt$

First, let us consider a function f that is continuous and nonnegative on the interval $[a, b]$ (see Figure 5.25, page 328). For any $x \in [a, b]$, we wish to express the area under f over $[a, x]$. In order to avoid confusion, we will let f be a function of t. Then the definite integral

$$\int_a^x f(t)\, dt$$

Isaac Barrow (1630–1677) began his career as a professor of Greek in 1660, but his interests were in the sciences, and he became a professor of mathematics in 1662. He was a prolific writer on mathematical and scientific subjects, and was among the first members elected to the British Royal Society. In 1664, he became the first Lucasian Professor of Mathematics at Trinity College, Cambridge, where he remained until 1669 when he resigned in favor of his pupil, Isaac Newton.

During this period, Barrow worked on the differential calculus, following the work of Fermat. In a sense, he can be called the inventor of calculus, because he was first to state the fundamental theorem in his *Geometrical Lectures* (1664, published 1670). Possibly because of his classical training, he expressed his ideas in the style of Euclid's geometry, which tended to obscure them. He himself seems not to have noticed the central importance of the fundamental theorem, for he made no important use of it.

Barrow left mathematics and science for a career in religious administration. After 5 years as a church functionary in Wales, he became Vice Chancellor of Trinity College, but died soon thereafter. He did not live long enough to see the flowering of the subject in which he so effectively laid the groundwork.

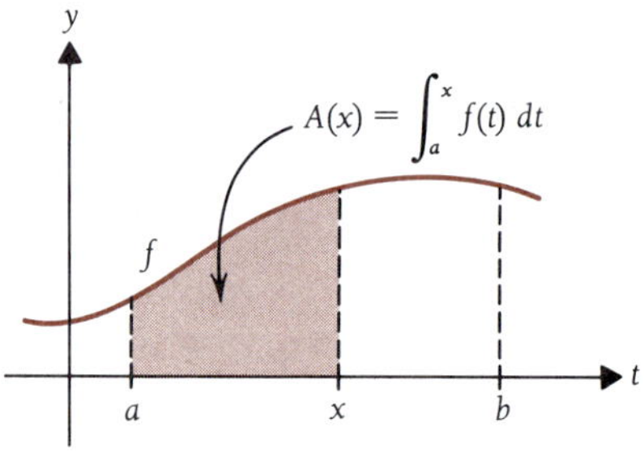

FIGURE 5.25 The area function

gives us the area of the region bounded by $t = a$, $t = x$, $y = 0$, and $y = f(t)$. This definite integral is actually a function of x, which we will call the **area function**, denoted by

$$A(x) = \int_a^x f(t)\,dt. \tag{1}$$

(Such functions are useful in several applications, notably in probability theory.)

Having constructed the area function, we consider its properties. We expect A to be a differentiable function, because area seems to be added at a continuous rate. We proceed to show that A is differentiable, as follows: By the definition of derivative, if $x \in (a, b)$, then

$$A'(x) = \lim_{h \to 0} \frac{A(x + h) - A(x)}{h},$$

provided the limit exists. We will show that it does. First, let $h > 0$. Then

$$\frac{A(x+h) - A(x)}{h} = \frac{\int_a^{x+h} f(t)\,dt - \int_a^x f(t)\,dt}{h}$$

$$= \frac{1}{h}\int_x^{x+h} f(t)\,dt.$$

See Figure 5.26 for a geometric interpretation. Now, by the mean value theorem for integrals, there is a number z between x and $x + h$ such that

$$\int_x^{x+h} f(t)\,dt = f(z)h.$$

Thus,

$$\frac{A(x+h) - A(x)}{h} = f(z)$$

for some $z \in [x, x+h]$. Now consider the limit as $h \to 0^+$. Since $z \in [x, x+h]$, $z \to x$ as $h \to 0^+$. Hence,

$$\lim_{h\to 0^+} \frac{A(x+h) - A(x)}{h} = \lim_{z\to x} f(z) = f(x),$$

because f is continuous. Similarly, we can show that

$$\lim_{h\to 0^-} \frac{A(x+h) - A(x)}{h} = f(x).$$

Thus, the limit exists, A is differentiable in (a, b), and, most importantly,

$$A'(x) = f(x) \tag{2}$$

for $x \in (a, b)$.

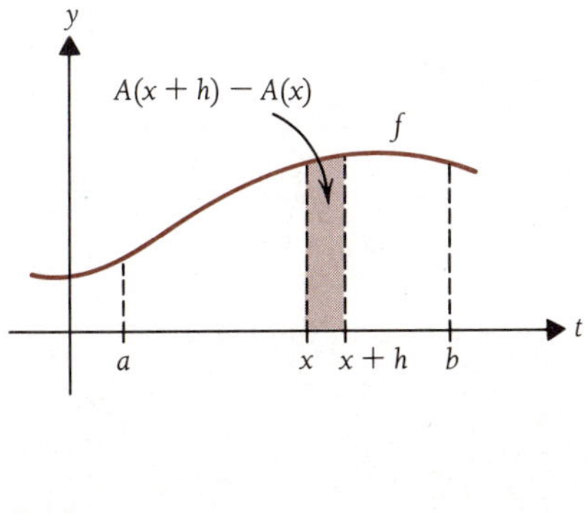

FIGURE 5:26 The difference in the area function

If $x = a$, then we cannot ask about the derivative of A, since A is not defined on both sides of a. However, we can show, as above, that

$$\lim_{h \to 0^+} \frac{A(a+h) - A(a)}{h} = f(a). \tag{3}$$

Similarly, we can show that

$$\lim_{h \to 0^-} \frac{A(b+h) - A(b)}{h} = f(b). \tag{4}$$

Whenever (2), (3), and (4) hold, we call A an **antiderivative of f on $[a, b]$**. It follows that A is continuous on $[a, b]$ (see Problem 54). All of this is also true if we do not require f to be nonnegative.

Before proceeding further, we pause to state formally what we have just proved.

THEOREM 5.11 If f is a continuous function on $[a, b]$, then the function

$$A(x) = \int_a^x f(t)\, dt$$

is an antiderivative of f on $[a, b]$.

Thus, we have proved that every function that is continuous on a closed interval has an antiderivative there, even though the antiderivative may be expressible only as a definite integral with variable upper limit. Moreover, Theorem 5.11 tells us how to find the derivative of a function defined as a definite integral with the independent variable as its upper limit: Simply evaluate the integrand at the upper limit. Formally stated, we have the following corollary.

COROLLARY If f is a continuous function on $[a, b]$ and

$$F(x) = \int_a^x f(t)\, dt,$$

then F is differentiable in (a, b), and $F'(x) = f(x)$.

■ **Example 5.17** Find $F'(x)$ if $F(x) = \int_1^x \sqrt{t^3 + 3}\, dt$.

Solution We just evaluate the integrand $\sqrt{t^3 + 3}$ at x, getting

$$F'(x) = \sqrt{x^3 + 3}.$$

■

The corollary will be generalized in the exercises.

Definite Integrals by Antidifferentiation

Since $A(x) = \int_a^x f(t)\,dt$ is an antiderivative of f on $[a, b]$, we have the differential equation

$$A'(x) = f(x).$$

From equation (1), we have the initial condition

$$A(a) = \int_a^a f(t)\,dt = 0.$$

Now, suppose g is any antiderivative of f on $[a, b]$. Then by Theorem 4.10, $A(x)$ and $g(x)$ differ by at most a constant, or

$$A(x) = g(x) + C.$$

Employing the initial condition $A(a) = 0$, we have

$$0 = A(a) = g(a) + C,$$

so that $C = -g(a)$. Hence,

$$A(x) = g(x) - g(a),$$

or, from equation (1),

$$\int_a^x f(t)\,dt = g(x) - g(a).$$

Letting $x = b$, we then have

$$\int_a^b f(t)\,dt = g(b) - g(a), \qquad (5)$$

and since t is a dummy variable, we may as well write (5) as

$$\int_a^b f(x)\,dx = g(b) - g(a). \qquad (6)$$

See what we have done! We have reduced the problem of evaluating the definite integral of f to the problem of finding an antiderivative of f and evaluating it at the limits of integration. This result is called the *fundamental theorem of calculus.*

THEOREM 5.12

The Fundamental Theorem of Calculus If the function f is continuous on $[a, b]$ and g is any antiderivative of f on $[a, b]$, then

$$\int_a^b f(x)\,dx = g(b) - g(a).$$

It is indeed remarkable that the process of evaluating a definite integral turns out to be solvable by finding an antiderivative! The discovery of this fundamental result signaled the beginning of the calculus as a discipline, and blended the two notions of *rate of change* and *area* into two sides of the same coin. Though they come from entirely different directions, differential calculus and integral calculus are shown to be part of the same whole by the fundamental theorem.

One notational convenience should be introduced before we go on. We use the symbol

$$g(x)\Big|_a^b \quad \text{or} \quad \Big[g(x)\Big]_a^b$$

for the difference $g(b) - g(a)$. For example,

$$x^2\Big|_2^3 = 3^2 - 2^2 = 5.$$

Now let's evaluate some definite integrals. First, recall the formulas of Section 4.6 for antiderivatives.

Example 5.18 Evaluate $\int_0^1 x^2\,dx$.

Solution Using formula (5) of Section 4.6, we have

$$\int_0^1 x^2\,dx = \frac{1}{3}x^3\Big|_0^1 = \frac{1}{3} - 0 = \frac{1}{3}.$$

(This is certainly easier than the method of Example 5.5 in Section 5.1!) ■

Example 5.19 Evaluate $\int_3^5 x^4\,dx$.

Solution

$$\int_3^5 x^4\,dx = \frac{1}{5}x^5\Big|_3^5 = \frac{1}{5}(5^5 - 3^5) = \frac{2882}{5}.$$

■

Example 5.20 Evaluate $\int_1^3 (x+1)\,dx$.

Solution

$$\int_1^3 (x+1)\,dx = \left[\frac{1}{2}x^2 + x\right]_1^3 = \left(\frac{9}{2} + 3\right) - \left(\frac{1}{2} + 1\right) = 6.$$

(Compare Example 5.9 in Section 5.2.) ■

Example 5.21 Evaluate $\int_0^2 (5x^3 - 4x^2 - 2x - 3)\,dx$.

Solution We first use the linearity properties of the definite integral:

$$\int_0^2 (5x^3 - 4x^2 - 2x - 3)\,dx = 5\int_0^2 x^3\,dx - 4\int_0^2 x^2\,dx - 2\int_0^2 x\,dx - 3\int_0^2 dx$$
$$= 5\left(\frac{1}{4}\right)x^4\Big|_0^2 - 4\left(\frac{1}{3}\right)x^3\Big|_0^2 - 2\left(\frac{1}{2}\right)x^2\Big|_0^2 - 3x\Big|_0^2$$
$$= \left[\frac{5}{4}x^4 - \frac{4}{3}x^3 - x^2 - 3x\right]_0^2$$
$$= 20 - \frac{32}{3} - 4 - 6 = -\frac{2}{3}.$$ ■

■ **Example 5.22** Evaluate $\int_1^3 \frac{x^3 + x}{x}\,dx$.

Solution

$$\int_1^3 \frac{x^3 + x}{x}\,dx = \int_1^3 (x^2 + 1)\,dx$$
$$= \left[\frac{1}{3}x^3 + x\right]_1^3$$
$$= (9 + 3) - \left(\frac{1}{3} + 1\right) = \frac{32}{3}.$$ ■

■ **Example 5.23** Evaluate $\int_1^3 \sqrt{x}\,dx$.

Solution

$$\int_1^3 \sqrt{x}\,dx = \int_1^3 x^{1/2}\,dx$$
$$= \frac{2}{3}x^{3/2}\Big|_1^3$$
$$= \frac{2}{3}(3\sqrt{3}) - \frac{2}{3}(1) = 2\sqrt{3} - \frac{2}{3}.$$ ■

■ **Example 5.24** Evaluate $\int_0^{\pi/2} \cos x\,dx$.

Solution

$$\int_0^{\pi/2} \cos x\,dx = \sin x\Big|_0^{\pi/2}$$
$$= \sin\frac{\pi}{2} - \sin 0 = 1.$$ ■

■ **Example 5.25** Evaluate $\int_{\pi/3}^{\pi/2} \sin(x/2)\,dx$.

Solution Using formula (15) of Section 4.6,

$$\int \sin ax\,dx = -\frac{1}{a}\cos ax + C,$$

we have

$$\begin{aligned}\int_{\pi/3}^{\pi/2} \sin \frac{x}{2}\,dx &= -2\cos \frac{x}{2}\Big|_{\pi/3}^{\pi/2}\\ &= -2\left(\cos\frac{\pi}{4} - \cos\frac{\pi}{6}\right)\\ &= -2\left(\frac{\sqrt{2}}{2} - \frac{\sqrt{3}}{2}\right) = \sqrt{3} - \sqrt{2}.\end{aligned}$$

■

Note that in these examples we are simplifying combinations of functions by freely using the linearity properties of the definite integral.

Theorem 5.12 says that we may choose *any* antiderivative of the integrand in order to evaluate a definite integral. This is because of the fact that if F and G are both antiderivatives of the integrand, then $F(x) = G(x) + C$, and

$$\begin{aligned}F(x)\Big|_a^b &= [G(x) + C]_a^b\\ &= [G(b) + C] - [G(a) + C]\\ &= G(b) - G(a)\\ &= G(x)\Big|_a^b.\end{aligned}$$

In the above examples, we have therefore chosen the simplest antiderivative in each case.

The process of evaluating a definite integral is called **integration**. Since it usually begins with antidifferentiation, most people also refer to antidifferentiation as integration. Thus, the word *integration* will be used with reference to either definite or indefinite integrals.

Exercises 5.4

Integrate.

1. $\int_0^1 x^6\,dx$
2. $\int_{-1}^1 x^{10}\,dx$
3. $\int_0^1 3x^{14}\,dx$
4. $\int_0^4 2x^3\,dx$
5. $\int_1^2 3x^{-4}\,dx$
6. $\int_{-1}^{-2} 2x^{-5}\,dx$
7. $\int_2^5 x^{-6}\,dx$
8. $\int_5^2 2x^{-3}\,dx$
9. $\int_0^2 (x - x^2)\,dx$
10. $\int_1^3 (1 + 2x)\,dx$
11. $\int_1^2 (x - x^{-2})\,dx$
12. $\int_1^4 (3x^2 - x^{-3})\,dx$
13. $\int_0^1 (x^6 - x^5 + 2x - 1)\,dx$
14. $\int_1^2 (3x^2 - 4x + 5)\,dx$

15. $\int_0^2 (4 - 3x - x^2)\,dx$

16. $\int_{-1}^1 (4x^7 + 3x^6 - x^3 + 6x^2 - 2x)\,dx$

17. $\int_0^1 \sqrt{x}\,dx$

18. $\int_1^2 3x\sqrt{x}\,dx$

19. $\int_1^4 x^{3/2}\,dx$

20. $\int_1^2 x^3\sqrt{x}\,dx$

21. $\int_1^8 x^{4/3}\,dx$

22. $\int_0^1 x^{7/5}\,dx$

23. $\int_1^9 x^{-3/2}\,dx$

24. $\int_{-2}^{-1} x^{-2/3}\,dx$

25. $\int_0^{\pi/4} \cos x\,dx$

26. $\int_0^{\pi/3} \sin x\,dx$

27. $\int_{-\pi/2}^{\pi/6} 3\cos x\,dx$

28. $\int_0^{\pi/4} (1 + \cos x)\,dx$

29. $\int_{-\pi/2}^{\pi/2} (\sin x + \cos x)\,dx$

30. $\int_0^{\pi} (2\sin x - \cos x)\,dx$

31. $\int_0^{\pi/2} (\cos x - \sin x)\,dx$

32. $\int_{-\pi/2}^{\pi/2} (4x - 2\cos x)\,dx$

33. $\int_0^{\pi/6} \cos 2x\,dx$

34. $\int_0^{\pi/4} \sin 3x\,dx$

35. $\int_{-\pi/12}^{0} 3\cos 4x\,dx$

36. $\int_{-\pi/12}^{-\pi/6} (-3\sin 2x)\,dx$

37. $\int_0^{\pi/4} \sec x \tan x\,dx$

38. $\int_{\pi/4}^{\pi/3} \csc x \cot x\,dx$

39. $\int_{-\pi/4}^{\pi/4} \sec^2 x\,dx$

40. $\int_{\pi/6}^{\pi/4} \csc^2 x\,dx$

41. $\int_0^{\pi/12} \sec 3x \tan 3x\,dx$

42. $\int_{\pi/12}^{\pi/8} \csc 2x \cot 2x\,dx$

43. $\int_{-\pi/6}^{0} \sec^2 2x\,dx$

44. $\int_{\pi/3}^{2\pi/3} \csc^2 \frac{x}{2}\,dx$

Find the derivative.

45. $A(x) = \int_0^x (t^3 + t)\,dt$

46. $A(x) = \int_1^x (t^3 + t^{-2})\,dt$

47. $G(x) = \int_0^x \cos 2t\,dt$

48. $F(x) = \int_0^x \frac{1}{t^2 + 1}\,dt$

49. Use the chain rule to show that the derivative of

$$F(x) = \int_a^{g(x)} f(t)\,dt$$

is $F'(x) = f(g(x))g'(x)$, given that g is a differentiable function.

Find the derivative.

50. $F(x) = \int_0^{x^2} \sqrt{t^3 + 1}\, dt$

51. $F(x) = \int_0^{\sqrt{x}} \frac{1}{t}\, dt$

52. $F(x) = \int_x^{3x} (t^2 + 4)^{-2/3}\, dt \left[\textit{Hint: } \int_x^{3x} = \int_a^{3x} - \int_a^x.\right]$

53. $F(x) = \int_{x^2}^{\sqrt{x}} \sin t^2\, dt$

54. Prove that an antiderivative of f on $[a, b]$ is continuous on $[a, b]$. [*Hint:* Use Theorem 3.1, Section 3.3.]

55. Show that if the function f is continuous and positive on $[a, b]$, then $F(x) = \int_a^x f(t)\, dt$ is an increasing function.

56. Given $F(x) = \int_8^{x^3} \sqrt{1 + t^2}\, dt$, show that F^{-1} exists, and find $D_x F^{-1}(x)|_{x=0}$.

57. The graph of the function f is given in the accompanying figure. Sketch the graph of $F(x) = \int_1^x f(t)\, dt$ over the interval $[1, 6]$.

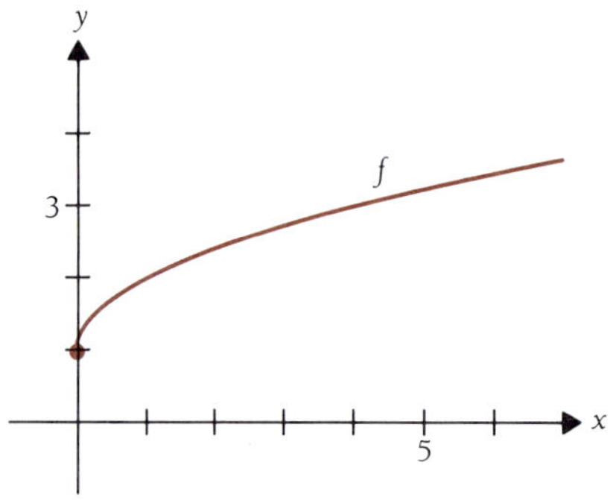

58. Repeat Problem 57 for the graph in the accompanying figure.

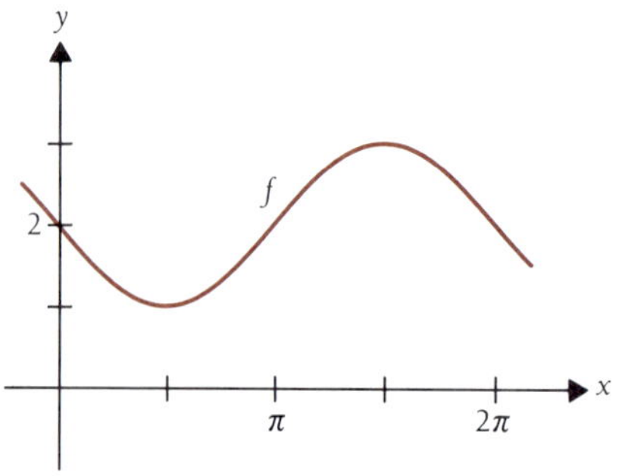

SECTION 5.5

Integration by Substitution

Even though the process of integration has been reduced to the finding of antiderivatives, our work has scarcely begun. The process of finding antiderivatives is largely one of trial and error, and no universal procedure exists for discovering antiderivatives. (Theorem 5.11 is universal, but gives only an integral form.) Various techniques do exist that remove some of the guesswork for many kinds of functions, and here, we introduce the simplest of these, called the **method of substitution**.

Any formula we have for antiderivatives may be applied directly to an integral if the integral has exactly the form of the formula. For example, the simplest antidifferentiation formula is

$$\int x^n \, dx = \frac{1}{n+1} x^{n+1} + C, \qquad n \neq -1. \tag{1}$$

(What to do if $n = -1$ is covered in Chapter 6.) The integral

$$\int x^8 \, dx$$

has exactly this form, so formula (1) applies directly:

$$\int x^8 \, dx = \frac{1}{9} x^9 + C.$$

On the other hand, the integral

$$\int (x+3)^8 \, dx$$

does not have exactly the form of (1), so formula (1) does not apply. However, this integral may be transformed by the method of substitution so that formula (1) will apply.

Substitution

The substitution method is given by the following theorem, which is a consequence of the chain rule.

THEOREM 5.13 If f is a differentiable function, g is a continuous function, and G is an antiderivative of g, then

$$\int g(f(x))f'(x) \, dx = G(f(x)) + C.$$

Proof By the chain rule,

$$\begin{aligned} D_x[G(f(x))] &= G'(f(x))f'(x) \\ &= g(f(x))f'(x), \end{aligned}$$

since G is an antiderivative of g. The indefinite integral form of this last statement is

$$\int g(f(x))f'(x)\,dx = G(f(x)) + C,$$

which is the statement of the theorem.

When Theorem 5.13 is used, the substitution $u = f(x)$, $du = f'(x)\,dx$, is usually employed. In so doing, the actual integration looks like the following:

$$\begin{aligned} \int g(f(x))f'(x)\,dx &= \int g(u)\,du \\ &= G(u) + C \\ &= G(f(x)) + C. \end{aligned}$$

It is from this u *substitution* that the method of substitution gets its name.

The method of substitution works only when the given integral is of the form stated in the theorem above. Finding the function u is the key step. Sometimes, we can easily recognize a function whose derivative is a factor of the integrand; sometimes, we can rewrite the integrand so that it has the necessary form.

Here are some simple examples to illustrate the technique.

■ **Example 5.26** Integrate $\int (x+3)^8\,dx$.

Solution If we let $u = x + 3$, then $du = dx$. Hence,

$$\int (x+3)^8\,dx = \int u^8\,du.$$

Now we can apply formula (1), after which we replace u by $x + 3$:

$$\begin{aligned} \int (x+3)^8\,dx &= \int u^8\,du \\ &= \frac{1}{9}u^9 + C \\ &= \frac{1}{9}(x+3)^9 + C. \end{aligned}$$ ■

■ **Example 5.27** Integrate $\int \sqrt{x^2+4}\,(2x)\,dx$.

Solution If we substitute $u = x^2 + 4$, then $du = 2x\,dx$. Hence,

$$\begin{aligned}\int \sqrt{x^2+4}\,(2x)\,dx &= \int \sqrt{u}\,du \\ &= \int u^{1/2}\,du \\ &= \frac{2}{3}u^{3/2} + C \\ &= \frac{2}{3}(x^2+4)^{3/2} + C.\end{aligned}$$

■

■ **Example 5.28** Integrate $\int x\sqrt{x^2+4}\,dx$.

Solution By analogy with Example 5.27, we substitute $u = x^2 + 4$; then $du = 2x\,dx$, and (dividing by 2) $\frac{1}{2}\,du = x\,dx$. Hence,

$$\begin{aligned}\int x\sqrt{x^2+4}\,dx &= \int \sqrt{x^2+4}(x)\,dx \\ &= \int \sqrt{u}\left(\frac{1}{2}\right)du \\ &= \frac{1}{2}\int \sqrt{u}\,du \\ &= \left(\frac{1}{2}\right)\left(\frac{2}{3}\right)u^{3/2} + C \\ &= \frac{1}{3}(x^2+4)^{3/2} + C.\end{aligned}$$

■

■ **Example 5.29** Integrate $\int (2x-1)^{2/3}\,dx$.

Solution The only evident substitution to make here is $u = 2x - 1$; then $du = 2\,dx$, and $\frac{1}{2}\,du = dx$. Hence,

$$\begin{aligned}\int (2x-1)^{2/3}\,dx &= \int u^{2/3}\left(\frac{1}{2}\right)du \\ &= \frac{1}{2}\int u^{2/3}\,du \\ &= \left(\frac{1}{2}\right)\left(\frac{3}{5}\right)u^{5/3} + C \\ &= \frac{3}{10}(2x-1)^{5/3} + C.\end{aligned}$$

■

For definite integrals, the substitution technique is similar. The version of Theorem 5.13 for definite integrals is as follows.

THEOREM 5.14 If f is a differentiable function and g is a continuous function, then

$$\int_a^b g(f(x))f'(x)\,dx = \int_{f(a)}^{f(b)} g(u)\,du.$$

Proof If G is an antiderivative of g, then by Theorem 5.13,

$$\begin{aligned}\int_a^b g(f(x))f'(x)\,dx &= G(f(x))\Big|_a^b\\ &= G(f(b)) - G(f(a))\\ &= G(u)\Big|_{f(a)}^{f(b)}\\ &= \int_{f(a)}^{f(b)} g(u)\,du.\end{aligned}$$

This theorem tells us that in the case of a definite integral, we can make the substitution $u = f(x)$, change the limits from a and b to $f(a)$ and $f(b)$ [when $x = a$, then $u = f(a)$; when $x = b$, then $u = f(b)$], and then forget about x altogether. It is not necessary to return to the original variable, as it is in the case of indefinite integrals. The following examples illustrate this technique.

Example 5.30 Evaluate $\int_2^3 x\sqrt{x^2-4}\,dx$.

Solution Letting $u = x^2 - 4$, we have $du = 2x\,dx$, or $\frac{1}{2}\,du = x\,dx$. Also, when $x = 2$, then $u = 0$; and when $x = 3$, then $u = 5$. Hence,

$$\begin{aligned}\int_2^3 x\sqrt{x^2-4}\,dx &= \int_0^5 \sqrt{u}\left(\frac{1}{2}\right)du\\ &= \frac{1}{2}\int_0^5 u^{1/2}\,du\\ &= \left(\frac{1}{2}\right)\left(\frac{2}{3}\right)u^{3/2}\Big|_0^5\\ &= \frac{1}{3}(5^{3/2} - 0) = \frac{5\sqrt{5}}{3}.\end{aligned}$$

■

Example 5.31 Evaluate $\int_3^{11} (5x-9)^{-3}\,dx$.

Solution If we let $u = 5x - 9$, then $du = 5\,dx$, or $\frac{1}{5}\,du = dx$. Also, when $x = 3$, then $u = 6$; and when $x = 11$, then $u = 46$. Hence,

$$\int_3^{11} (5x-9)^{-3}\,dx = \int_6^{46} u^{-3}\left(\frac{1}{5}\right) du$$
$$= \frac{1}{5}\int_6^{46} u^{-3}\,du$$
$$= \frac{1}{5}\left[\left(\frac{1}{-2}\right)u^{-2}\right]_6^{46}$$
$$= -\frac{1}{10}\,[(46)^{-2} - 6^{-2}] = \frac{13}{4761}.$$

■

■ Example 5.32 Integrate $\displaystyle\int_0^2 \frac{x^2}{\sqrt{x^3+1}}\,dx$.

Solution If we let $u = x^3 + 1$, then $du = 3x^2\,dx$, or $\frac{1}{3}\,du = x^2\,dx$. Also, $u = 1$ when $x = 0$; and $u = 9$ when $x = 2$. Hence,

$$\int_0^2 \frac{x^2}{\sqrt{x^3+1}}\,dx = \int_1^9 \frac{1}{\sqrt{u}}\left(\frac{1}{3}\right) du$$
$$= \frac{1}{3}\int_1^9 u^{-1/2}\,du$$
$$= \left(\frac{1}{3}\right)(2)u^{1/2}\Big|_1^9$$
$$= \left(\frac{2}{3}\right)(3-1) = \frac{4}{3}.$$

■

■ Example 5.33 Integrate $\int_0^{\pi/4} \sin^2 x \cos x\,dx$.

Solution Substituting $u = \sin x$, $du = \cos x\,dx$, we have

$$\int_0^{\pi/4} \sin^2 x \cos x\,dx = \int_0^{1/\sqrt{2}} u^2\,du$$
$$= \frac{1}{3}u^3\Big|_0^{1/\sqrt{2}}$$
$$= \left(\frac{1}{3}\right)\left(\frac{1}{\sqrt{2}}\right)^3 = \frac{1}{6\sqrt{2}}.$$

■

In evaluating definite integrals by the method of substitution, it is not necessary to return to the original variable, but we may. If we want to go back to the original variable, then we must also go back to the original limits. This is hinted at in the proof of Theorem 5.14, and illustrated by the following example.

■ Example 5.34 Rework the integration of Example 5.33 by returning to the original variable of integration.

Solution We proceed as in Example 5.33, except that we do not change the limits, but return to the variable x:

$$\begin{aligned}\int_0^{\pi/4} \sin^2 x \cos x\, dx &= \int_{x=0}^{x=\pi/4} u^2\, du \\ &= \frac{1}{3} u^3 \Big|_{x=0}^{x=\pi/4} \\ &= \frac{1}{3} \sin^3 x \Big|_0^{\pi/4} \\ &= \left(\frac{1}{3}\right)\left(\frac{1}{\sqrt{2}}\right)^3 = \frac{1}{6\sqrt{2}}.\end{aligned}$$ ■

The substitution technique is sometimes altered into what is called a **change of variable** technique. The procedure is only slightly different from substitution, in that we let x be a function of u, $x = g(u)$, and change $\int f(x)\, dx$ into $\int f(g(u))g'(u)\, du$. One way this often arises is that, having set u equal to a function of x, we solve for x in terms of u and proceed as above. The change of variable and substitution techniques are so closely related in this setting that little distinction is made between the two.

Consider the following simple example.

■ **Example 5.35** Integrate

$$\int \frac{x}{\sqrt{x+1}}\, dx.$$

Solution We let $u = x + 1$, so that $x = u - 1$. Then $dx = du$, and we get

$$\begin{aligned}\int \frac{x}{\sqrt{x+1}}\, dx &= \int \frac{u-1}{\sqrt{u}}\, du \\ &= \int (u^{1/2} - u^{-1/2})\, du \\ &= \frac{2}{3} u^{3/2} - 2u^{1/2} + C \\ &= \frac{2}{3} (x+1)^{3/2} - 2(x+1)^{1/2} + C.\end{aligned}$$ ■

As we remarked before, there is no guarantee that the method of substitution will work. For example, you will make no progress using substitution to integrate the integrals

$$\int (x^3 + 1)^{1/3}\, dx,$$

$$\int x(x^3 + 1)^{1/3}\, dx,$$

$$\int x^3(x^3+1)^{1/3}\,dx,$$

$$\int x^4(x^3+1)^{1/3}\,dx.$$

However, substitution works nicely on the integral

$$\int x^2(x^3+1)^{1/3}\,dx.$$

Can you explain why?

Exercises 5.5

Integrate.

1. $\int (x+1)^7\,dx$
2. $\int (x-6)^{1/5}\,dx$
3. $\int_{-1}^{1} (x+3)^4\,dx$
4. $\int_{0}^{3} (x-2)^{4/3}\,dx$
5. $\int (2x-1)^3\,dx$
6. $\int (3x+5)^{1/2}\,dx$
7. $\int (4x-3)^8\,dx$
8. $\int (1-x)^8\,dx$
9. $\int_{0}^{3} (2x+5)^3\,dx$
10. $\int_{-2}^{2} (4-x)^{1/3}\,dx$
11. $\int_{-1}^{2} (3x+1)^{4/5}\,dx$
12. $\int_{0}^{1} (5x+4)^{-1/2}\,dx$
13. $\int x(x^2+1)^{2/3}\,dx$
14. $\int x\sqrt{x^2+9}\,dx$
15. $\int 3x(x^2-4)^{3/2}\,dx$
16. $\int x(x^2+4)^{1/3}\,dx$
17. $\int_{0}^{4} x(1+4x^2)^{1/3}\,dx$
18. $\int_{-1}^{1} 2x(3x^2-8)^3\,dx$
19. $\int_{3}^{4} 4x\sqrt{x^2-9}\,dx$
20. $\int_{1}^{8} 3x(x^2+1)^{1/3}\,dx$
21. $\int_{1}^{3} \dfrac{x+1}{(x^2+2x)^2}\,dx$
22. $\int_{1}^{2} \dfrac{x-3}{\sqrt{1+6x-x^2}}\,dx$
23. $\int_{0}^{1} x(x^2+1)^{-1/2}\,dx$
24. $\int_{0}^{1} x^2\sqrt{x^3+1}\,dx$
25. $\int x^{-1/3}\sqrt{x^{2/3}+4}\,dx$
26. $\int x^{-2/5}(x^{3/5}+1)^{-3}\,dx$
27. $\int \sqrt{1+x^{-2/3}}\,dx$
28. $\int 5x^3(x^4-4)^{-1/2}\,dx$
29. $\int x(x-4)^{-1/2}\,dx$
30. $\int 3x\sqrt{x+5}\,dx$

31. $\int_0^1 x(x+1)^8\,dx$

32. $\int_1^2 x(x+4)^{-5}\,dx$

33. $\int \frac{x+2}{\sqrt{x-1}}\,dx$

34. $\int x(1-x)^{1/3}\,dx$

35. $\int x^2\sqrt{2x+3}\,dx$

36. $\int x(x+5)^{-3}\,dx$

37. $\int_0^{\pi/2} \sin x\cos^2 x\,dx$

38. $\int_0^{\pi/4} \tan x\sec^2 x\,dx$

39. $\int_0^{\pi/3} (1+\sin^2 x)\cos x\,dx$

40. $\int_{-\pi/6}^{\pi/6} \sin^3 x\cos x\,dx$

41. $\int x\sin x^2\,dx$

42. $\int \frac{\sin\sqrt{x}}{\sqrt{x}}\,dx$

43. $\int \sqrt{\tan x}\sec^2 x\,dx$

44. $\int \sec x(\sec x\tan x)\,dx$

Explain the difficulty in trying the method of substitution on each of the following integrals:

45. $\int \sqrt{x^2+4}\,dx$

46. $\int x^2\sqrt{x^2+4}\,dx$

47. $\int \frac{\sqrt{x^2+4}}{x}\,dx$

48. $\int \frac{x^2}{x^2+1}\,dx$

SECTION 5.6

Areas by Integration

As was noted in Section 5.2, the definite integral $\int_a^b f(x)\,dx$ of a nonnegative function f can be interpreted as the area under f over $[a, b]$. This fact enables us to find areas of many different regions whose boundaries are the graphs of functions. In this section, we shall develop this application of the definite integral.

Area between Two Curves

Suppose f and g are two integrable functions such that $0 \le g(x) \le f(x)$ for all $x \in [a, b]$ (see Figure 5.27). The region

$$\{(x, y) \mid a \le x \le b \text{ and } g(x) \le y \le f(x)\}$$

is called the **region between f and g over $[a, b]$**. The area of this region is called the **area between f and g over $[a, b]$**. Since the area between f and g over $[a, b]$ is equal to the area under f over $[a, b]$, minus the area under g over $[a, b]$, we have the formula

$$A = \int_a^b f(x)\,dx - \int_a^b g(x)\,dx = \int_a^b [f(x) - g(x)]\,dx. \tag{1}$$

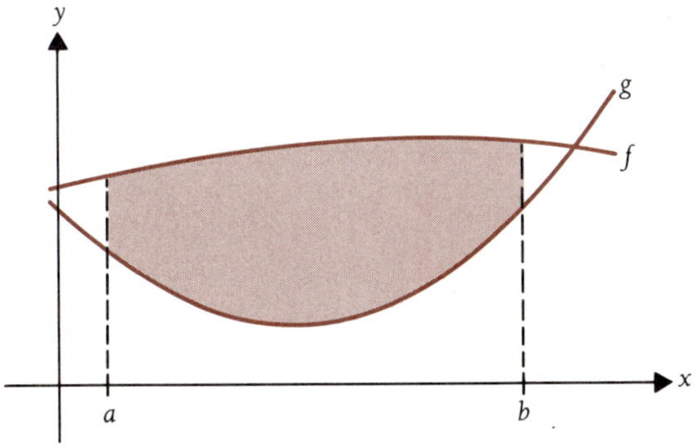

FIGURE 5.27 The region between f and g over $[a, b]$

Formula (1) holds in any situation where $g(x) \leq f(x)$ over $[a, b]$, whether $g(x) \geq 0$ or not. To see this, note that if g is integrable, then g is bounded below, or there is a constant c such that $g(x) + c \geq 0$ over $[a, b]$. The region between $f(x) + c$ and $g(x) + c$ (see Figure 5.28) has the same area as the region between f and g, so formula (1) applies.

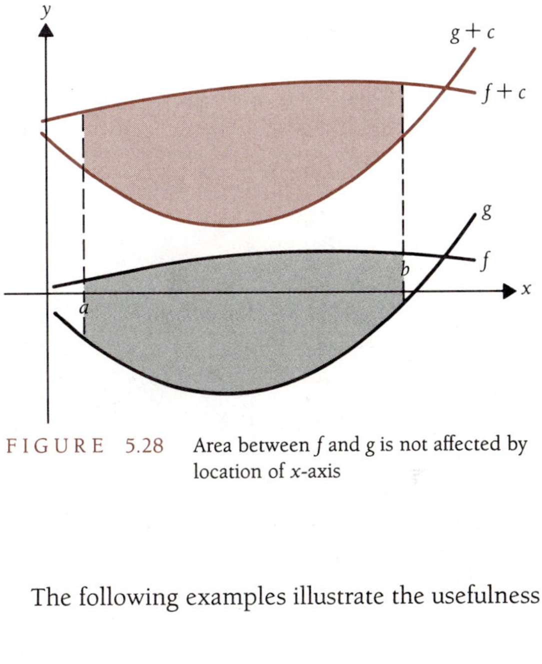

FIGURE 5.28 Area between f and g is not affected by location of x-axis

The following examples illustrate the usefulness of formula (1) in finding areas.

■ **Example 5.36** Find the area between $y = x$ and $y = x^2$ over $[0, 1]$.

Solution We see from the graphs of these curves (Figure 5.29) that $x^2 \leq x$ over $[0, 1]$. Hence, the area between $y = x$ and $y = x^2$ is

$$A = \int_0^1 (x - x^2)\, dx = \left[\frac{1}{2}x^2 - \frac{1}{3}x^3\right]_0^1$$
$$= \frac{1}{2} - \frac{1}{3} = \frac{1}{6}.$$

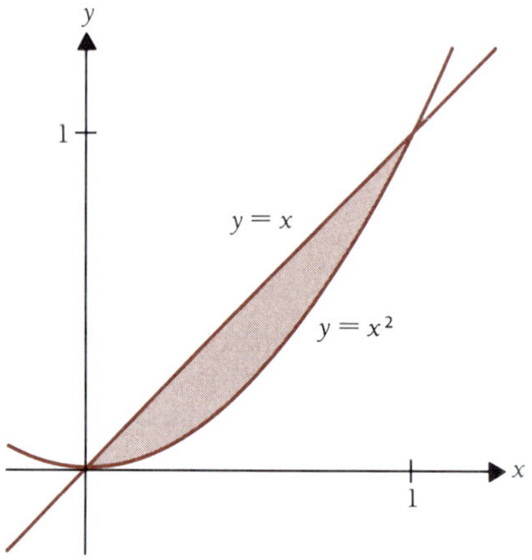

FIGURE 5.29 The region bounded by $y = x$ and $y = x^2$ ■

■ **Example 5.37** Find the area of the region bounded by the curve $y = x^2 - 4$ and the x-axis.

Solution By sketching the graph (Figure 5.30), we see that the region lies between $y = 0$ and $y = x^2 - 4$ over $[-2, 2]$. Thus, by formula (1), the area is

$$A = \int_{-2}^{2} [0 - (x^2 - 4)]\, dx$$
$$= \int_{-2}^{2} (4 - x^2)\, dx$$
$$= \left[4x - \frac{1}{3}x^3\right]_{-2}^{2}$$
$$= \left(8 - \frac{8}{3}\right) - \left(-8 + \frac{8}{3}\right) = \frac{32}{3}.$$

Note that, because of symmetry of this region about the x-axis, we could equally well have found the area of half the region and doubled that number. That is, we could have written

$$A = 2\int_0^2 [0 - (x^2 - 4)]\, dx.$$

This simplifies the calculations, because it is easy to evaluate the antiderivative at 0.

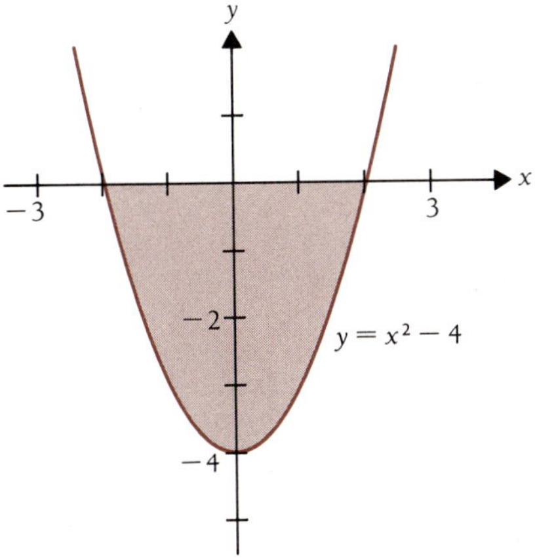

FIGURE 5.30 The region bounded by the x-axis and $y = x^2 - 4$ ■

■ **Example 5.38** Find the area of the region bounded by the curves $y = 4x - x^2$ and $y = 2x - 3$.

Solution We first sketch the graphs of these functions, as shown in Figure 5.31. To find the points of intersection, we solve the equations simultaneously:

$$\begin{aligned} 2x - 3 &= 4x - x^2, \\ x^2 - 2x - 3 &= 0, \\ (x + 1)(x - 3) &= 0, \\ x = -1 \qquad &\text{or} \qquad x = 3. \end{aligned}$$

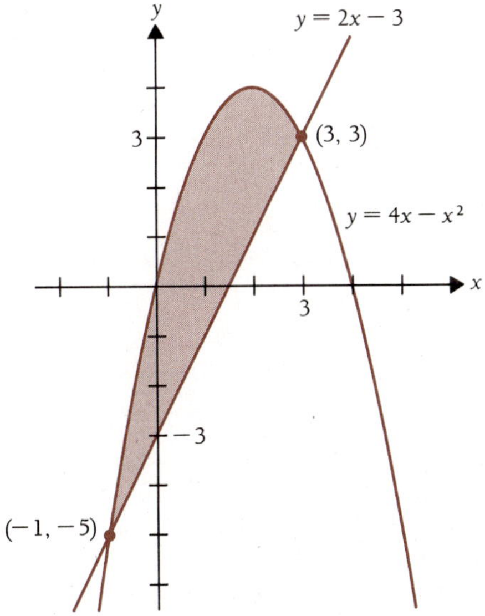

FIGURE 5.31 The region bounded by $y = 2x - 3$ and $y = 4x - x^2$

At $x=-1$, $y=-5$, and at $x=3$, $y=3$; thus, the points of intersection are $(-1,-5)$ and $(3,3)$. The area we seek is therefore the area between the curves over $[-1,3]$:

$$\begin{aligned} A &= \int_{-1}^{3} [(4x-x^2)-(2x-3)]\,dx \\ &= \int_{-1}^{3} (2x-x^2+3)\,dx \\ &= \left[x^2-\frac{1}{3}x^3+3x\right]_{-1}^{3} \\ &= (9-9+9)-\left(1+\frac{1}{3}-3\right)=\frac{32}{3}. \end{aligned}$$

■

■ **Example 5.39** Find the area of the region bounded by the curves $y=-x$, $y=\sqrt{x}$, and $y=1-2x$.

Solution We first sketch the graphs (Figure 5.32), and find the points of intersection as shown. We can then describe the area of the region enclosed as the sum of two areas: the area between $y=\sqrt{x}$ and $y=-x$ over $[0,\frac{1}{4}]$, plus the area between $y=1-2x$ and $y=-x$ over $[\frac{1}{4},1]$. Hence, the area is

$$\begin{aligned} A &= \int_{0}^{1/4} [\sqrt{x}-(-x)]\,dx + \int_{1/4}^{1} [1-2x-(-x)]\,dx \\ &= \int_{0}^{1/4} (\sqrt{x}+x)\,dx + \int_{1/4}^{1} (1-x)\,dx \end{aligned}$$

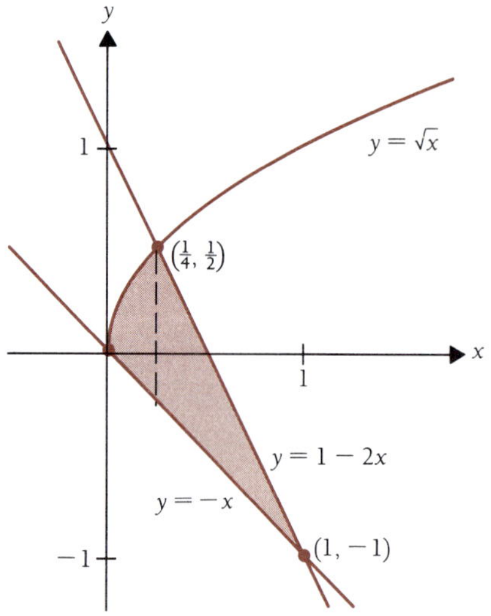

FIGURE 5.32 The region bounded by $y=\sqrt{x}$, $y=-x$, and $y=1-2x$

$$A = \left[\frac{2}{3}x^{3/2} + \frac{1}{2}x^2\right]_0^{1/4} + \left[x - \frac{1}{2}x^2\right]_{1/4}^1$$
$$= \left[\left(\frac{2}{24} + \frac{1}{32}\right) - 0\right] + \left[\left(1 - \frac{1}{2}\right) - \left(\frac{1}{4} - \frac{1}{32}\right)\right] = \frac{19}{48}.$$ ■

If two curves $y = f(x)$ and $y = g(x)$ cross each other in $[a, b]$, then the total area between f and g over $[a, b]$ will be the sum of two or more areas. An example should clarify this case.

■ **Example 5.40** Find the area between $y = x$ and $y = x^2$ over $[0, 3]$.

Solution By solving simultaneously, we find that these curves intersect at (0, 0) and (1, 1). Then, from the graph (Figure 5.33), we see that the total area between the curves is

$$A = \int_0^1 (x - x^2)\,dx + \int_1^3 (x^2 - x)\,dx$$
$$= \left[\frac{1}{2}x^2 - \frac{1}{3}x^3\right]_0^1 + \left[\frac{1}{3}x^3 - \frac{1}{2}x^2\right]_1^3$$
$$= \left[\left(\frac{1}{2} - \frac{1}{3}\right) - 0\right] + \left[\left(9 - \frac{9}{2}\right) - \left(\frac{1}{3} - \frac{1}{2}\right)\right] = \frac{29}{6}.$$

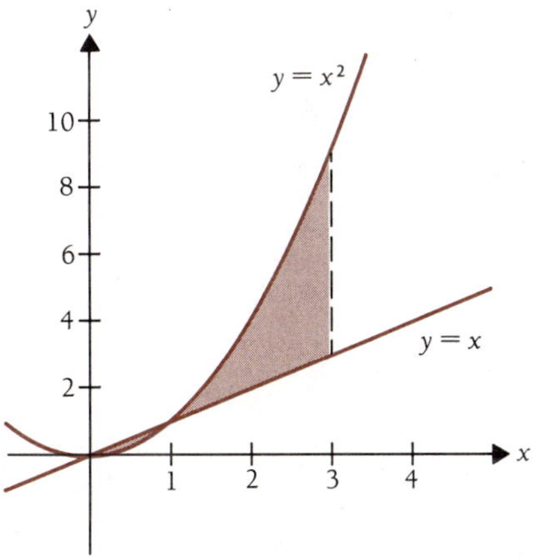

FIGURE 5.33 The region between $y = x^2$ and $y = x$ over $[0, 3]$ ■

In general, the area between two curves is given by the schematic formula

$$A = \int_a^b (\text{Upper curve} - \text{Lower curve})\,dx.$$

If the curves are given as functions of y rather than x, then we interchange the roles of x and y. This leads to the schematic formula

$$A = \int_c^d (\text{Right curve} - \text{Left curve})\, dy.$$

Example 5.41 Find the area bounded by the curves $x = 4 - y^2$ and $x = y^2$.

Solution Sketching these curves (Figure 5.34), we see that they intersect in two points. The coordinates are found by solving the equations simultaneously; we get $y = \sqrt{2}$ and $y = -\sqrt{2}$ when $x = 2$. Since the region bounded by the curves is symmetric with respect to the x-axis, we have

$$\begin{aligned} A &= 2\int_0^{\sqrt{2}} [(4 - y^2) - y^2]\, dy \\ &= 2\int_0^{\sqrt{2}} (4 - 2y^2)\, dy \\ &= 2\left[4y - \frac{2}{3}y^3\right]_0^{\sqrt{2}} \\ &= 2\left[4\sqrt{2} - \frac{4\sqrt{2}}{3}\right] = \frac{16\sqrt{2}}{3}. \end{aligned}$$

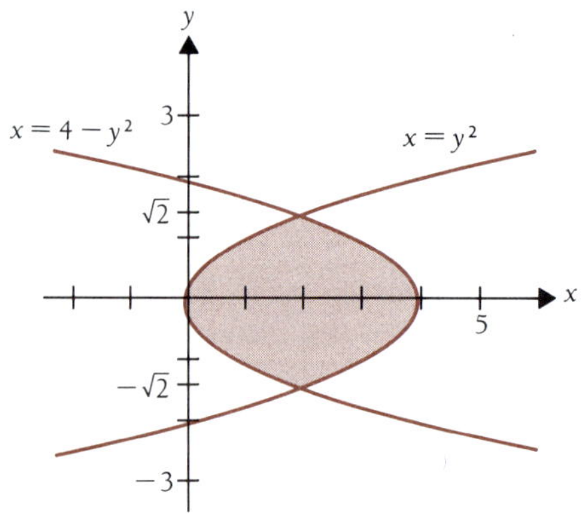

FIGURE 5.34 The region bounded by $x = y^2$ and $x = 4 - y^2$

Exercises 5.6

Find the area of the region enclosed by the given curves.

1. $y = 3 - 2x, \quad y = 0, \quad x = 0$
2. $y = 4 - x, \quad x = 0, \quad y = 0$
3. $y = x^2 - x, \quad y = 0$
4. $y = 0, \quad y = 9x - x^2$
5. $y = x, \quad y = x^2 - 4x + 6$
6. $y = x, \quad y = 2 - x^2$
7. $y = x - x^2, \quad y = 0$
8. $y = -1, \quad y = 4 - x^2$
9. $y = x^2 + 1, \quad y = 4 - x^2$
10. $y = x^2 - 4, \quad y = 2x - x^2$
11. $y = x^2 + 4, \quad y = 9 - x^2$
12. (A) $y = x^2 - 2x + 1, \quad y = 4 - x^2$

13. $y = 2x, \quad y = 4 - x, \quad y = -x/2$

14. $y = -x, \quad y = x + 2, \quad x = 3$

15. $y = x^2 - x, \quad y = x^2 - 9x + 16, \quad y = -x$

16. $y = x^2 - 1, \quad y = -x - 1, \quad y = x^2 - 9x + 15$

17. $x = y - y^2, \quad x = 0$

18. $x = 9y - y^2, \quad x = 0$

19. $x = 4 - y^2, \quad x = 0$

20. $x = 2y^2, \quad x = 2 - y^2$

21. $x = y - y^2, \quad y = x + 2$

22. $x = 4 - y^2, \quad x = y$

23. $x = y^3 + 1, \quad x = 3y - 1$

24. $x = y^2 - y, \quad x = 4y - y^2$

Find the area.

25. Under $y = |x^2 - 4|$, over $[-3, 1]$
26. Between $y = x^{2/3}$ and the x-axis, over $[-1, 1]$
27. Of $R = \{(x, y) | 1 \le x \le 4 \text{ and } -x \le y \le 1/\sqrt{x}\}$
28. Of $R = \{(x, y) | -1 \le x \le 2 \text{ and } 0 \le y \le |x^3|\}$
29. Under $y = \sin x$, over $[0, \pi/2]$
30. Under $y = \sin \pi x$, over $[0, \frac{1}{4}]$
31. Under $y = \sin x + \cos x$, over $[0, \pi/2]$
32. Under $y = \sec^2 x$, over $[-\pi/4, \pi/4]$
33. Between $y = \sin x$ and $y = \cos x$, over $[0, \pi/4]$
34. Between $y = \cos x$ and $y = \sec^2 x$, over $[0, \pi/3]$
35. Between $y = \cos(x/4)$ and $y = \frac{1}{2} \sin 2x$, over $[0, \pi]$
36. Between $y = x/(x^2 + 1)^2$ and $y = x^2 + 1$, over $[0, 1]$
37. Show that the area between $y = f(x)$ and the x-axis over $[a, b]$ is $A = \int_a^b |f(x)|\, dx$.
38. Show that the area between the curves $y = f(x)$ and $y = g(x)$ over $[a, b]$ is $A = \int_a^b |f(x) - g(x)|\, dx$.
39. When is the area between $y = f(x)$ and $y = 0$ over $[a, b]$ equal to $|\int_a^b f(x)\, dx|$?
40. The region bounded by $y = 4 - x^2$ and the x-axis is cut into two regions of equal area by the line $y = c$.
    (a) Find c by integrating with respect to y. (This puts c into the limits of the integral.)
    (b) Find c by integrating with respect to x. (This puts c into the integrand.)

SECTION 5.7

Average Values and Integrals of Rates of Change

To find the average $\bar{y}$ of a finite set of numbers $y_1, y_2, \ldots, y_n$, one simply adds the numbers and divides by n:

$$\bar{y} = \frac{1}{n} \sum_{i=1}^{n} y_i .$$

Using the definite integral, we can find the average of infinitely many numbers, given as the set of values (range) of a function. This will also lead us to integrals of rates of change.

Average Value of a Function

Let the function f be integrable on $[a, b]$, and let $x_0, x_1, \ldots, x_n$ be a regular subdivision of $[a, b]$. Then $\Delta x_i = (b-a)/n$. If $t_i \in [x_{i-1}, x_i]$, then the average of the numbers $f(t_1), f(t_2), \ldots, f(t_n)$ is

$$\frac{1}{n}\sum_{i=1}^{n} f(t_i).$$

This average can be rewritten as

$$\frac{1}{n\Delta x_i}\sum_{i=1}^{n} f(t_i)\Delta x_i = \frac{1}{b-a}\sum_{i=1}^{n} f(t_i)\Delta x_i,$$

which we recognize as a Riemann sum for the function

$$\frac{1}{b-a}f(x)$$

on $[a, b]$. This leads us to state the following definition.

DEFINITION 5.4

If the function f is integrable on $[a, b]$, then the **average value of f over $[a, b]$** is

$$\bar{y} = \frac{1}{b-a}\int_a^b f(x)\,dx.$$

Example 5.42 Find the average value of $f(x) = x^2$ over $[0, 2]$.

Solution By the definition, the average value is

$$\begin{aligned}\bar{y} &= \frac{1}{2-0}\int_0^2 x^2\,dx\\ &= \left(\frac{1}{2}\right)\left(\frac{1}{3}\right)x^3\Big|_0^2\\ &= \left(\frac{1}{2}\right)\left(\frac{8}{3}\right) = \frac{4}{3}.\end{aligned}$$

■

Example 5.43 Find the average value of $f(x) = \sin x$ over $[0, \pi]$.

Solution The average value is

$$\begin{aligned}\bar{y} &= \frac{1}{\pi-0}\int_0^\pi \sin x\,dx\\ &= \left(\frac{1}{\pi}\right)(-\cos x)\Big|_0^\pi\\ &= \left(\frac{1}{\pi}\right)[-\cos \pi - (-\cos 0)] = \frac{2}{\pi}.\end{aligned}$$

■

For a nonnegative function over $[a, b]$, there is a simple graphic interpretation for the average value. Since

$$\int_a^b f(x)\,dx = (b-a)\bar{y},$$

we see that the average value, $\bar{y}$, multiplied by the length of the interval, $b-a$, gives us the area under f over $[a, b]$. Thus, $\bar{y}$ is the height of a rectangle over $[a, b]$ whose area is the area under f (see Figure 5.35).

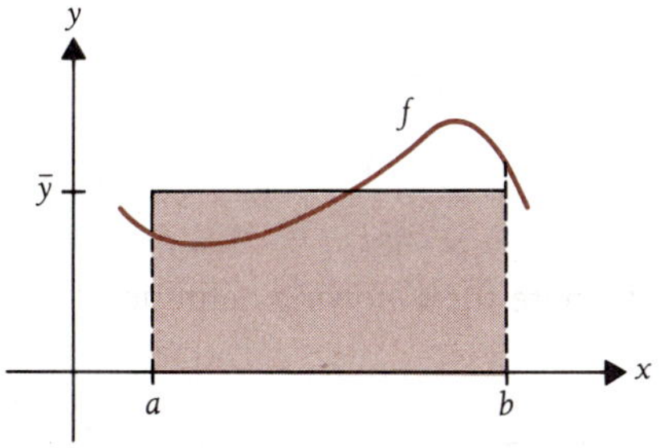

FIGURE 5.35 Geometric interpretation of the average value of f

In light of this definition of average value, the mean value theorem for integrals (Theorem 5.9, Section 5.3) says that a function that is continuous on $[a, b]$ takes on its average value at some point z.

Integrals of Rates of Change

The idea of average value has a useful interpretation when applied to a function expressing a rate of change. Suppose, for instance, that $v(t)$ represents the velocity of a moving object during the time interval $[a, b]$. Then the average value of v over $[a, b]$ is the **average velocity** during $[a, b]$,

$$v_{\text{ave}} = \frac{1}{b-a}\int_a^b v(t)\,dt.$$

Since $b-a$ is the elapsed time, it follows that $\int_a^b v(t)\,dt$ is the distance between the initial and final positions of the object. We call $\int_a^b v(t)\,dt$ the **displacement** of the object during $[a, b]$.

■ **Example 5.44** The velocity of an object moving along the x-axis is $v(t) = t^2 - t$. Find the displacement and the average velocity of the object during the time interval $[0, 2]$.

Solution The displacement is

$$\int_0^2 (t^2 - t)\,dt = \left[\frac{1}{3}t^3 - \frac{1}{2}t^2\right]_0^2 = \frac{8}{3} - 2 = \frac{2}{3}.$$

The average velocity is

$$v_{\text{ave}} = \frac{1}{2-0}\int_0^2 (t^2 - t)\,dt = \left(\frac{1}{2}\right)\left(\frac{2}{3}\right) = \frac{1}{3}.$$ ■

If the above considerations are applied to the speed $r(t) = |v(t)|$ of a moving object, then the **average speed** is

$$r_{\text{ave}} = \frac{1}{b-a}\int_a^b r(t)\,dt = \frac{1}{b-a}\int_a^b |v(t)|\,dt.$$

Here, $b - a$ is still the time elapsed, so the integral $\int_a^b |v(t)|\,dt$ is the **total distance traveled** by the object with velocity $v(t)$ during the time interval $[a, b]$. The distance traveled will not be the same as the displacement unless the velocity is always positive.

■ **Example 5.45** Find the distance traveled by the object in Example 5.44, and find its average speed.

Solution Since the velocity is $v(t) = t^2 - t$, the speed during $[0, 2]$ is $r(t) = |t^2 - t|$, so that

$$r(t) = \begin{cases} t - t^2 & \text{if } 0 \le t < 1 \\ t^2 - t & \text{if } 1 \le t \le 2. \end{cases}$$

Hence, the distance traveled is

$$\begin{aligned} \int_0^2 |t^2 - t|\,dt &= \int_0^1 (t - t^2)\,dt + \int_1^2 (t^2 - t)\,dt \\ &= \left[\frac{1}{2}t^2 - \frac{1}{3}t^3\right]_0^1 + \left[\frac{1}{3}t^3 - \frac{1}{2}t^2\right]_1^2 \\ &= \left(\frac{1}{2} - \frac{1}{3}\right) + \left[\left(\frac{8}{3} - 2\right) - \left(\frac{1}{3} - \frac{1}{2}\right)\right] = 1. \end{aligned}$$

The average speed is therefore

$$r_{\text{ave}} = \frac{1}{2-0}\int_0^2 |t^2 - t|\,dt = \left(\frac{1}{2}\right)(1) = \frac{1}{2}.$$ ■

The distance traveled found in Example 5.45 is not the same as the displacement found in Example 5.44 because the object retraced part of its path. The velocity during $[0, 1]$ was negative, so the object was moving to the left (see Figure 5.36); then the velocity during $[1, 2]$ was positive, so the object was moving to the right.

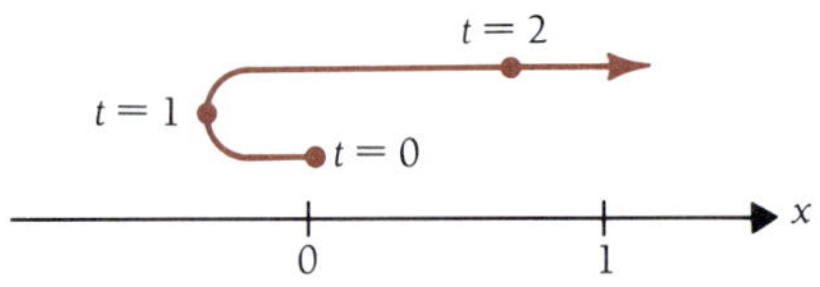

FIGURE 5.36 Path of a moving object

To generalize these ideas, we may say that if $f(x)$ represents the **rate of change** of a quantity with respect to x during the interval $[a, b]$, then the definite integral $\int_a^b f(x)\,dx$ represents the **net change** in the quantity as x changes from a to b. The integral of the absolute value, $\int_a^b |f(x)|\,dx$, represents the total fluctuation of the quantity as x changes from a to b.

Graphically, if the function f is positive over $[a, b]$, then the total change in the quantity is numerically equal to the area under f over $[a, b]$ (see Figure 5.37). More generally, the net change in the quantity is equal to the difference between the area under f above the x-axis and the area above f under the x-axis.

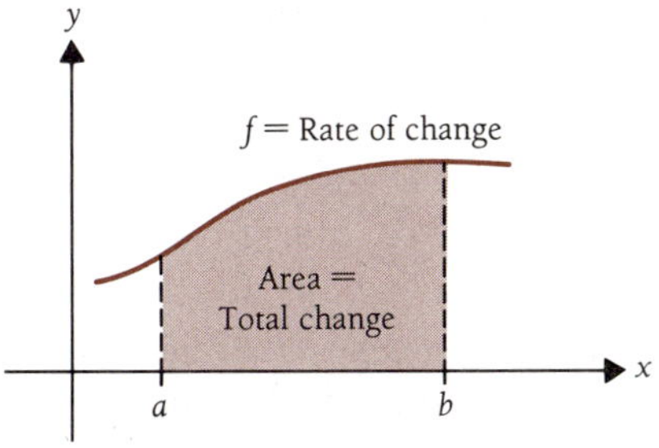

FIGURE 5.37 Relating rate of change to total change

Example 5.46 The growth rate of a certain business is

$$f(t) = 0.12 + 0.2t^{1/4},$$

measured in thousands of dollars of worth per year. How much does the business grow during its first year?

Solution The net growth is

$$\int_0^1 (0.12 + 0.2t^{1/4})\,dt = \left[0.12t + (0.2)\left(\frac{4}{5}\right)t^{5/4}\right]_0^1$$
$$= 0.12 + 0.16 = 0.28$$

thousands of dollars, or \$280. ■

Example 5.47 The growth rate of a certain business that experiences seasonal fluctuations is

$$g(t) = 0.12t - 0.005\cos\frac{\pi t}{6},$$

with t measured in months and g measured in thousands of dollars. What is the net growth of the business during its first year?

Solution The net growth during the first 12 months is

$$\int_0^{12}\left(0.12t - 0.005\cos\frac{\pi t}{6}\right)dt = \left[0.06t^2 - 0.005\left(\frac{6}{\pi}\right)\sin\frac{\pi t}{6}\right]_0^{12}$$
$$= 0.06(12)^2 - \left(\frac{0.03}{\pi}\right)\sin 2\pi$$
$$\approx 8.64,$$

or about \$8640. ■

By the fundamental theorem, we have the relationship

$$\int_a^b f'(x)\,dx = f(b) - f(a),$$

which can be rewritten as

$$f(b) = f(a) + \int_a^b f'(x)\,dx. \qquad (1)$$

Thus, if the value of f is known at $x = a$, and the rate of change of f is given, then we can compute the value of f at $x = b$ by using formula (1).

■ **Example 5.48** The velocity of an object is $v(t) = 2t - t^2$. If the object is at the origin when $t = 1$, where is it when $t = 4$?

Solution Let $s(t)$ be the position of the object at time t; then $s(1) = 0$. Hence, by formula (1),

$$s(4) = s(1) + \int_1^4 v(t)\,dt$$
$$= 0 + \int_1^4 (2t - t^2)\,dt$$
$$= \left[t^2 - \frac{1}{3}t^3\right]_1^4$$
$$= \left(16 - \frac{64}{3}\right) - \left(1 - \frac{1}{3}\right) = -6.$$
■

Exercises 5.7

Find the average value of the function f over the given interval.

1. $f(x) = x^2$, $[0, 3]$
2. $f(x) = 1 - x^2$, $[0, 3]$
3. $f(x) = \sqrt{x}$, $[1, 4]$
4. $f(x) = x^{2/3}$, $[1, 8]$
5. $f(x) = \cos x$, $[-\pi/2, \pi/2]$
6. $f(x) = \sin x$, $[0, \pi/2]$
7. $f(x) = \sin 2x + 2\cos x$, $[-\pi/4, \pi]$
8. $f(x) = x\cos x^2$, $[0, \pi]$
9. $f(x) = x\sqrt{x^2 + 1}$, $[1, 2]$
10. $f(x) = x\sqrt{1 - x^2}$, $[0, 1]$
11. $f(x) = \sqrt{2x + 1}$, $[\frac{2}{3}, 4]$
12. $f(x) = (1 - x)^{2/3}$, $[0, 9]$

An object moves along the x-axis with velocity $v(t)$. Find the displacement, average velocity, total distance traveled, and average speed during the given time interval.

13. $v(t) = 2$, $[0, 3]$
14. $v(t) = 3t$, $[1, 4]$
15. $v(t) = 3t$, $[0, 4]$
16. $v(t) = 3t$, $[3, 4]$
17. $v(t) = t^2$, $[-1, 2]$
18. $v(t) = t^2$, $[0, 2]$
19. $v(t) = -16t^2$, $[0, 5]$
20. $v(t) = -16t^2$, $[5, 10]$
21. $v(t) = t^2 - 4t$, $[1, 5]$
22. $v(t) = t^3 - t^2$, $[-1, 2]$
23. $v(t) = 3 \sin 2\pi t$, $[0, 1]$
24. $v(t) = 2 \cos(t/3)$, $[0, \pi]$
25. Find $s(2)$, given $s(0) = 2$ and $v(t) = s'(t) = 3 + t + t^2$.
26. Find $s(5)$, given $s(1) = 20$ and $v(t) = s'(t) = 80 - 16t^2$.
27. Find $s(1)$, given $s(0) = 2$ and $v(t) = s'(t) = 4 \cos t$.
28. Find $s(4)$, given $s(1) = -1$ and $v(t) = s'(t) = t^3 + 1$.
29. The value of a stock at $t = 0$ is $w(0) = \$36$, and its growth rate is $w'(t) = 0.02 + 0.05t$ dollars per month. What is its worth after 6 months?
30. What is the average value of the stock in Problem 29 over the first 6 months?
31. The growth rate of a ski manufacturing company is $g(t) = 0.18t + 0.06 \cos(\pi t/6)$ thousand dollars per month. What is the company's net growth during its first 15 months?
32. In a certain bacterial population, the growth rate is $P'(t) = 2t$ per hour. If $P(1) = 50{,}000$, find $P(3)$.
33. In Problem 32, what is the average growth rate during the time interval $[1, 3]$?
34. In Problem 32, what is the average population during the time interval $[1, 3]$?
35. Find the average width of the region bounded by the two curves $y = 4 - x^2$ and $y = 0$.
36. Find the average height of the region bounded by the two curves $y = 9 - x^2$ and $y = x + 3$.
37. Because of inflation, the buying power of a fixed income decreases at the rate of 12% per year. By how much has the buying power decreased at the end of 5 years?
38. A water tank is a cylinder of radius 3 feet, with its axis vertical. Water flows into the tank at the rate of 12 cubic feet per minute. What is the average depth of the water in the tank during the first 20 minutes?
39. Repeat Problem 38 if the tank is rectangular, 6 feet long, and 5 feet wide.
40. Repeat Problem 38 if the tank is conical, situated with its apex down, and has height 10 feet and top radius 6 feet.

For each interpretation of f, give an interpretation for the integral $\int_a^b f(x)\,dx$.

41. f is the acceleration of an object.
42. f is the growth rate of a population.
43. f is marginal cost.
44. f is marginal revenue.
45. f is the interest rate on an investment.
46. f is the interest rate on a mortgage.

47. f is the number of births per month.

48. f is the number of marriages per year.

49. The graphs in the accompanying figure give the growth rates of two stocks. Which stock appears to have grown more during the given time interval? Why?

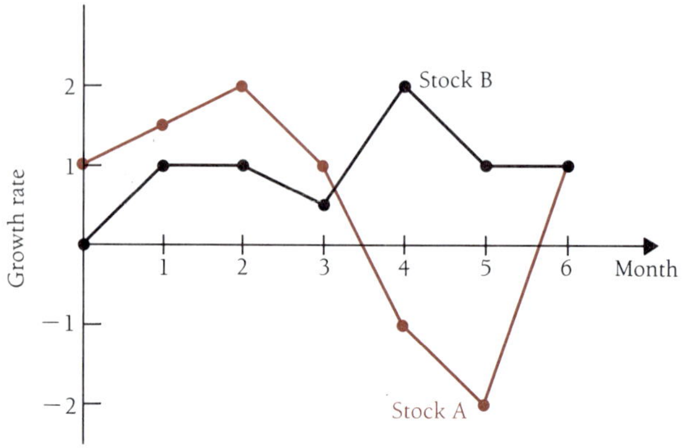

50. Use the fundamental theorem of calculus to show that if an object has position function $s(t)$, then its displacement during the time interval $[a, b]$ is just $s(b) - s(a)$.

SECTION 5.8

Chapter Review

Important Concepts

Method of exhaustion	Riemann sum	Change of variable
Region below a curve	Integrable function	Area between curves
Area under a curve	Definite integral	Average value of a function
Subinterval	Linearity property	Average velocity
Inner area	Mean value theorem for integrals	Displacement
Outer area	Antiderivative on an interval	Average speed
Summation notation	Fundamental theorem of calculus	Distance traveled
Subdivision	Integration	Rate of change
Norm of a subdivision	Integration by substitution	Net change

Review Exercises

Use the method of exhaustion to find the area under f over the given interval. Then check your work by integrating.

1. $f(x) = 9 - x^2$, $[0, 2]$

2. $f(x) = x^2 + 1$, $[-1, 1]$

Write each limit as a definite integral, and evaluate.

3. $\lim_{n\to\infty} \sum_{i=1}^{n} n^{-1/3}(n+i)^{-2/3}$

4. $\lim_{n\to\infty} \left(\frac{1}{8n^3}\right) \sum_{i=1}^{n} i^2$

Write each integral as the limit of a sum.

5. $\int_0^5 x^{2/3}\,dx$

6. $\int_0^4 \sqrt{x}\,dx$

Evaluate each definite integral by interpreting it as an area.

7. $\int_3^5 |x|\,dx$

8. $\int_0^3 |x|\,dx$

9. $\int_0^5 \sqrt{25-x^2}\,dx$

10. $\int_0^3 |x-2|\,dx$

11. $\int_0^3 (2x-1)\,dx$

12. $\int_{-4}^4 (-\sqrt{16-x^2})\,dx$

13. $\int_0^{2\pi} \sin x\,dx$

14. $\int_{-\pi/4}^{\pi/4} \tan x\,dx$

Evaluate each definite integral by using the fundamental theorem.

15. $\int_1^3 (x+x^{-2})\,dx$

16. $\int_1^4 (2x^2-3\sqrt{x})\,dx$

17. $\int_0^2 x^{2/3}\,dx$

18. $\int_1^{32} x^{-1/5}\,dx$

19. $\int_0^{\pi/6} \sec x \tan x\,dx$

20. $\int_0^{\pi} \sin \frac{x}{3}\,dx$

Integrate by substitution.

21. $\int (x+4)^{19}\,dx$

22. $\int_0^3 x\sqrt{9-x^2}\,dx$

23. $\int_0^2 \frac{3x}{(4x^2+1)^3}\,dx$

24. $\int \frac{x^2}{\sqrt{1-x}}\,dx$

25. $\int_0^{\pi/3} \tan^2 x \sec^2 x\,dx$

26. $\int x^{-1/2} \sin \sqrt{x}\,dx$

Differentiate.

27. $f(x) = \int_1^x \sqrt{3t^3+1}\,dt$

28. $f(x) = \int_x^{2x} (2t+3)^8\,dt$

29. $f(x) = \int_0^x \frac{1}{1-t^2}\,dt$

30. $f(x) = \int_x^{x^2} \frac{1}{1+t}\,dt$

Find the area.

31. Between $y = 25 - x^2$ and the x-axis, over $[0, 5]$

32. Between $y = 4 - x^3$ and $y = x^2$, over $[-1, 0]$

33. Between $y = x^2 + 4$ and $y = \sin x$, over $[-1, 1]$

34. Between $x = \sqrt{y}$ and $x = y - 2$, between $x = 0$ and $x = 4$

Let $R = \{(x, y) | 0 \le x \le 4$ and $0 \le y \le 8 - x^{2/3}\}$.

35. Find the area of R.

36. Find the average height of R.

37. Find the average width of R.

c38. Estimate c such that the line $x = c$ cuts R into two pieces of equal area. (Use Newton's method.)

39. A biology student can memorize species names at the rate of $2t - 3$ names per minute during a span of t minutes. How many names can he memorize in 10 minutes?

40. Find upper and lower bounds for

$$\int_0^1 \frac{1}{x^3 + 1}\, dx.$$

41. For $f(x) = 2x$ and $[a, b] = [1, 3]$, find $z \in [a, b]$ such that

$$f(z) = \frac{1}{b - a} \int_a^b f(x)\, dx.$$

42. Repeat Problem 41 for $f(x) = (2x + 5)^{1/3}$ and $[a, b] = [-2, \frac{3}{2}]$.

43. Explain why

$$\int_1^2 \frac{1}{2x - 1}\, dx \ge \int_1^2 \frac{1}{2x}\, dx.$$

Additional Exercises

1. Estimate

$$\frac{1}{\sqrt{n}\sqrt{n + 1}} + \frac{1}{\sqrt{n}\sqrt{n + 2}} + \cdots + \frac{1}{\sqrt{n}\sqrt{n + n}}.$$

Use Problem 40 of Section 5.3 to evaluate the following:

2. $\displaystyle\int_{-5}^{5} 3x(x^4 + 1)^8\, dx$

3. $\displaystyle\int_{-1}^{1} \frac{\sqrt{x^2 + 1}}{x^3}\, dx$

4. $\displaystyle\int_{-1}^{1} x \cos x\, dx$

5. $\displaystyle\int_{-2}^{2} (2x + 1)^4\, dx$

6. Show that

$$\int_{a+c}^{b+c} f(x - c)\, dx = \int_a^b f(x)\, dx.$$

7. Show that

$$\frac{1}{c} \int_{ac}^{bc} f\left(\frac{x}{c}\right) dx = \int_a^b f(x)\, dx.$$

8. Show that

$$\int_a^b (x - a)(b - x)\, dx = \frac{(b - a)^3}{6}.$$

9. Let f be a continuous function on $(-\infty, \infty)$, and let $A(x) = \int_a^x f(t)\, dt$. Show that the extrema of $A(x)$ occur only at zeros of f.

10. Let R be the region under $y = x^2$ over $[0, 1]$. Find a number m such that $y = mx$ cuts R into two pieces of equal area.

11. Let the function f be continuous on $[a, b]$. Find the area of the region

$$\left\{(x, y) | a \le x \le b \text{ and } \frac{f(x) - |f(x)|}{2} \le y \le \frac{f(x) + |f(x)|}{2}\right\}.$$

12. The rectangular region $\{(x, y) | 0 \le x \le a \text{ and } 0 \le y \le b\}$ is cut into two regions by the curve $y = ax^n$, $n \in \mathbb{N}$. What is the ratio of the areas of the two regions?

13. If $n \in \mathbb{N}$, find the area between the curves $y = x^n$ and $y = x^{n+1}$ over $[0, 1]$.

14. A function f defined on $[a, b]$ is **piecewise monotonic** on $[a, b]$ if and only if there is a subdivision $\{x_0, x_1, \ldots, x_n\}$ of $[a, b]$ such that f is monotonic on each open subinterval (x_{i-1}, x_i). Prove that a bounded, piecewise monotonic function on $[a, b]$ is integrable on $[a, b]$.

15. If function f is continuous on (a, b) and both the limits $\lim_{x\to a^+} f(x)$ and $\lim_{x\to b^-} f(x)$ exist, prove that for any $c \in (a, b)$, both the limits $\lim_{t\to a^+} \int_t^c f(x)\,dx$ and $\lim_{t\to b^-} \int_c^t f(x)\,dx$ exist.

16. Suppose f is a continuous function such that

$$[f(x)]^2 = 2\int_0^x f(t)\,dt.$$

Find f. [*Hint:* Differentiate.]

17. A function f defined on $(-\infty, \infty)$ is periodic with period p and is twice differentiable. Show that

$$\int_a^{a+p} f'(x)\,dx = 0.$$

18. Show that if function f is n times differentiable and periodic of period p, then

$$\int_a^{a+p} f^{(n-1)}(x)\,dx = 0.$$

19. Derive **Leibniz's rule**: If g and h are differentiable functions and the function f is continuous on $(-\infty, \infty)$, then the function $A(x) = \int_{g(x)}^{h(x)} f(t)\,dt$ is differentiable and has derivative

$$A'(x) = f(h(x))h'(x) - f(g(x))g'(x).$$

Differentiate.

20. $\int_x^{3x} \sqrt{t^2 + 1}\,dt$

21. $\int_{x^3}^{x^2} t(t^3 + 1)^5\,dt$

CHAPTER 6

Exponential, Logarithmic, and Other Transcendental Functions

INTRODUCTION

In this chapter, we develop the calculus of the exponential and logarithmic functions. We begin by introducing the natural logarithm as a function defined by a definite integral with variable upper limit. We develop its properties and those of its inverse function, the exponential function, and then go on to discuss other exponential and logarithmic functions. Then follows a section on applications of exponential functions, a discussion of L'Hôpital's rules, a section on the inverse trigonometric functions, and an optional section on the hyperbolic functions.

SECTION 6.1

The Natural Logarithm

Unlike the functions we have studied thus far, the natural logarithm is a function that depends on calculus for its very definition. Before we introduce it, let us discuss logarithmic functions in general.

Logarithmic Functions

We start with the definition of a logarithmic function.

DEFINITION 6.1

A **logarithmic function** L is a nonconstant, differentiable function defined on $(0, \infty)$, such that for any $x, y \in (0, \infty)$,

$$L(xy) = L(x) + L(y). \tag{1}$$

We can deduce some properties of logarithmic functions directly from the definition. For instance, if L is a logarithmic function, then

$$L(1) = L(1 \cdot 1) = L(1) + L(1) = 2L(1),$$

so that

$$L(1) = 0. \tag{2}$$

Also, for any $x \in (0, \infty)$, we have

$$0 = L(1) = L\left(x \cdot \frac{1}{x}\right) = L(x) + L\left(\frac{1}{x}\right),$$

so that

$$L\left(\frac{1}{x}\right) = -L(x). \tag{3}$$

Hence, it follows that for $x, y \in (0, \infty)$,

$$L\left(\frac{x}{y}\right) = L(x) - L(y). \tag{4}$$

[You will be asked to supply the two intermediate steps for the proof of property (4) in the exercises.]

Since a logarithmic function is differentiable, we have by definition

$$L'(x) = \lim_{h \to 0} \frac{L(x+h) - L(x)}{h}.$$

But by property (4),

$$L(x+h) - L(x) = L\left(\frac{x+h}{x}\right) = L\left(1 + \frac{h}{x}\right).$$

Therefore, using equation (2) and algebraic manipulation, we get

$$\begin{aligned} L'(x) &= \lim_{h \to 0} \frac{L(1 + h/x)}{h} \\ &= \lim_{h \to 0} \frac{L(1 + h/x) - L(1)}{h} \\ &= \lim_{h \to 0} \frac{1}{x} \cdot \frac{L(1 + h/x) - L(1)}{h/x} \\ &= \frac{1}{x} \lim_{h/x \to 0} \frac{L(1 + h/x) - L(1)}{h/x} \\ &= \frac{1}{x} L'(1). \end{aligned}$$

That is, for any logarithmic function L,

$$L'(x) = \frac{1}{x} L'(1). \tag{5}$$

Note that $L'(1)$ is not 0, since that would make L a constant function.

The Natural Logarithm

The most natural, or simplest, logarithmic function is one for which $L'(1) = 1$, because then $L'(x) = 1/x$. If $L'(x) = 1/x$, then $L(x)$ is an antiderivative of $1/x$. By Theorem 5.11, $\int_1^x (1/t)\, dt$ is also an antiderivative of $1/x$, so

$$L(x) = \int_1^x \frac{1}{t}\, dt + C.$$

Since $0 = L(1) = \int_1^1 (1/t)\, dt + C = C$, we get

$$L(x) = \int_1^x \frac{1}{t}\, dt.$$

Since the particular logarithmic function for which $L'(1) = 1$ is a natural choice, we give it the name *natural logarithm*.

DEFINITION 6.2

The **natural logarithm** of $x \in (0, \infty)$ is denoted by $\ln x$ and is defined by

$$\ln x = \int_1^x \frac{1}{t}\, dt.$$

We call the function $f(x) = \ln x$ the **natural logarithmic function.**

The function name, ln, is an abbreviation for the Latin *logarithmus naturalis.* Some writers use "log x" instead of "ln x."

To justify the name "logarithm" for $\ln x$, we must show that it is indeed a logarithmic function. First of all, $\ln x$ is defined for all $x \in (0, \infty)$. Secondly, $\ln x$ is differentiable, and

$$D_x \ln x = \frac{1}{x}. \tag{6}$$

It follows that $\ln x$ is not a constant function, since $1/x$ is never 0.

Finally, we must show that property (1) holds for $\ln x$. By definition, for $x, y \in (0, \infty)$,

$$\ln(xy) = \int_1^{xy} \frac{1}{t}\, dt$$

$$\begin{aligned}\ln(xy) &= \int_1^x \frac{1}{t}\,dt + \int_x^{xy} \frac{1}{t}\,dt \\ &= \ln x + \int_x^{xy} \frac{1}{t}\,dt.\end{aligned}$$

In this last integral, we make the change of variable $t = xu$; then $dt = x\,du$. Moreover, $u = 1$ when $t = x$, and $u = y$ when $t = xy$. Hence, we have

$$\begin{aligned}\ln(xy) &= \ln x + \int_1^y \frac{1}{xu}\,x\,du \\ &= \ln x + \int_1^y \frac{1}{u}\,du \\ &= \ln x + \ln y\end{aligned}$$

(u is a dummy variable). Thus, property (1) holds, and $\ln x$ is a logarithmic function.

The Graph of ln *x*

Now we describe the properties of the graph of $\ln x$, and work toward constructing the graph. From equation (6), we note that the derivative of $\ln x$ is positive for all $x \in (0, \infty)$; thus, $\ln x$ is increasing on $(0, \infty)$. From property (2), we know that $\ln 1 = 0$; hence, $\ln x > 0$ if $x > 1$ and $\ln x < 0$ if $0 < x < 1$. These facts can also be observed by interpreting $\ln x = \int_1^x (1/t)\,dt$ as the area under the curve $y = 1/t$ between 1 and x (see Figures 6.1 and 6.2).

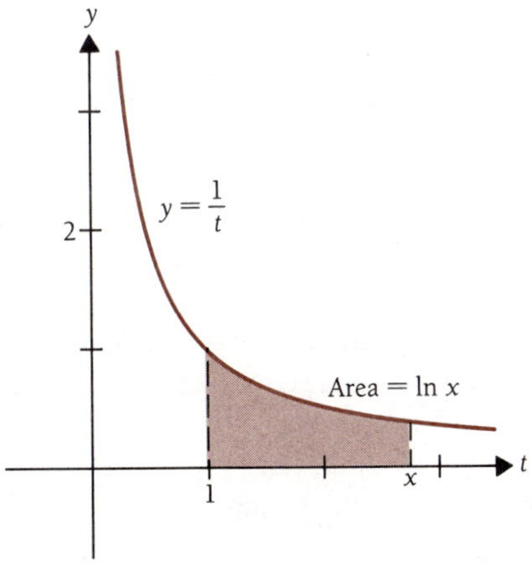

FIGURE 6.1 $\ln x > 0$ if $x > 1$

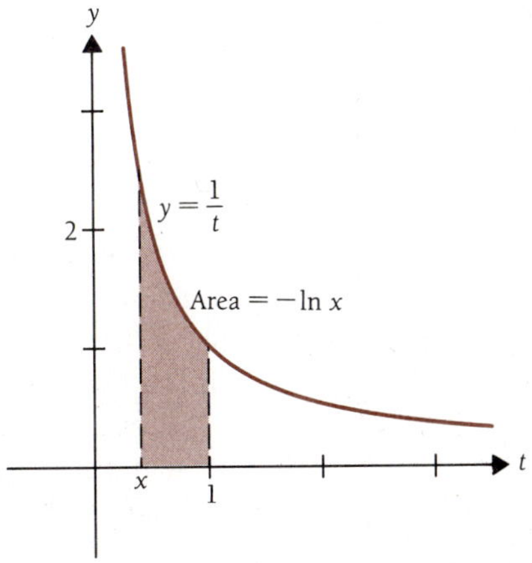

FIGURE 6.2 $\ln x < 0$ if $0 < x < 1$

These properties are evident in the graph of $y = \ln x$, shown in Figure 6.3 (page 366). The precise plotting of points depends on evaluating the integral $\int_1^x (1/t)\,dt$ for various values of x. This can be done to any appropriate degree of accuracy using an approximation based on formula (3) of Section 5.2 or a calculator.

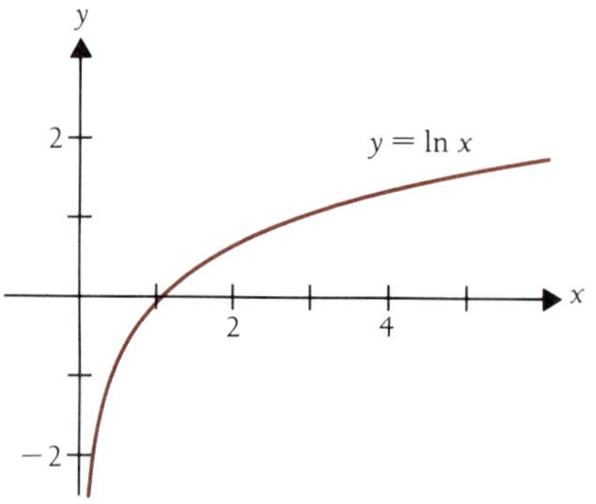

FIGURE 6.3 The graph of $y = \ln x$

One other property of the natural logarithmic function should be mentioned here. For any rational number n and for any $x \in (0, \infty)$,

$$\ln x^n = n \ln x. \tag{7}$$

To prove this fact, we start with the definition:

$$\ln x^n = \int_1^{x^n} \frac{1}{t}\, dt.$$

In this integral, we make the change of variable substitution $t = u^n$. Then $dt = nu^{n-1}\, du$; also, $u = 1$ when $t = 1$, and $u = x$ when $t = x^n$. Therefore,

$$\begin{aligned} \ln x^n &= \int_1^x \frac{1}{u^n} nu^{n-1}\, du \\ &= n \int_1^x \frac{1}{u}\, du \\ &= n \ln x. \end{aligned}$$

Some more properties of the natural logarithmic function are given in the following examples.

■ **Example 6.1** Show that $\lim_{x\to\infty} \ln x = \infty$ and $\lim_{x\to 0^+} \ln x = -\infty$.

Solution First, we show that $\ln x \to \infty$ as $x \to \infty$. To do this, we will let $x = 3^n$, and form a Riemann sum approaching $\int_1^{3^n} (1/t)\, dt$. Let $\{1, 3, 3^2, \ldots, 3^n\}$ be a subdivision of $[1, x]$. The ith subinterval $[3^{i-1}, 3^i]$ has length

$$\Delta x_i = 3^i - 3^{i-1} = 3^{i-1}(3 - 1) = 2 \cdot 3^{i-1}.$$

If we let $t_i = 3^i$, the right-hand endpoint of each subinterval, then the Riemann sum for $1/t$ over $[1, x]$ relative to this subdivision is

$$\sum_{i=1}^n \frac{1}{3^i} \Delta x_i = \sum_{i=1}^n \frac{1}{3^i} \cdot 2 \cdot 3^{i-1} = \sum_{i=1}^n \frac{2}{3} = \frac{2}{3} n.$$

This Riemann sum is also the sum of areas of rectangles that lie below the curve $y = 1/t$ (see Figure 6.4). At any rate, we have

$$\ln 3^n = \int_1^{3^n} \frac{1}{t}\, dt > \frac{2}{3} n. \tag{8}$$

Since $\frac{2}{3}n \longrightarrow \infty$ as $n \longrightarrow \infty$, it follows that $\ln x \longrightarrow \infty$ as $x \longrightarrow \infty$.

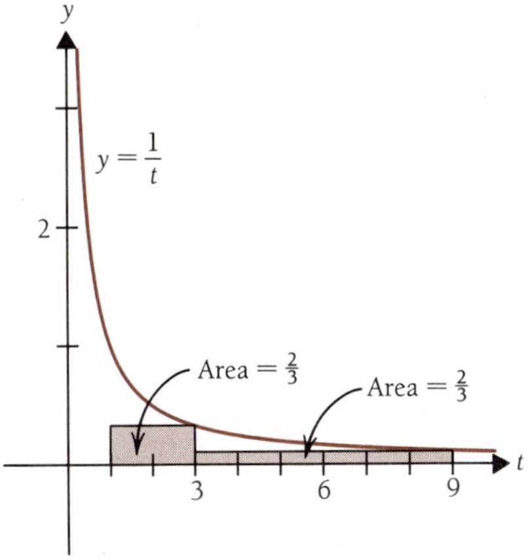

FIGURE 6.4 Rectangles below $y = 1/t$

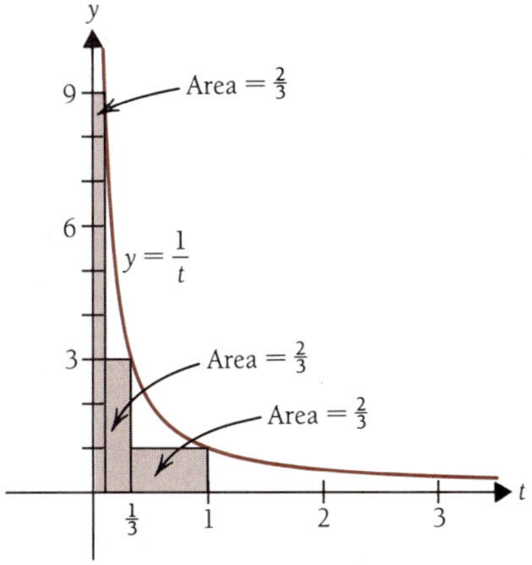

FIGURE 6.5 More rectangles below $y = 1/t$

Similarly, we can show that $\ln x \longrightarrow -\infty$ as $x \longrightarrow 0^+$. By property (3), we have from (8) the relation

$$-\ln\left(\frac{1}{3^n}\right) > \frac{2}{3} n.$$

Interpreting this in terms of rectangles under the curve $y = 1/t$ (see Figure 6.5), the result follows. ■

Differentiation of $\ln x$

The chain rule form of the differentiation formula (6) for the natural logarithm is

$$d(\ln u) = \frac{1}{u}\, du. \tag{9}$$

This formula can be generalized. First, we show that

$$D_x \ln|x| = \frac{1}{x}. \tag{10}$$

This is done by using the two cases $|x| = x$ if $x > 0$, and $|x| = -x$ if $x < 0$. In the first case, equation (10) reduces directly to equation (9). In the second case,

$$D_x \ln|x| = D_x \ln(-x) = \left(\frac{1}{-x}\right)(-1) = \frac{1}{x},$$

by (9). Now, the generalization we mentioned is the chain rule form of (10), namely,

$$d(\ln|u|) = \frac{1}{u}\,du. \tag{11}$$

An alternative form of (11) is

$$D_x \ln|f(x)| = \frac{f'(x)}{f(x)}. \tag{12}$$

Example 6.2 Find $D_x \ln|x^2 - 1|$.

Solution Using formula (12), we get

$$D_x \ln|x^2 - 1| = \frac{2x}{x^2 - 1}.$$

■

Example 6.3 Find $f'(x)$ if $f(x) = x^2 \ln|1 - x|$.

Solution Using the product rule and formula (12), we get

$$\begin{aligned} f'(x) &= 2x \ln|1 - x| + x^2\left(\frac{-1}{1-x}\right) \\ &= 2x \ln|1 - x| + \frac{x^2}{x-1}. \end{aligned}$$

■

Example 6.4 If $y = \ln|\sec x + \tan x|$, find y'.

Solution Again using formula (12), we have

$$\begin{aligned} y' &= \frac{\sec x \tan x + \sec^2 x}{\sec x + \tan x} \\ &= \frac{(\sec x)(\sec x + \tan x)}{\sec x + \tan x} = \sec x. \end{aligned}$$

■

Logarithmic Differentiation

The properties of the natural logarithm give us a technique of differentiation that simplifies many problems. This technique is called **logarithmic differentiation**, and involves implicitly differentiating—not the given function—but its natural logarithm. We will illustrate with some examples.

■ Example 6.5 Use logarithmic differentiation to find the derivative of $y = (x^2 + 4)^3$.

Solution Since y is positive for all values of x, we can use the properties of logarithms to write the natural logarithm of y as

$$\ln y = 3 \ln(x^2 + 4).$$

Then, differentiating implicitly, we get

$$\frac{1}{y} y' = 3\left(\frac{2x}{x^2 + 4}\right),$$

so that, solving for y', we obtain

$$\begin{aligned} y' &= \frac{6xy}{x^2 + 4} \\ &= \frac{6x(x^2 + 4)^3}{x^2 + 4} \\ &= 6x(x^2 + 4)^2. \end{aligned}$$

■

■ Example 6.6 Use logarithmic differentiation to find the derivative of

$$y = \sqrt{\left|\frac{x + 1}{x - 1}\right|}.$$

Solution We find the natural logarithm of y, simplify it using properties (7) and (4), and then differentiate implicitly:

$$\begin{aligned} \ln y &= \ln \sqrt{\left|\frac{x + 1}{x - 1}\right|} \\ &= \frac{1}{2} \ln\left|\frac{x + 1}{x - 1}\right| \\ &= \frac{1}{2} \ln|x + 1| - \frac{1}{2} \ln|x - 1|, \\ \frac{1}{y} y' &= \frac{1}{2} \cdot \frac{1}{x + 1} - \frac{1}{2} \cdot \frac{1}{x - 1}, \\ y' &= \frac{y}{2(x + 1)} - \frac{y}{2(x - 1)} \\ &= \frac{-y}{x^2 - 1}. \end{aligned}$$

■

■ Example 6.7 Differentiate:

$$y = (x + 3)^{3/2}(2x - 1)^{1/3}(x^2 + 5)^{-1/4}$$

Solution The absolute value of y is

$$|y| = |x+3|^{3/2}|2x-1|^{1/3}|x^2+5|^{-1/4}.$$

(Some of the absolute value signs are redundant, but there is no harm in using them anyway.) Hence, by properties (1) and (7), the natural logarithm of $|y|$ is

$$\ln|y| = \frac{3}{2}\ln|x+3| + \frac{1}{3}\ln|2x-1| - \frac{1}{4}\ln|x^2+5|.$$

Differentiating,

$$\frac{y'}{y} = \frac{3}{2}\cdot\frac{1}{x+3} + \frac{1}{3}\cdot\frac{2}{2x-1} - \frac{1}{4}\cdot\frac{2x}{x^2+5},$$

$$y' = \left[\frac{3}{2(x+3)} + \frac{2}{3(2x-1)} - \frac{x}{2(x^2+5)}\right]y.$$

This approach is simpler than using the product formula and power rules repeatedly. ■

We shall have further need of logarithmic differentiation later.

The Antiderivative of $1/u$

Formula (11) gives us the indefinite integral formula

$$\int \frac{1}{u}\,du = \ln|u| + C. \tag{13}$$

(This is the integral formula for u^n with $n = -1$.) We will illustrate the use of this formula with several examples.

■ **Example 6.8** Find

$$\int \frac{1}{x+3}\,dx.$$

Solution We substitute $u = x+3$, $du = dx$:

$$\int \frac{1}{x+3}\,dx = \int \frac{1}{u}\,du$$
$$= \ln|u| + C$$
$$= \ln|x+3| + C.$$ ■

■ **Example 6.9** Find $\int \tan x\,dx$.

Solution Since

$$\int \tan x\,dx = \int \frac{\sin x}{\cos x}\,dx,$$

we substitute $u = \cos x$, $du = -\sin x\,dx$:

$$\begin{aligned}\int \tan x\,dx &= -\int \frac{1}{u}\,du \\ &= -\ln|u| + C \\ &= -\ln|\cos x| + C.\end{aligned}$$

Using property (3), this expression also can be written as

$$-\ln|\cos x| + C = \ln\left|\frac{1}{\cos x}\right| + C = \ln|\sec x| + C.$$

Thus, we have the formula

$$\int \tan x\,dx = \ln|\sec x| + C. \qquad (14)$$ ■

■ **Example 6.10** Find

$$\int \frac{1}{x \ln x}\,dx.$$

Solution We substitute $u = \ln x$, $du = (1/x)\,dx$:

$$\begin{aligned}\int \frac{1}{x \ln x}\,dx &= \int \frac{1}{u}\,du \\ &= \ln|u| + C \\ &= \ln|\ln x| + C.\end{aligned}$$ ■

Formula (13) works equally well for definite integrals, as the next example shows.

■ **Example 6.11** Evaluate

$$\int_{-1}^{-3} \frac{1}{x}\,dx.$$

Solution

$$\begin{aligned}\int_{-1}^{-3} \frac{1}{x}\,dx &= \Big[\ln|x|\Big]_{-1}^{-3} \\ &= \ln|-3| - \ln|-1| \\ &= \ln 3.\end{aligned}$$

This integral is interpreted geometrically in Figure 6.6.

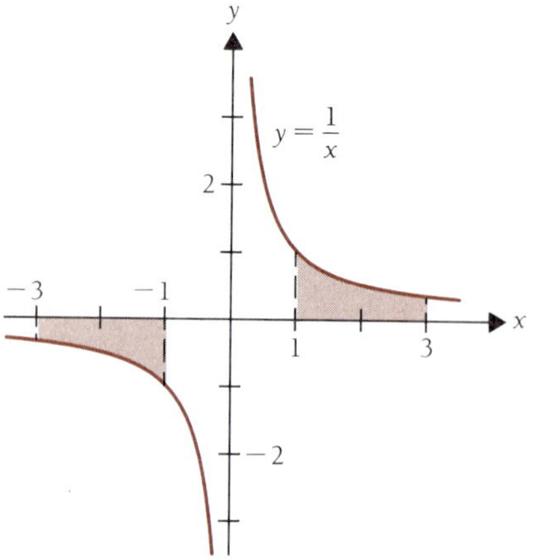

FIGURE 6.6 The areas are the same, so

$$\int_{-1}^{-3} \frac{1}{x}\,dx = \int_{1}^{3} \frac{1}{x}\,dx$$ ■

Exercises 6.1

1. Find an equation of the line tangent to the curve $y = \ln x$ at:

   (a) $x = 1$ (b) $x = 2$ (c) $x = 3$ (d) $x = 4$

2. Find the area under $y = 1/x$ over the interval:

   (a) $[1, 2]$ (b) $[2, 3]$ (c) $[\frac{1}{2}, 1]$ (d) $[\frac{1}{3}, 3]$

Differentiate.

3. $y = \ln 3x$
4. $y = \ln(3x + 5)$
5. $y = \ln(ax + b)$
6. $y = \ln(a - bx)$
7. $y = \ln\left(1 + \frac{1}{x}\right)$
8. $y = \ln\left(2 - \frac{1}{x}\right)$
9. $y = \ln(x^2 + 4)$
10. $y = \ln(1 - x^2)$
11. $y = \ln|4x^3 - 1|$
12. $y = \ln|x^4 + x^2 - 9|$
13. $y = \ln\sqrt{3x + 5}$
14. $y = \ln|1 - \sqrt{x}|$
15. $y = \ln\left|\frac{1}{x - 1}\right|$
16. $y = \ln\left|\frac{x^2}{2x + 1}\right|$
17. $y = \ln|\cos x|$
18. $y = \ln|\cot x + \csc x|$
19. $y = \ln|\csc x - \cot x|$
20. $y = \ln|\sin x + \cos x|$
21. $y = x^2\sqrt{x^3 + 5}$
22. $y = (x + 4)^{2/3}(x - 4)^{3/5}$
23. $y = \frac{\sqrt{x^2 + 4}}{\sqrt{(x^3 - 1)^5}}$
24. $y = (x + 5)^{-2}(4 - x)^{-1/2}$
25. $\ln(x^2 + y^2) = x - 2y$
26. $x + \ln y = 5xy$

Find the critical points and extrema of the following two functions. Also tell where each function increases or decreases, describe its concavity, and identify points of inflection.

27. $f(x) = \ln(1 + x^2)$
28. $f(x) = \ln(x + \sqrt{x^2 + 1})$

29. Show that for all $x \in (0, \infty)$, $\ln x < \sqrt{x}$.

C30. Show that the point on the curve $y = \ln x$ that is nearest the origin lies below the x-axis. Can you find the point?

Integrate.

31. $\int \frac{1}{x-3}\,dx$

32. $\int \frac{1}{1-x}\,dx$

33. $\int_1^2 \frac{1}{x-4}\,dx$

34. $\int \frac{2x}{x^2+1}\,dx$

35. $\int \frac{x}{3x^2+5}\,dx$

36. $\int \frac{1}{x \ln \sqrt{x}}\,dx$

37. $\int \frac{(\ln x)^2}{x}\,dx$

38. $\int_{-3}^{-5} \frac{x}{1-x^2}\,dx$

39. $\int \frac{\cos x}{1+\sin x}\,dx$

40. $\int \frac{\sec^2 x}{\tan x}\,dx$

41. $\int \tan 2x\,dx$

42. $\int_{-\pi/4}^{-\pi/3} \tan x\,dx$

Find the area.

43. Between $y = 1/x$ and $y = 2 + x$, over $[-1 + \sqrt{2}, 3]$

44. In the first quadrant, bounded by $y = 2$, $y = 1/x$, $x = 2$, and the coordinate axes

45. Explain why $\ln x_1 = \ln x_2$ implies $x_1 = x_2$.

46. Derive formula (4).

The next six problems illustrate that the natural logarithmic function is not the only function that can be defined as an integral.

47. Sketch the graph of $y = 1/(x^2 + 1)$.

48. Let

$$T(x) = \int_0^x \frac{1}{t^2+1}\,dt$$

for $x \in (-\infty, \infty)$. Show that $T(0) = 0$, $T(x) < 0$ if $x < 0$, and $T(x) > 0$ if $x > 0$.

49. Find $T'(x)$.

50. Show that $T(x)$ is increasing on $(-\infty, \infty)$.

51. Show that $D_x T^{-1}(x) = 1 + [T^{-1}(x)]^2$.

52. What familiar functions are $T(x)$ and $T^{-1}(x)$?

C53. Find both of the solutions of the equation

$$x^2 + \ln(\sin x) = 0$$

that lie in the interval $(0, \pi)$. [*Hint:* Use Newton's method.]

SECTION 6.2

The Exponential Function

In this section, we develop the exponential function as the inverse of the natural logarithmic function.

The Inverse of ln x

As was shown in the previous section, the natural logarithmic function is increasing, and hence, is one-to-one. Therefore, the function $f(x) = \ln x$ has an inverse function, which we will denote by $f^{-1}(x) = \exp(x)$. Thus, by definition of the inverse of a function,

$$y = \exp(x) \qquad \text{if and only if} \qquad x = \ln y. \tag{1}$$

Because $\ln x$ has domain $(0, \infty)$ and range $(-\infty, \infty)$, its inverse $\exp(x)$ has domain $(-\infty, \infty)$ and range $(0, \infty)$. Thus, $\exp(x)$ is positive for all x.

The most interesting property of $\exp(x)$ is that it is its own derivative. To show this, let $y = \exp(x)$. Then $\ln y = x$ by property (1). Differentiating implicitly, we get

$$\frac{1}{y}y' = 1,$$

$$y' = y.$$

Hence,

$$D_x \exp(x) = \exp(x). \tag{2}$$

Since $\exp(x)$ is positive, we see that $\exp(x)$ is an increasing function. Since $\ln 1 = 0$, we have $\exp(0) = 1$. Also, $\exp(x) \to \infty$ as $x \to \infty$, and $\exp(x) \to 0$ as $x \to -\infty$. The graph of $y = \exp(x)$ is shown in Figure 6.7.

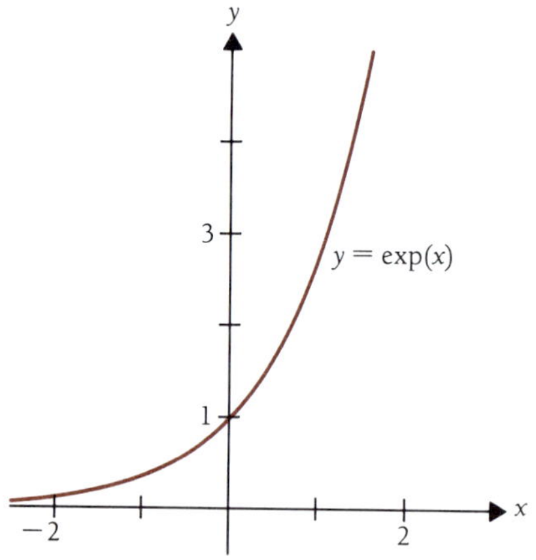

FIGURE 6.7 The graph of $y = \exp(x)$

The number $\exp(1)$ is of special importance, so we give it a special name.

DEFINITION 6.3 The number

$$e = \exp(1)$$

is called the **base of the natural logarithm**.

Since $e = \exp(1)$, we have

$$\ln e = \int_1^e \frac{1}{t}\,dt = 1. \tag{3}$$

Interpreted geometrically, equation (3) says that there is one unit of area under $y = 1/t$ over $[1, e]$ (see Figure 6.8). Clearly, $e > 2$.

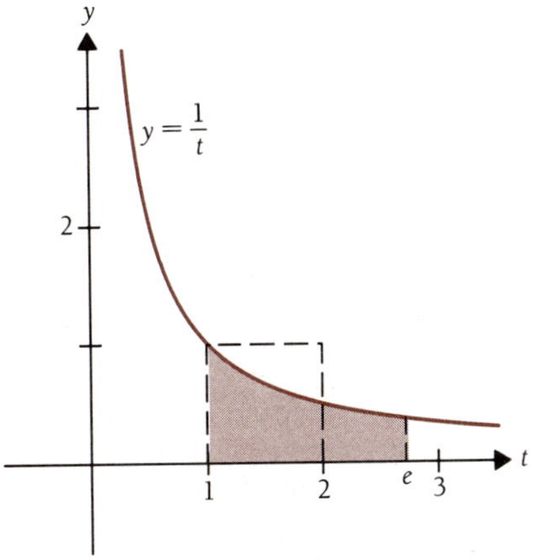

FIGURE 6.8 Graphic illustration of e

To get a better grasp of the value of e, we can express e as a limit.

THEOREM 6.1 If n represents an integer, then

$$e = \lim_{n\to\infty}\left(1 + \frac{1}{n}\right)^n.$$

Proof We use the properties of the natural logarithm to good advantage here. We begin with the natural logarithm of the limit, and use the fact that the logarithm is continuous to interchange it with the limit:

$$\ln\left[\lim_{n\to\infty}\left(1 + \frac{1}{n}\right)^n\right] = \lim_{n\to\infty} \ln\left(1 + \frac{1}{n}\right)^n$$

$$= \lim_{n\to\infty} n \ln\left(1 + \frac{1}{n}\right)$$

$$\lim_{n\to\infty} n \ln\left(1+\frac{1}{n}\right) = \lim_{n\to\infty} \frac{\ln(1+1/n)}{1/n}$$
$$= \lim_{1/n\to 0} \frac{\ln(1+1/n) - \ln 1}{1/n}.$$

This last expression is the derivative of $\ln x$ at $x = 1$, which is 1; so the natural logarithm of the limit is 1. Hence, the limit is $\exp(1) = e$, and the proof is done.

With this limit available, we may estimate e to a high degree of accuracy. For example, if $n = 10^6$, then

$$\left(1+\frac{1}{n}\right)^n \approx 2.71828.$$

This is the value of e correct to five decimal places.

The next theorem begins to show the usefulness of the number e.

THEOREM 6.2 If r is a rational number, then

$$\exp(r) = e^r.$$

Proof Let $y = \exp(r)$. Then

$$\ln y = r = r \cdot 1 = r \ln e = \ln e^r,$$

and hence, $y = e^r$ (see Problem 45 of Section 6.1).

Theorem 6.2 suggests that $\exp(x) = e^x$ for all real numbers x. However, we have not defined irrational exponents; e^x is meaningful only if x is rational. We therefore state the following definition.

DEFINITION 6.4 For any real number x,

$$e^x = \exp(x).$$

As an immediate consequence of this definition, we have, for any x,

$$\ln e^x = x, \tag{4}$$

by property (1). We also have, for any $x \in (0, \infty)$,

$$e^{\ln x} = x, \tag{5}$$

because if $y = \ln x$, then $x = \exp(y) = e^y = e^{\ln x}$.

We can now define an irrational power of any positive real number.

DEFINITION 6.5 If b is a positive real number, then for any real number x,

$$b^x = e^{(\ln b)x}.$$

Note that if x is rational, then Definition 6.5 holds, since

$$e^{(\ln b)x} = e^{\ln(b^x)} = b^x,$$

by property (5).

We can now prove that for any real number a,

$$\ln x^a = a \ln x. \tag{6}$$

By Definition 6.5, we have $x^a = e^{a \ln x}$; so $\ln x^a = \ln(e^{a \ln x}) = a \ln x$, by property (4).

The Exponential Function

DEFINITION 6.6 The function $y = \exp(x) = e^x$ is called **the exponential function with base *e***, or simply **the exponential function**.

By Definition 6.4, the exponential function is the inverse of the natural logarithmic function. Thus, statement (1) also can be written

$$y = e^x \qquad \text{if and only if} \qquad x = \ln y, \tag{7}$$

and we can restate formula (2) as

$$D_x e^x = e^x. \tag{8}$$

The chain rule form of equation (8) is

$$d(e^u) = e^u\, du, \tag{9}$$

which also can be written as

$$D_x e^{f(x)} = f'(x)e^{f(x)}. \tag{10}$$

The following examples illustrate the use of these differentiation formulas.

■ **Example 6.12** Find the derivative of $y = e^{3x}$.

Solution Using formula (10), we have

$$y' = 3e^{3x}.$$

■

■ **Example 6.13** Find y' when $y = e^x \ln x$.

Solution By formula (8) and the product rule,

$$y' = e^x \ln x + e^x\left(\frac{1}{x}\right)$$
$$= e^x\left(\ln x + \frac{1}{x}\right). \qquad \blacksquare$$

■ **Example 6.14** Find the derivative of $f(x) = \sqrt{1 + e^x}$.

Solution Using the chain rule, we write

$$f(x) = (1 + e^x)^{1/2},$$
$$f'(x) = \frac{1}{2}(1 + e^x)^{-1/2}(e^x)$$
$$= \frac{1}{2} \cdot \frac{e^x}{\sqrt{1 + e^x}}. \qquad \blacksquare$$

The Antiderivative of e^x

From formula (9), we have the antiderivative formula

$$\int e^u\, du = e^u + C. \tag{11}$$

■ **Example 6.15** Find $\int e^{3x}\, dx$.

Solution We substitute $u = 3x$, so $du = 3\, dx$, or $dx = \frac{1}{3}\, du$. Hence,

$$\int e^{3x}\, dx = \int e^u\left(\frac{1}{3}\right) du$$
$$= \frac{1}{3}\int e^u\, du$$
$$= \frac{1}{3} e^u + C = \frac{1}{3} e^{3x} + C. \qquad \blacksquare$$

■ **Example 6.16** Find $\int_0^1 xe^{1-x^2}\, dx$.

Solution Letting $u = 1 - x^2$, we have $du = -2x\, dx$, or $x\, dx = -(\frac{1}{2})\, du$. Therefore,

$$\int_0^1 xe^{1-x^2}\, dx = \int_1^0 e^u\left(-\frac{1}{2}\right) du$$
$$= \frac{1}{2}\int_0^1 e^u\, du$$
$$= \frac{1}{2} e^u\Big|_0^1 = \frac{1}{2}(e - 1). \qquad \blacksquare$$

Example 6.17 Evaluate

$$\int \frac{e^x}{1+e^x}\,dx.$$

Solution If $u = 1 + e^x$, then $du = e^x\,dx$, and

$$\begin{aligned}\int \frac{e^x}{1+e^x}\,dx &= \int \frac{du}{u}\\ &= \ln|u| + C\\ &= \ln|1+e^x| + C\\ &= \ln(1+e^x) + C\end{aligned}$$

(because $1 + e^x$ is already positive). ■

Exercises 6.2

Differentiate.

1. $y = e^{2x}$
2. $y = e^{x^2}$
3. $y = e^{1-x}$
4. $y = e^{ax+b}$
5. $y = xe^{-x}$
6. $y = x^2e^{2x}$
7. $y = (x^2+1)e^{3x}$
8. $y = e^{2x}\ln x$
9. $y = e^x\cos 2x$
10. $y = e^{2x}\cos 3x$
11. $y = e^{3x}\sin x$
12. $y = e^{\tan x}$
13. $y = (x^2+4)e^{-x}$
14. $y = e^{-x^2}$
15. $y = e^{1+\ln x}$
16. $y = \ln(1+e^x)$
17. $y = \sec e^x$
18. $y = \sin 2e^x$
19. $y = e^{2\tan x}$
20. $y = \ln(\cos e^x)$
21. $xy = e^{x+y}$
22. $\ln x + e^y = 2x - y$
23. $x + y = e^{xy}$
24. $\ln(x+2y) = x^2$

Find the critical points and the extrema of each function. Also tell where each function is increasing, decreasing, concave upward, and concave downward; and give the inflection points.

25. $f(x) = e^{-x}$
26. $f(x) = e^{x/10}$
27. $f(x) = e^{-x^2}$
28. $f(x) = \dfrac{1}{1-e^{-x}}$

Integrate.

29. $\int e^{2x}\,dx$
30. $\int e^{x+1}\,dx$
31. $\int_0^2 e^{4x}\,dx$
32. $\int_{-1}^1 e^{1-x}\,dx$
33. $\int_0^1 e^{-x}\,dx$
34. $\int e^{ax+b}\,dx$
35. $\int_1^4 (e^{\sqrt{x}})x^{-1/2}\,dx$
36. $\int_0^1 xe^{x^2}\,dx$
37. $\int x^2e^{x^3}\,dx$
38. $\int (e^{1/x})\,x^{-2}\,dx$
39. $\int \dfrac{e^x}{e^x+4}\,dx$
40. $\int e^x\sqrt{e^x+2}\,dx$

41. $\int e^x \cos(e^x)\, dx$

42. $\int_0^{1/2} \frac{e^{\ln(1+2x)}}{1+2x}\, dx$

43. Use logarithmic differentiation to show that

$$d(u^a) = au^{a-1}\, du$$

for any real number a, if u is a positive function of x.

Differentiate.

44. $y = x^\pi$

45. $y = x^{\sqrt{2}}$

46. $y = (x+1)^e$

47. $y = e^{2x} + x^{2e}$

48. Show that the tangent line to the curve $y = e^x$ at point (x_0, y_0) has slope y_0.

49. Find the area of the region between $y = e^{-x}$ and the tangent line to $y = e^{-x}$ at $x = 0$, over $[0, 1]$.

A50. Find the point of inflection of the curve

$$y = \frac{LP}{P + (L-P)e^{-kx}}.$$

(L, P, and k are constants.)

SECTION 6.3

Other Exponential and Logarithmic Functions

In this section we discuss the properties of real exponents, and develop other exponential and logarithmic functions besides e^x and $\ln x$.

Properties of Real Exponents

In Definition 6.5, we defined b^x for any positive real number b and any real number x. Now we wish to show that arbitrary real exponents obey the same rules as rational exponents. We start by proving some properties of the exponential function e^x.

THEOREM 6.3

For any real numbers x and y:

(a) $e^{x+y} = e^x e^y$,   (b) $e^{x-y} = \frac{e^x}{e^y}$,   (c) $(e^x)^y = e^{xy}$.

Proof

(a) Let $z = e^x$ and $w = e^y$. Then $x = \ln z$ and $y = \ln w$, so $x + y = \ln z + \ln w = \ln(zw)$. Therefore, $e^{x+y} = zw = e^x e^y$.

(b) This proof is similar to the proof of part (a), and is left as an exercise.

(c) Let $z = e^x$. Then $x = \ln z$, so $xy = y \ln z = \ln z^y$. Hence, $e^{xy} = z^y = (e^x)^y$.

Now we can prove the familiar rules of exponents for any real exponents.

THEOREM 6.4

For any real numbers x and y and for any positive real numbers a and b:

(a) $b^{x+y} = b^x b^y$, (b) $b^{x-y} = \dfrac{b^x}{b^y}$, (c) $(b^x)^y = b^{xy}$,

(d) $(ab)^x = a^x b^x$, (e) $\left(\dfrac{a}{b}\right)^x = \dfrac{a^x}{b^x}$.

Proof

(a) Using Definition 6.5 and Theorem 6.3,

$$\begin{aligned} b^{x+y} &= e^{(x+y)\ln b} \\ &= e^{x\ln b + y\ln b} \\ &= e^{x\ln b}e^{y\ln b} \\ &= b^x b^y. \end{aligned}$$

(b)–(c) The proofs of these parts are similar, and are left as exercises.

(d)
$$\begin{aligned} (ab)^x &= e^{x\ln(ab)} \\ &= e^{x(\ln a+\ln b)} \\ &= e^{x\ln a + x\ln b} \\ &= e^{x\ln a}e^{x\ln b} \\ &= a^x b^x. \end{aligned}$$

(e) This part is also left as an exercise.

The following curious example illustrates some of these rules of exponents.

Example 6.18

Find an irrational number a and an irrational exponent p such that a^p is rational.

Solution

Of many possible solutions, the following is probably simplest, and certainly one of the most interesting: Consider $(\sqrt{2})^{\sqrt{2}}$. If this number is rational, then we have a solution with $a = p = \sqrt{2}$. If this number is irrational, let $a = (\sqrt{2})^{\sqrt{2}}$ and $p = \sqrt{2}$; then

$$a^p = [(\sqrt{2})^{\sqrt{2}}]^{\sqrt{2}} = (\sqrt{2})^{\sqrt{2}\sqrt{2}} = (\sqrt{2})^2 = 2,$$

and the problem is solved.

Note that we have a solution, but we do not know for sure at what stage we have the solution! This is called an *existence proof,* because a solution is shown to exist without actually displaying a solution. (Here we displayed it, but we do not know which one it is.) An existence proof is an example of a *nonconstructive method.* In many parts of mathematics, we may be able to show that something with a certain property exists, but have no clue as to how to construct such a thing. ■

The Exponential Function with Base b

If we think of x as a variable, then b^x is a function with domain $(-\infty, \infty)$.

DEFINITION 6.7

If b is a positive real number, the **exponential function with base b** is the function

$$f(x) = b^x.$$

First, suppose $b \neq 1$. Since b^x is defined for all real numbers x, the domain of b^x is indeed $(-\infty, \infty)$. Since $b^x = e^{x \ln b}$, the range of b^x is the same as the range of e^x, namely $(0, \infty)$.

We can find the derivative of b^x using either the definition $b^x = e^{x \ln b}$ or logarithmic differentiation. Using the former, we have

$$\begin{aligned} D_x b^x &= D_x e^{x \ln b} \\ &= (\ln b) e^{x \ln b} \\ &= (\ln b) b^x. \end{aligned}$$

In chain rule form, this formula is

$$d(b^u) = (\ln b) b^u \, du. \tag{1}$$

■ **Example 6.19**

Find $f'(x)$ if $f(x) = 2^x$.

Solution

By formula (1), $f'(x) = (\ln 2)2^x$. ■

■ **Example 6.20**

Differentiate $y = x(10^x)$.

Solution

By the product rule and formula (1),

$$y' = 10^x + x(\ln 10)10^x.$$ ■

If $b = 1$, then the exponential function 1^x has derivative 0, since $\ln 1 = 0$. Thus, 1^x is constant (see Figure 6.9). If $b > 1$, then $\ln b > 0$, and b^x has a positive derivative. Thus, b^x is an increasing function if $b > 1$ (see, for example, Figure

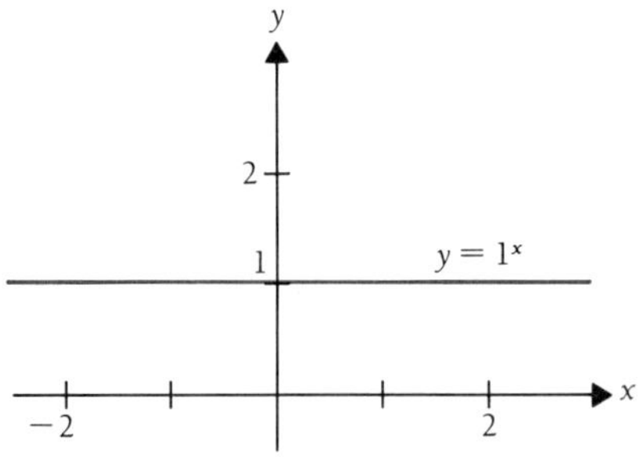

FIGURE 6.9 The exponential function with base 1 is constant

6.10). If $0 < b < 1$, then $\ln b < 0$, and b^x has a negative derivative. This makes b^x a decreasing function if $0 < b < 1$ (see, for example, Figure 6.11).

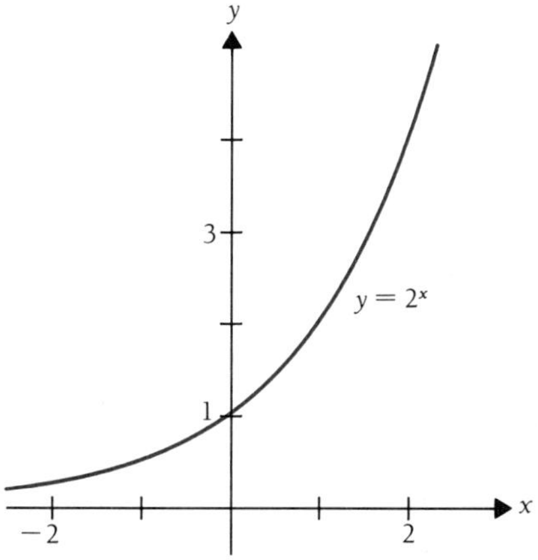

FIGURE 6.10 A typical exponential function with $b > 1$

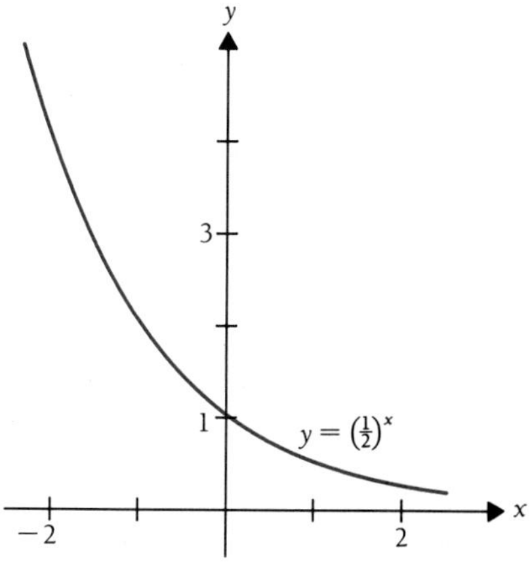

FIGURE 6.11 A typical exponential function with $0 < b < 1$

Formula (1) leads to the indefinite integral formula

$$\int b^u \, du = \frac{b^u}{\ln b} + C. \tag{2}$$

■ **Example 6.21** Find $\int 10^x \, dx$.

Solution

$$\int 10^x \, dx = \frac{10^x}{\ln 10} + C. \quad ■$$

■ **Example 6.22** Evaluate $\int e^x 2^{e^x} \, dx$.

Solution We substitute $u = e^x$, $du = e^x \, dx$:

$$\begin{aligned} \int e^x 2^{e^x} \, dx &= \int 2^u \, du \\ &= \frac{2^u}{\ln 2} + C \\ &= \frac{2^{e^x}}{\ln 2} + C. \end{aligned} \quad ■$$

The Logarithmic Function with Base b

If b is a positive real number different from 1, then the exponential function b^x is either increasing or decreasing, and so is one-to-one. Thus, b^x has an inverse function if $b \neq 1$. The inverses of the exponential functions are the logarithmic functions.

DEFINITION 6.8

If b is a positive real number and $b \neq 1$, the **logarithmic function with base b** is denoted by

$$f(x) = \log_b x,$$

and is defined by

$$y = \log_b x \qquad \text{if and only if} \qquad x = b^y.$$

Since $\log_b x$ is the inverse of b^x, we see that the domain of $\log_b x$ is $(0, \infty)$, and the range is $(-\infty, \infty)$.

Just as b^x can be expressed in terms of the exponential function with base e, or e^x, so can $\log_b x$ be expressed in terms of the logarithmic function with base e, or $\ln x$. The formula is

$$\log_b x = \frac{\ln x}{\ln b}, \tag{3}$$

which is established as follows: If $y = \log_b x$, then $x = b^y$ by Definition 6.8. Hence,

$$\ln x = \ln b^y = y \ln b,$$

so that

$$y = \frac{\ln x}{\ln b} = \log_b x.$$

With formula (3), we can easily find the derivative of $\log_b x$:

$$\begin{aligned}
\frac{d}{dx} \log_b x &= \frac{d}{dx} \frac{\ln x}{\ln b} \\
&= \frac{1}{\ln b} \frac{d}{dx} \ln x \\
&= \frac{1}{\ln b} \cdot \frac{1}{x} \\
&= \frac{1}{x \ln b}.
\end{aligned} \tag{4}$$

The chain rule form of this formula is

$$d(\log_b u) = \frac{du}{u \ln b}. \tag{5}$$

Example 6.23 Find $f'(x)$ if $f(x) = \log_2(3x+1)$.

Solution Using formula (4), we get

$$f'(x) = \frac{D_x(3x+1)}{(3x+1)\ln 2}$$
$$= \frac{3}{(3x+1)\ln 2}.$$

If $b > 1$, then $\ln b > 0$, and $\log_b x$ is an increasing function, by formula (4). Similarly, if $0 < b < 1$, then $\log_b x$ is a decreasing function (see, for example, Figures 6.12 and 6.13).

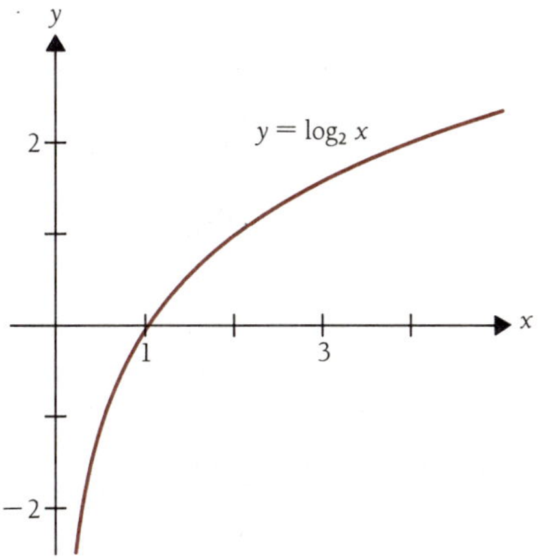

FIGURE 6.12 A typical logarithmic function with $b > 1$

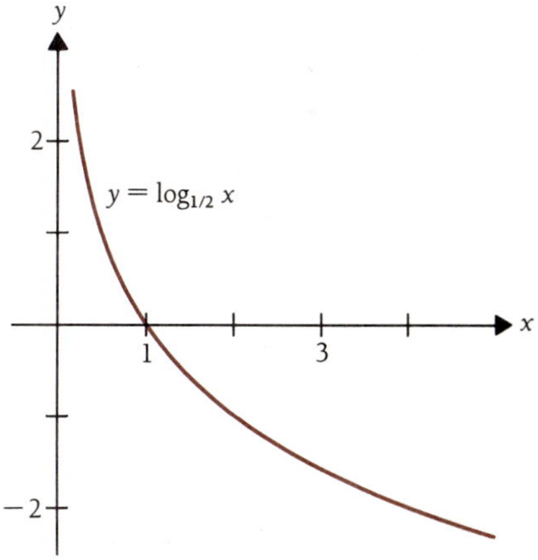

FIGURE 6.13 A typical logarithmic function with $0 < b < 1$

Since e is the base of the natural logarithm, it is also proper to write

$$\ln x = \log_e x.$$

Our notation $\ln x$ for $\log_e x$ is common, but not universal. Some people write $\log x$ for $\log_e x$, but others prefer to let $\log x$ stand for $\log_{10} x$, the **common logarithm** of x. In this book, we will continue to use $\ln x$ for $\log_e x$, and we will always specify the base in other cases.

It is easy to show that $\log_b x$ deserves to be called a logarithmic function, since it satisfies Definition 6.1. Even more than that, however, we can show that every logarithmic function is $\log_b x$ for some b. Recall from Section 6.1 that if $L(x)$ is any logarithmic function, then

$$L'(x) = \frac{1}{x} L'(1),$$

and $L'(1)$ cannot be 0. Suppose $L'(1) = m$; then

$$L'(x) = \frac{m}{x}.$$

But also the function $f(x) = m \ln x$ has derivative

$$f'(x) = \frac{m}{x},$$

so

$$L(x) = m \ln x + C$$

for some constant C. Since $L(1) = \ln 1 = 0$, we must have $C = 0$. Hence,

$$\begin{aligned} L(x) &= m \ln x \\ &= \frac{\ln x}{1/m} \\ &= \frac{\ln x}{\ln e^{1/m}} \\ &= \log_{e^{1/m}} x, \end{aligned}$$

as asserted. It is now clear that to say $L'(1) = 1$ is to say that $L(x) = \ln x$.

Variable Base and Exponent

Let f and g be differentiable functions with f positive. Then the function

$$y = [f(x)]^{g(x)}$$

is defined, and is differentiable. We find its derivative either by logarithmic differentiation or by using the identity

$$[f(x)]^{g(x)} = e^{g(x) \ln f(x)}.$$

Consider these examples.

■ **Example 6.24** Find the derivative of $y = x^{e^x}$, $x > 0$.

Solution We will use logarithmic differentiation:

$$\begin{aligned} \ln y &= \ln x^{e^x} \\ &= e^x \ln x, \\ \frac{1}{y} y' &= e^x \ln x + e^x \left(\frac{1}{x}\right), \\ y' &= e^x \left(\ln x + \frac{1}{x}\right) y. \end{aligned}$$

■

Example 6.25 Find $f'(x)$ if $f(x) = (x^2+1)^{x-1}$.

Solution We will set

$$f(x) = (x^2+1)^{x-1} = e^{(x-1)\ln(x^2+1)}.$$

Then

$$f'(x) = e^{(x-1)\ln(x^2+1)}\left[\ln(x^2+1) + (x-1)\frac{2x}{x^2+1}\right]$$
$$= (x^2+1)^{x-1}\left[\ln(x^2+1) + \frac{2x(x-1)}{x^2+1}\right]$$ ■

Exercises 6.3

Differentiate.

1. $y = 2^{-x}$
2. $y = 10^{3x}$
3. $y = (\frac{1}{2})^{x^2}$
4. $y = 4^{e^x}$
5. $y = \log_2(x^2+1)$
6. $y = \log_3(x-1)$
7. $y = \log_{10}(\sin x)$
8. $y = \log_{10}(\tan x)$
9. $y = x^x, \quad x > 0$
10. $y = (\ln x)^x, \quad x > 1$
11. $y = x^{\ln x}, \quad x > 0$
12. $y = (1+x)^{1/x}, \quad x > 0$
13. $y = 2^x + x^2 + (2x)^{x^2}, \quad x > 0$
14. $y^2 = x^{2x}, \quad x > 0$
15. $y = \ln(x + 3^y)$
16. $y = 2 + x^y$

Find the line tangent to the given curve at the given point.

17. $y = 3^x$; $(1, 3)$
18. $y = 10^{-x}$; $(2, 0.01)$
19. $y = x^x$; $(1, 1)$
20. $y = (\ln x)^{2x}$; $(e, 1)$

Find the critical points and extrema.

21. $f(x) = 2^{-x^2}$
22. $f(x) = \log_{10}(\sin x), \quad 0 < x < \pi$
23. $f(x) = x^{2x}$
C24. $f(x) = (x+4)^{4x}$

Integrate.

25. $\int 2^{-x}\,dx$
26. $\int 10^{5x}\,dx$
27. $\int 2^{\cos x}\sin x\,dx$
28. $\int e^x 2^x\,dx$
29. $\int e^x 2^{-x}\,dx$
30. $\int \frac{2^x}{1+2^x}\,dx$
31. $\int \frac{\log_2 x}{x}\,dx$
32. $\int \frac{1}{x\log_{10} x}\,dx$

Prove the following:

33. Theorem 6.3(b)
34. Theorem 6.4(b)
35. Theorem 6.4(c)
36. Theorem 6.4(e)
37. Show that $f(x) = \log_b x$ satisfies Definition 6.1.

38. Find a, p, and z such that $z = a^p$ with:
(a) a and p rational and z irrational **(b)** p and z rational and a irrational

39. Show that if $b^x = b^y$ and $b \neq 1$, then $x = y$.

40. Show that if $\log_b x = \log_b y$, then $x = y$.

SECTION 6.4

Exponential Growth and Decay

Exponential functions are very useful in modeling many real-world phenomena. Here, we present a simple example with many applications.

Exponential Growth

Let Q represent a physical quantity that grows at a rate proportional to Q itself. Then Q satisfies the differential equation

$$\frac{dQ}{dt} = kQ, \tag{1}$$

with k a positive constant. To express Q as a function of t, we solve equation (1) as follows: We rewrite equation (1) in differential notation, separate the variables, and then integrate:

$$\begin{aligned} \frac{1}{Q}\,dQ &= k\,dt, \\ \int \frac{1}{Q}\,dQ &= \int k\,dt, \\ \ln Q &= kt + C, \\ Q &= e^{kt+C}, \\ Q(t) &= Ae^{kt}, \end{aligned} \tag{2}$$

with $A = e^C$. This is the general solution of equation (1). We now suppose that at $t = 0$, $Q = Q_0$, the **initial value** of Q. Putting these values into equation (2), we get

$$Q_0 = Ae^0 = A,$$

and equation (2) becomes

$$Q(t) = Q_0 e^{kt}. \tag{3}$$

It is now evident why a quantity that grows at a rate proportional to the quantity itself is said to grow **exponentially**. Thus, equations (1) and (3) are called the **exponential growth model**. They are also called the **uninhibited growth model**, because $Q \to \infty$ as $t \to \infty$.

The typical example of a physical quantity that fits the exponential growth model is a natural population with unlimited room and food. The following examples illustrate exponential growth.

Example 6.26 Suppose that a certain bacterial culture grows at a rate proportional to the population itself. If there are 5000 bacteria initially, and there are 7500 bacteria after 1 hour, how many bacteria will there be after 10 hours?

Solution Let $Q(t)$ be the number of bacteria after t hours. Then $Q_0 = Q(0) = 5000$, so equation (3) is

$$Q(t) = 5000e^{kt}.$$

Now, $Q(1) = 7500$, so

$$7500 = 5000e^{k},$$
$$e^{k} = \frac{7500}{5000} = \frac{3}{2}.$$

(From this last equation, we can approximate the value of k with a calculator.) To answer the question, we now have

$$\begin{aligned} Q(10) &= 5000e^{10k} \\ &= 5000(e^{k})^{10} \\ &= 5000\left(\frac{3}{2}\right)^{10} \approx 288{,}000. \end{aligned}$$

Thus, after 10 hours, there will be about 288,000 bacteria. ■

Example 6.27 Suppose that the world's human population fits the exponential growth model, and is observed to increase by 4% in 1 year. At that rate, in how many years will the world's population double?

Solution By supposition, equation (3) applies, so the world population $Q(t)$ at time t is given by

$$Q(t) = Q_0e^{kt}.$$

Now, $Q(1) = (1.04)Q_0$, because the population increases by 4% in 1 year. Hence,

$$(1.04)Q_0 = Q_0e^{k},$$
$$e^{k} = 1.04,$$

and

$$Q(t) = Q_0(e^{k})^{t} = Q_0(1.04)^{t}.$$

We seek the value of t for which $Q(t) = 2Q_0$. This is found by setting $Q(t) = 2Q_0$ and solving for t:

$$\begin{aligned} 2Q_0 &= Q_0(1.04)^t, \\ 2 &= (1.04)^t, \\ \ln 2 &= t \ln 1.04, \\ t &= \frac{\ln 2}{\ln 1.04} \approx 17.67 \text{ years.} \end{aligned}$$

Thus, the doubling time would be less than 18 years. ■

■ **Example 6.28** If an amount A_0 of money is deposited in a savings account and earns interest at the rate r, compounded continuously, then the rate at which the amount in the account grows is equal to r times the amount, or

$$\frac{dA}{dt} = rA.$$

Hence, the amount at any time is given by

$$A = A_0 e^{rt}. \tag{4}$$

(For more on compound interest, see Problem 21.) If a bank advertises that an account will double your money in 10 years, what is the interest rate?

Solution The advertisement translates into the statement, "$A = 2A_0$ when $t = 10$." Thus, from equation (4), we have

$$\begin{aligned} 2A_0 &= A_0 e^{10r}, \\ 2 &= e^{10r}, \\ \ln 2 &= 10r, \\ r &= \frac{\ln 2}{10} \approx 0.0693. \end{aligned}$$

Thus, the interest rate is about 6.93%. (See also the *rule of 72* in Problem 12 in the Additional Exercises at the end of this chapter.) ■

Exponential Decay

If a physical quantity Q decreases at a rate proportional to Q, then equation (1) holds, but the constant k is negative. We emphasize that fact by writing

$$\frac{dQ}{dt} = -kQ, \tag{5}$$

with this k positive. We solve equation (5) just as before, and the solution is

$$Q(t) = Q_0 e^{-kt}. \tag{6}$$

We call equations (5) and (6) the **exponential decay model**.

A prime example of exponential decay is radioactive decay. Some isotopes (forms) of certain elements are unstable, and their atoms spontaneously break up, or decay, sending out subatomic particles that can be detected by appropriate instruments, such as a Geiger counter. The rate at which a radioactive substance decays (as measured by the number of clicks of a Geiger counter per unit time, for instance) is proportional to the quantity of the substance present. Since the isotope is changing atom by atom into other elements as it decays, the quantity of the original substance is decreasing.

The **half-life** of a radioactive isotope is the amount of time it takes for half of a given quantity of the isotope to decay. Stated another way, if h is the half-life of an element, then the probability is $\frac{1}{2}$ that a particular atom of it will decay in time h. The half-lives of different isotopes vary widely, as Table 6.1 shows.

TABLE 6.1

Element	Half-life
Radium, Ra-226	1622 years
Carbon, C-14	5760 years
Uranium, U-238	4.5×10^9 years
Polonium, Po-210	140 days
Krypton, Kr-91	10 seconds

Example 6.29 Of a sample of 150.0 milligrams of pure radium (Ra-226), how much remains after 100 years?

Solution Equation (6) applies, with $Q_0 = 150.0$. To determine k, we use the fact that $Q = \frac{1}{2}Q_0$ when $t = 1622$. This gives us

$$\frac{1}{2}Q_0 = Q_0 e^{-k(1622)},$$

$$\frac{1}{2} = e^{-1622k},$$

$$\ln\frac{1}{2} = -1622k,$$

$$k = \frac{-\ln\frac{1}{2}}{1622}$$

$$= \frac{\ln 2}{1622} \approx 0.000427.$$

Hence,

$$Q(t) = (150.0)e^{-(0.000427)t}$$

is the quantity present at time t. When $t = 100$, we have

$$Q(100) = (150.0)e^{-0.0427} \approx 143.7 \text{ milligrams.}$$ ■

Example 6.30 Uranium (U-238) decays into lead. Because of its low melting point, lead is removed from molten lava, so that a fresh lava flow contains no lead. Thus, the amount of lead present in an ancient lava flow indicates the age of the lava. In a certain sample of lava, it is found that there are 4 atoms of lead per 100 atoms of U-238. How much time has passed since the volcanic eruption that produced the lava?

Solution The presence of 4 atoms of lead per 100 atoms of U-238 indicates that $\frac{4}{104} \approx 3.85\%$ of the uranium has decayed since the eruption. That is, if Q_0 is the amount of uranium at the time of the eruption, then the present amount is $(0.9615)Q_0$. Thus, from equation (6), we have

$$\begin{aligned} 0.9615Q_0 &= Q_0 e^{-kt}, \\ 0.9615 &= e^{-kt}, \\ \ln 0.9615 &= -kt, \\ t &= -\left(\frac{1}{k}\right) \ln 0.9615. \end{aligned}$$

To determine k, note that in any radioactive decay problem, if h is the half-life, then

$$kh = \ln 2,$$

as we saw in Example 6.29. In the present case (from Table 6.1),

$$k = \frac{\ln 2}{4.5 \times 10^9} \approx 1.54 \times 10^{-10}.$$

Hence,

$$t = \frac{-\ln 0.9615}{1.54 \times 10^{-10}} \approx 255 \text{ million years.}$$

■

Example 6.31 Because of inflation, the buying power of a currency is decaying at the rate of 8% per year. What will be its relative buying power after 5 years?

Solution We assume the buying power B is decreasing exponentially, and so is given by

$$B = B_0 e^{-kt}.$$

When $t = 1$, $B = (0.92)B_0$, so

$$\begin{aligned} (0.92)B_0 &= B_0 e^{-k}, \\ \ln 0.92 &= -k. \end{aligned}$$

Thus,

$$B = B_0 e^{t \ln 0.92} = B_0 (0.92)^t.$$

In 5 years,

$$B = B_0 (0.92)^5 \approx (0.66)B_0.$$

That is, the buying power in 5 years will be about 66% of what it is at present. ■

Other Growth Models Involving Exponential Functions

We mention here some other models of growth that involve exponential functions, but differ in certain major aspects.

■ **Example 6.32** The **limited growth model** describes the growth of a quantity when the rate of growth is proportional to the difference between the quantity and a theoretical limit. If L is the maximum possible value of Q, then the equation of the limited growth model is

$$\frac{dQ}{dt} = k(L - Q). \tag{7}$$

Express Q in terms of t and the initial quantity Q_0.

Solution To solve equation (7), we separate the variables and integrate:

$$\frac{1}{L-Q}\,dQ = k\,dt,$$

$$\int \frac{1}{L-Q}\,dQ = \int k\,dt,$$

$$-\ln(L-Q) = kt + C,$$

$$\ln(L-Q) = -kt - C,$$

$$L - Q = e^{-kt-C},$$

$$Q = L - e^{-C}e^{-kt}.$$

Now, $Q(0) = Q_0$, so

$$Q_0 = L - e^{-C},$$

$$e^{-C} = L - Q_0.$$

Thus,

$$Q(t) = L - (L - Q_0)e^{-kt}. \tag{8}$$ ■

■ **Example 6.33** Discovery of a new drug is published in a medical journal, and the percentage of doctors that know of the drug fits the limited growth model. If 25% of the doctors know of the drug after 6 weeks, when will 90% of them know of it? (Assume $Q_0 = 0$.)

Solution We use equation (8) with $Q_0 = 0$ and $L = 1 (= 100\%)$. This gives us

$$Q = 1 - e^{-kt}.$$

Now, $Q = 0.25$ when $t = 6$, so

$$0.25 = 1 - e^{-6k},$$

$$e^{-6k} = 0.75,$$

$$k = -\frac{1}{6}\ln 0.75.$$

Finally, we set $Q = 0.90$ and solve for t:

$$\begin{aligned} 0.90 &= 1 - e^{-kt}, \\ e^{-kt} &= 0.10, \\ t &= -\frac{1}{k} \ln 0.10 \\ &= \frac{6 \ln 0.10}{\ln 0.75} \approx 48 \text{ weeks.} \end{aligned}$$

■

■ **Example 6.34** The **logistic model**, or **inhibited growth model**, describes population growth in a crowded environment, where the rate of growth varies jointly with the size of the population and the room for further growth. If L is the theoretical limit of the population, also called the **carrying capacity** of the environment, then the population P changes at the rate

$$\frac{dP}{dt} = kP(L - P). \tag{9}$$

Given the initial population P_0, find P as a function of t.

Solution We begin by solving the differential equation (9). First, we separate the variables,

$$\frac{1}{P(L-P)}\, dP = k\, dt,$$

and then integrate,

$$\int \frac{1}{P(L-P)}\, dP = \int k\, dt = kt + C_1.$$

To integrate the left-hand side, we use an algebraic identity:

$$\begin{aligned} \int \frac{1}{P(L-P)}\, dP &= \int \frac{1}{L}\left(\frac{1}{P} + \frac{1}{L-P}\right) dP \\ &= \frac{1}{L}\,[\ln P - \ln(L-P)] + C_2 \\ &= \frac{1}{L} \ln\left(\frac{P}{L-P}\right) + C_2. \end{aligned}$$

Hence,

$$\begin{aligned} \frac{1}{L} \ln\left(\frac{P}{L-P}\right) + C_2 &= kt + C_1, \\ \ln\left(\frac{P}{L-P}\right) &= kLt + C, \end{aligned}$$

with $C = L(C_1 - C_2)$. Now, when $t = 0$, $P = P_0$, so $C = \ln[P_0/(L - P_0)]$, and

$$\ln\left(\frac{P}{L-P}\right) = kLt + \ln\left(\frac{P_0}{L-P_0}\right),$$

$$\frac{P(L-P_0)}{P_0(L-P)} = e^{kLt}. \tag{10}$$

Now it is a straightforward but tedious task to solve for P; we eventually get

$$P(t) = \frac{LP_0}{P_0 + (L-P_0)e^{-kLt}}. \tag{11}$$ ■

■ **Example 6.35** Ten beavers are released in a watershed that can theoretically support 2000 beavers. Two years later, the beaver population is 30. According to the logistic model, when will the population reach 1000?

Solution The population is given by equation (11) with $P_0 = 10$ and $L = 2000$:

$$P = \frac{2000(10)}{10 + 1990e^{-2000kt}} = \frac{2000}{1 + 199e^{-2000kt}}.$$

We now determine k by using equation (10) and setting $P = 30$ and $t = 2$:

$$\frac{30(2000-10)}{10(2000-30)} = e^{2000k(2)},$$

$$k = \frac{1}{4000}\ln\frac{597}{197}.$$

Then setting $P = 1000$ in equation (10), we solve for t:

$$\frac{1000(2000-10)}{10(2000-1000)} = e^{2000kt},$$

$$199 = e^{(1/2)t\ln(597/197)},$$

$$\ln 199 = \frac{t}{2}\ln\frac{597}{197},$$

$$t = \frac{2\ln 199}{\ln\frac{597}{197}} \approx 9.55.$$

Hence, the beaver population will reach 1000 in about 9.55 years. ■

Exercises 6.4

1. Suppose a certain bacterial culture grows exponentially. If the initial 10,000 bacteria increase to 22,500 in 1 day, how many will be present after 4 days?
2. In a certain pond, the number of lily pads doubles each day. Starting with one lily pad, how many days will it take for 150 lily pads to appear?

3. Suppose the population of a certain state grows at a rate proportional to the population. If the 1960 population was 650,000 and the 1970 population was 720,000, when will the state reach 1,000,000 in population?

4. Suppose the rabbit population of Tulare County grows exponentially. If the number of rabbits was estimated at 30,000 in 1975 and at 45,000 in 1978, approximately what was the rabbit population in 1981?

5. If \$1000 is deposited in an account that earns interest at the rate of 6%, compounded continuously, how much will be in the account after 10 years? (This amount is called the **future value** of \$1000 in 10 years at 6% compounded continuously.)

6. Find the future value of \$100 in 5 years at 12%, compounded continuously. (See Problem 5.)

7. If the half-life of a radioactive substance is 8 days, how much of a 10 gram sample will be left after 6 days?

8. If the half-life of a radioactive substance is 8 days, how long will it take for 60% of a given sample to decay?

9. Of 3 milligrams of a certain radioactive element, 2 milligrams remained after 6 hours. What is its half-life?

10. A certain kind of moth ball evaporates at a rate proportional to the volume of the moth ball, and loses half its volume in 6 weeks. If the moth ball must have at least 10% of its initial volume to be effective, how long is this kind of moth ball effective?

11. Carbon-14 (C-14) is a radioactive isotope of carbon that is created in the earth's atmosphere through the bombardment of nitrogen by cosmic rays. The rates of creation and decay of C-14 are presumably the same, so that the concentration of C-14 in the atmosphere is constant. Carbon-14 atoms become a part of all living things along with normal carbon. When a living creature or plant dies, the C-14 decays, while normal carbon does not. Suppose that a piece of charcoal has only $\frac{2}{3}$ the concentration of C-14 that is found in the atmosphere. How old is the charcoal?

12. One theory of the earth's formation predicts that a certain amount of uranium (U-238) was formed in terrestrial rocks. At present, only about 52% of the predicted amount is actually found; the other 48% has presumably decayed. According to this theory, how old is the earth?

13. At the birth of a child, its parents wish to deposit an amount in a savings account earning 7.5%, compounded continuously, which will grow to \$5000 by the child's 18th birthday. What amount should be deposited? (This amount is called the **present value** of \$5000 due in 18 years at 7.5% compounded continuously.)

14. Find the present value of a government bond earning 4.75%, compounded continuously, which will be worth \$1000 when it matures in 6 years. (See Problem 13.)

15. The value of a new stock fits the limited growth model. If the maximum possible value is \$30, and the stock is worth \$12 after 3 months, how much is it worth after 1 year? Assume $Q_0 = 0$.

16. A new product is introduced in a certain area, and an advertising campaign promotes it. The percentage of the target population (potential purchasers) that has purchased the product after t days of advertising fits the limited growth model.
    (a) After 10 days of advertising, 25% of the target population has purchased the product. Give the buying percentage as a function of the number of days of advertising.

(b) The advertising campaign is costing \$500 per day, and the company figures that it makes a profit of \$7200 for each percent of the target population that buys the product. For how many days should the advertising campaign be run in order to maximize profit?

17. In a certain chemical solution, a compound is formed that acts as a catalyst to bring about increased formation of the compound. But formation of the compound also slows down the rate of formation, because the elements that make up the compound are removed from solution. Thus, the rate of formation of the compound fits the logistic model. At one point, 10% of the maximum possible amount of the compound is present; 20 minutes later, 40% of the possible amount is present. How much of the possible amount of the compound is present after 2 hours?

18. In Dull City, population 2000, the spread of a rumor fits the logistic model. If 10 people hear a rumor at 8:00 AM, and 50 people have heard the rumor by 9:00 AM, at what time will 1200 people have heard the rumor?

19. In the logistic model, show that the maximum growth occurs when $P = \frac{1}{2}L$.

20. The table gives the yearly population of pheasants on a certain island after their introduction there. What is the approximate carrying capacity of the island for pheasants?

Year	Number	Year	Number
1977	8	1981	641
1978	30	1982	1194
1979	81	1983	1898
1980	282	1984	2590

21. **Compound Interest** This problem develops the theory of compound interest. Suppose an amount A_0 is deposited in an account that earns interest at the rate r, compounded n times per year.

(a) Show that the amount A_1 in the account at the end of the first compounding period is

$$A_1 = A_0\left(1 + \frac{r}{n}\right).$$

(b) Show that the amount A_2 in the account at the end of the second compounding period is

$$A_2 = A_0\left(1 + \frac{r}{n}\right)^2.$$

(c) Show that the amount $A(t)$ in the account after t years is

$$A(t) = A_0\left(1 + \frac{r}{n}\right)^{nt}.$$

(d) Interest is said to be **compounded continuously** if the limit as $n \to \infty$ is taken. Show that under continuous compounding, the amount in the account after t years is $A(t) = A_0 e^{rt}$.

22. **Newton's Law of Cooling** If an object is placed in a cooler environment, it cools at a rate proportional to the difference between its temperature and the temperature of the environment. Its temperature therefore fits the limited growth model, but with negative constant. If an object at temperature 220°F is placed in a room at 60°F, the temperature of the object is found to be 160°F after 5 minutes. When will the temperature of the object be 100°F?

23. Sketch the graphs of the curves:

(a) $Q = L - (L - Q_0)e^{-kt}$ (b) $P = \dfrac{LP_0}{P_0 + (L - P_0)e^{-kLt}}$

SECTION 6.5

L'Hôpital's Rules

Differentiation can be used to evaluate limits under certain circumstances. The use of the exponential and logarithmic functions also aids in the process. In this section, we will discuss *limit forms* and introduce *L'Hôpital's rules.*

Limit Forms

If a function f is continuous at c, then $\lim_{x\to c} f(x)$ is found by evaluation: $\lim_{x\to c} f(x) = f(c)$. If the function is not defined at c, or if we are trying to find the limit as $x \to \infty$, then evaluation in the usual sense is not possible. However, we can arrive at what we call **limit forms** by a process similar to evaluation.

For limits as $x \to c$, $x \to c^+$, or $x \to c^-$, the form of a limit is obtained by substituting c for x, even though the substitution may not yield a number. For $x \to \infty$ or $x \to -\infty$, we also substitute ∞ for x, and then make simplifications based on the following rules:

$$\infty + \infty = \infty, \qquad \infty \cdot \infty = \infty,$$
$$\infty \cdot (-\infty) = -\infty, \qquad (-\infty) \cdot (-\infty) = \infty,$$
$$\infty \pm a = \infty, \qquad \infty \cdot (\pm a) = \pm\infty.$$

Various forms are illustrated in Examples 6.36–6.43; the assertions made about the limits will be justified following the examples.

Example 6.36 The limit

$$\lim_{x\to 0} \frac{ax}{x}$$

has form 0/0; the limit exists and equals a. ■

Example 6.37 The limit

$$\lim_{x\to 2} \frac{x-2}{(x-2)^2}$$

has form 0/0; the limit does not exist. ■

Example 6.38 The limit

$$\lim_{x\to\infty} \frac{ax+1}{x-1}, \qquad a>0,$$

has form ∞/∞; the value of the limit is a [see Example 2.30(a)]. We call any of the forms $\pm\infty/\pm\infty$ the ∞/∞ form. ∎

Example 6.39 The limit

$$\lim_{x\to\infty} x \ln\left(1+\frac{a}{x}\right)$$

has form $\infty \cdot 0$, since $1+(a/x) \to 1$ as $x \to \infty$, and $\ln 1 = 0$. The value of the limit is a, as we will see later. ∎

Example 6.40 The limit

$$\lim_{x\to\infty} (x - \sqrt{x^2 - 2ax})$$

has form $\infty - \infty$; the limit exists and has value a, as you can verify by rationalizing [see Example 2.30(d)]. ∎

Example 6.41 The limit

$$\lim_{x\to 0^+} \frac{1}{x}$$

has form $1/0$; the limit is ∞. ∎

Example 6.42 The limit

$$\lim_{x\to\infty} \frac{1}{x}$$

has form $1/\infty$; the limit equals 0. ∎

Example 6.43 The limit

$$\lim_{x\to 0^+} (1+ax)^{1/x}$$

has form 1^∞, since $1/x \to \infty$ as $x \to 0^+$. In this case, the limit exists and equals e^a, which we will verify later. ∎

Indeterminate Forms

As we saw in Examples 6.36 and 6.37, a limit with form 0/0 may have any value whatever, or fail to exist. That is, it is not possible to infer much from the form of the limit as to the value of the limit. The same thing is true of the forms ∞/∞ (Example

G. F. A. de L'Hôpital (1661–1704) was a French nobleman, mathematician, and writer. His *Analyse des infiniment petits* was the first calculus textbook, published in 1696, and had a dominant effect on the calculus of the 18th century. He paid Johann Bernoulli a regular salary in return for the right to use Bernoulli's discoveries. L'Hôpital also wrote a treatise on conic sections (published in 1707) that became the standard textbook on the subject for almost a century, because of its pedagogical quality.

6.38), $\infty \cdot 0$ (Example 6.39), $\infty - \infty$ (Example 6.40), 1^∞ (Example 6.43), and some others. For that reason, we call the limit forms listed below **indeterminate forms**:

$$\frac{0}{0}, \quad \frac{\infty}{\infty}, \quad 0 \cdot \infty, \quad \infty - \infty, \quad 1^\infty, \quad 0^\infty, \quad \infty^0, \quad 0^0.$$

Limits with the indeterminate forms $0/0$ and ∞/∞ lend themselves to techniques of evaluation that use differentiation. These techniques are known as **L'Hôpital's rules**, after G. F. A. de L'Hôpital, who published them in 1696. The rules were actually discovered, however, by his teacher, Johann Bernoulli.

L'Hôpital's 0/0 Rule

We will state L'Hôpital's 0/0 rule for limits as $x \to c^+$, but it also holds for limits as $x \to c$, $x \to c^-$, $x \to \infty$, and $x \to -\infty$.

THEOREM 6.5 Let functions f and g be differentiable in the interval (c, b), with g and g' nonzero in (c, b), such that $\lim_{x\to c^+} f(x) = \lim_{x\to c^+} g(x) = 0$. Then

$$\lim_{x\to c^+} \frac{f(x)}{g(x)} = \lim_{x\to c^+} \frac{f'(x)}{g'(x)},$$

in the sense that if the second limit exists, then the first limit also exists and is equal to the second limit.

Proof Define (or redefine) $f(c) = 0$ and $g(c) = 0$. For $z \in (c, x)$, define the function h on the interval $[c, z]$ by

$$h(x) = f(x) - \frac{f(z)}{g(z)}\, g(x). \tag{1}$$

Note that h is well-defined, because $g(z) \neq 0$. Since f and g are differentiable in (c, b), they are differentiable in (c, z) and continuous in $(c, z]$. Since $f(x) \to 0 = f(c)$ and $g(x) \to 0 = g(c)$ as $x \to c^+$, f and g are also continuous at c. Thus, h is continuous on $[c, z]$ and differentiable in (c, z). Moreover, $h(z) = h(c) = 0$ (verify this). Hence, by Rolle's theorem (Theorem 4.2), there is some $t \in (c, z)$ such that

$$h'(t) = f'(t) - \frac{f(z)}{g(z)} g'(t) = 0, \tag{2}$$

$$\frac{f(z)}{g(z)} = \frac{f'(t)}{g'(t)}. \tag{3}$$

We can rewrite equation (2) as (3), because $g'(t) \neq 0$. Now, $c < t < z < x$, so $t \to c^+$ as $z \to c^+$. Hence,

$$\lim_{x \to c^+} \frac{f(x)}{g(x)} = \lim_{z \to c^+} \frac{f(z)}{g(z)}$$
$$= \lim_{t \to c^+} \frac{f'(t)}{g'(t)},$$

as asserted. (We use the fact that t and x are dummy variables in the last step.)

Applications of the 0/0 Rule

■ **Example 6.44** Using L'Hôpital's 0/0 rule, verify that

$$\lim_{x \to 0} \frac{\sin x}{x} = 1.$$

Solution Since this limit has form 0/0, the rule applies. Thus,

$$\lim_{x \to 0} \frac{\sin x}{x} = \lim_{x \to 0} \frac{\cos x}{1} = \frac{\cos 0}{1} = 1.$$ ■

The tendency of mathematical talent to run in families is most clearly illustrated by the Bernoulli family of Switzerland.

Jakob (or Jacque) Bernoulli (1654–1705) founded the study of the calculus of variations. This is much like finding an extremum of a function, but instead of finding a number that maximizes or minimizes a function, one must find a function that maximizes or minimizes an integral. For example, one might ask for the curve of fixed length that encloses the greatest area. Jakob's other great contribution was *Ars Conjectandi* (1713), a book containing the first systematic approach to combinatorics.

Johann (or Jean) Bernoulli (1667 – 1748) learned calculus from his brother Jakob. Both brothers carried on an extensive correspondence with Leibniz. Johann's first professional position was tutor to the Marquis de L'Hôpital.

Johann's zeal for mathematics was perhaps exceeded only by his lack of tact, which kept him in constant controversy. Johann was one of the most vehement in arguing that Leibniz had priority over Newton. Johann and Jakob worked on the same ideas, and the letters between them bristled with accusations of plagiarism and theft. The bad feelings between the brothers became widely known, and Johann's sons Nikolaus and Daniel made special efforts to display brotherly kindness, to compensate.

Johann's generous treatment of Euler, whom he tutored free of charge, contrasts strangely with his treatment of his own son, Daniel (1700 – 1782). When father and son competed for the Paris prize in mathematics, Daniel's victory was rewarded by expulsion from the family home. Daniel spent some time in Russia, but disliking the climate, returned to Basel, Switzerland, as professor of anatomy. This required some medical research to be squeezed in among his mathematical researches, but finally, at the age of 50, he became professor of physics. His most important contributions relate to hydrodynamics.

The Bernoullis were largely responsible for the early spread of calculus in Europe as Leibniz formulated it, and they demonstrated conclusively the vast range of applications of the new subject.

Johann and Jakob were considered the two greatest mathematicians of their generation. The Bernoulli family produced a dozen members who achieved mathematical prominence during the 17th, 18th, and 19th centuries.

■ **Example 6.45** Evaluate

$$\lim_{x\to 0} \frac{x^3}{x - \sin x}.$$

Solution This limit has form 0/0, so we may apply L'Hôpital's rule:

$$\lim_{x\to 0} \frac{x^3}{x - \sin x} = \lim_{x\to 0} \frac{3x^2}{1 - \cos x},$$

provided the last limit exists. But it also has form 0/0, so we apply the rule again:

$$\lim_{x\to 0} \frac{3x^2}{1 - \cos x} = \lim_{x\to 0} \frac{6x}{\sin x},$$

provided this limit exists. One more application of L'Hôpital's rule is in order:

$$\lim_{x\to 0} \frac{6x}{\sin x} = \lim_{x\to 0} \frac{6}{\cos x} = \frac{6}{1} = 6.$$

Thus, the original limit has value 6. ■

■ **Example 6.46** Evaluate

$$\lim_{x\to 0} \frac{\ln(1+x)}{1+x}.$$

Solution This limit has form 0/1, which is not an indeterminate form; therefore, we may not apply L'Hôpital's rule. The value of the limit is 0, which is found by evaluation, since the function is continuous at 0. (Erroneous application of L'Hôpital's rule in this case would have given the incorrect value of 1 for the limit.) ■

L'Hôpital's ∞/∞ Rule

A second form of L'Hôpital's rule applies to limits of the form ∞/∞. Again, we will state the rule for limits as $x \to c^+$, but it also holds for all other limits as well.

THEOREM 6.6 Let functions f and g be differentiable in the interval (c, b), such that $\lim_{x\to c^+} f(x) = \lim_{x\to c^+} g(x) = \infty$. Then

$$\lim_{x\to c^+} \frac{f(x)}{g(x)} = \lim_{x\to c^+} \frac{f'(x)}{g'(x)}$$

in the sense that if the second limit exists, then the first limit also exists and is equal to the second limit.

We will omit the proof of Theorem 6.6, but its application is the same as for the L'Hôpital 0/0 rule.

■ **Example 6.47** Evaluate

$$\lim_{x\to\infty} \frac{x+3}{2x-5}.$$

Solution Here, we have the form ∞/∞, so we apply L'Hôpital's ∞/∞ rule:

$$\lim_{x\to\infty} \frac{x+3}{2x-5} = \lim_{x\to\infty} \frac{1}{2} = \frac{1}{2}.$$ ■

■ **Example 6.48** Find the limit:

$$\lim_{x\to\infty} \frac{2x^2+3}{x^2-2x}.$$

Solution Again, we have the form ∞/∞, so we apply Theorem 6.6—twice, as it turns out:

$$\lim_{x\to\infty} \frac{2x^2+3}{x^2-2x} = \lim_{x\to\infty} \frac{4x}{2x-2}$$

$$= \lim_{x\to\infty} \frac{4}{2} = 2.$$ ■

Other Indeterminate Forms

Usually, indeterminate forms other than 0/0 and ∞/∞ can be rewritten in those forms. Consider the following examples.

Example 6.49 Evaluate

$$\lim_{x\to 0^+}\left(\frac{1}{x}-\frac{1}{\sin x}\right).$$

Solution This limit has form $\infty - \infty$. Combining the terms into a single fraction puts it into the 0/0 form, to which we apply L'Hôpital's rule twice:

$$\begin{aligned}\lim_{x\to 0^+}\left(\frac{1}{x}-\frac{1}{\sin x}\right) &= \lim_{x\to 0^+}\frac{\sin x - x}{x\sin x}\\ &= \lim_{x\to 0^+}\frac{\cos x - 1}{x\cos x + \sin x}\\ &= \lim_{x\to 0^+}\frac{-\sin x}{\cos x - x\sin x + \cos x}\\ &= \frac{0}{2} = 0.\end{aligned}$$

■

Example 6.50 Find $\lim_{x\to 0^+} x\ln x$.

Solution Since $\ln x \to -\infty$ as $x \to 0^+$, this limit has form $0(-\infty)$. But it is easily rewritten in the ∞/∞ form, to which we apply L'Hôpital's rule:

$$\begin{aligned}\lim_{x\to 0^+} x\ln x &= \lim_{x\to 0^+}\frac{\ln x}{1/x}\\ &= \lim_{x\to 0^+}\frac{1/x}{-1/x^2}\\ &= \lim_{x\to 0^+}(-x) = 0.\end{aligned}$$

■

Example 6.51 Evaluate

$$\lim_{x\to\infty}\left(1+\frac{a}{x}\right)^x.$$

Solution This limit has form 1^∞. We will rewrite it using the identity $u^z = e^{z\ln u}$, and then use the continuity of the exponential function and L'Hôpital's 0/0 rule:

$$\lim_{x\to\infty}\left(1+\frac{a}{x}\right)^x = \lim_{x\to\infty} e^{x\ln(1+a/x)}$$

$$
\begin{aligned}
\lim_{x\to\infty} e^{x\ln(1+a/x)} &= \exp\left[\lim_{x\to\infty} x\ln\left(1+\frac{a}{x}\right)\right] \\
&= \exp\left[\lim_{x\to\infty} \frac{\ln\left(1+\dfrac{a}{x}\right)}{1/x}\right] \\
&= \exp\left(\lim_{x\to\infty} \frac{\dfrac{-a/x^2}{1+a/x}}{-1/x^2}\right) \\
&= \exp\left(\lim_{x\to\infty} \frac{a}{1+a/x}\right) \\
&= e^a.
\end{aligned}
$$

■

■ **Example 6.52** Evaluate $\lim_{x\to 0^+}(1+ax)^{1/x}$.

Solution This limit has form 1^∞. Proceeding as in Example 6.51, we have

$$
\begin{aligned}
\lim_{x\to 0^+}(1+ax)^{1/x} &= \lim_{x\to 0^+} \exp\left[\frac{1}{x}\ln(1+ax)\right] \\
&= \exp\left[\lim_{x\to 0^+} \frac{\ln(1+ax)}{x}\right] \\
&= \exp\left[\lim_{x\to 0^+} \frac{a/(1+ax)}{1}\right] \\
&= e^a.
\end{aligned}
$$

■

■ **Example 6.53** Evaluate $\lim_{x\to 0^+} x^x$.

Solution This limit has form 0^0. Again, we use the identity $u^z = e^{z\ln u}$ to rewrite it:

$$
\begin{aligned}
\lim_{x\to 0^+} x^x &= \lim_{x\to 0^+} e^{x\ln x} \\
&= \exp\left(\lim_{x\to 0^+} x\ln x\right) \\
&= e^0 = 1
\end{aligned}
$$

by Example 6.50. ■

■ **Example 6.54** Find

$$\lim_{x\to 0^+}\left(1+\frac{1}{x}\right)^x.$$

Solution Here, we have the form ∞^0. Rewriting it, we may use L'Hôpital's ∞/∞ rule:

$$\lim_{x\to 0^+}\left(1+\frac{1}{x}\right)^x = \lim_{x\to 0^+} e^{x\ln(1+1/x)}$$

$$\begin{aligned}\lim_{x\to 0^+} e^{x\ln(1+1/x)} &= \exp\left[\lim_{x\to 0^+} \frac{\ln(1+1/x)}{1/x}\right]\\ &= \exp\left[\lim_{x\to 0^+} \frac{(-1/x^2)/(1+1/x)}{-1/x^2}\right]\\ &= \exp\left(\lim_{x\to 0^+} \frac{x}{x+1}\right)\\ &= e^0 = 1.\end{aligned}$$

■

Exercises 6.5

Find the limit.

1. $\lim_{x\to 3} \dfrac{x^2-9}{x-3}$
2. $\lim_{x\to 3} \dfrac{\frac{1}{3}-1/x}{x-3}$
3. $\lim_{x\to 0} \dfrac{1-e^x}{\sin x}$
4. $\lim_{x\to 0} \dfrac{\sin x}{\sqrt{1-\cos x}}$
5. $\lim_{x\to\infty} \dfrac{\ln(1+1/x)}{\ln(1+2/x)}$
6. $\lim_{x\to\pi/2} \dfrac{x-\pi/2}{\cos x}$
7. $\lim_{x\to\infty} \dfrac{3x^2+4x-1}{x^2-2}$
8. $\lim_{x\to\infty} \dfrac{\ln x}{\sqrt{x}}$
9. $\lim_{x\to\infty} \dfrac{\ln(\ln x)}{\ln x}$
10. $\lim_{x\to\infty} \dfrac{\ln x}{x}$
11. $\lim_{x\to\infty} \dfrac{(\ln x)^2}{x}$
12. $\lim_{x\to\infty} \dfrac{(\ln x)^3}{x}$
13. $\lim_{x\to\infty} \dfrac{x^{10}}{e^x}$
14. $\lim_{x\to\infty} \dfrac{x}{\sqrt{1+x}}$
15. $\lim_{x\to 0^+} \dfrac{1}{x\ln x}$
16. $\lim_{x\to\infty} xe^{-x}$
17. $\lim_{x\to\infty} xe^{-\ln(x+1)}$
18. $\lim_{x\to\infty} e^{-x}\ln x$
19. $\lim_{x\to\pi/2} \left(x-\dfrac{\pi}{2}\right)\tan x$
20. $\lim_{x\to-\infty} e^x\ln(1-x)$
21. $\lim_{x\to 0^+} (\sin x)^x$
22. $\lim_{x\to 0^+} (\cos x)^{1/x}$
23. $\lim_{x\to 0^+} (1-x)^{1/x}$
24. $\lim_{x\to 0^-} (1+x)^{1/x}$
25. $\lim_{x\to\infty} (1+x)^{1/x}$
26. $\lim_{x\to 0^+} (2x+1)^{1/x}$
27. $\lim_{x\to 0^+} x^{1/x}$
28. $\lim_{x\to 0^+} (\sin x)^{1/x}$
29. $\lim_{x\to 0^+} (x^2+2x)^{1/\ln x}$
30. $\lim_{x\to 0^+} (x^2+ax)^{1/x}$
31. $\lim_{x\to\infty} (x+1)^{1/x^2}$
32. $\lim_{x\to\infty} \left(1+\dfrac{1}{x}\right)^{\ln x}$
33. $\lim_{x\to 3} \left(\dfrac{1}{x-3}\displaystyle\int_3^x \sin t^2\,dt\right)$
34. $\lim_{x\to 0} \left(\dfrac{1}{x}\displaystyle\int_0^x \sin t\,dt\right)$
35. Show that L'Hôpital's rules will not help with the limit

$$\lim_{x\to\infty} \frac{\sqrt{3x-1}}{\sqrt{x+1}}.$$

Find the limit another way.

36. Construct the graph of $y = xe^x$.
37. Construct the graph of $y = x^x$.
38. Let the tangent line to the curve $y = x^2$ at point (a, a^2) meet the y-axis at $(0, b)$. Find $\lim_{a\to\infty}(a/b)$.

SECTION 6.6

The Inverse Trigonometric Functions

Because the trigonometric functions are periodic, they are far from being one-to-one functions. Hence, they do not have inverses that are functions. However, the domains of these functions can be suitably restricted so that one-to-one functions are obtained.

To see how it is done, consider the sine function, whose graph is shown in Figure 6.14. The portion of the graph that lies over the interval $[-\pi/2, \pi/2]$ is the graph of a one-to-one function, by the horizontal line test. The same is true of other portions of the graph, but we choose the interval $[-\pi/2, \pi/2]$ to be what we call the **principal domain** of the sine function. Note that all possible values of the sine function are assumed over its principal domain (see Figure 6.15).

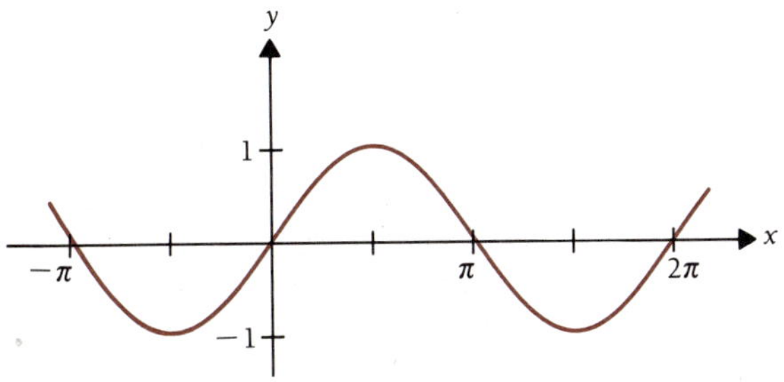

FIGURE 6.14 The graph of $y = \sin x$

The **inverse sine function** is now defined to be the inverse of the restriction of the sine function to its principal domain. The graph of the inverse sine function is shown in Figure 6.16.

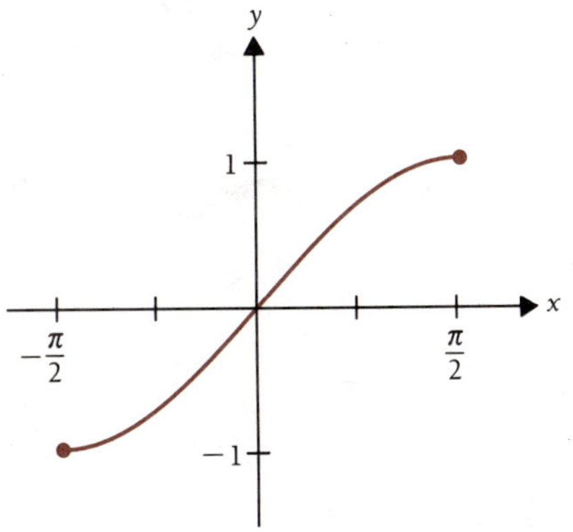

FIGURE 6.15 The graph of the sine function, restricted to its principal domain

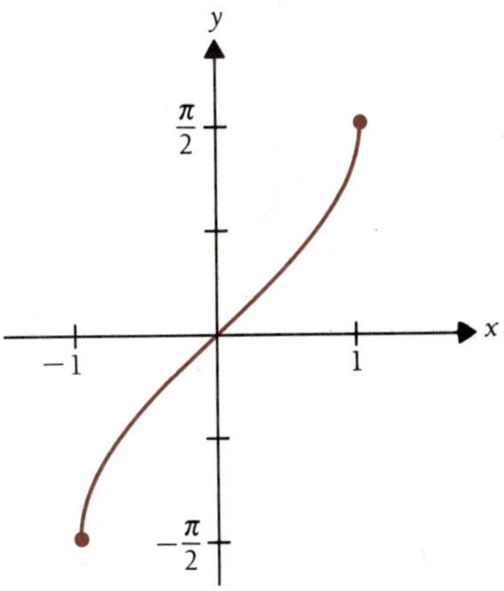

FIGURE 6.16 The graph of $y = \arcsin x$

Thus, the inverse sine function has domain $[-1, 1]$ and range $[-\pi/2, \pi/2]$. We will use either $\mathbf{sin^{-1}\,x}$ or $\mathbf{arcsin}\ \boldsymbol{x}$ to represent the inverse sine function. Summarizing, we can say that

$$y = \mathbf{arcsin}\ \boldsymbol{x} \quad \text{if and only if} \quad \sin y = x \quad \text{and} \quad y \in [-\pi/2, \pi/2]. \tag{1}$$

In a similar way, principal domains are chosen for the other trigonometric functions, and their inverses are defined by the following statements:

$$y = \mathbf{arccos}\ \boldsymbol{x} \quad \text{if and only if} \quad \cos y = x \quad \text{and} \quad y \in [0, \pi], \tag{2}$$

$$y = \mathbf{arctan}\ \boldsymbol{x} \quad \text{if and only if} \quad \tan y = x \quad \text{and} \quad y \in (-\pi/2, \pi/2), \tag{3}$$

$$y = \mathbf{arccot}\ \boldsymbol{x} \quad \text{if and only if} \quad \cot y = x \quad \text{and} \quad y \in (0, \pi), \tag{4}$$

$$y = \mathbf{arcsec}\ \boldsymbol{x} \quad \text{if and only if} \quad \sec y = x \quad \text{and} \quad y \in [-\pi, -\pi/2) \cup [0, \pi/2), \tag{5}$$

$$y = \mathbf{arccsc}\ \boldsymbol{x} \quad \text{if and only if} \quad \csc y = x \quad \text{and} \quad y \in (-\pi, -\pi/2] \cup (0, \pi/2]. \tag{6}$$

In each case, the range of the inverse function is the principal domain of the original function. The notations $\tan^{-1}x$ for arctan x, $\sec^{-1}x$ for arcsec x, etc., are also used. Some authors capitalize the initial letters of these functions. We should also note that there is not universal agreement as to what the principal domains should be in the cases of the inverse secant and inverse cosecant functions; some prefer $[0, \pi/2) \cup (\pi/2, \pi]$ for arcsec x and $[-\pi/2, 0) \cup (0, \pi/2]$ for arccsc x.

The graphs of the inverse trigonometric functions as we have defined them are shown in Figures 6.16–6.21.

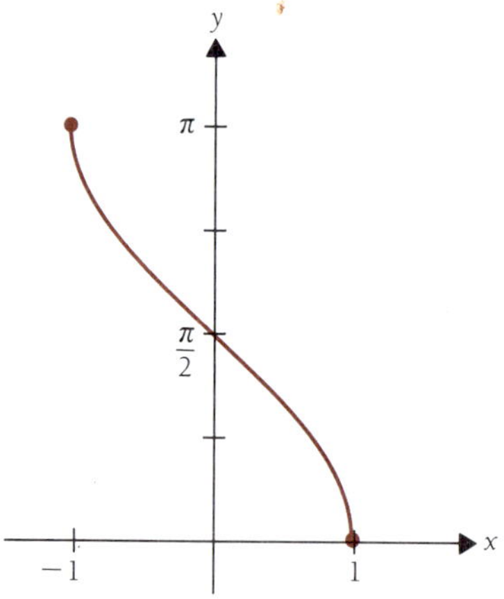

FIGURE 6.17 The graph of $y = \arccos x$

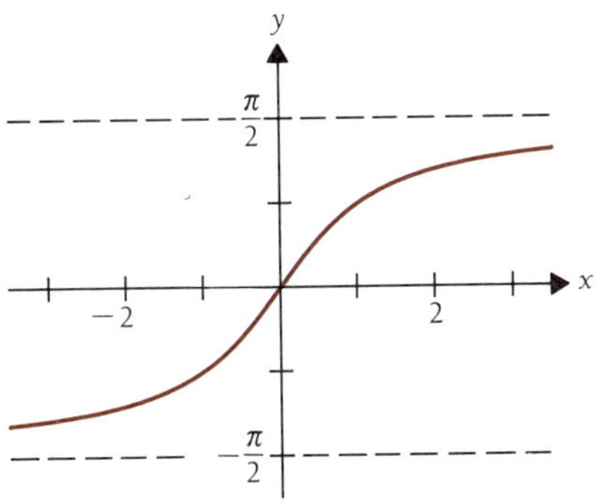

FIGURE 6.18 The graph of $y = \arctan x$

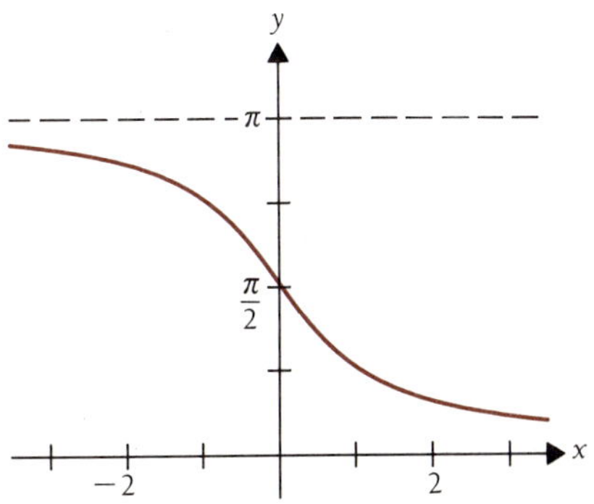

FIGURE 6.19 The graph of $y = \operatorname{arccot} x$

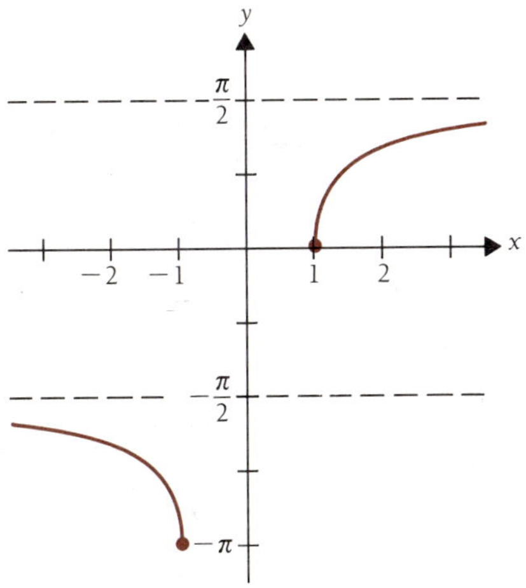

FIGURE 6.20 The graph of $y = \operatorname{arcsec} x$

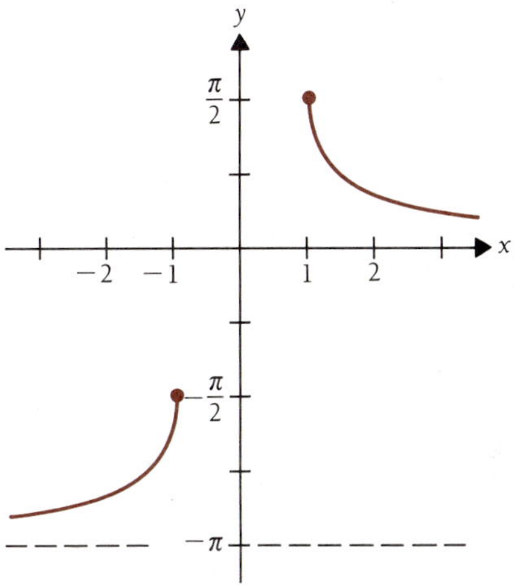

FIGURE 6.21 The graph of $y = \operatorname{arccsc} x$

Derivatives of the Inverse Trigonometric Functions

To find the derivatives of the inverse trigonometric functions, we use implicit differentiation (or the inverse function rule, Theorem 3.12). The results are summarized in the following theorem.

THEOREM 6.7 The inverse trigonometric functions are differentiable in the interiors of their domains, and

$$d(\arcsin u) = \frac{1}{\sqrt{1-u^2}}\,du, \tag{7}$$

$$d(\arccos u) = \frac{-1}{\sqrt{1-u^2}}\,du, \tag{8}$$

$$d(\arctan u) = \frac{1}{1+u^2}\,du, \tag{9}$$

$$d(\operatorname{arccot} u) = \frac{-1}{1+u^2}\,du, \tag{10}$$

$$d(\operatorname{arcsec} u) = \frac{1}{u\sqrt{u^2-1}}\,du, \tag{11}$$

$$d(\operatorname{arccsc} u) = \frac{-1}{u\sqrt{u^2-1}}\,du. \tag{12}$$

Proof To prove formula (7), let $y = \arcsin x$. Then $\sin y = x$, and implicit differentiation gives us

$$\begin{aligned}(\cos y)y' &= 1,\\ y' &= \frac{1}{\cos y}\\ &= \frac{1}{\sqrt{1-\sin^2 y}}\\ &= \frac{1}{\sqrt{1-x^2}}.\end{aligned}$$

Note that if $y = \arcsin x$, then y is between $-\pi/2$ and $\pi/2$; that is, y is in the first and fourth quadrants, where the cosine is positive. Thus, $\cos y = |\cos y| = \sqrt{\cos^2 y} = \sqrt{1-\sin^2 y}$, as was used above. Restatement of the above derivation in differential form gives us formula (7). The proof of formula (8) is similar to the above, and will be left as an exercise.

To prove formula (9), we let $y = \arctan x$, so that $\tan y = x$. Then we differentiate implicitly, and get

$$\begin{aligned}(\sec^2 y)y' &= 1,\\ y' &= \frac{1}{\sec^2 y}\\ &= \frac{1}{1+\tan^2 y}\\ &= \frac{1}{1+x^2},\end{aligned}$$

from which formula (9) follows. The proof of formula (10) is similar.

To prove formula (11), we let $y = \text{arcsec } x$, so that $\sec y = x$ and

$$(\sec y \tan y)y' = 1,$$
$$y' = \frac{1}{\sec y \tan y}.$$

Now, the principal domain of the secant is in Quadrants I and III, so that $\tan y > 0$. Hence, $\tan y = \sqrt{\sec^2 y - 1}$, and it is easy to see that formula (11) follows. This is the reason that the principal domain of the secant was chosen to be $[-\pi, -\pi/2) \cup [0, \pi/2)$—so that the derivative of the inverse secant will actually be the slope of the curve (see Figure 6.20 again). A similar approach establishes formula (12), and the theorem is proved.

The following examples use the differentiation formulas given in Theorem 6.7.

Example 6.55 Find the derivative:

(a) $y = \arcsin \frac{x}{2}$ **(b)** $y = x \arctan x$ **(c)** $y = \text{arcsec}(3 \tan x)$

Solution **(a)**

$$y' = \frac{1}{\sqrt{1 - (x/2)^2}} \cdot \frac{1}{2} = \frac{1}{\sqrt{4 - x^2}}.$$

(b) Using the product rule,

$$y' = 1 \arctan x + x \cdot \frac{1}{1 + x^2}$$
$$= \arctan x + \frac{x}{1 + x^2}.$$

(c)

$$y' = \frac{1}{3 \tan x \sqrt{(3 \tan x)^2 - 1}} (3 \sec^2 x)$$
$$= \frac{\sec^2 x}{\tan x \sqrt{9 \tan^2 x - 1}}.$$

■

Example 6.56 Show that $f(x) = \arctan x$ is increasing, and find its inflection points.

Solution Since

$$f'(x) = \frac{1}{1 + x^2},$$

the first derivative is always positive, so f is increasing. The second derivative is

$$f''(x) = \frac{-2x}{(1 + x^2)^2},$$

which changes sign at $x = 0$. Therefore, 0 is the only inflection point of f. ■

Integration Involving the Inverse Trigonometric Functions

The differentiation formulas (7)–(12) give us the following integration formulas:

$$\int \frac{du}{\sqrt{1-u^2}} = \arcsin u + C = -\operatorname{arccos} u + C, \tag{13}$$

$$\int \frac{du}{1+u^2} = \arctan u + C = -\operatorname{arccot} u + C, \tag{14}$$

$$\int \frac{du}{u\sqrt{u^2-1}} = \operatorname{arcsec} u + C = -\operatorname{arccsc} u + C. \tag{15}$$

These formulas can be generalized as follows (we assume that $a > 0$):

$$\int \frac{du}{\sqrt{a^2-u^2}} = \arcsin \frac{u}{a} + C, \tag{16}$$

$$\int \frac{du}{a^2+u^2} = \frac{1}{a} \arctan \frac{u}{a} + C, \tag{17}$$

$$\int \frac{du}{u\sqrt{u^2-a^2}} = \frac{1}{a} \operatorname{arcsec} \frac{u}{a} + C. \tag{18}$$

■ **Example 6.57** Integrate:

(a) $\displaystyle\int \frac{dx}{9+x^2}$ (b) $\displaystyle\int \frac{e^x\,dx}{\sqrt{1-e^{2x}}}$ (c) $\displaystyle\int_0^{1/2} \frac{dx}{\sqrt{1-x^2}}$

Solution (a) Using formula (17), we see immediately that

$$\int \frac{dx}{9+x^2} = \frac{1}{3} \arctan \frac{x}{3} + C.$$

(b) Substituting $u = e^x$, $du = e^x\,dx$, we get

$$\begin{aligned}\int \frac{e^x\,dx}{\sqrt{1-e^{2x}}} &= \int \frac{du}{\sqrt{1-u^2}}\\ &= \arcsin u + C\\ &= \arcsin e^x + C.\end{aligned}$$

(c) Using formula (13), we have

$$\begin{aligned}\int_0^{1/2} \frac{dx}{\sqrt{1-x^2}} &= \arcsin x \Big|_0^{1/2}\\ &= \arcsin \frac{1}{2} - \arcsin 0\\ &= \frac{\pi}{6}.\end{aligned}$$

■

Example 6.58 Solve the differential equation

$$y' - xy^2 = 4x,$$

subject to the condition $y = 2$ when $x = 0$.

Solution We first separate the variables, and then solve the equation:

$$y' = x(y^2 + 4),$$

$$\frac{dy}{y^2 + 4} = x\,dx,$$

$$\int \frac{dy}{y^2 + 4} = \int x\,dx,$$

$$\frac{1}{2}\arctan\frac{y}{2} = \frac{1}{2}x^2 + C.$$

Setting $x = 0$ and $y = 2$, we get

$$C = \frac{1}{2}\arctan 1 = \frac{\pi}{8}.$$

The solution is therefore

$$\frac{1}{2}\arctan\frac{y}{2} = \frac{1}{2}x^2 + \frac{\pi}{8},$$

$$y = 2\tan\left(x^2 + \frac{\pi}{4}\right).$$ ■

Exercises 6.6

Find the values of the following:

1. $\arcsin\frac{1}{2}$
2. $\arccos\frac{\sqrt{3}}{2}$
3. $\arcsin\left(-\frac{1}{2}\right)$
4. $\arccos\left(-\frac{\sqrt{3}}{2}\right)$
5. $\arctan\left(\sin\frac{\pi}{3}\right)$
6. $\operatorname{arcsec}\left(\tan\frac{3\pi}{4}\right)$
7. $\cos\left[\arctan\left(\sin\frac{\pi}{6}\right)\right]$
8. $\sin[\arctan(\cos 0)]$
9. Show that for any $x \in [-1, 1]$, $\arcsin x + \arccos x = \pi/2$.
10. Show that if $a \in [-1, 1]$, then the solution set of the equation $\cos x = a$ is $\{2n\pi \pm \arccos a \mid n \in \mathbb{Z}\}$.
11. Show that if $a \in [-1, 1]$, then the solution set of the equation $\sin x = a$ is $\{n\pi + (-1)^n \arcsin a \mid n \in \mathbb{Z}\}$.
12. Show that the solution set of the equation $\tan x = a$ is $\{n\pi + \arctan a \mid n \in \mathbb{Z}\}$.

Find the solution sets of the following equations:

13. $\cos x = 1$
14. $\sin x = -1$
15. $\cos^2 x = 1$
16. $\sin x \cos x = 1$
17. $\cos^2 x = \sin^2 x$
18. $\sin 2x = \cos x$

19. $\sin x = \cos x$ **20.** $\sin 2x + 2\cos^2 x = 0$ **21.** $\sec^2 x + 2\tan x = 0$

22. Explain why $\sin(\arcsin x) = x$ for all x, but $\arcsin(\sin x)$ need not equal x.

Find algebraic expressions in x for each of the following:

23. $\cos(\arcsin x)$ **24.** $\tan(\arcsin x)$ **25.** $\sin(\arctan x)$

26. $\cos(\arctan x)$ **27.** $\sin(\operatorname{arcsec} x)$ **28.** $\sec(\arctan x)$

29. $\tan(\operatorname{arcsec} x)$ **30.** $\cos(\operatorname{arcsec} x)$

31. Prove formula (8). **32.** Prove formula (10). **33.** Prove formula (12).

Find the derivative.

34. $y = \arcsin 3x$ **35.** $y = \arctan \dfrac{x}{2}$ **36.** $y = x \arccos x$

37. $y = \arcsin^2 x$ **38.** $y = \ln(\operatorname{arcsec} x)$ **39.** $y = e^{2 \arctan x}$

40. $xy = \arcsin x$ **41.** $\arcsin xy = 2x$

Find the second derivative.

42. $y = \arcsin x$ **43.** $y = \arctan 2x$

Integrate.

44. $\displaystyle\int \frac{dx}{4 + x^2}$ **45.** $\displaystyle\int \frac{dx}{\sqrt{4 - x^2}}$ **46.** $\displaystyle\int \frac{dx}{9 + x^2}$

47. $\displaystyle\int \frac{dx}{\sqrt{16 - x^2}}$ **48.** $\displaystyle\int_0^{\sqrt{3}/2} \frac{dx}{\sqrt{1 - x^2}}$ **49.** $\displaystyle\int_0^1 \frac{dx}{x^2 + 1}$

50. $\displaystyle\int_0^{\sqrt{3}} \frac{dx}{\sqrt{4 - x^2}}$ **51.** $\displaystyle\int_0^9 \frac{dx}{9 + x^2}$ **52.** $\displaystyle\int_{\sqrt{2}}^{2/\sqrt{3}} \frac{dx}{x\sqrt{x^2 - 1}}$

53. $\displaystyle\int_0^{1/2} \frac{e^{\arcsin x}\, dx}{\sqrt{1 - x^2}}$

Solve the differential equations.

54. $y' = (x - 1)(y^2 + 1)$ **55.** $xy' = \sqrt{1 - y^2}, \quad y(\frac{1}{2}) = 0$

56. The rate of growth of a certain stock is

$$g(t) = \frac{2000}{t^2 + 100}$$

dollars per month after t months. What is the value of the stock after 12 months?

57. Prove the identity

$$\frac{\pi}{4} = 4 \arctan \frac{1}{5} - \arctan \frac{1}{239}$$

by computing

$$\tan\left(2 \arctan \frac{1}{5}\right), \quad \tan\left(4 \arctan \frac{1}{5}\right), \quad \text{and} \quad \tan\left(4 \arctan \frac{1}{5} - \arctan \frac{1}{239}\right).$$

SECTION 6.7

Hyperbolic Functions (Optional)

The hyperbolic functions are special combinations of exponential functions that occur in certain applications. These functions have properties very similar to the properties of the circular functions. We shall define the hyperbolic functions here for the interested reader, and leave most of the properties as exercises. (There is no loss in continuity if this section is omitted.)

The Hyperbolic Sine and Cosine

DEFINITION 6.9

The **hyperbolic sine** function is defined by

$$\sinh x = \tfrac{1}{2}(e^x - e^{-x}).$$

The **hyperbolic cosine** function is defined by

$$\cosh x = \tfrac{1}{2}(e^x + e^{-x}).$$

The hyperbolic sine and cosine are obviously defined for all real numbers x, and are differentiable. Their derivatives are easily seen to be

$$D_x \sinh x = \cosh x, \tag{1}$$

$$D_x \cosh x = \sinh x. \tag{2}$$

From the definition, we see that $\cosh x$ is positive for all x. Hence, by formula (1), $\sinh x$ is increasing on $(-\infty, \infty)$. Since $\sinh 0 = 0$, it follows that $\sinh x < 0$ if $x < 0$ and $\sinh x > 0$ if $x > 0$. Thus, from formula (2), we see that $\cosh x$ is decreasing in $(-\infty, 0)$ and increasing in $(0, \infty)$. The graphs of $\sinh x$ and $\cosh x$ are shown in Figures 6.22 and 6.23.

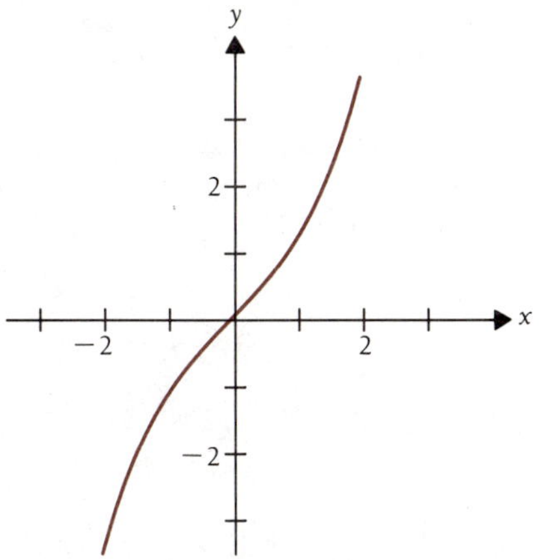

FIGURE 6.22 The graph of $y = \sinh x$

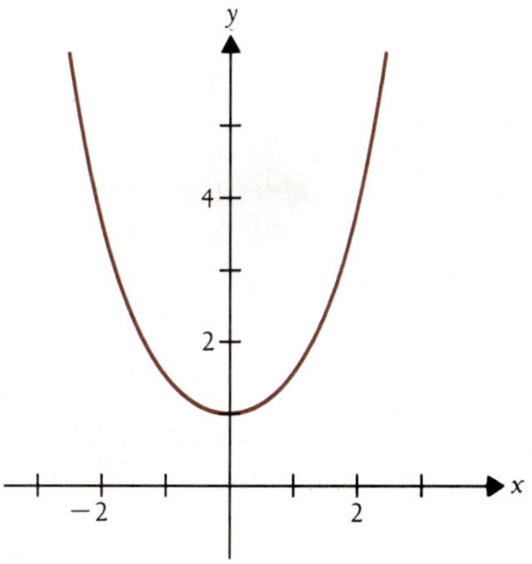

FIGURE 6.23 The graph of $y = \cosh x$

From Definition 6.9, we can also show that

$$\cosh^2 x - \sinh^2 x = 1. \tag{3}$$

Thus, for any number t, the point $(\cosh t, \sinh t)$ lies on the hyperbola $x^2 - y^2 = 1$ (see Figure 6.24). Moreover, the area of the shaded region in Figure 6.24 is $\frac{1}{2}t$. We thus see the similarity between the hyperbolic sine and cosine and the circular sine and cosine (see Figure 1.54). We also see why these functions are called *hyperbolic*.

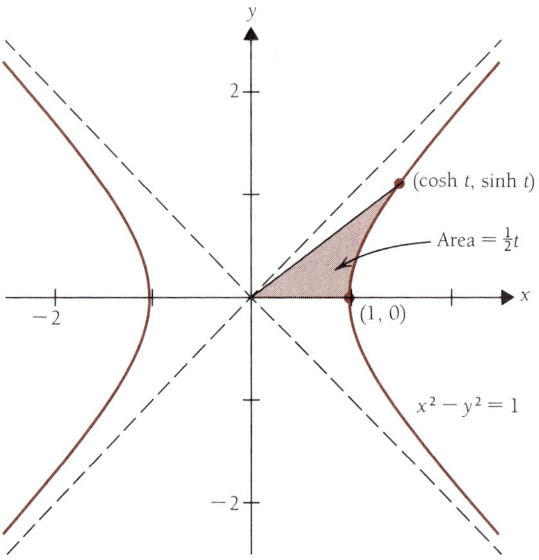

FIGURE 6.24 Development of the hyperbolic functions

Other Hyperbolic Functions

DEFINITION 6.10

The **hyperbolic tangent, hyperbolic cotangent, hyperbolic secant,** and **hyperbolic cosecant** functions are defined by

$$\tanh x = \frac{\sinh x}{\cosh x}, \qquad \coth x = \frac{\cosh x}{\sinh x},$$

$$\operatorname{sech} x = \frac{1}{\cosh x}, \qquad \operatorname{csch} x = \frac{1}{\sinh x}.$$

These functions have obvious similarities to the trigonometric functions, and many similar identities hold. These functions will be explored in the exercises.

Exercises 6.7

Establish each of the following statements:

1. $\cosh^2 x - \sinh^2 x = 1$
2. $1 - \tanh^2 x = \text{sech}^2 x$
3. $\coth^2 x - 1 = \text{csch}^2 x$
4. $D_x \sinh x = \cosh x$
5. $D_x \cosh x = \sinh x$
6. $D_x \tanh x = \text{sech}^2 x$
7. $D_x \coth x = -\text{csch}^2 x$
8. $D_x \text{sech}\, x = -\text{sech}\, x \tanh x$
9. $D_x \text{csch}\, x = -\text{csch}\, x \coth x$
10. $\int \sinh x\, dx = \cosh x + C$
11. $\int \cosh x\, dx = \sinh x + C$
12. $\int \tanh x\, dx = \ln(\cosh x) + C$
13. $\int \coth x\, dx = \ln|\sinh x| + C$
14. $\int \text{sech}\, x\, dx = 2 \arctan e^x + C$

Differentiate.

15. $y = \sinh 3x$
16. $y = \tanh(\cos x)$
17. $\sin y = \cosh(x + y)$
18. $y = (\sinh x)^{x+1}$

Integrate.

19. $\int \sinh 2x\, dx$
20. $\int \sin x \cosh(\cos x)\, dx$
21. $\int \frac{\cosh \sqrt{x+1}}{\sqrt{x+1}}\, dx$
22. $\int e^x \coth e^x\, dx$
23. Find the minimum value of $\cosh x$.
24. Describe the concavity of $\sinh x$.
25. Describe the concavity of $\cosh x$.
26. Tell where $f(x) = \tanh x$ is increasing and where it is decreasing, describe its concavity, and sketch its graph.
27. Sketch the graph of $f(x) = \text{sech}\, x$.
28. Show that $y = A \sinh bt + B \cosh bt$ is a solution of the differential equation

$$\frac{d^2y}{dt^2} = b^2 y.$$

29. Let the **inverse hyperbolic sine** function be defined by

$$y = \sinh^{-1} x \qquad \text{if and only if} \qquad x = \sinh y.$$

Show that

$$\sinh^{-1} x = \ln(x + \sqrt{x^2 + 1}).$$

(Since $\sinh x$ is defined in terms of the exponential function, it should be no surprise that the inverse may be expressed in terms of the natural logarithmic function.)

30. Let the **inverse hyperbolic cosine** function be defined by

$$y = \cosh^{-1} x \qquad \text{if and only if} \qquad x = \cosh y \qquad \text{and} \qquad x \geq 0.$$

Show that $\cosh^{-1} x = \ln(x + \sqrt{x^2 - 1})$.

31. Find $D_x \sinh^{-1}x$.

32. Find $D_x \cosh^{-1}x$.

33. Find $\displaystyle\int \frac{dx}{\sqrt{x^2+1}}$.

34. Find $\displaystyle\int \frac{dx}{\sqrt{x^2-1}}$.

35. Let the **inverse hyperbolic tangent** function be defined by

$$y = \tanh^{-1}x \qquad \text{if and only if} \qquad x = \tanh y.$$

Show that for $-1 < x < 1$,

$$\tanh^{-1}x = \frac{1}{2}\ln\frac{1+x}{1-x}.$$

36. Find $D_x \tanh^{-1}x$.

SECTION 6.8

Chapter Review

Important Concepts

Logarithmic function
Natural logarithm
Logarithmic differentiation
Base of the natural logarithm, e
Properties of real exponents
Exponential function
Base
Common logarithm
Exponential growth model
Uninhibited growth model
Exponential decay model
Limited growth model
Logistic model
Inhibited growth model
Future value
Present value
Compound interest
Limit form
Indeterminate forms
L'Hôpital's rules
Inverse trigonometric functions
Hyperbolic functions

Review Exercises

Find the limit, if it exists.

1. $\displaystyle\lim_{x\to 1}\frac{2^x-2}{x-1}$

2. $\displaystyle\lim_{x\to\infty}[\ln(x^2+1)-\ln(x^2-1)]$

3. $\displaystyle\lim_{x\to\infty} xe^{-2x}$

4. $\displaystyle\lim_{x\to 0^+}\frac{1-e^x}{\tan x}$

5. $\displaystyle\lim_{x\to\infty} x^{-1/3}\ln x$

6. $\displaystyle\lim_{x\to 0^+}\frac{x}{\ln x}$

7. $\displaystyle\lim_{x\to 1^-} x^{1/(x-1)}$

8. $\displaystyle\lim_{x\to \pi/2} (\tan x)^{x-\pi/2}$

Find dy/dx.

9. $y = e^x \ln(\cos x)$

10. $y = \ln(\arctan e^x)$

11. $y = e^{\arctan(\ln x)}$

12. $xe^y = \cos x$

13. $xy = \ln(x + y^2)$

14. $y = x^{x^2+1}$

15. $y = (x+3)^4(2-x)^3$

16. $y = \displaystyle\int_x^{x^2} \cos^3 2t\, dt$

17. $y = \arcsin 4x$

18. $y = 3 \arctan 5x$

19. $y = \cos x \arcsin 2x$

20. $y = \arcsin(3 \cos x)$

21. $y = \arctan \dfrac{x}{3}$

22. $y = \arctan|x|$

23. $y = \dfrac{1}{x} \arctan \dfrac{1}{x}$

24. $\arcsin(x + y) = y^2$

Integrate.

25. $\displaystyle\int e^{-4x}\, dx$

26. $\displaystyle\int_0^3 \frac{\ln(1+x)}{1+x}\, dx$

27. $\displaystyle\int_0^1 e^x \sin e^x\, dx$

28. $\displaystyle\int \frac{2x\, dx}{x^2+1}$

29. $\displaystyle\int \frac{2\, dx}{x^2+1}$

30. $\displaystyle\int \frac{e^x\, dx}{1-e^x}$

31. $\displaystyle\int_0^1 x^{1/3} \cos(x^{4/3})\, dx$

32. $\displaystyle\int \frac{dx}{e^{-x}-1}$

33. $\displaystyle\int \frac{x\, dx}{\sqrt{1-x^2}}$

34. $\displaystyle\int \frac{x\, dx}{x^2\sqrt{x^2-1}}$

Solve the differential equations.

35. $\dfrac{dy}{dx} = 3xy$

36. $\dfrac{dx}{dt} = k(1-x)$

37. $\dfrac{dr}{d\theta} = 2r \sin \theta$

38. $\dfrac{dy}{dx} = 2x(y^2+1)$

39. A certain radioactive substance loses 4% of its mass per day in radiation. What is its half-life?

40. The population of a certain nation doubles in 32 years. Assuming exponential growth, what is its growth rate per year?

41. A certain bacterial culture grows at a rate proportional to the population. After 2 hours of incubation, there are 10,000 bacteria; after 22 more hours, there are 16,000. What was the initial population?

42. The spread of a certain communicable disease fits the logistic model. If it takes 6 weeks for the infected portion of the population to increase from 1% to 10%, how long will it take for 90% of the population to become infected?

Additional Exercises

1. Find the points of inflection of $y = e^{-x^2}$.
2. Derive the inequality

$$\int_1^x \frac{1}{t}\, dt \le \int_1^x \frac{1}{\sqrt{t}}\, dt,$$

and use it to prove that

$$\lim_{x\to\infty} \frac{\ln x}{x} = 0.$$

3. Show that $x = Ae^{bt} + Be^{-bt}$ is a solution of the differential equation

$$\frac{d^2x}{dt^2} = b^2x.$$

4. A cylindrical barrel of radius 1 meter and height 2 meters has a leak in the bottom, through which water leaks at a rate proportional to the depth of water in the barrel. If the barrel is originally full and loses 5% of its volume in 1 hour, in how many hours will the barrel be half full?

5. An electric current with a resistance of R ohms and an inductance of L henries satisfies the equation

$$L\frac{di}{dt} + Ri = E,$$

where i is the current in amperes, E is the (constant) electromotive force in volts, and t is the time in seconds. Express i as a function of t.

6. The coroner arrives at the scene of a murder at 2:00 AM and finds the temperature of the corpse to be 76.5°F. One hour later, the temperature is 71.4°F. If the temperature of the surrounding environment is 60°F, when was the murder committed?

7. A certain object falling through the atmosphere encounters air resistance proportional to the velocity, so that the acceleration is $a(t) = g - kv(t)$, where g is the acceleration of gravity. Show that

$$v(t) = \frac{g}{k}(1 - e^{-kt}).$$

8. In Problem 7, find the limit of $v(t)$ as $t \to \infty$. (This limit is called the **terminal velocity** of the object.)

9. In Problem 7, how far does the object fall during the first t seconds?

10. A tank contains 20 gallons of water, in which 40 pounds of salt is dissolved. Pure water flows into the tank at the rate of 2 gallons per minute, and the mixture (kept uniform by stirring) flows out at the same rate. After how many minutes will there be 10 pounds of salt in the tank?

11. The growth rate of a certain business is

$$f(t) = e^{0.02t} + \sin\frac{\pi t}{6},$$

measured in thousands of dollars of net worth per month, at time t. How much growth does the business experience during its first year?

12. The **rule of 72** says that the amount of time it takes to double your money in a savings account with continuous compounding is approximately 72 divided by the interest rate. Find the exact doubling time, and explain where the rule of 72 comes from.

13. Does the rule of 72 apply to periodic compounding?

14. Does the rule of 72 apply to every case of exponential growth?

15. Does the rule of 72 apply to exponential decay?

16. A certain communicable disease spreads at a rate that fits the logistic model. Health officials become alarmed when 1% of the population has been infected, and issue public warnings 3 weeks later when 3% of the population has been infected. The warnings seem to have no effect on the spread of the disease, however. If an infected person is immobilized for the first 5 days he or she is infected, estimate the maximum percentage of the population that will be immobilized by this epidemic at any one time.

CHAPTER 7

Techniques of Integration

INTRODUCTION

At this point in our development, the process of integration is largely a trial-and-error process of guessing at antiderivatives. For some functions, we have formulas for finding the antiderivatives. The substitution technique introduced in Section 5.5 enables us to make many more functions fit the formulas. But for most functions, we do not—at this point—have formulas for integration.

Fortunately, several large classes of functions can be integrated by applying certain standard techniques. In this chapter, we will develop the techniques of integration by parts, algebraic manipulations, and various kinds of substitutions. We will thus be able to derive many general integration formulas.

Since some antiderivatives can be expressed only as integrals with variable limits, we cannot expect to integrate every function in terms of the elementary functions that we have already developed. For definite integrals, numerical approximation methods exist that work when all else fails. We will demonstrate some of these methods at the end of the chapter.

SECTION 7.1

Integration Formulas

The simplest integration technique is to apply a formula. This can be done when a formula exists that exactly fits the integral given. If no formula fits exactly, the integral often can be made to fit a known formula by the technique of substitution, as discussed in Section 5.5.

In this section, we will list the most basic integration formulas for reference. Then we will review the substitution technique and use it to derive several more formulas, some old and some new.

Basic Integration Formulas

We list here for reference the basic integration formulas. In these formulas, u and v represent functions of x, and a is a constant. All formulas will be given in indefinite integral form; the constant of integration is denoted by C.

$$\int du = u + C \tag{1}$$

$$\int au\, dx = a \int u\, dx \tag{2}$$

$$\int (u + v)\, dx = \int u\, dx + \int v\, dx \tag{3}$$

$$\int u^n\, du = \begin{cases} \dfrac{1}{n+1} u^{n+1} + C & \text{if } n \neq -1 \\ \ln|u| + C & \text{if } n = -1 \end{cases} \tag{4}$$

$$\int \sin u\, du = -\cos u + C \tag{5}$$

$$\int \cos u\, du = \sin u + C \tag{6}$$

$$\int \sec^2 u\, du = \tan u + C \tag{7}$$

$$\int \csc^2 u\, du = -\cot u + C \tag{8}$$

$$\int \sec u \tan u\, du = \sec u + C \tag{9}$$

$$\int \csc u \cot u\, du = -\csc u + C \tag{10}$$

$$\int \frac{du}{\sqrt{1 - u^2}} = \arcsin u + C \tag{11}$$

$$\int \frac{du}{1 + u^2} = \arctan u + C \tag{12}$$

$$\int \frac{du}{u\sqrt{u^2 - 1}} = \operatorname{arcsec} u + C \tag{13}$$

$$\int e^u\, du = e^u + C \tag{14}$$

The following examples review the use of the substitution technique.

Example 7.1 Integrate $\int (3x + 4)^{12}\, dx$.

Solution Formula (4) is similar to this integral, but does not fit exactly. Hence, we substitute $u = 3x + 4$, $du = 3\,dx$, to get

$$\begin{aligned}\int (3x+4)^{12}\,dx &= \frac{1}{3}\int (3x+4)^{12}3\,dx \\ &= \frac{1}{3}\int u^{12}\,du \\ &= \frac{1}{3}\left(\frac{1}{13}u^{13}+C\right) \qquad \text{[By formula (4)]} \\ &= \frac{1}{39}u^{13}+C \\ &= \frac{1}{39}(3x+4)^{13}+C.\end{aligned}$$

■

■ **Example 7.2** Integrate

$$\int \frac{3x\,dx}{x^2+4}.$$

Solution We note that the derivative of the denominator is in the numerator (up to the constant coefficient, which we can adjust). Hence, we let $u = x^2 + 4$, $du = 2x\,dx$. Thus,

$$\begin{aligned}\int \frac{3x\,dx}{x^2+4} &= \frac{3}{2}\int \frac{2x\,dx}{x^2+4} \\ &= \frac{3}{2}\int \frac{1}{u}\,du \\ &= \frac{3}{2}\ln|u|+C \qquad \text{[By formula (4)]} \\ &= \frac{3}{2}\ln|x^2+4|+C \\ &= \frac{3}{2}\ln(x^2+4)+C.\end{aligned}$$

Here, the absolute value symbol is redundant. ■

■ **Example 7.3** Integrate $\int \cos(2x+1)\,dx$.

Solution Formula (6) looks most applicable, so we let $u = 2x + 1$, $du = 2\,dx$. This gives us

$$\begin{aligned}\int \cos(2x+1)\,dx &= \frac{1}{2}\int \cos(2x+1)\,2\,dx \\ &= \frac{1}{2}\int \cos u\,du\end{aligned}$$

$$\frac{1}{2}\int \cos u \, du = \frac{1}{2} \sin u + C \qquad \text{[By formula (6)]}$$

$$= \frac{1}{2} \sin(2x + 1) + C.$$

■

■ **Example 7.4** Integrate $\int 2^x \, dx$.

Solution Since $2^x = e^{(\ln 2)x}$, formula (14) will apply. We substitute $u = (\ln 2)x$, $du = (\ln 2) \, dx$, or $dx = du/(\ln 2)$, and get

$$\int 2^x \, dx = \int e^{(\ln 2)x} \, dx$$

$$= \frac{1}{\ln 2} \int e^u \, du$$

$$= \frac{1}{\ln 2} e^u + C$$

$$= \frac{1}{\ln 2} e^{(\ln 2)x} + C$$

$$= \frac{1}{\ln 2} 2^x + C.$$

Note that we could have used formula (2) from Section 6.3 to obtain this result directly. ■

■ **Example 7.5** Integrate $\int \sec^2 x \tan x \, dx$.

Solution There are two ways to proceed in this integral. If we let $u = \tan x$, then $du = \sec^2 x \, dx$, and we get

$$\int \sec^2 x \tan x \, dx = \int u \, du$$

$$= \frac{1}{2} u^2 + C$$

$$= \frac{1}{2} \tan^2 x + C.$$

On the other hand, if we let $u = \sec x$, then $du = \sec x \tan x \, dx$, and we get

$$\int \sec^2 x \tan x \, dx = \int \sec x \sec x \tan x \, dx$$

$$= \int u \, du$$

$$= \frac{1}{2} u^2 + C$$

$$= \frac{1}{2} \sec^2 x + C.$$

At first glance, these answers appear to be different, but we know they actually are the same, because two antiderivatives of the same function differ only by a constant. Can you show that this is the case? ■

Integration Formulas for More General Cases

We will now list several other integration formulas that are almost as basic as formulas (1) – (14), and we will show how algebraic manipulation and the substitution technique can be used to derive them. In these formulas, a, b, and n are real numbers, $a \neq 0$.

$$\int (ax+b)^n\,dx = \begin{cases} \dfrac{1}{a(n+1)}(ax+b)^{n+1} + C & \text{if } n \neq -1 \\ \dfrac{1}{a}\ln|ax+b| + C & \text{if } n = -1 \end{cases} \tag{15}$$

$$\int \cos ax\,dx = \frac{1}{a}\sin ax + C \tag{16}$$

$$\int \sin ax\,dx = -\frac{1}{a}\cos ax + C \tag{17}$$

$$\int \tan u\,du = -\ln|\cos u| + C = \ln|\sec u| + C \tag{18}$$

$$\int \cot u\,du = \ln|\sin u| + C \tag{19}$$

$$\int \frac{dx}{\sqrt{a^2-x^2}} = \arcsin\frac{x}{a} + C, \quad a > 0 \tag{20}$$

$$\int \frac{dx}{x^2+a^2} = \frac{1}{a}\arctan\frac{x}{a} + C, \quad a > 0 \tag{21}$$

$$\int \frac{dx}{x\sqrt{x^2-a^2}} = \frac{1}{a}\operatorname{arcsec}\frac{x}{a} + C, \quad a > 0 \tag{22}$$

$$\int a^u\,du = \frac{1}{\ln a}a^u + C, \quad a > 0 \tag{23}$$

It should be evident that formula (15) is a general case of Example 7.1, and is derived from formula (4) as in Example 7.1. Formula (16) is derived from formula (6) as in Example 7.3, and the derivation of formula (17) is similar.

Formula (18) is derived by writing $\tan x$ as $(\sin x)/(\cos x)$, and substituting $u = \cos x$, $du = -\sin x\,dx$:

$$\begin{aligned} \int \tan x\,dx &= \int \frac{\sin x}{\cos x}\,dx \\ &= -\int \frac{1}{u}\,du \end{aligned}$$

$$
\begin{aligned}
-\int \frac{1}{u}\,du &= -\ln|u| + C \\
&= -\ln|\cos x| + C \\
&= \ln|\cos x|^{-1} + C \\
&= \ln|\sec x| + C.
\end{aligned}
$$

The derivation of formula (19) is similar.

Formula (20) is similar to formula (11), so we first try to rewrite it in that form. The first step is to factor a^2 from the radical:

$$\int \frac{dx}{\sqrt{a^2 - x^2}} = \int \frac{dx}{a\sqrt{1 - (x/a)^2}}.$$

Now, by comparison with formula (11), we let $u = x/a$, $du = (1/a)\,dx$:

$$
\begin{aligned}
\int \frac{dx}{a\sqrt{1 - (x/a)^2}} &= \int \frac{du}{\sqrt{1 - u^2}} \\
&= \arcsin u + C \\
&= \arcsin \frac{x}{a} + C.
\end{aligned}
$$

Formulas (21) and (22) are obtained similarly from formulas (12) and (13).

Finally, formula (23) is just formula (2) of Section 6.3.

The following examples illustrate the direct use of some of these formulas.

■ **Example 7.6** Find $\int \sin 5x\,dx$.

Solution By formula (17) with $a = 5$, we have

$$\int \sin 5x\,dx = -\frac{1}{5}\cos 5x + C.$$ ■

■ **Example 7.7** Find

$$\int \frac{dx}{x^2 + 9}.$$

Solution By formula (21) with $a = 3$,

$$\int \frac{dx}{x^2 + 9} = \frac{1}{3}\arctan \frac{x}{3} + C.$$ ■

■ **Example 7.8** Find

$$\int \frac{dx}{\sqrt{7 - x^2}}.$$

Solution By formula (20) with $a = \sqrt{7}$,

$$\int \frac{dx}{\sqrt{7 - x^2}} = \arcsin \frac{x}{\sqrt{7}} + C.$$

■

The foregoing has all been done in terms of indefinite integrals. For definite integrals, the work is the same, except that evaluation at the limits follows finding the antiderivatives. In case substitution is used, the limits are changed as was explained in Section 5.5.

Consider these examples.

■ **Example 7.9** Find

$$\int_0^1 \frac{dx}{x^2 + 4}.$$

Solution

$$\begin{aligned}\int_0^1 \frac{dx}{x^2 + 4} &= \left[\frac{1}{2} \arctan \frac{x}{2}\right]_0^1 \\ &= \frac{1}{2} \arctan \frac{1}{2} - \frac{1}{2} \arctan 0 \\ &= \frac{1}{2} \arctan \frac{1}{2}.\end{aligned}$$

■

■ **Example 7.10** Evaluate

$$\int_0^2 \frac{x\,dx}{x^2 + 4}.$$

Solution We use the substitution $u = x^2 + 4$ to make this integral fit formula (4). When $x = 0$, $u = 4$, and when $x = 2$, $u = 8$. Thus,

$$\begin{aligned}\int_0^2 \frac{x\,dx}{x^2 + 4} &= \frac{1}{2} \int_4^8 \frac{1}{u}\,du \\ &= \frac{1}{2} \Big[\ln|u|\Big]_4^8 \\ &= \frac{1}{2} (\ln 8 - \ln 4) \\ &= \frac{1}{2} \ln 2.\end{aligned}$$

■

■ **Example 7.11** Integrate

$$\int_0^{\pi/2} \tan \frac{x}{3}\,dx.$$

Solution

$$\int_0^{\pi/2} \tan\frac{x}{3}\,dx = 3\int_0^{\pi/6} \tan u\,du \qquad (u = x/3)$$
$$= 3\Big[\ln|\sec u|\Big]_0^{\pi/6}$$
$$= 3\ln\frac{2}{\sqrt{3}} - 3\ln 1$$
$$= 3\ln\frac{2}{\sqrt{3}}.$$

■

In this chapter, many more integration formulas will be developed, some of which are found in the table inside the back cover of this book. You may wish to construct your own personal table of integrals by adding new formulas to the list.

Exercises 7.1

Integrate.

1. $\int x^7\,dx$
2. $\int \frac{1}{t}\,dt$
3. $\int (2x-1)^3\,dx$
4. $\int_0^1 (4-3x)^5\,dx$
5. $\int \frac{1}{2-x}\,dx$
6. $\int \frac{dx}{2x+5}$
7. $\int_0^3 e^{2x}\,dx$
8. $\int_{-1}^1 e^{5x+6}\,dx$
9. $\int 10^x\,dx$
10. $\int 2^{3x}\,dx$
11. $\int (x^{-3}+e^x)\,dx$
12. $\int (x+xe^{x^2})\,dx$
13. $\int (4+e^x)^2\,dx$
14. $\int (e^x+e^{-x})\,dx$
15. $\int \sin 3x\,dx$
16. $\int_0^{\pi/4} \cos\left(x-\frac{\pi}{4}\right)dx$
17. $\int_{\pi/6}^{\pi/4} \cot 2x\,dx$
18. $\int \tan \pi x\,dx$
19. $\int (\sin x - \tan x)\,dx$
20. $\int \left(\frac{1}{x} - \sin x\right)dx$
21. $\int \sec x\,(\sec x + \tan x)\,dx$
22. $\int (\sec^2\theta - \csc^2\theta)\,d\theta$
23. $\int \sec^2 3x\,dx$
24. $\int \sec 2x \tan 2x\,dx$
25. $\int \sec^2 3x \tan 3x\,dx$
26. $\int \tan^2 x\,dx$
27. $\int_0^{1/2} \frac{dx}{\sqrt{1-x^2}}$
28. $\int_0^{\sqrt{3}} \frac{dx}{x^2+1}$
29. $\int_0^1 \frac{dx}{\sqrt{4-x^2}}$
30. $\int_1^{\sqrt{2}} \frac{dx}{x\sqrt{x^2-1}}$
31. $\int e^{\cos x}\sin x\,dx$
32. $\int e^{\tan x}\sec^2 x\,dx$
33. $\int \sin^2 x \cos x\,dx$
34. $\int \frac{\csc^2 x}{\cot x}\,dx$
35. $\int \frac{x\,dx}{3x^2+1}$
36. $\int \frac{x\,dx}{\sqrt{1-x^2}}$

37. $\displaystyle\int \frac{x+1}{x^2+1}\,dx$

38. $\displaystyle\int \frac{x+1}{\sqrt{1-x^2}}\,dx$

39. $\displaystyle\int \frac{e^x}{9-3e^x}\,dx$

40. $\displaystyle\int \frac{1}{2-e^{-x}}\,dx$

41. $\displaystyle\int \frac{e^x}{4+e^{2x}}\,dx$

42. $\displaystyle\int \frac{1}{4e^x+e^{-x}}\,dx$

Derive the following formulas:

43. Formula (17)

44. Formula (19)

45. Formula (21)

46. Formula (22)

47. A formula for $\displaystyle\int \tan ax\,dx$

48. A formula for $\displaystyle\int \cot ax\,dx$

49. The formula

$$\int \frac{dx}{a^2x^2+b^2} = \frac{1}{ab}\arctan\frac{ax}{b} + C$$

50. The formula

$$\int \frac{dx}{\sqrt{1-a^2x^2}} = \frac{1}{a}\sin^{-1}ax + C$$

Integrate.

51. $\displaystyle\int_0^1 \frac{dx}{3x^2+1}$

52. $\displaystyle\int \frac{dx}{2x^2+3}$

53. $\displaystyle\int \frac{x+3}{2x^2+12}\,dx$

54. $\displaystyle\int \frac{x+3}{x^2+4}\,dx$

55. $\displaystyle\int \frac{dx}{\sqrt{1-4x^2}}$

56. $\displaystyle\int \frac{dx}{\sqrt{1-9x^2}}$

57. $\displaystyle\int \frac{dx}{\sqrt{9-16x^2}}$

58. $\displaystyle\int \frac{dx}{\sqrt{3-4x^2}}$

Derive the following formulas:

59. $\displaystyle\int_0^{a/2} \frac{dx}{\sqrt{a^2-x^2}} = \frac{\pi}{6}$

60. $\displaystyle\int_0^a \frac{dx}{x^2+a^2} = \frac{\pi}{4a}$

SECTION 7.2

Integration by Parts

The differentiation formulas have integration counterparts that tell us properties of indefinite integrals. For example, if the addition formula

$$d(u+v) = du + dv$$

is thought of as a differential equation, then

$$\begin{aligned}\int (du+dv) &= \int d(u+v)\\ &= u+v+C\\ &= u+C_1+v+C_2\\ &= \int du + \int dv.\end{aligned}$$

That is,

$$\int (du + dv) = \int du + \int dv.$$

This is just a statement of the additive property of indefinite integrals, which we have seen before.

The Integration-by-Parts Formula

If the product rule

$$d(uv) = u\,dv + v\,du$$

is treated as a differential equation, we get

$$\begin{aligned}\int d(uv) &= \int (u\,dv + v\,du)\\ &= \int u\,dv + \int v\,du\end{aligned}$$

by the above additive property. Rearranging, we have

$$\int u\,dv = \int d(uv) - \int v\,du,$$

or

$$\int u\,dv = uv - \int v\,du, \tag{1}$$

where the constant of integration from $\int d(uv)$ is included in $\int v\,du$.

Formula (1) is called the **integration-by-parts formula**. It tells us that if an integrand can be expressed as the product of one function u and the derivative of another function v, then the integral can be rewritten in terms of the integral of the product of v and the derivative of u. Hopefully, the new integral $\int v\,du$ will be easier to evaluate than the original $\int u\,dv$.

The definite integral form of (1) is

$$\int_{x=a}^{x=b} u\,dv = uv\Big|_{x=a}^{x=b} - \int_{x=a}^{x=b} v\,du. \tag{2}$$

Integration by Parts

An integration that is done by using formula (1) or (2) is called an **integration by parts**. Integration by parts is a powerful technique of integration that is useful in many settings.

To carry out an integration by parts, we must find functions u and v so that the given integral is of the form $\int u\,dv$. Usually, we identify part of the integrand as u, and the rest of it, including dx, as dv. Then we compute du by differentiating and v by antidifferentiating. Finally, we use the expressions for u, v, du, and dv in the integration-by-parts formula, and thus rewrite the original integral in terms of a new one. The new integral may be attacked any way you like.

There is no set procedure for selecting u and dv, but we follow the general strategy of trying to find a new integral that is simpler than the original. We often select for u a function that is made simpler by differentiating. For dv, we must choose an expression whose antiderivative we can find; otherwise, we cannot compute v. We do not expect v to be simpler than dv very often; if v is no more complicated than dv, we consider ourselves lucky.

The following examples illustrate the simplifying strategy in integration by parts.

Example 7.12 Integrate $\int xe^x\,dx$.

Solution If the integrand did not contain the factor x, there would be no difficulty (that is, a known formula would apply). Since the derivative of x is constant, we try integrating by parts, letting

$$u = x \qquad \text{and} \qquad dv = e^x\,dx.$$

Then

$$du = dx \qquad \text{and} \qquad v = e^x.$$

(More precisely, $v = e^x + k$ with k a constant, but we may let $k = 0$ as will be shown shortly.) Thus, by formula (1), we have

$$\begin{aligned}\int xe^x\,dx &= xe^x - \int e^x\,dx \\ &= xe^x - (e^x + C) \\ &= xe^x - e^x + C.\end{aligned}$$

(Actually, it should be $-C$, but since $-C$ is still an arbitrary constant, we just write C.)

If we had written $v = e^x + k$, then formula (1) would have given us

$$\begin{aligned}\int xe^x\,dx &= x(e^x + k) - \int (e^x + k)\,dx \\ &= xe^x + kx - (e^x + kx + C) \\ &= xe^x + kx - e^x - kx + C \\ &= xe^x - e^x + C,\end{aligned}$$

as before. This demonstrates the general principle that in computing v from dv, we may let the constant of integration be any number that is convenient—in this case, $k = 0$. ∎

Example 7.13 Integrate $\int x^2 e^x\,dx$.

Solution Taking a hint from Example 7.12, we let

$$u = x^2 \qquad \text{and} \qquad dv = e^x\,dx.$$

Then

$$du = 2x\,dx \qquad \text{and} \qquad v = e^x,$$

so that

$$\int x^2 e^x\,dx = x^2 e^x - \int 2xe^x\,dx$$
$$= x^2 e^x - 2\int xe^x\,dx.$$

Now we can use Example 7.12 directly (or integrate by parts again) to obtain

$$\int x^2 e^x\,dx = x^2 e^x - 2(xe^x - e^x + C)$$
$$= x^2 e^x - 2xe^x + 2e^x + C.$$ ■

You might try Example 7.13 over again with $u = e^x$ and $dv = x^2\,dx$ to see why we did it the way we did.

Example 7.14 Find $\int x^3 \ln x\,dx$.

Solution We might be tempted to follow Examples 7.12 and 7.13 and let $u = x^3$ and $dv = \ln x\,dx$. But then we could not compute v (as yet). So we let

$$u = \ln x \qquad \text{and} \qquad dv = x^3\,dx.$$

Then

$$du = \frac{1}{x}\,dx \qquad \text{and} \qquad v = \frac{1}{4}x^4,$$

and

$$\int x^3 \ln x\,dx = \frac{1}{4}x^4 \ln x - \int \frac{1}{4}x^4 \frac{1}{x}\,dx$$
$$= \frac{1}{4}x^4 \ln x - \frac{1}{4}\int x^3\,dx$$
$$= \frac{1}{4}x^4 \ln x - \frac{1}{16}x^4 + C.$$ ■

■ Example 7.15 Integrate $\int \ln x \, dx$.

Solution We do not have a formula for the integral of ln x, but we can differentiate ln x. Also, in this integral, if we proceed by parts, there is little choice but to let

$$u = \ln x \qquad \text{and} \qquad dv = dx,$$

so

$$du = \frac{1}{x}\,dx \qquad \text{and} \qquad v = x,$$

so that

$$\begin{aligned}\int \ln x \, dx &= x \ln x - \int x \frac{1}{x}\, dx \\ &= x \ln x - \int dx \\ &= x \ln x - x + C.\end{aligned}$$

■

■ Example 7.16 Evaluate $\int_0^1 \arctan x \, dx$.

Solution Again, there is little choice but to let

$$u = \arctan x \qquad \text{and} \qquad dv = dx.$$

Then

$$du = \frac{dx}{x^2+1} \qquad \text{and} \qquad v = x,$$

so

$$\begin{aligned}\int_0^1 \arctan x \, dx &= \Big[x \arctan x\Big]_0^1 - \int_0^1 \frac{x\,dx}{x^2+1} \\ &= (\arctan 1 - 0) - \frac{1}{2}\int_1^2 \frac{1}{u}\,du \qquad (u = x^2+1) \\ &= \frac{\pi}{4} - \frac{1}{2}\Big[\ln|u|\Big]_1^2 \\ &= \frac{\pi}{4} - \frac{1}{2}\ln 2.\end{aligned}$$

■

In the foregoing examples, straightforward integration by parts, perhaps repeated, has sufficed to reduce the given integral to an integral we could evaluate. In some cases, a special twist will help.

■ Example 7.17 Integrate $\int e^x \sin x \, dx$.

Solution It appears that we may let u be either e^x or $\sin x$ and make progress. (We reject $u = e^x \sin x$, because its derivative is more complicated, not simpler.) If we let

$$u = \sin x \qquad \text{and} \qquad dv = e^x\,dx,$$

then

$$du = \cos x\,dx \qquad \text{and} \qquad v = e^x.$$

Hence,

$$\int e^x \sin x\,dx = e^x \sin x - \int e^x \cos x\,dx. \tag{3}$$

Now the new integral is not much simpler looking than the original, but it appears that we can integrate by parts again. Here it is easy to get into a "blind alley." If we let

$$u = e^x \qquad \text{and} \qquad dv = \cos x\,dx,$$

then

$$du = e^x\,dx \qquad \text{and} \qquad v = \sin x,$$

so that

$$\int e^x \cos x\,dx = e^x \sin x - \int e^x \sin x\,dx. \tag{4}$$

Then, putting (4) into equation (3), we get

$$\begin{aligned}\int e^x \sin x\,dx &= e^x \sin x - \left(e^x \sin x - \int e^x \sin x\,dx\right) \\ &= \int e^x \sin x\,dx.\end{aligned} \tag{5}$$

Well, equation (5) is certainly true, but it is not very helpful. So we back up, and integrate the last integral in equation (3) by letting

$$u = \cos x \qquad \text{and} \qquad dv = e^x\,dx.$$

Then

$$du = -\sin x\,dx \qquad \text{and} \qquad v = e^x,$$

so

$$\int e^x \cos x\,dx = e^x \cos x + \int e^x \sin x\,dx. \tag{6}$$

Now putting (6) into equation (3), we get

$$\int e^x \sin x\,dx = e^x \sin x - e^x \cos x - \int e^x \sin x\,dx. \tag{7}$$

Notice that both integrals in equation (7) are the same, so we just "solve" equation (7) for the original integral:

$$2\int e^x \sin x\,dx = e^x \sin x - e^x \cos x, \tag{8}$$

$$\int e^x \sin x\,dx = \frac{1}{2}e^x(\sin x - \cos x) + C. \tag{9}$$

The constant C in equation (9) is written explicitly, rather than being included in the last term of equation (7). (Remember to add on a constant whenever you solve for an indefinite integral.) ■

Reduction Formulas

Integration by parts can be used to derive formulas for reducing particular integrals to simpler integrals of the same type. These **reduction formulas** can then be used—repeatedly if necessary—to evaluate an integral of the particular form.

■ **Example 7.18** Derive the reduction formula

$$\int x^n e^{-x}\,dx = -x^n e^{-x} + n\int x^{n-1}e^{-x}\,dx. \tag{10}$$

Solution In this formula, the exponent on x is reduced by 1 in the new integral, so we will let

$$u = x^n \qquad \text{and} \qquad dv = e^{-x}\,dx.$$

Then

$$du = nx^{n-1}\,dx \qquad \text{and} \qquad v = -e^{-x},$$

so that

$$\begin{aligned}\int x^n e^{-x}\,dx &= -x^n e^{-x} + \int nx^{n-1}e^{-x}\,dx\\ &= -x^n e^{-x} + n\int x^{n-1}e^{-x}\,dx.\end{aligned}$$

Thus, we have the desired formula. ■

■ **Example 7.19** Integrate $\int x^4 e^{-x}\,dx$.

Solution We may integrate by parts several times, or we may apply the reduction formula (10). We will do the latter:

$$\begin{aligned}\int x^4 e^{-x}\,dx &= -x^4 e^{-x} + 4\int x^3 e^{-x}\,dx\\ &= -x^4 e^{-x} + 4\left(-x^3 e^{-x} + 3\int x^2 e^{-x}\,dx\right)\end{aligned}$$

$$= -x^4e^{-x} - 4x^3e^{-x} + 12\int x^2e^{-x}\,dx$$

$$= -x^4e^{-x} - 4x^3e^{-x} + 12\left(-x^2e^{-x} + 2\int xe^{-x}\,dx\right)$$

$$= -x^4e^{-x} - 4x^3e^{-x} - 12x^2e^{-x} + 24\int xe^{-x}\,dx$$

$$= -x^4e^{-x} - 4x^3e^{-x} - 12x^2e^{-x} + 24\left(-xe^{-x} + \int e^{-x}\,dx\right)$$

$$= -x^4e^{-x} - 4x^3e^{-x} - 12x^2e^{-x} - 24xe^{-x} + 24\int e^{-x}\,dx$$

$$= -x^4e^{-x} - 4x^3e^{-x} - 12x^2e^{-x} - 24xe^{-x} - 24e^{-x} + C$$

$$= -e^{-x}(x^4 + 4x^3 + 12x^2 + 24x + 24) + C.$$

■

Exercises 7.2

Integrate.

1. $\int x \sin x\,dx$
2. $\int x \cos x\,dx$
3. $\int xe^{2x}\,dx$
4. $\int xe^{-x}\,dx$
5. $\int_0^1 xe^{x^2}\,dx$
6. $\int_0^1 x^2e^{3x}\,dx$
7. $\int_0^{\pi/2} x^2 \cos x\,dx$
8. $\int x^2 \sin x\,dx$
9. $\int \sin x \cos 3x\,dx$
10. $\int_0^{\pi/2} \cos 2x \cos 5x\,dx$
11. $\int e^{2x} \sin x\,dx$
12. $\int e^{-x} \cos x\,dx$
13. $\int x \ln x\,dx$
14. $\int x \ln(x+1)\,dx$
15. $\int x\sqrt{x+1}\,dx$
16. $\int x^2\sqrt{x+4}\,dx$
17. $\int \arcsin x\,dx$
18. $\int \arctan 2x\,dx$
19. $\int e^{ax} \sin bx\,dx$
20. $\int e^{ax} \cos bx\,dx$
21. $\int \cos ax \cos bx\,dx$
22. $\int \sin ax \cos bx\,dx$
23. $\int xe^{4+3\ln x}\,dx$
24. $\int \sin(\ln x)\,dx$

Derive the following reduction formulas:

25. $\int (\ln x)^n\,dx = x(\ln x)^n - n\int (\ln x)^{n-1}\,dx, \quad n \neq -1$

26. $\int x^ne^{ax}\,dx = \frac{1}{a}x^ne^{ax} - \frac{n}{a}\int x^{n-1}e^{ax}\,dx$

27. $\int x^n\sqrt{ax+b}\,dx = \frac{2}{a(2n+3)}\left[x^n(ax+b)^{3/2} - bn\int x^{n-1}\sqrt{ax+b}\,dx\right]$

28. $\int x^m(ax+b)^n\,dx = \frac{1}{m+n+1}\left[x^{m+1}(ax+b)^n + bn\int x^m(ax+b)^{n-1}\,dx\right]$

Integrate.

29. $\int_1^e (\ln x)^2\, dx$

30. $\int (\ln x)^3\, dx$

31. $\int x^3 e^{2x}\, dx$

A32. $\int_0^1 x^4 e^{6x}\, dx$

33. $\int x^2 \sqrt{3x+5}\, dx$

34. $\int x^2(2x+1)^3\, dx$

35. $\int \ln(x^2+1)\, dx$

36. $\int x^3 e^{x^2}\, dx$

SECTION 7.3

Trigonometric Integrals

In this section, we discuss many integrals involving trigonometric functions. To integrate trigonometric functions, we may need to use algebraic manipulation or trigonometric identities, along with integration by formula, by substitution, by parts, and even by outright tricks. Formulas (5)–(10) and (16)–(19) of Section 7.1 will be used a great deal.

Integrals Involving $\sin^m x \cos^n x$

First, we will consider integrals of the form

$$\int \sin^m x \cos^n x\, dx. \tag{1}$$

If $n = 1$, then a simple substitution of $u = \sin x$ will do; if $m = 1$, the substitution $u = \cos x$ works.

Example 7.20 Integrate $\int \sin^{1/2} x \cos x\, dx$.

Solution We let $u = \sin x$:

$$\begin{aligned}\int \sin^{1/2} x \cos x\, dx &= \int u^{1/2}\, du \\ &= \tfrac{2}{3} u^{3/2} + C \\ &= \tfrac{2}{3} \sin^{3/2} x + C.\end{aligned}$$

■

If n is any odd positive integer in form (1), say $n = 2k + 1$, then we can write

$$\begin{aligned}\int \sin^m x \cos^n x\, dx &= \int \sin^m x \cos^{2k+1} x\, dx \\ &= \int \sin^m x \cos^{2k} x \cos x\, dx\end{aligned}$$

$$\begin{aligned}\int \sin^m x \cos^{2k}x \cos x\,dx &= \int \sin^m x\,(1-\sin^2 x)^k \cos x\,dx\\ &= \int u^m(1-u^2)^k\,du. \qquad (u=\sin x)\end{aligned}$$

This last integral can be expanded and integrated term by term.

Example 7.21 Integrate $\int \sin^5 x \cos^3 x\,dx$.

Solution

$$\begin{aligned}\int \sin^5 x \cos^3 x\,dx &= \int \sin^5 x \cos^2 x \cos x\,dx\\ &= \int \sin^5 x\,(1-\sin^2 x)\cos x\,dx\\ &= \int u^5(1-u^2)\,du \qquad (u=\sin x)\\ &= \int (u^5-u^7)\,du\\ &= \tfrac{1}{6}u^6 - \tfrac{1}{8}u^8 + C\\ &= \tfrac{1}{6}\sin^6 x - \tfrac{1}{8}\sin^8 x + C.\end{aligned}$$

■

If m is an odd positive integer in form (1), then we change all but one sine into cosines.

Example 7.22 Find $\int \sin^3 x \cos^2 x\,dx$.

Solution

$$\begin{aligned}\int \sin^3 x \cos^2 x\,dx &= \int \sin^2 x \cos^2 x \sin x\,dx\\ &= \int (1-\cos^2 x)\cos^2 x \sin x\,dx\\ &= -\int (1-u^2)u^2\,du \qquad (u=\cos x)\\ &= -\int (u^2-u^4)\,du\\ &= -\tfrac{1}{3}u^3 + \tfrac{1}{5}u^5 + C\\ &= \tfrac{1}{5}\cos^5 x - \tfrac{1}{3}\cos^3 x + C.\end{aligned}$$

■

Example 7.23 Find $\int \sin^5 x\,dx$.

Solution

$$\begin{aligned}\int \sin^5 x\,dx &= \int \sin^4 x \sin x\,dx\\ &= \int (1-\cos^2 x)^2 \sin x\,dx\end{aligned}$$

$$\begin{aligned}\int (1-\cos^2 x)^2 \sin x\, dx &= -\int (1-u^2)^2\, du \qquad (u=\cos x)\\ &= -\int (1-2u^2+u^4)\, du\\ &= -u+\tfrac{2}{3}u^3-\tfrac{1}{5}u^5+C\\ &= -\cos x+\tfrac{2}{3}\cos^3 x-\tfrac{1}{5}\cos^5 x+C.\end{aligned}$$

■

■ **Example 7.24** Evaluate

$$\int_{\pi/4}^{\pi/3} \frac{\cos^3 x}{\sin^2 x}\, dx.$$

Solution

$$\begin{aligned}\int_{\pi/4}^{\pi/3} \frac{\cos^3 x}{\sin^2 x}\, dx &= \int_{\pi/4}^{\pi/3} \frac{(1-\sin^2 x)\cos x}{\sin^2 x}\, dx\\ &= \int_{\sqrt{2}/2}^{\sqrt{3}/2} \frac{1-u^2}{u^2}\, du \qquad (u=\sin x)\\ &= \int_{\sqrt{2}/2}^{\sqrt{3}/2} (u^{-2}-1)\, du\\ &= \left[-\frac{1}{u}-u\right]_{\sqrt{2}/2}^{\sqrt{3}/2}\\ &= \frac{3\sqrt{2}}{2}-\frac{7\sqrt{3}}{6}.\end{aligned}$$

■

If both m and n are even integers in form (1), we make use of the identities

$$\sin^2\theta = \tfrac{1}{2}(1-\cos 2\theta), \qquad (2)$$

$$\cos^2\theta = \tfrac{1}{2}(1+\cos 2\theta), \qquad (3)$$

$$\sin\theta\cos\theta = \tfrac{1}{2}\sin 2\theta. \qquad (4)$$

Algebraic manipulation will usually be involved also.

■ **Example 7.25** Integrate $\int \sin^2 x\, dx$.

Solution

$$\begin{aligned}\int \sin^2 x\, dx &= \int \tfrac{1}{2}(1-\cos 2x)\, dx\\ &= \tfrac{1}{2}(x-\tfrac{1}{2}\sin 2x)+C\\ &= \tfrac{1}{2}x-\tfrac{1}{4}\sin 2x+C.\end{aligned}$$

■

■ **Example 7.26** Find $\int \sin^4 x\, dx$.

Solution Using identities (2) and (3), we have

$$\begin{aligned}\int \sin^4 x\,dx &= \int (\sin^2 x)^2\,dx\\ &= \int [\tfrac{1}{2}(1-\cos 2x)]^2\,dx\\ &= \int [\tfrac{1}{4}-\tfrac{1}{2}\cos 2x+\tfrac{1}{4}\cos^2 2x]\,dx\\ &= \int [\tfrac{1}{4}-\tfrac{1}{2}\cos 2x+\tfrac{1}{4}\cdot\tfrac{1}{2}(1+\cos 4x)]\,dx\\ &= \int [\tfrac{3}{8}-\tfrac{1}{2}\cos 2x+\tfrac{1}{8}\cos 4x]\,dx\\ &= \tfrac{3}{8}x-\tfrac{1}{4}\sin 2x+\tfrac{1}{32}\sin 4x+C.\end{aligned}$$

■

■ **Example 7.27** Evaluate $\int_0^{\pi/4} \sin^2 x \cos^2 x\,dx$.

Solution This time we use identity (4):

$$\begin{aligned}\int_0^{\pi/4} \sin^2 x\cos^2 x\,dx &= \int_0^{\pi/4} (\sin x\cos x)^2\,dx\\ &= \int_0^{\pi/4} \frac{1}{4}\sin^2 2x\,dx\\ &= \frac{1}{4}\int_0^{\pi/4} \frac{1}{2}(1-\cos 4x)\,dx\\ &= \left[\frac{1}{8}x-\frac{1}{32}\sin 4x\right]_0^{\pi/4}\\ &= \frac{\pi}{32}.\end{aligned}$$

■

■ **Example 7.28** Integrate

$$\int \frac{\sin^4 x}{\cos^2 x}\,dx.$$

Solution

$$\begin{aligned}\int \frac{\sin^4 x}{\cos^2 x}\,dx &= \int \frac{(\sin^2 x)^2}{\cos^2 x}\,dx\\ &= \int \frac{(1-\cos^2 x)^2}{\cos^2 x}\,dx\\ &= \int \frac{1-2\cos^2 x+\cos^4 x}{\cos^2 x}\,dx\end{aligned}$$

$$\int \frac{1 - 2\cos^2 x + \cos^4 x}{\cos^2 x}\,dx = \int \left[\frac{1}{\cos^2 x} - 2 + \cos^2 x\right] dx$$
$$= \int \left[\sec^2 x - 2 + \frac{1}{2}(1 + \cos 2x)\right] dx$$
$$= \int \left[\sec^2 x - \frac{3}{2} + \frac{1}{2}\cos 2x\right] dx$$
$$= \tan x - \frac{3}{2}x + \frac{1}{4}\sin 2x + C. \quad ■$$

Occasionally, in problems similar to these, integrals of the forms given below arise:

$$\int \sin ax \sin bx\,dx, \tag{5}$$
$$\int \sin ax \cos bx\,dx, \tag{6}$$
$$\int \cos ax \cos bx\,dx. \tag{7}$$

These can be integrated by parts, or the trigonometric identities below can be used to convert the products into sums:

$$\sin ax \sin bx = \tfrac{1}{2}[\cos(a - b)x - \cos(a + b)x], \tag{8}$$
$$\sin ax \cos bx = \tfrac{1}{2}[\sin(a + b)x + \sin(a - b)x], \tag{9}$$
$$\cos ax \cos bx = \tfrac{1}{2}[\cos(a + b)x + \cos(a - b)x]. \tag{10}$$

■ **Example 7.29** Integrate $\int \sin 4x \sin 2x\,dx$.

Solution Using identity (8),

$$\int \sin 4x \sin 2x\,dx = \int \tfrac{1}{2}[\cos 2x - \cos 6x]\,dx$$
$$= \tfrac{1}{2}[\tfrac{1}{2}\sin 2x - \tfrac{1}{6}\sin 6x] + C$$
$$= \tfrac{1}{4}\sin 2x - \tfrac{1}{12}\sin 6x + C. \quad ■$$

Integrals Involving $\sec^m x \tan^n x$

We next consider integrals of the form

$$\int \sec^m x \tan^n x\,dx. \tag{11}$$

If m is an even positive integer, say $m = 2k$, we write

$$\begin{aligned}\int \sec^m x \tan^n x\, dx &= \int \sec^{2k-2} x \tan^n x \sec^2 x\, dx \\ &= \int (\sec^2 x)^{k-1} \tan^n x \sec^2 x\, dx \\ &= \int (1 + \tan^2 x)^{k-1} \tan^n x \sec^2 x\, dx \\ &= \int (1 + u^2)^{k-1} u^n\, du,\end{aligned}$$

where we have substituted $u = \tan x$, $du = \sec^2 x\, dx$. This last integral can be expanded and integrated term by term. That is, we factor out $\sec^2 x$ and change the rest of the secants into tangents, after which we substitute $u = \tan x$.

Example 7.30 Find $\int \sec^6 x\, dx$.

Solution

$$\begin{aligned}\int \sec^6 x\, dx &= \int (\sec^2 x)^2 \sec^2 x\, dx \\ &= \int (1 + \tan^2 x)^2 \sec^2 x\, dx \\ &= \int (1 + u^2)^2\, du \qquad (u = \tan x) \\ &= \int (1 + 2u^2 + u^4)\, du \\ &= u + \tfrac{2}{3}u^3 + \tfrac{1}{5}u^5 + C \\ &= \tan x + \tfrac{2}{3}\tan^3 x + \tfrac{1}{5}\tan^5 x + C.\end{aligned}$$

If n in form (11) is an odd positive integer, we change all but one of the tangents into secants. That is, if $n = 2k + 1$, we write

$$\begin{aligned}\int \sec^m x \tan^n x\, dx &= \int \sec^{m-1} x \tan^{2k} x \sec x \tan x\, dx \\ &= \int \sec^{m-1} x\, (\sec^2 x - 1)^k \sec x \tan x\, dx \\ &= \int u^{m-1}(u^2 - 1)^k\, du. \qquad (u = \sec x)\end{aligned}$$

Example 7.31 Find $\int \sec^3 x \tan x\, dx$.

Solution

$$\begin{aligned}\int \sec^3 x \tan x\,dx &= \int \sec^2 x \sec x \tan x\,dx\\ &= \int u^2\,du \qquad (u = \sec x)\\ &= \tfrac{1}{3}u^3 + C\\ &= \tfrac{1}{3}\sec^3 x + C.\end{aligned}$$

■

If m is odd and n is even in form (11), then the tangents can all be changed into secants via the identity $\tan^2 x = \sec^2 x - 1$, yielding a sum of integrals of odd powers of the secant. An odd power of the secant is integrated by applying one of the following examples.

■ **Example 7.32** Show that $\int \sec x\,dx = \ln|\sec x + \tan x| + C$.

Solution This formula is obtained by sheer trickery, as follows:

$$\begin{aligned}\int \sec x\,dx &= \int \sec x\,\frac{\sec x + \tan x}{\sec x + \tan x}\,dx\\ &= \int \frac{\sec^2 x + \sec x \tan x}{\sec x + \tan x}\,dx\\ &= \int \frac{1}{u}\,du \qquad (u = \sec x + \tan x)\\ &= \ln|u| + C\\ &= \ln|\sec x + \tan x| + C.\end{aligned}$$

■

■ **Example 7.33** If $n > 1$, then show that

$$\int \sec^n x\,dx = \frac{1}{n-1}\sec^{n-2}x \tan x + \frac{n-2}{n-1}\int \sec^{n-2}x\,dx.$$

Solution This reduction formula is obtained by integrating by parts and using trigonometric identities. First, we write

$$\int \sec^n x\,dx = \int \sec^{n-2}x \sec^2 x\,dx. \qquad (12)$$

Next, we integrate by parts with

$$u = \sec^{n-2}x \qquad \text{and} \qquad dv = \sec^2 x\,dx.$$

Then

$$\begin{aligned}du &= (n-2)\sec^{n-3}x \sec x \tan x\,dx \qquad \text{and} \qquad v = \tan x.\\ &= (n-2)\sec^{n-2}x \tan x\,dx\end{aligned}$$

Hence,

$$\int \sec^{n-2}x \sec^2 x\, dx = \sec^{n-2}x \tan x - \int (n-2)\sec^{n-2}x \tan^2 x\, dx$$
$$= \sec^{n-2}x \tan x - (n-2)\int \sec^{n-2}x (\sec^2 x - 1)\, dx$$
$$= \sec^{n-2}x \tan x - (n-2)\int \sec^n x\, dx + (n-2)\int \sec^{n-2}x\, dx.$$

This, combined with equation (12), yields

$$(n-1)\int \sec^n x\, dx = \sec^{n-2}x \tan x + (n-2)\int \sec^{n-2}x\, dx,$$

from which the formula follows. ∎

Example 7.34 Find $\int \sec^3 x\, dx$.

Solution Using Example 7.33 and then Example 7.32, we have

$$\int \sec^3 x\, dx = \tfrac{1}{2}\sec x \tan x + \tfrac{1}{2}\int \sec x\, dx$$
$$= \tfrac{1}{2}\sec x \tan x + \tfrac{1}{2}\ln|\sec x + \tan x| + C.$$ ∎

Example 7.35 Find $\int \sec x \tan^2 x\, dx$.

Solution Changing the tangents to secants, we have

$$\int \sec x \tan^2 x\, dx = \int \sec x (\sec^2 x - 1)\, dx$$
$$= \int \sec^3 x\, dx - \int \sec x\, dx$$
$$= \tfrac{1}{2}\sec x \tan x + \tfrac{1}{2}\ln|\sec x + \tan x| - \ln|\sec x + \tan x| + C$$
$$= \tfrac{1}{2}\sec x \tan x - \tfrac{1}{2}\ln|\sec x + \tan x| + C.$$ ∎

Example 7.36 Find $\int \sec^5 x\, dx$.

Solution

$$\int \sec^5 x\, dx = \tfrac{1}{4}\sec^3 x \tan x + \tfrac{3}{4}\int \sec^3 x\, dx$$
$$= \tfrac{1}{4}\sec^3 x \tan x + \tfrac{3}{4}\left(\tfrac{1}{2}\sec x \tan x + \tfrac{1}{2}\ln|\sec x + \tan x|\right) + C$$
$$= \tfrac{1}{4}\sec^3 x \tan x + \tfrac{3}{8}\sec x \tan x + \tfrac{3}{8}\ln|\sec x + \tan x| + C.$$ ∎

Other Trigonometric Integrals

Integrals involving powers and products of cosecants and cotangents, of the form

$$\int \csc^m x \cot^n x \, dx, \tag{13}$$

are handled in the same way as form (11), but using the identity $1 + \cot^2 x = \csc^2 x$.

Example 7.37 Find $\int \csc^2 x \cot^2 x \, dx$.

Solution

$$\begin{aligned} \int \csc^2 x \cot^2 x \, dx &= -\int u^2 \, du \qquad (u = \cot x) \\ &= -\tfrac{1}{3}u^3 + C \\ &= -\tfrac{1}{3}\cot^3 x + C. \end{aligned}$$

The formula for the integral of the cosecant is

$$\int \csc x \, dx = \ln|\csc x - \cot x| + C. \tag{14}$$

It is derived in the same way as the formula of Example 7.32.

Since all the trigonometric functions can be expressed in terms of sines and cosines, any trigonometric integral can be expressed as an integral involving only sines and cosines. Sometimes, that will simplify an integral when nothing else works. Later, we will discuss a substitution that works (in theory) for any rational function of sines and cosines.

Exercises 7.3

Integrate.

1. $\int \sin^2 x \, dx$
2. $\int_0^{\pi/6} \cos^2 x \, dx$
3. $\int \sin^3 x \, dx$
4. $\int \cos^3 x \, dx$
5. $\int \sin^2 3x \cos 3x \, dx$
6. $\int \sin(x + 3) \cos^2(x + 3) \, dx$
7. $\int \sin x \cos^3 x \, dx$
8. $\int \cos^4 x \, dx$
9. $\int \cos^2 4x \sin^2 4x \, dx$
10. $\int \cos 2x \sin^3 2x \, dx$
11. $\int_0^{\pi/2} \sin^4 x \, dx$
12. $\int \sin^3 x \cos^6 x \, dx$
13. $\int \sin x \cos 2x \, dx$
14. $\int \sin 2x \cos x \, dx$

15. $\int \cos 2x \sin^2 x \, dx$

16. $\int \sin 2x \cos^3 x \, dx$

17. $\int \sin 5x \cos 2x \, dx$

18. $\int \cos 2x \cos 6x \, dx$

19. $\int \sec 2x \, dx$

20. $\int_0^{\pi} \sec \frac{x}{3} \, dx$

21. $\int \csc(x + 1) \, dx$

22. $\int \csc ax \, dx$

23. $\int \sec^2 3x \, dx$

24. $\int \sec 2x \tan 2x \, dx$

25. $\int \sec^2 2x \tan 2x \, dx$

26. $\int_0^{\pi/2} \sec \frac{x}{2} \tan^3 \frac{x}{2} \, dx$

27. $\int \sec^3 x \tan^3 x \, dx$

28. $\int \sec^3 x \tan^2 x \, dx$

29. $\int \sec^4 x \tan^3 x \, dx$

30. $\int \sec^{3/2} x \tan^3 x \, dx$

31. $\int \csc 2x \cot 2x \, dx$

32. $\int \csc x \cot^3 x \, dx$

33. $\int \frac{1}{\sin x} \, dx$

34. $\int \frac{1}{\cos x} \, dx$

35. $\int \frac{1}{\sec^2 x} \, dx$

36. $\int \sec^4 x \sqrt{\tan x} \, dx$

37. $\int \sin x \tan x \, dx$

38. $\int \frac{\tan^2 x}{\sec^2 x} \, dx$

39. $\int \frac{\tan x \, dx}{\sqrt{\sec x}}$

40. $\int \tan^2 x \, dx$

41. $\int \frac{\sin x}{1 - \cos x} \, dx$

42. $\int \frac{\cos x}{1 + \sin x} \, dx$

43. $\int \frac{\cos^2 x}{1 - \sin x} \, dx$

44. $\int \frac{\sin^2 x}{1 + \cos x} \, dx$

Derive the following formulas:

45. $\int \csc x \, dx = \ln|\csc x - \cot x| + C$

46. $\int \csc^n x \, dx = -\frac{1}{n-1} \csc^{n-2} x \cot x + \frac{n-2}{n-1} \int \csc^{n-2} x \, dx$

47. $\int \sin^n x \, dx = -\frac{1}{n} \sin^{n-1} x \cos x + \frac{n-1}{n} \int \sin^{n-2} x \, dx$

48. $\int \cos^n x \, dx = \frac{1}{n} \cos^{n-1} x \sin x + \frac{n-1}{n} \int \cos^{n-2} x \, dx$

49. $\int \tan^n x \, dx = \frac{1}{n-1} \tan^{n-1} x - \int \tan^{n-2} x \, dx$

SECTION 7.4

Trigonometric Substitution

Several types of algebraic functions can be integrated by using what is called a **trigonometric substitution** to change a sum or difference of squares into a single square. This is particularly useful in integrals involving such expressions as $\sqrt{a^2 - x^2}$, $\sqrt{a^2 + x^2}$, or $\sqrt{x^2 - a^2}$. We will deal with each of these types in turn. We assume $a > 0$ in each case.

Integrals Involving $\sqrt{a^2 - x^2}$

The expression $\sqrt{a^2 - x^2}$ is defined and real only for $-a \le x \le a$. By making the substitution $x = a \sin \theta$, or $\theta = \sin^{-1}(x/a)$, with $-\pi/2 \le \theta \le \pi/2$, we can simplify the expression as follows:

$$\begin{aligned}\sqrt{a^2 - x^2} &= \sqrt{a^2 - a^2 \sin^2\theta} \\ &= \sqrt{a^2(1 - \sin^2\theta)} \\ &= \sqrt{a^2 \cos^2\theta} \\ &= a \cos \theta,\end{aligned}$$

since $\cos \theta \ge 0$ for $-\pi/2 \le \theta \le \pi/2$.

Example 7.38 Evaluate $\int_0^a \sqrt{a^2 - x^2}\, dx$.

Solution We let $x = a \sin \theta$, so $dx = a \cos \theta\, d\theta$. Also, $\theta = 0$ if $x = 0$, and $\theta = \pi/2$ if $x = a$. Hence,

$$\begin{aligned}\int_0^a \sqrt{a^2 - x^2}\, dx &= \int_0^{\pi/2} a \cos \theta\, a \cos \theta\, d\theta \\ &= a^2 \int_0^{\pi/2} \cos^2\theta\, d\theta \\ &= \frac{a^2}{2} \int_0^{\pi/2} (1 + \cos 2\theta)\, d\theta \\ &= \frac{a^2}{2} \left[\theta + \frac{1}{2} \sin 2\theta\right]_0^{\pi/2} \\ &= \frac{a^2}{2} \left[\frac{\pi}{2} + 0 - 0 - 0\right] = \frac{\pi a^2}{4}.\end{aligned}$$

By interpreting the integral as the area of a quarter circle of radius a, we know this answer is correct. ■

Example 7.39 Evaluate

$$\int \frac{dx}{(4-x^2)^2}.$$

Solution We let $x = 2\sin\theta$ and $dx = 2\cos\theta\, d\theta$. Then $4 - x^2 = 4 - 4\sin^2\theta = 4(1-\sin^2\theta) = 4\cos^2\theta$, so $(4-x^2)^2 = 16\cos^4\theta$. Hence,

$$\begin{aligned}\int \frac{dx}{(4-x^2)^2} &= \int \frac{2\cos\theta}{16\cos^4\theta}\, d\theta \qquad (1)\\ &= \frac{1}{8}\int \sec^3\theta\, d\theta\\ &= \frac{1}{8}\left(\frac{1}{2}\sec\theta\tan\theta + \frac{1}{2}\ln|\sec\theta + \tan\theta|\right) + C.\end{aligned}$$

To return to the original variable x, we use such trigonometric identities as

$$\begin{aligned}\tan\theta &= \frac{\sin\theta}{\cos\theta} = \frac{\sin\theta}{\sqrt{1-\sin^2\theta}} \qquad (2)\\ &= \frac{x/2}{\sqrt{1-x^2/4}} = \frac{x}{\sqrt{4-x^2}},\\ \sec\theta &= \frac{1}{\cos\theta} = \frac{1}{\sqrt{1-\sin^2\theta}} \qquad (3)\\ &= \frac{1}{\sqrt{1-x^2/4}} = \frac{2}{\sqrt{4-x^2}}.\end{aligned}$$

These same identities may be obtained more readily by thinking of θ as an acute angle of a right triangle (see Figure 7.1). Since $x = 2\sin\theta$, $\sin\theta = x/2$, and we may give the side of the triangle opposite θ the length x and the hypotenuse length 2. Hence, the side adjacent to θ has length $\sqrt{4-x^2}$. Therefore, $\tan\theta$ and $\sec\theta$ can be read directly from this triangle. At any rate, putting identities (2) and (3) into equation (1), we get

$$\int \frac{dx}{(4-x^2)^2} = \frac{x}{8(4-x^2)} + \frac{1}{16}\ln\left|\frac{x+2}{\sqrt{4-x^2}}\right| + C.$$

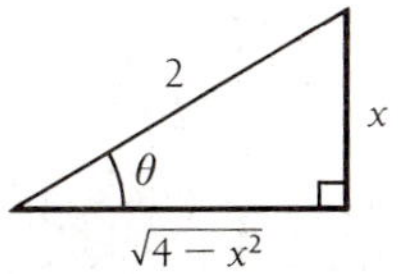

FIGURE 7.1 A triangle showing $\theta = \arcsin(x/2)$ ■

Integrals Involving $\sqrt{a^2 + x^2}$

The expression $\sqrt{a^2 + x^2}$ is defined for all x. The substitution $x = a \tan \theta$, or $\theta = \tan^{-1}(x/a)$, with $-\pi/2 < \theta < \pi/2$, will eliminate the radical:

$$\begin{aligned}\sqrt{a^2 + x^2} &= \sqrt{a^2 + a^2 \tan^2\theta} \\ &= \sqrt{a^2(1 + \tan^2\theta)} \\ &= \sqrt{a^2 \sec^2\theta} \\ &= a \sec \theta,\end{aligned}$$

because $\sec \theta > 0$ for $-\pi/2 < \theta < \pi/2$.

Example 7.40 Evaluate

$$\int_0^{\sqrt{3}} (x^2 + 1)^{-3/2}\, dx.$$

Solution We let $x = \tan \theta$ and $dx = \sec^2\theta\, d\theta$. Also, $\theta = 0$ when $x = 0$ and $\theta = \pi/3$ when $x = \sqrt{3}$. Hence,

$$\begin{aligned}\int_0^{\sqrt{3}} (x^2 + 1)^{-3/2}\, dx &= \int_0^{\pi/3} (\sec^2\theta)^{-3/2} \sec^2\theta\, d\theta \\ &= \int_0^{\pi/3} \frac{\sec^2\theta}{\sec^3\theta}\, d\theta \\ &= \int_0^{\pi/3} \cos \theta\, d\theta \\ &= \sin \theta \Big|_0^{\pi/3} = \frac{\sqrt{3}}{2}.\end{aligned}$$

■

Example 7.41 Find

$$\int \frac{dx}{(x^2 + 9)^2}.$$

Solution We let $x = 3 \tan \theta$ and $dx = 3 \sec^2\theta\, d\theta$. Then $x^2 + 9 = 9 \sec^2\theta$, so

$$\begin{aligned}\int \frac{dx}{(x^2 + 9)^2} &= \int \frac{3 \sec^2\theta}{81 \sec^4\theta}\, d\theta \\ &= \frac{1}{27} \int \cos^2\theta\, d\theta \\ &= \frac{1}{27} \cdot \frac{1}{2} \int (1 + \cos 2\theta)\, d\theta \\ &= \frac{1}{54} \left(\theta + \frac{1}{2} \sin 2\theta\right) + C \\ &= \frac{1}{54} (\theta + \sin \theta \cos \theta) + C,\end{aligned}$$

using the double-angle formula for the sine. Now, to get back to the original variable x, we again use a triangle. Let θ be an acute angle of a right triangle with opposite side x and adjacent side 3 (see Figure 7.2). Then $\tan\theta = x/3$, as given. It follows that the hypotenuse is $\sqrt{x^2+9}$, and we have

$$\sin\theta = \frac{x}{\sqrt{x^2+9}}, \qquad \cos\theta = \frac{3}{\sqrt{x^2+9}}.$$

Also, $\theta = \arctan(x/3)$. Hence, the integral becomes

$$\int \frac{dx}{(x^2+9)^2} = \frac{1}{54}\arctan\frac{x}{3} + \frac{x}{18(x^2+9)} + C.$$

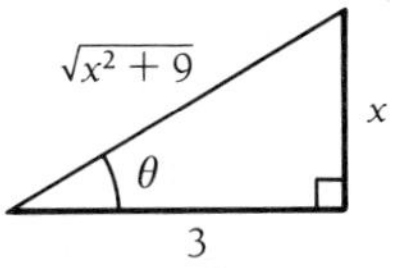

FIGURE 7.2 A triangle showing $\theta = \arctan(x/3)$ ■

Integrals Involving $\sqrt{x^2 - a^2}$

The expression $\sqrt{x^2-a^2}$ is defined for $x \geq a$ or $x \leq -a$. We make the substitution $x = a\sec\theta$, or $\theta = \sec^{-1}(x/a)$, with $0 \leq \theta < \pi/2$ if $x \geq a$ and $-\pi \leq \theta < -\pi/2$ if $x \leq -a$. This eliminates the radical:

$$\begin{aligned}\sqrt{x^2-a^2} &= \sqrt{a^2\sec^2\theta - a^2}\\ &= \sqrt{a^2(\sec^2\theta - 1)}\\ &= \sqrt{a^2\tan^2\theta}\\ &= a\tan\theta,\end{aligned}$$

since $\tan\theta > 0$ if θ is in Quadrant I or Quadrant III.

■ **Example 7.42** Find $\int_2^{2\sqrt{2}} x^3(x^2-4)^{3/2}\,dx$.

Solution We let $x = 2\sec\theta$ and $dx = 2\sec\theta\tan\theta\,d\theta$. Also, since $\theta = \text{arcsec}(x/2)$, $\theta = 0$ when $x = 2$ and $\theta = \pi/4$ when $x = 2\sqrt{2}$. Hence,

$$\begin{aligned}\int_2^{2\sqrt{2}} x^3(x^2-4)^{3/2}\,dx &= \int_0^{\pi/4} 8\sec^3\theta\; 8\tan^3\theta\; 2\sec\theta\tan\theta\,d\theta\\ &= \int_0^{\pi/4} 128\sec^4\theta\tan^4\theta\,d\theta\\ &= 128\int_0^{\pi/4}(1+\tan^2\theta)\tan^4\theta\sec^2\theta\,d\theta\\ &= 128\int_0^1 (1+u^2)u^4\,du \qquad (u = \tan\theta)\end{aligned}$$

$$128\int_0^1 (1+u^2)u^4\,du = 128\int_0^1 (u^4+u^6)\,du$$
$$= 128\left[\frac{u^5}{5}+\frac{u^7}{7}\right]_0^1$$
$$= 128\left(\frac{1}{5}+\frac{1}{7}\right) = \frac{1536}{35}.$$

■

■ **Example 7.43** Find

$$\int \frac{dx}{\sqrt{x^2-5}}.$$

Solution Letting $x=\sqrt{5}\sec\theta$, $dx=\sqrt{5}\sec\theta\tan\theta\,d\theta$, we find that $\sqrt{x^2-5}=\sqrt{5}\tan\theta$. Therefore,

$$\int \frac{dx}{\sqrt{x^2-5}} = \int \frac{\sqrt{5}\sec\theta\tan\theta\,d\theta}{\sqrt{5}\tan\theta}$$
$$= \int \sec\theta\,d\theta$$
$$= \ln|\sec\theta+\tan\theta|+C.$$

To get back to the variable x, we let θ be an acute angle of a right triangle with hypotenuse x and adjacent side $\sqrt{5}$ (see Figure 7.3). Thus, $\sec\theta = x/\sqrt{5}$, as needed. Now the opposite side is $\sqrt{x^2-5}$, and $\tan\theta=\sqrt{x^2-5}/\sqrt{5}$. Thus, the integral becomes

$$\int \frac{dx}{\sqrt{x^2-5}} = \ln\left|\frac{x}{\sqrt{5}}+\frac{\sqrt{x^2-5}}{\sqrt{5}}\right|+C$$
$$= \ln\left|\frac{x+\sqrt{x^2-5}}{\sqrt{5}}\right|+C.$$

This last expression may be written

$$\ln|x+\sqrt{x^2-5}|+C,$$

in which the constant $-\ln\sqrt{5}$ has been combined with C.

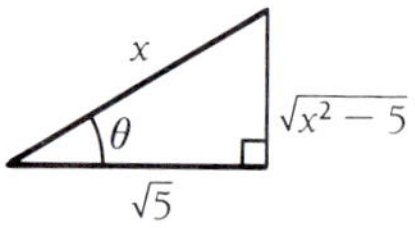

FIGURE 7.3 A triangle showing $\theta = \text{arcsec}(x/\sqrt{5})$

■

To summarize, trigonometric substitution can simplify expressions involving sums or differences of squares, changing algebraic integrals into trigonometric integrals. The substitutions to make are:

EXPRESSION	SUBSTITUTION	RESTRICTIONS
$a^2 - x^2$	$x = a \sin\theta$ $\theta = \sin^{-1}(x/a)$	$-a \le x \le a,\ -\pi/2 \le \theta \le \pi/2$
$a^2 + x^2$	$x = a \tan\theta$ $\theta = \tan^{-1}(x/a)$	$x \in \mathbb{R},\ -\pi/2 < \theta < \pi/2$
$x^2 - a^2$	$x = a \sec\theta$ $\theta = \sec^{-1}(x/a)$	$x \ge a$ and $0 \le \theta < \pi/2$, or $x \le -a$ and $-\pi \le \theta < -\pi/2$

Exercises 7.4

Integrate.

1. $\int \sqrt{1 - x^2}\, dx$
2. $\int_0^2 \sqrt{4 - x^2}\, dx$
3. $\int \sqrt{a^2 - x^2}\, dx$
4. $\int (9 - x^2)^{3/2}\, dx$
5. $\int \frac{dx}{3 - x^2}$
6. $\int_0^{a/2} \frac{x\, dx}{a^2 - x^2}$
7. $\int \frac{x\, dx}{(a^2 - x^2)^2}$
8. $\int (a^2 - x^2)^{-5/2}\, dx$
9. (A) $\int_0^{1/2} (1 - x^2)^{-3}\, dx$
10. $\int (5 - x^2)^{-3/2}\, dx$
11. $\int_0^a \sqrt{a^2 - x^2}\, dx$
12. $\int_{-1}^1 \frac{dx}{3 - x^2}$
13. $\int \sqrt{x^2 + 1}\, dx$
14. $\int_0^1 \sqrt{x^2 + 4}\, dx$
15. $\int x^3(3 + x^2)^{3/2}\, dx$
16. $\int_0^2 x^3(4 + x^2)^{3/2}\, dx$
17. $\int \frac{dx}{\sqrt{x^2 + 9}}$
18. $\int (5 + x^2)^{-1/2}\, dx$
19. $\int (x^2 + 1)^{-3/2}\, dx$
20. $\int (x^2 + 1)^{-5/2}\, dx$
21. $\int (4 + x^2)^{-2}\, dx$
22. (A) $\int (3 + x^2)^{-3}\, dx$
23. $\int_0^1 \sqrt{x^2 + 1}\, dx$
24. $\int_{-3}^0 \frac{dx}{\sqrt{x^2 + 9}}$
25. $\int \sqrt{x^2 - 1}\, dx$
26. $\int \sqrt{x^2 - 4}\, dx$
27. $\int_{\sqrt{3}}^{\sqrt{6}} \sqrt{x^2 - 3}\, dx$
28. $\int \sqrt{x^2 - a^2}\, dx$
29. $\int x\sqrt{x^2 - 1}\, dx$
30. $\int x^3\sqrt{x^2 - 1}\, dx$
31. $\int_1^3 \sqrt{x^2 - 1}\, dx$
32. $\int_a^{2a} \sqrt{x^2 - a^2}\, dx$
33. $\int \frac{dx}{x^2 - 4}$
34. $\int \frac{dx}{x^2 - 8}$
35. $\int \frac{dx}{x^2 - a^2}$
36. $\int \frac{x\, dx}{x^2 - 9}$
37. $\int \frac{x^2\, dx}{x^2 - 1}$
38. $\int \frac{x^2\, dx}{x^2 - a^2}$
39. $\int_5^8 \frac{dx}{x^2 - 16}$

A40. $\displaystyle\int_2^4 x^2\sqrt{x^2-4}\,dx$

41. $\displaystyle\int \frac{dx}{\sqrt{x^2-25}}$

42. $\displaystyle\int_{\sqrt{5}}^{2\sqrt{2}} \frac{dx}{\sqrt{x^2-4}}$

43. $\displaystyle\int x^2(x^2-a^2)^{-3/2}\,dx$

44. $\displaystyle\int x^3\sqrt{x^2-a^2}\,dx$

A45. $\displaystyle\int (x^2-a^2)^{-2}\,dx$

46. $\displaystyle\int \frac{x^3\,dx}{(x^2-a^2)^3}$

47. $\displaystyle\int \frac{dx}{x\sqrt{x^2-a^2}}$

48. $\displaystyle\int \frac{dx}{x\sqrt{a^2-x^2}}$

49. $\displaystyle\int \frac{dx}{x\sqrt{a^2+x^2}}$

50. $\displaystyle\int \frac{dx}{x^2\sqrt{a^2+x^2}}$

51. $\displaystyle\int \frac{dx}{1+a^2x^2}$

52. $\displaystyle\int \frac{dx}{\sqrt{a^2x^2-1}}$

53. $\displaystyle\int_0^a \sqrt{a^2x^2+1}\,dx$

54. $\displaystyle\int_0^{1/a} \sqrt{1-a^2x^2}\,dx$

55. Integrate the expression below using the substitution $x = a \sinh u$:

$$\int \frac{dx}{\sqrt{x^2+a^2}}$$

56. Integrate the expression below using the substitution $x = a \tanh u$:

$$\int \frac{dx}{\sqrt{a^2-x^2}}$$

57. Explain how to integrate an expression involving $\sqrt{a^2-x^2}$ using the substitution $x = a\cos\theta$, $0 \le \theta \le \pi$.

58. Integrate $\int \sqrt{x^2+a^2}\,dx$ using the substitution $x = a\cot\theta$, $0 < \theta < \pi$.

SECTION 7.5

Completing the Square

Here, we generalize the previous section in that we consider integrals involving expressions of the form

$$\sqrt{ax^2+bx+c} \tag{1}$$

with $b \neq 0$. The first step in simplifying such an expression is to **complete the square** — that is, write the expression as a sum or difference of squares.

Completing the Square

To complete the square in the quadratic trinomial

$$ax^2+bx+c, \tag{2}$$

it is convenient to first factor out the coefficient a, and then add and subtract the quantity that will yield the square of a binomial plus a constant:

$$ax^2+bx+c = a\left[x^2+\frac{b}{a}x+\frac{c}{a}\right] \tag{3}$$

$$= a\left[\left(x^2+\frac{b}{a}x+\frac{b^2}{4a^2}\right)+\frac{c}{a}-\frac{b^2}{4a^2}\right] \tag{4}$$

$$= a\left[\left(x+\frac{b}{2a}\right)^2+\frac{4ac-b^2}{4a^2}\right]. \tag{5}$$

The term added and subtracted in step (4) is the square of half the coefficient of x. The expression (5) is now a times a sum or difference of squares, depending on the sign of $b^2 - 4ac$.

Consider an example or two.

■ Example 7.44 Write $2x - x^2$ as a sum or difference of squares.

Solution

$$\begin{aligned} 2x - x^2 &= (-1)[x^2 - 2x] \\ &= (-1)[(x^2 - 2x + 1) - 1] \\ &= (-1)[(x-1)^2 - 1] \\ &= 1 - (x-1)^2. \end{aligned}$$

■

■ Example 7.45 Complete the square: $2x^2 + 5x - 7$.

Solution

$$\begin{aligned} 2x^2 + 5x - 7 &= 2[x^2 + \tfrac{5}{2}x - \tfrac{7}{2}] \\ &= 2[(x^2 + \tfrac{5}{2}x + \tfrac{25}{16}) - \tfrac{7}{2} - \tfrac{25}{16}] \\ &= 2[(x + \tfrac{5}{4})^2 - \tfrac{81}{16}] \\ &= 2[(x + \tfrac{5}{4})^2 - (\tfrac{9}{4})^2]. \end{aligned}$$

■

■ Example 7.46 Complete the square: $2x^2 + 5x + 7$.

Solution

$$\begin{aligned} 2x^2 + 5x + 7 &= 2[x^2 + \tfrac{5}{2}x + \tfrac{7}{2}] \\ &= 2[(x^2 + \tfrac{5}{2}x + \tfrac{25}{16}) + \tfrac{7}{2} - \tfrac{25}{16}] \\ &= 2[(x + \tfrac{5}{4})^2 + \tfrac{31}{16}] \\ &= 2[(x + \tfrac{5}{4})^2 + (\sqrt{31}/4)^2]. \end{aligned}$$

■

Integration Examples

Once we have completed the square, then the substitution

$$u = x + \frac{b}{2a} \tag{6}$$

gives us an integral we can treat as in the last section.

Consider a few examples.

Example 7.47 Integrate $\int \sqrt{2x - x^2}\, dx$.

Solution By Example 7.44, we have

$$
\begin{aligned}
\int \sqrt{2x-x^2}\,dx &= \int \sqrt{1-(x-1)^2}\,dx \\
&= \int \sqrt{1-u^2}\,du \qquad (u = x-1) \\
&= \int \sqrt{1-\sin^2\theta}\cos\theta\,d\theta \qquad (u = \sin\theta) \\
&= \int \cos^2\theta\,d\theta \\
&= \tfrac{1}{2}\int (1+\cos 2\theta)\,d\theta \\
&= \tfrac{1}{2}\left(\theta + \tfrac{1}{2}\sin 2\theta\right) + C \\
&= \tfrac{1}{2}\theta + \tfrac{1}{2}\sin\theta\cos\theta + C \\
&= \tfrac{1}{2}\arcsin u + \tfrac{1}{2}u\sqrt{1-u^2} + C \\
&= \tfrac{1}{2}\arcsin(x-1) + \tfrac{1}{2}(x-1)\sqrt{2x-x^2} + C.
\end{aligned}
$$

■

Example 7.48 Integrate $\int (2x^2 + 5x - 7)^{-1/2}\, dx$.

Solution By Example 7.45,

$$
\begin{aligned}
\int (2x^2+5x-7)^{-1/2}\,dx &= \int \left(2\left[\left(x+\frac{5}{4}\right)^2 - \left(\frac{9}{4}\right)^2\right]\right)^{-1/2} dx \\
&= 2^{-1/2}\int \left[u^2 - \left(\frac{9}{4}\right)^2\right]^{-1/2} du \qquad \left(u = x+\frac{5}{4}\right) \\
&= \frac{1}{\sqrt{2}}\int \left[\left(\frac{9}{4}\right)^2\sec^2\theta - \left(\frac{9}{4}\right)^2\right]^{-1/2} \frac{9}{4}\sec\theta\tan\theta\,d\theta \\
&\qquad\qquad \left(u = \frac{9}{4}\sec\theta\right) \\
&= \frac{1}{\sqrt{2}}\int \frac{\frac{9}{4}\sec\theta\tan\theta\,d\theta}{\frac{9}{4}\tan\theta} \\
&= \frac{1}{\sqrt{2}}\int \sec\theta\,d\theta \\
&= \frac{1}{\sqrt{2}}\ln|\sec\theta + \tan\theta| + C \\
&= \frac{1}{\sqrt{2}}\ln\left|\frac{4u}{9} + \frac{1}{9}\sqrt{16u^2-81}\right| + C
\end{aligned}
$$

$$= \frac{1}{\sqrt{2}} \ln\left|\frac{4x+5}{9} + \frac{1}{9}\sqrt{8(2x^2+5x-7)}\right| + C$$

$$= \frac{1}{\sqrt{2}} \ln|4x+5+\sqrt{8(2x^2+5x-7)}| + C.$$

[The constant $-(1/\sqrt{2})\ln 9$ is combined with C.] ■

■ **Example 7.49** Integrate $\int (2x^2+5x+7)^{-1/2}\,dx$.

Solution By Example 7.46, we have

$$\int (2x^2+5x+7)^{-1/2}\,dx = \int \left(2\left[\left(x+\frac{5}{4}\right)^2 + \left(\frac{\sqrt{31}}{4}\right)^2\right]\right)^{-1/2} dx$$

$$= 2^{-1/2}\int \left[u^2 + \left(\frac{\sqrt{31}}{4}\right)^2\right]^{-1/2} du \qquad \left(u = x + \frac{5}{4}\right)$$

$$= \frac{1}{\sqrt{2}}\int \left[\left(\frac{\sqrt{31}}{4}\right)^2 \tan^2\theta + \left(\frac{\sqrt{31}}{4}\right)^2\right]^{-1/2} \frac{\sqrt{31}}{4}\sec^2\theta\,d\theta$$

$$\left(u = \frac{\sqrt{31}}{4}\tan\theta\right)$$

$$= \frac{1}{\sqrt{2}}\int \frac{(\sqrt{31}/4)\sec^2\theta\,d\theta}{(\sqrt{31}/4)\sec\theta}$$

$$= \frac{1}{\sqrt{2}}\int \sec\theta\,d\theta$$

$$= \frac{1}{\sqrt{2}} \ln|4x+5+\sqrt{8(2x^2+5x+7)}| + C.$$

You can fill in the missing steps above; they are very similar to the steps in Example 7.48. ■

■ **Example 7.50** Integrate

$$\int \frac{dx}{x^2+x+1}.$$

Solution We first complete the square, getting

$$x^2+x+1 = x^2+x+\frac{1}{4}+\frac{3}{4}$$

$$= \left(x+\frac{1}{2}\right)^2 + \left(\frac{\sqrt{3}}{2}\right)^2.$$

Then

$$
\begin{aligned}
\int \frac{dx}{x^2+x+1} &= \int \frac{dx}{(x+\frac{1}{2})^2 + (\sqrt{3}/2)^2} \\
&= \int \frac{du}{u^2 + (\sqrt{3}/2)^2} \qquad \left(u = x + \frac{1}{2}\right) \\
&= \int \frac{(\sqrt{3}/2)\sec^2\theta \, d\theta}{\frac{3}{4}\sec^2\theta} \qquad \left(u = \frac{\sqrt{3}}{2}\tan\theta\right) \\
&= \frac{2}{\sqrt{3}} \int d\theta \\
&= \frac{2}{\sqrt{3}}\theta + C \\
&= \frac{2}{\sqrt{3}} \arctan \frac{2u}{\sqrt{3}} + C \\
&= \frac{2}{\sqrt{3}} \arctan \frac{2x+1}{\sqrt{3}} + C.
\end{aligned}
$$

■

Exercises 7.5

Integrate.

1. $\displaystyle\int \frac{dx}{\sqrt{2x-x^2}}$

2. $\displaystyle\int (4x-x^2)^{-3/2}\, dx$

A3. $\displaystyle\int \sqrt{x^2+3x}\, dx$

4. $\displaystyle\int \frac{dx}{\sqrt{x^2-x}}$

5. $\displaystyle\int_0^1 \sqrt{x^2+x+1}\, dx$

6. $\displaystyle\int (x^2+x+1)^{-2}\, dx$

7. $\displaystyle\int (x^2+2x+2)^{-1/2}\, dx$

8. $\displaystyle\int x(x^2+2x+2)^{-1/2}\, dx$

9. $\displaystyle\int_0^1 \sqrt{x-x^2}\, dx$

10. $\displaystyle\int_0^1 \frac{dx}{\sqrt{x^2+x+1}}$

11. $\displaystyle\int_0^2 \frac{dx}{x^2-x+1}$

12. $\displaystyle\int_{-2}^{-1} \sqrt{1-4x-x^2}\, dx$

13. $\displaystyle\int \frac{x+1}{\sqrt{x^2+2x+2}}\, dx$

14. $\displaystyle\int \frac{dx}{(x^2+2x+2)^2}$

15. $\displaystyle\int \frac{dx}{3x^2+4x+3}$

16. $\displaystyle\int \frac{dx}{3x^2+4x-3}$

A17. $\displaystyle\int_0^1 \sqrt{3x^2+4x+3}\, dx$

A18. $\displaystyle\int_1^2 \sqrt{3x^2+4x-3}\, dx$

Derive the following formulas:

A **19.** $\displaystyle\int \frac{dx}{\sqrt{ax^2+bx+c}} = \frac{1}{\sqrt{a}} \ln|2\sqrt{a}\sqrt{ax^2+bx+c} + 2ax + b| + C$

if $a > 0$ and $b^2 - 4ac > 0$.

A **20.** $\displaystyle\int \frac{dx}{\sqrt{ax^2+bx+c}} = -\frac{1}{\sqrt{-a}} \arcsin \frac{2ax+b}{\sqrt{b^2-4ac}} + C$

if $a < 0$ and $b^2 - 4ac > 0$.

SECTION 7.6

The Method of Partial Fractions

The **method of partial fractions** is a technique for integrating rational functions. It consists of algebraically rewriting a rational function as a sum of a polynomial and some fractions whose denominators are linear or quadratic polynomials (or powers thereof). For example, we can rewrite the rational function

$$\frac{2x^3+3x^2+x-3}{x^3-1} \quad \text{as} \quad 2 + \frac{1}{x-1} + \frac{2x+2}{x^2+x+1}.$$

This new expression has three parts that can be integrated one at a time by previous methods.

The focus of this section will be on the method of rewriting a rational function as a sum of partial fractions. We will rewrite a rational function as a sum of a polynomial and fractions of the form

$$\frac{A}{(ax+b)^k} \quad \text{or} \quad \frac{Ax+B}{(ax^2+bx+c)^k},$$

with A and B constants. Such a sum is called a **partial fraction decomposition** of the function.

The method will involve four steps: long division, factoring the denominator, decomposition, and determining the constants.

Step 1: Long Division

The first step in obtaining a partial fraction decomposition of the rational function $p(x)/q(x)$ is to carry out the indicated long division if the degree of $p(x)$ is not less than the degree of $q(x)$. In that case, the fraction is said to be **improper**. If the fraction is **proper**, meaning that the degree of $p(x)$ is less than the degree of $q(x)$, then we go on to the next step.

Thus, the first step reduces the given rational function $p(x)/q(x)$ to the sum of a polynomial $s(x)$ and a new, proper rational function $r(x)/q(x)$.

Example 7.51 Express the rational function $x^6/(x^4 - 1)$ as the sum of a polynomial and a proper rational function.

Solution The long division proceeds as follows:

$$\begin{array}{r} x^2 \\ x^4 - 1 \overline{)\, x^6 } \\ \underline{x^6 - x^2} \\ x^2 \end{array}$$

Thus,

$$\frac{x^6}{x^4 - 1} = x^2 + \frac{x^2}{x^4 - 1}.$$

Step 2: Factoring the Denominator

The second step is to completely factor the denominator into a product of linear and irreducible quadratic factors. That is, each factor of $q(x)$ will be either of the form $ax + b$ with a and b real numbers, or of the form $ax^2 + bx + c$ with a, b, and c real numbers and $b^2 - 4ac < 0$. [If $b^2 - 4ac \geq 0$, then $ax^2 + bx + c$ is not irreducible, but has zeros z_1 and z_2, and hence factors as $ax^2 + bx + c = a(x - z_1)(x - z_2)$.] The fact that $q(x)$ can be factored into linear and irreducible quadratic factors is a theorem proved in college algebra. The actual factorization may be difficult to find, but it always exists. Techniques for factoring include grouping, recognizing special forms, finding zeros, using the factor theorem, and synthetic division.

Example 7.52 Factor $x^4 - 1$ into linear and irreducible quadratic factors.

Solution This polynomial is a difference of squares, so the factorization is done easily:

$$\begin{aligned} x^4 - 1 &= (x^2 - 1)(x^2 + 1) \\ &= (x - 1)(x + 1)(x^2 + 1). \end{aligned}$$

The first two factors are linear and the third is irreducible and quadratic.

Example 7.53 Completely factor $x^4 - 4x^2 + 9x - 18$.

Solution We search for a zero. Possible rational zeros are $\pm 1, \pm 2, \pm 3, \pm 6, \pm 9$, and ± 18. Using synthetic division, we find that 2 is a zero:

$$\begin{array}{r|rrrrr} 2 & 1 & 0 & -4 & 9 & -18 \\ & & 2 & 4 & 0 & 18 \\ \hline & 1 & 2 & 0 & 9 & 0 \end{array}$$

Hence, $x-2$ is a factor, and $x^4-4x^2+9x-18=(x-2)(x^3+2x^2+9)$. Possible zeros of x^3+2x^2+9 are $\pm 1, \pm 3$, and ± 9. By trial and error, we find that -3 is a zero:

$$\begin{array}{r|rrrr} -3 & 1 & 2 & 0 & 9 \\ & & -3 & 3 & -9 \\ \hline & 1 & -1 & 3 & 0 \end{array}$$

Hence, $x^3+2x^2+9=(x+3)(x^2-x+3)$, so that

$$x^4-4x^2+9x-18=(x-2)(x+3)(x^2-x+3).$$

The last factor, x^2-x+3, is irreducible, because $(-1)^2-4(1)(3)<0$. Thus, the factorization is complete. ■

Step 3: Decomposition

The third step is to write the proper fraction $r(x)/q(x)$ as a sum of fractions of the desired form. There are four cases to consider.

Case 1. For each unrepeated linear factor $ax+b$ of $q(x)$, the sum will contain one term of the form

$$\frac{A}{ax+b},$$

with A a constant to be determined.

Case 2. For each repeated linear factor $(ax+b)^k$ of $q(x)$ with multiplicity k, the sum will contain the k terms

$$\frac{A_1}{ax+b}+\frac{A_2}{(ax+b)^2}+\cdots+\frac{A_k}{(ax+b)^k},$$

with constants $A_1, A_2, \ldots, A_k$ to be determined.

Case 3. For each unrepeated irreducible quadratic factor ax^2+bx+c of $q(x)$, the sum will contain one term of the form

$$\frac{Ax+B}{ax^2+bx+c},$$

with A and B constants to be determined.

Case 4. For each repeated irreducible quadratic factor $(ax^2+bx+c)^k$ of $q(x)$ with multiplicity k, the sum will contain the k terms

$$\frac{A_1x+B_1}{ax^2+bx+c}+\frac{A_2x+B_2}{(ax^2+bx+c)^2}+\cdots+\frac{A_kx+B_k}{(ax^2+bx+c)^k},$$

with $A_1, \ldots, A_k$ and $B_1, \ldots, B_k$ constants to be determined.

■ **Example 7.54** Rewrite $x^2/(x^4-1)$ in the form of a partial fraction decomposition.

Solution By Example 7.52, $x^4 - 1 = (x-1)(x+1)(x^2+1)$. Therefore, by Cases 1 and 3, the partial fraction decomposition has the form

$$\frac{x^2}{x^4-1} = \frac{A}{x-1} + \frac{B}{x+1} + \frac{Cx+D}{x^2+1},$$

with constants A, B, C, and D yet to be determined. ■

■ **Example 7.55** Give the form of the partial fraction decomposition of

$$\frac{x+3}{(x+2)^3(x^2+4)}.$$

Solution By Cases 2 and 3, the decomposition has the form

$$\frac{x+3}{(x+2)^3(x^2+4)} = \frac{A_1}{x+2} + \frac{A_2}{(x+2)^2} + \frac{A_3}{(x+2)^3} + \frac{Bx+C}{x^2+4},$$

with constants A_1, A_2, A_3, B, and C yet to be determined. ■

■ **Example 7.56** Give the form of the partial fraction decomposition of

$$\frac{1}{(x+1)(x-1)^2(x^2+x+1)^3}.$$

Solution Using Cases 1, 2, and 4, the decomposition must be of the form

$$\frac{1}{(x+1)(x-1)^2(x^2+x+1)^3} = \frac{A}{x+1} + \frac{B_1}{x-1} + \frac{B_2}{(x-1)^2} + \frac{C_1x+D_1}{x^2+x+1} + \frac{C_2x+D_2}{(x^2+x+1)^2} + \frac{C_3x+D_3}{(x^2+x+1)^3},$$

with the A's, B's, C's, and D's constants yet to be determined. ■

Step 4: Determining the Constants

The fourth step in obtaining the partial fraction decomposition is to determine the constants. Having written the form of the decomposition, we multiply both sides by $q(x)$ to clear the equation of fractions, thus obtaining a polynomial equation. This equation is to be an identity, so by either equating coefficients of like terms or substituting convenient values for x, we get a system of linear equations in which the constants to be determined are the unknowns.

■ **Example 7.57** Determine the constants in the decomposition

$$\frac{x^2}{x^4-1} = \frac{A}{x-1} + \frac{B}{x+1} + \frac{Cx+D}{x^2+1}.$$

Solution Multiplying this equation through by $x^4 - 1$, we get

$$x^2 = A(x+1)(x^2+1) + B(x-1)(x^2+1) + (Cx+D)(x-1)(x+1). \quad (1)$$

Expanding and collecting terms, we get

$$x^2 = (A+B+C)x^3 + (A-B+D)x^2 + (A+B-C)x + (A-B-D). \quad (2)$$

If equation (2) is to be an identity, then the coefficients of like powers of x must be the same (see the problems at the end of this section). The coefficient of x^3 on the left is 0 (there is no x^3 term), and on the right it is $A+B+C$. Hence,

$$A+B+C=0.$$

Similarly, comparing coefficients of the other terms,

$$A-B+D=1,$$
$$A+B-C=0,$$
$$A-B-D=0.$$

From the first and third of these equations, we have $A+B=0$; and from the second and fourth equations, we have $A-B=\frac{1}{2}$. Hence, $A=\frac{1}{4}$ and $B=-\frac{1}{4}$, and it follows that $C=0$ and $D=\frac{1}{2}$. Thus,

$$\frac{x^2}{x^4-1} = \frac{\frac{1}{4}}{x-1} - \frac{\frac{1}{4}}{x+1} + \frac{\frac{1}{2}}{x^2+1}.$$

We also could have used values of x to get a system of equations. For example, if $x=1$, then equation (1) gives us

$$1 = 4A;$$

and if $x=-1$, then equation (1) gives us

$$1 = -4B.$$

For $x=0$, we get

$$0 = A-B-D$$

again; and for $x=2$, we get

$$4 = 15A + 5B + 6C + 3D.$$

This system of equations is easily solved by substitution, and we get the same solution as before, of course. ■

■ **Example 7.58** Determine the constants in the decomposition found in Example 7.55.

Solution Multiplying through by $(x+2)^3(x^2+4)$, we get

$$x+3 = A_1(x+2)^2(x^2+4) + A_2(x+2)(x^2+4) + A_3(x^2+4) + (Bx+C)(x+2)^3.$$

When this is expanded, the terms collected, and the coefficients compared (all straightforward, but a bit tedious), we get the system of equations

$$\begin{aligned}
A_1 \qquad\qquad\qquad + B \qquad &= 0,\\
4A_1 + A_2 \qquad + 6B + C &= 0,\\
8A_1 + 2A_2 + A_3 + 12B + 6C &= 0,\\
16A_1 + 4A_2 \qquad + 8B + 12C &= 1,\\
16A_1 + 8A_2 + 4A_3 \qquad + 8C &= 3.
\end{aligned}$$

The solution of this system can be found by substitution or elimination, and is $A_1 = \frac{5}{64}$, $A_2 = \frac{3}{16}$, $A_3 = \frac{1}{8}$, $B = -\frac{5}{64}$, $C = -\frac{1}{32}$. Therefore, we may write

$$\frac{x+3}{(x+2)^3(x^2+4)} = \frac{\frac{5}{64}}{x+2} + \frac{\frac{3}{16}}{(x+2)^2} + \frac{\frac{1}{8}}{(x+2)^3} - \frac{\frac{5}{64}x + \frac{1}{32}}{x^2+4}. \qquad \blacksquare$$

Integration by Partial Fractions

Now we can use the method of partial fractions to integrate rational functions. Consider the following examples.

Example 7.59 Integrate

$$\int \frac{x^6}{x^4-1}\,dx.$$

Solution By Examples 7.51 and 7.57, we have

$$\begin{aligned}
\int \frac{x^6}{x^4-1}\,dx &= \int \left(x^2 + \frac{\frac{1}{4}}{x-1} - \frac{\frac{1}{4}}{x+1} + \frac{\frac{1}{2}}{x^2+1}\right) dx\\
&= \frac{1}{3}x^3 + \frac{1}{4}\ln|x-1| - \frac{1}{4}\ln|x+1| + \frac{1}{2}\arctan x + C. \qquad \blacksquare
\end{aligned}$$

Example 7.60 Integrate

$$\int \frac{x+3}{(x+2)^3(x^2+4)}\,dx.$$

Solution By Example 7.58, we can write

$$\begin{aligned}
\int \frac{x+3}{(x+2)^3(x^2+4)}\,dx &= \int \left[\frac{\frac{5}{64}}{x+2} + \frac{\frac{3}{16}}{(x+2)^2} + \frac{\frac{1}{8}}{(x+2)^3} - \frac{\frac{5}{64}x + \frac{1}{32}}{x^2+4}\right] dx\\
&= \frac{5}{64}\int \frac{dx}{x+2} + \frac{3}{16}\int (x+2)^{-2}\,dx + \frac{1}{8}\int (x+2)^{-3}\,dx\\
&\quad - \frac{5}{64}\int \frac{x\,dx}{x^2+4} - \frac{1}{32}\int \frac{dx}{x^2+4}
\end{aligned}$$

$$= \frac{5}{64}\ln|x+2| - \frac{3}{16(x+2)} - \frac{1}{16(x+2)^2} - \frac{5}{128}\ln(x^2+4) - \frac{1}{64}\arctan\frac{x}{2} + C. \quad \blacksquare$$

Example 7.61 Integrate

$$\int \frac{dx}{x^2-4}$$

by the method of partial fractions (instead of trigonometric substitution).

Solution We first find the partial fraction decomposition of the fraction $1/(x^2-4)$. Since this fraction is proper, no long division is necessary. The denominator factors as $x^2-4=(x-2)(x+2)$. Therefore, the decomposition has the form

$$\frac{1}{x^2-4} = \frac{A}{x-2} + \frac{B}{x+2}.$$

Multiplying by x^2-4, we get

$$1 = A(x+2) + B(x-2).$$

Now we can expand and compare coefficients, or we can select some values for x. Note that if $x=2$, then

$$1 = A(2+2) + B(2-2) = 4A,$$

and $A=\frac{1}{4}$. Also, if $x=-2$, we have

$$1 = A(-2+2) + B(-2-2) = -4B,$$

and $B=-\frac{1}{4}$. Hence,

$$\frac{1}{x^2-4} = \frac{\frac{1}{4}}{x-2} - \frac{\frac{1}{4}}{x+2}.$$

Taken altogether, we therefore have

$$\int \frac{dx}{x^2-4} = \int \left(\frac{\frac{1}{4}}{x-2} - \frac{\frac{1}{4}}{x+2}\right) dx$$
$$= \frac{1}{4}\ln|x-2| - \frac{1}{4}\ln|x+2| + C$$
$$= \frac{1}{4}\ln\left|\frac{x-2}{x+2}\right| + C. \quad \blacksquare$$

Example 7.62 Integrate

$$\int \frac{3x^4+6x^2+2x+3}{x^5+2x^3+x}\,dx.$$

Solution Again, we find the partial fraction decomposition of the integrand. The fraction is proper, so no long division is necessary. The denominator factors as

$$x^5 + 2x^3 + x = x(x^4 + 2x^2 + 1) = x(x^2 + 1)^2.$$

The decomposition thus has the form

$$\frac{3x^4 + 6x^2 + 2x + 3}{x^5 + 2x^3 + x} = \frac{A}{x} + \frac{B_1 x + C_1}{x^2 + 1} + \frac{B_2 x + C_2}{(x^2 + 1)^2}.$$

To determine the constants, we clear fractions by multiplying by $x(x^2 + 1)^2$, getting

$$\begin{aligned} 3x^4 + 6x^2 + 2x + 3 &= A(x^2 + 1)^2 + (B_1 x + C_1)x(x^2 + 1) + (B_2 x + C_2)x \\ &= (A + B_1)x^4 + C_1 x^3 + (2A + B_1 + B_2)x^2 + (C_1 + C_2)x + A. \end{aligned}$$

Comparing coefficients, we see that

$$\begin{aligned} A + B_1 &= 3, \\ C_1 &= 0, \\ 2A + B_1 + B_2 &= 6, \\ C_1 + C_2 &= 2, \\ A &= 3. \end{aligned}$$

This system has the solution $A = 3$, $B_1 = 0$, $C_1 = 0$, $B_2 = 0$, $C_2 = 2$, which is easily found by substitution. Thus,

$$\begin{aligned} \int \frac{3x^4 + 6x^2 + 2x + 3}{x^5 + 2x^3 + x}\,dx &= \int \left[\frac{3}{x} + \frac{2}{(x^2 + 1)^2}\right] dx \\ &= 3\int \frac{1}{x}\,dx + 2\int (x^2 + 1)^{-2}\,dx \\ &= 3\ln|x| + 2\int (\sec^2\theta)^{-2}\sec^2\theta\,d\theta \qquad (x = \tan\theta) \\ &= 3\ln|x| + 2\int \cos^2\theta\,d\theta \\ &= 3\ln|x| + \int (1 + \cos 2\theta)\,d\theta \\ &= 3\ln|x| + \theta + \frac{1}{2}\sin 2\theta + C \\ &= 3\ln|x| + \theta + \sin\theta\cos\theta + C \\ &= 3\ln|x| + \arctan x + \frac{x}{x^2 + 1} + C. \end{aligned}$$

■

■ **Example 7.63** Integrate

$$\int \frac{2x^3 + 3x^2 + x - 3}{x^3 - 1}\,dx.$$

Solution Seeking the partial fraction decomposition of the integrand, we first note that long division is necessary. Carrying out the long division, we find that

$$\frac{2x^3+3x^2+x-3}{x^3-1}=2+\frac{3x^2+x-1}{x^3-1}.$$

We next factor the denominator as

$$x^3-1=(x-1)(x^2+x+1).$$

Thus, the decomposition of the last fraction has the form

$$\frac{3x^2+x-1}{x^3-1}=\frac{A}{x-1}+\frac{Bx+C}{x^2+x+1}.$$

Clearing fractions, we get

$$\begin{aligned}3x^2+x-1&=A(x^2+x+1)+(Bx+C)(x-1)\\&=(A+B)x^2+(A-B+C)x+(A-C).\end{aligned}$$

Comparing coefficients, we get the system

$$\begin{aligned}A+B\qquad&=3,\\A-B+C&=1,\\A\qquad-C&=-1,\end{aligned}$$

which has solution $A=1$, $B=2$, $C=2$. Hence,

$$\frac{3x^2+x-1}{x^3-1}=\frac{1}{x-1}+\frac{2x+2}{x^2+x+1}.$$

Thus,

$$\begin{aligned}\int\frac{2x^3+3x^2+x-3}{x^3-1}\,dx&=\int\left(2+\frac{1}{x-1}+\frac{2x+2}{x^2+x+1}\right)dx\\&=2x+\ln|x-1|+\int\frac{2x+1}{x^2+x+1}\,dx+\int\frac{dx}{x^2+x+1}\\&=2x+\ln|x-1|+\ln(x^2+x+1)+\int\frac{dx}{(x+\frac{1}{2})^2+\frac{3}{4}}\\&=2x+\ln|x-1|+\ln(x^2+x+1)\\&\qquad+\frac{1}{\sqrt{3}/2}\arctan\frac{x+\frac{1}{2}}{\sqrt{3}/2}+C\\&=2x+\ln|x-1|+\ln(x^2+x+1)\\&\qquad+\frac{2}{\sqrt{3}}\arctan\frac{2x+1}{\sqrt{3}}+C.\quad\blacksquare\end{aligned}$$

One final remark should be made here: Do not forget about previous methods. For instance, $\int[3x^2/(x^3+8)]\,dx$ can be integrated by using partial fractions, but the simple substitution $u=x^3+8$ is much easier.

Exercises 7.6

Integrate.

1. $\int \frac{x^3\,dx}{x^2-1}$
2. $\int_2^3 \frac{2x-1}{x^2-1}\,dx$
3. $\int \frac{3x\,dx}{x^2-1}$
4. $\int \frac{x^2\,dx}{x^2-1}$
5. $\int \frac{dx}{x^2-x}$
6. $\int \frac{dx}{x^3-x}$
7. $\int \frac{x^3\,dx}{x^2+1}$
8. $\int_2^3 \frac{2x-1}{x^2+1}\,dx$
9. $\int \frac{dx}{x^3+x^2+x+1}$
10. $\int \frac{3x^2+2x+1}{x^3+x^2+x+1}\,dx$
11. $\int \frac{dx}{(x^2+1)^3}$
12. $\int \frac{x\,dx}{(x-1)^3}$
13. $\int \frac{3x^2+12x+11}{(x+1)(x+2)(x+3)}\,dx$
14. A $\int \frac{x^4+x^3+2x^2+x+2}{(x^2+1)^3}\,dx$
15. A $\int \frac{dx}{x^3+1}$
16. A $\int_0^a \frac{dx}{x^3+a^3}$
17. $\int \frac{dx}{(x+1)(2x^2+3x+1)}$
18. $\int \frac{dx}{(x+1)^2(x+2)^2}$
19. $\int \frac{2x}{x^3-8}\,dx$
20. A $\int_1^2 \frac{1-3x}{8x^3-1}\,dx$
21. $\int \frac{dx}{x^2-a^2}$
22. A $\int \frac{dx}{x^3-a^3}$
23. $\int \frac{dx}{x^4+1}$
24. A $\int \frac{dx}{x^6-1}$
25. $\int \frac{e^x\,dx}{e^{2x}+4}$
26. $\int_0^1 \frac{dx}{e^x+1}$
27. $\int \frac{dx}{x(1+\ln x)^2}$
28. $\int \frac{\cos x\,dx}{2-\sin^2 x}$
29. In Example 7.57, determine the constants by substituting $x = 1$, $x = -1$, and $x = i$ into equation (1).
30. Prove that if $f(x)$ and $g(x)$ are polynomial functions and $f(x) = g(x)$, then the coefficients of like powers of x in f and g are the same. [*Hint:* Substitute $x = 0$ to show that the constant terms are the same; then differentiate and let $x = 0$; and so forth.]

SECTION 7.7 Miscellaneous Methods; Integration Tables

There are many specialized integration techniques that apply to specific integrals. Also, a large number of formulas for integration have been developed and are readily available in published integration tables. Here, we mention a couple of substitution techniques and discuss briefly the use of tables, such as the one that appears inside the back cover of this book.

Rationalizing Substitutions

A substitution that aims to remove radicals or fractional powers is called a **rationalizing substitution.** Consider the following examples.

Example 7.64 Integrate $\int x^2(2x-1)^{2/5}\,dx$.

Solution To eliminate the fractional power, we let $u=(2x-1)^{1/5}$, or $u^5=2x-1$. [Note that $u=(2x-1)^{2/5}$ will not get rid of fractional powers when we solve for x.] Then $x=\frac{1}{2}(u^5+1)$, so that $dx=\frac{5}{2}u^4\,du$. Substituting, we get

$$\begin{aligned}\int x^2(2x-1)^{2/5}\,dx &= \int [\tfrac{1}{2}(u^5+1)]^2u^2\,\tfrac{5}{2}u^4\,du\\ &= \tfrac{5}{8}\int (u^5+1)^2u^6\,du\\ &= \tfrac{5}{8}\int (u^{16}+2u^{11}+u^6)\,du\\ &= \tfrac{5}{8}(\tfrac{1}{17}u^{17}+\tfrac{1}{6}u^{12}+\tfrac{1}{7}u^7)+C\\ &= \tfrac{5}{136}(2x-1)^{17/5}+\tfrac{5}{48}(2x-1)^{12/5}+\tfrac{5}{56}(2x-1)^{7/5}+C.\end{aligned}$$ ■

Example 7.65 Integrate $\int \sqrt{1+e^x}\,dx$.

Solution To remove the radical, we will let $u=\sqrt{1+e^x}$, or $u^2=1+e^x$. Then $e^x=u^2-1$, or $x=\ln(u^2-1)$, so that $dx=[2u/(u^2-1)]\,du$. Hence,

$$\begin{aligned}\int \sqrt{1+e^x}\,dx &= \int u\,\frac{2u}{u^2-1}\,du\\ &= \int \frac{2u^2}{u^2-1}\,du\\ &= \int \left(2+\frac{1}{u-1}-\frac{1}{u+1}\right)du\\ &= 2u+\ln\left|\frac{u-1}{u+1}\right|+C\\ &= 2\sqrt{1+e^x}+\ln\left|\frac{\sqrt{1+e^x}-1}{\sqrt{1+e^x}+1}\right|+C.\end{aligned}$$ ■

Example 7.66 Integrate

$$\int_0^8 \frac{x^{1/3}}{1+x^{1/2}}\,dx.$$

Solution To get rid of the fractional exponents $\frac{1}{3}$ and $\frac{1}{2}$ at the same time, we can let $u = x^{1/6}$; then $x = u^6$, $x^{1/3} = u^2$, $x^{1/2} = u^3$, and $dx = 6u^5\,du$. Upon substitution, the integral becomes

$$\int_0^8 \frac{x^{1/3}}{1 + x^{1/2}}\,dx = \int_0^{\sqrt{2}} \frac{u^2}{1+u^3}\,6u^5\,du$$
$$= \int_0^{\sqrt{2}} \frac{6u^7}{u^3+1}\,du.$$

The partial fraction decomposition of this integrand is

$$\frac{6u^7}{u^3+1} = 6u^4 - 6u - \frac{2}{u+1} + \frac{2u+2}{u^2-u+1},$$

which we will rewrite as

$$\frac{6u^7}{u^3+1} = 6u^4 - 6u - \frac{2}{u+1} + \frac{2u-1}{u^2-u+1} + \frac{3}{u^2-u+1}.$$

Integrating term by term, we get

$$\left[\frac{6}{5}u^5 - 3u^2 - 2\ln|u+1| + \ln|u^2-u+1|\right]_0^{\sqrt{2}} + 3\int_0^{\sqrt{2}} \frac{du}{u^2-u+1}$$
$$= \frac{24\sqrt{2}}{5} - 6 - 2\ln(1+\sqrt{2}) + \ln(3-\sqrt{2}) + 3\int_0^{\sqrt{2}} \frac{du}{u^2-u+1}.$$

This last integral becomes

$$3\int_0^{\sqrt{2}} \frac{du}{(u-\frac{1}{2})^2 + (\sqrt{3}/2)^2} = 3\int_{-1/2}^{\sqrt{2}-(1/2)} \frac{dt}{t^2 + (\sqrt{3}/2)^2} \qquad \left(t = u - \frac{1}{2}\right)$$
$$= 3\left[\frac{2}{\sqrt{3}}\arctan\frac{t}{\sqrt{3}/2}\right]_{-1/2}^{\sqrt{2}-(1/2)}$$
$$= 2\sqrt{3}\arctan\frac{2\sqrt{2}-1}{\sqrt{3}} - 2\sqrt{3}\arctan\left(\frac{-1}{\sqrt{3}}\right)$$
$$= 2\sqrt{3}\arctan\frac{2\sqrt{2}-1}{\sqrt{3}} - 2\sqrt{3}\left(-\frac{\pi}{6}\right).$$

Putting all this together, we have

$$\int_0^8 \frac{x^{1/3}}{1+x^{1/2}}\,dx = \frac{24\sqrt{2}}{5} - 6 - 2\ln(1+\sqrt{2}) + \ln(3-\sqrt{2})$$
$$+ 2\sqrt{3}\arctan\frac{2\sqrt{2}-1}{\sqrt{3}} + \frac{\pi}{\sqrt{3}}.$$ ∎

The Tangent Half-Angle Substitution

A rational function of $\sin x$ and $\cos x$ is like a rational function of x, except that $\sin x$ and/or $\cos x$ appear instead of x. The functions

$$\frac{\cos x + 1}{\cos x - 1}, \qquad \frac{1}{1 + \sin x}, \qquad \text{and} \qquad \frac{\cos^2 x}{\sin^2 x + 2\cos x}$$

are all rational functions of $\sin x$ and $\cos x$.

An integral of a rational function of $\sin x$ and $\cos x$ can be transformed into the integral of a rational function of z by the **tangent half-angle substitution**

$$z = \tan\frac{x}{2}. \tag{1}$$

Then $\arctan z = x/2$ and $x = 2\arctan z$, so that

$$dx = \frac{2}{1 + z^2}\,dz. \tag{2}$$

Furthermore,

$$\begin{aligned} \sin x &= 2\sin\frac{x}{2}\cos\frac{x}{2} \\ &= 2\,\frac{\sin(x/2)}{\cos(x/2)} \cdot \frac{1}{\sec^2(x/2)} \\ &= \frac{2\tan(x/2)}{1 + \tan^2(x/2)}, \end{aligned}$$

or

$$\sin x = \frac{2z}{1 + z^2}, \tag{3}$$

and

$$\begin{aligned} \cos x &= \cos^2\frac{x}{2} - \sin^2\frac{x}{2} \\ &= \cos^2\frac{x}{2}\left(1 - \tan^2\frac{x}{2}\right) \\ &= \frac{1 - \tan^2(x/2)}{\sec^2(x/2)}, \end{aligned}$$

or

$$\cos x = \frac{1 - z^2}{1 + z^2}. \tag{4}$$

Thus, substituting (2), (3), and (4) into the integral of a rational function of $\sin x$ and $\cos x$ will yield the integral of a rational function of z.

Consider some examples.

Example 7.67 Integrate

$$\int \frac{\sin x\, dx}{1+\sin x}.$$

Solution Using (2), (3), and (4), we get

$$\int \frac{\sin x\, dx}{1+\sin x} = \int \frac{\dfrac{2z}{1+z^2}\cdot\dfrac{2}{1+z^2}\,dz}{1+\dfrac{2z}{1+z^2}} = \int \frac{4z\, dz}{(z+1)^2(z^2+1)}.$$

The method of partial fractions yields

$$\frac{4z}{(z+1)^2(z^2+1)} = \frac{-2}{(z+1)^2} + \frac{2}{z^2+1},$$

so that

$$\begin{aligned}\int \frac{4z\, dz}{(z+1)^2(z^2+1)} &= -2\int (z+1)^{-2}\, dz + 2\int \frac{dz}{z^2+1}\\ &= \frac{2}{z+1} + 2\arctan z + C\\ &= \frac{2}{1+\tan(x/2)} + 2\arctan\left(\tan\frac{x}{2}\right) + C\\ &= \frac{2}{1+\tan(x/2)} + 2\left(\frac{x}{2}\right) + C\\ &= \frac{2}{1+\tan(x/2)} + x + C.\end{aligned}$$

■

Example 7.68 Find

$$\int_0^{\pi/2} \frac{dx}{1+\cos x}.$$

Solution Again using the tangent half-angle substitution, we have

$$\int_0^{\pi/2} \frac{dx}{1+\cos x} = \int_0^1 \frac{\dfrac{2}{1+z^2}\,dz}{1+\dfrac{1-z^2}{1+z^2}} = \int_0^1 dz = z\Big|_0^1 = 1.$$

■

■ **Example 7.69** Find

$$\int \frac{d\theta}{3-2\sin\theta}.$$

Solution Using $z = \tan(\theta/2)$, we get

$$\begin{aligned}
\int \frac{d\theta}{3-2\sin\theta} &= \int \frac{\dfrac{2}{1+z^2}\,dz}{3-2\left(\dfrac{2z}{1+z^2}\right)} \\
&= \int \frac{2\,dz}{3z^2-4z+3} \\
&= \frac{2}{3}\int \frac{dz}{z^2-\frac{4}{3}z+1} \\
&= \frac{2}{3}\int \frac{dz}{(z-\frac{2}{3})^2+\frac{5}{9}} \\
&= \frac{2}{3}\int \frac{du}{u^2+\frac{5}{9}} \qquad \left(u = z-\frac{2}{3}\right) \\
&= \frac{2}{3}\left(\frac{3}{\sqrt{5}}\arctan\frac{3u}{\sqrt{5}}\right)+C \\
&= \frac{2}{\sqrt{5}}\arctan\frac{3z-2}{\sqrt{5}}+C \\
&= \frac{2}{\sqrt{5}}\arctan\frac{3\tan(\theta/2)-2}{\sqrt{5}}+C.
\end{aligned}$$

■

Since all the trigonometric functions are rational functions of sines and cosines, the tangent half-angle substitution will also work with a rational function of any of the other trigonometric functions.

Integration Tables

The techniques of integration that we have discussed thus far can be applied to integrals of certain forms, with constants not specified, to yield integral formulas. For example, when the method of partial fractions is applied to the integral

$$\int \frac{dx}{x^2 - a^2},$$

it yields the integral formula

$$\int \frac{dx}{x^2 - a^2} = \frac{1}{2a} \ln\left|\frac{x-a}{x+a}\right| + C. \tag{5}$$

This formula holds for any positive constant a, so that, for instance,

$$\int \frac{dx}{x^2 - 36} = \frac{1}{12} \ln\left|\frac{x-6}{x+6}\right| + C.$$

This integration was performed by identifying a in equation (5) as 6.

Tables of integration formulas have been published that contain hundreds of formulas. In practice, if an integral is a bit complicated, it is often easier to find a formula that fits it than to use the various techniques necessary to integrate it. A short table of integrals is provided inside the back cover of this book. In using a formula in the table, care must be taken to see that the formula exactly fits the given integral; then the various constants are identified.

Example 7.70 Use the table of integrals in the back of this book to integrate the following:

(a) $\displaystyle\int x(5x-2)^3\, dx$ **(b)** $\displaystyle\int x^2 \cos 2x\, dx$ **(c)** $\displaystyle\int_0^1 \frac{dx}{x^2 - 5x + 6}$

Solution **(a)** This integral fits formula 17 in the table, with $a = 5$, $b = -2$, and $n = 3$. Hence,

$$\begin{aligned}\int x(5x-2)^3\, dx &= \frac{1}{5^2(3+2)}(5x-2)^{3+2} - \frac{-2}{5^2(3+1)}(5x-2)^4 + C\\ &= \frac{1}{125}(5x-2)^5 + \frac{1}{50}(5x-2)^4 + C.\end{aligned}$$

(b) For this integral, we use formula 34 with $a = 2$:

$$\begin{aligned}\int x^2 \cos 2x\, dx &= \frac{2}{2^2} x \cos 2x + \frac{1}{2^3}(2^2x^2 - 2)\sin 2x + C\\ &= \frac{1}{2} x \cos 2x + \frac{1}{4}(2x^2 - 1)\sin 2x + C.\end{aligned}$$

(c) To evaluate this integral, we first ignore the limits and factor the denominator, getting

$$\int \frac{dx}{x^2 - 5x + 6} = \int \frac{dx}{(x-2)(x-3)}.$$

This indefinite integral fits formula 20 in the table, with $a = c = 1$, $b = -2$, and $d = -3$. Therefore,

$$\int \frac{dx}{(x-2)(x-3)} = \frac{1}{(-3)-(-2)} \ln\left|\frac{x-2}{x-3}\right| + C$$
$$= \ln\left|\frac{x-3}{x-2}\right| + C.$$

Now we evaluate at the limits:

$$\int_0^1 \frac{dx}{x^2 - 5x + 6} = \left[\ln\left|\frac{x-3}{x-2}\right|\right]_0^1$$
$$= \ln 2 - \ln\frac{3}{2} = \ln\frac{4}{3}.$$

■

Exercises 7.7

Integrate by substitution methods.

1. $\int \frac{1}{x}\sqrt{2x+9}\,dx$

2. $\int \frac{1}{x}\sqrt{x-4}\,dx$

3. $\int_1^4 x^{-2}(4-x)^{1/2}\,dx$

4. $\int_7^{26} \frac{1}{x}(x+1)^{2/3}\,dx$

5. $\int \frac{x^2\,dx}{\sqrt{x+1}}$

6. $\int \frac{x\,dx}{\sqrt{2-x}}$

7. $\int \frac{\sqrt{x+5}}{x+1}\,dx$

8. $\int \frac{dx}{x^2\sqrt{x-4}}$

9. $\int \sqrt{\frac{x+1}{x}}\,dx$

10. $\int \sqrt{1-\frac{4}{x}}\,dx$

11. $\int \sqrt{4+e^x}\,dx$

12. $\int \frac{dx}{\sqrt{1+e^x}}$

13. $\int \frac{\sin x}{1-\sin x}\,dx$

14. $\int \frac{dx}{1-\cos x}$

15. $\int_0^{\pi/8} \frac{dx}{1-2\sin x}$

16. $\int_0^{\pi/3} \frac{dx}{1+\sin x}$

17. $\int_0^{\pi/3} \frac{dx}{1+\tan x}$

18. $\int_{\pi/6}^{\pi/3} \frac{1+\tan x}{\sin x}\,dx$

19. $\int \frac{dx}{\sin x + 2\cos x}$

20. $\int \frac{\cos x\,dx}{1+\sin^2 x}$

A21. $\displaystyle\int_0^{\pi/2} \frac{dx}{3+\sin^2 x}$

22. $\displaystyle\int \frac{\cos^2 x\,dx}{1+\cos^2 x}$

23. $\displaystyle\int_0^{1/2} \frac{\sqrt{x}\,dx}{1-\sqrt{x}}$

A24. $\displaystyle\int \frac{x^{1/3}\,dx}{1+8\sqrt{x}}$

25. $\displaystyle\int \frac{dx}{2\sqrt{x}+x^{1/4}}$

26. $\displaystyle\int_0^1 \frac{dx}{1+\sqrt{x}}$

27. $\displaystyle\int \frac{dx}{x^2\sqrt{4x^2+2x-1}}$

[*Hint:* Use $x = 1/u$.]

28. $\displaystyle\int \frac{dx}{x\sqrt{x^2+2x-1}}$

[*Hint:* See Problem 27.]

Integrate using the table of integrals.

29. $\displaystyle\int \tan(2x-1)\,dx$

30. $\displaystyle\int \frac{dx}{\sqrt{x^2+4}}$

31. $\displaystyle\int_0^2 \frac{x\,dx}{3x+2}$

32. $\displaystyle\int_0^2 \frac{dx}{x^3+8}$

33. $\displaystyle\int \sin^3 4x\,dx$

34. $\displaystyle\int \sin^2 3x\cos^2 3x\,dx$

35. $\displaystyle\int x\sin\frac{1}{2}x\,dx$

36. $\displaystyle\int \sqrt{1-\cos x}\,dx$

37. $\displaystyle\int x^4\ln x\,dx$

38. $\displaystyle\int x^4 e^x\,dx$

39. Using $\sin x = 2\sin(x/2)\cos(x/2)$, $\cos x = \cos^2(x/2) - \sin^2(x/2)$, and the figure below, show how to compute the tangent half-angle substitutions (3) and (4).

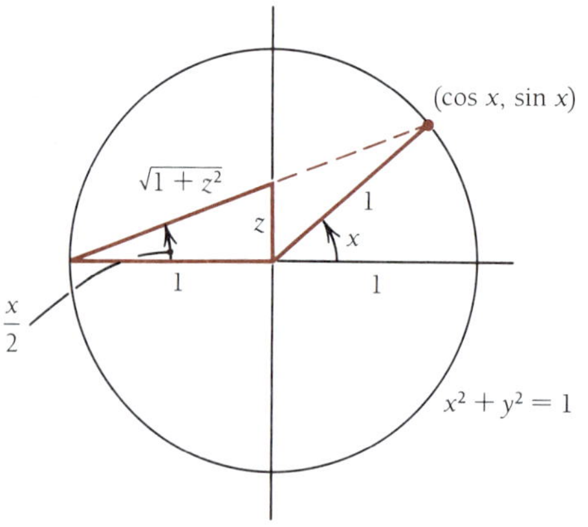

SECTION 7.8

Numerical Integration

Not every integrable function has an antiderivative that can be expressed in terms of elementary functions. That is, there exist integrals that cannot be "integrated," or written without the integral sign. Some examples are

$$\int (1 + x^3)^{-1/2}\, dx, \qquad \int e^{-x^2}\, dx, \qquad \int \sqrt{\sin x}\, dx.$$

(The second and third of these arise in certain statistical and engineering problems.) If you attempt to integrate the above integrals, you will see that all our techniques fail. In fact, no technique exists that will work.

For definite integrals with integrands of this sort, there exist methods of approximation known as **numerical integration** techniques. In this section, we will discuss three techniques of numerical integration.

Each technique starts out the same way. Given a definite integral

$$\int_a^b f(x)\, dx,$$

we first construct a regular subdivision

$$s = \{x_0, x_1, x_2, \ldots, x_n\}$$

of interval $[a, b]$ having n subintervals, each of length

$$\Delta x = \frac{b-a}{n}.$$

Therefore,

$$x_i = a + \left(\frac{b-a}{n}\right) i, \qquad i = 0, 1, \ldots, n.$$

For each i, we will let

$$y_i = f(x_i).$$

Riemann Sum Rules

Now, if $t_i \in [x_{i-1}, x_i]$, as was done in Section 5.2, the sum

$$\sum_{i=1}^{n} f(t_i) \left(\frac{b-a}{n}\right) \tag{1}$$

is a Riemann sum for f on $[a, b]$. Since

$$\int_a^b f(x)\, dx = \lim_{n\to\infty} \sum_{i=1}^{n} f(t_i) \left(\frac{b-a}{n}\right),$$

the sum (1) is an approximation to the integral. The larger the value of n, the closer we can expect the approximation to be. Expanding the sum, we may write

$$\int_a^b f(x)\, dx \approx \frac{b-a}{n} [f(t_1) + f(t_2) + \cdots + f(t_n)]. \tag{2}$$

Any approximation formula of the form (2) is called a **Riemann sum rule**.

Two particular Riemann sum rules are of special importance. If $t_i = x_{i-1}$, we have the **left endpoint rule**

$$\int_a^b f(x)\,dx \approx \frac{b-a}{n}(y_0 + y_1 + \cdots + y_{n-1}). \tag{3}$$

This rule approximates the area between f and the x-axis with rectangles as shown in Figure 7.4.

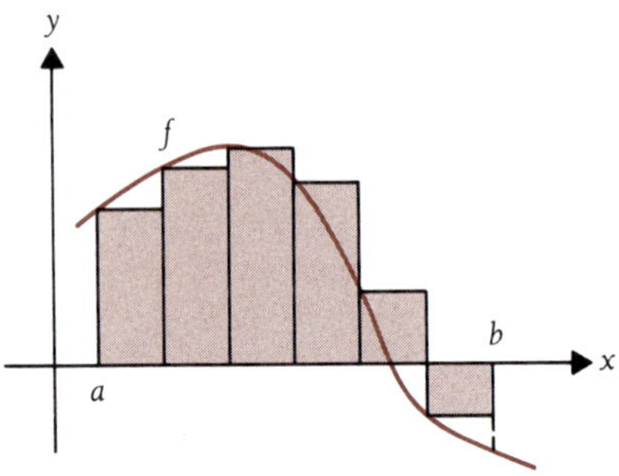

FIGURE 7.4 The left endpoint rule

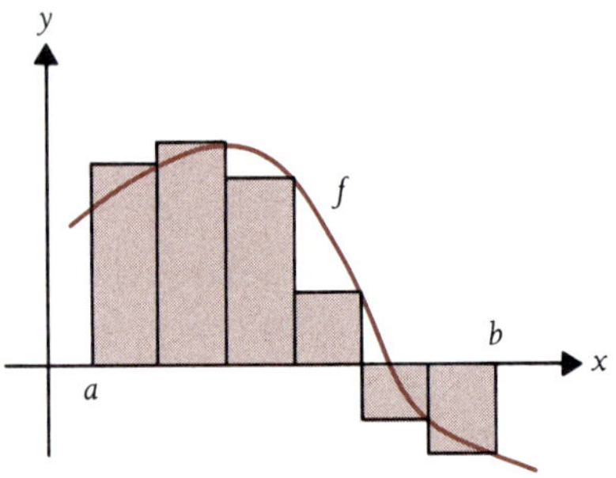

FIGURE 7.5 The right endpoint rule

If $t_i = x_i$, we have the **right endpoint rule**

$$\int_a^b f(x)\,dx \approx \frac{b-a}{n}(y_1 + y_2 + \cdots + y_n). \tag{4}$$

This rule approximates the area between f and the x-axis with rectangles as shown in Figure 7.5.

■ **Example 7.71** Approximate $\int_0^1 x^2\,dx$, using the left and right endpoint rules with $n = 5$.

Solution With $a = 0$, $b = 1$, $n = 5$, and $y_i = f(x_i) = x_i^2$, we have

$$\begin{aligned} x_0 &= 0, & y_0 &= 0, \\ x_1 &= \tfrac{1}{5}, & y_1 &= \tfrac{1}{25}, \\ x_2 &= \tfrac{2}{5}, & y_2 &= \tfrac{4}{25}, \\ x_3 &= \tfrac{3}{5}, & y_3 &= \tfrac{9}{25}, \\ x_4 &= \tfrac{4}{5}, & y_4 &= \tfrac{16}{25}, \\ x_5 &= 1, & y_5 &= 1. \end{aligned}$$

The left endpoint rule (3) gives us

$$\int_0^1 x^2\,dx \approx \frac{1-0}{5}\left(0 + \tfrac{1}{25} + \tfrac{4}{25} + \tfrac{9}{25} + \tfrac{16}{25}\right) = \tfrac{6}{25}.$$

The right endpoint rule (4) gives us

$$\int_0^1 x^2\,dx \approx \frac{1-0}{5}\left(\tfrac{1}{25}+\tfrac{4}{25}+\tfrac{9}{25}+\tfrac{16}{25}+1\right)=\tfrac{11}{25}.$$

The actual value, of course, is $\frac{1}{3}$, which is found by integrating in the usual way. ■

The question of how good an approximation a Riemann sum rule gives can be answered for some functions. We state the following theorem without proof.

THEOREM 7.1

If function f is differentiable and $|f'(x)| \le B$ on $[a, b]$, then the error in approximating $\int_a^b f(x)\,dx$ by a Riemann sum rule is not greater than

$$\frac{B}{2n}(b-a)^2.$$

■ **Example 7.72**

Approximate ln 3 with an error less than 0.001, using the right endpoint rule.

Solution

Recall that

$$\ln 3 = \int_1^3 \frac{1}{x}\,dx.$$

The right endpoint rule (4) applies with $a = 1$, $b = 3$, and $y_i = 1/x_i$. Since

$$x_i = 1 + \frac{2i}{n} = \frac{n+2i}{n},$$

we have

$$y_i = \frac{n}{n+2i}.$$

Then, by rule (4),

$$\int_1^3 \frac{1}{x}\,dx \approx \frac{3-1}{n}\left(\frac{n}{n+2}+\frac{n}{n+4}+\cdots+\frac{n}{n+2n}\right)$$

or

$$\ln 3 \approx 2\left(\frac{1}{n+2}+\frac{1}{n+4}+\cdots+\frac{1}{n+2n}\right). \tag{5}$$

To determine the size of n we should use, we apply Theorem 7.1. The integrand $f(x) = 1/x$ is differentiable, and

$$|f'(x)| = \left|-\frac{1}{x^2}\right| \le 1$$

on [1, 3]. Therefore, the error is not more than

$$\frac{1}{2n}(2)^2 = \frac{2}{n}.$$

Now we simply require that

$$\frac{2}{n} \leq 0.001,$$
$$n \geq 2000.$$

Hence, we may let $n = 2000$ in (5) and be assured of having the required accuracy. In other words, we must use fairly large values of n to get even a little accuracy. This shows that application of the Riemann sum rule can be quite lengthy if much accuracy is desired.

A computer evaluation of (5) with $n = 2000$ gives

$$\ln 3 \approx 1.098,$$

which is in error by about 0.0006. This approximation can be interpreted geometrically as the area of n rectangles inscribed under the graph of $y = 1/x$ (Figure 7.6). Thus, for increasing values of n, we expect increasing approximations—all less than the actual value so long as the right endpoint rule is used. The left endpoint rule would have given an overestimate.

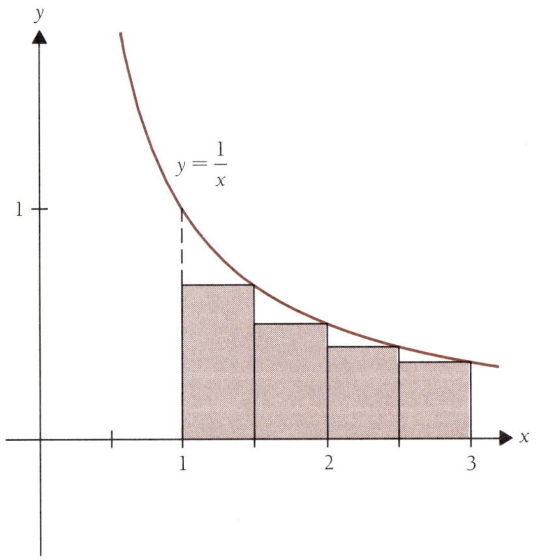

FIGURE 7.6 Estimating $\int_1^3 (1/x)\,dx$ with the right endpoint rule

■

In using a Riemann sum rule, we essentially approximate the integrand $f(x)$ by a function that is constant on each subinterval. We can get a better approximation for the same amount of work by using the next method.

The Trapezoidal Rule

If we approximate the integrand $f(x)$ by a function that is linear on each subinterval $[x_{i-1}, x_i]$ and joins the points (x_{i-1}, y_{i-1}) and (x_i, y_i), as shown in Figure 7.7, then the sum of the areas of the inscribed trapezoids approximates the area under the function. The trapezoid over $[x_{i-1}, x_i]$ has altitude (width) $(b - a)/n$ and bases y_{i-1} and y_i, so has area

$$\frac{1}{2}\left(\frac{b-a}{n}\right)(y_{i-1} + y_i).$$

Hence,

$$\int_a^b f(x)\, dx \approx \sum_{i=1}^{n} \frac{1}{2}\left(\frac{b-a}{n}\right)(y_{i-1} + y_i),$$

which simplifies to

$$\int_a^b f(x)\, dx \approx \frac{b-a}{2n}(y_0 + 2y_1 + 2y_2 + \cdots + 2y_{n-1} + y_n). \tag{6}$$

Formula (6) is called the **trapezoidal rule**.

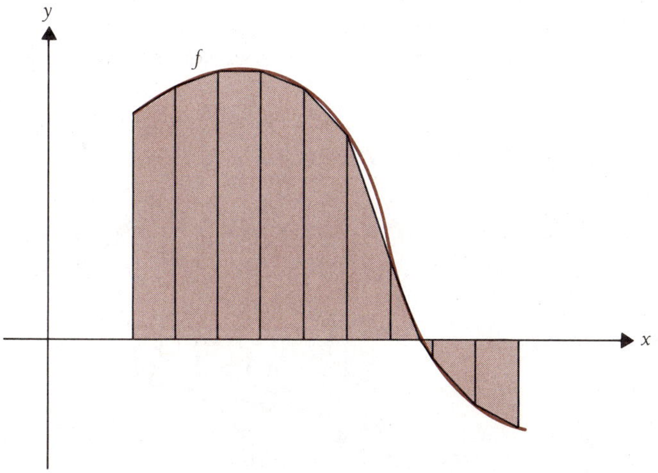

FIGURE 7.7 The trapezoidal rule

Example 7.73 Approximate

$$\int_{-1}^{1} \frac{dx}{x^2+1}$$

using the trapezoidal rule with $n = 4$.

Solution With $a = -1$, $b = 1$, $n = 4$, and $y_i = 1/(x_i^2 + 1)$, we have

$$\begin{aligned} x_0 &= -1, & y_0 &= \tfrac{1}{2}, \\ x_1 &= -\tfrac{1}{2}, & y_1 &= \tfrac{4}{5}, \\ x_2 &= 0, & y_2 &= 1, \\ x_3 &= \tfrac{1}{2}, & y_3 &= \tfrac{4}{5}, \\ x_4 &= 1, & y_4 &= \tfrac{1}{2}. \end{aligned}$$

Therefore, formula (6) gives

$$\begin{aligned} \int_{-1}^{1} \frac{dx}{x^2+1} &\approx \frac{1-(-1)}{2(4)}\left[\frac{1}{2} + 2\left(\frac{4}{5}\right) + 2(1) + 2\left(\frac{4}{5}\right) + \frac{1}{2}\right] \\ &= \frac{31}{20} = 1.55. \end{aligned}$$

The exact value, for comparison, is $\pi/2 \approx 1.571$. ■

The error in using the trapezoidal rule also can be determined for some functions, as is stated (without proof) by the following theorem.

THEOREM 7.2 If function f is twice differentiable and $|f''(x)| \le B$ on $[a, b]$, then the error in approximating $\int_a^b f(x)\,dx$ by the trapezoidal rule is not greater than

$$\frac{B}{12n^2}(b-a)^3.$$

Example 7.74 Approximate ln 3 with an error less than 0.0001, using the trapezoidal rule.

Solution Recalling that

$$\ln 3 = \int_1^3 \frac{1}{x}\,dx,$$

we apply formula (6) with $a = 1$, $b = 3$, and $y_i = n/(n + 2i)$. We get

$$\ln 3 \approx \frac{1}{n} + \frac{2}{n+2} + \frac{2}{n+4} + \cdots + \frac{2}{n+2n-2} + \frac{1}{n+2n}. \qquad (7)$$

To determine how large n must be to give an error less than 0.0001, we use Theorem 7.2. Since $f(x) = 1/x$, we have

$$|f''(x)| = \left|\frac{2}{x^3}\right| \leq 2$$

on [1, 3]. Accordingly, the error in (7) is not greater than

$$\frac{2}{12n^2}(2)^3 = \frac{4}{3n^2}.$$

Now we require

$$\frac{4}{3n^2} \leq 0.0001,$$

which is satisfied (noting that n is an integer) whenever

$$n \geq 116.$$

This value for n is still pretty large, but compared to Example 7.72, in which we used an endpoint formula, we get greater accuracy for a much smaller value of n. With $n = 116$, formula (7) yields

$$\ln 3 \approx 1.0986,$$

which is in error by about 0.00001. If the approximation (7) is interpreted geometrically (see Figure 7.8), it is seen that each approximation is larger than the actual value (the graph of $\ln x$ is concave upward), and the approximations decrease as n increases.

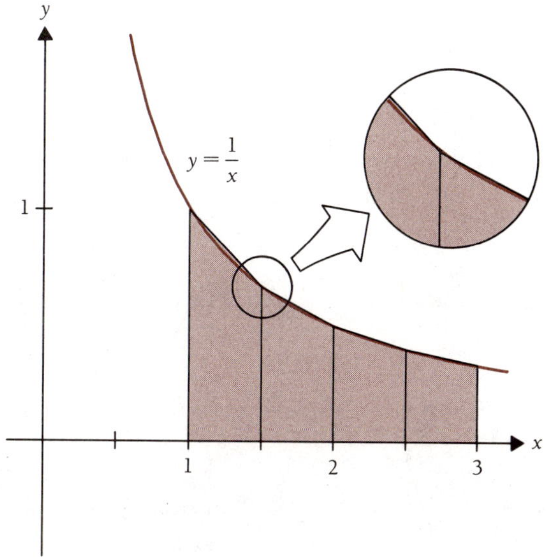

FIGURE 7.8 Estimating $\int_1^3 (1/x)\,dx$ with the trapezoidal rule

■

Simpson's Rule

Since approximating the integrand with linear functions (Example 7.74) instead of constant functions (Example 7.72) increased the accuracy while decreasing the amount of work, the next logical step is to approximate the integrand with quadratic functions, or portions of parabolas, over each subinterval.

To approximate $\int_a^b f(x)\,dx$ by this method, we subdivide interval $[a, b]$ into an *even* number of subintervals. Over the first *two* subintervals, or over $[x_0, x_2]$, we approximate f with the quadratic function

$$g_1(x) = a_1x^2 + b_1x + c_1$$

whose graph passes through the three points (x_0, y_0), (x_1, y_1), and (x_2, y_2), as shown in Figure 7.9. That there is a unique quadratic function g_1 passing through those points follows from the fact that the system of equations

$$\begin{aligned} a_1x_0^2 + b_1x_0 + c_1 &= y_0, \\ a_1x_1^2 + b_1x_1 + c_1 &= y_1, \\ a_1x_2^2 + b_1x_2 + c_1 &= y_2, \end{aligned}$$

uniquely determines a_1, b_1, and c_1. Then we use the fact that

$$\begin{aligned} \int_{x_0}^{x_2} g_1(x)\,dx &= \frac{x_2 - x_0}{6}\,[g_1(x_0) + 4g_1(x_1) + g_1(x_2)] \qquad (8) \\ &= \frac{b-a}{3n}\,(y_0 + 4y_1 + y_2). \end{aligned}$$

[The derivation of (8) is left as an exercise.] Hence,

$$\int_{x_0}^{x_2} f(x)\,dx \approx \int_{x_0}^{x_2} g_1(x)\,dx = \frac{b-a}{3n}\,(y_0 + 4y_1 + y_2).$$

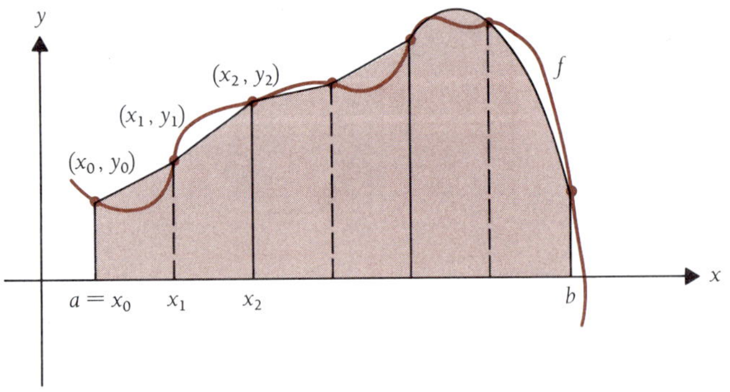

FIGURE 7.9 Simpson's rule

We next approximate f over the interval $[x_2, x_4]$ with the quadratic function $g_2(x)$ whose graph passes through the three points (x_2, y_2), (x_3, y_3), and (x_4, y_4). Then, as before,

$$\int_{x_2}^{x_4} f(x)\,dx \approx \frac{b-a}{3n}(y_2 + 4y_3 + y_4).$$

In general,

$$\int_{x_{2i-2}}^{x_{2i}} f(x)\,dx \approx \frac{b-a}{3n}(y_{2i-2} + 4y_{2i-1} + y_{2i}).$$

Thus,

$$\begin{aligned}\int_a^b f(x)\,dx &= \sum_{i=1}^{n/2} \int_{x_{2i-2}}^{x_{2i}} f(x)\,dx \\ &\approx \sum_{i=1}^{n/2} \frac{b-a}{3n}(y_{2i-2} + 4y_{2i-1} + y_{2i}),\end{aligned}$$

or

$$\int_a^b f(x)\,dx \approx \frac{b-a}{3n}(y_0 + 4y_1 + 2y_2 + 4y_3 + \cdots + 2y_{n-2} + 4y_{n-1} + y_n). \qquad (9)$$

This formula is known as **Simpson's rule**. Remember that in Simpson's rule, n must be even.

Example 7.75 Approximate $\int_0^\pi \sin x\,dx$ by Simpson's rule with $n = 6$.

Solution With $a = 0$, $b = \pi$, and $y_i = \sin x_i$, we have

$$\begin{aligned} x_0 &= 0, & y_0 &= 0, \\ x_1 &= \pi/6, & y_1 &= \tfrac{1}{2}, \\ x_2 &= \pi/3, & y_2 &= \sqrt{3}/2 \approx 0.8660, \\ x_3 &= \pi/2, & y_3 &= 1, \\ x_4 &= 2\pi/3, & y_4 &= \sqrt{3}/2 \approx 0.8660, \\ x_5 &= 5\pi/6, & y_5 &= \tfrac{1}{2}, \\ x_6 &= \pi, & y_6 &= 0. \end{aligned}$$

So by formula (9),

$$\begin{aligned}\int_0^\pi \sin x\,dx &\approx \frac{\pi - 0}{3(6)}\left[0 + 4\left(\frac{1}{2}\right) + 2\left(\frac{\sqrt{3}}{2}\right) + 4(1) + 2\left(\frac{\sqrt{3}}{2}\right) + 4\left(\frac{1}{2}\right) + 0\right] \\ &= \frac{\pi}{18}(8 + 2\sqrt{3}) \approx 2.0009.\end{aligned}$$

The actual value, of course, is 2. ■

There is also an error estimate for Simpson's rule, as given by Theorem 7.3.

THEOREM 7.3

If function f is four times differentiable and $|f^{(4)}(x)| \leq B$ on $[a, b]$, then the error in approximating $\int_a^b f(x)\, dx$ by Simpson's rule is not greater than

$$\frac{B}{180n^4}(b-a)^5.$$

Example 7.76 Approximate ln 3 with an error less than 0.00001, using Simpson's rule.

Solution To

$$\ln 3 = \int_1^3 \frac{1}{x}\, dx$$

we apply formula (9) with $a = 1$, $b = 3$, and $y_i = n/(n + 2i)$:

$$\ln 3 \approx \frac{2}{3n}\left(\frac{n}{n} + \frac{4n}{n+2} + \frac{2n}{n+4} + \cdots + \frac{4n}{n+2n-2} + \frac{n}{n+2n}\right),$$

or

$$\ln 3 \approx \frac{2}{3}\left(\frac{1}{n} + \frac{4}{n+2} + \frac{2}{n+4} + \cdots + \frac{4}{n+2n-2} + \frac{1}{n+2n}\right). \quad (10)$$

To determine what value of n to use in (10), we use Theorem 7.3. Since $f(x) = 1/x$, we have

$$|f^{(4)}(x)| = \left|\frac{24}{x^5}\right| \leq 24$$

on [1, 3]. Therefore, the error in (10) is no greater than

$$\frac{24}{180n^4}(2)^5 = \frac{64}{15n^4}.$$

We require

$$\frac{64}{15n^4} \leq 0.00001,$$

which is satisfied for

$$n \geq 26.$$

Using a calculator, we find that $n = 26$ in (10) yields

$$\ln 3 \approx 1.09861,$$

which is in error by about 0.000002. ■

Exercises 7.8

Approximate the following integrals using the left and right endpoint rules and the given value of n:

1. $\int_1^2 \frac{1}{x}\,dx, \quad n = 4$

2. $\int_1^2 \frac{1}{x}\,dx, \quad n = 6$

3. $\int_0^1 \frac{dx}{x^2+1}, \quad n = 6$

4. $\int_0^{1/2} (1 - x^2)^{-1/2}\,dx, \quad n = 4$

5.—8. Repeat Problems 1–4, using the trapezoidal rule.

9.–12. Repeat Problems 1–4, using Simpson's rule.

13.–16. For the integrals in Problems 1–4, which of the above methods seems to give the best approximation for the least amount of work?

Estimate:

17. $\int_1^e \frac{e^x\,dx}{x}$

18. $\int_1^2 \frac{\sin x\,dx}{x}$

19. $\int_0^1 \sqrt{4 + \cos^2 x}\,dx$

20. $\int_0^1 e^{-x^2/2}\,dx$

21. $\int_0^{\sqrt{\pi}} \sin x^2\,dx$

22. $\int_1^2 \cos(\ln x)\,dx$

23. How large must n be if $\int_{\pi/4}^{\pi/2} \sqrt{\sin x}\,dx$ is to be approximated to three decimal places by a Riemann sum rule?

24. Answer Problem 23 for the trapezoidal rule.

25. Answer Problem 23 for Simpson's rule.

26. For what type of integrand would you expect a Riemann sum rule to give an exact answer? [*Hint:* What does Theorem 7.1 say if the derivative of f is 0?]

27. Answer Problem 26 for the trapezoidal rule.

28. Answer Problem 26 for Simpson's rule.

29. Derive equation (8) by working both sides of the equation separately.

30. Show that the error in Theorem 7.1 is exactly $(B/2n)(b - a)^2$ for the function $f(x) = Bx$ on $[a, b]$. That is, the error estimate in Theorem 7.1 cannot be improved.

31. Can the error estimates in Theorems 7.2 and 7.3 be improved?

32. The **midpoint rule** is the Riemann sum rule obtained by setting $t_i = \frac{1}{2}(x_{i-1} + x_i)$, the midpoint of the ith subinterval. Derive a formula like (3) or (4) for the midpoint rule.

33. Show that the approximation given by the trapezoidal rule is the average of those given by the left and right endpoint rules.

34. For what kind of integrand is the midpoint rule the same as the trapezoidal rule?

SECTION 7.9

Chapter Review

Important Concepts

Integration formulas
Substitution
Integration by parts
Reduction formulas
Trigonometric substitution
Completing the square
Partial fraction decomposition
Comparing coefficients
Rationalizing substitution
Tangent half-angle substitution
Table of integrals
Numerical integration
Riemann sum rule
Left endpoint rule
Right endpoint rule
Trapezoidal rule
Simpson's rule
Error estimate

Review Exercises

Integrate without using tables.

1. $\int (3x+4)^8\,dx$

2. $\int \frac{2\,dx}{2x+5}$

3. $\int_0^1 \frac{2x-1}{x^2+1}\,dx$

4. $\int_0^{1/2} \frac{x+1}{\sqrt{1-x^2}}\,dx$

5. $\int xe^{-x^2}\,dx$

6. $\int 2^{x+1}\,dx$

7. $\int \frac{\arctan x}{x^2+1}\,dx$

8. $\int e^x \sin e^x\,dx$

9. $\int 2\cos(3x-5)\,dx$

10. $\int x\sqrt{x^2-1}\,dx$

11. $\int_0^{\pi} x^2 \sin 3x\,dx$

12. $\int_0^1 xe^{-2x}\,dx$

13. $\int x \arctan x\,dx$

14. $\int x \arcsin x\,dx$

15. $\int \operatorname{arcsec} 2x\,dx$

16. $\int x \operatorname{arcsec} x\,dx$

17. $\int_1^e x^2 \ln x\,dx$

18. $\int_1^e x^3 \ln x\,dx$

19. $\int \log_3 x\,dx$

20. $\int \frac{1}{x} \arctan(\ln x)\,dx$

21. $\int \sin^2 x \cos^3 x\,dx$

22. $\int \sec^2 x \tan^3 x\,dx$

23. $\int_0^{\pi/2} \cos^2 5x\,dx$

24. $\int_{-\pi/2}^{\pi/2} \sin^2 3x\,dx$

25. $\int \tan^2 4x\,dx$

26. $\int \sec^3 2x\,dx$

27. $\int \sec^3 x \tan x\,dx$

28. $\int \frac{\cos x\,dx}{1-\cos^2 x}$

29. $\int \sqrt{x^2+b^2}\,dx$

30. $\int (x^2+4)^{-3}\,dx$

31. $\int \frac{x^2}{(x^2+8)^{3/2}}\,dx$

32. $\int \frac{1}{x}\sqrt{x^2+4}\,dx$

33. $\int_1^2 \frac{1}{x}\sqrt{x^2+4}\,dx$

34. $\int_0^1 x(1-x^2)^{3/2}\,dx$

35. $\int x^2(1-x^2)^{-3/2}\,dx$

36. $\int \frac{\sqrt{1-x^2}}{x^2}\,dx$

37. $\int x^{-1}(1-x^2)^{-1/2}\,dx$

38. $\int (x^2-2x+2)^{-3/2}\,dx$

39. $\int (x+1)\sqrt{x^2+2x-3}\,dx$

A40. $\int \sqrt{3x^2-5}\,dx$

41. $\int (1-9x^2)^{-1/2}\,dx$

42. $\int (1+9x^2)^{-1/2}\,dx$

A43. $\int \frac{dx}{x^3+b^3}$

44. $\int \frac{dx}{(x+1)^3(x-1)}$

45. $\int_{-1}^{0} \frac{2x^5+20x^4+56x^3+16x^2-50x+28}{(x^2+4x-2)^3}\,dx$

46. $\int_0^1 \frac{x^4\,dx}{(x-2)^2}$

A47. $\int \frac{x^3+5}{x^2+5x-3}\,dx$

A48. $\int \frac{dx}{x^4-8x}$

49. $\int x^2 \arctan x\,dx$

50. $\int x^3 \arctan x\,dx$

51. $\int x^{-2} \arctan x\,dx$

52. $\int_1^2 \sqrt{x^3+1/x^2}\,dx$

53. $\int x\sqrt{2-x}\,dx$

54. $\int x(4x-1)^{3/8}\,dx$

55. $\int x^2(2x-1)^{-2/3}\,dx$

56. $\int e^x(1+e^{2x})^{-1/2}\,dx$

57. $\int \frac{\cos^2 x\,dx}{1+\sin x}$

58. $\int \frac{dx}{2-\cos^2 x}$

59. $\int \frac{dx}{1+\sin x+\cos x}$

60. $\int \frac{\tan x\,dx}{1-\cos x}$

61. $\int_1^e \ln x\,dx$

62. $\int_0^{\pi} 4\sin x\,dx$

Approximate by the left and right endpoint rules, the trapezoidal rule, and Simpson's rule.

63. $\int_1^5 \frac{1}{x}\,dx, \quad n=8$

64. $\int_0^1 \frac{dx}{1+x^2}, \quad n=4$

Approximate to three decimal places.

65. $\int_0^1 e^{-x^2/4}\,dx$

66. $\int_0^1 \sqrt{1+4\cos^2 x}\,dx$

Additional Exercises

Integrate.

1. $\int \frac{1}{x} 10^{\ln x}\,dx$

2. $\int x^2 \ln x^3\,dx$

3. $\int x\cos^2 x\,dx$

4. $\int e^x \sin^2 x\,dx$

5. $\int x^3(3-x)^{-3/4}\,dx$

6. $\int \left(\frac{x}{x+1}\right)^{1/2} dx$

7. $\int (1+\sqrt{x})^{1/2}\,dx$

8. $\int \left(1+\frac{2}{x}\right)^{1/2} dx$

9. $\int x^5 e^{-x^2}\,dx$

10. $\int x^3 e^{5x^2}\,dx$

11. $\int (1 + e^{2x})^{-1/2}\,dx$

12. $\int (1 - e^{4x})^{-1/2}\,dx$

13. $\int \frac{\sec x}{1 + \tan x}\,dx$

14. $\int x^a \ln x\,dx$

15. $\int x(ax + b)^{2/3}\,dx$

16. $\int x^2(ax + b)^{2/3}\,dx$

17. $\int \frac{xe^x\,dx}{(x + 1)^2}$

18. $\int x^5 e^{-x^3}\,dx$

19. $\int_0^{\pi/2} \sin^3 x\,dx$

20. $\int_0^{\pi/2} \cos^3 x\,dx$

21. $\int_0^{\pi/2} \sin ax \cos bx\,dx$

22. $\int_0^{\pi/2} \sin ax \sin bx\,dx$

23. $\int_0^{\pi/2} \cos ax \cos bx\,dx$

24. $\int_0^{2\pi} \sin ax \cos bx\,dx$

25. $\int_0^{2\pi} \sin^2 x\,dx$

26. $\int_0^{2\pi} \cos^2 x\,dx$

27. $\int_{-a}^{a} \sqrt{a^2 - x^2}\,dx$

28. $\int_0^{a} \sqrt{x^2 + a^2}\,dx$

29. $\int_0^{a} (a^2 - x^2)^{-1/2}\,dx$

30. $\int_0^{a} x^2(a^2 + x^2)^{-1/2}\,dx$

31. $\int \frac{\sin\theta\,d\theta}{1 - \cos\theta}$

32. $\int \frac{1 + \cos\theta}{\sin\theta}\,d\theta$

33. $\int \frac{dx}{\sin^2 x}$

34. $\int \frac{dx}{\sin^3 x}$

35. $\int \frac{dx}{\cot^3 x}$

36. $\int \ln(x + \sqrt{x^2 - 1})\,dx$

37. $\int \ln(x + \sqrt{x^2 + 1})\,dx$

38. $\int \ln(\sqrt{x} + \sqrt{x + 1})\,dx$

39. $\int x \sin x^2\,dx$

40. $\int x^2 \tan x\,dx$

41. $\int \frac{du}{u^2 + 4u + 3}$

42. $\int \frac{x^5\,dx}{\sqrt{x^2 + 1}}$

43. $\int \frac{x^5\,dx}{\sqrt{x + 1}}$

44. $\int e^x \sin^2 x\,dx$

45. $\int \frac{dx}{1 + x + \sqrt{x}}$

46. $\int \frac{e^x\,dx}{\sqrt[3]{1 + e^x}}$

47. $\int \ln(x + x^2)\,dx$

48. $\int \sqrt{4 + e^x}\,dx$

Using a numerical integration technique, approximate each of the following to four decimal places:

49. $\ln 5$ 50. π 51. $\arcsin \frac{1}{4}$ 52. $\sqrt{3}$

Factor the following polynomials into linear and irreducible quadratic factors:

53. $x^2 - 6$
54. $x^3 + 2$
55. $x^4 + x^2 + 1$
56. $x^4 - 16$
57. $x^3 + x^2 - 4x - 4$
58. $x^5 + x^4 + x + 1$

Derive the following formulas:

59. $\displaystyle\int \frac{du}{u^2 - a^2} = \frac{1}{2a} \ln\left|\frac{u-a}{u+a}\right| + C$

60. $\displaystyle\int \frac{du}{\sqrt{2au - u^2}} = \arccos\left(\frac{a-u}{a}\right) + C$

61. $\displaystyle\int u^n e^{au}\, du = \frac{1}{a} u^n e^{au} - \frac{n}{a} \int u^{n-1} e^{au}\, du$

62. $\displaystyle\int \csc^n u\, du = \frac{-1}{n-1} \csc^{n-2} u \cot u + \frac{n-2}{n-1} \int \csc^{n-2} u\, du$

63. Find the area between the curves $y = 4 - x^2$ and $y = 0$, outside the circle $x^2 + y^2 = 2y$.

64. Find the area inside both the circle $x^2 + y^2 = 1$ and the parabola $2y^2 = 2x - 1$.

65. Find the area inside both the circles $x^2 + y^2 = 1$ and $x^2 + y^2 - 4x + \frac{7}{4} = 0$.

66. A circle of radius 4 feet is cut by a chord that passes exactly 1 foot from the center of the circle. Find the areas of the two parts of the circle's interior created by the chord.

67. A circle of radius 2 feet has a chord 3 feet long. Find the area of the smaller part of the circle's interior cut off by the chord.

68. Find the area of the region bounded by $y = \cot(\pi x/4)$, $y = x$, and $y = 0$.

69. Circle C_1 has radius r_1 and circle C_2 has radius r_2; the distance between the centers of the circles is s. Derive formulas for the area:
(a) Inside both C_1 and C_2 (b) Inside C_1 and outside C_2

CHAPTER 8

Further Applications of Integration

INTRODUCTION

Because the definite integral is defined as the limit of a sum, it has many useful physical interpretations. Newton used the definite integral for areas and volumes, and to demonstrate certain properties of his theory of gravitation. Following his lead, physicists since then have found many applications for the definite integral, including displacements, moments, weights and masses, centers of mass, and inertia, as well as hosts of applications in electricity, heat, magnetism, and theoretical physics. Chemists, engineers, economists, statisticians, and most other scientists have likewise found myriads of uses for the definite integral.

In this chapter, we will sample some of the more common applications of the definite integral. We have already used the definite integral to find areas, average values, and changes and displacements in Sections 5.6 and 5.7. Here, we will discuss volumes, arc lengths, surface areas, centers of mass, and centroids, along with a few miscellaneous applications such as work, fluid pressure, population theory, and revenue streams.

SECTION 8.1

Volume: The Slicing Method

In this section and the next, we apply definite integrals to the finding of volumes. These methods apply to solids that can be described in certain ways. This section deals with what we will call *solids of known cross-section*. Such solids can be "sliced" into pieces whose volumes can be approximated. When the volumes of the slices are added together, we get a sum that approximates the volume of the solid. Then, taking a limit of this sum gives us a definite integral whose value is exactly the volume of the solid.

Solids of Revolution

We begin with the simplest of the solids of known cross-section, the *solids of revolution*, which are described as follows: Let R be a region in the plane, and let l be a line that does not cut through the interior of R. A **solid of revolution** is generated by revolving R through a full revolution in space about the line l as axis (see Figure 8.1). The solid consists of those points in space through which R passes as it revolves.

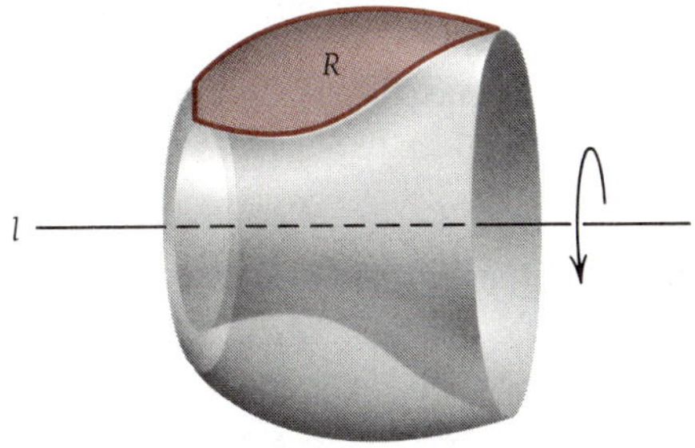

FIGURE 8.1 Generating a solid of revolution

For example, if a right triangular region is revolved about one of its legs, a right circular cone is generated, as shown in Figure 8.2(a). If a semicircular region is revolved about its diameter, a sphere is generated, as shown in Figure 8.2(b). If a

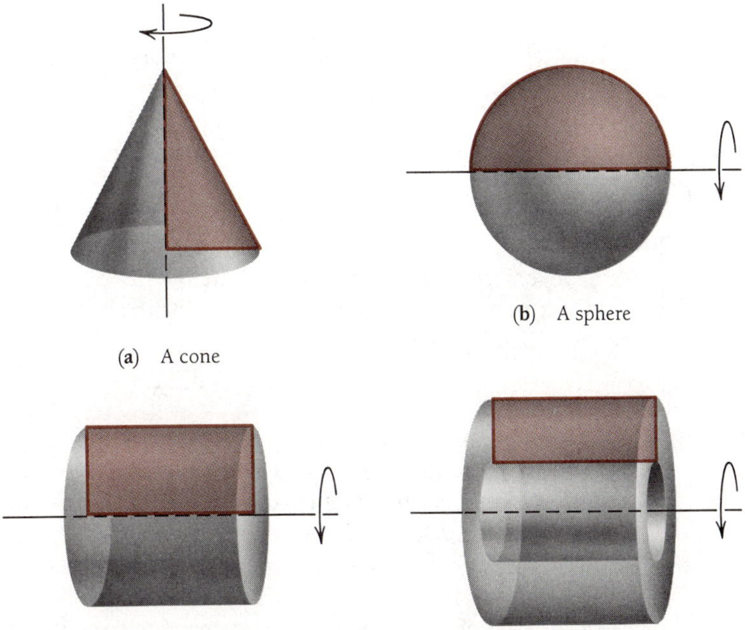

(a) A cone

(b) A sphere

(c) A cylinder

(d) A hollow cylinder

FIGURE 8.2 Generating solids of revolution

rectangular region is revolved about one of its sides, a right circular cylinder is generated, as shown in Figure 8.2(c); but if it is revolved about a line parallel to one of its sides, the resulting solid is a cylinder with a hole through it, as shown in Figure 8.2(d).

Slicing a Solid of Revolution

If we slice a solid of revolution by planes perpendicular to the axis of revolution, we get slices that are shaped like either discs or washers (Figure 8.3). The volume of such a slice is approximately the area of the slice times its thickness; the volume of the solid is then approximately the sum of the volumes of these slices.

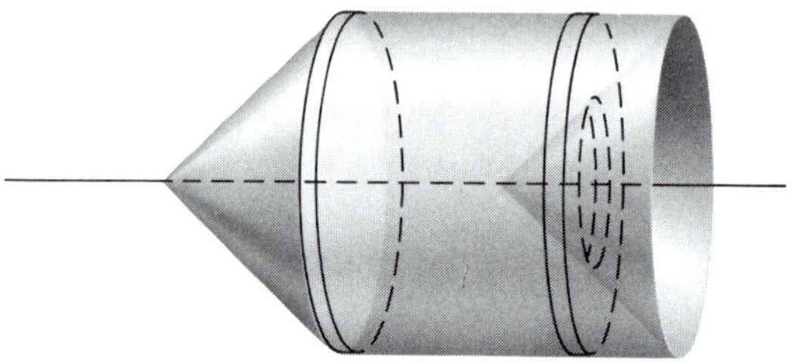

FIGURE 8.3 Slices of a solid of revolution

Let's take a specific example and be more precise. Let R be the region in the first quadrant bounded by the curves $y = x^2$, $y = 0$, and $x = 2$, as shown in Figure 8.4(a). We will revolve R about the x-axis to get a solid of revolution, as shown in Figure 8.4(b). To slice this solid, we will cut perpendicular to the x-axis at the points $x_1, x_2, \ldots, x_{n-1}$, with

$$0 < x_1 < x_2 < \cdots < x_{n-1} < 2.$$

For convenience, we will let $x_0 = 0$ and $x_n = 2$.

In this case, the slices we get are discs. The thickness of each slice is the distance between consecutive x_i's, and we will let

$$\Delta x_i = x_i - x_{i-1}$$

represent the thickness of the ith slice.

The area of each slice is the area of its circular face, which can be computed if we know its radius. Let t_i be a point in the interval $[x_{i-1}, x_i]$; then t_i^2 is approximately the radius of the ith disc. Thus, the volume of the ith slice is approximately

$$\Delta V_i = \pi (t_i^2)^2 \Delta x_i.$$

Then the sum of these volumes is approximately the volume of the solid; that is,

$$V \approx \sum_{i=1}^{n} \Delta V_i = \sum_{i=1}^{n} \pi t_i^4 \Delta x_i.$$

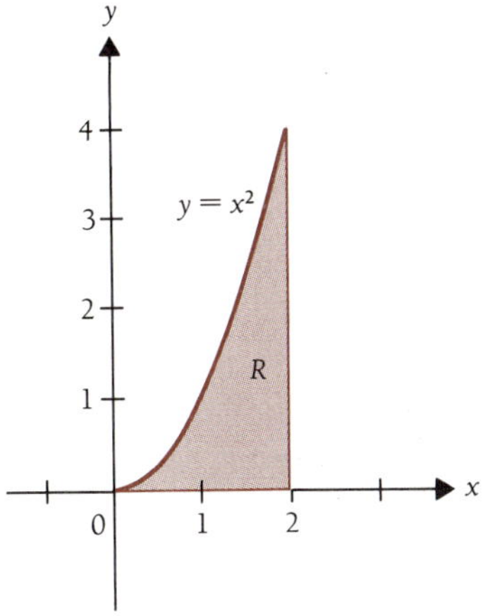

(a) The region R bounded by $y = 0$, $x = 2$, and $y = x^2$

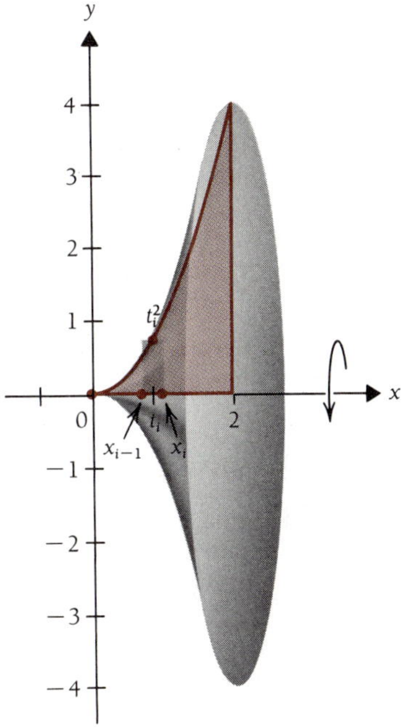

(b) The region R revolved about the x-axis

FIGURE 8.4

At this point, you should recognize this sum as a Riemann sum for the function $f(x) = \pi x^4$ over the interval $[0, 2]$. If we take the limit as the number of slices becomes infinite, then by the method of exhaustion, the volume of the solid is the definite integral of πx^4 over $[0, 2]$, namely

$$V = \int_0^2 \pi x^4 \, dx.$$

The value of this integral is $32\pi/5$, so the volume of this solid of revolution is $32\pi/5$.

Now suppose that our original region R shown in Figure 8.4(a) is revolved about the y-axis (Figure 8.5, page 496). This time, when we slice the solid of revolution perpendicular to the axis of revolution, we get slices shaped like washers. If we let

$$y_0 = 0 < y_1 < y_2 < \cdots < y_{n-1} < 4 = y_n,$$

then $\Delta y_i = y_i - y_{i-1}$ is the thickness of the ith slice. If we let t_i be a point in the ith subinterval $[y_{i-1}, y_i]$, then the inner radius of the washer-shaped slice is $\sqrt{t_i}$ and the outer radius is 2. The volume of the ith slice is therefore approximately

$$\begin{aligned}\Delta V_i &= \pi[(2)^2 - (\sqrt{t_i})^2]\Delta y_i \\ &= \pi(4 - t_i)\Delta y_i,\end{aligned}$$

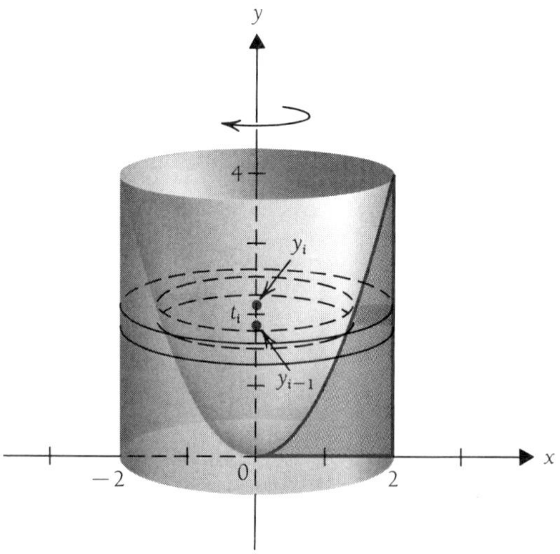

FIGURE 8.5 The region R revolved about the y-axis

and the volume of the solid is approximately

$$V \approx \sum_{i=1}^{n} \Delta V_i = \sum_{i=1}^{n} \pi(4 - t_i)\, \Delta y_i .$$

In the limit, then, we get the volume expressed as a definite integral, namely

$$V = \int_0^4 \pi(4 - y)\, dy,$$

whose value is 8π.

Thus, we see how the volume of a solid of revolution can be represented as a definite integral.

One case occurs often enough that we will develop a formula for it. Let f be a continuous function that is nonnegative over the interval $[a, b]$, and let R be the region below f,

$$R = \{(x, y) | a \le x \le b \text{ and } 0 \le y \le f(x)\}.$$

Let R be revolved about the x-axis to obtain a solid of revolution of volume V (see Figure 8.6). We find the volume V as follows: Let $s = \{x_0, x_1, \ldots, x_n\}$ be any subdivision of $[a, b]$. For $t_i \in [x_{i-1}, x_i]$, let R_i be the rectangular region with base $\Delta x_i = x_i - x_{i-1}$ and height $f(t_i)$. Let ΔV_i be the volume of the cylinder generated by revolving R_i about the x-axis. If Δx_i is small, this cylinder is a disc, somewhat like a thin slice of the solid perpendicular to the axis of rotation.

Since the volume of a disc is the thickness times the area of its face, we have

$$\Delta V_i = \pi[f(t_i)]^2 \Delta x_i .$$

Hence, the volume of the solid is approximately

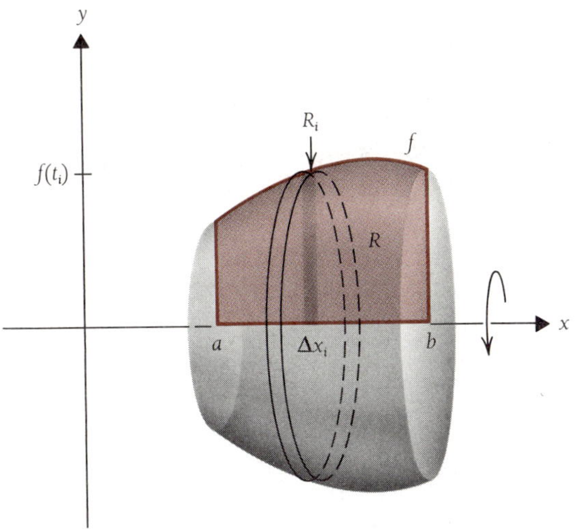

FIGURE 8.6 Revolving the region under a function

$$V \approx \sum_{i=1}^{n} \Delta V_i = \sum_{i=1}^{n} \pi[f(t_i)]^2 \Delta x_i .$$

This sum is a Riemann sum whose limit as $\|s\| \to 0$ is the integral

$$V = \int_a^b \pi[f(x)]^2 \, dx. \tag{1}$$

By the method of exhaustion, this integral is also the volume of the solid.

We can use formula (1) directly whenever we wish to compute the volume of a solid generated by revolving about the x-axis the region below a function.

■ **Example 8.1** The region below $y = \sin x$ over $[0, \pi]$ is revolved about the x-axis (Figure 8.7). Find the volume of the solid generated.

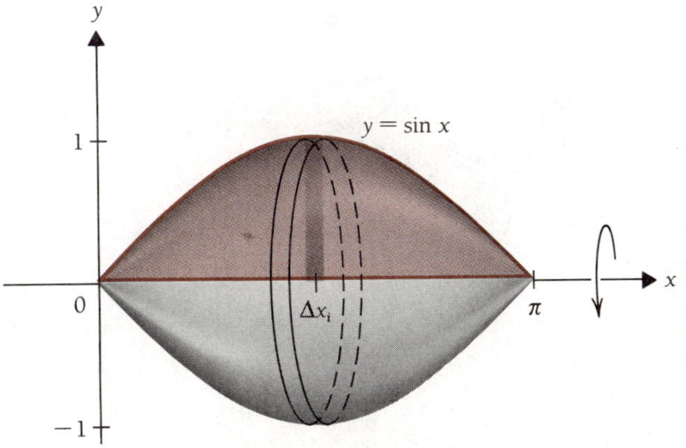

FIGURE 8.7 Revolving the region under $f(x) = \sin x$

Solution Using formula (1), we have

$$\begin{aligned} V &= \int_0^{\pi} \pi(\sin x)^2\, dx \\ &= \pi \int_0^{\pi} \frac{1}{2}(1 - \cos 2x)\, dx \\ &= \frac{\pi}{2}\left[x - \frac{1}{2}\sin 2x\right]_0^{\pi} \\ &= \frac{\pi^2}{2}. \end{aligned}$$

■

■ **Example 8.2** Find the volume of a sphere of radius r.

Solution We will let the sphere be generated by revolving the semicircular region below $f(x) = \sqrt{r^2 - x^2}$ over $[-r, r]$ about the x-axis (Figure 8.8). Then, by formula (1),

$$\begin{aligned} V &= \int_{-r}^{r} \pi\left(\sqrt{r^2 - x^2}\right)^2 dx \\ &= \pi \int_{-r}^{r} (r^2 - x^2)\, dx \\ &= \pi\left[r^2 x - \frac{1}{3}x^3\right]_{-r}^{r} \\ &= \frac{4}{3}\pi r^3. \end{aligned}$$

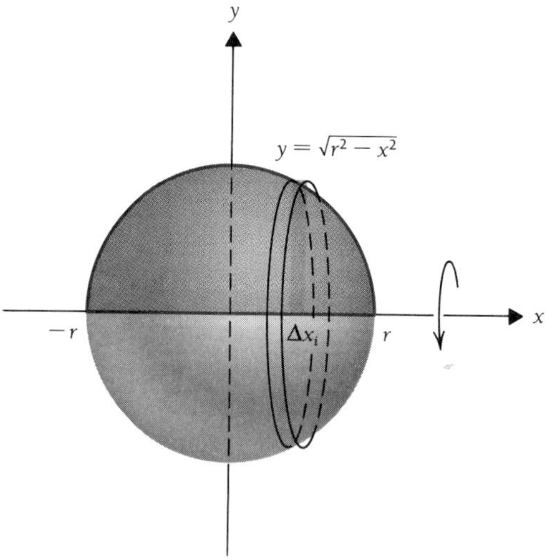

FIGURE 8.8 Generating a sphere of radius r

■

Example 8.3 Find the volume of a right circular cone of height h and base radius r.

Solution We will let the cone be generated by revolving a right triangular region with legs h and r about the leg of length h (Figure 8.9). This region can be described as the region below $f(x) = rx/h$ over $[0, h]$. Thus, using formula (1), we have

$$V = \int_0^h \pi \left(\frac{rx}{h}\right)^2 dx$$
$$= \frac{\pi r^2}{h^2} \int_0^h x^2\, dx$$
$$= \frac{\pi r^2}{h^2} \left[\frac{x^3}{3}\right]_0^h = \frac{1}{3}\pi r^2 h.$$

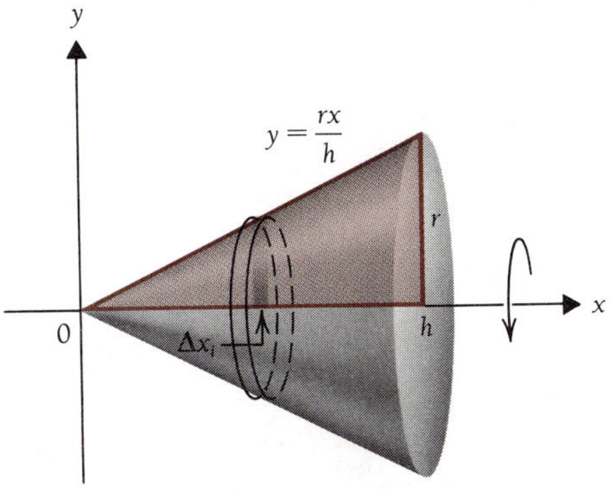

FIGURE 8.9 Generating a cone of height h and base radius r ■

In case the region revolved is not bounded below by the x-axis, formula (1) must be modified. Suppose the region between two nonnegative functions f and g is revolved about the x-axis (Figure 8.10). We can express the volume of the solid

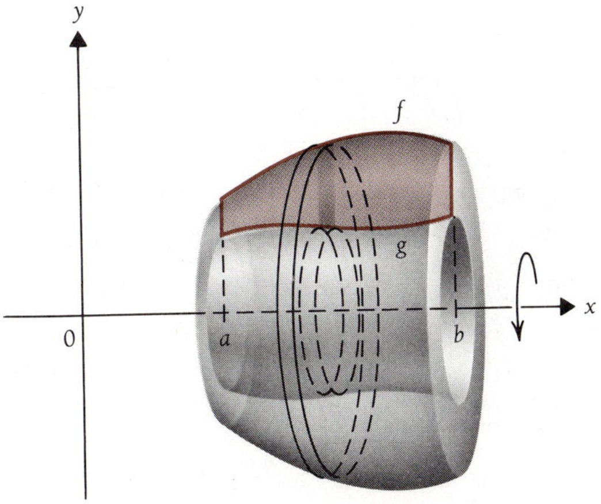

FIGURE 8.10 Revolving the region between two functions

generated as a difference of volumes, namely

$$V = \int_a^b \pi[f(x)]^2\,dx - \int_a^b \pi[g(x)]^2\,dx.$$

This can be rewritten as

$$V = \pi \int_a^b \{[f(x)]^2 - [g(x)]^2\}\,dx. \tag{2}$$

In this case, the slice is shaped like a washer, with outer radius $f(x)$ and inner radius $g(x)$ (see Figure 8.11).

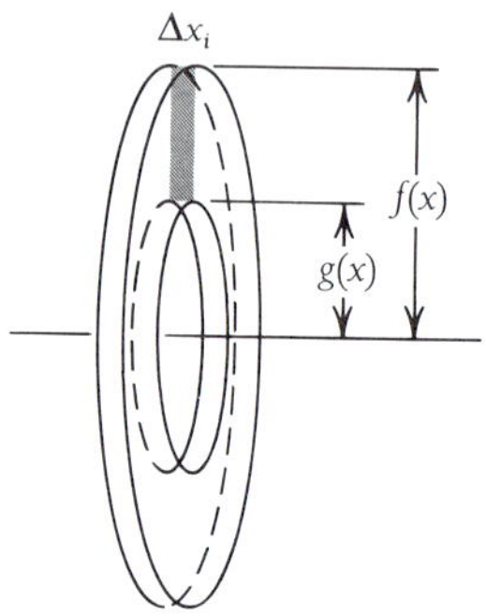

FIGURE 8.11 The washer-shaped slice in Figure 8.10

Example 8.4 The region between $y = x$ and $y = x^2$ over [0, 1] is revolved about the x-axis (Figure 8.12). Find the volume of the solid generated.

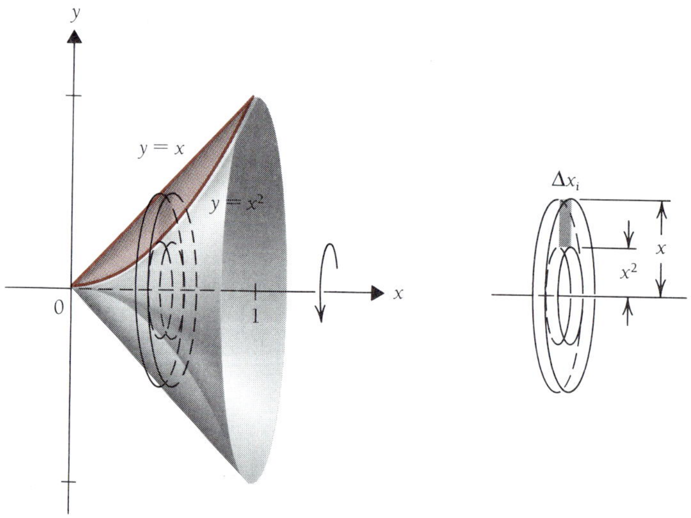

FIGURE 8.12 Revolving the region between $y = x$ and $y = x^2$ about the x-axis

Solution We use formula (2) with $f(x) = x$ and $g(x) = x^2$:

$$V = \pi \int_0^1 (x^2 - x^4)\, dx$$
$$= \pi \left[\frac{1}{3}x^3 - \frac{1}{5}x^5\right]_0^1 = \frac{2\pi}{15}.$$

■

■ **Example 8.5** The region in Example 8.4 is revolved about the line $y = 1$. Find the volume generated.

Solution If a cross-section of this solid is taken at x, parallel to the y-axis, the cross-section is a washer of radius $1 - x^2$ with a hole in it of radius $1 - x$ (Figure 8.13). Hence, we use formula (2) with $f(x) = 1 - x^2$ (the outside radius) and $g(x) = 1 - x$ (the inside radius):

$$V = \pi \int_0^1 [(1 - x^2)^2 - (1 - x)^2]\, dx$$
$$= \pi \int_0^1 (2x - 3x^2 + x^4)\, dx$$
$$= \pi \left[x^2 - x^3 + \frac{1}{5}x^5\right]_0^1 = \frac{\pi}{5}.$$

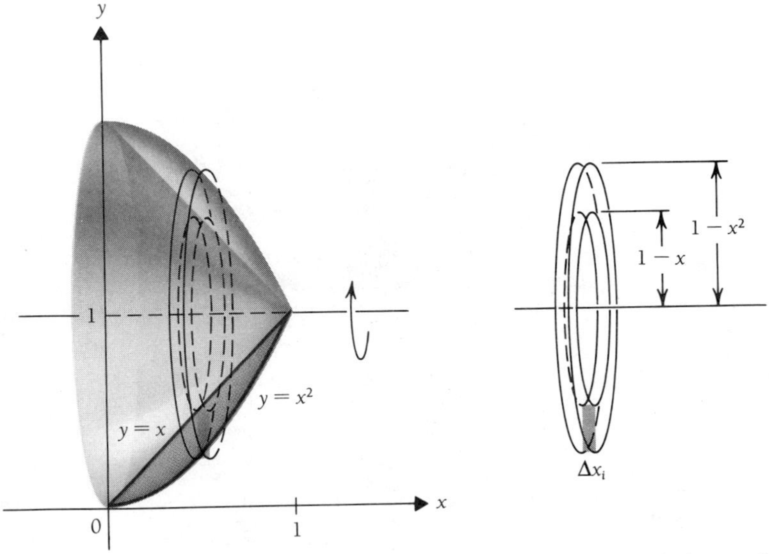

FIGURE 8.13 Revolving the region between $y = x$ and $y = x^2$ about the line $y = 1$ ■

Example 8.6 The region in Example 8.4 is revolved about the y-axis. Find the volume generated.

Solution Since we are now revolving about the y-axis, we will interchange the roles of x and y. We will view the region as bounded by the curves $x = y$ and $x = \sqrt{y}$, $0 \leq y \leq 1$ (see Figure 8.14). The formula (2) applies, but with x and y interchanged. The volume is

$$\begin{aligned} V &= \int_0^1 \pi[(\sqrt{y})^2 - (y)^2]\, dy \\ &= \pi \int_0^1 (y - y^2)\, dy \\ &= \pi \left[\frac{1}{2}y^2 - \frac{1}{3}y^3\right]_0^1 \\ &= \pi\left(\frac{1}{2} - \frac{1}{3}\right) = \frac{\pi}{6} \end{aligned}$$

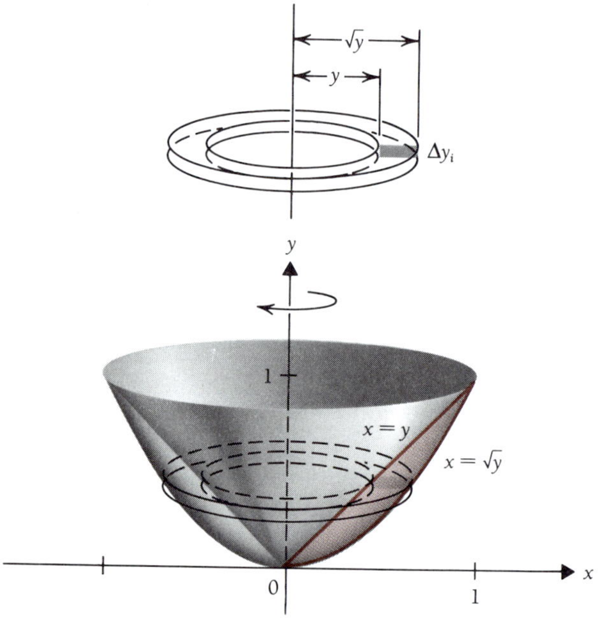

FIGURE 8.14 Revolving the region between $x = y$ and $x = \sqrt{y}$ about the y-axis

■

Example 8.7 The region in Example 8.4 is revolved about the line $x = 1$. Find the volume generated.

Solution This situation is similar to Example 8.5, except that again the roles of x and y are interchanged. We see from Figure 8.15 that the volume is

$$V = \int_0^1 \pi[(1-y)^2 - (1-\sqrt{y})^2]\, dy$$
$$= \pi \int_0^1 (-3y + y^2 + 2\sqrt{y})\, dy$$
$$= \pi \left[-\frac{3}{2}y^2 + \frac{1}{3}y^3 + \frac{4}{3}y^{3/2} \right]_0^1$$
$$= \pi \left(-\frac{3}{2} + \frac{1}{3} + \frac{4}{3} \right) = \frac{\pi}{6}.$$

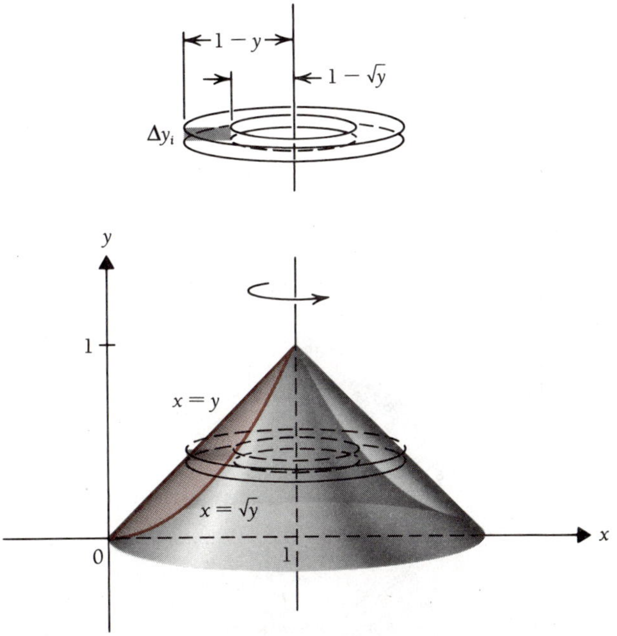

FIGURE 8.15 Revolving the region between $x = y$ and $x = \sqrt{y}$ about the line $x = 1$ ■

The Slicing Method for More General Solids

The slicing method can also be used to find volumes of solids other than solids of revolution. The key idea for solids of revolution is that a cross-section, or slice, always has a convenient form—it is circular. If a slice of a more general solid still has a convenient form, we may be able to find its volume using a definite integral. Such solids may be called *solids of known cross-section.*

Suppose such a solid extends from a to b along the x-axis (see Figure 8.16). Moreover, suppose that a cross-section, or slice, perpendicular to the x-axis at x is known in the sense that the area of the slice is given as a function of x. Let $A(x)$ be the area of the slice at x. Then the volume of the solid can be found as follows: Let $s = \{x_0, x_1, \ldots, x_n\}$ be a subdivision of the interval $[a, b]$, and for each i let $t_i \in [x_{i-1}, x_i]$. We approximate the volume of the portion of the solid between x_{i-1} and x_i by the product of $A(t_i)$ and $\Delta x_i = x_i - x_{i-1}$ (see Figure 8.16). Thus, the volume of the given solid is approximately

$$V \approx \sum_{i=1}^{n} A(t_i)\Delta x_i.$$

The limit of this Riemann sum as $\|s\| \to 0$ is the volume,

$$V = \int_a^b A(x)\, dx. \tag{3}$$

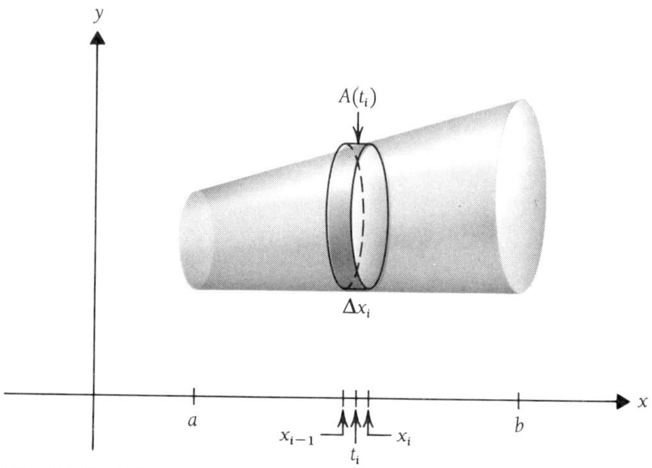

FIGURE 8.16 A solid of known cross-section

Now we see that formulas (1) and (2) are special cases of formula (3). The fact that formula (3) really is more general is illustrated by the following example.

■ **Example 8.8** The base of a certain solid is the region bounded by the curves $x = 4 - y^2$ and $x = 0$. A slice of the solid perpendicular to the x-axis is a square (Figure 8.17). Find the volume of the solid.

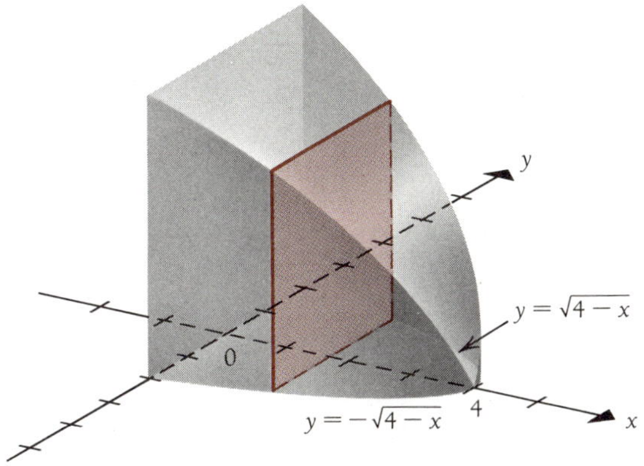

FIGURE 8.17 A solid with square cross-section

Solution Since a slice is a square with one side in the base of the solid, the side of the square is the distance from the curve $y = \sqrt{4-x}$ to the curve $y = -\sqrt{4-x}$, which is simply twice the value of $\sqrt{4-x}$. Thus, the slice at x has area

$$A(x) = [2\sqrt{4-x}]^2 = 4(4-x).$$

By formula (3), the volume of the solid is

$$V = \int_0^4 4(4-x)\,dx = 4\left[4x - \frac{1}{2}x^2\right]_0^4 = 32.$$ ■

■ **Example 8.9** Find the volume of a pyramid with square base of side s and having height h.

Solution The pyramid extends from 0 to h along a vertical axis. A horizontal slice at y on this axis is a square whose side is related to s as $h - y$ is related to h, by properties of similar triangles (see Figure 8.18). Hence, the side of the slice is $(s/h)(h-y)$, and

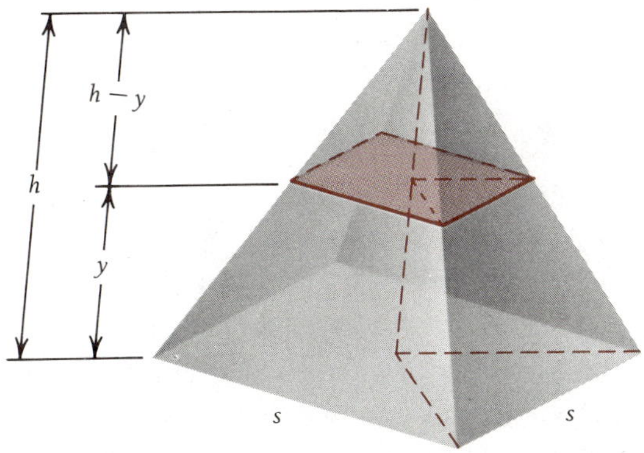

FIGURE 8.18 A square-based pyramid

the area of the slice is

$$A(y) = \frac{s^2}{h^2}(h - y)^2.$$

Therefore, by formula (3), the volume of the pyramid is

$$V = \int_0^h \frac{s^2}{h^2}(h^2 - 2hy + y^2)\,dy$$
$$= \frac{s^2}{h^2}\left[h^2y - y^2h + \frac{1}{3}y^3\right]_0^h = \frac{1}{3}s^2h.$$

■

Exercises 8.1

The given region is revolved about the x-axis. Find the volume of the solid generated.

1. Under $y = \cos x$ over $[0, \pi/2]$
2. Under $y = \tan x$ over $[0, \pi/4]$
3. Under $y = \sqrt{x}$ over $[0, 4]$
4. Under $y = x^2 - 4$ over $[0, 3]$
5. Bounded by $y = x^2 + 1$ and $y = 10$
6. Bounded by $y = 2x$ and $y = x^2$
7. Bounded by $y = 3x$, $y = 3$, and $x = 0$
8. Bounded by $y = \sqrt{x}$, $y = 4$, and $x = 0$
A9. Bounded by $y = (x - 2)^2$ and $y = 16 - (x - 2)^2$
10. Bounded by $y = -1 - x^2$ and $y = -2$
11. Under $y = e^{-x}$ over $[0, b]$
12. Under $y = 1/x$ over $[1, b]$

The following regions are revolved about the y-axis. Find the volume of the solid generated.

13. The region in Problem 3
14. The region in Problem 6
15. The region in Problem 7
16. The region in Problem 8
17. The region in Problem 11
18. The region in Problem 12

The following regions are revolved about the given line. Find the volume of the solid generated.

19. Under $y = \sqrt{1 - x^2}$; the line $y = -1$
20. The region of Problem 19; the line $x = 1$
21. $R = \{(x, y) | 0 \le x \le 2 \text{ and } 0 \le y \le 3\}$; the line $y = -2$
22. $R = \{(x, y) | 0 \le x \le a \text{ and } 0 \le y \le b\}$; the line $y = -r$ $(r > 0)$
23. $R = \{(x, y) | 0 \le x \le 4 \text{ and } 0 \le y \le \sqrt{16 - x^2}\}$; the line $x = 0$
24. $R = \{(x, y) | 0 \le x \le r \text{ and } 0 \le y \le \sqrt{r^2 - x^2}\}$; the line $x = 0$
25. $R = \{(x, y) | 0 \le y \le 1 \text{ and } 0 \le x \le y^2\}$; the line $y = 0$
26. $R = \{(x, y) | a \le x \le b \text{ and } g(x) \le y \le f(x)\}$, with $c \le g(x) \le f(x)$ over $[a, b]$; the line $y = c$

27. The base of a certain solid is the circular region $x^2 + y^2 \leq 1$. A plane section perpendicular to the y-axis is a square. Find the volume of the solid.

28. A certain solid has as base the circular region $x^2 + y^2 \leq 1$. A slice perpendicular to the x-axis is an equilateral triangle. Find the volume of the solid.

29. Find the volume of a frustum of a right circular cone of height h and base radii r_1 and r_2.

30. A ball of radius r is cut into two pieces by a plane. If the smaller piece is placed with its flat face down, its height is a. Find the volumes of the two pieces.

31. A certain oil barrel is a cylinder 6 feet long and 4 feet in diameter. If the barrel lies with its axis horizontal and oil fills the barrel to a depth of 3 feet, how many cubic feet of oil does the barrel contain?

32. A certain water tank is a hemispherical bowl of radius 5 feet. Water flows into the tank at the rate of 10 cubic feet per minute. How fast is the water level rising when the water is 3 feet deep at the center of the tank?

33. Repeat Problem 32, with the tank a parabolic bowl obtained by revolving the curve $y = x^2/2$ about the y-axis.

34. Find the volume of a pyramid of height h whose base is an equilateral triangle with side s.

SECTION 8.2

Volume: The Shell Method

In this section, we develop a second method for finding volumes of solids of revolution. It is called the **method of cylindrical shells**, because the volume is found by filling the solid with thin cylindrical shells, one inside another.

Let R be the region below a continuous nonnegative function f over the interval $[a, b]$, $a \geq 0$, and let R be revolved about the y-axis (Figure 8.19). Let V be the volume of the solid generated. Let $s = \{x_0, x_1, \ldots, x_n\}$ be a subdivision of $[a, b]$, and let t_i be the midpoint of $[x_{i-1}, x_i]$, namely $t_i = \frac{1}{2}(x_{i-1} + x_i)$. Let R_i be the rectangle over $[x_{i-1}, x_i]$ with height $f(t_i)$; the width of R_i is $\Delta x_i = x_i - x_{i-1}$.

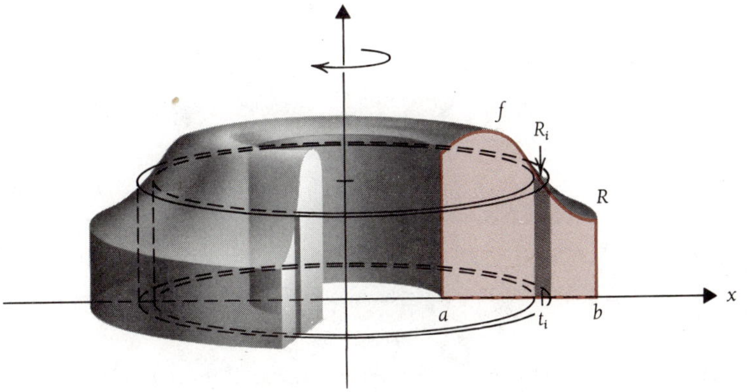

FIGURE 8.19 Revolving the region under f about the y-axis

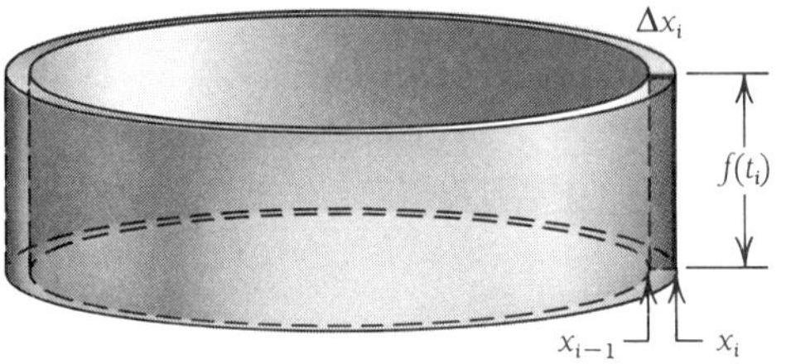

FIGURE 8.20 The cylindrical shell from Figure 8.19

■

As R_i revolves about the y-axis, it generates a cylindrical shell with inner radius x_{i-1}, outer radius x_i, and height $f(t_i)$, as shown in Figure 8.20. The volume of this cylindrical shell is the volume of the outer cylinder minus the volume of the hole, or

$$\begin{aligned}\Delta V_i &= \pi x_i^2 f(t_i) - \pi x_{i-1}^2 f(t_i)\\ &= \pi(x_i^2 - x_{i-1}^2) f(t_i)\\ &= \pi(x_i + x_{i-1}) f(t_i)(x_i - x_{i-1})\\ &= 2\pi\left[\frac{1}{2}(x_i + x_{i-1})\right] f(t_i)(x_i - x_{i-1})\\ &= 2\pi t_i f(t_i)\Delta x_i.\end{aligned}$$

This also can be seen by cutting the cylindrical shell and flattening it into a rectangular slab (see Figure 8.21). Since this is true for each i, the volume of the solid is approximately

$$V \approx \sum_{i=1}^{n} 2\pi t_i f(t_i)\Delta x_i.$$

This is a Riemann sum for $2\pi x f(x)$ over $[a, b]$. Also, the sum approaches V as $\|s\| \to 0$, and, in the limit,

$$V = \int_a^b 2\pi x f(x)\, dx. \tag{1}$$

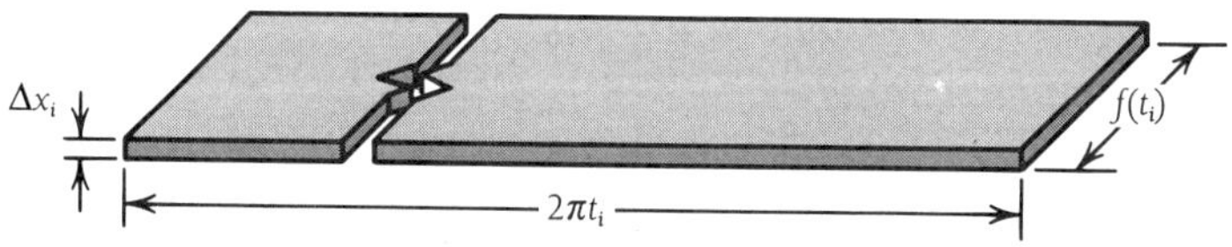

FIGURE 8.21 The cylindrical shell cut and flattened into a slab

■ **Example 8.10** Find the volume of the solid generated by revolving the region under $y = x^2$ over $[0, 1]$ about the y-axis (Figure 8.22).

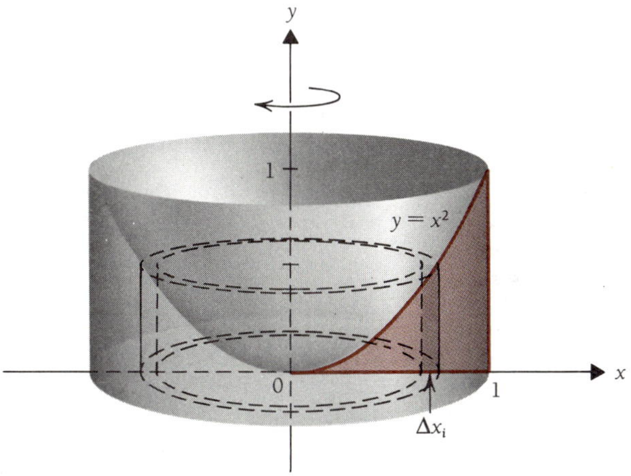

FIGURE 8.22 Revolving the region below $y = x^2$ about the y-axis

Solution By formula (1), the volume is

$$V = \int_0^1 2\pi x(x^2)\,dx$$
$$= 2\pi \int_0^1 x^3\,dx$$
$$= 2\pi \left.\frac{x^4}{4}\right|_0^1 = \frac{\pi}{2}.$$

■

■ **Example 8.11** Use the shell method to find the volume of a right circular cone of height h and radius r.

Solution We will revolve a right triangle of legs h and r about the leg of length h (see Figure 8.23, page 510). This region can be described as the region below

$$f(x) = h - \frac{h}{r}x$$

over $[0, r]$. Thus, by formula (1), we have

$$V = \int_0^r 2\pi x\left(h - \frac{h}{r}x\right)dx$$
$$= 2\pi h \int_0^r \left(x - \frac{1}{r}x^2\right)dx$$
$$= 2\pi h\left[\frac{1}{2}x^2 - \frac{1}{3r}x^3\right]_0^r = \frac{1}{3}\pi r^2 h.$$

Compare this approach with Example 8.3.

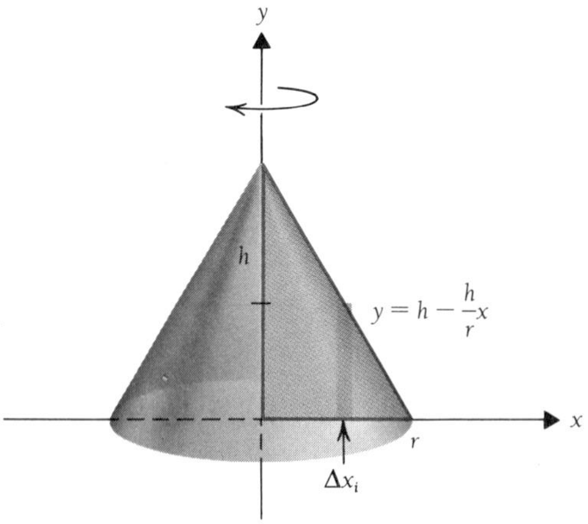

FIGURE 8.23 Generating a cone by revolving about the y-axis

■

■ **Example 8.12** Find the volume of the torus generated by revolving the region inside the circle $(x-a)^2 + y^2 = b^2$, $a > b$, about the y-axis.

Solution By symmetry (Figure 8.24), the top half of the torus is generated by revolving the region under $y = \sqrt{b^2 - (x-a)^2}$ over $[a-b, a+b]$ about the y-axis. Hence,

$$V = 2\int_{a-b}^{a+b} 2\pi x\sqrt{b^2 - (x-a)^2}\, dx$$

$$= 2\int_{-\pi/2}^{\pi/2} 2\pi(a + b\sin\theta)(b\cos\theta)(b\cos\theta)\, d\theta \qquad (x - a = b\sin\theta)$$

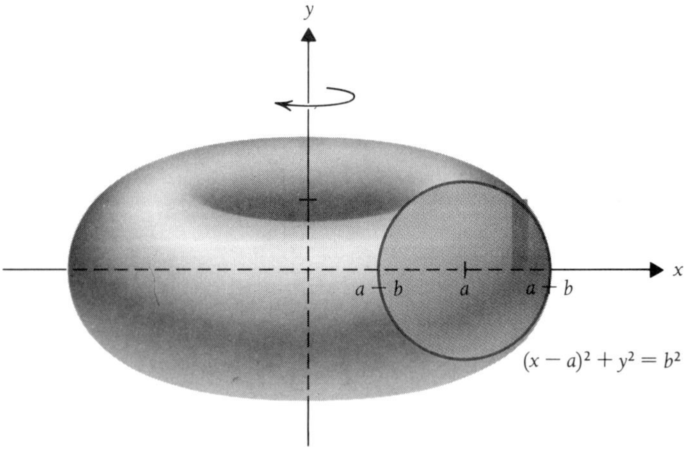

FIGURE 8.24 Generating a torus by revolving a disc

$$\begin{aligned} V &= 4\pi \int_{-\pi/2}^{\pi/2} ab^2 \cos^2\theta \, d\theta + 4\pi \int_{-\pi/2}^{\pi/2} b^3 \sin\theta \cos^2\theta \, d\theta \\ &= 2\pi ab^2 \int_{-\pi/2}^{\pi/2} (1 + \cos 2\theta)\, d\theta - 4\pi b^3 \int_0^0 u^2 \, du \qquad (u = \cos\theta) \\ &= 2\pi ab^2 \left[\theta + \frac{1}{2}\sin 2\theta\right]_{-\pi/2}^{\pi/2} = 2\pi^2 ab^2. \end{aligned}$$

The basic formula (1) can be modified to apply to other types of regions and other axes of rotation. The general schematic form of formula (1) is

$$V = \int_a^b 2\pi \begin{pmatrix}\text{Radius of}\\ \text{rotation}\end{pmatrix}\begin{pmatrix}\text{Height of}\\ \text{region at } x\end{pmatrix} dx. \tag{2}$$

For example, if region R between functions f and g is revolved about line $x = c$ ($c \le a$ or $c \ge b$), as shown in Figure 8.25, then the volume of the solid generated is given by

$$V = \int_a^b 2\pi|x - c|[f(x) - g(x)]\, dx. \tag{3}$$

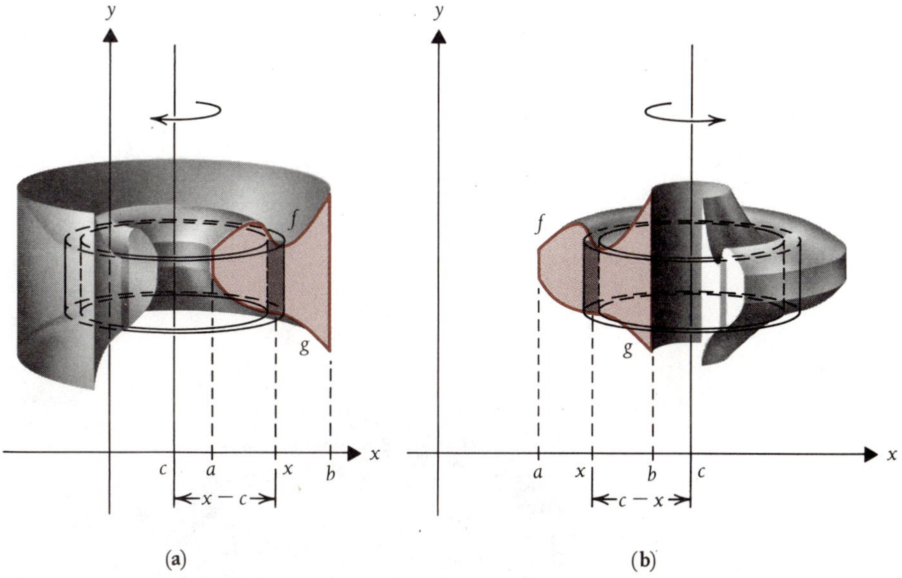

FIGURE 8.25 The region between f and g revolved about $x = c$

Example 8.13 The region between $y = x$ and $y = x^2$ over [0, 1] is revolved about the y-axis. Find the volume of the solid generated.

Solution From the graph of R (Figure 8.26) and formula (3), the volume is

$$\begin{aligned} V &= \int_0^1 2\pi|x - 0|(x - x^2)\,dx \\ &= 2\pi \int_0^1 (x^2 - x^3)\,dx \\ &= 2\pi \left[\frac{1}{3}x^3 - \frac{1}{4}x^4\right]_0^1 = \frac{\pi}{6}. \end{aligned}$$

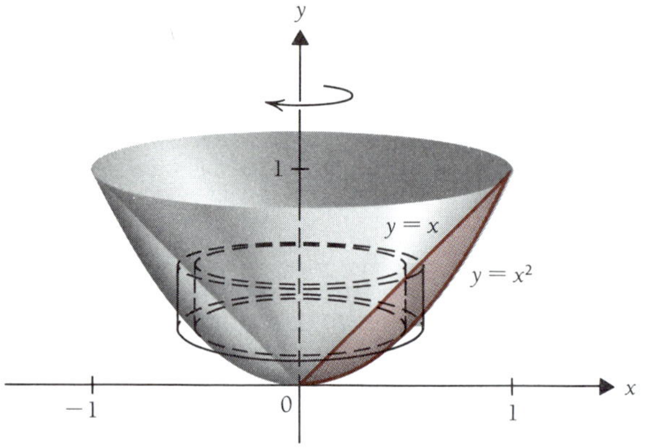

FIGURE 8.26 Revolving the region between $y = x$ and $y = x^2$ about the y-axis

■

Example 8.14 The region in Example 8.13 is revolved about the line $x = 3$. Find the volume of the solid generated.

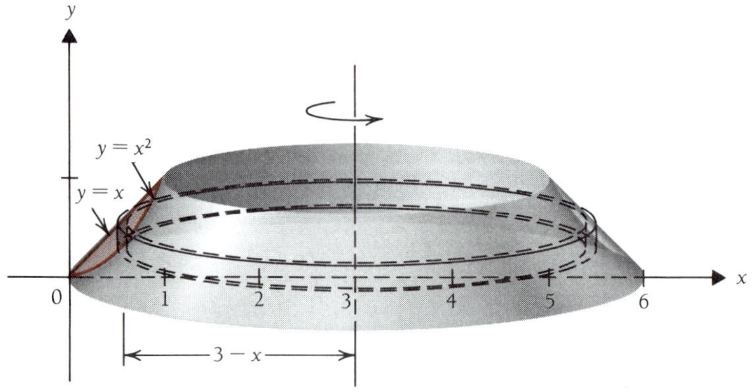

FIGURE 8.27 Revolving the region between $y = x$ and $y = x^2$ about the line $x = 3$

Solution Only the radius of rotation is changed (see Figure 8.27), so formula (3) gives

$$\begin{aligned} V &= \int_0^1 2\pi(3-x)(x-x^2)\,dx \\ &= 2\pi \int_0^1 (3x - 4x^2 + x^3)\,dx \\ &= 2\pi\left[\frac{3}{2}x^2 - \frac{4}{3}x^3 + \frac{1}{4}x^4\right]_0^1 = \frac{5\pi}{6}. \end{aligned}$$

■

■ Example 8.15 The region bounded by $y = 1$, $y = x + 1$, and $y = 4 - 2x$ (Figure 8.28) is revolved about the x-axis. Find the volume of the solid generated.

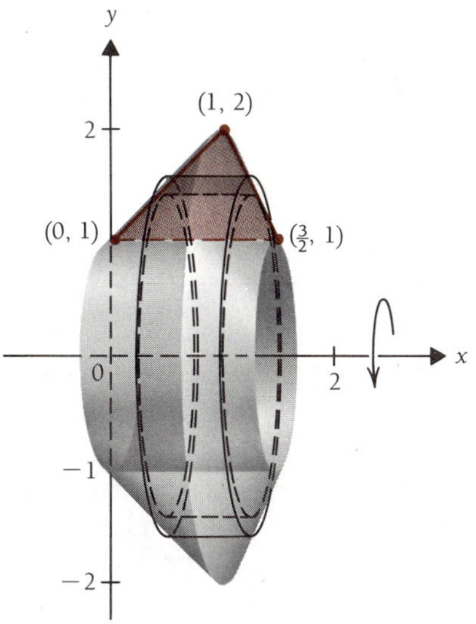

FIGURE 8.28 Revolving a triangular region about the x-axis

Solution We can use either the slicing method or the shell method to solve this problem. The slicing method would require two integrals, since two different functions bound the region above. So we use the shell method with y the variable of integration. First solving for x, the curves are written as $x = y - 1$ and $x = 2 - \frac{1}{2}y$. Then, from scheme (2), we get

$$\begin{aligned} V &= \int_1^2 2\pi y\left[\left(2 - \frac{1}{2}y\right) - (y-1)\right] dy \\ &= 2\pi \int_1^2 \left(3y - \frac{3}{2}y^2\right) dy \\ &= 2\pi\left[\frac{3}{2}y^2 - \frac{1}{2}y^3\right]_1^2 = 2\pi. \end{aligned}$$

■

■ **Example 8.16** The trapezoidal region with vertices (0, 0), (0, 1), (1, 1), and (2, 0) is revolved about the line $y = -x$ (see Figure 8.29). Find the volume generated.

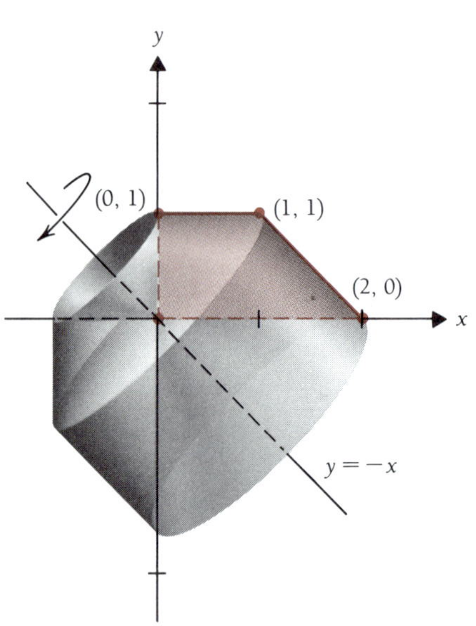

FIGURE 8.29 Revolving a trapezoidal region about the line $y = -x$

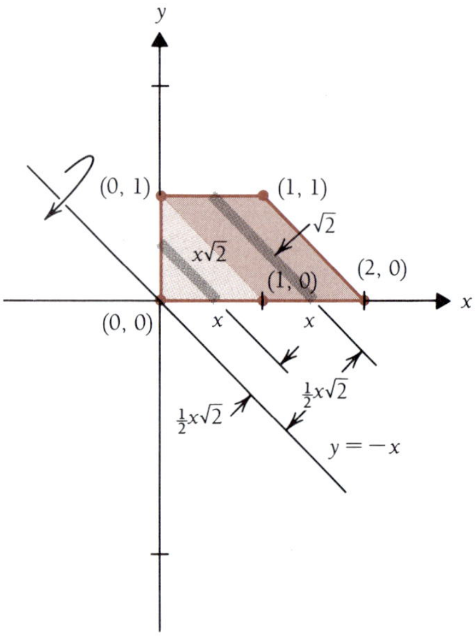

FIGURE 8.30 Subdividing the trapezoid

Solution This problem can be solved in several ways. We will divide it into two regions, the triangle with vertices (0, 0), (0, 1), and (1, 0), and the parallelogram with vertices (0, 1), (1, 1), (2, 0), and (1, 0) (see Figure 8.30). Using cylindrical shells, a line through the triangle parallel to $y = -x$ and meeting the x-axis at x has length $x\sqrt{2}$ and radius of rotation $\frac{1}{2}x\sqrt{2}$. Thus, the triangle generates the volume

$$\begin{aligned} V_1 &= \int_0^1 2\pi\left(\frac{1}{2}x\sqrt{2}\right)(x\sqrt{2})\,dx \\ &= 2\pi\int_0^1 x^2\,dx \\ &= \frac{2\pi}{3}. \end{aligned}$$

A line through the parallelogram parallel to $y = -x$ has length $\sqrt{2}$, and its radius of rotation is still $\frac{1}{2}x\sqrt{2}$, so the parallelogram generates the volume

$$V_2 = \int_1^2 2\pi\left(\frac{1}{2}x\sqrt{2}\right)(\sqrt{2})\,dx$$

$$V_2 = 2\pi \int_1^2 x\,dx$$
$$= 3\pi.$$

Thus, the total volume is

$$V = V_1 + V_2 = \frac{2\pi}{3} + 3\pi = \frac{11\pi}{3}.$$

■

Exercises 8.2

The given region is revolved about the y-axis. Use the shell method to find the volume of the solid generated.

1. Under $y = \cos x$ over $[0, \pi/2]$
2. Under $y = \sqrt{x}$ over $[0, 4]$
3. Bounded by $y = 2x$ and $y = x^2$
4. Bounded by $y = 3x$, $y = 3$, and $x = 0$
5. Bounded by $y = e^x$, $y = \ln x$, $x = 1$, and $x = e$
6. Bounded by $y = \cos 2x$ and $y = \sin 2x$ over $[0, \pi/8]$
7. $R = \{(x, y) | 0 \le x \le 1 \text{ and } -x \le y \le x^2\}$
8. $R = \{(x, y) | 1 \le x \le e \text{ and } \ln x \le y \le x\}$
9. $R = \{(x, y) | 0 \le x \le 2 \text{ and } 0 \le y \le 4 - x^2\}$
10. $R = \{(x, y) | 0 \le x \le 1 \text{ and } 1 - x^2 \le y \le e^x\}$
11. Bounded by $y = 2x$, $y = x^3$, and $x = 1$ over $[0, 1]$
12. Bounded by $y = 4$ and $y = x^2 - x + 4$

The given region is revolved about the given line. Find the volume of the solid generated. You may use either the slicing method or the shell method.

13. $R = \{(x, y) | 0 \le x \le 2 \text{ and } 0 \le y \le x^3\}$; $x = 3$
14. $R = \{(x, y) | 0 \le x \le 2 \text{ and } 0 \le y \le x^3\}$; $x = -1$
15. $R = \{(x, y) | 0 \le x \le 2 \text{ and } 0 \le y \le x^3\}$; $y = -1$
16. $R = \{(x, y) | 0 \le x \le 2 \text{ and } 0 \le y \le x^3\}$; $y = 10$
17. $R = \{(x, y) | 0 \le x \le \pi \text{ and } 0 \le y \le \sin x\}$; $x = -\pi/2$
18. $R = \{(x, y) | 0 \le x \le \pi \text{ and } 0 \le y \le \sin x\}$; $x = \pi$
19. $R = \{(x, y) | 0 \le x \le \pi \text{ and } 0 \le y \le \sin x\}$; $y = 1$
20. $R = \{(x, y) | 0 \le x \le \pi \text{ and } 0 \le y \le \sin x\}$; $y = -1$
21. $R = \{(x, y) | 0 \le x \le 1 \text{ and } x^2 + 1 \le y \le x + 1\}$; $x = 2$
22. $R = \{(x, y) | 0 \le x \le 1 \text{ and } x^2 + 1 \le y \le x + 1\}$; $x = 4$
23. $R = \{(x, y) | 3 \le x \le 5 \text{ and } \sqrt{x} \le y \le x^2\}$; $x = 1$
24. $R = \{(x, y) | 3 \le x \le 5 \text{ and } \sqrt{x} \le y \le x^2\}$; $x = 5$
25. $R = \{(x, y) | 3 \le x \le 5 \text{ and } \sqrt{x} \le y \le x^2\}$; $y = 1$
26. $R = \{(x, y) | 3 \le x \le 5 \text{ and } \sqrt{x} \le y \le x^2\}$; $y = \frac{3}{2}$
27. $R = \{(x, y) | 0 \le y \le 1 \text{ and } e^y \le x \le 2y + 1\}$; $y = 0$

28. $R = \{(x, y) | 0 \le y \le 1 \text{ and } e^y \le x \le 2y + 1\}$; $y = 1$
29. $R = \{(x, y) | 0 \le y \le 1 \text{ and } e^y \le x \le 2y + 1\}$; $x = 1$
30. $R = \{(x, y) | 0 \le y \le 1 \text{ and } e^y \le x \le 2y + 1\}$; $x = e$
31. $R = \{(x, y) | 1 \le y \le 2 \text{ and } \sqrt{y} \le x \le y\}$; $y = 0$
32. $R = \{(x, y) | 1 \le y \le 2 \text{ and } \sqrt{y} \le x \le y\}$; $y = 1$
33. $R = \{(x, y) | 1 \le y \le 2 \text{ and } \sqrt{y} \le x \le y\}$; $x = 1$
34. $R = \{(x, y) | 1 \le y \le 2 \text{ and } \sqrt{y} \le x \le y\}$; $x = 4$
35. $R = \{(x, y) | \pi/4 \le x \le 3\pi/4 \text{ and } \cos x \le y \le \sin x\}$; $x = 0$
36. $R = \{(x, y) | \pi/4 \le x \le 3\pi/4 \text{ and } \cos x \le y \le \sin x\}$; $x = 2\pi$
37. $R = \{(x, y) | \pi/4 \le x \le 3\pi/4 \text{ and } \cos x \le y \le \sin x\}$; $y = 1$
38. $R = \{(x, y) | \pi/4 \le x \le 3\pi/4 \text{ and } \cos x \le y \le \sin x\}$; $y = 2$
39. $R = \{(x, y) | -2 \le x \le 2 \text{ and } 0 \le y \le 4 - x^2\}$; $x = 5$
40. $R = \{(x, y) | a \le x \le b \text{ and } g(x) \le y \le f(x)\}$; $x = a - r, r > 0$
41. Find the volume of the solid generated by revolving an equilateral triangle with side s about:
    (a) A side
    (b) A line parallel to a side and a distance r from it
    (c) A line parallel to an altitude and through a vertex
    (d) A line parallel to an altitude and a distance r from the nearest vertex
    (e) A line parallel to a side and on the opposite vertex
    (f) A line parallel to a side and a distance r from the third vertex

SECTION 8.3

Arc Length

The length of a straight line segment in the coordinate plane can be found by using the distance formula, but the length of a curved segment is not so easily found. In this section, we define *arc length* and develop formulas for finding the lengths of certain curves. These formulas also aid in finding surface areas, as we will see in the next section.

Rectifiable Arcs

Let f be a differentiable function whose derivative is continuous on $[a, b]$. Such a function is called a **smooth** function. The graph of f, namely

$$C = \{(x, y) | y = f(x), a \le x \le b\},$$

is called a **smooth curve** or an **arc**.

We want to decide what is meant by the length of the arc C, and we start by making an approximation: Let $\sigma = \{x_0, x_1, \ldots, x_n\}$ be a subdivision of $[a, b]$, and for each $i = 0, 1, \ldots, n$, let P_i be the point $(x_i, f(x_i))$ on C. For each $i = 1, \ldots, n$, construct the straight line segment joining P_{i-1} to P_i (Figure 8.31). The broken line $P_0P_1P_2 \ldots P_n$ is called a **polygonal curve inscribed in** C.

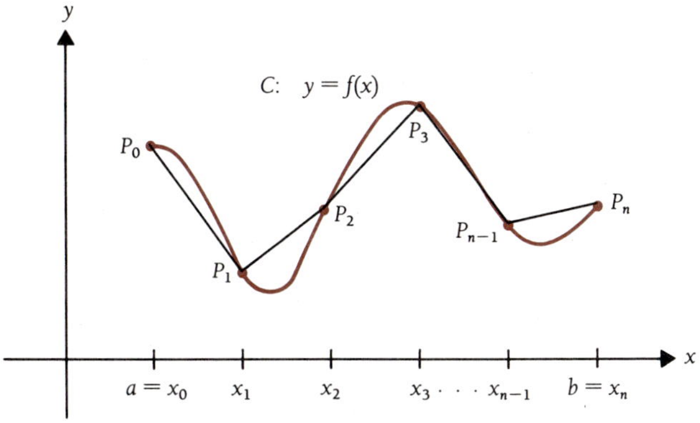

FIGURE 8.31 A polygonal arc inscribed in a curve

The polygonal curve inscribed in C is an approximation to C, and the length of the polygonal curve ought to be an approximation to what we want to regard as the length of C. We use the lengths of polygonal curves inscribed in C to define the length of arc C.

The length of each segment $P_{i-1}P_i$ of the polygonal curve inscribed in C is found by the distance formula to be

$$\begin{aligned} |P_{i-1}P_i| &= \sqrt{(x_i - x_{i-1})^2 + [f(x_i) - f(x_{i-1})]^2} \\ &= (x_i - x_{i-1}) \sqrt{1 + \left[\frac{f(x_i) - f(x_{i-1})}{x_i - x_{i-1}}\right]^2}. \end{aligned}$$

Hence, the length of the polygonal curve $P_0P_1 \ldots P_n$ is the sum

$$s_n = \sum_{i=1}^{n} |P_{i-1}P_i| = \sum_{i=1}^{n} (x_i - x_{i-1}) \sqrt{1 + \left[\frac{f(x_i) - f(x_{i-1})}{x_i - x_{i-1}}\right]^2} \tag{1}$$

We can simplify this sum by applying the mean value theorem to f on each subinterval. By that theorem, there exists $t_i \in [x_{i-1}, x_i]$ such that

$$f'(t_i) = \frac{f(x_i) - f(x_{i-1})}{x_i - x_{i-1}}.$$

We can also write $x_i - x_{i-1}$ as Δx_i; then formula (1) becomes

$$s_n = \sum_{i=1}^{n} \sqrt{1 + [f'(t_i)]^2}\, \Delta x_i. \tag{2}$$

Expression (2) is an underestimate of the length of the arc C, since each segment $P_{i-1}P_i$ is shorter than the actual curve. It is therefore evident that if s_n grows infinite as $n \to \infty$ and $\|\sigma\| \to 0$, then the arc C does not have length. On the other hand, if the limit of s_n does exist, the value of the limit ought to be the length of the arc C.

Since the sum (2) is also a Riemann sum, its limit as $n \to \infty$ and $\|\sigma\| \to 0$ is the definite integral

$$\int_a^b \sqrt{1 + [f'(x)]^2}\, dx, \tag{3}$$

if the limit exists. These considerations lead us to make the following definition.

DEFINITION 8.1

If the arc C is the graph of the function f over $[a, b]$ and the limit

$$\lim_{\|\sigma\| \to 0} \sum_{i=1}^{n} \sqrt{1 + [f'(t_i)]^2}\, \Delta x_i \tag{4}$$

exists, then C is said to **have length** or to be **rectifiable**, and the **length** of C is

$$s = \int_a^b \sqrt{1 + [f'(x)]^2}\, dx. \tag{5}$$

It can be shown that if f' is continuous, then the limit (4) exists, so that the arc C is rectifiable.

There are other approaches to the definition of arc length. One method is to define the arc to be rectifiable in case the set of lengths of all possible polygonal curves inscribed in the arc is bounded above. The length of the arc is then defined to be the least upper bound of the set of lengths, and the integral (3) is shown to be the length of the arc.

Notice that Definition 8.1 uses a definite integral not only as a formula for the length of an arc, but as the definition of arc length. Formula (5) can be applied directly in most cases to express the length of an arc as a definite integral. Often, the integral is difficult to evaluate, but it always can be approximated. Let's look at a few examples.

■ **Example 8.17** Find the length of the arc $y = x^{3/2}$ from $x = 0$ to $x = 4$.

Solution Here, $y = f(x) = x^{3/2}$, so $f'(x) = \frac{3}{2}x^{1/2}$, and by formula (5),

$$\begin{aligned} s &= \int_0^4 \sqrt{1 + \tfrac{9}{4}x}\, dx \\ &= \frac{4}{9}\int_1^{10} u^{1/2}\, du \qquad (u = 1 + \tfrac{9}{4}x) \\ &= \frac{4}{9} \cdot \frac{2}{3} u^{3/2}\Big|_1^{10} \\ &= \frac{8}{27}(10\sqrt{10} - 1). \end{aligned}$$

■

■ **Example 8.18** Find the length of a circular arc of radius r and central angle θ.

Solution First, suppose that $0 \le \theta \le \pi/2$. We regard the circular arc as the graph of the function $f(x) = \sqrt{r^2 - x^2}$ over the interval $[0, r \sin \theta]$, as shown in Figure 8.32. Then

$$f'(x) = \frac{-x}{(r^2 - x^2)^{1/2}},$$

so that

$$1 + [f'(x)]^2 = \frac{r^2}{r^2 - x^2}.$$

Therefore, by formula (5),

$$\begin{aligned} s &= \int_0^{r \sin \theta} \sqrt{1 + [f'(x)]^2}\, dx \\ &= \int_0^{r \sin \theta} \frac{r}{\sqrt{r^2 - x^2}}\, dx \\ &= \left[r \arcsin \frac{x}{r} \right]_0^{r \sin \theta} \\ &= r \arcsin(\sin \theta) = r\theta. \end{aligned}$$

It is easily shown that this formula holds for any positive value of θ. (This formula should be familiar to you from trigonometry.)

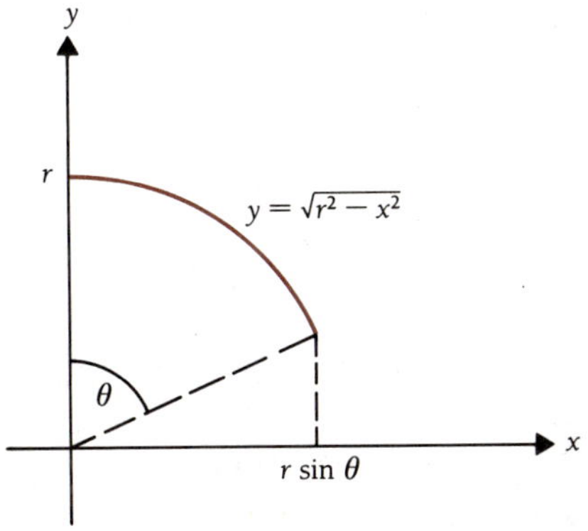

FIGURE 8.32 A circular arc ■

■ **Example 8.19** Find the length of the arc $y = 2\sqrt{x}$ from $x = 1$ to $x = 4$.

Solution In order to apply formula (5), we set $f(x) = 2\sqrt{x}$ and compute $f'(x) = 1/\sqrt{x}$, $1 + [f'(x)]^2 = 1 + (1/x)$. Then,

$$s = \int_1^4 \sqrt{1 + \frac{1}{x}}\, dx.$$

This integral requires a rationalizing substitution, $u = \sqrt{1 + (1/x)}$, or $x = 1/(u^2 - 1)$, $dx = -2u\,du/(u^2 - 1)^2$. Thus,

$$\begin{aligned} s &= \int_{\sqrt{2}}^{\sqrt{5}/2} u \frac{-2u\,du}{(u^2-1)^2} \\ &= \int_{\sqrt{2}}^{\sqrt{5}/2} \left[\frac{\frac{1}{2}}{u+1} - \frac{\frac{1}{2}}{(u+1)^2} - \frac{\frac{1}{2}}{u-1} - \frac{\frac{1}{2}}{(u-1)^2}\right] du \\ &= \left[\frac{1}{2}\ln\left|\frac{u+1}{u-1}\right| + \frac{1}{2(u+1)} + \frac{1}{2(u-1)}\right]_{\sqrt{2}}^{\sqrt{5}/2} \\ &= \frac{1}{2}\ln\left(\frac{\sqrt{5}+2}{\sqrt{5}-2}\right) + 2\sqrt{5} - \frac{1}{2}\ln\left(\frac{\sqrt{2}+1}{\sqrt{2}-1}\right) - \sqrt{2}. \end{aligned}$$

■

The Differential of Arc Length

Using formula (5), we can write the length of the arc $y = f(x)$ over $[a, x]$ as a function of x:

$$s(x) = \int_a^x \sqrt{1 + [f'(t)]^2}\,dt.$$

Differentiating this function, we get

$$\frac{ds}{dx} = \sqrt{1 + [f'(x)]^2},$$

or, in differential notation,

$$ds = \sqrt{1 + [f'(x)]^2}\,dx.$$

This also can be written

$$ds = \sqrt{1 + \left(\frac{dy}{dx}\right)^2}\,dx. \qquad (6)$$

Using the differentials dy and dx, we can rewrite equation (6) by putting dx inside the radical:

$$ds = \sqrt{(dx)^2 + (dy)^2}. \qquad (7)$$

Formula (7) may remind you of the Pythagorean theorem, and is illustrated geometrically in Figure 8.33.

Formula (7) also is useful in deriving an alternative form of formula (5). By factoring dy out of the radical, we obtain

$$ds = \sqrt{\left(\frac{dx}{dy}\right)^2 + 1}\,dy, \qquad (8)$$

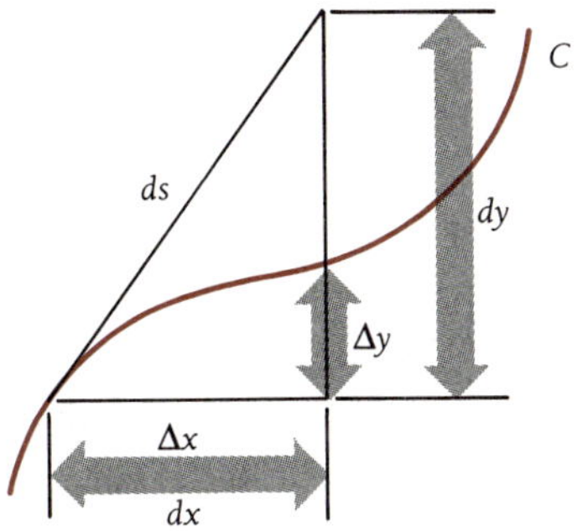

FIGURE 8.33 Relationship of differential of arc length to differentials of x and y

from which we get the formula

$$s = \int_c^d \sqrt{\left(\frac{dx}{dy}\right)^2 + 1}\, dy. \tag{9}$$

This formula is appropriate when an arc is given as the graph of a function $x = g(y)$ over $[c, d]$. In this case, formula (9) is also written

$$s = \int_c^d \sqrt{[g'(y)]^2 + 1}\, dy. \tag{10}$$

Example 8.20 Find the length of the curve $x = \frac{1}{4}y^2$ over $2 \le y \le 4$.

Solution Notice that this is the same arc as in Example 8.19. In order to use formula (9), we find $dx/dy = \frac{1}{2}y$, so $(dx/dy)^2 + 1 = \frac{1}{4}y^2 + 1$. Hence,

$$\begin{aligned}
s &= \int_2^4 \sqrt{\frac{1}{4}y^2 + 1}\, dy \\
&= \int_{\pi/4}^{\arctan 2} \sec\theta\, 2\sec^2\theta\, d\theta \qquad (y = 2\tan\theta) \\
&= 2\int_{\pi/4}^{\arctan 2} \sec^3\theta\, d\theta \\
&= \Big[\sec\theta\tan\theta + \ln|\sec\theta + \tan\theta|\Big]_{\pi/4}^{\arctan 2} \\
&= 2\sqrt{5} + \ln(\sqrt{5} + 2) - \sqrt{2} - \ln(\sqrt{2} + 1).
\end{aligned}$$

This answer is the same as the answer in Example 8.19, which you are invited to verify. Comparing this integration with that in Example 8.19, we see that the y integration is a little simpler. ■

An arc whose equation is given in the two variables x and y thus has two formulas for its length, (5) and (9). Convenience usually dictates which one should be used.

The length of an arc C can also be represented in the abstract form

$$s = \int_C ds, \tag{11}$$

where the differential ds is replaced by either formula (6) or (8), and the symbol $\int_C$ (read, "the integral over C") stands for the definite integral with appropriate limits corresponding to the ultimate variable of integration.

Example 8.21 Find the length of the arc

$$C = \{(x, y)|x^2 = y^3,\ 1 \leq x \leq 8,\ 1 \leq y \leq 4\}.$$

Solution The equation of the arc is $x^2 = y^3$. We can solve this equation for y and use formula (6), or solve for x and use formula (8). Trying both alternatives, we discover that solving for x is simpler. We get (since x is positive) $x = y^{3/2}$, $dx/dy = \frac{3}{2}y^{1/2}$, and $(dx/dy)^2 + 1 = \frac{9}{4}y + 1$. Hence, by formula (9),

$$\begin{aligned} s &= \int_1^4 \sqrt{\frac{9}{4}y + 1}\, dy \\ &= \frac{4}{9}\int_{13/4}^{10} u^{1/2}\, du \qquad (u = \tfrac{9}{4}y + 1) \\ &= \frac{4}{9} \cdot \frac{2}{3} u^{3/2}\Big|_{13/4}^{10} \\ &= \frac{8}{27}\left(10\sqrt{10} - \frac{13\sqrt{13}}{8}\right). \end{aligned}$$

■

Exercises 8.3

Find the length of each arc.

1. $C: y = x^2;\quad 0 \leq x \leq 2$
2. $C: y = 2x\sqrt{x};\quad 0 \leq x \leq 1$
3. $C: y = \ln(\cos x);\quad 0 \leq x \leq \pi/4$
A4. $C: y = \ln x;\quad 1 \leq x \leq 2$
5. $C: y = \frac{1}{4}x^2 - \frac{1}{2}\ln x;\quad 1 \leq x \leq 2$
6. $C: y = \frac{1}{2}(e^x + e^{-x});\quad 0 \leq x \leq 1$
7. $C: y = x\sqrt{8x};\quad 0 \leq x \leq 2$
A8. $C: y = 4e^x;\quad 0 \leq x \leq 1$
A9. $C: x = 1 + \ln y;\quad 1 \leq y \leq 3$
A10. $C: x = e^y;\quad 1 \leq y \leq 2$
11. $C: x = 4y\sqrt{y};\quad 1 \leq y \leq 9$
12. $C: 3x = (3y - 1)\sqrt{y};\quad 1 \leq y \leq 4$
A13. $C = \{(x, y)|y^2 - 8x = 4,\ 0 \leq x \leq 4,\ 2 \leq y \leq 6\}$
14. $C = \{(x, y)|e^x = \cos y,\ -\ln 2 \leq x \leq 0,\ 0 \leq y \leq \pi/3\}$
15. $C = \{(x, y)|x^2 + y^2 = 9,\ 0 \leq x \leq 3,\ -3 \leq y \leq 3\}$
16. $C = \{(x, y)|27y^2 = 4x^3,\ 0 \leq x \leq 192,\ 0 \leq y \leq 1024\}$
17. $C: x^{2/3} + y = 1;\quad$ from $x = \frac{1}{8}$ to $x = 1$
18. $C: x^{2/3} + y^{2/3} = a^{2/3},\ a > 0;\quad$ the entire curve

19. $C: y = \int_0^x \sqrt{\cos t}\, dt; \quad 0 \le x \le \pi/2$

20. $C: y = \int_0^x \sqrt{\sin t}\, dt; \quad 0 \le x \le \pi/2$ [*Hint:* Use $\sin x = \cos(\pi/2 - x)$.]

SECTION 8.4

Areas of Surfaces of Revolution

Let C be the graph of a smooth function $y = f(x)$ over the interval $[a, b]$, as shown in Figure 8.34. If C is revolved in space about the x-axis, a surface of revolution is generated (Figure 8.35). In this section, we show how to find the area of this surface and other surfaces of revolution.

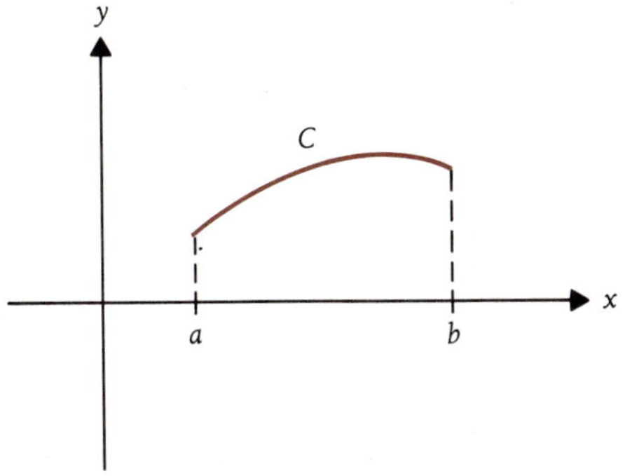

FIGURE 8.34 A curve in the plane

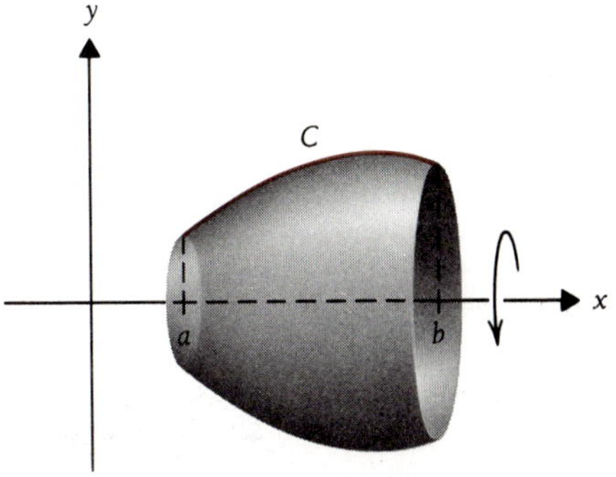

FIGURE 8.35 Revolving the curve about the x-axis to form a surface

Let $\sigma = \{x_0, x_1, \ldots, x_n\}$ be a subdivision of $[a, b]$, and let P_i be the point $(x_i, f(x_i))$. Then $P_0P_1 \ldots P_n$ is a polygonal curve inscribed in C. As each segment $P_{i-1}P_i$ is revolved about the x-axis, a surface is generated that is either a cylinder or a frustum of a cone (Figure 8.36). The entire polygonal curve generates a surface that approximates the surface generated by C.

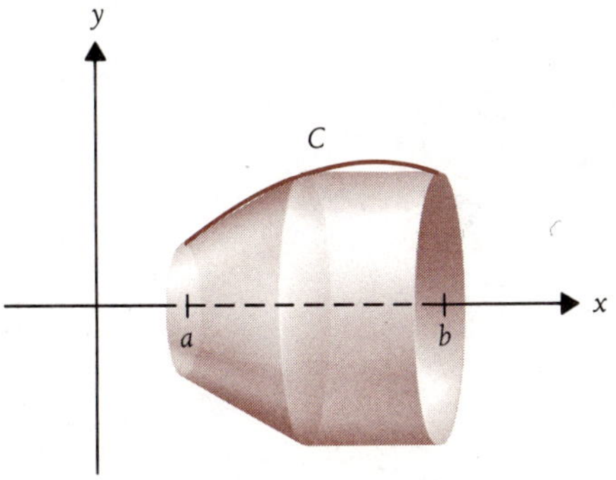

FIGURE 8.36 A polygonal arc generates cylinders and frustums of cones

To find the area of the surface generated by the polygonal curve $P_0P_1 \ldots P_n$, we first examine a frustum of a cone (Figure 8.37). If the frustum has slant height h and radii r_1 and r_2, we find its area by viewing it as a sector of an annular ring (Figure 8.38). From the figure and the formulas $s = r\theta$, $S = \frac{1}{2}r^2\theta$ for arc length and area of a circular sector, respectively, we get

$$s_1 = 2\pi r_1 = (R + h)\theta, \tag{1}$$

$$s_2 = 2\pi r_2 = R\theta, \tag{2}$$

$$S = \frac{1}{2}(R + h)^2\theta - \frac{1}{2}R^2\theta. \tag{3}$$

Putting equation (2) into (1), we get

$$2\pi r_1 = 2\pi r_2 + h\theta,$$

$$\theta = \frac{2\pi}{h}(r_1 - r_2).$$

Hence, from equation (2),

$$R = \frac{hr_2}{r_1 - r_2},$$

and from equation (3),

$$\begin{aligned} S &= \frac{1}{2}\left(\frac{hr_2}{r_1 - r_2} + h\right)^2 \frac{2\pi}{h}(r_1 - r_2) - \frac{1}{2}\left(\frac{hr_2}{r_1 - r_2}\right)^2 \frac{2\pi}{h}(r_1 - r_2) \qquad (4) \\ &= \pi(r_1 + r_2)h. \end{aligned}$$

This is the formula for the area of the slant surface of a frustum of a cone. Notice that it also holds for a cylinder, for which $r_1 = r_2$.

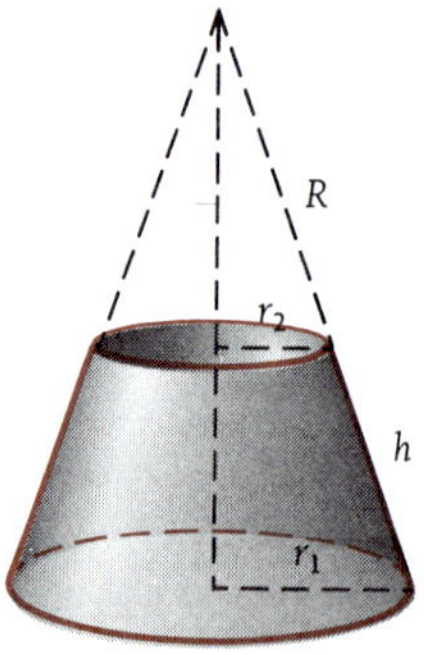

FIGURE 8.37 The frustum of a cone

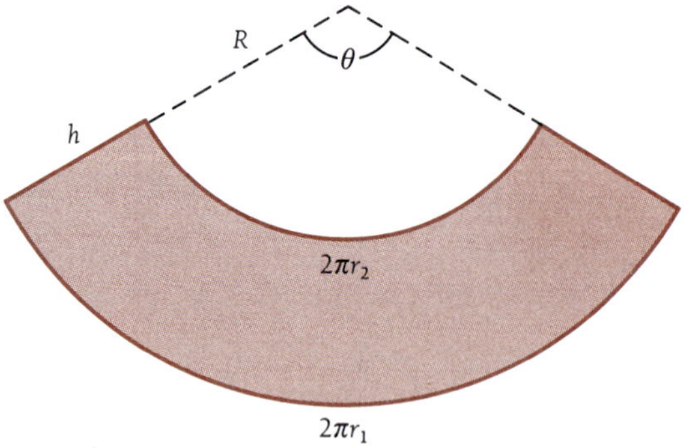

FIGURE 8.38 The frustum cut and flattened

As in Section 8.3, the segment $P_{i-1}P_i$ has length

$$|P_{i-1}P_i| = \sqrt{1 + [f'(t_i)]^2}\,\Delta x_i$$

for some $t_i \in [x_{i-1}, x_i]$. Hence, by formula (4), the segment $P_{i-1}P_i$ generates a surface of area

$$S_i = \pi[f(x_{i-1}) + f(x_i)]\sqrt{1 + [f'(t_i)]^2}\,\Delta x_i.$$

Therefore, the entire polygonal curve $P_0P_1 \ldots P_n$ generates a surface of area

$$S_n = \sum_{i=1}^{n} \pi[f(x_{i-1}) + f(x_i)]\sqrt{1 + [f'(t_i)]^2}\,\Delta x_i.$$

We expect the limit of S_n as $\|\sigma\| \to 0$ to yield the surface area. It can be shown (we will not digress to do so) that the limit of S_n as $n \to \infty$ is the integral

$$\int_a^b \pi[f(x) + f(x)]\sqrt{1 + [f'(x)]^2}\,dx = \int_a^b 2\pi f(x)\sqrt{1 + [f'(x)]^2}\,dx.$$

Thus, we are led to the following definition.

DEFINITION 8.2

If the rectifiable arc

$$C = \{(x, y) \mid y = f(x),\ a \le x \le b\}$$

is revolved about the x-axis, the area of the surface generated is

$$S = \int_a^b 2\pi f(x)\sqrt{1 + [f'(x)]^2}\,dx. \tag{5}$$

Formula (5) can be written in the more abstract form

$$S = \int_C 2\pi y\,ds, \tag{6}$$

where y is the *radius of rotation* and ds is the differential of arc length (see Figure 8.39). The symbol $\int_C$ is the integral over C, interpreted as in the previous section.

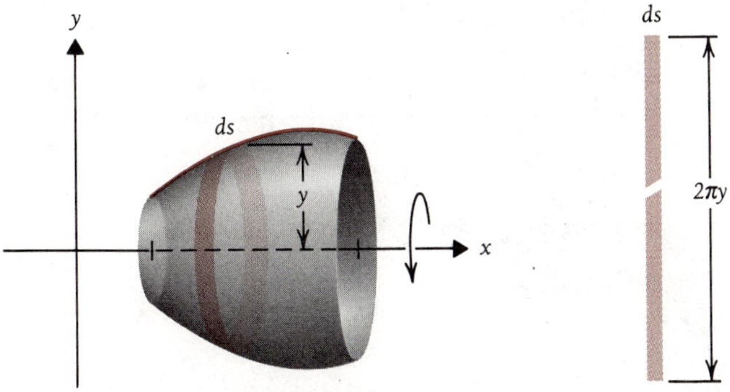

FIGURE 8.39 The ribbon generated by ds on the surface and then cut and flattened

If the arc is revolved about the y-axis, then formula (6) is changed to

$$S = \int_C 2\pi x \, ds. \tag{7}$$

More generally, if l is any line and ρ is a function of x or y that describes the distance between l and the point (x, y) on C (Figure 8.40), then $dS = 2\pi\rho \, ds$ and the area of the surface generated by revolving C about l is

$$S = \int_C 2\pi\rho \, ds. \tag{8}$$

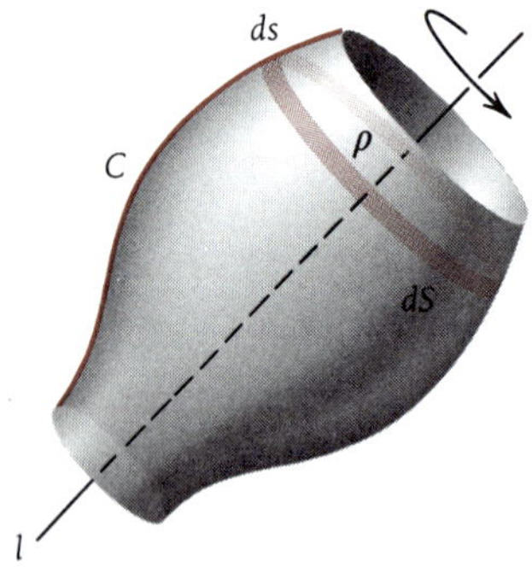

FIGURE 8.40 $dS = 2\pi\rho \, ds$

■ **Example 8.22** Find the surface area of a sphere of radius r.

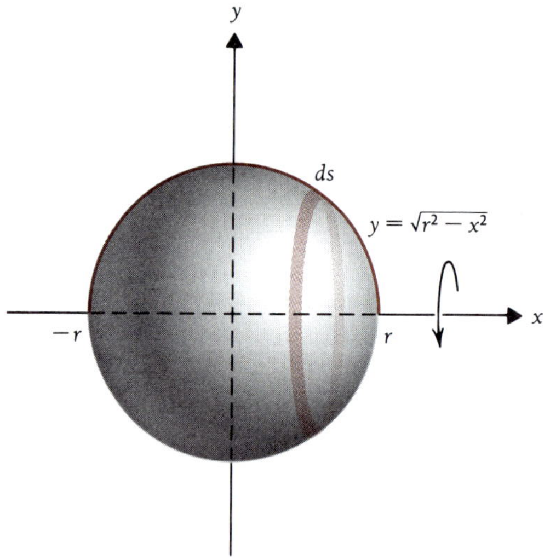

FIGURE 8.41 Generating a spherical surface

Solution We will describe the sphere as the surface generated by revolving the arc $x^2 + y^2 = r^2$, $y \geq 0$, about the x-axis (Figure 8.41). Then $dy/dx = -x/y$, so

$$ds = \sqrt{1 + \left(\frac{dy}{dx}\right)^2}\, dx = \sqrt{1 + \left(-\frac{x}{y}\right)^2}\, dx = \sqrt{\frac{r^2}{y^2}}\, dx = \frac{r}{y}\, dx,$$

and by formula (6),

$$S = \int_C 2\pi y\, ds = \int_{-r}^{r} 2\pi y \frac{r}{y}\, dx$$
$$= 2\pi r \int_{-r}^{r} dx = 4\pi r^2.$$

You probably recognize this formula as the one you've always used. ■

■ Example 8.23 The curve $y = 4 - x^2$, $x \in [0, 2]$, is revolved about the y-axis. Find the area of the surface generated (see Figure 8.42).

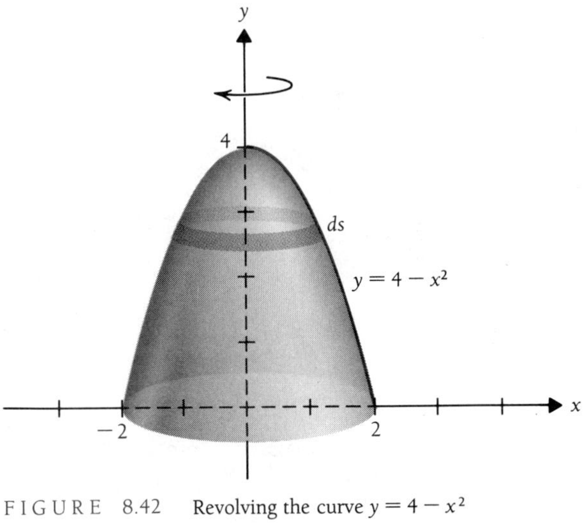

FIGURE 8.42 Revolving the curve $y = 4 - x^2$ about the y-axis

Solution We will use formula (7), with

$$ds = \sqrt{1 + \left(\frac{dy}{dx}\right)^2}\, dx.$$

Since

$$\frac{dy}{dx} = -2x,$$

$$1 + \left(\frac{dy}{dx}\right)^2 = 1 + 4x^2,$$

so formula (7) yields

$$S = \int_0^2 2\pi x\sqrt{1+4x^2}\,dx$$
$$= \frac{\pi}{4}\int_1^{17} u^{1/2}\,du \qquad (u = 1+4x^2)$$
$$= \frac{\pi}{4}\cdot\frac{2}{3}u^{3/2}\Big|_1^{17}$$
$$= \frac{\pi}{6}(17\sqrt{17}-1).$$

■

■ **Example 8.24** Find the surface area generated by revolving the curve $y = x^2$, $0 \le x \le 1$, about the line $x = -1$ (see Figure 8.43).

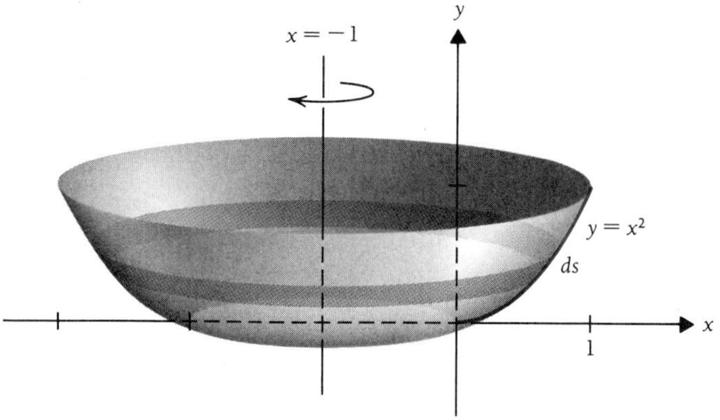

FIGURE 8.43 Revolving the curve $y = x^2$ about the line $x = -1$

Solution We will use formula (8) with $\rho = x + 1$. Since $y = x^2$, we get $ds = \sqrt{1+4x^2}\,dx$. Thus, the surface area is

$$S = \int_0^1 2\pi(x+1)\sqrt{1+4x^2}\,dx$$
$$= 2\pi\int_0^1 x\sqrt{1+4x^2}\,dx + 2\pi\int_0^1 \sqrt{1+4x^2}\,dx$$
$$= \frac{\pi}{4}\int_1^5 u^{1/2}\,du + \pi\int_0^{\tan^{-1}2} \sec^3\theta\,d\theta \qquad (u = 1+4x^2;\quad x = \tfrac{1}{2}\tan\theta)$$
$$= \frac{\pi}{6}[u^{3/2}]_1^5 + \pi\left[\frac{1}{2}\sec\theta\tan\theta + \frac{1}{2}\ln|\sec\theta+\tan\theta|\right]_0^{\tan^{-1}2}$$
$$= \frac{\pi}{6}(11\sqrt{5}-1) + \frac{\pi}{2}\ln(2+\sqrt{5}).$$

■

Exercises 8.4

The arc C is revolved about the given line. Find the area of the surface generated.

1. $C = \{(x, y)|x^2 + y^2 = 1, x \geq 0, y \geq 0\}$; $y = 0$
2. C as in Problem 1; $y = 1$
3. C as in Problem 1; $x = 2$
4. C as in Problem 1; $x = -2$
5. $C = \{(x, y)|y = 2x, 0 \leq x \leq 2\}$; $x = 2$
6. C as in Problem 5; $x = 5$
7. $C = \{(x, y)|x^2 = 2y^3, 0 \leq x \leq 4, 0 \leq y \leq 2\}$; $y = 0$

A8. $C = \{(x, y)|x + 1 = y^2, 1 \leq y \leq 3\}$; $x = 0$

9. $C = \{(x, y)|x^2 + y^2 = r^2, x \geq 0, y \geq 0\}$; $x = a, a > r$

A10. $C = \{(x, y)|b^2x^2 + a^2y^2 = a^2b^2, y \geq 0\}$; $y = 0$

A11. $C = \{(x, y)|y = 1/x, 1 \leq x \leq 3\}$; $y = 0$

12. A certain experimental apparatus is housed in a torus with outside diameter 50 feet and cross-sectional diameter 2 feet. The housing is to be painted with a special paint of which 1 gallon covers 450 square feet. How many gallons of paint will be needed?

SECTION 8.5

Moments and Centers of Mass

In this section, we discuss the physical quantities of *mass, moment,* and *center of mass.* The center of mass is a physical analog of the average value of a function. These topics also lead to the purely geometric idea of *centroid,* which is discussed in the next section.

First, note that the **mass** m of an object is the weight w of the object divided by the acceleration due to gravity g; that is, $m = w/g$. Thus, mass is a physical quantity that is independent of the gravitational system, while weight is not.

A **moment** is a product of a distance and a mass, similar to the product of a force and a distance used in computing leverage. The moment produced by a mass can be thought of as the leverage that would be produced by gravity acting on the mass.

One-Dimensional Moments and Centers of Mass

We will begin with a one-dimensional system. If a mass m is located at x on the real line (Figure 8.44), the **moment of m about the origin** is

$$M = xm.$$

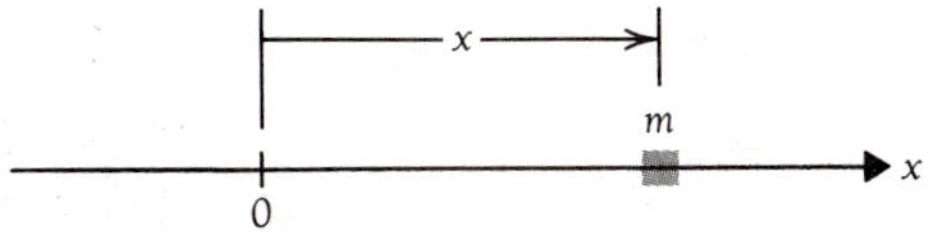

FIGURE 8.44 The moment about O is xm

If a system of masses is distributed along the line, say mass m_1 at x_1, m_2 at x_2, . . . , m_n at x_n, then the **total moment** of $m_1, m_2, \ldots, m_n$ about the origin is the sum of the individual moments,

$$M = x_1 m_1 + x_2 m_2 + \cdots + x_n m_n = \sum_{i=1}^{n} x_i m_i .$$

If the masses $m_1, m_2, \ldots, m_n$ were all combined into a single mass

$$m = m_1 + m_2 + \cdots + m_n = \sum_{i=1}^{n} m_i$$

and placed at point x, the moment produced would be xm. The point $\bar{x}$ at which m produces the same moment as the original system is called the **center of mass** of the system, and we have

$$\bar{x}m = M,$$

or

$$\bar{x} \sum_{i=1}^{n} m_i = \sum_{i=1}^{n} x_i m_i ,$$

so that

$$\bar{x} = \frac{\sum_{i=1}^{n} x_i m_i}{\sum_{i=1}^{n} m_i} . \tag{1}$$

Suppose that a solid rod lies along the real line, occupying the interval $[a, b]$, as shown in Figure 8.45. If the rod is thin, it may be thought of as a continuous distribution of mass throughout the interval $[a, b]$. The mass of the rod is expressed in terms of a **density function** δ, the mass per unit length of the rod. The density may vary from one point to another, so δ is a function of position x. If the density function is constant, the rod is said to be **uniform** or **homogeneous**.

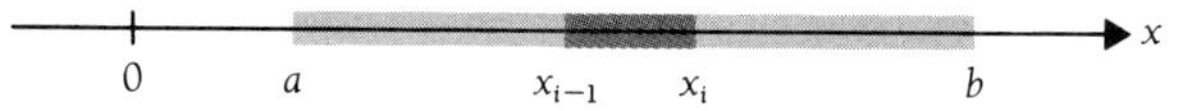

FIGURE 8.45 Finding the center of mass of a rod

We can approximate the moment of the rod about the origin by subdividing the interval $[a, b]$. Let $s = \{x_0, x_1, \ldots, x_n\}$ be a subdivision of $[a, b]$, and let the portion of the rod occupying the subinterval $[x_{i-1}, x_i]$ have mass Δm_i. If t_i is the midpoint of $[x_{i-1}, x_i]$, then Δm_i is approximately the density at t_i multiplied by the length $x_i - x_{i-1}$, or

$$\Delta m_i \approx \delta(t_i)(x_i - x_{i-1}) = \delta(t_i)\Delta x_i .$$

Thus, the moment produced by Δm_i about the origin is approximately

$$\Delta M_i \approx t_i \Delta m_i \approx t_i \delta(t_i) \Delta x_i.$$

Therefore, the total moment of the rod is approximately

$$M \approx \sum_{i=1}^{n} t_i \delta(t_i) \Delta x_i. \tag{2}$$

The sum in (2) is a Riemann sum for $x\delta(x)$ on $[a, b]$. The limit of the sum as $\|s\| \to 0$ is also intuitively the moment of the rod. Hence, we define the **moment** of the rod **about the origin** to be

$$M = \int_a^b x\delta(x)\, dx. \tag{3}$$

By similar reasoning, we conclude that the total **mass** of the rod is

$$m = \int_a^b \delta(x)\, dx. \tag{4}$$

Thus, the **center of mass** of the rod is

$$\bar{x} = \frac{\int_a^b x\delta(x)\, dx}{\int_a^b \delta(x)\, dx}. \tag{5}$$

This is the continuous analog of formula (1). The center of mass is also called the **center of gravity**; it can be interpreted as the point at which the rod will balance (Figure 8.46).

FIGURE 8.46 A rod will balance at its center of mass

■ **Example 8.25** A rod of length 2 has density that varies linearly from 1 at one end to 5 at the other end. Where is its center of mass?

Solution We first establish a coordinate system by placing the rod along the real line with its light end, say, at the origin and the other end at 2. Next, we must determine the density function δ. Since the density varies linearly, we know that δ has the form $\delta(x) = px + q$, for some p and q yet to be determined. By the placement of the rod, we have

$$1 = \delta(0) = p \cdot 0 + q,$$
$$5 = \delta(2) = p \cdot 2 + q.$$

Thus, $q = 1$ and $p = 2$, so

$$\delta(x) = 2x + 1.$$

Now we apply formula (5):

$$\bar{x} = \frac{\int_0^2 x(2x+1)\,dx}{\int_0^2 (2x+1)\,dx} = \frac{\int_0^2 (2x^2+x)\,dx}{\int_0^2 (2x+1)\,dx}$$
$$= \frac{[\frac{2}{3}x^3 + \frac{1}{2}x^2]_0^2}{[x^2+x]_0^2} = \frac{\frac{22}{3}}{6} = \frac{11}{9}.$$

Thus, the center of mass is located at point $\frac{11}{9}$, measured from the light end of the rod. ■

■ Example 8.26 Locate the center of mass of a uniform rod.

Solution For a uniform rod, density is constant. Let the rod have length L, with one end at the origin. Then formula (5) gives

$$\bar{x} = \frac{\int_0^L x\delta\,dx}{\int_0^L \delta\,dx} = \frac{x^2/2\Big|_0^L}{x\Big|_0^L} = \frac{L}{2}.$$

Thus the center of mass of a uniform rod is at the midpoint of the rod, as you would expect. ■

Two-Dimensional Moments and Centers of Mass

Now let's consider a two-dimensional coordinate system. If a mass m is located at point (x, y) in the plane (Figure 8.47), its **moment about the y-axis** is

$$M_y = xm,$$

and its **moment about the x-axis** is

$$M_x = ym.$$

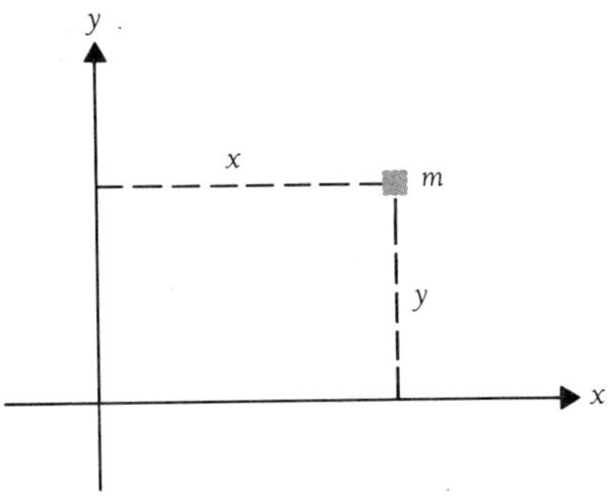

FIGURE 8.47 Moments about axes are $M_x = ym$ and $M_y = xm$

For a system of masses m_1 at (x_1, y_1), m_2 at (x_2, y_2), . . . , m_n at (x_n, y_n), the **total moment about the *y*-axis** is

$$M_y = \sum_{i=1}^{n} x_i m_i,$$

and the **total moment about the *x*-axis** is

$$M_x = \sum_{i=1}^{n} y_i m_i.$$

The center of mass of the system is the point $(\bar{x}, \bar{y})$ at which the total mass

$$m = \sum_{i=1}^{n} m_i$$

would produce the same total moments. Thus, $\bar{x}m = M_y$ and $\bar{y}m = M_x$, or

$$\bar{x} = \frac{\sum_{i=1}^{n} x_i m_i}{\sum_{i=1}^{n} m_i}, \qquad \bar{y} = \frac{\sum_{i=1}^{n} y_i m_i}{\sum_{i=1}^{n} m_i}. \tag{6}$$

Compare this two-dimensional version with the one-dimensional version (1).

Suppose now that a solid, two-dimensional plate occupies a region R in the plane. For the sake of simplicity, let R be the region under a nonnegative function f over $[a, b]$, as shown in Figure 8.48 (page 534). The mass of the plate is given in terms of a **density function** δ, which expresses mass per unit area, and is, in general, a function of both x and y. If density depends on both x and y, then the methods of multiple integration (Chapter 16) are necessary. Here, we will assume that density is a function of x only—that is, $\delta = \delta(x)$. If the density is constant, the plate is said to be **homogeneous** and is called a **lamina**.

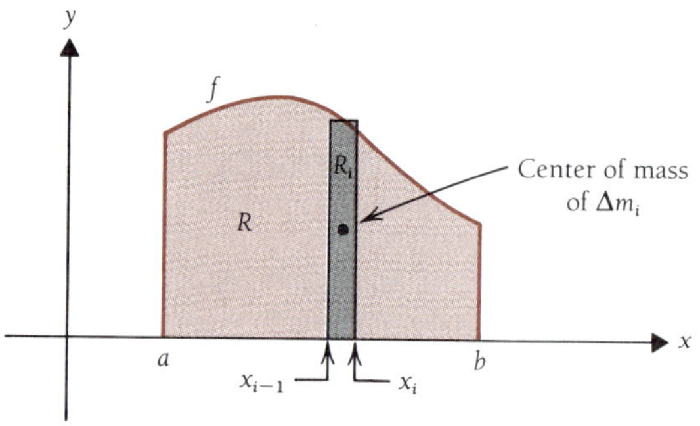

FIGURE 8.48 Finding the center of mass of a plate

Let $s = \{x_0, x_1, \ldots, x_n\}$ be a subdivision of $[a, b]$, and let R_i be the region under f over $[x_{i-1}, x_i]$. Let R_i have area ΔA_i, and let t_i be the midpoint of $[x_{i-1}, x_i]$. Since we have

$$\Delta A_i \approx f(t_i)\Delta x_i,$$

the mass of the portion of the plate lying over R_i is approximately

$$\Delta m_i \approx \delta(t_i)\Delta A_i \approx \delta(t_i)f(t_i)\Delta x_i.$$

Thus, the total mass of the plate is approximately

$$m \approx \sum_{i=1}^{n} f(t_i)\delta(t_i)\Delta x_i.$$

Now, if Δx_i is small, the center of mass of the strip of the plate over R_i, with mass Δm_i, is approximately $(t_i, \frac{1}{2}f(t_i))$, by Example 8.26. The moments produced by Δm_i about the x- and y-axes are $\frac{1}{2}f(t_i)\Delta m_i$ and $t_i\Delta m_i$, respectively. Thus, the total moments of the plate are approximately

$$M_y \approx \sum_{i=1}^{n} t_i f(t_i)\delta(t_i)\Delta x_i, \qquad M_x \approx \sum_{i=1}^{n} \frac{1}{2}[f(t_i)]^2\delta(t_i)\Delta x_i.$$

Therefore, we define the **mass** of the plate to be

$$m = \int_a^b f(x)\delta(x)\,dx, \tag{7}$$

and the **moments about the axes** to be

$$M_y = \int_a^b xf(x)\delta(x)\,dx, \qquad M_x = \int_a^b \frac{1}{2}[f(x)]^2\delta(x)\,dx. \tag{8}$$

The **center of mass** $(\bar{x}, \bar{y})$ is defined by

$$\bar{x} = \frac{\int_a^b x f(x)\delta(x)\,dx}{\int_a^b f(x)\delta(x)\,dx}, \qquad \bar{y} = \frac{\int_a^b \frac{1}{2}[f(x)]^2\delta(x)\,dx}{\int_a^b f(x)\delta(x)\,dx}. \tag{9}$$

These formulas should be compared with formula (5).

Example 8.27 Find the center of mass of a lamina covering the region bounded by $y = 4 - x^2$ and the x-axis (Figure 8.49).

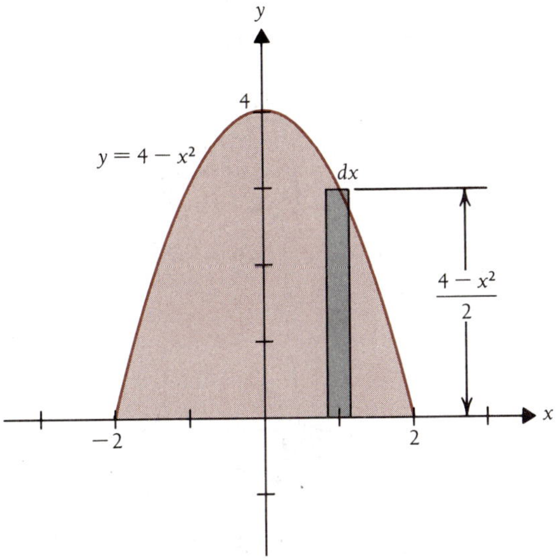

FIGURE 8.49 Finding the center of mass of a lamina covering the region bounded by $y = 4 - x^2$ and the x-axis

Solution We may use formulas (9) directly, with δ a constant:

$$\bar{x} = \frac{\int_{-2}^{2} x(4 - x^2)\delta\,dx}{\int_{-2}^{2} (4 - x^2)\delta\,dx} = \frac{\delta\int_{-2}^{2} (4x - x^3)\,dx}{\delta\int_{-2}^{2} (4 - x^2)\,dx} \qquad \bar{y} = \frac{\int_{-2}^{2} \frac{1}{2}(4 - x^2)^2\delta\,dx}{\int_{-2}^{2} (4 - x^2)\delta\,dx} = \frac{\delta\int_{-2}^{2} \frac{1}{2}(16 - 8x^2 + x^4)\,dx}{\delta\int_{-2}^{2} (4 - x^2)\,dx}$$

$$\bar{x} = \frac{[2x^2 - \frac{1}{4}x^4]_{-2}^2}{[4x - \frac{1}{3}x^3]_{-2}^2} = \frac{0}{\frac{32}{3}} = 0, \qquad \bar{y} = \frac{[8x - \frac{4}{3}x^3 + \frac{1}{10}x^5]_{-2}^2}{\frac{32}{3}} = \frac{\frac{256}{15}}{\frac{32}{3}} = \frac{8}{5}.$$

Thus, the center of mass is $(0, \frac{8}{5})$. The fact that $\bar{x} = 0$ illustrates the principle that if there is a line of symmetry for a lamina, then the center of mass lies on that line—the moments on opposite sides of the line "cancel out," so to speak (see Figure 8.49 again). ■

The formulas (9) may be generalized by considering dA to be an element of area of the region R. In the above development, dA has been a vertical strip, corresponding to ΔA_i. The element of mass of the plate may then be represented by $dm = \delta\, dA$. Let $(\hat{x}, \hat{y})$ be the center of mass of the element dm. Then

$$\bar{x} = \frac{\int_R \hat{x}\, dm}{\int_R dm} = \frac{\int_R \hat{x}\delta\, dA}{\int_R \delta\, dA} \tag{10}$$

and

$$\bar{y} = \frac{\int_R \hat{y}\, dm}{\int_R dm} = \frac{\int_R \hat{y}\delta\, dA}{\int_R \delta\, dA} \tag{11}$$

are the generalizations of (9). Here, the symbol $\int_R$ is the "integral over R," and stands for the definite integral with suitably chosen limits, depending on the form taken by dA.

■ **Example 8.28** The density of a semicircular plate varies directly as the distance from the diameter. Find its center of mass.

Solution Let the plate have radius r and occupy the region under $y = \sqrt{r^2 - x^2}$ over $[-r, r]$, as shown in Figure 8.50. Since the y-axis is a line of symmetry, not only geometrically but also with respect to density, we know that $\bar{x} = 0$. Since the distance from the diameter of the plate to point (x, y) is y, we may write the density function as

$$\delta(y) = ky,$$

k being the constant of variation. It will be necessary to take a horizontal strip for dA, since a vertical strip is not homogeneous and we therefore do not know $\hat{y}$. Solving $y = \sqrt{r^2 - x^2}$ for x, we get $x = \sqrt{r^2 - y^2}$ if $x > 0$, so that

$$dA = 2\sqrt{r^2 - y^2}\, dy.$$

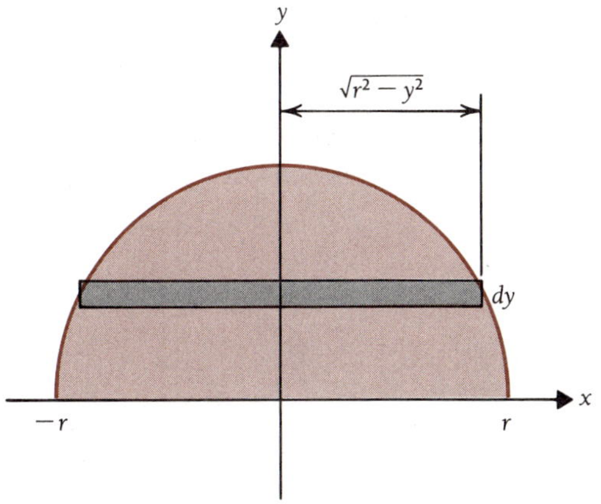

FIGURE 8.50 Finding the center of mass for the plate over the region below the semicircle $x^2 + y^2 = r^2$, $y \geq 0$, $dA = 2\sqrt{r^2 - y^2}\, dy$

Since the strip dA is taken at y, we have $\hat{y} = y$. Hence, formula (11) becomes

$$\bar{y} = \frac{\displaystyle\int_0^r yky2\sqrt{r^2 - y^2}\, dy}{\displaystyle\int_0^r ky2\sqrt{r^2 - y^2}\, dy} = \frac{2k\displaystyle\int_0^r y^2\sqrt{r^2 - y^2}\, dy}{2k\displaystyle\int_0^r y\sqrt{r^2 - y^2}\, dy}.$$

By the trigonometric substitution $y = r \sin \theta$, we find

$$\int_0^r y^2\sqrt{r^2 - y^2}\, dy = \frac{\pi r^4}{16};$$

and a simple substitution yields

$$\int_0^r y\sqrt{r^2 - y^2}\, dy = \frac{r^3}{3}.$$

Thus,

$$\bar{y} = \frac{\pi r^4/16}{r^3/3} = \frac{3\pi r}{16},$$

and the center of mass of the plate is on the axis of symmetry, a distance $3\pi/16$ of the radius from the center of the plate. ■

Example 8.29 A lamina covers the region between the curves $y = x$ and $y = x^2$. Find its center of mass.

Solution To find $\bar{x}$, we will let dA be a vertical strip at x (Figure 8.51). Then $\hat{x} = x$ and $dA = (x - x^2)\,dx$. Since δ is a constant, formula (10) gives us

$$\bar{x} = \frac{\int_0^1 x(x - x^2)\,dx}{\int_0^1 (x - x^2)\,dx} = \frac{\frac{1}{12}}{\frac{1}{6}} = \frac{1}{2}.$$

To find $\bar{y}$, we let dA be a horizontal strip at y (Figure 8.52). (We also could use vertical strips to find $\bar{y}$.) Then $\hat{y} = y$ and $dA = (\sqrt{y} - y)\,dy$, and formula (11) gives us

$$\bar{y} = \frac{\int_0^1 y(\sqrt{y} - y)\,dy}{\int_0^1 (\sqrt{y} - y)\,dy} = \frac{\frac{1}{15}}{\frac{1}{6}} = \frac{2}{5}.$$

Thus, the center of mass is $(\frac{1}{2}, \frac{2}{5})$.

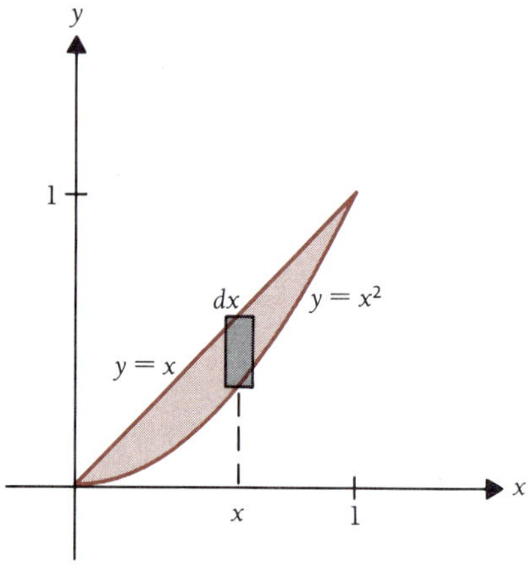

FIGURE 8.51 $dA = (x - x^2)\,dx$

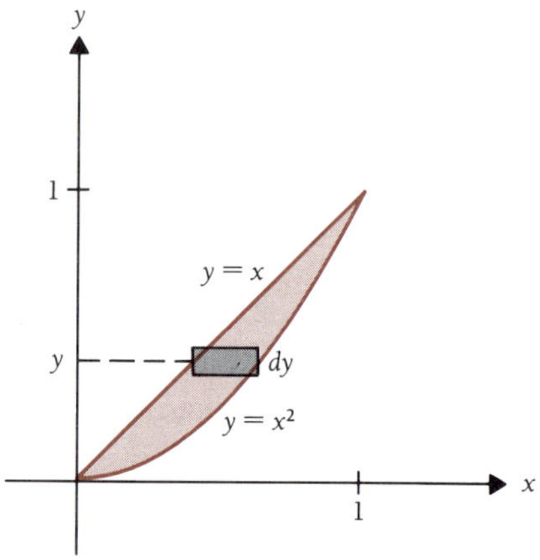

FIGURE 8.52 $dA = (\sqrt{y} - y)\,dy$ ■

As another type of two-dimensional system, a wire lying in the plane along an arc C may be considered (see Figure 8.53). In this case, the increment of mass Δm is related to the increment of arc length Δs by the density function δ of mass per unit length, according to the formula $\Delta m = \delta \Delta s$. If the increment Δm_i has center of mass (x_i, y_i), then the moments produced about the axes are

$$M_y \approx \sum_{i=1}^{n} x_i \Delta m_i, \qquad M_x \approx \sum_{i=1}^{n} y_i \Delta m_i.$$

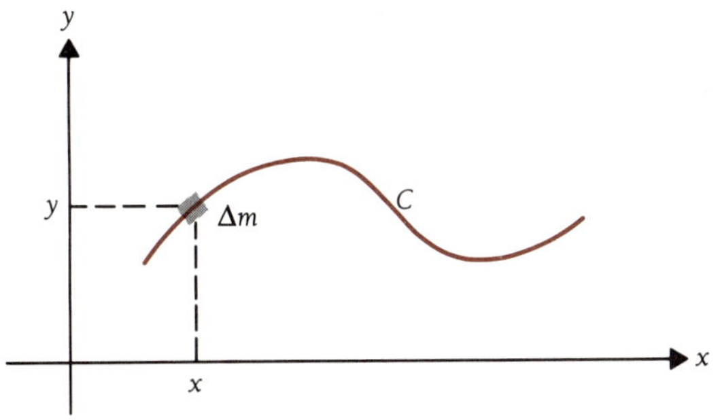

FIGURE 8.53 Finding the center of mass of a wire along C

Passing to the limit as before, we get the **moments**

$$M_y = \int_C x\delta \, ds, \qquad M_x = \int_C y\delta \, ds,$$

and the **mass**

$$m = \int_C \delta \, ds.$$

Hence, the **center of mass** of the wire is

$$\bar{x} = \frac{\int_C x\delta \, ds}{\int_C \delta \, ds}, \qquad \bar{y} = \frac{\int_C y\delta \, ds}{\int_C \delta \, ds}. \tag{12}$$

Here, $\int_C$, the integral over C, and ds, the differential of arc length, are the same as in Section 8.3.

Example 8.30 A homogeneous wire lies along the semicircle $y = \sqrt{4 - x^2}$. Find its center of mass.

Solution By symmetry (Figure 8.54, page 540), $\bar{x} = 0$. From $y = \sqrt{4 - x^2}$, we find that

$$ds = \frac{2 \, dx}{\sqrt{4 - x^2}},$$

so the numerator of $\bar{y}$ in formula (12) is

$$\int_C y\delta \, ds = \delta \int_{-2}^{2} \sqrt{4 - x^2} \cdot \frac{2 \, dx}{\sqrt{4 - x^2}}$$

$$= 2\delta \int_{-2}^{2} dx = 8\delta.$$

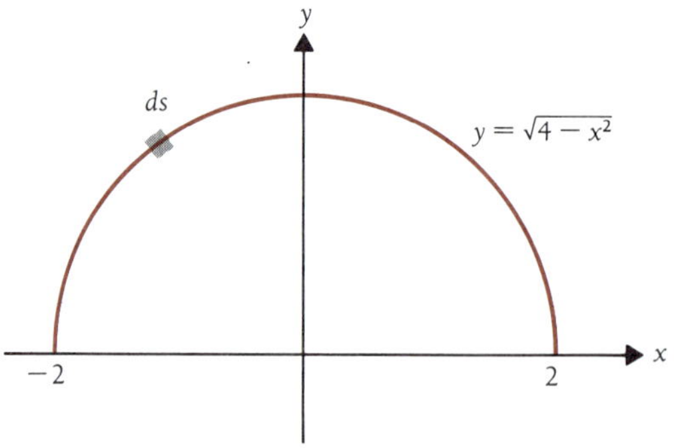

FIGURE 8.54 Finding the center of mass of a semicircular wire of radius 2

The denominator of $\bar{y}$ in formula (12) is

$$\int_C \delta\, ds = \delta \int_C ds = 2\pi\delta,$$

since $\int_C ds$ is just the length of arc C. Hence,

$$\bar{y} = \frac{8\delta}{2\pi\delta} = \frac{4}{\pi}.$$

The wire has center of mass $(0, 4/\pi)$. ■

■ **Example 8.31** A wire of constant density lies along the curve $y^2 = x^3$ between points (1, 1) and (4, 8), as shown in Figure 8.55. Find the integrals giving its center of mass.

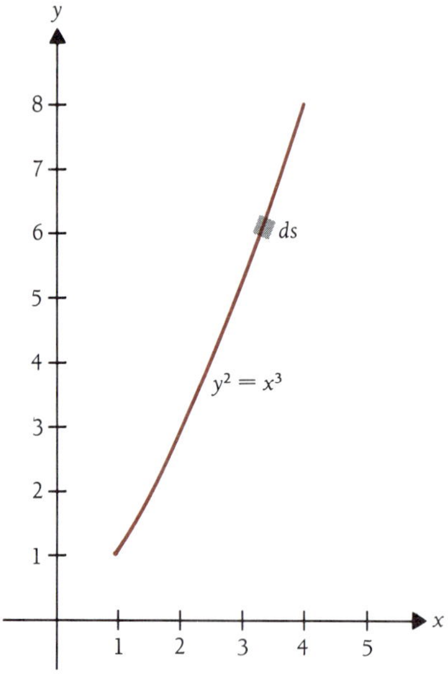

FIGURE 8.55 Finding the center of mass of a wire along the curve $y^2 = x^3$

Solution If we represent the given arc by $x = y^{2/3}$, $1 \le y \le 8$, then

$$\begin{aligned} ds &= \sqrt{\left(\frac{dx}{dy}\right)^2 + 1}\, dy \\ &= \sqrt{(\tfrac{2}{3}y^{-1/3})^2 + 1}\, dy \\ &= \sqrt{\tfrac{4}{9}y^{-2/3} + 1}\, dy \\ &= \tfrac{1}{3}y^{-1/3}\sqrt{4 + 9y^{2/3}}\, dy. \end{aligned}$$

The formulas (12) then give, after dividing out constants,

$$\bar{x} = \frac{\displaystyle\int_1^8 y^{1/3}\sqrt{4 + 9y^{2/3}}\, dy}{\displaystyle\int_1^8 y^{-1/3}\sqrt{4 + 9y^{2/3}}\, dy}, \qquad \bar{y} = \frac{\displaystyle\int_1^8 y^{2/3}\sqrt{4 + 9y^{2/3}}\, dy}{\displaystyle\int_1^8 y^{-1/3}\sqrt{4 + 9y^{2/3}}\, dy}. \tag{13}$$

On the other hand, if we represent the arc by $y = x^{3/2}$, $1 \le x \le 4$, then

$$\begin{aligned} ds &= \sqrt{1 + \left(\frac{dy}{dx}\right)^2}\, dx \\ &= \sqrt{1 + (\tfrac{3}{2}x^{1/2})^2}\, dx \\ &= \sqrt{1 + \tfrac{9}{4}x}\, dx \\ &= \tfrac{1}{2}\sqrt{4 + 9x}\, dx, \end{aligned}$$

and the formulas (12) yield

$$\bar{x} = \frac{\displaystyle\int_1^4 x\sqrt{4 + 9x}\, dx}{\displaystyle\int_1^4 \sqrt{4 + 9x}\, dx}, \qquad \bar{y} = \frac{\displaystyle\int_1^4 x^{3/2}\sqrt{4 + 9x}\, dx}{\displaystyle\int_1^4 \sqrt{4 + 9x}\, dx}. \tag{14}$$

We leave it to you to decide which set of integrals is easier to evaluate—(13) or (14)? I would probably use (14), maybe as the basis of a numerical approximation. ■

Three-Dimensional Moments and Centers of Mass

By now, you can guess what forms the formulas for center of mass of a three-dimensional solid will have. If a solid object occupies a region D in space, and its mass depends on density δ and volume V, then the element dm of mass is $dm = \delta\, dV$. If dm has center of mass $(\hat{x}, \hat{y}, \hat{z})$, then the center of mass $(\bar{x}, \bar{y}, \bar{z})$ of the object is given by

$$\bar{x} = \frac{\displaystyle\int_D \hat{x}\delta\, dV}{\displaystyle\int_D \delta\, dV}, \qquad \bar{y} = \frac{\displaystyle\int_D \hat{y}\delta\, dV}{\displaystyle\int_D \delta\, dV}, \qquad \bar{z} = \frac{\displaystyle\int_D \hat{z}\delta\, dV}{\displaystyle\int_D \delta\, dV}. \tag{15}$$

If a surface occupies a region E in space and its mass depends on density δ and surface area S, then the element of mass is $dm = \delta\, dS$. If dm has center of mass $(\hat{x}, \hat{y}, \hat{z})$, then the center of mass of the surface is given by

$$\bar{x} = \frac{\int_E \hat{x}\delta\, dS}{\int_E \delta\, dS}, \qquad \bar{y} = \frac{\int_E \hat{y}\delta\, dS}{\int_E \delta\, dS}, \qquad \bar{z} = \frac{\int_E \hat{z}\delta\, dS}{\int_E \delta\, dS}. \tag{16}$$

We will return to these formulas in Chapter 16. For now, we will let an example suffice.

Example 8.32 Locate the centers of mass of a homogeneous solid hemisphere and of a homogeneous hemispherical shell.

Solution First, we consider the solid hemisphere. Let it have radius r, and let it rest with its face on the xy-plane, with the z-axis as its axis of symmetry (Figure 8.56). By symmetry, $\bar{x} = \bar{y} = 0$. To find $\bar{z}$, let a horizontal slice of the hemisphere be taken at z. The radius of the slice is $\sqrt{r^2 - z^2}$, so $dV = \pi(r^2 - z^2)\, dz$. Since $\hat{z} = z$, formula (15) yields

$$z = \frac{\int_0^r z\delta\pi(r^2 - z^2)\, dz}{\int_0^r \delta\pi(r^2 - z^2)\, dz} = \frac{3r}{8}.$$

(You should be able to fill in the missing steps.) Thus, the center of mass is on the axis of symmetry, $\frac{3}{8}$ of the radius from the face of the hemisphere.

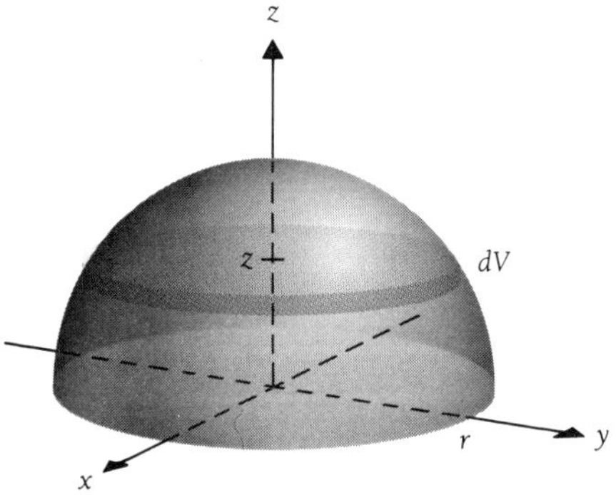

FIGURE 8.56 For the solid hemisphere, $dV = \pi(r^2 - z^2)\, dz$

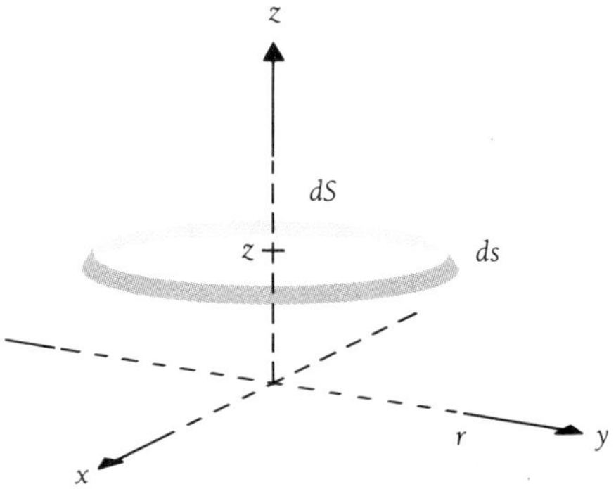

FIGURE 8.57 For the hemispherical shell, $dS = 2\pi\sqrt{r^2 - z^2}\, ds$

For the hemispherical shell, we use the same coordinate system. Again, by symmetry, $\bar{x} = \bar{y} = 0$. We will use formula (16) to compute $\bar{z}$. The element dS of surface area will be the thin band at height z (Figure 8.57). This element dS is generated by revolving the arc of length ds about the z-axis. Using the arc in the xz-plane, namely $x^2 + z^2 = r^2$, we have $z = \sqrt{r^2 - x^2}$, so $dz/dx = -x/\sqrt{r^2 - x^2}$ and $ds = r\,dx/\sqrt{r^2 - x^2}$. Then $dS = 2\pi\sqrt{r^2 - z^2}\,ds = 2\pi rx\,dx/\sqrt{r^2 - x^2}$. Since $\hat{z} = z$, formula (16) gives us

$$\bar{z} = \frac{\displaystyle\int_0^r z\delta 2\pi r(x/\sqrt{r^2 - x^2})\,dx}{\displaystyle\int_0^r \delta 2\pi r(x/\sqrt{r^2 - x^2})\,dx} = \frac{\displaystyle\int_0^r \sqrt{r^2 - x^2}(x/\sqrt{r^2 - x^2})\,dx}{\displaystyle\int_0^r (x/\sqrt{r^2 - x^2})\,dx} = \frac{\frac{1}{2}x^2\Big|_0^r}{-\sqrt{r^2 - x^2}\Big|_0^r} = \frac{r}{2}.$$

That is, the center of mass of a homogeneous hemispherical shell is on the axis of symmetry and midway between the base and top of the shell. Notice that the center of mass need not be in the shell (or solid) itself. ■

Exercises 8.5

Locate the center of mass of each object.

1. A uniform straight rod of length 3 meters
2. A circular disc in which density varies as the distance from the center
3. A uniform wire, bent into a square
4. A solid homogeneous cube
5. A straight rod, 2 feet long, whose density varies as the distance from one end
6. A straight rod, 2 feet long, whose density varies as the square of the distance from one end
7. A plate covering the region below $y = x^2$ over $[0, 2]$ whose density varies as the distance from the line $x = 3$
8. A semicircular plate of radius 3 whose density varies as the square of the distance from the diameter
9. A solid uniform cone of base radius r and height h
10. A solid cone of base radius r and height h whose density varies as the distance from the base
11. A cylinder of radius r and height h whose density varies as the distance from the base
12. A homogeneous pyramid of height h having a square base with side s
13. The homogeneous parabolic bowl obtained by revolving the curve $y = \frac{1}{2}x^2$, $x \in [0, 2]$, about the y-axis

14. An open conical shell of height h and base radius r, having constant density
15. Find the x-coordinate of the center of mass of a wire of uniform density lying along the curve $y = x\sqrt{x}$ between $(0, 0)$ and $(5, 5\sqrt{5})$.
16. Find the y-coordinate of the center of mass of the wire in Problem 15 by approximating an integral numerically.

SECTION 8.6

Centroids and the Theorems of Pappus

If a physical object has constant density, then the location of its center of mass depends only on geometric properties of the object. In that case, the center of mass is called the **centroid.** That is, the centroid of a geometrical object is the point that would be the center of mass of a homogeneous physical object of precisely the same shape and size.

For example, the centroid of a line segment is its midpoint, by Example 8.26. The centroid of a circular disc is its center, by symmetry. More generally, the centroid of an object lies on any line of symmetry of the object. If an object has two lines of symmetry, the centroid is their point of intersection (see Figure 8.58).

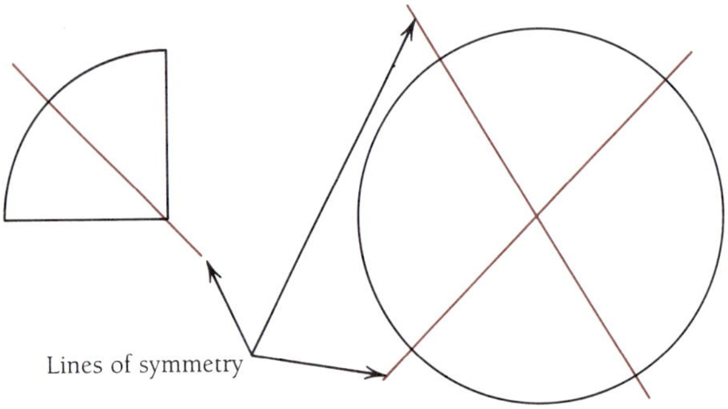

FIGURE 8.58 A centroid lies on any line of symmetry

By Example 8.30, the centroid of the semicircle $y = \sqrt{4 - x^2}$ is $(0, 4/\pi)$. More generally, the centroid of $y = \sqrt{r^2 - x^2}$ is $(0, 2r/\pi)$. By Example 8.32, the centroid of the solid hemisphere $0 \leq z \leq \sqrt{r^2 - x^2 - y^2}$ is $(0, 0, 3r/8)$. The centroid of the hollow hemispherical shell $z = \sqrt{r^2 - x^2 - y^2}$ is $(0, 0, r/2)$.

The centroid of a geometric object can be computed using the formulas of Section 8.5, with density constant.

The First Theorem of Pappus

There are two very pretty results for solids and surfaces of revolution, relating centroids, areas, and lengths to volume and surface area. These results are known as the **theorems of Pappus,** having been stated by Pappus of Alexandria about AD 320.

Pappus of Alexandria (flourished *c.* AD 320) was the last of the ancient Greek geometers. He wrote commentaries on the works of several of his predecessors, and his work *Collection* not only is an important mathematical treatise, but also provides many valuable insights into the history of Greek mathematics. Pappus presented many advances to the geometry of his day, and anticipated parts of analytic geometry and projective geometry. His theorems on volumes and surface areas of solids of revolution are an important topic in calculus today.

The first theorem of Pappus deals with solids of revolution. Let R be a region in the plane with area A and centroid P. Let l be a line that does not meet the interior of R (though it may meet the boundary of R). Let R be revolved about l to generate a solid with volume V (Figure 8.59).

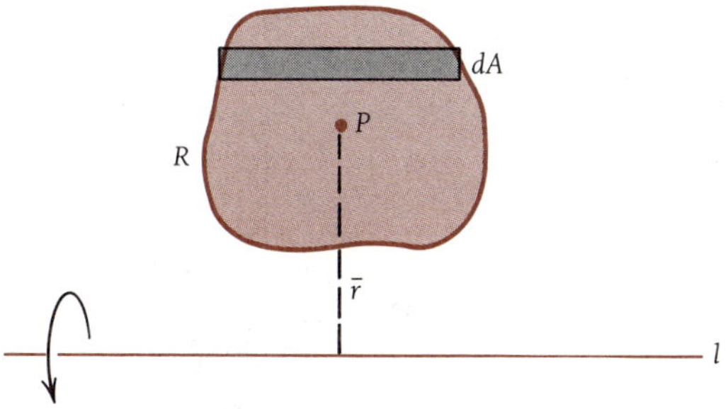

FIGURE 8.59 Deriving the first theorem of Pappus

We can express V in terms of an integral. If dA represents a strip of R parallel to l and a distance r from l, then

$$V = \int_R 2\pi r\, dA. \tag{1}$$

If the centroid P of R is a distance $\bar{r}$ from l, then we can express $\bar{r}$ in terms of integrals. If dA is again a strip parallel to l, then $\hat{r} = r$, and

$$\bar{r} = \frac{\int_R r\, dA}{\int_R dA} = \frac{1}{A} \int_R r\, dA. \tag{2}$$

Combining equations (1) and (2), we get

$$V = 2\pi \bar{r} A. \tag{3}$$

The restatement of equation (3) in words is the first theorem of Pappus.

THEOREM 8.1

First Theorem of Pappus Let a solid of revolution be generated by revolving a region about a line not meeting the interior of the region. The volume of the solid is the product of the area of the region and the length of the path described by the centroid of the region.

■ **Example 8.33**

Find the volume of the torus generated by revolving the circle $(x-a)^2+y^2=b^2$, $a \geq b$, about the y-axis.

Solution

(Compare with Example 8.12.) The centroid of the circular region is its center, $(a, 0)$, as can be seen in Figure 8.60. Thus, $\bar{r}$ in formula (3) is a, and the length of the path described by the centroid is $2\pi a$. The area of the circular region is πb^2. Therefore, by formula (3) or Theorem 8.1, the volume of the torus is

$$V=(2\pi a)(\pi b^2)=2\pi^2 ab^2.$$

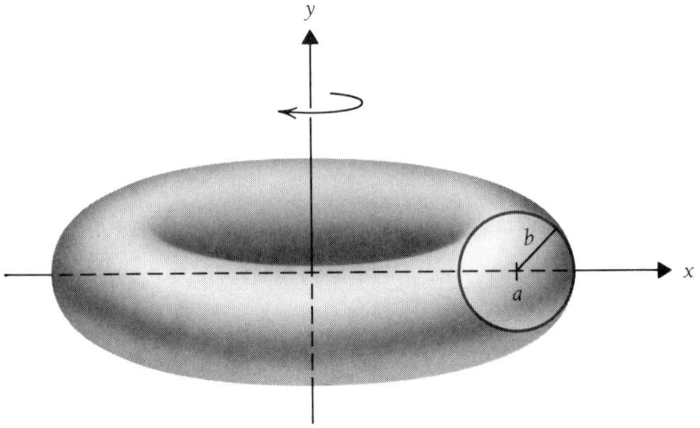

FIGURE 8.60 Revolving a circle to get a torus ■

■ **Example 8.34**

Find the centroid of the quarter-circular region

$$R=\{(x, y)|0 \leq x \leq r,\ 0 \leq y \leq \sqrt{r^2-x^2}\}.$$

Solution

By symmetry (see Figure 8.61), the centroid must lie on the line $y=x$, so $\bar{y}=\bar{x}$. If R is revolved about the x-axis, a solid hemisphere of volume $V=\frac{1}{2}(\frac{4}{3}\pi r^3)=\frac{2}{3}\pi r^3$ is generated. The area of the region is $A=\frac{1}{4}(\pi r^2)$. Hence, by formula (3),

$$\frac{2}{3}\pi r^3=2\pi\bar{y}\left(\frac{1}{4}\pi r^2\right),$$

so that

$$\bar{y}=\frac{4r}{3\pi}.$$

Thus, the centroid is the point $(4r/3\pi,\ 4r/3\pi)$.

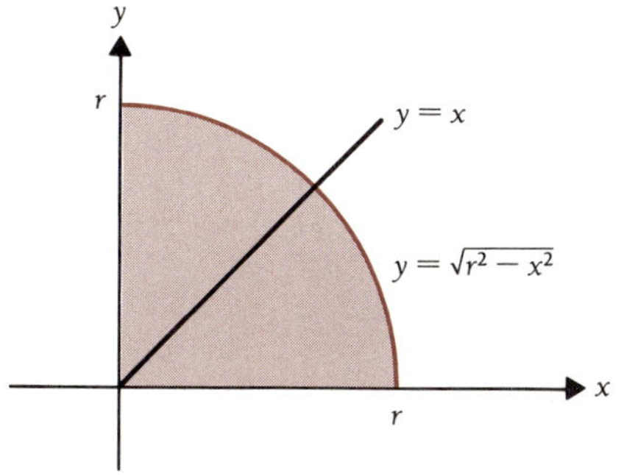

FIGURE 8.61 Revolving a quarter-circle to generate a hemisphere ■

The Second Theorem of Pappus

The second theorem of Pappus is analogous to the first, but applies to surfaces of revolution.

THEOREM 8.2

Second Theorem of Pappus Let a surface of revolution be generated by revolving an arc about a line that does not cut the arc. The area of the surface is the product of the length of the arc and the length of the path described by the centroid of the arc.

This theorem is illustrated in Figure 8.62. In symbols, it gives the surface area S as

$$S = 2\pi \bar{r} s, \tag{4}$$

where s is the arc length and $\bar{r}$ is the distance between the centroid of the arc and the axis of rotation. The justification for this theorem is very similar to the argument for Theorem 8.1. You will be invited to supply it in the exercises. The applications of this theorem are very much like those for Theorem 8.1.

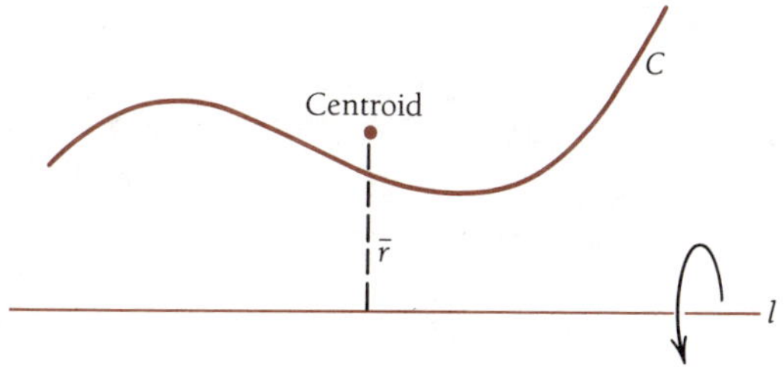

FIGURE 8.62 Deriving the second theorem of Pappus

■ **Example 8.35** Find the surface area of the torus that is generated by revolving the circle $(x-a)^2 + y^2 = b^2$, $a \geq b$, about the y-axis.

Solution The centroid of the circle is the point $(a, 0)$, so $\bar{r} = a$ in formula (4) (see Figure 8.60). The circumference of the circle is $s = 2\pi b$. Hence,

$$S = 2\pi a(2\pi b) = 4\pi^2 ab.$$

■

■ Example 8.36 Find the centroid of the arc $y = \sqrt{r^2 - x^2}$, $0 \le x \le r$ (Figure 8.63).

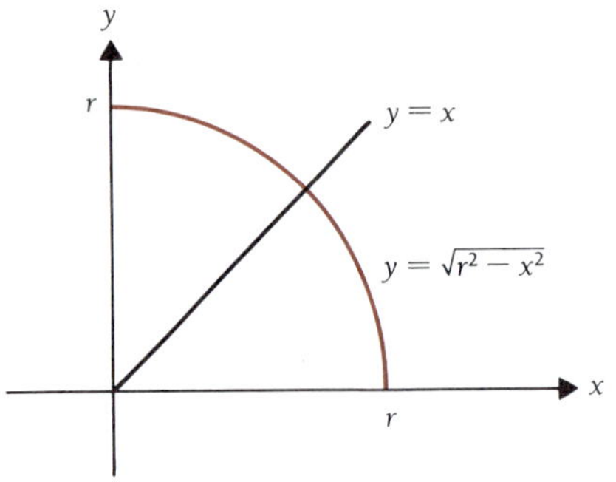

FIGURE 8.63 Revolving a circular arc to generate a hemispherical shell

Solution By symmetry, $\bar{x} = \bar{y}$. If the arc is revolved about the x-axis, a hemispherical shell is obtained, whose surface area is $S = \frac{1}{2}(4\pi r^2) = 2\pi r^2$. The length of the arc is $s = \frac{1}{4}(2\pi r) = \frac{1}{2}\pi r$. Hence, by formula (4),

$$2\pi r^2 = 2\pi\bar{y}\left(\frac{1}{2}\pi r\right),$$

$$\bar{y} = \frac{2r}{\pi}.$$

Thus, the centroid is $(2r/\pi, 2r/\pi)$. Note that this point is not on the arc, since $2/\pi < 1$. ■

Exercises 8.6

Find the volume of the solid generated by revolving R about the line l, using the first theorem of Pappus.

1. $R = \{(x, y) | -r \le x \le r, 0 \le y \le \sqrt{r^2 - x^2}\}$; $l: x = r$
2. $R = \{(x, y) | a \le x \le b, 0 \le y \le h\}$; $l: x = b + r, r \ge 0$
3. R is the interior of the isosceles triangle with base b and height h; l is a line perpendicular to the base and through a base vertex
4. $R = \left\{(x, y) | 0 \le x \le a, 0 \le y \le \frac{b}{a}(a - x)\right\}$; $l: x = a + r, r \ge 0$

Find the surface area of the surface generated by revolving C about the line l, using the second theorem of Pappus.

5. $C = \{(x, y) | y = mx, m > 0, 0 \le x \le h\}$; $l: y = -r - mh$
6. C is the square with vertices $(0, 0)$, $(0, s)$, (s, s), and $(s, 0)$; $l: x = r + s, r > 0$

7. $C = \{(x, y) | y = 4 - x^2, -2 \le x \le 2\}$; $l: x = 3$
8. $C = \{(x, y) | y = |x|, -1 \le x \le 1\}$; $l: x = -2$

Find the centroids of the following objects:

9. The region bounded by $y = x^3$, $y = 0$, and $x = 1$
10. The region $\{(x, y) | 0 \le x \le 1, 0 \le y \le 1 + x^2\}$
11. A solid cone of base radius r and height h
12. A solid frustum of a cone of height h and radii r_1 and r_2
13. The interior of a right triangle with legs a and b
14. A right triangle with legs a and b (not including the interior)
15. The interior of an isosceles triangle with base b and height h
16. An isosceles triangle with base b and height h
17. Show that the point of intersection of the medians of a triangle is the centroid of the interior of the triangle. (It is called the **centroid of the triangle**.) [*Hint:* Let a triangle have vertices $(0, 0)$, $(b, 0)$, and (a, h); show in two ways that the centroid is $((a + b)/3, h/3)$.]
18. Give an argument in support of Theorem 8.2.

SECTION 8.7

Miscellaneous Applications

In this section, we mention several more applications of the definite integral, any one of which is a whole topic in itself. The purpose of this section is to sample some of the many ways integration can be applied.

The Amount in an Annuity

An **annuity** is an account into which the owner periodically pays a given amount. The amount in an annuity at any time is the amount the owner has paid in, together with the interest earned by those payments.

Suppose an investor pays \$100 per month into an annuity that earns interest at the rate of 6% per annum, compounded continuously. How much is in the annuity after 5 years?

To answer the question, we start with the formula

$$P(t) = P_0 e^{rt},$$

which gives the value (principal plus interest) of an initial amount P_0 that earns interest at the periodic rate r, compounded continuously, after t time periods. In our example, the time period is 1 month, so $r = 0.06/12 = 0.005$. After 5 years, which is 60 months, the value of the \$100 paid at the beginning of the first month is $100e^{(0.005)60}$. The value of the \$100 paid at the beginning of the second month is

$100e^{(0.005)59}$, and so forth. Hence, the total amount in the annuity is

$$\begin{aligned} A &= 100e^{(0.005)60} + 100e^{(0.005)59} + \cdots + 100e^{(0.005)1} \\ &= \sum_{i=1}^{60} 100e^{(0.005)i}. \end{aligned}$$

This sum also can be written as

$$A = \sum_{i=1}^{60} 100e^{(0.005)i}\Delta t,$$

with $\Delta t = 1$ (see Figure 8.64). This sum is an approximation to the integral

$$\int_0^{60} 100e^{(0.005)t}\, dt.$$

Hence, the integral is approximately the amount in the annuity, and the answer to our question is

$$\begin{aligned} A &\approx \int_0^{60} 100e^{(0.005)t}\, dt \\ &= \frac{100}{0.005} e^{(0.005)t}\Big|_0^{60} \approx \$6997. \end{aligned}$$

(This compares with the exact value of \$7150; you can decide if the integral is a reasonable approximation.)

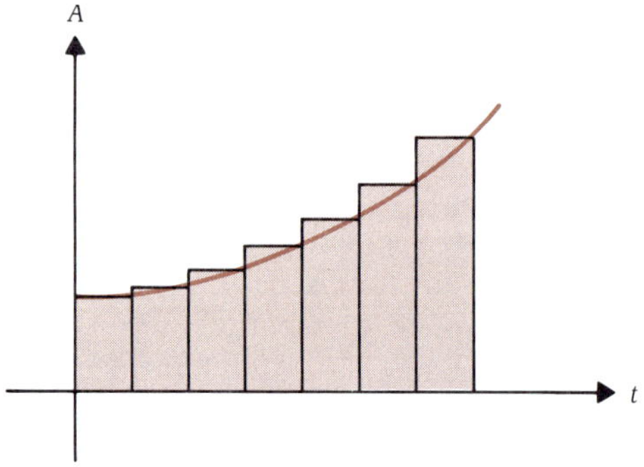

FIGURE 8.64 The amount in an annuity as a function of time

In general, if a periodic payment P is made into an annuity earning interest at the periodic rate r, compounded continuously, then the amount in the annuity after T time periods is approximately

$$A \approx \int_0^T Pe^{rt}\, dt. \tag{1}$$

Example 8.37 Each year, \$1000 is paid into an annuity earning interest at the rate of 8% per year, compounded continuously. What is the amount in the annuity after 10 years?

Solution The amount is approximately

$$A \approx \int_0^{10} 1000e^{0.08t}\,dt \approx \$15{,}319. \qquad \blacksquare$$

Imagine now that, instead of money being paid periodically into the annuity, money is paid continuously into the annuity at the varying rate of $P(t)$ dollars per unit time. Such a "flow" of money is called a **money stream** (in this case, we could call it a **payment stream**). Then the amount of money in the annuity at time T, under continuous compounding at interest rate r, is

$$A = \int_0^T P(t)e^{rt}\,dt. \tag{2}$$

This formula is exact if P is continuous. From this point of view, we see that the approximation (1) is obtained by treating the periodic payment as though it were a continuous payment stream at a constant rate.

Money streams occur in many phases of economics. An income stream is usually called a **revenue stream**.

Example 8.38 A certain rental property brings in income at a rate that reflects seasonal variations. If the revenue stream is

$$P(t) = 50\left(3 + 2\cos\frac{t}{2\pi}\right)$$

thousands of dollars per year, what will be the accrued value of this revenue stream at the end of 5 years? Assume continuous compounding at 10%.

Solution If we think of the revenue stream as a payment stream into an annuity, we see that formula (2) applies. Therefore, the total value after 5 years will be

$$\begin{aligned} A &= \int_0^5 50\left(3 + 2\cos\frac{t}{2\pi}\right)e^{0.10t}\,dt \\ &= \left[\frac{150}{0.10}e^{0.10t} + \frac{100e^{0.10t}}{(0.10)^2 + (1/2\pi)^2}\left(0.1\cos\frac{t}{2\pi} + \frac{1}{2\pi}\sin\frac{t}{2\pi}\right)\right]_0^5 \\ &\approx 1547. \end{aligned}$$

That is, the value will be \$1,547,000 in 5 years. ■

Hydrostatic Pressure

The pressure, or force per unit area, exerted by a fluid such as water depends only on the depth of the fluid, and is exerted equally in all directions. The pressure at a given point does not depend on the size or shape of the container, but only on the depth. The formula for the pressure p exerted by a fluid at depth h is $p = wh$, where w is a constant, the weight per unit volume of the fluid. For water, w is 62.5 pounds per cubic foot.

If a surface of area A is located at depth h, then the total force on the surface is the product of pressure and area,

$$F = pA = whA.$$

For a surface that is not all at the same depth, an integration is required to determine the total force.

Example 8.39 A gate in the face of a dam is a rectangle 3 feet wide and 2 feet high. It is situated vertically, with its upper edge 8 feet below the surface of the water (see Figure 8.65). What is the total force on the gate?

Solution Let dA be a horizontal strip across the face of the gate, of length 3 and width dh (Figure 8.65), located at depth h. The strip is approximately all at the same depth, so the force on the strip is approximately

$$dF = p\,dA = wh\,dA = wh3\,dh.$$

Hence, the total force on the gate is

$$F = \int_8^{10} 3wh\,dh = \frac{3}{2}wh^2\Big|_8^{10} = 54w.$$

Using $w = 62.5$, we get $F = 3375$ pounds.

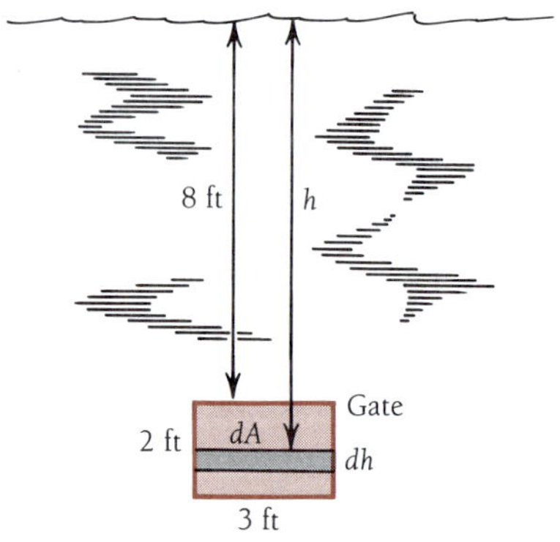

FIGURE 8.65 Finding the hydrostatic pressure on a gate ■

Work

If an object is moved along a straight line a distance s by a constant force F, the work done on the object by the force is the product $W = Fs$. In the case of a variable force, an integral is needed to compute the work done. If an object is moved from a to b along the x-axis by a force $F(x)$ that depends on the position x, then the work done is

$$W = \int_a^b F(x)\, dx.$$

Example 8.40 A particle on the x-axis is attracted toward the origin with a force equal to its distance from the origin. Find the work done in moving the particle from point 5 to point 3.

Solution If the particle is at point x, the force on it is $F(x) = -x$, where the minus sign indicates the direction "toward the origin." Therefore, the work done is

$$W = \int_5^3 (-x)\, dx = -\frac{1}{2} x^2 \Big|_5^3 = 8.$$

Example 8.41 A hemispherical tank that is 5 feet in diameter is filled with water. How much work is done in pumping the water into an adjacent tank whose rim is 2 feet above the top of the bowl?

Solution The amount of work done is determined by the weight of the water and the height it must be lifted. Consider the "slice" of water at level y below the top of the bowl (see Figure 8.66). If the slice has thickness dy, then its volume is approximately

$$dV = \pi[\sqrt{(2.5)^2 - y^2}]^2\, dy,$$

and its weight is

$$w\, dV = \pi w(6.25 - y^2)\, dy.$$

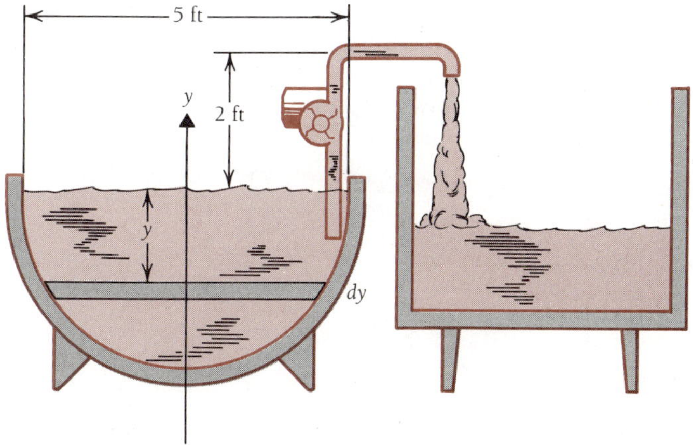

FIGURE 8.66 Pumping water from a hemispherical tank into a higher tank

The slice must be lifted a height $2 + y$, so that the work done on it is

$$dW = (2 + y)\pi w(6.25 - y^2)\, dy.$$

Hence, the work done is

$$\begin{aligned} W &= \int_0^{2.5} (2 + y)\pi w(6.25 - y^2)\, dy \\ &= \pi w \int_0^{2.5} (12.5 + 6.25y - 2y^2 - y^3)\, dy \\ &= 6008 \text{ foot-pounds.} \end{aligned}$$

■

Population Theory

Consider a population in which individuals both leave the population (die) and enter the population (are born). The number of individuals in the population at time t is a function $N(t)$. The rate at which individuals leave the population is described by a **survival function** $s(t)$, which gives the percentage of the population that was alive at $t = 0$ and survives at time t. Thus, $0 \leq s(t) \leq 1$, and s is necessarily a nonincreasing function. The rate at which individuals are added to the population is described by a **renewal function** $r(t)$, which gives the number of individuals added to the population at time t. Our task is to relate the functions N, s, and r.

Consider the time interval $[0, T]$, and let $\{t_0, t_1, \ldots, t_n\}$ be a subdivision of $[0, T]$. If $\bar{t}_i \in [t_{i-1}, t_i]$, then the number of individuals added to the population during $[t_{i-1}, t_i]$ is approximately $r(\bar{t}_i)(t_i - t_{i-1}) = r(\bar{t}_i)\Delta t_i$. But all of them do not necessarily survive to time T. The number that do survive is given by multiplying the number added during $[t_{i-1}, t_i]$ by the fraction that live to age $T - \bar{t}_i$, or

$$s(T - \bar{t}_i)r(\bar{t}_i)\Delta t_i.$$

Now the population at time T consists of the survivors of the original population at $t = 0$, together with the survivors of those added to the population during each time subinterval. Thus, we have approximately

$$N(T) \approx N(0)s(T) + \sum_{i=1}^{n} s(T - \bar{t}_i)r(\bar{t}_i)\Delta t_i.$$

This is a Riemann sum, whose limit gives us the formula for $N(T)$:

$$N(T) = N(0)s(T) + \int_0^T s(T - t)r(t)\, dt.$$

■ **Example 8.42** In a certain bacterial culture, the survival function is $s(t) = e^{-t/2}$ and the renewal function is $r(t) = 2^t$. Starting with 2000 bacteria, find the population after 12 hours.

Solution The formula gives us

$$
\begin{aligned}
N(12) &= 2000e^{-12/2} + \int_0^{12} e^{-(12-t)/2}2^t\,dt \\
&= 2000e^{-6} + \int_0^{12} e^{(1/2+\ln 2)t-6}\,dt \\
&\approx 3438.
\end{aligned}
$$

■

The foregoing examples are some of the types of additional uses for definite integrals. More applications are found in the exercises.

Exercises 8.7

1. A payment of \$50 per month is made into an annuity earning 7% interest per year, compounded continuously. Estimate the amount in the annuity after 10 years; 20 years; 30 years; 40 years.
2. A payment of \$500 per year is made into an annuity earning 10% interest per year, compounded continuously. Estimate the amount in the annuity after 8 years; 16 years; 24 years.
3. The proud new parents of a baby girl wish to pay into an annuity that will grow to \$10,000 in 20 years (to help pay for a college education). If they make equal monthly payments and the interest rate is 6% per year, compounded continuously, what monthly payment should be made?
4. A man aged 25 wishes to accumulate a retirement annuity of \$100,000 by age 65. He can invest his money at 8% per year, compounded continuously. What monthly payment should he make?
5. A Find the total value of the revenue stream $P(t) = 4t + 3\sin 2\pi t$ dollars per month after 1 year; 5 years; 10 years. Assume continuous compounding at 8%.
6. What constant revenue stream will result in a total value of \$1000 in 1 year under 5% interest, compounded continuously?
7. The face of a dam is a vertical rectangle 40 feet wide and 12 feet high. The water level is at the top of the dam. Find the force exerted on the dam. [For water, $w = 62.5$ pounds per cubic foot.]
8. A water tank is a cylinder 6 feet in diameter, and it stands on end. If water fills the tank to a depth of 4 feet, find the total force exerted on the sides of the tank.
9. A spillway valve is a circular plate 2 feet in diameter, situated vertically with its center 11 feet below the surface of the reservoir. What is the force on the valve?
10. The face of a dam is an isosceles trapezoid of height 20 feet and bases 10 feet and 40 feet. It is situated vertically, with the smaller base at the bottom. Find the force on the dam if the water level is 2 feet below the top of the dam.
11. The force acting on an object at position x is $F(x) = \sqrt{x^2 + 1}$. Find the work done in moving the object from $x = 1$ to $x = 5$.
12. Repeat Problem 11 for the force $F(x) = \sin \pi x$.
13. Find the work done in pumping the water from the tank in Problem 8 to another tank whose top is 8 feet above the bottom of the first tank.

14. The curve $y = 2x^2$, $x \in [0, 1]$, is revolved about the y-axis to form a tank that is 2 feet in diameter at the top. Water is pumped from a reservoir at the base of the tank into the tank. How much work is done in filling the tank?

15. In a certain population, the survival function is

$$s(t) = \begin{cases} 1 - t & \text{if } 0 \leq t \leq 1 \\ 0 & \text{if } t > 1, \end{cases}$$

and the renewal function is $r(t) = 2$. Starting with a population of 100, what is the population at time $t = 10$?

16. Rework Problem 15 if $s(t) = e^{-3t}$ and $r(t) = t^2$.

17. Rework Problem 15 if $s(t) = 1$ and $r(t) = e^t$.

18. Rework Problem 15 if

$$s(t) = \begin{cases} 1 & \text{if } 0 \leq t \leq 5 \\ 0 & \text{if } t > 5 \end{cases}$$

and $r(t) = 1 + \sin t$.

19. Explain what happens to a population if $s(t) = 1$.

20. **Demand for Resources** Suppose that the demand for a certain natural resource grows exponentially at the rate r per year and that the demand during last year was D_0. Show that under these conditions, the total demand for the resource for the next T years will be

$$D = \int_0^T D_0 e^{rt}\, dt.$$

21. Suppose that the demand for tin during 1980 was 5 million metric tons and that the demand is growing exponentially at the rate of 10% per year. Find the total demand for tin for the period 1980–2000.

22. Suppose that world reserves of tin amount to 250 million metric tons. Assuming no new reserves are discovered and that demand continues as described in Problem 21, when will world reserves of tin be depleted?

23. In 1973, the world use of copper was 7.7 million tons, and demand for copper was growing exponentially at the rate of 8% per year. World reserves of copper were 370 million tons. Assuming that the demand continues to grow at the same rate and that no new reserves are discovered, when will copper reserves be depleted?

24. **Capital Value** If a rental property brings in a revenue stream of $R(t)$ dollars per year, the **capital value** of the property for the next T years is

$$CV = \int_0^T R(t) e^{-rt}\, dt,$$

where r is the prevailing interest rate. Find the capital value of a property over the next 20 years if the revenue stream is constantly \$2400 per year and the prevailing interest rate is 9%.

25. A banker holds a mortgage on which the monthly payment is \$284. Find the capital value of the mortgage if it will be in force for 14 more years. Use 8% as the prevailing interest rate.

26. An investor holds a \$16,000 mortgage that will be paid off in 12 years, the payments being \$150 per month. If the mortgage is offered as part payment on a debt, what is a fair market value for the mortgage? Use 7.5% as the interest rate.

27. **Hooke's law** states that the force necessary to stretch a spring is proportional to the distance stretched. Thus, if a spring of length L is stretched to a length $L + x$, the force required is $F(x) = kx$, where k is a constant depending on the particular spring. Suppose a 12 inch spring is stretched to 14 inches by a 30 pound weight. Find the work done in stretching the spring from 12 to 16 inches.

28. A spring satisfying Hooke's law is stretched 0.5 inch by a 50 pound weight. Find the work done in stretching the spring an additional inch.

29. A certain chain is 8 feet long and weighs 6 pounds per foot. Find the work done in lifting it by one end until the chain just clears the ground.

30. Rework Problem 29 if the final position of the bottom end of the chain is 10 feet above the ground.

SECTION 8.8

Improper Integrals and Their Applications

The definite integral $\int_a^b f(x)\,dx$ was defined in Section 5.3 relative to a closed interval $[a, b]$. We also remarked there that an integrable function must be bounded. In this section, we consider integrals of continuous functions over infinite intervals and integrals of unbounded functions. In both cases, we call these integrals **improper** to distinguish them from definite integrals. Improper integrals may or may not represent real numbers; those that do can be used in various applications, as we will illustrate.

Improper Integrals with Infinite Limits

Suppose a function f is continuous on the infinite interval $[a, \infty)$. Then for any $b > a$, the definite integral $\int_a^b f(x)\,dx$ exists. Now consider the limit

$$\lim_{b\to\infty} \int_a^b f(x)\,dx.$$

If this limit exists, we call it an *improper integral.*

DEFINITION 8.3

If a function f is continuous on $[a, \infty)$, the **improper integral of f over $[a, \infty)$** is denoted by

$$\int_a^\infty f(x)\,dx$$

and is defined by

$$\int_a^\infty f(x)\,dx = \lim_{b\to\infty} \int_a^b f(x)\,dx.$$

If the limit exists, we say the improper integral **exists** or **converges**. If the limit does not exist, we say the improper integral **does not exist** or **diverges**.

Example 8.43 Determine whether $\int_1^\infty (1/x^2)\,dx$ converges.

Solution By Definition 8.3, we write

$$\begin{aligned}\int_1^\infty \frac{1}{x^2}\,dx &= \lim_{b\to\infty} \int_1^b \frac{1}{x^2}\,dx \\ &= \lim_{b\to\infty} \left[-\frac{1}{x}\right]_1^b \\ &= \lim_{b\to\infty} \left(-\frac{1}{b}+1\right) = 1.\end{aligned}$$

Since the limit exists, the improper integral converges and has value 1. ■

Example 8.44 Determine whether $\int_4^\infty (1/\sqrt{x})\,dx$ converges.

Solution Again by Definition 8.3,

$$\begin{aligned}\int_4^\infty \frac{1}{\sqrt{x}}\,dx &= \lim_{b\to\infty} \int_4^b \frac{1}{\sqrt{x}}\,dx \\ &= \lim_{b\to\infty} \left[2\sqrt{x}\right]_4^b \\ &= \lim_{b\to\infty} (2\sqrt{b}-4) = \infty.\end{aligned}$$

Since the limit does not exist, the improper integral diverges. ■

If the function f is continuous and positive over $[a, \infty)$ and $\int_a^\infty f(x)\,dx$ converges, then the value of the improper integral is the **area under f over $[a, \infty)$**. Thus, by Example 8.43, the area under $f(x) = 1/x^2$ over $[1, \infty)$ is 1 square unit (see Figure 8.67).

On the other hand, if f is continuous and positive over $[a, \infty)$ and $\int_a^\infty f(x)\,dx$ diverges, then the limit $\lim_{b\to\infty} \int_a^b f(x)\,dx$ is infinite. We therefore say that the **area under f over $[a, \infty)$ is infinite** also. For example, the area under $f(x) = 1/\sqrt{x}$ over $[4, \infty)$ is infinite, by Example 8.44 (see Figure 8.68).

An improper integral over an infinite interval of the form $(-\infty, b]$ is defined in a similar fashion.

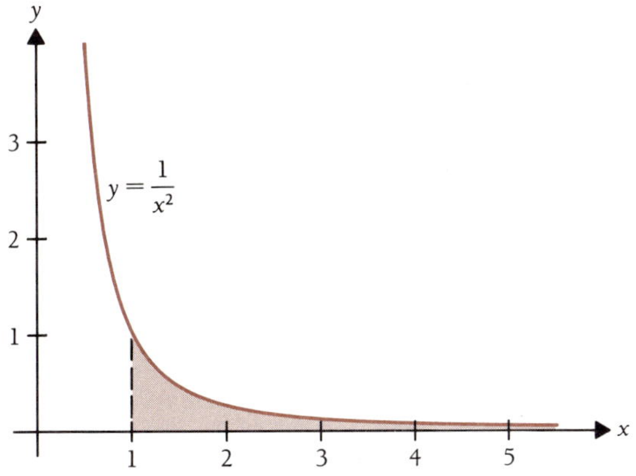

FIGURE 8.67 The area under $f(x) = 1/x^2$ over $[1, \infty)$ is finite

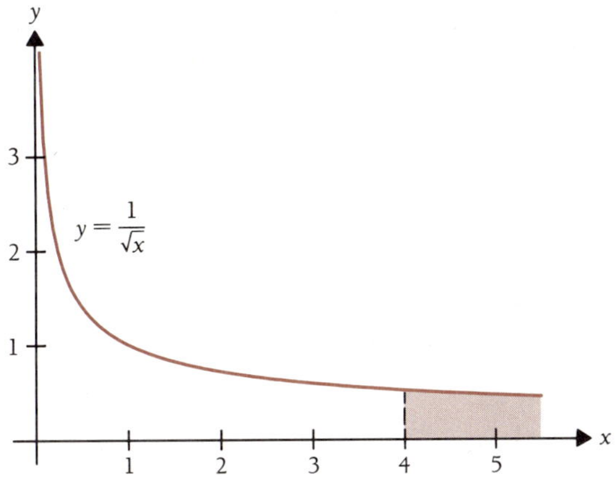

FIGURE 8.68 The area under $f(x) = 1/\sqrt{x}$ over $[4, \infty)$ is infinite

DEFINITION 8.4

If a function f is continuous on $(-\infty, b]$, the **improper integral of f over $(-\infty, b]$** is denoted by

$$\int_{-\infty}^{b} f(x)\, dx$$

and is defined by

$$\int_{-\infty}^{b} f(x)\, dx = \lim_{a \to -\infty} \int_{a}^{b} f(x)\, dx.$$

We say the improper integral **converges** or **diverges** according as the limit exists or does not exist.

Example 8.45 Determine whether $\int_{-\infty}^{-1} x^{-3}\,dx$ converges.

Solution Using Definition 8.4, we write

$$\begin{aligned}\int_{-\infty}^{-1} x^{-3}\,dx &= \lim_{a\to-\infty}\int_{a}^{-1} x^{-3}\,dx\\ &= \lim_{a\to-\infty}\left[-\frac{1}{2x^2}\right]_a^{-1}\\ &= \lim_{a\to-\infty}\left(-\frac{1}{2}+\frac{1}{2a^2}\right) = -\frac{1}{2}.\end{aligned}$$

Since the limit exists, the improper integral converges and has value $-\frac{1}{2}$. ■

We also can define an improper integral over the entire real line $(-\infty, \infty)$.

DEFINITION 8.5 If a function f is continuous on $(-\infty, \infty)$, the **improper integral of f over $(-\infty, \infty)$** is denoted by

$$\int_{-\infty}^{\infty} f(x)\,dx$$

and is defined by

$$\int_{-\infty}^{\infty} f(x)\,dx = \int_{-\infty}^{c} f(x)\,dx + \int_{c}^{\infty} f(x)\,dx$$

for any real number c. This improper integral converges if and only if *both* the integrals on the right-hand side converge.

It can be shown (see Problem 26) that if $\int_{-\infty}^{\infty} f(x)\,dx$ converges, its value does not depend on the number c in Definition 8.5.

Example 8.46 Determine whether the improper integral below converges:

$$\int_{-\infty}^{\infty} \frac{dx}{x^2+1}.$$

Solution Using Definition 8.5 with $c = 0$ (because it is convenient), we write

$$\begin{aligned}\int_{-\infty}^{\infty} \frac{dx}{x^2+1} &= \int_{-\infty}^{0} \frac{dx}{x^2+1} + \int_{0}^{\infty} \frac{dx}{x^2+1}\\ &= \lim_{a\to-\infty}\int_{a}^{0} \frac{dx}{x^2+1} + \lim_{b\to\infty}\int_{0}^{b} \frac{dx}{x^2+1}\end{aligned}$$

$$= \lim_{a \to -\infty} (-\tan^{-1} a) + \lim_{b \to \infty} \tan^{-1} b$$

$$= \frac{\pi}{2} + \frac{\pi}{2} = \pi.$$

Since both limits exist, the improper integral converges and has value π.

■ **Example 8.47** Determine whether $\int_{-\infty}^{\infty} x^{1/3}\,dx$ converges.

Solution Letting $c = 0$ again in Definition 8.5, we have

$$\int_{-\infty}^{\infty} x^{1/3}\,dx = \int_{-\infty}^{0} x^{1/3}\,dx + \int_{0}^{\infty} x^{1/3}\,dx$$

$$= \lim_{a \to -\infty} \int_{a}^{0} x^{1/3}\,dx + \lim_{b \to \infty} \int_{0}^{b} x^{1/3}\,dx$$

$$= \lim_{a \to -\infty} \left(-\frac{3}{4} a^{4/3}\right) + \lim_{b \to \infty} \left(\frac{3}{4} b^{4/3}\right).$$

Since neither of these limits is finite, the improper integral diverges. (Nonexistence of just one of the limits is enough to say the integral diverges.) ■

Improper Integrals with Unbounded Integrands

Suppose a function f is continuous on the interval $[a, b)$, but is unbounded there, say

$$\lim_{x \to b-} f(x) = \infty.$$

(See Figure 8.69.) For any $t \in [a, b)$, the definite integral $\int_a^t f(x)\,dx$ exists. We then consider the limit

$$\lim_{t \to b-} \int_a^t f(x)\,dx.$$

If this limit exists, we define the limit to be another improper integral.

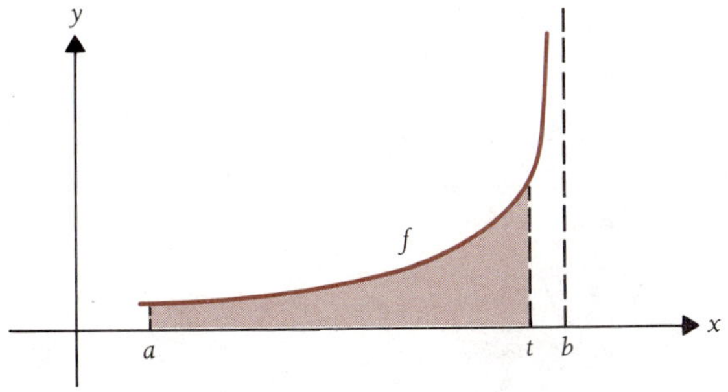

FIGURE 8.69 Toward the integral of an unbounded function

DEFINITION 8.6

If a function f is continuous on $[a, b)$ and $\lim_{x\to b^-} f(x) = \pm\infty$, the **improper integral of f over $[a, b)$** is denoted by

$$\int_a^b f(x)\,dx$$

and is defined by

$$\int_a^b f(x)\,dx = \lim_{t\to b^-} \int_a^t f(x)\,dx.$$

If the limit exists, we say the improper integral **exists** or **converges.** If the limit does not exist, we say the improper integral **does not exist** or **diverges.**

Note that in Definition 8.6, f need not be defined at b.

■ **Example 8.48**

Determine whether the improper integral below converges:

$$\int_0^1 \frac{dx}{x-1}.$$

Solution

Since

$$\lim_{x\to 1^-} \frac{1}{x-1} = -\infty,$$

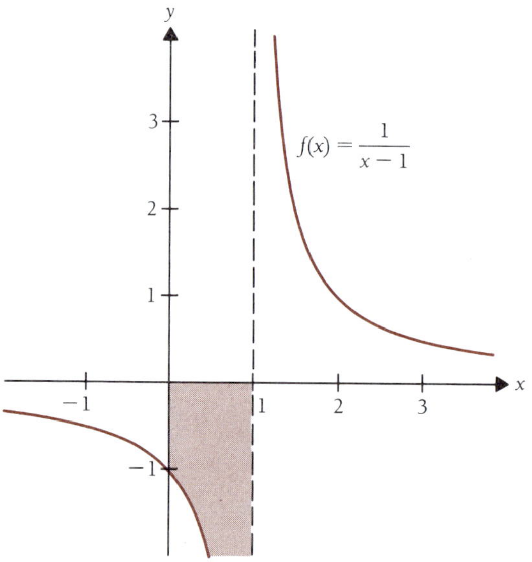

FIGURE 8.70 The graph of $f(x) = 1/(x-1)$

this integral is improper (see Figure 8.70). Using Definition 8.6, we write

$$\int_0^1 \frac{dx}{x-1} = \lim_{t\to 1^-} \int_0^t \frac{dx}{x-1}$$
$$= \lim_{t\to 1^-} \ln|t-1| = -\infty.$$

Since the limit does not exist, the improper integral diverges. ■

Improper integrals of unbounded functions also can be interpreted as areas. If f is continuous and positive on $[a, b)$ and $\int_a^b f(x)\,dx$ converges, even though f has an infinite discontinuity at b, then the improper integral represents the **area under f over $[a, b)$**. If the integral diverges, we say there is **infinite area under f over $[a, b)$**.

■ **Example 8.49** Find the area under $y = \sec^2 x$ over $[0, \pi/2)$.

Solution Since $\sec^2 x$ has an infinite discontinuity at $\pi/2$ (see Figure 8.71), we apply Definition 8.6:

$$\int_0^{\pi/2} \sec^2 x\,dx = \lim_{t\to \pi/2^-} \int_0^t \sec^2 x\,dx$$
$$= \lim_{t\to \pi/2^-} (\tan t) = \infty.$$

Since the limit is infinite, the area is also infinite.

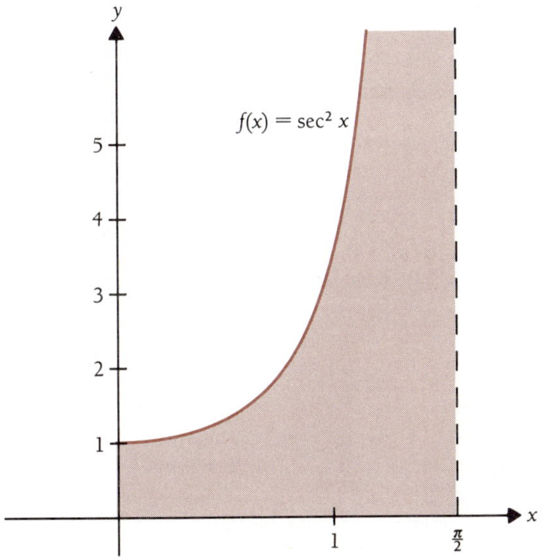

FIGURE 8.71 The graph of $f(x) = \sec^2 x$ over $[0, \pi/2)$ ■

Thus far, we have considered a function f over $[a, b)$ with an infinite discontinuity at b. If f has an infinite discontinuity at a, then the improper integral of f over $(a, b]$ is defined similarly.

DEFINITION 8.7

If a function f is continuous on $(a, b]$ and $\lim_{x\to a^+} f(x) = \pm\infty$, the **improper integral of f over $(a, b]$** is denoted by

$$\int_a^b f(x)\,dx$$

and is defined by

$$\int_a^b f(x)\,dx = \lim_{t\to a^+} \int_t^b f(x)\,dx.$$

We say the improper integral **converges** or **diverges** according as the limit exists or does not exist.

Example 8.50 Evaluate $\int_1^2 (x-1)^{-1/2}\,dx$.

Solution Since $f(x) = (x-1)^{-1/2}$ has an infinite discontinuity at 1 (see Figure 8.72), the integral is improper. By Definition 8.7, we have

$$\begin{aligned}\int_1^2 (x-1)^{-1/2}\,dx &= \lim_{t\to 1^+} \int_t^2 (x-1)^{-1/2}\,dx \\ &= \lim_{t\to 1^+} (2 - 2\sqrt{t-1}) = 2.\end{aligned}$$

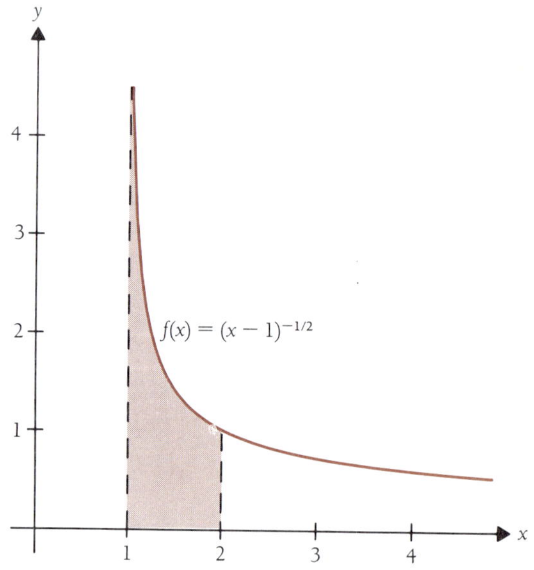

FIGURE 8.72 The graph of $f(x) = (x-1)^{-1/2}$

Thus, the improper integral converges and has value 2. We can also say that there are 2 square units of area under $f(x) = (x - 1)^{-1/2}$ over $(1, 2]$. ■

You have undoubtedly noticed that the improper integral of an unbounded function looks just like a regular definite integral. There is nothing in the notation to point out that the function is unbounded; you must notice that for yourself. One tip-off is that the function usually is not defined at one limit or the other. This was the case in Examples 8.49 and 8.50.

It also may happen that a function has an infinite discontinuity within the interval over which it is integrated. In that case, the integral is improper and is defined as the sum of two improper integrals.

DEFINITION 8.8

If a function f is continuous on $[a, c)$ and on $(c, b]$, and both the one-sided limits of f at c are infinite, then the **improper integral of f over $[a, c) \cup (c, b]$** is denoted by

$$\int_a^b f(x)\,dx$$

and defined by

$$\int_a^b f(x)\,dx = \int_a^c f(x)\,dx + \int_c^b f(x)\,dx.$$

We say the improper integral **converges** if and only if *both* the improper integrals on the right-hand side converge.

■ **Example 8.51** Evaluate $\int_{-1}^{2} x^{-4/3}\,dx$.

Solution Note that the integrand $x^{-4/3}$ has an infinite discontinuity at 0 (see Figure 8.73, page 566) which lies in the interval of integration. Hence, by Definition 8.8,

$$\begin{aligned}\int_{-1}^{2} x^{-4/3}\,dx &= \int_{-1}^{0} x^{-4/3}\,dx + \int_{0}^{2} x^{-4/3}\,dx \\ &= \lim_{t\to 0-} \int_{-1}^{t} x^{-4/3}\,dx + \lim_{t\to 0+} \int_{t}^{2} x^{-4/3}\,dx \\ &= \lim_{t\to 0-} (-3t^{-1/3} - 3) + \lim_{t\to 0+} [-3(2)^{-1/3} + 3t^{-1/3}].\end{aligned}$$

Since at least one of these limits fails to exist, the original improper integral diverges.

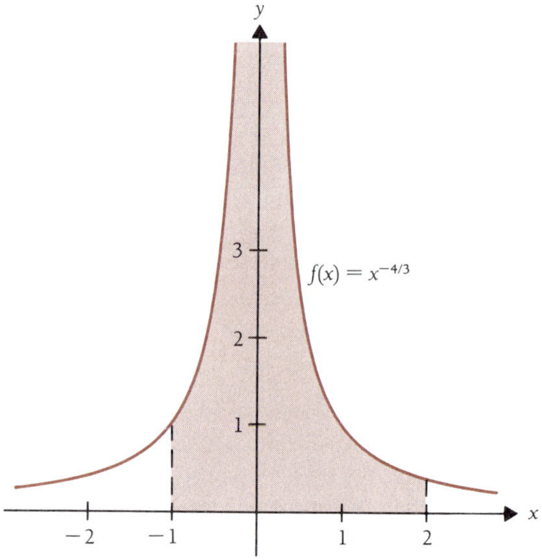

FIGURE 8.73 The graph of $f(x) = x^{-4/3}$

Because the integral in Example 8.51 is improper, the fundamental theorem of calculus does not apply. If we had failed to notice that the integral was improper and had mistakenly applied the fundamental theorem, we would have gotten

$$-3x^{-1/3}\Big|_{-1}^{2} = -3(1 + 2^{-1/3}),$$

a completely erroneous result.

Combinations of unbounded functions and unbounded intervals can occur, as in the following example.

Example 8.52 Evaluate $\int_0^\infty x^{-3}\,dx$.

Solution This integral is improper because the integrand is unbounded near 0 and also because the interval of integration is infinite. Choosing some convenient number in $(0, \infty)$, say 1, we rewrite the integral as two improper integrals:

$$\begin{aligned}\int_0^\infty x^{-3}\,dx &= \int_0^1 x^{-3}\,dx + \int_1^\infty x^{-3}\,dx \\ &= \lim_{t\to 0^+}\int_t^1 x^{-3}\,dx + \lim_{b\to\infty}\int_1^b x^{-3}\,dx \\ &= \lim_{t\to 0^+}\left(-\frac{1}{2}+\frac{1}{2t^2}\right) + \lim_{b\to\infty}\left(-\frac{1}{2b^2}+\frac{1}{2}\right).\end{aligned}$$

Since the first of these limits is infinite, the original integral diverges.

Exercises 8.8

Evaluate, if possible.

1. $\int_1^\infty x^{-5}\,dx$
2. $\int_2^\infty x^{-4}\,dx$
3. $\int_0^\infty \sqrt{x}\,dx$
4. $\int_{-\infty}^{-1} x^{-2}\,dx$
5. $\int_0^1 x^{-5}\,dx$
6. $\int_0^1 x^{-2/3}\,dx$
7. $\int_0^1 (x^2 - x^{-2})\,dx$
8. $\int_0^1 \frac{dx}{\sqrt{1-x^2}}$
9. $\int_{-\infty}^\infty x^{2/5}\,dx$
10. $\int_{-\infty}^\infty 2x\,dx$
11. $\int_0^2 (1-x)^{-2}\,dx$
12. $\int_0^1 (1-x)^{-1/2}\,dx$
13. $\int_{-\infty}^0 xe^{-x^2}\,dx$
14. $\int_{-\infty}^\infty e^{-x}\,dx$
15. $\int_{-\infty}^0 x^{-2}\,dx$
16. $\int_{-\infty}^\infty x^{-1/3}\,dx$
17. Show that $\int_1^\infty x^{-p}\,dx$ converges if $p > 1$ and diverges if $p \le 1$.
18. Show that $\int_0^1 x^{-p}\,dx$ converges if $p < 1$ and diverges if $p \ge 1$.
19. Find the area under $f(x) = e^{-kx}$ over $[0, \infty)$.
20. Find the area under $f(x) = 1/x^2$ over $[a, \infty)$, $a \ge 1$.
21. Show that there is infinite area under $f(x) = 1/x$ over $[1, \infty)$.
22. The region under $f(x) = 1/x$ over $[1, \infty)$ is revolved about the x-axis. Show that the volume of the solid generated is finite.
23. The curve $y = 1/x$ over $[1, \infty)$ is revolved about the x-axis. The surface generated is sometimes called **Gabriel's horn**. Show that it has infinite surface area. (But by Problem 22, the volume inside is finite; thus, it will not hold enough paint to paint the inside!)
24. Show that $\int_{-\infty}^\infty x^{1/3}\,dx$ diverges, but the limit $\lim_{t\to\infty} \int_{-t}^t x^{1/3}\,dx$ exists.
25. If the integral $\int_{-\infty}^\infty f(x)\,dx$ converges, is its value equal to $\lim_{t\to\infty} \int_{-t}^t f(x)\,dx$?
26. Show that the choice of c in Definition 8.5 is immaterial if the improper integral converges. [*Hint:* Let $f(x)$ have antiderivative $F(x)$, and show that $\int_{-\infty}^{c_1} f(x)\,dx + \int_{c_1}^\infty f(x)\,dx = \int_{-\infty}^{c_2} f(x)\,dx + \int_{c_2}^\infty f(x)\,dx$.]
27. A function f is a **probability density function** on $[a, b]$ if and only if $\int_a^b f(x)\,dx = 1$. Show that if $k > 0$, then $f(x) = ke^{-kx}$ is a probability density function on $[0, \infty)$.
28. If f is a probability density function on $[a, b]$, its **mean** or **expected value** is $\int_a^b xf(x)\,dx$. Find the mean of $f(x) = ke^{-kx}$ on $[0, \infty)$.
29. Show that the improper integral $\int_0^\infty \cos x\,dx$ diverges, but the limit $\lim_{t\to\infty} \int_0^t \cos x\,dx$ is not infinite.

SECTION 8.9

Chapter Review

Important Concepts

Volumes by slicing	Volumes by cylindrical shells	Polygonal curve
Solid of revolution	Smooth curve	Rectifiable arc
Solid of known cross-section	Arc	Differential of arc length

Surface of revolution	Lamina	Population theory
Mass	Centroid	Survival function
Moment	Theorems of Pappus	Renewal function
Center of mass	Annuity	Improper integral
Homogeneous	Revenue stream	Convergence
Density function	Pressure	Divergence
Center of gravity	Work	

Review Exercises

Let R be the region below $y = 8 - x^{3/2}$ over [0, 4]. Find the following:

1. The area of R 2. The centroid of R
3. The length of the curved part of the boundary of R
4. Integrals representing the centroid of the curved part of the boundary of R
5. The volume of the solid generated by revolving R about the line:
(a) $x = 0$ (b) $y = 0$ (c) $x = 4$ (d) $y = 8$
6. Integrals representing the area of the surface generated by revolving the curved part of the boundary of R about the line:
(a) $x = 0$ (b) $y = 0$ (c) $x = 4$ (d) $y = 8$
7. Integrals representing the center of mass of a plate covering R whose density varies as the distance from the line:
(a) $y = 0$ (b) $x = 4$
8. An earth-fill dam is 50 feet wide and 20 feet high, but the face of the dam is a rectangle inclined at an angle of 45°. If water fills the reservoir to a point 2 feet below the top of the dam, find the force on the dam. [For water, $w = 62.5$ pounds per cubic foot.]
9. The bottom of a cylindrical water tank is a hemispherical dome curving upward into the tank. If the tank is 2 feet in diameter and water fills the tank to a depth of 4 feet above the top of the dome, find an integral representing the force on the dome.
10. If \$100 per month is paid into an annuity earning 7% compounded continuously, how long will it take to accumulate \$10,000?
11. In population theory, explain what happens if $r(t) = 0$.
12. For a certain population,

$$s(t) = \begin{cases} 1 - t/10 & \text{if } 0 \le t \le 10 \\ 0 & \text{if } t > 10 \end{cases}$$

and $r(t) = e^{0.08t}$. If $N(0) = 5000$, find $N(20)$.

Evaluate, if possible.

13. $\int_0^1 x^{-2/3}\,dx$ 14. $\int_{-1}^1 x^{-2/3}\,dx$ 15. $\int_1^\infty x^{-4}\,dx$ 16. $\int_0^\infty x^{-4}\,dx$

Additional Exercises

1. Let R be the region under $y = x^2$ over [0, 1]. Find the number c such that the line $x = c$ cuts R into two pieces that generate the same volume when revolved about the:
(a) x-axis (b) y-axis
2. If a cone has height h and base of area B, show that its volume is $V = \frac{1}{3}Bh$, whatever the shape of its base may be.

3. **Cavalieri's Theorem** Two solid objects of equal height rest on the same plane, and any plane parallel to the base plane cuts the two solids in cross-sections of equal area. Prove that the two objects have equal volumes.

4. A solid object is made from a sphere by drilling a hole completely through the sphere along a diameter. If the hole is 6 inches long, find the volume of the solid.

5. Two cylinders of radius r intersect at right angles. Find the volume of their intersection.

6. A plane cuts a sphere into two pieces whose volumes are in the ratio of 2 to 1. How far from the center of the sphere does the plane pass?

7. Two parallel planes a distance h apart both meet a sphere of radius r. Show that the area of the portion of the surface of the sphere that lies between the planes is $2\pi rh$.

8. Find the length of the loop in the curve $9y^2 = x(x-3)^2$.

9. A certain buoy is a cylinder 6 feet long and 2 feet in diameter, and is weighted so that it floats vertically with half its length above the water. How much does it weigh?

10. The buoy in Problem 9 is forced down into the water until its top is just submerged. How much work is done in doing so?

11. The force of gravity exerted by the earth on an object of mass m at a distance r from the center of the earth is $F = km/r^2$, where k is a constant. If an object weighs 100 pounds at the surface ($F = 100$ if $r = 4000$ miles), find the work done in lifting the object to a height of 10 miles above the surface of the earth.

12. If g is the acceleration of gravity at the earth's surface and R is the radius of the earth, show that the work done in lifting an object of mass m to a height h above the earth's surface is $W = mgRh/(R+h)$.

13. Find the work done in lifting a 1000 pound payload from the earth's surface to a height 100 miles above the earth's surface.

14. How much work is done by gravity on an object of mass m that falls from infinity to the earth's surface?

15. Imagine a hole drilled straight through the center of the earth. An object of mass m in the hole, a distance r from the center of the earth, is attracted toward the center of the earth by a force $F = mgr/R$, where g is the acceleration of gravity at the earth's surface and R is the radius of the earth. How much work is done as the object falls from the surface of the earth to the center?

16. The winner of a sweepstakes puts \$10,000 into a savings account that earns interest at the rate of 6.5%, compounded continuously. He then withdraws \$100 at the end of each month. About how long can he continue to make these withdrawals before his account is depleted?

17. Can you apply population theory to the problem of the effect of inflation on a savings account?

18. Show by example that the centroid of a plane region may lie outside the region.

19. Prove that the force of hydrostatic pressure on one side of a vertical submerged plate is equal to the area of the plate multiplied by the pressure at its centroid.

20. A bag of sand is lifted at the constant rate of 10 feet per minute for 2 minutes. During that time, sand leaks out of the bag at a constant rate, reducing the weight of the sandbag from 100 pounds to 60 pounds. How much work is done in lifting the bag?

21. Find the area under the curve $x^2y = a^2(a-y)$.

22. Let a weight w be located at x on the x-axis. The **moment of inertia** of the weight **about the origin** is wx^2. Now, suppose a rod extends from a to b along the x-axis $(0 < a < b)$ and has density $\delta(x)$. Find an integral for its moment of inertia about the origin.

23. A wedge is cut from the end of a cylinder of radius r by a plane containing a diameter of the end of the cylinder and making an angle of $30°$ with the plane of the end. Find its volume.

Evaluate, if possible.

24. $\displaystyle\int_0^\infty \frac{dx}{x^2+4}$ **25.** $\displaystyle\int_0^\infty e^{-2x}\,dx$ **26.** $\displaystyle\int_1^\infty \frac{1}{x}\,dx$

27. $\displaystyle\int_0^1 \frac{1}{x}\,dx$ **28.** $\displaystyle\int_0^2 (4-x^2)^{-1/2}\,dx$ **29.** $\displaystyle\int_0^1 \frac{dx}{x\ln x}$

30. $\displaystyle\int_{-\infty}^\infty \frac{dx}{x^2+16}$ **31.** $\displaystyle\int_1^\infty \frac{dx}{x\sqrt{x^2-1}}$ **32.** $\displaystyle\int_1^\infty (x^2-1)^{3/2}\,dx$

33. The probability of getting a taxi on a certain corner within t minutes is

$$p(t) = \int_0^t \frac{1}{3}e^{-x/3}\,dx.$$

Find the probability:
(a) Of getting a taxi within 1 minute
(b) Of getting a taxi within 3 minutes
(c) Of not getting a taxi within 3 minutes
(d) Of having to wait at least 5 minutes for a taxi

34. What is the expected waiting time in Problem 33? (See Problem 28 of Section 8.8.)

CHAPTER 9

Sequences and Series

INTRODUCTION

In this chapter, we complete the development of the limit concept by studying sequences and series. We will concentrate on methods for determining whether a sequence or series converges. Then we look at series of functions, with the emphasis on power series. We will include Taylor series, term-by-term differentiation and integration of series, and the binomial series.

SECTION 9.1

Sequences of Real Numbers

In Section 2.1, we discussed limits from a computational point of view. For example, we showed that

$$\lim_{x \to 3} \frac{x^2 - 9}{x - 3} = 6,$$

and offered the following computations as evidence:

$$\begin{array}{rcccc} x = 2 & 2.9 & 2.99 & 2.999 & \ldots, \\ \dfrac{x^2-9}{x-3} = 5 & 5.9 & 5.99 & 5.999 & \ldots. \end{array}$$

What we were actually claiming was that since the x values 2, 2.9, 2.99, 2.999, . . . get closer and closer to 3, and the functional values 5, 5.9, 5.99, 5.999, . . . get closer and closer to 6, the limit of the function as $x \to 3$ must be 6.

The numbers 2, 2.9, 2.99, 2.999, . . . (the three dots mean the list does not terminate) form what is called a **sequence**. In this section, we will discuss sequences and their limits. At the end of this section, we will relate limits of sequences to limits of functions and examine the above claim more closely.

Sequences

We have seen several examples of sequences already. As mentioned above, the computational evidence for limits given in Section 2.1 involved sequences. The bisection method (Section 2.5) and Newton's method (Section 4.2) both involve sequences $\{x_0, x_1, x_2, \ldots\}$ in which each x_n is computed from the previous x_i's.

We will generally use the symbols

$$a_1, a_2, a_3, \ldots, a_n, \ldots$$

to denote the terms of a sequence. A sequence can be specified in various ways. The simplest is to give a formula, such as

$$a_n = n^2,$$

which is used like function notation. This equation tells us that

$$a_1 = 1^2, \qquad a_2 = 2^2, \qquad a_3 = 3^2,$$

and so forth. We also can list the terms of a sequence in braces; for example,

$$\{1^2, 2^2, 3^2, \ldots, n^2, \ldots\}.$$

Sometimes this notation is shortened to $\{n^2\}$, in which only the general term, or a formula for it, is given.

Sequences also can be given by **recursion**, in which each term is given in terms of previous terms. For example, the recursion formula

$$a_1 = 1, \qquad a_{n+1} = (n+1)a_n$$

defines the sequence

$$\{1, 2 \cdot 1, 3 \cdot 2 \cdot 1, 4 \cdot 3 \cdot 2 \cdot 1, 5 \cdot 4 \cdot 3 \cdot 2 \cdot 1, \ldots\}.$$

As another example, Newton's formula is a recursion formula for constructing a sequence of numbers that approach a zero of the function under consideration.

These ideas are stated succinctly in Definition 9.1.

DEFINITION 9.1

A **sequence** (of real numbers) is a function whose domain is the set of positive integers and whose range is a set of real numbers.

Thus, a_n is simply the image of n under the sequence a. The notation $\{a_1, a_2, a_3, \ldots\}$ is just a listing of the range of the sequence.

Some important types of sequences are introduced in Example 9.1.

Example 9.1

Write the first five terms of each sequence:

(a) $a_n = n$ (b) $a_n = \dfrac{1}{n}$ (c) $a_n = \dfrac{(-1)^n}{n}$ (d) $a_n = 2$ (e) $a_n = ar^{n-1}$

Solution

(a) If $a_n = n$, then the first five terms are $a_1 = 1, a_2 = 2, a_3 = 3, a_4 = 4$, and $a_5 = 5$. Thus, $\{n\} = \{1, 2, 3, 4, 5, \ldots\}$.

(b) If $a_n = 1/n$, then $\{1/n\} = \{1, \frac{1}{2}, \frac{1}{3}, \frac{1}{4}, \frac{1}{5}, \ldots\}$. This sequence is called the **harmonic sequence**.

(c) $\{(-1)^n/n\} = \{-1, \frac{1}{2}, -\frac{1}{3}, \frac{1}{4}, -\frac{1}{5}, \ldots\}$. This sequence is called the **alternating harmonic sequence**.

(d) $\{2\} = \{2, 2, 2, 2, 2, \ldots\}$. This is an example of a **constant sequence**.

(e) $\{ar^{n-1}\} = \{a, ar, ar^2, ar^3, ar^4, \ldots\}$. This sequence is called a **geometric sequence**. ■

A sequence can be represented geometrically by plotting its terms along the real line. The harmonic and alternating harmonic sequences are graphed in Figures 9.1 and 9.2. All their terms cannot be shown, of course, but the first few terms indicate the pattern.

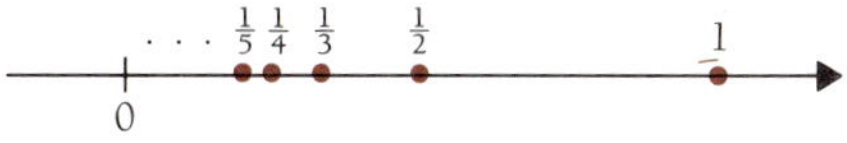

FIGURE 9.1 A few terms of the harmonic sequence

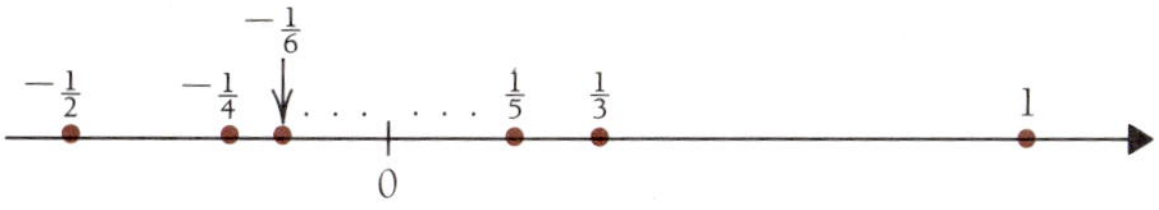

FIGURE 9.2 A few terms of the alternating harmonic sequence

The Limit of a Sequence

In Figures 9.1 and 9.2, the terms of the sequences seem to be getting successively closer to 0. In fact, if a distance ε is given, then all terms beyond a certain point will lie within ε of 0, in either sequence. This is the idea of the limit of a sequence.

DEFINITION 9.2

We say that the sequence $\{a_n\}$ **has limit L** or **converges to L**, written

$$\lim_{n\to\infty} a_n = L,$$

if and only if for any positive real number ε there exists a positive integer k such that if $n > k$, then $|a_n - L| < \varepsilon$. A sequence that has no limit (that is, for which the limit does not exist) is said to **diverge**.

If you compare this definition with Definition 2.4, you will see that it is just a modification to take into account that the domain of a sequence is the set of positive integers. Therefore, the properties of limits of sequences are similar to those of limits of functions at ∞, as we will discuss below.

We also use the notation $a_n \to L$ in place of $\lim_{n\to\infty} a_n = L$.

■ **Example 9.2** Find the limits, if they exist:

(a) $\left\{\frac{1}{n}\right\}$ (b) $\{n\}$ (c) $\{2\}$ (d) $\left\{\left(\frac{1}{2}\right)^n\right\}$ (e) $\{(-1)^n\}$ (f) $\{r^n\}$

Solution

(a) By analogy with $\lim_{x\to\infty}(1/x) = 0$, we have $\lim_{n\to\infty}(1/n) = 0$. (We could also prove this from Definition 9.2 by noting that if $\varepsilon > 0$ is given, then $1/n < \varepsilon$ whenever $n > 1/\varepsilon$.)

(b) $\lim_{n\to\infty} n = \infty$, so the sequence $\{n\}$ diverges. (We could also say it **diverges to** ∞.)

(c) Since $\lim_{n\to\infty} 2 = 2$, the sequence converges to 2.

(d) Because 2^n gets large as n gets large, we have $\lim_{n\to\infty}(\frac{1}{2})^n = 0$. [We could also prove this from Definition 9.2 by showing that $(\frac{1}{2})^n < \varepsilon$ whenever $k > \ln(1/\varepsilon)/\ln 2$.]

(e) This sequence may be partly written out as $\{1,-1,1,-1, \ldots\}$, and it obviously has no limit since there is no number L satisfying Definition 9.2. That is, $\lim_{n\to\infty}(-1)^n$ does not exist, and the sequence diverges. (This is an example of an **oscillating** sequence; that is, the terms oscillate between two different numbers. It is also an example of an **alternating** sequence, because the terms alternate in sign.)

(f) If $-1 < r < 1$, then $r^n \to 0$. [This can be proved by the same method used in part (d).] If $r = 1$, then $\{r^n\}$ is the constant sequence $\{1\}$ and has limit 1. If $r > 1$, then $\{r^n\}$ diverges to ∞. If $r \leq -1$, then $\{r^n\}$ simply diverges. ■

Limit Theorems for Sequences

The limit theorems of Section 2.3 also hold for limits of sequences. In Theorem 9.1, we will restate the main ones for limits of sequences in the notation introduced in this section.

THEOREM 9.1

If $a_n \to A$ and $b_n \to B$, then

$$a_n + b_n \to A + B,$$
$$a_n - b_n \to A - B,$$
$$a_n b_n \to AB,$$
$$\frac{a_n}{b_n} \to \frac{A}{B}. \qquad \text{(Provided each } b_n \text{ and } B \text{ are not 0)}$$

If c is a constant and $a_n \to A$, then

$$ca_n \to cA$$

If f is a continuous function and $a_n \to A$, then

$$f(a_n) \to f(A).$$

Finally, if $a_n \leq b_n \leq c_n$, $a_n \to L$, and $c_n \to L$, then $b_n \to L$.

L'Hôpital's rules can also be used with limits of sequences; we just treat n as though it were a continuous variable.

■ **Example 9.3** Find the limits, if they exist:

(a) $\left\{\frac{n+1}{n}\right\}$ (b) $\left\{\frac{n+1}{n^2}\right\}$ (c) $\left\{\frac{1}{n+2}\right\}$ (d) $\left\{\frac{10^{200}}{n}\right\}$

(e) $\left\{\frac{4^n-3^n}{5^n}\right\}$ (f) $\left\{\cos\frac{1}{n}\right\}$ (g) $\left\{\frac{\ln n}{n}\right\}$

Solution

(a) $\lim_{n\to\infty}\frac{n+1}{n}=\lim_{n\to\infty}\left(1+\frac{1}{n}\right)=1+0=1.$

(b) $\lim_{n\to\infty}\frac{n+1}{n^2}=\lim_{n\to\infty}\left(\frac{1}{n}+\frac{1}{n^2}\right)=0+0=0.$

(c) $\lim_{n\to\infty}\frac{1}{n+2}=\lim_{n\to\infty}\frac{1/n}{1+2/n}=\frac{0}{1+0}=0.$

(d) $\lim_{n\to\infty}\frac{10^{200}}{n}=10^{200}\lim_{n\to\infty}\frac{1}{n}=10^{200}(0)=0.$

(e) $\lim_{n\to\infty}\frac{4^n-3^n}{5^n}=\lim_{n\to\infty}\left[\left(\frac{4}{5}\right)^n-\left(\frac{3}{5}\right)^n\right]=0-0=0.$

(f) $\lim_{n\to\infty}\left(\cos\frac{1}{n}\right)=\cos\left(\lim_{n\to\infty}\frac{1}{n}\right)=\cos 0=1,$
by continuity of the cosine function.

(g) $\lim_{n\to\infty}\frac{\ln n}{n}=\lim_{n\to\infty}\frac{1/n}{1}=0$, by L'Hôpital's ∞/∞ rule. ■

Bounded Sequences

A number B is said to be an **upper bound** of a sequence $\{a_n\}$ if $a_n \le B$ for all n. A sequence with an upper bound is said to be **bounded above.** Similarly, a number b is said to be a **lower bound** of $\{a_n\}$ if $a_n \ge b$ for all n; a sequence with a lower bound is said to be **bounded below.** A sequence that is bounded above and bounded below is said to be **bounded.**

■ **Example 9.4** Determine whether each sequence is bounded above or bounded below:

(a) $\left\{\frac{1}{n}\right\}$ (b) $\{2^n\}$ (c) $\{-n\}$ (d) $\{(-n)^n\}$

Solution

(a) All the terms of $\{1/n\}$ are positive, so 0 is a lower bound. All the terms lie in (0, 1], so 1 is an upper bound. Thus, $\{1/n\}$ is bounded.

(b) All the terms of $\{2^n\}$ are positive, so 0 is a lower bound. Since $2^n \to \infty$, there is no upper bound. Thus, $\{2^n\}$ is bounded below but not bounded above.

(c) Each term of $\{-n\}$ is negative, so 0 is an upper bound. Since $-n \to -\infty$, there is no lower bound. Thus, $\{-n\}$ is bounded above but not bounded below.

(d) The terms of $\{(-n)^n\}$ are alternately positive and negative. Also, $n^n \to \infty$. Thus, $\{(-n)^n\}$ is neither bounded above nor bounded below. ■

Boundedness of a sequence is often related to its convergence. For example, we have Theorem 9.2.

THEOREM 9.2 Every convergent sequence is bounded.

Proof Let a_n be a sequence, and let $a_n \to L$. Let ε be a positive real number. Then there exists a positive integer k such that if $n > k$, then $|a_n - L| < \varepsilon$. That is, $L - \varepsilon < a_n < L + \varepsilon$ for $n > k$. Therefore, at worst, only the terms $a_1, a_2, \ldots, a_k$ do not lie in the interval $(L - \varepsilon, L + \varepsilon)$. It is now a simple matter to find both upper and lower bounds for $\{a_n\}$. (See Figure 9.3 for an example.) You will be invited to complete this proof in the exercises.

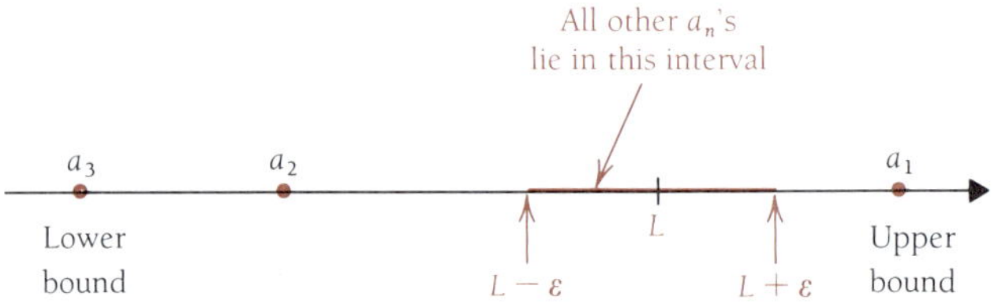

FIGURE 9.3 Illustrating that a convergent sequence is bounded

While every convergent sequence is bounded, it is not the case that every bounded sequence is convergent. For example, the sequence $\{(-1)^n\}$ lies in $[-1, 1]$ and is therefore bounded, but it diverges.

DEFINITION 9.3 A sequence $\{a_n\}$ is **increasing** if and only if $a_{n+1} \geq a_n$, and it is **decreasing** if and only if $a_{n+1} \leq a_n$ for each n. A sequence that is either increasing or decreasing is said to be **monotone**.

■ **Example 9.5** Determine whether each sequence is monotone:

(a) $\left\{\dfrac{1}{n}\right\}$ (b) $\{2n\}$ (c) $\{3\}$ (d) $\left\{\dfrac{(-1)^n}{n^2}\right\}$ (e) $\left\{\dfrac{n^2}{3^n}\right\}$

Solution

(a) Since $1/(n+1) < 1/n$, the sequence $\{1/n\}$ is decreasing.

(b) For all n, $2(n+1) \geq 2n$, so $\{2n\}$ is increasing.

(c) Since $3 \geq 3$ and $3 \leq 3$, the sequence $\{3\}$ is both increasing and decreasing.

(d) The first three terms of $\{(-1)^n/n^2\}$ are -1, $\frac{1}{4}$, and $-\frac{1}{9}$. Thus, this sequence is not monotone.

(e) The first five terms of $\{n^2/3^n\}$ are $\frac{1}{3}, \frac{4}{9}, \frac{1}{3}, \frac{16}{81}, \frac{25}{243}$, so the sequence is not monotone. Note that if the first term is omitted, the resulting sequence is decreasing. ■

Example 9.6 Show that if f is an increasing continuous function on $[1, \infty)$, then the sequence $\{f(n)\}$ is increasing.

Solution If f is increasing, then $n \leq n + 1$ implies that $f(n) \leq f(n + 1)$, so $\{f(n)\}$ is increasing. ■

Example 9.7 Let $\{a_n\}$ be a sequence with each $a_n > 0$. Show that $\{a_n\}$ is increasing if $a_{n+1}/a_n \geq 1$ for each n and decreasing if $a_{n+1}/a_n \leq 1$ for each n.

Solution If $a_{n+1}/a_n \geq 1$ and $a_n > 0$, then $a_{n+1} \geq a_n$, and $\{a_n\}$ is increasing. The decreasing case is similar. ■

As we noted before, not every bounded sequence is convergent; but if we add one condition, we get convergence.

THEOREM 9.3 Every bounded monotone sequence converges.

Proof Suppose a sequence $\{a_n\}$ is bounded and increasing. Then the set of terms of the sequence has an upper bound, and therefore by the completeness property it has a least upper bound. Let B be the least upper bound; we will show that $a_n \to B$. Let $\varepsilon > 0$ be given. Then $B - \varepsilon < B$, so that $B - \varepsilon$ is not an upper bound of $\{a_n\}$. Hence, some term a_k is greater than $B - \varepsilon$, and because $\{a_n\}$ is increasing, $a_n > B - \varepsilon$ for all $n > k$. Thus, $|a_n - B| < \varepsilon$ for $n > k$, and $a_n \to B$. Similarly, if $\{a_n\}$ is bounded and decreasing, it converges to its greatest lower bound.

Example 9.8 Show that $\{\arctan n\}$ is increasing and bounded, and find its limit.

Solution Let $f(x) = \arctan x$; then

$$f'(x) = \frac{1}{x^2 + 1},$$

which is always positive. Thus, f is an increasing continuous function on $[1, \infty)$, so $\{\arctan n\}$ is increasing by Example 9.6. The least upper bound of $\{\arctan n\}$ is $\pi/2$, because $\lim_{x\to\infty} \arctan x = \pi/2$, so $\pi/2$ is the limit of $\{\arctan n\}$. ■

A useful application of Theorem 9.3 is the following.

THEOREM 9.4

If $\{a_n\}$ is monotone and unbounded, then $\{1/a_n\}$ converges to 0.

Proof

Suppose $\{a_n\}$ is increasing (a similar proof holds for $\{a_n\}$ decreasing). Let $\varepsilon > 0$ be given. Then $1/\varepsilon > 0$, and since $\{a_n\}$ is increasing and unbounded, $1/\varepsilon$ is not an upper bound of $\{a_n\}$. Hence, some term a_k is greater than $1/\varepsilon$. It follows that if $n > k$, then $a_n > 1/\varepsilon$, or $1/a_n < \varepsilon$. Thus, $1/a_n \to 0$.

■ **Example 9.9**

Show that $\{1/(\ln n)\}$, $\{e^{-n}\}$, and $\{-1/n^n\}$ all converge to 0.

Solution

Each of the sequences $\{\ln n\}$, $\{e^n\}$, and $\{-n^n\}$ is monotone and unbounded, so the given sequences converge to 0 by Theorem 9.4. ■

One more useful theorem involves bounded sequences; we leave its proof as an exercise (see Problem 48).

THEOREM 9.5

If $\{a_n\}$ converges to 0 and $\{b_n\}$ is bounded, then $\{a_n b_n\}$ converges to 0.

■ **Example 9.10**

Show that the sequence below converges to 0:

$$\left\{\frac{1}{n}\cos n\right\}.$$

Solution

Since $1/n \to 0$ and $-1 \le \cos n \le 1$, the sequence $\{1/n\}$ converges to 0 and $\{\cos n\}$ is bounded. Therefore, applying Theorem 9.5, we can see that the given sequence converges to 0. ■

A Special Limit

One limit that will be useful to us is given in Theorem 9.6.

THEOREM 9.6

$\lim_{n\to\infty}(x^n/n!) = 0$ for any real number x.

Proof

Since

$$\frac{-|x|^n}{n!} \le \frac{x^n}{n!} \le \frac{|x|^n}{n!},$$

it is enough to show that $|x|^n/n! \to 0$ as $n \to \infty$. Let m be an integer such that $|x| < m$.

Since $n \to \infty$, we can assume that $n > m$; then

$$\frac{|x|^n}{n!} = \frac{|x|^n}{1 \cdot 2 \cdot \cdots \cdot m \cdot \underbrace{(m+1)(m+2) \cdot \cdots \cdot n}_{n-m \text{ terms}}}$$

$$\leq \frac{|x|^n}{m!m^{n-m}}$$

$$= \frac{m^m|x|^n}{m!m^n}$$

$$= \frac{m^m}{m!}\left(\frac{|x|}{m}\right)^n.$$

But $(|x|/m)^n \to 0$ [see Example 9.2(f)], so

$$0 \leq \frac{|x|^n}{n!} \leq \frac{m^m}{m!}\left(\frac{|x|}{m}\right)^n \quad \text{implies} \quad \frac{|x|^n}{n!} \to 0. \qquad \text{(By the squeeze play)}$$

Example 9.11 Find $\lim_{n\to\infty}(2^{n+1}/n!)$.

Solution $2^{n+1}/n! = 2 \cdot 2^n/n! \to 0$ by Theorem 9.6. ■

Sequential Limits by Computation

It is tempting to use a calculator to help determine the limit of a sequence. Often, if a_n is calculated for a "large" value of n, say $n = 10^{10}$, then the limit of the sequence is the number obtained, rounded off appropriately. However, care must be taken, because every calculator has limitations and cannot compute a_n for arbitrarily large n. Moreover, round-off errors become more significant in computations with large numbers. And, finally, the sequence simply may not lend itself to calculations. For example, the sequence $\{\ln(1 + n^p)/n^p\}$ converges to 0 (by L'Hôpital's rule), but if p is large (for example, $p = 10^{2500}$), no calculator will ever predict this result. Therefore, computational estimates should never be used to replace the theory of limits of sequences.

Sequential versus Functional Limits (Optional)

Because of the similarity of the definitions, we would naturally expect limits of sequences and limits of functions to be closely related. In one direction, the relationship is almost obvious. Suppose $f(x) \to L$ as $x \to c$, and suppose $\{x_n\}$ is a sequence converging to c. Then $\{f(x_n)\}$ is a sequence converging to L. The proof of this fact follows readily from the definitions.

We illustrated this in our opening example at the beginning of this section. If we let $x_n = 3 - 10^{1-n}$, or $\{x_n\} = \{2, 2.9, 2.99, 2.999, \ldots\}$, then with $f(x) = (x^2 - 9)/(x - 3)$,

$$\{f(x_n)\} = \left\{\frac{x_n^2 - 9}{x_n - 3}\right\} = \{x_n + 3\} = \{6 - 10^{1-n}\} = \{5, 5.9, 5.99, 5.999, \ldots\}.$$

Since $f(x) \to 6$ as $x \to 3$, and $x_n \to 3$, then $f(x_n) \to 6$.

The relationship in the other direction is not so simple. That is, if there is a sequence $\{x_n\}$ converging to c for which $\{f(x_n)\}$ converges to L, then it does not necessarily follow that $\lim_{x\to c} f(x) = L$, or even that the limit exists.

For example, the limit $\lim_{x\to 0}(|x|/x)$ fails to exist. However, with $f(x) = |x|/x$, we have

$$\frac{1}{n} \to 0 \quad \text{and} \quad f\left(\frac{1}{n}\right) \to 1,$$
$$2^{-n} \to 0 \quad \text{and} \quad f(2^{-n}) \to 1,$$
$$\left(\frac{3}{4}\right)^n \to 0 \quad \text{and} \quad f\left(\left(\frac{3}{4}\right)^n\right) \to 1,$$

suggesting that $f(x) \to 1$ as $x \to 0$. But of course, $f(x)$ does not have limit 1, since the limit fails to exist.

Thus, the existence of sequences for which the sequences of functional values converge is not enough to imply the existence of the limit of the function. Computational evidence for the limit of a function is *only* evidence, and is by no means a proof.

The question remains: Does convergence of sequences of functional values ever imply the existence of the functional limit? The answer is yes, and the condition is that *every* sequence $\{x_n\}$ converging to c must give rise to a sequence $\{f(x_n)\}$ that converges to L before we may conclude that $f(x) \to L$ as $x \to c$. Thus, we see that it is not possible to prove computationally that $\lim_{x\to c} f(x) = L$. To do so, we would have to test every possible sequence converging to c. Since there are infinitely many such sequences, the task is impossible.

We can *disprove* such a limit computationally, however. If we can find two sequences $\{x_n\}$ and $\{y_n\}$ converging to c such that the two sequences $\{f(x_n)\}$ and $\{f(y_n)\}$ do not have the same limit, then we know that $\lim_{x\to c} f(x)$ does not exist.

For example, with $f(x) = |x|/x$, both $\{1/n\}$ and $\{-1/n\}$ converge to 0, but $f(1/n) \to 1$ and $f(-1/n) \to -1$. Thus, $f(x)$ has no limit as $x \to 0$.

Exercises 9.1

Find the limit, if it exists.

1. $\left\{\frac{1}{2n}\right\}$
2. $\left\{\frac{2}{n}\right\}$
3. $\left\{\frac{1}{n+1}\right\}$
4. $\left\{\frac{4}{n+3}\right\}$
5. $\left\{1 + \frac{(-1)^n}{n}\right\}$
6. $\left\{1 - \frac{1}{n}\right\}$

7. $\left\{\frac{n+1}{n}\right\}$

8. $\left\{\frac{n^2+1}{n^2}\right\}$

9. $\{(-1)^n n\}$

10. $\left\{\frac{(-1)^n}{n^2}\right\}$

11. $\left\{\frac{n^2}{n+1}\right\}$

12. $\left\{\frac{n^2+1}{n+1}\right\}$

13. $\left\{\frac{\ln n}{n+1}\right\}$

14. $\left\{\frac{\sqrt{n}}{n+\ln n}\right]$

15. $\left\{\frac{\ln n}{n^2}\right\}$

16. $\left\{\frac{n^3}{e^n}\right\}$

17. $\left\{\tan\left(\frac{\ln n}{n}\right)\right\}$

18. $\{e^{-n}\sin n\}$

19. $\left\{\frac{\ln n\cos n}{n}\right\}$

20. $\left\{\frac{\sin e^n}{n}\right\}$

21. $\left\{\frac{1}{\ln(\ln n)}\right\}$

22. $\left\{\frac{\ln n}{\sqrt{n}}\right\}$

23. $\left\{\frac{\ln(\ln n)}{\sqrt{n}}\right\}$

24. $\{n^{-2/3}\ln n\}$

25. $\left\{\int_0^n e^{-x}\,dx\right\}$

26. $\left\{\int_e^n \frac{dx}{x\ln x}\right\}$

27. $\left\{\sum_{k=1}^{n} 2^{-k}\right\}$

28. $\left\{\sum_{k=1}^{n} e^{-rk}\right\}$

29. $\left\{\left(1+\frac{1}{n}\right)^n\right\}$

30. $\left\{\left(1-\frac{1}{n}\right)^n\right\}$

31. $\left\{n\sin\frac{1}{n}\right\}$

32. $\left\{n\tan\frac{1}{n}\right\}$

33. $\left\{\frac{1}{n!}\right\}$

34. $\{n!\}$

35. $\left\{\frac{3^n}{(n-1)!}\right\}$

36. $\left\{\frac{4^n}{(n+1)!}\right\}$

Using a calculator, compute a_n for $n = 10, 10^2, 10^3, 10^4, 10^5$.

37. $\left\{\frac{1}{n}\right\}$

38. $\left\{\frac{1}{n^2}\right\}$

39. $\left\{\frac{\sin n}{n}\right\}$

40. $\{e^n\}$

41. $\left\{\left(1+\frac{1}{n}\right)^n\right\}$

42. $\left\{\left(1-\frac{3}{n}\right)^n\right\}$

43. $\left\{\frac{1}{\ln n}\right\}$

44. $\{n\cos n\}$

45. Prove that if $|a_n| \to 0$, then $a_n \to 0$.

46. Find two sequences $\{a_n\}$ and $\{b_n\}$, neither of which converges, such that $\{a_n b_n\}$ converges.

47. Complete the proof of Theorem 9.2 by finding upper and lower bounds for $\{a_n\}$.

48. Prove Theorem 9.5.

49. Explain why the sequence $\{r^n\}$ diverges to ∞ if $r > 1$.

50. Explain why the geometric sequence $\{ar^{n-1}\}$ converges if $|r| < 1$.

51. Describe the bisection method (Section 2.5) in terms of sequences.

52. Describe Newton's method (Section 4.2) in terms of sequences.

53. Calculate four terms of the sequence $\{a_n\}$ for which $a_1 = 1$ and $a_{n+1} = a_n + \cos(a_n)$.

54. Draw a picture to suggest that the sequence in Problem 53 converges to $\pi/2$.

SECTION 9.2

Series of Real Numbers

It is sometimes desirable to consider the sum of infinitely many quantities. For example, we all learn in school that the decimal expansion of $\frac{1}{3}$ is 0.333. . . . Here, the three dots indicate that the digit 3 repeats *ad infinitum*. This decimal notation actually stands for an infinite sum—that is,

$$0.333\ldots = 0.3 + 0.03 + 0.003 + \cdots.$$

This sum is related to the number $\frac{1}{3}$, as will be shown below.

Infinite sums were first suggested in ancient times, notably by Zeno of Elea, who invented the following paradox: *Achilles, swiftest of runners, cannot overtake a crawling tortoise, slowest of creatures.* Suppose in a race, the tortoise is given a lead. However fast Achilles runs, it takes some time for him to reach the spot from which the tortoise started; and during that time the tortoise will have moved forward some (presumably smaller) distance. At that instant, it is as though the race were starting over again, with the contestants in new positions, but with the tortoise still ahead. Now the whole argument can be repeated—indefinitely. Thus, Achilles may get closer to the tortoise, but he can never overtake it.

The intuitive assumption that gives this argument its paradoxical quality is that the sum of infinitely many positive quantities must be infinite. As we shall see, when an infinite sum is treated with mathematical precision, it may turn out to be finite.

The basic tool for treating infinite sums with precision is a certain type of sequence.

Infinite Series

DEFINITION 9.4

Let $\{a_n\}$ be a sequence of real numbers. Let

$$s_1 = a_1,$$
$$s_2 = a_1 + a_2,$$
$$s_3 = a_1 + a_2 + a_3,$$

and, in general,

$$s_n = a_1 + a_2 + \cdots + a_n.$$

We call s_n the **nth partial sum** of $\{a_n\}$. The sequence $\{s_n\}$ of partial sums is called an **infinite series of real numbers**, and is also denoted by

$$\sum_{n=1}^{\infty} a_n = a_1 + a_2 + a_3 + \cdots.$$

The **terms** of the series are the numbers $a_1, a_2, \ldots$; the partial sums $s_1, s_2, \ldots$ are also called **partial sums** of the series.

Zeno of Elea (flourished *c.* 450 BC) was a philosopher who helped establish Athens as the Greek center of learning. He was motivated to construct his paradoxes by a philosophical dispute with Parmenides. Like the famous paradox of Achilles and the tortoise, all his arguments are based on the impossibility of a finite description of motion and the need for some notion of infinite sequence or limit to describe motion adequately. The paradoxes have had a long and vigorous life, having been debated continually through the centuries. Mathematicians generally feel that calculus and the fundamental mathematics on which it is built allows one to escape from the paradoxes, just as Achilles is able to pass the tortoise. In contrast, philosophers continue to argue about the paradoxes, a situation that is not likely to change.

Thus, for example, the decimal expansion 0.333. . . is the series

$$0.333. . . = \sum_{n=1}^{\infty} \frac{3}{10^n}.$$

In general, for any decimal expansion, the expression $0.a_1a_2a_3. . .$ actually means

$$\sum_{n=1}^{\infty} \frac{a_n}{10^n}.$$

Certain series of real numbers are given the names of the sequences from which they arise. For example, a **geometric series** has the form

$$\sum_{n=1}^{\infty} ar^{n-1} = a + ar + ar^2 + \cdots + ar^{n-1} + \cdots;$$

each term after the first is obtained by multiplying the preceding term by the number r, called the **ratio** of the series.

Another important series is the **harmonic series**,

$$\sum_{n=1}^{\infty} \frac{1}{n} = 1 + \frac{1}{2} + \frac{1}{3} + \cdots + \frac{1}{n} + \cdots.$$

Other special series will be introduced in this book as the need arises, and some of them will be given names.

■ **Example 9.12** The series

$$\sum_{n=1}^{\infty} \frac{1}{n(n+1)}$$

is an example of a **telescoping series**. Find its partial sums.

Solution The nth partial sum of this series is

$$s_n = \sum_{k=1}^{n} \frac{1}{k(k+1)}$$
$$= \frac{1}{1 \cdot 2} + \frac{1}{2 \cdot 3} + \frac{1}{3 \cdot 4} + \cdots + \frac{1}{n(n+1)}.$$

By the method of partial fractions, we can write the terms of this partial sum as

$$\frac{1}{k(k+1)} = \frac{1}{k} - \frac{1}{k+1}.$$

Therefore,

$$s_n = \left(1 - \frac{1}{2}\right) + \left(\frac{1}{2} - \frac{1}{3}\right) + \cdots + \left(\frac{1}{n} - \frac{1}{n+1}\right).$$

Dropping the parentheses and simplifying, we get

$$s_n = 1 - \frac{1}{n+1}.$$

It should now be evident where this type of series gets its name. ■

■ **Example 9.13** Find the partial sums of the geometric series

$$\sum_{n=1}^{\infty} ar^{n-1}.$$

Solution The nth partial sum of the geometric series is

$$s_n = a + ar + ar^2 + ar^3 + \cdots + ar^{n-1}.$$

Multiplying through this equation by the ratio r of the series, we get

$$rs_n = ar + ar^2 + ar^3 + ar^4 + \cdots + ar^n.$$

Now subtracting the second equation from the first, we have

$$s_n - rs_n = a - ar^n;$$

and solving for s_n, we get

$$s_n = \frac{a - ar^n}{1 - r}.$$

This, of course, assumes $r \neq 1$; if $r = 1$, then

$$s_n = a + a + a + \cdots + a = na.$$ ■

The Sum of a Series

Merely writing down an expression for an infinite sum does not give it meaning. We therefore must define what is meant by the sum of a series.

DEFINITION 9.5

Let $\Sigma_{n=1}^{\infty} a_n$ be a series. If the sequence $\{s_n\}$ of partial sums converges to s, we call s the **sum** of the series, and we write

$$\sum_{n=1}^{\infty} a_n = s.$$

If $\{s_n\}$ diverges, we say the series **diverges** or **has no sum.**

Example 9.14

Find the sum of the telescoping series of Example 9.12:

$$\sum_{n=1}^{\infty} \frac{1}{n(n+1)}.$$

Solution

By Example 9.12, the partial sums are

$$s_n = 1 - \frac{1}{n+1},$$

so $s_n \to 1$, and the sum of the series is 1; that is,

$$\sum_{n=1}^{\infty} \frac{1}{n(n+1)} = 1.$$

■

Example 9.15

Find the sum of the geometric series below, given that $|r| < 1$:

$$\sum_{n=1}^{\infty} ar^{n-1}.$$

Solution

By Example 9.13, the partial sums of this series are

$$s_n = \frac{a - ar^n}{1 - r} = \frac{a}{1-r} - \left(\frac{a}{1-r}\right) r^n.$$

Now if $|r| < 1$, then $1/|r| > 1$; and hence, $(1/|r|)^n$ is an increasing unbounded sequence. Therefore, $|r^n| \to 0$ by Theorem 9.4, and $r^n \to 0$ by Problem 45 of Section 9.1. It follows that $s_n \to a/(1-r)$, and thus, the geometric series with ratio r converges if $|r| < 1$. Its sum is the first term divided by $1 - r$; that is,

$$\sum_{n=1}^{\infty} ar^{n-1} = \frac{a}{1-r}.$$

■

Example 9.15 demonstrated the first part of the following theorem; the second part will be proved presently (see Example 9.19).

THEOREM 9.7

The geometric series

$$\sum_{n=1}^{\infty} ar^{n-1},$$

with first term a and ratio r, has sum

$$\frac{a}{1-r}$$

if $|r| < 1$, and diverges otherwise.

The following examples are some applications of geometric series.

Example 9.16

Suppose Achilles can run 100 times as fast as the tortoise. If the tortoise starts out 1000 meters ahead of Achilles, how far does Achilles run before he overtakes the tortoise?

Solution

By the time Achilles has run 1000 meters, the tortoise has gone 10 meters. By the time Achilles has run the additional 10 meters, the tortoise has gone an additional $\frac{1}{10}$ meter. By the time Achilles has run this additional $\frac{1}{10}$ meter, the tortoise has gone another $\frac{1}{1000}$ meter; and so forth. The total distance Achilles runs is therefore

$$\begin{aligned} 1000 + 10 + \frac{1}{10} + \frac{1}{1000} + \cdots &= 1000 + \frac{1000}{100} + \frac{1000}{(100)^2} + \frac{1000}{(100)^3} + \cdots \\ &= \sum_{n=1}^{\infty} \frac{1000}{(100)^{n-1}} \\ &= \sum_{n=1}^{\infty} 1000\left(\frac{1}{100}\right)^{n-1} \text{ meters.} \end{aligned}$$

This is a geometric series with first term 1000 and ratio $\frac{1}{100}$; its sum is

$$\frac{1000}{1-\frac{1}{100}} = \frac{100{,}000}{99}.$$

Thus, Achilles runs $100{,}000/99 = 1010\frac{10}{99}$ meters before overtaking the tortoise. ■

Example 9.17

Write the following decimal expansions as fractions:

(a) 0.333. . . **(b)** 0.151515. . .

Solution We rewrite each decimal as a geometric series and find the sum:

$$\text{(a)}\quad 0.333\ldots = \sum_{n=1}^{\infty} \frac{3}{10^n} = \sum_{n=1}^{\infty} \frac{3}{10}\left(\frac{1}{10}\right)^{n-1} = \frac{\frac{3}{10}}{1-\frac{1}{10}} = \frac{1}{3}.$$

$$\text{(b)}\quad 0.151515\ldots = \sum_{n=1}^{\infty} \frac{15}{(100)^n} = \frac{\frac{15}{100}}{1-\frac{1}{100}} = \frac{15}{99} = \frac{5}{33}.$$

■

■ **Example 9.18** A certain ball bounces half as high as it falls. If the ball is dropped from a height of 16 feet, how far does it travel before coming to rest?

Solution The ball falls the initial 16 feet and bounces 8 feet back up, so to the top of the first bounce the ball travels 24 feet. Then it falls 8 feet and bounces 4 feet, so it travels another 12 feet to the top of the second bounce. To the top of the third bounce, it travels another $4 + 2 = 6$ feet, and so on. Thus, the ball travels

$$24 + 12 + 6 + \cdots \text{ feet.}$$

This sum is an infinite geometric series with ratio $\frac{1}{2}$ and first term 24, so its sum is

$$\frac{24}{1-\frac{1}{2}} = 48.$$

Therefore, the ball travels 48 feet before coming to rest. ■

The Divergence Test

It is not always an easy task to determine whether a given series converges; and finding the sum when it exists is even harder. The next result is the first step toward determining whether a series converges.

By the definition of partial sum, it is clear that consecutive partial sums of a series differ by a single term of the series; specifically,

$$s_n - s_{n-1} = a_n.$$

If the series converges to a sum s, then both $s_n \to s$ and $s_{n-1} \to s$; hence, $a_n \to s - s = 0$. Thus, if a series $\sum_{n=1}^{\infty} a_n$ converges, then the sequence $\{a_n\}$ must converge to 0. We therefore have the **divergence test**.

THEOREM 9.8 **The Divergence Test** If the sequence $\{a_n\}$ does not converge to 0, then the series $\sum_{n=1}^{\infty} a_n$ diverges.

■ **Example 9.19** Show that the geometric series below diverges if $|r| \geq 1$:

$$\sum_{n=1}^{\infty} ar^{n-1}.$$

Solution If $|r| \geq 1$, then every term ar^{n-1} is at least as big as $|a|$ in absolute value. It is evident, then, that the sequence of terms $\{ar^{n-1}\}$ cannot converge to 0, so the series diverges by the divergence test. (This is the proof of the second part of Theorem 9.7.) ■

■ **Example 9.20** Show that the series below diverges:

$$\sum_{n=1}^{\infty}\left(1+\frac{1}{n}\right).$$

Solution The sequence $\{1 + 1/n\}$ converges to 1, not to 0, so the series diverges. ■

Divergence of the Harmonic Series

The next result allows us to prove that the harmonic series $\Sigma_{n=1}^{\infty}(1/n)$ diverges.

THEOREM 9.9 **The Cauchy Criterion** If the series $\Sigma_{n=1}^{\infty} a_n$ converges and ε is any positive real number, then there exists a positive integer k such that $|s_m - s_n| < \varepsilon$ whenever $m, n > k$.

Proof If the series $\Sigma_{n=1}^{\infty} a_n$ has sum s, then the sequence $\{s_n\}$ converges to s. Therefore, given $\varepsilon > 0$, there exists a positive integer k such that $|s_n - s| < \varepsilon/2$ if $n > k$. Hence, if both m and n are greater than k,

$$\begin{aligned}|s_m - s_n| &= |(s_m - s) + (s - s_n)|\\ &\leq |s_m - s| + |s_n - s|\\ &< \frac{\varepsilon}{2} + \frac{\varepsilon}{2} = \varepsilon.\end{aligned}$$

■ **Example 9.21** Show that the harmonic series $\Sigma_{n=1}^{\infty}(1/n)$ diverges.

Solution We will consider the partial sums s_n and s_{2n}; we have

$$\begin{aligned}s_{2n} - s_n &= \left(1 + \frac{1}{2} + \cdots + \frac{1}{n} + \frac{1}{n+1} + \cdots + \frac{1}{2n}\right) - \left(1 + \frac{1}{2} + \cdots + \frac{1}{n}\right)\\ &= \frac{1}{n+1} + \frac{1}{n+2} + \cdots + \frac{1}{2n}\\ &> \frac{1}{2n} + \frac{1}{2n} + \cdots + \frac{1}{2n} = \frac{1}{2}.\end{aligned}$$

Thus, if $\varepsilon < \frac{1}{2}$, we cannot satisfy the Cauchy criterion for s_n and s_{2n}. Therefore, the harmonic series must diverge. ■

Example 9.21 shows that the converse of the divergence test is false. That is, the harmonic sequence $\{1/n\}$ converges to 0, and the harmonic series $\Sigma_{n=1}^{\infty}(1/n)$ diverges. The fact that a sequence $\{a_n\}$ does converge to 0 is not enough to imply that the series $\Sigma_{n=1}^{\infty} a_n$ converges.

Combinations of Series

As with sequences, a certain amount of arithmetic is possible with series and their sums. The following theorems can be used to help determine whether a series converges and even to find sums on occasion.

THEOREM 9.10 If the series $\Sigma_{n=1}^{\infty} a_n$ and $\Sigma_{n=1}^{\infty} b_n$ are both convergent, say $\Sigma_{n=1}^{\infty} a_n = s$ and $\Sigma_{n=1}^{\infty} b_n = t$, then the series $\Sigma_{n=1}^{\infty}(a_n + b_n)$ is convergent and $\Sigma_{n=1}^{\infty}(a_n + b_n) = s + t$. If exactly one of $\Sigma_{n=1}^{\infty} a_n$ and $\Sigma_{n=1}^{\infty} b_n$ diverges, then $\Sigma_{n=1}^{\infty}(a_n + b_n)$ diverges.

The proof of this theorem is based on the limit properties of the sequences of partial sums, and is left as an exercise (see Problem 42).

Example 9.22 Evaluate

$$\sum_{n=1}^{\infty} \frac{3^n + 4^n}{12^n}.$$

Solution

$$\sum_{n=1}^{\infty} \frac{3^n + 4^n}{12^n} = \sum_{n=1}^{\infty} \left(\frac{1}{4^n} + \frac{1}{3^n}\right) = \frac{\frac{1}{4}}{1-\frac{1}{4}} + \frac{\frac{1}{3}}{1-\frac{1}{3}} = \frac{5}{6}.$$ ■

Example 9.23 Determine whether the series below converges:

$$\sum_{n=1}^{\infty} \frac{n + 2^n}{n2^n}.$$

Solution Since

$$\sum_{n=1}^{\infty} \frac{n + 2^n}{n2^n} = \sum_{n=1}^{\infty} \left(\frac{1}{n} + \frac{1}{2^n}\right)$$

and $\Sigma_{n=1}^{\infty}(1/n)$ diverges while $\Sigma_{n=1}^{\infty}(1/2^n)$ converges, the given series diverges. ■

THEOREM 9.11

If the series $\sum_{n=1}^{\infty} a_n$ converges to s and c is a constant, then the series $\sum_{n=1}^{\infty} ca_n$ converges to cs. If $\sum_{n=1}^{\infty} a_n$ diverges, then $\sum_{n=1}^{\infty} ca_n$ diverges if $c \neq 0$.

The proof is again omitted.

Example 9.24

Determine whether $\sum_{n=1}^{\infty}(1/2n)$ converges.

Solution

Since

$$\sum_{n=1}^{\infty} \frac{1}{2n} = \frac{1}{2} \sum_{n=1}^{\infty} \frac{1}{n}$$

and the harmonic series diverges, the given series diverges. ∎

THEOREM 9.12

If $a_n = b_n$ for all $n > k$, then the series $\sum_{n=1}^{\infty} a_n$ and $\sum_{n=1}^{\infty} b_n$ either both converge or both diverge.

Proof

Let s_n and t_n be the nth partial sums of $\sum_{n=1}^{\infty} a_n$ and $\sum_{n=1}^{\infty} b_n$, respectively. For $n > k$, we have

$$s_n = a_1 + \cdots + a_k + a_{k+1} + \cdots + a_n$$

and

$$t_n = b_1 + \cdots + b_k + b_{k+1} + \cdots + b_n,$$

so that

$$\begin{aligned} s_n - t_n &= (a_1 + \cdots + a_k) - (b_1 + \cdots + b_k) \\ &= s_k - t_k. \end{aligned}$$

Thus,

$$s_n = s_k - t_k + t_n$$

and

$$\lim_{n \to \infty} s_n = s_k - t_k + \lim_{n \to \infty} t_n.$$

That is, the limits of the partial sums either both exist or both fail to exist, and the theorem follows.

Theorem 9.12 states that if two series differ in only finitely many terms, they either both converge or both diverge. Stated another way, it says that a finite number of terms of a series may be changed without affecting its convergence. The sum, of course, would be affected.

Thus, the convergence of a series is not determined by any finite number of terms; convergence is determined by the infinite "tail end" of the series. This is picturesquely stated, "with infinite series, the tail wags the dog."

■ Example 9.25 Determine whether each series converges or diverges:

(a) $\sum_{n=1}^{\infty} \frac{1}{n+4}$ (b) $\sum_{n=1}^{\infty} a_n$, with $a_n = \begin{cases} \frac{1}{n} & \text{if } n < 1000 \\ \frac{1}{2^n} & \text{if } n \geq 1000 \end{cases}$

Solution

(a) This series is $\frac{1}{5} + \frac{1}{6} + \frac{1}{7} + \cdots$, which is the same as $0 + 0 + 0 + 0 + \frac{1}{5} + \frac{1}{6} + \frac{1}{7} + \cdots$. This is the same as the harmonic series for $n > 4$, so it diverges. The given series also could be written as $\sum_{n=5}^{\infty}(1/n)$.

(b) Even though this series is the same as the harmonic series for the first 1000 terms, it is geometric with ratio $\frac{1}{2}$ from that point on. Thus, the series converges. ■

Exercises 9.2

Find the sum of each series.

1. $\sum_{n=1}^{\infty} \frac{2}{10^n}$
2. $\sum_{n=1}^{\infty} \frac{1}{3e^n}$
3. $\sum_{n=0}^{\infty} \frac{(-1)^n}{2^n}$
4. $\sum_{n=0}^{\infty} \frac{(-1)^n}{3^n}$
5. $\sum_{n=1}^{\infty} 2^{1-2n}$
6. $\sum_{n=1}^{\infty} e^{3-2n}$
7. $\sum_{n=0}^{\infty} \frac{100}{2^n}$
8. $\sum_{n=0}^{\infty} \frac{1000}{7^n}$
9. $\sum_{n=1}^{\infty} 5 \cdot 2^n 3^{-n}$
10. $\sum_{n=1}^{\infty} (-1)^n 4^n 5^{-n}$
11. $\sum_{n=0}^{\infty} \frac{2^n + 3^n}{6^n}$
12. $\sum_{n=0}^{\infty} \frac{5^n - 2^n}{10^n}$
13. $\sum_{n=1}^{\infty} \frac{2}{n(n+1)}$
14. $\sum_{n=1}^{\infty} \frac{2}{n(n+2)}$
15. $\sum_{n=1}^{\infty} \frac{2}{(2n+1)(2n+3)}$
16. $\sum_{n=1}^{\infty} \frac{6}{n(n+2)}$

Determine whether each series converges or diverges.

17. $\sum_{n=1}^{\infty} \frac{n^2+1}{n^2}$
18. $\sum_{n=1}^{\infty} \frac{1-n^2}{2-3n^2}$
19. $\sum_{n=0}^{\infty} (1-e^{-n})^2$
20. $\sum_{n=0}^{\infty} \cos n$
21. $\sum_{n=2}^{\infty} \ln n$
22. $\sum_{n=1}^{\infty} \frac{1}{2+\ln n}$
23. $\sum_{n=1}^{\infty} \left(1+\frac{1}{n}\right)^n$
24. $\sum_{n=1}^{\infty} \left(\frac{4}{3}\right)^n$
25. $\sum_{n=1}^{\infty} (2^{-n}+3^{-n})$
26. $\sum_{n=1}^{\infty} (2^{-n}+n^{-1})$
27. $\sum_{n=1}^{\infty} \frac{4}{n}$
28. $\sum_{n=5}^{\infty} e^{-n}$
29. $\sum_{n=50}^{\infty} \frac{4}{n}$
30. $\sum_{n=900}^{\infty} \frac{1}{n(n+1)}$

Write each decimal expansion as a fraction.

31. 0.141414. . .

32. 0.027027027. . .

33. 1.176176176. . .

34. 0.571428571428571428. . .

35. A ball bounces $\frac{4}{5}$ of the height from which it falls. If the ball is dropped from a height of 20 centimeters, how far will it travel before coming to rest?

36. Suppose Achilles can run 20 times as fast as a tortoise and that the tortoise has a lead of 50 paces at the beginning of the race. How far will Achilles run in order to overtake the tortoise?

37. At what time after 3 o'clock does the minute hand overtake the hour hand? [Think of the minute hand as Achilles and the hour hand as the tortoise in Example 9.16.]

38. Suppose that any bank in the Federal Reserve System can reloan up to 80% of the money it borrows, thus increasing the money supply. By what factor is the money supply increased, potentially?

C**39.** For the harmonic series, use a computer to compute the partial sums s_{10}, s_{50}, s_{100}, and s_{200}.

C**40.** For the geometric series $\Sigma_{n=1}^{\infty}(\frac{4}{5})^n$, use a computer to compute the partial sums s_{10}, s_{20}, s_{50}, and s_{100}.

41. State the converse of Theorem 9.9.

42. Prove Theorem 9.10.

43. Prove that if $\Sigma_{n=1}^{\infty} a_n$ and $\Sigma_{n=1}^{\infty} b_n$ converge and have sums s and t, then $\Sigma_{n=1}^{\infty}(a_n - b_n)$ converges and has sum $s - t$.

44. Pat and Jan are walking toward each other along a straight road, each walking at the rate of 5 miles per hour. When they are 1 mile apart, Pat's dog runs at the rate of 10 miles per hour from Pat to Jan, then back to Pat, then to Jan again, and continues to run from one to another until Pat and Jan meet.

(a) Write the distance the dog runs as an infinite series, and find its sum.

(b) Find out how far the dog runs without using a series. [*Hint:* Find out how much time the dog spends running. The story is told that when the mathematician John von Neumann (1903–1957) was posed a similar problem, he answered almost immediately, doing the figuring in his head. When someone assumed that he had solved the problem the simple way because of the quickness of his answer, he explained that he had actually summed a series.]

45. Find the fallacy in the following argument: Let

$$s = 1 + 2 + 4 + 8 + \cdots .$$

Then

$$\begin{aligned} 2s &= 2 + 4 + 8 + 16 + \cdots \\ &= s - 1, \end{aligned}$$

so that

$$s = -1.$$

46. Under normal economic conditions, a nation spends about 90% of its income. That means that for \$100 income, about \$90 is spent first-hand; and then 90% of that \$90, or

$81, is spent second-hand; and so on. Thus, the $100 of income is actually worth a great deal more to the economy. The effect of this process of spending and respending is called the **multiplier effect**. If the rate of respending is 90%, what is the actual worth of $100 to the economy?

47. Show that for $|r| < 1$,

$$\sum_{n=0}^{\infty} r^n = \frac{1}{1-r}.$$

48. Show that for $|r| < 1$,

$$\sum_{n=0}^{\infty} (-1)^n r^n = \frac{1}{1+r}.$$

SECTION 9.3

Series of Positive Terms

A series $\sum_{n=1}^{\infty} a_n$ in which each term a_n is positive is called a **series of positive terms**. In this section, we will discuss several tests for convergence that apply to series of positive terms.

If $\sum_{n=1}^{\infty} a_n$ is a series of positive terms, then the sequence $\{s_n\}$ of partial sums is increasing. Thus, we have Theorem 9.13.

THEOREM 9.13 A series of positive terms converges if and only if the sequence of partial sums is bounded.

This theorem follows directly from Theorems 9.1 and 9.2. This fact is helpful in establishing a very useful set of tests for convergence.

The Integral Test

The first test relates a series to an improper integral.

THEOREM 9.14 **The Integral Test** If f is a positive, continuous, decreasing function on $[1, \infty)$ and $a_n = f(n)$, then the series $\sum_{n=1}^{\infty} a_n$ converges if and only if the improper integral $\int_1^{\infty} f(x)\, dx$ converges.

Proof First, suppose that the improper integral $\int_1^{\infty} f(x)\, dx$ converges. Let $\{s_n\}$ be the sequence of partial sums of $\sum_{n=1}^{\infty} a_n$. For each n, we may write

$$s_n = a_1 + \sum_{k=2}^{n} a_k.$$

The sum $\sum_{k=2}^{n} a_k = \sum_{k=2}^{n} f(k)$ is the area of the shaded rectangles lying under the curve $y = f(x)$ in Figure 9.4 (page 594). This area is less than the area under the curve on $[1, n]$, so $\{s_n\}$ must be bounded, and the series converges by Theorem 9.13.

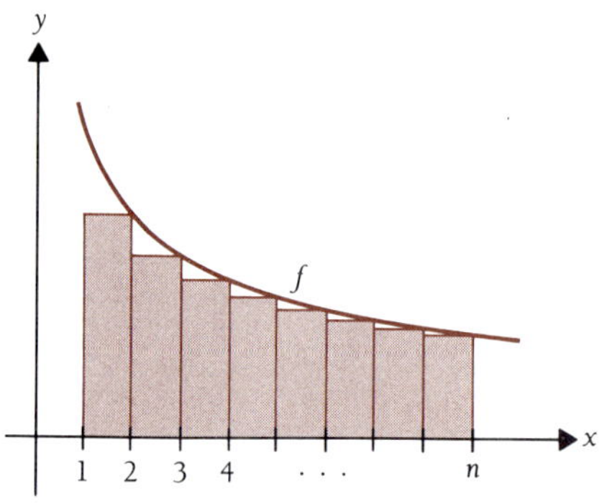

FIGURE 9.4 Rectangles enclosed in the region under f

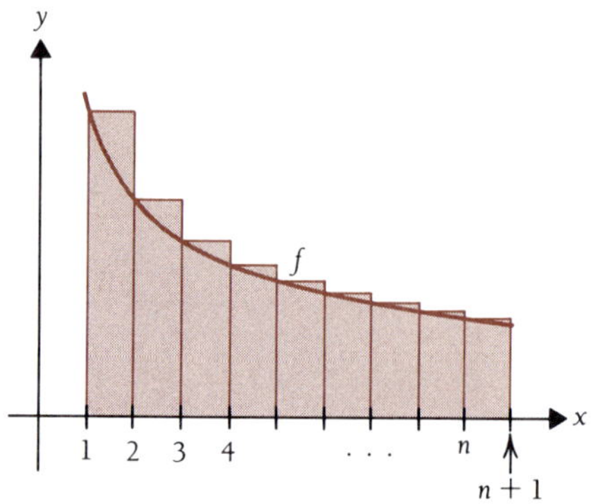

FIGURE 9.5 Rectangles enclosing the region under f

Conversely, suppose that the series $\sum_{n=1}^{\infty} a_n$ converges to s. Each partial sum $s_n = \sum_{k=1}^{n} a_k = \sum_{k=1}^{n} f(k)$ can be interpreted as the area of the shaded rectangles in Figure 9.5, which is greater than the area under the curve $y = f(x)$ over $[1, n + 1]$. Hence,

$$\int_1^{n+1} f(x)\,dx \leq s_n,$$

so that

$$\int_1^{\infty} f(x)\,dx = \lim_{n\to\infty} \int_1^{n+1} f(x)\,dx$$
$$\leq \lim_{n\to\infty} s_n = s,$$

so the integral converges.

■ **Example 9.26** Use the integral test (Theorem 9.14) to determine whether each of the following series converges:

(a) $\displaystyle\sum_{n=1}^{\infty} \frac{1}{n^2}$ (b) $\displaystyle\sum_{n=1}^{\infty} \frac{1}{n^2 + 1}$ (c) $\displaystyle\sum_{n=2}^{\infty} \frac{1}{n \ln n}$

Solution (a) Since the integral

$$\int_1^{\infty} \frac{dx}{x^2} = \lim_{t\to\infty} \int_1^{t} x^{-2}\,dx$$
$$= \lim_{t\to\infty} \left(1 - \frac{1}{t}\right) = 1$$

converges, the series $\sum_{n=1}^{\infty}(1/n^2)$ converges.

(b) Since the integral

$$\int_1^\infty \frac{dx}{x^2+1} = \lim_{t\to\infty} \int_1^t \frac{dx}{x^2+1}$$
$$= \lim_{t\to\infty} (\arctan t - \arctan 1) = \frac{\pi}{4}$$

converges, the series $\sum_{n=1}^{\infty}[1/(n^2+1)]$ converges.

(c) Since the integral

$$\int_2^\infty \frac{dx}{x \ln x} = \lim_{t\to\infty} \int_2^t \frac{dx}{x \ln x}$$
$$= \lim_{t\to\infty}[\ln(\ln t) - \ln(\ln 2)]$$

diverges, the series also diverges. ■

We should note here that in the case of a convergent series, the sum of the series and the value of the improper integral in Theorem 9.14 are not the same. (You might want to try to determine how they are related, if at all.)

A special type of positive-term series that we will find useful is the ***p*-series**, a series of the form

$$\sum_{n=1}^{\infty} \frac{1}{n^p} = 1 + \frac{1}{2^p} + \frac{1}{3^p} + \cdots.$$

For example, the series

$$\sum_{n=1}^{\infty} \frac{1}{\sqrt{n}} = \frac{1}{\sqrt{1}} + \frac{1}{\sqrt{2}} + \frac{1}{\sqrt{3}} + \cdots$$

is a p-series with $p = \frac{1}{2}$.

■ **Example 9.27** Show that a p-series converges if $p > 1$ and diverges if $p \le 1$.

Solution If $p = 1$, the p-series is just the harmonic series, which diverges. If $p \ne 1$, we use the integral test:

$$\int_1^\infty \frac{1}{x^p}\,dx = \lim_{t\to\infty} \int_1^t x^{-p}\,dx$$
$$= \lim_{t\to\infty} (1-p)^{-1}(t^{1-p} - 1).$$

Now if $p < 1$, then $1 - p > 0$, so the limit above does not exist and the series diverges. If $p > 1$, then the limit exists and the series converges. ■

Comparison Tests

Let $\sum_{n=1}^{\infty} a_n$ and $\sum_{n=1}^{\infty} b_n$ be series of positive terms. Series $\sum_{n=1}^{\infty} b_n$ is said to **dominate** the series $\sum_{n=1}^{\infty} a_n$ if and only if there is a positive integer k such that $a_n \le b_n$ for all $n > k$. That is, the terms of the dominating series are at least as large as the corresponding terms of the dominated series, beyond a certain point.

THEOREM 9.15

The Comparison Test Let $\sum_{n=1}^{\infty} a_n$ and $\sum_{n=1}^{\infty} b_n$ be series of positive terms such that $\sum_{n=1}^{\infty} b_n$ dominates $\sum_{n=1}^{\infty} a_n$. Then:

1. If $\sum_{n=1}^{\infty} b_n$ converges, then $\sum_{n=1}^{\infty} a_n$ converges.
2. If $\sum_{n=1}^{\infty} a_n$ diverges, then $\sum_{n=1}^{\infty} b_n$ diverges.

Proof

Let $s_n = \sum_{k=1}^{n} a_k$ and $t_n = \sum_{k=1}^{n} b_k$ be the partial sums of the series. Because the series have positive terms, both $\{s_n\}$ and $\{t_n\}$ are increasing sequences. Because $\sum_{n=1}^{\infty} b_n$ dominates $\sum_{n=1}^{\infty} a_n$, we have $a_n \le b_n$ for all $n > k$, so that for $n > k$, $s_n - s_k \le t_n - t_k$.

Now if $\sum_{n=1}^{\infty} b_n$ has sum t, then $t_n < t$ for all n. Hence,

$$s_n - s_k < t - t_k,$$

or

$$s_n < t - t_k + s_k < t + s_k$$

for all n. Thus, $\{s_n\}$ is bounded, and $\sum_{n=1}^{\infty} a_n$ converges. The contrapositive of this statement is also true, so the theorem is proved.

Following are some simple applications of the comparison test.

Example 9.28

Use the comparison test (Theorem 9.15) to determine whether each of the following series converges:

(a) $\displaystyle\sum_{n=1}^{\infty} \frac{1}{n^2+n}$ (b) $\displaystyle\sum_{n=0}^{\infty} \frac{1+3^n}{2^n}$ (c) $\displaystyle\sum_{n=2}^{\infty} \frac{1}{\ln n}$ (d) $\displaystyle\sum_{n=1}^{\infty} \frac{1}{2n+1}$

Solution

(a) The given series is dominated by the convergent p-series $\sum_{n=1}^{\infty}(1/n^2)$, so it converges.
(b) The given series dominates the divergent geometric series $\sum_{n=0}^{\infty}(\frac{3}{2})^n$, so it diverges.
(c) Since $\ln n < n$, we have $1/(\ln n) > 1/n$. The given series therefore dominates the harmonic series, so it diverges.
(d) Since $2n + 1 \le 3n$, it follows that $1/(2n + 1) \ge 1/3n$. Hence, the given series dominates the series $\sum_{n=1}^{\infty}(1/3n)$, which diverges, so it diverges. ■

In order to apply the comparison test, we must first have some idea about the convergence of the given series. If we believe it converges, then we try to find a convergent series that dominates it. If we believe it diverges, we look for a divergent series that it dominates. The series most often used for comparison are the geometric series and the p-series, including the harmonic series. But, of course, any series of positive terms may be used for comparison.

Theorem 9.16 is a refinement of the comparison test.

THEOREM 9.16 **The Limit Comparison Test** Let $\sum_{n=1}^{\infty} a_n$ and $\sum_{n=1}^{\infty} b_n$ be series of positive terms, and let $\lim_{n\to\infty}(a_n/b_n) = L$. If L is a positive real number, then the two series either both converge or both diverge.

Proof If $L > 0$, let $\varepsilon > 0$ be chosen so that $L - \varepsilon > 0$ (for example, $\varepsilon = L/4$). Then there exists a positive integer k such that $|a_n/b_n - L| < \varepsilon$ for all $n > k$. That is,

$$L - \varepsilon < \frac{a_n}{b_n} < L + \varepsilon,$$

$$(L - \varepsilon)b_n < a_n < (L + \varepsilon)b_n.$$

Now we can use the comparison test (Theorem 9.15). If $\sum_{n=1}^{\infty} a_n$ converges, then $\sum_{n=1}^{\infty}(L - \varepsilon)b_n$ also converges, so $\sum_{n=1}^{\infty} b_n$ converges by Theorem 9.10. If $\sum_{n=1}^{\infty} a_n$ diverges, then $\sum_{n=1}^{\infty}(L + \varepsilon)b_n$ also diverges, and $\sum_{n=1}^{\infty} b_n$ diverges.

The idea of the limit comparison test is that two series whose terms are of the same order of magnitude have the same behavior. Here are some examples.

■ **Example 9.29** Use the limit comparison test (Theorem 9.16) to determine whether each of the following series is convergent:

(a) $\displaystyle\sum_{n=1}^{\infty} \frac{\sqrt{n+1}}{n\sqrt{n}}$ (b) $\displaystyle\sum_{n=1}^{\infty} \frac{3}{2^n - 2n}$ (c) $\displaystyle\sum_{n=1}^{\infty} \sin\frac{2\pi}{n}$

(d) $\displaystyle\sum_{n=1}^{\infty} \frac{\sqrt{n}}{4n+9}$ (e) $\displaystyle\sum_{n=1}^{\infty} \frac{\sqrt{n}}{4n^2+9}$

Solution

(a) The "net degree" (the degree of the numerator minus the degree of the denominator) of each term of this series is about -1, so the series ought to behave like the harmonic series. Using the limit comparison test with the harmonic series, we get

$$\frac{\sqrt{n+1}\,/n\sqrt{n}}{1/n} = \frac{\sqrt{n+1}}{\sqrt{n}} \to 1,$$

and since the harmonic series diverges, so does this series.

(b) The portion of each term that predominates is the expression 2^n in the denominator (that is, if n is large, all other parts of each term are negligibly small). Therefore, the series is probably like the geometric series $\sum_{n=1}^{\infty}(\frac{1}{2})^n$, so we compare with that series:

$$\frac{3/(2^n - 2n)}{1/2^n} = \frac{3}{1 - (2n/2^n)} \to 3.$$

Since the geometric series converges, the given series also converges.

(c) For n large, $2\pi/n$ is small, and for small x, $\sin x$ behaves like x. Therefore, we compare this series with the harmonic series:

$$\frac{\sin(2\pi/n)}{1/n} = 2\pi\left[\frac{\sin(2\pi/n)}{2\pi/n}\right] \to 2\pi. \qquad \text{(By Theorem 2.15)}$$

Hence, the given series diverges.

(d) The terms of this series have "net degree" about $-\frac{1}{2}$, so we compare with the p-series in which $p = \frac{1}{2}$:

$$\frac{\sqrt{n}/(4n+9)}{1/n^{1/2}} = \frac{n}{4n+9} \to \frac{1}{4}.$$

Thus, since the p-series diverges if $p < 1$, the given series diverges.

(e) Here, the net degree is about $-\frac{3}{2}$, so we compare with the p-series in which $p = \frac{3}{2}$:

$$\frac{\sqrt{n}/(4n^2+9)}{1/n^{3/2}} = \frac{n^2}{4n^2+9} \to \frac{1}{4}.$$

Hence, the given series converges. ■

Exercises 9.3

Test each series for convergence. Give reasons for your conclusions.

1. $\sum_{n=1}^{\infty} \frac{100}{n^2}$
2. $\sum_{n=1}^{\infty} \frac{e}{n^3}$
3. $\sum_{n=1}^{\infty} \frac{5n}{n^2+5}$
4. $\sum_{n=1}^{\infty} \frac{5n}{n^3+5}$
5. $\sum_{n=2}^{\infty} \frac{1}{\sqrt{n}\ln n}$
6. $\sum_{n=2}^{\infty} \frac{\sqrt{n}}{\ln n}$
7. $\sum_{n=2}^{\infty} \frac{1}{n^2 \ln n}$
8. $\sum_{n=2}^{\infty} \frac{\ln n}{n}$
9. $\sum_{n=2}^{\infty} \frac{\ln n}{n\sqrt{n}}$
10. $\sum_{n=2}^{\infty} \frac{\ln n}{n^2}$
11. $\sum_{n=6}^{\infty} \frac{n+5}{n^2-5}$
12. $\sum_{n=2}^{\infty} \frac{n-5}{n^3-5}$
13. $\sum_{n=1}^{\infty} \frac{2n+3}{n}$
14. $\sum_{n=1}^{\infty} \frac{e^n}{n}$
15. $\sum_{n=1}^{\infty} \frac{\cos^2 n}{n^2}$
16. $\sum_{n=1}^{\infty} \frac{1}{\sin n}$
17. $\sum_{n=1}^{\infty} \frac{1}{n \tan n}$
18. $\sum_{n=1}^{\infty} \sin^2 \pi n$
19. $\sum_{n=1}^{\infty} \frac{e^n+1}{ne^n}$
20. $\sum_{n=2}^{\infty} \frac{\ln n}{n^{1+\varepsilon}}, \quad \varepsilon > 0$
21. Let function f be positive on $[1, \infty)$. Prove that the sequence $\{\int_1^n f(x)\,dx\}$ is bounded if and only if the improper integral $\int_1^\infty f(x)\,dx$ converges.

22. Let $\Sigma_{n=1}^{\infty} a_n$ and $\Sigma_{n=1}^{\infty} b_n$ be series of positive terms.
   (a) If $\lim_{n\to\infty}(a_n/b_n) = 0$ and $\Sigma_{n=1}^{\infty} b_n$ converges, prove that $\Sigma_{n=1}^{\infty} a_n$ converges.
   (b) If $\lim_{n\to\infty}(a_n/b_n) = \infty$ and $\Sigma_{n=1}^{\infty} b_n$ diverges, prove that $\Sigma_{n=1}^{\infty} a_n$ diverges.

23. Is the value of $\Sigma_{n=1}^{\infty} (\frac{1}{2})^n$ related to the value of $\int_1^{\infty} (\frac{1}{2})^x\, dx$? How?

SECTION 9.4

Series with Positive and Negative Terms

We now return to a discussion of infinite series whose terms may not all be positive. Some of this discussion will also apply to series of positive terms. We first consider series whose terms are alternately positive and negative.

Alternating Series

DEFINITION 9.6 An **alternating series** is an infinite series of real numbers whose terms are alternately positive and negative.

For example, if $\Sigma_{n=0}^{\infty} a_n$ is a series of positive terms, then $\Sigma_{n=0}^{\infty}(-1)^n a_n$ is an alternating series. (It is convenient to start the summation at $n = 0$, because that makes the first partial sum positive, $s_0 = a_0$.)

A particularly simple test for the convergence of an alternating series is the following result, due to Leibniz.

THEOREM 9.17 **The Alternating Series Test** If $\{a_n\}$ is a decreasing sequence of positive terms that converges to 0, then the alternating series $\Sigma_{n=0}^{\infty}(-1)^n a_n$ converges.

Proof Consider an even-numbered partial sum,

$$s_{2n} = (a_0 - a_1) + (a_2 - a_3) + \cdots + (a_{2n-2} - a_{2n-1}) + a_{2n}.$$

Note that s_{2n} is expressed as a sum of positive terms, so s_{2n} is positive. Also, the next even-numbered partial sum is

$$\begin{aligned} s_{2n+2} &= s_{2n} - a_{2n+1} + a_{2n+2} \\ &= s_{2n} - (a_{2n+1} - a_{2n+2}) \\ &< s_{2n}. \end{aligned}$$

Thus, the sequence $\{s_{2n}\}$ of even-numbered partial sums is decreasing and bounded below by 0, so $\{s_{2n}\}$ has a limit s, or $s_{2n} \to s$.

Now consider the odd-numbered partial sum

$$s_{2n+1} = s_{2n} - a_{2n+1}.$$

Since $a_{2n+1} \to 0$ and $s_{2n} \to s$ as $n \to \infty$, we must have $s_{2n+1} \to s$.

Clearly, then, for n even or odd, $s_n \to s$, and the series converges.

To apply the alternating series test, we must determine whether the absolute values of the terms of the given alternating series form a decreasing sequence that converges to 0.

Example 9.30 Show that the alternating harmonic series $\sum_{n=1}^{\infty}[(-1)^{n+1}/n]$ converges.

Solution The harmonic sequence decreases to 0, so the alternating harmonic series converges by the alternating series test (Theorem 9.17). ■

Example 9.31 Determine whether the series $\sum_{n=2}^{\infty}[(-1)^{n}/(\ln n)]$ converges.

Solution The sequence $\{1/(\ln n)\}$ decreases to 0, so the alternating series converges. ■

Example 9.32 Determine whether the series below converges:

$$\sum_{n=0}^{\infty} \frac{n+1}{2n+1}(-1)^n.$$

Solution The sequence $\{(n+1)/(2n+1)\}$ is decreasing but does not converge to 0. Therefore, the alternating series test does not apply, but the divergence test (Theorem 9.8) applies, indicating that the series diverges. ■

A result that is important for approximations is stated in Theorem 9.18.

THEOREM 9.18 If $\{a_n\}$ is a decreasing sequence of positive terms converging to 0, then the partial sum s_n of the alternating series $\sum_{n=0}^{\infty}(-1)^n a_n$ differs from the sum s of the series by less than the first unused term, a_{n+1}, or

$$|s_n - s| < a_{n+1}.$$

Proof The sum s exists by the alternating series test (Theorem 9.17). Because the sequence of even-numbered partial sums is decreasing (see the proof of the alternating series test), it follows that s is less than any even-numbered partial sum, or $s < s_{2n}$. Similarly, it can be shown that s is greater than any odd-numbered partial sum, or $s_{2n+1} < s$. But

$$s_{2n+1} = s_{2n} - a_{2n+1} < s \quad \text{implies} \quad s_{2n} - s < a_{2n+1},$$

and

$$s_{2n} = s_{2n-1} + a_{2n} > s \quad \text{implies} \quad s - s_{2n-1} < a_{2n}.$$

Thus, for any n, even or odd,

$$|s_n - s| < a_{n+1}.$$

Example 9.33 Show that the sum of the first 99 terms of the alternating harmonic series differs from the sum of the series by less than 0.01.

Solution By Theorem 9.18, $|s_{99} - s| < a_{100} = 0.01$. That is, to be sure that we have accuracy to two decimal places in adding up terms of the alternating harmonic series to approximate the sum, we would have to add up at least 99 terms. This is what we call "slow convergence." ■

Absolute Convergence

DEFINITION 9.7 A series $\sum_{n=0}^{\infty} a_n$ is **absolutely convergent** if and only if the series $\sum_{n=0}^{\infty} |a_n|$ converges. A convergent series that is not absolutely convergent is called **conditionally convergent.**

A convergent series of positive terms is automatically absolutely convergent. The alternating harmonic series is conditionally convergent, because its series of absolute values is the harmonic series, which diverges.

It would seem a contradiction in terms if an absolutely convergent series were not convergent; the next theorem states that this can never happen.

THEOREM 9.19 An absolutely convergent series converges.

Proof If $\sum_{n=0}^{\infty} |a_n|$ converges, then $\sum_{n=0}^{\infty} 2|a_n|$ converges. Since $-|a_n| \le a_n \le |a_n|$, we have $0 \le a_n + |a_n| \le 2|a_n|$, so by the comparison test, $\sum_{n=0}^{\infty}(a_n + |a_n|)$ converges. Consequently, $\sum_{n=0}^{\infty} a_n = \sum_{n=0}^{\infty}[(a_n + |a_n|) - |a_n|]$ converges.

■ **Example 9.34** Determine whether the series below converges:

$$1 - \frac{1}{2^2} - \frac{1}{3^2} + \frac{1}{4^2} - \frac{1}{5^2} - \frac{1}{6^2} + \frac{1}{7^2} - \frac{1}{8^2} - \cdots .$$

Solution The series of absolute values of these terms is the convergent p-series in which $p = 2$. Therefore, the given series is absolutely convergent, so it converges. ■

■ **Example 9.35** Determine whether the series below converges:

$$\sum_{n=1}^{\infty} (-1)^n n^{-3} \cos \frac{\pi n}{5}.$$

Solution We look at the corresponding series of absolute values,

$$\sum_{n=1}^{\infty} n^{-3} \left| \cos \frac{\pi n}{5} \right|.$$

Since $|\cos(\pi n/5)| \le 1$, this series is dominated by the p-series $\sum_{n=1}^{\infty} n^{-3}$. Thus, the given series is absolutely convergent, and it converges. ■

If a series is absolutely convergent, then no amount of rearranging of the terms of the series can affect the sum of the series. For a conditionally convergent series, however, the terms can be rearranged so that the new series has any preselected number whatever as its sum!

For example, we can rearrange the terms of the alternating harmonic series so that the new series has sum 100. We simply choose positive terms of the series until their sum first exceeds 100. We know this will eventually happen, because the series consisting of just the positive terms is divergent. Then we choose negative terms until the partial sum first dips below 100. Then we again choose positive terms, in order, until the partial sum is again greater than 100; then choose negative terms, in order; and so on. The sum of the series we create must be 100. The details necessary for a proof of this fact have been omitted here but are indicated in the exercises (see Problem 40).

The Ratio Test

The following test for convergence of a series is an important result, and will have much use in the next section.

THEOREM 9.20

The Ratio Test Let $\Sigma_{n=0}^{\infty} a_n$ be a series, and let

$$\rho = \lim_{n\to\infty} \left| \frac{a_{n+1}}{a_n} \right|.$$

1. If $\rho < 1$, then the series is absolutely convergent.
2. If $\rho > 1$ or $\rho = \infty$, then the series diverges.
3. If $\rho = 1$, the test fails (that is, gives us no information).

Proof To prove part 1, let $\rho < 1$ and choose $\varepsilon > 0$ so that $\rho + \varepsilon < 1$. Now because $|a_{n+1}/a_n| \to \rho$, there is a positive integer k such that if $n > k$, then

$$\rho - \varepsilon < \left| \frac{a_{n+1}}{a_n} \right| < \rho + \varepsilon.$$

In particular,

$$|a_{k+1}| < (\rho + \varepsilon)|a_k|,$$
$$|a_{k+2}| < (\rho + \varepsilon)|a_{k+1}| < (\rho + \varepsilon)^2|a_k|,$$

and so forth, so that in general,

$$|a_{k+j}| < (\rho + \varepsilon)^j|a_k|.$$

Thus, the geometric series $\Sigma_{n=0}^{\infty}(\rho + \varepsilon)^n|a_k|$ dominates the series $\Sigma_{n=0}^{\infty} |a_n|$; and since $\rho + \varepsilon < 1$, both series converge. Hence, $\Sigma_{n=0}^{\infty} a_n$ is absolutely convergent.

The proof of part 2 is similar. If $\rho > 1$, choose $\varepsilon > 0$ so that $\rho - \varepsilon > 1$. Then from $\rho - \varepsilon < |a_{n+1}/a_n|$, we can obtain

$$(\rho - \varepsilon)^j |a_k| < |a_{k+j}|.$$

Therefore, the series $\sum_{n=0}^{\infty} |a_n|$ dominates the series $\sum_{n=0}^{\infty} (\rho - \varepsilon)^n |a_k|$, which diverges. In particular, the terms a_n cannot approach 0, so the series $\sum_{n=0}^{\infty} a_n$ also diverges.

To prove part 3, we note that for any p-series, $\rho = 1$; but some p-series converge and some diverge. That is, we can make no conclusion about convergence in this case.

The limit ρ in the ratio test is called the **ratio** of the series. Note that this new use of an old term is the same as the old term in the case of a geometric series. Here are some applications of the ratio test.

Example 9.36 Use the ratio test (Theorem 9.20) to determine whether each of the following series converges or diverges:

(a) $\displaystyle\sum_{n=0}^{\infty} \frac{2^n}{n!}$ (b) $\displaystyle\sum_{n=1}^{\infty} \frac{n!}{10^{3n}}$ (c) $\displaystyle\sum_{n=1}^{\infty} \frac{n}{2^n}$ (d) $\displaystyle\sum_{n=1}^{\infty} \frac{n!}{2^{n^2}}$ (e) $\displaystyle\sum_{n=1}^{\infty} \frac{3n}{n^3+1}$

Solution (a) We first compute ρ:

$$\begin{aligned}\rho &= \lim_{n\to\infty} \left|\frac{a_{n+1}}{a_n}\right| \\ &= \lim_{n\to\infty} \frac{2^{n+1}/(n+1)!}{2^n/n!} \\ &= \lim_{n\to\infty} \frac{2}{n+1} = 0 < 1.\end{aligned}$$

Hence, the given series converges.

(b) Since

$$\begin{aligned}\rho &= \lim_{n\to\infty} \left|\frac{(n+1)!\,/10^{3n+3}}{n!/10^{3n}}\right| \\ &= \lim_{n\to\infty} \frac{n+1}{10^3} = \infty,\end{aligned}$$

the given series diverges.

(c) Since

$$\begin{aligned}\rho &= \lim_{n\to\infty} \left|\frac{(n+1)/2^{n+1}}{n/2^n}\right| \\ &= \lim_{n\to\infty} \frac{n+1}{2n} = \frac{1}{2} < 1,\end{aligned}$$

the series converges.

(d) The ratio of this series is

$$\rho = \lim_{n\to\infty} \left| \frac{(n+1)!/2^{(n+1)^2}}{n!/2^{n^2}} \right|$$

$$= \lim_{n\to\infty} \frac{n+1}{2^{2n+1}} = 0 < 1.$$

Hence, the series converges.

(e) The ratio is

$$\rho = \lim_{n\to\infty} \left| \frac{3(n+1)/[(n+1)^3+1]}{3n/(n^3+1)} \right|$$

$$= \lim_{n\to\infty} \frac{3(n+1)(n^3+1)}{3n(n^3+3n^2+3n+2)} = 1,$$

so the ratio test fails. However, we can show that this series converges by applying Theorem 9.16 (the limit comparison test) with $\Sigma_{n=1}^{\infty} (1/n^2)$. ■

The ratio test is particularly applicable when the terms involve factorials or powers of constants, as the above examples illustrate. For some series, though, the ratio test is not easy to apply, so we will discuss one more test that is useful in some cases.

The Root Test

THEOREM 9.21

The Root Test Let $\Sigma_{n=0}^{\infty} a_n$ be a series, and let

$$\rho = \lim_{n\to\infty} |a_n|^{1/n}.$$

1. If $\rho < 1$, then the series is absolutely convergent.
2. If $\rho > 1$, then the series diverges.
3. If $\rho = 1$, the test fails.

The proof of the root test is similar to that of the ratio test, but simpler, so we leave it as an exercise (see Problem 36). The limit ρ in the root test is called the **root** of the series. Consider some applications of the root test.

■ **Example 9.37** Use the root test to determine the convergence of the series:

(a) $\displaystyle\sum_{n=1}^{\infty} \left(\frac{2}{n}\right)^n$ (b) $\displaystyle\sum_{n=2}^{\infty} (-1)^n(\ln n)^{n/2}$ (c) $\displaystyle\sum_{n=1}^{\infty} \left(\frac{n-1}{n}\right)^n$

Solution

(a) We find that

$$\begin{aligned}\rho &= \lim_{n\to\infty} |a_n|^{1/n} \\ &= \lim_{n\to\infty} \left|\left(\frac{2}{n}\right)^n\right|^{1/n} \\ &= \lim_{n\to\infty} \frac{2}{n} = 0 < 1,\end{aligned}$$

so the series converges.

(b) Since

$$\begin{aligned}\rho &= \lim_{n\to\infty} |(-1)^n(\ln n)^{n/2}|^{1/n} \\ &= \lim_{n\to\infty} (\ln n)^{1/2} = \infty,\end{aligned}$$

the series diverges.

(c) The root is

$$\begin{aligned}\rho &= \lim_{n\to\infty} \left|\left(\frac{n-1}{n}\right)^n\right|^{1/n} \\ &= \lim_{n\to\infty} \frac{n-1}{n} = 1,\end{aligned}$$

so the root test fails. However,

$$\begin{aligned}\lim_{n\to\infty} a_n &= \lim_{n\to\infty} \left(\frac{n-1}{n}\right)^n \\ &= \lim_{n\to\infty} \left(1 - \frac{1}{n}\right)^n = e^{-1} \neq 0,\end{aligned}$$

so the series diverges by the divergence test (Theorem 9.8). ■

Summary

It should be mentioned that not every series will yield to the tests we have developed so far. More sophisticated tests for convergence can be found in advanced calculus texts.

Figure 9.6 on page 606 is a flowchart that you may find helpful in analyzing a series for convergence. The chart is no substitute for experience, but it may help you gain experience.

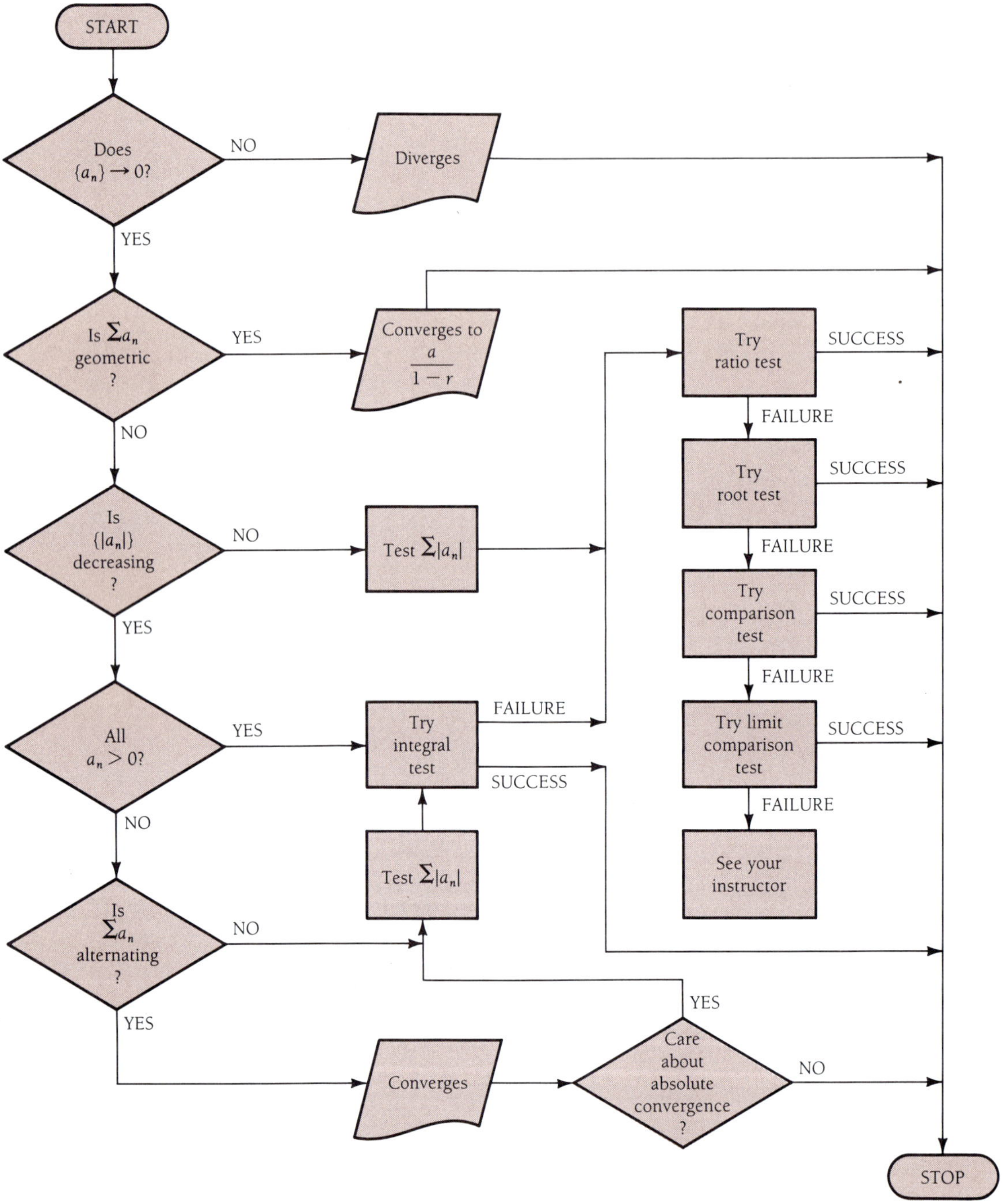

FIGURE 9.6 An algorithm to determine convergence of Σa_n

Exercises 9.4

Determine whether each series converges. If a series is convergent, determine whether the convergence is absolute or conditional.

1. $\sum_{n=1}^{\infty} \frac{(-1)^n}{n^2}$
2. $\sum_{n=1}^{\infty} \frac{1}{n^2+4}$
3. $\sum_{n=2}^{\infty} \frac{1}{n^2-1}$
4. $\sum_{n=3}^{\infty} \frac{(-1)^n}{n^2-2n}$
5. $\sum_{n=1}^{\infty} \frac{1}{n+1}$
6. $\sum_{n=1}^{\infty} \frac{(-1)^n}{n+1}$
7. $\sum_{n=0}^{\infty} \frac{n+1}{3^n}$
8. $\sum_{n=1}^{\infty} \frac{n}{10^{n-1}}$
9. $\sum_{n=1}^{\infty} \frac{n^2}{2^n}$
10. $\sum_{n=1}^{\infty} \frac{n^3}{4^n}$
11. $\sum_{n=1}^{\infty} \frac{n\sqrt{n}}{n^2+1}$
12. $\sum_{n=1}^{\infty} \frac{3^n}{2^n+1}$
13. $\sum_{n=1}^{\infty} \frac{n^{5/2}}{1+n^2}$
14. $\sum_{n=2}^{\infty} \frac{\ln n}{2n+3}$
15. $\sum_{n=1}^{\infty} \frac{n!}{(2n)!}$
16. $\sum_{n=1}^{\infty} \frac{5^{2n}}{(n-1)!}$
17. $\sum_{n=1}^{\infty} \left(1+\frac{1}{n}\right)^n$
18. $\sum_{n=1}^{\infty} \frac{1}{n^n}$
19. $\sum_{n=1}^{\infty} \frac{n^n}{(10n)^{2n}}$
20. $\sum_{n=1}^{\infty} \frac{n^n}{(n!)^2}$
21. $\sum_{n=1}^{\infty} \frac{1}{(n+1)(n+2)}$
22. $\sum_{n=1}^{\infty} \frac{1}{(n+1)(2n-1)}$
23. $\sum_{n=1}^{\infty} \frac{(-1)^n}{n^2+n}$
24. $\sum_{n=2}^{\infty} \frac{(-1)^n}{1+\ln n}$
25. $\sum_{n=1}^{\infty} \frac{n^2}{e^n}$
26. $\sum_{n=1}^{\infty} \frac{n^3}{e^n}$
27. $\sum_{n=1}^{\infty} \frac{1}{n(2n+1)}$
28. $\sum_{n=1}^{\infty} \left(1+\frac{1}{2n}\right)$
29. $\sum_{n=0}^{\infty} \left(\frac{1}{2^n}-\frac{1}{3^n}\right)$
30. $\sum_{n=1}^{\infty} \frac{(-1)^n(n^2+e^n)}{n^2e^n}$
31. $\sum_{n=1}^{\infty} \frac{1+\ln n}{n^2}$
32. $\sum_{n=1}^{\infty} \frac{1+\ln n}{n\sqrt{n}}$

c33. Compute the partial sums s_{10}, s_{20}, s_{30}, s_{40} for the alternating harmonic series $\sum_{n=1}^{\infty}[(-1)^{n+1}/n]$.

c34. Repeat Problem 33 for the series $\sum_{n=1}^{\infty}[(-1)^{n+1}/n^2]$.

35. Later, we will show that

$$\frac{1}{e}=\frac{1}{2!}-\frac{1}{3!}+\frac{1}{4!}-\frac{1}{5!}+\cdots.$$

How many terms of this series must be used to obtain an approximation to $1/e$ that is correct to within 10^{-3}?

36. Prove the root test (Theorem 9.21).

37. Refer to the flowchart in Figure 9.6. Verify that if $a_n \to 0$ and $\sum_{n=1}^{\infty} a_n$ is a geometric series, then

$$\sum_{n=1}^{\infty} a_n = \frac{a_1}{1-r}.$$

38. Refer to the flowchart in Figure 9.6. Verify the conclusion "Converges" if $\Sigma\, a_n$ is an alternating series.

39. Show that the methods discussed so far fail to determine the convergence of the series

$$\sum_{n=1}^{\infty} \frac{1 \cdot 3 \cdot 5 \cdot \cdots \cdot (2n-1)}{2 \cdot 4 \cdot 6 \cdot \cdots \cdot 2n}.$$

40. In the demonstration that the terms of the alternating harmonic sequence can be rearranged so that the new sum of the series is 100 (page 602), prove that the partial sums actually have 100 as their limit. [Use "first exceeds" strongly.]

41. Find the fallacy in the following argument: We know that the alternating harmonic series converges; let S be its sum. Then

$$S = 1 - \frac{1}{2} + \frac{1}{3} - \frac{1}{4} + \frac{1}{5} - \frac{1}{6} + \cdots,$$

so

$$2S = 2 - \frac{2}{2} + \frac{2}{3} - \frac{2}{4} + \frac{2}{5} - \frac{2}{6} + \cdots.$$

Rearranging terms, we have

$$\begin{aligned} 2S &= \left(2 - \frac{2}{2}\right) - \frac{2}{4} + \left(\frac{2}{3} - \frac{2}{6}\right) - \frac{2}{8} + \left(\frac{2}{5} - \frac{2}{10}\right) - \frac{2}{12} + \cdots \\ &= 1 - \frac{1}{2} + \frac{1}{3} - \frac{1}{4} + \frac{1}{5} - \frac{1}{6} + \cdots \\ &= S. \end{aligned}$$

That is,

$$\begin{aligned} 2S &= S, \\ 2 &= 1. \end{aligned}$$

SECTION 9.5

Power Series

Now that we have developed the properties of sequences and series of real numbers to an extent, we can consider sequences and series of functions. When we return to consider series of real numbers again later, we will be armed with a powerful tool, namely, power series.

Sequences and Series of Functions

A sequence of real numbers was defined as a function that assigns to each positive integer n a real number a_n. In a similar fashion, we define a **sequence of functions** to be a function that assigns to each positive integer n a function f_n.

A sequence $\{f_n\}$ of functions **converges at c** if and only if the sequence $\{f_n(c)\}$ of real numbers converges.

A series of real numbers was defined as a sequence of partial sums; we define a series of functions in a similar way. Let $\{f_n\}$ be a sequence of functions. The partial sums

$$S_n(x) = \sum_{k=1}^{n} f_k(x)$$

form a sequence $\{S_n\}$ which is also denoted by

$$\sum_{n=1}^{\infty} f_n(x),$$

and is called a **series of functions.** For any real number in the domain of each f_n, the series

$$\sum_{n=1}^{\infty} f_n(c)$$

is just a series of real numbers. We say the series $\sum_{n=1}^{\infty} f_n(x)$ **converges at** c, **converges absolutely at** c, or **diverges at** c, according as the series $\sum_{n=1}^{\infty} f_n(c)$ converges, converges absolutely, or diverges.

Power Series

Much of the subject of sequences and series of functions is beyond our scope. We will restrict our attention to series of functions of a certain type.

DEFINITION 9.8

A **power series in** x is a series of functions of the form

$$\sum_{n=0}^{\infty} a_n x^n = a_0 + a_1 x + a_2 x^2 + \cdots,$$

with each a_n a real number.

Convergence was a primary concern with series of real numbers. The same is true for power series, except that the question has now become: "For which values of x does $\sum_{n=0}^{\infty} a_n x^n$ converge?"

Consider the following examples, in which the ratio test (Theorem 9.20) plays a major role.

Example 9.38 Determine the values of x for which each power series converges:

(a) $\sum_{n=0}^{\infty} x^n$ (b) $\sum_{n=0}^{\infty} \frac{2^n}{n!} x^n$ (c) $\sum_{n=0}^{\infty} n^n x^n$

Solution **(a)** Viewing x as an unknown real number rather than as a variable, we may treat the power series just like a series of real numbers. Applying the ratio test to $\sum_{n=0}^{\infty} x^n$, we get

$$\rho = \lim_{n\to\infty} \left|\frac{a_{n+1}}{a_n}\right| = \lim_{n\to\infty} \left|\frac{x^{n+1}}{x^n}\right| = \lim_{n\to\infty} |x| = |x|.$$

Therefore, the series converges absolutely if $|x| = \rho < 1$; that is, if $-1 < x < 1$. It diverges if $|x| > 1$, and it diverges if $x = 1$ or $x = -1$ (by the divergence test, Theorem 9.8). Thus, the power series $\sum_{n=0}^{\infty} x^n$ converges absolutely in $(-1, 1)$ and diverges elsewhere. This series may be regarded as a geometric series with ratio x.

(b) We also apply the ratio test to this series, getting

$$\rho = \lim_{n\to\infty} \left|\frac{[2^{n+1}/(n+1)!]x^{n+1}}{(2^n/n!)x^n}\right| = \lim_{n\to\infty} \left|\frac{2x}{n+1}\right| = 0.$$

Since $\rho < 1$, this series converges for all x.

(c) The ratio test gives us

$$\rho = \lim_{n\to\infty} \left|\frac{(n+1)^{n+1}x^{n+1}}{n^n x^n}\right| = \lim_{n\to\infty} (n+1)\left(1+\frac{1}{n}\right)^n |x|,$$

and the limit is ∞ if $x \neq 0$. Therefore, the series $\sum_{n=0}^{\infty} n^n x^n$ converges only for $x = 0$. ∎

The method used in these examples can be used in general and leads to the following theorem. Note that the limit in this theorem is the reciprocal of the limit in the ratio test: $r = 1/\rho$.

THEOREM 9.22 If $\sum_{n=0}^{\infty} a_n x^n$ is a power series, let

$$r = \lim_{n\to\infty} \left|\frac{a_n}{a_{n+1}}\right|.$$

1. If $r = 0$, then the power series converges only for $x = 0$.
2. If $r = \infty$, then the power series converges absolutely for all x.
3. If $0 < r < \infty$, then the power series converges absolutely for $|x| < r$ and diverges for $|x| > r$. If $x = \pm r$, then the series may either converge or diverge.

Proof We apply the ratio test (Theorem 9.20), getting

$$\rho = \lim_{n\to\infty} \left|\frac{a_{n+1}x^{n+1}}{a_n x^n}\right| = \lim_{n\to\infty} \left|\frac{a_{n+1}}{a_n}\right| |x| = \frac{|x|}{r}.$$

Now if $r = 0$, then $\rho = \infty$ unless $x = 0$, so the series converges only if $x = 0$. This proves part 1. If $r = \infty$, then $\rho = 0$, so the series converges absolutely for all x. This proves part 2. Finally, suppose $0 < r < \infty$. If $\rho = |x|/r < 1$, then the series converges absolutely for $|x| < r$ and diverges for $|x| > r$. If $x = \pm r$, then $\rho = 1$, and anything may happen.

The number r in Theorem 9.22 is called the **radius of convergence** of the power series $\sum_{n=0}^{\infty} a_n x^n$. If $r = \infty$, we say the radius of convergence is infinite. The set of values of x for which the power series converges is called the **interval of convergence** of the power series and always extends from $-r$ to r, perhaps including one or both endpoints.

Example 9.39 Find the interval of convergence of each power series:

$$\text{(a)}\ \sum_{n=1}^{\infty} \frac{1}{n} x^n \qquad \text{(b)}\ \sum_{n=1}^{\infty} \frac{(-1)^{n+1}}{n(n+1)} x^n \qquad \text{(c)}\ \sum_{n=2}^{\infty} \frac{(-1)^n}{\ln n} x^n$$

Solution

(a) We first compute the radius of convergence:

$$r = \lim_{n\to\infty} \left| \frac{1/n}{1/(n+1)} \right| = \lim_{n\to\infty} \frac{n+1}{n} = 1.$$

Thus, the series converges at least in $(-1, 1)$. If $x = 1$, the series is the harmonic series, which diverges; if $x = -1$, we get the alternating harmonic series, which converges. Thus, $[-1, 1)$ is the interval of convergence of $\sum_{n=1}^{\infty}[(1/n)x^n]$.

(b) The radius of convergence is

$$r = \lim_{n\to\infty} \left| \frac{(-1)^{n+1}/n(n+1)}{(-1)^{n+2}/(n+1)(n+2)} \right| = \lim_{n\to\infty} \frac{n+2}{n} = 1.$$

If $x = 1$, the series is $\sum_{n=1}^{\infty}[(-1)^{n+1}/n(n+1)]$, which is absolutely convergent by comparison with the p-series with $p = 2$. If $x = -1$, the series is $-\sum_{n=1}^{\infty}[1/n(n+1)]$, which converges for the same reason. Therefore, $[-1, 1]$ is the interval of convergence.

(c) The radius of convergence is

$$r = \lim_{n\to\infty} \left| \frac{(-1)^n/(\ln n)}{(-1)^{n+1}/[\ln(n+1)]} \right| = \lim_{n\to\infty} \frac{\ln(n+1)}{\ln n} = 1. \quad \text{(By L'Hôpital's rule)}$$

If $x = 1$, the series is $\sum_{n=2}^{\infty}[(-1)^n/(\ln n)]$, which converges by the alternating series test (Theorem 9.17). If $x = -1$, the series is $\sum_{n=2}^{\infty}[1/(\ln n)]$, which diverges by comparison with the harmonic series. Hence, $(-1, 1]$ is the interval of convergence. ∎

A power series in x is also called a **power series about 0** or **centered at 0**, because 0 is the midpoint of its interval of convergence. If c is any real number, a **power series about c** or **centered at c** is a **power series in $(x - c)$**, and is of the form

$$\sum_{n=0}^{\infty} a_n(x-c)^n.$$

The radius of convergence of this series is still $r = \lim_{n\to\infty} |a_n/a_{n+1}|$, and if $r \neq \infty$, then the interval of convergence extends from $c - r$ to $c + r$, perhaps including one or both endpoints. Thus, c is the midpoint of the interval of convergence. If $r = \infty$, then the interval of convergence is $(-\infty, \infty)$.

More generally, if $g(x)$ is any expression in x, a **power series in $g(x)$** is a series of functions of the form

$$\sum_{n=0}^{\infty} a_n[g(x)]^n.$$

The interval of convergence of this series is computed by applying Theorem 9.22 to $g(x)$ instead of x.

■ **Example 9.40** Find the interval of convergence of each series:

(a) $\displaystyle\sum_{n=0}^{\infty} \frac{(-1)^n}{2^n}(x-5)^n$ (b) $\displaystyle\sum_{n=1}^{\infty} \frac{1}{n^2}(x-3)^n$ (c) $\displaystyle\sum_{n=0}^{\infty} \frac{1}{2^n}x^{2n}$ (d) $\displaystyle\sum_{n=0}^{\infty} e^n x^{3n}$

Solution (a) The radius of convergence is

$$r = \lim_{n\to\infty} \left| \frac{(-1)^n/2^n}{(-1)^{n+1}/2^{n+1}} \right| = 2.$$

Thus, this series converges at least in (3, 7). At $x = 3$, the series is

$$\sum_{n=0}^{\infty} \frac{(-1)^n}{2^n}(-2)^n = \sum_{n=0}^{\infty} 1;$$

and at $x = 7$, the series is

$$\sum_{n=0}^{\infty} \frac{(-1)^n}{2^n}(2)^n = \sum_{n=0}^{\infty} (-1)^n.$$

Both of these series diverge by the divergence test (Theorem 9.8). Thus, (3, 7) is the interval of convergence.

(b) The radius of convergence is easily found to be $r = 1$, so the series converges at least in (2, 4). At $x = 4$, the series becomes the p-series with $p = 2$; and at $x = 2$, the series becomes the alternating p-series with $p = 2$. Both these series converge, so the interval of convergence is [2, 4].

(c) This is a power series in x^2. The radius of convergence is $r = 2$, so the series converges for $x^2 \in (-2, 2)$. That is, the series converges for $x \in (-\sqrt{2}, \sqrt{2})$.

This is also the interval of convergence, since the series diverges at each endpoint.

(d) This is a power series in x^3. The radius of convergence is $r = e^{-1}$, so the series converges for $x^3 \in (-e^{-1}, e^{-1})$, or $x \in (-e^{-1/3}, e^{-1/3})$. At each endpoint, the series diverges, so this last interval is the interval of convergence. ■

In the next section we will begin to work our way toward finding a sum for a series.

Exercises 9.5

Find the interval of convergence of each power series.

1. $\sum_{n=1}^{\infty} \frac{10^n}{n} x^n$
2. $\sum_{n=0}^{\infty} \frac{x^n}{100^n}$
3. $\sum_{n=1}^{\infty} n^2 x^n$
4. $\sum_{n=1}^{\infty} n^3 x^n$
5. $\sum_{n=2}^{\infty} \frac{n(n-1)}{2} x^n$
6. $\sum_{n=1}^{\infty} \frac{(-1)^n}{n^2} x^n$
7. $\sum_{n=1}^{\infty} \frac{x^n}{n(2n+1)}$
8. $\sum_{n=0}^{\infty} \frac{x^n}{(n+1)(n+3)}$
9. $\sum_{n=1}^{\infty} \frac{n(n+3)x^n}{(n+1)(2n+1)}$
10. $\sum_{n=1}^{\infty} \frac{(-1)^n n}{2^n} x^n$
11. $\sum_{n=0}^{\infty} (1+n)x^n$
12. $\sum_{n=0}^{\infty} \frac{10^n}{n!} x^n$
13. $\sum_{n=1}^{\infty} \frac{n!}{n^{10}} x^n$
14. $\sum_{n=0}^{\infty} \frac{n!}{(2n)!} x^n$
15. $\sum_{n=0}^{\infty} \frac{3^n x^n}{2^n}$
16. $\sum_{n=0}^{\infty} e^{-n} x^n$
17. $\sum_{n=2}^{\infty} \frac{(x-2)^n}{\ln n}$
18. $\sum_{n=2}^{\infty} \frac{(-1)^n}{\ln n} (x+1)^n$
19. $\sum_{n=2}^{\infty} \frac{\ln n}{n} (x-1)^n$
20. $\sum_{n=2}^{\infty} \frac{\ln n}{n^2} (x-4)^n$
21. $\sum_{n=1}^{\infty} (1+\ln n)x^{2n}$
22. $\sum_{n=0}^{\infty} \frac{n^2+1}{2^n} x^{5n}$
23. $\sum_{n=0}^{\infty} (2-\sqrt{n})(x-1)^{3n}$
24. $\sum_{n=0}^{\infty} \frac{9^n \sqrt{n}}{1+n} (x+1)^{2n}$
25. Show that the radius of convergence of the power series $\sum_{n=0}^{\infty} a_n x^n$ is also $r = \lim_{n\to\infty} |1/a_n|^{1/n}$.

Find the interval of convergence of the following power series:

26. $\sum_{n=1}^{\infty} (nx)^n$
27. $\sum_{n=1}^{\infty} \left(\frac{x}{n}\right)^n$
28. $\sum_{n=0}^{\infty} (1+n)^n x^n$
29. $\sum_{n=1}^{\infty} \left(1+\frac{1}{n}\right)^{n^2} x^n$
30. $\sum_{n=1}^{\infty} \frac{(-10)^n}{\left(1+\frac{1}{n}\right)^n} (x+3)^n$
31. $\sum_{n=1}^{\infty} \frac{2^n}{\left(1+\frac{1}{n}\right)^n} (x-2)^n$
32. Show that $\sum_{n=0}^{\infty} x^n = 1/(1-x)$ for $x \in (-1, 1)$.

SECTION 9.6

Taylor Polynomials and Taylor Series

If a power series $\sum_{n=0}^{\infty} a_n x^n$ converges in the interval $(-r, r)$, then for each $x \in (-r, r)$, the series has a sum. If we denote the sum by $f(x)$, then the series converges to the function $f(x)$ on $(-r, r)$. In this section, we will find out how to construct a power series that converges to a given function. We will approach the topic from the point of view of polynomial approximations.

Polynomials and Transcendental Functions

First, we need to make some comments about polynomial functions and their relationship to transcendental functions. Throughout the history of calculus, polynomial functions have enjoyed a preferred status. The major reason for this "prejudice" against the transcendental functions is computational. With paper and pencil—and sufficient patience—anyone who can do arithmetic can compute the value of a given polynomial function at a given point. But for transcendental functions, tables or sophisticated computational methods are needed to find most function values. For example, to evaluate $p(x) = x^3 + 4x$ at $x = 1.5$ requires only arithmetic; but to evaluate $f(x) = \tan x$ at $x = 1.5$ requires a table or a calculator—it cannot be found by simple arithmetic. The computational problems caused by the transcendental functions have often led to the use of polynomial functions in applications, even when simple transcendental functions might give a better fit.

In recent years, advances in computing power have blurred these differences between polynomial and transcendental functions. The hand-held calculator, for instance, is making the use of tables and computations with logarithms obsolete. Today's student finds that calculating values for transcendental functions is just as easy as (or even easier than) computations with polynomials—both are reduced to pushing buttons on the calculator.

The fact remains that transcendental functions are fundamentally different from polynomial functions. For example, the sine function has higher derivatives that are always sine and cosine functions. For any polynomial, on the other hand, sufficiently high derivatives are all 0.

Polynomial Approximation

Formerly, difficulties in working with transcendental functions led to the idea of approximating a transcendental function by a polynomial function. We discuss the idea here, because it led to results that are much more significant than the computational problems. The aim is to construct a polynomial function of a given degree that has values as close as possible to the given transcendental function values in the vicinity of a given point. That is, we want the graphs of the two functions to look as much alike as possible near the given point. We accomplish this aim as well as it can be done by requiring the polynomial and its derivatives to have the same values as the given function and its derivatives at the given point.

Let's be more specific. Let f be a given function and let c be a given point. We will assume f has arbitrarily many derivatives at c (that is, we assume f is **infinitely differentiable** at c). Let $p_n(x)$ be a polynomial function of degree n which is to approximate f at c. If we write $p_n(x)$ in terms of powers of $x - c$, it will make our computations easier. Hence, we write

$$p_n(x) = a_0 + a_1(x-c) + a_2(x-c)^2 + \cdots + a_n(x-c)^n,$$

with the coefficients $a_0, a_1, \ldots, a_n$ yet to be determined.

We first require $p_n(c)$ to equal $f(c)$. Since $p_n(c) = a_0$, we see that

$$a_0 = f(c).$$

We next require the first derivatives to be equal; that is, $p_n'(c) = f'(c)$. Since $p_n'(c) = a_1$, we thus need

$$a_1 = f'(c).$$

Equating the second derivatives is the next step. Since $p_n''(c) = 2a_2$, we must have

$$a_2 = \frac{1}{2} f''(c).$$

Continuing in this fashion, we find that

$$a_3 = \frac{1}{2 \cdot 3} f'''(c),$$

$$a_4 = \frac{1}{2 \cdot 3 \cdot 4} f^{(4)}(c),$$

and, in general,

$$a_n = \frac{1}{n!} f^{(n)}(c).$$

Therefore, $p_n(x)$ can be written as

$$p_n(x) = f(c) + f'(c)(x-c) + \frac{1}{2} f''(c)(x-c)^2 + \cdots + \frac{1}{n!} f^{(n)}(c)(x-c)^n.$$

If we let the "zeroth derivative" $f^{(0)}$ of f stand for f itself, we can rewrite $p_n(x)$ in summation notation as follows:

$$p_n(x) = \sum_{k=0}^{n} \frac{f^{(k)}(c)}{k!} (x-c)^k. \qquad (1)$$

Taylor Polynomials

The polynomial (1) is called the **Taylor polynomial of degree n for f about c.** The British mathematician Brook Taylor (1685–1731) introduced this kind of polynomial in 1712.

■ **Example 9.41** Find the Taylor polynomial of degree 3 for the function $f(x) = \sin 2x$ about $\pi/2$.

Solution We must compute the first three derivatives of f at $\pi/2$:

$$\begin{aligned} f(x) &= \sin 2x, & f\left(\frac{\pi}{2}\right) &= 0; \\ f'(x) &= 2\cos 2x, & f'\left(\frac{\pi}{2}\right) &= -2; \\ f''(x) &= -4\sin 2x, & f''\left(\frac{\pi}{2}\right) &= 0; \\ f'''(x) &= -8\cos 2x, & f'''\left(\frac{\pi}{2}\right) &= 8. \end{aligned}$$

Hence, from polynomial (1), we get

$$\begin{aligned} p_3(x) &= 0 - 2\left(x - \frac{\pi}{2}\right) + 0 + \frac{8}{3!}\left(x - \frac{\pi}{2}\right)^3 \\ &= -2\left(x - \frac{\pi}{2}\right) + \frac{4}{3}\left(x - \frac{\pi}{2}\right)^3. \end{aligned}$$

This is the desired polynomial. To illustrate its use, suppose we approximate $\sin(3\pi/4)$ by using $p_3(x)$ with $x = 3\pi/8$. We get

$$\begin{aligned} p_3\left(\frac{3\pi}{8}\right) &= -2\left(\frac{3\pi}{8} - \frac{\pi}{2}\right) + \frac{4}{3}\left(\frac{3\pi}{8} - \frac{\pi}{2}\right)^3 \\ &= \frac{\pi}{4} - \frac{\pi^3}{384} \approx 0.7047. \end{aligned}$$

The actual value to four decimal places, for comparison, is 0.7071. ■

Taylor's Remainder

In order to say how good an approximation is given by a Taylor polynomial, we must have a measure of the error involved. If $p_n(x)$ is a Taylor polynomial for f, the difference

$$R_n(x) = f(x) - p_n(x)$$

is called **Taylor's remainder** for $p_n(x)$. That is,

$$f(x) = p_n(x) + R_n(x)$$

expresses $f(x)$ exactly in terms of the polynomial and the remainder. Thus, Taylor's remainder is the error in approximating $f(x)$ by $p_n(x)$.

The size of Taylor's remainder often can be determined by applying Theorem 9.23.

THEOREM 9.23

Taylor's Theorem Let a function f be infinitely differentiable in an interval (a, b) containing c. If

$$f(x) = \sum_{k=0}^{n} \frac{f^{(k)}(c)}{k!}(x-c)^k + R_n(x),$$

then

$$R_n(x) = \int_c^x \frac{f^{(n+1)}(t)}{n!}(x-t)^n\,dt$$

for all $x \in (a, b)$.

Proof

By the fundamental theorem of calculus,

$$f(x) - f(c) = \int_c^x f'(t)\,dt. \tag{2}$$

If we integrate equation (2) by parts, with $u = f'(t)$, $du = f''(t)\,dt$, $dv = dt$, and $v = t - x$ (x is a constant so far as t is concerned), we get

$$\begin{aligned}\int_c^x f'(t)\,dt &= f'(t)(t-x)\Big|_c^x - \int_c^x f''(t)(t-x)\,dt\\ &= f'(c)(x-c) + \int_c^x f''(t)(x-t)\,dt.\end{aligned}$$

Hence, from equation (2),

$$f(x) = f(c) + f'(c)(x-c) + \int_c^x f''(t)(x-t)\,dt.$$

The theorem therefore holds for $n = 1$. Proceeding inductively and using the same procedure as above, the theorem can be established for any positive integer n. We omit the details, since they are not difficult to supply.

The form of the remainder in Taylor's theorem (as well as the alternate form of the remainder found in the exercises; see Problem 38) enables us to put an upper bound on the remainder, at least for some functions. This allows us to estimate the error in using a Taylor polynomial approximation.

■ Example 9.42 Estimate the error in approximating $\sin 2x$ by $p_n(x)$ (see Example 9.41).

Solution By Taylor's theorem, the remainder is

$$R_n(x) = \int_{\pi/2}^x \frac{f^{(n+1)}(t)}{n!}(x-t)^n\,dt.$$

Now,

$$f^{(n+1)}(t) = \pm 2^{n+1} \sin 2t \qquad \text{or} \qquad \pm 2^{n+1} \cos 2t;$$

in any case,

$$|f^{(n+1)}(t)| \le 2^{n+1}.$$

Consequently,

$$\begin{aligned} |R_n(x)| &\le \left| \int_{\pi/2}^{x} \frac{2^{n+1}}{n!} (x-t)^n \, dt \right| \\ &= \left| \frac{2^{n+1}}{n!} \frac{(x-t)^{n+1}}{n+1} \Big|_{\pi/2}^{x} \right| \\ &= \frac{2^{n+1}}{(n+1)!} \left| x - \frac{\pi}{2} \right|^{n+1}. \end{aligned}$$

For example, if $n = 3$ and $x = 3\pi/8$, the error will not exceed

$$\frac{2^4}{4!} \left(\frac{\pi}{8}\right)^4 \approx 0.0159.$$

We saw in Example 9.41 that the error was substantially less than this amount. ■

Taylor Series

The Taylor polynomial $p_n(x)$ of degree n for f about c is the nth partial sum of the power series

$$\sum_{n=0}^{\infty} \frac{f^{(n)}(c)}{n!} (x-c)^n, \tag{3}$$

which is called the **Taylor series for f about c.** Since the difference between the value $f(x)$ and the nth partial sum $p_n(x)$ is Taylor's remainder $R_n(x)$, it follows that if the sequence $\{R_n(x)\}$ of remainders converges to 0, then the Taylor series (3) will converge to $f(x)$. Conversely, if the series (3) converges to $f(x)$, then the sequence of remainders converges to 0. We state this result as Theorem 9.24.

THEOREM 9.24 For all those values of x for which the sequence of Taylor's remainders $\{R_n(x)\}$ converges to 0, the Taylor series for f about c converges to $f(x)$, or

$$f(x) = \sum_{n=0}^{\infty} \frac{f^{(n)}(c)}{n!} (x-c)^n.$$

Example 9.43 Construct the Taylor series for $f(x) = \sin 2x$ about $\pi/2$, and prove that it converges to $\sin 2x$ for all x.

Solution The derivatives of $f(x) = \sin 2x$ fall into a repeating pattern:

$$f^{(4k)}(x) = 2^{4k} \sin 2x \qquad f^{(4k)}\left(\frac{\pi}{2}\right) = 0;$$

$$f^{(4k+1)}(x) = 2^{4k+1} \cos 2x, \qquad f^{(4k+1)}\left(\frac{\pi}{2}\right) = -2^{4k+1};$$

$$f^{(4k+2)}(x) = -2^{4k+2} \sin 2x, \qquad f^{(4k+2)}\left(\frac{\pi}{2}\right) = 0;$$

$$f^{(4k+3)}(x) = -2^{4k+3} \cos 2x, \qquad f^{(4k+3)}\left(\frac{\pi}{2}\right) = 2^{4k+3}.$$

Consequently, the Taylor series for $\sin 2x$ about $\pi/2$ is (after some simplification and reindexing of the summation)

$$\sum_{n=0}^{\infty} \frac{(-1)^{n+1}2^{2n+1}}{(2n+1)!}\left(x - \frac{\pi}{2}\right)^{2n+1} = -\frac{2}{1!}\left(x - \frac{\pi}{2}\right) + \frac{2^3}{3!}\left(x - \frac{\pi}{2}\right)^3 - \frac{2^5}{5!}\left(x - \frac{\pi}{2}\right)^5 + \cdots .$$

The sequence of Taylor's remainders, which was computed in Example 9.42, converges to 0 as $n \to \infty$ (because of Theorem 9.6). Therefore, the series converges to $\sin 2x$ for all x, and we may write

$$\sin 2x = -2\left(x - \frac{\pi}{2}\right) + \frac{2^3}{3!}\left(x - \frac{\pi}{2}\right)^3 - \frac{2^5}{5!}\left(x - \frac{\pi}{2}\right)^5 + \cdots .$$ ■

Example 9.44 Construct the Taylor series for $f(x) = e^x$ about 0.

Solution Since $f^{(n)}(x) = e^x$ for every n, we have $f^{(n)}(0) = e^0 = 1$. Hence, the Taylor series is

$$\sum_{n=0}^{\infty} \frac{1}{n!} x^n = 1 + x + \frac{x^2}{2!} + \frac{x^3}{3!} + \cdots .$$ ■

Example 9.45 Show that the Taylor series for e^x about 0 converges to e^x for all x.

Solution We must show that the Taylor's remainders tend to 0. For e^x about 0, the nth remainder is

$$R_n(x) = \int_0^x \frac{e^t}{n!}(x - t)^n \, dt.$$

For $x > 0$, if $0 < t < x$, then $e^t < e^x$. Hence,

$$\begin{aligned}|R_n(x)| &\leq \frac{1}{n!}\int_0^x e^x(x-t)^n\,dt \\ &= -\frac{e^x}{(n+1)!}(x-t)^{n+1}\Big|_0^x \\ &= \frac{e^x x^{n+1}}{(n+1)!}.\end{aligned}$$

Similarly, for $x < 0$, we get

$$|R_n(x)| \leq \frac{e^{|x|}|x|^{n+1}}{(n+1)!}.$$

In either case, the limit is 0 as $n \to \infty$ (by Theorem 9.6). Therefore, the Taylor series converges to e^x for all x, and

$$e^x = \sum_{n=0}^{\infty} \frac{x^n}{n!} = 1 + x + \frac{x^2}{2!} + \frac{x^3}{3!} + \cdots .$$

Figure 9.7 shows the graphs of $f(x)$ and the approximating Taylor polynomials p_0, p_1, p_2, and p_3.

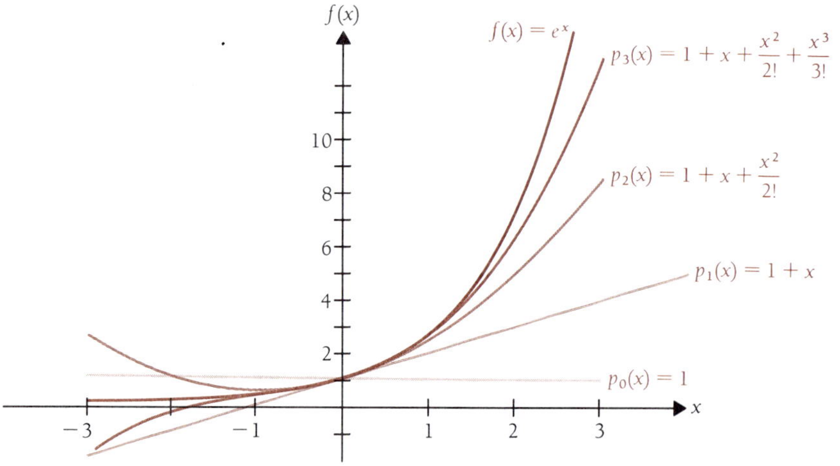

FIGURE 9.7 The graph of $f(x) = e^x$ and its first four Taylor polynomial approximations, centered at 0

Example 9.46 Construct the Taylor series for $f(x) = \ln x$ about 1. Determine for which x the series converges to $\ln x$.

Solution The derivatives of $f(x) = \ln x$ are

$$f'(x) = \frac{1}{x}, \quad f''(x) = -\frac{1}{x^2}, \quad f'''(x) = \frac{2}{x^3}, \quad \ldots, \quad f^{(k)}(x) = (-1)^{k+1}\frac{(k-1)!}{x^k}.$$

Hence,

$$f^{(k)}(1) = (-1)^{k+1}(k-1)!,$$

and the Taylor series is

$$\begin{aligned}\sum_{n=1}^{\infty} \frac{(-1)^{n+1}(n-1)!}{n!}(x-1)^n &= \sum_{n=1}^{\infty} \frac{(-1)^{n+1}}{n}(x-1)^n \\ &= (x-1) - \frac{1}{2}(x-1)^2 + \frac{1}{3}(x-1)^3 - \cdots.\end{aligned}$$

This power series has interval of convergence $(0, 2]$. To determine whether it converges to $\ln x$ there, we examine the remainders. For $\ln x$, we have

$$\begin{aligned}R_n(x) &= \int_1^x \frac{(-1)^{n+2}n!}{n!t^{n+1}}(x-t)^n\,dt \\ &= (-1)^n \int_1^x \frac{(x-t)^n}{t^{n+1}}\,dt.\end{aligned}$$

For $x \in [1, 2]$, we have $t^{n+1} \geq 1$, so $(x-t)^n/t^{n+1} \leq (x-t)^n$. Thus,

$$|R_n(x)| \leq \int_1^x (x-t)^n\,dt = \frac{(x-1)^{n+1}}{n+1} \to 0,$$

by comparison with a geometric sequence. For $x \in (0, 1)$, it gets a little tedious; we must break up the interval into several pieces. For any nonnegative integer k, if $2^{-(k+1)} \leq x < 2^{-k}$, we have

$$\begin{aligned}|R_n(x)| &= \int_x^1 \frac{(t-x)^n}{t^{n+1}}\,dt \\ &= \int_x^{1/2^k} \frac{(t-x)^n}{t^{n+1}}\,dt + \int_{1/2^k}^1 \frac{(t-x)^n}{t^{n+1}}\,dt \\ &\leq \int_x^{1/2^k} \frac{(1/2^k-x)^n}{t^{n+1}}\,dt + \int_{1/2^k}^1 \frac{(t-1/2^k)^n}{t^{n+1}}\,dt \\ &\leq \frac{1}{n}\left(\frac{1-2^kx}{2^kx}\right)^n[1-(2^kx)^n] + \frac{k}{n}\left(1-\frac{1}{2^n}\right) \to 0.\end{aligned}$$

Thus, for any $x \in (0, 2]$, $|R_n(x)| \to 0$, and

$$\ln x = (x-1) - \frac{1}{2}(x-1)^2 + \frac{1}{3}(x-1)^3 - \frac{1}{4}(x-1)^4 + \cdots . \quad \blacksquare$$

Maclaurin Series

A Taylor series about 0 is called a **Maclaurin series**, after the Scottish mathematician Colin Maclaurin (1698–1746). A Maclaurin series is therefore a power series in x and has the form

$$\sum_{n=0}^{\infty} \frac{f^{(n)}(0)}{n!} x^n.$$

Some particular Maclaurin series that are of great use are

$$e^x = 1 + x + \frac{x^2}{2!} + \frac{x^3}{3!} + \frac{x^4}{4!} + \cdots = \sum_{n=0}^{\infty} \frac{x^n}{n!}, \quad (4)$$

$$\sin x = x - \frac{x^3}{3!} + \frac{x^5}{5!} - \frac{x^7}{7!} + \cdots = \sum_{n=0}^{\infty} \frac{(-1)^n x^{2n+1}}{(2n+1)!}, \quad (5)$$

$$\cos x = 1 - \frac{x^2}{2!} + \frac{x^4}{4!} - \frac{x^6}{6!} + \cdots = \sum_{n=0}^{\infty} \frac{(-1)^n x^{2n}}{(2n)!}, \quad (6)$$

$$\ln(1+x) = x - \frac{x^2}{2} + \frac{x^3}{3} - \frac{x^4}{4} + \cdots = \sum_{n=1}^{\infty} \frac{(-1)^{n+1} x^n}{n}. \quad (7)$$

The series for e^x, $\sin x$, and $\cos x$ converge for all x, and the series for $\ln(1+x)$ converges for $x \in (-1, 1]$. The series (4) was derived in Example 9.44, and its convergence was proved in Example 9.45. It is easy to get the series (7) from the series in Example 9.46 by substituting $1 + x$ for x. Note that for $x = 1$ in (7) we get

$$\ln 2 = 1 - \frac{1}{2} + \frac{1}{3} - \frac{1}{4} + \cdots ,$$

the sum of the alternating harmonic series.

A Taylor series for a function is often called a **power series expansion** of the function. With power series expansions, approximate values of the functions may be computed by simple arithmetic. In this way, tables for functions can be constructed; it is not always the most efficient method, but it is always possible.

■ **Example 9.47** Approximate $\sin \frac{1}{2}$ to three decimal places using the Maclaurin series expansion of $\sin x$.

Solution From series (5), we have

$$\begin{aligned}\sin \frac{1}{2} &= \frac{1}{2} - \frac{1}{3!2^3} + \frac{1}{5!2^5} - \frac{1}{7!2^7} + \cdots \\ &\approx 0.5000 - 0.0208 + 0.0003 - \cdots \\ &= 0.4795 - \cdots ,\end{aligned}$$

so rounding to three decimal places gives us

$$\sin\frac{1}{2} \approx 0.479.$$

Looking at the remainder, we find from Taylor's theorem (Theorem 9.23) that for $f(x) = \sin x$,

$$|R_n(x)| \leq \frac{x^{n+1}}{(n+1)!},$$

so

$$\left|R_n\left(\frac{1}{2}\right)\right| \leq \frac{1}{(n+1)! \cdot 2^{n+1}}.$$

Thus, $|R_n(x)| \leq 0.001$ when $n \geq 4$, and our approximation above is correct. ■

Power series expansions also may provide a means of finding the sums of other series. We have already pointed out that the alternating harmonic series has sum ln 2, obtained from the Maclaurin series for $\ln(1 + x)$. Here are some more examples.

■ **Example 9.48** Find the sum of the series:

(a) $\displaystyle\sum_{n=0}^{\infty} \frac{2^n}{n!}$ (b) $\displaystyle\sum_{n=1}^{\infty} \frac{1}{n!}$ (c) $\displaystyle\sum_{n=0}^{\infty} \frac{(-1)^n 4^n}{(2n)!}$

Solution (a) Comparing this series with the Maclaurin series for e^x, we see that the sum is e^2.

(b) Comparing this series with the Maclaurin series for e^x, we note

$$e = \sum_{n=0}^{\infty} \frac{1}{n!} = 1 + \sum_{n=1}^{\infty} \frac{1}{n!},$$

so that

$$\sum_{n=1}^{\infty} \frac{1}{n!} = e - 1.$$

(c) Comparing this series with the Maclaurin series for $\cos x$, we see that

$$\sum_{n=0}^{\infty} \frac{(-1)^n 4^n}{(2n)!} = \sum_{n=0}^{\infty} \frac{(-1)^n 2^{2n}}{(2n)!} = \cos 2.$$ ■

Historical Note on Taylor Series

Taylor series were first used by the Scottish mathematician James Gregory (1638–1675), who described several examples in a letter to a friend in 1671. It is not clear whether Gregory ever wrote down the general formula, but he could find the series for virtually any function. The first publication on the subject is Johann Bernoulli's article of 1694, where the formula is expressed in terms of differentials. Brook Taylor, a disciple of Newton, published the now familiar form in 1715, and Colin Maclaurin partly based his treatment of calculus on Taylor series in his textbook of 1742. The remainder term, whose study is fundamental to a modern treatment of the subject, was first introduced by Lagrange about 1790.

Taylor series were vitally important to the history of calculus because all the elementary functions—polynomials, exponentials, logarithms, trigonometric functions and their inverses—could be expressed in terms of Taylor series and thus be treated in a uniform way. Taylor series do for functions what decimals do for real numbers: make them all look alike. This point of view was important in shaping the general concept of a function.

Exercises 9.6

Construct the Taylor polynomial $p_3(x)$ for each function $f(x)$ about the given point c.

1. $f(x) = \dfrac{1}{x+1}$; $c = 0$
2. $f(x) = \dfrac{1}{1-x}$; $c = 3$
3. $f(x) = \dfrac{1}{x^2+1}$; $c = 0$
4. $f(x) = \sqrt{x^2+1}$; $c = 0$
5. $f(x) = \cos 3x$; $c = \pi/3$
6. $f(x) = e^x \sin x$; $c = 0$
7. $f(x) = x^3 + 3x - 5$; $c = 0$
8. $f(x) = x^4 + 5x^3 - x^2 + x + 2$; $c = 0$

Obtain the Maclaurin series for each function, and give its interval of convergence.

9. $f(x) = e^{2x}$
10. $f(x) = e^{-x}$
11. $f(x) = \dfrac{1}{x+1}$
12. $f(x) = \dfrac{1}{x-1}$
13. $f(x) = \cosh x = \dfrac{1}{2}(e^x + e^{-x})$
14. $f(x) = \sinh x = \dfrac{1}{2}(e^x - e^{-x})$
15. $f(x) = e^{x^2}$
16. $f(x) = \cos(x^3)$

Use Taylor series or polynomials to approximate the following numbers to three decimal places:

17. $\cos 1$
18. e^2
19. $\cosh 2$
20. $\sinh(-\frac{1}{2})$

Find the sum of each series.

21. $\displaystyle\sum_{n=0}^{\infty} \frac{(-1)^n 2^n}{n!}$
22. $\displaystyle\sum_{n=0}^{\infty} \frac{(-1)^n}{n!}$
23. $\displaystyle\sum_{n=1}^{\infty} \frac{2^n}{n!}$
24. $\displaystyle\sum_{n=1}^{\infty} \frac{(-1)^{n+1}}{n2^n}$
25. $\displaystyle\sum_{n=1}^{\infty} \frac{1}{n2^n}$
26. $\displaystyle\sum_{n=2}^{\infty} \frac{1}{n2^n}$
27. $\displaystyle\sum_{n=0}^{\infty} \frac{2^{n+1}}{n!}$
28. $\displaystyle\sum_{n=1}^{\infty} \frac{1}{n2^{n-1}}$
29. $\displaystyle\sum_{n=0}^{\infty} \frac{1}{(2n)!}$
30. $\displaystyle\sum_{n=1}^{\infty} \frac{1}{(2n+1)!}$
31. $\displaystyle\sum_{n=1}^{\infty} \frac{1+(-1)^n}{(2n)!}$
32. $\displaystyle\sum_{n=1}^{\infty} \frac{(-1)^n 2^n}{(2n-1)!}$

33. Show that if f is a polynomial function of degree n, then f is its own Taylor polynomial of degree n about 0.

34. Let f be a polynomial function of degree n. How is f related to its Taylor polynomial of degree n about c?

35. Show that the Maclaurin series of a polynomial function terminates (that is, it is a polynomial).

36. Show that the Maclaurin series expansions of $\sin x$ and $\cos x$ converge to $\sin x$ and $\cos x$, respectively, for all x.

37. Suppose that f and g are continuous functions on $[a, b]$ and that g does not change sign in $[a, b]$. Prove that there is some $c \in (a, b)$ such that

$$\int_a^b f(x)g(x)\,dx = f(c)\int_a^b g(x)\,dx.$$

38. Show that Taylor's remainder $R_n(x)$ can be rewritten as

$$R_n(x) = \frac{f^{(n+1)}(\xi)}{(n+1)!}(x-c)^{n+1}$$

for some number ξ between x and c. (This form of the remainder is due to Joseph Louis Lagrange.)

39. Let $f(x) = e^{-1/x^2}$, $x \neq 0$, and $f(0) = 0$. Show that f is infinitely differentiable at 0. Construct the Maclaurin series for f, and show that it converges for all x, but not to $f(x)$.

SECTION 9.7

Operations on Power Series

A power series $\sum_{n=0}^{\infty} a_n x^n$ that converges on an interval (a, b) defines a function f on that interval. That is, a function is defined by

$$f(x) = \sum_{n=0}^{\infty} a_n x^n \qquad \text{for all } x \in (a, b).$$

We now investigate the differentiability and integrability of a function defined as the sum of a power series.

Differentiation of Power Series

The basic result concerning differentiation of power series is Theorem 9.25.

THEOREM 9.25

Let the power series $\sum_{n=0}^{\infty} a_n x^n$ have radius of convergence r, and let the function f be defined in $(-r, r)$ by $f(x) = \sum_{n=0}^{\infty} a_n x^n$. Then f is differentiable in $(-r, r)$, and

$$f'(x) = \sum_{n=0}^{\infty} n a_n x^{n-1}.$$

Proof Let g be the function defined by

$$g(x) = \sum_{n=0}^{\infty} na_n x^{n-1}.$$

Since the series $\Sigma_{n=0}^{\infty} na_n x^{n-1}$ also has radius of convergence r (as you may verify), g is defined on $(-r, r)$. Let $c \in (-r, r)$; then we wish to show that f is differentiable at c. To begin with, we compute

$$\begin{aligned}\left|\frac{f(x)-f(c)}{x-c} - g(c)\right| &= \left|\frac{\Sigma_{n=0}^{\infty} a_n x^n - \Sigma_{n=0}^{\infty} a_n c^n}{x-c} - \sum_{n=0}^{\infty} na_n c^{n-1}\right| \\ &= \left|\sum_{n=0}^{\infty} a_n \frac{x^n - c^n - (x-c)nc^{n-1}}{x-c}\right| \\ &= \left|\sum_{n=0}^{\infty} a_n \frac{x^n - nxc^{n-1} - (c^n - nc^n)}{x-c}\right|.\end{aligned}$$

(Rearrangement is permitted because of absolute convergence.)

At this point, we let $h(x) = x^n - nxc^{n-1}$. Since h is a polynomial function, we may apply the mean value theorem to h over $[c, x]$ or $[x, c]$. That is, there is a number t between x and c such that

$$\frac{h(x)-h(c)}{x-c} = h'(t) = nt^{n-1} - nc^{n-1}.$$

Hence,

$$\begin{aligned}\left|\frac{f(x)-f(c)}{x-c} - g(c)\right| &= \left|\sum_{n=0}^{\infty} a_n \frac{h(x)-h(c)}{x-c}\right| \\ &= \left|\sum_{n=0}^{\infty} a_n(nt^{n-1} - nc^{n-1})\right|.\end{aligned}$$

Now let $H(x) = nx^{n-1}$. Again by the mean value theorem, there is a number θ between t and c such that

$$\frac{H(t)-H(c)}{t-c} = H'(\theta) = n(n-1)\theta^{n-2},$$

or

$$H(t) - H(c) = (t-c)n(n-1)\theta^{n-2}.$$

Then we have

$$\begin{aligned}\left|\frac{f(x)-f(c)}{x-c} - g(c)\right| &= \left|\sum_{n=0}^{\infty} a_n[H(t) - H(c)]\right| \\ &= \left|\sum_{n=0}^{\infty} a_n(t-c)n(n-1)\theta^{n-2}\right| \\ &\le |x-c|\left|\sum_{n=0}^{\infty} a_n n(n-1)\theta^{n-2}\right|.\end{aligned}$$

Now the series $\sum_{n=0}^{\infty} a_n n(n-1)\theta^{n-2}$ has radius of convergence r (as you may easily show), and since $\theta \in (-r, r)$, the series converges to some number T. Then,

$$\left|\frac{f(x)-f(c)}{x-c} - g(c)\right| \leq |x-c||T|.$$

Therefore, as $x \to c$,

$$\left|\frac{f(x)-f(c)}{x-c} - g(c)\right| \to 0,$$

or

$$f'(c) = \lim_{x\to c} \frac{f(x)-f(c)}{x-c} = g(c).$$

This proves the theorem.

If we write out the series,

$$f(x) = a_0 + a_1 x + a_2 x^2 + a_3 x^3 + a_4 x^4 + \cdots,$$

then Theorem 9.25 says that the derivative of f is

$$f'(x) = a_1 + 2a_2 x + 3a_3 x^2 + 4a_4 x^3 + \cdots.$$

Thus, the derivative of a power series is obtained by differentiating term by term, or

$$\frac{d}{dx}\left(\sum_{n=0}^{\infty} a_n x^n\right) = \sum_{n=0}^{\infty} \frac{d}{dx}(a_n x^n).$$

By applying Theorem 9.25 to f', we see that f'' also exists, as well as f''', and all higher derivatives. That is, a function defined by a power series is infinitely differentiable in its interval of convergence.

The following examples illustrate some ways that differentiation of series can be used.

Example 9.49 Differentiate

$$f(x) = \sum_{n=1}^{\infty} \frac{x^n}{n^n}.$$

Solution Using Theorem 9.25 directly, we have

$$f'(x) = \sum_{n=1}^{\infty} \frac{nx^{n-1}}{n^n} = \sum_{n=1}^{\infty} \frac{x^{n-1}}{n^{n-1}}.$$ ■

Example 9.50 Use series expansions to verify that:

(a) $\frac{d}{dx} e^x = e^x$ (b) $\frac{d}{dx} \sin x = \cos x$

Solution **(a)** Using the Maclaurin series for e^x,

$$e^x = 1 + x + \frac{x^2}{2!} + \frac{x^3}{3!} + \cdots,$$

we find

$$\frac{d}{dx} e^x = 0 + 1 + \frac{2x}{2!} + \frac{3x^2}{3!} + \cdots$$
$$= 1 + x + \frac{x^2}{2!} + \cdots = e^x.$$

(b) Similarly,

$$\sin x = x - \frac{x^3}{3!} + \frac{x^5}{5!} - \frac{x^7}{7!} + \cdots,$$

$$\frac{d}{dx}\sin x = 1 - \frac{3x^2}{3!} + \frac{5x^4}{5!} - \frac{7x^6}{7!} + \cdots$$
$$= 1 - \frac{x^2}{2!} + \frac{x^4}{4!} - \frac{x^6}{6!} + \cdots = \cos x.$$ ■

■ **Example 9.51** Find the sum of the power series:

(a) $\sum_{n=1}^{\infty} nx^{n-1} = 1 + 2x + 3x^2 + 4x^3 + \cdots$

(b) $\sum_{n=0}^{\infty} \frac{x^{2n+1}}{2n+1} = x + \frac{x^3}{3} + \frac{x^5}{5} + \frac{x^7}{7} + \cdots$

Solution **(a)** We recognize this series as the derivative of the geometric series

$$1 + x + x^2 + \cdots = \frac{1}{1-x}.$$

Thus,

$$1 + 2x + 3x^2 + 4x^3 + \cdots = \frac{d}{dx}\left(\frac{1}{1-x}\right) = \frac{1}{(1-x)^2}.$$

(b) Let

$$f(x) = \sum_{n=0}^{\infty} \frac{x^{2n+1}}{2n+1}.$$

Then

$$f'(x) = \sum_{n=0}^{\infty} x^{2n} = 1 + x^2 + x^4 + \cdots = \frac{1}{1-x^2},$$

because this series is geometric with ratio x^2. Thus,

$$\begin{aligned} f(x) &= \int f'(x)\,dx \\ &= \int \frac{dx}{1-x^2} \\ &= \frac{1}{2}\ln\left|\frac{1+x}{1-x}\right| + C. \end{aligned}$$

To evaluate the constant C, we let $x = 0$. Then $f(0) = 0$, and $C = 0$. That is,

$$\sum_{n=0}^{\infty} \frac{x^{2n+1}}{2n+1} = \frac{1}{2}\ln\left|\frac{1+x}{1-x}\right|.$$

Note that this series converges in $(-1, 1)$. ■

Differentiation of power series enables us to make another observation relating Taylor series to other power series.

THEOREM 9.26 If function f is defined by the power series

$$f(x) = \sum_{n=0}^{\infty} a_n(x-c)^n$$

on $(c-r, c+r)$, then that same power series is the Taylor series for f about c.

Proof The kth derivative of f is

$$\begin{aligned} f^{(k)}(x) &= \sum_{n=0}^{\infty} a_n n(n-1)\cdot\cdots\cdot(n-k+1)(x-c)^{n-k} \\ &= a_k k! + \sum_{n=k+1}^{\infty} a_n n(n-1)\cdot\cdots\cdot(n-k+1)(x-c)^{n-k}. \end{aligned}$$

It follows that

$$f^{(k)}(c) = a_k k!,$$

so that

$$a_k = \frac{f^{(k)}(c)}{k!}.$$

It is now obvious that $\sum_{n=0}^{\infty} a_n(x-c)^n$ is the Taylor series for f.

Theorem 9.26 says that any series we obtain for a function is the Taylor series on its interval of convergence. As an application of this theorem, consider Example 9.52 on the next page.

Example 9.52 Find the Maclaurin series for $f(x) = x^2e^{3x}$.

Solution Starting with the Maclaurin series for e^x,

$$e^x = 1 + x + \frac{x^2}{2!} + \frac{x^3}{3!} + \frac{x^4}{4!} + \cdots,$$

we first substitute $3x$ for x:

$$\begin{aligned} e^{3x} &= 1 + 3x + \frac{(3x)^2}{2!} + \frac{(3x)^3}{3!} + \frac{(3x)^4}{4!} + \cdots \\ &= 1 + 3x + \frac{3^2x^2}{2!} + \frac{3^3x^3}{3!} + \frac{3^4x^4}{4!} + \cdots. \end{aligned}$$

Then we multiply by x^2:

$$x^2e^{3x} = x^2 + 3x^3 + \frac{3^2x^4}{2!} + \frac{3^3x^5}{3!} + \frac{3^4x^6}{4!} + \cdots.$$

In summation notation,

$$x^2e^{3x} = \sum_{n=0}^{\infty} \frac{3^n}{n!} x^{n+2}.$$

This is a power series about 0, so it is the Maclaurin series for x^2e^{3x}, by Theorem 9.26. ■

Integration of Power Series

A function that is differentiable in (a, b) is continuous in (a, b), so it is integrable inside (a, b). This is also true of a function defined by a power series.

THEOREM 9.27 Let $\sum_{n=0}^{\infty} a_nx^n$ be a power series with radius of convergence r, and let function f be defined on $(-r, r)$ by

$$f(x) = \sum_{n=0}^{\infty} a_nx^n.$$

If $[a, b] \subseteq (-r, r)$, then f is integrable on $[a, b]$, and

$$\int f(x)\,dx = \sum_{n=0}^{\infty} \frac{a_n}{n+1} x^{n+1} + C.$$

Proof Let

$$g(x) = \sum_{n=0}^{\infty} \frac{a_n}{n+1} x^{n+1}.$$

Then g has radius of convergence r (as is easily verified), so g is differentiable in $(-r, r)$. Then

$$g'(x) = \sum_{n=0}^{\infty} a_n x^n = f(x),$$

so

$$\int f(x)\, dx = \int g'(x)\, dx = g(x) + C.$$

Thus, the integral of a power series is found by integrating term by term. In the case of a definite integral, we have

$$\int_a^b \left(\sum_{n=0}^{\infty} a_n x^n\right) dx = \sum_{n=0}^{\infty} \left(\int_a^b a_n x^n\, dx\right).$$

The following examples illustrate the use of term-by-term integration of power series.

Example 9.53 Establish the series

$$\arctan x = x - \frac{x^3}{3} + \frac{x^5}{5} - \frac{x^7}{7} + \cdots = \sum_{n=0}^{\infty} \frac{(-1)^n x^{2n+1}}{2n+1}.$$

Solution Starting with the geometric series

$$\frac{1}{1-x} = 1 + x + x^2 + x^3 + \cdots = \sum_{n=0}^{\infty} x^n,$$

we replace x by $-x^2$:

$$\frac{1}{1+x^2} = 1 - x^2 + x^4 - x^6 + \cdots = \sum_{n=0}^{\infty} (-1)^n x^{2n}.$$

Now we integrate term by term, to get

$$\begin{aligned}\arctan x &= \int \frac{dx}{1+x^2}\\ &= x - \frac{x^3}{3} + \frac{x^5}{5} - \frac{x^7}{7} + \cdots + C\\ &= \sum_{n=0}^{\infty} \frac{(-1)^n x^{2n+1}}{2n+1} + C.\end{aligned}$$

To evaluate C, we set $x = 0$, getting $\arctan 0 = 0 + C$, or $C = 0$, and the desired result follows. The arctangent series converges for $x \in (-1, 1]$. ■

Example 9.54 Find the sum of the power series:

(a) $\sum_{n=0}^{\infty} \frac{x^{n+1}}{n+1}$ (b) $\sum_{n=0}^{\infty} \frac{x^{2n}}{n+1}$

Solution (a) We recognize this series as the integral of the geometric series

$$\sum_{n=0}^{\infty} x^n = \frac{1}{1-x}.$$

Hence,

$$\sum_{n=0}^{\infty} \frac{x^{n+1}}{n+1} = \int \frac{1}{1-x}\,dx = -\ln|1-x| + C.$$

Letting $x = 0$, we find $C = 0$, so

$$\sum_{n=0}^{\infty} \frac{x^{n+1}}{n+1} = \ln\left|\frac{1}{1-x}\right|.$$

This series converges on $[-1, 1)$.

(b) Let

$$f(x) = \sum_{n=0}^{\infty} \frac{x^{2n}}{n+1} = \sum_{n=0}^{\infty} \frac{(x^2)^n}{n+1}.$$

Multiplying by x^2, we get

$$x^2 f(x) = \sum_{n=0}^{\infty} \frac{(x^2)^{n+1}}{n+1},$$

which we recognize from part (a) as

$$x^2 f(x) = \ln\left|\frac{1}{1-x^2}\right|.$$

Therefore,

$$f(x) = \frac{1}{x^2}\ln\left|\frac{1}{1-x^2}\right|.$$

This series converges in $(-1, 1)$. ■

With our increased ability to sum power series, we also have increased ability to sum series of real numbers. Consider a couple of examples.

Example 9.55 Establish the identity

$$\ln\frac{4}{3} = \frac{1}{4} + \frac{1}{4 \cdot 2 \cdot 2^2} + \frac{1}{4 \cdot 3 \cdot 2^4} + \frac{1}{4 \cdot 4 \cdot 2^6} + \frac{1}{4 \cdot 5 \cdot 2^8} + \cdots.$$

Solution From Example 9.54(b), we have

$$\ln\left|\frac{1}{1-x^2}\right| = x^2 \sum_{n=0}^{\infty} \frac{x^{2n}}{n+1}.$$

If we let $x = \frac{1}{2}$ in this identity, we get

$$\begin{aligned}\ln\frac{4}{3} &= \frac{1}{4}\sum_{n=0}^{\infty}\frac{(\frac{1}{2})^{2n}}{n+1}\\ &= \frac{1}{4}\left(1 + \frac{1}{2\cdot 2^2} + \frac{1}{3\cdot 2^4} + \frac{1}{4\cdot 2^6} + \cdots\right),\end{aligned}$$

and the desired result follows. ■

■ **Example 9.56** Sum the series

$$\sum_{n=0}^{\infty} \frac{(-1)^n}{(2n+1)3^n\sqrt{3}}.$$

Solution First we write

$$3^n\sqrt{3} = 3^{n+1/2} = 3^{(2n+1)/2} = (\sqrt{3})^{2n+1},$$

so that

$$\sum_{n=0}^{\infty} \frac{(-1)^n}{(2n+1)3^n\sqrt{3}} = \sum_{n=0}^{\infty} \frac{(-1)^n}{(2n+1)}\left(\frac{1}{\sqrt{3}}\right)^{2n+1}.$$

Comparing this series with the arctangent series of Example 9.53, we see that

$$\sum_{n=0}^{\infty} \frac{(-1)^n}{(2n+1)3^n\sqrt{3}} = \arctan\frac{1}{\sqrt{3}} = \frac{\pi}{6}.$$ ■

Term-by-term integration of a series also can be helpful in estimating definite integrals for which we can find no antiderivatives.

■ **Example 9.57** Estimate $\int_0^1 e^{-x^2}\,dx$ to four decimal places.

Solution We first construct a power series expansion of e^{-x^2}. If we replace x in the Maclaurin series for e^x with $-x^2$, we get

$$e^{-x^2} = 1 - x^2 + \frac{x^4}{2!} - \frac{x^6}{3!} + \frac{x^8}{4!} - \cdots = \sum_{n=0}^{\infty} \frac{(-1)^n x^{2n}}{n!}.$$

Now we integrate term by term:

$$\begin{aligned}\int_0^1 e^{-x^2}\,dx &= \int_0^1 \left(\sum_{n=0}^{\infty}\frac{(-1)^n x^{2n}}{n!}\right)dx = \sum_{n=0}^{\infty}\frac{(-1)^n}{n!}\left(\int_0^1 x^{2n}\,dx\right)\\ &= \sum_{n=0}^{\infty}\frac{(-1)^n}{n!}\left(\frac{x^{2n+1}}{2n+1}\Big|_0^1\right) = \sum_{n=0}^{\infty}\frac{(-1)^n}{(2n+1)n!}.\end{aligned}$$

To get a four-place approximation, we recall that a partial sum of an alternating series differs from the sum by less than the first unused term. If we want $1/(2n+1)n! < 0.0001$, then we must have $(2n+1)n! > 10^4$, which is true for $n \geq 7$. Then,

$$\int_0^1 e^{-x^2}\,dx \approx 1 - \frac{1}{3} + \frac{1}{5 \cdot 2!} - \frac{1}{7 \cdot 3!} + \frac{1}{9 \cdot 4!} - \frac{1}{11 \cdot 5!} + \frac{1}{13 \cdot 6!} - \frac{1}{15 \cdot 7!} + \cdots$$

$$\approx 1 - 0.33333 + 0.1 - 0.02381 + 0.00463 - 0.00076 + 0.00011 - 0.00001 + \cdots$$

$$\approx 0.7468.$$

A numerical integration such as the trapezoidal rule would involve much more work. ∎

Exercises 9.7

Find the sum of each power series, and give its interval of convergence.

1. $\sum_{n=2}^{\infty} n(n-1)x^{n-2}$
2. $\sum_{n=3}^{\infty} n(n-1)(n-2)x^{n-3}$
3. $\sum_{n=1}^{\infty} (-1)^n n x^{n-1}$
4. $\sum_{n=1}^{\infty} (-1)^n n x^n$
5. $\sum_{n=2}^{\infty} (-1)^n n(n-1)x^{n-2}$
6. $\sum_{n=2}^{\infty} (-1)^n n(n-1)x^n$
7. $\sum_{n=0}^{\infty} \frac{1}{n+1} x^n$
8. $\sum_{n=0}^{\infty} \frac{(-1)^n x^{n+1}}{n+1}$
9. $\sum_{n=0}^{\infty} \frac{(-1)^n}{n+1} x^{2n}$
10. $\sum_{n=0}^{\infty} \frac{(-1)^n}{n+1} x^{n+2}$
11. $\sum_{n=0}^{\infty} \frac{x^{n+2}}{(n+1)(n+2)}$
12. $\sum_{n=0}^{\infty} \frac{(-1)^n x^{n+2}}{(n+1)(n+2)}$
13. $\sum_{n=1}^{\infty} (-1)^n 2n x^{2n-1}$
14. $\sum_{n=1}^{\infty} (-1)^n 2n(2n-1)x^{2n}$
15. $\sum_{n=1}^{\infty} n(x-1)^{n-1}$
16. $\sum_{n=0}^{\infty} (-1)^n \frac{(x+1)^{n+1}}{n+1}$
17. $\sum_{n=1}^{\infty} \frac{n x^{2n-2}}{2^n}$
18. $\sum_{n=1}^{\infty} \frac{2n}{n+1} x^{2n-1}$

Establish the following identities:

19. $\frac{\pi}{4} = 1 - \frac{1}{3} + \frac{1}{5} - \frac{1}{7} + \frac{1}{9} - \cdots$ [*Hint:* $\pi/4 = \arctan 1$.]
20. $\pi\sqrt{3} = \sum_{n=0}^{\infty} \frac{(-1)^n 6}{3^n(2n+1)}$
21. $\ln 2 = \frac{1}{1 \cdot 2} + \frac{1}{2 \cdot 2^2} + \frac{1}{3 \cdot 2^3} + \frac{1}{4 \cdot 2^4} + \cdots$

22. $\ln 3 = \frac{2}{1 \cdot 3} + \frac{2^2}{2 \cdot 3^2} + \frac{2^3}{3 \cdot 3^3} + \frac{2^4}{4 \cdot 3^4} + \cdots$

23. $\int e^{x^2}\,dx = \sum_{n=0}^{\infty} \frac{x^{2n+1}}{(2n+1)n!} + C$

24. $\int \arctan x\,dx = \sum_{n=0}^{\infty} \frac{(-1)^n x^{2n+2}}{(2n+1)(2n+2)} + C$

25. $\int x^2 \sin x\,dx = \sum_{n=0}^{\infty} \frac{(-1)^n x^{2n+4}}{(2n+4)(2n+1)!} + C$

26. $\int \frac{e^x}{x}\,dx = \ln x + \sum_{n=1}^{\infty} \frac{x^n}{n \cdot n!} + C$

Find power series expansions for the following functions, and give their intervals of convergence:

27. $\frac{1}{1-x^3}$

28. $\frac{1}{1+x^3}$

29. $\frac{1}{(1+x)^2}$

30. $\ln(1+x^2)$

31. $\ln|1+x^3|$

32. $\arctan 3x^2$

Sum the series.

33. $1 - \frac{2}{2} + \frac{3}{4} - \frac{4}{8} + \frac{5}{16} - \cdots$

34. $\frac{1}{2} - \frac{2}{4} + \frac{3}{8} - \frac{4}{16} + \frac{5}{32} - \cdots$

35. $1 + \frac{1}{2 \cdot 2} + \frac{1}{3 \cdot 2^2} + \frac{1}{4 \cdot 2^3} + \frac{1}{5 \cdot 2^4} + \cdots$

36. $\frac{1 \cdot 2}{1} + \frac{2 \cdot 3}{3} + \frac{3 \cdot 4}{3^2} + \frac{4 \cdot 5}{3^3} + \frac{5 \cdot 6}{3^4} + \cdots$

37. Compare the rates of convergence of the two series given below:

$$\ln 2 = 1 - \frac{1}{2} + \frac{1}{3} - \frac{1}{4} + \cdots,$$

$$\ln 2 = \frac{1}{1 \cdot 2} + \frac{1}{2 \cdot 2^2} + \frac{1}{3 \cdot 2^3} + \frac{1}{4 \cdot 2^4} + \cdots.$$

38. Estimate $\int_0^1 x^2 \sin x\,dx$ to four decimal places.

SECTION 9.8

The Binomial Series

Precalculus mathematics usually includes the study of a result known as the **binomial theorem**, which gives the expansion of a positive integral power of a binomial. The theorem states that

$$(a+b)^n = a^n + na^{n-1}b + \frac{n(n-1)}{2}a^{n-2}b^2 + \cdots + b^n$$

for any numbers a and b and any positive integer n. The binomial theorem can be generalized to include the case that n is any real number, but in that case a series is obtained instead of a finite expansion. The convergence of the resulting series to the power of the binomial was proved by Isaac Newton. Here, we will give a proof of his result and apply it to the analysis of series.

THEOREM 9.28 If $x \in (-1, 1)$ and r is any real number, then

$$(1+x)^r = 1 + rx + \frac{r(r-1)}{2!}x^2 + \frac{r(r-1)(r-2)}{3!}x^3 + \cdots + \frac{r(r-1)\cdot \cdots \cdot(r-n+1)}{n!}x^n + \cdots.$$

Before proving this theorem, we wish to make some remarks. If r is a positive integer, then the binomial series in Theorem 9.28 actually terminates, because all the terms beyond the $(r+1)$th term are 0. That is, the original binomial theorem is included as a special case of Theorem 9.28.

The coefficients in the binomial series are often abbreviated by a special symbol,

$$\binom{r}{n} = \frac{r(r-1)(r-2)\cdot \cdots \cdot(r-n+1)}{n!},$$

with $\binom{r}{0}$ defined to be 1. For example,

$$\binom{5}{2} = \frac{5\cdot 4}{2!} = 10,$$
$$\binom{5}{7} = \frac{5\cdot 4\cdot 3\cdot 2\cdot 1\cdot 0\cdot(-1)}{7!} = 0,$$
$$\binom{\frac{1}{3}}{2} = \frac{\frac{1}{3}(-\frac{2}{3})}{2!} = -\frac{1}{9}.$$

Using this notation, the binomial series may be written in the more compact form

$$(1+x)^r = \sum_{n=0}^{\infty}\binom{r}{n}x^n. \tag{1}$$

The symbol $\binom{r}{n}$ is called a **binomial coefficient** and is read as "r choose n." This rather strange phrase comes from the fact that if r is a positive integer, then $\binom{r}{n}$ is the number of combinations of r objects, chosen n at a time. Some useful identities involving the binomial coefficients are

$$n\binom{r}{n} = (r-n+1)\binom{r}{n-1}, \tag{2}$$

$$\binom{r}{n+1} = \frac{r-n}{n+1}\binom{r}{n}. \tag{3}$$

These are actually two versions of the same identity and follow from the definition of the binomial coefficients. You should be able to derive identity (2) from the definition and identity (3) from (2). (Try it.)

Proof of Theorem 9.28

The plan of the proof is to obtain equation (1) by solving a certain differential equation. We start by setting

$$y = \sum_{n=0}^{\infty} \binom{r}{n} x^n.$$

Using identity (3), we show that y has radius of convergence 1:

$$\lim_{n\to\infty} \left| \frac{\binom{r}{n}}{\binom{r}{n+1}} \right| = \lim_{n\to\infty} \left| \frac{\binom{r}{n}}{\frac{r-n}{n+1}\binom{r}{n}} \right| = \lim_{n\to\infty} \left| \frac{n+1}{r-n} \right| = 1.$$

Thus, y is defined and differentiable in $(-1, 1)$. Now,

$$\begin{aligned} \frac{dy}{dx} &= \sum_{n=1}^{\infty} n \binom{r}{n} x^{n-1} \\ &= \sum_{n=1}^{\infty} (r-n+1) \binom{r}{n-1} x^{n-1}, \end{aligned}$$

by identity (2). Multiplying this equation by $(1+x)$, we get

$$\begin{aligned} (1+x)\frac{dy}{dx} &= (1+x) \sum_{n=1}^{\infty} (r-n+1) \binom{r}{n-1} x^{n-1} \\ &= \sum_{n=1}^{\infty} (r-n+1) \binom{r}{n-1} x^{n-1} + \sum_{n=1}^{\infty} (r-n+1) \binom{r}{n-1} x^{n} \\ &= r + \sum_{n=2}^{\infty} (r-n+1) \binom{r}{n-1} x^{n-1} + \sum_{n=1}^{\infty} n \binom{r}{n} x^{n}, \end{aligned}$$

by identity (2) again. Now,

$$\sum_{n=2}^{\infty} (r-n+1) \binom{r}{n-1} x^{n-1} = \sum_{n=1}^{\infty} (r-n) \binom{r}{n} x^{n},$$

adjusting the index. Therefore,

$$\begin{aligned} (1+x)\frac{dy}{dx} &= r + \sum_{n=1}^{\infty} (r-n) \binom{r}{n} x^{n} + \sum_{n=1}^{\infty} n \binom{r}{n} x^{n} \\ &= r + \sum_{n=1}^{\infty} r \binom{r}{n} x^{n} \\ &= r \sum_{n=0}^{\infty} \binom{r}{n} x^{n} = ry. \end{aligned}$$

Thus, y satisfies the differential equation

$$(1+x)\frac{dy}{dx} = ry. \tag{4}$$

Now we solve this separable differential equation in the usual way:

$$\begin{aligned}
\frac{1}{y}\,dy &= \frac{r}{1+x}\,dx, \\
\ln y &= r\ln(1+x) + C \\
&= \ln(1+x)^r + C, \\
y &= e^{\ln(1+x)^r + C} \\
&= (1+x)^r e^C.
\end{aligned}$$

Since $y = 1$ when $x = 0$, we have $e^C = 1$, so that

$$y = (1+x)^r.$$

This establishes equation (1), and the proof is done.

Following are some applications of the binomial series.

Example 9.58 Establish the series:

(a) $$\frac{1}{\sqrt{1+x}} = 1 - \frac{1}{2}x + \frac{1\cdot 3}{2^2 2!}x^2 - \frac{1\cdot 3\cdot 5}{2^3 3!}x^3 + \frac{1\cdot 3\cdot 5\cdot 7}{2^4 4!}x^4 - \cdots$$

(b) $$\frac{1}{\sqrt{1-x^2}} = 1 + \frac{1}{2}x^2 + \frac{1\cdot 3}{2^2 2!}x^4 + \frac{1\cdot 3\cdot 5}{2^3 3!}x^6 + \cdots$$

(c) $$\arcsin x = x + \frac{1}{3\cdot 2}x^3 + \frac{1\cdot 3}{5\cdot 2^2 2!}x^5 + \frac{1\cdot 3\cdot 5}{7\cdot 2^3 3!}x^7 + \cdots$$

Solution

(a) Since $1/\sqrt{1+x} = (1+x)^{-1/2}$, we use the binomial series with $r = -\frac{1}{2}$. A typical binomial coefficient is

$$\begin{aligned}
\binom{-\frac{1}{2}}{n} &= \frac{(-\frac{1}{2})(-\frac{1}{2}-1)(-\frac{1}{2}-2)\cdot\cdots\cdot(-\frac{1}{2}-n+1)}{n!} \\
&= \frac{(-\frac{1}{2})(-\frac{3}{2})(-\frac{5}{2})\cdot\cdots\cdot[-(2n-1)/2]}{n!} \\
&= \frac{(-1)^n 1\cdot 3\cdot 5\cdot\cdots\cdot(2n-1)}{2^n n!}.
\end{aligned}$$

The desired series readily follows. You can verify that the radius of convergence is 1, just as Theorem 9.28 asserts.

(b) In part (a), replace x by $-x^2$.

(c) In part (b), integrate term by term. The constant of integration will be 0. ∎

■ **Example 9.59** Find power series expansions for:

(a) $\sqrt{1+x}$ (b) $(4-x)^{-2}$

Solution (a)

$$\begin{aligned}\sqrt{1+x} &= (1+x)^{1/2}\\ &= 1+\frac{1}{2}x+\frac{(\frac{1}{2})(-\frac{1}{2})}{2!}x^2+\frac{(\frac{1}{2})(-\frac{1}{2})(-\frac{3}{2})}{3!}x^3+\cdots\\ &= 1+\frac{1}{2}x-\frac{1}{2^2 2!}x^2+\frac{1\cdot 3}{2^3 3!}x^3-\frac{1\cdot 3\cdot 5}{2^4 4!}x^4+\cdots .\end{aligned}$$

(b)

$$(4-x)^{-2} = 4^{-2}\left(1-\frac{x}{4}\right)^{-2};$$

consequently, we start with the series

$$\begin{aligned}(1+x)^{-2} &= 1+(-2)x+\frac{(-2)(-3)}{2!}x^2+\frac{(-2)(-3)(-4)}{3!}x^3+\cdots\\ &= 1-2x+3x^2-4x^3+\cdots .\end{aligned}$$

Then we replace x by $-x/4$ and multiply by $\frac{1}{16}$:

$$\begin{aligned}(4-x)^{-2} &= \frac{1}{16}\left[1-2\left(-\frac{x}{4}\right)+3\left(-\frac{x}{4}\right)^2-4\left(-\frac{x}{4}\right)^3+\cdots\right]\\ &= \frac{1}{2^4}+\frac{2}{2^6}x+\frac{3}{2^8}x^2+\frac{4}{2^{10}}x^3+\cdots .\end{aligned}$$

■

■ **Example 9.60** Find a series converging to $\sqrt{5}$.

Solution It is tempting to use the series in Example 9.59(a) with $x=4$, but that series only converges for $x \in (-1, 1)$. So we write

$$\sqrt{5}=\sqrt{4+1}=\sqrt{4(1+\tfrac{1}{4})}=2\sqrt{1+\tfrac{1}{4}}.$$

Now we use the series in Example 9.59(a) with $x=\frac{1}{4}$:

$$\begin{aligned}\sqrt{5} &= 2\left[1+\frac{1}{2}\left(\frac{1}{4}\right)-\frac{1}{2^2 2!}\left(\frac{1}{4}\right)^2+\frac{1\cdot 3}{2^3 3!}\left(\frac{1}{4}\right)^3-\frac{1\cdot 3\cdot 5}{2^4 4!}\left(\frac{1}{4}\right)^4+\cdots\right]\\ &= 2+\frac{1}{2^2}-\frac{1}{2^5 2!}+\frac{1\cdot 3}{2^8 3!}-\frac{1\cdot 3\cdot 5}{2^{11}4!}+\cdots .\end{aligned}$$

■

Exercises 9.8

Find a power series expansion for each of the following functions, and give the interval of convergence for each one:

1. $(1+x)^{2/3}$
2. $(1+x)^{-1/3}$
3. $(1-x)^{1/3}$
4. $(1-x)^{1/5}$
5. $(x+1)^{-1}$
6. $(x+1)^{-3}$
7. $x(1+x)^{1/3}$
8. $\dfrac{x}{1-x}$
9. $\ln|1-x|$
10. $\sqrt{x+3}$
11. $\sqrt{a^2+x^2}$
12. $x\sqrt{1+x^2}$

Establish the identities.

13. $\dfrac{\pi}{2} = 1 + \dfrac{1}{3 \cdot 2} + \dfrac{1 \cdot 3}{5 \cdot 2^2 2!} + \dfrac{1 \cdot 3 \cdot 5}{7 \cdot 2^3 3!} + \cdots$

14. $\dfrac{\pi}{6} = \dfrac{1}{2} + \dfrac{1}{3 \cdot 2^4} + \dfrac{1 \cdot 3}{5 \cdot 2^7 2!} + \dfrac{1 \cdot 3 \cdot 5}{7 \cdot 2^{10} 3!} + \cdots$

15. $\dfrac{1}{\sqrt{2}} = 1 - \dfrac{1}{2} + \dfrac{1 \cdot 3}{2^2 2!} - \dfrac{1 \cdot 3 \cdot 5}{2^3 3!} + \dfrac{1 \cdot 3 \cdot 5 \cdot 7}{2^4 4!} - \cdots$

16. $4 = 1 + \dfrac{2}{2} + \dfrac{3}{4} + \dfrac{4}{8} + \dfrac{5}{16} + \cdots$

17. Show that the series

$$y = \sum_{n=0}^{\infty} \frac{x^n}{n^2 + 1}$$

is a solution of the differential equation

$$x^2 y'' + xy' + y = \frac{1}{1 - x}.$$

SECTION 9.9

Chapter Review

Important Concepts

Sequence
Recursion
Harmonic sequence
Geometric sequence
Limit
Convergence
Divergence
Bounded sequence
Upper bound
Lower bound
Increasing sequence
Decreasing sequence
Monotone sequence
Partial sum
Series
Geometric series
Harmonic series
Telescoping series
Sum
Divergence test
Cauchy criterion
Positive-term series
Integral test
p-series
Dominate
Comparison test
Limit comparison test
Alternating series
Alternating series test
Absolute convergence
Conditional convergence
Ratio test
Root test
Power series
Radius of convergence
Interval of convergence
Center of a series
Polynomial approximation
Infinitely differentiable
Taylor polynomial
Taylor's remainder
Taylor's theorem
Taylor series
Maclaurin series
Power series expansion
Binomial theorem
Binomial coefficient

Review Exercises

Determine whether the sequence converges or diverges. If it converges, find the limit.

1. $\left\{\dfrac{6}{n}\right\}$

2. $\left\{\dfrac{1}{3n}\right\}$

3. $\left\{\dfrac{\cos n^2}{n}\right\}$

4. $\left\{\dfrac{\sin^2 n}{n^2}\right\}$

5. $\left\{\dfrac{\sqrt{n}}{n+1}\right\}$

6. $\left\{\dfrac{1}{1+\sqrt{n}}\right\}$

7. $\left\{\dfrac{(-1)^n \ln n}{2+n}\right\}$ 8. $\left\{\left(-\dfrac{1}{2}\right)^n\right\}$ 9. $\{(-2)^n\}$

10. $\left\{\dfrac{10^n}{n^2}\right\}$ 11. $\left\{\dfrac{2n^2+7}{n^2-4}\right\}$ 12. $\left\{\dfrac{n^3+4}{3n-n^3}\right\}$

Determine whether the series converges or diverges. Give the sum if possible.

13. $\sum_{n=0}^{\infty} e^{-n}$ 14. $\sum_{n=0}^{\infty} \left(\dfrac{3}{4}\right)^n$ 15. $\sum_{n=0}^{\infty} \left(\dfrac{e}{2}\right)^n$

16. $\sum_{n=1}^{\infty} (-1)^n$ 17. $\sum_{n=1}^{\infty} \left(1+\dfrac{1}{n}\right)$ 18. $\sum_{n=0}^{\infty} \dfrac{n}{2^n}$

19. $\sum_{n=1}^{\infty} \dfrac{10^n}{n!}$ 20. $\sum_{n=1}^{\infty} \dfrac{n}{1+n\ln n}$ 21. $\sum_{n=1}^{\infty} n^{-5/2}$

22. $\sum_{n=1}^{\infty} n^{-2/5}$ 23. $\sum_{n=1}^{\infty} n^{1/n}$ 24. $\sum_{n=1}^{\infty} (1+\ln n)$

25. $\sum_{n=1}^{\infty} \dfrac{14}{100^n}$ 26. $\sum_{n=1}^{\infty} \dfrac{201}{1000^n}$ 27. $\sum_{n=1}^{\infty} \dfrac{12}{10^n}$

28. $\sum_{n=1}^{\infty} 3e^{-n}$ 29. $\sum_{n=1}^{\infty} \left[\dfrac{1}{2^n}+\dfrac{(-1)^n}{n}\right]$ 30. $\sum_{n=1}^{\infty} \left(\dfrac{1}{n^2}+\dfrac{1}{n}\right)$

31. $\sum_{n=1}^{\infty} \dfrac{(-1)^n 3n}{100^n}$ 32. $\sum_{n=1}^{\infty} \dfrac{(-1)^{n+1}}{n(n+1)}$ 33. $\sum_{n=1}^{\infty} \dfrac{(-1)^n n}{(n+1)^2}$

34. $\sum_{n=1}^{\infty} \dfrac{\sqrt{n}}{n^3+n^2}$ 35. $\sum_{n=1}^{\infty} \left(\dfrac{n}{n^2+1}\right)^n$ 36. $\sum_{n=1}^{\infty} \left(\dfrac{n^2}{n^2+4}\right)^n$

Give the interval of convergence of the power series, and find the sum if possible.

37. $\sum_{n=0}^{\infty} \left(\dfrac{x}{10}\right)^n$ 38. $\sum_{n=1}^{\infty} \dfrac{x^n}{2^n}$ 39. $\sum_{n=1}^{\infty} \dfrac{(-1)^n}{n^2} x^n$

40. $\sum_{n=0}^{\infty} \dfrac{3^n x^n}{5^n}$ 41. $\sum_{n=1}^{\infty} \dfrac{n}{e^n} x^{n-1}$ 42. $\sum_{n=1}^{\infty} \dfrac{n}{2^n} x^{n-1}$

43. $\sum_{n=1}^{\infty} \dfrac{nx^n}{n+1}$ 44. $\sum_{n=1}^{\infty} \dfrac{n}{2^n} x^{2n}$ 45. $\sum_{n=0}^{\infty} \dfrac{1}{n!} x^{n+1}$

46. $\sum_{n=1}^{\infty} \dfrac{(-1)^{n+1}}{n!} x^{n+2}$ 47. $\sum_{n=0}^{\infty} \dfrac{5^n}{n!} x^n$ 48. $\sum_{n=1}^{\infty} \dfrac{n}{(n+1)2^n} x^{n+1}$

Construct a Maclaurin series for each function, and give the interval of convergence.

49. e^{3x} 50. e^{-3x} 51. $\sin 5x$ 52. $\cos \frac{1}{2} x$

53. $\dfrac{1}{x}\ln(1+x^2)$ 54. $x \arctan x$ 55. $(1+x)^{-1/3}$ 56. $\sqrt{x^2+25}$

Construct a Taylor series for each function about c, and give the interval of convergence.

57. e^{-3x}; $c=1$ 58. e^{-3x}; $c=-2$

Find a series for the integral.

59. $\displaystyle\int_0^1 x^2 \arctan x \, dx$ 60. $\displaystyle\int \dfrac{\sin x}{x} \, dx$

Additional Exercises

Determine whether the sequence converges or diverges. If it converges, find the limit.

1. $\left\{\frac{10^n}{n!}\right\}$
2. $\left\{\frac{e^n}{1+10^n}\right\}$
3. $\{2^{1/n}\}$
4. $\left\{\frac{n^2+1}{(\ln n)^3}\right\}$
5. $\left\{\left(1-\frac{a}{n}\right)^n\right\}$
6. $\left\{\frac{\ln n}{\ln(\ln n)}\right\}$

Find the values of x for which the sequence converges.

7. $\{x^{2n}\}$
8. $\{x^{-n}\}$
9. $\left\{\frac{n}{1+nx}\right\}$
10. $\left\{\frac{nx}{1+n}\right\}$
11. $\{2^n x^n\}$
12. $\left\{\left(\frac{x}{10}\right)^n\right\}$

Determine whether the series converges or diverges.

13. $\sum_{n=1}^{\infty} (n+1)^n e^{-n}$
14. $\sum_{n=1}^{\infty} \left(1+\frac{1}{n}\right)^n 10^{-n}$
15. $\sum_{n=2}^{\infty} \frac{n(n-1)}{2^{n-2}}$
16. $\sum_{n=0}^{\infty} \frac{1}{(n+1)2^{n+1}}$
17. $\sum_{n=1}^{\infty} \frac{(-1)^{n+1} 1 \cdot 3 \cdot 5 \cdot \cdots \cdot (2n-1)}{2 \cdot 4 \cdot 6 \cdot \cdots \cdot 2n}$
18. $\sum_{n=1}^{\infty} \frac{1 \cdot 3 \cdot 5 \cdot \cdots \cdot (2n-1)}{n!}$
19. $\sum_{n=1}^{\infty} \left[\frac{n!}{(2n)!}\right]^n$
20. $\sum_{n=1}^{\infty} \frac{n^{1/n}}{n!}$

Give the interval of convergence.

21. $\sum_{n=1}^{\infty} \frac{(-1)^n 1 \cdot 3 \cdot 5 \cdot \cdots \cdot (2n-1)}{2^n n!} x^n$
22. $\sum_{n=1}^{\infty} \frac{1 \cdot 3 \cdot 5 \cdot \cdots \cdot (2n-1)}{2 \cdot 4 \cdot 6 \cdot \cdots \cdot (2n)} x^n$
23. $\sum_{n=1}^{\infty} n^{1/n} x^n$
24. $\sum_{n=1}^{\infty} \left(\frac{n+1}{n!}\right)^n x^n$
25. $\sum_{n=0}^{\infty} \frac{5^n}{n!}(x-3)^n$
26. $\sum_{n=1}^{\infty} \frac{n^2}{6^n}(x+1)^n$
27. $\sum_{n=0}^{\infty} \frac{(-1)^n x^{2n}}{(n!)^2 2^{2n}}$
28. $\sum_{n=0}^{\infty} \frac{(-1)^n x^{2n+1}}{n!(n+1)! 2^{2n+1}}$
29. $\sum_{n=0}^{\infty} \frac{(2n)!}{(n!)^2} x^{2n}$
30. $\sum_{n=2}^{\infty} \frac{5^{3n}}{\sqrt{n}} x^n$

31. Let function f be continuous on (a, b) and let $f(c) > 0$ at some point $c \in (a, b)$. Prove that there is an interval (r, s) containing c such that $f(x) > 0$ for every $x \in (r, s)$.
32. In summing the alternating p-series

$$\sum_{n=1}^{\infty} \frac{(-1)^n}{n^2},$$

how many terms are necessary to give a partial sum within 0.0001 of the sum?

33. Show that the sequence

$$a_n = \left(1 + \frac{1}{2} + \frac{1}{3} + \cdots + \frac{1}{n}\right) - \ln n$$

is bounded and decreasing. (Its limit is **Euler's constant,** $\gamma = 0.57721. \ldots$)

34. The functions

$$J_0(x) = \sum_{n=0}^{\infty} \frac{(-1)^n x^{2n}}{(n!)^2 2^{2n}}, \qquad J_1(x) = \sum_{n=0}^{\infty} \frac{(-1)^n x^{2n+1}}{n!(n+1)! 2^{2n+1}}$$

are called **Bessel functions.** Show that:

(a) $J_0'(x) = -J_1(x)$ **(b)** $\dfrac{d}{dx}[xJ_1(x)] = xJ_0(x)$

(c) $y = J_0(x)$ is a solution of $xy'' + y' + xy = 0$.

(d) $y = J_1(x)$ is a solution of $x^2y'' + xy' + (x^2 - 1)y = 0$.

CHAPTER 10

Plane Analytic Geometry

INTRODUCTION

Calculus often can be applied to geometry. In this chapter, plane analytic geometry is revisited and extended, applying calculus where appropriate. After considering some additional topics concerning lines, we develop the conic sections. Translations and rotations of coordinate systems are discussed in the process. Then we look at quadratic curves in general. Some additional topics will be considered in Chapter 11.

In our time, it is difficult for us to realize how completely separate algebra and geometry were at the time of Descartes. His great achievement lay in demonstrating that the two subjects are interrelated in a richly productive way. With the invention of analytic geometry, the concepts and methods of each subject became applicable to the other, and as a result, a much more profound understanding of each became possible. The algorithmic processes of algebra suggested new ways of attacking geometric problems, and geometric concepts helped to organize algebra and suggested fruitful directions for research. The influence of these cross connections extends far beyond the subject now called analytic geometry; it is fundamental to all of modern mathematics. Without this interplay of geometric and algebraic ideas, calculus would never have developed at all.

It should also be borne in mind that analytic geometry is an extremely broad and rich body of knowledge. We have space here only to skim the surface.

SECTION 10.1

More on Lines in the Plane; Translation of Axes

In this section, we use the inclination of a line to express the angle between two lines. We also use a translation of the coordinate system to obtain a formula for the distance between a point and a line.

Inclination of a Line

Let line l in the plane meet the x-axis at point A (Figure 10.1). Let B be a point on the x-axis to the right of A, and let C be a point on l that is above the x-axis. In both possible cases shown in the figure, the **angle of inclination**, or simply **inclination**, of l is the measure ϕ of angle BAC. If line l is horizontal, we define its inclination to be 0.

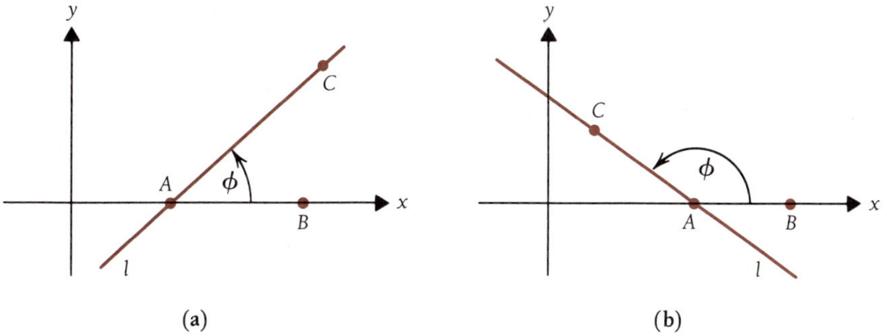

FIGURE 10.1 Angle of inclination of line l

From the definition of slope, we can see (Figure 10.2) that the slope of a line is the tangent of the angle of inclination, or

$$m = \tan \phi. \tag{1}$$

Equation (1) might even be said to hold for a vertical line, which has no slope, since $\tan(\pi/2)$ is undefined. It clearly holds for a horizontal line.

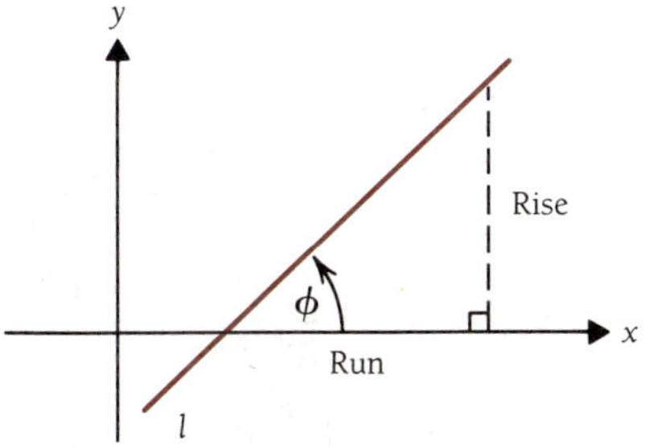

FIGURE 10.2 Slope is the tangent of the angle of inclination, $m = \tan \phi$

Angle between Two Lines

Given two intersecting lines in the plane, we can use their inclinations to determine the angle between them. Let l_1 and l_2 be two distinct, intersecting lines with inclinations ϕ_1 and ϕ_2 (Figure 10.3, page 646). Let l_1 and l_2 intersect at point P, and let θ be the measure of the angle between lines l_1 and l_2, measured in the counterclockwise direction from l_1 to l_2.

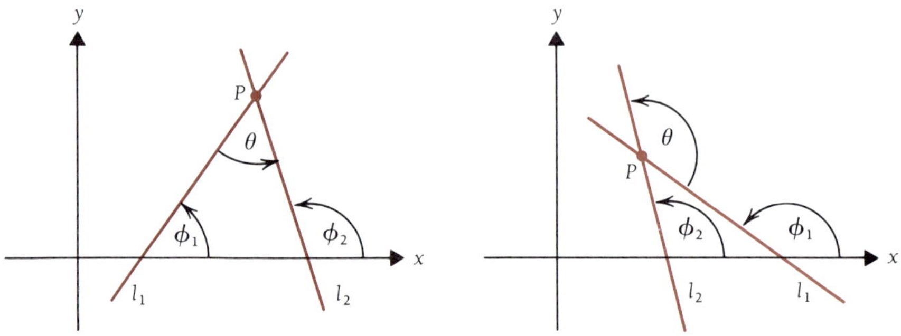

(a) (b)

FIGURE 10.3 Relating the angle between two lines to the angles of inclination

Now, using the fact that the measures of the angles of a triangle add up to π, we get one of the relationships $\theta = \phi_2 - \phi_1$ [from Figure 10.3(a)] or $\theta = \pi + \phi_2 - \phi_1$ [from Figure 10.3(b)]. In either case (since $\tan\theta$ has period π), we then have

$$\begin{aligned}\tan\theta &= \tan(\phi_2 - \phi_1) \\ &= \frac{\tan\phi_2 - \tan\phi_1}{1 + \tan\phi_1 \tan\phi_2}.\end{aligned} \tag{2}$$

If neither line l_1 nor l_2 is vertical, we can now express the tangent of the angle between them in terms of their slopes. If the slopes are $m_1 = \tan\phi_1$ and $m_2 = \tan\phi_2$, then equation (2) becomes

$$\tan\theta = \frac{m_2 - m_1}{1 + m_1 m_2}. \tag{3}$$

■ **Example 10.1** Find the angle between the lines $x + 3y = 3$ and $x + y = 2$.

Solution The lines have slopes $m_1 = -\frac{1}{3}$ and $m_2 = -1$, so the angle between the lines satisfies

$$\tan\theta = \frac{-1 - (-\frac{1}{3})}{1 + (-\frac{1}{3})(-1)} = -\frac{1}{2}.$$

Hence, $\theta = \pi + \arctan(-\frac{1}{2}) \approx 2.68 \approx 153°26'$. ■

If either line l_1 or l_2 is vertical, then formula (3) does not hold. You can readily verify that if l_2 is vertical, then $\theta = \pi - \phi_1 \pm \pi/2$, and if l_1 is vertical, then $\theta = \phi_2 \pm \pi/2$ (see Figure 10.4).

If lines l_1 and l_2 are parallel, then equation (3) tells us that the angle between them is 0. That is our convention; it is equivalent to giving a horizontal line an inclination of 0. If lines l_1 and l_2 are perpendicular, then the angle between them is $\pi/2$. In that case, both sides of equation (3) are undefined. Thus, equation (3) can be considered to hold even for perpendicular lines.

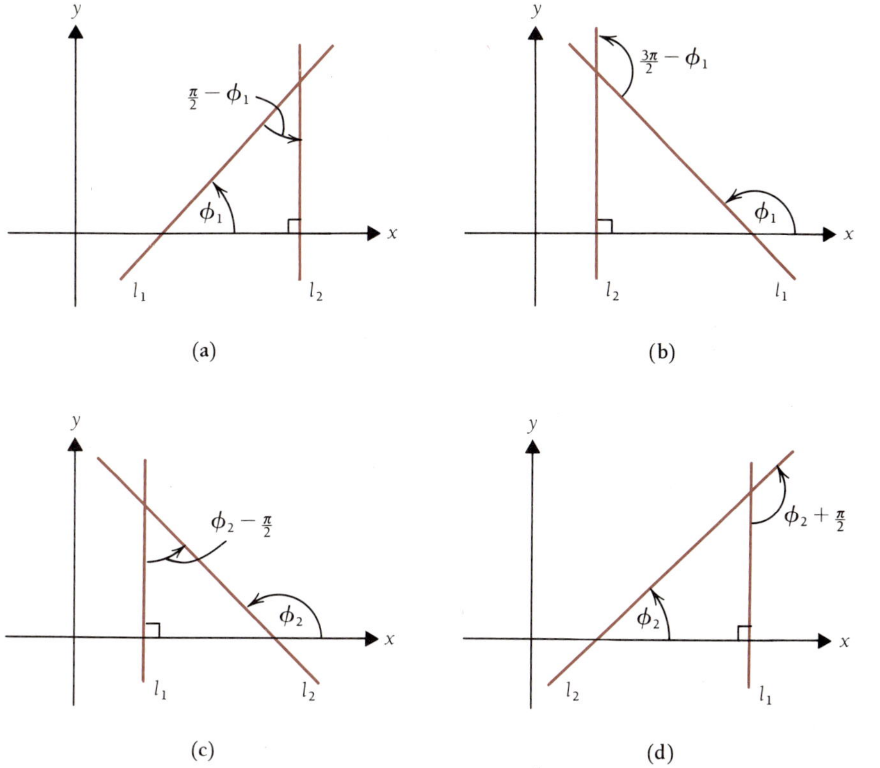

FIGURE 10.4 The angle between lines when one line is vertical

Distance between the Origin and a Line

Let line l with equation $ax + by + c = 0$ lie in the plane with point F the foot of the perpendicular from the origin to l (Figure 10.5). The distance r between the origin and F is the distance between the origin and l. If $b \neq 0$, then line l has slope $m = -a/b$, so the slope of any line perpendicular to l is $-1/m = b/a$. Therefore, the line joining F to the origin has equation

$$y = \frac{b}{a}x.$$

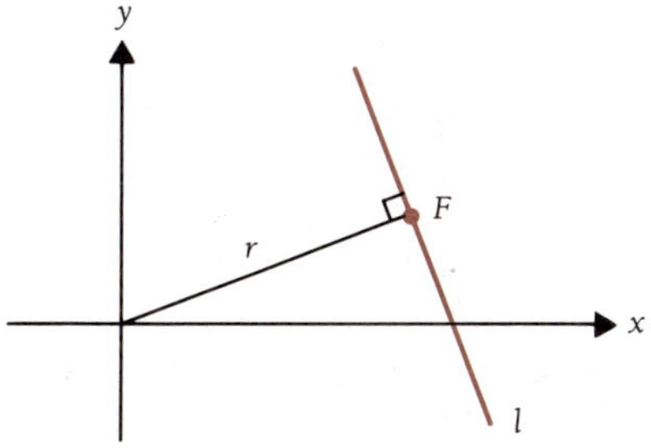

FIGURE 10.5 Distance from a line to the origin

Solving this equation simultaneously with the equation for l, we find the coordinates of point F to be

$$\left(\frac{-ac}{a^2+b^2}, \frac{-bc}{a^2+b^2}\right).$$

Finally, r is the distance between F and the origin. Using the distance formula, we get

$$r = \frac{|c|}{\sqrt{a^2+b^2}}. \tag{4}$$

If $b = 0$, then line l is vertical (Figure 10.6), and its distance from the origin is $|c/a|$, so that equation (4) also holds when $b = 0$.

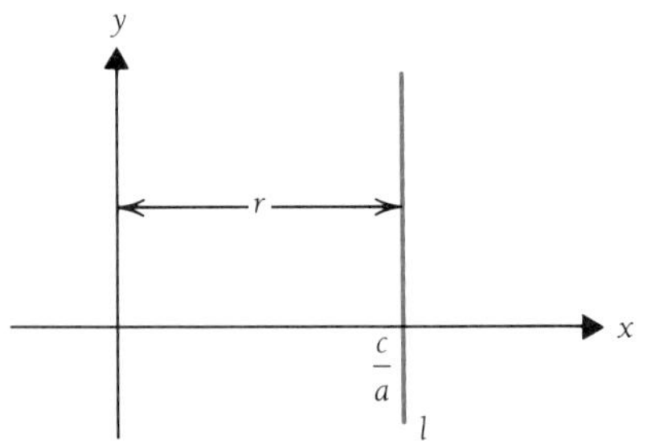

FIGURE 10.6 Distance from a vertical line to the origin

■ **Example 10.2** Find the distance between the origin and the line $3x - 6y + 5 = 0$.

Solution By formula (4), the distance is

$$r = \frac{|5|}{\sqrt{3^2+(-6)^2}} = \frac{\sqrt{5}}{3}. \qquad ■$$

In order to derive a formula for the distance between an arbitrary point and a line, we first consider a change in the coordinate system.

Translation of Axes

In an xy-coordinate system, let point P have coordinates (h, k), as shown in Figure 10.7. Through the point P, we draw a horizontal line x' and a vertical line y'. By treating x' as a new horizontal axis (the x'-axis) and y' as a new vertical axis (the y'-axis), we superimpose a new coordinate system on the plane. The origin of the $x'y'$-coordinate system is the point P. We call the $x'y'$-coordinate system a **translation** of the xy-coordinate system.

Now, any point A will have coordinates (x, y) in the xy system and coordinates (x', y') in the new $x'y'$ system. From Figure 10.7, we can see that the old and new

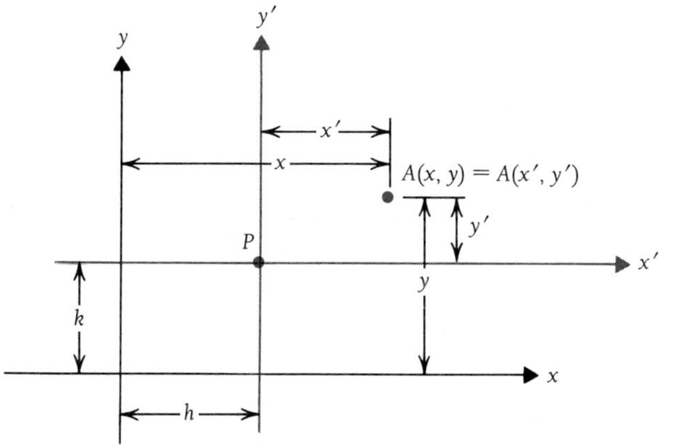

FIGURE 10.7 Deriving the equations of a translation

coordinates of A are related by the equations

$$x' = x - h, \qquad y' = y - k, \tag{5}$$

or, equivalently,

$$x = x' + h, \qquad y = y' + k. \tag{6}$$

We call equations (5) and (6) the **equations of the translation.**

To illustrate the usefulness of a translation of axes, we return to the problem of distance between a point and a line.

Distance between a Point and a Line

Let $P(h, k)$ be a point and let l: $ax + by + c = 0$ be a line in the plane. Let a translation of axes be performed, so that the origin of the new $x'y'$ system is the point P. The equation of l is given in the original xy system. To find its equation in the new $x'y'$ system, we use equations (6):

$$\begin{aligned} ax + by + c &= 0, \\ a(x' + h) + b(y' + k) + c &= 0, \\ ax' + by' + (ah + bk + c) &= 0. \end{aligned} \tag{7}$$

Now we may use equation (4) to find the distance between P and l, relative to the new coordinate system:

$$r = \frac{|ah + bk + c|}{\sqrt{a^2 + b^2}}. \tag{8}$$

Notice that we had to use equation (7) in applying the formula. But now notice that equation (8) is the same in either coordinate system, since distances do not change under translation. That is, equation (8) gives the distance between any point $P(h, k)$ and any line l: $ax + by + c = 0$.

Example 10.3 Find the distance between the point (2, 5) and the line $3x - 6y + 4 = 0$.

Solution Using equation (8) directly, we find

$$r = \frac{|3(2) - 6(5) + 4|}{\sqrt{3^2 + (-6)^2}} = \frac{4\sqrt{5}}{3}.$$

Exercises 10.1

Find the angle between the two lines.

1. $x + 2y - 3 = 0, \quad 3x + y - 4 = 0$
2. $5x - y = 2, \quad y = 2x + 1$
3. $2x - 3y - 5 = 0, \quad 3x + 2y + 5 = 0$
4. $x + 4y = 1, \quad 8y = 1 - 2x$

*The **angle at which two curves intersect** is the angle between their tangent lines at the point of intersection. Find the angle between the curves at the given point.*

5. $y = x^2, \quad y = \sqrt{x}; \quad (1, 1)$
6. $y = x^2, \quad y = 1/x^2; \quad (-1, 1)$
7. $x^2 + y^2 = 6, \quad y = x^2; \quad (\sqrt{2}, 2)$
8. $y = \sin x, \quad y = \cos x; \quad (\pi/4, 1/\sqrt{2})$

Find the distance between the point and the line.

9. $(0, 0), \quad 4x - y - 1 = 0$
10. $(0, 0), \quad y = 3x - 7$
11. $(1, -1), \quad y = 3x + 2$
12. $(4, -1), \quad x = 12$

A translation of axes places the new origin at the point $(4, -1)$. Find the equations of the following curves in the new $x'y'$ system:

13. $2x - y - 3 = 0$
14. $y = 4x + 1$
15. $x^2 + 4 = 3y$
16. $x^2 + y^2 = 1$

SECTION 10.2

The Parabola

In this section we begin the development of the conic sections.

Conic Sections

The circle, parabola, ellipse, and hyperbola are the most studied curves in geometry. They are collectively called **conic sections**, or simply **conics**, because they can be obtained by the intersection of a double-napped cone and a plane (Figure 10.8).

Conics arise naturally in physical situations where an inverse square law of attraction or repulsion governs. For example, the trajectory of a baseball is a parabola, the orbit of a planet is an ellipse, and the path of an electron past a fixed negative charge is a hyperbola. Certain geometric properties of the conics also make them useful; a parabolic reflector is just one example that will be described later.

We turn first to a study of the parabola.

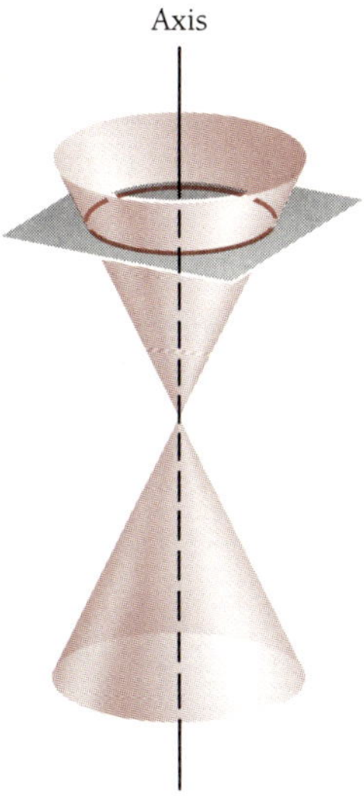

(a) Circle – the plane is perpendicular to the axis

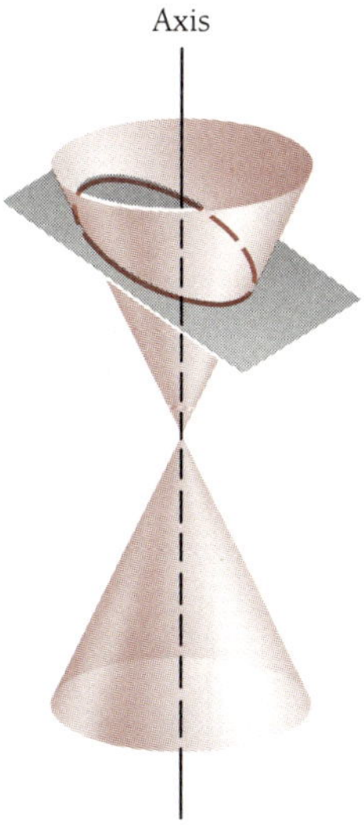

(b) Ellipse

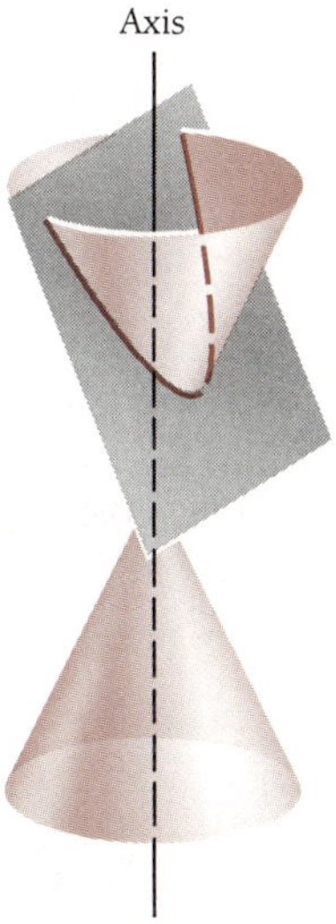

(c) Parabola – the plane is parallel to the side of the cone

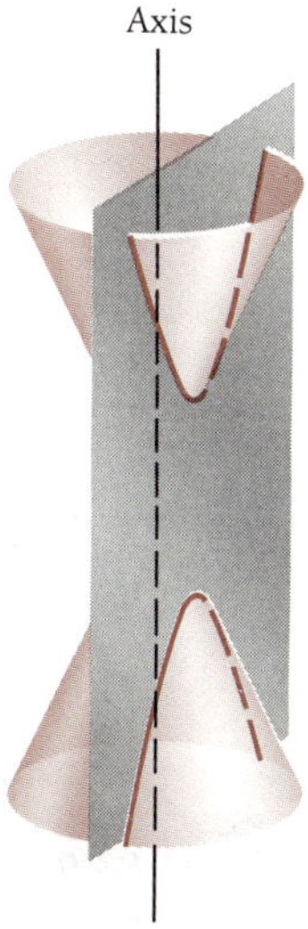

(d) Hyperbola – the plane is parallel to the axis

FIGURE 10.8 Sections of a cone by a plane — the conic sections

Definition of Parabola

DEFINITION 10.1

A **parabola** is the set of points in the plane equidistant from a fixed point (called the **focus**) and a fixed line (called the **directrix**).

That is, if F is a fixed point and d is a fixed line (Figure 10.9), the parabola with focus F and directrix d is the set of points P such that the distance $|PF|$ is equal to the distance $|Pd|$.

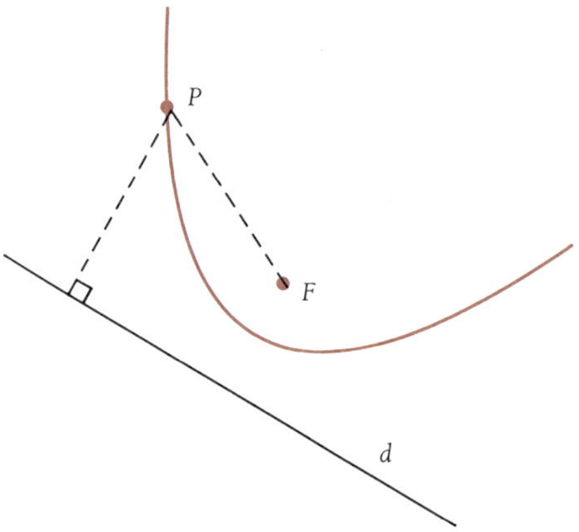

FIGURE 10.9 For a parabola, $|PF| = |Pd|$

Equation of a Parabola

Our first goal is to derive an equation for a parabola in the coordinate plane. Suppose that the focus given is the point $F(0, c)$ and the directrix d is the line $y = -c$ (see Figures 10.10 and 10.11). If $P(x, y)$ is any point, then

$$|PF| = \sqrt{x^2 + (y - c)^2}$$

and

$$|Pd| = |y + c|.$$

The parabola with focus F and directrix d consists of the points P such that $|PF| = |Pd|$, or

$$\sqrt{x^2 + (y - c)^2} = |y + c|.$$

We simplify this equation by squaring and combining terms:

$$x^2 + (y - c)^2 = (y + c)^2,$$
$$x^2 + y^2 - 2cy + c^2 = y^2 + 2cy + c^2,$$
$$x^2 = 4cy. \tag{1}$$

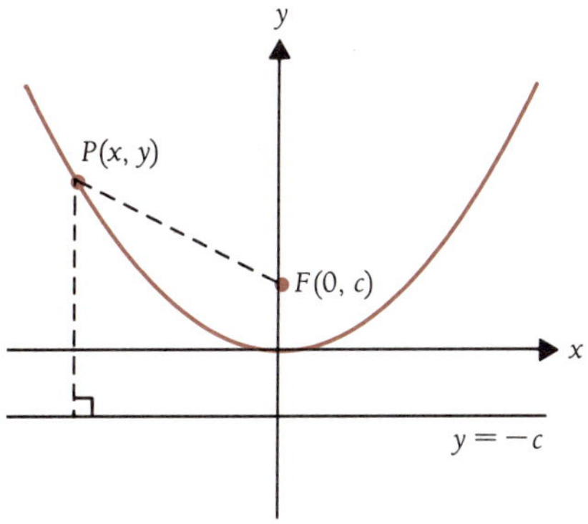

FIGURE 10.10 A parabola that opens upward

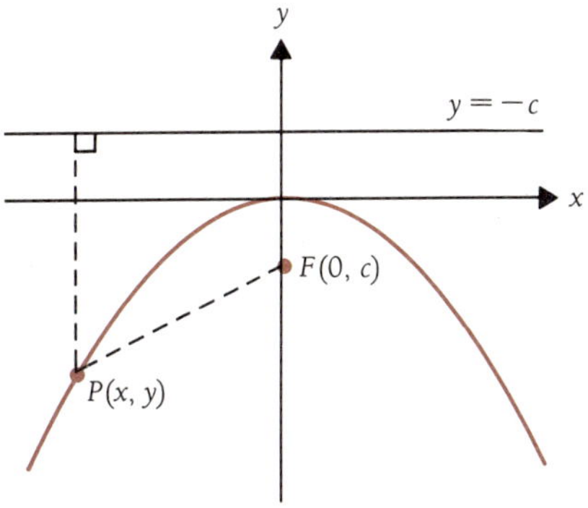

FIGURE 10.11 A parabola that opens downward

We call (1) the **equation of the parabola**, because the point $P(x, y)$ belongs to the parabola if and only if the coordinates of P satisfy equation (1).

If c is positive in equation (1), the parabola opens upward, as shown in Figure 10.10. If c is negative, the parabola opens downward, as in Figure 10.11.

The line passing through the focus of a parabola and perpendicular to the directrix is an axis of symmetry of the parabola (in the case of Figures 10.10 and 10.11, the y-axis), and is called the **axis** of the parabola. The axis meets the parabola at a point called the **vertex** (in the case of Figures 10.10 and 10.11, the vertex is the origin).

Obviously, different choices for the focus and directrix of a parabola give rise to different equations. For example, if the point $F(c, 0)$ is the focus and the line $x = -c$ is the directrix (Figures 10.12 and 10.13), then the equation of the parabola is

$$y^2 = 4cx. \tag{2}$$

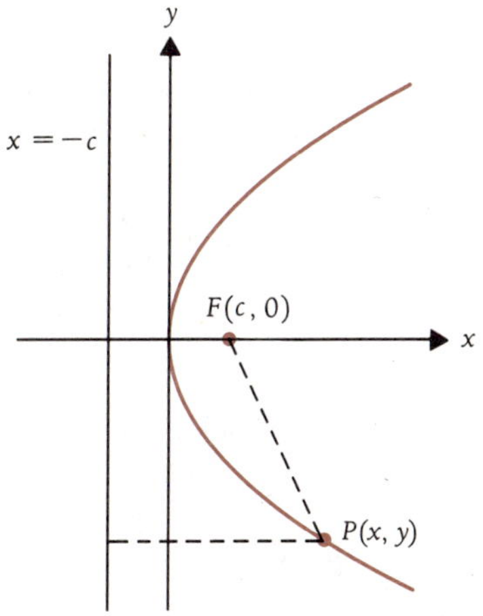

FIGURE 10.12 A parabola that opens to the right

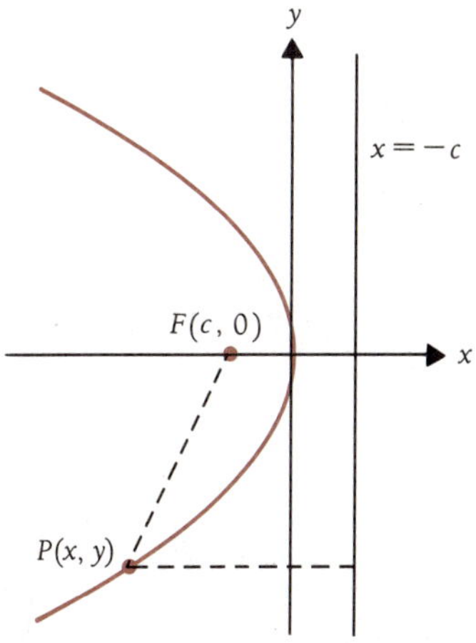

FIGURE 10.13 A parabola that opens to the left

This equation is derived in the same way as equation (1), or we could have simply interchanged x and y in equation (1). If c is positive in equation (2), the parabola opens to the right, as shown in Figure 10.12. If c is negative, the parabola opens to the left, as in Figure 10.13. In each case, the axis of the parabola is the x-axis, and the vertex is the origin.

A parabola whose equation is of the form of equation (1) or (2) is said to be in **standard position**, and equations (1) and (2) are **standard forms** of the equation of a parabola. A parabola in standard position will always have its vertex at the origin and one of the coordinate axes as its axis. Several examples of parabolas in standard position are given in Figure 10.14.

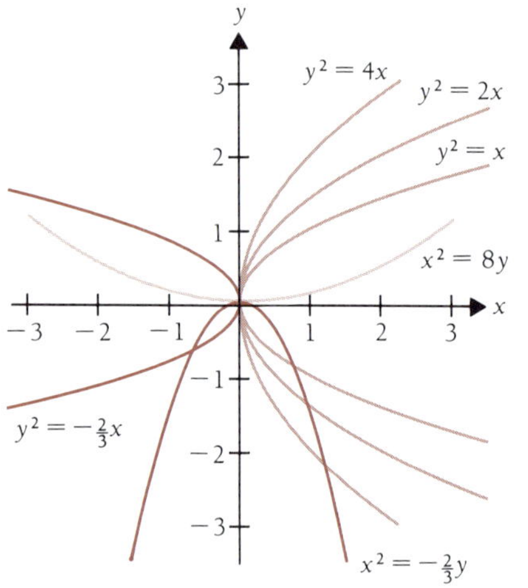

FIGURE 10.14 Parabolas in standard position

As you may suspect, a parabola that is not in standard position has a more complicated equation. If the vertex of the parabola is not at the origin, but its axis is parallel to a coordinate axis, then we may use a translation of axes to get its equation.

Suppose that the axis of a parabola is vertical and that its vertex is at the point (h, k), as shown in Figure 10.15. Let a new $x'y'$-coordinate system be chosen with its origin at (h, k). The equations of the translation are

$$x' = x - h, \qquad y' = y - k. \tag{3}$$

In the new system, our parabola has an equation of the form

$$x'^2 = 4cy', \tag{4}$$

by comparison with Figure 10.10. We find its equation in the original coordinate system by substituting equations (3) into equation (4):

$$(x - h)^2 = 4c(y - k). \tag{5}$$

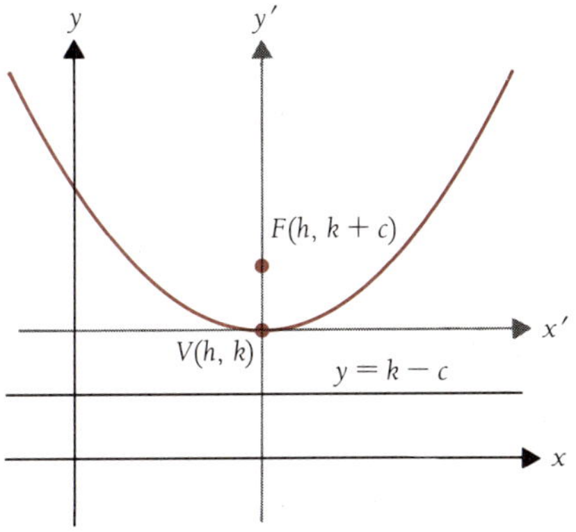

FIGURE 10.15 A parabola with vertex $V(h, k)$

If the axis of the parabola had been horizontal, its equation would have had the form

$$(y - k)^2 = 4c(x - h). \tag{6}$$

For parabolas with equations of the forms (5) and (6), it is helpful to remember that the vertex is midway between the focus and the directrix, and the distance between the focus and the vertex is $|c|$.

Example 10.4 Find the equation of the parabola whose focus is $F(2, 3)$ and whose vertex is $V(4, 3)$.

Solution The distance from the focus to the vertex is 2, and since the focus is to the left of the vertex (Figure 10.16), $c = -2$. Also, since the vertex is $(4, 3)$, we have $h = 4$ and $k = 3$, and the equation of the parabola is of the form (6), namely,

$$(y - 3)^2 = -8(x - 4).$$

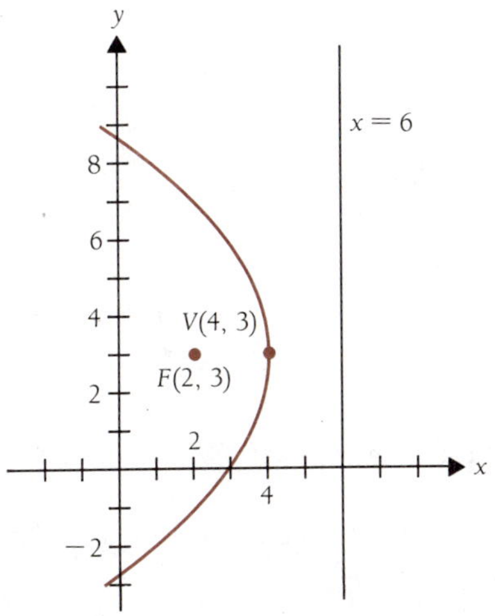

FIGURE 10.16 The parabola $(y - 3)^2 = -8(x - 4)$ ■

We often may locate the graph of a parabola by writing its equation in the form (5) or (6). This is done by completing the square.

Example 10.5 Describe and sketch the graph of

$$x^2 + 8x - 4y - 2 = 0.$$

Solution We first collect the x terms and the y terms separately, and then complete the square:

$$x^2 + 8x = 4y + 2,$$
$$x^2 + 8x + 16 = 4y + 18,$$
$$(x + 4)^2 = 4\left(y + \frac{9}{2}\right).$$

Comparing this equation with equation (5), we see that it represents a parabola with vertex $V(-4, -\frac{9}{2})$. Since $c = 1$, the focus is $F(-4, -\frac{7}{2})$ and the directrix is d: $y = -\frac{11}{2}$. The graph is sketched in Figure 10.17.

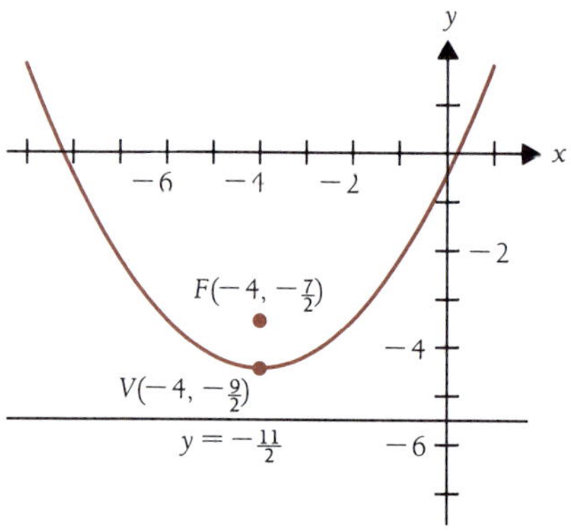

FIGURE 10.17 The parabola $x^2 + 8x - 4y - 2 = 0$ ■

A Parabola in General Position

A parabola whose directrix is neither horizontal nor vertical has an equation that is even more complicated. Consider Example 10.6.

Example 10.6 Find the equation of the parabola with focus $F(1, 2)$ and directrix $d: x - y + 3 = 0$.

Solution If $P(x, y)$ is any point, then the distance between P and F is

$$|PF| = \sqrt{(x-1)^2 + (y-2)^2}.$$

The distance between P and d is

$$|Pd| = \frac{|x - y + 3|}{\sqrt{1^2 + (-1)^2}} = \frac{\sqrt{2}}{2}|x - y + 3|.$$

Hence, the equation of the parabola is

$$\sqrt{(x-1)^2 + (y-2)^2} = \frac{\sqrt{2}}{2}|x - y + 3|,$$

$$(x-1)^2 + (y-2)^2 = \frac{1}{2}(x - y + 3)^2,$$

which eventually simplifies to

$$x^2 + 2xy + y^2 - 10x - 2y + 1 = 0.$$

This is the equation of the parabola; the curve is sketched in Figure 10.18.

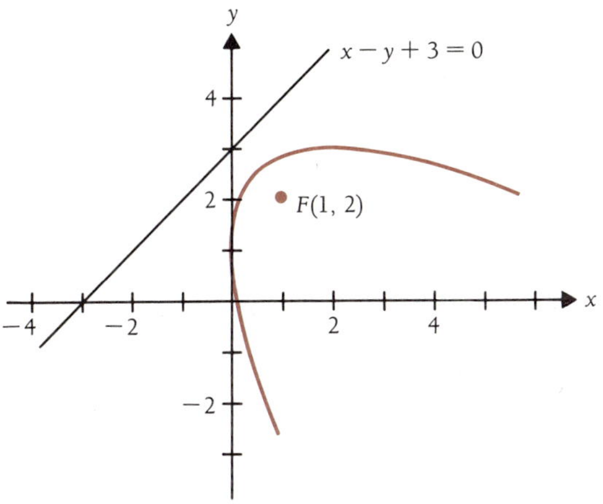

FIGURE 10.18 The parabola $x^2 + 2xy + y^2 - 10x - 2y + 1 = 0$

■

Applications of the Parabola

The principle that makes a parabolic reflector work is a reflection property of parabolas. Let T be a point on a parabola, and let line t be tangent to the parabola at T (Figure 10.19, page 658). Then the line FT makes the same angle with t as the line through T that is parallel to the axis of the parabola (see Problem 17). In terms of the reflector, this means that light rays emanating from the focus of a parabola are all reflected in a direction parallel to the axis. Or, reversing the process, light rays coming into the parabolic reflector from a direction parallel to the axis are all focused at one point—the focus. Spotlights, radar antennae, reflecting telescopes, solar reflectors, and parabolic microphones are a few of the applications of parabolic reflectors.

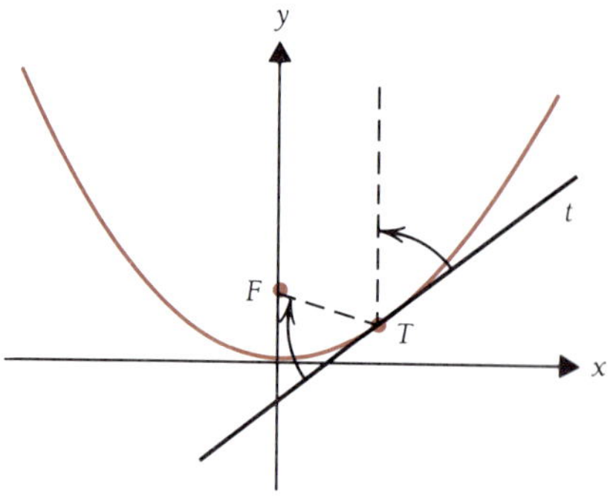

FIGURE 10.19 The reflective property of a parabola

Parabolas also occur in natural contexts. The path of a projectile in the earth's gravitational system is a parabola, and the cables supporting a suspension bridge follow a parabolic arc.

Exercises 10.2

Find an equation for the parabola:

1. With focus $F(0, 2)$ and directrix d: $y = -2$
2. With focus $F(3, 0)$ and directrix d: $x = -3$
3. With focus $F(0, 0)$ and vertex $V(0, 1)$
4. With vertex $V(2, -3)$ and focus $F(0, -3)$
5. With focus $F(1, 2)$ and directrix d: $x + y + 1 = 0$
6. With focus $F(0, 0)$ and directrix d: $y = 2x + 3$

Give the vertex, focus, and directrix of each parabola and sketch its graph.

7. $x^2 = 4y$
8. $y^2 = -2x$
9. $x^2 + 6x = -3y$
10. $y^2 = 4x - 8$
11. $x^2 - 2x + 4y - 6 = 0$
12. $x - 5y^2 + 10y - 25 = 0$

Let point $T(x_0, y_0)$ be a point on the parabola $x^2 = 4cy$, and let line t be tangent to the parabola at T.

13. Find the slope of t.
14. Find an equation for t.
15. Let F be the focus of the parabola, and find the angle between lines t and TF.
16. Find the angle between t and the vertical line through T.
17. Let t meet the axis of the parabola at point A. Show that triangle TFA is isosceles.

SECTION 10.3

The Ellipse

In this section, we develop the properties of the ellipse. As in the case of the parabola, we first give a purely geometric definition and then find equations for ellipses in various positions. From the equations, we learn the properties of the ellipse.

Definition of Ellipse

DEFINITION 10.2 An **ellipse** is the set of points, the sum of whose distances from two fixed points (called the **foci**) is a constant.

That is, if F_1 and F_2 are two fixed points and k is a positive constant, then the ellipse determined by F_1, F_2, and k (see Figure 10.20) is the set of points P such that

$$|PF_1| + |PF_2| = k.$$

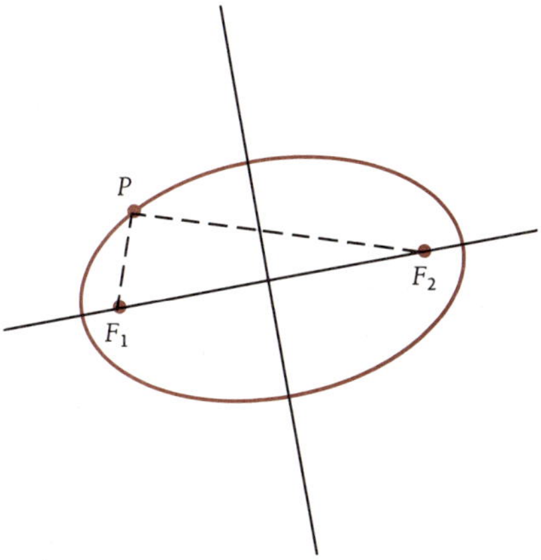

FIGURE 10.20 An ellipse with foci F_1 and F_2

Note that the constant k must be at least as large as the distance between F_1 and F_2 if the ellipse is to have any points at all. Also, if $k = |F_1F_2|$, the ellipse consists only of the line segment joining F_1 and F_2. We will therefore consider only the case $k > |F_1F_2|$.

Equations of an Ellipse

Suppose that foci $F_1(c, 0)$ and $F_2(-c, 0)$ are given (Figure 10.21, page 660) and constant k is specified. Since $k > |F_1F_2| = 2c$, we will set $k = 2a$ with $a > c$. If $P(x, y)$ is any point, then

$$|PF_1| = \sqrt{(x-c)^2 + y^2}$$

and

$$|PF_2| = \sqrt{(x+c)^2 + y^2}.$$

Hence, the ellipse has equation

$$\sqrt{(x-c)^2 + y^2} + \sqrt{(x+c)^2 + y^2} = 2a,$$

which we simplify by rearranging and squaring a couple of times:

$$\begin{aligned}
\sqrt{(x-c)^2+y^2} &= 2a-\sqrt{(x+c)^2+y^2},\\
(x-c)^2+y^2 &= 4a^2-4a\sqrt{(x+c)^2+y^2}+(x+c)^2+y^2,\\
x^2-2cx+c^2 &= 4a^2-4a\sqrt{(x+c)^2+y^2}+x^2+2cx+c^2,\\
a\sqrt{(x+c)^2+y^2} &= a^2+cx,\\
a^2x^2+2a^2cx+a^2c^2+a^2y^2 &= a^4+2a^2cx+c^2x^2,\\
(a^2-c^2)x^2+a^2y^2 &= a^2(a^2-c^2),\\
\frac{x^2}{a^2}+\frac{y^2}{a^2-c^2} &= 1. \qquad (1)
\end{aligned}$$

Since $a > c$, the number $a^2 - c^2$ in equation (1) is positive; so, for convenience, we let

$$b^2 = a^2 - c^2. \qquad (2)$$

Then we can write the equation of the ellipse in the form

$$\frac{x^2}{a^2}+\frac{y^2}{b^2} = 1. \qquad (3)$$

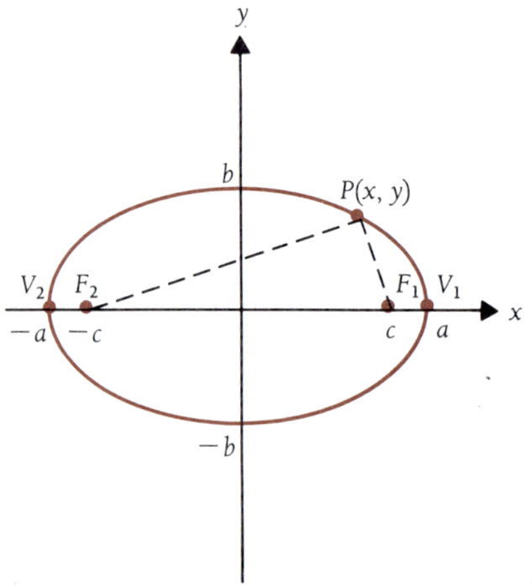

FIGURE 10.21 The ellipse $\frac{x^2}{a^2}+\frac{y^2}{b^2}=1$

The line passing through both foci of an ellipse is an axis of symmetry of the ellipse and meets the ellipse at two points called **vertices**. The line segment joining the vertices is called the **major axis** of the ellipse. In the case of Figure 10.21, the major axis lies along the x-axis and has length $2a$. For that reason, the number a is called the **semimajor axis** of the ellipse.

The perpendicular bisector of the major axis of an ellipse is also an axis of symmetry of the ellipse and meets the ellipse at two points. The line segment joining those points is called the **minor axis** of the ellipse. In the case of Figure 10.21, the minor axis lies along the y-axis and has length $2b$; thus, b is called the **semiminor axis** of the ellipse.

The **center** of an ellipse is the intersection of its two axes. In the case of Figure 10.21, the center is at the origin.

If the foci of an ellipse are given as $F_1(0, c)$ and $F_2(0, -c)$, then the equation of the ellipse will have the form

$$\frac{y^2}{a^2} + \frac{x^2}{b^2} = 1. \tag{4}$$

This equation is derived in the same manner as equation (3); the ellipse is sketched in Figure 10.22. Again, its center is at the origin and its axes lie along the coordinate axes.

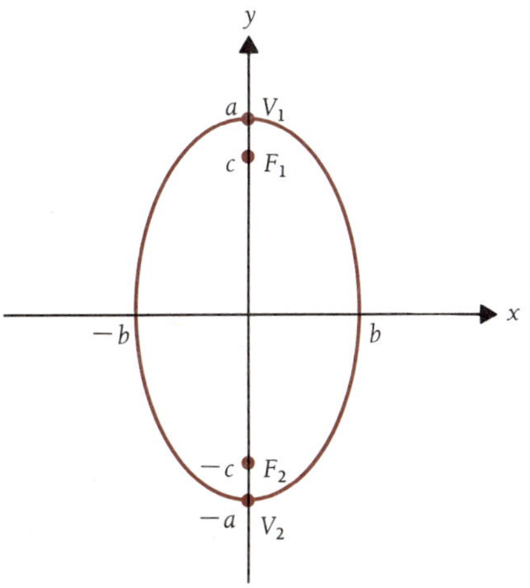

FIGURE 10.22 The ellipse $\frac{x^2}{b^2} + \frac{y^2}{a^2} = 1$

An ellipse centered at the origin and with its axes lying along the coordinate axes is said to be in **standard position**. Equations (3) and (4) are **standard forms** of the equation of an ellipse. Additional examples of ellipses in standard position are shown in Figure 10.23 on page 662.

If an ellipse is not in standard position, its equation is more complicated than the standard form. If its axes are still parallel to the coordinate axes, a translation can be used to get its equation.

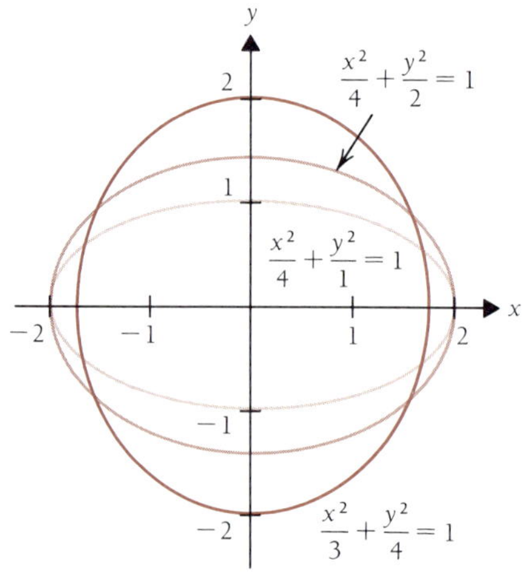

FIGURE 10.23 Ellipses in standard position

Ellipses and Translations

Suppose an ellipse has center $C(h, k)$ and foci $F_1(h + c, k)$ and $F_2(h - c, k)$, as shown in Figure 10.24. Let a new $x'y'$-coordinate system be chosen with origin (h, k). In the new system, the ellipse will have an equation of the form

$$\frac{x'^2}{a^2} + \frac{y'^2}{b^2} = 1.$$

Hence, in the original system, the ellipse has equation

$$\frac{(x-h)^2}{a^2} + \frac{(y-k)^2}{b^2} = 1. \tag{5}$$

In this case, the vertices are the points $V_1(h + a, k)$ and $V_2(h - a, k)$.

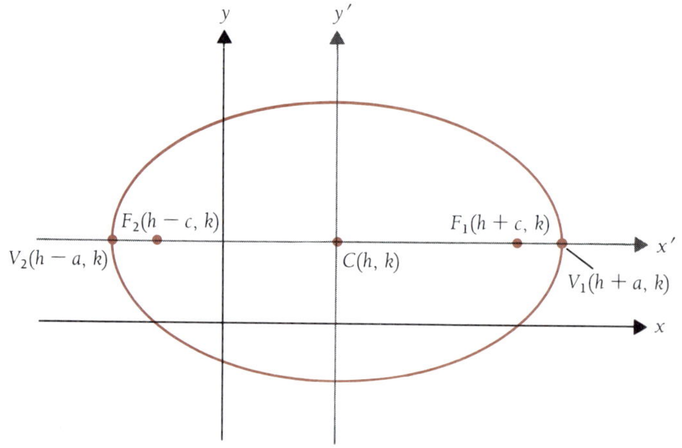

FIGURE 10.24 An ellipse with center $C(h, k)$

The analog to equation (5) for an ellipse with major axis vertical is

$$\frac{(y-k)^2}{a^2} + \frac{(x-h)^2}{b^2} = 1. \tag{6}$$

In this case, the center is $C(h, k)$, the foci are $F_1(h, k+c)$ and $F_2(h, k-c)$, and the vertices are $V_1(h, k+a)$ and $V_2(h, k-a)$. (As always for an ellipse, $b^2 = a^2 - c^2$.)

■ **Example 10.7** Find the equation of the ellipse with semimajor axis $a = 6$ and foci $F_1(2, -1)$ and $F_2(2, 2)$.

Solution The center of the ellipse must be midway between F_1 and F_2, so it is the point $C(2, \frac{1}{2})$. Thus, $c = \frac{3}{2}$, and

$$b^2 = a^2 - c^2 = 36 - \frac{9}{4} = \frac{135}{4}.$$

From equation (6), we find the equation to be

$$\frac{(y-\frac{1}{2})^2}{36} + \frac{(x-2)^2}{\frac{135}{4}} = 1.$$

■

■ **Example 10.8** Give the center, foci, vertices, and semiaxes of the ellipse

$$2x^2 + y^2 - 4x + 6y - 1 = 0.$$

Solution We first write the equation in the form of equation (5) or (6) by rearranging and completing the square:

$$2x^2 - 4x + y^2 + 6y = 1,$$
$$2(x^2 - 2x + 1) + (y^2 + 6y + 9) = 1 + 2 + 9,$$
$$2(x-1)^2 + (y+3)^2 = 12,$$
$$\frac{(x-1)^2}{6} + \frac{(y+3)^2}{12} = 1.$$

This equation is of the form shown in equation (6) with $h = 1$, $k = -3$, $a = \sqrt{12}$, and $b = \sqrt{6}$. Hence, $c = \sqrt{6}$. This tells us that the center is $C(1, -3)$, the foci are $F_1(1, -3+\sqrt{6})$ and $F_2(1, -3-\sqrt{6})$, the vertices are $V_1(1, -3+\sqrt{12})$ and $V_2(1, -3-\sqrt{12})$, the semimajor axis is $\sqrt{12}$, and the semiminor axis is $\sqrt{6}$. The ellipse is drawn in Figure 10.25 on page 664. ■

An Ellipse in General Position

If the axes of an ellipse are not parallel to the coordinate axes, the ellipse has an even more complicated equation. Consider Example 10.9.

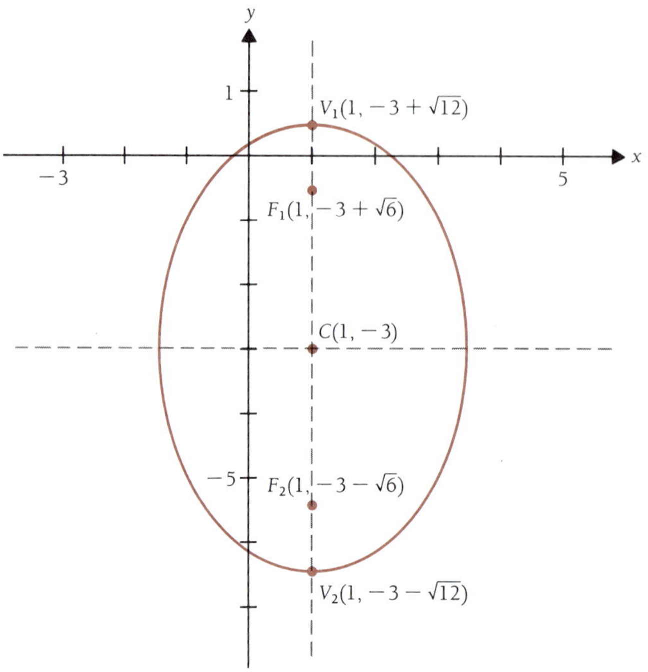

FIGURE 10.25 The ellipse $2x^2 + y^2 - 4x + 6y - 1 = 0$

Example 10.9 Find the equation of the ellipse with foci $F_1(1, 2)$ and $F_2(3, 1)$ and major axis of length 6.

Solution Let $P(x, y)$ be any point. Then the equation of the ellipse will be

$$\begin{aligned}
|PF_1| + |PF_2| &= 6, \\
|PF_1| &= 6 - |PF_2|, \\
|PF_1|^2 &= 36 - 12|PF_2| + |PF_2|^2, \\
(x-1)^2 + (y-2)^2 &= 36 - 12\sqrt{(x-3)^2 + (y-1)^2} \\
&\quad + (x-3)^2 + (y-1)^2, \\
12\sqrt{(x-3)^2 + (y-1)^2} &= 41 - 4x + 2y, \\
144[(x-3)^2 + (y-1)^2] &= 1681 + 16x^2 + 4y^2 - 328x \\
&\quad + 164y - 16xy,
\end{aligned}$$

$$128x^2 + 16xy + 140y^2 - 536x - 452y - 241 = 0.$$

This is the equation of the ellipse; it is sketched in Figure 10.26.

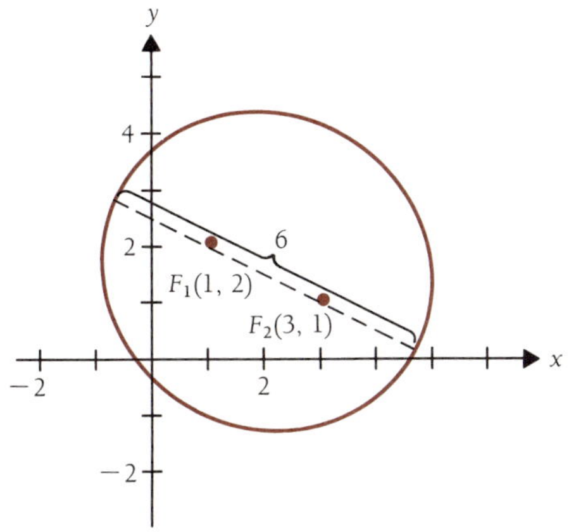

FIGURE 10.26 The ellipse $128x^2 + 16xy + 140y^2 - 536x - 452y - 241 = 0$ ■

Applications of an Ellipse

An ellipse also has an interesting and useful reflection property. Let T be any point on an ellipse with foci F_1 and F_2, and let t be the tangent line to the ellipse at T. Then the lines TF_1 and TF_2 make the same angle with t (Figure 10.27). This means, for example, that if an auditorium is elliptical in shape and a speaker stands at one focus, the sound waves emanating from the speaker will be reflected toward the other focus. Moreover, since every path from one focus to a point on the ellipse to the other focus is the same length, the sound waves all converge at the other focus at the same time. Thus, a whisper at one focus can be heard at the other focus, perhaps a considerable distance away.

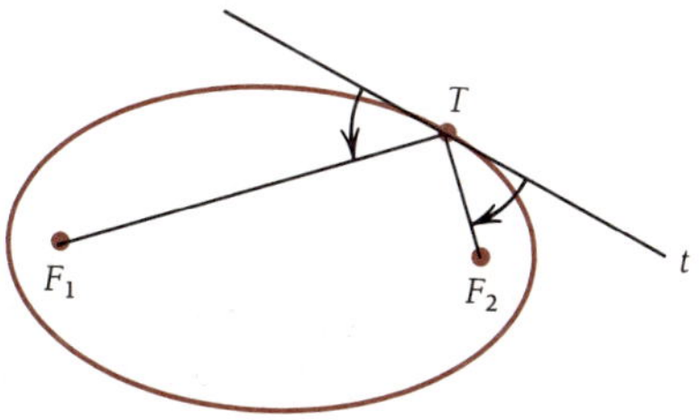

FIGURE 10.27 The reflective property of an ellipse

Exercises 10.3

Find the equation of the ellipse:

1. With foci (0, 1) and (0, −1), and vertex (0, 4)
2. With foci (3, 0) and (−3, 0), and vertex (4, 0)
3. With foci (0, 0) and (4, 0), and vertex (5, 0)
4. With center (1, 2), vertex (5, 2), and minor axis 6

A5. With vertices (0, 0) and (4, 3), and focus $(1, \frac{3}{4})$

6. With semiaxes 10 and 2, center (0, 0), and vertex (8, −6)

Give the center, foci, vertices, and semiaxes of each ellipse, and sketch its graph.

7. $4x^2 + 9y^2 = 36$
8. $4x^2 + y^2 = 4$
9. $x^2 + 2y^2 + 4y = 4$
10. $6x^2 - 18x + y^2 + 4y = \frac{19}{2}$
11. $x^2 + 8x + 4y^2 = 0$
12. $25x^2 + 16y^2 - 200x - 64y + 64 = 0$

Let R be the region bounded by an ellipse with semiaxes a and b.

13. Find the area of R.

14. Locate the centroid of R.

15. Find the volume generated by revolving R about:
    (a) The major axis (b) The minor axis
    (c) The line tangent to the ellipse at a vertex
    (d) The line tangent to the ellipse at the end of the minor axis

16. Express the perimeter of R as an integral.

Let $T(x_0, y_0)$ be a point on the ellipse $b^2x^2 + a^2y^2 = a^2b^2$, and let t be the line tangent to the ellipse at T.

17. Find the slope of t.

A18. If the ellipse has foci F_1 and F_2, show that lines TF_1 and TF_2 meet t at the same angle.

19. Suppose a swimming pool is elliptic in shape. Explain why a person diving in at one focus will produce a splash at the other focus a few moments later.

20. Discuss the special features of an elliptic billiard table.

SECTION 10.4

The Hyperbola

DEFINITION 10.3

A **hyperbola** is the set of points in the plane, the difference of whose distances from two fixed points (called the **foci**) is a constant.

That is, given two points F_1 and F_2 and a positive constant k, the hyperbola with foci F_1 and F_2 determined by k is the set of points P such that

$$||PF_1| - |PF_2|| = k$$

(see Figure 10.28). In this case, we must have $k < |F_1F_2|$ in order for the hyperbola to exist.

Equation of a Hyperbola

We obtain an equation for a hyperbola in much the same way as for an ellipse. Given fixed points $F_1(c, 0)$ and $F_2(-c, 0)$, we let $k = 2a$ with $a < c$. The equation of the

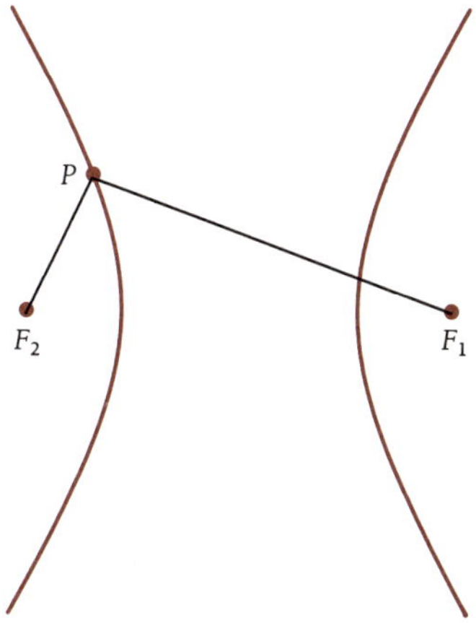

FIGURE 10.28 A hyperbola with foci F_1 and F_2

hyperbola so determined is

$$\left|\sqrt{(x-c)^2+y^2}-\sqrt{(x+c)^2+y^2}\right|=2a.$$

This equation can be simplified to

$$\frac{x^2}{a^2}-\frac{y^2}{c^2-a^2}=1. \tag{1}$$

Since $c > a$, we may set

$$b^2=c^2-a^2, \tag{2}$$

and thereby rewrite equation (1) as

$$\frac{x^2}{a^2}-\frac{y^2}{b^2}=1. \tag{3}$$

This hyperbola is symmetric about each coordinate axis, as shown in Figure 10.29. The axis of symmetry that passes through the foci is called the **transverse axis** and

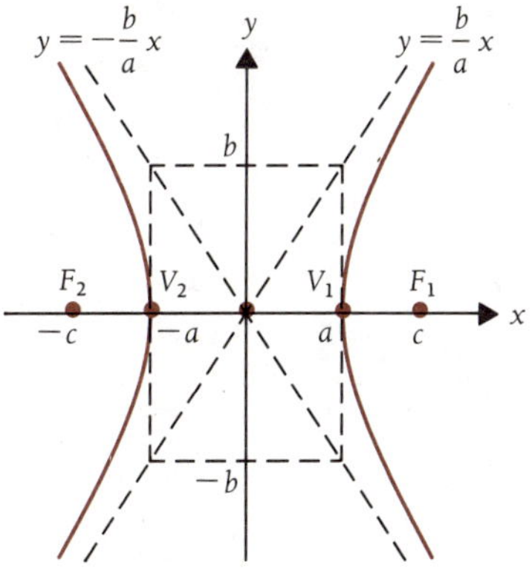

FIGURE 10.29 The hyperbola $\frac{x^2}{a^2}-\frac{y^2}{b^2}=1$

meets the hyperbola at two points called **vertices**. The perpendicular bisector of the segment joining the foci is called the **conjugate axis** and meets the transverse axis at the **center** of the hyperbola.

If we solve equation (3) for y, we can write

$$y = \pm \frac{b}{a} x \sqrt{1 - \frac{a^2}{x^2}}.$$

Then, as $x \to \pm\infty$, $y \to \pm(b/a)x$. That is, the curve approaches the lines

$$y = \frac{b}{a} x, \qquad y = -\frac{b}{a} x,$$

which are called **asymptotes** of the hyperbola. The asymptotes meet at the center of the hyperbola. The rectangle of width $2a$ and height $2b$, centered at the center of the hyperbola, has the asymptotes as its diagonals.

If the foci are chosen to be on the y-axis, say $F_1(0, c)$ and $F_2(0, -c)$, then the equation of the hyperbola is

$$\frac{y^2}{a^2} - \frac{x^2}{b^2} = 1. \tag{4}$$

This hyperbola is illustrated in Figure 10.30.

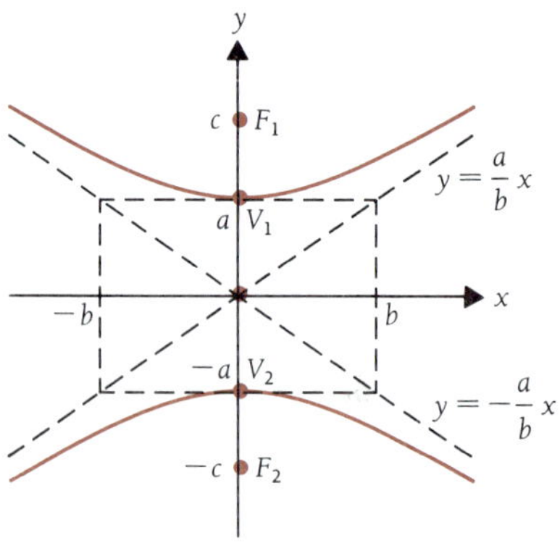

FIGURE 10.30 The hyperbola $\frac{y^2}{a^2} - \frac{x^2}{b^2} = 1$

The hyperbolas shown in Figures 10.29 and 10.30 are in **standard position.** Equations (3) and (4) are **standard forms** of the equation of a hyperbola. Other hyperbolas in standard position are shown in Figure 10.31.

Suppose a hyperbola has center $C(h, k)$ and foci $F_1(h + c, k)$ and $F_2(h - c, k)$, as in Figure 10.32. By using a translation of the coordinate system as in the preceding sections, we can show that the equation of the hyperbola is

$$\frac{(x-h)^2}{a^2} - \frac{(y-k)^2}{b^2} = 1. \tag{5}$$

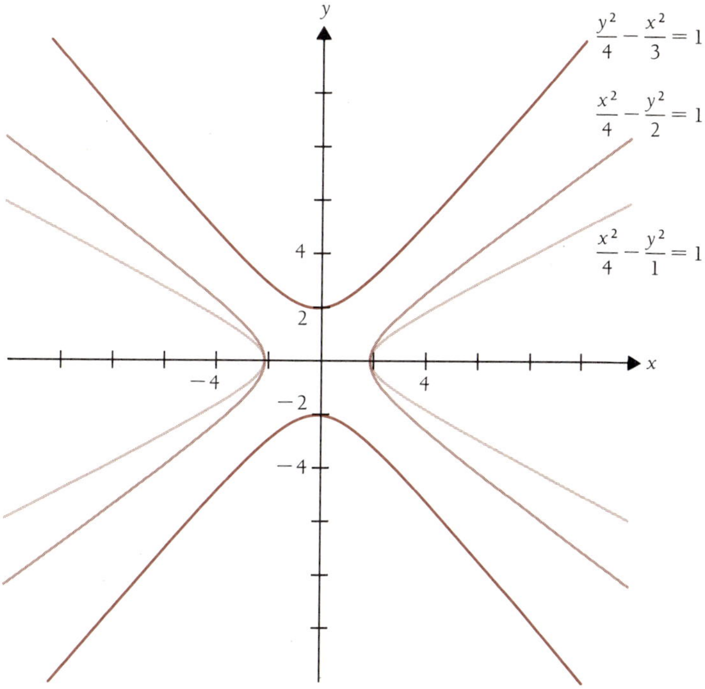

FIGURE 10.31 Hyperbolas in standard position

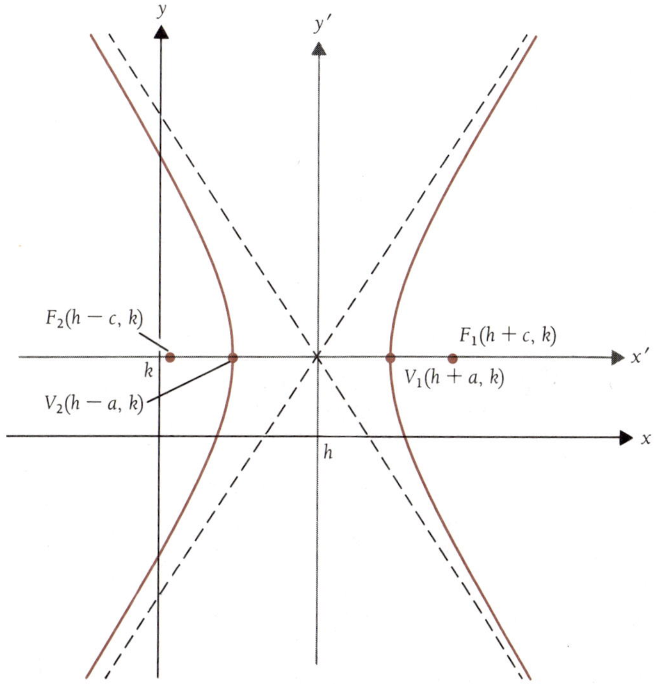

FIGURE 10.32 A hyperbola with center (h, k)

The vertices of this hyperbola are $V_1(h+a, k)$ and $V_2(h-a, k)$, and the asymptotes are

$$\frac{y-k}{b} = \pm\frac{x-h}{a}.$$

If a hyperbola has a vertical transverse axis, its equation is of the form

$$\frac{(y-k)^2}{a^2} - \frac{(x-h)^2}{b^2} = 1. \tag{6}$$

Here, the center is $C(h, k)$, the foci are $F_1(h, k+c)$ and $F_2(h, k-c)$, the vertices are $V_1(h, k+a)$ and $V_2(h, k-a)$, and the asymptotes are

$$\frac{y-k}{a} = \pm\frac{x-h}{b}.$$

Example 10.10 Find the equation of the hyperbola with center $C(1, 2)$, focus $F(1, 4)$, and vertex $V(1, 3)$.

Solution We have $c = 2$ and $a = 1$, so $b = \sqrt{3}$. Also, $h = 1$ and $k = 2$. Since the transverse axis is vertical, we use the form (6), and get the equation

$$\frac{(y-2)^2}{1} - \frac{(x-1)^2}{3} = 1.$$

This hyperbola is sketched in Figure 10.33.

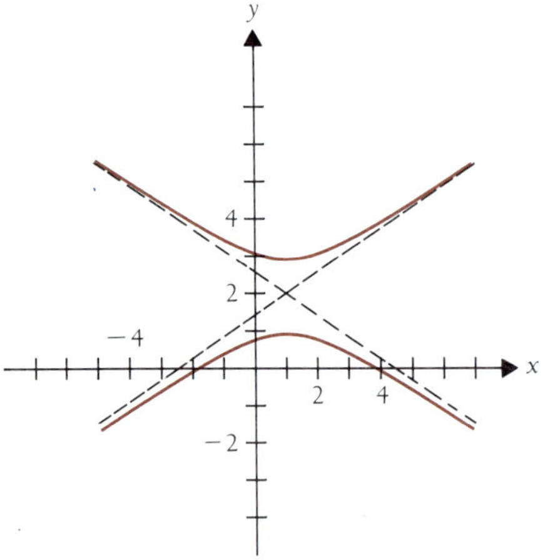

FIGURE 10.33 The graph of $\frac{(y-2)^2}{1} - \frac{(x-1)^2}{3} = 1$

Example 10.11 Give the center, foci, vertices, and asymptotes of the hyperbola

$$4x^2 - y^2 + 16x + 6y - 9 = 0,$$

and sketch its graph.

Solution First, we complete the square:

$$\begin{aligned} 4(x^2 + 4x) - (y^2 - 6y) &= 9, \\ 4(x^2 + 4x + 4) - (y^2 - 6y + 9) &= 9 + 16 - 9, \\ 4(x + 2)^2 - (y - 3)^2 &= 16, \\ \frac{(x + 2)^2}{4} - \frac{(y - 3)^2}{16} &= 1. \end{aligned}$$

This hyperbola, which is drawn in Figure 10.34, has center $C(-2, 3)$, foci $F_1(-2 + 2\sqrt{5}, 3)$ and $F_2(-2 - 2\sqrt{5}, 3)$, vertices $V_1(0, 3)$ and $V_2(-4, 3)$, and asymptotes

$$\frac{y - 3}{4} = \pm\frac{x + 2}{2},$$

or $y = 2x + 7$ and $y = -2x - 1$.

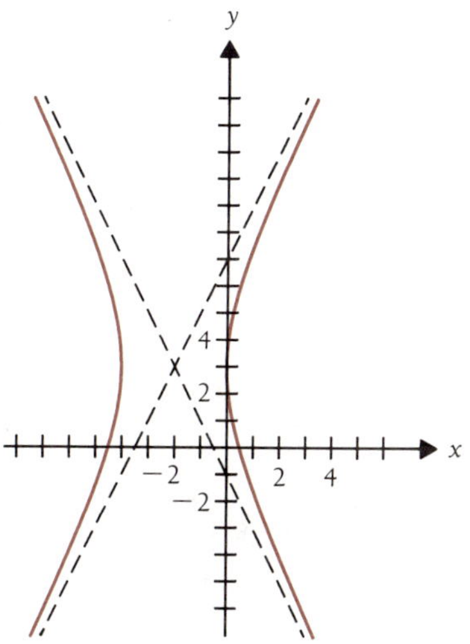

FIGURE 10.34 The graph of $4x^2 - y^2 + 16x + 6y - 9 = 0$

Hyperbolas in General Position

If the axes of a hyperbola are not parallel to the coordinate axes, its equation is not of the form (5) or (6). Consider Example 10.12.

■ **Example 10.12** Find the equation of the hyperbola with foci $(2, -2)$ and $(-2, 2)$ determined by the constant $k = 4$.

Solution If F_1 and F_2 are the foci, then point $P(x, y)$ belongs to this hyperbola if and only if

$$\begin{aligned} ||PF_1| - |PF_2|| &= 4, \\ |PF_1|^2 - 2|PF_1||PF_2| + |PF_2|^2 &= 16, \\ |PF_1|^2 + |PF_2|^2 - 16 &= 2|PF_1||PF_2|, \\ (|PF_1|^2 + |PF_2|^2 - 16)^2 &= 4|PF_1|^2|PF_2|^2. \end{aligned}$$

Now substituting

$$|PF_1|^2 = (x - 2)^2 + (y + 2)^2$$

and

$$|PF_2|^2 = (x + 2)^2 + (y - 2)^2$$

and carrying out the tedious but straightforward simplification, we eventually get the equation

$$xy = -2.$$

This hyperbola is sketched in Figure 10.35.

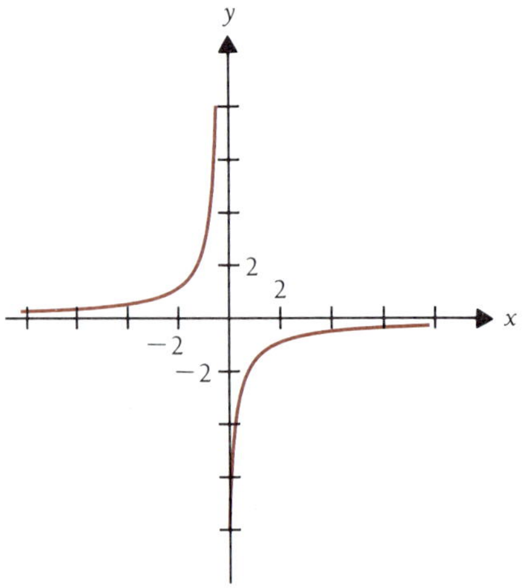

FIGURE 10.35 The graph of $xy = -2$ ■

Applications of Hyperbolas

There are at least two important practical applications of hyperbolas. The first is based on the definition and is used to determine position. For example, suppose that soldiers are trying to pinpoint the location of an enemy gun hidden in the jungle. Observers at positions P and Q record the instant that they hear the sound of the gun. Using the speed of sound and the difference in the times recorded, a hyperbola with foci P and Q can be determined, on which the gun must lie. If the observers make another observation from new positions, or employ a third observer, a second hyperbola is obtained. The gun is at one of the intersections of the two hyperbolas (see Figure 10.36).

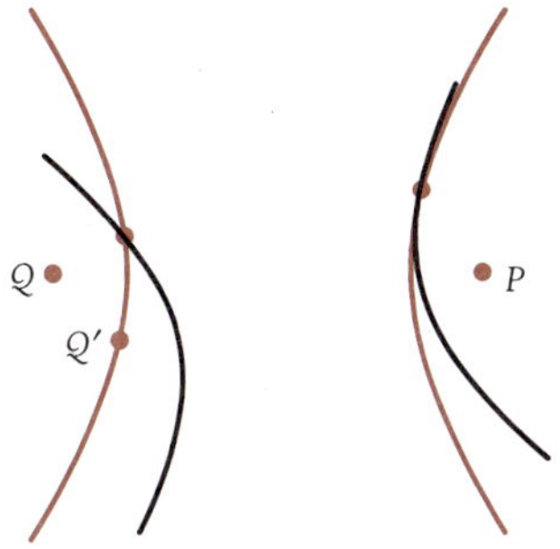

FIGURE 10.36 Locating a gun in the jungle

The second application is based on a reflective property of the hyperbola. Let T be any point on the hyperbola, and let t be the line tangent to the hyperbola at T (see Figure 10.37). If F_1 and F_2 are the foci, then the two lines TF_1 and TF_2 make the same angle with t. Thus, a beam moving toward one focus is reflected toward the other focus. This principle in combination with the parabolic reflector is applied to the construction of compound reflecting telescopes (see Figure 10.38).

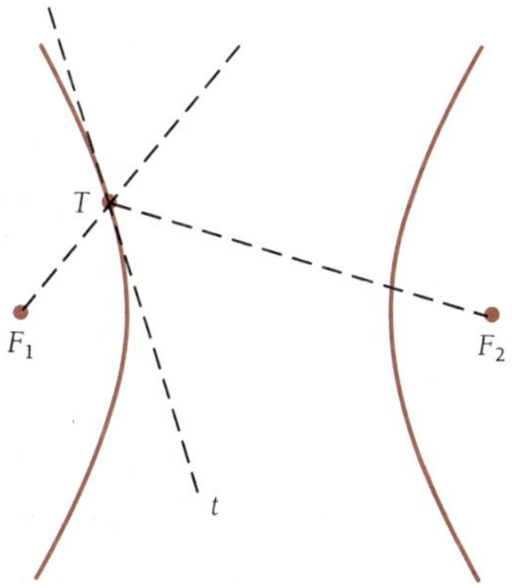

FIGURE 10.37 The reflective property of a hyperbola

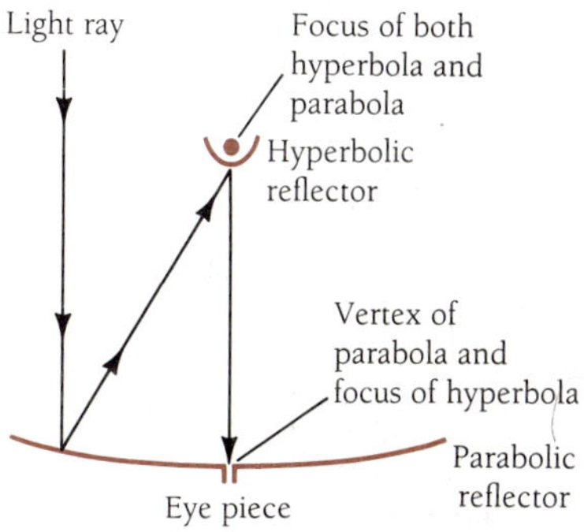

FIGURE 10.38 Compound reflecting telescope

Exercises 10.4

Find the equation of the hyperbola:

1. With foci (1, 0) and (−1, 0), and vertex $(\frac{3}{4}, 0)$
2. With foci (0, 2) and (0, −2), and vertex (0, −1)

3. With foci $(0, 0)$ and $(0, 3)$, and vertex $(0, 1)$
4. With vertices $(2, -2)$ and $(6, -2)$, and focus $(1, -2)$
5. With vertices $(1, 1)$ and $(-1, -1)$, and focus $(\sqrt{2}, \sqrt{2})$
6. With foci $(1, 1)$ and $(5, 5)$, and vertex $(2, 2)$

Give the center, foci, vertices, and asymptotes of each hyperbola.

7. $4x^2 - 9y^2 = 36$
8. $y^2 - 4x^2 = 4$
9. $x^2 - y^2 = 1$
10. $5y^2 - 16x^2 = 80$
11. $x^2 + 8x - 4y^2 = 0$
12. $2y^2 - x^2 + 4y = 4$

In Problems 13–16, let $T(x_0, y_0)$ be a point on the hyperbola $b^2x^2 - a^2y^2 = a^2b^2$, and let t be the line tangent to the hyperbola at T.

13. Find the slope of t.
14. If the hyperbola has foci F_1 and F_2, show that lines TF_1 and TF_2 make the same angle with t.
15. Explain, using Figure 10.38, how the compound reflecting telescope works.
A 16. Let line t meet the asymptotes of the hyperbola at points G_1 and G_2. Show that T is the midpoint of the segment joining G_1 and G_2.
A,C 17. When observer B is 1000 meters due east of observer A, he hears the sound of a gun 1 second before A does. If A remains stationary and B moves to a position 1000 meters due south of A, he hears the gun 2 seconds later than A does. Sound travels 330 meters per second. Use A's position as the origin of a coordinate system, and obtain equations of the two hyperbolas on which the gun lies. Then sketch the graphs to estimate the gun's position. (Finding the gun's coordinates requires solving the equations simultaneously, which is best done numerically by a computer in this case.)

SECTION 10.5

Rotation of Axes

Thus far we have seen that a conic section that has an axis parallel to one of the coordinate axes has an equation of a certain convenient form. In particular, the equation does not involve an xy term. On the other hand, the examples of conics in more general position have all had equations involving an xy term (see Examples 10.6, 10.9, and 10.12).

Conics whose equations involve no xy term can be analyzed as in the preceding sections; nothing more complicated than a translation of the coordinate system is required. On the other hand, a conic whose equation involves an xy term has no axis of symmetry parallel to a coordinate axis. The xy term may be eliminated by choosing a new coordinate system, one of whose axes is parallel to an axis of the conic. This involves a rotation of the coordinate system.

Rotation of the Coordinate System

In the xy-coordinate system, let a new $x'y'$ system be chosen with the same origin, but with the x'-axis having an inclination α (see Figure 10.39). If the original axes

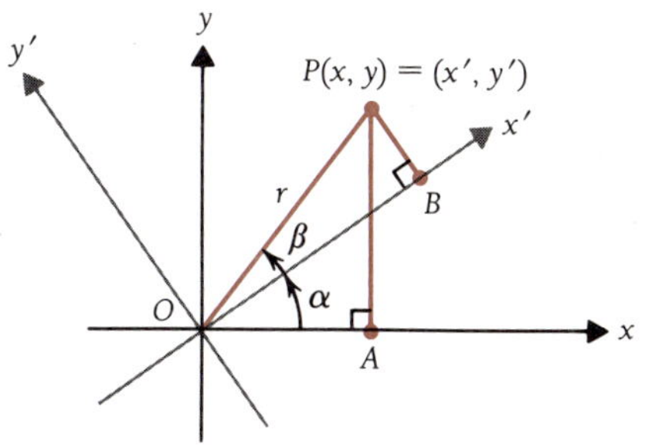

FIGURE 10.39 Rotation of axes

were rotated through an angle α, they would coincide with the new axes, so the new system is called a **rotation** of the old one.

To discover how coordinates of the old and new systems are related, let point P have coordinates (x, y) in the old system and (x', y') in the new (rotated) system. Let O be the origin, and let β be the inclination of the line OP in the $x'y'$ system (Figure 10.39). Let PA be perpendicular to the x-axis, and let PB be perpendicular to the x'-axis. Let r be the distance between O and P.

Now, in triangle OAP, we have

$$\begin{aligned} x &= |OA| \\ &= r\cos(\alpha+\beta) \\ &= r\cos\beta\cos\alpha - r\sin\beta\sin\alpha, \qquad (1) \\ y &= |AP| \\ &= r\sin(\alpha+\beta) \\ &= r\sin\beta\cos\alpha + r\cos\beta\sin\alpha. \qquad (2) \end{aligned}$$

In triangle OBP, we have

$$\begin{aligned} x' &= |OB| \\ &= r\cos\beta, \\ y' &= |BP| \\ &= r\sin\beta. \end{aligned}$$

Substituting x' and y' into equations (1) and (2), we get

$$\begin{cases} x = x'\cos\alpha - y'\sin\alpha \\ y = x'\sin\alpha + y'\cos\alpha. \end{cases} \qquad (3)$$

These equations are called **equations of the rotation**. If the system (3) is solved for x' and y', we get the alternative set of equations

$$\begin{cases} x' = x\cos\alpha + y\sin\alpha \\ y' = -x\sin\alpha + y\cos\alpha. \end{cases} \qquad (4)$$

Eliminating the xy Term

Now we return to the problem of eliminating the xy term from the equation of a conic. If we rotate the coordinate system so that a coordinate axis is parallel to an axis of the conic, then the new equation of the conic in the rotated system will have no $x'y'$ term.

Let's be more specific. Let the given conic have equation

$$ax^2 + bxy + cy^2 + dx + ey + f = 0, \tag{5}$$

with $b \neq 0$. (If $b = 0$, all this is unnecessary.) We will rotate the axes through an angle α, although we do not yet know what value of α will give the desired result. However, the equation of the conic in the rotated system is found by substituting system (3) into equation (5). The new equation is

$$Ax'^2 + Bx'y' + Cy'^2 + Dx' + Ey' + F = 0, \tag{6}$$

with

$$\begin{cases} A = a\cos^2\alpha + b\sin\alpha\cos\alpha + c\sin^2\alpha \\ B = b(\cos^2\alpha - \sin^2\alpha) - 2(a-c)\sin\alpha\cos\alpha \\ C = a\sin^2\alpha - b\sin\alpha\cos\alpha + c\cos^2\alpha \\ D = d\cos\alpha + e\sin\alpha \\ E = e\cos\alpha - d\sin\alpha \\ F = f. \end{cases} \tag{7}$$

(These equations are found by simple, but tedious, algebraic manipulations; there are no tricks involved at all. You are asked to do this in Problem 20, at the end of this section.)

Our goal of eliminating the xy term will be realized if we choose the value of α that makes B in system (7) equal to 0. To find out what value of α that is, we set $B = 0$ and solve for α:

$$\begin{aligned} b(\cos^2\alpha - \sin^2\alpha) - 2(a-c)\sin\alpha\cos\alpha &= 0, \\ b\cos 2\alpha - (a-c)\sin 2\alpha &= 0, \\ b\cot 2\alpha &= a - c, \\ \cot 2\alpha &= \frac{a-c}{b}. \end{aligned} \tag{8}$$

That is, we choose a value of α that satisfies equation (8). Once α is chosen, we obtain the new equation (6), which can be treated as in previous sections relative to the $x'y'$ system. We will illustrate the procedure with some examples.

Example 10.13 Sketch the graph of

$$x^2 + 2xy + y^2 + 12\sqrt{2}x - 12\sqrt{2}y = 0.$$

Solution Since an xy term is present, a rotation of axes is necessary. Comparing this equation with the general form (5), we note that $a = 1$, $b = 2$, and $c = 1$. Hence, equation (8) gives

$$\cot 2\alpha = \frac{1-1}{2} = 0,$$

$$2\alpha = \cot^{-1} 0 = \frac{\pi}{2},$$

$$\alpha = \frac{\pi}{4}.$$

This is the angle through which we will rotate the axes.

Putting $\alpha = \pi/4$ into system (3), we find the equations of the rotation to be

$$\begin{cases} x = \dfrac{\sqrt{2}}{2}(x' - y') \\ y = \dfrac{\sqrt{2}}{2}(x' + y'). \end{cases}$$

Now we substitute these expressions for x and y in our original equation, getting

$$\left[\frac{\sqrt{2}}{2}(x'-y')\right]^2 + 2\left[\frac{\sqrt{2}}{2}(x'-y')\right]\left[\frac{\sqrt{2}}{2}(x'+y')\right] + \left[\frac{\sqrt{2}}{2}(x'+y')\right]^2$$
$$+ 12\sqrt{2}\left[\frac{\sqrt{2}}{2}(x'-y')\right] - 12\sqrt{2}\left[\frac{\sqrt{2}}{2}(x'+y')\right] = 0.$$

Simplifying, we get the equation

$$x'^2 = 12y',$$

which we recognize as the equation of a parabola with focus (0, 3) and directrix $y' = -3$ in the $x'y'$ system. Its graph is shown in Figure 10.40.

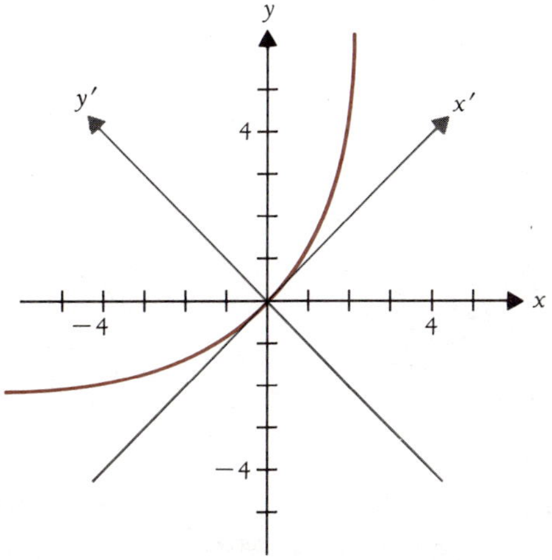

FIGURE 10.40 The graph of $x'^2 = 12y'$ ■

Example 10.14 Sketch the graph of

$$23x^2 + 26\sqrt{3}xy - 3y^2 - 16x + 16\sqrt{3}y = -128.$$

Solution Using equation (8) with $a = 23$, $b = 26\sqrt{3}$, and $c = -3$, we find α:

$$\cot 2\alpha = \frac{23 - (-3)}{26\sqrt{3}} = \frac{\sqrt{3}}{3},$$

$$2\alpha = \frac{\pi}{3},$$

$$\alpha = \frac{\pi}{6}.$$

This value of α gives us the rotation equations

$$\begin{cases} x = \frac{1}{2}(x'\sqrt{3} - y') \\ y = \frac{1}{2}(x' + y'\sqrt{3}). \end{cases}$$

Putting these expressions for x and y into the equation of our conic and simplifying, we get

$$36x'^2 - 16y'^2 + 32y' + 128 = 0.$$

Having eliminated the xy term, we complete the square to get

$$\frac{(y'-1)^2}{9} - \frac{x'^2}{4} = 1.$$

We recognize this equation as that of a hyperbola with center (0, 1) in the $x'y'$ system, with $a = 3$, $b = 2$, and $c = \sqrt{13}$. The curve is sketched in Figure 10.41.

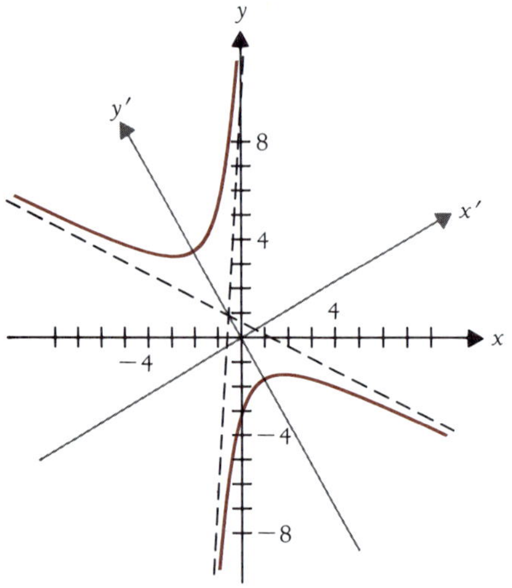

FIGURE 10.41 The graph of $\frac{(y'-1)^2}{9} - \frac{x'^2}{4} = 1$

Example 10.15 Sketch the graph of

$$5x^2 + 4xy + 4y^2 - 3 = 0.$$

Solution As before, we first look for the angle of rotation necessary, using equation (8) with $a = 5$, $b = 4$, and $c = 4$:

$$\cot 2\alpha = \frac{5-4}{4} = \frac{1}{4},$$

$$\alpha = \frac{1}{2} \operatorname{arccot} \frac{1}{4}.$$

Thus, $\alpha \approx 0.663$. However, approximations for $\sin \alpha$ and $\cos \alpha$ are not entirely satisfactory in the rotation equations. We need exact values. From $\cot 2\alpha = \frac{1}{4}$, we get

$$\sin 2\alpha = 4 \cos 2\alpha.$$

Then, from the identity $\sin^2 2\alpha + \cos^2 2\alpha = 1$, we get

$$16 \cos^2 2\alpha + \cos^2 2\alpha = 1,$$

$$\cos 2\alpha = \frac{\sqrt{17}}{17}.$$

We therefore find

$$\sin \alpha = \left(\frac{1 - \cos 2\alpha}{2}\right)^{1/2} = \left(\frac{17 - \sqrt{17}}{34}\right)^{1/2},$$

$$\cos \alpha = \left(\frac{1 + \cos 2\alpha}{2}\right)^{1/2} = \left(\frac{17 + \sqrt{17}}{34}\right)^{1/2}.$$

These might not look like the nicest numbers to work with, but if we stifle our prejudices and use them in the rotation equations (3) and then in the original equation of our conic, we eventually get

$$(9 + \sqrt{17})x'^2 + (9 - \sqrt{17})y'^2 = 6.$$

This can be further transformed into

$$\frac{y'^2}{3(9 + \sqrt{17})/32} + \frac{x'^2}{3(9 - \sqrt{17})/32} = 1,$$

which is an ellipse with semiaxes

$$a = \left[\frac{3(9 + \sqrt{17})}{32}\right]^{1/2} \approx 1.11,$$

$$b = \left[\frac{3(9 - \sqrt{17})}{32}\right]^{1/2} \approx 0.68.$$

The graph is sketched in Figure 10.42.

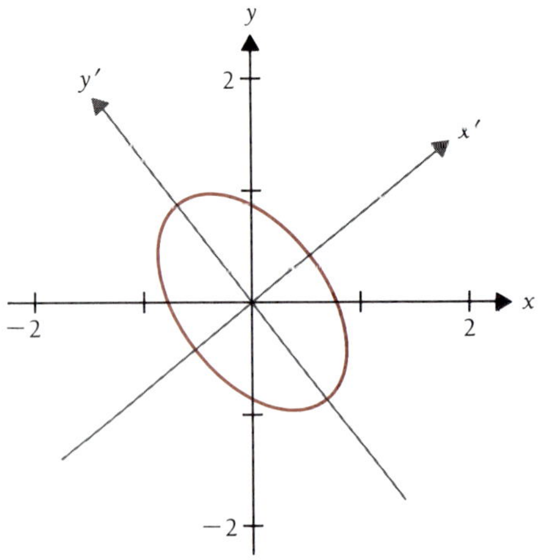

FIGURE 10.42 The graph of $5x^2 + 4xy + 4y^2 - 3 = 0$, or

$$\frac{y'^2}{3(9+\sqrt{17})/32} + \frac{x'^2}{3(9-\sqrt{17})/32} = 1$$

Exercises 10.5

Find an angle α of rotation that will eliminate the xy term.

1. $xy = 9$
2. $xy + y - 4x = 0$
3. $x^2 + xy = 2$
4. $2xy + 2y^2 - 5 = 0$
5. $x^2 - xy + (1 + \sqrt{3})y^2 + 2x + 4y - 6 = 0$
6. $6x^2 + \sqrt{3}xy + 5y^2 - 4x + y - 9 = 0$
7. $2x^2 + 3xy - x + 5y - 1 = 0$
8. $9x^2 - 9xy - 5y^2 = 1$

In Problems 9–16, eliminate the xy term and then sketch the graph.

A9. $x^2 + xy = 2$

10. $xy = 9$

A11. $x^2 - 3xy + (1 + \sqrt{3})y^2 = 4$

12. $6x^2 + \sqrt{3}xy + 5y^2 = 9$

A13. $x^2 + 24xy - 6y^2 = 1$

A14. $x^2 + 2xy - 3y^2 = 16$

15. $x^2 + \sqrt{3}xy + 2y^2 + 2x - 5 = 0$

16. $23x^2 + 26\sqrt{3}xy - 3y^2 + (32 - 36\sqrt{3})x - (36 + 32\sqrt{3})y = 172$

17. Show that
$$\frac{y^2}{a^2} + \frac{x^2}{b^2} = 1$$
is transformed into
$$\frac{x'^2}{a^2} + \frac{y'^2}{b^2} = 1$$
by a rotation through $\pi/2$.

18. Show that
$$b^2y^2 - a^2x^2 = a^2b^2$$
is transformed into
$$b^2x'^2 - a^2y'^2 = a^2b^2$$
by a rotation through $\pi/2$.

19. Derive equations (4) from (3).

20. Derive equations (7).

SECTION 10.6

Quadratic Curves

We have seen thus far that the parabola, ellipse, and hyperbola have equations of the form

$$ax^2 + bxy + cy^2 + dx + ey + f = 0. \tag{1}$$

Any equation of this form is called a **quadratic equation in *x* and *y***, and its graph is called a **quadratic curve**. Thus, all the conic sections are quadratic curves.

In this section, we will examine the possible types of quadratic curves, together with some means of identifying and classifying them. We will also show how the parabola, ellipse, and hyperbola can be described in a unified fashion.

Types of Quadratic Curves

Let us begin with a quadratic equation of the form shown in equation (1). We assume that not all of the coefficients a, b, and c are 0, because otherwise the equation would not be quadratic. If $b \neq 0$, we can eliminate the xy term by a suitable rotation of axes, as discussed in Section 10.5, and obtain an equation of the form

$$Ax'^2 + Cy'^2 + Dx' + Ey' + F = 0. \tag{2}$$

If $b = 0$, equation (1) already has the form shown in equation (2). Thus, equation (2) is all we need to consider.

Now, we cannot have both A and C equal to 0 (can you tell why?), so we have essentially two cases to consider:

Case 1. If one of A and C is 0, say $C = 0$, then a translation changes equation (2) into an equation of the form

$$Ax''^2 + Ey' + F' = 0. \tag{3}$$

(a) If $E \neq 0$ in equation (3), then we have a parabola, since the equation can be written as

$$Ax''^2 = -E\left(y' + \frac{F'}{E}\right).$$

(b) If $E = 0$ in equation (3), then the equation can be written as

$$x''^2 = F''. \tag{4}$$

(i) If F'' is positive, say $F'' = k^2$, then equation (4) can be factored into

$$(x'' + k)(x'' - k) = 0,$$

whose graph is the pair of parallel lines $x'' = k$ and $x'' = -k$.

(ii) If $F'' = 0$, then equation (4) is just $x''^2 = 0$, whose graph is the single line $x'' = 0$.

(iii) If F'' is negative, then no point satisfies equation (4), and the graph is nonexistent; we say the graph is **imaginary**.

Thus, in Case 1, the graph of the equation is a parabola, two parallel lines, a single line, or no graph at all.

Case 2. If both A and C are not 0 in equation (2), then a translation changes (2) into the form

$$Ax''^2 + Cy''^2 + F' = 0. \tag{5}$$

Now we consider the possible distribution of signs of the coefficients in equation (5).

(a) If A and C have the same sign, we can write equation (5) as

$$A'^2x''^2 + C'^2y''^2 = F''. \tag{6}$$

(i) If F'' is positive, we have an ellipse; in the special case that $A' = C'$ in equation (6), we have a circle.

(ii) If $F'' = 0$, we have as the graph only the point $x'' = 0, y'' = 0$.

(iii) If F'' is negative, the curve has no graph at all, or is imaginary.

(b) If A and C in equation (5) have opposite signs, we can write (5) in the form

$$A'^2x''^2 - C'^2y''^2 = F''. \tag{7}$$

(i) If $F'' \neq 0$, the graph is a hyperbola.

(ii) If $F'' = 0$, then equation (7) can be written in factored form as

$$(A'x'' + C'y'')(A'x'' - C'y'') = 0,$$

and its graph is the pair of intersecting lines

$$A'x'' + C'y'' = 0, \qquad A'x'' - C'y'' = 0.$$

Thus, in Case 2, the graph will be an ellipse, a circle, a point, imaginary, a hyperbola, or two intersecting lines.

If a quadratic curve consists of two lines, one line, or a single point, it is called a **singular conic.** If it is a parabola, ellipse, circle, or hyperbola, it is called a **nonsingular conic.**

Let's look at some examples.

■ Example 10.16 Describe the curve

$$x^2 - y^2 + 2y - 1 = 0.$$

Solution We describe the conic by writing its equation in one of the forms discussed above. Since no xy term is present, no rotation is necessary. Completing the square, we get

$$x^2 - (y-1)^2 = 0,$$
$$[x + (y-1)][x - (y-1)] = 0.$$

This fits into Case 2(b)(ii); the conic is singular, consisting of the two lines

$$x + y - 1 = 0, \qquad x - y + 1 = 0.$$

■

Example 10.17 Describe the curve

$$x^2 + 4y^2 - 4x + 8y + 8 = 0.$$

Solution Again, no rotation is necessary. Completing the square, we get

$$(x - 2)^2 + 4(y + 1)^2 = 0.$$

This curve fits into Case 2(a)(ii); the graph is the single point $(2, -1)$. ■

Example 10.18 Describe the curve

$$4x^2 - 20xy + 25y^2 + 12x - 30y + 9 = 0.$$

Solution In this case, we must perform a rotation of axes through an angle α satisfying

$$\cot 2\alpha = \frac{21}{20}.$$

We find that

$$\sin \alpha = \frac{2}{\sqrt{29}}, \qquad \cos \alpha = \frac{5}{\sqrt{29}},$$

and eventually we transform our original equation into the form

$$\left(y' - \frac{3}{\sqrt{29}}\right)^2 = 0.$$

This curve fits Case 1(b)(ii), and is the single line

$$y' = \frac{3}{\sqrt{29}}.$$

Or, going back to the original coordinates via equations (4) of Section 10.5, the equation of the line is

$$2x - 5y + 3 = 0.$$

Now we can see in retrospect that the original equation is just

$$(2x - 5y + 3)^2 = 0.$$

■

Invariants: Discriminant and Determinant

It would be nice to have some means of describing a conic without having to go through the process of rotating and translating axes. Luckily, such means exist.

Given any quadratic equation in x and y, we can classify its graph by using certain combinations of the coefficients. These combinations are called **invariants**, because their values do not change under translations or rotations.

The first invariant we consider is the discriminant. The **discriminant** of the quadratic equation

$$ax^2 + bxy + cy^2 + dx + ey + f = 0$$

is

$$\Delta = b^2 - 4ac.$$

It can be shown that the discriminant is indeed invariant under translations and rotations. Moreover, considering the various cases as before, we find that:

1. If $\Delta < 0$, then the quadratic curve is an ellipse, a circle, a point, or imaginary.
2. If $\Delta = 0$, then the quadratic curve is a parabola, two parallel lines, a single line, or imaginary.
3. If $\Delta > 0$, then the quadratic curve is a hyperbola or two intersecting lines.

The second invariant we consider is the determinant. The **determinant** of the quadratic equation

$$ax^2 + bxy + cy^2 + dx + ey + f = 0$$

is

$$\delta = \begin{vmatrix} 2a & b & d \\ b & 2c & e \\ d & e & 2f \end{vmatrix}$$
$$= 8acf + 2bde - 2ae^2 - 2cd^2 - 2fb^2.$$

It also can be shown that the determinant is invariant under translations and rotations, though the computations become a bit tedious. Again, by considering cases, it can be shown that:

1. If $\delta = 0$, then the quadratic curve is singular.
2. If $\delta \neq 0$, then the quadratic curve is nonsingular.

Using the discriminant and the determinant together, we can classify a conic according to Table 10.1.

TABLE 10.1

	$\delta \neq 0$	$\delta = 0$
$\Delta < 0$	Ellipse, circle, or imaginary	Point
$\Delta = 0$	Parabola	Two parallel lines, one line, or imaginary
$\Delta > 0$	Hyperbola	Two intersecting lines

The following examples illustrate the use of the invariants.

Example 10.19 Classify the following conics:

(a) $x^2 + 2xy + y^2 - x - y = 0$ (b) $2x^2 + xy + y^2 + 5 = 0$
(c) $x^2 + 4xy + 3y^2 - 2x + y - 6 = 0$

Solution (a) The discriminant is

$$\Delta = 2^2 - 4 \cdot 1 \cdot 1 = 0,$$

and the determinant is

$$\delta = \begin{vmatrix} 2 & 2 & -1 \\ 2 & 2 & -1 \\ -1 & -1 & 0 \end{vmatrix} = 0.$$

This conic is therefore two parallel lines, one line, or imaginary. Since the point (0, 0) satisfies the equation, we can quickly eliminate the imaginary case. further study reveals that the equation can be factored as

$$(x + y)^2 - (x + y) = 0,$$
$$(x + y)(x + y - 1) = 0.$$

Therefore, the conic consists of two parallel lines.

(b) The discriminant is

$$\Delta = 1^2 - 4 \cdot 2 \cdot 1 = -7 < 0,$$

and the determinant is

$$\delta = \begin{vmatrix} 4 & 1 & 0 \\ 1 & 2 & 0 \\ 0 & 0 & 10 \end{vmatrix} = 70 \neq 0.$$

Therefore, the conic is an ellipse, a circle, or imaginary. Because there is an xy term, it cannot be a circle. If we solve the equation for y, we get

$$y = \frac{1}{2}(-x \pm \sqrt{-7x^2 - 20}),$$

which is an imaginary number for every real number x. Thus, the conic is imaginary.

(c) The discriminant is

$$\Delta = 4^2 - 4 \cdot 1 \cdot 3 = 4 > 0,$$

and the determinant is

$$\delta = \begin{vmatrix} 2 & 4 & -2 \\ 4 & 6 & 1 \\ -2 & 1 & -12 \end{vmatrix} = 6 \neq 0,$$

so the conic is a hyperbola. ∎

Eccentricity, Focus, and Directrix

We conclude this section with a discussion of the eccentricity of a nonsingular conic and a description of ellipses and hyperbolas in terms of focus and directrix properties.

First, we consider the case of an ellipse. Let the coordinate system be chosen so that the ellipse has equation

$$\frac{x^2}{a^2} + \frac{y^2}{b^2} = 1.$$

The number

$$e = \frac{c}{a} = \frac{1}{a}\sqrt{a^2 - b^2}$$

is called the **eccentricity of the ellipse**. Since $a > c > 0$, it follows that

$$0 < e < 1.$$

Figure 10.43 shows ellipses with the same foci and various eccentricities.

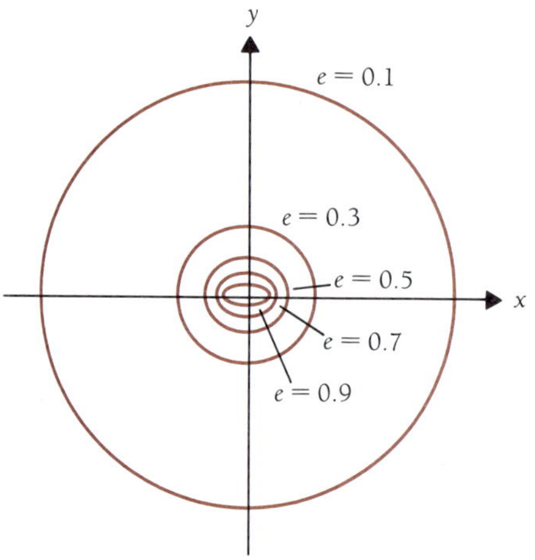

FIGURE 10.43 Ellipses with the same foci and various eccentricities

Now, let one focus of the ellipse be $F_1 = (c, 0)$, and let line d_1 have equation

$$x = \frac{a^2}{c}.$$

Then we can prove Theorem 10.1.

THEOREM 10.1 Point P is on the ellipse

$$\frac{x^2}{a^2} + \frac{y^2}{b^2} = 1$$

if and only if the distance $|PF_1|$ between P and $F_1(c, 0)$ and the distance $|Pd_1|$ between P and line $d_1: x = a^2/c$ satisfy the equation

$$|PF_1| = e|Pd_1|,$$

where e is the eccentricity and $e = c/a$.

The proof of this theorem will be left as an exercise (see Problem 17).

We call the line d_1 the **directrix** of the ellipse corresponding to the focus F_1. Thus, the ellipse can be defined in terms of a focus and a directrix, using the eccentricity. The ellipse also has a second focus–directrix pair, namely the focus $F_2(-c, 0)$ and the directrix $d_2: x = -a^2/c$, in terms of which the ellipse can be defined (see Figure 10.44).

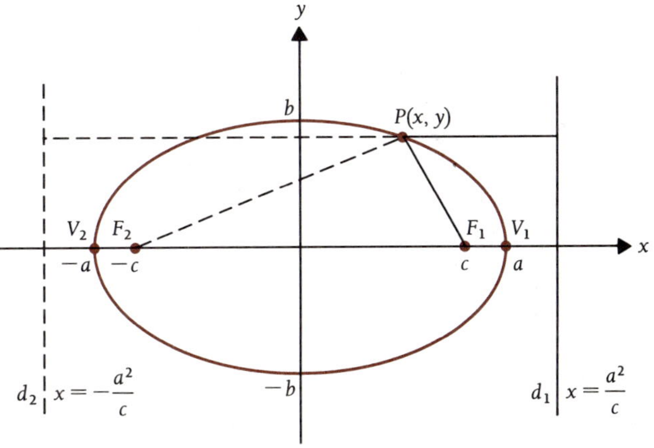

FIGURE 10.44 $|PF_i| = e|Pd_i|$, where $e = c/a$

Next, consider the hyperbola

$$\frac{x^2}{a^2} - \frac{y^2}{b^2} = 1.$$

The number

$$e = \frac{c}{a} = \frac{1}{a}\sqrt{a^2 + b^2}$$

is called the **eccentricity of the hyperbola.** Since $c > a > 0$, we have

$$e > 1.$$

Figure 10.45 (on page 688) shows hyperbolas with the same foci and various eccentricities.

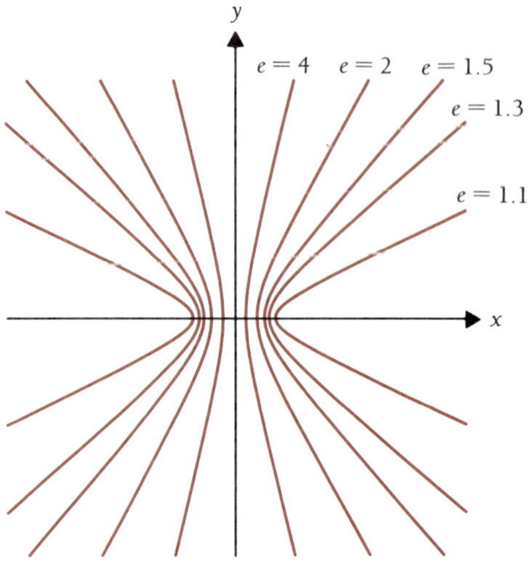

FIGURE 10.45 Hyperbolas with the same foci and various eccentricities

The hyperbola also can be defined in terms of a focus–directrix pair. The foci are $F_1(c, 0)$ and $F_2(-c, 0)$, and the corresponding directrices are $d_1: x = a^2/c$ and $d_2: x = -a^2/c$ (see Figure 10.46). The hyperbola consists of those points P such that

$$|PF_i| = e|Pd_i|.$$

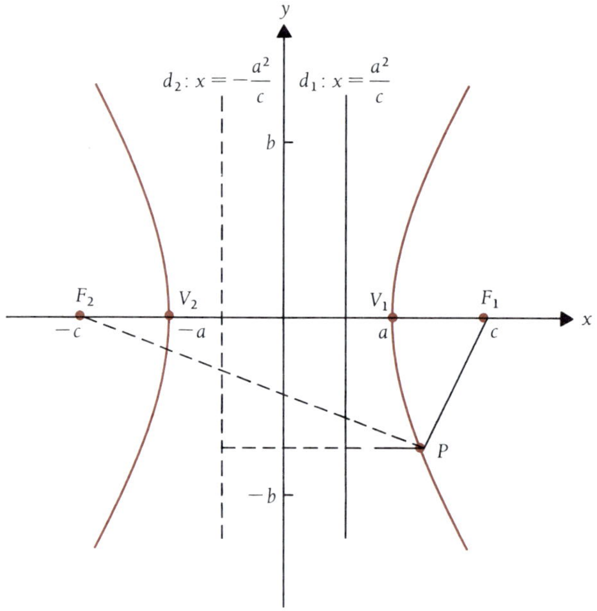

FIGURE 10.46 $|PF_i| = e|Pd_i|$, where $e = c/a$

We can now describe the parabola, ellipse, and hyperbola in the same terms. Let F be a fixed point, and let d be a fixed line. The locus of points P such that

$$|PF| = e|Pd|$$

is an ellipse if $0 < e < 1$, a parabola if $e = 1$, and a hyperbola if $e > 1$. Because of this relationship, we say that a parabola always has eccentricity 1.

Exercises 10.6

Without using translations or rotations, classify each conic.

1. $x^2 - 2xy + y^2 + y = 0$
2. $4x^2 + xy + 3y^2 = 0$
3. $xy + y^2 - 3 = 0$
4. $2x^2 + 3xy - 9y^2 - 9x - 18y + 7 = 0$
5. $x^2 - 2xy + y^2 - 1 = 0$
6. $x^2 - 4xy + 4y^2 = 0$
7. $2x^2 + 4xy + 2y^2 + 3x + 3y - 5 = 0$
8. $4x^2 + 4xy + y^2 + 3 = 0$

Give the eccentricity of each curve.

9. $4x^2 + 9y^2 = 36$
10. $9x^2 - 4y^2 = 36$
11. $4x^2 + 9y = 36$
12. $6x^2 + 5y^2 - 30y + 15 = 0$

A 13. Show that the discriminant Δ is invariant under a translation of axes.

14. Show that the discriminant Δ is invariant under a rotation of axes.

15. Show that the determinant δ is invariant under a translation of axes.

A 16. (a) Suppose the discriminant Δ of a conic is nonzero. Show that there exist numbers h and k such that the translation $x' = x - h$, $y' = y - k$ transforms the equation of the conic into the form $Ax'^2 + Bx'y' + Cy'^2 + F = 0$.
(b) Given that $\Delta \neq 0$, show that δ is invariant under a rotation of axes.
(c) Given that $\Delta = 0$, show that δ is invariant under a rotation of axes.

17. Prove Theorem 10.1.

18. State and prove a theorem for hyperbolas, similar to Theorem 10.1.

SECTION 10.7

Chapter Review

Important Concepts

Angle of inclination
Angle between lines
Translation of axes
Equations of translation
Conic section
Parabola
Focus
Directrix
Axis
Vertex
Standard form
General position
Ellipse
Foci
Vertices
Major axis
Semimajor axis
Minor axis
Semiminor axis
Center
Hyperbola
Foci
Transverse axis
Vertices
Conjugate axis
Center
Asymptotes
Rotation of axes
Equations of rotation
Quadratic curve
Imaginary curve
Singular conic
Nonsingular conic
Invariant
Discriminant
Determinant
Eccentricity

Review Exercises

Give an equation for each conic.

1. The parabola with focus (1, 2) and directrix $y = -1$
2. The parabola with focus (1, 2) and vertex (3, 2)
3. An ellipse with center (4, 3) and vertices (6, 3) and (2, 3)
4. An ellipse with focus (0, 1) and vertex (0, 3)
5. The hyperbola with foci (2, −1) and (2, 1) and vertex $(2, \frac{1}{2})$
6. A hyperbola with asymptotes $y = \pm 3x + 2$ and vertex (1, 2)
7. The conic with focus (1, 3), directrix $x = 4$, and eccentricity 2
8. The conic with focus (2, −1), directrix $y = 4$, and eccentricity $\frac{2}{3}$

Sketch each curve.

9. $y^2 = 8x$
10. $x^2 + 2x = 4y$
11. $x^2 + 2xy + y^2 = 0$
12. $x^2 - 2xy + y^2 + 4x - 4y = 16$
13. $4x^2 + 9y^2 = 36$
14. $4x^2 - 9y^2 = 36$
15. $x^2 + 4y^2 - 2x - 8y + 3 = 0$
16. $3x^2 - 2y^2 + 12x + 12y - 6 = 0$
17. $xy + 2x - 3y = 5$
18. [A] $4x^2 - xy + y^2 = 1$
19. $x^2 - y^2 + 2y - 1 = 0$
20. $16x^2 + 16y^2 + 4x - 8y = 1$

Additional Exercises

1. Find the angle at which the curves $y = mx$ and $x^2 + y^2 = 2x$ intersect.
2. Two circles (or other curves) are **orthogonal** if they meet at right angles. Show that any circle tangent to the x-axis at the origin is orthogonal to any circle tangent to the y-axis at the origin.
3. A **latus rectum** of a conic is a chord of the conic that is perpendicular to the axis of the conic at a focus. Find the length of the latus rectum of:
(a) $x^2 = 4cy$ (b) $b^2x^2 + a^2y^2 = a^2b^2$ (c) $b^2x^2 - a^2y^2 = a^2b^2$

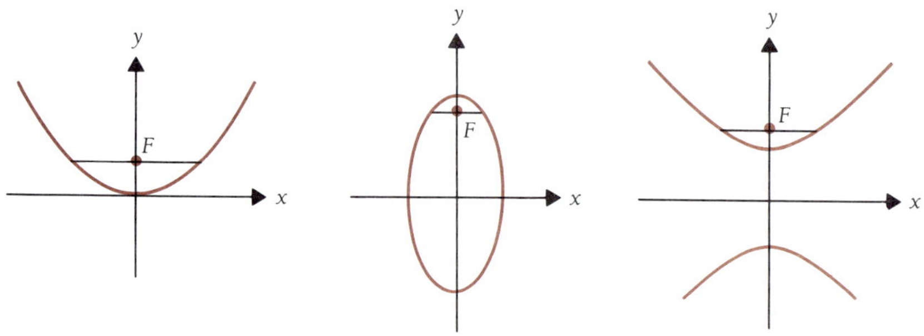

4. Let two circles of radii a and b, $a > b$, be centered at the origin. Let a common radius of the circles meet the larger circle at P and the smaller circle at Q. Prove that the vertical line through P and the horizontal line through Q meet on the ellipse $b^2x^2 + a^2y^2 = a^2b^2$.
5. Explain how Problem 4 leads to a point-by-point construction of an ellipse using a ruler and compass.

[A]6. Two observers listen for an enemy gun hidden in the jungle. On one observation, B is located 1000 yards south of A and hears the gun 2 seconds after A hears it. For the second observation, B moves 500 yards north and A moves 500 yards east. This time, B hears the gun 1 second earlier than A does. If the speed of sound is 360 yards per second, approximately where is the gun located?

7. What point on a conic lies nearest the focus?

8. The earth's orbit is an ellipse with eccentricity $e \approx \frac{1}{60}$ and with the sun at one focus. If the major axis of the ellipse is 185.5 million miles, how close does the earth get to the sun?

9. A certain earth satellite has an elliptic orbit with eccentricity $\frac{1}{3}$ and the center of the earth at one focus. If the satellite's nearest approach to the earth's surface is 300 miles, what is its greatest distance above the earth's surface? Use 4000 miles as the earth's radius.

[A]10. Devise formulas for the equations of the tangent lines to a conic that pass through a given point not on the conic.

CHAPTER 11

Parametric Equations and Polar Coordinates

INTRODUCTION

This chapter deals with two more topics in analytic geometry — parametric equations and polar coordinates. Using parametric equations, we can give a more complete treatment of curves in the plane. For example, an object moving in the plane may trace out a curve that crosses itself or has other features that cannot be described in terms of graphs of functions or even graphs of equations. With parametric equations, the coordinates of a point are described independently in terms of a new independent variable, called a **parameter.** This gives us much greater freedom in describing motion, and enables us to discuss many more curves than we could previously.

The polar coordinate system describes the location of a point in terms of distance and direction, rather than in terms of two distances in perpendicular directions, as the rectangular coordinate system does. With polar coordinates, we meet a whole new family of curves. Derivatives and integrals have different interpretations in polar coordinates than in rectangular coordinates.

In this chapter, we will develop the calculus of parametric equations and apply it to polar coordinates. We then develop the applications of derivatives and integrals that are appropriate for the polar coordinate system.

SECTION 11.1

Plane Curves and Parametric Equations

Thus far, we have regarded a curve lying in the plane as the graph of a function (defined explicitly or implicitly). This point of view is a little restrictive and not always convenient. In this section, we will define what we mean by a curve, using parameters.

Parametrized Curves

Let t be an independent variable, and let f and g be functions that are continuous on an interval $[a, b]$. Corresponding to each value $t \in [a, b]$, there is a point $(f(t), g(t))$ in the plane. The set of these points is a **parametrized curve**, or simply **curve**, denoted

$$C = \{(f(t), g(t)) \mid a \leq t \leq b\}.$$

We can think of the curve C as being traced out by the point $(f(t), g(t))$ as t varies from a to b.

The curve C also can be described as the set of points (x, y) such that

$$\begin{cases} x = f(t) \\ y = g(t), \quad t \in [a, b]. \end{cases} \tag{1}$$

The equations (1) are called **parametric equations** for C, with t as **parameter.**

This definition of curve includes the graphs of continuous functions. If $y = f(x)$ is a continuous function on $[a, b]$, we can let

$$\begin{cases} x = t \\ y = f(t), \quad t \in [a, b]. \end{cases}$$

This is a set of parametric equations for the graph of $y = f(x)$. Thus, the graphs we have been calling curves are still curves in this new sense.

Often, a parametrized curve can be described as the graph of a function. This is done by eliminating the parameter from the parametric equations of the curve, and thus, expressing y directly or implicitly as a function of x, or x as a function of y.

Some examples will illustrate these ideas.

■ Example 11.1 Sketch the curve

$$\begin{cases} x = 2t \\ y = 1 - t, \quad t \in (-\infty, \infty). \end{cases}$$

Solution We may either plot points to get the curve or eliminate the parameter and obtain a function whose graph we know. We will use the latter method. We can easily solve $x = 2t$ for t, getting $t = x/2$, and put this expression into $y = 1 - t$:

$$y = 1 - \frac{x}{2}.$$

The graph of this equation is the line with slope $-\frac{1}{2}$ and intercept 1 (Figure 11.1, page 694).

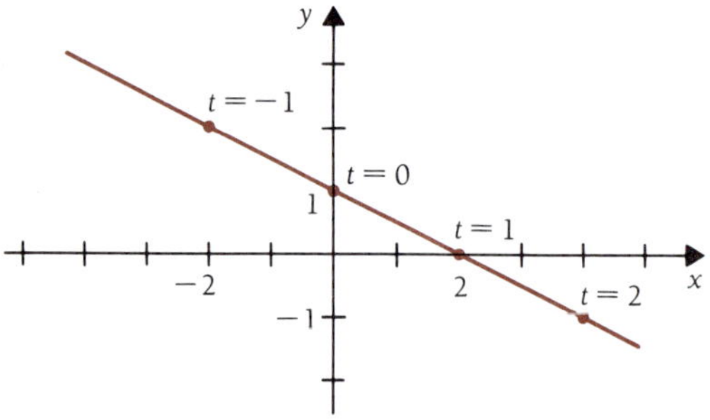

FIGURE 11.1 The graph of $\begin{cases} x = 2t \\ y = 1 - t \end{cases}$

Example 11.2 Sketch the curve

$$\begin{cases} x = a \cos t \\ y = b \sin t, \quad t \in [0, 2\pi]. \end{cases}$$

Solution In order to eliminate the parameter in this case, we write

$$\cos t = \frac{x}{a}, \qquad \sin t = \frac{y}{b}.$$

Then we use the identity $\cos^2 t + \sin^2 t = 1$, and get

$$\left(\frac{x}{a}\right)^2 + \left(\frac{y}{b}\right)^2 = 1.$$

The curve is therefore an ellipse (Figure 11.2). As t varies from 0 to 2π, the point $(x, y) = (a \cos t, b \sin t)$ traces out the entire ellipse in the counterclockwise direction.

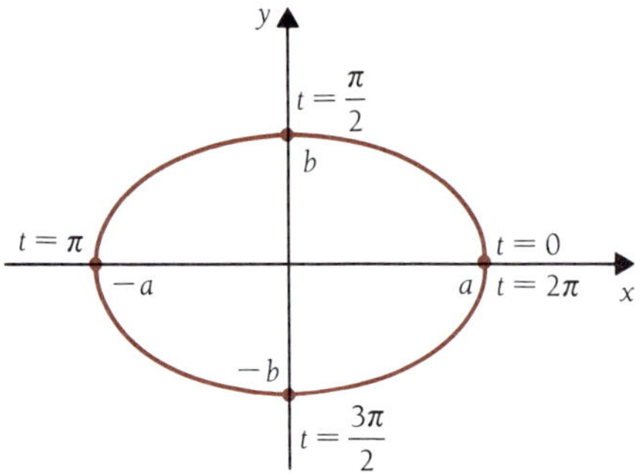

FIGURE 11.2 The graph of $\begin{cases} x = a \cos t \\ y = b \sin t \end{cases}$

Example 11.3 Sketch the curve

$$\begin{cases} x = \sin^2 t \\ y = \cos t. \end{cases}$$

Solution Since no restrictions are placed on t, we assume $t \in (-\infty, \infty)$. To eliminate the parameter, we note that

$$\begin{aligned} y^2 &= \cos^2 t \\ &= 1 - \sin^2 t \\ &= 1 - x \\ &= -(x - 1). \end{aligned}$$

The graph of this equation is a parabola, as shown in Figure 11.3. But the parametric equations actually represent only part of the parabola; since $0 \leq \sin^2 t \leq 1$ and $-1 \leq \cos t \leq 1$, the curve is only the solid portion of the parabola in Figure 11.3. As t varies from $-\infty$ to ∞, the point (x, y) traces this portion of the parabola infinitely many times.

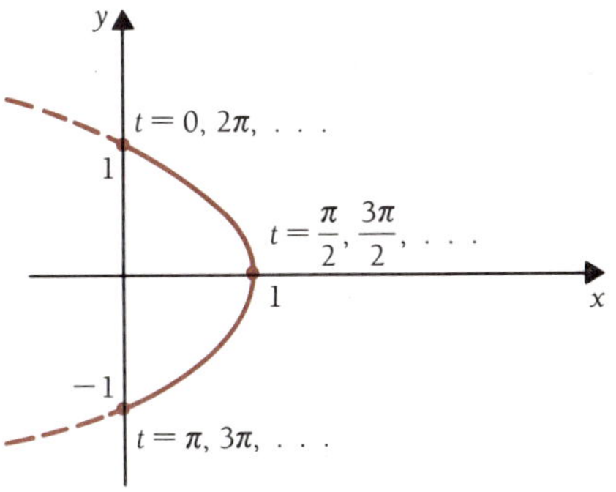

FIGURE 11.3 The graph of $\begin{cases} x = \sin^2 t \\ y = \cos t \end{cases}$ ■

A curve may have many different sets of parametric equations, or **parametrizations**. The precise way in which the point $(f(t), g(t))$ traces out the curve is determined by the parametrization.

■ **Example 11.4** Show that

$$\begin{cases} x = \cos t \\ y = \sin t \end{cases} \tag{2}$$

and

$$\begin{cases} x = \sin t \\ y = -\cos t, \quad t \in [0, 2\pi] \end{cases} \tag{3}$$

are both parametrizations of the unit circle.

Solution We note that in (2),

$$x^2 + y^2 = \cos^2 t + \sin^2 t = 1,$$

and in (3),

$$x^2 + y^2 = \sin^2 t + (-\cos t)^2 = 1,$$

so that the unit circle contains the curve in each case. The equations (2) trace out the circle in the counterclockwise direction infinitely many times, while the equations (3) trace out the circle only once, in the clockwise direction, starting and ending at the point $(0, -1)$. ■

It is not always convenient or even possible to eliminate the parameter from a set of parametric equations. In those cases, the curve is sketched by plotting points.

■ **Example 11.5** Sketch the curve

$$\begin{cases} x = t - 2 \sin t \\ y = 1 - 2 \cos t, \quad t \in [-\pi, \pi]. \end{cases}$$

Solution Since eliminating the parameter is unproductive in this case, we plot points. From the following table of values, we plot points and sketch the curve, as shown in Figure 11.4.

$t =$	$-\pi$	-2	-1	0	1	2	π
$x =$	$-\pi$	-0.18	0.68	0	-0.68	0.18	π
$y =$	3	1.83	-0.08	-1	-0.08	1.83	3

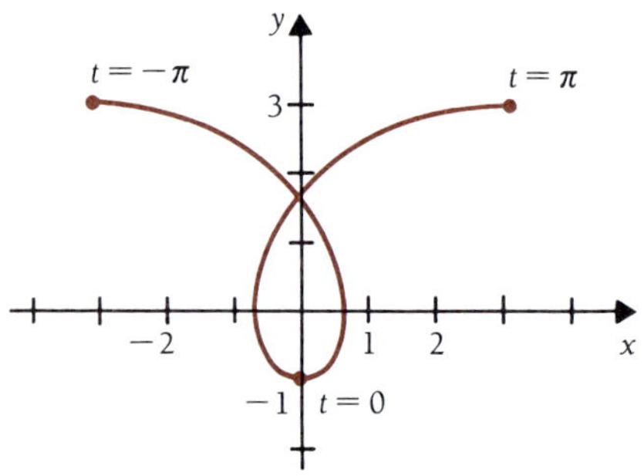

FIGURE 11.4 The graph of $\begin{cases} x = t - 2 \sin t \\ y = 1 - 2 \cos t, \quad -\pi \le t \le \pi \end{cases}$ ■

Tangents to Parametrized Curves

If a curve is the graph of a differentiable function, we can find the slope of the curve by the usual differentiation methods. If it is not possible to eliminate the parameter, we must have a technique for finding the derivative dy/dx. Theorem 11.1 tells us how.

THEOREM 11.1 **Parametric Form of the Chain Rule** If $x = f(t)$ and $y = g(t)$ are differentiable functions and $f'(t) \neq 0$, then dy/dx exists and

$$\frac{dy}{dx} = \frac{dy/dt}{dx/dt} = \frac{g'(t)}{f'(t)}.$$

Proof If dy/dx exists, then by the chain rule,

$$\frac{dy}{dt} = \frac{dy}{dx}\frac{dx}{dt}.$$

We solve this equation for dy/dx to get the desired formula.

To show that dy/dx exists, let $P(f(t), g(t))$ and $Q(f(t+h), g(t+h))$ be two points on the curve (Figure 11.5). Then the slope of the line PQ is

$$m_h = \frac{g(t+h) - g(t)}{f(t+h) - f(t)} = \frac{[g(t+h) - g(t)]/h}{[f(t+h) - f(t)]/h}.$$

The slope of the curve at P is then the limit of m_h as $h \to 0$, which exists and equals $g'(t)/f'(t)$.

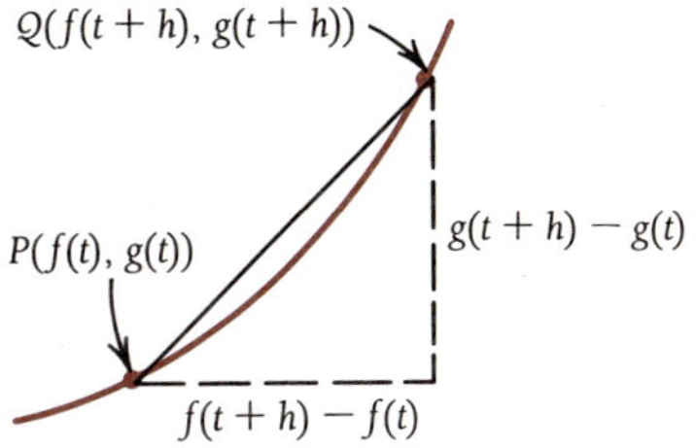

FIGURE 11.5 The slope of a parametrized curve

■ **Example 11.6** Find dy/dx:

(a) $\begin{cases} x = 2t + 3 \\ y = t^2 - 1 \end{cases}$ (b) $\begin{cases} x = t - \sin t \\ y = 1 - \cos t \end{cases}$

Solution **(a)** By Theorem 11.1,

$$\frac{dy}{dx} = \frac{dy/dt}{dx/dt} = \frac{2t}{2} = t.$$

(b) Again by the parametric form of the chain rule,

$$\frac{dy}{dx} = \frac{dy/dt}{dx/dt} = \frac{\sin t}{1 - \cos t}.$$

■

■ **Example 11.7** The curve

$$\begin{cases} x = t - 2\sin t \\ y = 1 - 2\cos t, \quad t \in [-\pi, \pi] \end{cases}$$

of Example 11.5 crosses itself at the point whose coordinates are approximately (0, 1.638), as shown in Figure 11.4. [This point is found by estimating the solutions of the equation $t - 2 \sin t = 0$, which are approximately ± 1.895, and using those values in the expression for y. In other words, if $\pm z$ are the solutions of $t - 2 \sin t = 0$, then $y = 1 - 2 \cos(\pm z) = 1 - 2\sqrt{1 - \sin^2 z} = 1 - \sqrt{4 - z^2}$.] Find equations of the tangent lines to the curve at that point.

Solution The slope of the curve is

$$\frac{dy}{dx} = \frac{2 \sin t}{1 - 2 \cos t}.$$

When $t = z$, the slope is

$$\frac{dy}{dx} = \frac{2 \sin z}{1 - 2 \cos z} = \frac{z}{1 + \sqrt{4 - z^2}},$$

which leads to the tangent line

$$y = \left(\frac{z}{1 + \sqrt{4 - z^2}}\right) x + 1 + \sqrt{4 - z^2}.$$

When $t = -z$, we get the tangent line

$$y = \left(\frac{-z}{1 + \sqrt{4 - z^2}}\right) x + 1 + \sqrt{4 - z^2}.$$ ∎

In the parametric form of the chain rule, if $dx/dt = 0$, the derivative dy/dx does not exist. If dy/dt exists and is not 0, then the tangent line to the curve must be vertical, because

$$\frac{dx}{dy} = \frac{dx/dt}{dy/dt} = 0.$$

▪ Example 11.8 Find the points on the curve of Example 11.7 at which the tangent lines are vertical.

Solution We seek the points for which dx/dt is 0, so we set $dx/dt = 0$ and solve for t:

$$\frac{dx}{dt} = 1 - 2 \cos t = 0,$$

$$\cos t = \frac{1}{2},$$

$$t = \pm\frac{\pi}{3}.$$

When $t = -\pi/3$, we get the point $(-\pi/3 + \sqrt{3}, 0)$; and when $t = \pi/3$, we get the point $(\pi/3 - \sqrt{3}, 0)$. Since $dy/dt = 2 \sin t \neq 0$ when $t = \pm\pi/3$, these points are those at which the tangent lines are vertical (see Figure 11.4). ∎

Smooth Curves

A **smooth curve** is a curve that has a tangent line at every point. Thus, the graph of a differentiable function is a smooth curve. A parametrized curve is smooth whenever its parametric equations are differentiable.

The Second Derivative in Parametric Form

The parametric form of the chain rule applies not only for the first derivative, but for all derivatives. However, one slight quirk in the process makes it necessary to discuss and illustrate the finding of the second derivative.

The second derivative is

$$\frac{d^2y}{dx^2} = \frac{d}{dx}\left(\frac{dy}{dx}\right).$$

Let us denote dy/dx by y'; then

$$\frac{d^2y}{dx^2} = \frac{d}{dx}(y') = \frac{dy'}{dx}.$$

By the parametric form of the chain rule, we therefore have

$$\frac{d^2y}{dx^2} = \frac{dy'}{dx} = \frac{\dfrac{dy'}{dt}}{\dfrac{dx}{dt}} = \frac{\dfrac{d}{dt}\left(\dfrac{dy}{dx}\right)}{\dfrac{dx}{dt}}. \tag{4}$$

That is, to find d^2y/dx^2, we differentiate dy/dx with respect to t and then divide by dx/dt. [*Note:* It is very easy to forget to divide by dx/dt!]

■ **Example 11.9** Find d^2y/dx^2:

(a) $\begin{cases} x = 3t \\ y = t^3 \end{cases}$ (b) $\begin{cases} x = t - \sin t \\ y = 1 - \cos t \end{cases}$

Solution (a) First, we find dy/dx:

$$\frac{dy}{dx} = \frac{dy/dt}{dx/dt} = \frac{3t^2}{3} = t^2.$$

Then we use equation (4):

$$\frac{d^2y}{dx^2} = \frac{\dfrac{d}{dt}\left(\dfrac{dy}{dx}\right)}{\dfrac{dx}{dt}} = \frac{2t}{3}.$$

(b) From Example 11.6(b), we have

$$\frac{dy}{dx} = \frac{\sin t}{1 - \cos t}.$$

Hence,

$$\begin{aligned}\frac{d}{dt}\left(\frac{dy}{dx}\right) &= \frac{(1 - \cos t)(\cos t) - (\sin t)(\sin t)}{(1 - \cos t)^2} \\ &= \frac{\cos t - 1}{(1 - \cos t)^2}.\end{aligned}$$

So, by equation (4),

$$\frac{d^2y}{dx^2} = \frac{\cos t - 1}{(1 - \cos t)^3} = \frac{-1}{(1 - \cos t)^2}.$$ ■

The computation done in this last example can be done in general; it gives us a form of equation (4) involving derivatives of x and y with respect to t. We have

$$\begin{aligned}\frac{d}{dt}\left(\frac{dy}{dx}\right) &= \frac{d}{dt}\left(\frac{\dfrac{dy}{dt}}{\dfrac{dx}{dt}}\right) \\ &= \frac{\left(\dfrac{dx}{dt}\right)\left(\dfrac{d^2y}{dt^2}\right) - \left(\dfrac{dy}{dt}\right)\left(\dfrac{d^2x}{dt^2}\right)}{\left(\dfrac{dx}{dt}\right)^2}\end{aligned}$$

by the quotient rule, so that by equation (4),

$$\frac{d^2y}{dx^2} = \frac{\left(\dfrac{dx}{dt}\right)\left(\dfrac{d^2y}{dt^2}\right) - \left(\dfrac{dy}{dt}\right)\left(\dfrac{d^2x}{dt^2}\right)}{\left(\dfrac{dx}{dt}\right)^3}. \tag{5}$$

Arc Length of Parametrized Curves

From Section 8.3, we recall that the differential of arc length is expressed in terms of the differentials of x and y by the formula

$$ds = \sqrt{(dx)^2 + (dy)^2}.$$

If x and y are given parametrically as $x = f(t)$ and $y = g(t)$, then $dx = f'(t)\,dt$ and $dy = g'(t)\,dt$, and

$$\begin{aligned} ds &= \sqrt{[f'(t)\,dt]^2 + [g'(t)\,dt]^2} \\ &= \sqrt{[f'(t)]^2 + [g'(t)]^2}\,dt \\ &= \sqrt{\left(\frac{dx}{dt}\right)^2 + \left(\frac{dy}{dt}\right)^2}\,dt. \end{aligned}$$

Therefore, the curve

$$C: \quad \begin{cases} x = f(t) \\ y = g(t), \quad t \in [a, b] \end{cases}$$

has length

$$s = \int_a^b \sqrt{\left(\frac{dx}{dt}\right)^2 + \left(\frac{dy}{dt}\right)^2}\,dt. \tag{6}$$

■ **Example 11.10** Find the length of the curve

$$C: \quad \begin{cases} x = t - \sin t \\ y = 1 - \cos t, \quad t \in [0, 2\pi]. \end{cases}$$

Solution Using equation (6) directly, we get

$$\begin{aligned} s &= \int_0^{2\pi} \sqrt{(1 - \cos t)^2 + (\sin t)^2}\,dt \\ &= \int_0^{2\pi} \sqrt{2 - 2\cos t}\,dt \\ &= \int_0^{2\pi} \sqrt{4\sin^2 \frac{t}{2}}\,dt \\ &= \int_0^{2\pi} 2\sin \frac{t}{2}\,dt \\ &= -4\cos \frac{t}{2}\Big|_0^{2\pi} = 8. \end{aligned}$$

■

The Cycloid

Parametric equations can be used to describe certain curves when no other analytic description exists. Example 11.11 illustrates an important case.

■ **Example 11.11** As a wheel rolls without slipping along a level road, a fixed point on the rim of the wheel describes a curve called a **cycloid** (Figure 11.6) Find parametric equations for the cycloid described by a circle of radius a rolling along the x-axis. Assume that the point on the circle that traces out the cycloid starts out at the origin.

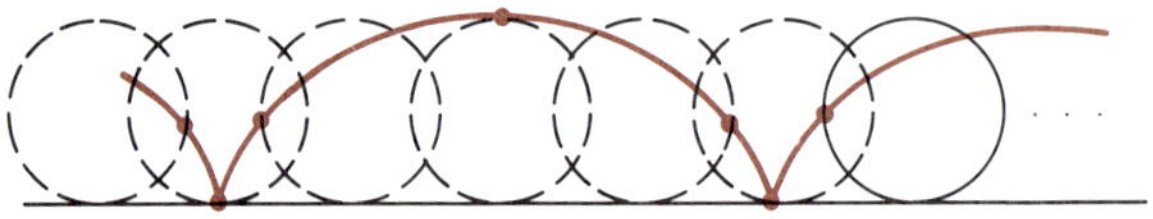
FIGURE 11.6 A point on a rolling circle generates a cycloid

Solution We start with our circle touching the origin (Figure 11.7). Let P be the point on the circle that is initially at the origin. As the circle rolls along the x-axis in the positive direction, the point P rises and falls, tracing out the cycloid. We will stop the circle at some point Q and compute the coordinates of P at that position.

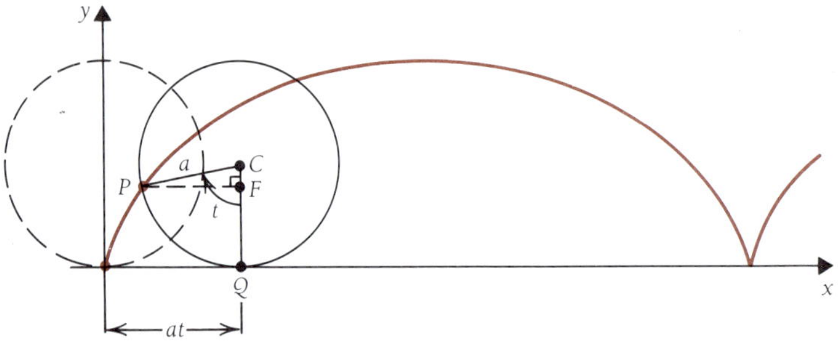

FIGURE 11.7 Finding parametric equations of the cycloid

Let C be the center of the circle, and let t be the angle (in radians) through which CP has rotated. Since the circle rolls without slipping, the distance that the circle has moved horizontally is the same as the length of the circular arc $PQ = at$. Let F be the foot of the perpendicular from P to the line CQ. Since $CP = a$ and the measure of angle FCP is t, we have

$$|PF| = a \sin t.$$

Thus, the x-coordinate of P is

$$x = at - a \sin t, \tag{7}$$

keeping in mind the sign of $\sin t$ in the various quadrants. Similarly, we have

$$|CF| = a \cos t,$$

so that the y-coordinate of P is

$$y = a - a \cos t. \tag{8}$$

Therefore, the cycloid has parametric equations

$$\begin{cases} x = a(t - \sin t) \\ y = a(1 - \cos t). \end{cases} \tag{9}$$

One arch of the cycloid is traced out as t varies from 0 to 2π ■

The cycloid is a curve with some interesting properties. Suppose an arch of the cycloid is inverted, with highest point A and lowest point B (Figure 11.8). Now suppose the curve is a wire, on which a bead slides without friction. The bead is released at A and slides to B under the influence of gravity. Of all curves joining A and B, the cycloid allows the bead to slide from A to B in the least time. For this reason, the cycloid is called a **brachistochrone** (least-time curve). Moreover, if the bead is released at any point between A and B, it always takes the same amount of time to slide to B. For this reason, the cycloid is called a **tautochrone** (same-time curve).

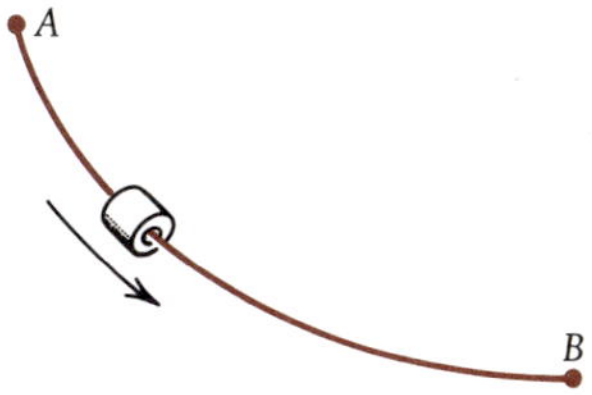

FIGURE 11.8 A bead sliding on a cycloidal wire

Exercises 11.1

Eliminate the parameter and sketch the curve.

1. $\begin{cases} x = 3t + 5 \\ y = 2 - t \end{cases}$

2. $\begin{cases} x = 1 - t \\ y = 1 + t \end{cases}$

3. $\begin{cases} x = t^2 \\ y = 2 - t \end{cases}$

4. $\begin{cases} x = 2t + 1 \\ y = 1 - t^2 \end{cases}$

5. $\begin{cases} x = 2 \sin 3t \\ y = 2 \cos 3t \end{cases}$

6. $\begin{cases} x = 3 \cos 2t \\ y = 2 \sin 2t \end{cases}$

7. $\begin{cases} x = e^t \\ y = 3 - t \end{cases}$

8. $\begin{cases} x = \arctan t \\ y = 2 \ln t \end{cases}$

Find dy/dx.

9. $\begin{cases} x = 4t + 2 \\ y = 2 - 3t \end{cases}$

10. $\begin{cases} x = t^2 + 4 \\ y = 2t - 3 \end{cases}$

11. $\begin{cases} x = t^3 + 1 \\ y = 1/t \end{cases}$

12. $\begin{cases} x = t^2 - t \\ y = t^2 + 2t \end{cases}$

13. $\begin{cases} x = \sin t \\ y = 2 \cos t \end{cases}$

14. $\begin{cases} x = e^t \cos t \\ y = t \cos t \end{cases}$

15. $\begin{cases} x = a(t - \sin t) \\ y = a(1 - \cos t) \end{cases}$

16. $\begin{cases} x = t \ln t \\ y = t \cos t \end{cases}$

Find d^2y/dx^2.

17. $\begin{cases} x = t^2 \\ y = 1 - t \end{cases}$

18. $\begin{cases} x = t^3 + 1 \\ y = 1/t \end{cases}$

19. $\begin{cases} x = \cos t \\ y = \sin t \end{cases}$

20. $\begin{cases} x = 3 \sin 4t \\ y = 2 \cos 4t \end{cases}$

21. $\begin{cases} x = \sec t \\ y = \tan t \end{cases}$

22. $\begin{cases} x = a(t - \sin t) \\ y = a(1 - \cos t) \end{cases}$

Find the length of each curve C.

23. $C = \{(t, t^2) | 0 \leq t \leq 3\}$

24. $C = \{(t^2, 2t) | -1 \leq t \leq 1\}$

25. $C = \{(t^2, t^3) | -1 \leq t \leq 1\}$

26. $C = \{(3t^2, 2t^3) | 0 \leq t \leq 8\}$

27. $C = \{(a(t - \sin t), a(1 - \cos t)) | 0 \leq t \leq 2\pi\}$

28. $C = \{(2 \sin t, 2 \cos t) | 0 \leq t \leq \pi/2\}$

29. Show that equation (5) can be written as

$$\frac{d^2y}{dx^2} = \begin{vmatrix} x'(t) & x''(t) \\ y'(t) & y''(t) \end{vmatrix} [x'(t)]^{-3}.$$

30. Find an expression similar to equation (5) for d^3y/dx^3.

31. Find the area under one arch of the cycloid

$$\begin{cases} x = a(t - \sin t) \\ y = a(1 - \cos t). \end{cases}$$

[*Hint:* Use $A = \int_{t=0}^{t=2\pi} y\, dx$, with $y = a(1 - \cos t)$ and $dx = a(1 - \cos t)\, dt$.]

32. One arch of the cycloid is revolved about the x-axis. Find the area of the surface generated.

33. Repeat Problem 32 for rotation about the y-axis.

34. The region under one arch of the cycloid is revolved about the x-axis. Find the volume of the solid generated.

35. Repeat Problem 34 for rotation about the y-axis.

SECTION 11.2

Polar Coordinates

The rectangular coordinate system can be thought of as a sort of "city block" system. To reach the point (x, y) from the origin, one travels x "blocks" east or west and y "blocks" north or south. Such a system is appropriate in a city where one is usually constrained to follow the streets.

In a different setting, however, a rectangular coordinate system might be inconvenient to use. At sea, for example, a ship would not normally sail 2 knots east and then $2\sqrt{3}$ knots north to reach a certain point, but rather 4 knots in the direction 30° east of north (see Figure 11.9). You can probably think of other situations where an "as-the-crow-flies" coordinate system would be more convenient.

In calculus, some applications are simplified by such an alternative system of coordinates. And some natural phenomena are difficult to describe in any other coordinate system.

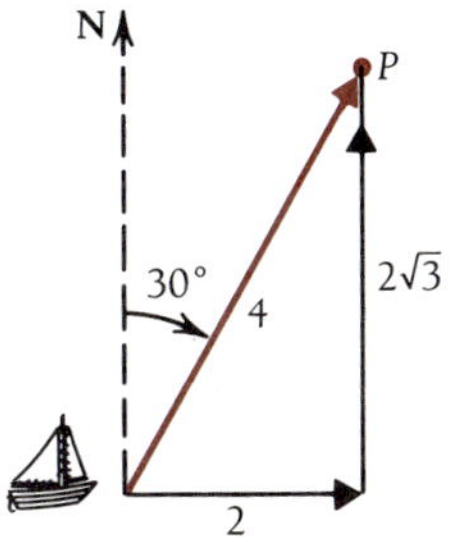

FIGURE 11.9 Straight-line route is shorter

The Polar Coordinate System

A coordinate system in which points are specified in terms of distance and direction is called a **polar coordinate system**. Distance is measured from a fixed point, called the **pole**, or **origin**, and direction is given as an angle relative to a fixed ray or half-line emanating from the pole, called the **polar axis** (see Figure 11.10).

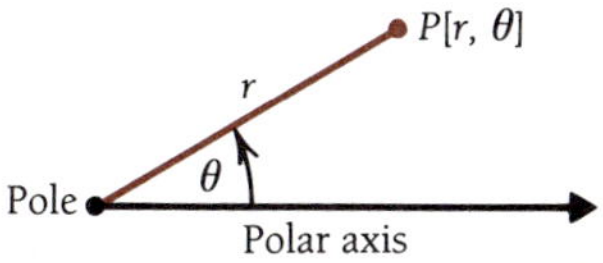

FIGURE 11.10 The polar coordinate system

If a point P lies a distance r from the pole, and the ray from the pole through P forms an angle of measure θ with the polar axis, we give P the **polar coordinates** $[r, \theta]$. We customarily select the polar axis to be horizontal, extending to the right, and take θ to be positive in the counterclockwise direction. We usually measure θ in radians.

In the polar coordinate system, a point does not have just one pair of coordinates, as in the rectangular system. For example, the pole is represented as $[0, \theta]$ with θ any value whatsoever. Also, since the angles θ and $\theta + 2n\pi$ (n an integer) are coterminal, the points $[r, \theta]$ and $[r, \theta + 2n\pi]$ are the same (see Figure 11.11). Moreover, the point $[r, \theta]$ may be written as $[-r, \theta + \pi]$; that is, one may reach the point $[r, \theta]$ by facing in the direction $\theta + \pi$ (the opposite direction to θ) and backing up a distance r (see Figure 11.12). Thus, each point has infinitely many pairs of polar coordinates.

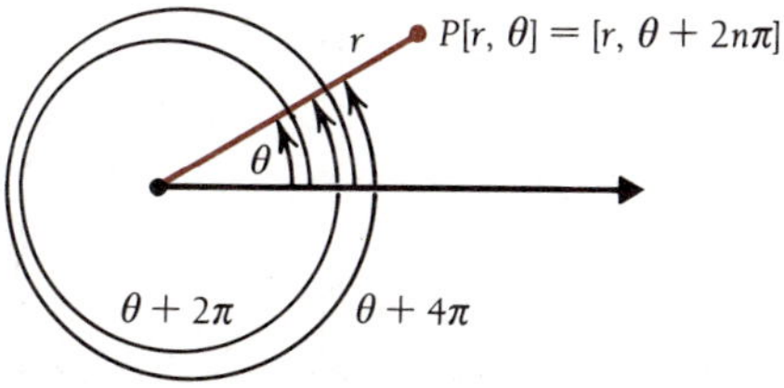

FIGURE 11.11 Angles are not unique

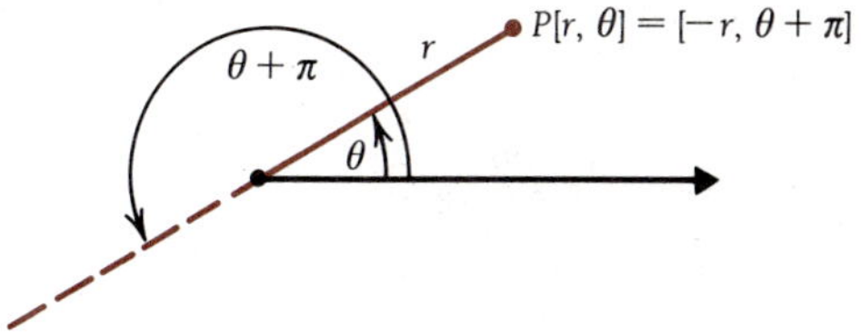

FIGURE 11.12 Reversing direction changes angles

Example 11.12 Write six other pairs of coordinates for the point $[2, \pi/4]$.

Solution We could write $[2, \pi/4]$ as $[2, 9\pi/4]$, $[2, 17\pi/4]$, $[2, -7\pi/4]$, $[-2, 5\pi/4]$, $[-2, 13\pi/4]$, or $[-2, -3\pi/4]$. ■

Example 11.13 Give the polar coordinates of a point on the polar axis.

Solution A point on the polar axis has coordinates of the form $[r, 2n\pi]$ or $[-r, \pi + 2n\pi]$ with $r \geq 0$ and n an integer. ■

Relation of Polar Coordinates to Rectangular Coordinates

To compare the rectangular and polar coordinate systems, we will superimpose a polar coordinate system on a rectangular system. We will choose the origin as the pole, and the positive half of the x-axis as the polar axis. This is depicted in Figure 11.13. If P is a point with rectangular coordinates (x, y) and polar coordinates $[r, \theta]$, then the right-triangle properties of the trigonometric functions tell us that

$$\begin{cases} x = r\cos\theta \\ y = r\sin\theta. \end{cases} \tag{1}$$

These equations give the rectangular coordinates (x, y) in terms of the polar coordinates $[r, \theta]$.

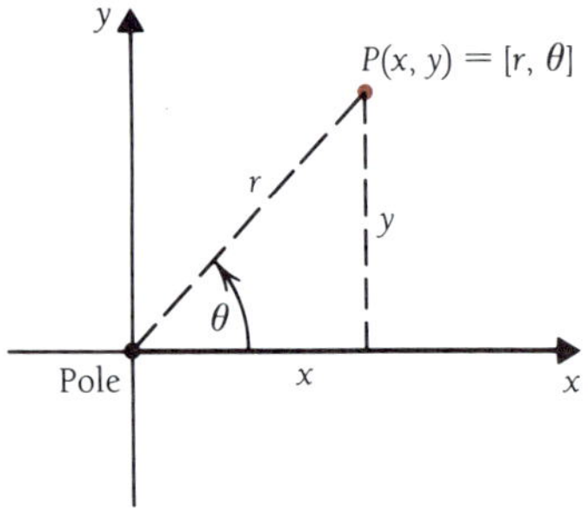

FIGURE 11.13 Relationship between rectangular and polar coordinates

Example 11.14 Give the rectangular coordinates of the points:

(a) $[1, \pi/6]$ **(b)** $[-1, \pi/2]$ **(c)** $[2, 1]$

Solution Using equations (1) directly, we have:

(a) $(x, y) = (1\cos(\pi/6), 1\sin(\pi/6)) = (\sqrt{3}/2, \frac{1}{2})$.
(b) $(x, y) = (-1\cos(\pi/2), -1\sin(\pi/2)) = (0, -1)$.
(c) $(x, y) = (2\cos 1, 2\sin 1) \approx (1.08, 1.68)$. ■

From Figure 11.13, we also can obtain the relationships

$$\begin{cases} r^2 = x^2 + y^2 \\ \theta = \arctan(y/x) = \arcsin(y/r) = \arccos(x/r). \end{cases} \tag{2}$$

These equations enable us to find polar coordinates for a point whose rectangular coordinates are given. Note that θ must be chosen according to the quadrant in which (x, y) lies.

Example 11.15 Give polar coordinates for the points:

(a) $(-1, \sqrt{3})$ **(b)** $(-2, 2)$ **(c)** $(4, 3)$

Solution Using equations (2), we have:

(a) $[r, \theta] = [\sqrt{(-1)^2 + (\sqrt{3})^2}, \arctan(-\sqrt{3})] = [2, 2\pi/3]$.

(b) $[r, \theta] = [\sqrt{(-2)^2 + (2)^2}, \arccos(-2/2\sqrt{2})] = [2\sqrt{2}, 3\pi/4]$.

(c) $[r, \theta] = [\sqrt{4^2 + 3^2}, \arctan \frac{3}{4}] = [5, \arctan \frac{3}{4}] \approx [5, 0.64]$. ■

Equations in Polar Coordinates; Transformations

An equation in the two variables x and y defines a curve in the rectangular coordinate system that is called the graph of the equation. In exactly the same way, an equation in the two variables r and θ defines a curve in the polar coordinate system. The equations (1) and (2) are called **transformation equations** and provide a means of transforming an equation in one system to an equation in the other.

Example 11.16 Find an equation in polar coordinates for the line $y = mx$.

Solution Substituting equations (1) directly into $y = mx$, we get

$$r \sin \theta = mr \cos \theta.$$

Then solving for θ, we get

$$\theta = \arctan m.$$

That is, θ is a constant and r is any number (see Figure 11.14).

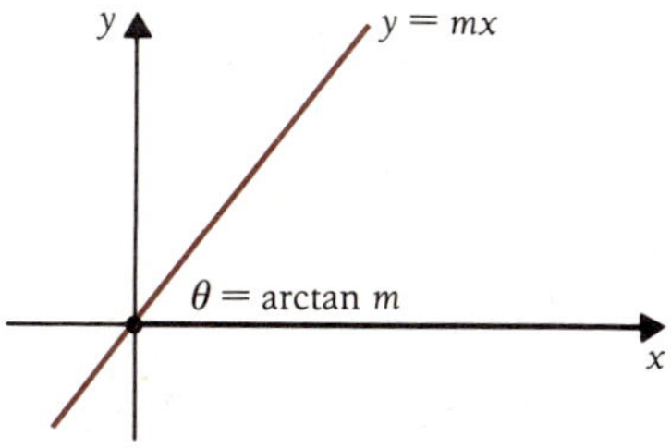

FIGURE 11.14 The equation of a line through the origin is $\theta =$ constant ■

Example 11.17 Find an equation in rectangular coordinates for the curve $r = k$, k a constant.

Solution Using equations (2), we have

$$\sqrt{x^2 + y^2} = k,$$
$$x^2 + y^2 = k^2.$$

The curve is therefore a circle of radius $|k|$, centered at the origin (see Figure 11.15).

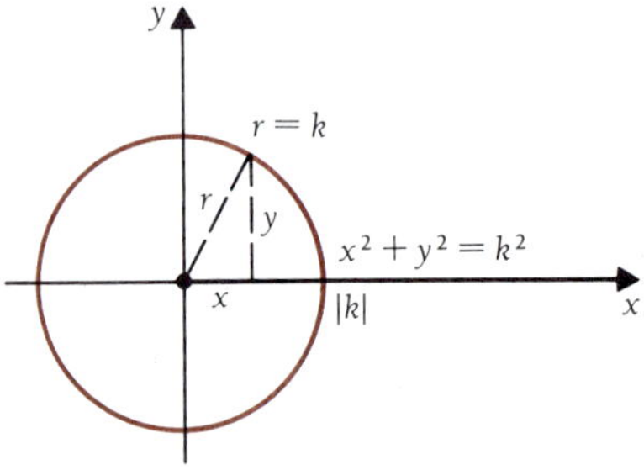

FIGURE 11.15 The equation of a circle centered at the origin is $r =$ constant

Example 11.18 Transform the equation $x^2 + y^2 = 2x$ into polar coordinates.

Solution Using equations (1) and (2), we get

$$r^2 = 2r \cos \theta,$$
$$r = 2 \cos \theta.$$

In order to divide by r, we had to assume $r \neq 0$; but the pole, which is the only point for which $r = 0$, is a solution of $r = 2 \cos \theta$ when written as $[0, \pi/2]$. Thus, no points are lost by dividing by r.

Example 11.19 Transform the equation $r = \sin 2\theta$ into rectangular coordinates.

Solution We first expand $\sin 2\theta$ by the double-angle formula, and then use equations (1) and (2):

$$r = 2 \sin \theta \cos \theta,$$
$$r = 2\left(\frac{y}{r}\right)\left(\frac{x}{r}\right),$$
$$r^3 = 2xy,$$
$$(x^2 + y^2)^{3/2} = 2xy,$$
$$(x^2 + y^2)^3 = 4x^2y^2.$$

This equation defines the same curve as $r = \sin 2\theta$, but in a different coordinate system.

Exercises 11.2

Give four additional pairs of coordinates for each point, two with r positive and two with r negative. Then plot the point in a polar coordinate system.

1. $[1, \pi/2]$
2. $[4, \pi/4]$
3. $[-2, 0]$
4. $[-1, -\pi/3]$

Give rectangular coordinates for each point.

5. $[0, 2]$
6. $[-1, 0]$
7. $[2, 0]$
8. $[3, \pi/2]$
9. $[2, \pi/3]$
10. $[2, -\pi/6]$
11. $[1, \pi/8]$
12. $[4, \pi/12]$
13. $[2, -2]$
14. $[3, 5]$
15. $[\sqrt{2}, \pi/4]$
16. $[\pi, \pi]$

Give polar coordinates for each point.

17. $(0, 1)$
18. $(1, 0)$
19. $(2, 2)$
20. $(-1, -1)$
21. $(0, \sqrt{3})$
22. $(-1, -\sqrt{3})$
23. $(\sqrt{3}, -1)$
24. $(a, -a\sqrt{3})$
25. $(7, 6)$
26. $(-3, 5)$
27. $(\pi, -\pi)$
28. $(\sqrt{2}, -\sqrt{2})$

Transform each equation into polar coordinates.

29. $y = x$
30. $y = x + 1$
31. $x^2 + y^2 = 4$
32. $x^2 + y^2 = 2y$
33. $2xy = 1$
34. $x = 3$
35. $y = \sqrt{1 - x^2}$
36. $y = |x|$

Transform each equation into rectangular coordinates.

37. $\theta = \pi$
38. $r = 2$
39. $r = \sin\theta$
40. $r = -2\cos\theta$
41. $r = \sec\theta$
42. $r = \tan\theta$
43. $r^2 = \sin\theta$
44. $r = 1 - \sin\theta$

SECTION 11.3

Curves in Polar Coordinates

This section has two purposes. First, we will develop the equations in polar coordinates of certain specific curves, namely lines, circles, and the eccentric conic sections. Second, we will discuss a variety of curves defined by equations in polar coordinates.

Lines

Let's begin by finding the polar equation of a line. We have already seen in Example 11.16 that the equation of a line passing through the pole (Figure 11.16, page 710) is

$$\theta = k,$$

where the constant k is the arctangent of the slope of the line.

If a line l does not pass through the pole (Figure 11.17), let the perpendicular from the pole to l meet l at $F[a, \alpha]$. If $P[r, \theta]$ is any point on l, then from the right triangle OPF we have

$$r\cos(\theta - \alpha) = a. \tag{1}$$

This is the equation of l in polar coordinates.

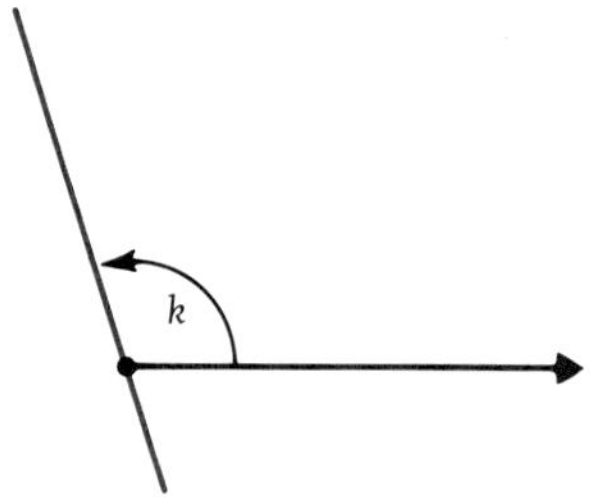

FIGURE 11.16
The line $\theta = k$

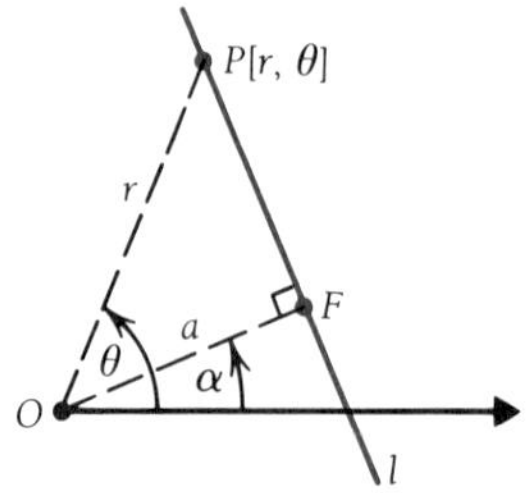

FIGURE 11.17
The line $r\cos(\theta - \alpha) = a$

■ **Example 11.20** Discuss the equations of vertical and horizontal lines.

Solution If line l is vertical and lies to the right of the pole (Figure 11.18), then $\alpha = 0$ in equation (1), and the equation of l is

$$r\cos\theta = a.$$

(That is just $x = a$ in rectangular coordinates.) If l is vertical and lies to the left of the pole (Figure 11.19), then $\alpha = \pi$, and the equation of l is

$$r\cos\theta = -a.$$

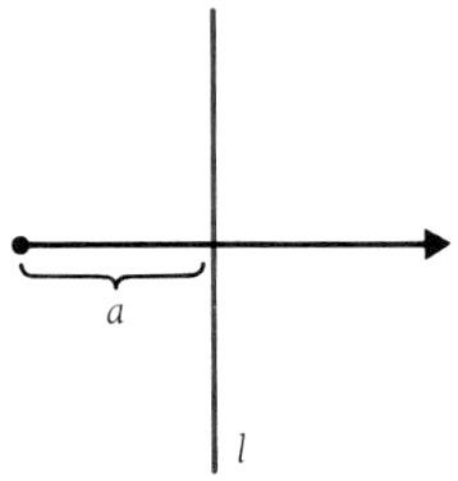

FIGURE 11.18
The line $r\cos\theta = a$

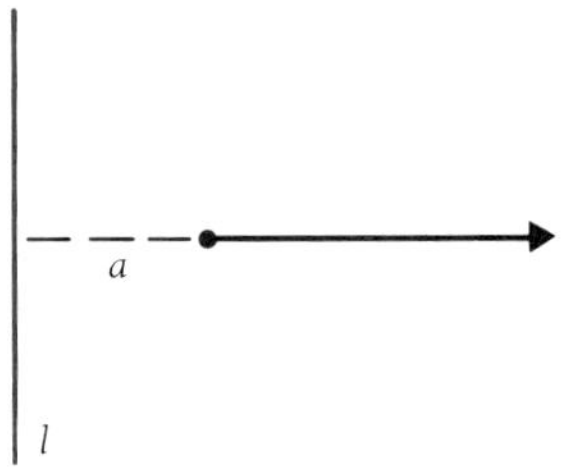

FIGURE 11.19
The line $r\cos\theta = -a$

If l is horizontal and lies above the pole (Figure 11.20), then $\alpha = \pi/2$, and the equation of l is

$$r\sin\theta = a.$$

If l is horizontal and lies below the pole (Figure 11.21), then $\alpha = -\pi/2$, and the equation of l is

$$r\sin\theta = -a.$$

FIGURE 11.20 The line $r \sin \theta = a$

FIGURE 11.21 The line $r \sin \theta = -a$ ■

Circles

Next, we consider circles. Suppose a circle has center $C[c, \alpha]$ and radius a (Figure 11.22). If $P[r, \theta]$ is any point on the circle, we may apply the law of cosines to the triangle OPC to obtain

$$a^2 = r^2 + c^2 - 2rc \cos(\theta - \alpha),$$
$$r^2 = a^2 - c^2 + 2rc \cos(\theta - \alpha), \tag{2}$$

which is the most general form of the equation of a circle.

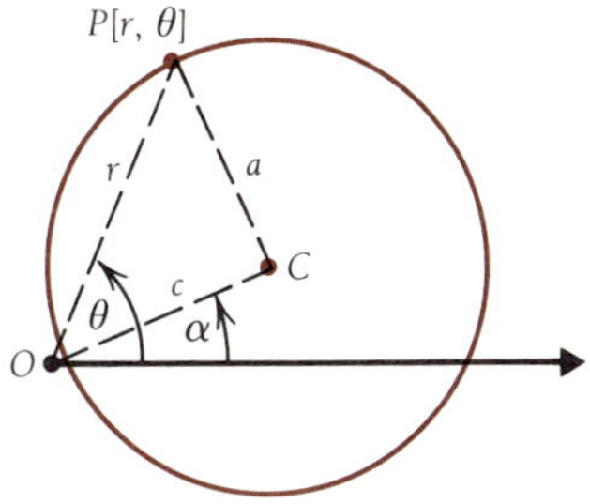

FIGURE 11.22 The circle $r^2 = a^2 - c^2 + 2rc \cos(\theta - \alpha)$

■ **Example 11.21** Discuss the equations of the following circles:

(a) With center the pole **(b)** Passing through the pole
(c) Passing through the pole and having center on the x-axis or y-axis

Solution **(a)** If the center of the circle is the pole (Figure 11.23), then $c = 0$ in equation (2), and the equation of the circle is $r^2 = a^2$, or simply

$$r = a.$$

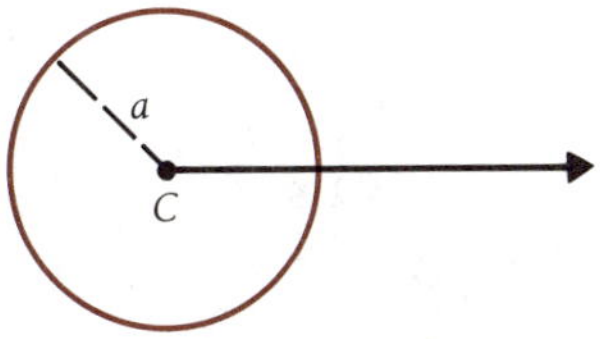

FIGURE 11.23
The circle $r = a$

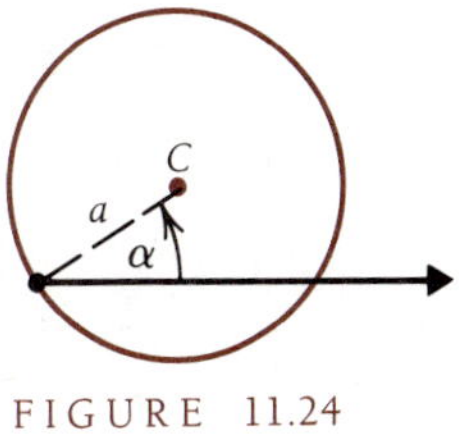

FIGURE 11.24
The circle $r = 2a \cos(\theta - \alpha)$

(b) If the circle passes through the pole (Figure 11.24), then $c = a$ in equation (2), and the equation of the circle is $r^2 = 2ra\cos(\theta - \alpha)$, or simply

$$r = 2a\cos(\theta - \alpha). \tag{3}$$

(c) If in addition to passing through the pole, the circle has its center on one of the coordinate axes, then there are four cases to consider. If the center is on the x-axis and to the right of the pole (Figure 11.25), then $\alpha = 0$ in equation (3), and the circle has equation

$$r = 2a\cos\theta.$$

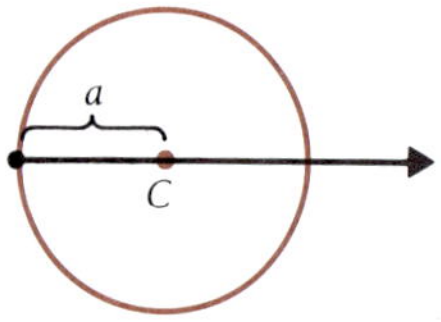

FIGURE 11.25
The circle $r = 2a\cos\theta$

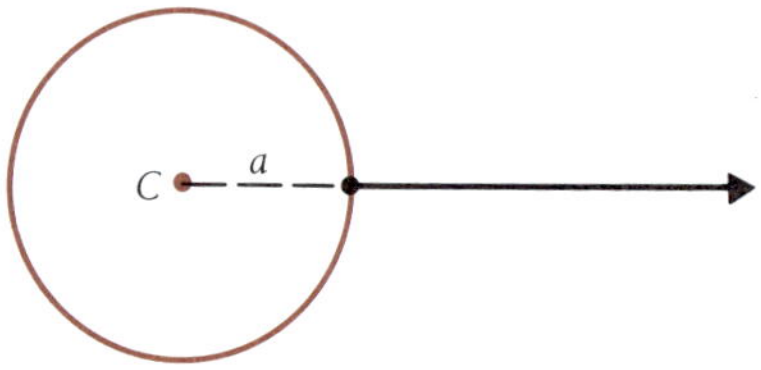

FIGURE 11.26
The circle $r = -2a\cos\theta$

If the center is to the left of the pole (Figure 11.26), then $\alpha = \pi$ in equation (3), and the circle has equation

$$r = -2a\cos\theta.$$

If the center of the circle lies on the y-axis and above the pole (Figure 11.27), then $\alpha = \pi/2$ in equation (3), and the circle has equation

$$r = 2a\sin\theta.$$

If the center lies below the pole (Figure 11.28), then $\alpha = -\pi/2$ in equation (3), and the equation of the circle is

$$r = -2a\sin\theta.$$

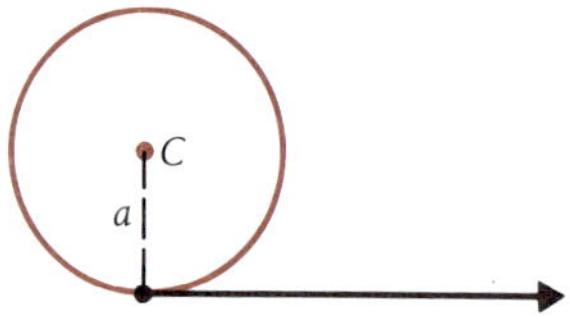

FIGURE 11.27
The circle $r = 2a\sin\theta$

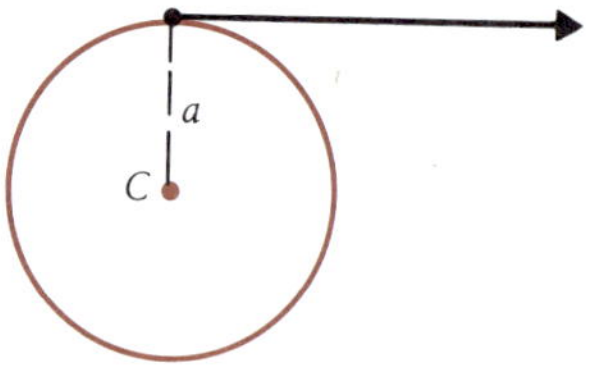

FIGURE 11.28
The circle $r = -2a\sin\theta$

■

The Eccentric Conic Sections

Now we look at the eccentric conic sections. Suppose a conic has eccentricity e, focus F at the pole, and directrix the line given by equation (1), $d: r\cos(\theta - \alpha) = a$ (see Figure 11.29). Then if $P[r, \theta]$ is any point on the conic,

$$|PF| = e|Pd|.$$

From Figure 11.29, we see that $|PF| = r$ and $|Pd| = a + r\sin(\alpha - \theta - \pi/2) = a - r\cos(\theta - \alpha)$. Hence,

$$r = e[a - r\cos(\theta - \alpha)].$$

Solving this equation for r, we get

$$r = \frac{ea}{1 + e\cos(\theta - \alpha)}. \tag{4}$$

This is the equation of the conic.

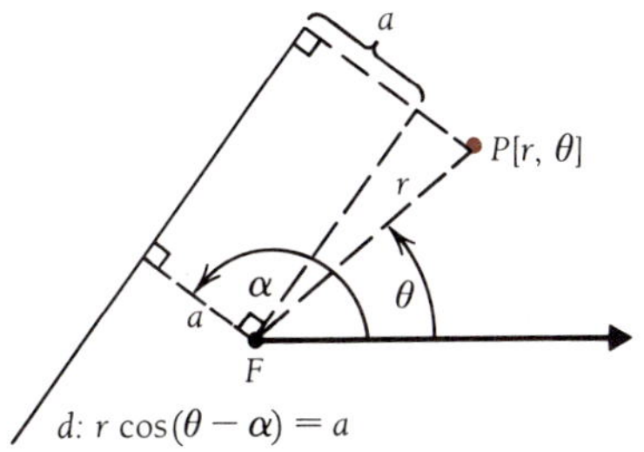

FIGURE 11.29 Deriving the equation of a conic section

Example 11.22 Discuss the equations of the conics with:

(a) Directrix vertical and lying to the left of the pole
(b) Directrix horizontal and lying below the pole

Solution

(a) If the directrix is vertical and to the left of the pole, then $\alpha = \pi$ in equation (4), and the equation is

$$r = \frac{ea}{1 - e\cos\theta}.$$

(See Figure 11.30, page 714; the ellipse shown is merely representative.)

(b) If the directrix is horizontal and lies below the pole (see Figure 11.31), then $\alpha = -\pi/2$ in equation (4), and the equation is

$$r = \frac{ea}{1 - e\sin\theta}. \tag{5}$$

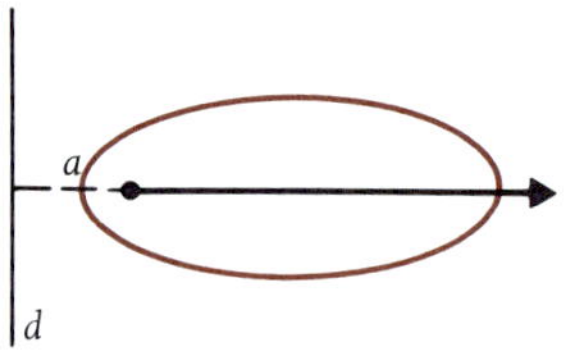

FIGURE 11.30
Directrix vertical, left of pole

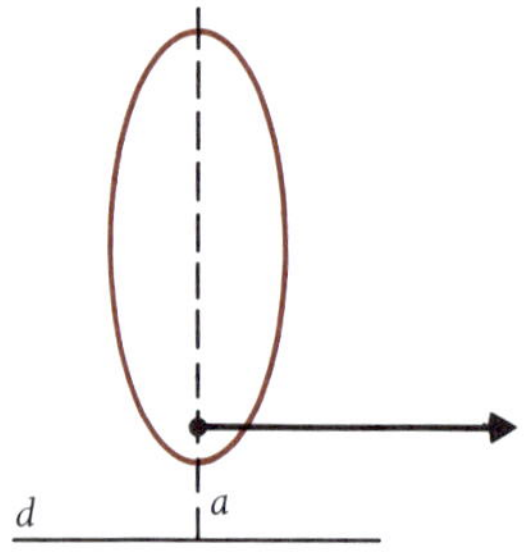

FIGURE 11.31
Directrix horizontal, below pole ■

■ **Example 11.23** Identify the graphs of the equations:

(a) $r = \dfrac{4}{2 + \cos\theta}$ **(b)** $r = \dfrac{2}{1 - \sin\theta}$

Solution **(a)** We will rewrite the equation as

$$r = \frac{(\frac{1}{2})4}{1 + (\frac{1}{2})\cos\theta}.$$

Comparing this with equation (4), we see that the graph is a conic section with focus the pole, directrix the line $r\cos\theta = 4$, and eccentricity $e = \frac{1}{2}$. The graph is thus an ellipse (Figure 11.32).

(b) Comparing the given equation with equation (5), we see that the graph is a conic with focus the pole, directrix the line $r\sin\theta = -2$, and eccentricity $e = 1$. The graph is thus a parabola (Figure 11.33).

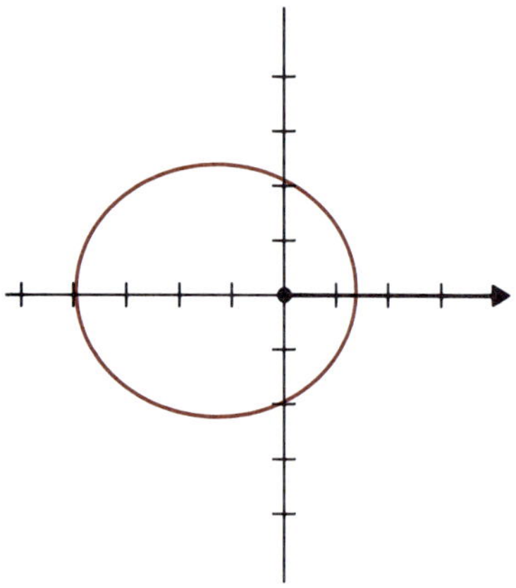

FIGURE 11.32 The graph of $r = \dfrac{4}{2 + \cos\theta}$

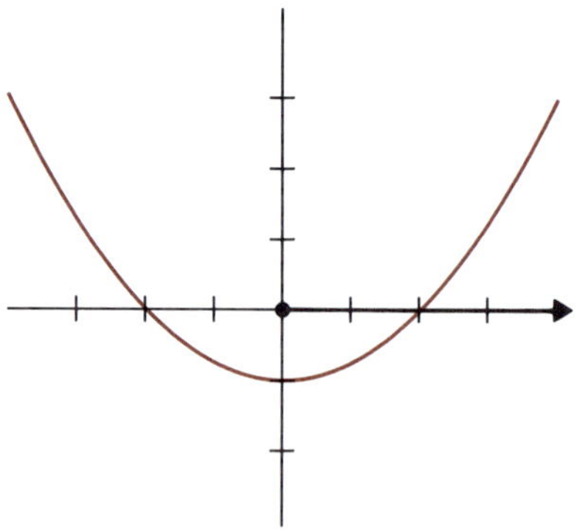

FIGURE 11.33 The graph of $r = \dfrac{2}{1 - \sin\theta}$ ■

Other Polar Curves

We consider now the graphs of some other equations in polar coordinates, mostly of the form $r = f(\theta)$. The equations listed below are simply representative.

If f is a constant, the equation is

$$r = c,$$

whose graph is a circle centered at the pole [see Examples 11.17 and 11.21(a)].

If f is of the form $f(\theta) = k\theta$, k a constant, then the equation is

$$r = k\theta.$$

The graph of this equation (Figure 11.34) is called a **spiral of Archimedes.**

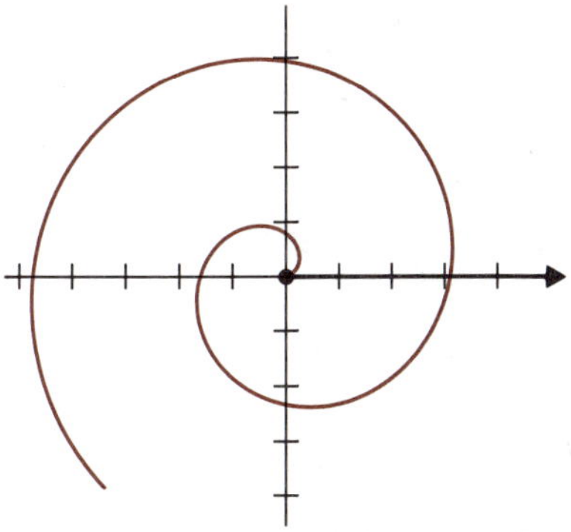

FIGURE 11.34 The graph of $r = k\theta$

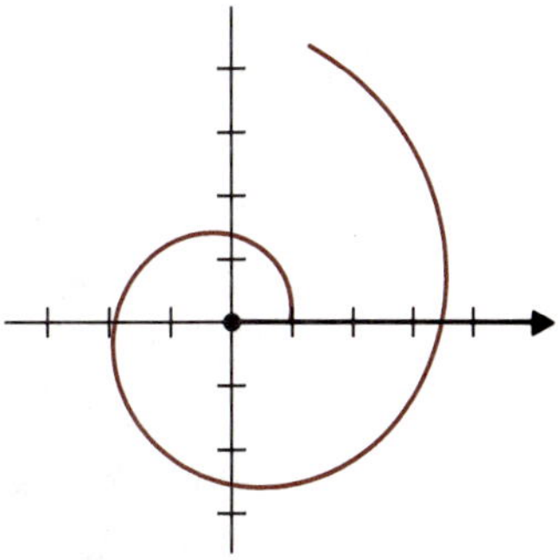

FIGURE 11.35 The graph of $r = e^{k\theta}$

If f is an exponential function, the equation is of the form

$$r = e^{k\theta},$$

whose graph (Figure 11.35) is called a **logarithmic spiral**. Such curves occur in nature—for example, in the face of a sunflower and in the chambered nautilus (Figures 11.36 and 11.37, page 716).

If f is a simple trigonometric function, say $f(\theta) = k \sin \theta$ or $f(\theta) = k \cos \theta$, the graph is a circle [see Example 11.21(c)]. If $f(\theta) = k \sec \theta$ or $f(\theta) = k \csc \theta$, the graph is a line (see Example 11.20).

Equations of the forms

$$r = a + b \cos \theta \qquad \text{and} \qquad r = a + b \sin \theta$$

have graphs (Figure 11.38, page 716) that are called **limaçons.** The special cases in which $a = b$ give equations of the forms

$$r = a(1 + \cos \theta) \qquad \text{and} \qquad r = a(1 + \sin \theta),$$

whose graphs (Figure 11.39, page 717) are called **cardioids.**

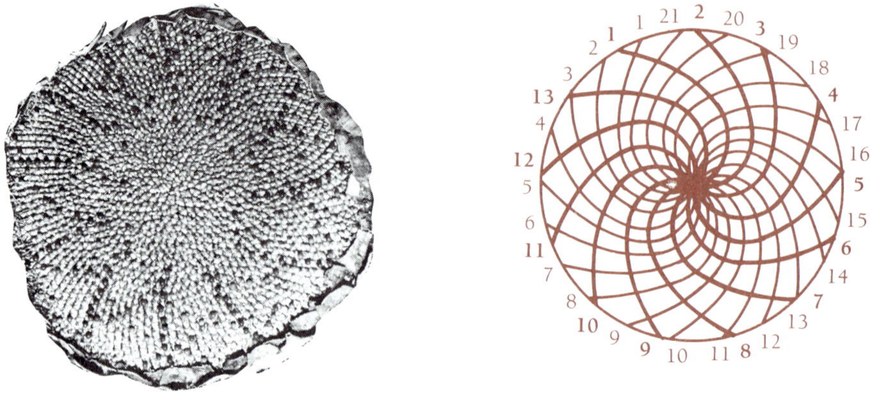

FIGURE 11.36 A logarithmic spiral occurs in the face of a sunflower

FIGURE 11.37 A logarithmic spiral in the chambered nautilus

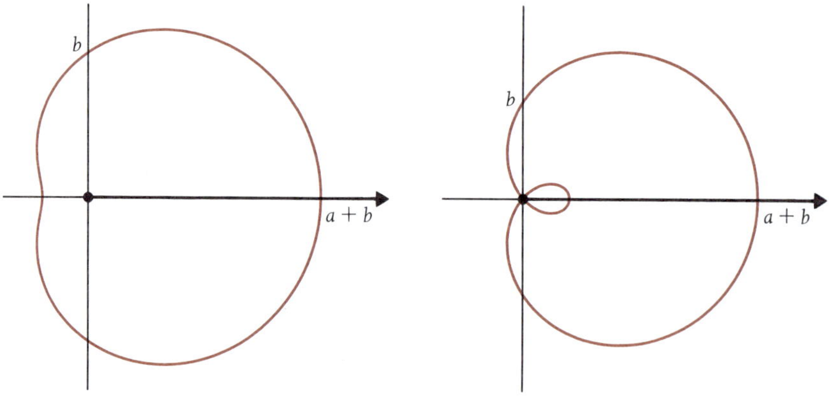

(a) $a > b$ (b) $a < b$

FIGURE 11.38 Limaçons, $r = a + b\cos\theta$

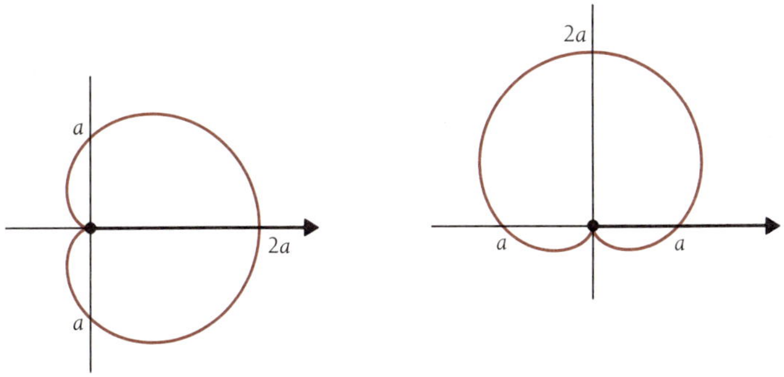

(a) $r = a(1 + \cos \theta)$ (b) $r = a(1 + \sin \theta)$

FIGURE 11.39 Cardioids

Equations of the forms

$$r = a \cos n\theta \qquad \text{and} \qquad r = a \sin n\theta$$

have graphs that are called **rose curves**. For example, if $n = 2$, the graph is a **four-leaved rose** (Figure 11.40); if $n = 3$, the graph is a **three-leaved rose** (Figure 11.41). Other examples of rose curves (with slightly different shapes of the petals) are equations of the forms

$$r^2 = a^2 \cos n\theta \qquad \text{and} \qquad r^2 = a^2 \sin n\theta.$$

Here, if $n = 2$, the curve is also called a **lemniscate** (Figure 11.42, page 718).

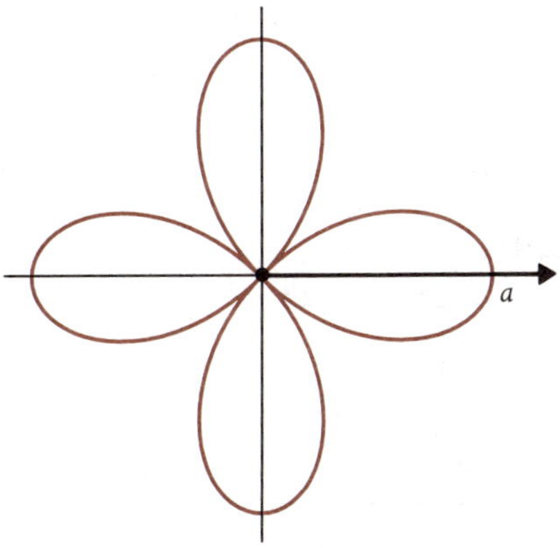

FIGURE 11.40 The rose curve $r = a \cos 2\theta$

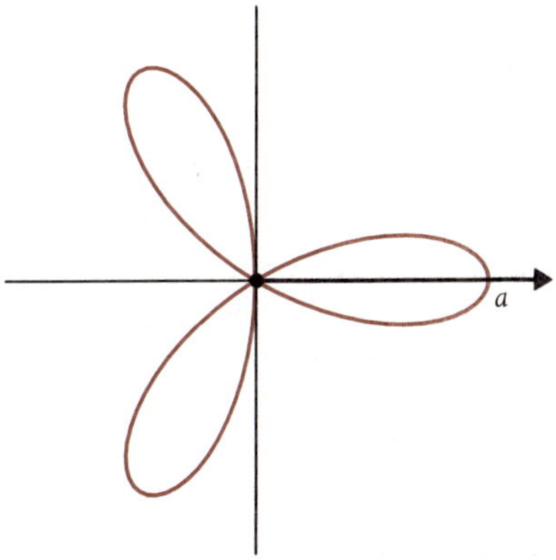

FIGURE 11.41 The rose curve $r = a \cos 3\theta$

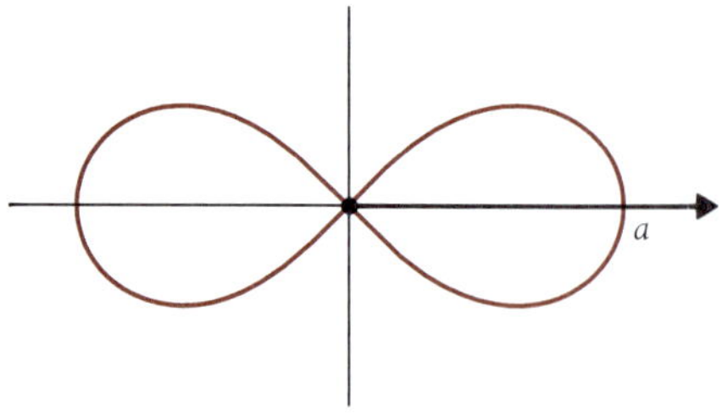

FIGURE 11.42 The lemniscate $r^2 = a^2 \cos 2\theta$

Graphing Polar Curves

Some general remarks about graphing an equation in polar coordinates are in order. When confronted with an equation that is not one of the specific types mentioned above, we can do two things to make sketching its graph a little easier. One is to consider symmetry, which we discuss below. The other is to apply the derivative and use slopes, which we discuss in the next section. (Unfortunately, the derivative often turns out to be of too little practical use.)

If, in the equation $r = f(\theta)$, a substitution of $-\theta$ for θ leaves the equation unchanged, then the curve has symmetry about the x-axis. This is because the points $[r, \theta]$ and $[r, -\theta]$ are symmetrically located relative to the x-axis (see Figure 11.43).

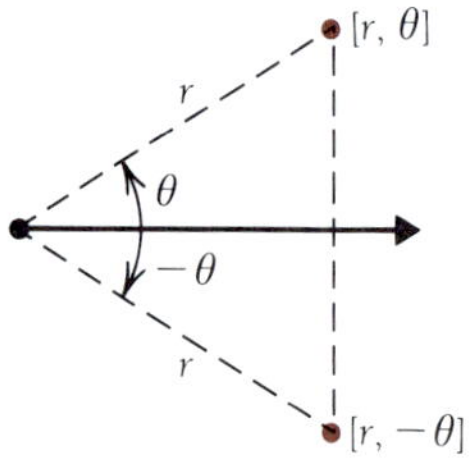

FIGURE 11.43 Symmetry about the x-axis

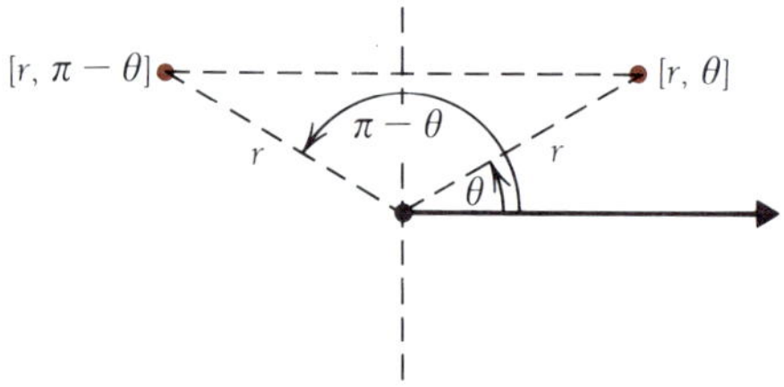

FIGURE 11.44 Symmetry about the y-axis

If the equation $r = f(\theta)$ is unchanged when $\pi - \theta$ is substituted for θ, then the y-axis will be an axis of symmetry (see Figure 11.44).

If the equation $r = f(\theta)$ is unchanged when r is replaced by $-r$, then the curve has central symmetry (see Figure 11.45).

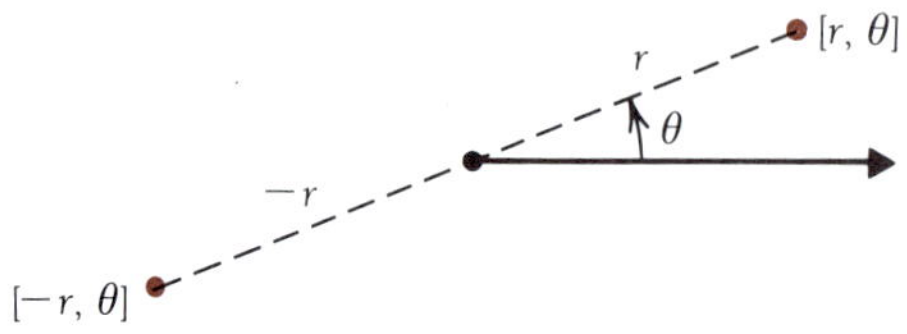

FIGURE 11.45 Symmetry about the origin

The conditions above are sufficient to imply symmetry, but are not necessary. It sometimes happens that a curve has one of these kinds of symmetry, but the corresponding condition does not hold for its equation. As a simple example, note that the circle $r = a$ has all three types of symmetry, but replacing r by $-r$ yields $-r = a$, a different equation. The new equation still has the same graph, however. The fact that points in polar coordinates have different sets of coordinates allows for this complication.

While symmetry considerations are often helpful in sketching graphs, the plotting of points is still necessary in most cases. Point plotting is simplified by thinking of the polar axis as rotating about the pole, with a point sliding along it tracing out the curve. Consider an example.

Example 11.24 Sketch the graph of the equation

$$r = 1 - \sin 3\theta.$$

Solution To make the sketching as simple as possible, we first check for symmetry. Since $\sin 3(\pi - \theta) = \sin 3\theta$, we do have symmetry about the y-axis. There appears to be no other kind of symmetry, however. Next, we note that the sine function is periodic of period 2π, so that r will be periodic of period $2\pi/3$. That is, once we plot the curve for $0 \le \theta \le 2\pi/3$, then we simply repeat that curve for $2\pi/3 \le \theta \le 4\pi/3$ and for $4\pi/3 \le \theta \le 2\pi$.

Now we begin plotting points. At $\theta = 0$, $r = 1$. As θ varies from 0 to $\pi/6$, 3θ varies from 0 to $\pi/2$, so $\sin 3\theta$ varies from 0 to 1. Hence, r varies from 1 to 0. This much of the curve is plotted in Figure 11.46(a) on page 720.

Next, we let θ vary from $\pi/6$ to $\pi/3$. Then 3θ varies from $\pi/2$ to π, so $\sin 3\theta$ varies from 1 to 0 and r varies from 0 to 1. This is added to the curve, as shown in Figure 11.46(b).

Finally, we let θ vary from $\pi/3$ to $\pi/2$. As it does, 3θ varies from π to $3\pi/2$, so $\sin 3\theta$ varies from 0 to -1. Thus, r varies from 1 to 2. This is added to the picture, as shown in Figure 11.46(c).

At this point, symmetry and periodicity give us the rest of the curve. By symmetry, we can copy what we have so far by reflecting it in the y-axis, extending the curve to $0 \le \theta \le \pi$, as shown in Figure 11.46(d). By periodicity, the portion of the curve for $0 \le \theta \le 2\pi/3$ can be repeated for $2\pi/3 \le \theta \le 4\pi/3$ and for $4\pi/3 \le \theta \le 2\pi$. This gives us the entire curve shown in Figure 11.47. ■

Intersections of Polar Curves

The fact that points in polar coordinates do not have unique representations causes some difficulty in finding the points of intersection of two curves. In rectangular coordinates, we find the points of intersection of two curves by solving their equations simultaneously. In polar coordinates, that method alone often does not yield all the points of intersection. Consider Example 11.25.

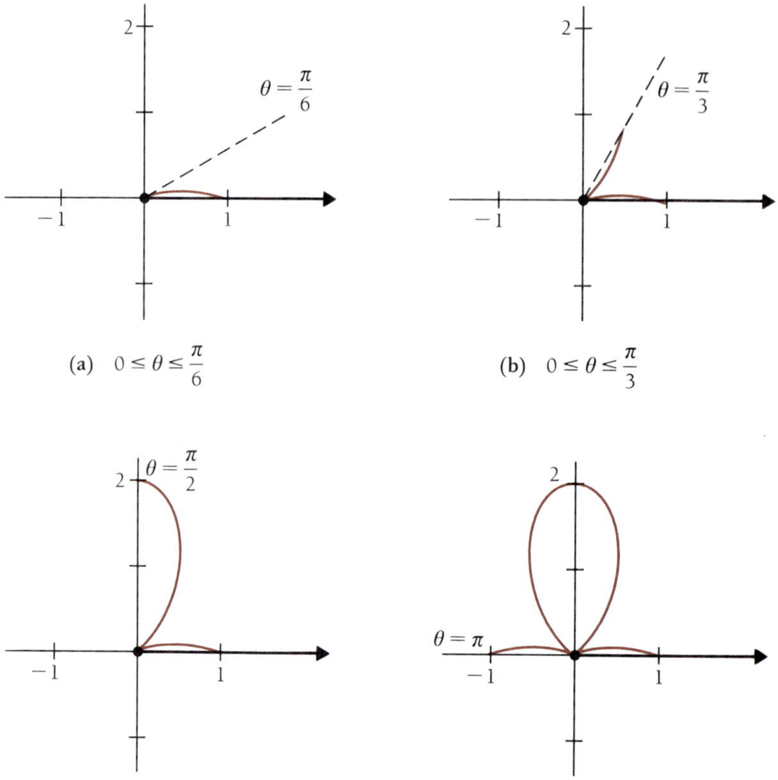

(a) $0 \le \theta \le \frac{\pi}{6}$

(b) $0 \le \theta \le \frac{\pi}{3}$

(c) $0 \le \theta \le \frac{\pi}{2}$

(d) $0 \le \theta \le \pi$

FIGURE 11.46 Plotting the graph of $r = 1 - \sin 3\theta$

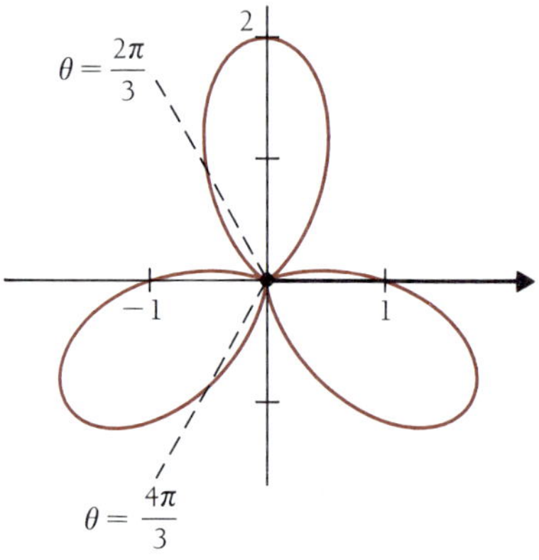

FIGURE 11.47 The graph of $r = 1 - \sin 3\theta$

Example 11.25 Find the points of intersection of the circle $r = \cos\theta$ and the cardioid $r = 1 - \cos\theta$.

Solution We start by solving the equations simultaneously. If we set

$$\cos\theta = 1 - \cos\theta,$$

then

$$\cos\theta = \frac{1}{2},$$

and

$$\theta = \pm\frac{\pi}{3} + 2n\pi.$$

Hence, the points $[\frac{1}{2}, \pi/3]$ and $[\frac{1}{2}, -\pi/3]$ are on both curves, as we see from their graphs (Figure 11.48). But the curves also appear to intersect at the origin. How did we miss it?

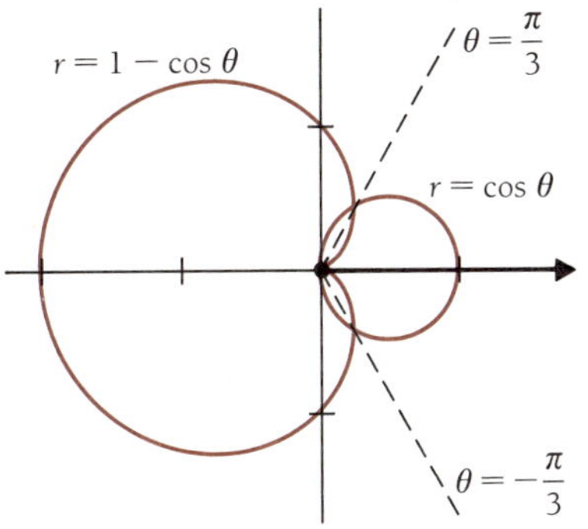

FIGURE 11.48 Intersections of a circle and a cardioid

The answer is that the origin satisfies the equation of the circle $r = \cos\theta$ when represented as $[0, \pi/2]$; on the other hand, the origin satisfies the equation $r = 1 - \cos\theta$ when represented as $[0, 0]$. Thus, the origin is on both curves without being a simultaneous solution of their equations. If we think of two points sliding along the polar axis as it rotates about the pole to trace out the two curves, the points collide at points that are simultaneous solutions of the two equations, such as $[\frac{1}{2}, \pi/3]$ and $[\frac{1}{2}, -\pi/3]$. Both curves also go through the origin, but for different values of θ. ■

When searching for points of intersection of two curves in polar coordinates, it is essential to sketch their graphs. In that way, points of intersection that are not simultaneous solutions usually can be found.

Exercises 11.3

Give equations in polar coordinates for the following curves:

1. The line through [1, 0] and [1, $\pi/2$]
2. The line through [1,0] and [$\sqrt{2}$, $\pi/4$]
3. The circle of radius 3, centered at the pole
4. The circle of radius $\sqrt{2}$, centered at the pole
5. The circle with center [2, $\pi/2$] and passing through the pole
6. The circle with center [4, $5\pi/4$] and radius 4
7. The conic with focus the pole, directrix $r \cos \theta = -2$, and eccentricity $\frac{1}{2}$
8. The conic with focus the pole, directrix $r \sin \theta = 2$, and eccentricity 1

Sketch the curves.

9. $\theta = \dfrac{3\pi}{4}$
10. $\theta = 5$
11. $r \cos \theta = 1$
12. $r \cos \theta = -3$
13. $r \sin \theta = 2$
14. $r \sin \theta = -4$
15. $r \cos\left(\theta - \dfrac{\pi}{4}\right) = 1$
16. $r\left(\dfrac{1}{2} \cos \theta + \dfrac{\sqrt{3}}{2} \sin \theta\right) = 2$
17. $r = 2$
18. $r = \pi$
19. $r = 2 \cos \theta$
20. $r = -4 \sin \theta$
21. $r = 4 \cos\left(\theta - \dfrac{\pi}{3}\right)$
22. $r = 3 \cos\left(\theta + \dfrac{\pi}{4}\right)$
23. $r^2 = 9 + 8r \cos \theta$
24. $r^2 = 1 + 2r \sin \theta$
25. $r = \dfrac{3}{1 + \cos \theta}$
26. $r = \dfrac{2}{2 - \sin \theta}$
27. $r = \dfrac{6}{3 - 2 \cos \theta}$
28. $r = \dfrac{4}{1 + 3 \sin \theta}$
29. $r = 2\theta$
30. $r = e^{2\theta}$
31. $r = 1 + \cos \theta$
32. $r = 2 - 2 \sin \theta$
33. $r = 2 + 3 \sin \theta$
34. $r = 4 - 3 \cos \theta$
35. $r = 3 \sin 2\theta$
36. $r = 2 \sin 3\theta$
37. $r = \cos 5\theta$
38. $r = \cos 4\theta$
39. $r^2 = \cos 2\theta$
40. $r^2 = r \sin 2\theta$
41. $r = 2 \tan \theta$
42. $r = \cot \theta$
43. $r = 1 + \sin 2\theta$
44. $r = 1 + \cos 4\theta$
45. Show that $r = a \cos n\theta$ is an n-leaved rose if n is odd and a $2n$-leaved rose if n is even.
46. Show that $r^2 = a^2 \cos n\theta$ is an n-leaved rose if n is even and a $2n$-leaved rose if n is odd.

Find the points of intersection of the curves.

47. $r = 4; \quad r = 2(1 - \sin \theta)$
48. $r = -2; \quad r = 4 \cos \theta$
49. $r = -\sin \theta; \quad r = \cos \theta$
50. $r = -\sin \theta; \quad r = \frac{1}{2}$
51. $r^2 = 4 \cos 2\theta; \quad r = \cos \theta$
52. $r = \sin 3\theta; \quad r = 1$

c53. $r = 2\theta$; $r = 1 - \sin\theta$

54. $r = 1 + \cos\theta$; $r^2 = \cos 2\theta$

55. Find the five points of intersection of the two curves $r = 1 - \sin\theta$ and $r = 1 + 2\cos\theta$.

56. Rewrite the equation $r\cos(\theta - \alpha) = a$ in rectangular coordinates. Compare your result with Problem 29 in the Additional Exercises at the end of Chapter 1.

SECTION 11.4

The Derivative in Polar Coordinates

In the rectangular coordinate system, the derivative dy/dx of a function $y = f(x)$ is the slope of the graph of the function. In the polar coordinate system, the derivative $dr/d\theta$ of a function $r = f(\theta)$ does not have the same interpretation. In this section, we discuss the geometric significance of $dr/d\theta$ and its relationship to the slope of the curve $r = f(\theta)$.

Slope of a Polar Curve

Suppose $r = f(\theta)$ is a differentiable function of θ in polar coordinates. Let line t be tangent to the graph of $r = f(\theta)$ at point $[r, \theta]$, as shown in Figure 11.49. Let m be the slope of t, and let ϕ be the inclination of t. Then

$$m = \tan\phi.$$

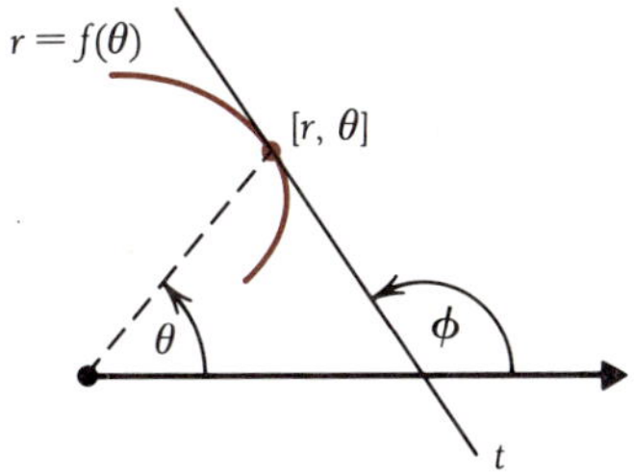

FIGURE 11.49 The inclination of the tangent line

To relate the slope of t, and hence of the curve, to the derivative, we will use rectangular coordinates. From Section 11.2, we recall that

$$\begin{cases} x = r\cos\theta \\ y = r\sin\theta. \end{cases}$$

Since $r = f(\theta)$, we therefore have

$$\begin{cases} x = f(\theta)\cos\theta \\ y = f(\theta)\sin\theta. \end{cases}$$

These are parametric equations of the curve $r = f(\theta)$ in rectangular coordinates, where θ is the parameter.

Now, using the parametric form of the chain rule, we can write the slope as

$$m = \frac{dy}{dx} = \frac{dy/d\theta}{dx/d\theta}$$

$$= \frac{f'(\theta)\sin\theta + f(\theta)\cos\theta}{f'(\theta)\cos\theta - f(\theta)\sin\theta} \quad (1)$$

$$= \frac{f'(\theta)\tan\theta + f(\theta)}{f'(\theta) - f(\theta)\tan\theta} \quad (2)$$

$$= \frac{(dr/d\theta)\tan\theta + r}{(dr/d\theta) - r\tan\theta}. \quad (3)$$

This gives us three formulas for the slope—we can use whichever form is most convenient.

Example 11.26 Find the slope of the cardioid $r = 1 + \cos\theta$ at the point $[1, \pi/2]$.

Solution Since $r = 1 + \cos\theta$, $dr/d\theta = -\sin\theta$, and the slope is, by equation (1),

$$m = \frac{(-\sin\theta)\sin\theta + (1 + \cos\theta)\cos\theta}{(-\sin\theta)\cos\theta - (1 + \cos\theta)\sin\theta}.$$

At $\theta = \pi/2$, we get

$$m = \frac{(-1)(1) + (1+0)(0)}{(-1)(0) - (1+0)(1)} = 1.$$

Therefore, the slope is 1 at $[1, \pi/2]$ (see Figure 11.50).

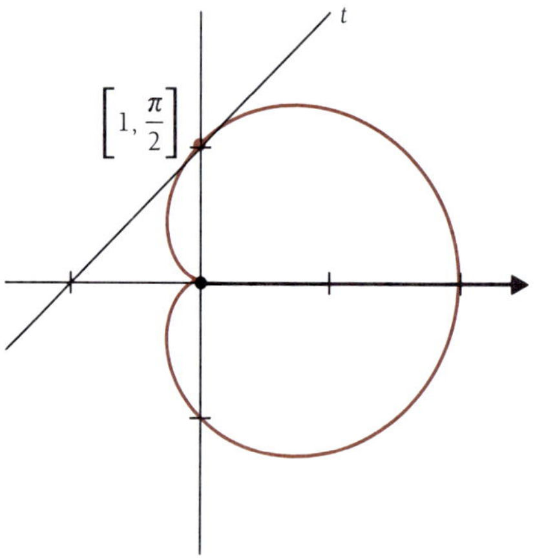

FIGURE 11.50 The cardioid $r = 1 + \cos\theta$ with tangent line at $[1, \pi/2]$ ■

Example 11.27 Find the points on the curve $r^2 = \sin 2\theta$ at which the tangent line is horizontal.

Solution If $r^2 = \sin 2\theta$, then we differentiate implicitly to get

$$\frac{dr}{d\theta} = \frac{1}{r}\cos 2\theta.$$

Then, from equation (3), we find the slope to be

$$\begin{aligned} m &= \frac{(1/r)\cos 2\theta \tan\theta + r}{(1/r)\cos 2\theta - r\tan\theta} \\ &= \frac{\cos 2\theta \tan\theta + r^2}{\cos 2\theta - r^2 \tan\theta} \\ &= \frac{\cos 2\theta \tan\theta + \sin 2\theta}{\cos 2\theta - \sin 2\theta \tan\theta} \\ &= \frac{\cos 2\theta \sin\theta + \sin 2\theta \cos\theta}{\cos 2\theta \cos\theta - \sin 2\theta \sin\theta} \\ &= \frac{\sin 3\theta}{\cos 3\theta}. \end{aligned}$$

The tangent line is horizontal when $m = 0$, or $\sin 3\theta = 0$. This equation has solutions $\theta = 0, \pi/3, 2\pi/3, \pi, 4\pi/3, 5\pi/3$. Since no point on the graph (Figure 11.51) corresponds to $\theta = 2\pi/3$ or $\theta = 5\pi/3$, we discard those solutions. Thus, the lemniscate $r^2 = \sin 2\theta$ has horizontal tangent lines at the pole and at points $[(\sqrt{3}/2)^{1/2}, \pi/3]$ and $[(\sqrt{3}/2)^{1/2}, 4\pi/3]$.

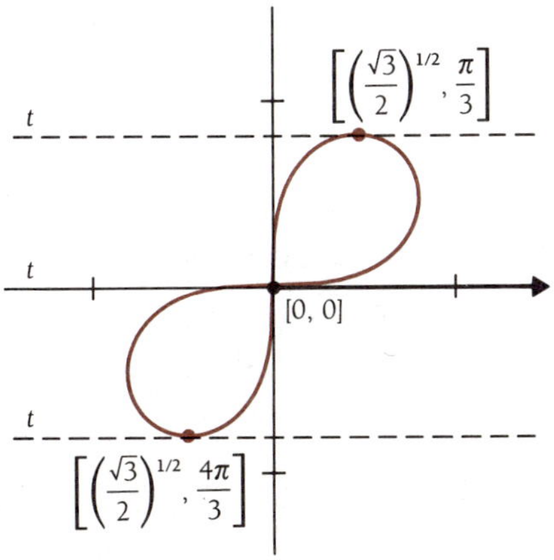

FIGURE 11.51 The lemniscate $r^2 = \sin 2\theta$ with horizontal tangent lines at $[0, 0]$, $[(\sqrt{3}/2)^{1/2}, \pi/3]$, and $[(\sqrt{3}/2)^{1/2}, 4\pi/3]$ ■

In general, the curve $r = f(\theta)$ has a horizontal tangent line whenever $m = 0$, or from equation (3), whenever

$$\tan\theta = \frac{-r}{dr/d\theta}. \tag{4}$$

Similarly, the curve $r = f(\theta)$ will have a vertical tangent line whenever the denominator of equation (3) is 0, or

$$\tan\theta = \frac{dr/d\theta}{r}. \tag{5}$$

The Angle ψ

Consider the graph of $r = f(\theta)$ shown in Figure 11.52. The ray extending from the pole through the point $[r, \theta]$ is called the **radial ray**. The measure of the angle between the radial ray and the tangent line, measured in the positive (counterclockwise) direction, is denoted by ψ. The angle ψ is related to the slope and to the derivative $dr/d\theta$, and is often useful.

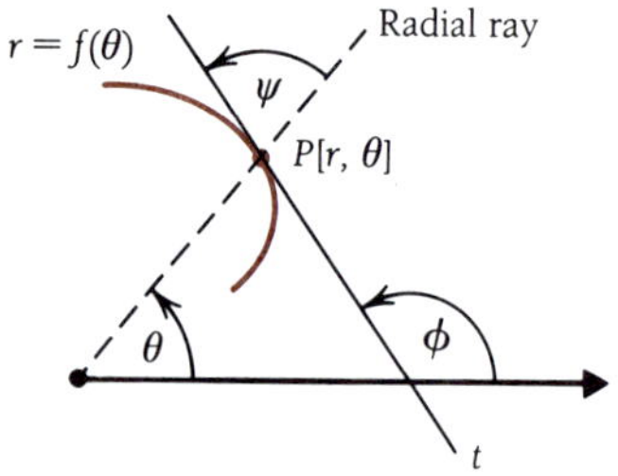

FIGURE 11.52 The angle ψ

To find a formula involving ψ, we note that if the inclination of the tangent line is ϕ, then

$$\psi = \phi - \theta.$$

We therefore have

$$\begin{aligned}
\tan\psi &= \tan(\phi - \theta) \\
&= \frac{\tan\phi - \tan\theta}{1 + \tan\phi\tan\theta} \\
&= \frac{m - \tan\theta}{1 + m\tan\theta} \\
&= \frac{\dfrac{(dr/d\theta)\tan\theta + r}{(dr/d\theta) - r\tan\theta} - \tan\theta}{1 + \dfrac{(dr/d\theta)\tan\theta + r}{(dr/d\theta) - r\tan\theta}\tan\theta} \\
&= \frac{r(1 + \tan^2\theta)}{(dr/d\theta)(1 + \tan^2\theta)} \\
&= \frac{r}{dr/d\theta}.
\end{aligned} \tag{6}$$

An alternative form of (6) is

$$\cot \psi = \frac{1}{r}\frac{dr}{d\theta}. \tag{7}$$

Notice that when a tangent line is horizontal, equations (4) and (6) together yield

$$\tan \theta = -\tan \psi.$$

The reason for this can be seen in Figure 11.53.

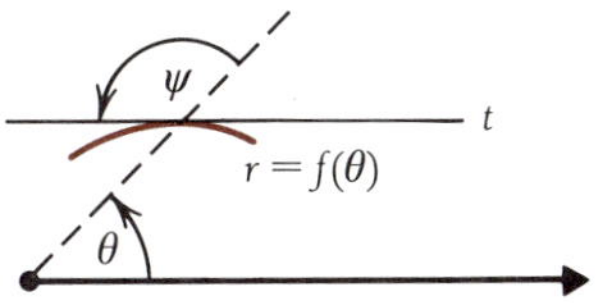

FIGURE 11.53 $\theta + \psi = \pi$

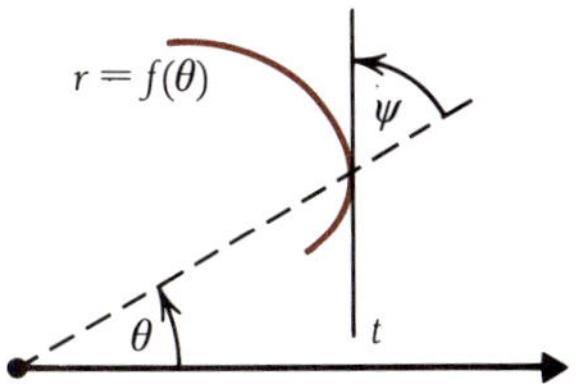

FIGURE 11.54 $\theta + \psi = \pi/2$

When a tangent line is vertical, equations (5) and (7) tell us

$$\tan \theta = \cot \psi.$$

Figure 11.54 offers the explanation.

■ Example 11.28 Find the points on the curve $r = 1 - \sin\theta$ at which the tangent line is perpendicular to the radial ray.

Solution We seek the points on the curve for which $\psi = \pi/2$. That is, we seek points $[r, \theta]$ such that

$$\cot \psi = \frac{1}{r}\frac{dr}{d\theta} = 0.$$

Now, $dr/d\theta = -\cos\theta$, so we need

$$\frac{-\cos\theta}{1 - \sin\theta} = 0.$$

This holds for $\theta = 3\pi/2$ (it is undefined for $\theta = \pi/2$). Hence, only at the point $[2, 3\pi/2]$ is the tangent line perpendicular to the radial ray (see Figure 11.55).

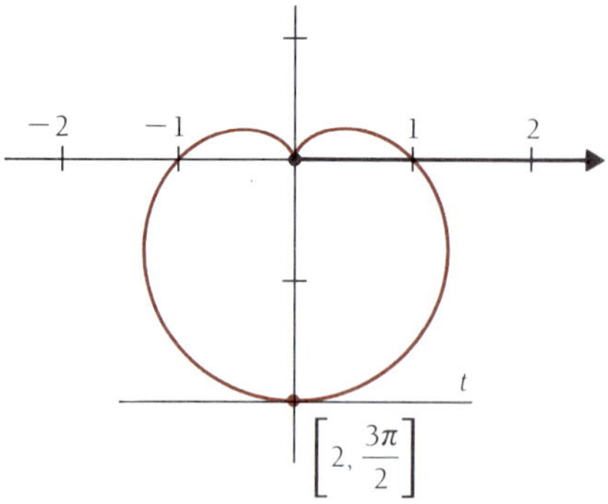

FIGURE 11.55 The graph of $r = 1 - \sin\theta$ with tangent line at $[2, 3\pi/2]$

Example 11.29 For the logarithmic spiral $r = e^{k\theta}$, show that ψ is constant. (For this reason, $r = e^{k\theta}$ is also called the **equiangular spiral**.)

Solution By equation (6), we have

$$\tan\psi = \frac{r}{dr/d\theta} = \frac{e^{k\theta}}{ke^{k\theta}} = \frac{1}{k},$$

which is independent of θ. Thus, ψ is constant (see Figure 11.56).

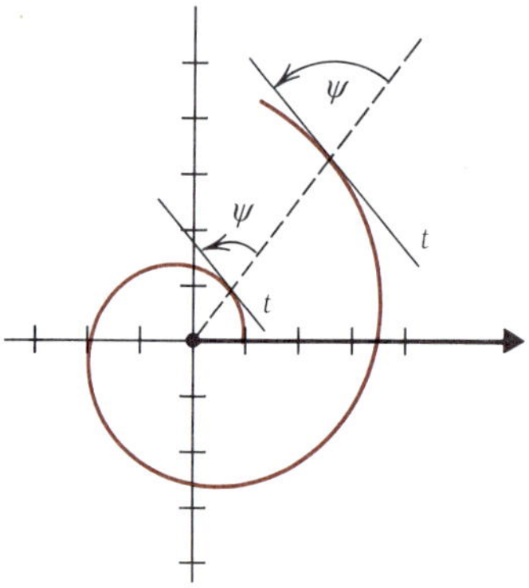

FIGURE 11.56 The equiangular spiral $r = e^{k\theta}$

Exercises 11.4

Find the slope of the tangent line and the angle ψ at the given point.

1. $r = a$; $[a, \pi/3]$
2. $r = a$; $[a, \theta]$
3. $r = 2\sin\theta$; $[1, \pi/6]$
4. $r = 2\sin\theta$; $[2, \pi/2]$
5. $r = -3\cos\theta$; $[-\frac{3}{2}, \pi/3]$
6. $r = -3\cos\theta$; $[3\sqrt{3}/2, 5\pi/6]$
7. $r = 3\theta$; $[3\pi, \pi]$
8. $r = 3\theta$; $[6\pi, 2\pi]$
9. $r = \sin 2\theta$; $[\sqrt{3}/2, \pi/6]$
10. $r = \sin 2\theta$; $[-\sqrt{3}/2, 2\pi/3]$
11. $r^2 = 4\cos 2\theta$; $[2, 0]$
12. $r^2 = 4\cos 2\theta$; $[0, \pi/4]$

Find the points at which the tangent lines are horizontal or vertical.

13. $r = 1 + \cos\theta$

14. $r = \sin\theta - 1$

15. $r = \cos 2\theta$

16. $r = 4\sin 3\theta$

17. $r = e^{2\theta}$

18. $r^2 = \cos 2\theta$

19. Find the maximum value of r for any point on the limaçon $r = 3 + 4\cos\theta$. Does this suggest another interpretation for $dr/d\theta$?

SECTION 11.5

Area and Arc Length in Polar Coordinates

The familiar techniques for finding area and arc length need to be modified slightly for polar coordinates. First we consider area.

Area

The technique for finding areas bounded by curves in polar coordinates is similar to the case of rectangular coordinates, but gives rise to a different basic formula.

Let R be the region of the plane bounded by the lines $\theta = \alpha$ and $\theta = \beta$ and the curve $r = f(\theta)$, with f nonnegative over the interval $[\alpha, \beta]$ (see Figure 11.57). Let $s = \{\theta_0, \theta_1, \ldots, \theta_n\}$ be a subdivision of $[\alpha, \beta]$, and let $t_i \in [\theta_{i-1}, \theta_i]$ for each i. This subdivision divides R into sectors, as shown in Figure 11.58 (page 730). The area of the ith sector is approximately the same as the area of the circular sector with central angle $\Delta\theta_i = \theta_i - \theta_{i-1}$ and radius $f(t_i)$ (see Figure 11.59). Thus, the area of R is approximately

$$A \approx \sum_{i=1}^{n} \frac{1}{2}[f(t_i)]^2\,\Delta\theta_i.$$

The actual area is the limit of this Riemann sum, which is

$$A = \int_\alpha^\beta \frac{1}{2}[f(\theta)]^2\,d\theta = \int_\alpha^\beta \frac{1}{2}r^2\,d\theta. \tag{1}$$

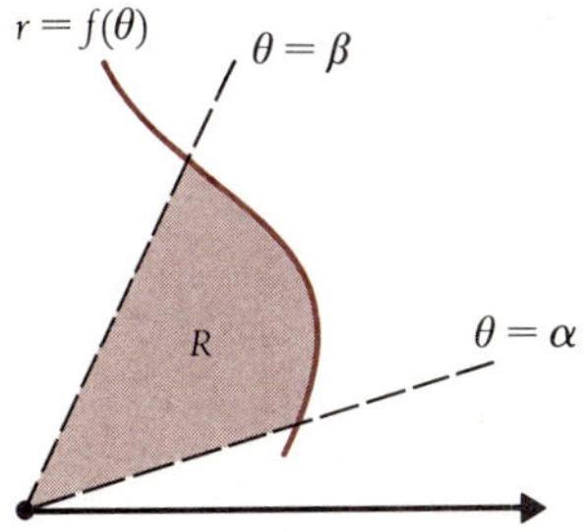

FIGURE 11.57 A polar region bounded by $r = f(\theta)$, $\theta = \alpha$, and $\theta = \beta$

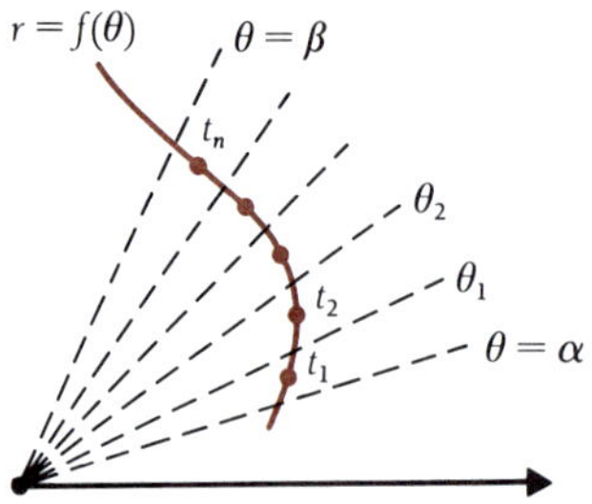

FIGURE 11.58 A subdivision of the interval $[\alpha, \beta]$

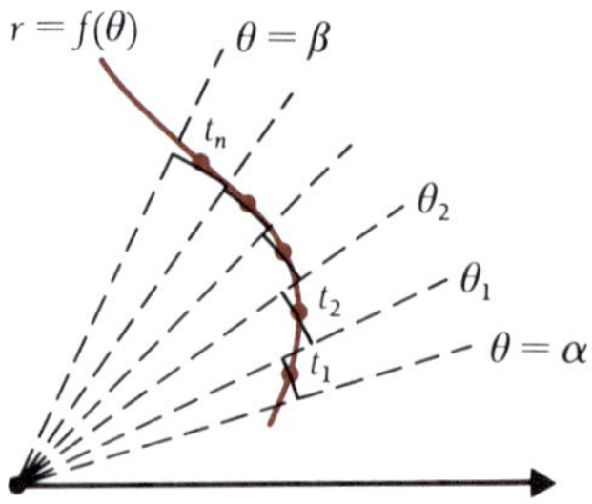

FIGURE 11.59 The polar region approximated by circular sectors

Example 11.30 Find the area enclosed by the cardioid

$$r = a(1 - \cos \theta).$$

Solution The cardioid is traced out as θ varies from 0 to 2π (see Figure 11.60); also, r is nonnegative for all θ, so we may use equation (1):

$$\begin{aligned}
A &= \int_0^{2\pi} \frac{1}{2} r^2 \, d\theta \\
&= \int_0^{2\pi} \frac{1}{2} [a(1 - \cos \theta)]^2 \, d\theta \\
&= a^2 \int_0^{2\pi} \frac{1}{2} (1 - 2 \cos \theta + \cos^2\theta) \, d\theta \\
&= a^2 \int_0^{2\pi} \left[\frac{1}{2} - \cos \theta + \frac{1}{4} (1 + \cos 2\theta)\right] d\theta \\
&= a^2 \int_0^{2\pi} \left(\frac{3}{4} - \cos \theta + \frac{1}{4} \cos 2\theta\right) d\theta \\
&= a^2 \left[\frac{3}{4} \theta - \sin \theta + \frac{1}{8} \sin 2\theta\right]_0^{2\pi} \\
&= \frac{3}{2} \pi a^2.
\end{aligned}$$

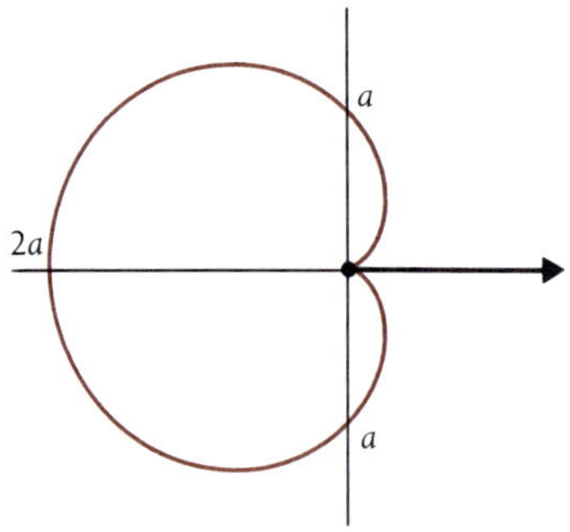

FIGURE 11.60 The cardioid $r = a(1 - \cos \theta)$ ■

To find the area between two curves, we use formula (1) twice, finding a difference. Other variations on this theme are treated in much the same way as for rectangular coordinates.

Example 11.31 Find the area inside the lemniscate $r^2 = 4\cos 2\theta$ and outside the circle $r = 1$ (Figure 11.61).

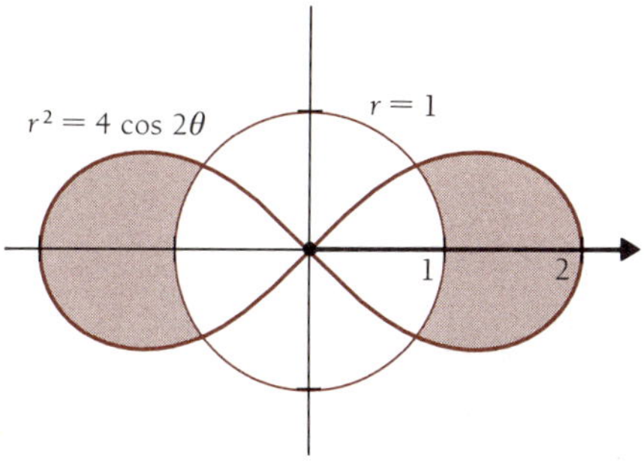

FIGURE 11.61 The region inside the lemniscate and outside the circle

Solution The circle and the lemniscate meet at the point $[1, \frac{1}{2}\cos^{-1}\frac{1}{4}]$ and three other points (see Figure 11.62). We will find the area between the curves for $0 \le \theta \le \frac{1}{2}\cos^{-1}\frac{1}{4}$ (the shaded portion of Figure 11.62) and then multiply by 4 to find the total area. That is,

$$\begin{aligned} A &= 4\left[\int_0^{(1/2)\cos^{-1}(1/4)} \frac{1}{2}(4\cos 2\theta)\,d\theta - \int_0^{(1/2)\cos^{-1}(1/4)} \frac{1}{2}(1)^2\,d\theta\right] \\ &= \int_0^{(1/2)\cos^{-1}(1/4)} (8\cos 2\theta - 2)\,d\theta \\ &= \Big[4\sin 2\theta - 2\theta\Big]_0^{(1/2)\cos^{-1}(1/4)} \\ &= \sqrt{15} - \cos^{-1}\frac{1}{4}. \end{aligned}$$

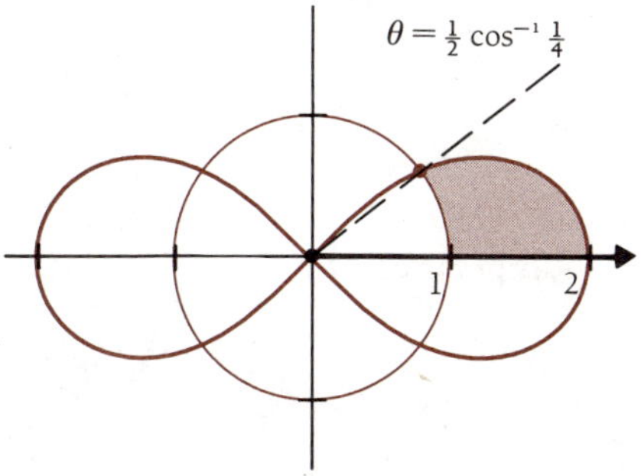

FIGURE 11.62 Finding the area ■

Arc Length

To find the length of an arc in polar coordinates, we represent the curve parametrically in rectangular coordinates. If the curve is $r = f(\theta)$, $\theta \in [\alpha, \beta]$, then the parametric equations of the curve are

$$\begin{cases} x = f(\theta)\cos\theta \\ y = f(\theta)\sin\theta, \quad \theta \in [\alpha, \beta]. \end{cases}$$

The differential of arc length is

$$\begin{aligned} (ds)^2 &= \left[\left(\frac{dx}{d\theta}\right)^2 + \left(\frac{dy}{d\theta}\right)^2\right](d\theta)^2 \\ &= \{[f'(\theta)\cos\theta - f(\theta)\sin\theta]^2 + [f'(\theta)\sin\theta + f(\theta)\cos\theta]^2\}(d\theta)^2 \\ &= \{[f'(\theta)]^2 + [f(\theta)]^2\}(d\theta)^2 \\ &= \left[r^2 + \left(\frac{dr}{d\theta}\right)^2\right](d\theta)^2, \end{aligned}$$

$$ds = \sqrt{r^2 + (dr/d\theta)^2}\, d\theta. \tag{2}$$

Therefore, the arc length is

$$s = \int_\alpha^\beta \sqrt{r^2 + (dr/d\theta)^2}\, d\theta. \tag{3}$$

Example 11.32 Find the circumference of the cardioid

$$r = a(1 - \cos\theta).$$

Solution The cardioid is traced out as θ varies from 0 to 2π (see Figure 11.60). By equation (3), we have

$$\begin{aligned} s &= \int_0^{2\pi} \sqrt{a^2(1-\cos\theta)^2 + a^2\sin^2\theta}\, d\theta \\ &= a\int_0^{2\pi} \sqrt{2 - 2\cos\theta}\, d\theta \\ &= a\int_0^{2\pi} 2\sin\frac{\theta}{2}\, d\theta \\ &= -4a\cos\frac{\theta}{2}\Big|_0^{2\pi} \\ &= 8a. \end{aligned}$$

■

Formula (2) is useful for surface areas, centroids, and the like, whenever the curves involved are given in polar coordinates.

Example 11.33 Find the surface area generated by revolving the top half of the cardioid $r = a(1 - \cos\theta)$ about the x-axis (Figure 11.63).

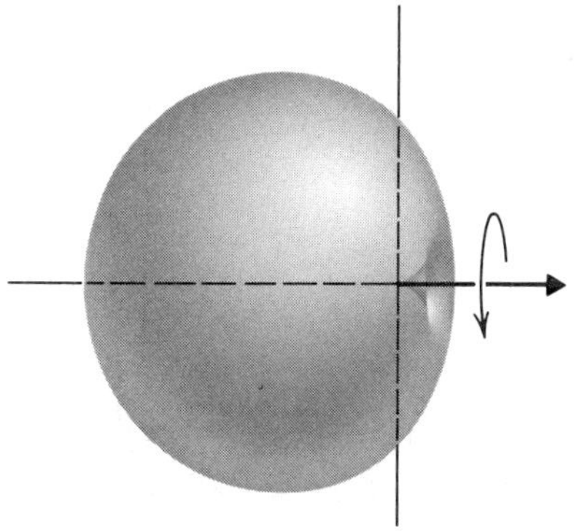

FIGURE 11.63 The cardioid revolved about the x-axis

Solution The top half of the cardioid is traced out as θ varies from 0 to π. The surface area S is therefore

$$\begin{aligned}
S &= \int_C 2\pi y\, ds \\
&= \int_0^\pi 2\pi(r \sin \theta)\sqrt{r^2 + (dr/d\theta)^2}\, d\theta \\
&= \int_0^\pi 2\pi a(1 - \cos \theta) \sin \theta \sqrt{2a^2(1 - \cos \theta)}\, d\theta \\
&= 2\pi a^2\sqrt{2} \int_0^\pi (1 - \cos \theta)^{3/2} \sin \theta\, d\theta \\
&= 2\pi a^2\sqrt{2} \int_0^2 u^{3/2}\, du \qquad (u = 1 - \cos \theta) \\
&= 2\pi a^2\sqrt{2}\, \frac{2}{5}\, u^{5/2}\Big|_0^2 \\
&= \frac{32\pi a^2}{5}.
\end{aligned}$$

■

Exercises 11.5

Find the area enclosed by each curve.

1. $r = 6 \cos \theta$
2. $r = 3 \sin \theta$
3. $r^2 = 9 \cos 2\theta$
4. $r^2 = 16 \sin 2\theta$
5. $r = 2 - \cos \theta$
6. $r = 4 + 3 \sin \theta$
7. $r = \sin 3\theta$
8. $r = 4 \cos 3\theta$
9. $r = \cos 6\theta$
10. $r = a \sin 2\theta$

[A]11. $r = \dfrac{2}{2 - \sin \theta}$

[A]12. $r = \dfrac{3}{2 + \cos \theta}$

Find the area of each region.

13. Inside $r = 2 \sin 3\theta$ and outside $r = 1$
14. Inside $r = 3 \cos 2\theta$ and outside $r = 2$
15. Inside both curves $r^2 = 4 \cos 2\theta$ and $r = 1$
16. Inside both curves $r^2 = 9 \sin 2\theta$ and $r = 2$

Find the length of each curve.

17. $r = 3 \cos \theta$
18. $r = 2 \sin \theta$
19. $r = a(1 + \cos \theta)$
20. $r = a(1 - \sin \theta)$
21. $r = e^{\theta}, \quad 0 \le \theta \le 2\pi$
22. $r = 4\theta, \quad 0 \le \theta \le \pi/2$

Find the surface area generated by revolving each curve.

23. $r = 5(1 - \cos \theta), \quad 0 \le \theta \le \pi/2,$ about the x-axis
24. $r^2 = a^2 \cos 2\theta, \quad 0 \le \theta \le \pi/4,$ about the x-axis

Find the volume generated by revolving each region.

A25. Inside the cardioid $r = a(1 - \cos \theta), \quad 0 \le \theta \le \pi,$ about the x-axis

[*Hint:* $V = \int_{\theta=\pi}^{\theta=\pi/3} \pi y^2 \, dx - \int_{\theta=0}^{\theta=\pi/3} \pi y^2 \, dx$; get x and y in terms of θ.]

A26. Inside one loop of the lemniscate $r^2 = a^2 \cos 2\theta,$ about the y-axis

Find the centroid of each curve.

A27. $r = a(1 - \cos \theta), \quad 0 \le \theta \le \pi$
A28. $r = e^{\theta}, \quad 0 \le \theta \le 2\pi$

SECTION 11.6

Chapter Review

Important Concepts

Parameter
Parametrized curve
Parametric equations
Parametrizations
Parametric form of the chain rule
Smooth curve
Second derivative in parametric form
Arc length
Cycloid
Polar coordinates
Pole
Polar axis
Transformation equations
Curves in polar coordinates
Spiral of Archimedes
Logarithmic spiral
Limaçon
Cardioid
Rose curves
Lemniscate
Derivative in polar coordinates
Slope of a polar curve
Radial ray
Angle ψ
Area in polar coordinates
Arc length in polar coordinates

Review Exercises

Give an equation for the conic section:

1. With focus the pole, directrix $r \cos \theta = -3$, and eccentricity $\frac{1}{2}$
2. With focus the pole, directrix $r \sin \theta = 2$, and eccentricity 1

Sketch each curve.

3. $\begin{cases} x = 3t + 1 \\ y = t^2 + 1 \end{cases}$
4. $\begin{cases} x = 3t - 3 \sin t \\ y = 3 - 3 \cos t \end{cases}$

5. $\begin{cases} x = 4\cos t \\ y = 3\sin t \end{cases}$

6. $\begin{cases} x = 4\tan t \\ y = 2\sec t \end{cases}$

7. $r = 2\theta$

8. $r = 3e^{2\theta}$

9. $\theta = \dfrac{3\pi}{4}$

10. $r = \dfrac{\sqrt{2}}{2}\sec\left(\theta - \dfrac{\pi}{4}\right)$

11. $r\sin\theta = 5$

12. $\theta = 0$

13. $r = 2\cos\theta$

14. $r = 2\cos 2\theta$

15. $r = 2\cos 3\theta$

16. $r = 2\cos 5\theta$

17. $r = 1 + 2\cos\theta$

18. $r = 2 + \cos\theta$

19. $r = 1 + \cos 2\theta$

20. $r^2 = \cos\theta$

21. $r^2 = 4\cos 2\theta$

22. $r^2 = 4\cos 3\theta$

23. $r = \dfrac{1}{1 + \cos\theta}$

24. $r = \dfrac{2}{1 + \cos\theta}$

25. $r = \dfrac{1}{2 + \cos\theta}$

26. $r = \dfrac{2}{1 - \cos\theta}$

27. $r = \dfrac{3}{2 + 5\cos\theta}$

28. $r = \dfrac{5}{3 - 2\cos\theta}$

29. $r = \dfrac{1}{\theta}$

30. $(r - 1)^2 = 4\theta$

Find the points at which each curve has horizontal or vertical tangent lines.

31. $r = 4\sin\theta$

32. $r = -9\cos\theta$

33. $r = \dfrac{1}{1 + \cos\theta}$

34. $r = \dfrac{3}{1 - \sin\theta}$

35. $r = e^{3\theta}$

36. $r = 2\theta$

37. $r = 3 - 2\cos\theta$

38. $r^2 = 4\cos 2\theta$

C39. $r = \sin 3\theta$

40. $r = \cos 2\theta$

Find the area.

41. Inside the limaçon $r = a - b\cos\theta$, $0 < b < a$

42. Inside the outer loop and outside the inner loop of $r = a - b\cos\theta$, $0 < a < b$

43. Inside both circles $r = 2a\cos\theta$ and $r = 2b\sin\theta$

44. Inside the lemniscate $r^2 = a^2\sin 2\theta$

45. Inside the lemniscate $r^2 = a^2\cos 2\theta$ and outside the circle $r = b$, $0 < b < a$

46. Inside the circle $r = 2a\cos\theta$ and outside the lemniscate $r^2 = a^2\cos 2\theta$

47. Inside one leaf of $r^2 = a^2\sin 4\theta$

48. Inside one leaf of $r = a\sin 3\theta$

Find an integral representing the length of each arc, and evaluate it if possible.

49. $\begin{cases} x = 2t \\ y = t^2 + 1, \quad t \in [0, 1] \end{cases}$

50. $\begin{cases} x = e^t\sin t \\ y = e^t\cos t, \quad t \in [0, \pi/2] \end{cases}$

51. $r = a\cos(\theta - \pi/4)$

52. $r = a - b\cos\theta$

53. $r = e^{k\theta}, \quad \theta \in [\alpha, \beta]$

A54. $r^2 = a^2 \cos 2\theta$

Additional Exercises

Find the angle at which the two curves intersect.

1. $y = mx - 1$ and $\begin{cases} x = t - \sin t \\ y = 1 - \cos t \end{cases}$

2. $r = e^{2\theta}$ and $r = a$

3. $r^2 = a^2 \cos 2\theta$ and $r = b, \quad 0 < b < a$

4. As a wheel rolls along a level road, a point on a spoke of the wheel describes a curve known as a **curtate cycloid**, or **trochoid** (see the figure). If the wheel has radius a and the point lies at distance b from the center of the wheel, show that the parametric equations for the curtate cycloid are

$$\begin{cases} x = at - b\sin t \\ y = a - b\cos t, \quad a > b. \end{cases}$$

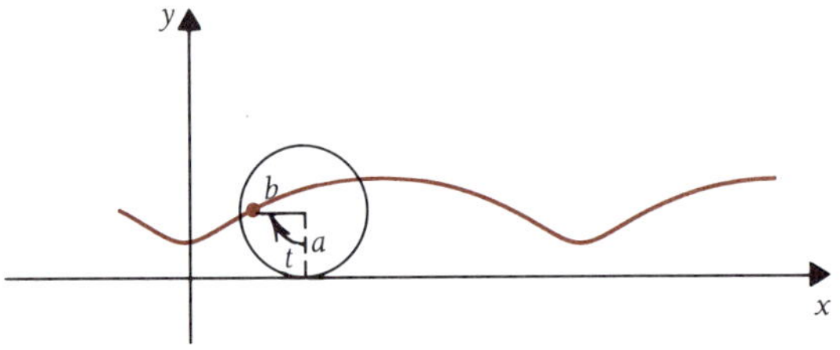

5. If $b > a$ in Problem 4, the curve is called a **prolate cycloid**. Draw its graph.

6. If a circle of radius b rolls around the outside of a fixed circle of radius a, a point on the first circle traces out a curve called an **epicycloid** (see the figure). Show that the parametric equations for the epicycloid are

$$\begin{cases} x = (a+b)\cos t - b\cos \dfrac{a+b}{b}t \\ y = (a+b)\sin t - b\sin \dfrac{a+b}{b}t. \end{cases}$$

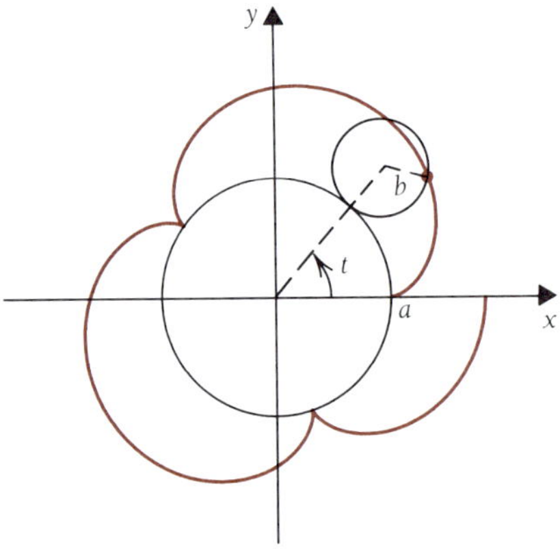

7. If the circle in Problem 6 rolls around the inside of the fixed circle, a point on the moving circle traces out a **hypocycloid** (see the figure). Show that the parametric equations for the hypocycloid are

$$\begin{cases} x = (a-b)\cos t + b\cos\dfrac{a-b}{b}t \\ y = (a-b)\sin t - b\sin\dfrac{a-b}{b}t. \end{cases}$$

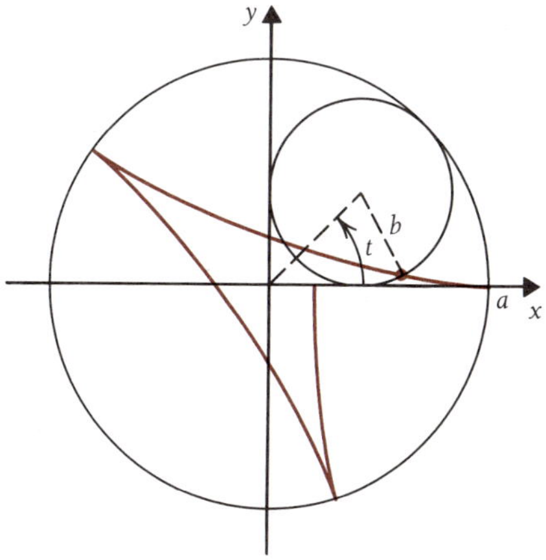

8. In Problem 7, if $a = nb$, n an integer, the hypocycloid is said to have ***n* cusps.** From the parametric equations for the hypocycloid of 4 cusps (also called an **astroid**), use the identities

$$3\cos t + \cos 3t = 4\cos^3 t,$$
$$3\sin t - \sin 3t = 4\sin^3 t$$

to eliminate the parameter.

9. A man stands at the origin, and his dog stands at the point $(0, a)$, straining at the leash. Then the man walks along the positive x-axis, dragging the reluctant dog along. The curve traced out by the dog is called a **tractrix** (see the figure). Using the distance the man walks as parameter, show that the parametric equations for the tractrix are

$$\begin{cases} x = t - a\tanh\dfrac{t}{a} \\ y = a\,\mathrm{sech}\,\dfrac{t}{a}. \end{cases}$$

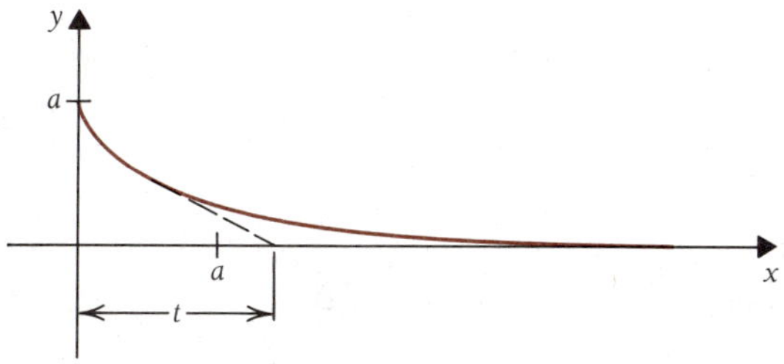

10. The **involute of a circle** is a curve described as the path of the end of a string that is unwound from the circle and held taut. If Γ is a fixed circle with radius a (see the figure), we assume the end of the string starts at point $A(a, 0)$. If P is a point on the involute, then the tangent PT to Γ has the same length as arc TA. Let the angle t be the parameter, and show that the parametric equations of the involute are

$$\begin{cases} x = a\cos t + at\sin t \\ y = a\sin t - at\cos t. \end{cases}$$

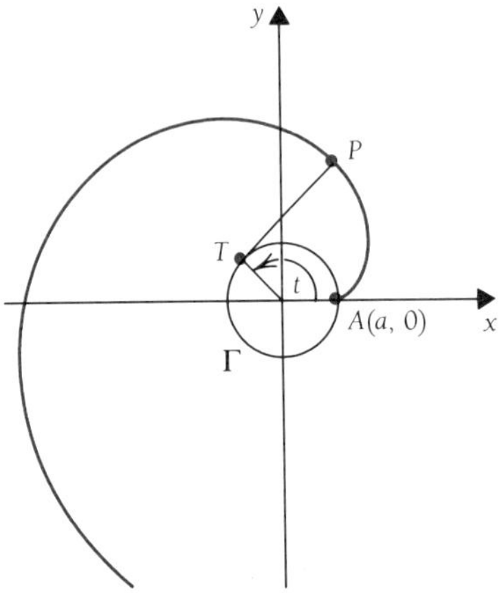

11. The **witch of Agnesi** is a curve described as follows: Let Γ be a fixed circle of diameter d, resting on the x-axis at the origin O (see the figure). Let Q be any point on the line $y = d$, and let the line segment OQ meet Γ at R. Then the horizontal line through R and the vertical line through Q meet at P, a point on the witch. Using the angle t as parameter, show that the parametric equations of the witch of Agnesi are

$$\begin{cases} x = d\cot t \\ y = d\sin^2 t. \end{cases}$$

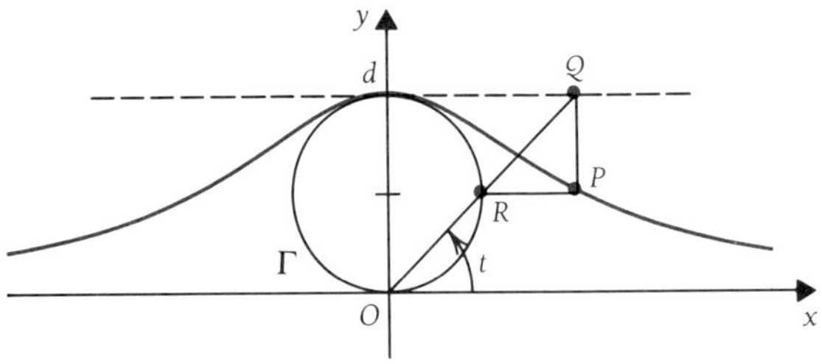

12. Find the area enclosed by $r = a\cos n\theta$. [*Hint:* Consider the two cases n even and n odd.]

13. Find the area enclosed by $r^2 = a^2\cos n\theta$.

14. The inclination ϕ of the tangent line to a curve $y = f(x)$ is $\phi(x) = \arctan(dy/dx)$. The **curvature** of the curve $y = f(x)$ is $d\phi/ds$, the derivative of ϕ with respect to arc length.

For a curve given parametrically, show that the curvature is

$$\frac{d\phi}{ds} = \frac{(dx/dt)(d^2y/dt^2) - (dy/dt)(d^2x/dt^2)}{[\sqrt{(dx/dt)^2 + (dy/dt)^2}]^3}.$$

15. Refer to Problem 14. Show that the curvature of a straight line is 0 and that the curvature of a circle is a nonzero constant.
16. Refer to Problem 14. Find a formula for the curvature of a curve given in polar coordinates.
17. Find a formula for the distance between points $[r_1, \theta_1]$ and $[r_2, \theta_2]$.
18. Find the area enclosed by the inner loop of the limaçon $r = 3 - 6 \cos \theta$.
19. A goat is tethered to the corner of a barn with a rope 20 feet long. If the dimensions of the barn are 12 feet by 16 feet, what area of grass can the goat reach?
20. This time, the goat is tethered to a tree of diameter 6 feet with a rope 8 feet long. What area of grass can the goat reach?

CHAPTER 12

Solid Geometry and Vectors

INTRODUCTION

Having studied some of the geometry of the plane, we now expand our study into space. Solid analytic geometry is the study of geometric objects in three-dimensional space, and it uses the tools of algebra and calculus where appropriate.

Vectors represent quantities that have both magnitude and direction. Many physical phenomena can be described using vectors, so that vectors are a tool of the scientist. But many theoretical problems also can be solved using vectors, even though the problems may not involve vectors in the physical sense. In particular, many geometric ideas are simplified by the use of vectors.

After developing the mathematical concept of vector as an abstraction of the physical concept of vector, we study lines, planes, and surfaces in space. We also examine some alternative coordinate systems.

SECTION 12.1

Rectangular Coordinates in Three Dimensions

In the plane, we form a rectangular coordinate system by selecting two perpendicular coordinate axes in order to assign a pair of coordinates to each point in the plane. In space, we form a rectangular coordinate system by selecting three mutually perpendicular lines as coordinate axes, and we use them to assign a triple of coordinates to each point in space.

Coordinate Axes

In three-dimensional space, let x, y, and z be three mutually perpendicular lines passing through the point O (see Figure 12.1). Let each of these lines be positioned so that O is the origin on each line. We call x, y, and z the **coordinate axes** in a **rectangular coordinate system** with **origin** O.

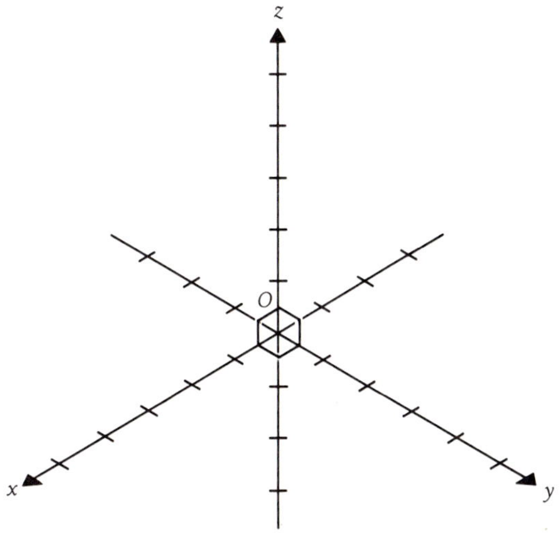

FIGURE 12.1 Three mutually perpendicular lines form a three-dimensional coordinate system

In our graphs, we will usually picture only the positive half of each axis for the sake of simplicity. We will use the so-called **right-handed** coordinate system (Figure 12.2) in which the x-axis points "east," the y-axis points "north," and the z-axis points "up." (You are viewing Figure 12.2 from a point above and "northeast" of the origin.) This system is called right-handed because if the index finger of the right hand points east and the second finger points north, then the thumb will point up. If the x- and y-axes are interchanged, we obtain a **left-handed** system (Figure 12.3).

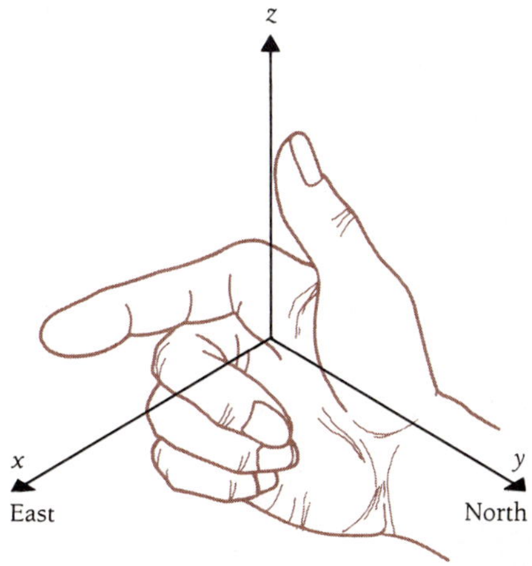

FIGURE 12.2 A right-handed coordinate system

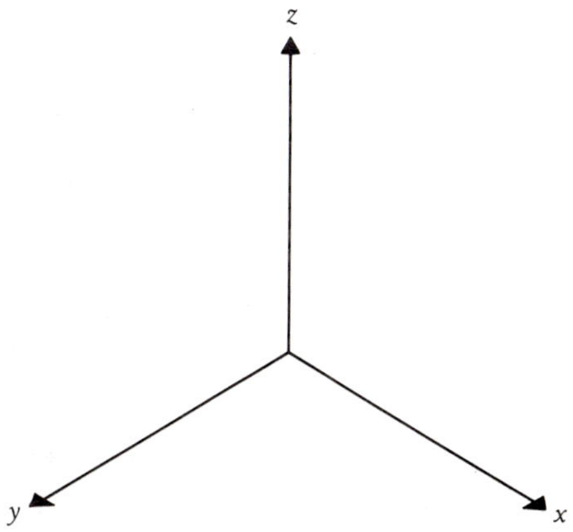

FIGURE 12.3 A left-handed coordinate system

Coordinate Planes

The coordinate axes, taken in pairs, determine three planes called the **coordinate planes**, as pictured in Figure 12.4. The coordinate planes are named for the axes they contain; the xy-plane contains the x- and y-axes, etc.

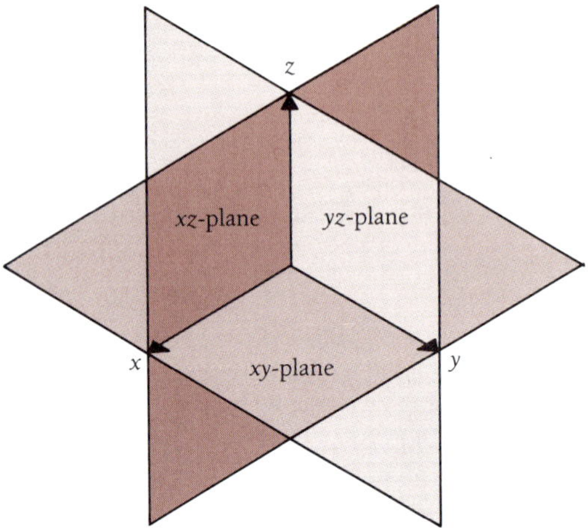

FIGURE 12.4 The coordinate planes

The coordinate planes divide space into eight parts, called **octants**. The **first octant** is the octant bordering the positive halves of all three axes. The other seven octants are not numbered.

Coordinates

We now can assign a triple of coordinates to each point in space. If P is a point in space, let the three planes through P and parallel to the coordinate planes meet the x-axis at point a, the y-axis at point b, and the z-axis at point c, respectively (see Figure 12.5). Then we give P the **coordinates** (a, b, c). In this way, each point is

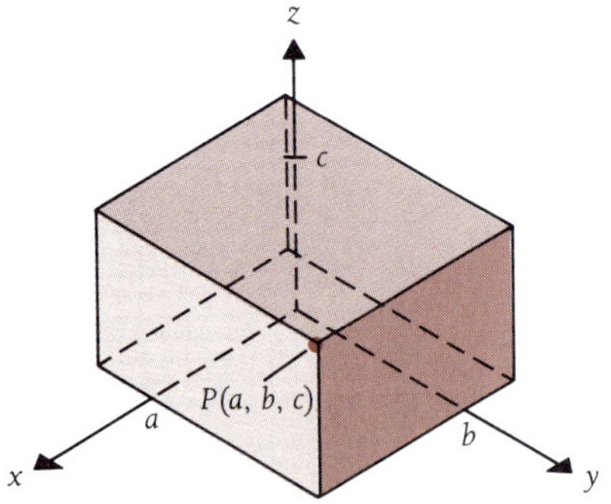

FIGURE 12.5 The coordinates of point P

assigned a triple of coordinates. Conversely, every triple (a, b, c) of real numbers represents the coordinates of some point in space. The three planes perpendicular to the three axes at the points a, b, and c, respectively, meet at a point P whose coordinate triple is (a, b, c).

Example 12.1 Plot the points (2, 1, 3), (−1, 1, −2), and (−4, −2, 1).

Solution See Figure 12.6.

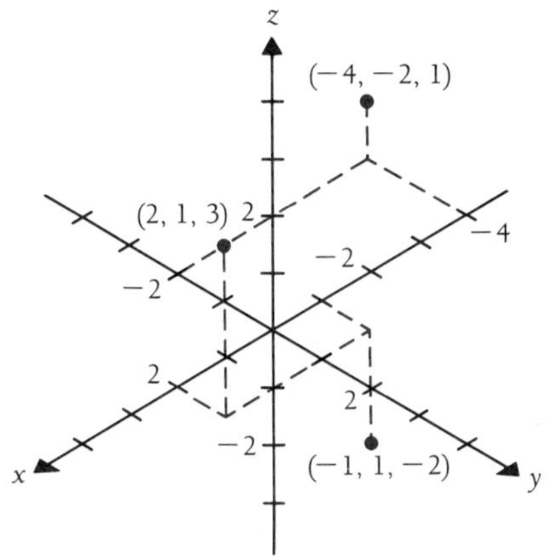

FIGURE 12.6 Some points in space ■

It should be evident that points in the same octant have coordinates of the same signs. Points on the x-axis have coordinates of the form $(a, 0, 0)$, those on the y-axis have coordinates of the form $(0, b, 0)$, and those on the z-axis have coordinates of the form $(0, 0, c)$. The origin, of course, is $(0, 0, 0)$.

Any point in the xy-plane has coordinates of the form $(a, b, 0)$. That is, the xy-plane consists of the points (x, y, z) such that $z = 0$. We therefore use $z = 0$ as the equation of the xy-plane; that is, we say "the xy-plane" or "the plane $z = 0$" interchangeably. Similarly, the xz-plane is the plane $y = 0$, and the yz-plane is the plane $x = 0$.

The Distance Formula

The distance formula for the plane is easily generalized to obtain a distance formula for space. Let $P_1(x_1, y_1, z_1)$ and $P_2(x_2, y_2, z_2)$ be two points in space, and let d be the distance between them (Figure 12.7). Let Π be the plane through P_1 that is parallel to the xy-plane, and let the vertical line through P_2 meet Π at Q. Then Q has coordinates (x_2, y_2, z_1). If s is the distance between P_1 and Q, then by the distance formula for the plane,

$$s^2 = (x_2 - x_1)^2 + (y_2 - y_1)^2.$$

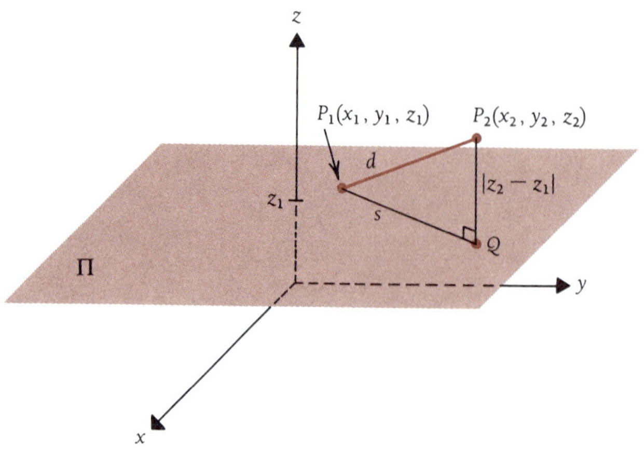

FIGURE 12.7 The distance formula:
$d = \sqrt{(x_2 - x_1)^2 + (y_2 - y_1)^2 + (z_2 - z_1)^2}$

Since the distance between P_2 and Q is $|z_2 - z_1|$, the Pythagorean theorem gives us

$$d^2 = s^2 + (z_2 - z_1)^2.$$

Combining these two equations, we get

$$d = \sqrt{(x_2 - x_1)^2 + (y_2 - y_1)^2 + (z_2 - z_1)^2}.$$

This is the distance formula for three-dimensional space.

■ **Example 12.2** Find the distance between points $(1, 2, -1)$ and $(3, -5, 2)$.

Solution By the distance formula,

$$d = \sqrt{(3 - 1)^2 + (-5 - 2)^2 + [2 - (-1)]^2}$$
$$= \sqrt{62}.$$

■

Equation of a Sphere

A **sphere** is the set of points in space a given distance (the **radius**) from a given point (the **center**). We can use the distance formula to derive an equation for a sphere.

Suppose a sphere has radius r and center (a, b, c). If $P(x, y, z)$ is an arbitrary point, then P belongs to the sphere if and only if the distance between P and (a, b, c) is r, or

$$(x - a)^2 + (y - b)^2 + (z - c)^2 = r^2.$$

This is the **equation of the sphere.**

■ **Example 12.3** Find the equation of the sphere of radius 5 with center $(4, -3, -1)$.

Solution From the above equation, we have

$$(x-4)^2+(y+3)^2+(z+1)^2=25. \quad \blacksquare$$

Example 12.4 Find the equation of the sphere of radius r with center at the origin.

Solution Since the origin is (0, 0, 0), the equation is

$$x^2+y^2+z^2=r^2. \quad \blacksquare$$

Some Special Planes

The set of points 3 units above the xy-plane is another plane parallel to the xy-plane. Each point in that plane has coordinates of the form $(a, b, 3)$; thus, $z=3$ is the equation of that plane. More generally, the plane $z=c$ is a plane parallel to the xy-plane and meets the z-axis at c (see Figure 12.8). Similarly, the plane $y=b$ is parallel to the xz-plane and meets the y-axis at b (Figure 12.9), and the plane $x=a$ is parallel to the yz-plane and meets the x-axis at a (Figure 12.10).

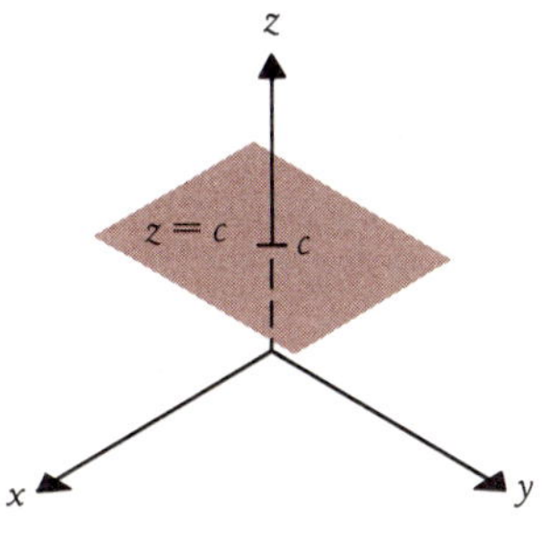

FIGURE 12.8 The plane $z=c$

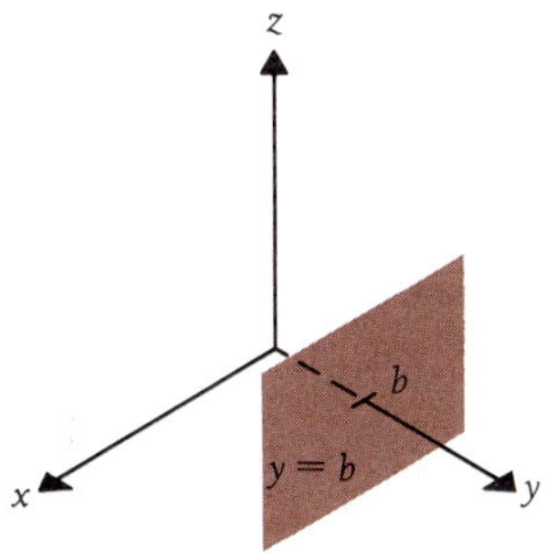

FIGURE 12.9 The plane $y=b$

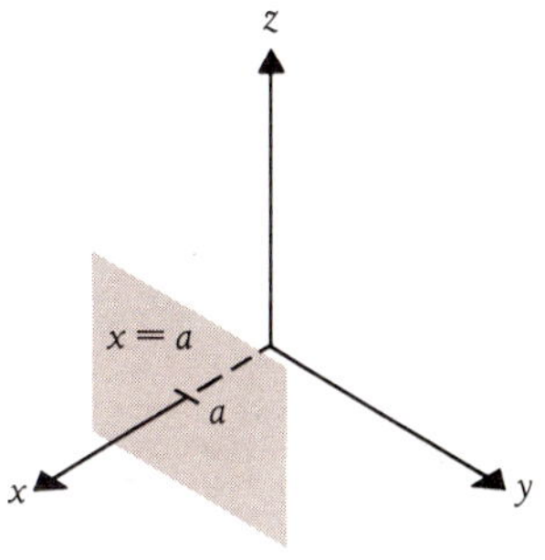

FIGURE 12.10 The plane $x=a$

In each of these special cases, the plane is represented by a single linear equation. We will see later that any linear equation in x, y, and z is the equation of a plane, and that every plane has such an equation.

Exercises 12.1

Find the distance between the points.

1. $(2, 0, -3)$; $(5, 1, 1)$
2. $(-4, 1, -1)$; $(6, 3, 2)$
3. $(\frac{1}{3}, \frac{2}{3}, \frac{2}{3})$; $(0, 0, 0)$
4. $(1, 1, -1)$; $(-1, 1, 1)$
5. Use the distance formula to determine whether the points $A(1, 1, 2)$, $B(3, 5, -2)$, and $C(0, 1, 1)$ are the vertices of a right triangle.
6. How long a rod may be packed in a rigid box that measures 2 feet by 3 feet by 18 inches?

Find an equation for the sphere:

7. With radius 4 and center the origin
8. With radius $\sqrt{3}$ and center the origin
9. With radius 1 and center (2, 1, 0)
10. With radius 2 and center (−1, 4, 3)
11. With center (1, 2, −2) and passing through the origin
12. With center at the origin and tangent to the plane $z = 5$

Sketch each plane.

13. $z = 1$
14. $x = -2$
15. $y = 4$
16. $z = 5$
17. Show that the midpoint of the segment joining (a, b, c) and (a', b', c') is

$$\left(\frac{a+a'}{2}, \frac{b+b'}{2}, \frac{c+c'}{2}\right).$$

18. A rectangular solid has its faces parallel to the coordinate planes. If two of its vertices are (−1, 1, 3) and (2, 8, −1), give its dimensions.

SECTION 12.2

Vectors

In physical applications, we often deal with quantities that have both magnitude and direction. For example, the velocity of a moving object has both magnitude (the speed of the object) and direction (the direction of travel). If the motion takes place along a line, two directions are possible (Figure 12.11); in the plane or in space, infinitely many directions are possible (Figure 12.12).

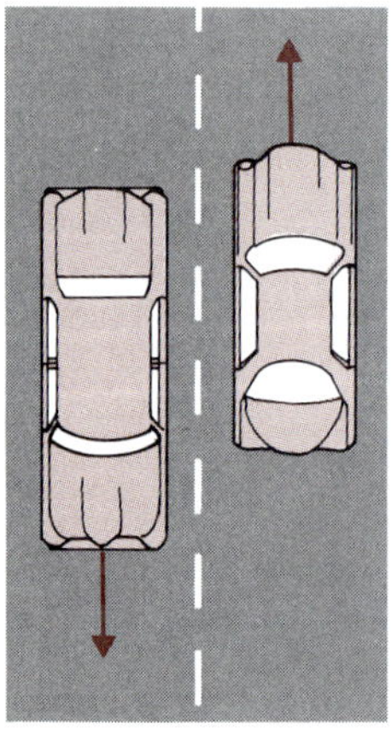

FIGURE 12.11

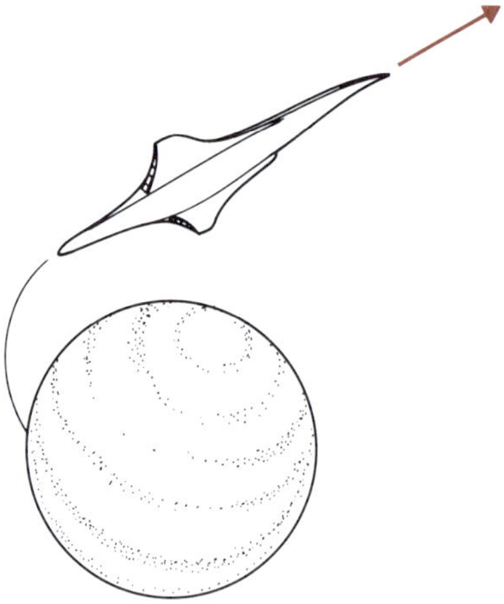

FIGURE 12.12

Vectors and Arrows

A quantity with both magnitude and direction is called a **vector quantity**, or simply, a **vector**. Examples of physical quantities that are vectors are velocity, force, and displacement. We denote a vector by a boldface letter, such as $\mathbf{v}$.

Geometrically, a vector is represented by a directed line segment, or **arrow**. The length of the arrow (relative to the scale of the coordinate system) represents the magnitude of the vector, and the direction of the arrow (relative to the coordinate system) represents the direction of the vector. For example, if an object is displaced from point P to point Q in a coordinate system, then the displacement is logically represented by an arrow with its tail at P and its head at Q (Figure 12.13). We denote the arrow by $\overrightarrow{PQ}$, and we denote the vector it represents by $\mathbf{v}(\overrightarrow{PQ})$.

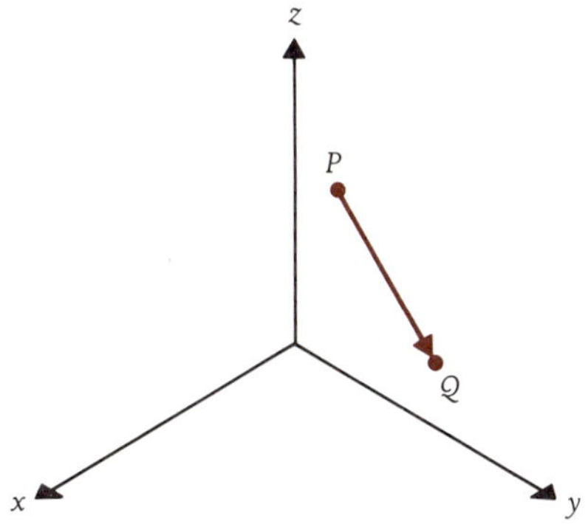

FIGURE 12.13 An arrow represents a vector

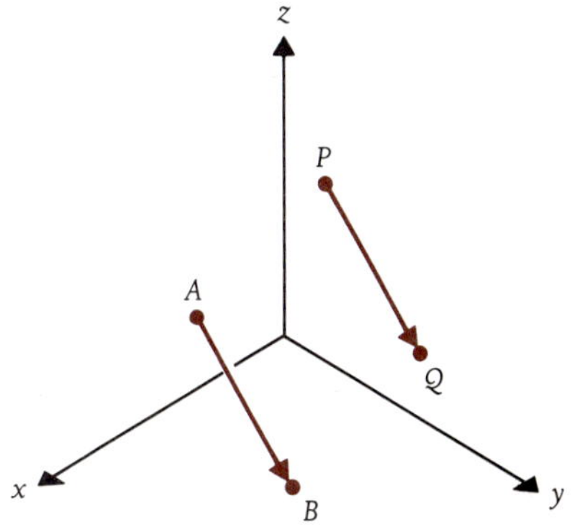

FIGURE 12.14 Arrows with the same length and direction represent the same vector

If $\overrightarrow{AB}$ is another arrow (Figure 12.14) with the same direction and length as $\overrightarrow{PQ}$, then $\overrightarrow{AB}$ is also an arrow representing $\mathbf{v}(\overrightarrow{PQ})$. That is, if an object is displaced from A to B, its displacement is the same as that of the object that was moved from P to Q. In symbols, $\mathbf{v}(\overrightarrow{AB}) = \mathbf{v}(\overrightarrow{PQ})$. Thus, we conclude that arrows with the same length and direction represent the same vector. Or, put another way, the location of an arrow does not matter.

Of all the arrows representing a given vector $\mathbf{v}$, there is exactly one that has its tail at the origin O of the coordinate system (Figure 12.15, page 748). This arrow is called the **position arrow of v**. If $\overrightarrow{OP}$ is the position arrow of $\mathbf{v}$, then $\mathbf{v} = \mathbf{v}(\overrightarrow{OP})$; any vector is uniquely represented by its position arrow.

Coordinates for Vectors

Because a vector $\mathbf{v}$ has just one position arrow $\overrightarrow{OP}$, $\mathbf{v}$ can be uniquely specified by naming the point P. That is, the coordinates of P can be used to designate $\mathbf{v}$. If point

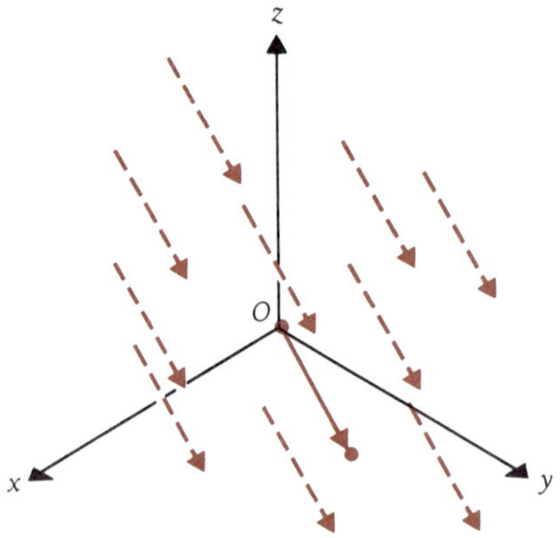

FIGURE 12.15 The tail of a position arrow is at the origin

P has coordinates (a, b, c), we write

$$\mathbf{v}(\overrightarrow{OP}) = \langle a, b, c\rangle$$

and call a, b, and c the **coordinates** of $\mathbf{v}(\overrightarrow{OP})$. For a point P in the plane with coordinates (a, b), we write $\mathbf{v}(\overrightarrow{OP}) = \langle a, b\rangle$, and thus a vector in two dimensions has only two coordinates.

The use of coordinates for vectors is an algebraic representation and is the most convenient form to use. We will develop the properties of vectors in terms of coordinates—that is, algebraically—using physical and geometric representations to motivate and clarify the development.

The Algebra of Vectors

It is now time to give a formal definition.

DEFINITION 12.1

A **vector** in three dimensions is an ordered triple of real numbers,

$$\mathbf{v} = \langle v_1, v_2, v_3\rangle.$$

Two vectors $\mathbf{u} = \langle u_1, u_2, u_3\rangle$ and $\mathbf{v} = \langle v_1, v_2, v_3\rangle$ are **equal** if and only if $u_1 = v_1$, $u_2 = v_2$, and $u_3 = v_3$.

Since the vector $\mathbf{v} = \langle v_1, v_2, v_3\rangle$ is represented by the position arrow from the origin to the point (v_1, v_2, v_3), the magnitude of $\mathbf{v}$ is defined algebraically as stated in Definition 12.2.

DEFINITION 12.2

The **magnitude**, or **norm**, or **absolute value**, of the vector $\mathbf{v} = \langle v_1, v_2, v_3\rangle$ is the nonnegative real number

$$\|\mathbf{v}\| = \sqrt{v_1^2 + v_2^2 + v_3^2}.$$

A vector with magnitude 1 is called a **unit vector**.

Example 12.5 Find the magnitudes of the vectors $\mathbf{u} = \langle \frac{1}{3}, -\frac{2}{3}, \frac{2}{3} \rangle$ and $\mathbf{v} = \langle 4, -1, \frac{1}{2} \rangle$.

Solution By Definition 12.2,

$$\|\mathbf{u}\| = \sqrt{(\tfrac{1}{3})^2 + (-\tfrac{2}{3})^2 + (\tfrac{2}{3})^2} = 1,$$
$$\|\mathbf{v}\| = \sqrt{(4)^2 + (-1)^2 + (\tfrac{1}{2})^2} = \tfrac{1}{2}\sqrt{69}.$$

The only vector with magnitude 0 is the **zero vector**,

$$\mathbf{0} = \langle 0, 0, 0 \rangle.$$

We do not specify the direction of the zero vector.

The unit vectors

$$\mathbf{i} = \langle 1, 0, 0, \rangle, \qquad \mathbf{j} = \langle 0, 1, 0 \rangle, \qquad \mathbf{k} = \langle 0, 0, 1 \rangle$$

are called the **standard unit vectors** (Figure 12.16). These vectors will play an important role in our work.

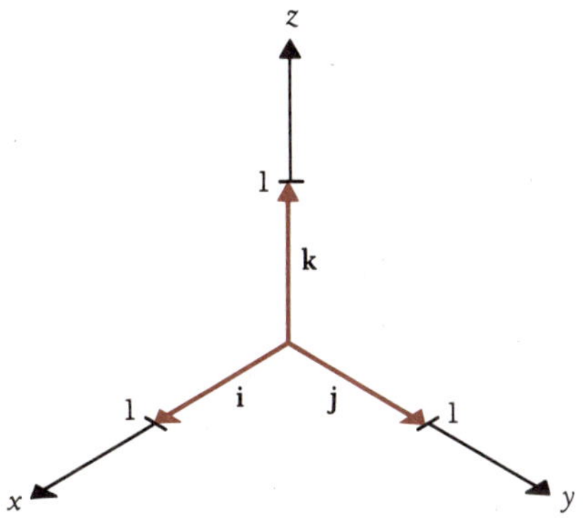

FIGURE 12.16 Standard unit vectors

We next define addition of vectors. To give a concrete illustration of the idea, we use displacements. Suppose an object is displaced from point A to point B, and then displaced again from point B to point C (see Figure 12.17). The net result is a displacement from A to C. We say that the **sum** of $\mathbf{v}(\overrightarrow{AB})$ and $\mathbf{v}(\overrightarrow{BC})$ is $\mathbf{v}(\overrightarrow{AC})$. The sum of two vectors is also called the **resultant**.

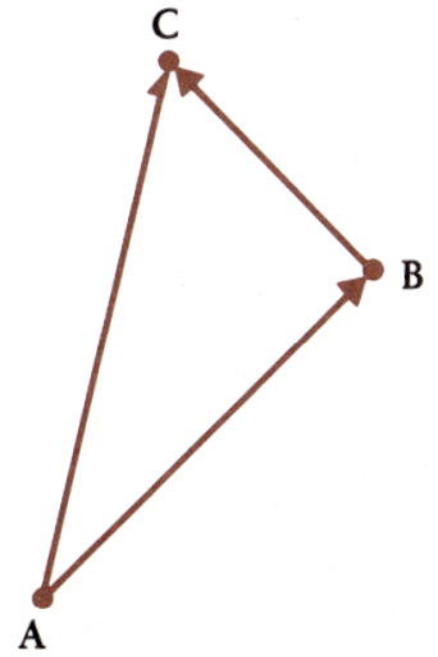

FIGURE 12.17 The resultant of two vectors

Using position arrows, we can obtain a formula for the sum in terms of coordinates. Let $\mathbf{v}(\overrightarrow{AB}) = \mathbf{v}(\overrightarrow{OP})$ with $P = (u_1, u_2, u_3)$, and let $\mathbf{v}(\overrightarrow{BC}) = \mathbf{v}(\overrightarrow{OQ})$ with $Q = (v_1, v_2, v_3)$ [see Figure 12.18(a)]. If $\mathbf{v}(\overrightarrow{AC}) = \mathbf{v}(\overrightarrow{OR})$, then we can see by Figure 12.18(b) that R is the point $(u_1 + v_1, u_2 + v_2, u_3 + v_3)$. We therefore have Definition 12.3.

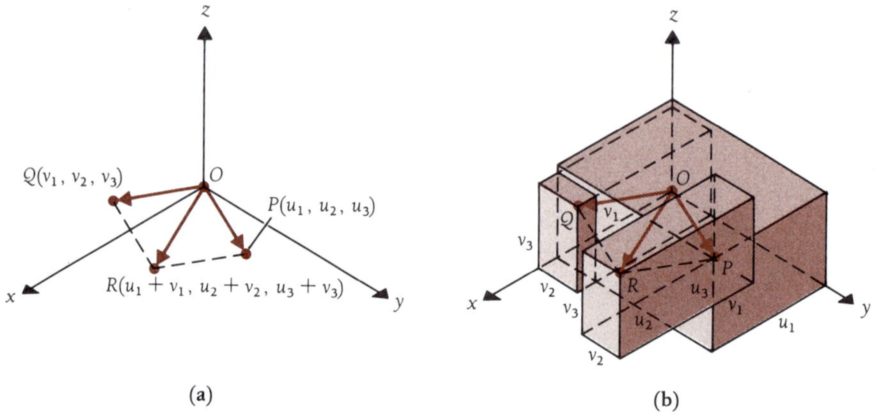

FIGURE 12.18 Coordinates for a resultant

DEFINITION 12.3

The **sum**, or **resultant**, of the two vectors $\mathbf{u} = \langle u_1, u_2, u_3 \rangle$ and $\mathbf{v} = \langle v_1, v_2, v_3 \rangle$ is the vector

$$\mathbf{u} + \mathbf{v} = \langle u_1 + v_1, u_2 + v_2, u_3 + v_3 \rangle.$$

Figure 12.18(a) illustrates the reason that this definition of vector addition is often called the **parallelogram law**. The resultant of two vectors is represented by the diagonal of the parallelogram of which the position arrows of the two vectors are adjacent sides.

■ **Example 12.6** Find the sum of the vectors $\mathbf{u} = \langle 6, 1, -3 \rangle$ and $\mathbf{v} = \langle 0, -1, 2 \rangle$.

Solution By Definition 12.3, we have

$$\mathbf{u} + \mathbf{v} = \langle 6, 0, -1 \rangle. \quad ■$$

Several useful properties of vector addition follow from the definition. For instance, vector addition is associative and commutative, and the zero vector is an additive identity. That is, for any vectors $\mathbf{u}$, $\mathbf{v}$, and $\mathbf{w}$, we have

$$\mathbf{u} + (\mathbf{v} + \mathbf{w}) = (\mathbf{u} + \mathbf{v}) + \mathbf{w}, \tag{1}$$

$$\mathbf{u} + \mathbf{v} = \mathbf{v} + \mathbf{u}, \tag{2}$$

$$\mathbf{v} + \mathbf{0} = \mathbf{v}. \tag{3}$$

The magnitude of a sum of vectors is never greater than the sum of the magnitudes of the vectors. In symbols,

$$\|\mathbf{u}+\mathbf{v}\| \le \|\mathbf{u}\| + \|\mathbf{v}\|. \tag{4}$$

This statement is called the **triangle inequality**, taking its name from the fact that the length of one side of a triangle is less than the sum of the lengths of the other two sides (see Figure 12.19). We have equality in (4) when **u** and **v** are in the same direction.

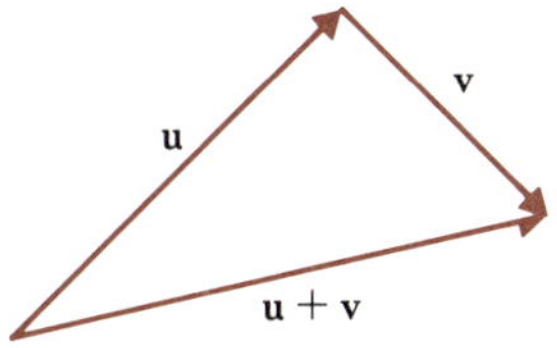

FIGURE 12.19 The triangle inequality

In contrast to vectors, real numbers are often called **scalars**, and quantities that can be represented by single numbers are called **scalar quantities**. We can define a scalar multiple of a vector as stated in Definition 12.4.

DEFINITION 12.4 The **scalar multiple** of a vector $\mathbf{v} = \langle v_1, v_2, v_3 \rangle$ by a scalar a is the vector

$$a\mathbf{v} = \langle av_1, av_2, av_3 \rangle.$$

To illustrate this idea, suppose an object has velocity **v**, represented by the first arrow in Figure 12.20. The velocities of objects traveling in a direction parallel to **v** can be described by comparing them with **v**. An object going in the same direction, but twice as fast, has velocity 2**v**. An object with velocity $\frac{1}{3}\mathbf{v}$ is also traveling in the same direction, but $\frac{1}{3}$ as fast. An object traveling in the opposite direction, but with the same speed, has velocity $(-1)\mathbf{v}$, which is usually written $-\mathbf{v}$. An object with velocity $-\frac{4}{3}\mathbf{v}$ is traveling in the opposite direction at $\frac{4}{3}$ the rate. Arrows representing all these velocities are shown in Figure 12.20. In each case, the velocity is given as a scalar multiple of **v**.

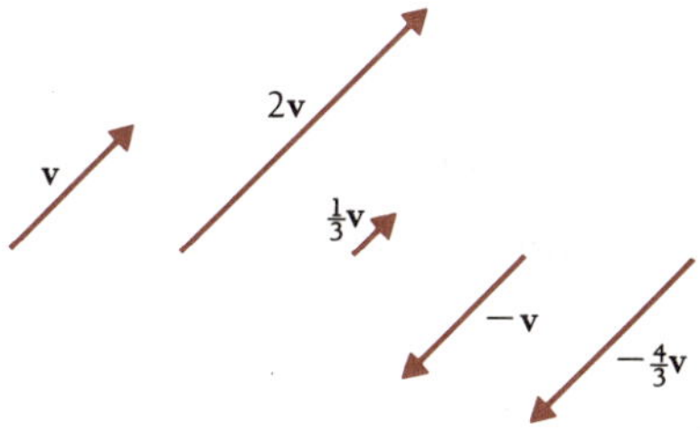

FIGURE 12.20 Scalar multiples of **v**

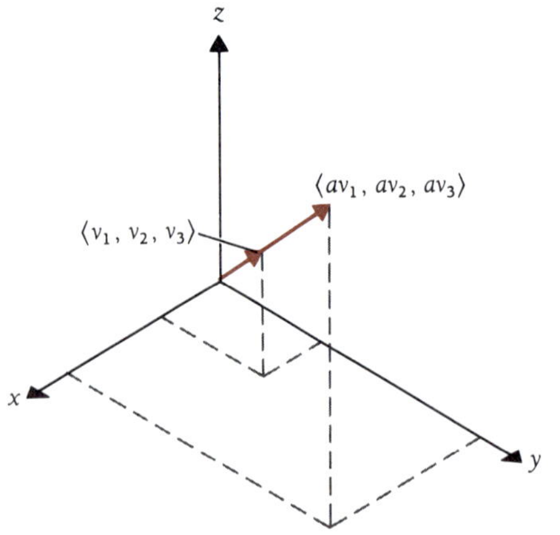

FIGURE 12.21 A scalar multiple

In terms of coordinates, it is easy to see that the vectors $\langle v_1, v_2, v_3\rangle$ and $\langle av_1, av_2, av_3\rangle$ have the same or opposite direction and that the second is $|a|$ times as long as the first (see Figure 12.21).

Example 12.7 For the vector $\mathbf{v} = \langle 2, -1, 3\rangle$, find the scalar multiples $2\mathbf{v}$, $\frac{1}{3}\mathbf{v}$, $-\mathbf{v}$, $-3\mathbf{v}$, and $0\mathbf{v}$.

Solution By Definition 12.4, we have

$$\begin{aligned}
2\mathbf{v} &= \langle 2(2), 2(-1), 2(3)\rangle = \langle 4, -2, 6\rangle,\\
\frac{1}{3}\mathbf{v} &= \left\langle \frac{1}{3}(2), \frac{1}{3}(-1), \frac{1}{3}(3)\right\rangle = \left\langle \frac{2}{3}, -\frac{1}{3}, 1\right\rangle,\\
-\mathbf{v} &= (-1)\mathbf{v} = \langle -2, 1, -3\rangle,\\
-3\mathbf{v} &= \langle (-3)(2), (-3)(-1), (-3)(3)\rangle = \langle -6, 3, -9\rangle,\\
0\mathbf{v} &= \langle 0(2), 0(-1), 0(3)\rangle = \langle 0, 0, 0\rangle = \mathbf{0}.
\end{aligned}$$

■

Several properties of scalar multiples are now evident. For example,

$$1\mathbf{v} = \mathbf{v}, \tag{5}$$
$$0\mathbf{v} = \mathbf{0}. \tag{6}$$

Also, it is easily shown that for any vectors $\mathbf{u}$ and $\mathbf{v}$ and for any scalars a and b,

$$a(\mathbf{u} + \mathbf{v}) = a\mathbf{u} + a\mathbf{v}, \tag{7}$$
$$(a + b)\mathbf{v} = a\mathbf{v} + b\mathbf{v}, \tag{8}$$
$$(ab)\mathbf{v} = a(b\mathbf{v}), \tag{9}$$
$$\|a\mathbf{v}\| = |a|\,\|\mathbf{v}\|. \tag{10}$$

As a result of property (10), we observe that if $\mathbf{v}$ is not the zero vector, then the scalar multiple

$$\mathbf{u} = \left(\frac{1}{\|\mathbf{v}\|}\right)\mathbf{v} \tag{11}$$

is a unit vector in the direction of **v**. To verify this, note that the scalar $1/\|\mathbf{v}\|$ is positive, so that **u** is in the direction of **v**. Moreover,

$$\|\mathbf{u}\| = \left\|\left(\frac{1}{\|\mathbf{v}\|}\right)\mathbf{v}\right\| = \left|\frac{1}{\|\mathbf{v}\|}\right|\|\mathbf{v}\| = \frac{1}{\|\mathbf{v}\|}\|\mathbf{v}\| = 1,$$

so that **u** is a unit vector.

Example 12.8 Find a unit vector in the direction of $\mathbf{v} = \langle 2, -1, 3\rangle$.

Solution The magnitude of **v** is

$$\|\mathbf{v}\| = \sqrt{2^2 + (-1)^2 + 3^2} = \sqrt{14},$$

so that by equation (11),

$$\mathbf{u} = \left(\frac{1}{\sqrt{14}}\right)\mathbf{v} = \left\langle \frac{2}{\sqrt{14}}, \frac{-1}{\sqrt{14}}, \frac{3}{\sqrt{14}}\right\rangle$$

is the unit vector in the direction of **v**. ■

The vector $-\mathbf{v} = (-1)\mathbf{v}$ is called the **additive inverse**, or **negative**, of **v**, and satisfies the equation

$$\mathbf{v} + (-\mathbf{v}) = \mathbf{0}. \tag{12}$$

We also define subtraction of vectors in terms of the additive inverse, as follows:

DEFINITION 12.5 The **difference** of vectors $\mathbf{u} = \langle u_1, u_2, u_3\rangle$ and $\mathbf{v} = \langle v_1, v_2, v_3\rangle$ is the vector

$$\begin{aligned}\mathbf{u} - \mathbf{v} &= \mathbf{u} + (-\mathbf{v})\\ &= \langle u_1 - v_1, u_2 - v_2, u_3 - v_3\rangle.\end{aligned}$$

Example 12.9 Find the difference of the vectors $\mathbf{u} = \langle 6, -1, 3\rangle$ and $\mathbf{v} = \langle 0, 1, -2\rangle$.

Solution By Definition 12.5, the difference is

$$\mathbf{u} - \mathbf{v} = \langle 6, -2, 5\rangle.$$ ■

Geometrically, if arrows representing **u** and **v** are adjacent sides of a parallelogram, then the diagonals represent the sum $\mathbf{u} + \mathbf{v}$ [see Figure 12.18(a)] and the difference $\mathbf{u} - \mathbf{v}$ (see Figure 12.22). Note also that $(\mathbf{u} - \mathbf{v}) + \mathbf{v} = \mathbf{u}$.

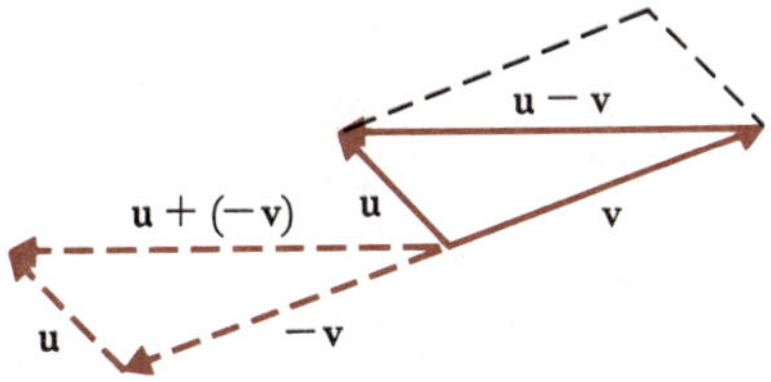

FIGURE 12.22 Vector subtraction

Vector subtraction enables us to solve the problem posed in Example 12.10.

Example 12.10 Given points $P = (x_1, y_1, z_1)$ and $Q = (x_2, y_2, z_2)$, find coordinates for the vector $\mathbf{v}(\overrightarrow{PQ})$.

Solution Let O be the origin. Then from Figure 12.23, we have

$$\begin{aligned}\mathbf{v}(\overrightarrow{PQ}) &= \mathbf{v}(\overrightarrow{OQ}) - \mathbf{v}(\overrightarrow{OP}) \\ &= \langle x_2, y_2, z_2\rangle - \langle x_1, y_1, z_1\rangle \\ &= \langle x_2 - x_1, y_2 - y_1, z_2 - z_1\rangle. \end{aligned} \tag{13}$$

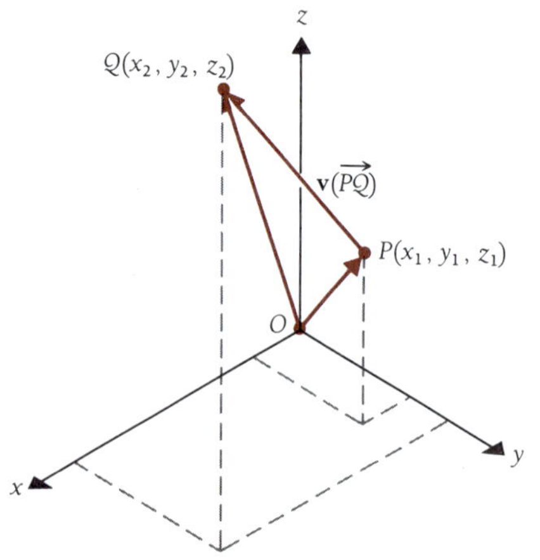

FIGURE 12.23 $\mathbf{v}(\overrightarrow{PQ}) = \langle x_2 - x_1, y_2 - y_1, z_2 - z_1\rangle$ ■

Example 12.11 Find coordinates for the vector represented by the arrow with its tail at $P = (-4, -4, 3)$ and its head at $Q = (7, 2, 0)$.

Solution By equation (13), we have

$$\mathbf{v}(\overrightarrow{PQ}) = \langle 11, 6, -3\rangle.$$ ■

A sum of scalar multiples of two or more vectors is called a **linear combination** of the vectors. For example, an arbitrary vector $\mathbf{v} = \langle v_1, v_2, v_3\rangle$ can be written as a linear combination of the standard unit vectors $\mathbf{i} = \langle 1, 0, 0\rangle$, $\mathbf{j} = \langle 0, 1, 0\rangle$, and $\mathbf{k} = \langle 0, 0, 1\rangle$, as follows:

$$\begin{aligned}\mathbf{v} &= \langle v_1, v_2, v_3\rangle \\ &= \langle v_1, 0, 0\rangle + \langle 0, v_2, 0\rangle + \langle 0, 0, v_3\rangle \\ &= v_1\langle 1, 0, 0\rangle + v_2\langle 0, 1, 0\rangle + v_3\langle 0, 0, 1\rangle \\ &= v_1\mathbf{i} + v_2\mathbf{j} + v_3\mathbf{k}. \end{aligned} \tag{14}$$

The representation (14) provides an alternative notation for vectors in terms of the standard unit vectors. In this notation, the sum and scalar multiple rules are

$$(u_1\mathbf{i} + u_2\mathbf{j} + u_3\mathbf{k}) + (v_1\mathbf{i} + v_2\mathbf{j} + v_3\mathbf{k}) = (u_1 + v_1)\mathbf{i} + (u_2 + v_2)\mathbf{j} + (u_3 + v_3)\mathbf{k}, \quad (15)$$

$$a(v_1\mathbf{i} + v_2\mathbf{j} + v_3\mathbf{k}) = av_1\mathbf{i} + av_2\mathbf{j} + av_3\mathbf{k}. \quad (16)$$

■ **Example 12.12** Carry out the operations:

(a) $(2\mathbf{i} - \mathbf{j} + 3\mathbf{k}) + (\mathbf{i} - 2\mathbf{k})$ (b) $4(\mathbf{i} + 2\mathbf{k})$
(c) $(-\mathbf{i} - \mathbf{j} + 3\mathbf{k}) - (2\mathbf{i} + 3\mathbf{j} - 4\mathbf{k})$

Solution (a) By equation (15), we have

$$(2\mathbf{i} - \mathbf{j} + 3\mathbf{k}) + (\mathbf{i} - 2\mathbf{k}) = 3\mathbf{i} - \mathbf{j} + \mathbf{k}.$$

(b) By equation (16), we have

$$4(\mathbf{i} + 2\mathbf{k}) = 4\mathbf{i} + 8\mathbf{k}.$$

(c) By the definition of subtraction and equations (15) and (16), we have

$$(-\mathbf{i} - \mathbf{j} + 3\mathbf{k}) - (2\mathbf{i} + 3\mathbf{j} - 4\mathbf{k}) = -3\mathbf{i} - 4\mathbf{j} + 7\mathbf{k}.$$ ■

Exercises 12.2

Let $\mathbf{u} = \langle -1, 0, 3 \rangle$ *and* $\mathbf{v} = \langle 2, 4, 1 \rangle$. *Find the following:*

1. $\|\mathbf{u}\|$
2. $\|\mathbf{v}\|$
3. $\mathbf{u} + \mathbf{v}$
4. $3\mathbf{v}$
5. $-2\mathbf{u}$
6. $\mathbf{u} - \mathbf{v}$
7. $\|-2\mathbf{u}\|$
8. $\|\mathbf{u} + \mathbf{v}\|$
9. A unit vector in the direction of $\mathbf{u}$
10. A unit vector in the direction of $\mathbf{v}$

Let $\mathbf{u} = \langle -3, 1, 5 \rangle$ *and* $\mathbf{v} = \langle -3, -1, -5 \rangle$. *Find the following:*

11. $\mathbf{u} + \mathbf{v}$
12. $\mathbf{v} - \mathbf{u}$
13. $-\mathbf{v} - 3\mathbf{u}$
14. $7\mathbf{u} + 3\mathbf{v}$
15. Scalars a and b such that $a\mathbf{u} + b\mathbf{v} = \mathbf{i}$
16. Scalars a and b such that $a\mathbf{u} + b\mathbf{v} = \mathbf{0}$
17. A unit vector in the direction of $\mathbf{u} - \mathbf{v}$
18. A unit vector opposite to $\mathbf{v}$

Develop the following formulas for vectors in two dimensions as special cases of the properties of vectors in three dimensions:

19. $\|\langle v_1, v_2 \rangle\| = \sqrt{v_1^2 + v_2^2}$
20. $\langle u_1, u_2 \rangle + \langle v_1, v_2 \rangle = \langle u_1 + v_1, u_2 + v_2 \rangle$
21. $a\langle v_1, v_2 \rangle = \langle av_1, av_2 \rangle$
22. $\langle u_1, u_2 \rangle - \langle v_1, v_2 \rangle = \langle u_1 - v_1, u_2 - v_2 \rangle$

Let $\mathbf{u} = \langle 2, -1 \rangle$ *and* $\mathbf{v} = \langle -4, 3 \rangle$ *be vectors in two dimensions. Find the following:*

23. $\mathbf{u} + \mathbf{v}$
24. $\mathbf{u} - \mathbf{v}$
25. $3\mathbf{u}$
26. $-\mathbf{v}$
27. $4\mathbf{u} + 2\mathbf{v}$
28. $5\mathbf{u} - 6\mathbf{v}$
29. $\|\mathbf{u}\|$
30. $\|\mathbf{v}\|$
31. Scalars a and b such that $a\mathbf{u} + b\mathbf{v} = \langle 1, 0 \rangle$
32. A unit vector in the direction of $\mathbf{v}$

Find the following:

33. $(2\mathbf{i} + \mathbf{j}) + (\mathbf{i} - 2\mathbf{j} + \mathbf{k})$
34. $(-\mathbf{i} + 2\mathbf{j} - \mathbf{k}) + (3\mathbf{i} + 2\mathbf{k})$
35. $(3\mathbf{i} - 2\mathbf{j} - 4\mathbf{k}) - (4\mathbf{i} + 3\mathbf{j} - 4\mathbf{k})$
36. $(3\mathbf{j} - \mathbf{k}) - (\mathbf{i} + \mathbf{j} + 3\mathbf{k})$
37. $5(\mathbf{i} + 2\mathbf{j}) - 3(2\mathbf{i} - \mathbf{j})$
38. $5(\mathbf{i} + \mathbf{j} + \mathbf{k}) + 4(\mathbf{i} + 2\mathbf{j} - \mathbf{k})$
39. $\|\mathbf{i} - \mathbf{j} + 2\mathbf{k}\|$
40. $\|\frac{5}{13}\mathbf{i} - \frac{12}{13}\mathbf{k}\|$

Express $\langle 4, 0, -3\rangle$ as a linear combination of the following sets of vectors:

41. $\mathbf{i}$, $\mathbf{j}$, and $\mathbf{k}$
42. $\langle 1, 0, 1\rangle$, $\langle 1, 1, 0\rangle$, and $\langle 0, 0, 1\rangle$
43. $\langle 3, 0, 1\rangle$ and $\langle 1, 0, 2\rangle$
44. $\langle 1, 2, 0\rangle$, $\langle 0, 1, 2\rangle$, and $\langle 1, 0, 2\rangle$
45. Do scalars a and b exist such that
$$a\langle 1, 2, 0\rangle + b\langle 0, 3, 0\rangle = \langle 1, 2, 1\rangle?$$
46. Do scalars a, b, and c exist such that
$$a\langle 1, 2, 0\rangle + b\langle 0, 3, 0\rangle = c\langle 1, 2, 1\rangle?$$
47. Show that $\|\mathbf{v}(\overrightarrow{PQ})\|$ is the distance between P and Q.
48. Show that for any three points P, Q, and R,
$$\mathbf{v}(\overrightarrow{PQ}) + \mathbf{v}(\overrightarrow{QR}) + \mathbf{v}(\overrightarrow{RP}) = 0.$$

Prove the identities. Use $\mathbf{u} = u_1\mathbf{i} + u_2\mathbf{j} + u_3\mathbf{k}$, etc.

49. $a(\mathbf{u} + \mathbf{v}) = a\mathbf{u} + a\mathbf{v}$
50. $(a + b)\mathbf{v} = a\mathbf{v} + b\mathbf{v}$
51. $\|a\mathbf{v}\| = |a|\,\|\mathbf{v}\|$
52. $\|\mathbf{u} - \mathbf{v}\| \geq \|\mathbf{u}\| - \|\mathbf{v}\|$

SECTION 12.3

Products of Vectors

Besides adding, subtracting, and finding scalar multiples of vectors, there are two useful ways to form products of vectors: the dot product and the cross product.

The Dot Product

The dot product of two vectors is useful in many applications. We will first give the definition and then some interpretations and uses.

DEFINITION 12.6

The **dot product** of the vectors $\mathbf{u} = \langle u_1, u_2, u_3\rangle$ and $\mathbf{v} = \langle v_1, v_2, v_3\rangle$ is the scalar
$$\mathbf{u} \cdot \mathbf{v} = u_1v_1 + u_2v_2 + u_3v_3.$$
The dot product is also called the **inner product** or the **scalar product** of the vectors.

Example 12.13 Find the dot product of the vectors $\mathbf{u} = \langle 1, 4, -2\rangle$ and $\mathbf{v} = \langle 3, -3, -5\rangle$.

Solution By Definition 12.6, we have

$$\mathbf{u} \cdot \mathbf{v} = (1)(3) + (4)(-3) + (-2)(-5) = 1.$$

Angle between Vectors

A geometric interpretation of the dot product is possible in terms of the angle between position arrows. If vectors **u** and **v** have position arrows $\overrightarrow{OP}$ and $\overrightarrow{OQ}$, respectively, then the **angle between u** and **v** is the measure of angle POQ, measured so that the angle is between 0 and π (see Figure 12.24).

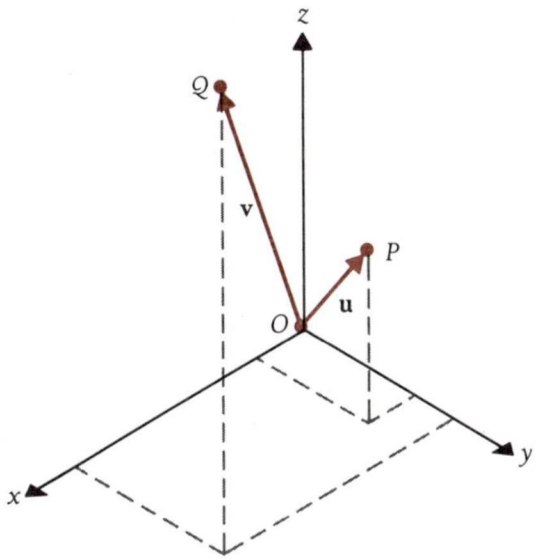

FIGURE 12.24

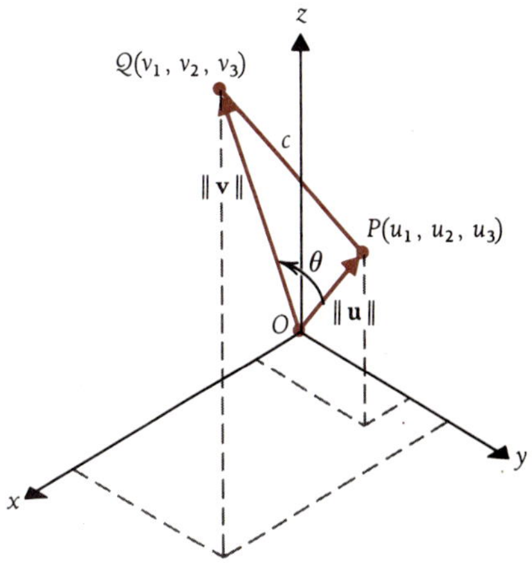

FIGURE 12.25 Using the law of cosines to relate the angle between vectors to the dot product

Now let $\mathbf{u} = \langle u_1, u_2, u_3 \rangle$ and $\mathbf{v} = \langle v_1, v_2, v_3 \rangle$ be two nonzero vectors. Then P is the point (u_1, u_2, u_3) and Q is the point (v_1, v_2, v_3) (see Figure 12.25), so that the distance c between P and Q satisfies

$$\begin{aligned} c^2 &= (u_1 - v_1)^2 + (u_2 - v_2)^2 + (u_3 - v_3)^2 \\ &= u_1^2 + u_2^2 + u_3^2 - 2(u_1v_1 + u_2v_2 + u_3v_3) + v_1^2 + v_2^2 + v_3^2 \\ &= \|\mathbf{u}\|^2 - 2(\mathbf{u} \cdot \mathbf{v}) + \|\mathbf{v}\|^2. \end{aligned} \tag{1}$$

If we let θ be the angle between **u** and **v** and apply the law of cosines to the triangle OPQ, we get

$$c^2 = \|\mathbf{u}\|^2 + \|\mathbf{v}\|^2 - 2\|\mathbf{u}\|\,\|\mathbf{v}\| \cos\theta. \tag{2}$$

Combining equations (1) and (2), we get

$$\mathbf{u} \cdot \mathbf{v} = \|\mathbf{u}\|\,\|\mathbf{v}\| \cos\theta. \tag{3}$$

That is, the dot product of two vectors is the product of their magnitudes, multiplied by the cosine of the angle between them.

If we solve formula (3) for $\cos\theta$, we get a formula for the angle θ between vectors **u** and **v**:

$$\cos\theta = \frac{\mathbf{u}\cdot\mathbf{v}}{\|\mathbf{u}\|\,\|\mathbf{v}\|}. \tag{4}$$

Example 12.14 Find the angle between the vectors:

(a) $\langle 3, 4, -5\rangle$ and $\langle 2, 1, 1\rangle$ (b) $\langle 3, 4, -5\rangle$ and $\langle 2, 1, 2\rangle$
(c) $\langle 3, 4, -5\rangle$ and $\langle 6, 8, -10\rangle$

Solution We use formula (4) in each case.

(a) $\cos\theta = \dfrac{\langle 3, 4, -5\rangle\cdot\langle 2, 1, 1\rangle}{\|\langle 3, 4, -5\rangle\|\,\|\langle 2, 1, 1\rangle\|} = \dfrac{5}{(5\sqrt{2})(\sqrt{6})} = \dfrac{1}{\sqrt{12}}$;

therefore, $\theta \approx 1.28$.

(b) $\cos\theta = \dfrac{\langle 3, 4, -5\rangle\cdot\langle 2, 1, 2\rangle}{\|\langle 3, 4, -5\rangle\|\,\|\langle 2, 1, 2\rangle\|} = \dfrac{0}{(5\sqrt{2})(3)} = 0$;

therefore, $\theta = \pi/2$.

(c) $\cos\theta = \dfrac{\langle 3, 4, -5\rangle\cdot\langle 6, 8, -10\rangle}{\|\langle 3, 4, -5\rangle\|\,\|\langle 6, 8, -10\rangle\|} = \dfrac{100}{(5\sqrt{2})(10\sqrt{2})} = 1$;

therefore, $\theta = 0$. ■

If the angle between two vectors is 0, we say the vectors are **parallel**. This means that arrows representing the vectors are in the same direction. If the angle between two vectors is π, we say they are **antiparallel**, and their arrows are in opposite directions. If the angle between two vectors is $\pi/2$, we say the vectors are **orthogonal.** This means that the position arrows of the two vectors are perpendicular. You can verify that vectors **u** and **v** are orthogonal if and only if

$$\mathbf{u}\cdot\mathbf{v} = 0. \tag{5}$$

Components and Projections

Further applications of the dot product can be obtained by viewing equation (3) in the right light. In triangle OPQ, let line FQ be drawn perpendicular to line OP (Figure 12.26). The distance $|OF|$, by elementary right-triangle relationships, is

$$|OF| = \|\mathbf{v}\|\cos\theta,$$

which by equation (3) can be rewritten as

$$|OF| = \frac{\mathbf{u}\cdot\mathbf{v}}{\|\mathbf{u}\|}. \tag{6}$$

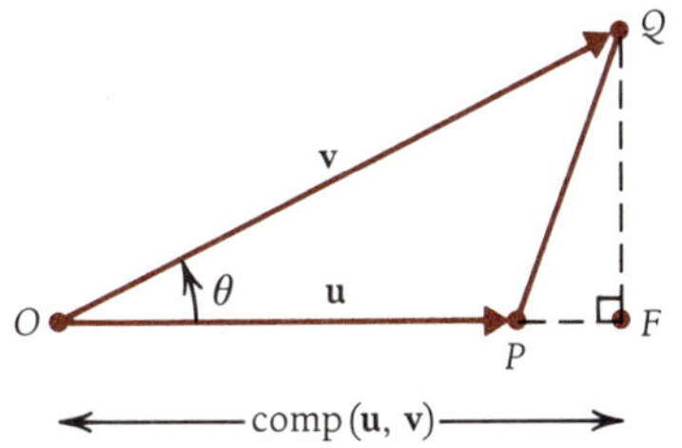

FIGURE 12.26 The component of **v** in the direction of **u**

This distance is called the **component of v in the direction of u**, which we will also denote by comp(**u**, **v**). By similar reasoning, the quantity

$$\text{comp}(\mathbf{v}, \mathbf{u}) = \frac{\mathbf{u} \cdot \mathbf{v}}{\|\mathbf{v}\|} \tag{7}$$

is called the **component of u in the direction of v**.

Example 12.15 Find the components of the vector $\mathbf{v} = \langle 2, -1, 3\rangle$ in the directions of **i**, **j**, and **k**.

Solution First, to find the component of **v** in the direction of **i**, we let **i** play the role of **u** in equation (6):

$$\text{comp}(\mathbf{i}, \mathbf{v}) = \frac{\mathbf{i} \cdot \mathbf{v}}{\|\mathbf{i}\|} = \frac{\langle 1, 0, 0\rangle \cdot \langle 2, -1, 3\rangle}{1} = 2.$$

Similarly, the component of **v** in the direction of **j** is

$$\text{comp}(\mathbf{j}, \mathbf{v}) = \frac{\mathbf{j} \cdot \mathbf{v}}{\|\mathbf{j}\|} = \frac{\langle 0, 1, 0\rangle \cdot \langle 2, -1, 3\rangle}{1} = -1;$$

and the component of **v** in the direction of **k** is

$$\text{comp}(\mathbf{k}, \mathbf{v}) = \frac{\mathbf{k} \cdot \mathbf{v}}{\|\mathbf{k}\|} = \frac{\langle 0, 0, 1\rangle \cdot \langle 2, -1, 3\rangle}{1} = 3.$$ ■

Example 12.15 illustrates a general principle. If **u** is any unit vector, then

$$\text{comp}(\mathbf{u}, \mathbf{v}) = \mathbf{u} \cdot \mathbf{v}.$$

Thus, the components of $\mathbf{v} = \langle v_1, v_2, v_3\rangle$ in the directions of **i**, **j**, and **k** are

$$\mathbf{i} \cdot \mathbf{v} = v_1, \qquad \mathbf{j} \cdot \mathbf{v} = v_2, \qquad \mathbf{k} \cdot \mathbf{v} = v_3.$$

For this reason, the coordinates of **v** are also called the **components of v**.

Now, let **u** and **v** be arbitrary vectors. If we multiply the scalar comp(**u**, **v**) by a unit vector in the direction of **u**, we obtain a vector called the **projection of v on u**, denoted by proj(**u**, **v**). Since the length of proj(**u**, **v**) is just comp(**u**, **v**), as shown in Figure 12.27 on page 760, the projection of **v** on **u** is

$$\text{proj}(\mathbf{u}, \mathbf{v}) = \left(\frac{\mathbf{u} \cdot \mathbf{v}}{\|\mathbf{u}\|^2}\right)\mathbf{u}. \tag{8}$$

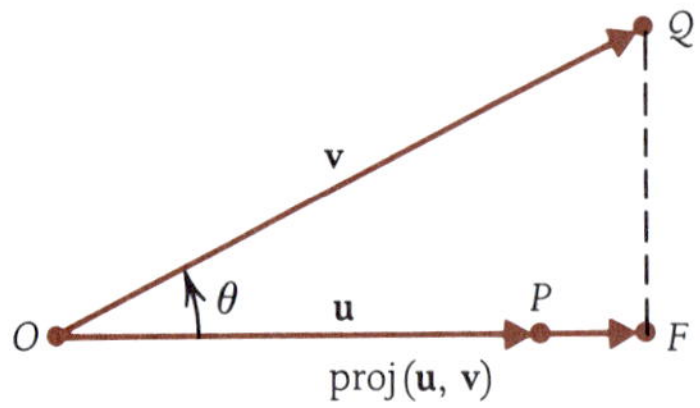

FIGURE 12.27 The projection of **v** on **u**

■ **Example 12.16** Find the projections of the vector $\mathbf{v} = \langle 2, -1, 3\rangle$ on **i**, **j**, and **k**.

Solution Letting **i**, **j**, and **k**, in turn, play the role of **u** in equation (8), we find the projections to be

$$\text{proj}(\mathbf{i}, \mathbf{v}) = \left(\frac{\mathbf{i}\cdot\mathbf{v}}{\|\mathbf{i}\|^2}\right)\mathbf{i} = \left(\frac{\langle 1, 0, 0\rangle \cdot \langle 2, -1, 3\rangle}{1}\right)\mathbf{i} = 2\mathbf{i},$$
$$\text{proj}(\mathbf{j}, \mathbf{v}) = \left(\frac{\mathbf{j}\cdot\mathbf{v}}{\|\mathbf{j}\|^2}\right)\mathbf{j} = \left(\frac{\langle 0, 1, 0\rangle \cdot \langle 2, -1, 3\rangle}{1}\right)\mathbf{j} = -\mathbf{j},$$
$$\text{proj}(\mathbf{k}, \mathbf{v}) = \left(\frac{\mathbf{k}\cdot\mathbf{v}}{\|\mathbf{k}\|^2}\right)\mathbf{k} = \left(\frac{\langle 0, 0, 1\rangle \cdot \langle 2, -1, 3\rangle}{1}\right)\mathbf{k} = 3\mathbf{k}.$$ ■

Example 12.16 also illustrates a general principle. If **u** is any unit vector, then

$$\text{proj}(\mathbf{u}, \mathbf{v}) = (\mathbf{u}\cdot\mathbf{v})\mathbf{u}.$$

It then follows that any vector $\mathbf{v} = \langle v_1, v_2, v_3\rangle = v_1\mathbf{i} + v_2\mathbf{j} + v_3\mathbf{k}$ is the sum of its projections on **i**, **j**, and **k**.

Other properties of the dot product include the following:

$$\mathbf{u}\cdot\mathbf{v} = \mathbf{v}\cdot\mathbf{u} \tag{9}$$
$$(a\mathbf{u})\cdot\mathbf{v} = \mathbf{u}\cdot(a\mathbf{v}) = a(\mathbf{u}\cdot\mathbf{v}) \tag{10}$$
$$\mathbf{u}\cdot(\mathbf{v}+\mathbf{w}) = \mathbf{u}\cdot\mathbf{v} + \mathbf{u}\cdot\mathbf{w} \tag{11}$$
$$\mathbf{v}\cdot\mathbf{v} = \|\mathbf{v}\|^2 \tag{12}$$

The Cross Product

The cross product will have even more applications than the dot product, but its definition is also more complicated.

DEFINITION 12.7 The **cross product** of the vectors $\mathbf{u} = \langle u_1, u_2, u_3\rangle$ and $\mathbf{v} = \langle v_1, v_2, v_3\rangle$ is the vector

$$\mathbf{u}\times\mathbf{v} = \langle u_2v_3 - u_3v_2,\ u_3v_1 - u_1v_3,\ u_1v_2 - u_2v_1\rangle.$$

The cross product can be written in two other forms that you might find easier to remember. Both use the idea of determinants (see Appendix A). One is

$$\mathbf{u} \times \mathbf{v} = \left\langle \begin{vmatrix} u_2 & u_3 \\ v_2 & v_3 \end{vmatrix}, \begin{vmatrix} u_3 & u_1 \\ v_3 & v_1 \end{vmatrix}, \begin{vmatrix} u_1 & u_2 \\ v_1 & v_2 \end{vmatrix} \right\rangle, \tag{13}$$

and the other is

$$\mathbf{u} \times \mathbf{v} = \begin{vmatrix} \mathbf{i} & \mathbf{j} & \mathbf{k} \\ u_1 & u_2 & u_3 \\ v_1 & v_2 & v_3 \end{vmatrix}. \tag{14}$$

Example 12.17 Find the cross product of the vectors, $\mathbf{u} = \langle 3, 4, -5 \rangle$ and $\mathbf{v} = \langle 2, -1, 3 \rangle$.

Solution By formula (13), we have

$$\begin{aligned}\mathbf{u} \times \mathbf{v} &= \left\langle \begin{vmatrix} 4 & -5 \\ -1 & 3 \end{vmatrix}, \begin{vmatrix} -5 & 3 \\ 3 & 2 \end{vmatrix}, \begin{vmatrix} 3 & 4 \\ 2 & -1 \end{vmatrix} \right\rangle \\ &= \langle 7, -19, -11 \rangle.\end{aligned}$$

■

Geometrically, the cross product of two vectors is represented by a position arrow that is perpendicular to the position arrows of both vectors (see Figure 12.28, page 762). This is because $\mathbf{u} \times \mathbf{v}$ is orthogonal to both $\mathbf{u}$ and $\mathbf{v}$. To show that $\mathbf{u} \times \mathbf{v}$ is orthogonal to $\mathbf{v}$, we compute $(\mathbf{u} \times \mathbf{v}) \cdot \mathbf{v}$ and use equation (5):

$$\begin{aligned}(\mathbf{u} \times \mathbf{v}) \cdot \mathbf{v} &= \langle u_2v_3 - u_3v_2, u_3v_1 - u_1v_3, u_1v_2 - u_2v_1 \rangle \cdot \langle v_1, v_2, v_3 \rangle \\ &= (u_2v_3 - u_3v_2)v_1 + (u_3v_1 - u_1v_3)v_2 + (u_1v_2 - u_2v_1)v_3 \\ &= u_2v_3v_1 - u_3v_2v_1 + u_3v_1v_2 - u_1v_3v_2 + u_1v_2v_3 - u_2v_1v_3 \\ &= 0.\end{aligned}$$

We can show that $\mathbf{u} \times \mathbf{v}$ is orthogonal to $\mathbf{u}$ in a similar manner.

The direction of the cross product is determined geometrically by the **right-hand rule.** If one grasps the position arrow of $\mathbf{u} \times \mathbf{v}$ with the right hand, with the thumb pointing in the direction of $\mathbf{u} \times \mathbf{v}$, then the fingers curl in the direction of the smallest angle from the position arrow of $\mathbf{u}$ to the position arrow of $\mathbf{v}$ (see Figure 12.29). The vector $\mathbf{v} \times \mathbf{u}$ is in the opposite direction from $\mathbf{u} \times \mathbf{v}$, since

$$\mathbf{v} \times \mathbf{u} = -(\mathbf{u} \times \mathbf{v}) \tag{15}$$

(which you are asked to verify in Problem 53).

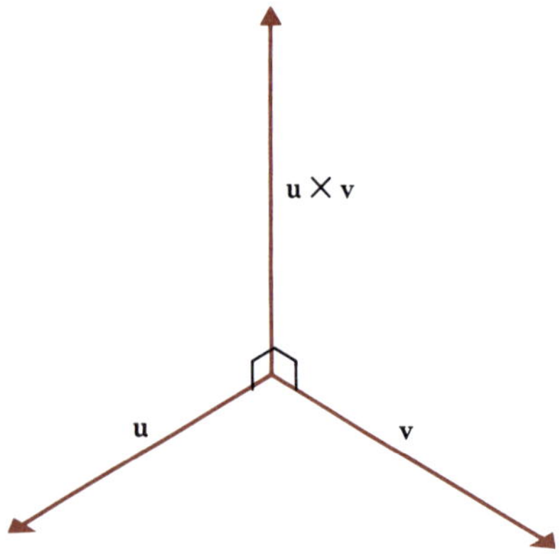

FIGURE 12.28 The cross product

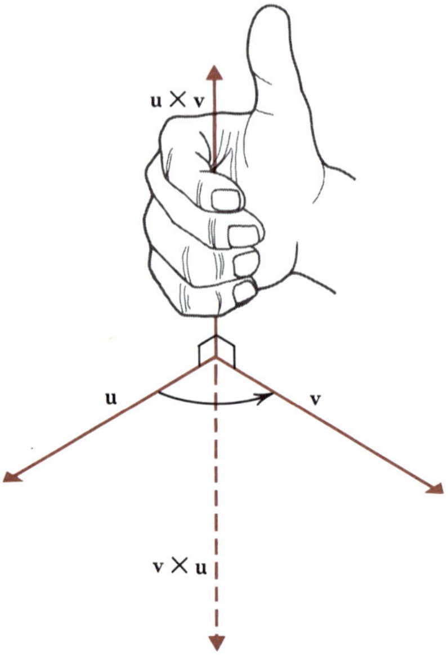

FIGURE 12.29 The right-hand rule gives the direction of the cross product

Further geometric interpretation of $\mathbf{u} \times \mathbf{v}$ comes from computing its magnitude. We have

$$\begin{aligned}
\|\mathbf{u} \times \mathbf{v}\|^2 &= (u_2v_3 - u_3v_2)^2 + (u_3v_1 - u_1v_3)^2 + (u_1v_2 - u_2v_1)^2 \\
&= (u_1^2 + u_2^2 + u_3^2)(v_1^2 + v_2^2 + v_3^2) - (u_1v_1 + u_2v_2 + u_3v_3)^2 \\
&\qquad\qquad \text{(Verify this equation.)} \\
&= \|\mathbf{u}\|^2\|\mathbf{v}\|^2 - (\mathbf{u} \cdot \mathbf{v})^2 \\
&= \|\mathbf{u}\|^2\|\mathbf{v}\|^2 - \|\mathbf{u}\|^2\|\mathbf{v}\|^2 \cos^2\theta \\
&= \|\mathbf{u}\|^2\|\mathbf{v}\|^2(1 - \cos^2\theta) \\
&= \|\mathbf{u}\|^2\|\mathbf{v}\|^2 \sin^2\theta,
\end{aligned}$$

where θ is the angle between $\mathbf{u}$ and $\mathbf{v}$. Hence,

$$\|\mathbf{u} \times \mathbf{v}\| = \|\mathbf{u}\|\,\|\mathbf{v}\| \sin \theta; \tag{16}$$

that is, the magnitude of the cross product of two vectors is the product of their magnitudes, multiplied by the sine of the angle between them. Thus, we see from Figure 12.30 that the magnitude of the cross product $\mathbf{u} \times \mathbf{v}$ is the area of the parallelogram of which arrows representing $\mathbf{u}$ and $\mathbf{v}$ are adjacent sides.

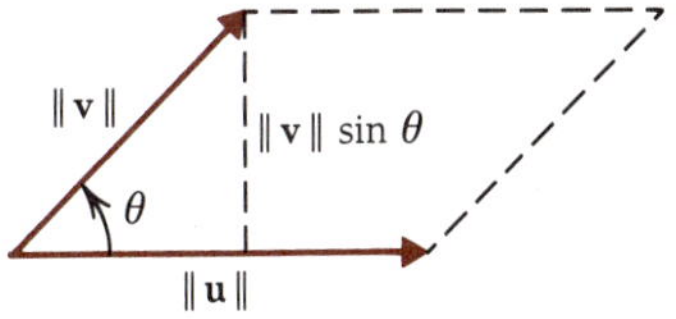

FIGURE 12.30 Geometric interpretation of the magnitude of the cross product

Example 12.18 Find the area of the parallelogram with vertices $O(0, 0, 0)$, $P(2, 1, -4)$, $Q(5, 2, -1)$, and $R(7, 3, -5)$.

Solution Note that $\mathbf{v}(\overrightarrow{OP}) + \mathbf{v}(\overrightarrow{OQ}) = \mathbf{v}(\overrightarrow{OR})$, so the given points are indeed the vertices of a parallelogram. By the above paragraph, the area is

$$\begin{aligned}\|\mathbf{v}(\overrightarrow{OP}) \times \mathbf{v}(\overrightarrow{OQ})\| &= \|\langle 2, 1, -4\rangle \times \langle 5, 2, -1\rangle\| \\ &= \|\langle 7, -18, -1\rangle\| \\ &= \sqrt{49 + 324 + 1} = \sqrt{374}.\end{aligned}$$

■

Example 12.19 Find the area of the triangle with vertices $A(-1, 1, 3)$, $B(4, 1, -1)$, and $C(0, 4, 3)$.

Solution A triangle is half a parallelogram (see Figure 12.31), so the area is

$$\begin{aligned}\frac{1}{2}\|\mathbf{v}(\overrightarrow{AB}) \times \mathbf{v}(\overrightarrow{AC})\| &= \frac{1}{2}\|\langle 5, 0, -4\rangle \times \langle 1, 3, 0\rangle\| \\ &= \frac{1}{2}\|\langle 12, -4, 15\rangle\| \\ &= \frac{1}{2}\sqrt{144 + 16 + 225} = \frac{1}{2}\sqrt{385}.\end{aligned}$$

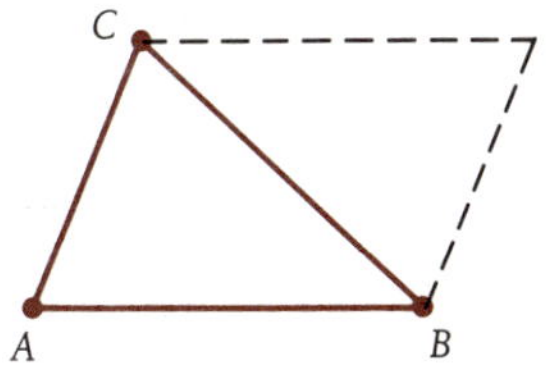

FIGURE 12.31 A triangle is half of a parallelogram ■

Equation (15) shows that the cross product is a noncommutative operation. Other properties of the cross product are

$$(a\mathbf{u}) \times \mathbf{v} = \mathbf{u} \times (a\mathbf{v}) = a(\mathbf{u} \times \mathbf{v}) \tag{17}$$

$$\mathbf{u} \times (\mathbf{v} + \mathbf{w}) = (\mathbf{u} \times \mathbf{v}) + (\mathbf{u} \times \mathbf{w}) \tag{18}$$

$$(\mathbf{u} + \mathbf{v}) \times \mathbf{w} = (\mathbf{u} \times \mathbf{w}) + (\mathbf{v} \times \mathbf{w}) \tag{19}$$

$$\mathbf{v} \times \mathbf{v} = \mathbf{0} \tag{20}$$

The Triple Product

Combinations of the dot and cross products of the form

$$\mathbf{u} \cdot \mathbf{v} \times \mathbf{w} \qquad \text{and} \qquad \mathbf{u} \times \mathbf{v} \cdot \mathbf{w}$$

are called **triple products.** Because the dot product yields a scalar and the cross product yields a vector, the triple products make sense only when the cross product is computed first. Hence, grouping symbols are not needed in these expressions.

■ **Example 12.20** Compute the triple product $\langle 2, 2, -3\rangle \cdot \langle 1, -3, 0\rangle \times \langle 2, 0, 1\rangle$.

Solution Since $\langle 1, -3, 0\rangle \times \langle 2, 0, 1\rangle = \langle -3, -1, 6\rangle$, we get

$$\begin{aligned}\langle 2, 2, -3\rangle \cdot \langle 1, -3, 0\rangle \times \langle 2, 0, 1\rangle &= \langle 2, 2, -3\rangle \cdot \langle -3, -1, 6\rangle \\ &= -26.\end{aligned}$$ ■

There is a handy computational device for the triple product in terms of determinants. If $\mathbf{u} = \langle u_1, u_2, u_3\rangle$, $\mathbf{v} = \langle v_1, v_2, v_3\rangle$, and $\mathbf{w} = \langle w_1, w_2, w_3\rangle$, then it can be shown by expanding both sides that

$$\mathbf{u} \cdot \mathbf{v} \times \mathbf{w} = \begin{vmatrix} u_1 & u_2 & u_3 \\ v_1 & v_2 & v_3 \\ w_1 & w_2 & w_3 \end{vmatrix}. \tag{21}$$

It then follows that

$$\mathbf{u} \times \mathbf{v} \cdot \mathbf{w} = \mathbf{u} \cdot \mathbf{v} \times \mathbf{w}, \tag{22}$$

since

$$\begin{aligned}\mathbf{u} \times \mathbf{v} \cdot \mathbf{w} &= \mathbf{w} \cdot \mathbf{u} \times \mathbf{v} && \text{[By equation (9)]} \\ &= \begin{vmatrix} w_1 & w_2 & w_3 \\ u_1 & u_2 & u_3 \\ v_1 & v_2 & v_3 \end{vmatrix} && \text{[By equation (21)]} \\ &= \begin{vmatrix} u_1 & u_2 & u_3 \\ v_1 & v_2 & v_3 \\ w_1 & w_2 & w_3 \end{vmatrix} && \text{(See Appendix A.)} \\ &= \mathbf{u} \cdot \mathbf{v} \times \mathbf{w}. && \text{[By equation (21)]}\end{aligned}$$

A geometric interpretation of the triple product is obtained by combining the interpretations of the dot and cross products. Let vectors **u**, **v**, and **w** be represented by arrows as in Figure 12.32, and consider the parallelepiped formed as shown in the figure. The base (or top face) of the parallelepiped has area $\|\mathbf{v} \times \mathbf{w}\|$. The height of the parallelepiped is the absolute value of the component of **u** in the direction of

$\mathbf{v} \times \mathbf{w}$, which by equation (7) is

$$\text{comp}(\mathbf{v} \times \mathbf{w}, \mathbf{u}) = \frac{\mathbf{u} \cdot \mathbf{v} \times \mathbf{w}}{\|\mathbf{v} \times \mathbf{w}\|}.$$

Hence, the volume of the parallelepiped is the area of the base times the height, which is $|\mathbf{u} \cdot \mathbf{v} \times \mathbf{w}|$, the absolute value of the triple product.

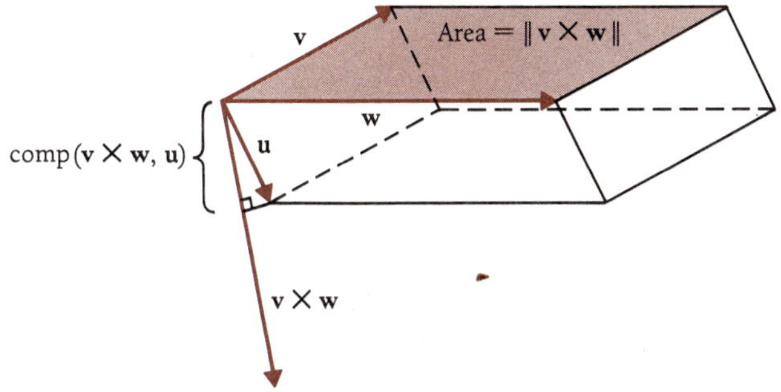

FIGURE 12.32 The volume of the parallelepiped is $|\mathbf{u} \cdot \mathbf{v} \times \mathbf{w}|$

Historical Note on Vectors

The concept of vectors has a rather complex history. In the 1840's, the Irish mathematician Sir William Rowan Hamilton (1805–1865) invented the concept of a *quaternion,* which somewhat resembles a number plus a vector. Hamilton was able to define dot and cross products, and eventually won a few converts to his theory. But in spite of certain nice algebraic properties, quaternions are clumsy for use in applications.

Simultaneously, in Germany, Heinrich Grassmann (1809–1877)invented the *Ausdehnungslehre,* a very general algebraic system for describing geometric objects, and from this theory the dot and cross products also emerged. Because Grassmann's work was too advanced for its time, vectors did not make much headway in Germany either, although certain related concepts involving determinants became widely used.

In the 1890's, the American mathematician Josiah Willard Gibbs (1839–1903) decided to remove vectors from the surrounding mix of quaternions and more general concepts; he presented them along with the dot and cross products as a self-contained theory. Gibbs's book, and similar work in Germany, created vector analysis as an independent discipline. The simplicity and obvious utility of these methods quickly gained widespread acceptance.

Exercises 12.3

For Problems 1–32, let $\mathbf{u} = \langle -3, 4, 1 \rangle$, $\mathbf{v} = \langle 0, 2, 3 \rangle$, *and* $\mathbf{w} = \langle 1, 1, -1 \rangle$. *Find the following:*

1. $\mathbf{u} \cdot \mathbf{v}$
2. $\mathbf{u} \cdot \mathbf{w}$
3. $\mathbf{v} \cdot \mathbf{w}$
4. $\mathbf{w} \cdot \mathbf{u}$
5. $\mathbf{u} \cdot (\mathbf{v} + \mathbf{w})$
6. $\mathbf{v} \cdot (\mathbf{u} + \mathbf{w})$
7. $\mathbf{u} \cdot (\mathbf{v} - \mathbf{w})$
8. $\mathbf{w} \cdot (\mathbf{u} - \mathbf{v})$

9. $(2\mathbf{u}) \cdot \mathbf{v}$
10. $(3\mathbf{u}) \cdot (4\mathbf{v})$
11. $(2\mathbf{w}) \cdot (3\mathbf{v})$
12. $\mathbf{w} \cdot (-\mathbf{w})$
13. The angle between $\mathbf{u}$ and $\mathbf{v}$
14. The angle between $\mathbf{u}$ and $\mathbf{v} - \mathbf{w}$
15. The angle between $\mathbf{v}$ and $3\mathbf{v}$
16. The angle between $\mathbf{u}$ and $-\mathbf{u}$
17. The component of $\mathbf{u}$ in the direction of $\mathbf{v}$
18. The component of $\mathbf{v}$ in the direction of $\mathbf{w}$
19. The projection of $\mathbf{v}$ on $\mathbf{u}$
20. The projection of $\mathbf{w}$ on $\mathbf{u}$
21. $\mathbf{u} \times \mathbf{v}$
22. $\mathbf{v} \times \mathbf{u}$
23. $\mathbf{u} \times \mathbf{w}$
24. $\mathbf{w} \times \mathbf{v}$
25. $\mathbf{u} \times (\mathbf{v} \times \mathbf{w})$
26. $\mathbf{u} \times (\mathbf{u} \times \mathbf{v})$
27. $(\mathbf{u} + \mathbf{v}) \times \mathbf{w}$
28. $(\mathbf{u} \times \mathbf{w}) + (\mathbf{v} \times \mathbf{w})$
29. $\|\mathbf{u} \times \mathbf{w}\|$
30. $\|\mathbf{v} \times \mathbf{w}\|$
31. $\mathbf{u} \cdot \mathbf{v} \times \mathbf{w}$
32. $\mathbf{v} \cdot \mathbf{u} \times \mathbf{w}$
33. The area of the parallelogram with vertices $(0, 0, 0)$, $(-3, 4, 1)$, $(0, 2, 3)$, and $(-3, 6, 4)$
34. The area of the parallelogram with vertices $(0, 0, 0)$, $(0, 2, 3)$, $(1, 1, -1)$, and $(1, 3, 2)$
35. The area of the triangle with vertices $(0, 0, 0)$, $(0, 2, 3)$, and $(1, 1, -1)$
36. The area of the triangle with vertices $(0, 0, 0)$, $(-3, 4, 1)$, and $(1, 1, -1)$

Carry out the indicated operations.

37. $(5\mathbf{i} + 3\mathbf{j}) \cdot (2\mathbf{i} - 3\mathbf{j} + \mathbf{k})$
38. $(\mathbf{i} + \mathbf{j} + \mathbf{k}) \cdot (4\mathbf{i} + 5\mathbf{j} - \mathbf{k})$
39. $(7\mathbf{i} - 3\mathbf{k}) \times (-\mathbf{i} - \mathbf{j} + \mathbf{k})$
40. $(-\mathbf{j} - 3\mathbf{k}) \times (\mathbf{i} + 10\mathbf{j})$
41. $\mathbf{i} \cdot \mathbf{j}$
42. $\mathbf{i} \cdot \mathbf{k}$
43. $\mathbf{j} \cdot \mathbf{k}$
44. $(\mathbf{i} + \mathbf{j}) \cdot \mathbf{k}$
45. $\mathbf{i} \times \mathbf{j}$
46. $\mathbf{j} \times \mathbf{k}$
47. $\mathbf{i} \times \mathbf{k}$
48. $\mathbf{k} \times \mathbf{k}$
49. $(2\mathbf{i} + 2\mathbf{j} - \mathbf{k}) \cdot (-\mathbf{i} + 3\mathbf{j}) \times (\mathbf{j} + 2\mathbf{k})$
50. $(\mathbf{i} - \mathbf{j} - 4\mathbf{k}) \times (\mathbf{i} + \mathbf{j} + \mathbf{k}) \cdot (2\mathbf{i} - 5\mathbf{k})$

Establish the following identities:

51. $(a\mathbf{u}) \cdot \mathbf{v} = a(\mathbf{u} \cdot \mathbf{v})$
52. $\mathbf{v} \cdot \mathbf{v} = \|\mathbf{v}\|^2$
53. $\mathbf{v} \times \mathbf{u} = -(\mathbf{u} \times \mathbf{v})$
54. $(a\mathbf{u}) \times \mathbf{v} = a(\mathbf{u} \times \mathbf{v})$
55. $(\mathbf{u} + \mathbf{v}) \cdot (\mathbf{u} - \mathbf{v}) = \|\mathbf{u}\|^2 - \|\mathbf{v}\|^2$
56. $(\mathbf{u} + \mathbf{v}) \times (\mathbf{u} - \mathbf{v}) = 2(\mathbf{v} \times \mathbf{u})$
57. Let **u**, **v**, and **w** be vectors whose position arrows all lie in the xy-plane. Show that $\mathbf{u} \cdot \mathbf{v} \times \mathbf{w} = 0$.
58. Vectors **u**, **v**, and **w** are said to be **coplanar** if and only if $\mathbf{u} \cdot \mathbf{v} \times \mathbf{w} = 0$. Show that any linear combination of **u** and **v** is coplanar with **u** and **v**.
59. Refer to Problem 58. If **u**, **v**, and **w** are coplanar, show that **w** is a linear combination of **u** and **v**.
60. Refer to Problem 58. Determine whether the vectors $\langle 2, -5, 3 \rangle$, $\langle 0, 1, 4 \rangle$, and $\langle 2, -2, 15 \rangle$ are coplanar.
61. If a force **F** acts on an object to produce a displacement **r**, then the **work** done by the force is defined to be $W = \|\mathbf{F}\|\,\|\mathbf{r}\|$. This assumes that the force and displacement are in the same direction. If the force and displacement are not in the same direction, show that the work done is $W = |\mathbf{F} \cdot \mathbf{r}|$.
62. Refer to Problem 61. A force $\mathbf{F} = \langle 20, 50, 80 \rangle$ acts on an object and produces a displacement $\mathbf{r} = \langle 1, 2, -1 \rangle$. Find the work done.
63. When a particle with electric charge q travels with velocity **v** through a magnetic field with flux **B**, the force **F** exerted by the magnetic field on the particle is $\mathbf{F} = q(\mathbf{v} \times \mathbf{B})$,

where $\|\mathbf{F}\|$ is measured in newtons, q in coulombs, $\|\mathbf{v}\|$ in meters per second, and $\|\mathbf{B}\|$ in webers per square meter. If a charged particle with velocity $\mathbf{v} = 10^6\mathbf{i}$ experiences a force $\mathbf{F} = 8 \times 10^{-23}\mathbf{j}$ when traveling through a magnetic field with flux $\mathbf{B} = 5 \times 10^{-10}\mathbf{k}$, what is the charge on the particle?

64. Find a unit vector that is orthogonal to both $\mathbf{u} = \langle 4, 1, -1\rangle$ and $\mathbf{v} = \langle 0, -1, 1\rangle$.

SECTION 12.4

Lines in Space

We now return to solid analytic geometry. Our first study is the description of lines in space using vectors.

Elementary properties of lines in space include the facts that two points determine a line, and two lines may either intersect (Figure 12.33) or be parallel (Figure 12.34) if they lie in the same plane. Lines not lying in the same plane are called **skew lines** (Figure 12.35).

FIGURE 12.33
Intersecting lines

FIGURE 12.34
Parallel lines

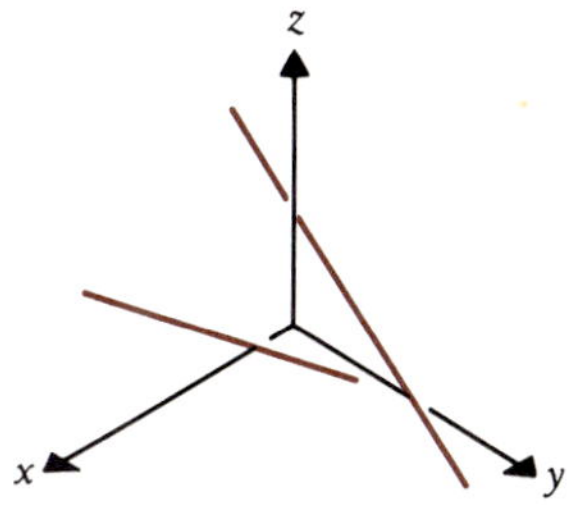

FIGURE 12.35
Skew lines

In the plane, we usually describe a line in terms of its slope, which tells us its direction. In space, the direction of a line is described in terms of vectors.

Direction Vectors, Direction Numbers, and Direction Cosines

Let l be a line in space. If P and Q are two points on l, then the vector $\mathbf{v}(\overrightarrow{PQ})$ is called a **direction vector** of line l (Figure 12.36). Thus, if (x_1, y_1, z_1) and (x_2, y_2, z_2) are two points on l, then the vector

$$\langle x_2 - x_1, y_2 - y_1, z_2 - z_1\rangle \tag{1}$$

is a direction vector of l.

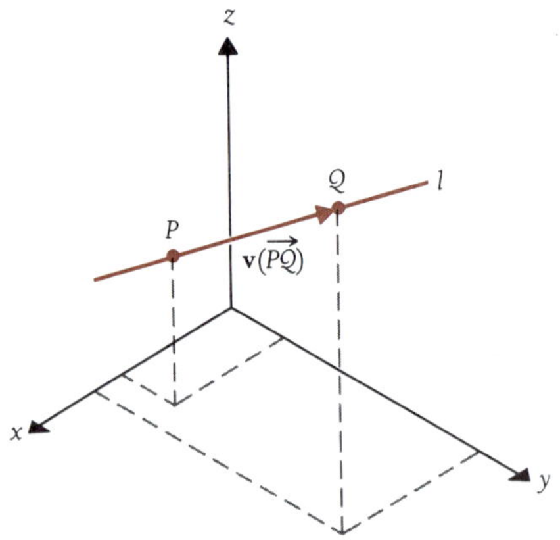

FIGURE 12.36 A direction vector of l

Example 12.21 Find a direction vector for the line through the points:

(a) $(2, -1, 5)$ and $(4, 4, -1)$ **(b)** $(3, -2, -1)$ and the origin

Solution **(a)** Using expression (1), a direction vector for this line is

$$\mathbf{u} = \langle 2-4, -1-4, 5+1 \rangle = \langle -2, -5, 6 \rangle.$$

(b) Similarly, a direction vector is

$$\mathbf{v} = \langle 3-0, -2-0, -1-0 \rangle = \langle 3, -2, -1 \rangle.$$ ■

Note that the direction vector gives the line a direction only, so that any scalar multiple of a direction vector of l is another direction vector of l. Also, parallel lines have the same direction vectors.

We define the **angle between two lines** to be the angle between their direction vectors, whether or not the lines lie in the same plane. Thus, if vectors $\mathbf{v}_1 = \langle a_1, b_1, c_1 \rangle$ and $\mathbf{v}_2 = \langle a_2, b_2, c_2 \rangle$ are direction vectors of the lines l_1 and l_2, then from equation (4) of Section 12.3 we know that the angle θ between l_1 and l_2 satisfies

$$\cos\theta = \frac{\mathbf{v}_1 \cdot \mathbf{v}_2}{\|\mathbf{v}_1\|\,\|\mathbf{v}_2\|} = \frac{a_1a_2 + b_1b_2 + c_1c_2}{\sqrt{a_1^2 + b_1^2 + c_1^2}\,\sqrt{a_2^2 + b_2^2 + c_2^2}}.$$

Orthogonal lines are lines whose direction vectors are orthogonal, and **perpendicular lines** are orthogonal lines that intersect.

Example 12.22 Find the angle between the two lines in Example 12.21.

Solution The direction vectors are $\mathbf{u} = \langle -2, -5, 6\rangle$ and $\mathbf{v} = \langle 3, -2, -1\rangle$, so the angle between them satisfies

$$\cos\theta = \frac{\mathbf{u}\cdot\mathbf{v}}{\|\mathbf{u}\|\,\|\mathbf{v}\|} = \frac{-2}{\sqrt{65}\sqrt{14}}.$$

Thus, the angle between the lines is $\theta = \arccos(-2/\sqrt{65}\sqrt{14}) \approx 1.64$. ■

The coordinates of any direction vector of line l are called **direction numbers** of l. If a direction vector of l is a unit vector, its coordinates are called **direction cosines** of l. To see the significance of the direction cosines of a line, suppose that line l has unit direction vector $\mathbf{u} = \langle a, b, c\rangle$. Then the angles α, β, and γ between l and the coordinate axes (see Figure 12.37) are such that

$$\cos\alpha = \mathbf{u}\cdot\mathbf{i} = a,$$
$$\cos\beta = \mathbf{u}\cdot\mathbf{j} = b,$$
$$\cos\gamma = \mathbf{u}\cdot\mathbf{k} = c.$$

That is, the direction cosines of a line are the cosines of the angles between the line and the coordinate axes; these angles are called **direction angles**.

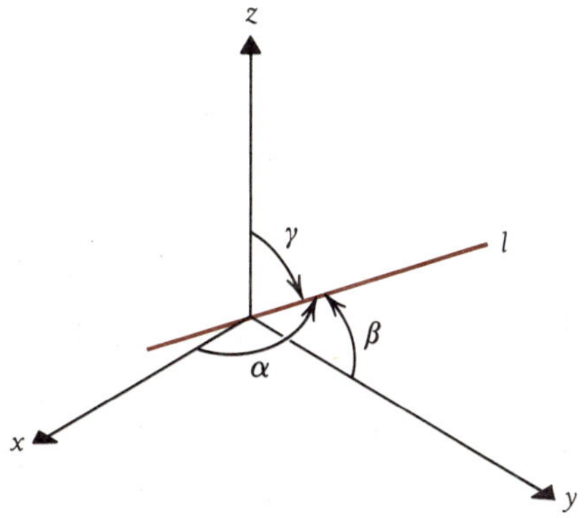

FIGURE 12.37 Direction angles

Because the direction cosines of a line are the coordinates of a unit vector, it follows that the direction angles α, β, and γ of a line satisfy the condition

$$\cos^2\alpha + \cos^2\beta + \cos^2\gamma = 1.$$

■ **Example 12.23** Find direction cosines and direction angles of the line with direction vector $\mathbf{d} = \langle 2, -3, 6\rangle$.

Solution To find direction cosines, we need to find a unit vector in the direction of **d**, the direction vector. Since

$$\|\mathbf{d}\| = \sqrt{4+9+36} = \sqrt{49} = 7,$$

a unit vector in the direction of **d** is

$$\mathbf{u} = \langle \tfrac{2}{7}, -\tfrac{3}{7}, \tfrac{6}{7} \rangle.$$

Thus, the direction cosines are $\frac{2}{7}$, $-\frac{3}{7}$, and $\frac{6}{7}$. It follows that the direction angles are

$$\alpha = \arccos \tfrac{2}{7} \approx 1.28,$$
$$\beta = \arccos(-\tfrac{3}{7}) \approx 2.01,$$
$$\gamma = \arccos \tfrac{6}{7} \approx 0.54.$$

■

Equations of a Line

Let l be a line with direction vector $\mathbf{d} = \langle a, b, c \rangle$ and passing through the point $P_1(x_1, y_1, z_1)$. Only one line has these properties. If $P(x, y, z)$ is any point on the line l, then there is a scalar t such that $\mathbf{v}(\overrightarrow{P_1P}) = t\mathbf{d}$ (see Figure 12.38). All possible points on l are described as t ranges over all possible real numbers. If O is the origin, then

$$\mathbf{v}(\overrightarrow{OP}) = \mathbf{v}(\overrightarrow{OP_1}) + \mathbf{v}(\overrightarrow{P_1P}),$$

or, equivalently,

$$\langle x, y, z \rangle = \langle x_1, y_1, z_1 \rangle + t\langle a, b, c \rangle. \quad (2)$$

This can be rewritten as

$$\langle x, y, z \rangle = \langle x_1 + at, y_1 + bt, z_1 + ct \rangle. \quad (3)$$

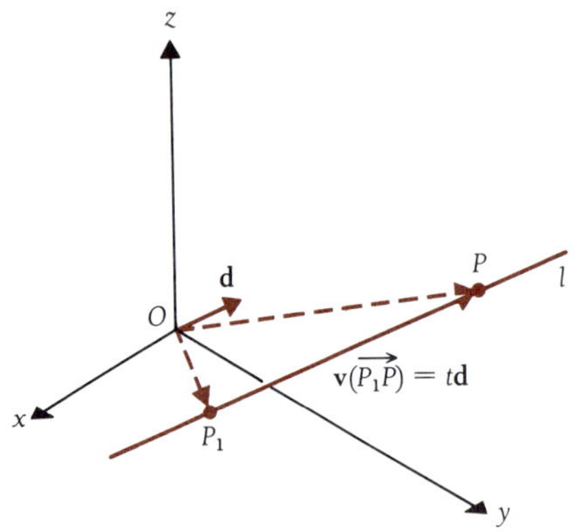

FIGURE 12.38 Finding the equation of a line in space

Each of the equations (2) and (3) is called the **vector equation** of the line l, with $t \in (-\infty, \infty)$. Equation (3) also can be written as a set of parametric equations:

$$\begin{cases} x = x_1 + at \\ y = y_1 + bt \\ z = z_1 + ct, \quad t \in (-\infty, \infty). \end{cases} \tag{4}$$

These equations are called **parametric equations** of l.

Example 12.24 Find vector and parametric equations of each line:

(a) Through point $(7, 1, -3)$ and with direction vector $\mathbf{d} = \langle 2, 0, -3 \rangle$
(b) Through the two points $(4, -1, 5)$ and $(0, 2, 1)$

Solution **(a)** Using equations (3) and (4) directly, we get the vector equation

$$\langle x, y, z \rangle = \langle 7 + 2t, 1, -3 - 3t \rangle$$

and the parametric equations

$$\begin{cases} x = 7 + 2t \\ y = 1 \\ z = -3 - 3t. \end{cases}$$

(b) First, we must compute the direction vector of the line. Using the two points on the line, we find that $\mathbf{d} = \langle 4, -3, 4 \rangle$ is a direction vector. Then, using the point $(4, -1, 5)$ and equations (3) and (4) as above, we get

$$\langle x, y, z \rangle = \langle 4 + 4t, -1 - 3t, 5 + 4t \rangle$$

and

$$\begin{cases} x = 4 + 4t \\ y = -1 - 3t \\ z = 5 + 4t. \end{cases}$$

If we use the point $(0, 2, 1)$ instead, we get

$$\langle x, y, z \rangle = \langle 4t, 2 - 3t, 1 + 4t \rangle$$

and

$$\begin{cases} x = 4t \\ y = 2 - 3t \\ z = 1 + 4t. \end{cases}$$

Here, it is understood that $t \in (-\infty, \infty)$ unless otherwise specified. ■

Example 12.24(b) illustrates the case of a line being specified by two points. In general, if (x_1, y_1, z_1) and (x_2, y_2, z_2) are two points, then the line through them has direction vector

$$\mathbf{d} = \langle x_2 - x_1, y_2 - y_1, z_2 - z_1 \rangle$$

and parametric equations

$$\begin{cases} x = x_1 + (x_2 - x_1)t \\ y = y_1 + (y_2 - y_1)t \\ z = z_1 + (z_2 - z_1)t. \end{cases} \tag{5}$$

Note that for specific values of t in equation (5), we get specific points on the line. For example, if $t = 0$, we get the point (x_1, y_1, z_1); if $t = 1$, we get the point (x_2, y_2, z_2). If $0 < t < 1$, we get a point between these two (see Figure 12.39); if $t > 1$, we get a point beyond (x_2, y_2, z_2); and if $t < 0$, we get a point beyond (x_1, y_1, z_1).

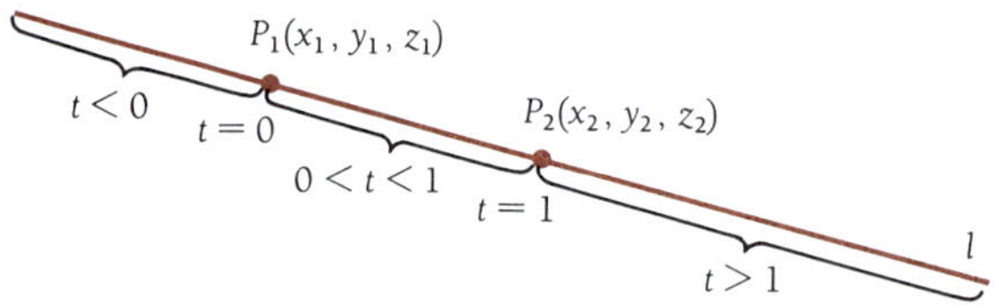

FIGURE 12.39 Various values of the parameter give various points on the line

Example 12.25 Find parametric equations for the line through the two points $(-3, 0, 4)$ and $(2, 7, -1)$.

Solution Setting $(x_1, y_1, z_1) = (-3, 0, 4)$ and $(x_2, y_2, z_2) = (2, 7, -1)$, we find from equations (5) that the parametric equations of the line are

$$\begin{cases} x = -3 + 5t \\ y = 7t \\ z = 4 - 5t. \end{cases}$$

■

Some Special Lines

Thus far, we have tacitly assumed that our lines have not had any particular position relative to the coordinate planes; all that we have said applies to all lines. We now want to consider separately some lines lying in or parallel to the coordinate planes.

To begin, consider the coordinate axes. The x-axis has direction vector $\mathbf{i} = \langle 1, 0, 0 \rangle$, so the vector equation of the x-axis is

$$\langle x, y, z \rangle = \langle t, 0, 0 \rangle.$$

Similarly, the y-axis has equation

$$\langle x, y, z\rangle = \langle 0, t, 0\rangle,$$

and the z-axis has equation

$$\langle x, y, z\rangle = \langle 0, 0, t\rangle.$$

A line parallel to the x-axis (see Figure 12.40) must meet the yz-plane at some point $(0, y_1, z_1)$, and $\mathbf{i}$ is still a direction vector, so a vector equation of such a line is

$$\langle x, y, z\rangle = \langle t, y_1, z_1\rangle.$$

Similar equations can be found for lines parallel to the y-axis and the z-axis (see Figures 12.41 and 12.42).

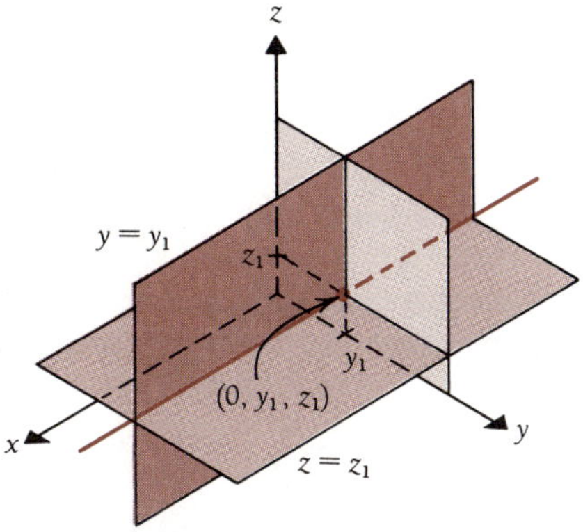

FIGURE 12.40 The line $x = t$, $y = y_1$, $z = z_1$

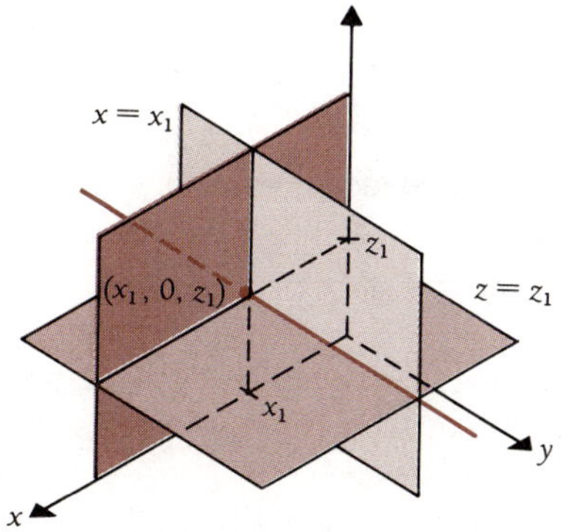

FIGURE 12.41 The line $x = x_1$, $y = t$, $z = z_1$

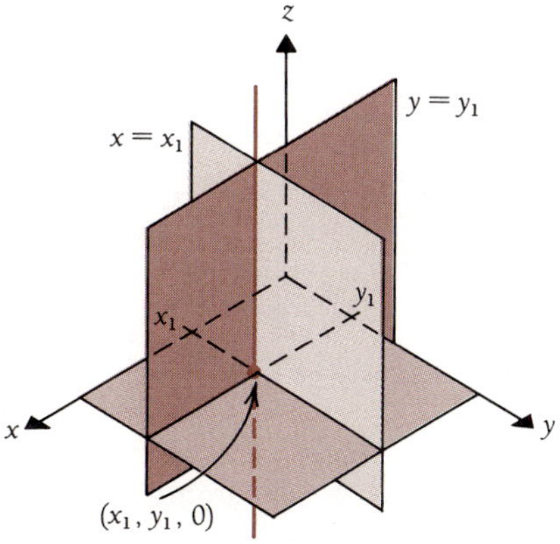

FIGURE 12.42 The line $x = x_1$, $y = y_1$, $z = t$

Now consider a line lying in the xy-plane (see Figure 12.43). Since the angle between the line and the z-axis is $\pi/2$, the third direction cosine of the line must be 0, and therefore the third direction number must always be 0. The same is true of any line parallel to the xy-plane. Conversely, any line with its third direction number 0 must be in or parallel to the xy-plane.

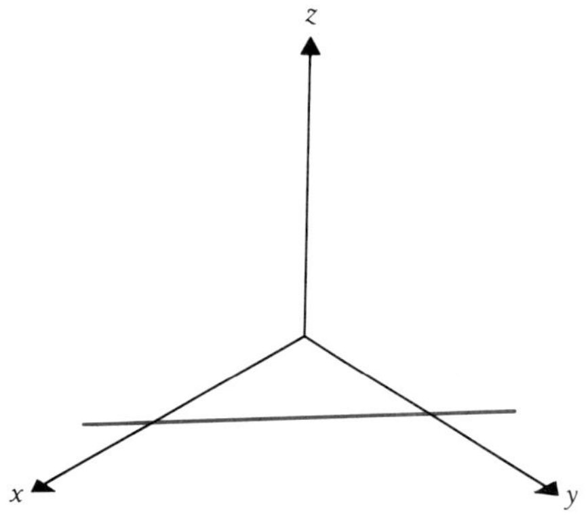

FIGURE 12.43 A line in the xy-plane

Similarly, a line is in or parallel to the xz-plane if and only if its second direction number is 0, and a line is in or parallel to the yz-plane if and only if its first direction number is 0. Can you tell why no line can have all three direction numbers equal to 0?

Symmetric Forms

Let line l have equations (4). If none of the direction numbers a, b, c is 0, then we may solve for t in each equation, getting

$$\frac{x - x_1}{a} = \frac{y - y_1}{b} = \frac{z - z_1}{c} = t.$$

The equations (actually two independent equations)

$$\frac{x - x_1}{a} = \frac{y - y_1}{b} = \frac{z - z_1}{c} \tag{6}$$

are called the **symmetric form** of the equations of l. Equations (6) can be thought of as *point–direction* symmetric equations. The *two-point* symmetric form is obtained from (5):

$$\frac{x - x_1}{x_2 - x_1} = \frac{y - y_1}{y_2 - y_1} = \frac{z - z_1}{z_2 - z_1}. \tag{7}$$

If one of the direction numbers of l is 0, say $a = 0$, then the equations of l can be put into the modified symmetric form

$$x = x_1, \qquad \frac{y - y_1}{b} = \frac{z - z_1}{c}.$$

If two of the direction numbers are 0, we have the special case of a line parallel to a coordinate axis. If, for example, the line is parallel to the z-axis, then the equations take the form

$$x = x_1, \qquad y = y_1.$$

Example 12.26 Find equations in symmetric form for the lines:

(a) Through (4, 0, 1), with direction numbers 3, -1, and 1
(b) Through (2, 6, -5) and (3, 8, 1)
(c) Through (2, 3, 5) and (0, 3, -1)
(d) Through (1, 4, 1) and (1, -2, 1)

Solution **(a)** From equations (6), we get

$$\frac{x-4}{3} = \frac{y}{-1} = \frac{z-1}{1}.$$

(b) From equations (7), we get

$$\frac{x-2}{3-2} = \frac{y-6}{8-6} = \frac{z+5}{1+5},$$

$$\frac{x-2}{1} = \frac{y-6}{2} = \frac{z+5}{6}.$$

(c) The direction numbers of this line are $2 - 0 = 2$, $3 - 3 = 0$, and $5 + 1 = 6$. Hence, the equations are

$$\frac{x-2}{2} = \frac{z-5}{6}, \qquad y = 3.$$

(d) The direction numbers are 0, 6, and 0, so the equations of this line are

$$x = 1, \qquad z = 1.$$

■

Lines and the Cross Product

The properties of the cross product enable us to construct lines orthogonal to each of two lines. If $\mathbf{v}_1$ and $\mathbf{v}_2$ are direction vectors of the lines l_1 and l_2, then any line with direction vector $\mathbf{v}_1 \times \mathbf{v}_2$ is orthogonal to both l_1 and l_2 (see Figure 12.44).

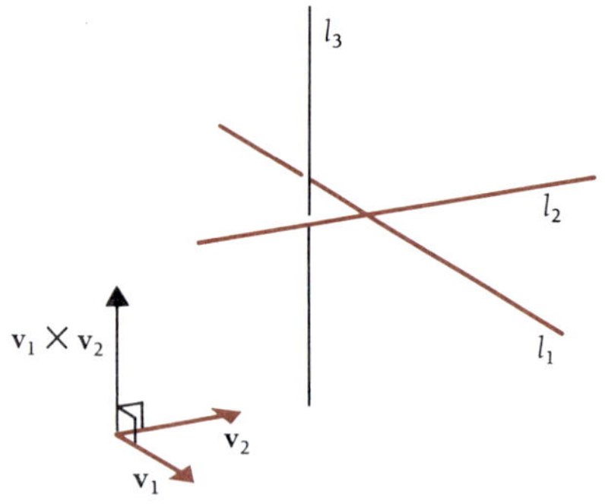

FIGURE 12.44 Line l_3 is orthogonal to both lines l_1 and l_2

Example 12.27 Find equations for the line through the point (4, −1, 3) and orthogonal to both the lines

$$\begin{cases} x = 1 + 3t \\ y = 2t \\ z = -1 - t \end{cases} \quad \text{and} \quad \begin{cases} x = 1 - 2t \\ y = -t \\ z = -1 + 2t. \end{cases}$$

Solution Direction vectors of the two given lines are $\mathbf{u} = \langle 3, 2, -1 \rangle$ and $\mathbf{v} = \langle -2, -1, 2 \rangle$. Hence, the vector $\mathbf{u} \times \mathbf{v} = \langle 3, -4, 1 \rangle$ is orthogonal to both lines and is a direction vector of the desired line. The solution is the line through (4, −1, 3) with direction numbers 3, −4, and 1; its parametric equations are

$$\begin{cases} x = 4 + 3t \\ y = -1 - 4t \\ z = 3 + t. \end{cases}$$

■

Exercises 12.4

Find the direction vector for each line.

1. $x = t, \quad y = 2t, \quad z = -3t$
2. $x = 1 + 4t, \quad y = 1, \quad z = 2 + 2t$
3. The line through (2, 2, 3) and (4, 4, 1)
4. The line through (2, 8, −3) and (−7, 1, 1)

Find the angle between the given lines.

5. $x = t, \quad y = t, \quad z = -t; \quad x = 2 + 2t, \quad y = 3 + 3t, \quad z = -1$
6. $\dfrac{x+1}{4} = \dfrac{y-1}{2} = \dfrac{z-1}{-1}$; the z-axis
7. $x = 4t, \quad y = 1 - 3t, \quad z = 2 + t; \quad x = 2 - 3t, \quad y = 1 - 4t, \quad z = 7$
8. The line through (4, 3, −3) and (2, 0, 1); the line through (4, 3, −3) and (1, 2, −1)

Find equations for the lines:

9. Through $(2, -1, 3)$ and parallel to the y-axis
10. Through $(2, -1, 3)$ and parallel to the z-axis
11. Through $(1, 4, -1)$ and the origin
12. Through $(2, 0, -3)$ and the origin
13. Through the origin, with direction numbers 4, 2, and -1
14. Through the origin, with direction numbers 0, 3, and 5
15. Through $(4, -1, 1)$, with direction numbers $1, -1$, and 1
16. Through $(1, \frac{1}{2}, \frac{1}{3})$, with direction cosines $\frac{1}{3}, -\frac{2}{3}$, and $\frac{2}{3}$
17. Through the origin, parallel to
$$\frac{x+2}{5} = \frac{y-1}{1} = \frac{z+1}{6}$$
18. Through $(0, 1, -1)$, parallel to
$$\frac{x+3}{2} = \frac{y-3}{4} = \frac{z+1}{5}$$
19. Through $(1, -1, 0)$ and $(4, 3, 1)$
20. Through $(6, 3, 1)$ and $(1, -2, 1)$
21. Through $(1, 2, -5)$ and $(4, 2, -5)$
22. Through $(0, 1, 1)$ and $(1, 0, 1)$
23. Through $(4, 3, 2)$ and parallel to $x = -1, y = 2$
24. Through $(4, 3, 2)$ and parallel to $x = 1 + 4t, y = 2, z = 3t$
25. Find direction cosines for the line $x = 4t, y = 3t, z = 2$.
26. Find direction cosines for
$$\frac{x-1}{-1} = \frac{y+2}{2} = \frac{z}{-2}.$$
27. Find the points of intersection of the line $x = 3t, y = -5t, z = 2t$ with the unit sphere $x^2 + y^2 + z^2 = 1$.
28. Explain why there is one and only one line through a given point having a given set of direction numbers.
29. Let l be the line $x = 2t, y = t, z = 6t$. Identify the point or portion of l corresponding to the following values of t:
    (a) $t = 0$ (b) $t = 1$ (c) $t = \frac{1}{2}$
    (d) $t = -1$ (e) $t \in [0, 1]$ (f) $t < 0$
30. Repeat Problem 29 for the line $x = -2t, y = -t, z = -6t$.
31. Repeat Problem 29 for the line $x = 6t, y = 3t, z = 18t$.
32. Repeat Problem 29 for the line $x = x_1 + (x_2 - x_1)t, y = y_1 + (y_2 - y_1)t$, $z = z_1 + (z_2 - z_1)t$.
33. Let lines l_1 and l_2 intersect at the origin O and have direction cosines λ_1, μ_1, ν_1 and λ_2, μ_2, ν_2, respectively. Let $P_1 = (\lambda_1, \mu_1, \nu_1)$ and $P_2 = (\lambda_2, \mu_2, \nu_2)$, and let θ represent the angle P_1OP_2. By applying the law of cosines to triangle P_1OP_2, show that
$$\cos\theta = \lambda_1\lambda_2 + \mu_1\mu_2 + \nu_1\nu_2.$$
34. Show that the angle between lines l_1 and l_2, having direction numbers a_1, b_1, c_1 and a_2, b_2, c_2, respectively, is $\pi/2$ if and only if $a_1a_2 + b_1b_2 + c_1c_2 = 0$.
35. Let Π be a plane in space, and let (x_1, y_1, z_1) be a point in Π. Let a line perpendicular to Π have direction numbers a, b, and c. Show that point $P(x, y, z)$ lies in Π if and only if
$$a(x - x_1) + b(y - y_1) + c(z - z_1) = 0.$$

SECTION 12.5

Planes

Planes are most easily described in terms of vectors that are perpendicular to the planes. A line l is defined to be **perpendicular** to a plane Π in case every line lying in Π is orthogonal to l. If l is perpendicular to plane Π, then any direction vector of l is said to be **normal** to Π.

Equation of a Plane

We can easily obtain the equation of a plane if we know a point in the plane and a vector normal to the plane. Suppose point $P_1(x_1, y_1, z_1)$ lies in the plane Π, and vector $\mathbf{u} = \langle a, b, c\rangle$ is normal to Π. Let $P(x, y, z)$ be an arbitrary point in Π (see Figure 12.45). Then $\mathbf{v}(\overrightarrow{P_1P})$ is orthogonal to $\mathbf{u}$, or

$$\mathbf{u} \cdot \mathbf{v}(\overrightarrow{P_1P}) = 0.$$

Since $\mathbf{v}(\overrightarrow{P_1P}) = \langle x - x_1, y - y_1, z - z_1\rangle$, we then have

$$a(x - x_1) + b(y - y_1) + c(z - z_1) = 0. \tag{1}$$

This is the equation of Π, since it is satisfied by all the points in Π and no others.

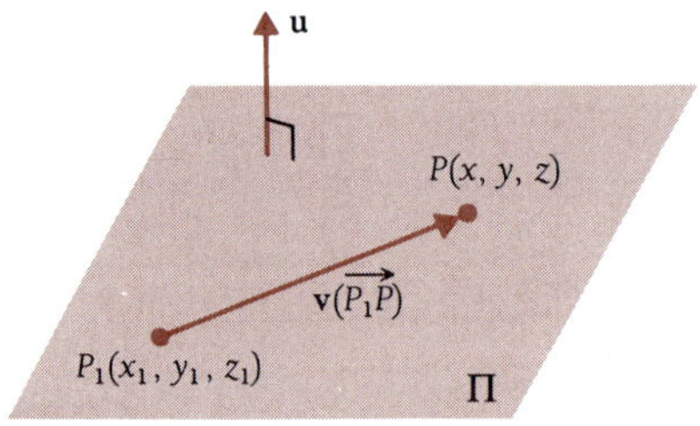

FIGURE 12.45 Finding the equation of a plane

■ **Example 12.28** Find an equation for the plane through the point $(1, 1, -3)$ and with normal vector $\mathbf{v} = \langle 4, -1, -1\rangle$.

Solution Using equation (1), we get

$$4(x - 1) - 1(y - 1) - 1(z + 3) = 0,$$

which simplifies to

$$4x - y - z - 6 = 0. \quad ■$$

■ **Example 12.29** Show that the equation $x + 4y + 3z - 5 = 0$ is the equation of a plane, and find a vector normal to it.

Solution The given equation can be rewritten in many ways, one of which is

$$(x - 1) + 4(y - 1) + 3(z - 0) = 0.$$

This is the equation of the plane through the point (1, 1, 0) and having normal vector $\mathbf{v} = \langle 1, 4, 3 \rangle$. Any other way we rewrite the equation in the same form, we get a different point, but always a normal vector that is a scalar multiple of $\mathbf{v}$. ■

From equation (1) we can conclude that every plane has an equation of the form

$$ax + by + cz + d = 0. \tag{2}$$

Conversely, every equation of this form is the equation of a plane. The coefficients a, b, and c are sometimes called **attitude numbers** of the plane; they are coordinates of a vector normal to the plane and also direction numbers of any line perpendicular to the plane (see Figure 12.46).

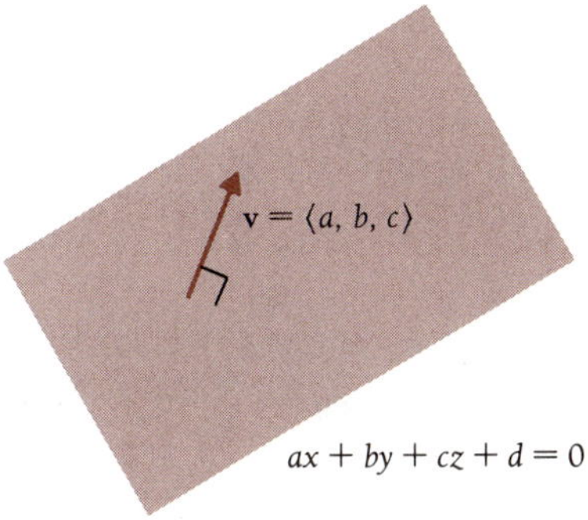

FIGURE 12.46 Equation of a plane with normal vector $\mathbf{v}$

A plane is determined not only by a point and a normal vector, but also by three points not on a line (Figure 12.47), by a line and a point not on it (Figure 12.48), or by two intersecting lines (Figure 12.49). To find the equation of the plane in each case, we begin by finding a normal vector.

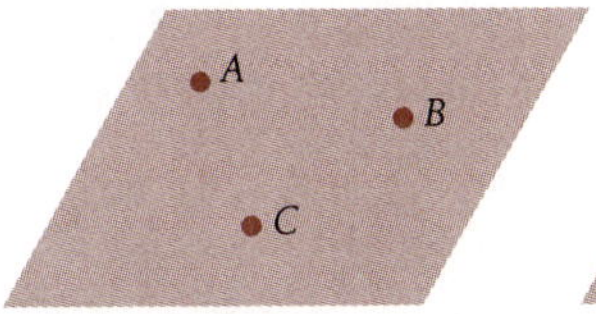

FIGURE 12.47 Three noncollinear points determine a plane

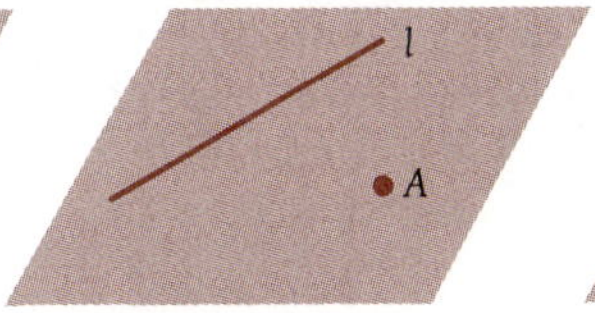

FIGURE 12.48 A line and a point not on the line determine a plane

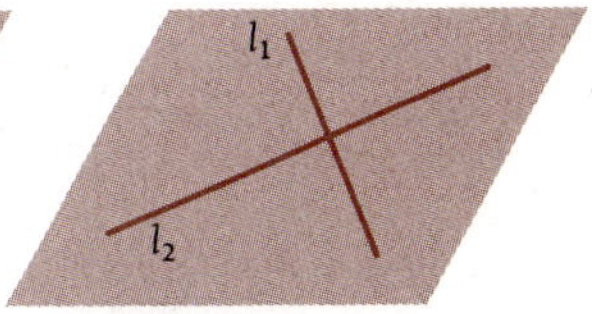

FIGURE 12.49 Two intersecting lines determine a plane

■ **Example 12.30** Find an equation of the plane containing the three points $A(1, 3, -1)$, $B(-2, -1, 1)$, and $C(6, 3, 2)$.

Solution The vectors $\mathbf{v}(\overrightarrow{AB})$ and $\mathbf{v}(\overrightarrow{AC})$ are direction vectors of lines in the plane (see Figure 12.50), so their cross product will be normal to the plane. Let

$$\begin{aligned}\mathbf{u} &= \mathbf{v}(\overrightarrow{AB}) \times \mathbf{v}(\overrightarrow{AC}) \\ &= \langle -3, -4, 2\rangle \times \langle 5, 0, 3\rangle \\ &= \langle -12, 19, 20\rangle.\end{aligned}$$

Using $\mathbf{u}$ and point A, we get the equation

$$-12(x-1) + 19(y-3) + 20(z+1) = 0$$

from equation (1), which simplifies to

$$-12x + 19y + 20z - 25 = 0.$$

To check our work, we can verify that each of the points A, B, and C satisfies this equation.

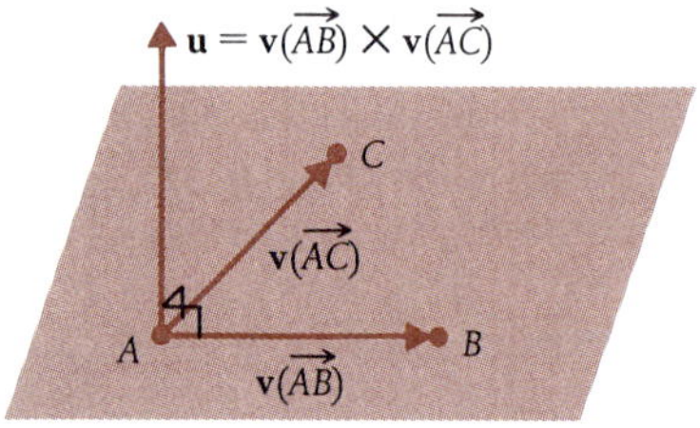

FIGURE 12.50 Using the cross product to find a normal vector ■

■ **Example 12.31** Find the equation of the plane containing the two intersecting lines

$$l_1: \frac{x+2}{7} = \frac{y+2}{5} = \frac{z-9}{9},$$

$$l_2: \frac{x+2}{5} = \frac{y+2}{8} = \frac{z-9}{1}.$$

Solution The vectors $\mathbf{u} = \langle 7, 5, 9\rangle$ and $\mathbf{v} = \langle 5, 8, 1\rangle$ are direction vectors of the given lines (see Figure 12.51), so their cross product $\mathbf{u} \times \mathbf{v} = \langle -67, 38, 31\rangle$ is a normal vector to the plane we seek. Using the common point $(-2, -2, 9)$ of the lines, equation (1) gives us

$$-67(x+2) + 38(y+2) + 31(z-9) = 0,$$

which simplifies to

$$-67x + 38y + 31z - 337 = 0.$$

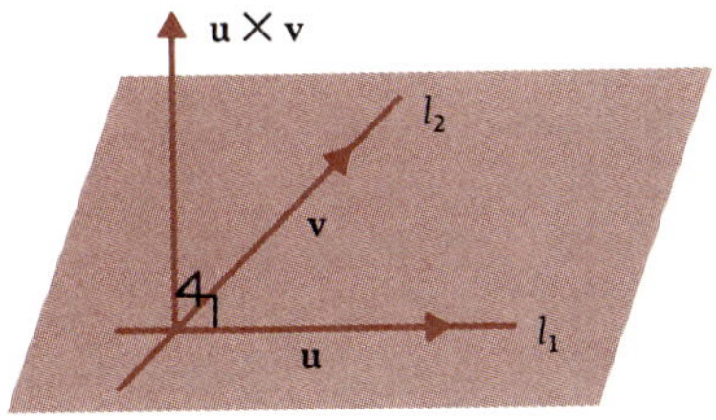

FIGURE 12.51 ■

■ **Example 12.32** Find an equation of the plane containing the point $(-4, 5, 2)$ and the line

$$\begin{cases} x = 1 + t \\ y = t \\ z = 4. \end{cases}$$

Solution The vector $\mathbf{u} = \langle 1, 1, 0 \rangle$ is the direction vector of the given line, and the point $P(1, 0, 4)$ is on that line. Let $Q(-4, 5, 2)$ be the given point. Then the vector $\mathbf{v}(\overrightarrow{PQ}) = \langle -5, 5, -2 \rangle$ is also a direction vector of a line lying in the plane (see Figure 12.52). Therefore, the cross product $\mathbf{u} \times \mathbf{v}(\overrightarrow{PQ}) = \langle -2, 2, 10 \rangle$ is a normal vector to the plane. This gives us the equation

$$-2(x+4) + 2(y-5) + 10(z-2) = 0,$$

which reduces to

$$-x + y + 5z - 19 = 0.$$

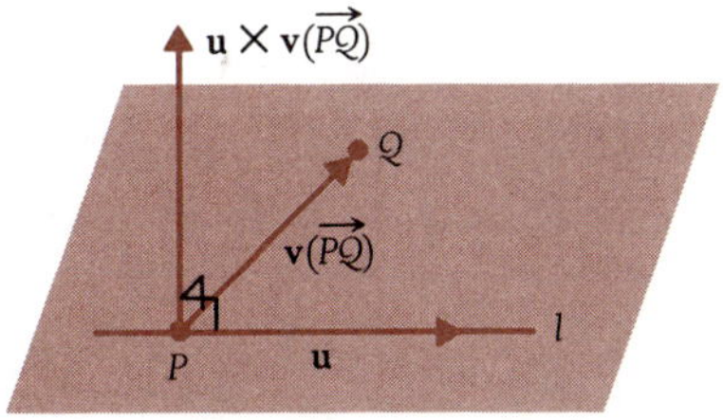

FIGURE 12.52 ■

Intersecting and Parallel Planes

The **angle between two planes** is defined to be the acute angle between their normal vectors (see Figure 12.53, page 782). If the angle between two planes is 0, then the planes are **parallel** (see Figure 12.54). If the angle between two planes is $\pi/2$, then the planes are **perpendicular** (see Figure 12.55).

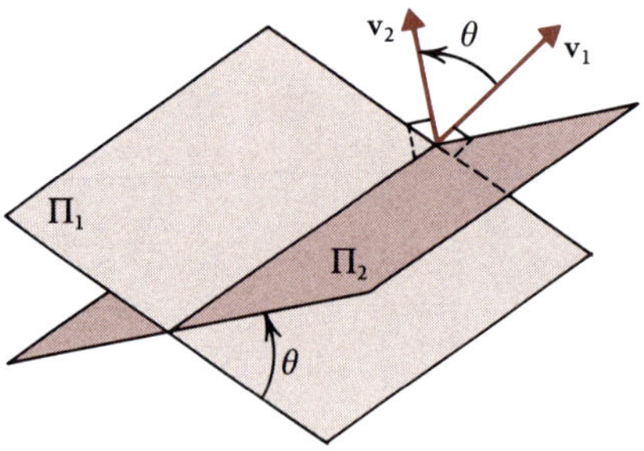

FIGURE 12.53 The angle between two planes

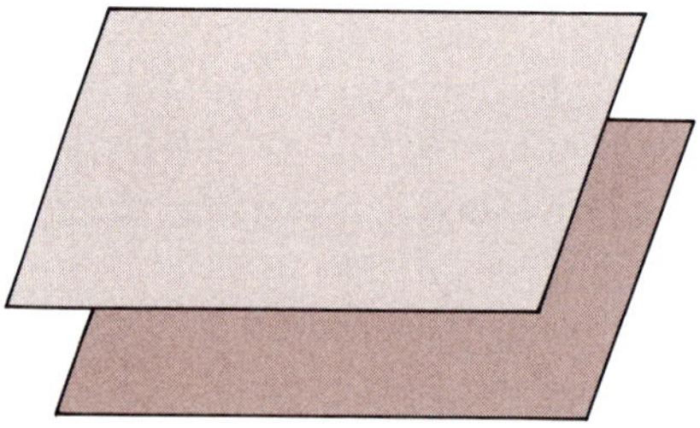

FIGURE 12.54 Parallel planes

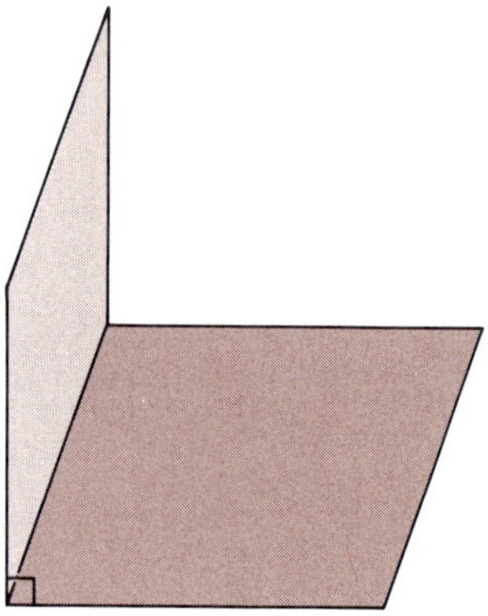

FIGURE 12.55 Perpendicular planes

If two planes are not parallel, then they intersect in a line. To find equations of the line of intersection, we may proceed in two ways, as the next two examples show.

■ **Example 12.33** Find equations of the line of intersection of the two planes

$$3x + 6y - z - 5 = 0 \qquad \text{and} \qquad 9x + y + 5z - 18 = 0.$$

Solution Using algebra, we may solve the equations of the planes simultaneously for x and y in terms of z. From

$$\begin{cases} 3x + 6y = z + 5 \\ 9x + y \;\;= -5z + 18, \end{cases}$$

we get

$$x = -\frac{31}{51}z + \frac{103}{51}, \qquad y = \frac{8}{17}z - \frac{3}{17}.$$

(You may fill in the details of this solution.)

Now we let $z = t$ be the parameter, and we have these equations of the line:

$$\begin{cases} x = \frac{103}{51} - \frac{31}{51}t \\ y = -\frac{3}{17} + \frac{8}{17}t \\ z = t. \end{cases}$$

■

■ **Example 12.34** Find the line of intersection of the planes

$$x + 3y - z = 0 \qquad \text{and} \qquad 2x + 5z = 7.$$

Solution This time we will use vector methods. Let l be the line of intersection. The vectors $\mathbf{u} = \langle 1, 3, -1\rangle$ and $\mathbf{v} = \langle 2, 0, 5\rangle$ are the normals to the planes, so each is orthogonal to l. Thus, $\mathbf{u} \times \mathbf{v} = \langle 15, -7, -6\rangle$ is a direction vector of l (see Figure 12.56). Now all we need is a point on l, which is simply any point in both planes. For example, if we let $y = 0$ and solve the equations of the planes simultaneously, we get $z = x = 1$. Thus, (1, 0, 1) is a point on l, and therefore,

$$\begin{cases} x = 1 + 15t \\ y = -7t \\ z = 1 - 6t \end{cases}$$

are equations for l.

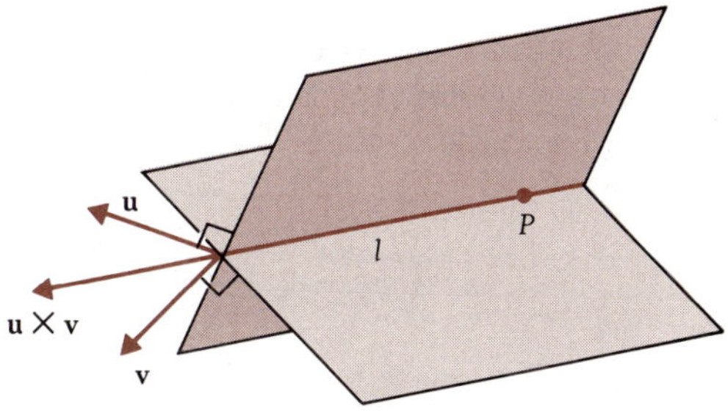

FIGURE 12.56 The line of intersection of two planes

■

As a slightly different kind of problem, consider Example 12.35.

■ **Example 12.35** Find the two parallel planes that contain the two skew lines

$$\frac{x-5}{4} = \frac{y+1}{1} = \frac{z-3}{-2} \qquad \text{and} \qquad \frac{x+1}{2} = \frac{y+1}{3} = \frac{z-1}{5}.$$

Solution The vectors $\mathbf{u} = \langle 4, 1, -2\rangle$ and $\mathbf{v} = \langle 2, 3, 5\rangle$ are direction vectors for the lines. Hence, the cross product $\mathbf{u} \times \mathbf{v} = \langle 11, -24, 10\rangle$ is a vector orthogonal to both lines (see Figure 12.57). Since the two planes we seek are parallel, they have the

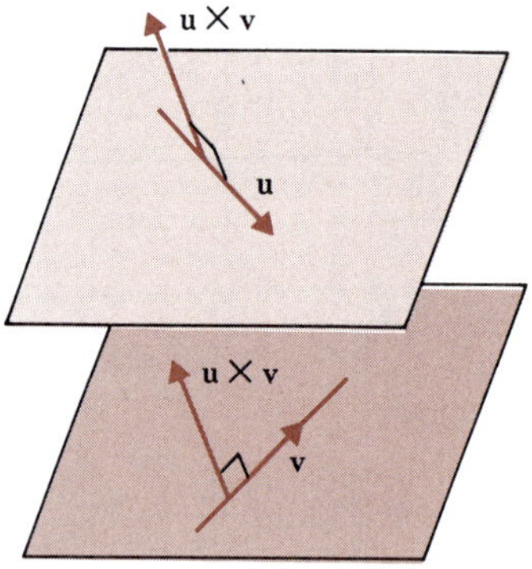

FIGURE 12.57 Parallel planes containing given skew lines

same normal vector; and a vector normal to the two planes must be orthogonal to both lines. Therefore, $\mathbf{u} \times \mathbf{v}$ is normal to the two planes we seek. The equations of the two planes are therefore

$$11(x-5) - 24(y+1) + 10(z-3) = 0,$$
$$11(x+1) - 24(y+1) + 10(z-1) = 0.$$ ■

Distance from a Point to a Plane

In Section 10.1 we found the distance from a point to a line in two dimensions. We can find the distance from a point to a plane in three dimensions in very much the same fashion. We start with the origin.

Let Π be the plane

$$ax + by + cz + d = 0,$$

and let O be the origin. The vector $\mathbf{v} = \langle a, b, c\rangle$ is normal to the plane Π, and therefore the line $x = at, y = bt, z = ct$ passes through the origin and is perpendicular to Π (see Figure 12.58). The intersection of this line with Π is found by substituting the parametric expressions for x, y, and z into the equation of Π and solving for t:

$$a^2t + b^2t + c^2t + d = 0,$$
$$t = \frac{-d}{a^2 + b^2 + c^2}.$$

Thus, the point of intersection is

$$\left(\frac{-ad}{a^2+b^2+c^2}, \frac{-bd}{a^2+b^2+c^2}, \frac{-cd}{a^2+b^2+c^2}\right).$$

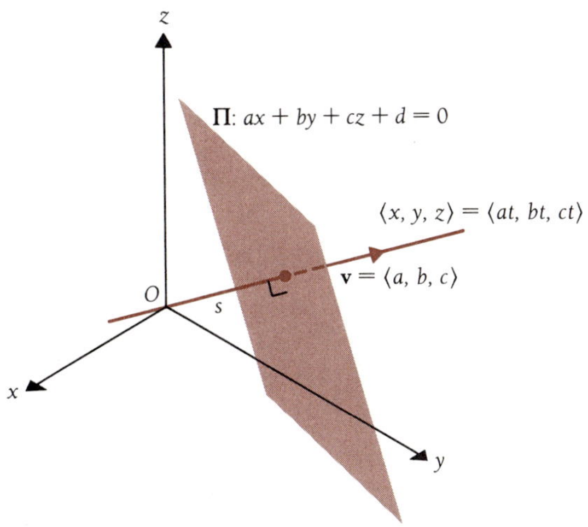

FIGURE 12.58 The distance from the origin to a plane

The distance s from the origin to this point is then

$$s = \frac{|d|}{\sqrt{a^2 + b^2 + c^2}}. \tag{3}$$

(The computations are left as an exercise; see Problem 45.) That is, equation (3) is the formula for the distance from the origin to the plane $ax + by + cz + d = 0$.

Example 12.36 Find the distance from the origin to the plane $2x - 3y + 2z + 8 = 0$.

Solution Using equation (3), we find that the distance s from the origin to this plane is

$$s = \frac{|8|}{\sqrt{2^2 + (-3)^2 + 2^2}} = \frac{8}{\sqrt{17}}.$$ ■

Now let P be the point (x_1, y_1, z_1); we want to find the distance from the point P to the plane Π: $ax + by + cz + d = 0$. Using the idea of translation of axes that was used in Section 10.1, we find the distance s to be

$$s = \frac{|ax_1 + by_1 + cz_1 + d|}{\sqrt{a^2 + b^2 + c^2}}. \tag{4}$$

Example 12.37 Find the distance from the point $(4, -2, 1)$ to the plane $x - 3y + 2z - 5 = 0$.

Solution By equation (4), the distance s is

$$s = \frac{|1(4) + (-3)(-2) + 2(1) + (-5)|}{\sqrt{1^2 + (-3)^2 + 2^2}} = \frac{7}{\sqrt{14}}.$$ ■

Formula (4) also can be used to find the distance between two skew lines. After a little thought, it is apparent that there must be two points, one on each of two skew lines, that are a minimum distance apart. The distance between them is the same as the distance between parallel planes that contain the two lines. Thus, to find the distance between skew lines, we find the distance from a point on one of them to the parallel plane containing the other.

Example 12.38 Find the distance between the skew lines in Example 12.35.

Solution From Example 12.35, we recall that the parallel planes containing the two skew lines are

$$11(x-5)-24(y+1)+10(z-3)=0,$$
$$11(x+1)-24(y+1)+10(z-1)=0.$$

A point in the second plane (in fact, on the second line) is $(-1, -1, 1)$, and its distance s from the first plane, $11x-24y+10z-109=0$, is

$$s=\frac{|11(-1)+(-24)(-1)+10(1)-109|}{\sqrt{11^2+(-24)^2+10^2}}=\frac{86}{\sqrt{797}}.$$

This is the distance between the skew lines. ■

Exercises 12.5

Find an equation of the given plane.

1. With normal vector $\langle -1, 5, 8\rangle$ and through $(2, 0, 3)$
2. With normal vector $\langle -2, 1, 0\rangle$ and through $(0, 3, 5)$
3. With normal vector $7\mathbf{i}+\frac{1}{2}\mathbf{j}+2\mathbf{k}$ and through $(1, \sqrt{3}, -1)$
4. With normal vector $3\mathbf{k}$ and through $(1, 2, -1)$
5. Parallel to $2x-y+z=5$ and through $(0, 3, 4)$
6. Perpendicular to the line
$$\frac{x+6}{2}=\frac{y+1}{1}=\frac{z}{3}$$
at $(-6, -1, 0)$
7. Containing the points $(2, -1, 0)$, $(4, 6, 3)$, and $(-1, -1, -1)$
8. Containing the points $(3, 1, 0)$, $(1, 0, 3)$, and $(0, 3, 1)$
9. Containing the points $(0, 0, 0)$, $(4, 1, -2)$, and $(-3, 0, 8)$
10. Containing the points $(-1, 0, 0)$, $(0, 4, 0)$, and $(0, 0, 3)$
11. Containing the lines
$$\frac{x-2}{5}=\frac{y+3}{7}=\frac{z-1}{2} \quad\text{and}\quad \frac{x-2}{3}=\frac{y+3}{1}, \quad z=1$$
12. Containing the lines
$$\frac{x+6}{2}=\frac{z}{3}, \quad y=-1 \quad\text{and}\quad \frac{x+6}{4}=\frac{y+1}{1}=\frac{z}{3}$$

13. Containing the lines $x = 1 - 2t$, $y = 3 + 5t$, $z = -1 + 6t$ and $x = -2t$, $y = 5t$, $z = 1 + 6t$

14. Containing the lines $x = -4 - t$, $y = -3t$, $z = -1 + 2t$ and $x = -5 - t$, $y = -2 - 3t$, $z = 2t$

15. Containing the line $x = 3t$, $y = 4 - t$, $z = 5$ and the point $(7, -1, 2)$

16. Containing the line

$$\frac{x+2}{5} = \frac{y-1}{4} = \frac{z+6}{7}$$

and the origin

17. Containing the line $x = 3 + t$, $y = 5 - 3t$, $z = 2 + 2t$ and parallel to:
(a) The x-axis (b) The y-axis (c) The z-axis

18. Containing the points $(-1, 1, 3)$ and $(2, 0, 2)$ and parallel to:
(a) The x-axis (b) The y-axis (c) The z-axis

Find equations of the given line.

19. Common to $x + y - 3z = 2$ and $2x - y + z = 9$

20. Common to $3x + 8y + z = 6$ and $-2x + 5y - 4z = 7$

21. Common to $x - 5y + z = -4$ and:
(a) $x = 0$ (b) $y = 0$ (c) $z = 0$

22. Common to $ax + by + cz = d$ and:
(a) $x = 0$ (b) $y = 0$ (c) $z = 0$

23. Common to $2x + 3z = 0$ and $x + y = 1$

24. Common to $2x + 2y + 3z = 4$ and $7x - 8y + 2z = 3$

25. Perpendicular to $x - 3y - z = 4$ at $(2, 1, -5)$

26. Perpendicular to $12x + 8y + 6z = 24$ and through the origin

27. Perpendicular to $x - 3y + 4z + 2 = 0$ and passing through the origin

28. Parallel to both $x + y = 0$ and $3x - 4y + z = 1$ and through $(0, 0, 2)$

29. Parallel to both of the planes $x - y = 1$ and $2x + 4z + 1 = 0$ and passing through $(1, 0, 0)$

30. Parallel to both $3x + z = 4$ and $y + 2z = -5$ and through $(1, 1, 0)$

31. Through the origin and perpendicular to the line $x = 4 + t$, $y = 5 + t$, $z = -t$

32. Perpendicular to both the lines $x = 2t$, $y = -1 - t$, $z = 7 + 2t$ and $x = 4$, $y = 1 + 4t$, $z = 3t$

Find equations for parallel planes containing the given skew lines.

33. $x = y = -z$ and $\dfrac{x+2}{2} = \dfrac{y+3}{3}$, $z = -1$

34. $x = 2 + 5t$, $y = 1 + 4t$, $z = 6 + 7t$ and $x = 3t$, $y = 5t$, $z = 3 + 3t$

Find the angle between the given planes.

35. $x + y + 2z = 1$ and $3x - y - z = 0$

36. $4x + 5y + z = 9$ and $2x - y - 3z = 12$

Find the point at which the given line meets the given plane.

37. $x = 3 + 5t, \quad y = -1 - t, \quad z = 2t; \quad 3x - y - z = 5$

38. $\dfrac{x+1}{6} = \dfrac{y-2}{2} = \dfrac{z+7}{1}; \quad 4x + 5z = 9$

39. The intersections of a plane with the coordinate axes are called the **intercepts** of the plane. Show that the plane with intercepts $(a, 0, 0)$, $(0, b, 0)$, and $(0, 0, c)$ has equation

$$\frac{x}{a} + \frac{y}{b} + \frac{z}{c} = 1.$$

40. Find the intercepts of $Ax + By + Cz = D$.

41. Show that the equation of the plane containing points (x_1, y_1, z_1), (x_2, y_2, z_2), and (x_3, y_3, z_3) can be written in the determinant form

$$\begin{vmatrix} x & y & z & 1 \\ x_1 & y_1 & z_1 & 1 \\ x_2 & y_2 & z_2 & 1 \\ x_3 & y_3 & z_3 & 1 \end{vmatrix} = 0.$$

42. Show that the line $x = x_1 + at$, $y = y_1 + bt$, $z = z_1 + ct$ lies in the plane $Ax + By + Cz = D$ if and only if $aA + bB + cC = 0$ and $Ax_1 + By_1 + Cz_1 = D$.

43. Show that the line $x = x_1 + at$, $y = y_1 + bt$, $z = z_1 + ct$ is parallel to plane $Ax + By + Cz = D$ if and only if $aA + bB + cC = 0$ and $Ax_1 + By_1 + Cz_1 \neq D$.

44. Show that the lines $x = x_1 + a_1t$, $y = y_1 + b_1t$, $z = z_1 + c_1t$ and $x = x_2 + a_2t$, $y = y_2 + b_2t$, $z = z_2 + c_2t$ lie in the same plane if and only if

$$\begin{vmatrix} a_1 & a_2 & x_1 - x_2 \\ b_1 & b_2 & y_1 - y_2 \\ c_1 & c_2 & z_1 - z_2 \end{vmatrix} = 0.$$

45. Show that the distance s between the origin and the plane $ax + by + cz + d = 0$ is

$$s = \frac{|d|}{\sqrt{a^2 + b^2 + c^2}}$$

by carrying out the computations omitted in the text.

46. Show that the distance s between the point (x_1, y_1, z_1) and the plane $ax + by + cz + d = 0$ is

$$s = \frac{|ax_1 + by_1 + cz_1 + d|}{\sqrt{a^2 + b^2 + c^2}}.$$

47. Show that the distance s between the parallel planes $ax + by + cz + d_1 = 0$ and $ax + by + cz + d_2 = 0$ is

$$s = \frac{|d_1 - d_2|}{\sqrt{a^2 + b^2 + c^2}}.$$

48. Find a formula for the distance between the skew lines $x = x_1 + a_1t$, $y = y_1 + b_1t$, $z = z_1 + c_1t$ and $x = x_2 + a_2t$, $y = y_2 + b_2t$, $z = z_2 + c_2t$.

49. The **angle between a line l and a plane Π** is defined to be the complement of the angle between the direction vector of l and the normal vector of Π (see the figure). Show that the angle θ between the line $x = x_1 + at$, $y = y_1 + bt$, $z = z_1 + ct$ and the plane $Ax + By + Cz + D = 0$ satisfies

$$\sin\theta = \frac{aA + bB + cC}{\sqrt{a^2 + b^2 + c^2}\sqrt{A^2 + B^2 + C^2}}.$$

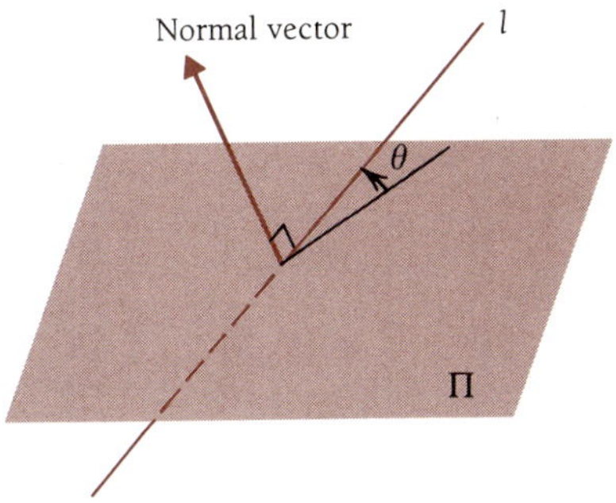

SECTION 12.6

Surfaces

The graph of a single equation in x, y, and z is called a **surface** in three dimensions. In this section, we will study briefly some of the more common or useful surfaces.

Planes

A linear equation in x, y, and z has as its graph a plane. We have already studied planes to a great extent, and need say no more about them here.

Cylinders

A **cylinder** is a surface generated by moving a line in space without changing the direction of the line. That is, a cylinder consists of the points lying on a family of parallel lines. Each of the parallel lines is called a **generator** of the cylinder.

For example, as shown on the next page, a **right circular cylinder** can be described as the set of lines perpendicular to a certain plane and meeting a circle lying in that plane (Figure 12.59). An **elliptic cylinder** is a similar surface, in which an ellipse replaces the circle (Figure 12.60). **Parabolic** and **hyperbolic cylinders** are similarly defined (Figures 12.61 and 12.62).

A cylinder can be based on any plane curve. Simply pick one line not in the plane and meeting the curve, and take as the generators of the cylinder all the lines that meet the curve and are parallel to the line first picked.

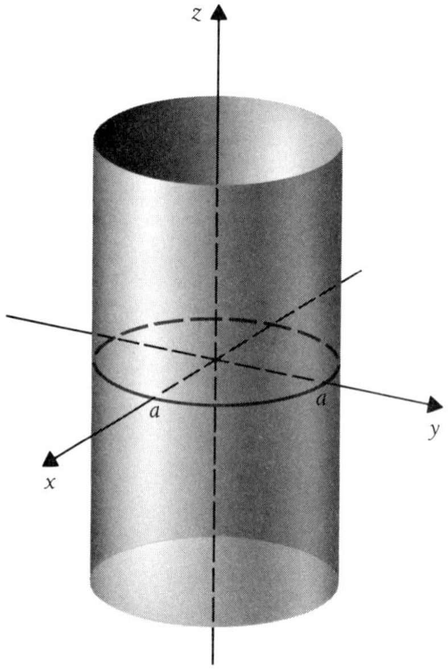

FIGURE 12.59 The right circular cylinder $x^2 + y^2 = a^2$

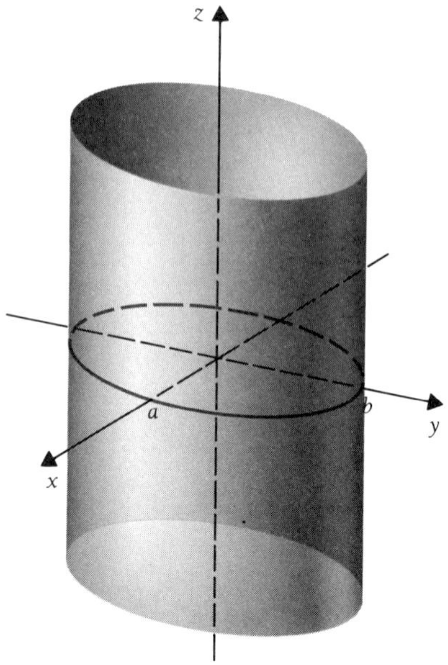

FIGURE 12.60 The elliptic cylinder $\frac{x^2}{a^2} + \frac{y^2}{b^2} = 1$

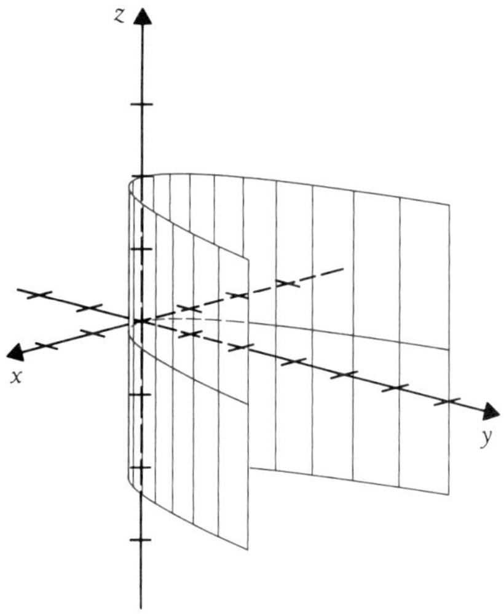

FIGURE 12.61 The parabolic cylinder $y = x^2$

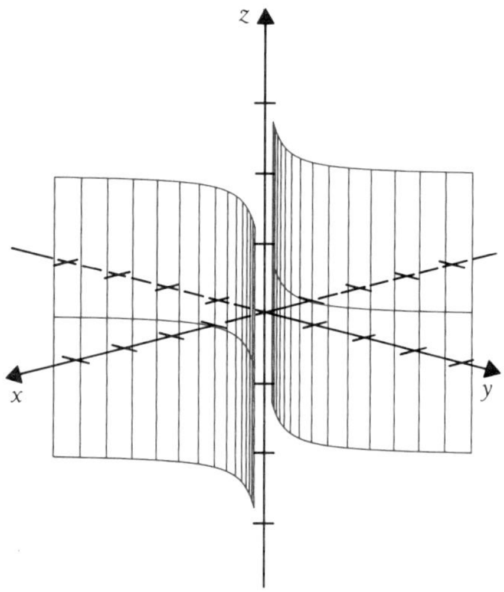

FIGURE 12.62 The hyperbolic cylinder $x^2 - y^2 = 1$

Some equations in x, y, and z can be easily recognized as equations of cylinders. If one of the variables x, y, or z is missing from the equation, then the graph is a cylinder. For example, if z does not occur and (x_1, y_1, z_1) is a point on the graph, then (x_1, y_1, z) is on the graph for any z. That is, the line $x = x_1, y = y_1$ is contained in the graph, and the graph consists of lines parallel to the z-axis. Similarly, if x is missing, the graph consists of lines parallel to the x-axis; and if y is missing, the graph consists of lines parallel to the y-axis.

Example 12.39 Graph the cylinders:

(a) $z = \sin y$ (b) $z = e^y$ (c) $z + x^3 = 0$

Solution (a) The cylinder $z = \sin y$ is the wavy sheet pictured in Figure 12.63.

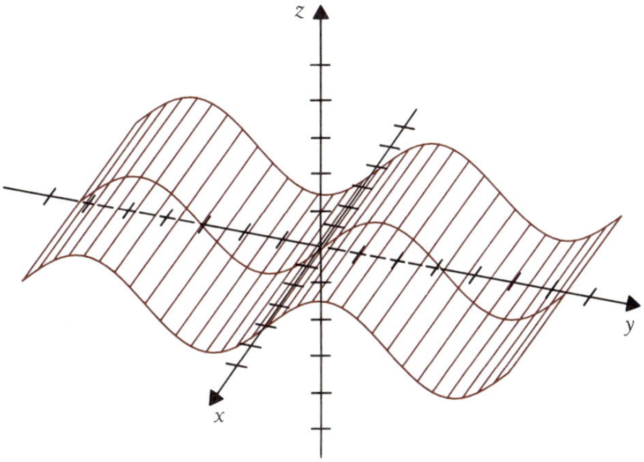

FIGURE 12.63 The cylinder $z = \sin y$

(b) $z = e^y$ is the exponential cylinder shown in Figure 12.64 (page 792). Note that the surface is always above the xy-plane.
(c) The graph of $z + x^3 = 0$ is the cubic cylinder shown in Figure 12.65 (page 792). Note that the scale on the z-axis is compressed, making the graph appear more flattened than it actually is. ■

Surfaces of Revolution

Recall that a surface of revolution is a surface generated by revolving a curve in space about an axis. Our study of surfaces of revolution in Chapter 8 should have made you familiar with many examples. The key feature of a surface of revolution is that plane sections perpendicular to the axis of revolution are circles. This fact enables us to identify a surface of revolution from its equation at times.

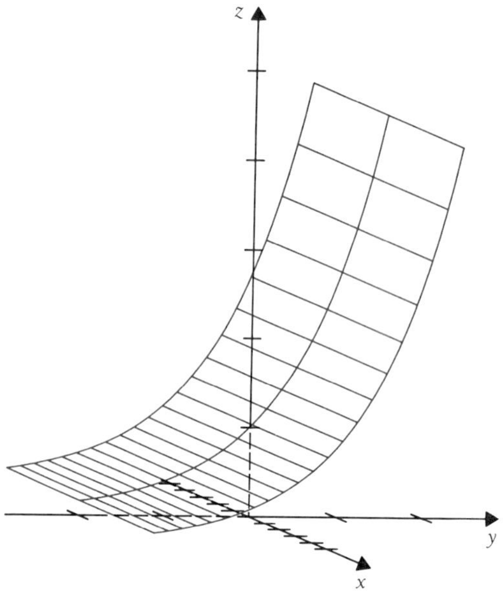

FIGURE 12.64 The cylinder $z = e^y$

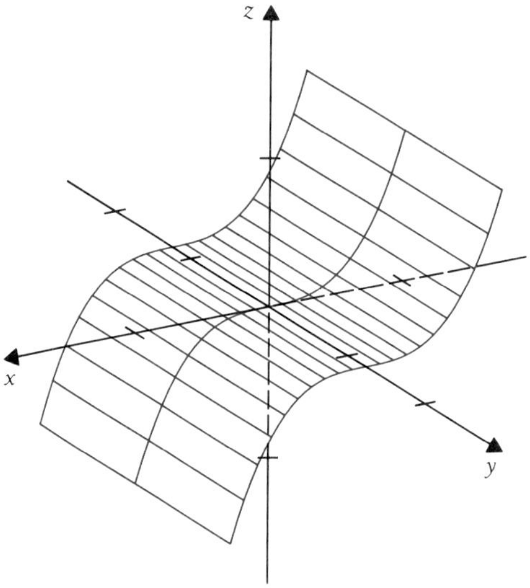

FIGURE 12.65 The cylinder $z + x^3 = 0$

■ **Example 12.40** Describe the surface $x^2 + y^2 - \cos z = 2$.

Solution Any plane $z = k$ meets the surface in the circle $x^2 + y^2 = 2 + \cos k$. The surface is generated by revolving the curve $x = \sqrt{2 + \cos z}$ (in the xz-plane) about the z-axis to obtain a spindle, as shown in Figure 12.66.

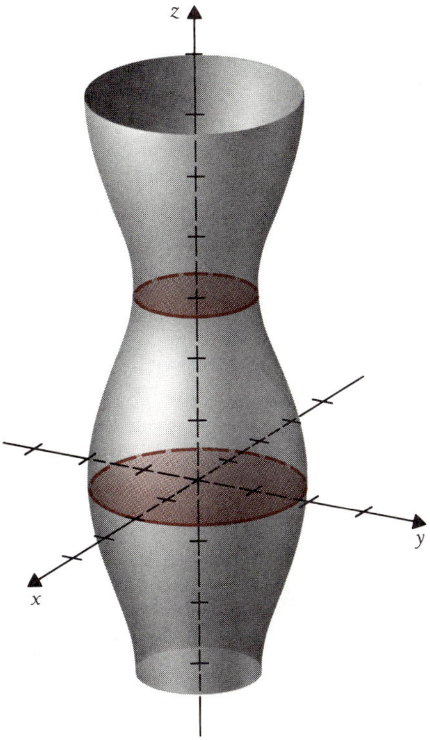

FIGURE 12.66 The surface of revolution $x^2 + y^2 - \cos z = 2$ ■

Quadric Surfaces

A quadratic equation in x, y, and z has as its graph a **quadric surface**. As with quadratic curves in the plane, there are a number of singular cases, but we will consider only the nonsingular quadric surfaces.

When working with conic sections, we tried to choose the coordinate system so that the conic had an equation in standard form. We will do the same for quadric surfaces and discuss only the "nicest" forms of their equations.

First consider a quadratic equation of the form

$$Ax^2 + By^2 + Cz^2 = D. \tag{1}$$

If one of the variables in equation (1) is missing, we have a cylinder, as already discussed.

If all the coefficients in equation (1) have the same sign, the equation can be written in the form

$$\frac{x^2}{a^2} + \frac{y^2}{b^2} + \frac{z^2}{c^2} = 1. \tag{2}$$

The graph of this equation is called an **ellipsoid**. The intersection of the ellipsoid with the plane $z = k$ is an ellipse

$$\frac{x^2}{a^2} + \frac{y^2}{b^2} = 1 - \frac{k^2}{c^2}$$

for $|k| < c$, a point for $|k| = c$, and empty if $|k| > c$. The same reasoning is true for planes $x = k$ or $y = k$. Thus, the numbers a, b, and c are the semiaxes of the ellipsoid (Figure 12.67).

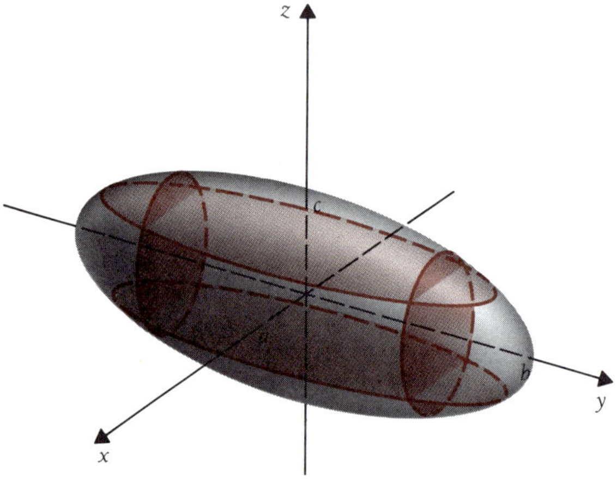

FIGURE 12.67 The ellipsoid $\frac{x^2}{a^2} + \frac{y^2}{b^2} + \frac{z^2}{c^2} = 1$

If two of the numbers a, b, c in equation (2) are equal, the ellipsoid is a surface of revolution and is called a **spheroid**. If the third number is greater than the other two, the spheroid is **prolate** (Figure 12.68); if the third number is less, the spheroid is **oblate** (Figure 12.69). If the three numbers a, b, and c in equation (2) are equal, the ellipsoid is a **sphere** (Figure 12.70).

If the coefficients in equation (1) have different signs, there are two possibilities. One case is the **hyperboloid of one sheet** with equation of the form

$$\frac{x^2}{a^2} + \frac{y^2}{b^2} - \frac{z^2}{c^2} = 1, \qquad (3)$$

as shown in Figure 12.71 on page 796. Sections by planes $z = k$ are ellipses, and sections by planes $x = k$ or $y = k$ are hyperbolas or pairs of intersecting lines.

If $a = b$ in equation (3), the surface is a **hyperboloid of revolution** obtained by revolving a hyperbola about its conjugate axis (Figure 12.72, page 796).

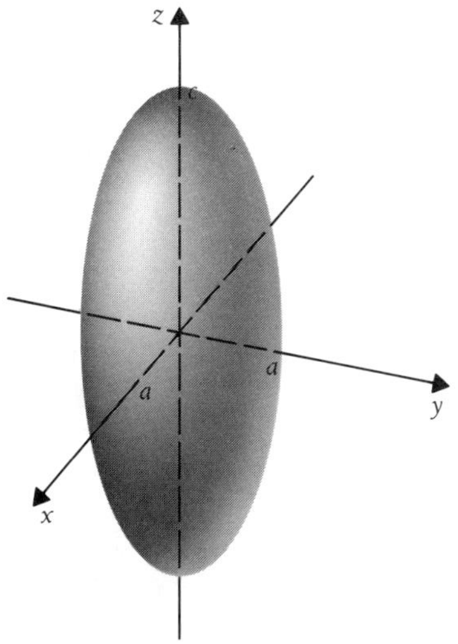

FIGURE 12.68
The prolate spheroid
$\frac{x^2}{a^2}+\frac{y^2}{a^2}+\frac{z^2}{c^2}=1\ (a<c)$

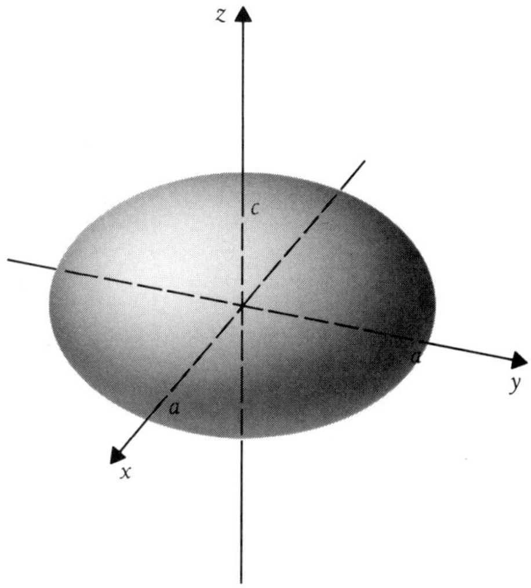

FIGURE 12.69
The oblate spheroid
$\frac{x^2}{a^2}+\frac{y^2}{a^2}+\frac{z^2}{c^2}=1\ (a>c)$

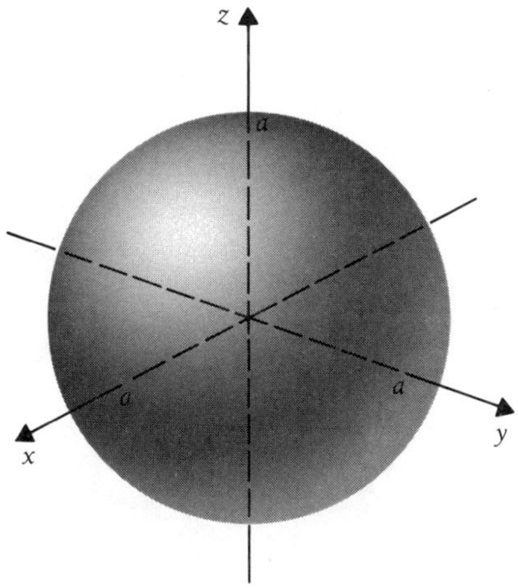

FIGURE 12.70 The sphere
$x^2+y^2+z^2=a^2$

The other case, when the coefficients in equation (1) differ in sign, is the **hyperboloid of two sheets**, with equation of the form

$$\frac{x^2}{a^2}+\frac{y^2}{b^2}-\frac{z^2}{c^2}=-1, \tag{4}$$

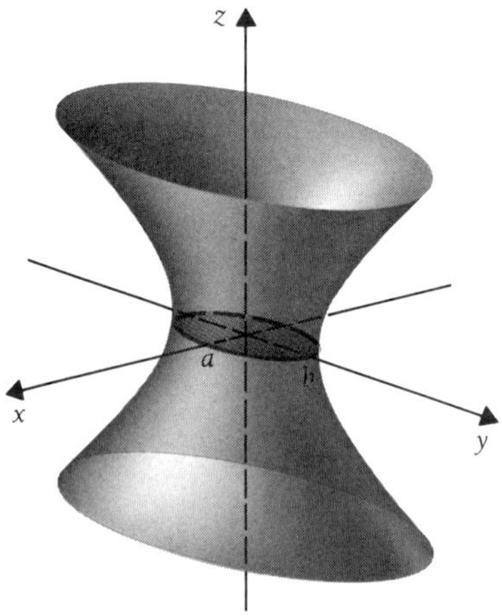

FIGURE 12.71 The hyperboloid of one sheet

$$\frac{x^2}{a^2}+\frac{y^2}{b^2}-\frac{z^2}{c^2}=1$$

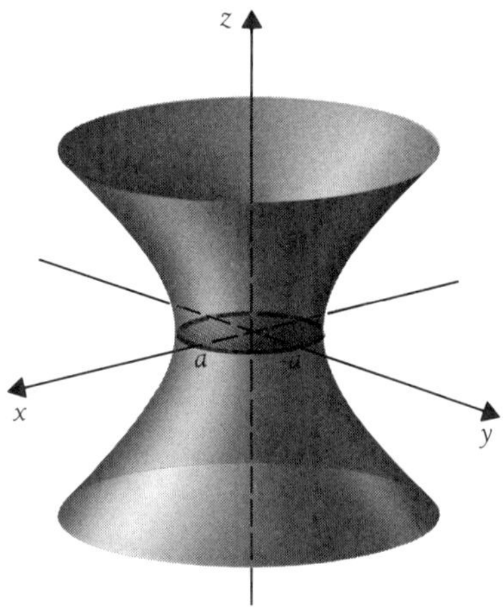

FIGURE 12.72 The circular hyperboloid

$$\frac{x^2}{a^2}+\frac{y^2}{a^2}-\frac{z^2}{c^2}=1$$

as shown in Figure 12.73. Sections by planes $z=k$ are ellipses when $|k|>c$, points when $|k|=c$, and empty when $|k|<c$. Sections by planes $x=k$ or $y=k$ are hyperbolas.

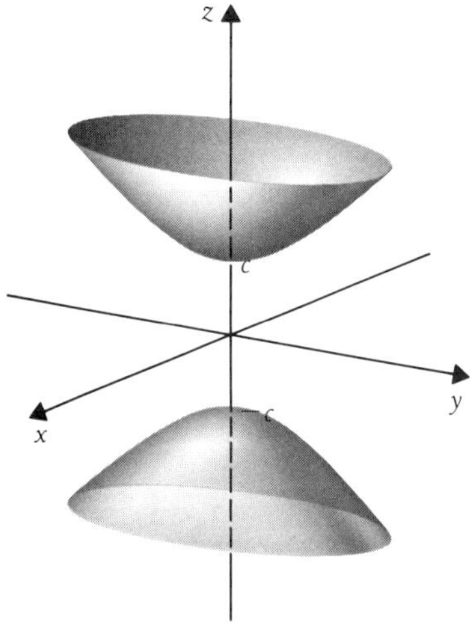

FIGURE 12.73 The hyperboloid of two sheets

$$\frac{x^2}{a^2}+\frac{y^2}{b^2}-\frac{z^2}{c^2}=-1$$

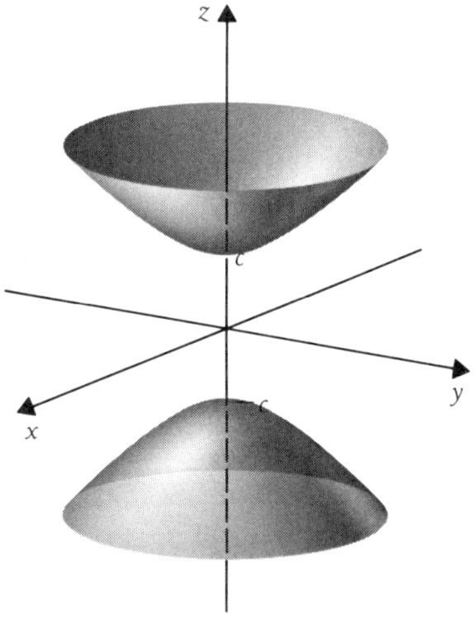

FIGURE 12.74 The circular hyperboloid of two sheets

$$\frac{x^2}{a^2}+\frac{y^2}{a^2}-\frac{z^2}{c^2}=-1$$

If $a = b$ in equation (4), we have the **hyperboloid of revolution** generated by revolving a hyperbola about its transverse axis (Figure 12.74).

If the constant term in equation (1) is 0, the only nonsingular case is the **elliptic cone**

$$\frac{x^2}{a^2} + \frac{y^2}{b^2} - \frac{z^2}{c^2} = 0, \tag{5}$$

which is graphed in Figure 12.75. Horizontal plane sections are either ellipses or a single point. Vertical plane sections are hyperbolas or pairs of lines. If $a = b$ in equation (5), we have a circular cone.

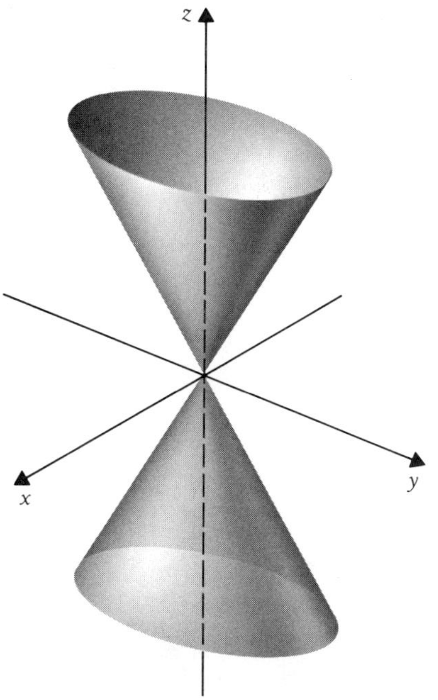

FIGURE 12.75 The elliptic cone $\frac{x^2}{a^2} + \frac{y^2}{b^2} - \frac{z^2}{c^2} = 0$

Next, consider a quadratic equation of the form

$$Ax^2 + By^2 = Cz. \tag{6}$$

If A or B is 0 in equation (6), the surface is a parabolic cylinder.

If A and B in equation (6) have the same sign, the surface is an **elliptic paraboloid,** with equation of the form

$$z = \frac{x^2}{a^2} + \frac{y^2}{b^2}. \tag{7}$$

This surface is shown in Figure 12.76. Sections by planes $z = k$ are ellipses if $k > 0$, a point if $k = 0$, and empty if $k < 0$. Sections by vertical planes are parabolas. The case $a = b$ in equation (7) yields a **circular paraboloid** generated by revolving a parabola about its axis (Figure 12.77).

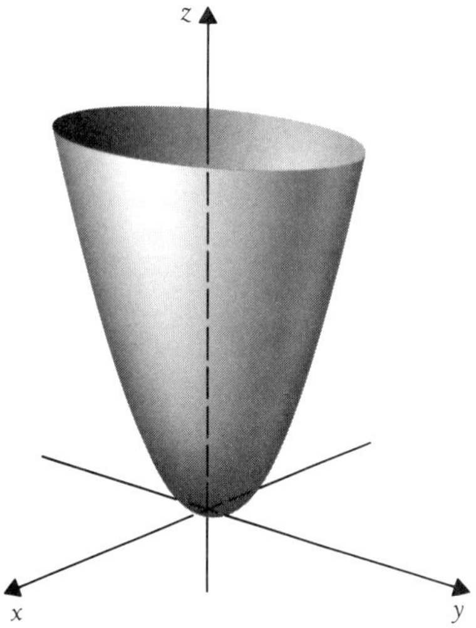

FIGURE 12.76 The elliptic paraboloid $z = \frac{x^2}{a^2} + \frac{y^2}{b^2}$

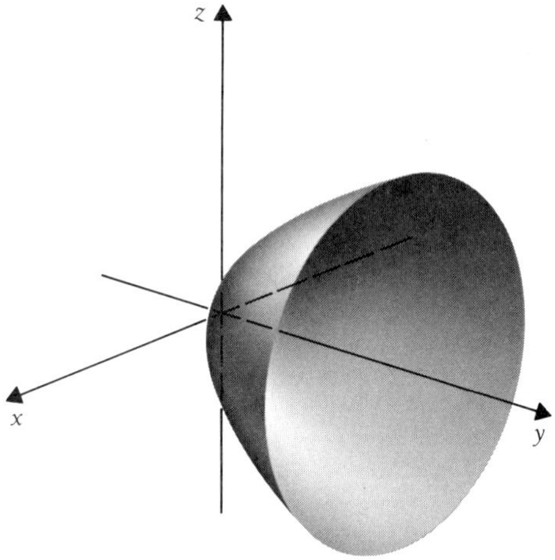

FIGURE 12.77 The circular paraboloid $x^2 + z^2 = a^2y$

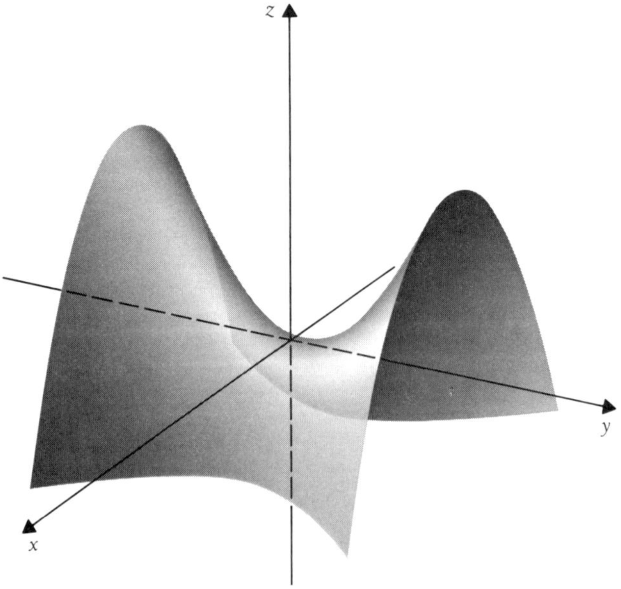

FIGURE 12.78 The hyperbolic paraboloid $z = \frac{x^2}{a^2} - \frac{y^2}{b^2}$

If A and B in equation (6) have different signs, the surface is a **hyperbolic paraboloid**, with equation of the form

$$z = \frac{x^2}{a^2} - \frac{y^2}{b^2}. \tag{8}$$

This surface is shown in Figure 12.78. Sections by planes $z = k$ are hyperbolas if $k \neq 0$ and a pair of intersecting lines if $k = 0$. Sections by planes $x = k$ are parabolas opening upward, and sections by planes $y = k$ are parabolas opening downward. The hyperbolic paraboloid is also called a **saddle surface**; the origin in Figure 12.78 is called a **saddle point.**

All other quadratic equations in three variables have graphs of the above types or are singular and represent planes, points, or "imaginary" graphs.

■ **Example 12.41** Name the quadric surfaces:

(a) $4x^2 - 8y^2 - 32z = 0$ (b) $4x^2 - 8y^2 - 32z^2 = 1$
(c) $4x^2 - 8y^2 - 32z^2 = 0$ (d) $4x^2 - 8y^2 + 32z^2 = 1$
(e) $4x^2 + 8y^2 + 32z^2 = 1$

Solution (a) Rewriting the equation as

$$z = \frac{x^2}{8} - \frac{y^2}{4},$$

we see by comparison with equation (8) that the graph is a hyperbolic paraboloid.

(b) With the equation written in the form

$$8y^2 + 32z^2 - 4x^2 = -1,$$

we see that we have a hyperboloid of two sheets by comparing with equation (4).

(c) We write

$$\frac{y^2}{4} + z^2 - \frac{x^2}{8} = 0$$

to see that the equation represents an elliptic cone.

(d) Interchanging the y and z terms, we compare with equation (3) and see that the graph is a hyperboloid of one sheet.

(e) Comparison with equation (2) shows that we have an ellipsoid. ■

Many of the quadric surfaces have a number of interesting properties, most of which, unfortunately, are beyond our scope. One property that is mentioned in the Additional Exercises at the end of this chapter is that some of the surfaces are **ruled surfaces,** so called because every point on the surface lies on a line that is contained in the surface. Among the ruled surfaces are the planes, cylinders, and cones, of course, but surprisingly, also the hyperboloid of one sheet and the hyperbolic paraboloid (see Figure 12.79, page 800).

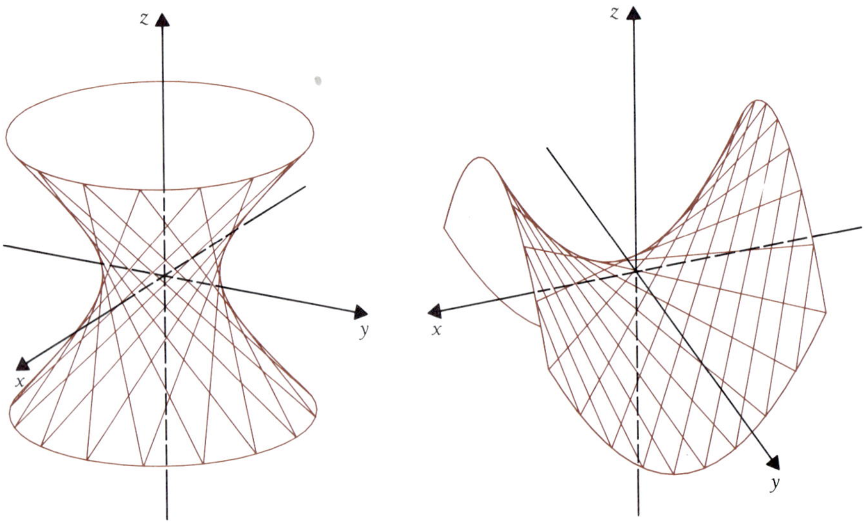

(a) Hyperboloid of one sheet (b) Hyperbolic paraboloid

FIGURE 12.79 Ruled surfaces

Other surfaces will be considered in Chapter 14, where we study the graphs of functions of two variables.

Exercises 12.6

Identify and sketch the graphs of the following surfaces:

1. $x^2 + y^2 = 9$
2. $x^2 - 4y = 0$
3. $z = y^3$
4. $x^2 - z^3 = 0$
5. $2x^2 + 2y^2 = e^z$
6. $x^2 - \ln y + z^2 = 0$
7. $x^2 + y^2 + z^2 = 9$
8. $x^2 + y^2 + 4z^2 = 4$
9. $x^2 + y^2 - 4z^2 = 4$
10. $x^2 + y^2 + 4 = 4z^2$
11. $x^2 + y^2 - z = 0$
12. $x^2 + y^2 - z^2 = 0$
13. $x^2 - y^2 - z^2 = 0$
14. $x^2 - y^2 - z = 0$
15. $x^2 - y^2 = 0$
16. $x - y - z^2 = 0$

Describe the following singular quadric surfaces:

17. $x^2 = 0$
18. $x^2 + y^2 = 0$
19. $x^2 + y^2 + z^2 = 0$
20. $x^2 - 9y^2 = 0$
21. $x^2 + y^2 + 1 = 0$
22. $x^2 + y^2 + z^2 + 1 = 0$

SECTION 12.7

Cylindrical and Spherical Coordinate Systems

In space, as in the plane, a rectangular coordinate system is not the only coordinate system possible. There are two other coordinate systems in common use, both of which are generalizations of the polar coordinate system in the plane. Each of these is more convenient in certain settings than rectangular coordinates.

Cylindrical Coordinates

A **cylindrical coordinate system** is a combination of polar coordinates in the xy-plane with the rectangular z-coordinate (Figure 12.80). The point with rectangular coordinates (x, y, z) has **cylindrical coordinates** $[r, \theta, z]$ with

$$\begin{cases} x = r\cos\theta \\ y = r\sin\theta \\ z = z. \end{cases} \tag{1}$$

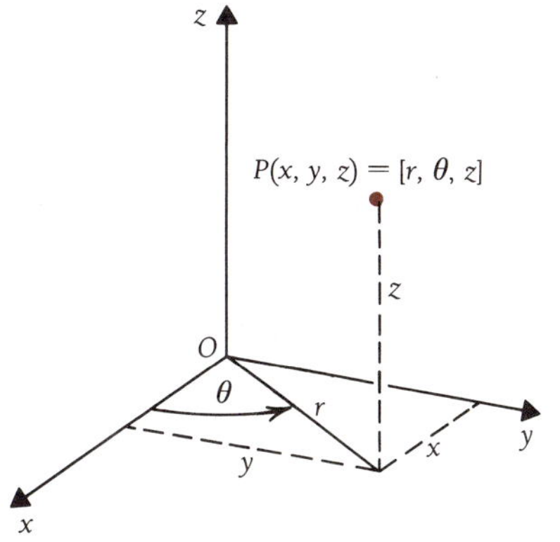

FIGURE 12.80 The cylindrical coordinate system

■ **Example 12.42** Find cylindrical coordinates for the points with rectangular coordinates:

(a) $(1, 1, 3)$ **(b)** $(0, 2, -1)$ **(c)** $(0, 0,1)$

Solution

(a) Since $r^2 = x^2 + y^2 = 2$ and $\sin\theta = 1/\sqrt{2}$ and $\cos\theta = 1/\sqrt{2}$, the cylindrical coordinates are $[\sqrt{2}, \pi/4, 3]$.

(b) Here, $\theta = \pi/2$ and $r = 2$, so the cylindrical coordinates are $[2, \pi/2, -1]$.

(c) The cylindrical coordinates are $[0, \theta, 1]$, where θ may be any value. ■

The cylindrical coordinate system gets its name from the fact that the equation $r = k$, $k \neq 0$, has as its graph a circular cylinder around the z-axis (Figure 12.81, page 802). The cylindrical coordinate system is often more convenient than rectangular coordinates for describing surfaces with symmetry about the z-axis.

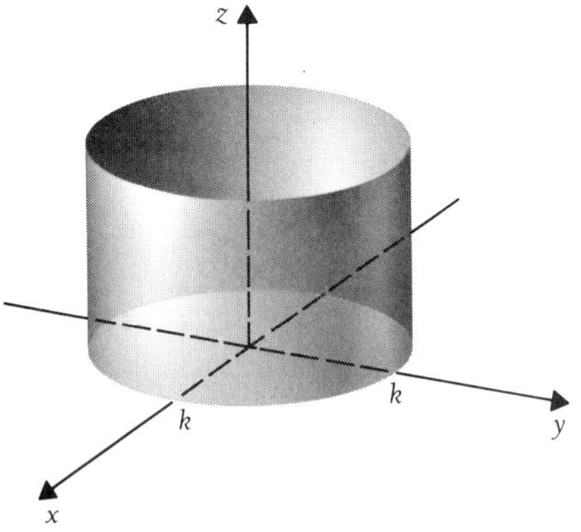

FIGURE 12.81 The cylinder $r = k$

■ **Example 12.43** Find an equation in cylindrical coordinates for the cone

$$z^2 = \frac{1}{4}(x^2 + y^2).$$

Solution Since $x^2 + y^2 = r^2$, the equation is $z^2 = \frac{1}{4}r^2$, which is equivalent to $z = \frac{1}{2}r$. ■

■ **Example 12.44** Find an equation in rectangular coordinates for the surface $z = 4r^2$, and describe the surface.

Solution The rectangular form of the equation is $z = 4x^2 + 4y^2$, which represents a circular paraboloid. ■

Spherical Coordinates

The **spherical coordinate system** is more nearly the three-dimensional version of polar coordinates. Let P be the point with rectangular coordinates (x, y, z), as shown in Figure 12.82. Let O be the origin of the coordinate system, and let F be the point at which the vertical line through P meets the xy-plane. Then point P has **spherical coordinates** $[\rho, \theta, \phi]$, where

$$\rho = \text{Distance } |OP|,$$
$$\theta = \text{Angle between positive } x\text{-axis and ray } \overrightarrow{OF},$$
$$\phi = \text{Angle between positive } z\text{-axis and ray } \overrightarrow{OP}.$$

Note that θ is the same as in cylindrical coordinates, and that $r = \rho \sin\phi$. Thus,

$$\begin{cases} x = \rho \sin\phi \cos\theta \\ y = \rho \sin\phi \sin\theta \\ z = \rho \cos\phi. \end{cases} \qquad (2)$$

For convenience, we let $0 \le \phi \le \pi$ and $0 \le \theta \le 2\pi$.

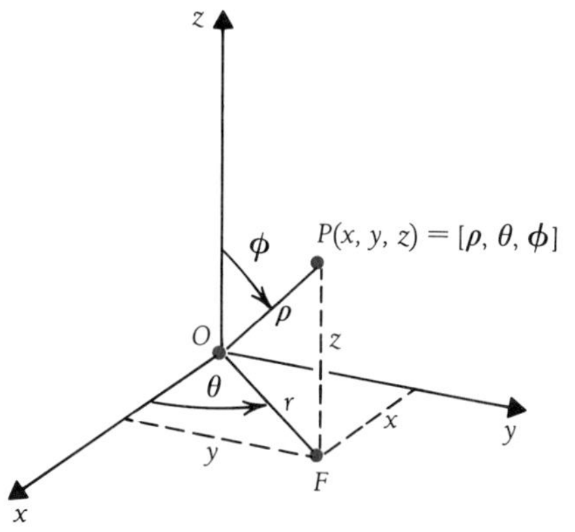

FIGURE 12.82 The spherical coordinate system

■ **Example 12.45** Find spherical coordinates for the points with rectangular coordinates:

(a) (1, 1, 1) **(b)** (0, 0, 2) **(c)** (0, 0, 0)

Solution

(a) It is evident that $\theta = \pi/4$ and $\phi = \pi/4$. From the distance formula, we get $\rho = \sqrt{3}$. Hence, the spherical coordinates are $[\sqrt{3}, \pi/4, \pi/4]$.

(b) Since this point is on the z-axis and 2 units above the origin, it has spherical coordinates $[2, \theta, 0]$, where θ is any value.

(c) The origin has spherical coordinates $[0, \theta, \phi]$ with θ and ϕ any values. ■

The spherical coordinate system gets its name from the fact that the equation $\rho = k$, $k > 0$, represents a sphere centered at the origin (Figure 12.83). The spherical coordinate system is often more convenient than rectangular or cylindrical coordinates for describing surfaces that are symmetric about the origin.

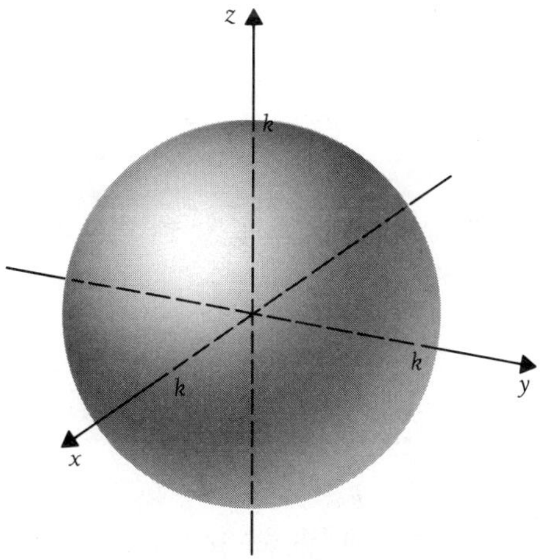

FIGURE 12.83 The sphere $\rho = k$

Example 12.46 Find an equation in spherical coordinates for the hyperboloid $x^2 + y^2 - z^2 = 1$.

Solution Using equations (2), this equation becomes

$$\rho^2 \sin^2\phi \cos^2\theta + \rho^2 \sin^2\phi \sin^2\theta - \rho^2 \cos^2\phi = 1,$$

which simplifies to

$$\rho^2 \cos 2\phi = -1.$$ ■

Example 12.47 Find an equation in rectangular coordinates for the surface $\rho \sin \phi = \cos \theta$, and describe the surface.

Solution We first multiply the equation through by $\rho \sin \phi$, getting

$$\rho^2 \sin^2\phi = \rho \sin \phi \cos \theta.$$

From equations (2), we can derive the identity $x^2 + y^2 = \rho^2 \sin^2\phi$, so the equation in rectangular coordinates is

$$x^2 + y^2 = x.$$

The graph is a circular cylinder. ■

Exercises 12.7

Find cylindrical and spherical coordinates for the points with the given rectangular coordinates.

1. $(1, \sqrt{3}, 3)$
2. $(2, -\sqrt{2}, \sqrt{2})$
3. $(0, 2, -2)$
4. $(-1, 1, 0)$

Find the rectangular and spherical coordinates for the points with the given cylindrical coordinates.

5. $[2, \pi/2, 3]$
6. $[1, -\pi/3, -1]$
7. $[-1, \pi/2, 2]$
8. $[\sqrt{2}, \pi/4, -1]$

Find rectangular and cylindrical coordinates for the points with the given spherical coordinates.

9. $[1, \pi/2, \pi/2]$
10. $[3, \pi/3, \pi/6]$
11. $[3, 0, \pi/2]$
12. $[4, -\pi/2, 2\pi/3]$

Find equations in cylindrical and spherical coordinates for the following surfaces, and describe each one in words or by drawing a picture.

13. $x^2 + y^2 = 4$
14. $x^2 + y^2 + z^2 = 4$
15. $x^2 + y^2 = 4z^2$
16. $z = x^2 + y^2$
17. $z = 4 - x^2 - y^2$
18. $z = x^2 - y^2$

Find equations in rectangular coordinates for the following surfaces, and describe each one verbally or graphically.

19. $r = 16$
20. $\theta = \pi/4$
21. $z = 3r$
22. $z = 9 - r$
23. $r^2 + z^2 = 1$
24. $z = 2r \cos \theta$
25. $\rho = 16$
26. $\phi = \pi/4$
27. $\phi = \pi/2$
28. $\phi = \pi$
29. $\rho = \cos \phi$
30. $\rho = \sec \phi$
31. $\rho = \cos \theta$
32. $\rho(1 - \cos \theta) = \cos \phi$
33. Find equations for r, θ, z in terms of x, y, z.

34. Find equations for ρ, θ, ϕ in terms of r, θ, z.

35. Find equations for ρ, θ, ϕ in terms of x, y, z.

36. Find equations for r, θ, z in terms of ρ, θ, ϕ.

SECTION 12.8

Chapter Review

Important Concepts

Right-handed coordinate system
Coordinate planes
Octant
Distance formula
Sphere
Vector
Arrow
Position arrow
Coordinates of a vector
Magnitude
Norm
Unit vector
Zero vector
Standard unit vector
Sum of vectors
Resultant
Parallelogram law
Triangle inequality
Scalar
Scalar multiple
Additive inverse
Negative vector
Difference of vectors
Linear combination
Dot product
Angle between vectors
Parallel vectors
Antiparallel vectors
Orthogonal vectors
Component
Projection
Cross product
Right-hand rule
Triple product
Skew lines
Direction vector
Angle between lines
Orthogonal lines
Perpendicular lines
Direction numbers
Direction cosines
Direction angles
Vector equation of a line
Parametric equations
Symmetric equations
Plane
Normal vector
Attitude numbers
Angle between planes
Parallel planes
Perpendicular planes
Distance from a point to a plane
Intercepts of a plane
Surface
Cylinder
Surface of revolution
Quadric surface
Ellipsoid
Spheroid
Hyperboloid
Cone
Paraboloid
Saddle surface
Ruled surface
Cylindrical coordinate system
Spherical coordinate system

Review Exercises

Let $\mathbf{u} = \langle 3, 5, -2 \rangle$ *and* $\mathbf{v} = \langle -4, -3, 5 \rangle$. *Find the following:*

1. $\mathbf{u} + \mathbf{v}$
2. $\mathbf{u} - \mathbf{v}$
3. $2\mathbf{u} - 3\mathbf{v}$
4. $5\mathbf{v} - 3\mathbf{u}$
5. $\|\mathbf{u}\|$
6. $\|\mathbf{v}\|$
7. $\|\mathbf{u} + \mathbf{v}\|$
8. $\|\mathbf{u} - \mathbf{v}\|$
9. A unit vector in the direction of $\mathbf{u}$
10. A unit vector in the direction of $\mathbf{u} - \mathbf{v}$
11. $\mathbf{u} \cdot \mathbf{v}$
12. $\mathbf{j} \cdot \mathbf{u}$
13. The angle between $\mathbf{u}$ and $\mathbf{v}$
14. The angle between $\mathbf{u} + \mathbf{v}$ and $\mathbf{u} - \mathbf{v}$
15. The component of $\mathbf{u}$ in the direction of $\mathbf{v}$

16. The projection of **u** on **v**

17. $\mathbf{u} \times \mathbf{v}$ 18. $\mathbf{v} \times \mathbf{u}$ 19. $\mathbf{u} \cdot \mathbf{u} \times \mathbf{v}$ 20. $(\mathbf{u} + \mathbf{v}) \cdot \mathbf{u}$

Simplify.

21. $(4\mathbf{i} - \mathbf{j} + 3\mathbf{k}) + (2\mathbf{j} - \mathbf{k})$
22. $(4\mathbf{i} - \mathbf{j} + 3\mathbf{k}) - (2\mathbf{j} - \mathbf{k})$
23. $(4\mathbf{i} - \mathbf{j} + 3\mathbf{k}) \cdot (2\mathbf{j} - \mathbf{k})$
24. $(4\mathbf{i} - \mathbf{j} + 3\mathbf{k}) \times (2\mathbf{j} - \mathbf{k})$

Given points $P(-1, 0, 3)$, $Q(1, 2, 0)$, and $R(4, 4, 5)$, find the following:

25. $\mathbf{v}(\overrightarrow{PQ})$
26. $\mathbf{v}(\overrightarrow{QP})$
27. $\mathbf{v}(\overrightarrow{QR}) + \mathbf{v}(\overrightarrow{QP})$
28. $\mathbf{v}(\overrightarrow{PQ}) \cdot \mathbf{v}(\overrightarrow{PR})$
29. $\mathbf{v}(\overrightarrow{RP}) \times \mathbf{v}(\overrightarrow{RQ})$
30. Equations for the line PQ
31. The equation of the plane PQR
32. The area of triangle PQR
33. Equations of the line through R and perpendicular to plane PQR
34. The equation of the plane through R and perpendicular to line PQ
35. The angle between lines PQ and PR
36. The angle between plane PQR and the line $x = 1 + 3t$, $y = 1 - 3t$, $z = 4t$

Find equations (in rectangular coordinates) for the following:

37. The line through points $(2, 1, -3)$ and $(4, -2, 2)$
38. The line through points $(-1, 1, -1)$ and $(4, 0, 3)$
39. The line through the point $(2, 0, 3)$ with direction numbers 2, 1, and -1
40. The line through the point $(2, 0, 3)$ and parallel to the line $x = 1 + 4t$, $y = -1 + 2t$, $z = 1$
41. The sphere of radius 4 and with center $(1, 0, 0)$
42. The sphere of radius $\sqrt{2}$ and with center $(1, 1, 0)$
43. The circular cylinder of radius 2 with axis the z-axis
44. The circular cylinder of radius 3 with axis the x-axis

Describe the following surfaces:

45. $x = 1$
46. $y = -2$
47. $x^2 = y^2 + z^2$
48. $x^2 = y^2 - z^2$
49. $y = 9 - x^2 - z^2$
50. $z = x^2 + 4y^2$
51. $x^2 + y^2 + z^2 = 9$
52. $x^2 - y^2 + z^2 = 9$
53. $x^2 - y^2 - z^2 = 9$
54. $x^2 - y^2 - z = 9$
55. $x^2 = 1 + y^2 + z^2$
56. $x^2 - y^2 = 4$
57. $z = 4y^2$
58. $z^2 + 9y^2 = 9$

Write each equation in cylindrical and spherical coordinates.

59. $z = 4$
60. $z = 4 - x^2 - y^2$
61. $x^2 + y^2 + z^2 = 25$
62. $z = x^2 - y^2$
63. $36x^2 = 9y^2 + 4z^2$
64. $4x^2 + 4y^2 - z^2 = 4$

Write each equation in rectangular coordinates.

65. $z = r \sin\theta$
66. $z^2 = r^2 \sin 2\theta$
67. $\rho \sin\phi = \sin\theta$
68. $\theta = \pi$
69. $\cos\theta = r$
70. $\rho = \sin\theta$

Find direction cosines for the given line.

71. $x = 2 - t,\quad y = 1 + 3t,\quad z = \frac{1}{2}t$
72. $\frac{x+1}{1} = \frac{y}{3} = \frac{z-2}{-4}$

73. Find the point of intersection of the two lines $x = 1$, $y = 2 + t$, $z = -1 + t$ and $x = 2 - t$, $y = 5 - 3t$, $z = 4 - 5t$.

74. Find the points of intersection of the line $x = 4t$, $y = -t$, $z = 2t$ and the sphere $x^2 + y^2 + z^2 = 16$.

Additional Exercises

1. For any vectors **u** and **v**, show that the vectors $\|\mathbf{v}\|\mathbf{u} + \|\mathbf{u}\|\mathbf{v}$ and $\|\mathbf{v}\|\mathbf{u} - \|\mathbf{u}\|\mathbf{v}$ are orthogonal.
2. Under what conditions is $\mathbf{u} \times (\mathbf{v} \times \mathbf{w}) = (\mathbf{u} \times \mathbf{v}) \times \mathbf{w}$?
3. Find the unit vector **u** such that if θ is the angle between **v** and **w**, then $\mathbf{v} \times \mathbf{w} = \|\mathbf{v}\|\,\|\mathbf{w}\|(\sin\theta)\mathbf{u}$.
4. If $\mathbf{u} \times \mathbf{v} = \mathbf{u} \times \mathbf{w}$, show that **u** and $\mathbf{v} - \mathbf{w}$ are parallel.
5. Show that $\mathbf{u} \times (\mathbf{v} \times \mathbf{w}) = (\mathbf{u} \cdot \mathbf{w})\mathbf{v} - (\mathbf{u} \cdot \mathbf{v})\mathbf{w}$.
6. Show that $(\mathbf{u} \times \mathbf{v}) \times \mathbf{w} = (\mathbf{u} \cdot \mathbf{w})\mathbf{v} - (\mathbf{v} \cdot \mathbf{w})\mathbf{u}$.
7. Show that $\mathbf{u} \times (\mathbf{v} \times \mathbf{w}) + \mathbf{v} \times (\mathbf{w} \times \mathbf{u}) + \mathbf{w} \times (\mathbf{u} \times \mathbf{v}) = \mathbf{0}$.
8. Use vectors to prove the following:
   (a) The diagonals of a parallelogram bisect one another.
   (b) The sum of the squares of the sides of a parallelogram is equal to the sum of the squares of the diagonals.
   (c) The sum of the distances from a point inside an equilateral triangle to the three sides is a constant.
9. Find the distance from the origin to the plane
$$\frac{x}{a} + \frac{y}{b} + \frac{z}{c} = 1.$$
10. Do the points $(1, -1, 1)$, $(5, 1, 1)$, and $(9, 2, 1)$ lie on a single line?
11. Devise a general method for determining whether three points $(x_1, y_1\ z_1)$, (x_2, y_2, z_2), and (x_3, y_3, z_3) lie on a single line.
12. The system
$$\begin{cases} a_1x + b_1y = c_1 \\ a_2x + b_2y = c_2 \\ a_3x + b_3y = c_3 \end{cases}$$
of three equations in two unknowns has a solution if and only if
$$\begin{vmatrix} a_1 & b_1 & c_1 \\ a_2 & b_2 & c_2 \\ a_3 & b_3 & c_3 \end{vmatrix} = 0.$$

Show that the two lines

$$\begin{cases} x = x_1 + a_1 t \\ y = y_1 + b_1 t \\ z = z_1 + c_1 t \end{cases} \quad \text{and} \quad \begin{cases} x = x_2 + a_2 t \\ y = y_2 + b_2 t \\ z = z_2 + c_2 t \end{cases}$$

intersect if and only if

$$\begin{vmatrix} a_1 & a_2 & x_1 - x_2 \\ b_1 & b_2 & y_1 - y_2 \\ c_1 & c_2 & z_1 - z_2 \end{vmatrix} = 0.$$

13. Devise a method for determining whether two lines are perpendicular.

14. Show that for any real number s, the line

$$\begin{cases} x = s\sqrt{1 + c^2}\, t \\ y = \sqrt{c^2 - (1 - c^2)s^2}\, t \\ z = t \end{cases}$$

lies entirely on the cone $c^2z^2 = x^2 + y^2$. (Each line is a generator of the cone.)

15. Show that all the generators of the cone $c^2z^2 = x^2 + y^2$ pass through the origin. (The point of intersection of the generators is called the **vertex** of the cone.)

16. Show that exactly one generator of the cone $c^2z^2 = x^2 + y^2$ passes through each point of the cone, other than the vertex.

17. **Ruled Surfaces** Show that for any real number s, the hyperboloid of one sheet $x^2 + y^2 - z^2 = 1$ entirely contains the lines

$$\begin{cases} x = \dfrac{2s}{1 + s^2} - (1 - s^2)t \\ y = \dfrac{1 - s^2}{1 + s^2} + 2st \\ z = (1 + s^2)t \end{cases} \quad \text{and} \quad \begin{cases} x = \dfrac{2s}{1 + s^2} - (1 - s^2)t \\ y = -\dfrac{1 - s^2}{1 + s^2} - 2st \\ z = (1 + s^2)t. \end{cases}$$

Also show that for any real number s, the hyperbolic paraboloid $cz = x^2 - y^2$ entirely contains the lines

$$\begin{cases} x = \dfrac{1}{2s} + sct \\ y = \dfrac{-1}{2s} + sct \\ z = 2t \end{cases} \quad \text{and} \quad \begin{cases} x = \dfrac{-1}{2s} - sct \\ y = \dfrac{-1}{2s} + sct \\ z = 2t. \end{cases}$$

[Each family of lines generated by letting s vary over $(-\infty, \infty)$ is called a **regulus**; each surface described above contains two reguli. Thus, these surfaces are ruled surfaces.]

18. Refer to Problem 17. Show that on a ruled surface, each line of one regulus meets every line of the other regulus, but no other line of its own regulus.

19. Refer to Problem 17. Show that through each point on a ruled surface, there passes one line of each regulus.

20. Find the volume of the ellipsoid

$$a^2x^2 + b^2y^2 + b^2z^2 = a^2b^2.$$

21. Find the volume of the solid enclosed by the paraboloid $z = a^2 - x^2 - y^2$ and the plane $z = 0$.

22. Find the volume enclosed by the hyperboloid $x^2 - y^2 + z^2 = 1$ and the planes $y = 0$ and $y = 4$.

23. Develop a formula for the distance between a point and a line in space.

CHAPTER 13

Vector Functions and Calculus

INTRODUCTION

We are now ready to introduce the concept of a vector-valued function. In this chapter, we develop the properties of vector functions and use the tools of calculus, as well as the tools of vector analysis, to describe curves in the plane and in space. We also discuss curvilinear motion and the notion of curvature.

SECTION 13.1

Vector Functions and Space Curves

The functions we have met thus far are all real-valued functions, meaning that to each real number x in the domain of a function there is assigned a real number $f(x)$. In contrast, a vector function assigns to each real number in its domain a vector; that is, it is vector-valued.

Vector-Valued Functions

DEFINITION 13.1

Let D be a set of real numbers. A **vector-valued function** on D is a function $\mathbf{f}$ that assigns to each $t \in D$ a vector $\mathbf{f}(t) = \langle x(t), y(t), z(t) \rangle$. The functions x, y, and z are called the **components of f.**

We will call a vector-valued function a **vector function** for short. An alternative representation for a vector function $\mathbf{f}(t) = \langle x(t), y(t), z(t) \rangle$ in terms of the standard unit vectors is

$$\mathbf{f}(t) = x(t)\mathbf{i} + y(t)\mathbf{j} + z(t)\mathbf{k}.$$

Because a real-valued function has scalars (real numbers) as its values, we will call it a **scalar function.** Thus, the components of a vector function are scalar functions.

Example 13.1 Give the domain of the vector function

$$\mathbf{f}(t) = \left\langle t^2 + 1, \sqrt{1+t}, \frac{e^t}{t-3} \right\rangle.$$

Solution The vector function **f** is defined for those values of t for which each component of **f** is defined. Thus, the domain of **f** is $[-1, 3) \cup (3, \infty)$. ■

Space Curves

A vector function is represented geometrically by a variable position arrow. That is, if $\mathbf{f}(t) = \langle x(t), y(t), z(t) \rangle$ is a vector function and t_0 is in the domain of **f**, we represent the vector $\mathbf{f}(t_0) = \langle x(t_0), y(t_0), z(t_0) \rangle$ by its position arrow. Thus, as t varies, the position arrow varies, and the tip of the arrow traces out a curve in space (see Figure 13.1). In general, this curve does not lie in a plane, so it is called a **space curve.** This curve is what is meant by the **graph of f.**

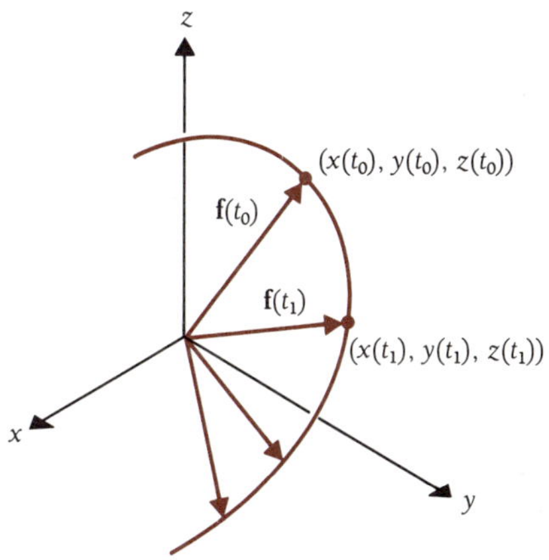

FIGURE 13.1 The graph of a vector function is a space curve

The graph of a vector function $\mathbf{f}(t) = \langle x(t), y(t), z(t) \rangle$ also can be thought of as the curve represented parametrically by the equations

$$\begin{cases} x = x(t) \\ y = y(t) \\ z = z(t). \end{cases}$$

We take the largest possible domain for t unless otherwise specified.

■ **Example 13.2** Describe the graphs of the vector functions:

(a) $\mathbf{f}(t) = \langle 1 + t, 1 - t, 3t \rangle$ (b) $\mathbf{f}(t) = \langle \cos t, \sin t, 0 \rangle$
(c) $\mathbf{f}(t) = \langle a \cos t, a \sin t, bt \rangle$

Solution (a) Using the parametric interpretation, the graph of this vector function is

$$\begin{cases} x = 1 + t \\ y = 1 - t \\ z = 3t, \end{cases}$$

which is the line through the point $(1, 1, 0)$ with direction vector $\langle 1, -1, 3 \rangle$.

(b) Since the point $(\cos t, \sin t, 0)$ lies on the unit circle in the xy-plane (see Example 11.4), the graph of $\mathbf{f}$ is just that circle (Figure 13.2).

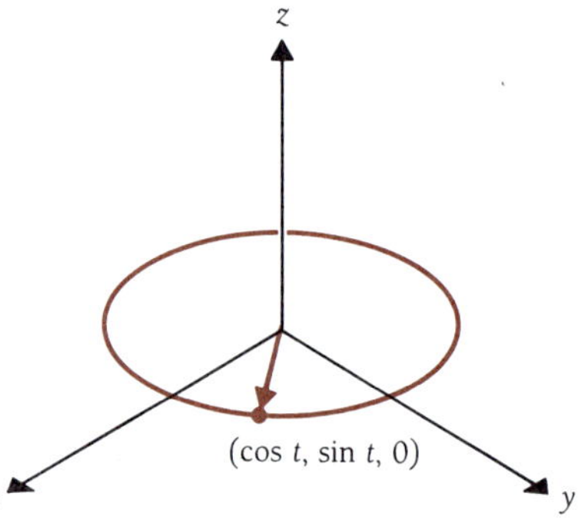

FIGURE 13.2
The circle $\mathbf{f}(t) = \langle \cos t, \sin t, 0 \rangle$

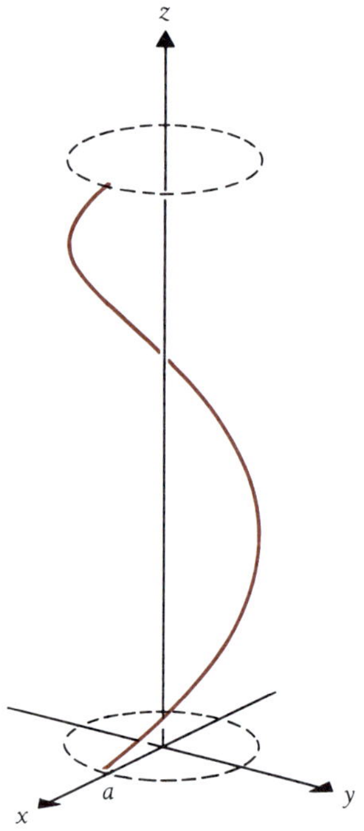

FIGURE 13.3
The circular helix
$\mathbf{f}(t) = \langle a \cos t, a \sin t, bt \rangle$

(c) Since $x = a \cos t$ and $y = a \sin t$, the graph of this function lies on the cylinder $x^2 + y^2 = a^2$. It is a curve that winds around the cylinder (Figure 13.3), resembling a stretched spring. Such a curve is called a **circular helix**. ■

We will return to a study of space curves later.

Combinations of Vector and Scalar Functions

We can form sums, scalar multiples, and dot and cross products with vector and scalar functions, just as with vectors and scalars. These combinations are defined as follows: If **f** and **g** are vector functions and u is a scalar function, then for each t, we have

$$(\mathbf{f}+\mathbf{g})(t)=\mathbf{f}(t)+\mathbf{g}(t), \tag{1}$$

$$(u\mathbf{f})(t)=u(t)\mathbf{f}(t), \tag{2}$$

$$(\mathbf{f}\cdot\mathbf{g})(t)=\mathbf{f}(t)\cdot\mathbf{g}(t), \tag{3}$$

$$(\mathbf{f}\times\mathbf{g})(t)=\mathbf{f}(t)\times\mathbf{g}(t). \tag{4}$$

■ **Example 13.3** Given $\mathbf{f}(t)=\langle \cos t, \sin t, t\rangle$, $\mathbf{g}(t)=\langle \cos t, t, 0\rangle$, and $u(t)=t^2+1$, find:

(a) $(\mathbf{f}+\mathbf{g})(t)$ **(b)** $(u\mathbf{f})(t)$ **(c)** $(\mathbf{f}\cdot\mathbf{g})(t)$ **(d)** $(\mathbf{f}\times\mathbf{g})(t)$

Solution **(a)** By equation (1), we have

$$\begin{aligned}(\mathbf{f}+\mathbf{g})(t)&=\mathbf{f}(t)+\mathbf{g}(t)\\&=\langle \cos t, \sin t, t\rangle+\langle \cos t, t, 0\rangle\\&=\langle 2\cos t, t+\sin t, t\rangle.\end{aligned}$$

(b) By equation (2), we have

$$\begin{aligned}(u\mathbf{f})(t)&=u(t)\mathbf{f}(t)\\&=(t^2+1)\langle \cos t, \sin t, t\rangle\\&=\langle (t^2+1)\cos t, (t^2+1)\sin t, t^3+t\rangle.\end{aligned}$$

(c) By equation (3), we get

$$\begin{aligned}(\mathbf{f}\cdot\mathbf{g})(t)&=\mathbf{f}(t)\cdot\mathbf{g}(t)\\&=\langle \cos t, \sin t, t\rangle\cdot\langle \cos t, t, 0\rangle\\&=\cos^2 t+t\sin t.\end{aligned}$$

(d) Using equation (4), we get

$$\begin{aligned}(\mathbf{f}\times\mathbf{g})(t)&=\mathbf{f}(t)\times\mathbf{g}(t)\\&=\langle \cos t, \sin t, t\rangle\times\langle \cos t, t, 0\rangle\\&=\langle -t^2, t\cos t, (t-\sin t)\cos t\rangle.\end{aligned}$$ ■

The **magnitude** of a vector function **f** is the scalar function $\|\mathbf{f}\|$ defined by

$$\|\mathbf{f}\|(t)=\|\mathbf{f}(t)\|. \tag{5}$$

Thus, if $\mathbf{f}(t)=\langle x(t), y(t), z(t)\rangle$, then

$$\|\mathbf{f}\|(t)=\sqrt{[x(t)]^2+[y(t)]^2+[z(t)]^2}. \tag{6}$$

Example 13.4 Find the magnitude of the vector function $\mathbf{f}(t) = \langle \cos t, \sin t, t \rangle$.

Solution Using equation (6), we have

$$\|\mathbf{f}\|(t) = \sqrt{\cos^2 t + \sin^2 t + t^2} = \sqrt{1 + t^2}.$$

The Calculus of Vector Functions

Since a vector function is given in terms of the scalar functions that are its components, the calculus of scalar functions can be extended to vector functions. Let us be precise.

The **limit** of a vector function $\mathbf{f}(t) = \langle x(t), y(t), z(t) \rangle$ as $t \to c$ is defined by

$$\lim_{t \to c} \mathbf{f}(t) = \left\langle \lim_{t \to c} x(t), \lim_{t \to c} y(t), \lim_{t \to c} z(t) \right\rangle. \tag{7}$$

Thus, the limit of a vector function exists only when the limits of all its components exist. It follows from this definition that $\lim_{t \to c} \mathbf{f}(t) = \mathbf{L}$ if and only if for every $\varepsilon > 0$ there exists $\delta > 0$ such that if $0 < |t - c| < \delta$, then $\|\mathbf{f}(t) - \mathbf{L}\| < \varepsilon$. Note that $\|\mathbf{f}(t) - \mathbf{L}\|$ is the magnitude of a vector. Thus, limits of vector functions are treated in very much the same way as limits of scalar functions.

Example 13.5 Find $\lim_{t \to 0} \mathbf{f}(t)$, given that

$$\mathbf{f}(t) = \left\langle \frac{\tan t}{t}, \frac{1 - \cos t}{t}, \frac{e^t - 1}{t} \right\rangle.$$

Solution Since

$$\lim_{t \to 0} \frac{\tan t}{t} = 1, \qquad \lim_{t \to 0} \frac{1 - \cos t}{t} = 0, \qquad \lim_{t \to 0} \frac{e^t - 1}{t} = 1,$$

we get $\lim_{t \to 0} \mathbf{f}(t) = \langle 1, 0, 1 \rangle$.

We say a vector function $\mathbf{f}$ is **continuous at c** if and only if each component of $\mathbf{f}$ is continuous at c. The graph of a continuous vector function is a space curve that is continuous in the same sense that a plane curve is continuous.

A vector function $\mathbf{f}$ is **differentiable at c** if and only if each component of $\mathbf{f}$ is differentiable at c. The derivative of $\mathbf{f}(t) = \langle x(t), y(t), z(t) \rangle$ is defined to be

$$\frac{d\mathbf{f}}{dt} = \mathbf{f}'(t) = \langle x'(t), y'(t), z'(t) \rangle \tag{8}$$

if $\mathbf{f}$ is differentiable. We can also show that if $\mathbf{f}$ is differentiable at t, then

$$\mathbf{f}'(t) = \lim_{h \to 0} \frac{1}{h} [\mathbf{f}(t + h) - \mathbf{f}(t)]. \tag{9}$$

Example 13.6 If $\mathbf{f}(t) = \langle t - \cos t, t \sin t, 2t \rangle$, find $\mathbf{f}'(t)$ and $\mathbf{f}''(t)$.

Solution By equation (8), we find

$$\mathbf{f}'(t) = \langle 1 + \sin t, \sin t + t \cos t, 2 \rangle.$$

Applying equation (8) again, we get

$$\mathbf{f}''(t) = \langle \cos t, 2 \cos t - t \sin t, 0 \rangle.$$

■

The geometric interpretation of $\mathbf{f}'(t)$ is interesting and useful. Let C be the curve that is the graph of $\mathbf{f}$ (Figure 13.4). For small h, the vector

$$\frac{1}{h}[\mathbf{f}(t + h) - \mathbf{f}(t)]$$

is approximately equal to $\mathbf{f}'(t)$. As $h \to 0$, $\mathbf{f}(t + h) \to \mathbf{f}(t)$, so that the arrow representing $\mathbf{f}(t + h) - \mathbf{f}(t)$ approaches the direction of the tangent line to C. Thus, $\mathbf{f}'(t)$ is a direction vector for the tangent line to C at $\mathbf{f}(t)$, and we call $\mathbf{f}'(t)$ a **tangent vector** to C.

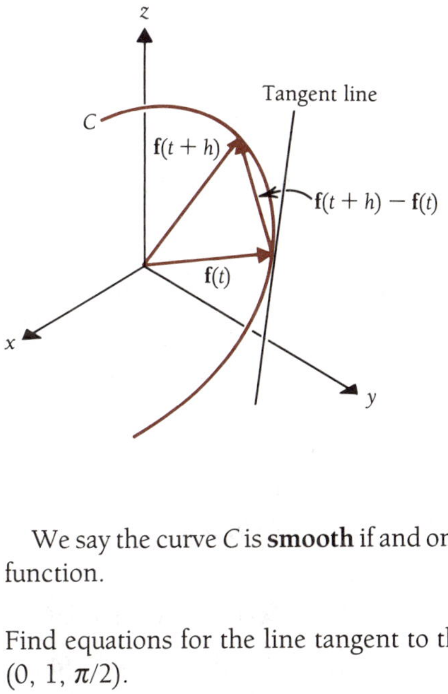

FIGURE 13.4 The derivative is a tangent vector

We say the curve C is **smooth** if and only if it is the graph of a differentiable vector function.

Example 13.7 Find equations for the line tangent to the helix $\mathbf{f}(t) = \langle \cos t, \sin t, t \rangle$ at the point $(0, 1, \pi/2)$.

Solution The point $(0, 1, \pi/2)$ is on the graph of $\mathbf{f}$ at the point $t = \pi/2$ (Figure 13.5). Thus, the tangent line at that point is parallel to $\mathbf{f}'(\pi/2)$. Since $\mathbf{f}'(t) = \langle -\sin t, \cos t, 1 \rangle$, we

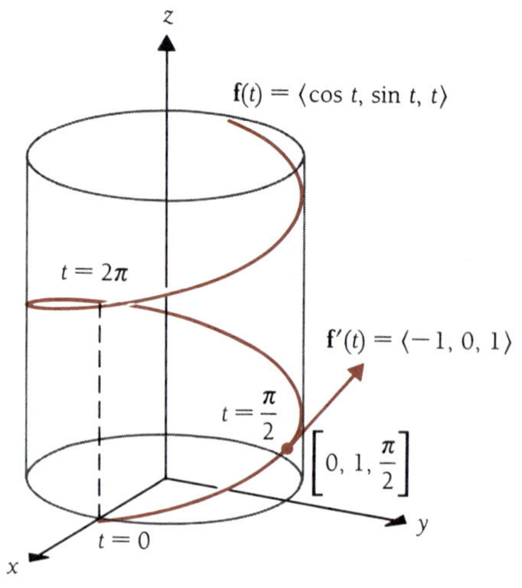

FIGURE 13.5

have $\mathbf{f}'(\pi/2) = \langle -1, 0, 1 \rangle$. This is a direction vector for the tangent line, so its equations are

$$x = -t, \qquad y = 1, \qquad z = \pi/2 + t.$$

Note that we can write these equations in vector form as

$$\langle -t, 1, \pi/2 + t \rangle = \langle 0, 1, \pi/2 \rangle + t\langle -1, 0, 1 \rangle.$$

■

In general, the **line tangent to the curve** that is the graph of $\mathbf{f}$ at $t = c$ is the line

$$\mathbf{f}(c) + t\mathbf{f}'(c).$$

The differentiation formulas for vector functions are similar to those for scalar functions. It is not difficult to show that if $\mathbf{f}$ and $\mathbf{g}$ are vector functions and u is a scalar function, then

$$(\mathbf{f} + \mathbf{g})'(t) = \mathbf{f}'(t) + \mathbf{g}'(t), \tag{10}$$

$$(u\mathbf{f})'(t) = u'(t)\mathbf{f}(t) + u(t)\mathbf{f}'(t), \tag{11}$$

$$(\mathbf{f} \cdot \mathbf{g})'(t) = \mathbf{f}'(t) \cdot \mathbf{g}(t) + \mathbf{f}(t) \cdot \mathbf{g}'(t), \tag{12}$$

$$(\mathbf{f} \times \mathbf{g})'(t) = \mathbf{f}'(t) \times \mathbf{g}(t) + \mathbf{f}(t) \times \mathbf{g}'(t). \qquad \text{(Order is important!)} \tag{13}$$

■ **Example 13.8**

Given $\mathbf{f}(t) = \langle \cos t, \sin t, t \rangle$, $\mathbf{g}(t) = \langle \cos t, t, 0 \rangle$, and $u(t) = t^2 + 1$, find:

(a) $(\mathbf{f} + \mathbf{g})'(t)$ **(b)** $(u\mathbf{f})'(t)$ (c) $(\mathbf{f} \cdot \mathbf{g})'(t)$ **(d)** $(\mathbf{f} \times \mathbf{g})'(t)$

Solution

We could refer to Example 13.3 and simply differentiate the vector functions found there. But to illustrate formulas (10)–(13), we will use $\mathbf{f}'(t) = \langle -\sin t, \cos t, 1 \rangle$,

$\mathbf{g}'(t) = \langle -\sin t, 1, 0\rangle$, and $u'(t) = 2t$, as follows:

(a) $$\begin{aligned}(\mathbf{f}+\mathbf{g})'(t) &= \mathbf{f}'(t) + \mathbf{g}'(t)\\ &= \langle -\sin t, \cos t, 1\rangle + \langle -\sin t, 1, 0\rangle\\ &= \langle -2\sin t, 1+\cos t, 1\rangle.\end{aligned}$$

(b) $$\begin{aligned}(u\mathbf{f})'(t) &= u'(t)\mathbf{f}(t) + u(t)\mathbf{f}'(t)\\ &= 2t\langle \cos t, \sin t, t\rangle + (t^2+1)\langle -\sin t, \cos t, 1\rangle\\ &= \langle 2t\cos t - (t^2+1)\sin t, 2t\sin t + (t^2+1)\cos t, 3t^2+1\rangle.\end{aligned}$$

(c) $$\begin{aligned}(\mathbf{f}\cdot\mathbf{g})'(t) &= \mathbf{f}'(t)\cdot\mathbf{g}(t) + \mathbf{f}(t)\cdot\mathbf{g}'(t)\\ &= \langle -\sin t, \cos t, 1\rangle \cdot \langle \cos t, t, 0\rangle\\ &\qquad + \langle \cos t, \sin t, t\rangle \cdot \langle -\sin t, 1, 0\rangle\\ &= -\sin t\cos t + t\cos t - \sin t\cos t + \sin t\\ &= -2\sin t\cos t + t\cos t + \sin t.\end{aligned}$$

(d) $$\begin{aligned}(\mathbf{f}\times\mathbf{g})'(t) &= \mathbf{f}'(t)\times\mathbf{g}(t) + \mathbf{f}(t)\times\mathbf{g}'(t)\\ &= \langle -\sin t, \cos t, 1\rangle \times \langle \cos t, t, 0\rangle\\ &\qquad + \langle \cos t, \sin t, t\rangle \times \langle -\sin t, 1, 0\rangle\\ &= \langle -t, \cos t, -t\sin t - \cos^2 t\rangle + \langle -t, -t\sin t, \cos t + \sin^2 t\rangle\\ &= \langle -2t, \cos t - t\sin t, \cos t - t\sin t - \cos^2 t + \sin^2 t\rangle.\end{aligned}$$ ■

Note that in applying formula (13), care must be taken to keep the vectors in proper order, since the cross product is not commutative.

A vector function $\mathbf{f}$ is **integrable over $[a, b]$** if and only if each component of $\mathbf{f}$ is integrable over $[a, b]$. The integral of $\mathbf{f}(t) = \langle x(t), y(t), z(t)\rangle$ is defined by

$$\int_a^b \mathbf{f}(t)\,dt = \left\langle \int_a^b x(t)\,dt, \int_a^b y(t)\,dt, \int_a^b z(t)\,dt \right\rangle. \tag{14}$$

A similar relationship holds for indefinite integrals.

■ **Example 13.9** Given

$$\mathbf{f}(t) = \left\langle \cos t, \frac{1}{t^2+1}, e^t \right\rangle,$$

find $\int_0^1 \mathbf{f}(t)\,dt$.

Solution By equation (14),

$$\begin{aligned}\int_0^1 \mathbf{f}(t)\,dt &= \left\langle \int_0^1 \cos t\,dt, \int_0^1 \frac{1}{t^2+1}\,dt, \int_0^1 e^t\,dt \right\rangle\\ &= \left\langle \sin t\Big|_0^1, \arctan t\Big|_0^1, e^t\Big|_0^1 \right\rangle\\ &= \langle \sin 1, \pi/4, e-1\rangle.\end{aligned}$$ ■

Example 13.10 Find $\mathbf{f}(t)$, given that $\mathbf{f}'(t) = \langle 2t, \cos t, 1/t \rangle$ and $\mathbf{f}(1) = \langle 2, 1, e \rangle$.

Solution If $\mathbf{f}'(t) = \langle 2t, \cos t, 1/t \rangle$, then

$$\mathbf{f}(t) = \langle t^2 + C_1, \sin t + C_2, \ln|t| + C_3 \rangle.$$

Then $\mathbf{f}(1) = \langle 1 + C_1, \sin 1 + C_2, C_3 \rangle = \langle 2, 1, e \rangle$, so that $C_1 = 1, C_2 = 1 - \sin 1$, and $C_3 = e$. Thus,

$$\mathbf{f}(t) = \langle t^2 + 1, \sin t + 1 - \sin 1, \ln|t| + e \rangle.$$

■

Arc Length

Arc length will be important in the next section, when we study motion. We will first consider arc length in the plane. Recall from Section 11.1 that if s represents the arc length of the curve

$$C: \quad \begin{cases} x = x(t) \\ y = y(t), \end{cases}$$

then

$$\frac{ds}{dt} = \sqrt{\left(\frac{dx}{dt}\right)^2 + \left(\frac{dy}{dt}\right)^2}.$$

In terms of vectors in two dimensions, if $\mathbf{f}(t) = \langle x(t), y(t) \rangle$, then

$$\frac{ds}{dt} = \|\mathbf{f}'(t)\|. \tag{15}$$

Thus, if C is the graph of the vector function $\mathbf{f}$ over $a \le t \le b$, then the length of C is

$$s = \int_a^b \|\mathbf{f}'(t)\|\, dt. \tag{16}$$

If we now regard $\mathbf{f}$ as a vector function in three dimensions, the equations (15) and (16) still hold. Therefore, if curve C is the graph of $\mathbf{f}(t) = \langle x(t), y(t), z(t) \rangle$ over $[a, b]$, then the length of C is

$$s = \int_a^b \sqrt{\left(\frac{dx}{dt}\right)^2 + \left(\frac{dy}{dt}\right)^2 + \left(\frac{dz}{dt}\right)^2}\, dt. \tag{17}$$

Example 13.11 Find the length of the helix $\mathbf{f}(t) = \langle \cos t, \sin t, t \rangle$ over $0 \le t \le 2\pi$.

Solution Using either equation (16) or (17), we get

$$\begin{aligned} s &= \int_0^{2\pi} \sqrt{(-\sin t)^2 + (\cos t)^2 + 1^2}\, dt \\ &= \int_0^{2\pi} \sqrt{2}\, dt = 2\pi\sqrt{2}. \end{aligned}$$

■

Exercises 13.1

Give the domain of **f**.

1. $\mathbf{f}(t) = \left\langle \frac{1}{t}, t^2, \frac{t}{t-1} \right\rangle$

2. $\mathbf{f}(t) = \langle (1-t^2)^{1/2}, (4-t^2)^{1/2}, t^{-1} \rangle$

3. $\mathbf{f}(t) = t\mathbf{i} - (\ln t)\mathbf{j} + e^t\mathbf{k}$

4. $\mathbf{f}(t) = \left(1 + \frac{1}{t}\right)\mathbf{i} - \left(\frac{1}{\sqrt{t}}\right)\mathbf{j}$

Describe the graph of **f**.

5. $\mathbf{f}(t) = \langle t, t+1, t-1 \rangle$

6. $\mathbf{f}(t) = \langle a_1t + b_1, a_2t + b_2, a_3t + b_3 \rangle$

7. $\mathbf{f}(t) = \langle \cos t, \sin t, -t \rangle$

8. $\mathbf{f}(t) = \langle 2 \cos t, 3 \sin t, 4t \rangle$

9. $\mathbf{f}(t) = t\mathbf{i} + t\mathbf{j} + t^2\mathbf{k}$

10. $\mathbf{f}(t) = t\mathbf{i} + t^2\mathbf{j} + t^3\mathbf{k}$

For the given **f** *and* **g**, *find:*
(a) $\mathbf{f} + \mathbf{g}$ *(b)* $\mathbf{f} \cdot \mathbf{g}$ *(c)* $\mathbf{f} \times \mathbf{g}$ *(d)* $\mathbf{f}'$
(e) $\mathbf{g}'$ *(f)* $(\mathbf{f} + \mathbf{g})'$ *(g)* $(\mathbf{f} \cdot \mathbf{g})'$ *(h)* $(\mathbf{f} \times \mathbf{g})'$

11. $\mathbf{f}(t) = \langle 2t, t^2, t-1 \rangle$; $\mathbf{g}(t) = \langle t^2, t^3, 2 \rangle$
12. $\mathbf{f}(t) = \langle e^t, t, 1 \rangle$; $\mathbf{g}(t) = \langle 2t, e^t, -t \rangle$
13. $\mathbf{f}(t) = \langle t, \ln t, \tan t \rangle$; $\mathbf{g}(t) = \langle 0, t^2, \ln t \rangle$
14. $\mathbf{f}(t) = \langle t - \cos t, 1 + \sin t, 0 \rangle$; $\mathbf{g}(t) = \langle 1 + \sin t, \cos t, 0 \rangle$

For the given **f** *and u, find:*
(a) $u\mathbf{f}$ *(b)* u' *(c)* $\mathbf{f}'$ *(d)* $(u\mathbf{f})'$

15. $\mathbf{f}(t) = \langle 1 + 3t, 1 - 3t, t^2 \rangle$; $u(t) = t^3$
16. $\mathbf{f}(t) = \langle t, 3t, t - 3 \rangle$; $u(t) = 1/t$
17. $\mathbf{f}(t) = \langle \cos t, 2 \sin t, t \rangle$; $u(t) = \sin 2t$
18. $\mathbf{f}(t) = \langle e^{2t}, e^{-t}, \ln(1 + t) \rangle$; $u(t) = t^2$

Find equations of the line tangent to the graph of **f** *at the given point.*

19. $\mathbf{f}(t) = \langle \cos t, \sin t, 2t \rangle$; $t = \pi$
20. $\mathbf{f}(t) = \langle t - \cos t, 1 + \sin t, 0 \rangle$; $t = \pi/2$
21. $\mathbf{f}(t) = \langle t \sin t, 1, 1 - \cos t \rangle$; $t = 0$
22. $\mathbf{f}(t) = \langle e^t + e^{-t}, e^t - e^{-t}, 2t \rangle$; $t = 0$

Find $\mathbf{f}(t)$, *given* $\mathbf{f}'(t)$ *and the boundary condition.*

23. $\mathbf{f}'(t) = \langle 2, 1, 3 \rangle$; $\mathbf{f}(0) = \langle 1, 0, 3 \rangle$
24. $\mathbf{f}'(t) = \langle t, 2t, 3t \rangle$; $\mathbf{f}(0) = \langle -1, -1, -2 \rangle$
25. $\mathbf{f}'(t) = \langle e^t, 1/t, e^{-t} \rangle$; $\mathbf{f}(1) = \langle e, 1, 1/e \rangle$
26. $\mathbf{f}'(t) = \langle 20, 30, -16t \rangle$; $\mathbf{f}(0) = \langle 0, 0, 0 \rangle$

Find the length of the graph of **f** *over the given interval.*

27. $\mathbf{f}(t) = \langle 3 \sin t, 2, 3 \cos t \rangle$; $[0, \pi]$
28. $\mathbf{f}(t) = \langle t - \cos t, 1 + \sin t, 0 \rangle$; $[0, 2\pi]$
29. $\mathbf{f}(t) = \langle 1, t, \ln(\sec t) \rangle$; $[0, \pi/4]$
30. $\mathbf{f}(t) = \langle e^t, 2e^t, 1 - 3e^t \rangle$; $[0, 1]$

Derive the following formulas:

31. $(\mathbf{f} + \mathbf{g})'(t) = \mathbf{f}'(t) + \mathbf{g}'(t)$
32. $(u\mathbf{f})'(t) = u'(t)\mathbf{f}(t) + u(t)\mathbf{f}'(t)$
33. $(\mathbf{f} \cdot \mathbf{g})'(t) = \mathbf{f}'(t) \cdot \mathbf{g}(t) + \mathbf{f}(t) \cdot \mathbf{g}'(t)$
34. $(\mathbf{f} \times \mathbf{g})'(t) = \mathbf{f}'(t) \times \mathbf{g}(t) + \mathbf{f}(t) \times \mathbf{g}'(t)$
35. Can you find a vector function (other than one with constant components) such that $\mathbf{f} \cdot \mathbf{f}' = 0$?
36. Does there exist a vector function $\mathbf{f}$ such that $\mathbf{f} \times \mathbf{f}' = \mathbf{0}$?
37. Show that if the individual limits exist, then:

    (a) $\lim_{t \to c} [u(t)\mathbf{f}(t)] = \left[\lim_{t \to c} u(t)\right]\left[\lim_{t \to c} \mathbf{f}(t)\right]$

    (b) $\lim_{t \to c} [\mathbf{f}(t) \times \mathbf{g}(t)] = \left[\lim_{t \to c} \mathbf{f}(t)\right] \times \left[\lim_{t \to c} \mathbf{g}(t)\right]$
38. A **normal plane** to a curve C at point P is a plane through P such that the tangent vector to C at P is normal to the plane. Find the equation of the normal plane to the graph of $\mathbf{f}(t) = \langle t, e^t, te^{-t} \rangle$ at $t = 0$.
39. Show that $(\mathbf{f}' \cdot \mathbf{f})' = \|\mathbf{f}'\|^2 + \mathbf{f} \cdot \mathbf{f}''$.
40. Show that

$$\frac{d}{dt}\left[\frac{\mathbf{f}(t)}{u(t)}\right] = \frac{u(t)\mathbf{f}'(t) - u'(t)\mathbf{f}(t)}{[u(t)]^2}.$$

41. Let $\theta(t)$ be the measure of the angle between $\mathbf{f}(t)$ and $\mathbf{g}(t)$. Find $\theta'(t)$.
42. Establish the vector form of the chain rule,

$$\frac{d}{dt}\mathbf{f}(u(t)) = u'(t)\mathbf{f}'(u(t)).$$

SECTION 13.2

Curvilinear Motion

Motion that takes place along a straight line was discussed in Section 4.7. Here, we consider **curvilinear motion**—that is, motion that takes place along a curved path in space.

Motion in the Plane

We first consider curvilinear motion that takes place in a plane, because it can be described as taking place in a two-dimensional coordinate system. For example, although the earth moves through space as it goes around the sun, its orbit lies in a plane that passes through the sun.

An object moving in the plane has at any time t a position $(x(t), y(t))$. We can therefore describe the position of the object by a two-dimensional vector function

$$\mathbf{r}(t) = \langle x(t), y(t) \rangle = x(t)\mathbf{i} + y(t)\mathbf{j}, \tag{1}$$

which is called the **position vector** of the object. The position vector is most conveniently represented geometrically by the position arrow, and the graph of $\mathbf{r}(t)$ is the **path** along which the object moves (see Figure 13.6).

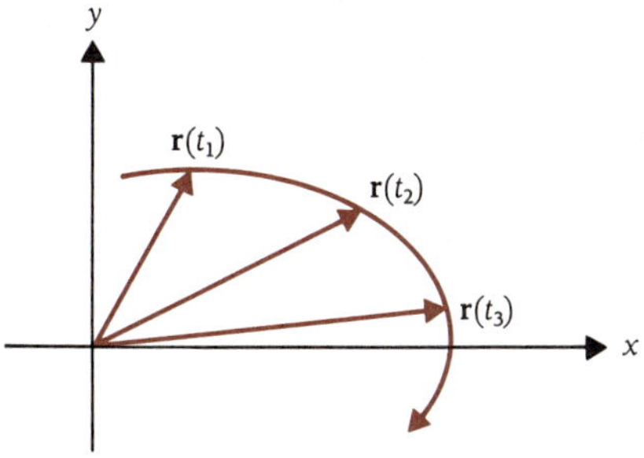

FIGURE 13.6 The position of a moving object

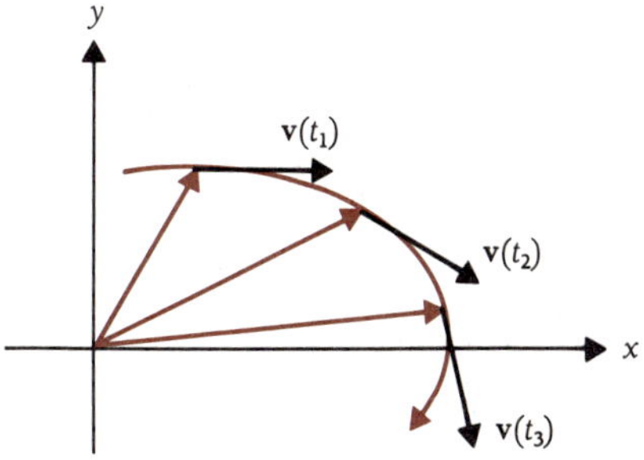

FIGURE 13.7 The tangential velocity of a moving object

The vector function

$$\mathbf{v}(t) = \mathbf{r}'(t) = \langle x'(t), y'(t) \rangle \tag{2}$$

is called the **velocity** of the object. Because $\mathbf{r}'(t)$ is parallel to the tangent to the path, $\mathbf{v}(t)$ is also called the **tangential velocity** of the object. The velocity of the object at any time t is in the direction of motion (see Figure 13.7).

If $s(t)$ is the length of the path of the object from some initial point $\mathbf{r}(t_0)$ to the point $\mathbf{r}(t)$, then by equation (15) of Section 13.1, we have

$$\frac{ds}{dt} = \|\mathbf{r}'(t)\| = \|\mathbf{v}(t)\|. \tag{3}$$

Thus, $\|\mathbf{v}(t)\|$ is the rate of change of length of path (distance traveled) with respect to time, and so is the **speed** of the object. That is, speed is the magnitude of the velocity.

The vector function

$$\mathbf{a}(t) = \mathbf{v}'(t) = \mathbf{r}''(t) = \langle x''(t), y''(t) \rangle \tag{4}$$

is called the **acceleration** of the object. This is the change in velocity with respect to time.

Example 13.12 For an object with position vector $\mathbf{r}(t) = a(\cos \omega t)\mathbf{i} + a(\sin \omega t)\mathbf{j}$ $(a > 0)$, find:

(a) The velocity **(b)** The speed **(c)** The acceleration
(d) The angle between velocity and acceleration

Solution

(a) The velocity is

$$\mathbf{v}(t) = \mathbf{r}'(t) = -a\omega(\sin \omega t)\mathbf{i} + a\omega(\cos \omega t)\mathbf{j}.$$

Note that $\mathbf{v}(t)$ is orthogonal to $\mathbf{r}(t)$ in this case.

(b) The speed is

$$\begin{aligned}\|\mathbf{v}(t)\| &= \sqrt{(-a\omega \sin \omega t)^2 + (a\omega \cos \omega t)^2} \\ &= \sqrt{a^2\omega^2(\sin^2\omega t + \cos^2\omega t)} = a|\omega|.\end{aligned}$$

(c) The acceleration is

$$\begin{aligned}\mathbf{a}(t) = \mathbf{v}'(t) &= -a\omega^2(\cos \omega t)\mathbf{i} - a\omega^2(\sin \omega t)\mathbf{j} \\ &= -\omega^2[a(\cos \omega t)\mathbf{i} + a(\sin \omega t)\mathbf{j}] \\ &= -\omega^2\mathbf{r}(t).\end{aligned}$$

(d) The dot product of $\mathbf{a}(t)$ and $\mathbf{v}(t)$ is

$$\begin{aligned}\mathbf{a}(t) \cdot \mathbf{v}(t) &= [-a\omega^2(\cos \omega t)\mathbf{i} - a\omega^2(\sin \omega t)\mathbf{j}] \cdot [-a\omega(\sin \omega t)\mathbf{i} + a\omega(\cos \omega t)\mathbf{j}] \\ &= a^2\omega^3 \cos \omega t \sin \omega t - a^2\omega^3 \sin \omega t \cos \omega t = 0.\end{aligned}$$

Thus, the angle between $\mathbf{a}(t)$ and $\mathbf{v}(t)$ is $\pi/2$. ■

The object in Example 13.12 has as its path a circle of radius a (see Figure 13.8). Since the speed of the object is constant, we call the motion **uniform circular motion**. Note that the acceleration has direction opposite to the position vector, so it is directed toward the center of the circle (see Figure 13.9); for this reason the acceleration is called **centripetal acceleration**. The magnitude of the centripetal acceleration is called the **centrifugal acceleration**, and in the case of Example 13.12 is $a\omega^2$. The object makes one complete circuit of its path in a time interval of length $2\pi/\omega$, because the functions $\cos \omega t$ and $\sin \omega t$ are periodic with period $2\pi/\omega$. The **angular velocity** of the object is $\omega/2\pi$ revolutions, or ω radians, per unit time.

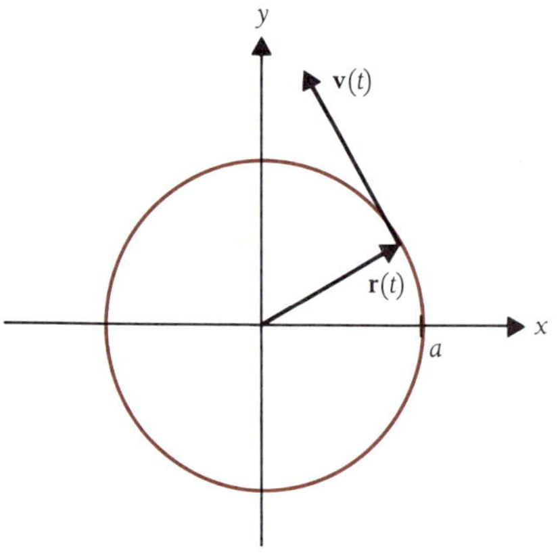

FIGURE 13.8 Position and velocity in uniform circular motion

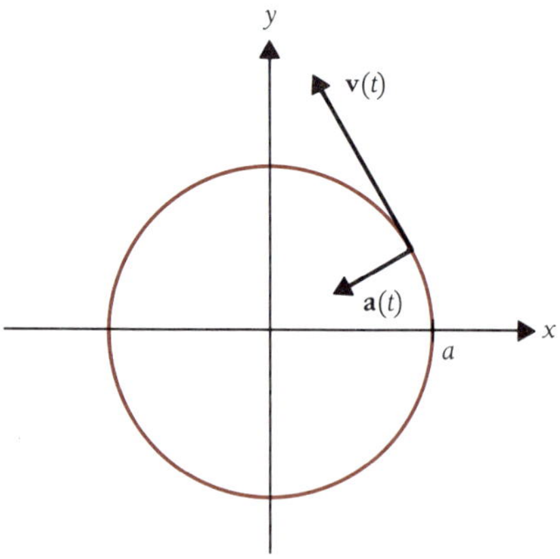

FIGURE 13.9 Velocity and acceleration in uniform circular motion

Example 13.13 A centrifuge has a radius of 5 feet, and it is desired to produce a centrifugal acceleration twice that of gravity. What angular velocity, in revolutions per minute (rpm), will produce the desired acceleration?

Solution Because 5 is the radius, the centrifugal acceleration is $5\omega^2$, where ω is the angular velocity in radians per unit time. The acceleration of gravity is $g \approx 32$ feet per second/second $= 32(60)^2$ feet per minute/minute. We thus need ω such that $5\omega^2 = 2g$, or

$$\omega = \sqrt{\tfrac{2}{5}g} = 96\sqrt{5} \text{ radians per minute.}$$

The desired angular velocity is therefore

$$\frac{\omega}{2\pi} = \frac{48\sqrt{5}}{\pi} \approx 34.16 \text{ revolutions per minute.}$$

■

Example 13.14 A projectile is fired with an initial speed of v and at an angle of θ radians from the horizontal. Assuming only the acceleration g of gravity acts on the projectile, find the equation of its trajectory.

Solution The trajectory (path) of the projectile is a curve in a vertical plane. We will let the distance downrange be represented by x and the altitude by y, and we will assume that the projectile was fired from the origin (see Figure 13.10). Since gravity acts only in the vertical direction and downward, we can represent the acceleration of the projectile by the vector function

$$\mathbf{a}(t) = \langle 0, -g \rangle.$$

That is,

$$\mathbf{v}'(t) = \langle 0, -g \rangle.$$

Integrating, we find

$$\mathbf{v}(t) = \langle C_1, -gt + C_2 \rangle.$$

To evaluate the constants C_1 and C_2, we use the initial condition that the speed is v and the angle of elevation is θ when $t = 0$. Thus, the horizontal component of velocity must be $v \cos \theta$, and the vertical component must be $v \sin \theta$. That is,

$$\mathbf{v}(0) = \langle v \cos \theta, v \sin \theta \rangle.$$

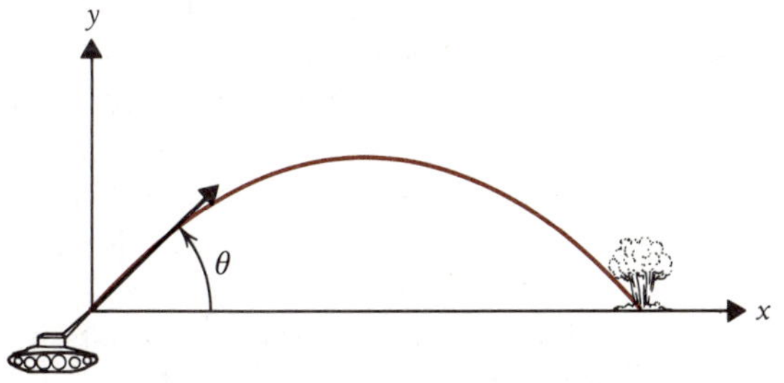

FIGURE 13.10 Trajectory of a projectile

Thus, $C_1 = v \cos\theta$ and $C_2 = v \sin\theta$, so that

$$\mathbf{v}(t) = \langle v \cos\theta,\ v \sin\theta - gt \rangle.$$

Therefore,

$$\mathbf{r}'(t) = \langle v \cos\theta,\ v \sin\theta - gt \rangle.$$

Integrating again, we get

$$\mathbf{r}(t) = \left\langle (v \cos\theta)t + C_3,\ (v \sin\theta)t - \frac{gt^2}{2} + C_4 \right\rangle.$$

To evaluate C_3 and C_4, we recall that $\mathbf{r}(0) = \langle 0, 0 \rangle$. Hence, $C_3 = C_4 = 0$, and

$$\mathbf{r}(t) = \left\langle (v \cos\theta)t,\ (v \sin\theta)t - \frac{gt^2}{2} \right\rangle.$$

The trajectory therefore has parametric equations

$$\begin{cases} x = (v \cos\theta)t \\ y = (v \sin\theta)t - \dfrac{gt^2}{2}, \end{cases}$$

or, eliminating the parameter t,

$$\begin{aligned} y &= (v \sin\theta)\frac{x}{v\cos\theta} - \frac{1}{2}g\left(\frac{x}{v\cos\theta}\right)^2 \\ &= (\tan\theta)x - \left(\frac{g}{2v^2\cos^2\theta}\right)x^2. \end{aligned} \tag{5}$$

The trajectory is a parabola. ■

■ **Example 13.15** Assume the projectile of Example 13.14 is fired over level ground. With what angle of elevation will the projectile achieve its maximum range?

Solution When the projectile lands, the altitude will be 0. Hence, from equation (5), we must have

$$\begin{aligned} (\tan\theta)x - \left(\frac{g}{2v^2\cos^2\theta}\right)x^2 &= 0, \\ x(2v^2 \sin\theta\cos\theta - gx) &= 0. \end{aligned}$$

Thus, the altitude is 0 when $x = 0$ (when the projectile is fired) and when $x = 2(v^2/g)\sin\theta\cos\theta$ (when the projectile lands). This second value of x is the range. We thus seek the value of θ that will maximize the function

$$x = 2\left(\frac{v^2}{g}\right)\sin\theta\cos\theta = \left(\frac{v^2}{g}\right)\sin 2\theta.$$

Since $dx/d\theta = 2(v^2/g)\cos 2\theta$ is 0 when $\theta = \pi/4$, the second derivative test tells us that the range is maximum when the angle of elevation θ is $\pi/4$, or 45°. The maximum range turns out to be v^2/g. ■

Motion in Space

The same ideas governing curvilinear motion in the plane can be extended to motion in space by using three-dimensional vectors. That is, the position vector of an object moving in space is a function of the form

$$\mathbf{r}(t) = \langle x(t), y(t), z(t) \rangle.$$

As before, the velocity and acceleration are $\mathbf{v}(t) = \mathbf{r}'(t)$ and $\mathbf{a}(t) = \mathbf{v}'(t) = \mathbf{r}''(t)$, and the speed is $ds/dt = \|\mathbf{r}'(t)\| = \|\mathbf{v}(t)\|$.

Example 13.16 A particle has position vector

$$\mathbf{r}(t) = \langle a \sin t, bt, a \cos t \rangle.$$

Find its velocity, speed, and acceleration, and describe its path.

Solution The path of the particle is a helix of radius a, winding around the y-axis (Figure 13.11). The velocity is

$$\mathbf{v}(t) = \langle a \cos t, b, -a \sin t \rangle,$$

the speed is

$$\begin{aligned} \|\mathbf{v}(t)\| &= \sqrt{a^2 \cos^2 t + b^2 + a^2 \sin^2 t} \\ &= \sqrt{a^2 + b^2}, \end{aligned}$$

and the acceleration is

$$\mathbf{a}(t) = \langle -a \sin t, 0, -a \cos t \rangle.$$

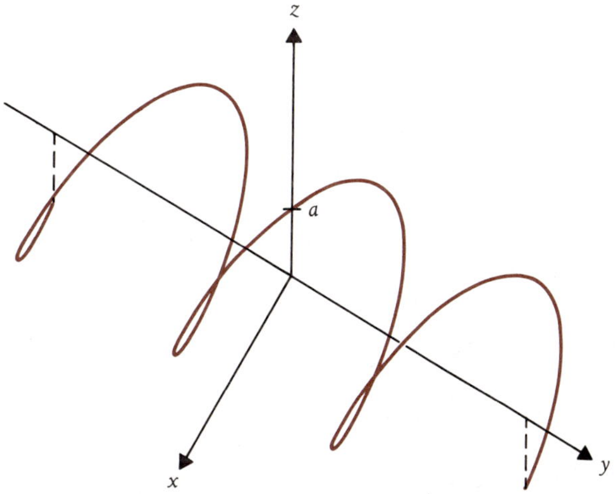

FIGURE 13.11 The helical path $\mathbf{r}(t) = \langle a \sin t, bt, a \cos t \rangle$ ■

Exercises 13.2

Given the position vector $\mathbf{r}$*, find the velocity, speed, and acceleration.*

1. $\mathbf{r}(t) = \langle t, 1-t, 2-3t \rangle$
2. $\mathbf{r}(t) = \langle t^2, 1/t, t^3 \rangle$
3. $\mathbf{r}(t) = \langle 3t, 2\sin t, 2\cos t \rangle$
4. $\mathbf{r}(t) = \langle 3\cos 2t, 3\sin 2t \rangle$
5. $\mathbf{r}(t) = (\sin 3t)\mathbf{i} - (\cos 3t)\mathbf{j} + e^t\mathbf{k}$
6. $\mathbf{r}(t) = 2t\mathbf{j} + t^2\mathbf{k}$

Given the velocity or acceleration and the boundary conditions, find the position vector.

7. $\mathbf{v}(t) = \langle t/2, 2, 1/t \rangle$; $\mathbf{r}(1) = \langle 0, 0, 1 \rangle$
8. $\mathbf{v}(t) = (t+1)\mathbf{i} - \mathbf{j} - t^2\mathbf{k}$; $\mathbf{r}(0) = \mathbf{i} + \mathbf{j}$
9. $\mathbf{a}(t) = \langle 0, 1, 2 \rangle$; $\mathbf{v}(0) = \langle 2, 1, 0 \rangle$; $\mathbf{r}(0) = \langle 0, 1, 0 \rangle$
10. $\mathbf{a}(t) = (\cos 2t)\mathbf{j} + (\sin 2t)\mathbf{k}$; $\mathbf{v}(0) = 3\mathbf{i} + \frac{1}{2}\mathbf{k}$; $\mathbf{r}(0) = \frac{1}{4}\mathbf{j}$
11. A centrifuge of radius 2 feet develops a centrifugal acceleration of 6g, where $g \approx 32$ feet per second/second. What is its angular velocity?
12. A man throws a ball at a speed of 80 miles per hour. If his arm (length 32 inches) is in uniform circular motion at the time the ball is released, what is its angular velocity?
13. A grindstone of radius 6 centimeters turns at the rate of 250 revolutions per minute. What is the centrifugal acceleration at the rim?
14. A tractor with tire diameter 5 feet travels down the road at 20 miles per hour. Mud is being thrown off the tire tangentially. With what speed (relative to the road) does the mud leave the tire? [*Hint:* The speed will be a function of the point from which the mud leaves the tire. The forward motion of the tractor must be taken into account.]
15. In Problem 14, how high into the air may the mud be thrown?
16. Find the maximum altitude of the projectile in Example 13.14.
17. If a person throws a ball vertically into the air at a speed of 80 miles per hour, how far does it rise?
18. If a ballplayer can throw a ball 80 miles per hour, how far can he throw it across a level field to another player?
19. Prove that

$$\mathbf{a}(t) = \mathbf{v}(t)\,\frac{d\mathbf{v}}{ds}.$$

20. An object with mass m and velocity $\mathbf{v}(t)$ has **momentum** $\mathbf{p}(t) = m\mathbf{v}(t)$. Using $\mathbf{F}(t) = m\mathbf{a}(t)$, show that $\mathbf{p}'(t) = \mathbf{F}(t)$.
21. If an object of mass m moves at velocity $\mathbf{v}(t)$ under a force $\mathbf{F}(t)$, the **power** expended by the force is $\mathbf{F}(t) \cdot \mathbf{v}(t)$, and the **kinetic energy** of the object is $\frac{1}{2}m\|\mathbf{v}(t)\|^2$. Show that the power is the derivative (with respect to t) of the kinetic energy.
22. A force $\mathbf{F}(t)$ acting on an object of mass m with position $\mathbf{r}(t)$ produces a **torque** $\boldsymbol{\tau}(t) = \mathbf{r}(t) \times \mathbf{F}(t)$ and an **angular momentum** $\mathbf{L}(t) = \mathbf{r}(t) \times m\mathbf{v}(t)$. Show that $\mathbf{L}'(t) = \boldsymbol{\tau}(t)$.
23. Refer to Problem 22. Show that a force that keeps an object in uniform circular motion produces no torque.
24. Show that motion with 0 acceleration is rectilinear.
25. Show that if the speed of an object is constant, then the velocity and acceleration vectors are orthogonal.

SECTION 13.3

Curvature

To describe the shape of a curve in space, we must develop the notion of curvature.

Curvature

Let curve C be the graph of the vector function $\mathbf{f}(t) = \langle x(t), y(t), z(t)\rangle$, and let P be the point $(x(t), y(t), z(t))$ on C. Since the derivative $\mathbf{f}'(t)$ is a direction vector (in the direction of increasing t) of the tangent line to C at P, the unit vector

$$\mathbf{T}(t) = \frac{\mathbf{f}'(t)}{\|\mathbf{f}'(t)\|} \tag{1}$$

is called the **unit tangent vector** of C at P. Note that if $s(t)$ is the arc length of C, then $ds/dt = \|\mathbf{f}'(t)\|$ implies by the chain rule that

$$\mathbf{f}'(t) = \frac{d\mathbf{f}}{dt} = \left(\frac{ds}{dt}\right)\left(\frac{d\mathbf{f}}{ds}\right) = \|\mathbf{f}'(t)\|\frac{d\mathbf{f}}{ds},$$

or

$$\mathbf{T}(t) = \frac{d\mathbf{f}}{ds}. \tag{2}$$

That is, the unit tangent vector is the rate of change of $\mathbf{f}$ with respect to arc length.

Now the unit tangent vector $\mathbf{T}(t)$ does not change in magnitude, but only in direction. The rate of change of $\mathbf{T}$ with respect to arc length is called the **curvature vector** of C, and is denoted by the Greek letter kappa,

$$\boldsymbol{\kappa}(t) = \frac{d\mathbf{T}}{ds}. \tag{3}$$

The curvature vector indicates how the curve is changing direction and turns out to be orthogonal to the unit tangent vector, as we will see later. The magnitude of the curvature vector is called the **curvature** of C and is denoted by

$$\kappa(t) = \|\boldsymbol{\kappa}(t)\|. \tag{4}$$

The curvature tells how sharply the curve is curving.

To compute the curvature vector, we first find $\mathbf{T}(t)$ using equation (1). Then by equation (3) and the chain rule, we have

$$\boldsymbol{\kappa}(t) = \frac{d\mathbf{T}}{ds} = \frac{d\mathbf{T}/dt}{ds/dt} = \frac{\mathbf{T}'(t)}{\|\mathbf{f}'(t)\|}. \tag{5}$$

■ **Example 13.17** Show that the line

$$\mathbf{f}(t) = \langle x_1 + at, y_1 + bt, z_1 + ct\rangle$$

has 0 curvature.

Solution We first find

$$\mathbf{f}'(t) = \langle a, b, c \rangle,$$

so that

$$\|\mathbf{f}'(t)\| = \sqrt{a^2 + b^2 + c^2}.$$

Thus, $\mathbf{T}(t)$ is the constant vector

$$\mathbf{T}(t) = \frac{\langle a, b, c \rangle}{\sqrt{a^2 + b^2 + c^2}}$$

so that $\mathbf{T}'(t) = \mathbf{0}$ and $\boldsymbol{\kappa}(t) = \mathbf{0}$. It follows that the curvature $\kappa(t)$ is 0. ■

■ **Example 13.18** Show that the circle

$$\mathbf{f}(t) = \langle a \cos t, a \sin t, 0 \rangle$$

of radius a has curvature $1/a$.

Solution We find

$$\mathbf{f}'(t) = \langle -a \sin t, a \cos t, 0 \rangle,$$

so that

$$\|\mathbf{f}'(t)\| = \sqrt{a^2 \sin^2 t + a^2 \cos^2 t} = a.$$

Hence,

$$\mathbf{T}(t) = \langle -\sin t, \cos t, 0 \rangle.$$

Then

$$\mathbf{T}'(t) = \langle -\cos t, -\sin t, 0 \rangle,$$

so that

$$\boldsymbol{\kappa}(t) = \frac{1}{a} \langle -\cos t, -\sin t, 0 \rangle.$$

Hence,

$$\kappa(t) = \frac{1}{a} \sqrt{\cos^2 t + \sin^2 t} = \frac{1}{a}.$$ ■

Because of the result found in Example 13.18, if the curvature of C at P is not 0, we call its reciprocal the **radius of curvature** of C at P and denote it by the Greek letter rho,

$$\rho(t) = \frac{1}{\kappa(t)}. \tag{6}$$

Thus, the radius of curvature at P is the radius of a circle whose curvature is the same as the curvature of C at P.

Now let Q be the point such that

$$\mathbf{v}(\overrightarrow{PQ}) = [\rho(t)]^2\boldsymbol{\kappa}(t) = \rho(t)\,\frac{\boldsymbol{\kappa}(t)}{\|\boldsymbol{\kappa}(t)\|}. \tag{7}$$

We call Q the **center of curvature** of C at P. The **circle of curvature** of C at P is the circle with center Q and radius $\rho(t)$ that lies in the plane containing Q and the line tangent to C at P (see Figure 13.12). Thus, the tangent line to C at P is also tangent to the circle of curvature. In a sense, the circle of curvature is the circle that best approximates C at P. The plane containing the circle of curvature is called the **osculating plane** of C at P.

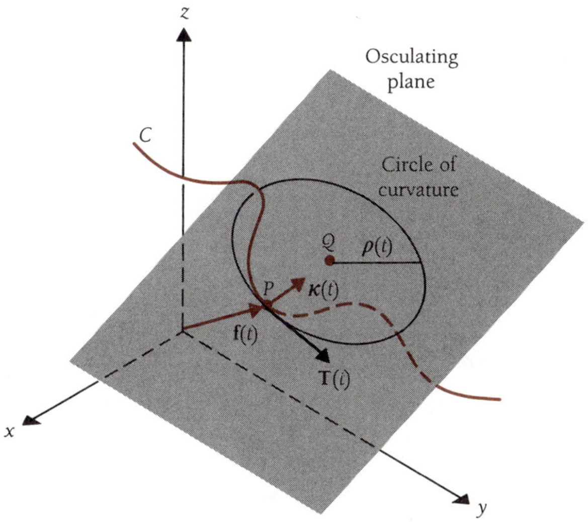

FIGURE 13.12 Radius and circle of curvature

■ **Example 13.19** For the helix

$$\mathbf{f}(t) = \langle a\cos t,\ a\sin t,\ bt\rangle,$$

find the curvature, the radius of curvature, the center of curvature, and the osculating plane at the point $t = 0$.

Solution To find the curvature, we compute as follows:

$$\mathbf{f}'(t) = \langle -a\sin t,\ a\cos t,\ b\rangle,$$

$$\|\mathbf{f}'(t)\| = \sqrt{a^2\sin^2 t + a^2\cos^2 t + b^2} = \sqrt{a^2 + b^2},$$

$$\mathbf{T}(t) = \frac{1}{\sqrt{a^2+b^2}}\langle -a\sin t,\ a\cos t,\ b\rangle,$$

$$\mathbf{T}'(t) = \frac{1}{\sqrt{a^2 + b^2}} \langle -a \cos t, -a \sin t, 0 \rangle,$$

$$\boldsymbol{\kappa}(t) = \frac{1}{a^2 + b^2} \langle -a \cos t, -a \sin t, 0 \rangle,$$

$$\kappa(t) = \frac{1}{a^2 + b^2} \sqrt{a^2 \cos^2 t + a^2 \sin^2 t} = \frac{|a|}{a^2 + b^2}.$$

This is the curvature. The radius of curvature is thus

$$\rho(t) = \frac{a^2 + b^2}{|a|}.$$

At $t = 0$, P is the point $(a, 0, 0)$ and $\boldsymbol{\kappa}(0) = \langle -a/(a^2 + b^2), 0, 0 \rangle$ is the curvature vector. Let the center of curvature be $Q(x, y, z)$. Then $\mathbf{v}(\overrightarrow{PQ}) = \langle x - a, y, z \rangle$, and by equation (7),

$$\begin{aligned} \mathbf{v}(\overrightarrow{PQ}) &= [\rho(0)]^2 \boldsymbol{\kappa}(0) \\ &= \frac{(a^2 + b^2)^2}{a^2} \left\langle -\frac{a}{a^2 + b^2}, 0, 0 \right\rangle \\ &= \left\langle -\frac{a^2 + b^2}{a}, 0, 0 \right\rangle. \end{aligned}$$

Hence, $x - a = -(a^2 + b^2)/a$, $y = 0$, and $z = 0$, so Q is the point $(-b^2/a, 0, 0)$. This is the center of curvature. Finally, the osculating plane contains the line tangent to the helix at P and the line PQ, so the plane has normal vector

$$\begin{aligned} \mathbf{v}(\overrightarrow{PQ}) \times \mathbf{f}'(0) &= \left\langle -\frac{a^2 + b^2}{a}, 0, 0 \right\rangle \times \langle 0, a, b \rangle \\ &= \left\langle 0, \frac{b(a^2 + b^2)}{a}, -(a^2 + b^2) \right\rangle \\ &= \frac{a^2 + b^2}{a} \langle 0, b, -a \rangle. \end{aligned}$$

Thus, $0, b, -a$ are attitude numbers of the plane, and the equation of the osculating plane is

$$\begin{aligned} 0(x - a) + b(y - 0) - a(z - 0) &= 0, \\ by - az &= 0. \end{aligned}$$

■

Tangential and Normal Components of Acceleration

To develop other properties of the curvature vector, first note that if $\mathbf{u}$ is any unit vector, then $\mathbf{u} \cdot \mathbf{u} = 1$. In particular,

$$\mathbf{T}(t) \cdot \mathbf{T}(t) = 1.$$

Differentiating with respect to t, we get

$$\mathbf{T}'(t)\cdot\mathbf{T}(t)+\mathbf{T}(t)\cdot\mathbf{T}'(t)=0,$$

which says that

$$\mathbf{T}'(t)\cdot\mathbf{T}(t)=0.$$

Hence,

$$\boldsymbol{\kappa}(t)\cdot\mathbf{T}(t)=\frac{1}{\|\mathbf{f}'(t)\|}\mathbf{T}'(t)\cdot\mathbf{T}(t)=0,$$

or the curvature vector is orthogonal to the unit tangent vector.

The unit vector

$$\mathbf{N}(t)=\frac{\boldsymbol{\kappa}(t)}{\kappa(t)}=\rho(t)\boldsymbol{\kappa}(t) \tag{8}$$

is therefore orthogonal to the unit tangent vector and is called the **unit normal vector**. Components of any vector along $\mathbf{T}(t)$ and $\mathbf{N}(t)$ are called the **tangential** and **normal components** of that vector.

These ideas are particularly interesting when applied to curvilinear motion. We will derive the tangential and normal components of acceleration.

Let $\mathbf{r}(t)$ be the position vector. Then the speed of the object is $ds/dt=\|\mathbf{r}'(t)\|$, so by equation (1),

$$\mathbf{r}'(t)=\frac{ds}{dt}\mathbf{T}(t). \tag{9}$$

Differentiating this equation with respect to t, we get

$$\mathbf{r}''(t)=\frac{d^2s}{dt^2}\mathbf{T}(t)+\frac{ds}{dt}\mathbf{T}'(t). \tag{10}$$

Now, from equation (5), we have

$$\begin{aligned}\mathbf{T}'(t)&=\frac{ds}{dt}\boldsymbol{\kappa}(t)\\&=\frac{ds}{dt}\frac{1}{\rho(t)}\mathbf{N}(t)\qquad\text{[By equation (8)]}\end{aligned}$$

Putting this into equation (10), we get

$$\mathbf{r}''(t)=\frac{d^2s}{dt^2}\mathbf{T}(t)+\left(\frac{ds}{dt}\right)^2\frac{1}{\rho(t)}\mathbf{N}(t). \tag{11}$$

Now the component of acceleration in the direction of $\mathbf{T}(t)$ is

$$\begin{aligned}\mathbf{r}''(t)\cdot\mathbf{T}(t)&=\frac{d^2s}{dt^2}\mathbf{T}(t)\cdot\mathbf{T}(t)+\left(\frac{ds}{dt}\right)^2\frac{1}{\rho(t)}\mathbf{N}(t)\cdot\mathbf{T}(t)\\&=\frac{d^2s}{dt^2},\end{aligned} \tag{12}$$

since $\mathbf{T}(t)$ is a unit vector and $\mathbf{N}(t)$ is orthogonal to $\mathbf{T}(t)$. That is, the tangential component of acceleration is the rate of change of the speed with respect to time. For example, it is the tangential component of acceleration that pushes you back in your seat as an airplane takes off.

Similarly, the normal component of acceleration is

$$\begin{aligned}\mathbf{r}''(t)\cdot\mathbf{N}(t) &= \frac{d^2s}{dt^2}\mathbf{T}(t)\cdot\mathbf{N}(t)+\left(\frac{ds}{dt}\right)^2\frac{1}{\rho(t)}\mathbf{N}(t)\cdot\mathbf{N}(t)\\ &= \left(\frac{ds}{dt}\right)^2\frac{1}{\rho(t)},\end{aligned} \tag{13}$$

which is the square of the speed divided by the radius of curvature. For example, it is the normal component of acceleration that pushes you against the car door as you drive around a curve. Note that if the speed is high or the radius of curvature is small, the normal component of acceleration is large. This is the theoretical equivalent of the obvious fact that to negotiate a sharp turn, one must usually reduce the speed.

Next, we express the tangential and normal components of acceleration in terms of the velocity and acceleration alone. Combining equations (9) and (11) by the dot product, we get

$$\begin{aligned}\mathbf{r}'(t)\cdot\mathbf{r}''(t) &= \frac{ds}{dt}\left(\frac{d^2s}{dt^2}\right)\mathbf{T}(t)\cdot\mathbf{T}(t)+\left(\frac{ds}{dt}\right)^3\frac{1}{\rho(t)}\mathbf{N}(t)\cdot\mathbf{T}(t)\\ &= \frac{ds}{dt}\left(\frac{d^2s}{dt^2}\right),\end{aligned}$$

because $\mathbf{N}$ and $\mathbf{T}$ are orthogonal. Hence, the tangential component of acceleration is

$$\frac{d^2s}{dt^2}=\frac{\mathbf{r}'(t)\cdot\mathbf{r}''(t)}{ds/dt}=\frac{\mathbf{r}'(t)\cdot\mathbf{r}''(t)}{\|\mathbf{r}'(t)\|}. \tag{14}$$

Similarly, combining equations (9) and (11) by the cross product, we have

$$\begin{aligned}\mathbf{r}'(t)\times\mathbf{r}''(t) &= \frac{ds}{dt}\left(\frac{d^2s}{dt^2}\right)\mathbf{T}(t)\times\mathbf{T}(t)+\left(\frac{ds}{dt}\right)^3\frac{1}{\rho(t)}\mathbf{T}(t)\times\mathbf{N}(t)\\ &= \left(\frac{ds}{dt}\right)^3\frac{1}{\rho(t)}[\mathbf{T}(t)\times\mathbf{N}(t)].\end{aligned}$$

Since $\mathbf{T}$ and $\mathbf{N}$ are orthogonal unit vectors, we have $\|\mathbf{T}(t)\times\mathbf{N}(t)\|=1$, so that

$$\|\mathbf{r}'(t)\times\mathbf{r}''(t)\|=\left(\frac{ds}{dt}\right)^3\frac{1}{\rho(t)}.$$

Hence, the normal component of acceleration is

$$\left(\frac{ds}{dt}\right)^2\frac{1}{\rho(t)}=\frac{\|\mathbf{r}'(t)\times\mathbf{r}''(t)\|}{ds/dt}=\frac{\|\mathbf{r}'(t)\times\mathbf{r}''(t)\|}{\|\mathbf{r}'(t)\|}. \tag{15}$$

■ Example 13.20 The position vector of an object is $\mathbf{r}(t) = \langle e^t, t, t^2 \rangle$. Find the tangential and normal components of acceleration.

Solution We have

$$\begin{aligned}
\mathbf{r}'(t) &= \langle e^t, 1, 2t \rangle, \\
\|\mathbf{r}'(t)\| &= \sqrt{e^{2t} + 1 + 4t^2}, \\
\mathbf{r}''(t) &= \langle e^t, 0, 2 \rangle.
\end{aligned}$$

Preparing to use equations (14) and (15), we compute

$$\begin{aligned}
\mathbf{r}'(t) \cdot \mathbf{r}''(t) &= e^{2t} + 4t, \\
\mathbf{r}'(t) \times \mathbf{r}''(t) &= \langle 2, 2e^t(t-1), -e^t \rangle, \\
\|\mathbf{r}'(t) \times \mathbf{r}''(t)\| &= \sqrt{4 + 4e^{2t}(t-1)^2 + e^{2t}} \\
&= \sqrt{4 + e^{2t}(4t^2 - 8t + 5)}.
\end{aligned}$$

Therefore, the tangential component of acceleration is

$$\frac{e^{2t} + 4t}{\sqrt{e^{2t} + 4t^2 + 1}}$$

and the normal component is

$$\sqrt{\frac{4 + e^{2t}(4t^2 - 8t + 5)}{e^{2t} + 4t^2 + 1}}.$$

■

We can reinterpret formula (15) so as to obtain an alternate formula for curvature. Since $ds/dt = \|\mathbf{f}'(t)\|$ and $1/\rho(t) = \kappa(t)$, formula (15) gives us

$$\kappa(t) = \frac{\|\mathbf{f}'(t) \times \mathbf{f}''(t)\|}{\|\mathbf{f}'(t)\|^3}. \tag{16}$$

This is also expressible as

$$\kappa(t) = \frac{\|\mathbf{v}(t) \times \mathbf{a}(t)\|}{\|\mathbf{v}(t)\|^3}. \tag{17}$$

■ Example 13.21 Show that the curvature of the plane curve $\mathbf{f}(t) = \langle x(t), y(t) \rangle$ is

$$\kappa(t) = \frac{|x'(t)y''(t) - y'(t)x''(t)|}{(\sqrt{[x'(t)]^2 + [y'(t)]^2})^3}.$$

Solution We will use equation (16) with $\mathbf{f}(t) = \langle x(t), y(t), 0 \rangle$. Since $\mathbf{f}'(t) = \langle x'(t), y'(t), 0 \rangle$ and $\mathbf{f}''(t) = \langle x''(t), y''(t), 0 \rangle$, we have

$$\mathbf{f}'(t) \times \mathbf{f}''(t) = \langle 0, 0, x'(t)y''(t) - y'(t)x''(t) \rangle.$$

Thus,

$$\|\mathbf{f}'(t) \times \mathbf{f}''(t)\| = x'(t)y''(t) - y'(t)x''(t).$$

Also,

$$\|\mathbf{f}'(t)\| = \sqrt{[x'(t)]^2 + [y'(t)]^2},$$

so that the desired formula follows directly from equation (16). ■

■ Example 13.22 Show that the curvature of the plane curve $y = f(x)$ is

$$\kappa(x) = \frac{|f''(x)|}{(\sqrt{1 + [f'(x)]^2})^3}.$$

Solution We will let the curve be parametrized by setting $x = t$, $y = f(t)$. Then by Example 13.21,

$$\kappa(x) = \kappa(t) = \frac{|(1)f''(t) - f'(t)(0)|}{(\sqrt{[1]^2 + [f'(t)]^2})^3},$$

which reduces to the desired formula. ■

Exercises 13.3

For each curve, compute the curvature, the radius of curvature, the center of curvature, and the osculating plane at the given point.

1. $\mathbf{f}(t) = \langle t, 0, t^2 \rangle$; $t = 2$
2. $\mathbf{f}(t) = \langle 0, 3 \sin t, 3 \cos t \rangle$; $t = \pi/2$
3. $\mathbf{f}(t) = \langle \cos 2t, \sin 2t, t \rangle$; $t = \pi/6$
4. $\mathbf{f}(t) = \langle e^t + e^{-t}, e^t - e^{-t}, 4t \rangle$; $t = 0$

Find the tangential and normal components of acceleration.

5. $\mathbf{r}(t) = \langle \frac{3}{2}t^3, \frac{3}{2}t^2, t \rangle$
6. $\mathbf{r}(t) = \langle 3t, 2 \cos t, 2 \sin t \rangle$
7. $\mathbf{r}(t) = \langle 1 - t, t^2, 1 + t^2 \rangle$
8. $\mathbf{r}(t) = \langle \ln t, 2, t^2 \rangle$

Find the curvature of each plane curve.

9. $x = t^2$, $y = t^3$
10. $x = 1 + t$, $y = \sqrt{1 + t^2}$
11. $y = x^3$
12. $y = \sqrt{x^2 - 1}$
13. Show that the cross product of two orthogonal unit vectors is a unit vector.
14. The vector $\mathbf{B}(t) = \mathbf{T}(t) \times \mathbf{N}(t)$ is called the **unit binormal vector** to C at P. Find the binormal component of acceleration.

SECTION 13.4

Chapter Review

Important Concepts

Vector-valued function	Tangent vector	Position vector
Scalar function	Smooth curve	Tangential velocity
Space curve	Arc length	Acceleration
Helix	Curvilinear motion	Uniform circular motion

Centripetal acceleration	Curvature	Osculating plane
Centrifugal acceleration	Radius of curvature	Unit normal vector
Angular velocity	Center of curvature	Tangential component
Unit tangent vector	Circle of curvature	Normal component
Curvature vector		

Review Exercises

Given $\mathbf{f}(t) = \langle 2t, \sin t, e^t\rangle$ *and* $\mathbf{g}(t) = \langle t^2, \cos t, e^{-t}\rangle$, *find the following:*

1. $(\mathbf{f}+\mathbf{g})(t)$
2. $(\mathbf{f}-\mathbf{g})(t)$
3. $(\mathbf{f}\cdot\mathbf{g})(t)$
4. $(\mathbf{f}\times\mathbf{g})(t)$
5. $(\mathbf{f}+\mathbf{g})'(t)$
6. $(\mathbf{f}-\mathbf{g})'(t)$
7. $(\mathbf{f}\cdot\mathbf{g})'(t)$
8. $(\mathbf{f}\times\mathbf{g})'(t)$
9. $t^3\mathbf{f}(t)$
10. $\dfrac{d}{dt}[t^3\mathbf{f}(t)]$
11. $\|\mathbf{f}'(t)\|$
12. $\dfrac{d}{dt}\|\mathbf{f}(t)\|$

Given $\mathbf{r}(t) = \langle 1, t^{1/2}, t^{3/2}\rangle$, *find the following:*

13. $\mathbf{r}'(t)$
14. $\|\mathbf{r}'(t)\|$
15. $\mathbf{r}''(t)$
16. $\mathbf{T}(t)$
17. $\mathbf{T}'(t)$
18. $\boldsymbol{\kappa}(t)$
19. $\kappa(t)$
20. $\rho(t)$
21. $\mathbf{N}(t)$
22. $\mathbf{r}''(t)\cdot\mathbf{T}(t)$
23. $\mathbf{r}''(t)\cdot\mathbf{N}(t)$
24. $\displaystyle\int_0^1 \|\mathbf{r}'(t)\|\,dt$
25. Given $\mathbf{r}''(t) = \langle t, 4t, 1-t\rangle$, $\mathbf{r}'(0) = \langle 1, 0, -1\rangle$, and $\mathbf{r}(0) = \langle 1, 0, -2\rangle$, find $\mathbf{r}(t)$.
26. If a football is thrown at a speed of 80 feet per second, how many yards can it travel?

Additional Exercises

1. The angle at which two curves intersect is the angle between their tangent vectors at the point of intersection. Find the angle between the curves $\mathbf{f}(t) = \langle t, \sin \pi t, \cos \pi t\rangle$ and $\mathbf{g}(t) = \langle e^t, 2t, 1-2e^t\rangle$ at $(1, 0, -1)$.
2. The angle between a curve and a plane is the angle between the tangent line to the curve and the plane at their point of intersection. Find the angle between the helix $\mathbf{f}(t) = \langle a\cos\omega t, a\sin\omega t, bt\rangle$ and the plane $z = k$.
3. The **tractrix** is the plane curve with parametric equations $x = t - a\tanh(t/a)$, $y = a\,\mathrm{sech}(t/a)$. Show that the tangent line to the curve at t meets the x-axis at t.
4. Show that
$$\frac{d}{dt}[u(t)\mathbf{f}(t)\times v(t)\mathbf{g}(t)] = u(t)v(t)\frac{d}{dt}[\mathbf{f}(t)\times\mathbf{g}(t)] + [\mathbf{f}(t)\times\mathbf{g}(t)]\frac{d}{dt}[u(t)v(t)].$$
5. For $\mathbf{r}(t) = (1+t^2)\mathbf{i} + t\mathbf{j}$, find the points at which $\mathbf{r}(t)$ and $\mathbf{r}'(t)$ are:
(a) Orthogonal **(b)** In the same direction **(c)** In opposite directions
6. Show that the tangent line to the curve $\mathbf{f}(t)$ passes through the origin if and only if $\mathbf{f}(t)$ and $\mathbf{f}'(t)$ are parallel or antiparallel.
7. Show that
$$(\mathbf{f}\cdot\mathbf{g}\times\mathbf{h})'(t) = (\mathbf{f}'\cdot\mathbf{g}\times\mathbf{h})(t) + (\mathbf{f}\cdot\mathbf{g}'\times\mathbf{h})(t) + (\mathbf{f}\cdot\mathbf{g}\times\mathbf{h}')(t).$$
8. A frictionless block slides down a plane inclined 10° from the horizontal. How far does it slide in 5 seconds, starting from rest?
9. A cannon with muzzle velocity 400 feet per second fires a shell 2500 feet. Find the two possible angles of elevation.

10. What minimum initial speed must an arrow have in order to travel 1200 feet?

11. A satellite moves in a circular orbit of radius 4800 miles about the center of the earth. Find its speed.

12. A satellite moves in a circular orbit at a speed of 17,750 miles per hour. What is the radius of its orbit?

13. A bucket of water is swung in a circle of diameter 5 feet in a vertical plane. What minimum angular velocity (in revolutions per minute) is necessary to keep the water in the bucket?

14. If an object's velocity and acceleration vectors are always unit vectors, show that the curvature of its path is constantly 1.

15. Show that $\left\| \int_a^b \mathbf{f}(t)\, dt \right\| \le \int_a^b \|\mathbf{f}(t)\|\, dt.$

16. In a polar coordinate system, let

$$\mathbf{u} = (\cos\theta)\mathbf{i} + (\sin\theta)\mathbf{j}, \qquad \mathbf{v} = -(\sin\theta)\mathbf{i} + (\cos\theta)\mathbf{j}.$$

Let $\mathbf{f}(t) = r(t)\mathbf{u} + \theta(t)\mathbf{v}$. Show that

$$\mathbf{f}'(t) = r'(t)\mathbf{u} + r(t)\theta'(t)\mathbf{v}$$

and

$$\mathbf{f}''(t) = [r''(t) - r(t)\theta''(t)]\mathbf{u} + [r(t)\theta''(t) + 2r'(t)\theta'(t)]\mathbf{v}.$$

17. Show that the curvature of the polar curve $r = f(\theta)$ is

$$\kappa(\theta) = \frac{|2[f'(\theta)]^2 - f(\theta)f''(\theta) + [f(\theta)]^2|}{(\sqrt{[f'(\theta)]^2 + [f(\theta)]^2})^3}.$$

18. If a curve has parametric equations

$$r = r(t), \qquad \theta = \theta(t), \qquad z = z(t)$$

in cylindrical coordinates, show that the derivative of arc length is

$$\frac{ds}{dt} = \sqrt{\left(\frac{dr}{dt}\right)^2 + r^2\left(\frac{d\theta}{dt}\right)^2 + \left(\frac{dz}{dt}\right)^2}.$$

19. If a curve has parametric equations

$$\rho = \rho(t), \qquad \theta = \theta(t), \qquad \phi = \phi(t)$$

in spherical coordinates, show that the derivative of arc length is

$$\frac{ds}{dt} = \sqrt{\left(\frac{d\rho}{dt}\right)^2 + \rho^2 \sin^2\phi \left(\frac{d\theta}{dt}\right)^2 + \rho^2\left(\frac{d\phi}{dt}\right)^2}.$$

20. The **evolute** of a plane curve is the locus of centers of curvature of the curve. Show that the curve $\mathbf{f}(t) = \langle x(t), y(t)\rangle$ has evolute

$$\mathbf{E}(t) = \left\langle x(t) - \frac{\rho(t)y'(t)}{\sqrt{[x'(t)]^2 + [y'(t)]^2}},\ y(t) + \frac{\rho(t)x'(t)}{\sqrt{[x'(t)]^2 + [y'(t)]^2}} \right\rangle.$$

21. Referring to the normal component of acceleration, explain how banking a curve enables a car to negotiate the curve at a greater speed.

22. How might nonuniform circular motion be described?
[*Hint:* Consider $\mathbf{r}(t) = \langle a \cos \omega(t), a \sin \omega(t), 0 \rangle$.]

23. Suppose a projectile is fired with initial speed v and at an angle of elevation θ up a hillside that is inclined at an angle α. Find an expression for the distance up the hillside that the projectile lands.

24. Suppose the projectile in Example 13.14 experiences a head wind that produces an acceleration a in the horizontal direction. Describe where the projectile lands.

25. Rework Problem 24 for the case of a crosswind.

C H A P T E R 14

Partial Differentiation

INTRODUCTION

In this chapter, we introduce functions of several variables and describe some of their applications. We develop the ideas of limits, continuity, and partial derivatives; and these concepts are related to the gradient and differentiability and applied to directional derivatives and extrema. The proper theoretical development of these topics belongs to advanced calculus, so we will skip many of the proofs and use an intuitive approach for the most part.

SECTION 14.1

Functions of Several Variables

Up to this point we have considered only functions of one variable. A function of the form $y = f(x)$ assigns to each real number x in its domain a real number y. Because the independent variable x takes on real values and the dependent variable y also has real values, we call this function a "real-valued function of a single real variable." In Chapter 13, we discussed vector functions. A vector function of the form $\mathbf{f}(t) = \langle x(t), y(t), z(t)\rangle$ assigns to each real number t a vector $\mathbf{f}(t)$. Thus, we may call a vector function a "vector-valued function of a single real variable."

In this chapter, we introduce functions of several variables, or more precisely, "real-valued functions of several real variables." We will start with functions of two variables and work our way up.

Functions of Two Variables

Before we state any formal definitions, let's look at some familiar examples. The formula for the area of a rectangle of length l and width w is

$$A(l, w) = lw.$$

Here, A is expressed as a function of the two independent variables l and w. Because of the physical interpretation attached to this function, it is defined only for l and w nonnegative numbers.

As another example, consider a manufacturer's profit, which is given in terms of the total cost C and the total revenue R by

$$P(C, R) = R - C.$$

As yet another example, consider a weather map on which the barometric pressure is given as a function of location—that is, of the two variables latitude and longitude.

Now let's look at a formal definition.

DEFINITION 14.1

Let D be a set of ordered pairs (x, y) of real numbers. (Think of D as a set of points in the plane.) A **function of two variables** with **domain** D is a rule f that assigns to every ordered pair $(x, y) \in D$ exactly one real number $f(x, y)$, called the **image** of (x, y) under f. The **range** of f is the set of real numbers z such that $z = f(x, y)$ for $(x, y) \in D$. The **graph** of f is the set of points (x, y, z) in space such that $z = f(x, y)$.

If the domain of a function of two variables is not specified, it is customary to use the largest possible domain. The graph of a function of two variables is generally a surface in space, just as the graph of a function of one variable is a curve in the plane.

■ **Example 14.1**

Let function f be defined by

$$f(x, y) = x^2 + 2y^2.$$

Find the images of $(0, 0)$, $(1, -2)$, $(3, 4)$, and $(0, 3)$. Also give the domain and range of f, and describe its graph.

Solution

We first find the images:

$$\begin{aligned} f(0, 0) &= 0^2 + 2(0)^2 = 0, \\ f(1, -2) &= 1^2 + 2(-2)^2 = 9, \\ f(3, 4) &= 3^2 + 2(4)^2 = 41, \\ f(0, 3) &= 0^2 + 2(3)^2 = 18. \end{aligned}$$

The domain of f is the entire plane since $x^2 + 2y^2$ is defined for any numbers x and y. The range of f is the set of all nonnegative numbers, because $x^2 + 2y^2$ is never negative and $f(\sqrt{z}, 0) = z$ for any nonnegative number z. The graph of f is the elliptic paraboloid $z = x^2 + 2y^2$ shown in Figure 14.1.

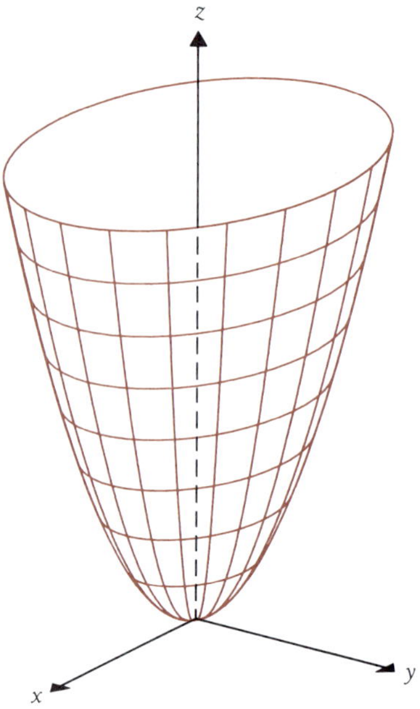

FIGURE 14.1 The graph of $z = x^2 + 2y^2$, an elliptic paraboloid

Example 14.2 Give the domain and range of the function $g(x, y) = \sqrt{169 - x^2 - y^2}$.

Solution The domain is the set of points (x, y) such that

$$169 - x^2 - y^2 \geq 0,$$
$$x^2 + y^2 \leq 13^2.$$

This is the circular disc of radius 13 centered at the origin (Figure 14.2). It is easy to see that the range of g is the interval $[0, 13]$. The graph of g is the hemisphere $z = \sqrt{13^2 - x^2 - y^2}$ (Figure 14.3).

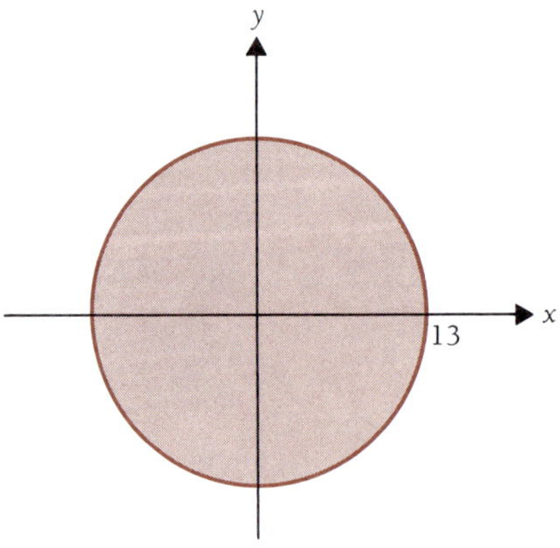

FIGURE 14.2 The domain is a disc of radius 13

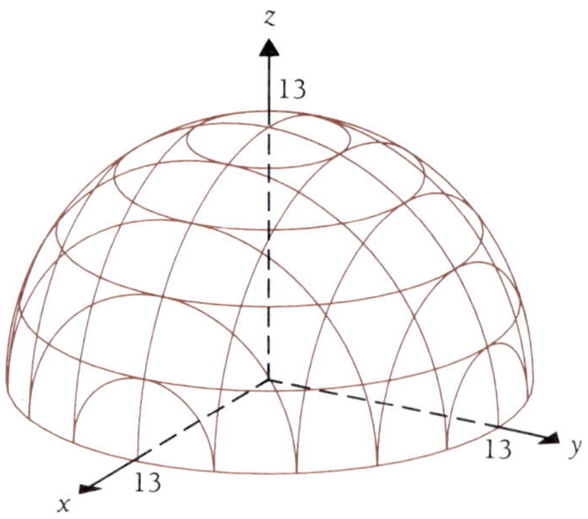

FIGURE 14.3 The graph of $z = \sqrt{13^2 - x^2 - y^2}$, a hemisphere

If P is the point (x, y) or $\mathbf{v}$ is the vector $\langle x, y \rangle$, we may use $f(P)$ or $f(\mathbf{v})$ in place of $f(x, y)$. In the latter notation, f can be regarded as a "real-valued function of a single vector variable."

■ **Example 14.3** If $\mathbf{v} = \langle x, y \rangle$, write $f(\mathbf{v}) = \|\mathbf{v}\|$ as a function of two variables.

Solution If $\mathbf{v} = \langle x, y \rangle$, then $\|\mathbf{v}\| = \sqrt{x^2 + y^2}$. Hence, we may write f as

$$f(x, y) = \sqrt{x^2 + y^2}. \qquad ■$$

Level Curves

A useful tool in graphing a function of two variables is the notion of level curves. The idea is the same as that of a topographical map, in which contours of constant elevation on a flat map impart information about the shape of the land. Figure 14.4 shows a topographical map in which adjacent contour lines represent constant differences in altitude. Where contour lines are close together, the land is relatively steep; where they are far apart, the land is almost flat.

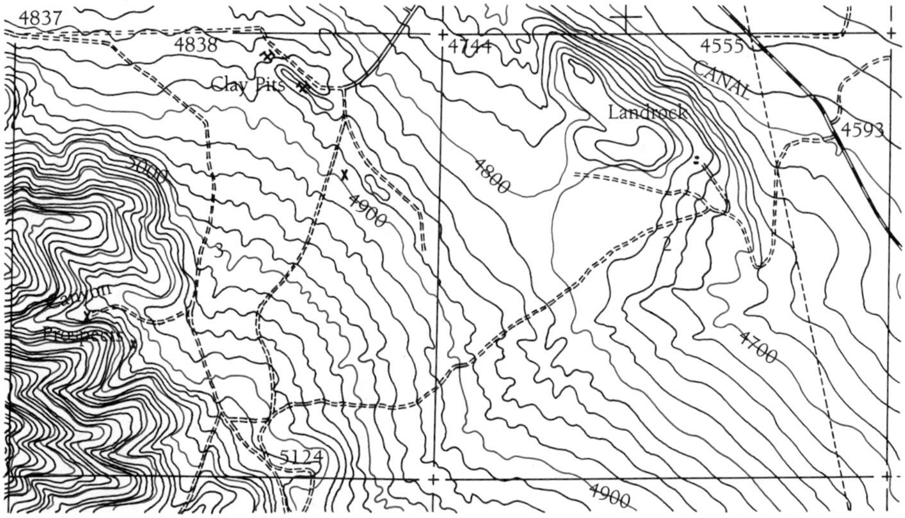

FIGURE 14.4 A topographical map

Let $z = f(x, y)$ define a function of two variables. The graph of f is a surface in space. The intersection of that surface with the horizontal plane $z = c$ is a curve, and the projection of that curve into the xy-plane is a curve with equation

$$f(x, y) = c.$$

This curve is called the **level curve at c** or the **c-level curve** of f (see Figure 14.5, page 842). Level curves are also called **contour curves.** By selecting several values of c and sketching the corresponding level curves, we obtain a **contour map** of f, similar to Figure 14.4.

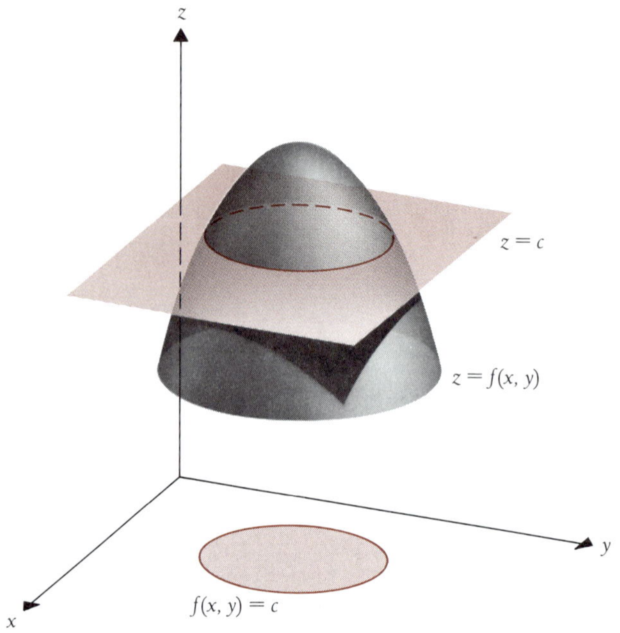

FIGURE 14.5 A level curve is the projection of an intersection of the surface with a horizontal plane

■ **Example 14.4** Describe several level curves of the function $f(x, y) = 4x^2 - 9y^2$.

Solution The level curves of f are all of the form

$$4x^2 - 9y^2 = c.$$

If $c = 0$, we get the two intersecting lines $y = \frac{2}{3}x$ and $y = -\frac{2}{3}x$. If $c \neq 0$, we get a hyperbola. Several level curves of f are sketched in Figure 14.6. The graph of f is the hyperbolic paraboloid $z = 4x^2 - 9y^2$ (Figure 14.7). ■

■ **Example 14.5** Describe the level curves of the function $f(x, y) = y - x^2$.

Solution Each level curve of f is of the form

$$y - x^2 = c,$$
$$y = x^2 + c,$$

which is a parabola. Several level curves of f are sketched in Figure 14.8. The graph of f is the parabolic cylinder $z = y - x^2$ (Figure 14.9). ■

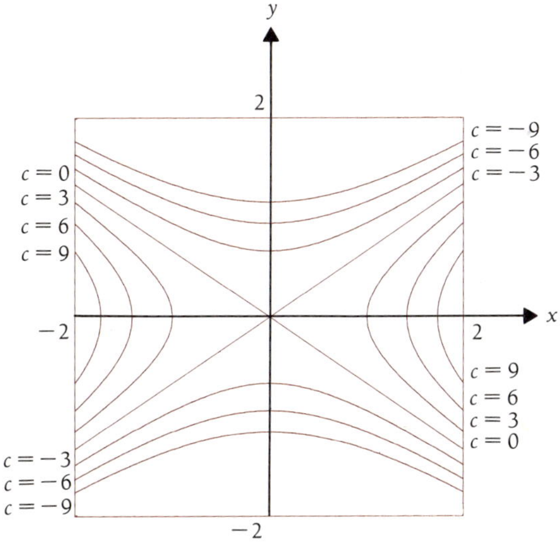

FIGURE 14.6 The contour map of $z = 4x^2 - 9y^2$

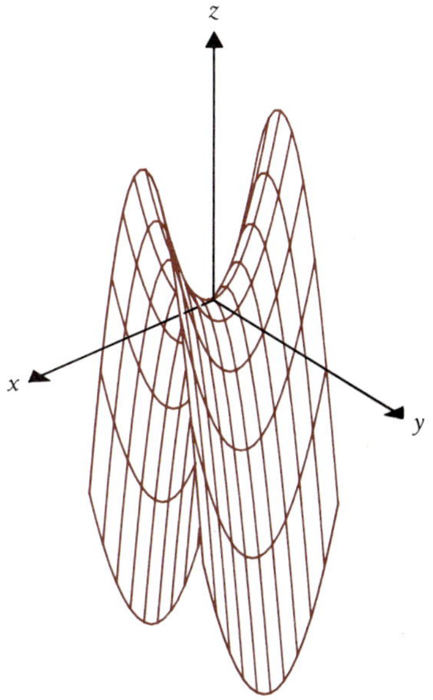

FIGURE 14.7 The graph of $z = 4x^2 - 9y^2$, a hyperbolic paraboloid

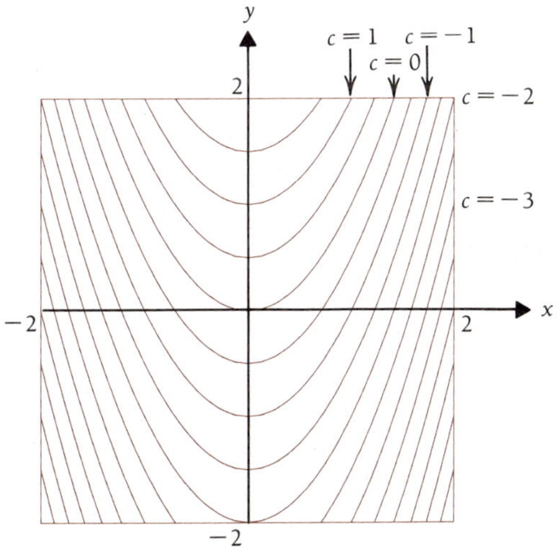

FIGURE 14.8 The contour map of $z = y - x^2$

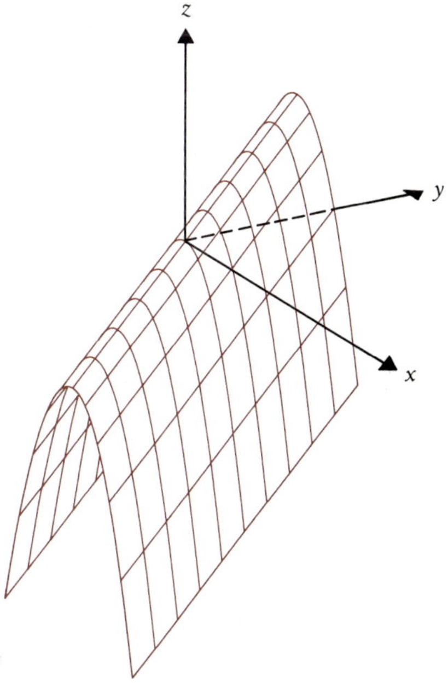

FIGURE 14.9 The graph of $z = y - x^2$, a parabolic cylinder

Functions of Three Variables

Now let's consider functions of three variables. As a simple example, consider the formula for the area of a trapezoid with bases b_1 and b_2 and height h (see Figure 14.10),

$$A(b_1, b_2, h) = \frac{1}{2}(b_1 + b_2)h.$$

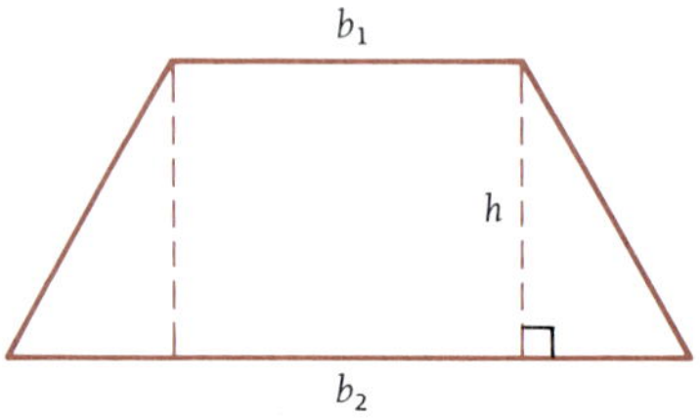

FIGURE 14.10 A trapezoid

As another example, consider a temperature function $T(x, y, z)$, which gives the temperature at point (x, y, z). This can be used to describe a solid object, a liquid in a container, or even the atmosphere.

DEFINITION 14.2 Let D be a set of ordered triples (x, y, z) of real numbers. (Think of D as a set of points in space.) A **function of three variables** with **domain** D is a rule f that assigns to every ordered triple $(x, y, z) \in D$ exactly one real number $f(x, y, z)$, called the **image** of (x, y, z) under f. The **range** of f is the set of real numbers $w = f(x, y, z)$ for $(x, y, z) \in D$.

We do not usually discuss the graph of a function of three variables, since it is a "surface" in four-dimensional space and is impossible to visualize.

■ **Example 14.6** Describe the domain and range of the function $f(x, y, z) = \sqrt{z - x^2 - 4y^2}$.

Solution The domain of f is the set of points (x, y, z) in space such that

$$z - x^2 - 4y^2 \geq 0,$$
$$z \geq x^2 + 4y^2.$$

Graphically, the domain is the elliptic paraboloid $z = x^2 + 4y^2$, together with its interior (Figure 14.11). To find the range of f, we note that when $x = y = 0$, then $f(x, y, z) = \sqrt{z}$, which can take on any nonnegative value. Therefore, the range of f is $[0, \infty)$.

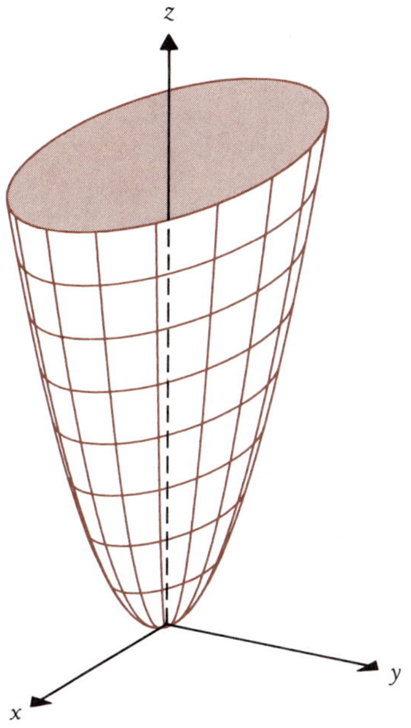

FIGURE 14.11 The graph of $z = x^2 + 4y^2$, an elliptic paraboloid ■

If P is the point (x, y, z) or $\mathbf{v}$ is the vector $\langle x, y, z\rangle$, we may write $f(P)$ or $f(\mathbf{v})$ for $f(x, y, z)$.

■ **Example 14.7** Write the function $f(\mathbf{v}) = \mathbf{v} \cdot \mathbf{v}$ as a function of three variables.

Solution If $\mathbf{v} = \langle x, y, z\rangle$, then $\mathbf{v} \cdot \mathbf{v} = x^2 + y^2 + z^2$. Hence, f may be written as

$$f(x, y, z) = x^2 + y^2 + z^2.$$ ■

Level Surfaces

An idea similar to the notion of level curves can be applied to a function of three variables. If $w = f(x, y, z)$ defines a function of three variables, we set $w = c$ and obtain

$$f(x, y, z) = c.$$

This is the equation of a surface in space called the **c-level surface** of f. The level surfaces of f are about as close to the graph of f as we can get.

As a physical example, the level surfaces of a temperature function $T(x, y, z)$ are surfaces of constant temperature. For instance, if a heat source is located at a point P inside a homogeneous substance, we would expect the level surfaces of the temperature function to be spheres centered at P.

Example 14.8 Describe the level surfaces of the function $f(x, y, z) = x^2 + y^2 - z^2$.

Solution Each level surface of f is of the form

$$x^2 + y^2 - z^2 = c.$$

If $c = 0$, the surface is a cone. If $c > 0$, the surface is a hyperboloid of one sheet. If $c < 0$, the surface is a hyperboloid of two sheets. By analogy with Example 14.4, we might guess that the graph of f is a four-dimensional version of a hyperbolic paraboloid. ■

Functions of Many Variables

We now give a general definition of a real-valued function of n variables, where n is any positive integer.

DEFINITION 14.3 Let n be a positive integer, and let D be a set of ordered n-tuples $(x_1, x_2, \ldots, x_n)$. A **function of *n* variables** with **domain** D is a rule f that assigns to every ordered n-tuple $(x_1, x_2, \ldots, x_n) \in D$ exactly one real number $f(x_1, x_2, \ldots, x_n)$, called the **image** of $(x_1, x_2, \ldots, x_n)$ under f. The **range** of f is the set of real numbers $y = f(x_1, x_2, \ldots, x_n)$ for $(x_1, x_2, \ldots, x_n) \in D$.

Note that if $n = 1$, 2, or 3, this definition agrees with our previous definitions. If $n > 3$, the usual geometric interpretations of the domain and graph of f are lost. For that reason, we will devote most of our attention to functions of two or three variables.

As a sample illustration of a function of many variables, consider the arithmetic mean (i.e., average) of the n numbers $x_1, x_2, \ldots, x_n$, which is defined by

$$\mu(x_1, x_2, \ldots, x_n) = \frac{1}{n}(x_1 + x_2 + \cdots + x_n).$$

Combinations of Functions of Several Variables

Functions of several variables can be combined to form sums, differences, products, quotients, and composites, just as for functions of a single variable, but now we have many more possible combinations.

Example 14.9 Let $f(x, y) = 2xy$, $g(x, y) = x + y$, and $h(x) = x^2 + 1$. Find $(f + g)(x, y)$, $(fg)(x, y)$, $(h \circ f)(x, y)$, and $g(f(x, y), h(y))$.

Solution We have

$$\begin{aligned}(f+g)(x, y) &= f(x, y) + g(x, y)\\ &= 2xy + x + y,\\ (fg)(x, y) &= f(x, y)g(x, y)\\ &= 2xy(x + y),\\ (h \circ f)(x, y) &= h(f(x, y))\\ &= h(2xy) = (2xy)^2 + 1,\\ g(f(x, y), h(y)) &= f(x, y) + h(y)\\ &= 2xy + y^2 + 1.\end{aligned}$$

■

Some terminology that will be helpful later is as follows:

A **monomial** in two variables x, y is a function of the form cx^iy^j, with c a real number and i and j nonnegative integers. The given monomial is of **degree i in x**, of **degree j in y**, and of **total degree $i + j$**. For example, the monomial $3x^5y^2$ is of total degree 7.

A **polynomial** in two variables x, y is a sum of monomials in x and y, each of which is called a **term**. The **degree** of a polynomial in two variables is the greatest of the total degrees of its terms. For example, the polynomial $3x^2 - 5xy^2 + xy$ is of degree 3.

A **rational function** in two variables x, y is the ratio of two polynomials in x and y. For example,

$$\frac{2xy}{x^2 + y^2}$$

is a rational function.

It should be obvious how these definitions generalize to more than two variables.

Exercises 14.1

Describe the domain of each function.

1. $f(x, y) = x^2 + 3xy + y^2$
2. $f(x, y) = x^3 - 3xy^2 + 5$
3. $f(x, y) = \sqrt{9 - x^2 - y^2}$
4. $f(x, y) = \sqrt{25 - 4x^2 - y^2}$
5. $f(x, y) = \sqrt{x^2 + y^2 - 9}$
6. $f(x, y) = \sqrt{4x^2 + y^2 - 25}$
7. $f(x, y) = x\sqrt{x^2 + y^2 - 9}$
8. $f(x, y) = y\sqrt{x^2 + y^2 - 9}$
9. $f(x, y) = \dfrac{1}{y}\sqrt{x^2 + y^2 - 9}$
10. $f(x, y) = \dfrac{y}{\sqrt{x^2 + y^2 - 9}}$
11. $f(x, y) = x \ln y$
12. $f(x, y) = y - \ln x$
13. $f(x, y) = \ln(x + y)$
14. $f(x, y) = \ln x + \ln y$
15. $f(x, y, z) = \sqrt{1 - x^2 - y^2 - z^2}$
16. $f(x, y, z) = \sqrt{x^2 + y^2 + z^2 - 4}$

17. $f(x, y, z) = \dfrac{x - 2y}{z^2}$

18. $f(x, y, z) = \dfrac{2xyz}{x^2 - y^2 + z}$

19. Match the following functions with their graphs, which are shown in the accompanying figures: $f_1(x, y) = x^2 + y^2$, $f_2(x, y) = x^2 - y^2$, $f_3(x, y) = xy$, $f_4(x, y) = 4 - x^2 - y^2$, $f_5(x, y) = x^2$.

(a)

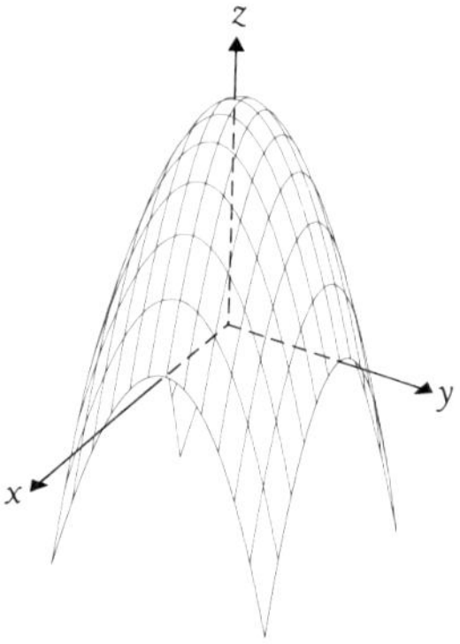

(b)

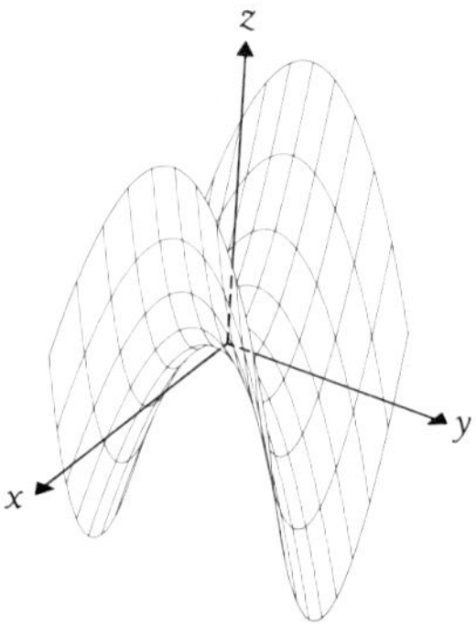

(c)

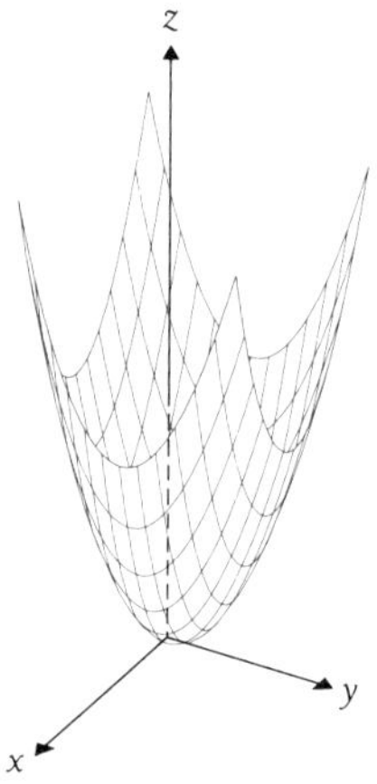

(d)

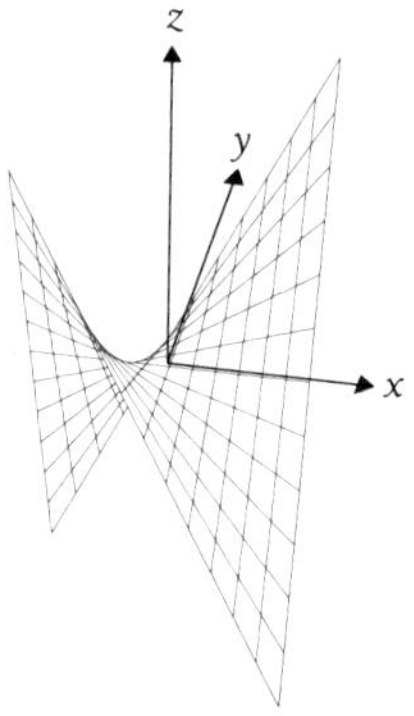

(e)

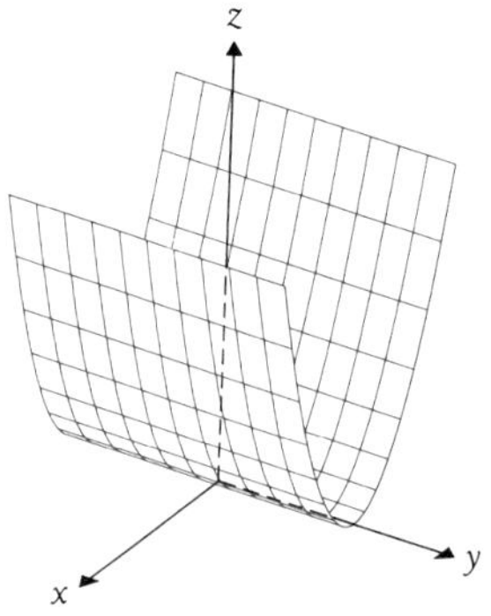

20. Match the functions in Problem 19 with their contour maps, which are shown in the accompanying figures.

(a)

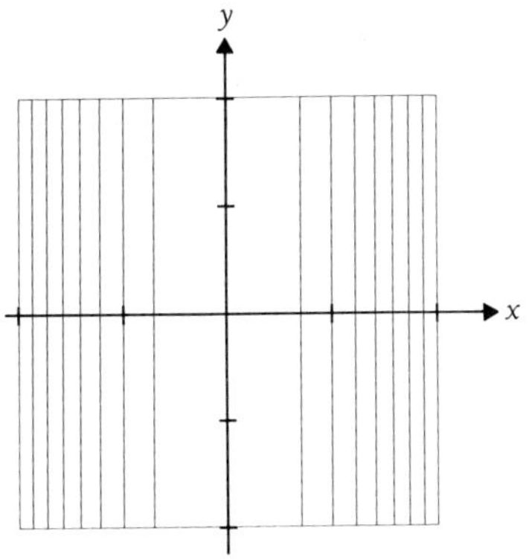

(b)

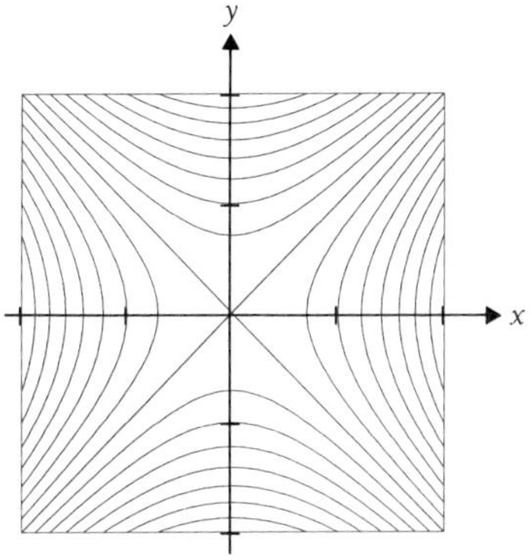

(c)

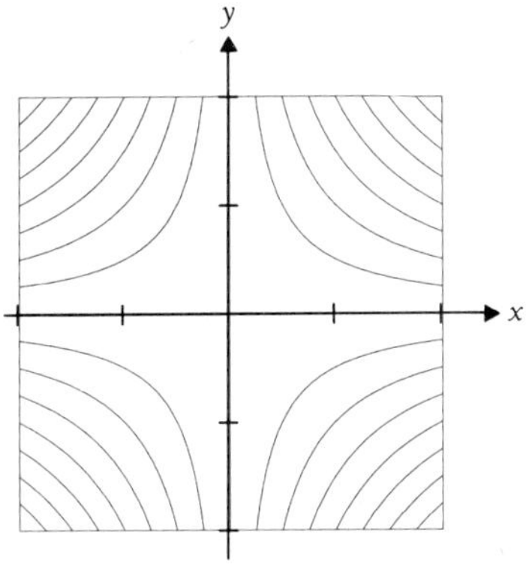

(d)

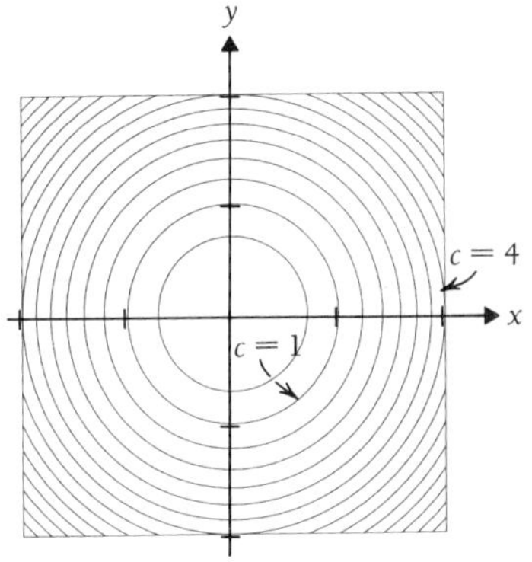

(e)

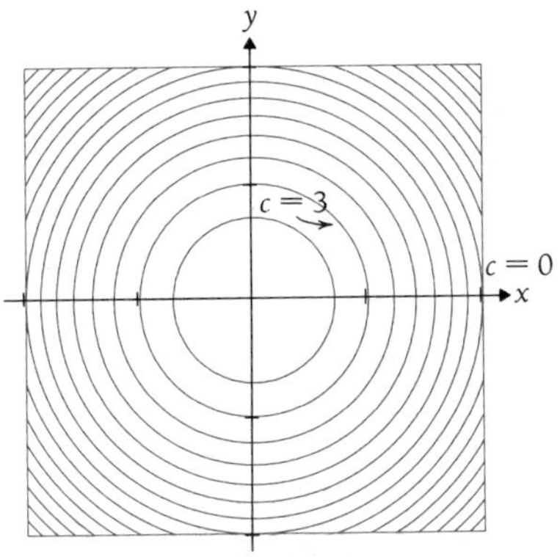

21. Match the following functions with their graphs, which are shown in the accompanying figures: $f_1(x, y) = x - \frac{1}{3}x^3 - \frac{1}{2}y^2$, $f_2(x, y) = x^2y^2 - x^2y^3 - \frac{1}{4}x^4$, $f_3(x, y) = x^2 \cos y$, $f_4(x, y) = e^{-x} \cos y$, $f_5(x, y) = e^{-x^2} \cos y$.

(a)

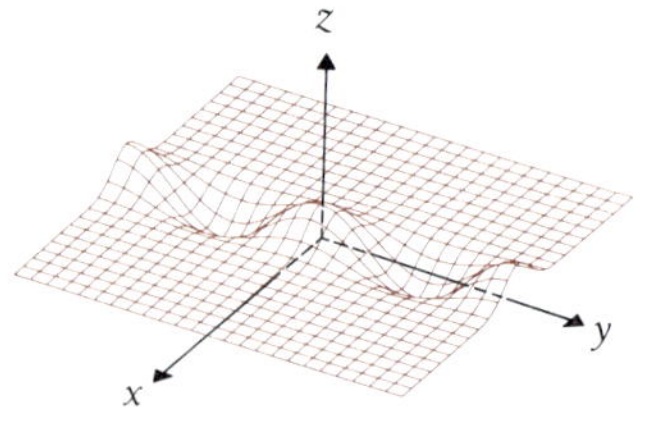

(b)

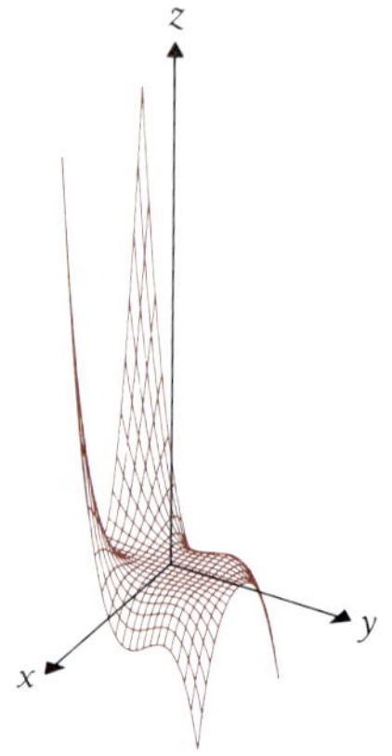

(c)

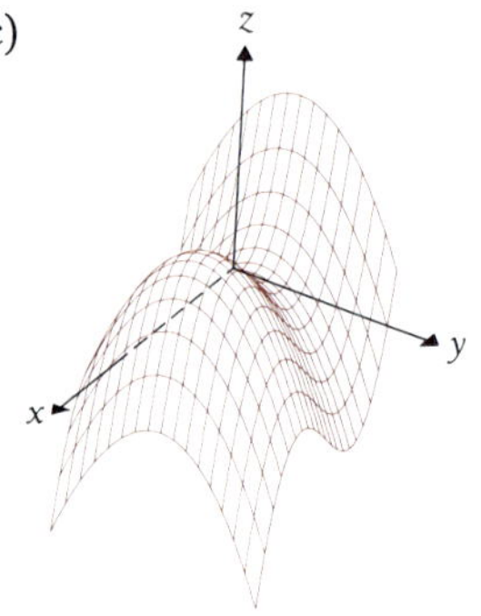

(d)

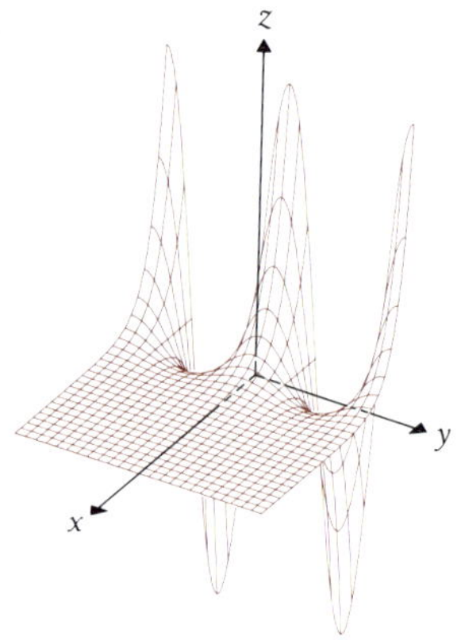

(e)

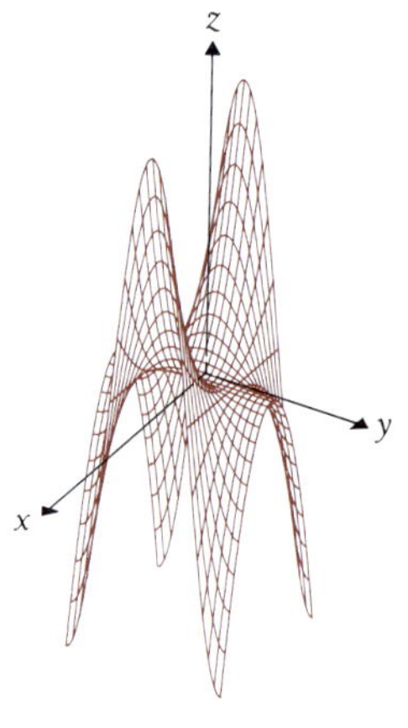

22. Match the functions in Problem 21 with their contour maps, which are shown in the accompanying figures.

(a)

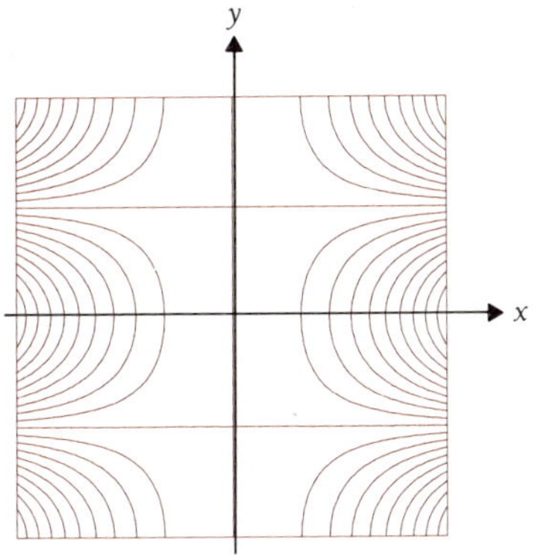

(b)

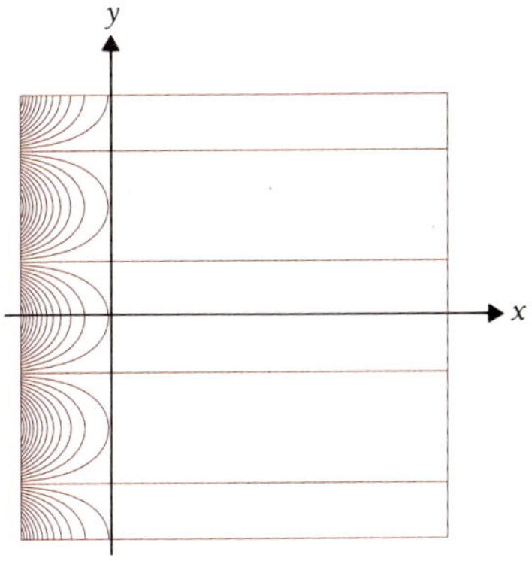

(c)

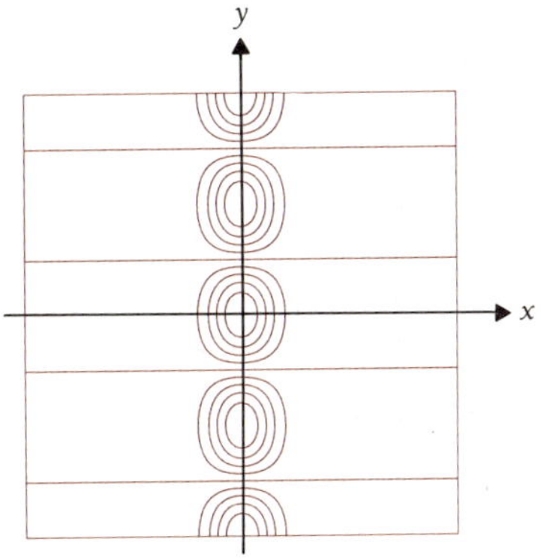

(d)

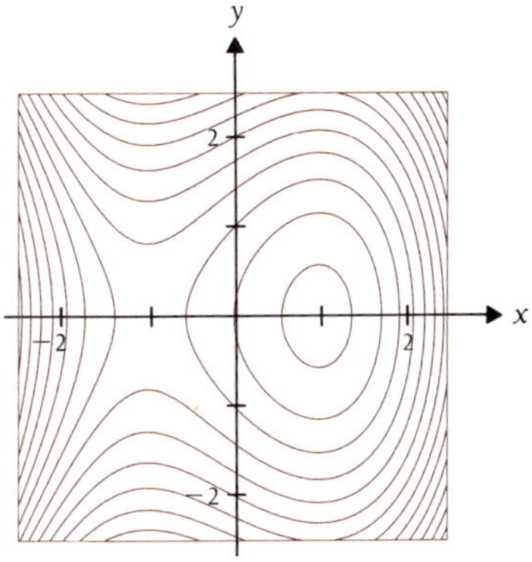

(e)

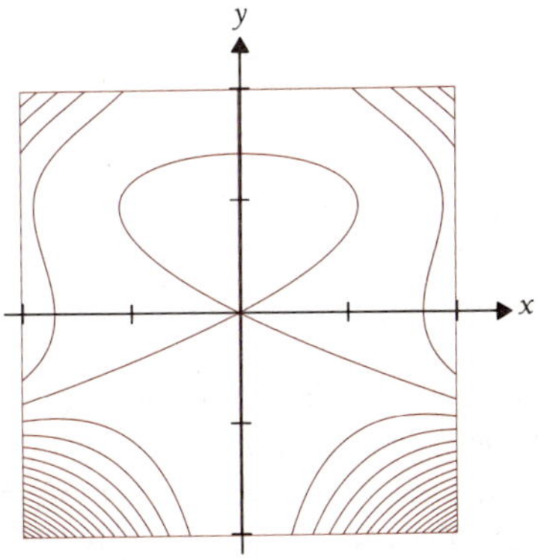

Describe the graph of each function.

23. $f(x, y) = x^2 + y^2 - 4$
24. $f(x, y) = x^2 - y^2 - 4$
25. $f(x, y) = 2x - y - 1$
26. $f(x, y) = -4x + y + 2$
27. $f(x, y) = -x^2 - y^2$
28. $f(x, y) = x^2 + 9y^2$
29. $f(x, y) = \sqrt{4 - x^2 - y^2}$
30. $f(x, y) = \sqrt{x^2 + 4y^2 - 4}$

Sketch several level curves for each function.

31. $f(x, y) = \ln(1 + x^2 + y^2)$
32. $f(x, y) = \dfrac{1}{\sqrt{x^2 + y^2}}$
33. $f(x, y) = x^2 - 4y$
34. $f(x, y) = 3x + 12y^2$
35. $f(x, y) = e^{x-y}$
36. $f(x, y) = x + 3\ln(1 + y)$
37. $f(x, y) = x - 2\cos y$
38. $f(x, y) = y \sec x$
39. $f(x, y) = \sin(x + y)$
40. $f(x, y) = y - \tan^{-1}x$

Describe the level surfaces of each function.

41. $f(x, y, z) = x^2 - y^2 + z^2$
42. $f(x, y, z) = x^2 - y^2 + z$
43. $f(x, y, z) = x^2 + y^2 - z$
44. $f(x, y, z) = 2x + y + 5z$

Given $f(x, y) = 2x - y$, $g(x, y) = x^2 - y^2$, and $h(x, y, z) = xy - z^2$, find the following:

45. $(f + g)(x, y)$
46. $(f - g)(x, y)$
47. $(fg)(x, y)$
48. $\left(\dfrac{f}{g}\right)(x, y)$
49. $f(2, 2)$
50. $g(1, 3)$
51. $h(1, 1, 3)$
52. $f(x, x)$
53. $h(x, y, x^2)$
54. $f(g(x, y), 1)$
55. $h(f(x, y), g(x, y), z)$
56. $f(g(x, y), h(x, y, z))$

A function of two variables is **homogeneous** *of* **index** *n if and only if for any constant k,*

$$f(kx, ky) = k^n f(x, y).$$

Determine which of the following functions are homogeneous, and give the index:

57. $f(x, y) = 3x^2 + 4xy - 2y^2$
58. $f(x, y) = x^3 + xy - y + 1$
59. $f(x, y) = \dfrac{x + y}{x - y}$
60. $f(x, y) = \cos\dfrac{x}{y}$
61. Prove that a polynomial in two variables is homogeneous if and only if each term has the same total degree.

SECTION 14.2

Limits and Continuity

In this section, the concept of limit is developed for functions of several variables. We begin by reviewing the definition of limit for functions of a single variable, and then expand it to functions of two, three, or more variables. We conclude with a brief discussion of continuity.

Limits of Functions of Two Variables

For a function f of a single variable, the limit $\lim_{x \to c} f(x)$ was defined to be a number L such that for any $\varepsilon > 0$, there exists $\delta > 0$ such that if $0 < |x - c| < \delta$, then $|f(x) - L| < \varepsilon$. Moreover, it was assumed that f was defined in an open interval (a, b) containing c, except possibly at c itself (see Definition 2.1).

To define the limit of a function of two variables, we must translate the above terms into their two-dimensional analogs. Instead of x approaching c, we have a point (x, y) approaching a point (a, b), which we will denote by $(x, y) \to (a, b)$. The distance $|x - c|$ between x and c is replaced by the distance $\sqrt{(x-a)^2 + (y-b)^2}$ between (x, y) and (a, b). As the two-dimensional analog of an open interval, we will use an **open disc**, which is the interior of a circle in the plane (see Figure 14.12).

FIGURE 14.12 An open disc

We can now state the definition of limit for a function of two variables.

DEFINITION 14.4

Let f be a function of two variables that is defined in an open disc containing the point (a, b), except possibly at (a, b). Then

$$\lim_{(x, y) \to (a, b)} f(x, y) = L$$

if and only if for any $\varepsilon > 0$ there exists $\delta > 0$ such that

$$|f(x, y) - L| < \varepsilon \qquad \text{when} \qquad 0 < \sqrt{(x-a)^2 + (y-b)^2} < \delta.$$

Note that the set of points (x, y) such that $0 < \sqrt{(x-a)^2 + (y-b)^2} < \delta$ is an open disc of radius δ, centered at (a, b), with the point (a, b) deleted. Thus, Definition 14.4 says that the limit of $f(x, y)$ as (x, y) approaches (a, b) is L whenever, for any positive ε, we can find an open disc D_ε centered at (a, b) such that if $(x, y) \in D_\varepsilon$ and $(x, y) \neq (a, b)$, then $f(x, y)$ differs from L by less than ε (see Figure 14.13). Intuitively, the portion of the surface $z = f(x, y)$ lying over D_ε is between the planes $z = L - \varepsilon$ and $z = L + \varepsilon$. Thus, the intuitive interpretation of limits in two variables is about the same as for the one-variable case.

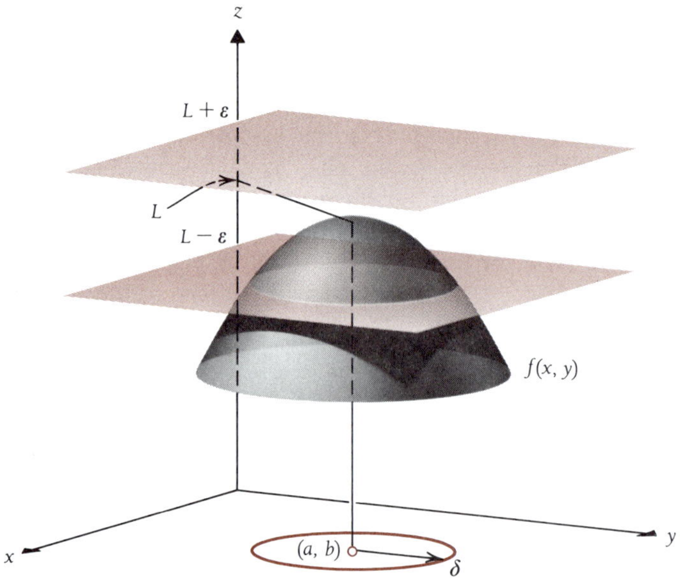

FIGURE 14.13 The definition of limit

Note also that an open disc of radius δ centered at (a, b) is the set of vectors $\langle x, y\rangle$ such that $\|\langle x, y\rangle - \langle a, b\rangle\| < \delta$.

The limit theorems of Sections 2.2 and 2.3 hold with little change for functions of two variables. That is, the limit of a sum is the sum of the limits, the limit of a product is the product of the limits, and so forth. The limit of a polynomial function is found by evaluation; that is,

$$\lim_{(x, y)\to(a, b)} p(x, y) = p(a, b)$$

if p is a polynomial in two variables.

Example 14.10 Find

$$\lim_{(x, y)\to(1, 2)} (x^2 + 3xy).$$

Solution

$$\lim_{(x, y)\to(1, 2)} (x^2 + 3xy) = 1^2 + 3(1)(2) = 7.$$

The composite limit theorem (Theorem 2.10) can be stated in the following form for a function of two variables.

THEOREM 14.1

If f is a function of two variables, g is a continuous function of a single variable, and $\lim_{(x,y)\to(a,b)} f(x, y) = L$, then

$$\lim_{(x, y)\to(a, b)} g(f(x, y)) = g(L).$$

The proof is omitted, since it is very similar to the one-variable case.

■ **Example 14.11** Find

$$\lim_{(x, y)\to(1, 2)} \sqrt{\frac{2x}{x^2 + y}}.$$

Solution

$$\lim_{(x, y)\to(1, 2)} \sqrt{\frac{2x}{x^2 + y}} = \sqrt{\frac{2(1)}{1^2 + 2}} = \sqrt{\frac{2}{3}}.$$ ■

■ **Example 14.12** Find

$$\lim_{(x, y)\to(2, 0)} \frac{e^{3x}}{\cos xy}.$$

Solution

$$\lim_{(x, y)\to(2, 0)} \frac{e^{3x}}{\cos xy} = \frac{e^6}{\cos 0} = e^6.$$ ■

Approaches

For functions of one variable, we have one-sided limits, in which x approaches c from the right or from the left. In the plane, it is possible for (x, y) to approach (a, b) from infinitely many directions. To talk about approaching (a, b) from one direction only, we will use the following new concept: A set S of points in the plane is called an **approach to (a, b)** if and only if every open disc containing (a, b) contains infinitely many points of S.

For example, any continuous curve passing through (a, b) is an approach to (a, b). An approach also may be more general. As an approach to the origin, we may use the upper half-plane, the interior of an acute angle with vertex at the origin, or the set of points $\{(1/n, 0) | n \text{ a positive integer}\}$.

The two-variable version of the definition of a one-sided limit is stated in terms of approaches.

DEFINITION 14.5

Let f be a function of two variables defined in an open disc containing the point (a, b), except possibly at (a, b), and let S be an approach to (a, b). The **limit of $f(x, y)$**

as (x, y) approaches (a, b) along S is L, written

$$\lim_{\substack{(x, y)\to(a, b)\\(x, y)\in S}} f(x, y) = L,$$

if and only if for any $\varepsilon > 0$ there exists $\delta > 0$ such that if $(x, y) \in S$ and $0 < \sqrt{(x-a)^2 + (y-b)^2} < \delta$, then $|f(x, y) - L| < \varepsilon$.

Let's look at some examples.

Example 14.13 Find the limit of $xy/(x^2 + y^2)$ as (x, y) approaches the origin along the x-axis.

Solution If S is the set of points on the x-axis, then the condition $(x, y) \in S$ is the same as the condition $y = 0$. Therefore,

$$\begin{aligned}\lim_{\substack{(x, y)\to(0, 0)\\(x, y)\in S}} f(x, y) &= \lim_{\substack{(x, y)\to(0, 0)\\y=0}} f(x, y)\\ &= \lim_{(x, y)\to(0, 0)} f(x, 0)\\ &= \lim_{x\to 0} f(x, 0)\\ &= \lim_{x\to 0} \frac{0}{x^2 + 0} = 0.\end{aligned}$$

In this case, selecting the particular approach reduced the limit to the limit of a function of a single variable. ■

Example 14.14 Find the limit of $xy/(x^2 + y^2)$ as (x, y) approaches the origin along the line $y = x$.

Solution

$$\begin{aligned}\lim_{\substack{(x, y)\to(0, 0)\\y=x}} \frac{xy}{x^2 + y^2} &= \lim_{(x, y)\to(0, 0)} \frac{x^2}{x^2 + x^2}\\ &= \lim_{x\to 0} \frac{1}{2} = \frac{1}{2}.\end{aligned}$$ ■

Example 14.15 Find the limit of $xy/(x^2 + y^2)$ as (x, y) approaches the origin along the curve $y = x^2$.

Solution

$$\begin{aligned}\lim_{\substack{(x, y)\to(0, 0)\\y=x^2}} \frac{xy}{x^2 + y^2} &= \lim_{x\to 0} \frac{x^3}{x^2 + x^4}\\ &= \lim_{x\to 0} \frac{x}{1 + x^2} = 0.\end{aligned}$$ ■

Just as for one-sided limits in the one-variable case, the values of limits along various approaches tell us something about the existence of the limit. The following theorem is an extension of Theorem 2.16.

THEOREM 14.2

If

$$\lim_{(x,y)\to(a,b)} f(x, y) = L$$

and S is any approach to (a, b), then

$$\lim_{\substack{(x,y)\to(a,b)\\(x,y)\in S}} f(x, y) = L.$$

The proof of this theorem is an immediate consequence of the definition of limit.

As an immediate application of Theorem 14.2, note that if a function has different limits along different approaches, then the (unrestricted) limit of the function does not exist.

Example 14.16 Show that

$$\lim_{(x,y)\to(0,0)} \frac{xy}{x^2 + y^2}$$

does not exist.

Solution Since the limits along different approaches (Examples 14.13 and 14.14) are different, the limit cannot exist by Theorem 14.2. ■

Limits of Functions of Several Variables

The process of generalizing the definition of limit to two variables easily extends to three or more variables. The only questionable point concerns distance in space of dimension more than three. We therefore define the "distance" between "points" $P(x_1, x_2, \ldots, x_n)$ and $Q(y_1, y_2, \ldots, y_n)$ in the space of dimension n to be

$$|PQ| = \sqrt{(x_1 - y_1)^2 + (x_2 - y_2)^2 + \cdots + (x_n - y_n)^2}.$$

Note that this "distance formula" reduces to the usual distance formulas if $n = 1, 2,$ or 3.

We also define the "sphere" of radius r, centered at A in n-dimensional space, to be the set of points P such that $|AP| = r$. For $n = 1$, we get a set of two points; for $n = 2$, we get a circle of radius r; and for $n = 3$, we get an ordinary sphere of radius r (see Figure 14.14). We then define an **open ball** in n-dimensional space to be the interior of a "sphere." For $n = 3$, an open ball is actually the interior of an ordinary

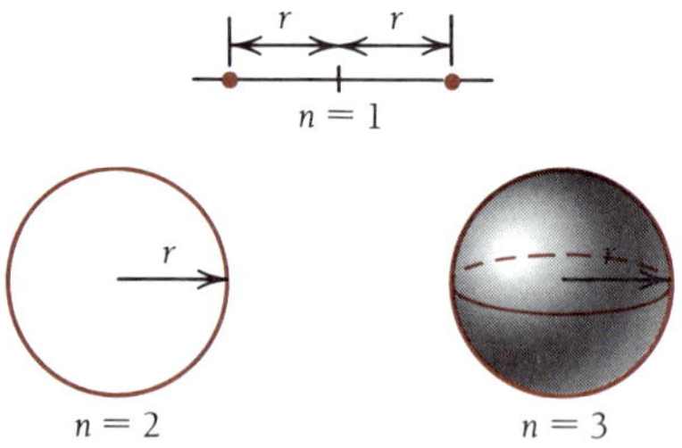

FIGURE 14.14 "Spheres" in 1, 2, and 3 dimensions

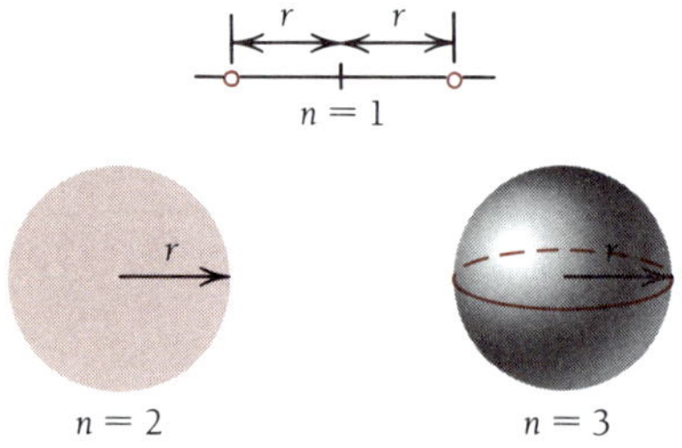

FIGURE 14.15 "Open balls" in 1, 2, and 3 dimensions

sphere; for $n = 1$ and $n = 2$, an open ball is an open interval and an open disc, respectively (see Figure 14.15).

We can now state the definition of limit for a function of any number of variables.

DEFINITION 14.6

Let f be a function of n variables defined in an open ball containing point A, except possibly at A. Then

$$\lim_{P \to A} f(P) = L$$

if and only if for any $\varepsilon > 0$ there exists $\delta > 0$ such that $|f(P) - L| < \varepsilon$ whenever P is in the domain of f and $0 < |AP| < \delta$.

Notice that Definition 14.6 reduces to the previous definitions for $n = 1$ or 2.

Continuity

The definition of continuity for functions of several variables is essentially the same as for functions of one variable.

DEFINITION 14.7

Let f be a function of n variables defined in an open ball containing point A. We say f is **continuous at A** if and only if

$$\lim_{P \to A} f(P) = f(A).$$

We say f is **continuous on set D** if and only if f is continuous at each point of D.

A function f may fail to be continuous at A for three reasons: f may not be defined at A, $\lim_{P \to A} f(P)$ may not exist, or $\lim_{P \to A} f(P)$ may exist but fail to be equal to $f(A)$.

A polynomial function of several variables is continuous at all points, and a rational function is continuous wherever it is defined. Combinations of continuous functions of several variables are again continuous.

Open, Closed, and Compact Sets

For functions of a single variable, a continuous function on a closed interval has some important special properties that we have studied already. We want to consider similar properties of functions of two or more variables, but we first have to develop the higher-dimensional analog of the closed interval.

We first consider the two-dimensional case. Recall that an open disc in the plane is the interior of a circle. A set S of points in the plane is called **open** if and only if every point of S lies in some open disc that lies entirely in S. That is, S is an open set if and only if for every point $P \in S$ there exists an open disc D containing P and lying entirely in S (see Figure 14.16). For example, an open disc, the first quadrant (see Figure 14.17), and the entire plane are all open sets.

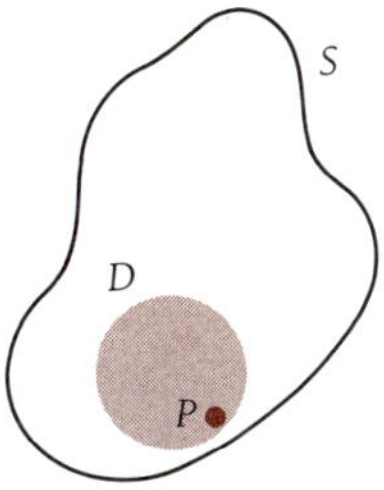

FIGURE 14.16
An open set

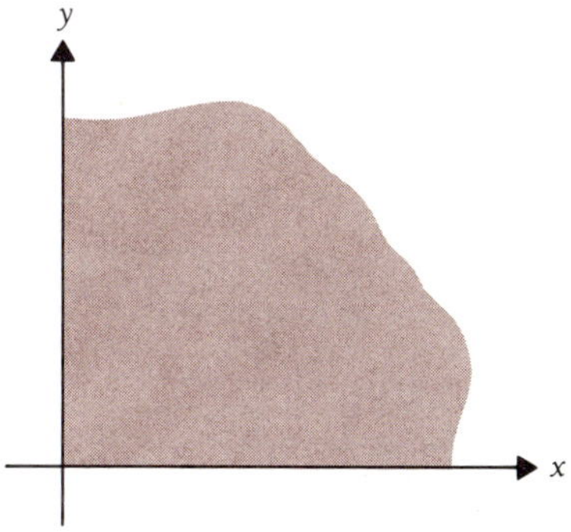

FIGURE 14.17
Quadrant I is an open set

A set S of points in the plane is called **closed** if and only if the complement of S (the set of points not in S) is open. That is, S is a closed set if and only if for every point $P \notin S$ there exists an open disc D containing P and lying entirely outside of S (see Figure 14.18). For example, a line segment with endpoints is closed, since any point not belonging to the segment can be enclosed in an open disc that misses the segment (see Figure 14.19). On the other hand, the interior of a circle is not closed, since an open disc containing a point on the circle must contain points of the interior (see Figure 14.20). The entire plane is also a closed set.

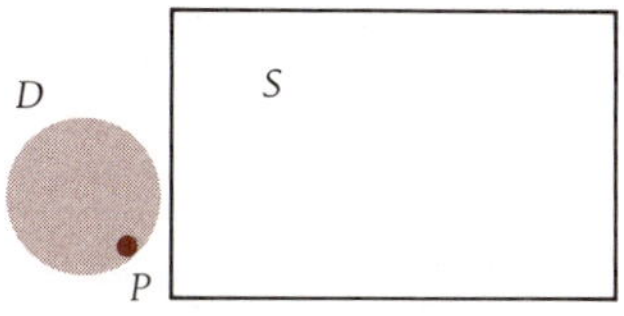

FIGURE 14.18 A closed set

FIGURE 14.19
A line segment in the plane is closed

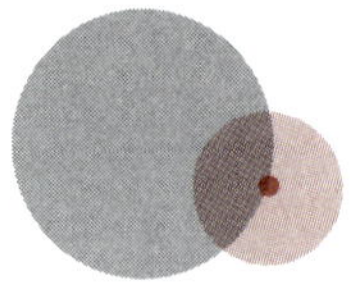

FIGURE 14.20
The interior of a circle is not closed

A set S of points in the plane is called **bounded** if and only if there is a circle whose interior entirely contains S. That is, S is a bounded set if and only if there exists an open disc D entirely containing S. Thus, a bounded set cannot extend to infinity.

The definitions are the same for higher dimensions, but using open balls instead of open discs. A set S is **open** if every point of S lies in an open ball contained in S. A set S is **closed** if its complement is open. A set S is **bounded** if it is contained in an open ball.

Now we can define a compact set.

DEFINITION 14.8

A set is **compact** if and only if it is both closed and bounded.

With this definition, we can state a multidimensional version of the extreme value theorem.

THEOREM 14.3

Extreme Value Theorem If f is a continuous function defined on a compact set D, then f has both an absolute maximum and an absolute minimum on D.

For example, if the temperature function on a sphere in three dimensions is continuous, then there must be a point on the sphere at which the temperature is a maximum and there must be another point at which it is a minimum, because the sphere is a compact set. This would imply, for instance, that there are points on earth at which minimum and maximum temperatures exist at any given time, provided the temperature function is continuous.

You should be able to see the similarity between Theorem 14.3 and the earlier version of the extreme value theorem (Theorem 2.21), noting that a closed interval is a compact set.

Exercises 14.2

1. Let

$$f(x, y) = \frac{xy}{x^2 + y^2} \qquad \text{and} \qquad S = \left\{ \left(\frac{1}{n}, \frac{2}{n} \right) \middle| n \text{ a positive integer} \right\}.$$

Compute $f(1/n, 2/n)$ for several values of n, and guess the value of

$$\lim_{\substack{(x, y) \to (0, 0) \\ (x, y) \in S}} f(x, y).$$

2. Repeat Problem 1 for

$$S = \left\{ \left(\frac{1}{n}, \frac{1}{2n} \right) \middle| n \text{ a positive integer} \right\}.$$

3. Repeat Problem 1 for

$$f(x, y) = \frac{\arctan xy}{x + y}.$$

4. Repeat Problem 3 for

$$S = \left\{ \left(\frac{1}{n}, \frac{1}{n} \right) \middle| n \text{ a positive integer} \right\}.$$

Find the limit, if it exists.

5. $\lim_{(x, y) \to (1, 1)} (3xy^2 - x^2y)$

6. $\lim_{(x, y) \to (-1, 2)} (4x + xy^2 - 1)$

7. $\lim_{(x, y) \to (2, 1)} \frac{x - y}{x^2 + y}$

8. $\lim_{(x, y) \to (0, 2)} \frac{2x + y}{xy - 1}$

9. $\lim_{(x, y) \to (0, 0)} \cos \sqrt{\frac{x + 1}{y + 4}}$

10. $\lim_{(x, y) \to (0,1)} \frac{\cos x + \sin y}{x^2 + y^2}$

11. $\lim_{(x, y) \to (0, 0)} \frac{2xy}{x^2 + y^2}$

12. $\lim_{(x, y) \to (0, 0)} \frac{x^2 + y^2}{3xy}$

13. $\lim_{(x, y, z) \to (0, 0, 1)} \frac{xyz}{x^2 + y^2 + z}$

14. $\lim_{(x, y, z) \to (0, 0, 0)} \frac{xyz}{x^2 + y^2 + z^2}$

15. $\lim_{\substack{(x, y) \to (0, 0) \\ y = x}} \frac{2x^2 - y^2}{x^2 + y^2}$

16. $\lim_{\substack{(x, y) \to (0, 0) \\ y = mx}} \frac{2x^2 - y^2}{x^2 + y^2}$

17. $\lim_{\substack{(x, y) \to (0, 0) \\ y = mx}} \frac{xy - y^2}{x^2 + y^2}$

18. $\lim_{\substack{(x, y) \to (0, 0) \\ x = y^2}} \frac{xy - y^2}{x^2 + y^2}$

19. $\lim_{\substack{(x, y, z) \to (0, 0, 0) \\ z = x^2 - y^2}} \frac{xy}{x^2 + y^2 + z}$

20. $\lim_{\substack{(x, y, z) \to (0, 0, 0) \\ y = x,\ z = 0}} \frac{xy}{x^2 + y^2 + z}$

21. $\lim_{\substack{(x, y, z) \to (0, 0, 0) \\ x = y = z}} \frac{2x - z}{x + y + z}$

22. $\lim_{\substack{(x, y, z) \to (0, 0, 0) \\ x = y = 2z}} \frac{2x - z}{x + y + z}$

23. Given $\lim_{(x, y) \to (a, b)} f(x, y) = L$, show that

$$\lim_{x \to a} f(x, b) = L \quad \text{and} \quad \lim_{y \to b} f(a, y) = L.$$

24. Is the converse of Problem 23 true?

Determine whether the function is continuous at the given point.

25. $f(x, y) = \dfrac{xy}{x^2 + y^2}$; $(1, 0)$

26. $f(x, y) = \dfrac{\arctan xy}{x^2 + y^2}$; $(1, 1)$

27. $f(x, y) = \begin{cases} (x^2 - y^2)/(x^2 + y^2) & \text{if } (x, y) \neq (0, 0) \\ 0 & \text{if } (x, y) = (0, 0); \end{cases}$ $(0, 0)$

28. $f(x, y) = \begin{cases} (x-y)/(x^2-y^2) & \text{if } x \neq y \\ 1/x^2 & \text{if } x = y; \end{cases}$ (2, 2)

29. Show that a circle in the plane, together with its interior, is a compact set.

30. Show that an ellipsoid in space is a compact set.

31. Is the union of two compact sets a compact set?

32. Is the intersection of two compact sets a compact set?

33. Show that the intersection of a compact set in space with a plane is another compact set.

SECTION 14.3 Partial Derivatives

Here we begin the study of differential calculus of functions of several variables. We start by differentiating such functions one variable at a time. This leads to partial derivatives, which we define first for functions of two variables.

Partial Derivatives

DEFINITION 14.9

Let f be a function of two variables x and y. The **partial derivative of f with respect to x** is the function

$$f_x(x, y) = \lim_{h \to 0} \frac{f(x+h, y) - f(x, y)}{h},$$

and the **partial derivative of f with respect to y** is the function

$$f_y(x, y) = \lim_{h \to 0} \frac{f(x, y+h) - f(x, y)}{h}$$

(provided the limits exist).

■ **Example 14.17** Given $f(x, y) = x^2 + 2xy$, find f_x and f_y.

Solution From the definition, we have

$$\begin{aligned} f_x(x, y) &= \lim_{h \to 0} \frac{f(x+h, y) - f(x, y)}{h} \\ &= \lim_{h \to 0} \frac{(x+h)^2 + 2(x+h)y - x^2 - 2xy}{h} \\ &= \lim_{h \to 0} \frac{x^2 + 2xh + h^2 + 2xy + 2hy - x^2 - 2xy}{h} \\ &= \lim_{h \to 0} (2x + h + 2y) = 2x + 2y, \end{aligned}$$

$$
\begin{aligned}
f_y(x, y) &= \lim_{h \to 0} \frac{f(x, y+h) - f(x, y)}{h} \\
&= \lim_{h \to 0} \frac{x^2 + 2x(y+h) - x^2 - 2xy}{h} \\
&= \lim_{h \to 0} 2x = 2x.
\end{aligned}
$$

■

Notice that the computation of f_x involves only x in the actual differentiation. That is, if we consider y to be a constant and let $F(x) = f(x, y)$, then

$$
\begin{aligned}
F'(x) &= \lim_{h \to 0} \frac{F(x+h) - F(x)}{h} \\
&= \lim_{h \to 0} \frac{f(x+h, y) - f(x, y)}{h} \\
&= f_x(x, y).
\end{aligned}
$$

Hence, to compute f_x, we may simply regard y as a constant and differentiate $f(x, y)$ with respect to x, using the usual techniques. Similarly, to compute f_y, we simply hold x constant and differentiate with respect to y. You can quickly check the results of this procedure with the results found in Example 14.17.

■ **Example 14.18** Compute f_x and f_y for:

(a) $f(x, y) = xy^2 - 2x \cos y$ **(b)** $f(x, y) = \dfrac{xy}{x^2 + y^2}$

Solution **(a)** Holding y constant and differentiating with respect to x, we get

$$f_x(x, y) = y^2 - 2 \cos y.$$

Holding x fixed and differentiating with respect to y, we get

$$f_y(x, y) = 2xy + 2x \sin y.$$

(b) Using the quotient rule, we get

$$
\begin{aligned}
f_x(x, y) &= \frac{(x^2 + y^2)y - xy(2x)}{(x^2 + y^2)^2} = \frac{y^3 - x^2 y}{(x^2 + y^2)^2}, \\
f_y(x, y) &= \frac{(x^2 + y^2)x - xy(2y)}{(x^2 + y^2)^2} = \frac{x^3 - xy^2}{(x^2 + y^2)^2}.
\end{aligned}
$$

Note that if x and y are interchanged, $f(x, y)$ is not changed. Therefore, by symmetry, we could have obtained f_y from f_x by just interchanging x and y.

■

Other forms of notation for partial derivatives include

$$f_1, \quad D_x f, \quad \frac{\partial f}{\partial x}; \qquad \text{and} \qquad f_2, \quad D_y f, \quad \frac{\partial f}{\partial y}.$$

To represent the value of f_x at point (a, b), we use any of the symbols

$$f_x(a, b), \quad f_1(a, b), \quad D_x f(a, b), \quad \left.\frac{\partial f}{\partial x}\right|_{(a, b)};$$

and similarly for f_y. The process of finding partial derivatives is called **partial differentiation**.

The definition of partial derivative is easily extended to functions of three or more variables.

DEFINITION 14.10

Let f be a function of three variables x, y, and z. The **partial derivatives of f with respect to x, y, and z** are, respectively,

$$f_x(x, y, z) = \lim_{h \to 0} \frac{f(x+h, y, z) - f(x, y, z)}{h},$$

$$f_y(x, y, z) = \lim_{h \to 0} \frac{f(x, y+h, z) - f(x, y, z)}{h},$$

$$f_z(x, y, z) = \lim_{h \to 0} \frac{f(x, y, z+h) - f(x, y, z)}{h}.$$

DEFINITION 14.11

Let f be a function of n variables $x_1, x_2, \ldots, x_n$. The **partial derivative of f with respect to x_i** is

$$f_i = \lim_{h \to 0} \frac{f(x_1, \ldots, x_{i-1}, x_i + h, x_{i+1}, \ldots, x_n) - f(x_1, \ldots, x_n)}{h}$$

Other notations for f_z are

$$f_3, \quad D_z f, \quad \frac{\partial f}{\partial z};$$

and other notations for f_i are

$$D_i f, \quad \frac{\partial f}{\partial x_i}.$$

■ Example 14.19

Find f_x, f_y, and f_z for the function

$$f(x, y, z) = x^2 y \sin z.$$

Solution

Holding two variables constant at a time, we get

$$f_x(x, y, z) = 2xy \sin z,$$
$$f_y(x, y, z) = x^2 \sin z,$$
$$f_z(x, y, z) = x^2 y \cos z.$$

■

Higher-Order Partial Derivatives

The partial derivatives of a function of several variables are other functions of several variables, so they may be differentiated, in turn. This leads to **higher-order** partial derivatives.

If f is a function of two variables x and y, we call f_x and f_y the **first-order** partial derivatives. The **second-order** partial derivatives are

$$f_{xx} = \frac{\partial^2 f}{\partial x^2}, \qquad f_{xy} = \frac{\partial^2 f}{\partial y\,\partial x}, \qquad f_{yx} = \frac{\partial^2 f}{\partial x\,\partial y}, \qquad f_{yy} = \frac{\partial^2 f}{\partial y^2}.$$

Here,

$$\frac{\partial^2 f}{\partial x^2} \quad \text{means} \quad \frac{\partial}{\partial x}\left(\frac{\partial f}{\partial x}\right), \qquad \frac{\partial^2 f}{\partial y\,\partial x} \quad \text{means} \quad \frac{\partial}{\partial y}\left(\frac{\partial f}{\partial x}\right), \qquad \text{etc.}$$

Some of the **third-order** partial derivatives would be f_{xxx}, f_{xxy}, f_{xyx}, and so forth.

For a function of three variables x, y, and z, there are three first-order partial derivatives, f_x, f_y, and f_z, and nine second-order partial derivatives, f_{xx}, f_{xy}, f_{xz}, f_{yx}, f_{yy}, f_{yz}, f_{zx}, f_{zy}, and f_{zz}.

Example 14.20 Find all the first- and second-order partial derivatives of $f(x, y) = x^3 + 4xy^2 + 3y$.

Solution The partial derivatives are

$$f_x = 3x^2 + 4y^2, \qquad f_y = 8xy + 3,$$
$$f_{xx} = 6x, \qquad f_{xy} = 8y, \qquad f_{yx} = 8y, \qquad f_{yy} = 8x.$$

■

Example 14.21 Find all the first-order and second-order partial derivatives of $f(x, y, z) = 2x^2y + e^x \sin z$.

Solution We find

$$\begin{aligned} &f_x = 4xy + e^x \sin z, && f_y = 2x^2, && f_z = e^x \cos z,\\ &f_{xx} = 4y + e^x \sin z, && f_{xy} = 4x, && f_{xz} = e^x \cos z,\\ &f_{yx} = 4x, && f_{yy} = 0, && f_{yz} = 0,\\ &f_{zx} = e^x \cos z, && f_{zy} = 0, && f_{zz} = -e^x \sin z. \end{aligned}$$

■

The second-order partial derivatives f_{xy}, f_{yx}, f_{xz}, f_{zx}, f_{yz}, and f_{zy} are called **mixed partials**. Note that in Example 14.20, $f_{xy} = f_{yx}$; and in Example 14.21, $f_{xy} = f_{yx}$, $f_{xz} = f_{zx}$, and $f_{yz} = f_{zy}$. This is not an accident. Second-order mixed partials involving the same variables are often equal. The governing theorem (proved in advanced calculus) is that if f_x, f_y, f_{xy}, and f_{yx} are all continuous, then $f_{xy} = f_{yx}$. This result extends to any number of variables and to higher-order mixed partials. For an example of a function whose mixed second-order partials are not equal, see Problem 27 at the end of this section.

Interpretations of Partial Derivatives

For a function of two variables, a geometric interpretation of the partial derivatives is possible. Consider the graph of the function $z = f(x, y)$, and let $P(a, b, f(a, b))$ be a point on the graph (Figure 14.21). The partial derivative $f_x(a, b)$ is found by holding y constant, differentiating with respect to x, and then setting $x = a$ and $y = b$. The same result will be obtained if we let $y = b$, then differentiate with respect to x, and let $x = a$. To set $y = b$ is to intersect the surface $z = f(x, y)$ with the plane $y = b$; the intersection is the curve $z = f(x, b)$ in the plane $y = b$. Now the partial derivative $f_x(a, b)$ is seen to be the slope of the curve $z = f(x, b)$ at point P in the plane $y = b$ (see Figure 14.21). Thus, $f_x(a, b)$ can be thought of as the slope of the surface $z = f(x, y)$ "in the x direction" at P.

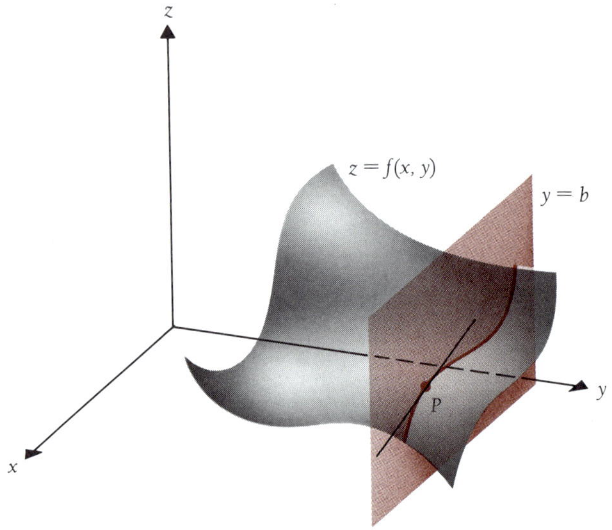

FIGURE 14.21 Interpretation of f_x as a slope at $P(a, b, f(a, b))$

Similarly, the partial derivative $f_y(a, b)$ can be thought of as the slope of the surface $z = f(x, y)$ "in the y direction" at P (see Figure 14.22).

An interpretation as a rate of change is also possible for a partial derivative. If f is a function of several variables and x is one of them, then f_x is the rate of change of f with respect to x alone (all other variables being held constant).

■ **Example 14.22** Find the rate of change of the volume of a cylinder:

(a) With respect to its radius **(b)** With respect to its height

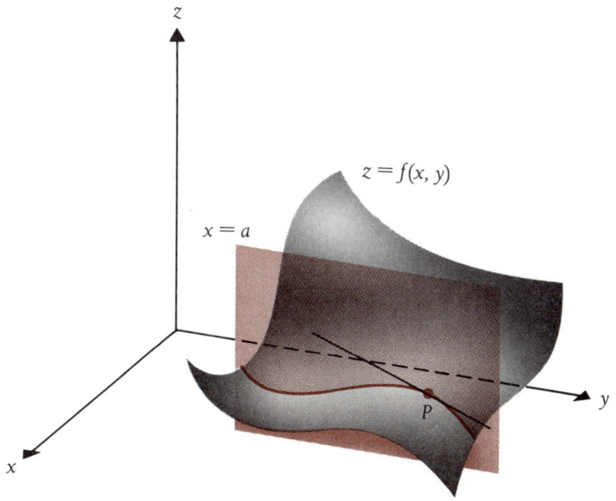

FIGURE 14.22 Interpretation of f_y as a slope at $P(a, b, f(a, b))$

Solution The volume of a cylinder of radius r and height h is $V = \pi r^2 h$.

(a) Therefore, the rate of change of V with respect to r is

$$\frac{\partial V}{\partial r} = 2\pi r h,$$

which just happens to be the lateral surface area. (Imagine painting the sides of a cylinder with thick paint; the rate at which the volume increases is proportional to the lateral surface area.)

(b) The rate of change of V with respect to h is

$$\frac{\partial V}{\partial h} = \pi r^2,$$

which happens to be the area of the base. (Imagine filling a bucket with water; the rate at which the volume increases is proportional to the area of the base.) ■

Exercises 14.3

Find all the first- and second-order partial derivatives.

1. $f(x, y) = 4x^3 - 3xy^2 + y^4$
2. $f(x, y) = x^4 + 3x^3y + 7x^2y^2 + 5y^3 - 6$
3. $f(x, y) = xy \ln(x^2 + y^2)$
4. $f(x, y) = (x^2 + y^2) \cos xy$
5. $f(x, y) = \sin^{-1}(x + 2y)$
6. $f(x, y) = 3x\sqrt{x^2 + y^2}$
7. $f(x, y) = e^{x^2+y^2}$
8. $f(x, y) = x \ln y - ye^x$

9. $f(x, y, z) = xy \sin z$

10. $f(x, y, z) = e^{xyz}$

11. $f(x, y, z) = x^2 + 2xy + 2yz + 3xz + y^2 + z^2$

12. $f(x, y, z) = \dfrac{xy}{z}$

Evaluate f_x and f_y at the point given.

13. $f(x, y) = x^2 - xy + 3y^2$; $(1, 1)$

14. $f(x, y) = e^x \cos y$; $(0, \pi)$

15. $f(x, y) = \dfrac{x+1}{y^2+1}$; $(3, 0)$

16. $f(x, y) = \sin x \ln y$; $(\pi/2, e^2)$

17. Find the rate of change of the volume of a cone with respect to:
(a) Its height (b) Its base radius

18. If an electric current of I amperes flows through a circuit of R ohms resistance for t seconds, the amount of heat produced is

$$H(I, t) = 0.24RI^2t \text{ calories.}$$

Find the rate of heat production in calories per second.

19. If $f(x, y, z) = \sin xyz$, find f_{xyz}.

20. If $f(x, y, z) = x^4y^3z - 3y^2e^{xz}$, show that

$$f_{xxy} = f_{xyx} = f_{yxx}.$$

21. Show that $f(x, y) = Ax^4 + 2Bx^2y^2 + Cy^4$ satisfies the partial differential equation

$$x\frac{\partial f}{\partial x} + y\frac{\partial f}{\partial y} = 4f.$$

22. Show that $f(x, y) = (y - 2x)^3 - \sqrt{y - 2x}$ satisfies the equation

$$f_{xx} - 4f_{yy} = 0.$$

23. Show that $f(x, y) = x^2y^2/(x + y)$ satisfies the equation

$$x\frac{\partial f}{\partial x} + y\frac{\partial f}{\partial y} = 3f.$$

[A]24. **Laplace's equation** is the partial differential equation

$$\frac{\partial^2 f}{\partial x^2} + \frac{\partial^2 f}{\partial y^2} = 0.$$

Show that the following functions satisfy Laplace's equation:

(a) $f(x, y) = \ln(x^2 + y^2)$

(b) $f(x, y) = \arctan \dfrac{y}{x}$

(c) $f(x, y) = e^x \sin y + e^y \cos x$

(d) $f(x, y) = \arctan \dfrac{2xy}{x^2 - y^2}$

25. Let curve C be the intersection of the surface $z = 9 - x^2 - y^2$ with the plane $y = 2$. Find the parametric equations of the line tangent to C at the point $(2, 2, 1)$.

26. Find f_x and f_y for $f(x, y) = \int_x^y (1 + t^3)^{-1/2}\, dt$.

27. Let the function f be defined by

$$f(x, y) = \begin{cases} (x^3y - xy^3)/(x^2 + y^2) & \text{if } (x, y) \neq (0, 0) \\ 0 & \text{if } (x, y) = (0, 0). \end{cases}$$

(a) Find $f_x(0, y)$ and $f_x(0, 0)$.

(b) Find $f_y(x, 0)$ and $f_y(0, 0)$.

(c) Find $f_{xy}(0, 0) = \lim_{h \to 0} \dfrac{f_x(0, h) - f_x(0, 0)}{h - 0}$.

(d) Find $f_{yx}(0, 0) = \lim_{h \to 0} \dfrac{f_y(h, 0) - f_y(0, 0)}{h - 0}$.

(e) Find $f_{xy}(x, 0)$ for $x \neq 0$, and verify that f_{xy} is not continuous at $(0, 0)$.

Rectangular coordinates (x, y) and polar coordinates $[r, \theta]$ are related by the equations $x = r\cos\theta$, $y = r\sin\theta$, $r = \sqrt{x^2 + y^2}$, and $\theta = \arctan(y/x)$. Find the following partial derivatives:

28. $\dfrac{\partial x}{\partial r}$ 29. $\dfrac{\partial x}{\partial \theta}$ 30. $\dfrac{\partial y}{\partial r}$ 31. $\dfrac{\partial y}{\partial \theta}$

32. $\dfrac{\partial r}{\partial x}$ 33. $\dfrac{\partial r}{\partial y}$ 34. $\dfrac{\partial \theta}{\partial x}$ 35. $\dfrac{\partial \theta}{\partial y}$

36. An equation of the form

$$\frac{\partial^2 u}{\partial x^2} = c^2 \frac{\partial^2 u}{\partial t^2},$$

with c a constant, is called a **wave equation.** Show that the function

$$u = \sin ax \sin bt$$

satisfies the wave equation with $c = a/b$.

37. Show that the function

$$z = e^{-ay} \cos ax$$

satisfies the equation

$$\frac{\partial^2 z}{\partial x^2} = a \frac{\partial z}{\partial y}.$$

38. Let R be the rectangle $\{(x, y) | a \leq x \leq b \text{ and } c \leq y \leq d\}$. Let f be a continuous function of two variables with domain R, and suppose f_x is also continuous. Then $\int_c^d f(x, y)\, dy$ is a differentiable function of x and

$$D_x \int_c^d f(x, y)\, dy = \int_c^d f_x(x, y)\, dy.$$

Use this fact to find the derivative with respect to x of:

(a) $\displaystyle\int_1^{\sqrt{3}} \ln(x^2 + y^2)\, dy$ (b) $\displaystyle\int_1^2 \frac{1}{y} \arctan \frac{x}{y}\, dy$

39. A bacterium is roughly the shape of a cylinder of radius r and height h, capped on each end by a hemisphere. The volume of the bacterium is

$$V = \pi r^2 \left(h + \frac{4r}{3} \right),$$

and the surface area is

$$S = 2\pi r(h + 2r).$$

The rate R at which a chemical substance can be absorbed into the bacterium is

$$R = c \left(\frac{S}{V} \right),$$

with c a positive constant. Show that $\partial R/\partial r < 0$ and $\partial R/\partial h < 0$. (This shows that an increase in either r or h results in a decrease in the rate of chemical absorption.)

SECTION 14.4

Differentiability and the Chain Rule

A function of a single variable is called differentiable if its first derivative exists. For functions of several variables, differentiability involves more than the existence of the first-order partial derivatives. For example, it is not possible to prove a chain rule for functions of two variables if we know only that the partial derivatives exist. In this section, we will define differentiability for functions of several variables and develop the chain rule.

Differentiability

In order to define differentiability of a function of two variables, we will use an extended form of the mean value theorem (Theorem 4.3) for functions of one variable. We will state and prove the theorem here.

THEOREM 14.4

Extended Mean Value Theorem Let f be a function of one variable, and suppose the derivative f' exists in an interval $(a - \delta, a + \delta)$ containing a. Then there exists a function η defined on the interval $(-\delta, \delta)$ such that $\eta(h) \to 0$ as $h \to 0$, and for every h in $(-\delta, \delta)$,

$$f(a + h) - f(a) = f'(a)h + \eta(h)h.$$

Proof If we let

$$\eta(h) = \frac{1}{h}\,[f(a + h) - f(a)] - f'(a),$$

then the final conclusion of the theorem is clearly true (just multiply by h). Moreover,

$$\lim_{h\to 0} \eta(h) = \lim_{h\to 0} \left\{\frac{1}{h}[f(a+h) - f(a)] - f'(a)\right\}$$
$$= f'(a) - f'(a) = 0,$$

and the theorem is proved.

From this theorem we can see that since $\eta(h)$ is small when h is small, the product $\eta(h)h$ is very small. Thus, for small h,

$$f(a+h) - f(a) \approx f'(a)h.$$

If we let h be an increment in x (that is, $h = \Delta x$), then $f(a+h) - f(a)$ is the corresponding increment in f, and

$$\Delta f(a) \approx f'(a)\Delta x. \tag{1}$$

This is just a restatement of the fact that the increment of the function f is approximately the differential of f, which is the derivative of f times the increment of the independent variable.

Now we are ready to state the version of this theorem for two variables.

THEOREM 14.5 Let f be a function of two variables x and y, and suppose the first-order partial derivatives f_x and f_y exist and are continuous in an open disc D of radius δ centered at point $A(a, b)$. Then there exist functions η_1 and η_2 defined on the disc of radius δ centered at $(0, 0)$ such that $\eta_1(h, k) \to 0$ and $\eta_2(h, k) \to 0$ as $(h, k) \to (0, 0)$, and for every $(a+h, b+k)$ in D,

$$f(a+h, b+k) - f(a, b) = f_x(a, b)h + f_y(a, b)k + \eta_1(h, k)h + \eta_2(h, k)k.$$

Proof First, we write

$$f(a+h, b+k) - f(a, b) = f(a+h, b+k) - f(a, b+k) + f(a, b+k) - f(a, b)$$

and let

$$F(x) = f(x, b+k).$$

Then

$$F'(x) = f_x(x, b+k).$$

Since f_x exists in D, we may apply Theorem 14.4 to F. Hence, there exists a function η such that $\eta(h) \to 0$ as $h \to 0$ and

$$\begin{aligned} f(a+h, b+k) - f(a, b+k) &= F(a+h) - F(a) \\ &= F'(a)h + \eta(h)h \\ &= f_x(a, b+k)h + \eta(h)h. \end{aligned}$$

Similarly, if $G(y) = f(a, y)$, then $G'(y) = f_y(a, y)$, and by Theorem 14.4 again, there exists a function η^* such that $\eta^*(k) \to 0$ as $k \to 0$ and

$$\begin{aligned} f(a, b+k) - f(a, b) &= G(b+k) - g(b) \\ &= G'(b)k + \eta^*(k)k \\ &= f_y(a, b)k + \eta^*(k)k. \end{aligned}$$

Now, since f_x is continuous in D, $f_x(a, b+k) \to f_x(a, b)$ as $k \to 0$. Let

$$\eta^{**}(k) = f_x(a, b+k) - f_x(a, b).$$

Then let $\eta_1(h, k) = \eta(h) + \eta^{**}(k)$ and let $\eta_2(h, k) = \eta^*(k)$. Then

$$\begin{aligned} f(a+h, b+k) - f(a, b) &= f(a+h, b+k) - f(a, b+k) + f(a, b+k) - f(a, b) \\ &= f_x(a, b+k)h + \eta(h)h + f_y(a, b)k + \eta^*(k)k \\ &= [f_x(a, b) + \eta^{**}(k)]h + \eta(h)h + f_y(a, b)k + \eta^*(k)k \\ &= f_x(a, b)h + \eta_1(h, k)h + f_y(a, b)k + \eta_2(h, k)k. \end{aligned}$$

We also have $\eta_1(h, k) \to 0$ and $\eta_2(h, k) \to 0$ as $(h, k) \to (0, 0)$, so the theorem is proved.

In Theorem 14.5, if we let h and k be increments of x and y (that is, $h = \Delta x$ and $k = \Delta y$), then $f(a+h, b+k) - f(a, b)$ is the corresponding increment of f,

$$\Delta f(a, b) = f(a + \Delta x, b + \Delta y) - f(a, b).$$

The conclusion of Theorem 14.5 can then be written as

$$\Delta f(a, b) = f_x(a, b)\Delta x + f_y(a, b)\Delta y + \eta_1 \Delta x + \eta_2 \Delta y.$$

Since η_1 and η_2 both approach 0 as Δx and Δy approach 0, the products $\eta_1 \Delta x$ and $\eta_2 \Delta y$ are very small, and

$$\Delta f(a, b) \approx f_x(a, b)\Delta x + f_y(a, b)\Delta y. \tag{2}$$

This is the two-variable version of equation (1) and can be interpreted geometrically in a similar way. The approximation in equation (1) is along a tangent line instead of along the graph of f (see Section 3.7); and the approximation in equation (2) is along a tangent plane instead of along the graph of f. We will have more to say about the tangent plane later.

The form of Theorem 14.5 that holds for any number of variables is stated in Theorem 14.6 without proof.

THEOREM 14.6 Let f be a function of n variables, and suppose that each first-order partial derivative f_{x_i} exists and is continuous in an open ball B of radius δ centered at point $A(a_1, \ldots, a_n)$. Then there exist functions $\eta_1, \ldots, \eta_n$ defined on the ball of radius δ centered at $(0, \ldots, 0)$ such that for each i, $\eta_i(h_1, \ldots, h_n) \to 0$ as $(h_1, \ldots, h_n) \to (0, \ldots, 0)$, and for every $(a_1 + h_1, \ldots, a_n + h_n)$ in B,

$$f(a_1 + h_1, \ldots, a_n + h_n) - f(a_1, \ldots, a_n) = f_{x_1}(a_1, \ldots, a_n)h_1 + \cdots + f_{x_n}(a_1, \ldots, a_n)h_n + \eta_1(h_1, \ldots, h_n)h_1 + \cdots + \eta_n(h_1, \ldots, h_n)h_n.$$

Stated in terms of increments, the conclusion of this theorem is that

$$\Delta f(a_1, \ldots, a_n) = \sum_{i=1}^{n} f_{x_i}(a_1, \ldots, a_n)\Delta x_i + \sum_{i=1}^{n} \eta_i \Delta x_i.$$

We can now give the definition of differentiability for a function of several variables. We will first treat two variables and then many.

DEFINITION 14.12

A function f of two variables x and y is **differentiable at point** $A(a, b)$ if and only if $f_x(a, b)$ and $f_y(a, b)$ exist, and there exist functions η_1 and η_2 of Δx and Δy such that $\eta_1 \to 0$ and $\eta_2 \to 0$ as $(\Delta x, \Delta y) \to (0, 0)$ and

$$\Delta f(a, b) = f_x(a, b)\Delta x + f_y(a, b)\Delta y + \eta_1 \Delta x + \eta_2 \Delta y.$$

DEFINITION 14.13

A function f of n variables $x_1, \ldots, x_n$ is **differentiable at** $A(a_1, \ldots, a_n)$ if and only if each partial derivative $f_{x_i}(a_1, \ldots, a_n)$ exists, and there exist functions $\eta_1, \ldots, \eta_n$ of $\Delta x_1, \ldots, \Delta x_n$ such that for each i, $\eta_i \to 0$ as the $\Delta x_i \to 0$ and

$$\Delta f(a_1, \ldots, a_n) = \sum_{i=1}^{n} f_{x_i}(a_1, \ldots, a_n)\Delta x_i + \sum_{i=1}^{n} \eta_i \Delta x_i.$$

If f is differentiable at A and all the first-order partial derivatives of f are continuous at A, then f is said to be **continuously differentiable at A**. We will say that a function is **differentiable** if it is differentiable at each point of its domain.

We see by Theorem 14.6 that a function whose first-order partial derivatives all exist and are continuous in an open ball containing A is continuously differentiable at A. It is possible for a function to be differentiable at A without its partial derivatives being continuous at A; thus, continuity of the partial derivatives is sufficient but not necessary for differentiability. It is also possible for a function to have partial derivatives at A but fail to be differentiable at A; that is, existence of the first-order partial derivatives is necessary but not sufficient for differentiability.

The Chain Rule

We will now apply the definition of differentiability to establish the chain rule. As is the case for functions of one variable, the chain rule tells how to find derivatives of composite functions of several variables. We will prove the chain rule for functions of two variables, and then give some general and special cases without proof.

THEOREM 14.7

The Chain Rule Let f be a differentiable function of two variables u and v, and let u and v each be differentiable functions of two variables x and y. Then the partial derivatives of f with respect to x and y exist, and

$$\frac{\partial f}{\partial x} = \frac{\partial f}{\partial u}\frac{\partial u}{\partial x} + \frac{\partial f}{\partial v}\frac{\partial v}{\partial x}, \tag{3}$$

$$\frac{\partial f}{\partial y} = \frac{\partial f}{\partial u}\frac{\partial u}{\partial y} + \frac{\partial f}{\partial v}\frac{\partial v}{\partial y}. \tag{4}$$

Proof By Definition 14.12, there exist η_1 and η_2 such that $\eta_1 \to 0$ and $\eta_2 \to 0$ as $(\Delta u, \Delta v) \to (0, 0)$, and

$$\Delta f = f_u \Delta u + f_v \Delta v + \eta_1 \Delta u + \eta_2 \Delta v.$$

Let Δx be an increment in x, and let

$$\Delta u = u(x + \Delta x, y) - u(x, y),$$
$$\Delta v = v(x + \Delta x, y) - v(x, y)$$

be the corresponding increments in u and v. Then by Definition 14.9, we have

$$\begin{aligned}
\frac{\partial f}{\partial x} &= \lim_{\Delta x \to 0} \frac{f(u(x + \Delta x, y), v(x + \Delta x, y)) - f(u(x, y), v(x, y))}{\Delta x} \\
&= \lim_{\Delta x \to 0} \frac{f(u + \Delta u, v + \Delta v) - f(u, v)}{\Delta x} \\
&= \lim_{\Delta x \to 0} \frac{\Delta f}{\Delta x} \\
&= \lim_{\Delta x \to 0} \frac{1}{\Delta x}(f_u \Delta u + f_v \Delta v + \eta_1 \Delta u + \eta_2 \Delta v) \\
&= \lim_{\Delta x \to 0} \left(f_u \frac{\Delta u}{\Delta x} + f_v \frac{\Delta v}{\Delta x} + \eta_1 \frac{\Delta u}{\Delta x} + \eta_2 \frac{\Delta v}{\Delta x}\right).
\end{aligned} \tag{5}$$

Now,

$$\frac{\Delta u}{\Delta x} = \frac{u(x + \Delta x, y) - u(x, y)}{\Delta x},$$

so that

$$\lim_{\Delta x \to 0} \frac{\Delta u}{\Delta x} = \frac{\partial u}{\partial x}.$$

Similarly,

$$\lim_{\Delta x \to 0} \frac{\Delta v}{\Delta x} = \frac{\partial v}{\partial x}.$$

Also, $\Delta u \to 0$ and $\Delta v \to 0$ as $\Delta x \to 0$, so $\eta_1 \to 0$ and $\eta_2 \to 0$ as $\Delta x \to 0$. Hence, equation (5) yields

$$\frac{\partial f}{\partial x} = f_u \frac{\partial u}{\partial x} + f_v \frac{\partial v}{\partial x},$$

which is equation (3). The proof of (4) is similar.

Let's look at an example.

Example 14.23 Let $f(u, v) = e^{uv}$, $u(x, y) = x \sin y$, and $v(x, y) = y \sin x$. Find $\partial f/\partial x$ and $\partial f/\partial y$.

Solution By the chain rule, we get

$$\begin{aligned} \frac{\partial f}{\partial x} &= \frac{\partial f}{\partial u}\frac{\partial u}{\partial x} + \frac{\partial f}{\partial v}\frac{\partial v}{\partial x} \\ &= ve^{uv} \sin y + ue^{uv} y \cos x, \\ \frac{\partial f}{\partial y} &= \frac{\partial f}{\partial u}\frac{\partial u}{\partial y} + \frac{\partial f}{\partial v}\frac{\partial v}{\partial y} \\ &= ve^{uv} x \cos y + ue^{uv} \sin x. \end{aligned}$$

■

The most general form of the chain rule is stated in Theorem 14.8.

THEOREM 14.8 If f is a differentiable function of n variables $u_1, \ldots, u_n$ and each u_j is a differentiable function of m variables $x_1, \ldots, x_m$, then $\partial f/\partial x_i$ exists for each i, and

$$\frac{\partial f}{\partial x_i} = \sum_{j=1}^{n} \frac{\partial f}{\partial u_j}\frac{\partial u_j}{\partial x_i}.$$

The special case of Theorem 14.8 in which each u_j is a function of the single variable x gives a formula for the "total" derivative df/dx, since in that case f is a function of x alone:

$$\frac{df}{dx} = \sum_{j=1}^{n} \frac{\partial f}{\partial u_j}\frac{du_j}{dx}. \tag{6}$$

Example 14.24 Let $f(x, y, z) = xy^2 - xz + y^2 z$, and let $x = \cos t$, $y = \tan t$, and $z = e^t$. Find df/dt.

Solution By equation (6), with x, y, and z playing the roles of the u_j's, we have

$$\begin{aligned}\frac{df}{dt} &= \frac{\partial f}{\partial x}\frac{dx}{dt} + \frac{\partial f}{\partial y}\frac{dy}{dt} + \frac{\partial f}{\partial z}\frac{dz}{dt}\\ &= (y^2 - z)(-\sin t) + (2xy + 2yz)\sec^2 t + (-x + y^2)e^t.\end{aligned}$$

■

■ **Example 14.25** A certain gas obeys the law $PV = kT$, which relates pressure P, volume V, and temperature T, with k a constant. If the pressure is increasing at the rate of 0.05 pound per square inch/second and the volume is decreasing at the rate of 0.25 cubic inch/second, how is the temperature changing when $P = 12$ pounds per square inch and $V = 120$ cubic inches?

Solution From the given equation, we have

$$T = \frac{1}{k}PV;$$

therefore,

$$\begin{aligned}\frac{dT}{dt} &= \frac{\partial T}{\partial P}\frac{dP}{dt} + \frac{\partial T}{\partial V}\frac{dV}{dt}\\ &= \frac{V}{k}\frac{dP}{dt} + \frac{P}{k}\frac{dV}{dt}.\end{aligned}$$

When $P = 12$, $V = 120$, we have

$$\frac{dT}{dt} = \frac{120}{k}(0.05) + \frac{12}{k}(-0.25) = \frac{3}{k},$$

so the temperature is increasing at the rate of $3/k$ degrees per second. ■

Exercises 14.4

Find $\partial f/\partial x$ and $\partial f/\partial y$.

1. $f(u, v) = u + 3v + 5; \quad u = x^2 + y^2, \quad v = x^2 - xy + y^2$
2. $f(u, v) = e^u \ln v; \quad u = x + \dfrac{1}{y}, \quad v = x \tan y$
3. $f(u, v, w) = uv - w^2; \quad u = x^2 - y^2, \quad v = 2xy, \quad w = y^2 - x^2$
4. $f(u, v, w) = w \sin uv; \quad u = e^x, \quad v = e^{-y}, \quad w = \ln(x + y)$
5. $f(r, s, t, u) = r^2 + s^2 + t^2 + u^2; \quad r = \sin^{-1}xy, \quad s = x + y, \quad t = x - y, \quad u = xy$
6. $f(r, s, t, u) = rs + tu - rt - su; \quad r = 2x - y, \quad s = 2y - x, \quad t = x^2, \quad u = xy$
7. $f(u_1, u_2, u_3) = (u_1 + u_2 + u_3)^2; \quad u_1 = xy, \quad u_2 = x^2y^2, \quad u_3 = x^3y^3$
8. $f(u_1, \ldots, u_n) = u_1 + \cdots + u_n; \quad u_i = x^i + y^i \ (i = 1, \ldots, n)$

Find $\partial f/\partial x$, $\partial f/\partial y$, and $\partial f/\partial z$.

9. $f(u, v) = \sin^2 u - \sin^2 v; \quad u = x + y + z, \quad v = x - y + z$
10. $f(u, v) = \sqrt{u + v}; \quad u = \sqrt{1 + x^2}, \quad v = \sqrt{1 - y^2}$

11. $f(u, v, w) = uv - \sin w$; $u = \dfrac{x+y}{2}$, $v = \dfrac{x-y}{2}$, $w = \dfrac{1}{z}$

12. $f(u, v, w) = (1 + u)e^{vw}$; $u = yz \ln x$, $v = xz \ln y$, $w = xy \ln z$

Find df/dt.

13. $f(x, y) = \sqrt{x^2 + y^2 - 9}$; $x = 2t^2$, $y = \dfrac{1}{t}$

14. $f(x, y) = \arctan(x + y)$; $x = \cos t$, $y = \sin t$

15. $f(u, v) = \dfrac{uv}{u+v}$; $u = \dfrac{1}{t}$, $v = \dfrac{1+t}{1-t}$

16. $f(r, s) = e^{rs}$; $r = t + \ln t$, $s = t \cos t$

17. The base of a triangle is increasing at the rate of 2 inches per minute, and the height is decreasing at the rate of 0.5 inch per minute. How is the area of the triangle changing when the base is 20 inches and the altitude is 12 inches?

18. The height of a right circular cone is increasing at the rate of 0.025 meter per second, and the base radius is decreasing at the rate of 0.05 meter per second. At a certain instant, the height and radius are both 1 meter. How is the volume changing at that instant?

19. Let f be a differentiable function of two variables, and let t be a differentiable function of x. Show that

$$\frac{\partial}{\partial x} f(t, x) = f_1(t, x) \frac{dt}{dx} + f_2(t, x).$$

20. If f is a differentiable function of u and v, and u and v are differentiable functions of x and y, show that

$$\frac{\partial f}{\partial u} = \frac{\dfrac{\partial f}{\partial x}\dfrac{\partial v}{\partial y} - \dfrac{\partial f}{\partial y}\dfrac{\partial v}{\partial x}}{\dfrac{\partial u}{\partial x}\dfrac{\partial v}{\partial y} - \dfrac{\partial u}{\partial y}\dfrac{\partial v}{\partial x}}, \qquad \frac{\partial f}{\partial v} = \frac{\dfrac{\partial f}{\partial y}\dfrac{\partial u}{\partial x} - \dfrac{\partial f}{\partial x}\dfrac{\partial u}{\partial y}}{\dfrac{\partial u}{\partial x}\dfrac{\partial v}{\partial y} - \dfrac{\partial u}{\partial y}\dfrac{\partial v}{\partial x}}.$$

21. If $z = f(u, v)$, $u = x + y$, $v = x - y$, and f has continuous partial derivatives, show that

$$\frac{\partial z}{\partial x}\frac{\partial z}{\partial y} = \left(\frac{\partial f}{\partial u}\right)^2 - \left(\frac{\partial f}{\partial v}\right)^2.$$

22. If $z = f(u)$, $u = ax + by$, and f is a differentiable function of u, show that

$$a\left(\frac{\partial z}{\partial y}\right) = b\left(\frac{\partial z}{\partial x}\right).$$

23. If $z = f(x, y)$, f has continuous partial derivatives, and we substitute polar coordinates $x = r \cos \theta$, $y = r \sin \theta$, show that

$$\frac{\partial z}{\partial r} = f_x \cos \theta + f_y \sin \theta$$

and

$$\frac{1}{r}\frac{\partial z}{\partial \theta} = -f_x \sin \theta + f_y \cos \theta.$$

24. Solve the equations in Problem 23 for f_x and f_y in terms of r, θ, $\partial z/\partial r$, and $\partial z/\partial\theta$.

25. In Problem 23, show that

$$\left(\frac{\partial z}{\partial r}\right)^2 + \frac{1}{r^2}\left(\frac{\partial z}{\partial \theta}\right)^2 = f_x^2 + f_y^2.$$

26. Show that if a function f of two variables is differentiable at point $A(a, b)$, then f is continuous at A. [*Hint:* Use the definition of differentiability and show that $f(x, y) \to f(a, b)$ as $(x, y) \to (a, b)$.]

27. Let

$$f(x, y) = \begin{cases} xy/(x^2+y^2) & \text{if } (x, y) \neq (0, 0) \\ 0 & \text{if } (x, y) = (0, 0) \end{cases}$$

Show that the first-order partial derivatives of f exist at (0, 0), but f is not differentiable at (0, 0). [*Hint:* Use Problem 26.]

28. Let

$$f(x, y) = \begin{cases} x & \text{if } |x| \leq |y| \\ -x & \text{if } |x| > |y| \end{cases}$$

Show that f is continuous at (0, 0) and its first-order partial derivatives exist at (0, 0), but f is not differentiable at (0, 0).

29. Show that the function $f(x, y) = \sqrt{x^2 + y^2}$ is not differentiable at the origin.

30. Let

$$f(x, y) = \begin{cases} (x^2+y^2)\sin(1/\sqrt{x^2+y^2}) & \text{if } (x, y) \neq (0, 0) \\ 0 & \text{if } (x, y) = (0, 0) \end{cases}$$

Show that f is differentiable at the origin, but its first-order partial derivatives are not continuous at (0, 0).

31. Determine whether the function

$$f(x, y) = \begin{cases} 1 - x^2 - y^2 & \text{if } x \neq y \\ 0 & \text{if } x = y \end{cases}$$

is differentiable at the origin.

SECTION 14.5

Gradients and Directional Derivatives

In this section, we introduce the gradient of a differentiable function and use it to develop the directional derivative of a function of two or three variables.

The Gradient

Let f be a differentiable function of three variables. By the definition of differentiability, there exist functions η_1, η_2, and η_3 that all approach 0 as $(\Delta x, \Delta y, \Delta z) \to (0, 0, 0)$, such that

$$\Delta f = f_x\Delta x + f_y\Delta y + f_z\Delta z + \eta_1\Delta x + \eta_2\Delta y + \eta_3\Delta z. \tag{1}$$

We will rewrite equation (1) in vector notation by setting

$$\begin{aligned} \mathbf{h} &= \langle \Delta x, \Delta y, \Delta z \rangle, \\ \boldsymbol{\eta} &= \langle \eta_1, \eta_2, \eta_3 \rangle, \\ \nabla f &= \langle f_x, f_y, f_z \rangle. \end{aligned} \tag{2}$$

Then equation (1) becomes

$$\Delta f = \nabla f \cdot \mathbf{h} + \boldsymbol{\eta} \cdot \mathbf{h}. \tag{3}$$

Therefore, f is differentiable if and only if there exist ∇f and $\boldsymbol{\eta}$ such that $\boldsymbol{\eta} \to \mathbf{0}$ as $\mathbf{h} \to \mathbf{0}$ and equation (3) holds.

The vector (2) is called the **gradient of f**, and is also written as

$$\nabla f = f_x\mathbf{i} + f_y\mathbf{j} + f_z\mathbf{k}, \tag{4}$$

where **i**, **j**, and **k** are the standard unit vectors. The gradient of f is a function that assigns to each point a vector; this means that the gradient of f is a vector-valued function of three variables. Such a function is often called a **vector field**.

The symbol ∇ is read, "del," and is sometimes treated as a vector differential operator, being written as

$$\nabla = \left\langle \frac{\partial}{\partial x}, \frac{\partial}{\partial y}, \frac{\partial}{\partial z} \right\rangle. \tag{5}$$

From this point of view, ∇ operates on f, giving

$$\nabla f = \left\langle \frac{\partial f}{\partial x}, \frac{\partial f}{\partial y}, \frac{\partial f}{\partial z} \right\rangle.$$

To indicate that the gradient of f is evaluated at point $P(a, b, c)$, we can write

$$\nabla f(P) = \langle f_x(a, b, c), f_y(a, b, c), f_z(a, b, c) \rangle.$$

For a function of two variables, the gradient is

$$\nabla f = \langle f_x, f_y \rangle. \tag{6}$$

The following properties of the gradient are easily verified:

$$\nabla(f + g) = \nabla f + \nabla g \tag{7}$$

$$\nabla(af) = a\nabla f \tag{8}$$

$$\nabla(fg) = f\,\nabla g + g\nabla f \tag{9}$$

■ Example 14.26 Let $f(x, y) = x^2 + 3xy^2$; find ∇f and $\nabla f(-1, 3)$.

Solution Since $f_x = 2x + 3y^2$ and $f_y = 6xy$, the gradient is

$$\nabla f = \langle 2x + 3y^2, 6xy \rangle.$$

Therefore,

$$\nabla f(-1, 3) = \langle 2(-1) + 3(3)^2, 6(-1)(3)\rangle = \langle 25, -18\rangle. \qquad \blacksquare$$

Example 14.27 Find the gradient of the function $f(x, y, z) = e^{xy} + \cos yz$, and evaluate it at $(0, 1, \pi/2)$.

Solution The gradient is

$$\nabla f = \langle ye^{xy}, xe^{xy} - z \sin yz, -y \sin yz\rangle,$$

so that

$$\nabla f\left(0, 1, \frac{\pi}{2}\right) = \left\langle e^0, 0 - \frac{\pi}{2} \sin \frac{\pi}{2}, -\sin \frac{\pi}{2}\right\rangle = \left\langle 1, -\frac{\pi}{2}, -1\right\rangle. \qquad \blacksquare$$

Directional Derivatives

Now we will leave gradients for a moment in order to define a directional derivative. Let f be a function of three variables. The partial derivatives of f, namely

$$f_x(x, y, z) = \lim_{h\to 0} \frac{f(x + h, y, z) - f(x, y, z)}{h},$$

$$f_y(x, y, z) = \lim_{h\to 0} \frac{f(x, y + h, z) - f(x, y, z)}{h},$$

$$f_z(x, y, z) = \lim_{h\to 0} \frac{f(x, y, z + h) - f(x, y, z)}{h},$$

can be written in vector form as

$$f_x(\mathbf{v}) = \lim_{h\to 0} \frac{f(\mathbf{v} + h\mathbf{i}) - f(\mathbf{v})}{h},$$

$$f_y(\mathbf{v}) = \lim_{h\to 0} \frac{f(\mathbf{v} + h\mathbf{j}) - f(\mathbf{v})}{h},$$

$$f_z(\mathbf{v}) = \lim_{h\to 0} \frac{f(\mathbf{v} + h\mathbf{k}) - f(\mathbf{v})}{h}.$$

Thus, each partial derivative of f is of the form

$$\lim_{h\to 0} \frac{f(\mathbf{v} + h\mathbf{u}) - f(\mathbf{v})}{h}, \tag{10}$$

with $\mathbf{u}$ one of the unit vectors $\mathbf{i}$, $\mathbf{j}$, or $\mathbf{k}$. If $\mathbf{u}$ is allowed to be any unit vector, then $\mathbf{u}$ specifies a direction and we get a derivative of f in the direction of $\mathbf{u}$.

DEFINITION 14.14

Let f be a function of two or three variables, and let **u** be a unit vector. The **directional derivative** of f **in the direction of u** is the limit

$$f_{\mathbf{u}}(\mathbf{v}) = \lim_{h \to 0} \frac{f(\mathbf{v} + h\mathbf{u}) - f(\mathbf{v})}{h},$$

provided the limit exists.

If **w** is any vector, the directional derivative of f **in the direction of w** is $f_{\mathbf{u}}(\mathbf{v})$ with **u** a unit vector in the direction of **w**.

Thus, we see that the partial derivatives of f are special directional derivatives, namely

$$f_x = f_{\mathbf{i}}, \qquad f_y = f_{\mathbf{j}}, \qquad f_z = f_{\mathbf{k}}.$$

Definition 14.14 does not give the handiest formula for evaluating directional derivatives. For differentiable functions, a better formula is given by Theorem 14.9.

THEOREM 14.9

If f is a differentiable function of two or three variables and **u** is a unit vector, then the directional derivative of f in the direction of **u** is given by

$$f_{\mathbf{u}}(\mathbf{v}) = \mathbf{u} \cdot \nabla f(\mathbf{v}).$$

Proof

$$\begin{aligned} f(\mathbf{v} + h\mathbf{u}) - f(\mathbf{v}) &= \Delta f \\ &= \nabla f(\mathbf{v}) \cdot h\mathbf{u} + \boldsymbol{\eta} \cdot h\mathbf{u} && \text{[By equation (3)]} \\ &= h(\mathbf{u} \cdot \nabla f(\mathbf{v}) + \boldsymbol{\eta} \cdot \mathbf{u}). \end{aligned}$$

Hence,

$$\frac{f(\mathbf{v} + h\mathbf{u}) - f(\mathbf{v})}{h} = \mathbf{u} \cdot \nabla f(\mathbf{v}) + \boldsymbol{\eta} \cdot \mathbf{u}.$$

We know that $h\mathbf{u} \to \mathbf{0}$ as $h \to 0$, so that $\boldsymbol{\eta} \to \mathbf{0}$ as $h \to 0$. Therefore,

$$f_{\mathbf{u}}(\mathbf{v}) = \lim_{h \to 0} [\mathbf{u} \cdot \nabla f(\mathbf{v}) + \boldsymbol{\eta} \cdot \mathbf{u}] = \mathbf{u} \cdot \nabla f(\mathbf{v}).$$

■ **Example 14.28**

Find the directional derivative of $f(x, y) = x^2 - y^2$ in the direction of $\mathbf{u} = \langle 1/\sqrt{10}, -3/\sqrt{10} \rangle$ at the point $(1, -2)$.

Solution

To use Theorem 14.9, we note that **u** is a unit vector and find the gradient of f:

$$\nabla f(x, y) = \langle 2x, -2y \rangle.$$

Thus, in general, the directional derivative is

$$\begin{aligned} f_{\mathbf{u}} &= \mathbf{u} \cdot \nabla f \\ &= \left\langle \frac{1}{\sqrt{10}}, \frac{-3}{\sqrt{10}} \right\rangle \cdot \langle 2x, -2y \rangle \\ &= \frac{2x + 6y}{\sqrt{10}}. \end{aligned}$$

Hence,

$$f_{\mathbf{u}}(1, -2) = \frac{2(1) + 6(-2)}{\sqrt{10}} = -\sqrt{10}.$$

■

■ **Example 14.29** Find the directional derivative of $f(x, y, z) = x^2 + y^2 - z$ in the direction of $\mathbf{w} = \langle 1, 1, -1 \rangle$ at the point $(3, -1, 0)$.

Solution First, we find the gradient of f:

$$\begin{aligned} \nabla f &= \langle 2x, 2y, -1 \rangle, \\ \nabla f(3, -1, 0) &= \langle 6, -2, -1 \rangle. \end{aligned}$$

Next, we find a unit vector in the direction of $\mathbf{w}$:

$$\mathbf{u} = \frac{\mathbf{w}}{\|\mathbf{w}\|} = \left\langle \frac{1}{\sqrt{3}}, \frac{1}{\sqrt{3}}, \frac{-1}{\sqrt{3}} \right\rangle.$$

Finally, we use Theorem 14.9:

$$\begin{aligned} f_{\mathbf{u}}(3, -1, 0) &= \mathbf{u} \cdot \nabla f(3, -1, 0) \\ &= \left\langle \frac{1}{\sqrt{3}}, \frac{1}{\sqrt{3}}, \frac{-1}{\sqrt{3}} \right\rangle \cdot \langle 6, -2, -1 \rangle \\ &= \frac{5}{\sqrt{3}}. \end{aligned}$$

■

If θ is the angle between a unit vector $\mathbf{u}$ and the gradient ∇f, then

$$\begin{aligned} f_{\mathbf{u}}(\mathbf{v}) &= \mathbf{u} \cdot \nabla f(\mathbf{v}) \\ &= \|\mathbf{u}\| \|\nabla f(\mathbf{v})\| \cos \theta \\ &= \|\nabla f(\mathbf{v})\| \cos \theta. \end{aligned}$$

It follows that the directional derivative is a maximum when $\cos \theta = 1$, that is, when $\theta = 0$. Therefore, *the directional derivative is greatest in the direction of the gradient.* Put another way, *the gradient points in the direction of steepest ascent.* Also, the magnitude of the gradient is the maximum value of the directional derivative. On the other hand, the minimum value of the directional derivative is the negative of the magnitude of the gradient and occurs in the direction opposite to the gradient. Symbolically,

$$\max\{f_{\mathbf{u}}(\mathbf{v})\} = \|\nabla f(\mathbf{v})\|, \qquad \mathbf{u} \text{ in the direction of } \nabla f(\mathbf{v}),$$
$$\min\{f_{\mathbf{u}}(\mathbf{v})\} = -\|\nabla f(\mathbf{v})\|, \qquad \mathbf{u} \text{ in the direction of } -\nabla f(\mathbf{v}).$$

Example 14.30 Find the maximum and minimum directional derivatives of $f(x, y) = x^2 - y^2$ at the point $(1, -2)$.

Solution We first find the magnitude of the gradient of f at $(1, -2)$:

$$\nabla f(x, y) = \langle 2x, -2y \rangle,$$
$$\nabla f(1, -2) = \langle 2, 4 \rangle,$$
$$\|\nabla f(1, -2)\| = \|\langle 2, 4 \rangle\| = 2\sqrt{5}.$$

Therefore, the maximum directional derivative is $2\sqrt{5}$, and the minimum directional derivative is $-2\sqrt{5}$. ■

For a function of two variables, there is a simple geometric interpretation of the directional derivative. Let f be a function of two variables, and let (a, b) be a point in the domain of f. Let $c = f(a, b)$; then (a, b, c) is a point on the graph of f. Let $\mathbf{u}$ be a given unit vector. The vertical plane through (a, b, c) that is parallel to $\mathbf{u}$ cuts the graph of f in a curve C (see Figure 14.23). The directional derivative $f_{\mathbf{u}}(a, b)$ is the slope of the line in that plane that is tangent to C at (a, b, c). (You should compare Figure 14.23 with Figures 14.21 and 14.22 on pages 866 and 867.)

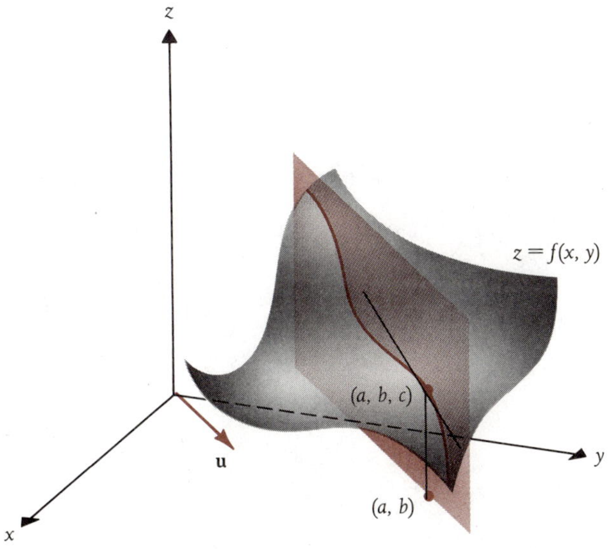

FIGURE 14.23 The directional derivative as a slope

In this geometric interpretation, the gradient of a function of two variables is a vector pointing in the direction of steepest ascent of the graph of the function. The negative of the gradient points in the direction of steepest descent.

For a function of three variables, we can get a similar interpretation for the directional derivative, but in terms of rate of change. The directional derivative $f_{\mathbf{u}}(a, b, c)$ is the rate of change of f at (a, b, c) in the direction of $\mathbf{u}$. Thus, the gradient points in the direction of the greatest rate of increase, and its negative points in the direction of the greatest rate of decrease of f.

Path of a Heat-Seeking Particle

Our last interpretation of the gradient makes possible an interesting application. Suppose that a function f of three variables gives the temperature at each point in a solid. Then the gradient always points in the direction of greatest rate of increase of temperature. A **heat-seeking particle** is a particle that always moves in the direction of most rapid increase of heat—that is, in the direction of the gradient. The path traveled by a heat-seeking particle has tangent vector parallel to the temperature gradient at each point. This fact allows us to determine the path of a heat-seeking particle.

Example 14.31 The temperature of a certain solid is $f(x, y, z) = x^2 + 2y - z$ at point (x, y, z). Find the path of a heat-seeking particle that starts at the point $(1, 0, 0)$.

Solution Let the path of the particle be the graph of the vector function $\mathbf{r}(t) = \langle x(t), y(t), z(t) \rangle$. Then the tangent vector $\mathbf{r}'(t) = \langle x'(t), y'(t), z'(t) \rangle$ at each point is in the direction of the gradient $\nabla f(x, y, z) = \langle 2x, 2, -1 \rangle$ (see Figure 14.24). Hence, for some positive constant k,

$$\mathbf{r}'(t) = k\nabla f(x, y, z),$$

which implies that

$$x'(t) = 2kx(t), \qquad y'(t) = 2k, \qquad z'(t) = -k.$$

These differential equations are separable (see Section 4.6). Solving them individually, we get

$$x(t) = Ae^{2kt}, \qquad y(t) = 2kt + B, \qquad z(t) = -kt + C.$$

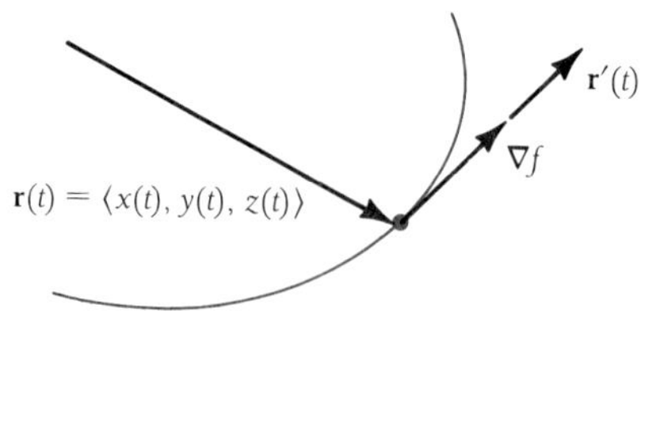

FIGURE 14.24 Along the path of a heat-seeking particle, the tangent vector is parallel to the gradient at each point

Since the particle starts at point (1, 0, 0), we have

$$x(0) = 1, \qquad y(0) = 0, \qquad z(0) = 0.$$

Therefore,

$$A = 1, \qquad B = 0, \qquad C = 0.$$

The path of the particle is thus

$$\mathbf{r}(t) = \langle e^{2kt}, 2kt, -kt \rangle.$$

■

A two-dimensional setting for the same type of problem is handled in the same way, except that the path is a curve in the plane.

■ **Example 14.32**

The temperature of a plate at point (x, y) is $T(x, y) = 20 - x^2 + 2y^2$. An ant is dropped onto the plate at the point (1, 3). If the ant flees from the heat by moving in the direction of greatest rate of temperature decrease, what is its path?

Solution

Let $\mathbf{r}(t) = \langle x(t), y(t) \rangle$ be the path. Then the tangent vector $\mathbf{r}'(t) = \langle x'(t), y'(t) \rangle$ of the path is always in the direction opposite to the gradient $\nabla T(x, y) = \langle -2x, 4y \rangle$. That is, for some positive constant k,

$$\mathbf{r}'(t) = -k\nabla T(x, y),$$

or

$$x'(t) = 2kx(t), \qquad y'(t) = -4ky(t).$$

These differential equations have solutions

$$x(t) = Ae^{2kt}, \qquad y(t) = Be^{-4kt}.$$

Using the initial condition $\mathbf{r}(0) = \langle 1, 3 \rangle$, we get $A = 1$ and $B = 3$. Therefore, the path is

$$\mathbf{r}(t) = \langle e^{2kt}, 3e^{-4kt} \rangle.$$

If the parameter t is eliminated from the equations

$$x = e^{2kt}, \qquad y = 3e^{-4kt},$$

we get the equation

$$y = \frac{3}{x^2}.$$

This is the ant's path, traversed downward and to the right from the point (1, 3).

■

Exercises 14.5

Find the gradient, and evaluate it at the given point.

1. $f(x, y) = 3x - 4y$; $(3, -1)$
2. $f(x, y) = 6x + 2y$; $(1, -1)$
3. $f(x, y) = x^2 + y^2$; $(4, 0)$
4. $f(x, y) = x^2 + 3y^2$; $(0, 2)$

5. $f(r, s) = e^{3rs}$; $(2, -1)$
6. $f(p, q) = \ln(p^2 + q^2)$; $(2, 0)$
7. $f(x, y) = \tan(x/y)$; $(\pi, 4)$
8. $f(x, y) = \sin x - x \cos y$; $(0, \pi/2)$
9. $f(x, y, z) = x^2 - y + xz^2$; $(1, 1, -1)$
10. $f(x, y, z) = z - xy - x^2 - y^2$; $(1, 1, 0)$
11. $f(x, y, z) = e^x \cos y \ln z$; $(0, \pi, e)$
12. $f(x, y, z) = e^{-x^2-y^2} \ln z$; $(0, 0, 1)$

Find the directional derivative of f in the direction of **v** *at the given point.*

13. $f(x, y) = 2x + y$; $\mathbf{v} = \langle -\frac{1}{2}, \sqrt{3}/2 \rangle$; $(1, 1)$
14. $f(x, y) = x^2 + y^2$; $\mathbf{v} = \mathbf{i}$; $(2, 3)$
15. $f(x, y) = x^2 + y^2$; $\mathbf{v} = \langle 1, 1 \rangle$; $(2, 3)$
16. $f(x, y) = x^2 - y^2$; $\mathbf{v} = \langle 1, -1 \rangle$; $(1, 2)$
17. $f(x, y, z) = 3x + 5y - 4z$; $\mathbf{v} = \langle 1, 0, 1 \rangle$; $(1, 1, -1)$
18. $f(x, y, z) = 3z - xz - xy + x^2 - y^2$; $\mathbf{v} = \langle 0, 1, 2 \rangle$; $(0, 0, 5)$
19. $f(x, y, z) = 2e^{xyz}$; $\mathbf{v} = \nabla f$; $(0, 0, 1)$
20. $f(x, y, z) = z \sin xy$; $\mathbf{v} = -\nabla f$; $(\pi, \frac{1}{2}, 3)$
21. The temperature at point (x, y) in a plate is $f(x, y) = 16 - x^2 - 2y^2$. Find the following:
    (a) The rate of temperature change at point $(1, 4)$ in the direction of (i) $\langle 1, 1 \rangle$; (ii) greatest heating
    (b) The path of a heat-seeking particle that starts at the point $(1, 4)$
22. Repeat Problem 21 if the temperature is $f(x, y) = 100 - e^{-x} + e^{-y}$.
23. The temperature at point (x, y, z) in a solid is $f(x, y, z) = 100 - e^x - e^{2y} - e^{3z}$. Find the following:
    (a) The rate of temperature change at point $(0, 1, 1)$ in the direction of (i) greatest cooling; (ii) $\mathbf{j} - \mathbf{i}$
    (b) The path of a heat-fleeing particle that starts at the point $(0, 1, 1)$
24. Repeat Problem 23 for the temperature function $f(x, y, z) = x^2 - 3x + y - z^2$.

Find a unit vector **u** *such that* $f_{\mathbf{u}}$ *is 0 at the given point.*

25. $f(x, y) = 9 - x^2 - y^2$; $(\frac{3}{2}, -2)$
26. $f(x, y) = x^2 - y^2$; $(2, 3)$
27. $f(x, y, z) = x^2 + y^2 - z^2$; $(2, 2, -1)$
28. $f(x, y, z) = 6xy - x^3 - y^2 + z$; $(1, 1, 1)$

Establish the following formulas:

29. $\nabla(f + g) = \nabla f + \nabla g$
30. $\nabla(af) = a\nabla f$
31. $\nabla(fg) = f\nabla g + g\nabla f$
32. $\nabla(f^n) = nf^{n-1}\nabla f$

c33. A mountain climber is standing on the side of a mountain, studying a contour map. He notes that the y-axis of his coordinate system points north and the x-axis points east. If Riemann were present, he could show that the altitude of any point on the climber's map is

$$A(x, y) = 2 + \frac{1}{3}(2x^2 + x^3 - x^4 - y^2) \text{ kilometers above sea level}$$

and that the climber is standing at the point (1.44, −0.80) on the map (measured in kilometers).

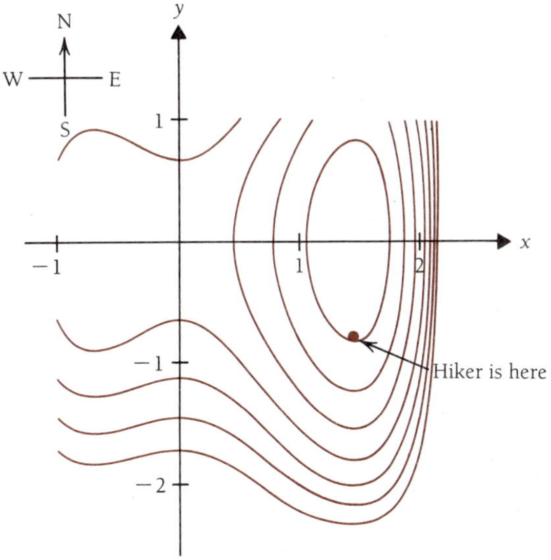

(a) In which (compass) direction should the climber set off if he wants the steepest uphill path? How steep is it?
(b) In which direction is the steepest downhill path?
(c) How steep is the path that heads northwest?
(d) Can you tell in which direction(s) the climber should set off if he wants to walk around the mountain, maintaining his current altitude?

SECTION 14.6

Normals and Tangents

In this section, we deal with normal vectors to curves and surfaces, tangent lines to curves in the plane, and tangent planes to surfaces in space. As you may suspect, the gradient will prove to be a useful tool.

Normal Vectors

Let C be a smooth curve, and let P be a point on C. Let $\mathbf{t}$ be the tangent vector to C at P. A vector $\mathbf{v}$ is **normal** to C at P if and only if $\mathbf{v}$ is orthogonal to $\mathbf{t}$ (see Figure 14.25).

Now let S be a surface in space, and let P be a point on S. A vector $\mathbf{v}$ is **normal** to S at P if and only if, for every smooth curve C lying on S and passing through P, $\mathbf{v}$ is normal to C at P (see Figure 14.26).

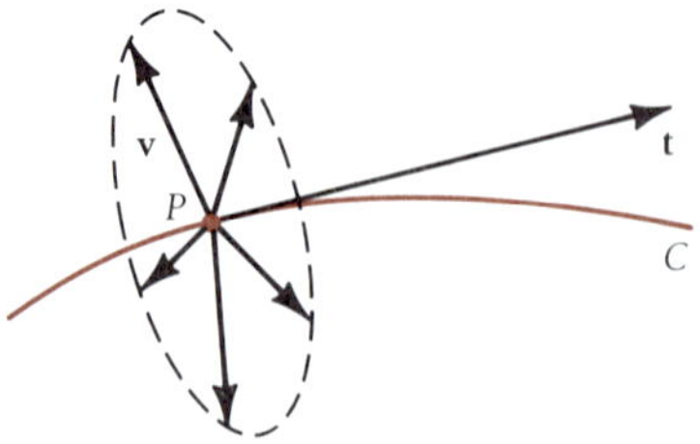

FIGURE 14.25 A smooth curve in space has many normal vectors at each point

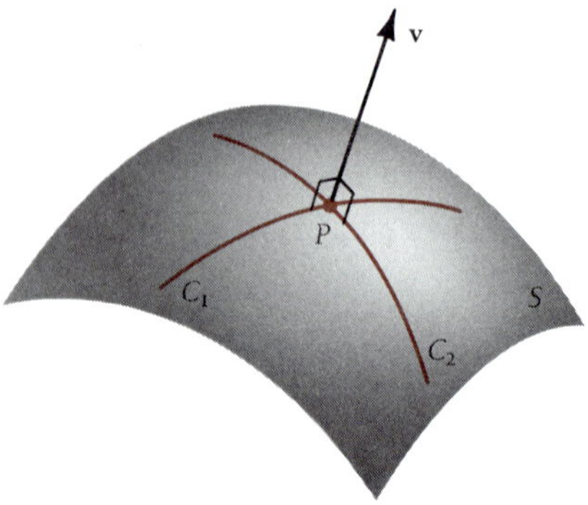

FIGURE 14.26 A normal vector to a surface S at a point P is normal to every curve in S through P

The Gradient as Normal Vector

The gradient of a function can be interpreted as a normal vector. For a function of two variables, the gradient is normal to the level curves, as we will now prove.

THEOREM 14.10 Let function f be a continuously differentiable function of two variables, and let $C: f(x, y) = c$ be any level curve of f. If P is any point on C, then $\nabla f(P)$ is normal to C at P (see Figure 14.27).

Proof Let $\mathbf{r}(t) = \langle x(t), y(t) \rangle$ be a parametrization for C. Since f is continuously differentiable, it follows that C is smooth, so x and y are differentiable. Let $P = (x(t_0), y(t_0))$ be any point on C. The tangent vector to C at P is $\mathbf{r}'(t_0) = \langle x'(t_0), y'(t_0) \rangle$. Now C is a c-level curve of f, so

$$f(\mathbf{r}(t)) = c.$$

Hence,

$$\frac{d}{dt} f(\mathbf{r}(t)) = 0.$$

But by the chain rule,

$$\begin{aligned}\frac{d}{dt} f(r(t)) &= \frac{d}{dt} f(x(t), y(t)) \\ &= f_x \frac{dx}{dt} + f_y \frac{dy}{dt} \\ &= \nabla f \cdot \mathbf{r}'(t).\end{aligned}$$

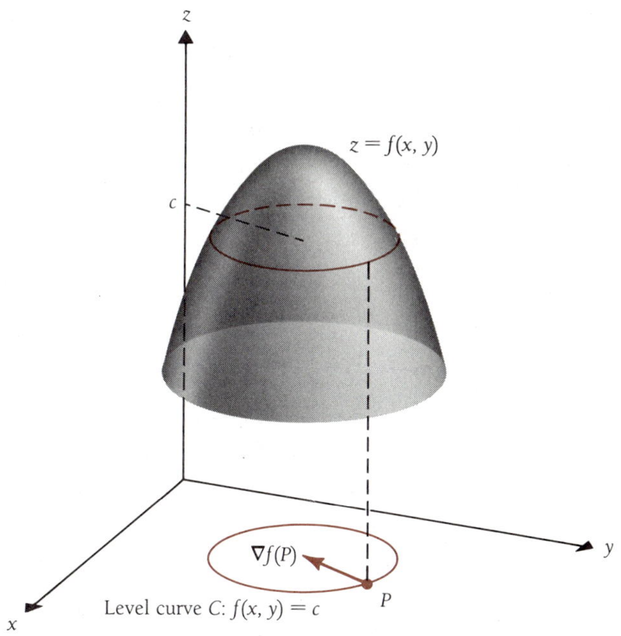

FIGURE 14.27 The gradient of f at any point P on a level curve is normal to the level curve

Therefore,

$$\nabla f(P) \cdot \mathbf{r}'(t_0) = 0,$$

and $\nabla f(P)$ is normal to C at P.

As a consequence of this result, we note that the direction of steepest ascent or descent on a surface at any point is normal to the level curve through that point. For example, if a path around a hillside has a constant elevation, then to go "straight up" the hill, one would have to leave the path at a right angle. Similarly, a skier coming down the hill along a path of steepest descent would cross the level path at right angles.

The corresponding result for a function of three variables is that the gradient at any point is normal to the level surface through that point.

THEOREM 14.11 Let f be a continuously differentiable function of three variables, and let $S: f(x, y, z) = c$ be any level surface of f. If P is any point on S, then $\nabla f(P)$ is normal to S at P (see Figure 14.28).

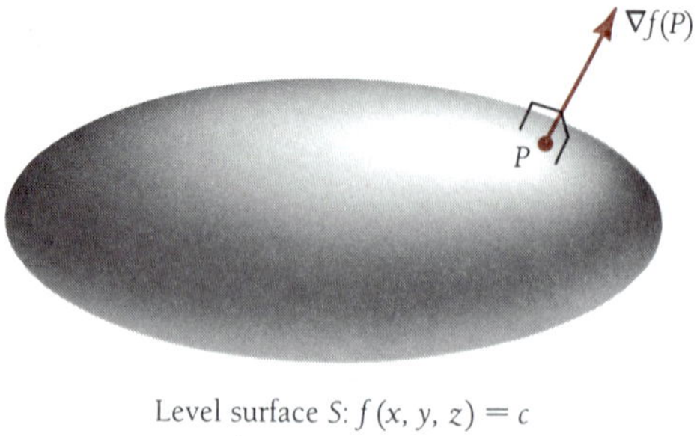

FIGURE 14.28 The gradient of f at any point P on a level surface is normal to the level surface

Since the proof of this theorem is almost identical to that of Theorem 14.10, we will leave it as an exercise (see Problem 28).

A level surface of a temperature distribution is called an **isotherm.** A physical interpretation of Theorem 14.11 is that the path of a heat-seeking particle is normal to any isotherm.

Normal Lines

Though we have discussed normal lines to a curve in the plane before (Section 3.1), here is another formulation. Let $C: f(x, y) = c$ be a smooth curve in the plane, and let $P = (a, b)$ be a point on C. Then

$$\nabla f(P) = \langle f_x(a, b), f_y(a, b) \rangle$$

is normal to C, as shown in Figure 14.29(a). The line normal to C at P has slope $m = f_y(a, b)/f_x(a, b)$, and the normal line has equation

$$f_y(a, b)(x - a) - f_x(a, b)(y - b) = 0. \tag{1}$$

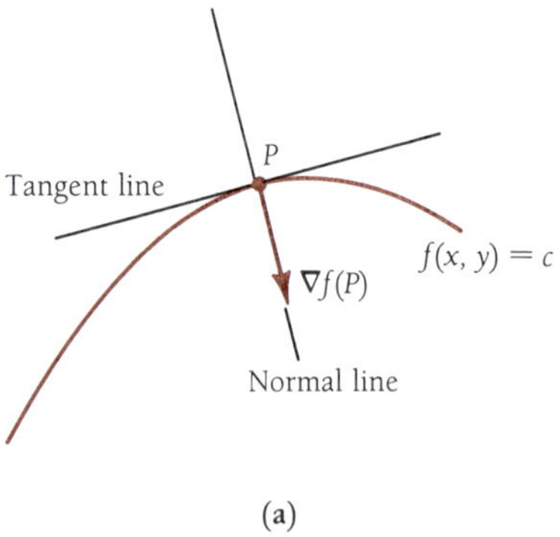

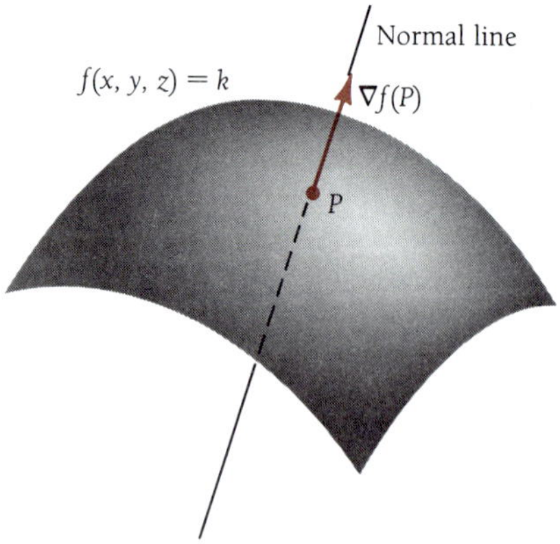

FIGURE 14.29 The gradient gives the normal line

For a surface S in space, a line is said to be **normal** to S at point P if and only if the line passes through P and has direction vector normal to S at P. If surface S has equation $f(x, y, z) = k$ and $P = (a, b, c)$ is a point on S, then $\nabla f(P)$ is a direction vector for the line normal to S at P, as shown in Figure 14.29(b). The normal line therefore has equations

$$\begin{cases} x = a + f_x(a, b, c)t \\ y = b + f_y(a, b, c)t \\ z = c + f_z(a, b, c)t. \end{cases} \tag{2}$$

Example 14.33 Find equations for the line normal to the cone $z^2 = x^2 + y^2$ at the point $(4, -3, 5)$.

Solution We must first describe the surface as a level surface of some function. If we let

$$f(x, y, z) = z^2 - x^2 - y^2,$$

then our surface is the zero-level surface of f. The gradient of f is

$$\nabla f = \langle -2x, -2y, 2z \rangle,$$

so that the vector

$$\nabla f(4, -3, 5) = \langle -8, 6, 10 \rangle$$

is a direction vector for the normal line. The normal line therefore has equations

$$\begin{cases} x = 4 - 8t \\ y = -3 + 6t \\ z = 5 + 10t. \end{cases}$$

■

A surface that is the graph of a function of two variables is a special but common case, which we can treat in the same way. Suppose S: $z = f(x, y)$ is such a surface. Then S is also the zero-level surface of the function

$$F(x, y, z) = z - f(x, y).$$

The gradient of F is

$$\nabla F = \langle -f_x, -f_y, 1 \rangle.$$

Therefore, if $(a, b, f(a, b))$ is a point on the surface, the normal line at that point is

$$\begin{cases} x = a - f_x(a, b)t \\ y = b - f_y(a, b)t \\ z = f(a, b) + t. \end{cases} \tag{3}$$

Example 14.34 Find equations for the line normal to the surface $z = x^2 + 3y^2$ at the point $(1, 2, 13)$.

Solution With $f(x, y) = x^2 + 3y^2$, we have

$$f_x = 2x, \qquad f_y = 6y.$$

Then, using $a = 1$ and $b = 2$, equations (3) give us the normal line

$$\begin{cases} x = 1 - 2t \\ y = 2 - 12t \\ z = 13 + t. \end{cases}$$

■

Tangent Lines

We have already studied the problem of finding the tangent line to a plane curve, but Theorem 14.9 gives us yet another way to do it. Let $C\colon f(x, y) = c$ be a smooth curve in the plane, and let $P = (a, b)$ be a point on C. Since $\nabla f(P) = \langle f_x(a, b), f_y(a, b)\rangle$ is normal to C at P, a tangent vector to C at P must be $\langle f_y(a, b), -f_x(a, b)\rangle$ (because $\langle u, v\rangle$ and $\langle -v, u\rangle$ are always orthogonal). Therefore, the slope of C at P is $-f_x(a, b)/f_y(a, b)$, and the equation of the tangent line to C at P is

$$f_x(a, b)(x - a) + f_y(a, b)(y - b) = 0. \tag{4}$$

Tangent Planes

A plane Π is said to be **tangent** to a surface S at point P if and only if Π passes through P and the normal vector to S at P is also normal to Π. If surface S has equation $f(x, y, z) = k$ and $P = (a, b, c)$ is a point on S, then the gradient

$$\nabla f(P) = \langle f_x(a, b, c), f_y(a, b, c), f_z(a, b, c)\rangle$$

is a normal vector to the tangent plane to S at P (see Figure 14.30). The equation of the tangent plane is therefore

$$f_x(a, b, c)(x - a) + f_y(a, b, c)(y - b) + f_z(a, b, c)(z - c) = 0. \tag{5}$$

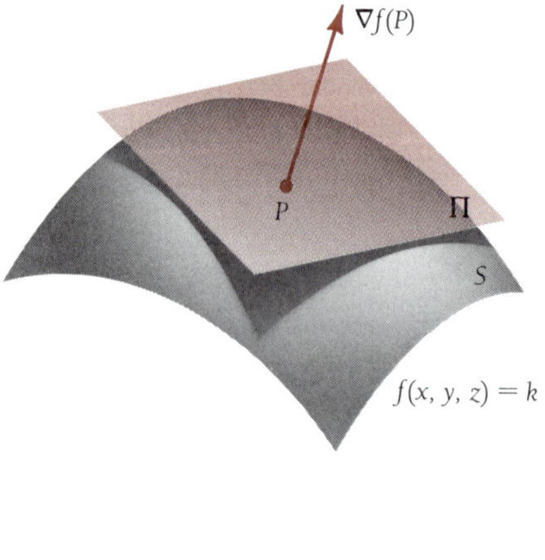

FIGURE 14.30 The gradient of f at P is normal to the tangent plane to S at P

Example 14.35 Find an equation of the plane tangent to the surface $z = x^2 + y^2$ at the point $(1, 2, 5)$.

Solution We first describe the surface as a level surface of a function of three variables. If we let $f(x, y, z) = z - x^2 - y^2$, then the surface given is the zero-level surface of f. Therefore, the gradient of f is normal to the surface and can be used in the formula (5). We have

$$\nabla f = \langle -2x, -2y, 1 \rangle,$$
$$\nabla f(1, 2, 5) = \langle -2, -4, 1 \rangle.$$

Hence, by formula (5), the tangent plane is

$$-2(x-1) - 4(y-2) + 1(z-5) = 0,$$
$$-2x - 4y + z + 5 = 0.$$

■

Example 14.36 Find a point on the surface

$$x^2 + y^2 - 2x + 4y + z + 1 = 0$$

at which the tangent plane is horizontal.

Solution Let $f(x, y, z) = x^2 + y^2 - 2x + 4y + z + 1$; then the given surface is the zero-level surface of f. The tangent plane to the surface will be horizontal (parallel to the xy-plane) at any point where the gradient of f is parallel to the standard unit vector $\mathbf{k}$. The gradient is

$$\nabla f = \langle 2x - 2, 2y + 4, 1 \rangle,$$

and will be parallel to $\mathbf{k} = \langle 0, 0, 1 \rangle$ if and only if

$$2x - 2 = 0, \qquad 2y + 4 = 0.$$

This happens for $x = 1, y = -2, z = 4$, so there is a horizontal tangent plane to the surface at the point $(1, -2, 4)$, as shown in Figure 14.31 on page 894. ■

Example 14.37 Find the form of equation (5) that holds for a surface whose equation is of the form $z = f(x, y)$.

Solution The surface $z = f(x, y)$ is the zero-level surface of $g(x, y, z) = z - f(x, y)$. The gradient of g is

$$\nabla g = \langle -f_x, -f_y, 1 \rangle.$$

Using this in equation (5) gives us the tangent plane

$$-f_x(a, b)(x-a) - f_y(a, b)(y-b) + (z-c) = 0, \tag{6}$$

or

$$z = f(a, b) + f_x(a, b)(x-a) + f_y(a, b)(y-b). \tag{7}$$

■

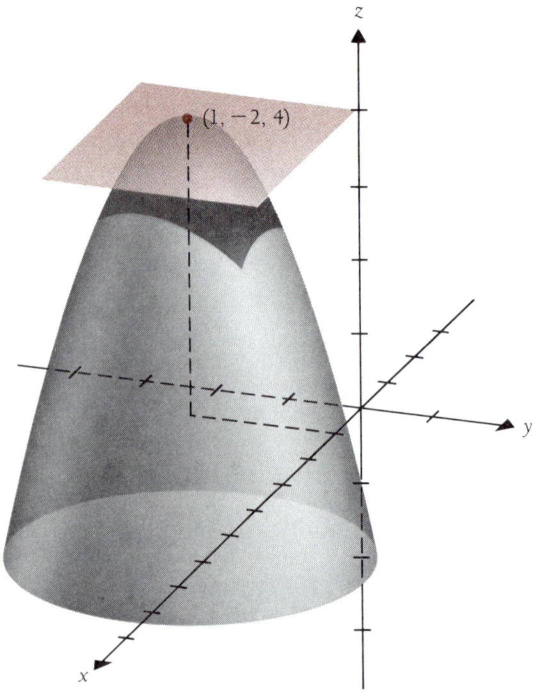

FIGURE 14.31 A horizontal tangent plane to the surface $x^2 + y^2 - 2x + 4y + z + 1 = 0$

Exercises 14.6

Find equations for the normal line and tangent plane to the given surface at the given point.

1. $z = x^2 + y^2$; $(0, 0, 0)$
2. $z = x^2 - y^2$; $(0, 0, 0)$
3. $z = 3x^2 + 5y^2$; $(1, 2, 23)$
4. $z = 3x^2 + 5y^2$; $(2, -1, 17)$
5. $x^2 + y^2 + 2z^2 = 2$; $(0, 0, 1)$
6. $x^2 + y^2 + 2z^2 = 2$; $(1, 1, 0)$
7. $2x^2 - y^2 + 3z^2 = 6$; $(\sqrt{3}, 0, 0)$
8. $2x^2 - y^2 + 3z^2 = 6$; $(\sqrt{2}, 1, 1)$

Find the point(s) at which the tangent plane to the given surface is horizontal.

9. $z = x^2 + 2y^2$
10. $z = x^2 - 2y^2$
11. $x^2 + y^2 - z^2 - 1 = 0$
12. $x^2 + xyz = 5$
13. $x^2 - 3xy + y^2 + 4y + z^2 + 4 = 0$
14. $4x^2 + 9y^2 - z^2 + 36 = 0$
15. Show that the tangent plane to the ellipsoid

$$\frac{x^2}{a^2} + \frac{y^2}{b^2} + \frac{z^2}{c^2} = 1$$

at point (x_1, y_1, z_1) has equation

$$\frac{xx_1}{a^2} + \frac{yy_1}{b^2} + \frac{zz_1}{c^2} = 1.$$

16. If a curve C meets a surface S at point P, the **angle between C and S** at P is defined to be the angle between the tangent line to C at P and the tangent plane to S at P. If $\mathbf{T}$ is the unit tangent vector to C at P and $\mathbf{n}$ is the unit normal vector to S at P, show that the angle θ between C and S at P satisfies $\sin\theta = \mathbf{n} \cdot \mathbf{T}$.

Find the angle at which the given curve meets the given surface. (See Problem 16.)

17. C: $\mathbf{r}(t) = \langle t, 2t, t+3 \rangle$; S: $2x + 3y - 5z = 6$
18. C: $\mathbf{r}(t) = \langle t, t^2, 2t \rangle$; S: $z = x^2 - y^2$; at $(0, 0, 0)$
19. C: $\mathbf{r}(t) = \langle \cos t, \sin t, t \rangle$; S: $x^2 + y^2 + z^2 = 4$
20. C: $\mathbf{r}(t) = \langle t, t, t^3 \rangle$; S: $z = x - x^2 - y^2$
21. Find all the points on the ellipsoid

$$\frac{x^2}{a^2} + \frac{y^2}{b^2} + \frac{z^2}{c^2} = 1$$

at which the normal line passes through the origin.

22. If two surfaces meet at point P, the **angle between the surfaces** at P is defined to be the angle between their tangent planes at P. If surfaces S_1 and S_2 have unit normal vectors $\mathbf{n}_1$ and $\mathbf{n}_2$ at P, show that the angle θ between S_1 and S_2 satisfies $\cos\theta = \mathbf{n}_1 \cdot \mathbf{n}_2$.

Find the angle between the given surfaces at P. (See Problem 22.)

23. $z = x^2 + y^2$; $x^2 + y^2 + z^2 = 2$; $P(0, 1, 1)$
24. $x^2 + 4y^2 + 4z^2 = 4$; $x + y = 0$; $P(0, 0, 1)$
25. $z = x^2 - y^2$; $z = 4$; $P(2, 0, 4)$
26. $z = x^2 - y^2$; $z = 4$; $P(3, \sqrt{5}, 4)$
27. If f is a function of two variables and the directional derivative $f_{\mathbf{u}}(a, b) = 0$, show that $\mathbf{u}$ is tangent to the level curve of f through (a, b).
28. Prove Theorem 14.11
29. Let f be a continuously differentiable function of three variables, and let C be any smooth curve that lies on a level surface of f. Show that ∇f is normal to C at each point of C.

SECTION 14.7

Maxima and Minima

We have already noted (Theorem 14.3) that a function of several variables defined on a compact set has absolute extrema there. We also can extend the notion of relative extrema to functions of several variables. Our aim in this section is to be able to locate both relative and absolute extrema of functions of several variables. We will start with functions of two variables, for which there are graphical interpretations. Then we will extend these ideas to more than two variables.

Relative Extrema

Let f be a function of two variables. We say f has a **relative maximum** at point (a, b) in case there is an open disc D centered at (a, b) such that $f(x, y) \leq f(a, b)$ for every point $(x, y) \in D$. Similarly, f has a **relative minimum** at (a, b) in case there is an open disc D centered at (a, b) such that $f(x, y) \geq f(a, b)$ for every point $(x, y) \in D$. We say f has an **absolute maximum** at (a, b) in case $f(x, y) \leq f(a, b)$ for all points (x, y) in the domain of f. And, finally, we say f has an **absolute minimum** at (a, b) in case $f(x, y) \geq f(a, b)$ for all points (x, y) in the domain of f.

For a function of one variable, an extremum can occur only where the graph has either a horizontal tangent line or no tangent line at all. For a function of two variables, an extremum can occur only where the graph has either a horizontal tangent plane or no tangent plane at all. To see this, in the case of a differentiable function f, suppose f has a relative maximum at (a, b), as shown in Figure 14.32. Then the curve $z = f(x, b)$, $y = b$, must have a relative maximum at $x = a$. Therefore, its slope, $f_x(x, y)$, must be 0 at (a, b); that is, $f_x(a, b) = 0$. Similarly, the curve $z = f(a, y)$, $x = a$, must have a horizontal tangent at (a, b), so $f_y(a, b) = 0$. Therefore, the gradient $\nabla f(a, b) = \mathbf{0}$, and the tangent plane to $z = f(x, y)$ at (a, b) is horizontal by equation (6) in Section 14.6. Seen another way, the tangent plane contains two horizontal tangent lines, so it must be horizontal. A similar argument applies to a minimum.

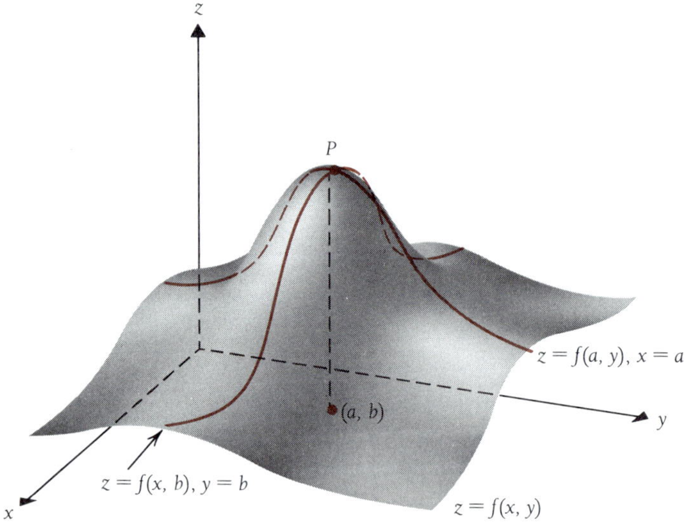

FIGURE 14.32

Critical Points

If f is a function of two variables, a point at which the gradient of f is the zero vector or does not exist at all is called a **critical point** or a **stationary point** of f. We can now extend Theorem 4.1 to functions of two variables, but first, we must define the two-dimensional analog of an endpoint of the domain of f: A point P is called a **boundary point** of the set S if and only if every open ball containing P contains a point in S and a point not in S. The set of boundary points of the set S is called the **boundary** of S.

THEOREM 14.12 If f is a differentiable function of two variables and f has a relative extremum at point (a, b), then (a, b) is either a boundary point of the domain of f or a critical point of f.

As is the case with functions of one variable, a function of two variables can have a critical point without having an extremum there. For example, the function $f(x, y) = x^2 - y^2$ has as its graph the hyperbolic paraboloid shown in Figure 14.33. At the origin, the function has a critical point and a horizontal tangent plane, but no extremum. Such a point is called a **saddle point**; in some ways, this is the two-variable version of an inflection point.

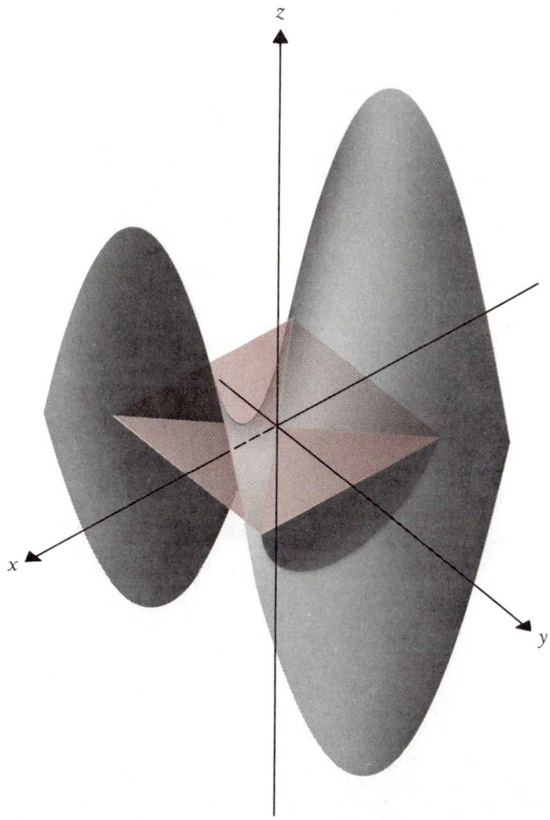

FIGURE 14.33 The graph of $z = x^2 - y^2$, with a horizontal tangent plane at the origin

The process of finding relative (or absolute) extrema of a function of two variables begins with finding the critical points. Then the function is examined at each critical point to determine whether it has an extremum there. We may also have to examine the function on boundary points of its domain, if any. Let's look at some examples.

Example 14.38 Find the extrema of the function

$$f(x, y) = 9x^2 + 4y^2 - 18x - 24y + 41.$$

Solution We begin by searching for critical points. We find the gradient, and then find the points at which it is the zero vector. We have

$$\nabla f(x, y) = \langle 18x - 18, 8y - 24 \rangle,$$

which is the zero vector when

$$\begin{aligned} 18x - 18 &= 0 \qquad \text{and} \qquad 8y - 24 = 0, \\ x &= 1 \qquad \text{and} \qquad y = 3. \end{aligned}$$

Thus, $(1, 3)$ is the only critical point of f. In this case, to determine whether f has an extremum at $(1, 3)$, we can rewrite f by completing the squares in the form

$$f(x, y) = 9(x - 1)^2 + 4(y - 3)^2 - 4.$$

Thus, $f(1, 3) = -4$, and if $(x, y) \neq (1, 3)$, then $f(x, y) > -4$. Therefore, f has an absolute minimum of -4 at $(1, 3)$. ■

Example 14.39 Find the extrema of $f(x, y) = xy^2 - y + 1$.

Solution The gradient of f is

$$\nabla f(x, y) = \langle y^2, 2xy - 1 \rangle,$$

which is the zero vector when

$$y^2 = 0 \qquad \text{and} \qquad 2xy = 1.$$

However, no point satisfies both these equations, so f has no critical points and therefore has no extrema. ■

Example 14.40 Find the extrema of $f(x, y) = x^3 - xy + y + 1$.

Solution The gradient of f is

$$\nabla f(x, y) = \langle 3x^2 - y, -x + 1 \rangle,$$

which is the zero vector when

$$3x^2 - y = 0 \qquad \text{and} \qquad -x + 1 = 0.$$

The only solution of this system is $x = 1, y = 3$, so $(1, 3)$ is the only critical point of f. The value of f at $(1, 3)$ is $f(1, 3) = 2$. We may determine whether 2 is an extremum of f by examining points near $(1, 3)$. If h and k are numbers close to 0, then point $(1 + h, 3 + k)$ is close to $(1, 3)$, and

$$\begin{aligned} f(1 + h, 3 + k) &= (1 + h)^3 - (1 + h)(3 + k) + (3 + k) + 1 \\ &= h^3 + 3h^2 - hk + 2. \end{aligned}$$

We now try to determine whether the expression

$$h^3 + 3h^2 - hk = h^2(h + 3) - hk \tag{1}$$

is positive or negative for various choices of h and k. First, note that if h is positive and k is negative, then expression (1) is positive; thus, 2 is not a relative maximum, since points arbitrarily near $(1, 3)$ give greater values for f. Thus, the only question remaining is whether we can choose h and k arbitrarily small but so that expression (1) is negative. After a bit of manipulation, we see that if h is negative and $k < h(h + 3)$, then expression (1) is negative. Thus, 2 is not a minimum, so f must have a saddle point at $(1, 3)$. Therefore, f has no extrema. ■

■ **Example 14.41** Find the extrema of $f(x, y) = 4 - x^2 - 2y^2$.

Solution By inspection, we see that $f(0, 0) = 4$ and $f(x, y) < 4$ for any $(x, y) \neq (0, 0)$. Thus, f has an absolute maximum of 4 at $(0, 0)$. To verify it using the gradient, and also to find other extrema, if any, we proceed as in the previous examples. The gradient is

$$\nabla f(x, y) = \langle -2x, -4y \rangle,$$

which is the zero vector only at $(0, 0)$. Thus, there are no other extrema besides the maximum of 4 at $(0, 0)$. ■

■ **Example 14.42** Find the extrema of $f(x, y) = (y - x^2)(2y - x^2)$.

Solution Expanding, we can write

$$f(x, y) = 2y^2 - 3x^2y + x^4,$$

so that

$$\nabla f(x, y) = \langle -6xy + 4x^3, 4y - 3x^2 \rangle.$$

The gradient is the zero vector for

$$-6xy + 4x^3 = 0 \qquad \text{and} \qquad 4y - 3x^2 = 0.$$

From the second equation, $y = \frac{3}{4}x^2$. Putting this expression into the first equation, we get $x^3 = 0$, so $x = y = 0$, and $(0, 0)$ is the only critical point. To determine the behavior of f near $(0, 0)$, we plot the zero-level curves of f shown in Figure 14.34. In the shaded region between the parabolas, the values of f are negative; elsewhere, f has positive values. Thus, f has a saddle point at the origin, and no extrema.

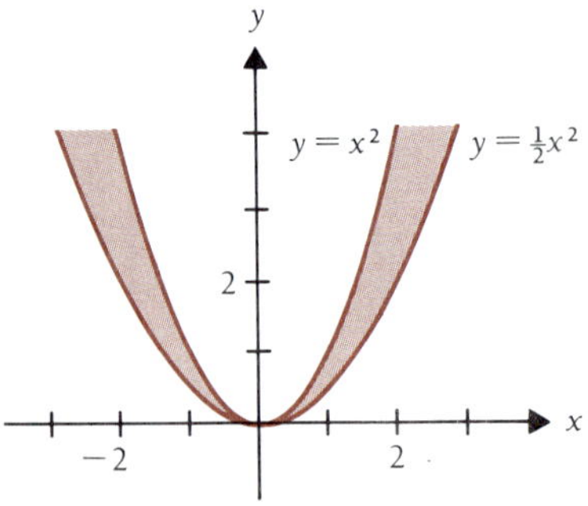

FIGURE 14.34 The zero-level curves of $z = (y - x^2)(2y - x^2)$ ■

■ **Example 14.43** Find the absolute maximum and minimum values of the function

$$f(x, y) = 12 - 4y - x^2 - 2y^2$$

over the domain $D = \{(x, y) | 0 \leq x \leq 4 \text{ and } 0 \leq y \leq 16 - x^2\}$.

Solution Here, we have a differentiable (hence, continuous) function defined over a compact set, so we know by the extreme value theorem that absolute extrema exist. We begin by locating the critical points of f. The gradient is

$$\nabla f = \langle -2x, -4 - 4y \rangle,$$

which is the zero vector only at the point $(0, -1)$. Since this point is not in the domain D, our function has no critical points inside D. It remains to examine f over the boundary points of D.

The boundary of D consists of the segment

$$S_1 = \{(x, y) | x = 0 \text{ and } 0 \leq y \leq 16\},$$

the segment

$$S_2 = \{(x, y) | y = 0 \text{ and } 0 \leq x \leq 4\},$$

and the curve

$$C = \{(x, y) | 0 \leq x \leq 4 \text{ and } y = 16 - x^2\}.$$

Over S_1, f has values $f(0, y) = 12 - 4y - 2y^2$, $0 \leq y \leq 16$. This function of y has its only extreme value at $y = -1$, as you may check by the methods of Chapter 4. Since -1 is not between 0 and 16, only the endpoints of the y interval $[0, 16]$ need to be checked. For $y = 0$, $f(0, 0) = 12$; for $y = 16$, $f(0, 16) = -564$.

Over S_2, f has values $f(x, 0) = 12 - x^2$, $0 \leq x \leq 4$. The only extremum of this function occurs at $x = 0$, and $f(0, 0) = 12$. Checking the endpoints of the x interval $[0, 4]$, we find $f(0, 0) = 12$ (which we found before) and $f(4, 0) = -4$.

Finally, over the curve C, f has the values

$$f(x, 16 - x^2) = -564 + 67x^2 - 2x^4, \qquad 0 \leq x \leq 4.$$

The critical points of this function are at $x = 0$ and $x = \pm\sqrt{67}/2$. We found that $f(0, 16) = -564$ previously, and we can ignore the latter two values, because they are outside the interval [0, 4].

Summarizing, the only possible extrema are

$$f(0, 0) = 12,$$
$$f(0, 16) = -564,$$
$$f(4, 0) = -4.$$

Obviously, 12 is the absolute maximum value and -564 is the absolute minimum. ■

The Second Derivative Test

As you have seen in these examples, it may or may not be easy to tell whether a function of two variables has an extremum at a critical point. One result that is helpful is the following two-variable version of the second derivative test (Theorem 4.8). It is based on the intuitive idea that if the second derivatives f_{xx} and f_{yy} are both negative, then the situation in Figure 14.32 might occur and f would have a maximum. Similarly for f_{xx} and f_{yy} both positive and f would have a minimum. However, the actual situation is not that simple, since the mixed partial derivatives f_{xy} and f_{yx} also must be taken into account.

THEOREM 14.13

The Second Derivative Test Let f be a function of two variables whose second-order partial derivatives exist and are continuous, and let (a, b) be a critical point of f. Let

$$D(x, y) = \begin{vmatrix} f_{xx}(x, y) & f_{xy}(x, y) \\ f_{yx}(x, y) & f_{yy}(x, y) \end{vmatrix}$$
$$= f_{xx}(x, y)f_{yy}(x, y) - [f_{xy}(x, y)]^2.$$

Then:

1. If $D(a, b) > 0$ and $f_{xx}(a, b) < 0$, then f has a relative maximum at (a, b).
2. If $D(a, b) > 0$ and $f_{xx}(a, b) > 0$, then f has a relative minimum at (a, b).
3. If $D(a, b) < 0$, then f has a saddle point at (a, b).
4. If $D(a, b) = 0$, the test fails.

The proof of this theorem is quite involved and can be found in more advanced texts. To say that the test fails when $D(a, b) = 0$ means that other methods must be used to determine the extrema. The functions $f(x, y) = x^4 + y^4$, $g(x, y) = -x^4 - y^4$, and $h(x, y) = x^4 - y^4$ (Figure 14.35) illustrate that anything can happen when $D = 0$.

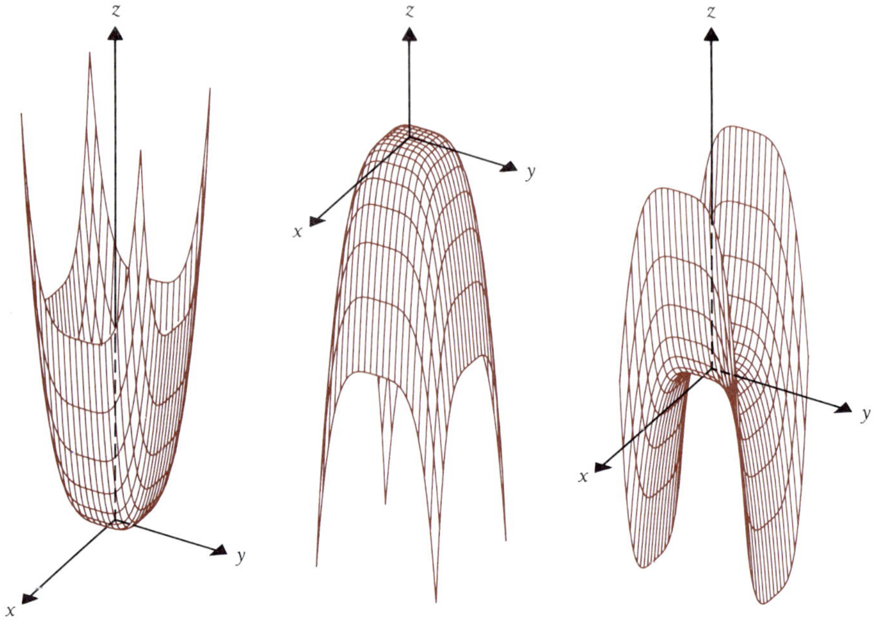

(a) The graph of $f(x, y) = x^4 + y^4$

(b) The graph of $g(x, y) = -x^4 - y^4$

(c) The graph of $h(x, y) = x^4 - y^4$

FIGURE 14.35

Now we look at some applications of the second derivative test.

Example 14.44 Find the extrema of the function

$$f(x, y) = x - xy + y^4 + 4.$$

Solution We first locate the critical points. The gradient is

$$\nabla f(x, y) = \langle 1 - y, -x + 4y^3 \rangle,$$

which is the zero vector for $y = 1$, $x = 4$. Thus, $(4, 1)$ is the only critical point. The second-order partial derivatives of f are

$$f_{xx} = 0, \qquad f_{xy} = -1, \qquad f_{yy} = 12y^2,$$

so that

$$D(x, y) = \begin{vmatrix} 0 & -1 \\ -1 & 12y^2 \end{vmatrix} = -1.$$

Thus, $D(4, 1) = -1 < 0$, and f has a saddle point at $(4, 1)$ and no extrema. ■

Example 14.45 Find the extrema of $f(x, y) = \sin x + \cos y$.

Solution The gradient of f is

$$\nabla f(x, y) = \langle \cos x, -\sin y \rangle,$$

which is the zero vector for $x = (2n + 1)\pi/2$, $y = m\pi$, where m and n are both integers. Thus, f has infinitely many critical points. The second-order partial derivatives are

$$f_{xx} = -\sin x, \qquad f_{xy} = 0, \qquad f_{yy} = -\cos y,$$

so that

$$D(x, y) = \begin{vmatrix} -\sin x & 0 \\ 0 & -\cos y \end{vmatrix} = \sin x \cos y.$$

Hence,

$$\begin{aligned} D\left((2n+1)\frac{\pi}{2}, m\pi\right) &= \sin\left[(2n+1)\frac{\pi}{2}\right]\cos m\pi \\ &= \sin\left(n\pi + \frac{\pi}{2}\right)\cos m\pi \\ &= \cos n\pi \cos m\pi \\ &= (-1)^n(-1)^m. \end{aligned}$$

Therefore, if m and n are both even, or the critical point is of the form $(\pi/2 + 2k\pi, 2l\pi)$, where k and l are both integers, then $D > 0$ and $f_{xx} < 0$, so f has a relative maximum of 2. If m and n are both odd, or the critical point is of the form $(3\pi/2 + 2k\pi, \pi + 2l\pi)$, then $D > 0$ and $f_{xx} > 0$, so f has a relative minimum of -2. By considering the function f, we see that the maximum and minimum values we found are actually absolute extreme values. If m and n have opposite parity, then $D < 0$ and f has a saddle point. ■

Extremum Problems

Now we consider some applications of these techniques for finding extrema.

Example 14.46 A rectangular box without a lid is to be constructed so that it will contain a volume of 32 cubic feet. Find the dimensions of the box that requires the least material.

Solution Let the box have length x, width y, and depth z. The amount of material used in constructing the box is proportional to the surface area, which is

$$A = xy + 2xz + 2yz.$$

Thus, we wish to choose x, y, and z so that A is a minimum. We also must have

$$xyz = 32,$$
$$z = \frac{32}{xy}.$$

Thus, we may rewrite A as

$$A = xy + \frac{64}{y} + \frac{64}{x},$$

which is a function of two variables. Now we minimize A. The gradient is

$$\nabla A = \left\langle y - \frac{64}{x^2}, x - \frac{64}{y^2} \right\rangle,$$

which is the zero vector when

$$y = \frac{64}{x^2}, \qquad x = \frac{64}{y^2}.$$

Eliminating y from these two equations, we get

$$x^4 = 64x,$$

and since $x \neq 0$, we must have $x = 4$. It follows that $y = 4$ also. You can verify by the second derivative test that A has a relative minimum at (4, 4). This minimum turns out to be absolute, as you will have a chance to show in the exercises (see Problem 36). We now find $z = 32/xy = 2$. The box should be 4 feet square and 2 feet deep. ■

For functions of more than two variables, the finding of extrema is a similar process. A higher-dimensional version of the second derivative test may be stated, but it is much more complicated. In most of the problems we encounter, it will be evident that the function has an extremum at the critical point we find.

■ **Example 14.47** A customer may buy quantities w, x, y, and z of four different commodities. The utility (value derived) of the purchase to the customer is

$$U = 500 - w^{5/2} - x^2 - y^{3/2} - z.$$

The commodities cost, respectively, \$20, \$10, \$6, and \$1 per unit, and the customer has \$300 to spend. How many units of each commodity should be bought to maximize the customer's utility?

Solution The cost constraint is

$$20w + 10x + 6y + z = 300,$$
$$z = 300 - 20w - 10x - 6y.$$

We may therefore rewrite U as a function of three variables,

$$\begin{aligned} U &= 500 - w^{5/2} - x^2 - y^{3/2} - (300 - 20w - 10x - 6y) \\ &= 200 - w^{5/2} + 20w - x^2 + 10x - y^{3/2} + 6y. \end{aligned}$$

The gradient of U is

$$\nabla U = \left\langle -\frac{5}{2} w^{3/2} + 20, -2x + 10, -\frac{3}{2} y^{1/2} + 6 \right\rangle,$$

which is the zero vector when $w = 4$, $x = 5$, and $y = 16$. Hence,

$$z = 300 - 20(4) - 10(5) - 6(16) = 74.$$

Therefore, the customer should buy 4, 5, 16, and 74 units of the commodities, respectively. ■

The Method of Least Squares

One measure of the distance from a point (x_1, y_1) in the plane to a line $y = mx + b$ is the square of the vertical distance (see Figure 14.36),

$$(mx_1 + b - y_1)^2.$$

For a collection of points $(x_1, y_1), \ldots, (x_n, y_n)$, the sum of such squared vertical distances is

$$S = \sum_{i=1}^{n} (mx_i + b - y_i)^2.$$

This sum is a function of the slope m and intercept b of the line. If the line is chosen so that the sum S is a minimum, we call the line the **least squares line** for the data points $(x_1, y_1), \ldots, (x_n, y_n)$. The least squares line is used to "smooth" data, particularly when looking for trends. The process of finding the least squares line is called **linear regression analysis**.

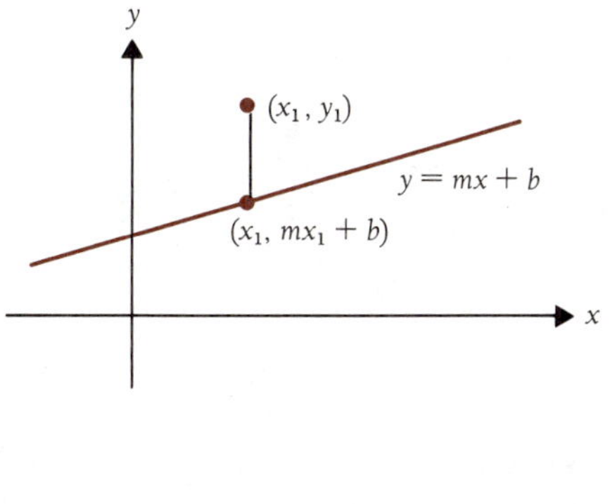

FIGURE 14.36

The function S can be expanded and rewritten in the form

$$S(m, b) = m^2\left(\sum_{i=1}^{n} x_i^2\right) - 2m\left(\sum_{i=1}^{n} x_i y_i\right) + \sum_{i=1}^{n} y_i^2 + 2mb\left(\sum_{i=1}^{n} x_i\right) - 2b\left(\sum_{i=1}^{n} y_i\right) + nb^2.$$

To find m and b for the least squares line, we can treat S like any other function of two variables to be minimized. We must set the partial derivatives S_m and S_b equal to 0, which leads us to the pair of equations

$$\begin{cases} \left(\sum_{i=1}^{n} x_i^2\right) m + \left(\sum_{i=1}^{n} x_i\right) b = \sum_{i=1}^{n} x_i y_i \\ \left(\sum_{i=1}^{n} x_i\right) m + nb = \sum_{i=1}^{n} y_i \end{cases} \tag{2}$$

These equations can be solved as they are for general formulas (to program into your computer, for instance), or the coefficients can be computed in each case and then the system can be solved. If you have a calculator with linear regression built in, then you need only enter the data points and push the right buttons. (By all means use the calculating power you have, if you have it!)

■ **Example 14.48** Find the least squares line for the data points (1, 2.5), (2, 2.7), (3, 3.1), and (4, 3.3).

Solution Here, we have four data points, so $n = 4$. By adding up the x-coordinates, their squares, the y-coordinates, and the products of the x- and y-coordinates, we find

$$\sum_{i=1}^{4} x_i = 10, \qquad \sum_{i=1}^{4} x_i^2 = 30, \qquad \sum_{i=1}^{4} y_i = 11.6, \qquad \sum_{i=1}^{4} x_i y_i = 30.4.$$

The above system of equations (2) then becomes

$$\begin{cases} 30m + 10b = 30.4 \\ 10m + 4b = 11.6. \end{cases}$$

The solution of this system is $m = 0.28$, $b = 2.2$; the least squares line is therefore (see Figure 14.37)

$$y = 0.28x + 2.2.$$

■

The method of least squares is also used in numerical analysis to find least squares curves of other kinds, such as parabolas and logarithmic curves.

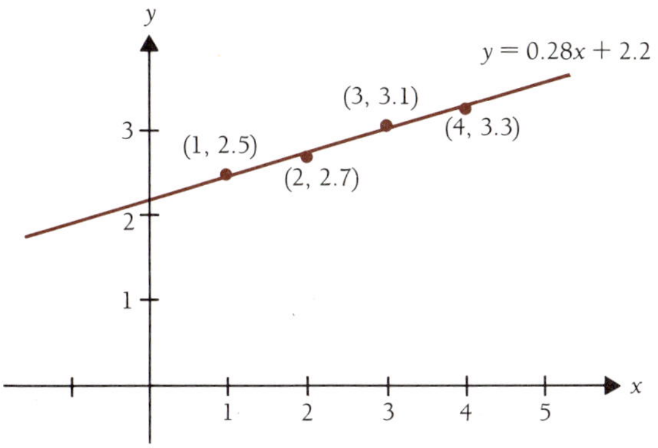

FIGURE 14.37

Exercises 14.7

Find the extrema.

1. $f(x, y) = x^2 + 4y^2$
2. $f(x, y) = 1 - 2x^2 - y^2$
3. $f(x, y) = x^3 + xy + 5$
4. $f(x, y) = x^3 + 2xy - y - 1$
5. $f(x, y) = x - \frac{1}{3}x^3 - \frac{1}{2}y^2$
6. $f(x, y) = x^2y^2 - x^2y^3 - \frac{1}{4}x^4$
7. $f(x, y) = 3xy - x^3 - y^3$
8. $f(x, y) = \frac{3}{2}x - \frac{1}{2}x^3 - xy^2$
9. $f(x, y) = x^4 - 2x^2 + y^2$
10. $f(x, y) = 2y^3 - 3x^4 - 6x^2y$
11. $f(x, y) = y^4 - x^4 - 2y^2 + 2x^2$
12. $f(x, y) = y^4 - 4xy + 2x^2$
13. $f(x, y) = \cos(x + y)$
14. $f(x, y) = x^2 \cos y$
15. $f(x, y) = e^{-x^2} \cos y$
16. $f(x, y) = e^{-x} \cos y$
17. $f(x, y) = e^{-x^2-y^2}$
18. $f(x, y) = \sin x + \sin y$
19. $f(x, y, z) = x^2 + 2yz$
20. $f(x, y, z) = 4 - x^2 - y^2 - z^2$
21. Find the dimensions of the rectangular box without a lid that uses 48 square feet of material and has maximum volume.
22. A window is in the shape of a rectangle surmounted by an isosceles triangle. If the perimeter is 360 inches, find the dimensions that give the window the maximum area.
23. A rectangular solid inscribed in the hemisphere $z = \sqrt{4 - x^2 - y^2}$ has what maximum possible volume?
24. Find the maximum volume of a rectangular solid inscribed in the ellipsoid

$$\frac{x^2}{a^2} + \frac{y^2}{b^2} + \frac{z^2}{c^2} = 1.$$

25. A rectangular solid has three faces in the coordinate planes and its remaining vertex on the plane $x + y + z = a$. Find the maximum volume of the solid.
26. A rectangular solid has its base in the xy-plane and its upper vertices on the paraboloid $z = a^2 - b^2x^2 - c^2y^2$. Find its maximum possible volume.
27. Find the minimum distance between the lines $x = t$, $y = 2 + t$, $z = 1 + 3t$ and $x = t$, $y = -1 + 2t$, $z = -1 - t$. [*Hint:* The distance is minimum precisely when the square of the distance is minimum.]

A **28.** Find the point on the surface $x^2 + 3y^2 - 2x - 6y + z + 3 = 0$ that is nearest to the origin.

29. A manufacturer produces cassette players and television sets at a cost of \$50 per cassette player and \$60 per television set. If the selling price is \$$x$ per cassette player and \$$y$ per television set, then the numbers of cassette players and television sets that can be sold per week are

$$N_{CP} = 250(y - x) \qquad \text{and} \qquad N_{TV} = 32{,}000 + 250(x - 2y),$$

respectively. What selling prices will maximize the profit?

30. A manufacturer produces thumbtacks at a cost of 8¢ per box and paper clips at a cost of 6¢ per box. If thumbtacks sell for x¢ per box and paper clips sell for y¢ per box, then $4{,}000{,}000/xy$ boxes of thumbtacks and $8{,}000{,}000/xy$ boxes of paper clips can be sold per week. What selling prices will maximize profit?

31. A company finds that if it spends \$$A$ million on advertising and charges \$$S$ for each item it produces, then the company's profit, in millions of dollars, is

$$P = 2AS + 80S - 1.5S^2 - (0.1)AS^2 - 100.$$

Find the maximum of P and the values of A and S that yield the maximum.

A **32.** If a consumer purchases quantities x, y, and z of three commodities, the utility to the consumer is

$$U = x^{1/2}y^{1/3}z^{1/4}.$$

If the commodities cost \$9, \$4, and \$3 per unit, respectively, and the consumer has \$117 to spend, how many units of each commodity should be bought to maximize the utility?

A **33.** Show that of all triangles with a given perimeter, the equilateral triangle has maximum area. [*Hint:* If p is the perimeter and a, b, and c are the sides, the area is $A = \frac{1}{4}\sqrt{p(p - 2a)(p - 2b)(p - 2c)}$.]

34. Show that of all rectangular solids with a given surface area, the cube has the greatest volume.

35. A triangle is inscribed in a circle of radius r so that its perimeter is as large as possible. What kind of triangle is it, and what is its perimeter?

A **36.** Show that

$$xy + \frac{64}{y} + \frac{64}{x} \geq 48$$

for all positive x and y (see Example 14.46). [*Hint:* Assume otherwise, and complete the square in y; then the expression in x on the other side of the inequality must be positive, and that leads to a contradiction.]

37. Can you rework Example 14.47 so as to minimize the utility? (This is a problem in "nonlinear programming." The minimum occurs on a boundary.)

Find the least squares line for each set of data points.

38. $(2, -1)$, $(3, 0)$, $(5, 2)$

39. $(0, 6)$, $(2, 4.5)$, $(4, 3)$

40. $(0, 12.5)$, $(1, 10.7)$, $(2, 11.3)$, $(3, 12.5)$

41. $(1, 2.3)$, $(2, 1.8)$, $(3, 1.4)$, $(4, 1.1)$, $(5, 0.7)$

A42. Show that the least squares line for two data points is the line passing through the two points.

43. If $y = mx + b$ is the least squares line for a set of data points and $\hat{x}$ is another value for x, the value $\hat{y} = m\hat{x} + b$ is called the **linear estimate** for $\hat{x}$, relative to the data points. (Linear estimates are often used for forecasting, interpolating, and extrapolating.)
    (a) Find the linear estimate for $\hat{x} = 4$ relative to the data points in Problem 38.
    (b) Find the linear estimate for $\hat{x} = 6$ relative to the data points in Problem 41.

44. Find equations similar to system (2) for the coefficients a, b, and c of a least squares parabola $y = ax^2 + bx + c$ for the set of data points $(x_1, y_1), \ldots, (x_n, y_n)$.

SECTION 14.8

Extrema with Constraints

Many of the problems in Section 14.7 asked us to find an extremum of a function, subject to a side condition, or **constraint**. For instance, in Example 14.46, we wished to find the minimum of the function

$$A(x, y, z) = xy + 2xz + 2yz$$

subject to the constraint

$$xyz = 32.$$

In that example, we used the constraint to eliminate one of the variables. In this section, we introduce a more general approach, called the **method of Lagrange**.

The Method of Lagrange

Let f be a differentiable function of two variables, and let $g(x, y) = 0$ be a constraint on x and y. We assume that g is differentiable and ∇g is not the zero vector. The key idea of Lagrange's method is that if f has an extremum (relative or absolute) at point (a, b) on the curve $g(x, y) = 0$, then $\nabla f(a, b)$ is parallel to $\nabla g(a, b)$. To show this, let the curve $g(x, y) = 0$ be parametrized by the vector function $\mathbf{r}(t)$, with $\mathbf{r}(t_0) = \langle a, b \rangle$. Then the function $f(\mathbf{r}(t))$ has an extremum at t_0, so

$$\left.\frac{d}{dt} f(\mathbf{r}(t))\right|_{t=t_0} = \nabla f(a, b) \cdot \mathbf{r}'(t_0) = 0,$$

and $\nabla f(a, b)$ is normal to the curve $g(x, y) = 0$ at (a, b). But $\nabla g(a, b)$ is also normal to $g(x, y) = 0$ at (a, b), because the gradient is always normal to a level curve. Since $\nabla f(a, b)$ and $\nabla g(a, b)$ are both normal to the same plane curve at (a, b), they must be parallel.

Therefore, there is a scalar λ such that

$$\nabla f(a, b) = \lambda \nabla g(a, b). \tag{1}$$

This equation can be generalized to any number of variables. We call λ a **Lagrange multiplier**.

Joseph Louis Lagrange (1736–1813), of mixed French and Italian parentage, was forced to make his own way in the world when his father squandered the family fortune. He began his career at the age of 16 as Professor of Mathematics at the Royal Artillery School at Turin. (At that time, gunnery officers were among the major consumers of mathematics.) Even at this early age, Lagrange had mastered the available material on calculus and had begun to make original contributions. Some of these he showed to Euler, who had been thinking along similar lines, but Euler generously suppressed his own work in favor of Lagrange. Euler recommended Lagrange for a position at the Berlin Scientific Academy, and Lagrange remained there for more than 20 years. When Frederick the Great died, the atmosphere in Berlin became uncongenial and Lagrange moved to Paris just as the French Revolution broke out. When the new French government had achieved some measure of stability, it was felt that an improved system of weights and measures was necessary. Lagrange, a member of the committee chosen to make recommendations, is generally credited with the design of the metric system.

While still a youth in Turin, Lagrange had dreamed of a unified approach to mathematics; he realized his dream in his masterpiece, the *Mechanique Analytique* of 1788. The publisher was unsure of how the book would be received and required a financial guarantee. That precaution turned out to be entirely unnecessary — the book is still in print today!

The procedure for finding relative extrema of a function f subject to a constraint $g(\mathbf{v}) = 0$ is to write

$$\nabla f = \lambda \nabla g, \tag{2}$$

or equivalently,

$$\nabla(f - \lambda g) = \mathbf{0}. \tag{3}$$

The solution points are the critical points of f satisfying $g(\mathbf{v}) = 0$. Note that $f - \lambda g$ is a function in not only the variables of f and g but also the Lagrange multiplier λ.

Consider some examples.

■ **Example 14.49** Rework Example 14.46 using the method of Lagrange.

Solution We wish to minimize the function

$$A = xy + 2xz + 2yz$$

subject to the constraint

$$xyz = 32.$$

We set $g(x, y, z) = xyz - 32$, and construct the function $F = A - \lambda g$, or

$$F(x, y, z, \lambda) = xy + 2xz + 2yz - \lambda(xyz - 32).$$

Then we require the partial derivatives of F to be 0, that is,

$$\begin{aligned} F_x(x, y, z, \lambda) &= y + 2z - \lambda yz = 0, \\ F_y(x, y, z, \lambda) &= x + 2z - \lambda xz = 0, \\ F_z(x, y, z, \lambda) &= 2x + 2y - \lambda xy = 0, \\ F_\lambda(x, y, z, \lambda) &= -(xyz - 32) = 0. \end{aligned}$$

Solving these equations simultaneously, we eventually get $x = y = 4$, $z = 2$, the same solution as before. ■

■ **Example 14.50** The temperature at point (x, y, z) on the sphere $x^2 + y^2 + z^2 = 4$ is $T(x, y, z) = xy^2z$. Find the absolute maximum temperature and the points at which it occurs.

Solution We must find the maximum of $T(x, y, z) = xy^2z$, subject to the constraint

$$g(x, y, z) = x^2 + y^2 + z^2 - 4 = 0. \tag{4}$$

The gradients are

$$\nabla T(x, y, z) = \langle y^2z, 2xyz, xy^2 \rangle$$

and

$$\nabla g(x, y, z) = \langle 2x, 2y, 2z \rangle.$$

Therefore, equation (2) becomes

$$\langle y^2z, 2xyz, xy^2 \rangle = \lambda \langle 2x, 2y, 2z \rangle,$$

or

$$y^2z = 2\lambda x, \tag{5}$$
$$2xyz = 2\lambda y, \tag{6}$$
$$xy^2 = 2\lambda z. \tag{7}$$

We now solve these equations simultaneously with the constraint (4). From equation (6) we get $y(\lambda - xz) = 0$, so that either $y = 0$ or $\lambda = xz$. If $y = 0$, then $T = 0$, which is clearly not a maximum. If $\lambda = xz$, then equations (5) and (7) become $z(y^2 - 2x^2) = 0$ and $x(y^2 - 2z^2) = 0$. Again, if $z = 0$ or $x = 0$, then $T = 0$, which is not the maximum. Therefore, $y^2 = 2x^2 = 2z^2$, so that $x^2 = z^2$. Then from equation (4), we get $x^2 + 2x^2 + x^2 = 4$, or $x = \pm 1$. It follows that $y = \pm\sqrt{2}$ and $z = \pm 1$, and the critical points are $(\pm 1, \pm\sqrt{2}, \pm 1)$, with signs chosen independently.

At each of the points $(1, \pm\sqrt{2}, 1)$ and $(-1, \pm\sqrt{2}, -1)$, we have $T = 2$; and at each of the points $(1, \pm\sqrt{2}, -1)$ and $(-1, \pm\sqrt{2}, 1)$, we have $T = -2$. Thus, the maximum temperature is 2, and it occurs at $(1, \pm\sqrt{2}, 1)$ and $(-1, \pm\sqrt{2}, -1)$. ■

■ **Example 14.51** Find the points on the curve $x^2 + xy = 4$ that are nearest the origin.

Solution The distance is minimum when the square of the distance is minimum, so to avoid radicals we will minimize the function

$$f(x, y) = x^2 + y^2$$

subject to the constraint

$$g(x, y) = x^2 + xy - 4 = 0.$$

We will let

$$\begin{aligned} F(x, y, \lambda) &= f(x, y) - \lambda g(x, y) \\ &= x^2 + y^2 - \lambda(x^2 + xy - 4), \end{aligned}$$

from which we find

$$\begin{aligned} \nabla(f - \lambda g) &= \nabla F(x, y, \lambda) \\ &= \langle 2x - 2x\lambda - y\lambda, 2y - x\lambda, -(x^2 + xy - 4) \rangle. \end{aligned}$$

The gradient is the zero vector if and only if

$$2x - 2x\lambda - y\lambda = 0, \tag{8}$$

$$2y - x\lambda = 0, \tag{9}$$

$$x^2 + xy - 4 = 0. \tag{10}$$

We will solve these equations simultaneously. [Notice that equation (10) is just the given constraint—we get it back automatically.] From equation (9), we have $y = \frac{1}{2}x\lambda$, which, substituted into equation (8), yields $x(4 - 4\lambda - \lambda^2)$. If $x = 0$, then also $y = 0$ by equation (9), but then equation (10) is not satisfied. Hence, $x \neq 0$ and $\lambda = -2 \pm 2\sqrt{2}$. Then $y = (-1 \pm \sqrt{2})x$, and equation (10) yields $x = \pm 2^{3/4}$ when $y = (-1 + \sqrt{2})x$, so $y = \pm(-1 + \sqrt{2})2^{3/4}$. Thus, the points $(2^{3/4}, (-1 + \sqrt{2})2^{3/4})$ and $(-2^{3/4}, (1 - \sqrt{2})2^{3/4})$ are critical points. Since the given curve is a hyperbola, these points must be the points we seek. ■

The method of Lagrange also allows for more than one constraint. To find an extremum of a function f subject to the set of constraints $g_1(\mathbf{v}) = 0, \ldots, g_n(\mathbf{v}) = 0$, we require that

$$\nabla(f - \lambda_1 g_1 - \lambda_2 g_2 - \cdots - \lambda_n g_n) = \mathbf{0} \tag{11}$$

in order to find the critical points. This involves solving simultaneously a system of equations in the variables of f and the Lagrange multipliers $\lambda_1, \ldots, \lambda_n$.

■ Example 14.52 Find the extreme values of z on the curve of intersection of the plane $x + y + 8z = 8$ and the ellipsoid $x^2 + y^2 + 4z^2 = 16$.

Solution We must find the extreme values of the function

$$f(x, y, z) = z,$$

subject to the constraints

$$g_1(x, y, z) = x + y + 8z - 8 = 0$$

and

$$g_2(x, y, z) = x^2 + y^2 + 4z^2 - 16 = 0.$$

We thus form

$$\begin{aligned} F(x, y, z, \lambda_1, \lambda_2) &= f - \lambda_1 g_1 - \lambda_2 g_2 \\ &= z - \lambda_1(x + y + 8z - 8) - \lambda_2(x^2 + y^2 + 4z^2 - 16). \end{aligned}$$

Setting each component of the gradient of F equal to 0, we get the system of equations

$$\lambda_1 + 2x\lambda_2 = 0, \tag{12}$$
$$\lambda_1 + 2y\lambda_2 = 0, \tag{13}$$
$$8\lambda_1 + 8z\lambda_2 = 1, \tag{14}$$
$$x + y + 8z = 8, \tag{15}$$
$$x^2 + y^2 + 4z^2 = 16. \tag{16}$$

Solving this system simultaneously, we have from equations (12) and (13) the relation $(x - y)\lambda_2 = 0$. If $\lambda_2 = 0$, then $\lambda_1 = 0$ from equation (12), contradicting equation (14). Hence, $x = y$. Then equation (15) becomes $2x + 8z = 8$, or $x = 4(1 - z)$, and equation (16) becomes $2x^2 + 4z^2 = 16$, or $x^2 + 2z^2 = 8$. Substituting for x, we get $z = (8 \pm 2\sqrt{7})/9$. These values of z correspond to critical points of f, subject to the constraints. Thus, $f(x, y, z) = z$ has maximum value $(8 + 2\sqrt{7})/9$ and minimum value $(8 - 2\sqrt{7})/9$, subject to the constraints. Both extrema are absolute, since they are the only extrema over a compact set. ■

Exercises 14.8

Find the critical points of f, subject to the given constraints.

[A]1. $f(x, y) = x^2 + 24xy + 8y^2$; $x^2 + y^2 = 25$

[A]2. $f(x, y) = x^2 + 24xy + 8y^2$; $x^2 - 2y^2 = 2$

3. $f(x, y, z) = 2x^2 + y^2 + z^2$; $x + 3y + 2z = 1$

4. $f(x, y, z) = 2x^2 + y^2 + z^2$; $x - 2y - z = 2$

5. $f(x, y, z) = x^2 + y^2 + z^2$; $x^2yz = 1$

6. $f(x, y, z) = x^2 + y^2 + z^2$; $xyz = 12$

7. $f(x, y, z) = xyz$; $x^2 + y^2 = 1$

8. $f(x, y, z) = xyz$; $x + z^2 = y$

[A]9. $f(x, y, z) = x^2 + y^2 + z^2$; $x + 2y + z = 1$ and $2x + y - 3z = 4$

10. $f(x, y, z) = x^2 + y^2 + z^2$; $x - y = 5$ and $y - z = 2$

[A]11. $f(x, y, z, w) = x^2 + y^2 + z^2 + w^2$; $x + y - z + 2w = 2$ and $2x - y + z + 2w = 4$

12. $f(x, y, z, w) = x^2 + y^2 + z^2 + w^2$; $x + y + z + w = 4$ and $y = z$

13. Find the minimum distance between the origin and the curve $y^2 = x + 1$.

14. Find the point on the curve $xy - 2x - y = -1$ that is nearest the origin.

[A]15. The plane $x + y + 2z = 0$ meets the ellipsoid $x^2 + y^2 + 3z^2 = 6$ in a curve C. Find the point(s) on C nearest the origin.

[A]16. The surfaces $z = 4 - x^2 - y^2$ and $(x - 1)^2 + y^2 + z^2 = 4$ meet in a curve C. Find the point(s) on C nearest the origin.

17. The temperature on the circle $x^2 + y^2 = 4$ is $T(x, y) = 16 + x^2 - y$. Find the extremes of temperature.

18. The temperature on the surface of the sphere $x^2 + y^2 + z^2 = 16$ is given by $T(x, y, z) = 2xy + 5z^2$. Find the extremes of temperature.

19. Find the maximum of $f(x, y, z) = x^2 + y^2$ on the surface $x^2 + y^2 + 4z^2 = 12$.

20. The density of a plate covering the region $x^2 + y^2 \leq 1$ is $\delta(x, y) = 3xy$. Find the point of greatest density.

21. Three numbers are chosen so that their sum is S and their product is a maximum. Show that the numbers are all the same.

[A]22. A plane passes through point (1, 2, 3) and creates with the coordinate planes a solid tetrahedron of minimal volume. Find the intercepts of the plane with the coordinate axes. [*Hint:* The volume of the tetrahedron is one-third the product of the height and the area of the base.]

23. Find the point P such that the sum of the squares of the distances from P to the vertices of a quadrilateral is a minimum. [*Hint:* Give the vertices coordinates, and find P in terms of the coordinates.]

24. A company can produce quantities x, y, and z of three products, subject to the constraint

$$x^2 + 2y^2 + 4z^2 = 3.4 \times 10^9,$$

and thereby realize a profit $P = 2x + 8y + 20z$. Find the maximum profit.

SECTION 14.9

Chapter Review

Important Concepts

Function of several variables
Level curve
Contour
Level surface
Open disc
Approach
Open ball
Open set
Closed set
Bounded set
Compact set
Extreme value theorem
Partial derivative
Partial differentiation
Higher-order partial derivatives
Second-order mixed partials
Differentiable
Chain rule
Gradient
Vector field
Directional derivative
Heat-seeking particle
Normal vector
Isotherm
Normal line
Tangent line
Tangent plane
Relative extrema
Absolute extrema
Critical points
Boundary points
Saddle point
Second derivative test
Least squares method
Least squares line
Linear regression analysis
Constraint
Method of Lagrange
Lagrange multiplier

Review Exercises

Express the given statement as a function of several variables.

1. The area A of a trapezoid is the product of the height h with the average of the bases b_1 and b_2.
2. The cost C of painting the walls in a rectangular room of dimensions l and w and height h is the product of the area of the walls and the cost c per gallon of paint, divided by the area s that 1 gallon of paint will cover.

Give the domain of each f, and describe its graph by means of level curves or level surfaces.

3. $f(x, y) = 2x + y - 3$
4. $f(x, y) = y^2 - 4x$
5. $f(x, y) = \dfrac{x^2}{y}$
6. $f(x, y) = x \ln y$
7. $f(x, y, z) = 3x + 4y - z + 1$
8. $f(x, y, z) = \dfrac{x^2 + y^2}{z}$
9. Describe the domains of the functions $f + g$ and fg in terms of the domains of the functions f and g of two variables.
10. If f is a function of two variables and g is a function of one variable, describe the domain of $g \circ f$ in terms of the domains of f and g.

Find the limits, if they exist.

11. $\displaystyle\lim_{(x, y) \to (0, 0)} \frac{x^2 + y^2}{x^2 - y^2}$
12. $\displaystyle\lim_{(x, y) \to (0, 0)} \frac{x^2 + xy + y^2}{2xy}$
13. $\displaystyle\lim_{\substack{(x, y,) \to (1, 2) \\ x+y=3}} (x + y)$
14. $\displaystyle\lim_{(x, y, z) \to (0, 0, 0)} \frac{xyz}{x^2 + yz}$

Discuss the continuity of each function.

15. $f(x, y) = \sin xy$
16. $f(x, y) = \tan xy$
17. $f(x, y) = \tan xy; \quad x^2 + y^2 \le 1$
18. $f(x, y) = x \tan y; \quad y^2 \le 1$

Find the first-order partial derivatives.

19. $f(x, y) = x\sqrt{y} - y\sqrt{x}$
20. $f(x, y) = 3x^2y + \dfrac{1}{xy}$
21. $f(x, y) = x^y$
22. $f(x, y) = ye^{xy}$
23. $f(x, y, z) = y^{x+z}$
24. $f(x, y, z) = (x + y)^z$

Find the second-order partial derivatives.

25. $f(x, y) = x^2 + y^2$
26. $f(x, y) = e^x \cos y$
27. $f(x, y) = e^{x+y} + e^{x-y}$
28. $f(x, y) = \arctan xy$
29. $f(x, y, z) = xyz$
30. $f(x, y, z) = \cos x \sin y \tan z$

Find $\partial f/\partial x$.

31. $f(u, v) = u^2 - v^2; \quad u = 2x - y, \quad v = 2x + 3y$
32. $f(u, v) = \sin uv; \quad u = \cos x, \quad v = y \sin x$

33. $f(u, v, w) = e^{uvw}$; $u = \tan xy$, $v = e^y$, $w = x + y$

34. $f(u, v, w) = u^2 + v^2 - w$; $u = x - y + z$, $v = x + y + z$, $w = x^2 + y^2 - z$

Find df/dx.

35. $f(u, v) = e^u \arcsin v$; $u = \cos x$, $v = e^x$

36. $f(u, v) = \ln(u + v)$; $u = 1 + x^2$, $v = 1 - x^3$

37. $f(u, v, w) = u^2 + v^2 + 3w^2$; $u = \cos x$, $v = \sin x$, $w = \tan x$

38. $f(u, v, w, t) = uvw + t^2$; $u = x^2 + y^2$, $v = x^2 - y^2$, $w = xy$, $t = x + y$, $y = 1/x$

Find the gradient.

39. $f(x, y) = \sin x \cos y$

40. $f(x, y) = e^{x-y}$

41. $f(x, y, z) = \sin xyz$

42. $f(x, y, z) = (x + y)^{2z}$

Find the directional derivative of each f at the given point in the direction of **v**.

43. $f(x, y) = (1 + x)^y$; $(0, 2)$; $\mathbf{v} = \mathbf{i} + 2\mathbf{j}$

44. $f(x, y) = e^{x \cos y}$; $(1, 0)$; $\mathbf{v} = \mathbf{i} - \mathbf{j}$

45. $f(x, y) = x^2y^2$; $(2, -1)$; $\mathbf{v} = \nabla f(2, -1)$

46. $f(x, y) = x^2 + y^2$; $(3, 1)$; $\mathbf{v} = \nabla f(1, -1)$

47. $f(x, y, z) = (x + y)^{\cos z}$; $(1, 0, 0)$; $\mathbf{v} = \langle \frac{1}{3}, \frac{2}{3}, \frac{1}{3} \rangle$

48. $f(x, y, z) = x^2 - y^2 - z^2$; $(2, 1, 1)$; $\mathbf{v} = \nabla f(1, 1, 2)$

Find equations of the normal line and tangent plane to each surface at the given point.

49. $z = 5x^2 + 4y^2$; $(1, 1, 9)$

50. $z^2 = 5x^2 + 4y^2$; $(1, 1, 3)$

51. $z = e^{x+y}$; $(0, 0, 1)$

52. $z = e^{x+y}$; $(0, 1, e)$

Find the extrema, if any.

53. $f(x, y) = x^2 + y - y^3$

54. $f(x, y) = x^2 + y^4$

55. $f(x, y) = 2x - \frac{2}{3}x^3 - y^3$

56. $f(x, y) = 4 - x + x^2 - x^3 - y^3$

57. $f(x, y, z) = x + y^2 - z^2$

58. $f(x, y, z) = 1 - x^2 - y^2z^2$

Find the minimum distance between the origin and the given curve.

A59. $y = 2 - (x - 1)^2$

C60. $y = e^x$

C61. $y = \cos x$

A62. $x^2 + y^2 + 6x - 2y - 4 = 0$

63. Show that two spheres meet at the same angle at each point of intersection.

Additional Exercises

1. Find

$$\lim_{(x, y)\to(1, \pi/2)} (x \sin y).$$

2. Find

$$\lim_{(x, y)\to(1/\sqrt{2}, 1/\sqrt{2})} (x^2 + y^2).$$

Find the first-order partial derivatives.

3. $f(x, y) = y^{\cos xy}$

4. $f(x, y) = \dfrac{1}{x} \arctan \dfrac{1}{y}$

5. $f(x, y, z, w) = xy \cos zw$

6. $f(x, y, z, w) = \dfrac{ax + by + cz + dw}{ax - by - cz + dw}$

7. Show that the function below is not differentiable at the origin:

$$f(x, y) = \begin{cases} xy/(x^2+y^2) & \text{if } (x, y) \neq (0, 0) \\ 0 & \text{if } (x, y) = (0, 0). \end{cases}$$

8. Let f be a function of u and v whose second-order partial derivatives are continuous, and let u and v be continuously differentiable functions of x and y. Show that:

(a) $$\frac{\partial^2 f}{\partial x^2} = \frac{\partial^2 f}{\partial u^2}\left(\frac{\partial u}{\partial x}\right)^2 + 2\frac{\partial^2 f}{\partial u\,\partial v}\frac{\partial u}{\partial x}\frac{\partial v}{\partial x} + \frac{\partial^2 f}{\partial v^2}\left(\frac{\partial v}{\partial x}\right)^2 + \frac{\partial f}{\partial u}\frac{\partial^2 u}{\partial x^2} + \frac{\partial f}{\partial v}\frac{\partial^2 v}{\partial x^2}$$

(b) $$\frac{\partial^2 f}{\partial y\,\partial x} = \frac{\partial^2 f}{\partial u^2}\frac{\partial u}{\partial x}\frac{\partial u}{\partial y} + \frac{\partial^2 f}{\partial u\,\partial v}\left(\frac{\partial u}{\partial x}\frac{\partial v}{\partial y} + \frac{\partial u}{\partial y}\frac{\partial v}{\partial x}\right) + \frac{\partial^2 f}{\partial v^2}\frac{\partial v}{\partial x}\frac{\partial v}{\partial y} + \frac{\partial f}{\partial u}\frac{\partial^2 u}{\partial y\,\partial x} + \frac{\partial f}{\partial v}\frac{\partial^2 v}{\partial y\,\partial x}$$

9. Find the minimum distance between the origin and the surface:

(a) $z = 4 - x^2 - 2y^2$ (b) $x = 6 - y^2 - z^2$

10. Find the minimum distance between the curves $x^2 + y^2 = 1$ and $y = (x - 3)^2$.

11. Find the minimum distance between the two surfaces $x^2 + y^2 + z^2 = 4$ and $16x^2 - 9(y - 6)^2 - 9z^2 = 144$.

12. A certain solar collector is a flat plate with area A. If θ is the angle between the collector and the sun's rays, then the **effective area** of the collector is

$$A_e = A \sin \theta.$$

The orientation of the collector is given by a vector $\mathbf{n}$ normal to the collector, relative to a coordinate system whose x-axis points south, y-axis points east, and z-axis is normal to the earth. The position of the sun is given by the **hour angle** h, the compass bearing in the xy-plane to the sun measured from the positive x-axis, and the **elevation** α, the angle at which the sun's rays meet the xy-plane. Find a formula for the effective area of the collector in terms of $\mathbf{n}$, h, and α.

13. Two manufacturers are in competition. Let x and y be their respective outputs, and let P and Q be their respective profits. Suppose $P = 50x - x^2 - y^2$ and $Q = 40y - \frac{1}{2}x^2 - y^2$. Find the profits if each manufacturer works independently to maximize profit. Then find the maximum value of $P + Q$ if the manufacturers cooperate.

14. Let θ be the angle between $\mathbf{v}$ and $\mathbf{i}$. Show that the directional derivative of $f(x, y)$ in the direction of $\mathbf{v}$ is $f_x \cos\theta + f_y \sin\theta$.

15. Find the point interior to a given triangle such that the sum of the squares of the distances to the sides of the triangle is a minimum.

16. A quadrilateral has sides of lengths x, y, z, and w. Determine the quadrilateral with maximum area.

17. Find the curve on the surface $x^2 + y^2 = 36$ whose tangent vector is constantly 45° above the horizontal and which passes through the point (6, 0, 0).

18. Find the points on the curve of intersection of $x^2 + y^2 + z = 16$ and $x + 2y + 3z = 3$ that are nearest and farthest from the origin.

19. A surface is said to be **smooth** if and only if it has a continuous unit normal vector $\mathbf{n}(x, y, z)$ at each point (x, y, z) of the surface. Show that the surface $z = f(x, y)$ is smooth if and only if the function f is continuously differentiable.

20. Find the outward-pointing unit normal vector for the surface $x^2 + y^2 + z^2 = a^2$.

CHAPTER 15

Differentials and Path Integrals

INTRODUCTION

Differentials of functions of several variables can be quite useful, and they also suggest a several-variable version of the antiderivative. In this chapter, we examine the total differential of a function of several variables. Then we look at differential forms and study path integrals and their application to work.

SECTION 15.1

Total Differentials

Before introducing the differential of a function of several variables, we will review the idea for a function of one variable (see Section 3.7).

Review of Differentials in One Variable

If $y = f(x)$ is a differentiable function of a single variable x, the differential dx of x is another independent variable, and the differential dy of y is a new dependent variable defined by

$$dy = f'(x)\ dx.$$

Thus, dy, also denoted by df, is a function of the two variables x and dx, or

$$df(x, dx) = f'(x)\ dx.$$

Recall that the differential df is approximately equal to the increment Δf, or

$$df \approx \Delta f = f(x + dx) - f(x).$$

Differentials in Two Variables

For a function of two variables, the differential is analogous to the single-variable case.

DEFINITION 15.1

Let $z = f(x, y)$ be a differentiable function of two variables. The **differentials** dx and dy of x and y are new independent variables. The **total differential** of f (or of z) is denoted by df (or dz) and defined by

$$df = f_x\,dx + f_y\,dy. \tag{1}$$

If df is evaluated at particular values of x, y, dx, and dy, we write

$$df(x, y, dx, dy) = f_x(x, y)\,dx + f_y(x, y)\,dy.$$

We can rewrite equation (1) in vector notation. If we let $\mathbf{r} = \langle x, y\rangle$ and $d\mathbf{r} = \langle dx, dy\rangle$, then equation (1) can be written as

$$df = \nabla f \cdot d\mathbf{r}. \tag{2}$$

Recall from Section 14.5 that if f is a differentiable function of two variables and $\mathbf{h} = \langle \Delta x, \Delta y\rangle$, then the increment

$$\Delta f(x, y) = f(x + \Delta x, y + \Delta y) - f(x, y)$$

may be written as

$$\Delta f(x, y) = \nabla f(x, y) \cdot \mathbf{h} + \boldsymbol{\eta} \cdot \mathbf{h}, \tag{3}$$

with $\boldsymbol{\eta} \to \mathbf{0}$ as $\mathbf{h} \to \mathbf{0}$. If we now set $dx = \Delta x$ and $dy = \Delta y$, then $d\mathbf{r} = \mathbf{h}$ and equation (3) becomes

$$\Delta f(x, y) = \nabla f(x, y) \cdot d\mathbf{r} + \boldsymbol{\eta} \cdot \mathbf{h}. \tag{4}$$

Hence,

$$\Delta f(x, y) \to \nabla f(x, y) \cdot d\mathbf{r} = df \qquad \text{as} \qquad \mathbf{h} \to \mathbf{0}.$$

Thus, df may be used as an approximation to $\Delta f(x, y)$, or

$$df \approx f(x + dx, y + dy) - f(x, y). \tag{5}$$

Equivalently,

$$\begin{aligned} f(x + dx, y + dy) &\approx f(x, y) + df \\ &= f(x, y) + f_x(x, y)\,dx + f_y(x, y)\,dy. \end{aligned} \tag{6}$$

■

In the case of one variable, df is the change along the tangent line instead of the graph; in the case of two variables, df is the change along the tangent plane instead of along the graph (see Figure 15.1).

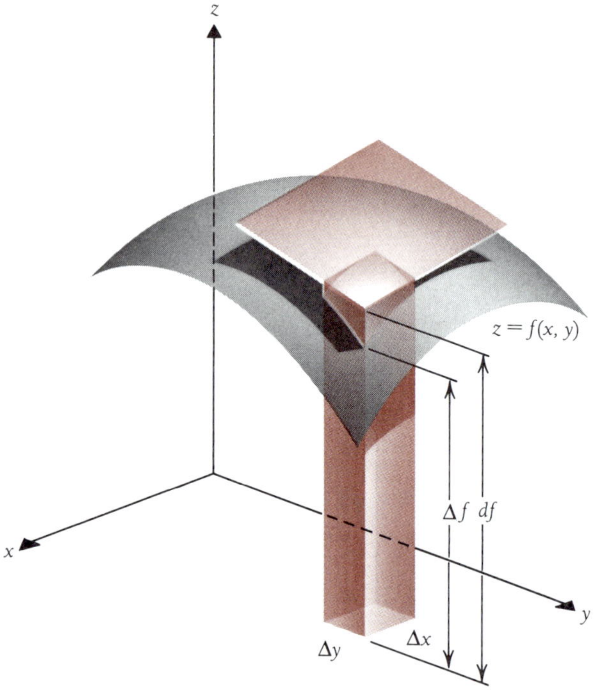

FIGURE 15.1 The differential of f is the change along the tangent plane

Example 15.1 Use differentials to approximate $(25)^{2/3} + \sqrt{65}$.

Solution We will approximate the function

$$f(x, y) = x^{2/3} + \sqrt{y}$$

at the point (25, 65). We can easily evaluate f at the point (27, 64), so we let

$$x = 27, \qquad dx = -2,$$
$$y = 64, \qquad dy = 1.$$

We will now approximate $f(x + dx, y + dy)$ using equation (6). We find first that

$$f_x(x, y) = \frac{2}{3}x^{-1/3} \quad \text{and} \quad f_y(x, y) = \frac{1}{2\sqrt{y}}.$$

Hence,

$$\begin{aligned}(25)^{2/3} + \sqrt{65} &= f(x + dx, y + dy) \\ &\approx f(x, y) + f_x(x, y)\,dx + f_y(x, y)\,dy \\ &= x^{2/3} + \sqrt{y} + \frac{2}{3}x^{-1/3}\,dx + \frac{1}{2\sqrt{y}}\,dy\end{aligned}$$

$$= (27)^{2/3} + \sqrt{64} + \frac{2}{3}(27)^{-1/3}(-2) + \frac{1}{2\sqrt{64}}(1)$$

$$= 9 + 8 - \frac{4}{9} + \frac{1}{16} \approx 16.618.$$

(The actual value to three decimal places is 16.612.) ■

■ **Example 15.2** Estimate the change in the volume of a cone brought about by increasing the base radius r by 2% and increasing the height h by 1%.

Solution The volume of the cone is

$$V = \frac{1}{3}\pi r^2 h.$$

We will let $dr = (0.02)r$ and $dh = (0.01)h$. Then the differential of V is

$$\begin{aligned} dV &= V_r\,dr + V_h\,dh \\ &= \frac{2}{3}\pi rh(0.02)r + \frac{1}{3}\pi r^2(0.01)h \\ &= (0.04)\frac{1}{3}\pi r^2 h + (0.01)\frac{1}{3}\pi r^2 h \\ &= (0.05)V. \end{aligned}$$

Thus, the change in V is approximately a 5% increase. ■

Differentials in Several Variables

It should be clear how Definition 15.1 can be extended to functions of any number of variables. For three variables, we have

$$df = f_x\,dx + f_y\,dy + f_z\,dz. \tag{7}$$

For n variables, we have

$$df = \sum_{i=1}^{n} f_i\,dx_i. \tag{8}$$

In fact, formula (2) is valid for any number of variables. If $\mathbf{r} = \langle x_1, \ldots, x_n\rangle$, then $d\mathbf{r} = \langle dx_1, \ldots, dx_n\rangle$ and $df = \nabla f \cdot d\mathbf{r}$.

■ **Example 15.3** If the length l and width w of a rectangular box are each increased by 3% and the height h is decreased by 2%, then approximately how much does the volume change?

Solution The volume is

$$V = lwh.$$

We set $dl = 0.03l$, $dw = 0.03w$, and $dh = -0.02h$. Then the change in volume is approximately

$$\begin{aligned} dV &= V_l\,dl + V_w\,dw + V_h\,dh \\ &= wh(0.03l) + lh(0.03w) + lw(-0.02h) \\ &= 0.04lwh \\ &= 0.04V. \end{aligned}$$

That is, the volume increases by about 4%. ■

■ **Example 15.4** A set of numbers $x_1, \ldots, x_n$ has **mean**

$$m = \frac{1}{n} \sum_{i=1}^{n} x_i$$

and **standard deviation**

$$s = \left[\frac{1}{n} \sum_{i=1}^{n} (x_i - m)^2\right]^{1/2}.$$

Approximate the effect on m and s of increasing each number x_i by an amount c_i.

Solution First, we calculate the effect on m. If each x_i is replaced by $x_i + c_i$, the new mean is

$$M = \frac{1}{n} \sum_{i=1}^{n} (x_i + c_i) = \frac{1}{n} \sum_{i=1}^{n} x_i + \frac{1}{n} \sum_{i=1}^{n} c_i.$$

The expression $(1/n) \sum_{i=1}^{n} c_i$ is the mean of the numbers $c_1, \ldots, c_n$, which we will denote by c. Hence,

$$M = m + c.$$

Now we will calculate the effect on s. We find the new standard deviation by replacing x_i by $x_i + c_i$ and m by $m + c$:

$$s = \left[\frac{1}{n} \sum_{i=1}^{n} (x_i + c_i - m - c)^2\right]^{1/2}.$$

This expression is not easily simplified by algebraic methods, so we will use differentials to approximate the change in s. Since

$$\begin{aligned} \frac{\partial s}{\partial x_i} &= \frac{1}{2}\left[\frac{1}{n} \sum_{i=1}^{n} (x_i - m)^2\right]^{-1/2} \frac{1}{n}\, 2(x_i - m) \\ &= \frac{1}{ns}(x_i - m), \end{aligned}$$

we let $dx_i = c_i$ and get

$$\begin{aligned} ds &= \sum_{i=1}^{n} \frac{\partial s}{\partial x_i}\, dx_i \\ &= \sum_{i=1}^{n} \frac{1}{ns}(x_i - m)c_i \\ &= \frac{1}{s}\left(\frac{1}{n}\sum_{i=1}^{n} x_i c_i - \frac{m}{n}\sum_{i=1}^{n} c_i\right) \\ &= \frac{1}{s}\left(\frac{1}{n}\sum_{i=1}^{n} x_i c_i - mc\right) \end{aligned}$$

This is the approximate change in s. This quantity is related to what statisticians call the **covariance** of the x_i's and the c_i's, which measures to some extent the correlation between them. ■

Exercises 15.1

Identify the values of x, y, dx, and dy to be used in estimating the value of each function at the given point, and then estimate the function by use of the total differential.

1. $f(x, y) = \sin(x^2 + y^2)$; $(0.02, -0.03)$
2. $f(x, y) = e^{2xy}$; $(1.05, 0.49)$
3. $f(x, y) = \sqrt{xy + y^2}$; $(-0.033, 2.025)$
4. $f(x, y) = \ln(x^2 - y^2)$; $(0.998, 0.098)$

Identify the values of x, y, z, dx, dy, and dz to be used in estimating the value of each function at the given point, and then estimate the function by use of the total differential.

5. $f(x, y, z) = \sqrt{x^2 + y^2 + z^2}$; $(2.04, 0.13, -0.05)$
6. $f(x, y, z) = \cos(\pi xyz)$; $(0.51, 1.99, 1.03)$
7. $f(x, y, z) = \dfrac{1}{z}\ln(x^2 + y^2)$; $(-0.02, 0.97, 0.98)$
8. $f(x, y, z) = \cos x \sin y \tan z$; $(0.12, 1.56, 0.75)$

Use the total differential to estimate the number given.

9. $e^{0.01}\sqrt{0.99}$
10. $\sqrt{50}\cos 1.5$
11. $(7.6)^{1/3}(16.1)^{-1/4}e^{-0.02}$
12. $(8.1)^{-1/3}\sqrt{(8.8)(15.9)}$
13. What is the approximate effect on the volume of a cylinder if the radius is increased by 3% and the height is decreased by 5%?
14. A machine turns out a metal cylinder whose length and diameter are within tolerances of 0.001 inch. Within what tolerances are the volume and the total surface area?
15. An ellipse has axes of lengths 6 and 8. If each axis is increased by 0.1, what is the approximate increase in the area of the ellipse? (If a and b are the semiaxes, the area is $A = \pi ab$.)
16. [A] If each side of a triangle is decreased by 2%, approximately how does the area change? [If the sides are a, b, and c, and s is the semiperimeter, i.e., $s = \frac{1}{2}(a + b + c)$, then the area is $A = \sqrt{s(s - a)(s - b)(s - c)}$.]
17. If the national work force has decreased by 0.5% but average take-home pay has increased by 3%, how has the nation's total take-home pay changed?

18. How are the mean and standard deviation of several numbers changed if each number is increased by the same constant c?
19. How are the mean and standard deviation of several numbers changed if each number is multiplied by the same constant c?
20. How are the mean and standard deviation of several numbers changed if each number is multiplied by a weighting factor w_i?
21. The **geometric mean** g of the three positive numbers x_1, x_2, and x_3 is the cube root of their product, $g = \sqrt[3]{x_1 x_2 x_3}$. If the three numbers are each decreased by 5%, approximately how is their geometric mean changed?
22. The geometric mean g of n positive numbers $x_1, \ldots, x_n$ is the nth root of their product, $g = \sqrt[n]{x_1 x_2 \cdot \cdots \cdot x_n}$. If each x_i is multiplied by a weighting factor w_i, approximately how will the geometric mean change?

SECTION 15.2

Exact Differential Forms

If f is a differentiable function of two variables, the total differential of f is

$$df = f_x(x, y)\, dx + f_y(x, y)\, dy.$$

More generally, if P and Q are any functions of two variables, the expression

$$P(x, y)\, dx + Q(x, y)\, dy$$

is called a **differential form**. A differential form is **exact** if and only if it is the total differential of some differentiable function.

In this section, we discuss two problems: how to recognize an exact differential form; and how to reconstruct a function from its total differential, or equivalently, from its gradient. These ideas will be useful in subsequent work.

Exact Differential Forms

The following theorem solves the first problem by telling us when a differential form is exact, and it will be useful in solving the second problem of reconstructing the function. Its proof involves details that are treated in advanced calculus, but we will sketch the main outline of the proof below.

THEOREM 15.1 If P and Q are continuously differentiable functions of two variables x and y, then the differential form $P\, dx + Q\, dy$ is exact if and only if $P_y = Q_x$.

Proof First, suppose that the differential form is exact. Then there is a function f such that $f_x = P$ and $f_y = Q$. Since P and Q are continuously differentiable, the second-order mixed partials of f are equal, so that $P_y = f_{xy} = f_{yx} = Q_x$.

Conversely, suppose that $P_y = Q_x$. Since P is differentiable, it is continuous. In particular, P is continuous with respect to x and therefore integrable with respect to x. Similarly, Q is integrable with respect to y. Hence, for any convenient point (a, b), we may construct the function

$$f(x, y) = \int_a^x P(t, b)\,dt + \int_b^y Q(x, t)\,dt. \tag{1}$$

Now, applying the fundamental theorem of calculus (Theorem 5.12) and a more advanced rule allowing us to differentiate under the second integral sign, we differentiate equation (1) with respect to x:

$$f_x(x, y) = P(x, b) + \int_b^y Q_x(x, t)\,dt. \tag{2}$$

Then we can rewrite equation (2) as

$$\begin{aligned} f_x(x, y) &= P(x, b) + \int_b^y P_y(x, t)\,dt \\ &= P(x, b) + P(x, t)\Big|_b^y \\ &= P(x, b) + P(x, y) - P(x, b) \\ &= P(x, y). \end{aligned}$$

If equation (1) is differentiated with respect to y, we get

$$f_y(x, y) = 0 + Q(x, y).$$

Thus, $P\,dx + Q\,dy = df$, and the differential form is exact.

Example 15.5 Determine whether these differential forms are exact:

(a) $(x^2 + 2xy)\,dx + 4y\,dy$ **(b)** $(2x + 3y^2)\,dx + (6xy + 1)\,dy$

Solution

(a) We let $P(x, y) = x^2 + 2xy$ and $Q(x, y) = 4y$. Since $P_y = 2x$ and $Q_x = 0$, the differential form is not exact by Theorem 15.1.

(b) This time we let $P(x, y) = 2x + 3y^2$ and $Q(x, y) = 6xy + 1$. Then P and Q are continuously differentiable, since they are polynomials, and $P_y = 6y = Q_x$. Hence, the differential form is exact. ■

In three variables, a differential form is an expression like

$$P(x, y, z)\,dx + Q(x, y, z)\,dy + R(x, y, z)\,dz.$$

The three-variable version of Theorem 15.1 is stated in Theorem 15.2.

THEOREM 15.2 If P, Q, and R are continuously differentiable functions of three variables, then the differential form $P(x, y, z)\,dx + Q(x, y, z)\,dy + R(x, y, z)\,dz$ is exact if and only if $P_y = Q_x$, $P_z = R_x$, and $Q_z = R_y$.

■ **Example 15.6** Determine whether the differential form $(2x + yz^2)\,dx + (1 + xz^2)\,dy + 2xyz\,dz$ is exact.

Solution We let $P(x, y, z) = 2x + yz^2$, $Q(x, y, z) = 1 + xz^2$, and $R(x, y, z) = 2xyz$. Then we find that

$$\begin{aligned} P_y &= z^2 = Q_x, \\ P_z &= 2yz = R_x, \\ Q_z &= 2xz = R_y. \end{aligned}$$

Therefore, the given differential form is exact. ■

Reconstructing a Function from Its Gradient

We now turn to the second problem mentioned earlier—that of reconstructing a function from its total differential or gradient. Since the total differential df and the gradient ∇f are closely related by the formula $df = \nabla f \cdot d\mathbf{r}$, we can obtain either one from the other. The reconstruction can therefore begin with either the total differential or the gradient.

In finding the function f, given its total differential df, equation (1) in the proof of Theorem 15.1 offers a solution. Consider some examples.

■ **Example 15.7** Find the functions f for which each expression is the total differential:

(a) $(2x + 3y^2)\,dx + (6xy + 1)\,dy$ **(b)** $\ln(1 + y)\,dx + \dfrac{x}{1 + y}\,dy$

Solution **(a)** We will form the function f using equation (1), with $P(x, y) = 2x + 3y^2$ and $Q(x, y) = 6xy + 1$. We already know from Example 15.5(b) that the given differential form is exact. We get

$$\begin{aligned} f(x, y) &= \int_a^x P(t, b)\,dt + \int_b^y Q(x, t)\,dt \\ &= \int_a^x (2t + 3b^2)\,dt + \int_b^y (6xt + 1)\,dt \\ &= \Big[t^2 + 3b^2t\Big]_a^x + \Big[3xt^2 + t\Big]_b^y \\ &= x^2 + 3b^2x - a^2 - 3ab^2 + 3xy^2 + y - 3xb^2 - b \\ &= x^2 + 3xy^2 + y - (a^2 + 3ab^2 + b). \end{aligned}$$

Since the point (a, b) is arbitrary, the expression $-(a^2 + 3ab^2 + b)$ is an arbitrary constant, which we will denote by C. Hence,

$$f(x, y) = x^2 + 3xy^2 + y + C$$

are the functions we seek. Each of these functions works, as you may easily check.

(b) This time we let $P(x, y) = \ln(1 + y)$ and $Q(x, y) = x/(1 + y)$. We check that $P_y = Q_x = 1/(1 + y)$, so the given differential form is indeed exact. Again using equation (1), we have

$$\begin{aligned} f(x, y) &= \int_a^x P(t, b)\, dt + \int_b^y Q(x, t)\, dt \\ &= \int_a^x \ln(1 + b)\, dt + \int_b^y \frac{x}{1 + t}\, dt \\ &= t \ln(1 + b)\Big|_a^x + x \ln(1 + t)\Big|_b^y \\ &= (x - a) \ln(1 + b) + x \ln(1 + y) - x \ln(1 + b) \\ &= x \ln(1 + y) + C. \end{aligned}$$

This solution works. ■

An alternative method of reconstructing the function is illustrated in the next example.

■ **Example 15.8** Find f, given $\nabla f = \langle 2 + ye^{xy}, xe^{xy} \rangle$.

Solution We check that

$$\frac{\partial}{\partial y}(2 + ye^{xy}) = e^{xy} + xye^{xy} = \frac{\partial}{\partial x}(xe^{xy}),$$

so the function f exists. Since $f_x = 2 + ye^{xy}$, we find the antiderivative with respect to x, getting

$$f(x, y) = 2x + e^{xy} + C(y). \tag{3}$$

Here, $C(y)$ is the "constant" of integration—it may involve y, but it is constant with respect to x. Since $f_y = xe^{xy}$, we find the antiderivative with respect to y, getting

$$f(x, y) = e^{xy} + D(x), \tag{4}$$

with the "constant" $D(x)$ perhaps involving x, but not y. Combining equations (3) and (4), we have

$$\begin{aligned} 2x + e^{xy} + C(y) &= e^{xy} + D(x), \\ C(y) &= D(x) - 2x. \end{aligned} \tag{5}$$

Now the left-hand side of equation (5) does not involve x, and the right-hand side does not involve y. This means that neither side involves either x or y, so each is a constant. That is,

$$C(y) = K = D(x) - 2x, \qquad K \text{ a constant.}$$

Putting either $C(y) = K$ into equation (3) or $D(x) = K + 2x$ into equation (4), we get

$$f(x, y) = e^{xy} + 2x + K.$$

This is the solution to the problem. ■

For functions of three variables, formula (1) becomes

$$f(x, y, z) = \int_a^x P(t, b, c)\,dt + \int_b^y Q(x, t, c)\,dt + \int_c^z R(x, y, t)\,dt. \tag{6}$$

This enables us to reconstruct f from its gradient $\nabla f = \langle P, Q, R\rangle$.

■ **Example 15.9**

Find the functions f, given

$$df = (2x + yz^2)\,dx + (1 + xz^2)\,dy + 2xyz\,dz.$$

Solution

By Example 15.6, we know the function f exists. Using equation (6), we have

$$\begin{aligned} f(x, y, z) &= \int_a^x (2t + bc^2)\,dt + \int_b^y (1 + xc^2)\,dt + \int_c^z 2xyt\,dt \\ &= \Big[t^2 + bc^2 t\Big]_a^x + \Big[(1 + xc^2)t\Big]_b^y + \Big[xyt^2\Big]_c^z \\ &= x^2 + bc^2x - a^2 - bc^2a + (1 + xc^2)(y - b) + xyz^2 - xyc^2 \\ &= x^2 + y + xyz^2 + (-a^2 - b - abc^2) \\ &= x^2 + y + xyz^2 + C. \end{aligned}$$

These are clearly solutions to the problem. ■

A function of three variables also can be constructed from its gradient by the method of Example 15.8, but the use of formula (6) is much simpler.

Exercises 15.2

Determine which differential forms are exact, and construct the corresponding functions, where possible.

1. $x\,dx + y\,dy$
2. $xy\,dx - x^2\,dy$
3. $2xy\,dx + x^2\,dy$
4. $\dfrac{1}{y}\,dx - \dfrac{x}{y^2}\,dy$
5. $e^{x+2y}\,dx + e^{x+2y}\,dy$
6. $-\dfrac{1}{x^2}\,dx - \dfrac{1}{y^2}\,dy$

7. $\ln y\, dx + \dfrac{x + 2y^2}{y}\, dy$

8. $\dfrac{dx}{x + 3y} + \dfrac{dy}{x + 3y}$

9. $x\, dx + y\, dy + z\, dz$

10. $yz\, dx + xz\, dy + xy\, dz$

11. $dx + e^y\, dy - \sin z\, dz$

12. $2xe^y \sin z\, dx + x^2 e^y \sin z\, dy + x^2 e^y \cos z\, dz$

Find the function f, given its gradient.

13. $\nabla f = \langle 2x, -2y \rangle$

14. $\nabla f = \langle e^x \sin y, e^x \cos y \rangle$

15. $\nabla f = \langle 3x^2 + 2xy - y^2, x^2 - 2xy - 3y^2 \rangle$

16. $\nabla f = \left\langle \dfrac{2y}{2x + z}, \ln(2x + z), \dfrac{y}{2x + z} \right\rangle$

17. $\nabla f = \langle 2x - 3y - 7z, -3x + 10y + 11z, -7x + 11y - 26z \rangle$

18. $\nabla f = \langle 2xy + yz + z^2, x^2 + xz, xy + 2xz \rangle$

SECTION 15.3

Path Integrals

Just as area was used in Chapter 5 to introduce the definite integral, we will use the concept of work to introduce the idea of the path integral.

Work

If **F** is a constant force that acts along a straight line and moves an object a distance x along the line (see Figure 15.2), then the **work** done is defined to be

$$W = \|\mathbf{F}\| x.$$

Thus, if **F** is a constant force and moves an object from point A to point B along a straight line, the work done is

$$W = \mathbf{F} \cdot \mathbf{v}(\overrightarrow{AB}).$$

If **F** does not act along the line, we multiply x not by $\|\mathbf{F}\|$ but by the component of **F** along the line (see Figure 15.3).

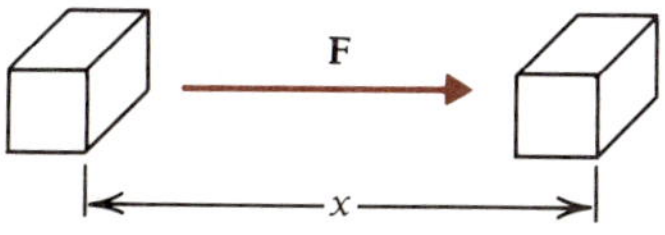

FIGURE 15.2 $W = \|\mathbf{F}\| x$

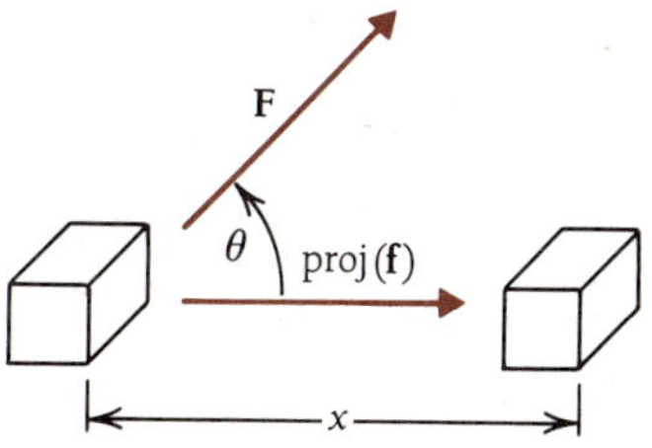

FIGURE 15.3 $W = \|\mathbf{F}\|(\cos \theta) x$

Vector Fields

Now, suppose the object does not move along a straight line, but along a curve C whose equation is

$$\mathbf{r}(t) = \langle x(t), y(t), z(t) \rangle, \qquad t \in [a, b].$$

We will suppose C is a smooth curve, meaning that $\mathbf{r}$ is differentiable. We will also suppose the force $\mathbf{F}$ is not necessarily constant, but varies from place to place. That is, we will let $\mathbf{F}$ be a vector-valued function of three variables,

$$\mathbf{F}(x, y, z) = \langle P(x, y, z), Q(x, y, z), R(x, y, z) \rangle.$$

Such a function is called a **vector field**. A vector field interpreted as a force is called a **force field** (physical examples are gravitational fields and magnetic fields). We will assume that $\mathbf{F}$ is continuous, which by definition means that each component of $\mathbf{F}$ is a continuous function.

On the curve C, the force field $\mathbf{F}$ becomes

$$\mathbf{F}(\mathbf{r}(t)) = \langle P(\mathbf{r}(t)), Q(\mathbf{r}(t)), R(\mathbf{r}(t)) \rangle.$$

We will use an integral to represent the work done by $\mathbf{F}$ in moving an object along C. Let $s = \{a = t_0, t_1, t_2, \ldots, t_n = b\}$ be a subdivision of $[a, b]$. If A_i is the point $(x(t_i), y(t_i), z(t_i))$, then

$$\mathbf{v}(\overrightarrow{A_{i-1}A_i}) = \mathbf{r}(t_i) - \mathbf{r}(t_{i-1}).$$

If $t_i^* \in [t_{i-1}, t_i]$, then $\mathbf{F}(\mathbf{r}(t))$ over $[t_{i-1}, t_i]$ is approximately equal to $\mathbf{F}(\mathbf{r}(t_i^*))$. Therefore, the work done in moving the object from A_{i-1} to A_i along C is approximately (see Figure 15.4)

$$\begin{aligned}\Delta W_i &\approx \mathbf{F}(\mathbf{r}(t_i^*)) \cdot \mathbf{v}(\overrightarrow{A_{i-1}A_i}) \\ &= \mathbf{F}(\mathbf{r}(t_i^*)) \cdot [\mathbf{r}(t_i) - \mathbf{r}(t_{i-1})].\end{aligned}$$

Now,

$$\mathbf{r}(t_i) - \mathbf{r}(t_{i-1}) = \langle x(t_i) - x(t_{i-1}), y(t_i) - y(t_{i-1}), z(t_i) - z(t_{i-1}) \rangle.$$

Since x, y, and z are differentiable, we may apply the mean value theorem to each of them on $[t_{i-1}, t_i]$. That is, there exist $\xi_i, \psi_i, \zeta_i \in [t_{i-1}, t_i]$ such that

$$\begin{aligned}x(t_i) - x(t_{i-1}) &= x'(\xi_i)(t_i - t_{i-1}), \\ y(t_i) - y(t_{i-1}) &= y'(\psi_i)(t_i - t_{i-1}), \\ z(t_i) - z(t_{i-1}) &= z'(\zeta_i)(t_i - t_{i-1}).\end{aligned}$$

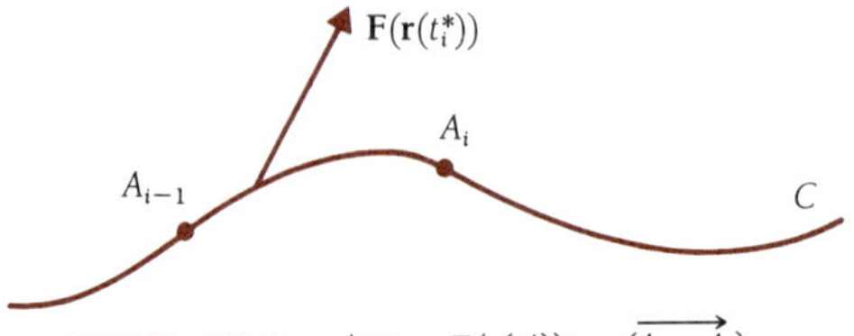

FIGURE 15.4 $\Delta W_i \approx \mathbf{F}(\mathbf{r}(t_i^*)) \cdot \mathbf{v}(\overrightarrow{A_{i-1}A_i})$

Hence,

$$\mathbf{r}(t_i) - \mathbf{r}(t_{i-1}) = (t_i - t_{i-1})\langle x'(\xi_i), y'(\psi_i), z'(\zeta_i)\rangle.$$

Therefore, letting $t_i - t_{i-1} = \Delta t_i$, we have

$$\begin{aligned}\Delta W_i &\approx \mathbf{F}(\mathbf{r}(t_i^*)) \cdot [\mathbf{r}(t_i) - \mathbf{r}(t_{i-1})]\\ &= \langle P(\mathbf{r}(t_i^*)), Q(\mathbf{r}(t_i^*)), R(\mathbf{r}(t_i^*))\rangle \cdot \langle x'(\xi_i), y'(\psi_i), z'(\zeta_i)\rangle \Delta t_i\\ &= [P(\mathbf{r}(t_i^*))x'(\xi_i) + Q(\mathbf{r}(t_i^*))y'(\psi_i) + R(\mathbf{r}(t_i^*))z'(\zeta_i)]\Delta t_i.\end{aligned}$$

Summing this last expression over i, we get a Riemann sum whose limit is the definite integral

$$W = \int_a^b [P(\mathbf{r}(t))x'(t) + Q(\mathbf{r}(t))y'(t) + R(\mathbf{r}(t))z'(t)]\, dt. \tag{1}$$

This is the work done by $\mathbf{F}$ in moving an object along C from $t = a$ to $t = b$.

We can write equation (1) in the briefer form

$$W = \int_a^b \mathbf{F}(\mathbf{r}(t)) \cdot \mathbf{r}'(t)\, dt, \tag{2}$$

or, using $d\mathbf{r} = \mathbf{r}'(t)\, dt$, in the form

$$W = \int_C \mathbf{F} \cdot d\mathbf{r}. \tag{3}$$

Here, $\int_C$ means the integral along the curve C whose equation is $\mathbf{r}$, from the beginning point of C to the endpoint of C (as determined by the parametrization). We also can write equation (3) in the form

$$W = \int_C [P(x, y, z)\, dx + Q(x, y, z)\, dy + R(x, y, z)\, dz], \tag{4}$$

which is usually abbreviated to

$$W = \int_C (P\, dx + Q\, dy + R\, dz). \tag{5}$$

Example 15.10 Find the work done by the force field $\mathbf{F}(x, y, z) = \langle x + y, x - y, z^2\rangle$ in moving an object along the curve C: $\mathbf{r}(t) = \langle e^t, e^{-t}, t\rangle$ from $t = 0$ to $t = 1$.

Solution We can use formula (1) directly:

$$\begin{aligned}W &= \int_0^1 [(e^t + e^{-t})e^t - (e^t - e^{-t})e^{-t} + t^2]\, dt\\ &= \int_0^1 (e^{2t} + e^{-2t} + t^2)\, dt\end{aligned}$$

$$W = \left[\frac{1}{2}e^{2t} - \frac{1}{2}e^{-2t} + \frac{1}{3}t^3\right]_0^1$$
$$= \frac{1}{2}(e^2 - e^{-2}) + \frac{1}{3}.$$

Path Integrals

In general, the integral of a differential form over a curve,

$$\int_C (P\,dx + Q\,dy + R\,dz),$$

is called a **path integral** or a **line integral**. The curve C is called the **path of integration**, and the differential form $P\,dx + Q\,dy + R\,dz$ is called the **integrand**. We see by equation (5) that work is given as a path integral.

To evaluate a path integral, we find a suitable parametrization $\mathbf{r}(t)$ of the curve C, and rewrite the path integral as a definite integral of the form (1). Let's consider some examples.

Example 15.11 Evaluate $\int_C (x^2\,dx - y^2\,dy + z\,dz)$, with C the path $\mathbf{r}(t) = \langle \sin t, \cos t, t\rangle$, $t \in [0, \pi]$ (Figure 15.5).

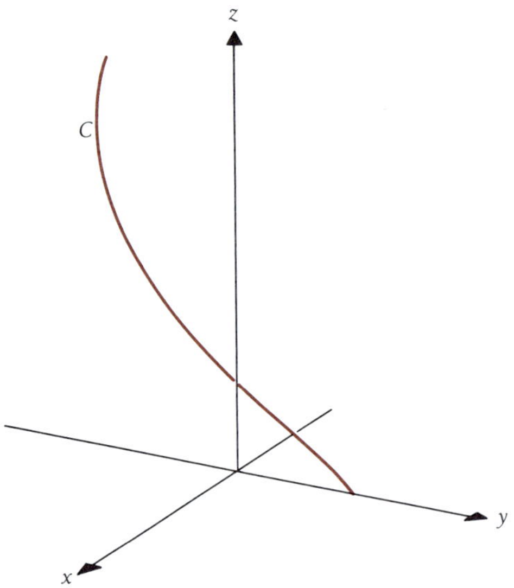

FIGURE 15.5 The path $\mathbf{r}(t) = \langle \sin t, \cos t, t\rangle$, $0 \le t \le \pi$

Solution The parametrization is already given, so we may use equation (1) directly. We get

$$\int_C (x^2\,dx - y^2\,dy + z\,dz) = \int_0^\pi (\sin^2 t\cos t + \cos^2 t\sin t + t)\,dt$$
$$= \left[\frac{1}{3}\sin^3 t - \frac{1}{3}\cos^3 t + \frac{1}{2}t^2\right]_0^\pi$$
$$= \frac{2}{3} + \frac{\pi^2}{2}.$$ ■

■ **Example 15.12** Evaluate $\int_C [xy\,dx + z\,dy + (x-y)\,dz]$, for C the portion of the parabola $y = x^2$ in the xy-plane from the origin to the point (1, 1, 0) (Figure 15.6).

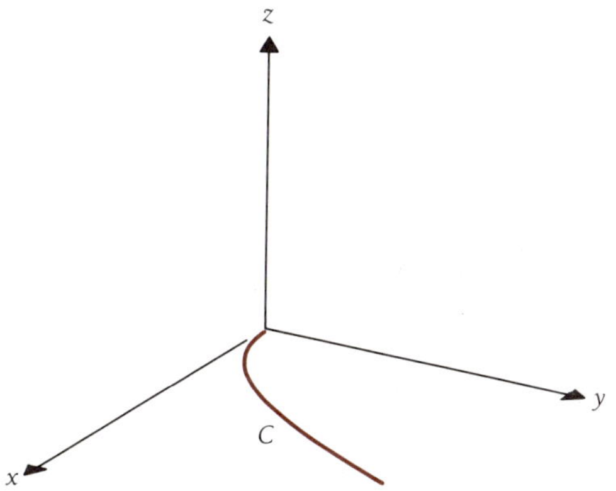

FIGURE 15.6 The path $\mathbf{r}(t) = \langle t, t^2, 0\rangle,\ 0 \le t \le 1$

Solution Our first task is to find a parametrization for C (see Section 11.1). If we let $x = t$, $y = t^2$, $z = 0$, then $\mathbf{r}(t) = \langle x, y, z\rangle$ traces out C as t varies from 0 to 1. Then equation (1) gives us

$$\int_C [xy\,dx + z\,dy + (x-y)\,dz] = \int_0^1 (t^3 + 0 + 0)\,dt = \frac{1}{4}.$$ ■

The two-dimensional version of a path integral is the same as the three-dimensional case, but with $\mathbf{F}(x, y) = \langle P(x, y), Q(x, y)\rangle$ and $\mathbf{r}(t) = \langle x(t), y(t)\rangle$. Thus, all that we have done also applies to the two-variable case.

■ **Example 15.13** Find the work done by the force $\mathbf{F}(x, y) = \langle 2xy, 3y^2\rangle$ in moving an object along the unit circle from (1, 0) to (0, 1) in the counterclockwise direction.

Solution The path of integration (Figure 15.7) can be parametrized by $\mathbf{r}(t) = \langle \cos t, \sin t \rangle$, $0 \le t \le \pi/2$. Hence, by formula (1) with $R = 0$, the work done is

$$W = \int_0^{\pi/2} [2 \cos t \sin t\,(-\sin t) + 3 \sin^2 t \cos t]\, dt$$
$$= \int_0^{\pi/2} \sin^2 t \cos t\, dt$$
$$= \frac{1}{3} \sin^3 t \Big|_0^{\pi/2} = \frac{1}{3}.$$

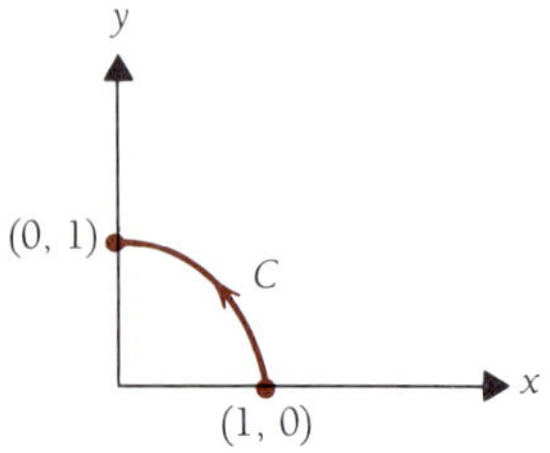

FIGURE 15.7 ■

If C_1 and C_2 are smooth curves such that the endpoint of C_1 is the beginning point of C_2, we denote by $C_1 + C_2$ the union of the two curves. The resulting curve may not be smooth, but it is called **piecewise smooth**. In general, a curve C is piecewise smooth if it can be written as a sum of a number of curves, $C = C_1 + \cdots + C_n$, with each C_i smooth.

Path integrals over piecewise smooth curves are defined as sums of path integrals over the smooth pieces:

$$\int_{C_1+C_2+\cdots+C_n} = \int_{C_1} + \int_{C_2} + \cdots + \int_{C_n}.$$

■ **Example 15.14** Evaluate $\int_C (x^2\, dx + 2xy\, dy)$ for C the triangle from $(1, 0)$ to $(0, 2)$ to $(3, 1)$ to $(1, 0)$ (see Figure 15.8).

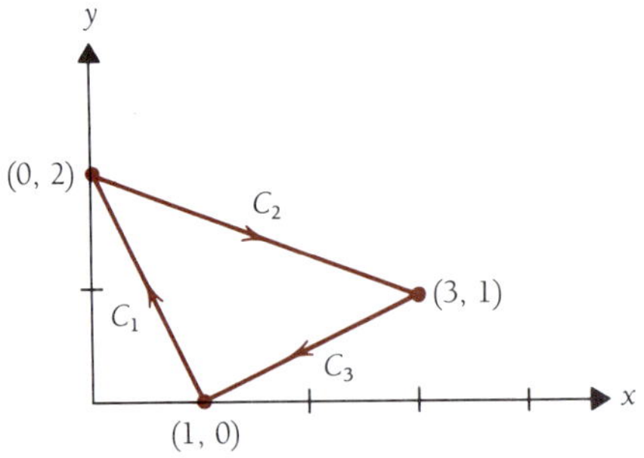

FIGURE 15.8

Solution Let $C = C_1 + C_2 + C_3$, as in Figure 15.8. We find parametrizations for these three line segments (parametric equations for the lines will do) as follows:

$$\begin{aligned} C_1&: \mathbf{r}_1(t) = \langle 1-t, 2t\rangle, && 0 \le t \le 1,\\ C_2&: \mathbf{r}_2(t) = \langle 3t, 2-t\rangle, && 0 \le t \le 1,\\ C_3&: \mathbf{r}_3(t) = \langle 3-2t, 1-t\rangle, && 0 \le t \le 1. \end{aligned}$$

Then

$$\begin{aligned} \int_C (x^2\,dx + 2xy\,dy) &= \int_0^1 [(1-t)^2(-1) + 2(1-t)2t(2)]\,dt \\ &\quad + \int_0^1 [(3t)^2(3) + 2(3t)(2-t)(-1)]\,dt \\ &\quad + \int_0^1 [(3-2t)^2(-2) + 2(3-2t)(1-t)(-1)]\,dt \\ &= \int_0^1 (12t^2 + 32t - 25)\,dt = -5. \end{aligned}$$

■

Since a path integral is equal to a definite integral, the properties of definite integrals extend to path integrals. For instance,

$$\begin{aligned} \int_C (P_1\,dx + Q_1\,dy + R_1\,dz) &+ \int_C (P_2\,dx + Q_2\,dy + R_2\,dz) \\ &= \int_C [(P_1 + P_2)\,dx + (Q_1 + Q_2)\,dy + (R_1 + R_2)\,dz] \end{aligned}$$

and

$$\int_C (aP\,dx + aQ\,dy + aR\,dz) = a\int_C (P\,dx + Q\,dy + R\,dz).$$

If $-C$ denotes the same curve as C, but traversed in the reverse direction, then

$$\int_{-C} (P\,dx + Q\,dy + R\,dz) = -\int_C (P\,dx + Q\,dy + R\,dz).$$

It is shown in advanced calculus that the value of a path integral does not depend on the parametrization used for the path C of integration.

Exercises 15.3

Find the work done by the force **F** *in moving an object along the curve C.*

1. $\mathbf{F}(x, y) = \langle 3x, -2y\rangle$; C: $\mathbf{r}(t) = \langle t, 2t+3\rangle$, $t \in [0, 2]$
2. $\mathbf{F}(x, y) = \langle xy, x+y\rangle$; C: $\mathbf{r}(t) = \langle \sin t, \cos t\rangle$, $t \in [0, \pi]$
3. $\mathbf{F}(x, y, z) = \langle yz, xz, xy\rangle$; C: $\mathbf{r}(t) = \langle t, t, -t\rangle$, $0 \le t \le 1$
4. $\mathbf{F}(x, y, z) = \langle x^2, y, -z\rangle$; C: $\mathbf{r}(t) = \langle 1-t, t, 2t\rangle$, $t \in [0, 1]$
5. $\mathbf{F}(x, y, z) = \langle z, 2x, y\rangle$; C: $x = 0$, $y = z^2$ from $(0, 0, 0)$ to $(0, 4, 2)$
6. $\mathbf{F}(x, y, z) = \langle xy, xz, yz\rangle$; C: one counterclockwise revolution of the circle $x^2 + z^2 = 9$, $y = 2$, starting at $(3, 2, 0)$

Evaluate $\int_C (y^2\,dx - 2xy\,dy)$ *for each path C.*

7. C is one clockwise circuit of the unit circle, starting at $(1, 0)$.
8. C is counterclockwise around the unit circle from $(1, 0)$ to $(-1, 0)$.
9. C is the line segment from $(1, 0)$ to $(-1, 0)$.
10. C is the polygonal line from $(1, 0)$ to $(0, 2)$ to $(-1, 0)$.
11. C is the square from $(1, 0)$ to $(0, 1)$ to $(-1, 0)$ to $(0, -1)$ to $(1, 0)$.
12. C is the line segment from $(1, 0)$ to $(2, 0)$, then the semicircle $y = \sqrt{4 - x^2}$ from $(2, 0)$ to $(-2, 0)$, and then the line segment from $(-2, 0)$ to $(-1, 0)$.

Evaluate $\int_C (z^2\,dx + 2\,dy + 2xz\,dz)$ *for each path C.*

13. C is the line segment from $(0, 0, 0)$ to $(1, 0, 1)$.
14. C is the curve $z = x^2$, $y = 0$ from $(0, 0, 0)$ to $(1, 0, 1)$.
15. C is the polygonal line from $(0, 0, 0)$ to $(0, 1, 0)$ to $(0, 1, 1)$ to $(1, 0, 1)$.
16. C is the triangle from $(0, 1, 0)$ to $(0, 0, 1)$ to $(1, 0, 0)$ to $(0, 1, 0)$.
17. Recalling from Section 13.1 that $ds = \|\mathbf{r}'(t)\|\,dt$, evaluate the following:
   (a) $\int_C xy\,ds$; C: $\mathbf{r}(t) = \langle \sin t, \cos t \rangle$, $t \in [0, \pi/2]$
   (b) $\int_C (x^2 + yz)\,ds$; C: $\mathbf{r}(t) = \langle t, -t, 2t \rangle$, $t \in [0, 3]$
18. Refer to Problem 17. If a wire in space has density $\delta(x, y, z)$, show that the mass of the wire is $m = \int_C \delta(x, y, z)\,ds$, where C is the curve along which the wire lies.
19. For the wire in Problem 18, show that the moments in the x, y, and z directions are $M_x = \int_C x\delta(x, y, z)\,ds$, $M_y = \int_C y\delta(x, y, z)\,ds$, and $M_z = \int_C z\delta(x, y, z)\,ds$.
20. Find the center of mass of the helical wire of constant density lying on C: $\mathbf{r}(t) = \langle \cos t, \sin t, t \rangle$, $t \in [0, 2\pi]$.
21. If $\mathbf{T}(t)$ is the unit tangent vector to curve C: $\mathbf{r}(t) = \langle x(t), y(t), z(t) \rangle$, show that the work done by force $\mathbf{F}$ in moving an object along C is

$$W = \int_C \mathbf{F} \cdot \mathbf{T}\,ds.$$

22. Let curve C_1 be parametrized by

$$\mathbf{r}_1(t) = \langle t, t, t \rangle, \qquad t \in [0, 1]$$

and let curve C_2 be parametrized by

$$\mathbf{r}_2(t) = \langle \sin t, \sin t, \sin t \rangle, \qquad t \in [0, \pi/2].$$

Evaluate

$$\int_{C_1} (xy\,dx + z\,dy + y^2\,dz) \qquad \text{and} \qquad \int_{C_2} (xy\,dx + z\,dy + y^2\,dz).$$

Are the results the same? Why or why not?

23. A person weighing 130 pounds carries a bucket of water weighing 40 pounds up a spiral staircase. The path of the person is the helix

$$\mathbf{r}(t) = \langle 4\cos t, 4\sin t, 8t/\pi \rangle, \qquad t \in [0, 4\pi].$$

Find the amount of work done by gravity on the person and the bucket of water.

24. In Problem 23, suppose that the bucket leaks at a steady rate, losing half of its contents during the person's ascent. Find the amount of work done in this case.

SECTION 15.4

Path-Independent Path Integrals

The path integral $\int_C (P\,dx + Q\,dy + R\,dz)$ has some special properties in case the differential form $P\,dx + Q\,dy + R\,dz$ is exact. The basic theorem is that the value of the path integral depends only on the endpoints of the curve C. This result is sometimes called the **fundamental theorem for path integrals**, because of its similarity to the fundamental theorem of calculus (Theorem 5.12).

THEOREM 15.3

Fundamental Theorem for Path Integrals If $P\,dx + Q\,dy + R\,dz$ is an exact differential form, f is a function such that $df = P\,dx + Q\,dy + R\,dz$, and C is any piecewise smooth curve from point (x_1, y_1, z_1) to point (x_2, y_2, z_2), then

$$\int_C (P\,dx + Q\,dy + R\,dz) = f(x_2, y_2, z_2) - f(x_1, y_1, z_1).$$

Proof Let $C = C_1 + \cdots + C_n$, with each C_i smooth. Suppose C_i runs from point A_{i-1} to point A_i. Let C_i be parametrized by a function $\mathbf{r}_i(t)$, $a_i \le t \le b_i$, with $A_{i-1} = \mathbf{r}_i(a_i)$ and $A_i = \mathbf{r}_i(b_i)$. Since $df = P\,dx + Q\,dy + R\,dz$, we have

$$\frac{df}{dt} = P\frac{dx}{dt} + Q\frac{dy}{dt} + R\frac{dz}{dt}.$$

Hence,

$$\begin{aligned}\int_{C_i} (P\,dx + Q\,dy + R\,dz) &= \int_{a_i}^{b_i} \left[P(\mathbf{r}_i(t))\frac{dx}{dt} + Q(\mathbf{r}_i(t))\frac{dy}{dt} + R(\mathbf{r}_i(t))\frac{dz}{dt}\right] dt \\ &= \int_{a_i}^{b_i} \frac{d}{dt} f(\mathbf{r}_i(t))\,dt \\ &= f(\mathbf{r}_i(t))\Big|_{a_i}^{b_i} \\ &= f(\mathbf{r}_i(b_i)) - f(\mathbf{r}_i(a_i)) \\ &= f(A_i) - f(A_{i-1}).\end{aligned}$$

Therefore,

$$\begin{aligned}\int_C (P\,dx + Q\,dy + R\,dz) &= \sum_{i=1}^{n} \int_{C_i} (P\,dx + Q\,dy + R\,dz) \\ &= \sum_{i=1}^{n} [f(A_i) - f(A_{i-1})] \\ &= f(A_n) - f(A_0).\end{aligned}$$

But by the way the parametrizations are set up, $A_n = (x_2, y_2, z_2)$ and $A_0 = (x_1, y_1, z_1)$. The theorem is thus proved.

Path-Independence

A path integral whose value depends only on the endpoints of the path is called **path-independent.** Thus, Theorem 15.3 says that *a path integral of an exact differential is path-independent.* As a consequence, the path integral

$$\int_C \nabla f \cdot d\mathbf{r}$$

is path-independent, because the differential form

$$\nabla f \cdot d\mathbf{r} = f_x\, dx + f_y\, dy + f_z\, dz$$

is exact.

Path-independent path integrals have special physical significance when interpreted in terms of work. A force field is called **conservative** in case the work done in moving an object depends only on the initial and final positions of the object, and not on the path of motion. Thus, if the vector field **F** is a **gradient field,**

$$\mathbf{F} = \nabla f, \tag{1}$$

then **F** is a conservative force field. The function f in equation (1) is called the **potential function** of **F**.

Example 15.15 Evaluate $\int_C [(1 + ye^x)\, dx + e^x\, dy]$, with C the curve $y = 4 - x^2$ from $(-2, 0)$ to $(0, 4)$.

Solution Let $P = 1 + ye^x$ and $Q = e^x$. Then $P_y = e^x = Q_x$, so the differential form is exact and the integral is path-independent. It is not difficult to see that the integrand is the total differential of $f(x, y) = x + ye^x$, so

$$\begin{aligned}\int_C [(1 + ye^x)\, dx + e^x\, dy] &= f(0, 4) - f(-2, 0) \\ &= 4 - (-2) = 6.\end{aligned}$$

■

Example 15.16 Show that a constant force field is conservative.

Solution Let $\mathbf{F} = \langle a, b, c \rangle$ be a constant force field. Then the potential of **F** is

$$f(x, y, z) = ax + by + cz,$$

because **F** is the gradient of f. Thus,

$$W = \int_C \mathbf{F} \cdot d\mathbf{r} = \int_C \nabla f \cdot d\mathbf{r}$$

is path-independent, and **F** is conservative. ■

Example 15.17 Over small distances, the earth's gravitational field is approximately constant, namely

$$\mathbf{F} = \langle 0, 0, -g \rangle.$$

Find the work done by gravity on an object that falls a distance h.

Solution If the initial position of the object is (x_1, y_1, z_1), then the final position is $(x_1, y_1, z_1 - h)$. A potential function of $\mathbf{F}$ is $f(x, y, z) = -gz$. Hence, the work done is

$$\begin{aligned} W &= \int_C \nabla f \cdot d\mathbf{r} \\ &= f(x_1, y_1, z_1 - h) - f(x_1, y_1, z_1) \\ &= -g(z_1 - h) + gz_1 = gh. \end{aligned}$$

Closed-Path Integrals

A path that starts and ends at the same point is called a **closed path**. A path integral over a closed path is called a **closed-path integral**, and the special integral symbol $\oint$ is sometimes used to denote it.

A closed-path integral of an exact differential form must be 0, by Theorem 15.3. That is,

$$\oint_C \nabla f \cdot d\mathbf{r} = 0.$$

Occasionally, this fact can be used to simplify a problem.

Example 15.18 Evaluate $\int_C (e^x \cos 2y\, dx - 2e^x \sin 2y\, dy)$, with C one clockwise circuit of the unit circle, starting at $(1, 0)$.

Solution Let $P = e^x \cos 2y$ and $Q = -2e^x \sin 2y$. Then $P_y = -2e^x \sin 2y = Q_x$, so the integrand is exact and the closed-path integral is 0.

Example 15.19 Evaluate $\int_C [e^x \cos 2y\, dx + (x - 2e^x \sin 2y)\, dy]$, with C as in Example 15.18.

Solution We will rewrite the original integral, taking a cue from Example 15.18, as

$$\int_C [e^x \cos 2y\, dx + (x - 2e^x \sin 2y)\, dy] = \int_C (e^x \cos 2y\, dx - 2e^x \sin 2y\, dy) + \int_C x\, dy.$$

The first integral is 0 by Example 15.18, so we need only evaluate the last integral. Let C be parametrized by $\mathbf{r}(t) = \langle \cos t, -\sin t \rangle$, $t \in [0, 2\pi]$. Then $x = \cos t$, $dy = -\cos t\, dt$, and

$$\int_C x\, dy = -\int_0^{2\pi} \cos^2 t\, dt$$
$$= -\frac{1}{2}\left(t + \frac{1}{2}\sin 2t\right)\Bigg|_0^{2\pi} = -\pi.$$

You can see that it would be next to impossible to evaluate the original integral without splitting off the exact part. ■

Exercises 15.4

Problems 1–30 refer to the following paths:

C_1: *The curve* $y = \sqrt{1 - x^2}$ *from* $(1, 0)$ *to* $(0, 1)$
C_2: *One counterclockwise circuit of the unit circle, beginning at* $(1, 0)$
C_3: *The polygonal line from* $(1, 0)$ *to* $(0, 0)$ *to* $(0, 1)$
C_4: *The line segment from* $(0, 1)$ *to* $(1, 0)$
C_5: *The polygonal line from* $(0, 0, 0)$ *to* $(1, 1, 1)$ *to* $(0, 0, 1)$
C_6: *The line segment from* $(0, 0, 1)$ *to* $(0, 0, 0)$
C_7: *The helix* $x = \cos t$, $y = \sin t$, $z = t$ *from* $(1, 0, 0)$ *to* $(1, 0, 2\pi)$; *then the line segment from* $(1, 0, 2\pi)$ *to* $(1, 0, 0)$

Evaluate the path integrals.

1. $\int_{C_1} (y\, dx + x\, dy)$
2. $\int_{C_3} (y\, dx + x\, dy)$
3. $\int_{-C_4} (y\, dx + x\, dy)$
4. $\int_{C_1 + C_4} (y\, dx + x\, dy)$
5. $\int_{C_4 + C_1} (y\, dx + x\, dy)$
6. $\int_{C_3 + C_4} (y\, dx + x\, dy)$
7. $\int_{C_1} (2x\, dx - 2y\, dy)$
8. $\int_{C_3} (2x\, dx - 2y\, dy)$
9. $\int_{C_1} [(2x - 2y - 3)\, dx + (-2x + 2y + 5)\, dy]$
10. $\int_{C_3} [(2x - 2y - 3)\, dx + (-2x + 2y + 5)\, dy]$
11. $\int_{C_4} [(3x^2 + 2xy - y^2)\, dx + (x^2 - 2xy - 3y^2)\, dy]$
12. $\int_{C_3} [(3x^2 + 2xy - y^2)\, dx + (x^2 - 2xy - 3y^2)\, dy]$
13. $\int_{C_3 + C_4} (y^2\, dx - xy\, dy)$
14. $\int_{C_1 - C_3} (4y\, dx - x\, dy)$
15. $\int_{C_5} (x\, dx + y\, dy + z\, dz)$
16. $\int_{C_5} (y\, dx + x\, dy + dz)$

17. $\int_{C_5+C_6} (x\,dx + 2y\,dy - z\,dz)$

18. $\int_{C_6} (y\,dx + x\,dy + dz)$

19. $\int_{C_7} (\cos x\,dx + \cos y\,dy - \sin z\,dz)$

20. $\int_{C_7} (x\,dy + z\,dz)$

21. $\int_{C_5} [(2x - 3y - 7z)\,dx + (10y - 3x + 11z)\,dy + (11y - 7x - 26z)\,dz]$

22. $\int_{C_6} [(2x - 3y - 7z)\,dx + (10y - 3x + 11z)\,dy + (11y - 7x - 26z)\,dz]$

Find the work done by each force **F** *in moving an object along the given path.*

23. $\mathbf{F}(x, y) = \langle -y, -x \rangle;\quad C_1$

24. $\mathbf{F}(x, y) = \langle x^2, y^2 \rangle;\quad C_1$

25. $\mathbf{F}(x, y) = \langle \cos x, y^2 \rangle;\quad C_2$

A26. $\mathbf{F}(x, y) = \langle y \cos x, y \sin x \rangle;\quad C_2$

27. $\mathbf{F}(x, y, z) = \langle 1, 0, -z \rangle;\quad C_6$

28. $\mathbf{F}(x, y, z) = \langle 1, 0, -z \rangle;\quad C_7$

29. $\mathbf{F}(x, y, z) = \langle y, 0, -z \rangle;\quad C_5$

30. $\mathbf{F}(x, y, z) = \langle y, 0, -z \rangle;\quad C_7$

Determine whether each force field is conservative, and if so, find its potential.

31. $\mathbf{F}(x, y) = \langle -y, -x \rangle$

32. $\mathbf{F}(x, y) = \langle y, -x \rangle$

33. $\mathbf{F}(x, y, z) = \langle 0, 0, 1 \rangle$

34. $\mathbf{F}(x, y, z) = \langle 1, 0, 1/z \rangle$

35. $\mathbf{F}(x, y, z) = \langle 1, 1, -4z \rangle$

36. $\mathbf{F}(x, y, z) = \langle 2x, z, -1 \rangle$

37. An electrical charge concentrated at a point exerts an attractive force on oppositely charged particles with magnitude k/r^2, where k is a constant and r is the distance to the particle. Find an expression for the force field of the charge. Is it conservative?

38. A force field $\mathbf{F}(x, y, z)$ at each point is directed away from the origin and has magnitude inversely proportional to the distance from the origin. Show that **F** is conservative, and find its potential.

SECTION 15.5

Chapter Review

Important Concepts

Total differential	Force field	Path-independent
Mean	Path integral	Conservative force field
Standard deviation	Line integral	Gradient field
Differential form	Path of integration	Potential function
Exact differential form	Piecewise smooth curve	Closed path
Work	Fundamental theorem for path integrals	Closed-path integral
Vector field		

Review Exercises

1. Use differentials to approximate:
   (a) $\sqrt{27\sqrt[3]{1010}}$ **(b)** $(25)^{1/3}(99)^{1/2}(1.01)^{-1}$

2. If the height and base radius of a cone are both increased by 5%, estimate the change in:
   (a) Volume **(b)** Lateral surface area

3. If each edge of a rectangular solid is increased by 1%, estimate the change in:
   (a) Volume **(b)** Surface area

4. The base radii of the frustum of a cone are both decreased by 3%, and the height is increased by 2%. How is the volume changed?
5. A rectangle 10 feet by 12 feet is to have its perimeter increased by 2 feet. What should the new dimensions be so that the area is increased as much as possible?

Determine which differential forms are exact, and construct the corresponding functions, where possible.

6. $dx + dy - 4\,dz$
7. $x\,dx - z\,dy - y\,dz$
8. $e^x\,dx + e^y\,dy - dz$
9. $e^y\,dx + xe^y\,dy + dz$
10. $\cos z\,dx - dy - x\sin z\,dz$
11. $\sqrt{x^2+y^2}\,dx + y\ln\left(\frac{x+\sqrt{x^2+y^2}}{y}\right)dy$
12. $\sqrt{x^2-y^2}\,dx - y\ln\left(\frac{x+\sqrt{x^2-y^2}}{y}\right)dy$
13. $\frac{2y}{x^2-y^2}\,dx + \frac{2x}{x^2-y^2}\,dy$
14. $\frac{y}{\sqrt{1-x^2y^2}}\,dx + \frac{x}{\sqrt{1-x^2y^2}}\,dy$

Find the functions, given their gradients.

15. $\nabla f(x, y) = \left\langle \frac{-2y}{(x-y)^2}, \frac{2x}{(x-y)^2} \right\rangle$
16. $\nabla f(x, y) = \left\langle \frac{y}{1+x^2y^2}, \frac{x}{1+x^2y^2} \right\rangle$
17. $\nabla f(x, y, z) = \left\langle 2x + \tan^{-1}yz, \frac{xz}{1+y^2z^2}, \frac{xy}{1+y^2z^2} \right\rangle$
18. $\nabla f(x, y, z) = \langle 3x^2y + yz, x^3 + xz - z\sin yz, xy - y\sin yz \rangle$

Evaluate the path integrals.

19. $\int_C (x^2y\,dx + xy^2\,dy)$; C: $\mathbf{r}(t) = \langle t, t^2 - 1\rangle$, $0 \le t \le 1$
20. $\int_C [x^2y\,dx + (x+y)\,dy]$; C: $\mathbf{r}(t) = \langle t^2, t^3\rangle$, $-1 \le t \le 1$
21. $\int_C (\sin y\,dx + \tan y\,dy)$; C: $\mathbf{r}(t) = \langle t + \pi, 2t\rangle$, $0 \le t \le \pi/6$
22. $\int_C (\sec x\,dx - \cos y\,dy)$; C: $\mathbf{r}(t) = \langle \pi/2 - t, \frac{1}{2}t\rangle$, $\pi/4 \le t \le \pi/3$
23. $\int_C (xy\,dx + x^2\,dy)$; C: from (1, 1) to (0, 0) along $y = x^2$, then to (1, 0) along the x-axis
24. $\int_C \left(\frac{y}{x^2-y^2}\,dx + \frac{x}{x^2-y^2}\,dy\right)$; C: the path from (1, 0) to (2, −1) along:
    (a) The straight line ᴬ**(b)** The curve $y = \frac{1}{2}(x - x^2)$
25. $\int_C (xy\,dx + x^2\,dy)$; C: the line segment from (1, 1) to (1, 0)

26. $\int_C (dx + dz)$; C: the path:

(a) $\mathbf{r}(t) = \langle 2, \cos t, \sin t \rangle$, $0 \le t \le 2\pi$
(b) $\mathbf{r}(t) = \langle \sin t, t, \cos t \rangle$, $0 \le t \le \pi$

27. $\int_C (dx + y\,dy + z^2\,dz)$; C: the path:

(a) $\mathbf{r}(t) = \langle 2, \cos t, \sin t \rangle$, $0 \le t \le \pi$
(b) From (2, 1, 0) to (2, −1, 0) along the straight line

28. Find the work done by the force field $\mathbf{F}(x, y) = \langle x^2 + y^2, xy \rangle$ in moving an object from (0, 0) to (1, 1) along the curve:
(a) $y = x$ (b) $y = x^2$ (c) $y = x^3$ (d) $y = \sqrt{x}$

29. Find the work done by the force field $\mathbf{F}(x, y, z) = \langle e^x, e^y, e^z \rangle$ in moving an object from (0, 0, 0) to (1, 0, 0) along the curve:
(a) $y = 0$, $z = 0$ (b) $y = x^2 - x$, $z = 0$ (c) $z = x - x^2$, $y = 0$
(d) $\mathbf{r}(t) = \langle t, t^2 - t, t^3 - t \rangle$, $0 \le t \le 1$

30. A force field **F** is directed toward the origin and has magnitude proportional to the distance from the origin. Show that this force field is conservative, and find its potential function.

31. A certain magnetic field is given by

$$\mathbf{F}(x, y, z) = \left\langle -\frac{ky}{x^2 + y^2}, \frac{kx}{x^2 + y^2}, 0 \right\rangle.$$

Find the work done in moving a particle once around the circle $x^2 + y^2 = a^2$, $z = c$ in the counterclockwise direction, starting at $(a, 0, c)$.

32. Repeat Problem 31 for the curve $\mathbf{r}(t) = \langle \cos t, \sin t, t \rangle$, $0 \le t \le 2\pi$.

Additional Exercises

1. A wire of density $\delta(x, y, z) = x^2y/z$ lies along the curve C: $\mathbf{r}(t) = \langle t, t^2, 3t \rangle$, $0 \le t \le 1$. Find:
(a) The mass (b) The moments (c) The center of mass
(d) The moments of inertia $\int_C x^2\delta\,ds$, $\int_C y^2\delta\,ds$, and $\int_C z^2\delta\,ds$
(Use numerical integration.)

2. Find the length of the wire in Problem 1.

3. A certain magnetic field has flux $\mathbf{B}(x, y, z) = \langle 0, 0, k \rangle$. A charged particle enters the field with initial velocity $\mathbf{v}(0) = \langle -v, 0, 0 \rangle$ and initial position $\mathbf{r}(0) = \langle 0, b, c \rangle$. The force produced on the particle is proportional to the cross product of the velocity and the flux, or $\mathbf{F} = q(\mathbf{v} \times \mathbf{B})$, where q is the charge on the particle. Using the fact that the acceleration of the particle is $\mathbf{a}(t) = (1/m)\mathbf{F}(t)$, where m is the mass of the particle, and the solution of the differential equation $u'' = -h^2u$ is $u = A\cos(ht + \phi)$, where A and ϕ are constants, determine the path of the particle.

4. The vector field

$$\mathbf{F}(x, y, z) = \frac{K}{(x^2 + y^2 + z^2)^{n/2}} \langle x, y, z \rangle$$

is called a **central force field.** Show that a central force field is conservative, and find its potential. [*Hint:* The case $n = 2$ must be treated separately.]

5. Find the work done against gravity in lifting a payload weighing P pounds to a height H miles above the earth's surface. Use 4000 miles as the earth's radius and the fact that the earth's gravitational field is a central force field with $n = 2$.

6. Find the work done by gravity on a 200 pound satellite in a circular orbit of radius 6000 miles during:
   (a) $\frac{1}{2}$ revolution (b) $\frac{3}{4}$ revolution

CHAPTER 16

Multiple Integrals

INTRODUCTION

In Chapter 5, we introduced the definite integral, which can also be described as the integral of a function of one variable over an interval on the real line. The integral of a function of two variables over a region in the plane is called a **double integral**; the integral of a function of three variables over a region in space is called a **triple integral**. In general, these are called **multiple integrals**. In this chapter, we develop the theory and applications of multiple integrals.

SECTION 16.1

Double Integrals

In this section, we develop the double integral as the two-variable version of the definite integral. First of all, the definite integral $\int_a^b f(x)\,dx$ is defined relative to a closed interval $[a, b]$. The two-variable version of a closed interval is a region in the plane; we therefore digress briefly to define a region in the plane.

Regions in the Plane

Recall from Chapter 14 that an open set S is such that every point of S lies in an open disc that lies entirely in S, and a closed set is the complement of an open set. Recall also that a bounded set is one that lies entirely in some open disc, and a compact set is one that is closed and bounded. Finally, a boundary point of a set S is a point P such that every open disc that contains P contains a point in S and a point not in S.

A point of a set S that is not a boundary point of S is called an **interior point** of S; the set of interior points of S is called the **interior** of S. It follows that the interior of any set is an open set. It is also true that any closed set contains all its boundary points, while an open set contains none of its boundary points.

We can now define a region.

DEFINITION 16.1 A **region** in the plane is a compact set with nonempty interior.

For example, a polygon together with its interior is a region called a **polygonal region** (Figure 16.1). A circle and its interior—that is, a closed disc—is a **circular region** (Figure 16.2). On the other hand, a line segment is compact but has empty interior, and so fails to be a region.

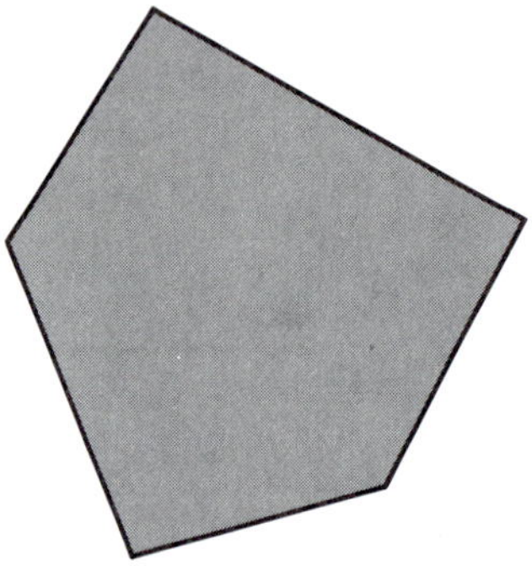

FIGURE 16.1 A polygonal region

FIGURE 16.2 A circular region

Unfortunately, difficulties arise if the boundary of a region is too irregular. To avoid this, we will consider regions of only two types. A region of **Type I** is a set of points of the form

$$\{(x, y) | a \leq x \leq b \text{ and } g(x) \leq y \leq h(x)\},$$

and a region of **Type II** is of the form

$$\{(x, y) | g(y) \leq x \leq h(y) \text{ and } c \leq y \leq d\},$$

with g and h continuous functions in each case. Regions of these types are shown in Figures 16.3 and 16.4. Every region we shall consider is either one of these types or a finite union of regions of these types.

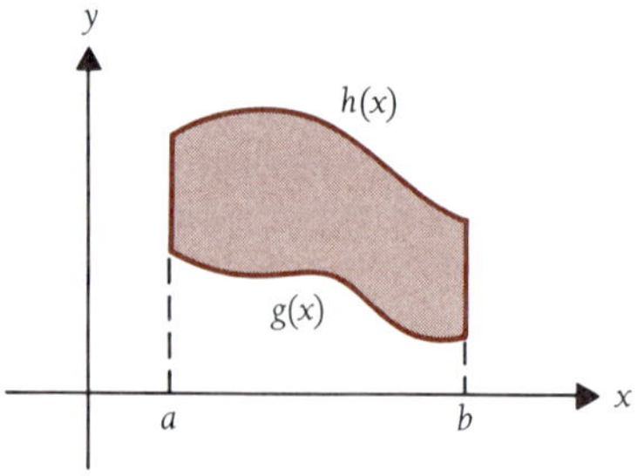

FIGURE 16.3 A Type I region

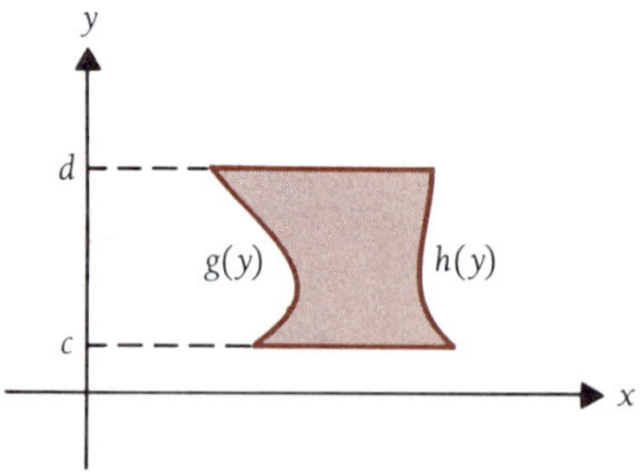

FIGURE 16.4 A Type II region

Double Integrals over Rectangular Regions

We now define a double integral of a function of two variables over a rectangular region.

Let f be a function of two variables, and let

$$\begin{aligned} R &= \{(x, y) \mid a \le x \le b \text{ and } c \le y \le d\} \\ &= [a, b] \times [c, d] \end{aligned}$$

be the rectangular region lying in the domain of f (see Figure 16.5). Let $s_1 = \{a = x_0, x_1, \ldots, x_i, \ldots, x_n = b\}$ be a partition of $[a, b]$, and let $s_2 = \{c = y_0, y_1, \ldots, y_j, \ldots, y_m = d\}$ be a partition of $[c, d]$. We call

$$s = s_1 \times s_2 = \{(x_i, y_j) \mid x_i \in s_1 \text{ and } y_j \in s_2\}$$

a **partition of** R. The points (x_{i-1}, y_{j-1}), (x_i, y_{j-1}), (x_i, y_j), and (x_{i-1}, y_j) are vertices of the i, jth **rectangular subregion** of R, which we denote by R_{ij}, as shown in Figure 16.5. The length and width of R_{ij} are $\Delta x_i = x_i - x_{i-1}$ and $\Delta y_j = y_j - y_{j-1}$, so the area of R_{ij} is

$$\Delta A_{ij} = \Delta x_i \Delta y_j .$$

The partition s of R determines mn subrectangular regions. The **norm** of s is the maximum dimension of any rectangular subregion, namely

$$\|s\| = \max\{\Delta x_i, \Delta y_j\}.$$

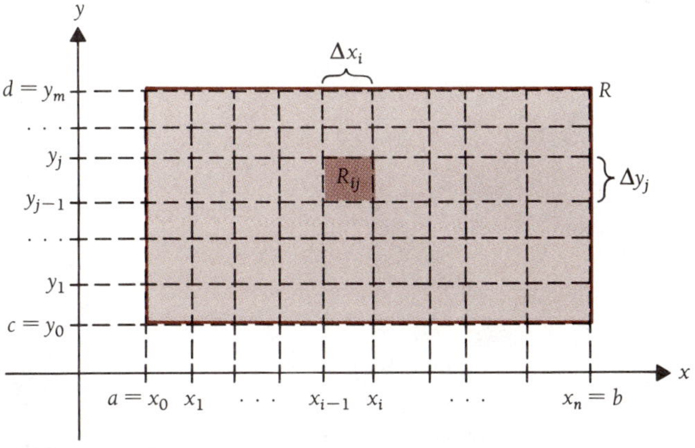

FIGURE 16.5 Subdivision of a rectangular region R

Now let $(\bar{x}_i, \bar{y}_j)$ be any point of R_{ij}. The double sum

$$\sum_{i=1}^{n} \sum_{j=1}^{m} f(\bar{x}_i, \bar{y}_j) \Delta A_{ij}$$

is called a **double Riemann sum** of f over R. The value of the double Riemann sum depends on the partition s and on the choices of the points $(\bar{x}_i, \bar{y}_j)$. However, if the double Riemann sums so obtained have a limit as the norms of the partitions approach 0, independent of the choices of the points $(\bar{x}_i, \bar{y}_j)$, we define the limit to be the **double integral of f over R**, and denote it by

$$\iint_R f(x, y)\, dA. \tag{1}$$

That is,

$$\iint_R f(x, y)\, dA = \lim_{\|s\| \to 0} \sum_{i=1}^{n} \sum_{j=1}^{m} f(\bar{x}_i, \bar{y}_j)\Delta A_{ij}.$$

In the double integral (1), we call $f(x, y)$ the **integrand**, and we refer to dA as the **differential of area**. We read (1) as the **double integral of $f(x, y)$ over R with respect to area**.

Just as the definite integral can be interpreted as an area, the double integral can be interpreted as a volume. If $f(x, y)$ is positive throughout R, then $f(\bar{x}_i, \bar{y}_j)\Delta A_{ij}$ is approximately the volume beneath the graph of f over R_{ij} (see Figure 16.6). Therefore, the limit of the double sum of such expressions is the volume under the graph of f over R, or

$$V = \iint_R f(x, y)\, dA.$$

(See Figure 16.7.)

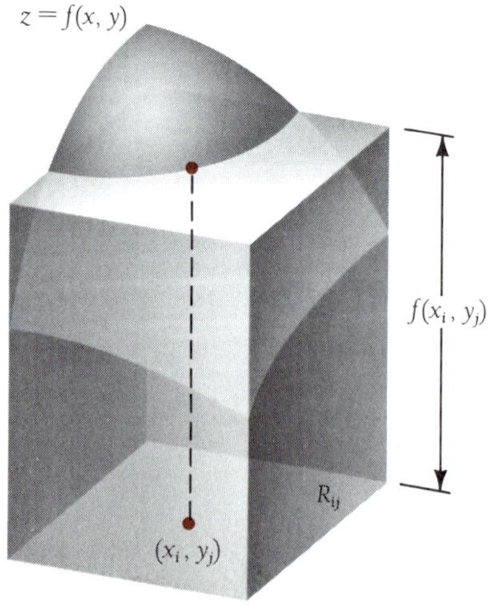

FIGURE 16.6 The prism over R_{ij}

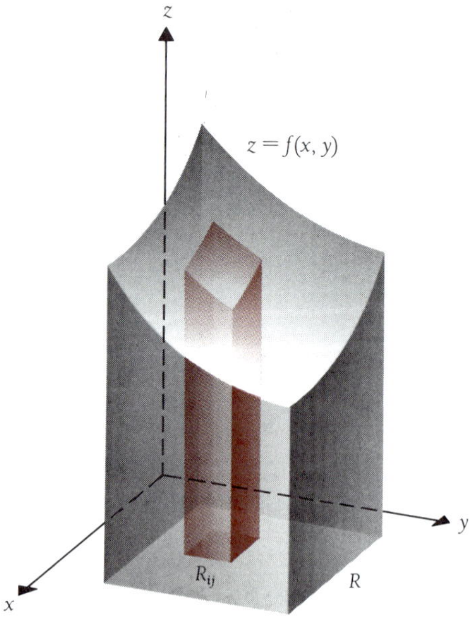

FIGURE 16.7

As a special case, we have

$$\iint_R dA = \text{Area of } R.$$

This is the volume of a solid of height 1 over R (Figure 16.8).

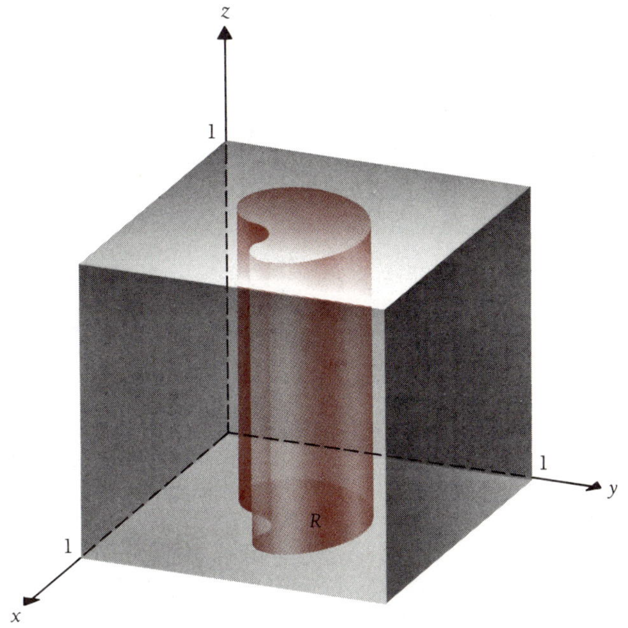

FIGURE 16.8 The special case $\iint_R dA = \text{Area of } R$

Example 16.1 Find the double integral $\iint_R k\, dA$, k constant, over $R = [a, b] \times [c, d]$ (see Figure 16.9, page 950).

Solution Since the double integral is the volume, and the constant function $f(x, y) = k$ defines a rectangular solid over R, the volume is

$$\iint_R k\, dA = k(b - a)(d - c).$$

This assumes k is positive. If k is negative, then we interpret the double integral as the negative of the volume of the solid, and the result is the same as before.

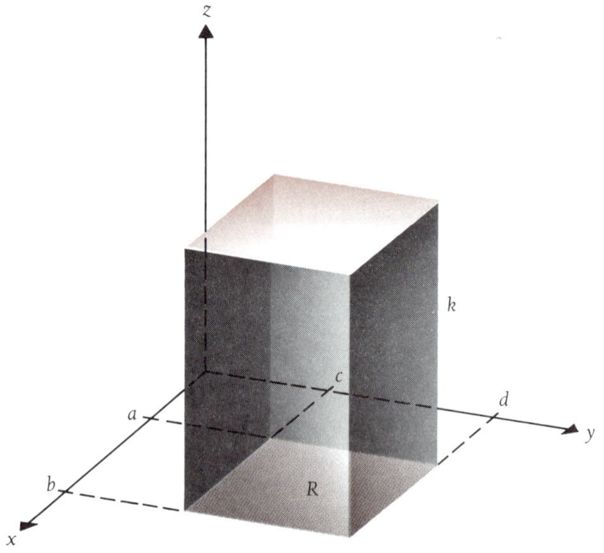

FIGURE 16.9 ■

Double Integrals over Other Regions

Let f be a function of two variables, and let R be a region (not necessarily of Type I or Type II) lying in the domain of f. If R is not a rectangular region, we can find a rectangular region R' that entirely contains R, because R is bounded. Then we partition R' into subrectangular regions R'_{ij} (see Figure 16.10). For each subrectangular region R'_{ij} that lies *entirely* in R, we select a point $(\bar{x}_i, \bar{y}_j) \in R'_{ij}$, and form the sum

$$\sum_{R'_{ij} \subseteq R} f(\bar{x}_i, \bar{y}_j) \Delta A_{ij}.$$

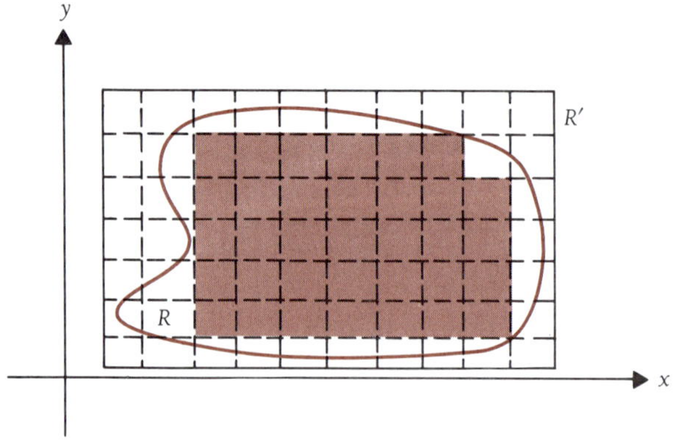

FIGURE 16.10 Subdivision of a plane region R

If the limit of such sums exists as the norms of the subdivisions approach 0, we say f is **integrable over R**, and we define the double integral $\iint_R f(x, y)\, dA$ to be that limit. We may interpret this integral as a volume just as we did when the region was a rectangle.

Properties of Double Integrals

The basic properties of double integrals are very much like the basic properties of definite integrals. We will state these properties as theorems, but because they are intuitively clear and geometrically evident, we will omit the proofs.

THEOREM 16.1 If functions f and g are integrable over R, then so is their sum, and

$$\iint_R [f(x, y) + g(x, y)]\, dA = \iint_R f(x, y)\, dA + \iint_R g(x, y)\, dA.$$

THEOREM 16.2 If f is integrable over R and c is a constant, then cf is integrable over R, and

$$\iint_R cf(x, y)\, dA = c \iint_R f(x, y)\, dA.$$

THEOREM 16.3 If R_1 and R_2 are regions whose intersection has empty interior, and function f is integrable over R_1 and over R_2, then f is integrable over $R_1 \cup R_2$, and

$$\iint_{R_1 \cup R_2} f(x, y)\, dA = \iint_{R_1} f(x, y)\, dA + \iint_{R_2} f(x, y)\, dA.$$

Some additional properties of double integrals are found in the exercises at the end of this section. (Compare Theorem 16.3 with Problem 26.)

We should mention that not just any function is integrable over any region; however, *any continuous function is integrable over any region of Type I or II.* We will be concerned only with these kinds of problems.

Evaluating a Double Integral

The definition of a double integral does not tell us how to evaluate it, unless we can interpret the double integral as a known volume. The usual method of evaluating a double integral is to transform it into a pair of "nested" definite integrals, called an **iterated integral**.

Suppose R is a region of Type I (see Figure 16.3). Let f be a continuous function of two variables defined on R, and for the sake of illustration, suppose f is positive over R. Let S be the solid bounded below by R and above by the graph of f (see Figure 16.11). For any fixed point t between a and b, let $A(t)$ be the area obtained by intersecting S with the plane $x = t$ (shaded in Figure 16.11). We can express $A(t)$ as an integral, since it is the area in the plane $x = t$ below the curve $z = f(t, y)$ with $g(t) \leq y \leq h(t)$. Hence,

$$A(t) = \int_{g(t)}^{h(t)} f(t, y)\, dy.$$

Now, by the slicing method of Section 8.1, the volume of S is

$$V = \int_a^b A(t)\, dt,$$

or, because t is a dummy variable,

$$V = \int_a^b A(x)\, dx.$$

But the volume of S is also the double integral,

$$V = \iint_R f(x, y)\, dA.$$

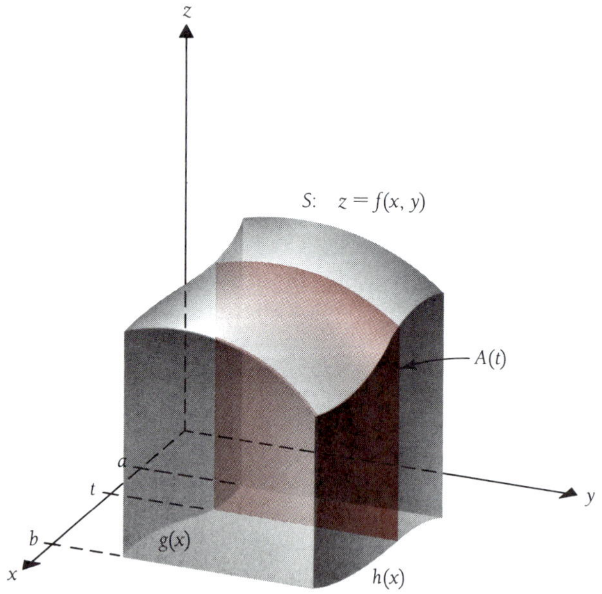

FIGURE 16.11 Volume interpretation of an iterated integral

We therefore have

$$\iint_R f(x, y)\, dA = \int_a^b A(x)\, dx$$
$$= \int_a^b \left[\int_{g(x)}^{h(x)} f(x, y)\, dy\right] dx,$$

which we usually write as

$$\iint_R f(x, y)\, dA = \int_a^b \int_{g(x)}^{h(x)} f(x, y)\, dy\, dx. \tag{2}$$

We have thus transformed the double integral into an iterated integral.

By an analogous argument, if R is a region of Type II, then

$$\iint_R f(x, y)\, dA = \int_c^d \int_{g(y)}^{h(y)} f(x, y)\, dx\, dy. \tag{3}$$

Equations (2) and (3) were published in 1909 by the Italian mathematician Guido Fubini (1879–1943).

We illustrate the use of these formulas with some examples.

Example 16.2 Let R be the region bounded by the curves $y = x^2$ and $y = 4$ (Figure 16.12). Find

$$\iint_R (x^2 + 2xy)\, dA.$$

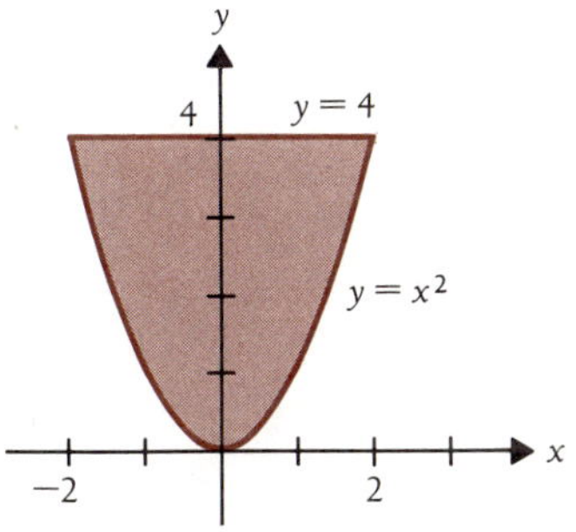

FIGURE 16.12

Solution Since $R = \{(x, y) | -2 \le x \le 2 \text{ and } x^2 \le y \le 4\}$, it is a region of Type I, and we use formula (2):

$$\iint_R (x^2 + 2xy)\, dA = \int_{-2}^{2} \int_{x^2}^{4} (x^2 + 2xy)\, dy\, dx$$
$$= \int_{-2}^{2} \left[x^2y + xy^2\right]_{x^2}^{4} dx$$

$$= \int_{-2}^{2} (4x^2 + 16x - x^4 - x^5)\,dx$$

$$= \left[\frac{4}{3}x^3 + 8x^2 - \frac{1}{5}x^5 - \frac{1}{6}x^6\right]_{-2}^{2} = \frac{128}{15}.$$

Alternatively, we can describe R as

$$R = \{(x, y) | -\sqrt{y} \le x \le \sqrt{y} \text{ and } 0 \le y \le 4\},$$

which is a region of Type II. Hence, by formula (3), we get

$$\iint_R (x^2 + 2xy)\,dA = \int_0^4 \int_{-\sqrt{y}}^{\sqrt{y}} (x^2 + 2xy)\,dx\,dy$$

$$= \int_0^4 \left[\frac{1}{3}x^3 + x^2 y\right]_{-\sqrt{y}}^{\sqrt{y}} dy$$

$$= \int_0^4 \frac{2}{3}y^{3/2}\,dy$$

$$= \left[\frac{4}{15}y^{5/2}\right]_0^4 = \frac{128}{15}.$$

■

■ **Example 16.3** Find

$$\iint_R (x - 2y)\,dA$$

if $R = \{(x, y) | 0 \le x \le 4 \text{ and } 0 \le y \le \sqrt{x}\}$ (Figure 16.13).

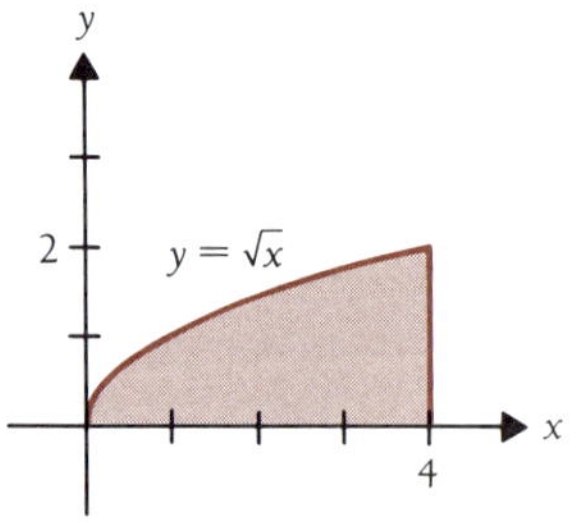

FIGURE 16.13

Solution As given, R is of Type I, so by formula (2), we have

$$\iint_R (x - 2y)\,dA = \int_0^4 \int_0^{\sqrt{x}} (x - 2y)\,dy\,dx$$

$$= \int_0^4 \left[xy - y^2\right]_0^{\sqrt{x}} dx$$

$$= \int_0^4 (x^{3/2} - x)\, dx$$

$$= \left[\frac{2}{5}x^{5/2} - \frac{1}{2}x^2\right]_0^4 = \frac{24}{5}.$$

Example 16.4 Find

$$\iint_R x \cos y \, dA$$

for $R = \{(x, y) | -\pi/2 \le y \le \pi/2 \text{ and } 1 - y^2 \le x \le 2\}$ (Figure 16.14).

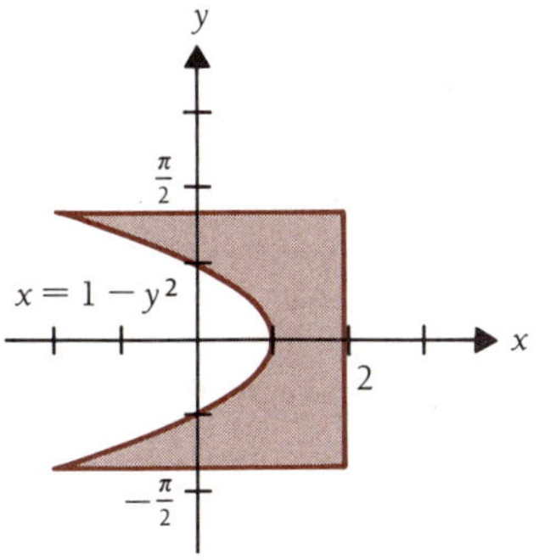

FIGURE 16.14

Solution This region is described as a Type II region, so by formula (3), we have

$$\iint_R x \cos y \, dA = \int_{-\pi/2}^{\pi/2} \int_{1-y^2}^{2} x \cos y \, dx\, dy$$

$$= \int_{-\pi/2}^{\pi/2} \left[\frac{1}{2}x^2 \cos y\right]_{1-y^2}^{2} dy$$

$$= \int_{-\pi/2}^{\pi/2} \left(\frac{3}{2}\cos y + y^2 \cos y - \frac{1}{2}y^4 \cos y\right) dy.$$

This integral can be evaluated by integrating by parts on the second and third terms, eventually getting

$$\iint_R x \cos y \, dA = -25 + \frac{7}{2}\pi^2 - \frac{1}{16}\pi^4.$$

Changing the Order of Integration

As we have seen, an iterated integral represents a double integral. Sometimes, given an iterated integral, we must identify the region over which the corresponding double integral is defined in order to evaluate the iterated integral. In such problems, the region can be described as either Type I or Type II, and the choice of description makes a lot of difference in the solution of the problem. Consider an example.

Example 16.5 Evaluate the iterated integral

$$\int_{-2}^{2}\int_{y^2}^{4} ye^{x^2}\,dx\,dy.$$

Solution We see from the limits given that $y^2 \le x \le 4$ and $-2 \le y \le 2$, so the given iterated integral is the same as

$$\iint_R ye^{x^2}\,dA,$$

with R the region between $x = y^2$ and $x = 4$ (Figure 16.15). Thus, the given iterated integral treats R as a Type II region. But we cannot evaluate the given iterated integral, because we cannot find an antiderivative of e^{x^2}. Instead, we will describe R as a Type I region:

$$R = \{(x, y) | 0 \le x \le 4 \text{ and } -\sqrt{x} \le y \le \sqrt{x}\}.$$

Then

$$\begin{aligned}\int_{-2}^{2}\int_{y^2}^{4} ye^{x^2}\,dx\,dy &= \iint_R ye^{x^2}\,dA \\ &= \int_0^4\int_{-\sqrt{x}}^{\sqrt{x}} ye^{x^2}\,dy\,dx \\ &= \int_0^4 \frac{y^2}{2}e^{x^2}\Bigg|_{y=-\sqrt{x}}^{y=\sqrt{x}}\,dx \\ &= \int_0^4 0\,dx = 0.\end{aligned}$$

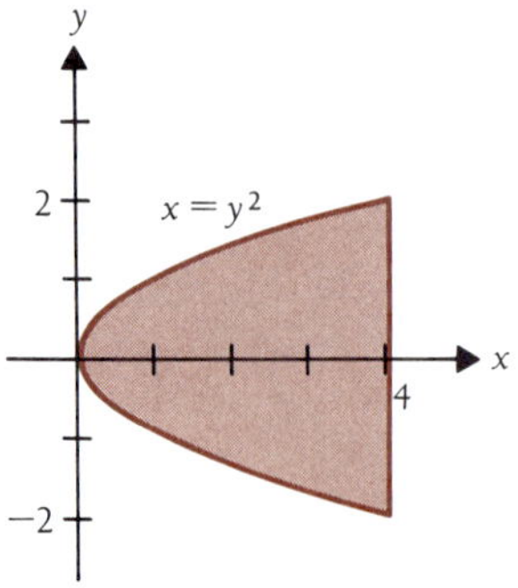

FIGURE 16.15 ■

The process of changing the description of a region from one type to the other is called **changing the order of integration.**

Exercises 16.1

Evaluate each double integral. First sketch the region R in each case.

1. $\iint_R xy\, dA$; $R = [0, 2] \times [1, 5]$
2. $\iint_R (x + y)\, dA$; $R = [-1, 2] \times [0, 1]$
3. $\iint_R (x^2 + y^2)\, dA$; $R = \{(x, y) | 1 \le x \le 3 \text{ and } x \le y \le x^2\}$
4. $\iint_R (x + 3y + 5)\, dA$; $R = \{(x, y) | 0 \le x \le 1, x^2 \le y \le x\}$
5. $\iint_R \sin x\, dA$; $R = \{(x, y) | 0 \le x \le \pi \text{ and } \frac{1}{2}x \le y \le 2x\}$
6. $\iint_R e^{x+y}\, dA$; $R = \{(x, y) | 1 \le x \le y \text{ and } 1 \le y \le 2\}$
7. $\iint_R e^{x^2}\, dA$; R bounded by $x = 1$, $y = 0$, $y = x$ (Choose type carefully!)
8. $\iint_R \sin y^2\, dA$; R bounded by $x = 0$, $y = x$, $y = 1$

Let $R_1 = [1, 2] \times [1, 3]$ and $R_2 = [2, 3] \times [1, 2]$. Evaluate each double integral.

9. $\iint_{R_1} xe^y\, dA$
10. $\iint_{R_1} y \cos \pi x\, dA$
11. $\iint_{R_2} xe^y\, dA$
12. $\iint_{R_2} y \cos \pi x\, dA$
13. $\iint_{R_1 \cup R_2} xe^y\, dA$
14. $\iint_{R_1 \cup R_2} y \cos \pi x\, dA$
15. $\iint_{R_1} (xe^y + y \cos \pi x)\, dA$
16. $\iint_{R_2} (xe^y - y \cos \pi x)\, dA$

Evaluate the following double integrals by interpreting them as known volumes.

17. $\displaystyle\iint_R \sqrt{4 - x^2 - y^2}\, dA$; $R = \{(x, y) | x^2 + y^2 \le 4\}$
18. $\displaystyle\iint_R \sqrt{9 - x^2 - y^2}\, dA$; $R = \{(x, y) | 0 \le x \le 3 \text{ and } 0 \le y \le \sqrt{9 - x^2}\}$
19. $\displaystyle\iint_R (2 - y)\, dA$; $R = [0, 1] \times [0, 2]$
20. $\displaystyle\iint_R \left(h - \frac{h}{r}\sqrt{x^2 + y^2}\right) dA$; $R = \{(x, y) | x^2 + y^2 \le r^2\}$

Evaluate each iterated integral by changing the order of integration. First sketch the region in each case.

21. $\displaystyle\int_{-1}^{1} \int_0^{\pi} x\sqrt{\sin y}\, dy\, dx$
22. $\displaystyle\int_0^1 \int_x^1 e^{y^2}\, dy\, dx$
23. $\displaystyle\int_0^{\sqrt{\pi}} \int_y^{\sqrt{\pi}} \cos x^2\, dx\, dy$
24. $\displaystyle\int_0^1 \int_{\arcsin y}^{\pi/2} \sqrt{4 - \cos^2 x}\, dx\, dy$
25. Explain why $\iint_R dA$ is just the area of R.
26. Show that if R_1 and R_2 are any regions and f is integrable, then

$$\iint_{R_1 \cup R_2} f(x, y)\, dA = \iint_{R_1} f(x, y)\, dA + \iint_{R_2} f(x, y)\, dA - \iint_{R_1 \cap R_2} f(x, y)\, dA.$$

27. Show that if f and g are integrable and $f(x, y) \leq g(x, y)$ over R, then

$$\iint_R f(x, y)\, dA \leq \iint_R g(x, y)\, dA.$$

28. If region R has area A and f is a function integrable over R such that $m \leq f(x, y) \leq M$, show that

$$mA \leq \iint_R f(x, y)\, dA \leq MA.$$

29. If $R = [a, b] \times [c, d]$, show that

$$\iint_R f(x)g(y)\, dA = \left[\int_a^b f(x)\, dx\right]\left[\int_c^d g(y)\, dy\right].$$

30. Prove that a set of points in the plane is closed if and only if it contains all its boundary points.
31. Prove that the boundary of any set of points in the plane is a closed set.
32. Prove that the interior of any set of points in the plane is an open set.
33. Prove that a set contains none of its boundary points if and only if it is an open set.
34. How might a double integral over an unbounded closed set be related to "improper" iterated integrals?
35. Decide whether the following "improper" double integrals exist, and if so, find their values.
 (a) $\iint_R x^{-2}\, dA$; $R = \{(x, y) | x \geq 1 \text{ and } 0 \leq y \leq 1\}$
 (b) $\iint_R e^{-y}\, dA$; $R = \{(x, y) | x \geq 0 \text{ and } x - 1 \leq y \leq x + 1\}$

SECTION 16.2

Applications of Double Integrals

As we saw in Section 16.1, a double integral can be interpreted as a volume. That is, double integrals can be used to find volumes of solids. They also can be used to find areas, masses, centers of mass, and centroids. We will discuss these applications in this section.

Volume

In Chapter 8, we used definite integrals to find volumes of solids of revolution. With double integrals, we can find volumes of more general solids.

First, suppose that a function f of two variables is positive over a region R in the plane. The **solid below f** is the solid whose base is the region R and whose upper surface is the graph of $z = f(x, y)$.

When we considered areas in Chapter 5, we noted that not every positive function defines a region having area. The same is true here; that is, not every positive function of two variables defines a solid having volume. In Chapter 5, we proved that a continuous function has area below it. We also can prove that if a positive function of two variables is continuous, then the solid below it has volume. We will use this fact without proving it here.

If f is a continuous function of two variables that is positive over the region R, then

$$\iint_R f(x, y)\, dA$$

is the volume of the solid below f.

The following examples show how using this interpretation enables us to find volumes of various solid objects.

Example 16.6 Find the volume enclosed by the planes $x = 0$, $y = 0$, $x + y = 1$, $z = 0$, and the paraboloid $z = 4 - x^2 - y^2$.

Solution Sketching the solid, as shown in Figure 16.16(a), we see that the volume is $\iint_R (4 - x^2 - y^2)\, dA$ with $R = \{(x, y) | 0 \le x \le 1$ and $0 \le y \le 1 - x\}$ [Figure 16.16(b)].

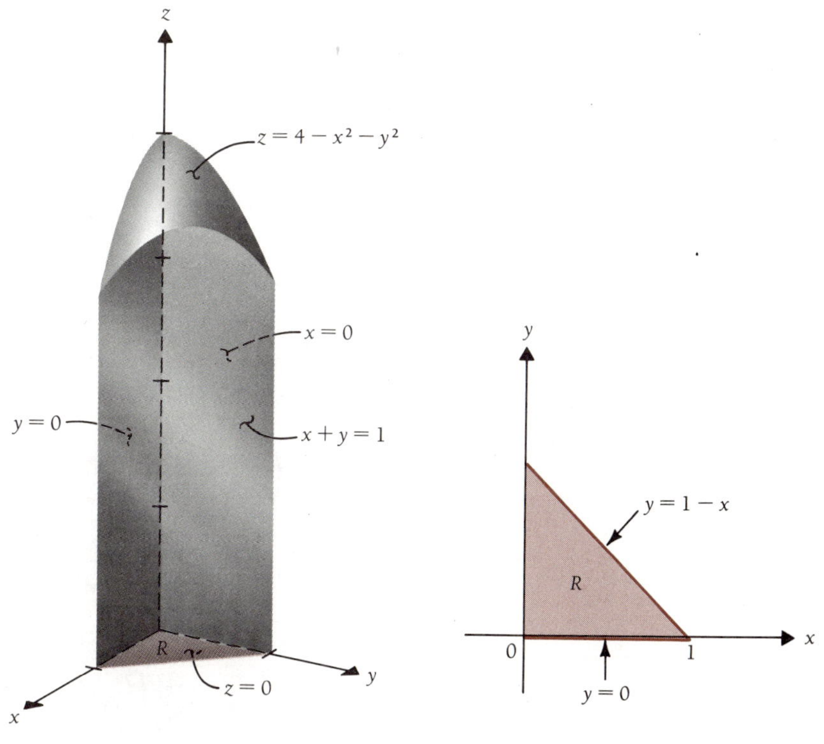

(a) A solid **(b)** The region over which it lies

FIGURE 16.16

Hence, the volume is

$$\begin{aligned} V &= \iint_R (4 - x^2 - y^2)\, dA \\ &= \int_0^1 \int_0^{1-x} (4 - x^2 - y^2)\, dy\, dx \\ &= \int_0^1 \left[4y - x^2y - \frac{1}{3}y^3 \right]_0^{1-x} dx \\ &= \int_0^1 \left(\frac{11}{3} - 3x - 2x^2 + \frac{4}{3}x^3 \right) dx \\ &= \left[\frac{11}{3}x - \frac{3}{2}x^2 - \frac{2}{3}x^3 + \frac{1}{3}x^4 \right]_0^1 = \frac{11}{6}. \end{aligned}$$

■

■ **Example 16.7** Find the volume inside the cylinder $x^2 + y^2 = 4$, above the plane $z = 1 - x/2$, and below the paraboloid $z = 4 + x^2 + y^2$.

Solution Again we sketch the solid described, as shown in Figure 16.17(a), and we see that the entire solid lies above the region $R = \{(x, y) | x^2 + y^2 \leq 4\}$ [Figure 16.17(b)]. The volume can be expressed as the difference of two volumes. From the volume under $z = 4 + x^2 + y^2$ over R, we subtract the volume under the plane $z = 1 - x/2$ over R. Hence, the volume of the solid is

$$\begin{aligned} V &= \iint_R (4 + x^2 + y^2)\, dA - \iint_R \left(1 - \frac{x}{2}\right) dA \\ &= \iint_R \left(3 + x^2 + y^2 + \frac{x}{2}\right) dA. \end{aligned}$$

Now we will describe R as a Type II region,

$$R = \{(x, y) | -\sqrt{4 - y^2} \leq x \leq \sqrt{4 - y^2} \text{ and } -2 \leq y \leq 2\}.$$

Therefore,

$$\begin{aligned} V &= \int_{-2}^{2} \int_{-\sqrt{4-y^2}}^{\sqrt{4-y^2}} \left(3 + x^2 + y^2 + \frac{x}{2}\right) dx\, dy \\ &= \int_{-2}^{2} \left[3x + \frac{1}{3}x^3 + y^2x + \frac{1}{4}x^2 \right]_{-\sqrt{4-y^2}}^{\sqrt{4-y^2}} dy \\ &= \int_{-2}^{2} \frac{2}{3}(13 + 2y^2)\sqrt{4 - y^2}\, dy \\ &= \frac{26}{3} \int_{-2}^{2} \sqrt{4 - y^2}\, dy + \frac{4}{3} \int_{-2}^{2} y^2\sqrt{4 - y^2}\, dy. \end{aligned}$$

The first integral is the area of a semicircle of radius 2, so it has value $\frac{1}{2}\pi(2)^2 = 2\pi$. The second integral can be evaluated by a trigonometric substitution, eventually

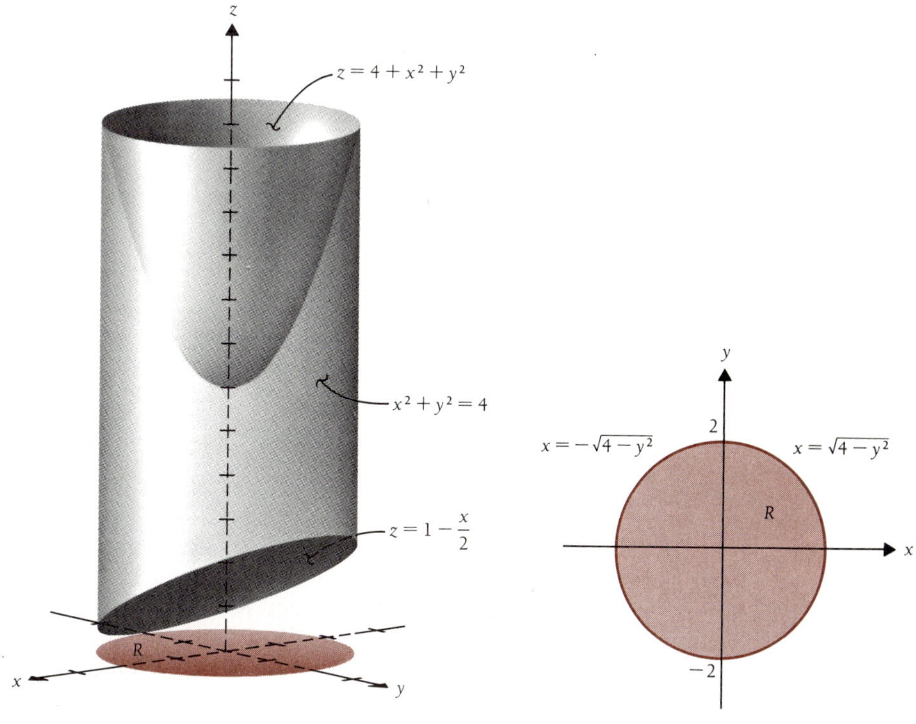

(a) A solid

(b) The region over which it lies

FIGURE 16.17

getting 2π as its value. Therefore, the volume is

$$V = \frac{26}{3}(2\pi) + \frac{4}{3}(2\pi) = 20\pi.$$

■ **Example 16.8** Find the volume of the solid bounded on the sides by the planes $x = 0, x = 1, y = 0, y = 1$, below by the plane $x + y + z = 1$, and above by the paraboloid $z = 4 - x^2 - y^2$.

Solution By sketching the solid, we see that it lies over (and partly below) the region $R = [0, 1] \times [0, 1]$, as shown in Figure 16.18 (page 962). The volume is the double integral of a difference of functions, much like the previous example:

$$V = \iint_R [(4 - x^2 - y^2) - (1 - x - y)]\, dA$$

$$= \iint_R (3 - x^2 - y^2 + x + y)\, dA$$

$$\begin{aligned} V &= \int_0^1 \int_0^1 (3 - x^2 - y^2 + x + y)\, dy\, dx \\ &= \int_0^1 \left[3y - x^2 y - \frac{1}{3} y^3 + xy + \frac{1}{2} y^2\right]_0^1 dx \\ &= \int_0^1 \left(\frac{19}{6} - x^2 + x\right) dx \\ &= \left[\frac{19}{6} x - \frac{1}{3} x^3 + \frac{1}{2} x^2\right]_0^1 = \frac{10}{3}. \end{aligned}$$

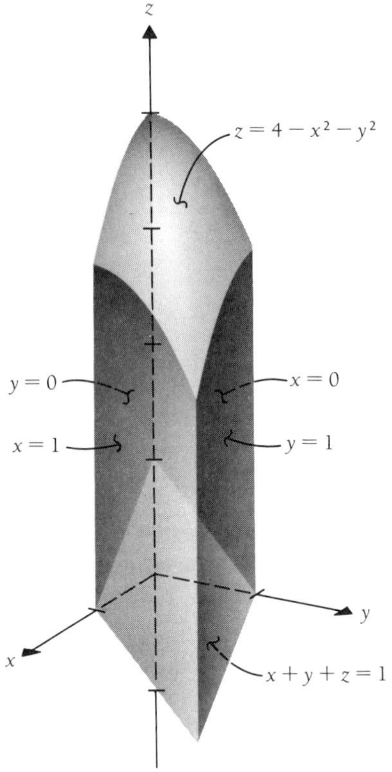

FIGURE 16.18 ■

■ **Example 16.9** Find the volume between the paraboloids $z = x^2 + y^2$ and $z = 4 - x^2 - y^2$.

Solution The two surfaces (Figure 16.19) intersect in the circle $x^2 + y^2 = 2$, $z = 2$. Hence, the solid lies above the circular region $R = \{(x, y) | x^2 + y^2 \le 2\}$, and its volume is the double integral of a difference,

$$V = \iint_R [4 - x^2 - y^2 - (x^2 + y^2)]\, dA$$
$$= \iint_R (4 - 2x^2 - 2y^2)\, dA.$$

By symmetry of both R and the solid, we can find the volume over the region

$$R' = \{(x, y) | 0 \le x \le \sqrt{2} \text{ and } 0 \le y \le \sqrt{2 - x^2}\}$$

and multiply by 4. Thus,

$$V = 4 \iint_{R'} (4 - 2x^2 - 2y^2)\, dA$$
$$= 4 \int_0^{\sqrt{2}} \int_0^{\sqrt{2-x^2}} (4 - 2x^2 - 2y^2)\, dy\, dx$$
$$= 4 \int_0^{\sqrt{2}} \left[4y - 2x^2y - \frac{2}{3}y^3 \right]_0^{\sqrt{2-x^2}} dx$$

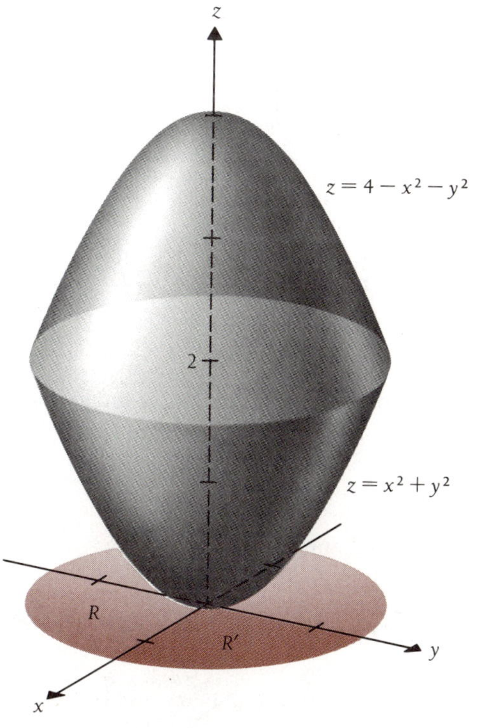

FIGURE 16.19

$$\begin{aligned} V &= 4\int_0^{\sqrt{2}} \frac{4}{3}(2 - x^2)^{3/2}\,dx \\ &= \frac{64}{3}\int_0^{\pi/2} \sin^4\theta\,d\theta && (x = \sqrt{2}\cos\theta) \\ &= 4\pi. && \text{(See Example 7.26.)} \end{aligned}$$

■

Area

Under our current assumptions, $\iint_R f(x, y)\,dA$ is the volume under $z = f(x, y)$ over the region R. Hence, the integral $\iint_R dA$ is the volume of a solid over R of height 1. That is, $\iint_R dA$ is numerically equal to the area of R. Thus, a double integral can be used to find the area of a region in the plane.

■ **Example 16.10** Find the area of a region of Type I,

$$R = \{(x, y) | a \le x \le b \text{ and } g(x) \le y \le h(x)\}.$$

Solution The area is

$$\begin{aligned} A &= \iint_R dA \\ &= \int_a^b \int_{g(x)}^{h(x)} dy\,dx \\ &= \int_a^b y\Big|_{g(x)}^{h(x)} dx \\ &= \int_a^b [h(x) - g(x)]\,dx. \end{aligned}$$

This is formula (1) of Section 5.6. ■

Example 16.10 illustrates that when a double integral is used to find the area of a region, the first integration reduces the problem to a definite integral, the same as used in Chapter 5.

■ **Example 16.11** Find the area of the region bounded by the curves $y = \ln x$, $y = 1 - x$, and $y = -2$.

Solution The given region (Figure 16.20) can most easily be described as a Type II region,

$$R = \{(x, y) | e^y \le x \le 1 - y \text{ and } -2 \le y \le 0\}.$$

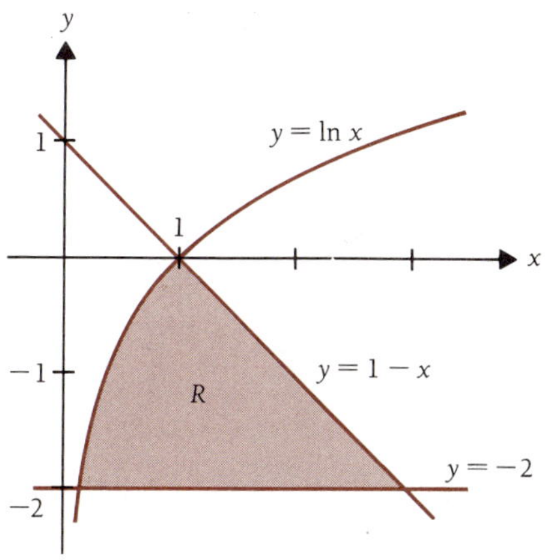

FIGURE 16.20

Hence, the area is

$$\begin{aligned} A &= \iint_R dA \\ &= \int_{-2}^{0} \int_{e^y}^{1-y} dx\, dy \\ &= \int_{-2}^{0} (1 - y - e^y)\, dy \\ &= \left[y - \frac{1}{2} y^2 - e^y \right]_{-2}^{0} = 3 + e^{-2}. \end{aligned}$$

■

Mass

Suppose a thin plate covers a region R of the plane, and suppose the density of the plate is a function $\delta(x, y)$ of two variables. We will show that the mass of the plate is

$$m = \iint_R \delta(x, y)\, dA. \tag{1}$$

We partition the region R and let $(\bar{x}_i, \bar{y}_j)$ be a point in the subrectangular region R_{ij}. If the area of R_{ij} is ΔA_{ij}, then the mass of the portion of the plate covering R_{ij} is approximately $\delta(\bar{x}_i, \bar{y}_j)\Delta A_{ij}$. A double sum of such terms is an approximation to the mass of the plate, and in the limit, the mass is given by formula (1).

■ **Example 16.12** A plate covers the rectangular region $R = [0, 1] \times [0, 2]$ and has density $\delta(x, y) = 1 + x + 2xy$. Find its mass.

Solution By formula (1), the mass is

$$m = \iint_R \delta(x, y)\, dA = \int_0^1 \int_0^2 (1 + x + 2xy)\, dy\, dx = \int_0^1 (2 + 6x)\, dx = 5.$$

■

■ **Example 16.13** A semicircular plate of radius a has density varying as the distance from the diameter. Find its mass.

Solution We will place the plate on the region

$$R = \{(x, y) | -a \leq x \leq a \text{ and } 0 \leq y \leq \sqrt{a^2 - x^2}\}$$

(Figure 16.21). Then the density function is $\delta(x, y) = ky$, k a constant of proportionality. Hence, the mass is

$$m = \iint_R \delta(x, y)\, dA = \int_{-a}^{a} \int_0^{\sqrt{a^2-x^2}} ky\, dy\, dx = \int_{-a}^{a} \frac{k}{2}(a^2 - x^2)\, dx = \frac{2}{3} ka^3.$$

(Compare with Example 8.28.)

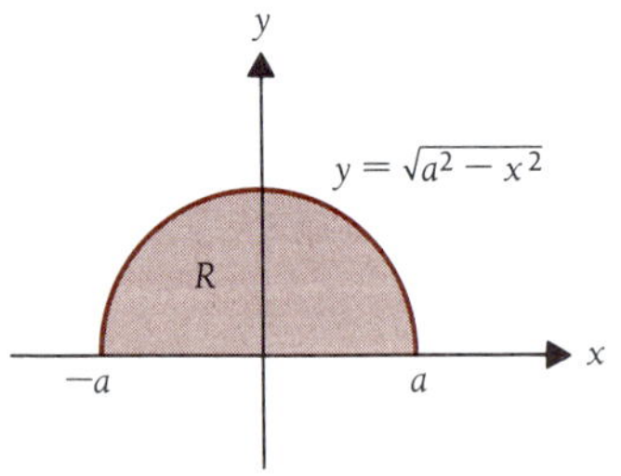

FIGURE 16.21

■

Centers of Mass

Double integrals can be used to find the center of mass of a flat plate. If R is the region occupied by the plate and δ is the density function, then formulas (10) and

(11) of Section 8.5 can be rewritten as

$$\bar{x} = \frac{\iint_R x\delta(x, y)\, dA}{\iint_R \delta(x, y)\, dA}, \qquad \bar{y} = \frac{\iint_R y\delta(x, y)\, dA}{\iint_R \delta(x, y)\, dA}. \tag{2}$$

These formulas give the coordinates of the center of mass of the plate.

Example 16.14 A plate with density $\delta(x, y) = 1 + xy$ covers the region

$$R = \{(x, y) | 0 \leq x \leq 2 \text{ and } x^2 \leq y \leq 4\}$$

(Figure 16.22). Find its center of mass.

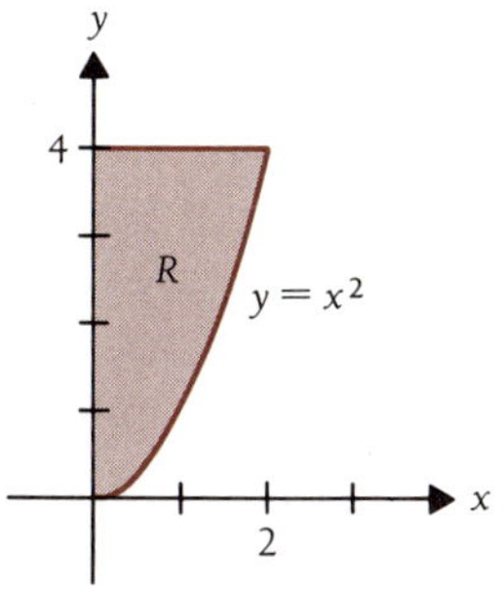

FIGURE 16.22

Solution The mass is

$$\begin{aligned} m &= \iint_R \delta(x, y)\, dA \\ &= \int_0^2 \int_{x^2}^4 (1 + xy)\, dy\, dx \\ &= \int_0^2 \left(4 + 8x - x^2 - \frac{1}{2}x^5\right) dx = 16. \end{aligned}$$

The moments are

$$\begin{aligned} M_y &= \iint_R x\delta(x, y)\, dA \\ &= \int_0^2 \int_{x^2}^4 x(1 + xy)\, dy\, dx \\ &= \int_0^2 \left(4x + 8x^2 - x^3 - \frac{1}{2}x^6\right) dx = \frac{340}{21}, \end{aligned}$$

$$\begin{aligned}
M_x &= \iint_R y\delta(x, y)\, dA \\
&= \int_0^2 \int_{x^2}^4 y(1 + xy)\, dy\, dx \\
&= \int_0^2 \left(8 - \frac{64}{3}x - \frac{1}{2}x^4 + \frac{1}{3}x^7\right) dx = \frac{224}{5}.
\end{aligned}$$

Hence, by formulas (2), we have

$$(\bar{x}, \bar{y}) = \left(\frac{85}{84}, \frac{14}{5}\right).$$

■

Centroids

If the density of a plate is constant, the plate is called **homogeneous**; and a homogeneous plate is also called a **lamina.** The center of mass of a lamina is called a **centroid**. The **centroid of a region** R is the centroid of a lamina covering R. In general, the center of mass of a nonhomogeneous plate will not coincide with the centroid of the region it covers. The centroid of R always lies on an axis of symmetry of R.

The centroid of a region R can be found using the formulas (2), and since the density δ is constant, it divides out. That leaves us with the formulas

$$\bar{x} = \frac{\iint_R x\, dA}{\iint_R dA}, \qquad \bar{y} = \frac{\iint_R y\, dA}{\iint_R dA}. \tag{3}$$

Equivalently, since $\iint_R dA$ is just the area A of R,

$$\iint_R x\, dA = \bar{x}A, \qquad \iint_R y\, dA = \bar{y}A. \tag{4}$$

The formulas (4) sometimes can be used to good advantage in evaluating double integrals. If the integrand is a linear function of x and y, and the centroid $(\bar{x}, \bar{y})$ and the area A of R are known, then from (4) the formula

$$\iint_R (ax + by + c)\, dA = (a\bar{x} + b\bar{y} + c)A \tag{5}$$

enables us to evaluate the double integral without integrating.

■ **Example 16.15** Evaluate $\iint_R (2x - y + 3)\, dA$ for R the circular region $\{(x, y) | x^2 + (y - 1)^2 \le 9\}$.

Solution The centroid of a circular region is the center of the circle. For this region, the centroid is the point $(\bar{x}, \bar{y}) = (0, 1)$. Also, the area of R is $A = \pi(3)^2 = 9\pi$. Thus, from formula (5), we get

$$\iint_R (2x - y + 3)\, dA = (2\bar{x} - \bar{y} + 3)A$$
$$= [2(0) - 1 + 3](9\pi) = 18\pi.$$

■

Exercises 16.2

Find the volume of the solid bounded by the given surfaces.

1. $y = 0,\ \ x + y = 1,\ \ x + 2y = 2,\ \ z = 0,\ \ z = 3x + 4y + 5$
2. $x = 0,\ \ x + y = 1,\ \ 2x + y = 2,\ \ z = 0,\ \ z = 2x + 3y + 3$
3. $x = 0,\ \ y = 0,\ \ x + y = 1,\ \ z = 0,\ \ z = x + y - 4$
4. $x = 0,\ \ y = 0,\ \ z = 0,\ \ \frac{x}{a} + \frac{y}{b} + \frac{z}{c} = 1\ \ (a, b, c > 0)$
5. $x^2 + y^2 = 1,\ \ z = 0,\ \ z = x + y + 2$
6. $x^2 + y^2 = a^2,\ \ z = 0,\ \ z = px + qy + r\ \ (r > 0)$
7. $z = 0,\ \ z = a^2 - x^2 - y^2$
8. $z = 4,\ \ z = 2x^2 + 3y^2$
9. $z = a^2x^2 + b^2y^2,\ \ z = c\ \ (c > 0)$
10. $\frac{x^2}{a^2} + \frac{y^2}{b^2} + \frac{z^2}{c^2} = 1$

For each region R and density function δ, find:
(a) The area of R
(b) The mass of a plate covering R and having density δ
(c) The center of mass of the plate
(d) The centroid of R

11. $R = \{(x, y) | x^2 + y^2 \leq 1\};\ \ \delta(x, y) = y + 1$
12. $R = \{(x, y) | 0 \leq x \leq 1 \text{ and } 0 \leq y \leq x\};\ \ \delta(x, y) = x + 2$
13. $R = \{(x, y) | -2 \leq x \leq 2 \text{ and } x^2 \leq y \leq 4\};\ \ \delta(x, y) = x + y + 6$
14. $R = \{(x, y) | 0 \leq y \leq \sqrt{2}/2,\ y \leq x \leq \sqrt{1 - y^2}\};\ \ \delta(x, y) = x^2 + y^2$
15. $R = [0, a] \times [0, b];\ \ \delta(x, y) = x^2 + y$
16. $R = \{(x, y) | 0 \leq x \leq y^2 \text{ and } 0 \leq y \leq 1\};\ \ \delta(x, y) = 2e^{x/y}$
17. The **moment of inertia** of a plate covering region R and having density δ is

$$I = \iint_R (x^2 + y^2)\delta(x, y)\, dA.$$

Find the moment of inertia of the plate in:
(a) Problem 11 **(b)** Problem 12

18. If function f is continuous over region R, the **average value of f** over R is defined to be

$$\bar{f} = \frac{1}{\text{Area of } R} \iint_R f(x, y)\, dA.$$

Find the average value of:
(a) $f(x, y) = x^2 + y^2$ over $R = [0, a] \times [0, b]$
(b) $f(x, y) = ax + by + c$ over $R = \{(x, y) | x^2 + y^2 \leq r^2\}$

SECTION 16.3

Double Integrals in Polar Coordinates

In this section, we give the definition of a double integral over a region defined in polar coordinates. Then we show how a change from rectangular coordinates to polar coordinates can simplify some double integrals.

Regions in Polar Coordinates

We will start with the simplest type of region in polar coordinates. A **polar rectangle** is a region in the polar coordinate plane of the form

$$R = \{[r, \theta] | a \le r \le b \text{ and } \alpha \le \theta \le \beta\},$$

as shown in Figure 16.23. Here, it is assumed that $a \ge 0$ and $\beta \le \alpha + 2\pi$.

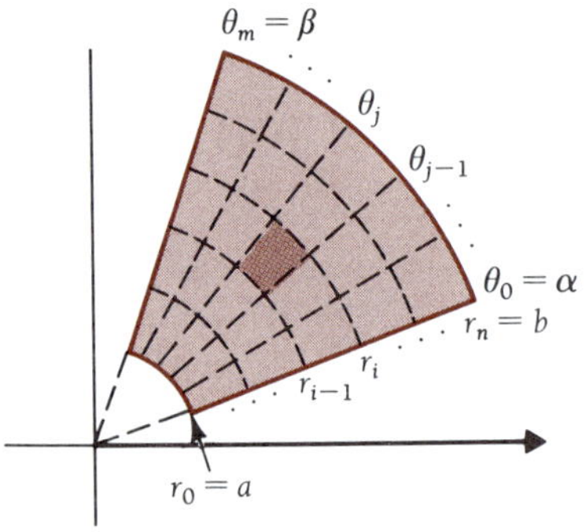

FIGURE 16.23 A polar rectangle with R_{ij} shaded

If f is a continuous function of the variables r and θ, and f is defined on R, the double integral of f over R with respect to area is defined as the limit of a double Riemann sum. To form the Riemann sum, we let $s_1 = \{a = r_0, r_1, \ldots, r_n = b\}$ be a partition of the interval $[a, b]$, and we let $s_2 = \{\alpha = \theta_0, \theta_1, \ldots, \theta_m = \beta\}$ be a partition of the interval $[\alpha, \beta]$. Then

$$s = s_1 \times s_2 = \{[r_i, \theta_j] | 0 \le i \le n \text{ and } 0 \le j \le m\}$$

is a partition of R into polar subrectangles, with the i, jth subrectangle R_{ij} bounded by $r = r_{i-1}$, $r = r_i$, $\theta = \theta_{j-1}$, and $\theta = \theta_j$ (see Figure 16.23). The area of R_{ij} is

$$\Delta A_{ij} = \frac{1}{2} r_i^2 (\theta_j - \theta_{j-1}) - \frac{1}{2} r_{i-1}^2 (\theta_j - \theta_{j-1})$$

$$= \frac{1}{2} (r_i + r_{i-1})(r_i - r_{i-1})(\theta_j - \theta_{j-1}).$$

If we set $\bar{r}_i = \frac{1}{2}(r_i + r_{i-1})$, $\Delta r_i = r_i - r_{i-1}$, and $\Delta\theta_j = \theta_j - \theta_{j-1}$, then (see Figure 16.24)

$$\Delta A_{ij} = \bar{r}_i \Delta r_i \Delta\theta_j.$$

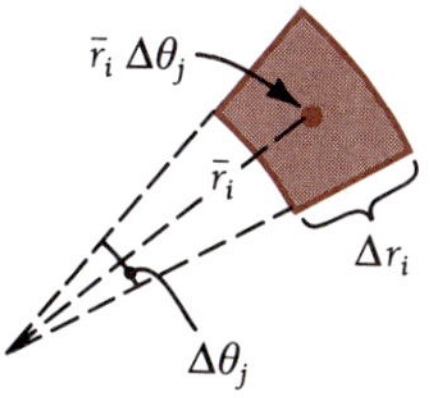

FIGURE 16.24 $\Delta A_{ij} = \bar{r}_i \Delta r_i \Delta \theta_j$

If $\overline{\theta}_j \in [\theta_{j-1}, \theta_j]$, then $[\bar{r}_i, \overline{\theta}_j] \in R_{ij}$, and we form the double Riemann sum

$$\sum_{i=1}^{n} \sum_{j=1}^{m} f(\bar{r}_i, \overline{\theta}_j)\Delta A_{ij} = \sum_{i=1}^{n} \sum_{j=1}^{m} f(\bar{r}_i, \overline{\theta}_j)\bar{r}_i \Delta r_i \Delta \theta_j .$$

The limit (if it exists) of such double Riemann sums as the norms of the partitions approach 0 is the **double integral**

$$\iint_R f(r, \theta)\, dA.$$

This definition is extended to regions that are not polar rectangles in the same way as for rectangular coordinates.

To evaluate the double integral above, we transform it into an iterated integral. If (see Figure 16.25)

$$R = \{[r, \theta] | g(\theta) \le r \le h(\theta) \text{ and } \alpha \le \theta \le \beta\},$$

then we have

$$\iint_R f(r, \theta)\, dA = \int_\alpha^\beta \int_{g(\theta)}^{h(\theta)} f(r, \theta) r\, dr\, d\theta. \tag{1}$$

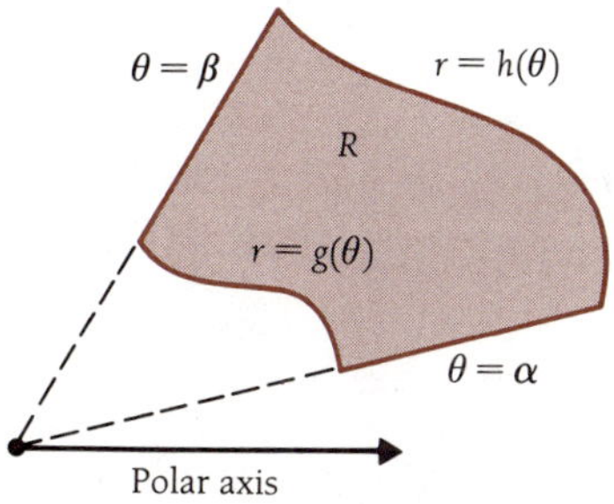

FIGURE 16.25 A Type I polar region

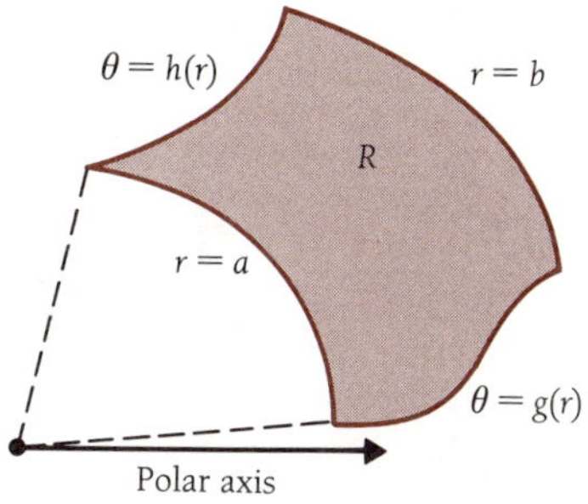

FIGURE 16.26 A Type II polar region

If (see Figure 16.26)

$$R = \{[r, \theta] | a \le r \le b \text{ and } g(r) \le \theta \le h(r)\},$$

then we have

$$\iint_R f(r,\theta)\,dA = \int_a^b \int_{g(r)}^{h(r)} f(r,\theta) r\,d\theta\,dr. \tag{2}$$

The proofs of these formulas are similar to the corresponding proofs in Section 16.1. Note that in rectangular coordinates, dA becomes $dy\,dx$ or $dx\,dy$; in polar coordinates, dA becomes $r\,dr\,d\theta$ or $r\,d\theta\,dr$.

Example 16.16 Evaluate $\iint_R r\cos\theta\,dA$ with

$$R = \{[r,\theta] | 0 \le r \le 1-\sin\theta \text{ and } 0 \le \theta \le \pi/2\}.$$

Solution The region R is shown in Figure 16.27. By formula (1), we have

$$\begin{aligned}\iint_R r\cos\theta\,dA &= \int_0^{\pi/2}\int_0^{1-\sin\theta} r\cos\theta\, r\,dr\,d\theta \\ &= \int_0^{\pi/2}\int_0^{1-\sin\theta} r^2\cos\theta\,dr\,d\theta \\ &= \int_0^{\pi/2} \frac{1}{3} r^3 \cos\theta \Big|_0^{1-\sin\theta} d\theta \\ &= \int_0^{\pi/2} \frac{1}{3}(1-\sin\theta)^3\cos\theta\,d\theta \\ &= \frac{1}{12}.\end{aligned}$$

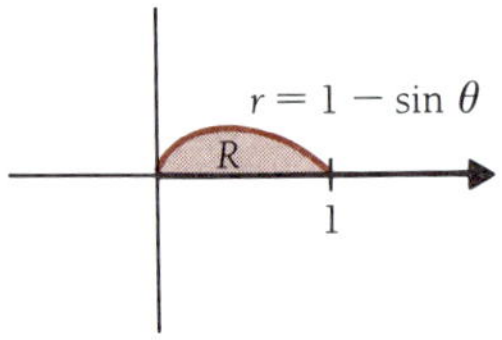

FIGURE 16.27 ■

The applications of double integrals in polar coordinates are about the same as in rectangular coordinates. Consider some examples.

Example 16.17 Find the volume below the surface $z = 1 + r^2$ over the region

$$R = \{[r,\theta] | 0 \le r \le 3 \text{ and } 0 \le \theta \le 2\pi\}.$$

Solution The region R is shown in Figure 16.28. The volume is

$$V = \iint_R (1 + r^2)\, dA$$
$$= \int_0^{2\pi} \int_0^3 (1 + r^2) r\, dr\, d\theta$$
$$= \int_0^{2\pi} \frac{99}{4}\, d\theta = \frac{99\pi}{2}.$$

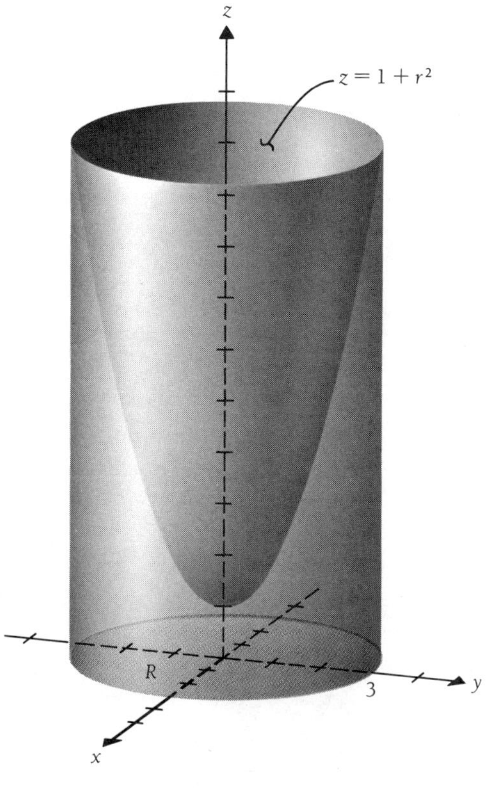

FIGURE 16.28 ■

■ **Example 16.18** Find the volume of the solid inside both the sphere $r^2 + z^2 = 4$ and the cone $z = r\sqrt{3}$, $r \geq 0$ (Figure 16.29, page 974).

Solution The cone and the sphere meet in the circle $r = 1$, $z = \sqrt{3}$, so the solid lies over the region $R = \{[r, \theta] | 0 \leq r \leq 1 \text{ and } 0 \leq \theta \leq 2\pi\}$. The upper surface of the solid is

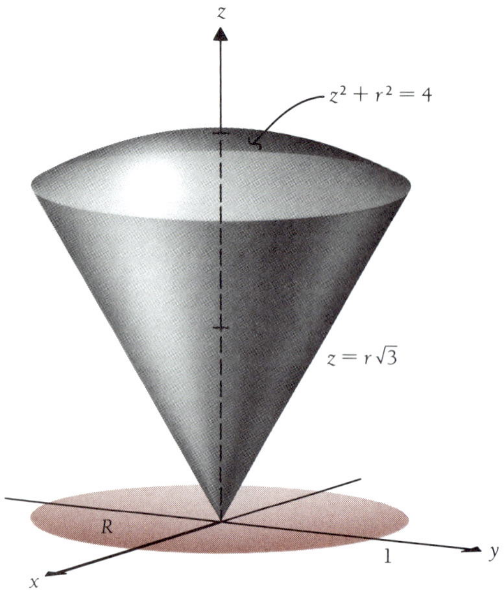

FIGURE 16.29

$z = \sqrt{4 - r^2}$, so the volume is

$$\begin{aligned} V &= \iint\limits_R (\sqrt{4 - r^2} - r\sqrt{3})\, dA \\ &= \int_0^{2\pi} \int_0^1 (\sqrt{4 - r^2} - r\sqrt{3})r\, dr\, d\theta \\ &= \int_0^{2\pi} \left[-\frac{1}{3}(4 - r^2)^{3/2} - \frac{\sqrt{3}}{3} r^3 \right]_0^1 d\theta \\ &= \frac{1}{3}(8 - 4\sqrt{3}) \int_0^{2\pi} d\theta = \frac{2\pi}{3}(8 - 4\sqrt{3}). \end{aligned}$$

■

■ **Example 16.19** Find the area outside the limaçon $r = 3 + 2\cos\theta$ and inside the circle $r = 6\cos\theta$ (see Figure 16.30).

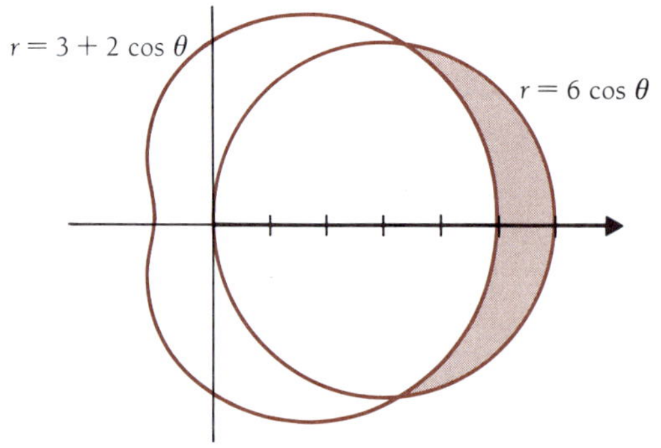

FIGURE 16.30

Solution The two curves meet at the points $[\frac{9}{2}, \cos^{-1}\frac{3}{4}]$ and $[\frac{9}{2}, -\cos^{-1}\frac{3}{4}]$. Hence, the area is just $\iint_R dA$ with

$$R = \{[r, \theta] | 3 + 2\cos\theta \le r \le 6\cos\theta \text{ and } -\cos^{-1}\tfrac{3}{4} \le \theta \le \cos^{-1}\tfrac{3}{4}\}.$$

Thus,

$$\begin{aligned} A &= \int_{-\cos^{-1}(3/4)}^{\cos^{-1}(3/4)} \int_{3+2\cos\theta}^{6\cos\theta} r\,dr\,d\theta \\ &= \int_{-\cos^{-1}(3/4)}^{\cos^{-1}(3/4)} \left[\frac{1}{2}r^2\right]_{3+2\cos\theta}^{6\cos\theta} d\theta \\ &= \int_{-\cos^{-1}(3/4)}^{\cos^{-1}(3/4)} \left(8\cos 2\theta - 6\cos\theta + \frac{7}{2}\right) d\theta \\ &= \left[4\sin 2\theta - 6\sin\theta + \frac{7}{2}\theta\right]_{-\cos^{-1}(3/4)}^{\cos^{-1}(3/4)} \\ &= 7\cos^{-1}\frac{3}{4}. \end{aligned}$$

■

■ **Example 16.20** A plate covers the region bounded by the cardioid $r = 1 + \sin\theta$ and has density $\delta(r, \theta) = 2r$. Find its mass.

Solution The region inside the cardioid is

$$R = \{[r, \theta] | 0 \le r \le 1 + \sin\theta \text{ and } 0 \le \theta \le 2\pi\}$$

(Figure 16.31, page 976). Hence, the mass is

$$\begin{aligned} m &= \iint_R \delta(r, \theta)\,dA \\ &= \int_0^{2\pi} \int_0^{1+\sin\theta} (2r)r\,dr\,d\theta \\ &= \int_0^{2\pi} \frac{2}{3}(1 + \sin\theta)^3\,d\theta \\ &= \int_0^{2\pi} \frac{2}{3}(1 + 3\sin\theta + 3\sin^2\theta + \sin^3\theta)\,d\theta \\ &= \int_0^{2\pi} \left[\frac{2}{3} + 2\sin\theta + 2\left(\frac{1-\cos 2\theta}{2}\right) + \frac{2}{3}(1 - \cos^2\theta)\sin\theta\right] d\theta \\ &= \int_0^{2\pi} \frac{2}{3}\,d\theta + 2\int_0^{2\pi} \sin\theta\,d\theta + \int_0^{2\pi} (1 - \cos 2\theta)\,d\theta - \frac{2}{3}\int_1^1 (1 - u^2)\,du \\ &= \frac{2}{3}(2\pi) + 0 + 2\pi + 0 = \frac{10\pi}{3}. \end{aligned}$$

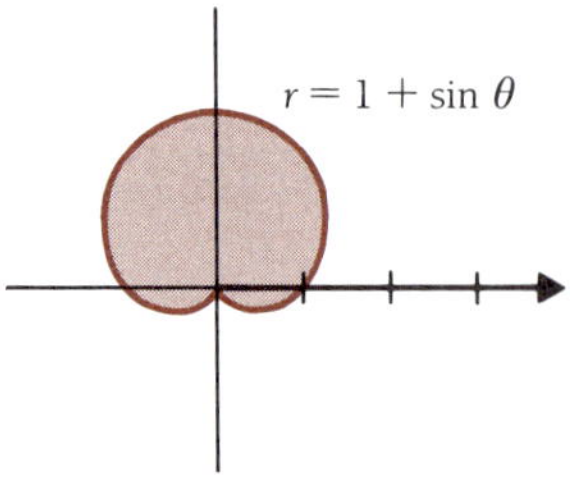

FIGURE 16.31

■ **Example 16.21** Find the center of mass of the plate in Example 16.20.

Solution The center of mass is given by formula (2) in Section 16.2. Using $x = r\cos\theta$ and $y = r\sin\theta$, we get

$$\begin{aligned}\iint_R x\delta\,dA &= \int_0^{2\pi}\int_0^{1+\sin\theta} r\cos\theta\,(2r)r\,dr\,d\theta\\ &= \int_0^{2\pi}\frac{1}{2}(1+\sin\theta)^4\cos\theta\,d\theta\\ &= 0,\\ \iint_R y\delta\,dA &= \int_0^{2\pi}\int_0^{1+\sin\theta} r\sin\theta\,(2r)r\,dr\,d\theta\\ &= \int_0^{2\pi}\frac{1}{2}(1+\sin\theta)^4\sin\theta\,d\theta\\ &= \frac{7\pi}{2}.\end{aligned}$$

(We have left out several steps.) Therefore, using the result of Example 16.20, we find the center of mass to be $(\bar{x}, \bar{y}) = (0, \frac{21}{20})$, which is $[\frac{21}{20}, \pi/2]$ in polar coordinates. ■

Switching to Polar Coordinates

Sometimes a double integral in rectangular coordinates can be more easily evaluated by switching to polar coordinates (or cylindrical coordinates, from the spatial point of view). The switch to polar coordinates is made in the double integral $\iint_R f(x, y)\,dA$ by redescribing R in polar coordinates and substituting $x = r\cos\theta$ and $y = r\sin\theta$.

Consider an example.

■ **Example 16.22** Evaluate the iterated integral below by switching to polar coordinates:

$$\int_{-4}^{4}\int_{-\sqrt{16-x^2}}^{\sqrt{16-x^2}} \sqrt{4x^2+4y^2+1}\,dy\,dx.$$

Solution This integral is the same as

$$\iint_R \sqrt{4x^2 + 4y^2 + 1}\, dA,$$

with

$$R = \{(x, y) | -4 \le x \le 4 \text{ and } -\sqrt{16 - x^2} \le y \le \sqrt{16 - x^2}\}$$

(Figure 16.32). That is, R is the disc $x^2 + y^2 \le 16$, which is described in polar coordinates as

$$R = \{[r, \theta] | 0 \le r \le 4 \text{ and } 0 \le \theta \le 2\pi\}.$$

With $x = r\cos\theta$ and $y = r\sin\theta$, $\sqrt{4x^2 + 4y^2 + 1}$ becomes $\sqrt{4r^2 + 1}$. Hence, the original iterated integral becomes

$$\begin{aligned}\int_0^4 \int_0^{2\pi} \sqrt{4r^2 + 1}\, r\, d\theta\, dr &= 2\pi \int_0^4 \sqrt{4r^2 + 1}\, r\, dr \\ &= 2\pi \int_1^{65} \frac{1}{8} u^{1/2}\, du \qquad (u = 4r^2 + 1) \\ &= \frac{\pi}{6}(65^{3/2} - 1).\end{aligned}$$

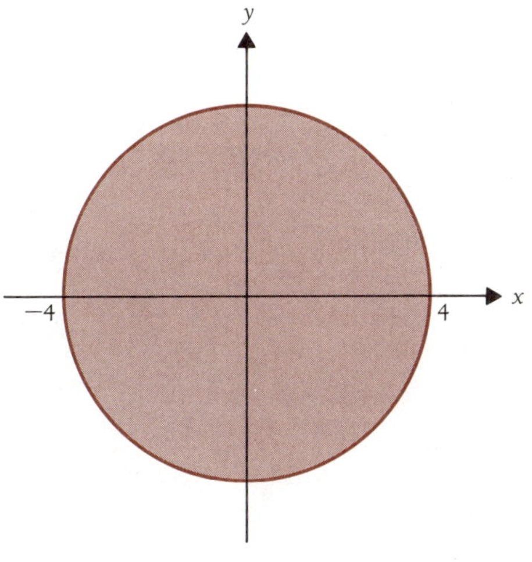

FIGURE 16.32 ■

Switching to Other Coordinate Systems

The double integral $\iint_R f(x, y)\, dA$ in rectangular coordinates is equal to an iterated integral in which dA is replaced by $dy\, dx$ or $dx\, dy$. In switching to polar coordinates, we get an iterated integral in which $f(x, y)$ is replaced by $f(r\cos\theta, r\sin\theta)$ and dA is

The Jacobian determinant was named for **C. G. J. Jacobi** (1804–1851), the first prominent Jewish mathematician of modern times. He was born in Königsberg (now Kalinengrad) and was a professor at the university there all his adult life. His work was in elliptic functions, number theory, and mechanics, to all of which he made major contributions.

replaced by $r\,dr\,d\theta$ or $r\,d\theta\,dr$. The extra factor r can be obtained from the equations $x = r\cos\theta$, $y = r\sin\theta$ by using a formula known as the **Jacobian.**

In general terms, if the xy-coordinate system and a uv-coordinate system are related by the equations $x = g(u, v)$, $y = h(u, v)$, then the **Jacobian of x and y relative to u and v** is the determinant of partial derivatives,

$$\frac{\partial(x, y)}{\partial(u, v)} = \begin{vmatrix} \dfrac{\partial x}{\partial u} & \dfrac{\partial x}{\partial v} \\ \dfrac{\partial y}{\partial u} & \dfrac{\partial y}{\partial v} \end{vmatrix}$$

For $x = r\cos\theta$, $y = r\sin\theta$, the Jacobian is

$$\begin{aligned} \frac{\partial(x, y)}{\partial(r, \theta)} &= \begin{vmatrix} \dfrac{\partial x}{\partial r} & \dfrac{\partial x}{\partial \theta} \\ \dfrac{\partial y}{\partial r} & \dfrac{\partial y}{\partial \theta} \end{vmatrix} \\ &= \begin{vmatrix} \cos\theta & -r\sin\theta \\ \sin\theta & r\cos\theta \end{vmatrix} \\ &= r\cos^2\theta + r\sin^2\theta = r. \end{aligned}$$

In general, if the double integral $\iint_R f(x, y)\,dA$ is equal to an iterated integral in uv-coordinates, then $f(x, y)$ is replaced by $f(g(u, v), h(u, v))$ and dA is replaced by

$$\frac{\partial(x, y)}{\partial(u, v)}\,du\,dv \qquad \text{or} \qquad \frac{\partial(x, y)}{\partial(u, v)}\,dv\,du.$$

■ **Example 16.23** Compute the Jacobian of x and y relative to u and v if $x = u^2 - v^2$ and $y = u^2 + v^2$.

Solution Using the definition, we find

$$\begin{aligned} \frac{\partial(x, y)}{\partial(u, v)} &= \begin{vmatrix} \dfrac{\partial x}{\partial u} & \dfrac{\partial x}{\partial v} \\ \dfrac{\partial y}{\partial u} & \dfrac{\partial y}{\partial v} \end{vmatrix} \\ &= \begin{vmatrix} 2u & -2v \\ 2u & 2v \end{vmatrix} \\ &= 4uv + 4uv = 8uv. \end{aligned}$$

■

Exercises 16.3

Problems 1–18 use the following regions:

$$R_1 = \{[r, \theta] | 0 \le r \le 1 \text{ and } 0 \le \theta \le 2\pi\},$$
$$R_2 = \{[r, \theta] | 0 \le r \le 2 \cos \theta \text{ and } 0 \le \theta \le \pi/2\},$$
$$R_3 = \{[r, \theta] | 0 \le r \le 1 \text{ and } 0 \le \theta \le 2r\},$$
$$R_4 = \{[r, \theta] | 1 - \cos \theta \le r \le 1 + \cos \theta \text{ and } 0 \le \theta \le \pi/2\}.$$

Evaluate the following double integrals:

1. $\iint_{R_1} r^2 \cos^2\theta \, dA$
2. $\iint_{R_1} r^2 \, dA$
3. $\iint_{R_2} r \sin \theta \, dA$
4. $\iint_{R_2} \sin 2\theta \, dA$
5. $\iint_{R_3} r\theta \, dA$
6. $\iint_{R_3} (r^3 + \cos \theta) \, dA$
7. $\iint_{R_4} \sin \theta \, dA$
8. $\iint_{R_4} \sqrt{r} \, dA$

Find the following:

9. The area of R_3
10. The area of R_4
11. The centroid of R_4
12. The area of $R_1 \cap R_2$
13. The centroid of $R_1 \cap R_2$
14. The volume inside $f(r, \theta) = r \cos \theta$ over R_1

Find the mass and center of mass of each plate:

15. Covering R_1; $\delta(r, \theta) = k\theta$
16. Covering R_2; $\delta(r, \theta) = r^2$
17. Covering R_3; $\delta(r, \theta) = 1 - r$
18. Covering R_4; $\delta(r, \theta) = \cos \theta$

Find the area:

19. Of one leaf of $r = 3 \sin 2\theta$
20. Inside the lemniscate $r^2 = 4 \cos \theta$ and outside the circle $r = 1$

Find the volume:

21. Inside the cylinder $(x - 1)^2 + y^2 = 1$, above the xy-plane, and below the cone $z^2 = x^2 + y^2$
22. Inside the cylinder $x^2 + y^2 = 1$ and inside the sphere $x^2 + y^2 + z^2 = 4$

Evaluate the iterated integrals by switching to polar coordinates.

23. $\displaystyle\int_0^1 \int_x^{\sqrt{2-x^2}} e^{x^2+y^2} \, dy \, dx$
24. $\displaystyle\int_{-1}^1 \int_{-\sqrt{1-x^2}}^{\sqrt{1-x^2}} \cos(x^2 + y^2) \, dy \, dx$
25. $\displaystyle\int_0^1 \int_0^{\sqrt{1-y^2}} \sqrt{1 - x^2 - y^2} \, dx \, dy$
26. $\displaystyle\int_0^2 \int_0^{\sqrt{4-x^2}} \sqrt{9x^2 + 9y^2 + 1} \, dy \, dx$
27. Evaluate the improper double integral $\iint_R e^{-(x^2+y^2)} \, dA$ for $R = \{(x, y) | x \ge 0\}$ by switching to polar coordinates.
28. Compute the Jacobian for x and y relative to u and v if $x = 2uv$, $y = u^2 - v^2$.
29. For the equations $r = \sqrt{x^2 - y^2}$, $\theta = \arctan(y/x)$, compute the Jacobian of r and θ relative to x and y.

SECTION 16.4

Triple Integrals

Now that you have seen double integrals, triple integrals will not be too difficult for you to understand. A region in space is similar to a region in the plane, except that it is solid. For example, a circular region in the plane has a spherical region as its analog in space, and an open disc in the plane becomes an open ball in space. As was the case with regions in the plane, we will consider only regions of certain very special forms.

Regions in Space and Triple Integrals

The simplest region in space is a **rectangular region**, having the form

$$\begin{aligned} R &= [a, b] \times [c, d] \times [e, f] \\ &= \{(x, y, z) \mid a \leq x \leq b,\ c \leq y \leq d,\ e \leq z \leq f\}. \end{aligned} \tag{1}$$

(See Figure 16.33.) Such a region can be partitioned into subregions R_{ijk} (see Figure 16.34), which have volumes $\Delta V_{ijk} = \Delta x_i \Delta y_j \Delta z_k$. If f is a function of three variables defined on R, a **triple Riemann sum** of f over R has the form

$$\sum_{i=1}^{n} \sum_{j=1}^{m} \sum_{k=1}^{p} f(\bar{x}_i, \bar{y}_j, \bar{z}_k) \Delta V_{ijk},$$

with $(\bar{x}_i, \bar{y}_j, \bar{z}_k)$ a point of R_{ijk}. If the limit of such triple Riemann sums exists as the norms of the subdivisions approach 0, we say f is **integrable over R**, and we denote the limit by the **triple integral**

$$\iiint\limits_R f(x, y, z)\, dV.$$

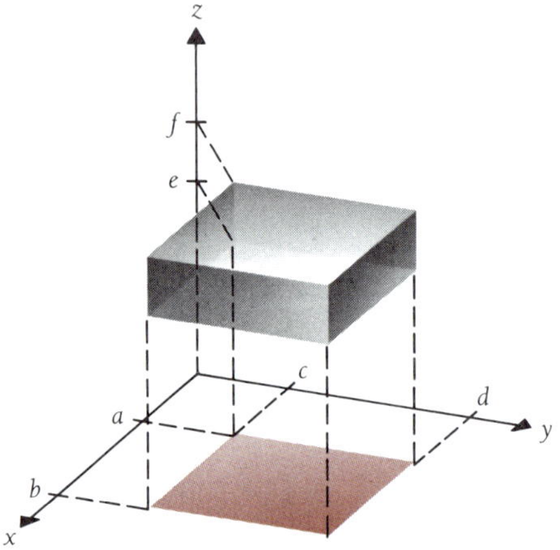

FIGURE 16.33 A rectangular region in space

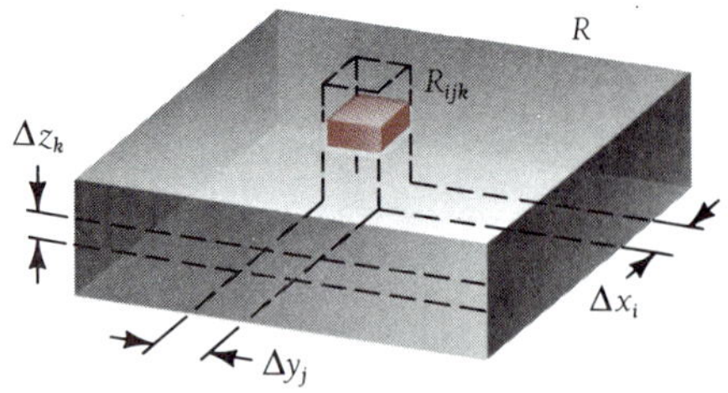

FIGURE 16.34 Subregion R_{ijk}

We generalize to triple integrals over nonrectangular regions as in the two-dimensional case.

We will consider regions such as

$$R = \{(x, y, z) | a \le x \le b,\ g(x) \le y \le h(x),\ G(x, y) \le z \le H(x, y)\}. \tag{2}$$

This region is bounded below and above by the surfaces $z = G(x, y)$ and $z = H(x, y)$, left and right by the cylinders $y = g(x)$ and $y = h(x)$, and front and back by the planes $x = b$ and $x = a$ (see Figure 16.35). We will also allow for a permutation of the variables in such regions, giving us six types of regions in all; these correspond to the various possible orientations of the region in Figure 16.35.

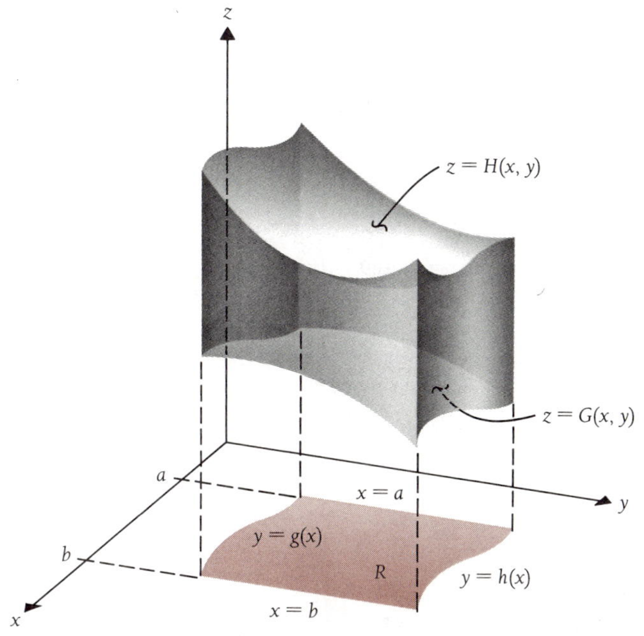

FIGURE 16.35 A region in space

It can be shown that any continuous function is integrable over such a region.

Threefold Iterated Integrals

To evaluate a triple integral, we transform it into a threefold iterated integral. For example, if R is the region (2) shown in Figure 16.35, then

$$\iiint_R f(x, y, z)\, dV = \int_a^b \int_{g(x)}^{h(x)} \int_{G(x, y)}^{H(x, y)} f(x, y, z)\, dz\, dy\, dx. \tag{3}$$

Other types of regions give rise to similar iterated integrals, nested in other orders. Here are some examples.

Example 16.24 Find $\iiint_R dV$ for $R = [a, b] \times [c, d] \times [e, f]$.

Solution Comparing equations (1), (2), and (3), we see that

$$\begin{aligned}\iiint\limits_R dV &= \int_a^b \int_c^d \int_e^f dz\, dy\, dx \\ &= \int_a^b \int_c^d (f-e)\, dy\, dx \\ &= \int_a^b (d-c)(f-e)\, dx \\ &= (b-a)(d-c)(f-e),\end{aligned}$$

which is the volume of R. ■

Example 16.25 Let R be the region bounded by the coordinate planes and the plane $x + y + z = 1$ (see Figure 16.36). Evaluate $\iiint_R x^2\, dV$.

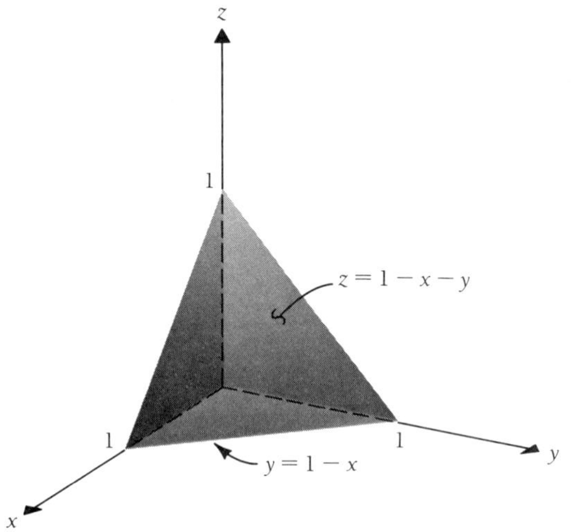

FIGURE 16.36

Solution Our first task is to describe R. Using the equation of the plane, we may describe R as

$$R = \{(x, y, z) | 0 \leq x \leq 1,\ 0 \leq y \leq 1 - x,\ 0 \leq z \leq 1 - x - y\}.$$

Using formula (3), we have

$$\iiint_R x^2\,dV = \int_0^1 \int_0^{1-x} \int_0^{1-x-y} x^2\,dz\,dy\,dx$$
$$= \int_0^1 \int_0^{1-x} x^2(1-x-y)\,dy\,dx$$
$$= \int_0^1 x^2\left[(1-x) - x(1-x) - \frac{1}{2}(1-x)^2\right] dx$$
$$= \int_0^1 \left(\frac{1}{2}x^2 - x^3 + \frac{1}{2}x^4\right) dx = \frac{1}{60}.$$

■

Applications of Triple Integrals

If R is a region in the plane, $\iint_R dA$ is just the area of R. Similarly, if R is a region in space, then $\iiint_R dV$ is the volume of R. If a solid object in space occupies the region R and has density $\delta(x, y, z)$, then $\iiint_R \delta(x, y, z)\,dV$ is the mass of the object. The center of mass of the object is the point $(\bar{x}, \bar{y}, \bar{z})$ given by

$$\bar{x} = \frac{\iiint_R x\delta\,dV}{\iiint_R \delta\,dV}, \qquad \bar{y} = \frac{\iiint_R y\delta\,dV}{\iiint_R \delta\,dV}, \qquad \bar{z} = \frac{\iiint_R z\delta\,dV}{\iiint_R \delta\,dV}. \tag{4}$$

■ **Example 16.26** Let R be the region bounded by $y = x^2 + z^2$, the xy-plane, and $y = 2$ (Figure 16.37). A solid object occupying R has density that varies as the distance from the xy-plane. Find the volume, mass, center of mass, and centroid of the object.

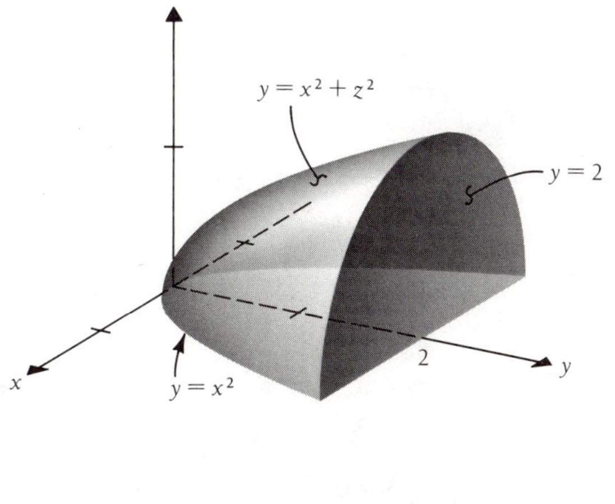

FIGURE 16.37

Solution We can describe R as

$$R = \{(x, y, z) | -\sqrt{2} \le x \le \sqrt{2},\ 0 \le z \le \sqrt{2 - x^2},\ x^2 + z^2 \le y \le 2\}.$$

Hence, the volume of the solid is

$$\begin{aligned} V = \iiint_R dV &= \int_{-\sqrt{2}}^{\sqrt{2}} \int_0^{\sqrt{2-x^2}} \int_{x^2+z^2}^{2} dy\, dz\, dx \\ &= \int_{-\sqrt{2}}^{\sqrt{2}} \int_0^{\sqrt{2-x^2}} (2 - x^2 - z^2)\, dz\, dx \\ &= \int_{-\sqrt{2}}^{\sqrt{2}} \frac{2}{3} (2 - x^2)^{3/2}\, dx = \pi. \end{aligned}$$

The density of the solid is $\delta(x, y, z) = kz$, so its mass is

$$\begin{aligned} m = \iiint_R \delta\, dV &= \int_{-\sqrt{2}}^{\sqrt{2}} \int_0^{\sqrt{2-x^2}} \int_{x^2+z^2}^{2} kz\, dy\, dz\, dx \\ &= \int_{-\sqrt{2}}^{\sqrt{2}} \int_0^{\sqrt{2-x^2}} kz(2 - x^2 - z^2)\, dz\, dx \\ &= \int_{-\sqrt{2}}^{\sqrt{2}} \frac{k}{4} (2 - x^2)^2\, dx = \frac{16k\sqrt{2}}{15}. \end{aligned}$$

To find the center of mass, we use formulas (4). By symmetry, we know $\bar{x} = 0$. We find

$$\begin{aligned} \bar{y} = \frac{\iiint_R y\delta\, dV}{\iiint_R \delta\, dV} &= \frac{15}{16k\sqrt{2}} \int_{-\sqrt{2}}^{\sqrt{2}} \int_0^{\sqrt{2-x^2}} \int_{x^2+z^2}^{2} ykz\, dy\, dz\, dx \\ &= \frac{15}{16\sqrt{2}} \int_{-\sqrt{2}}^{\sqrt{2}} \int_0^{\sqrt{2-x^2}} \left(2z - \frac{1}{2} x^4 z - x^2 z^3 - \frac{1}{2} z^5\right) dz\, dx \\ &= \frac{15}{16\sqrt{2}} \int_{-\sqrt{2}}^{\sqrt{2}} \left(\frac{4}{3} - x^2 + \frac{1}{12} x^6\right) dx = \frac{10}{7}, \end{aligned}$$

$$\begin{aligned} \bar{z} = \frac{\iiint_R z\delta\, dV}{\iiint_R \delta\, dV} &= \frac{15}{16k\sqrt{2}} \int_{-\sqrt{2}}^{\sqrt{2}} \int_0^{\sqrt{2-x^2}} \int_{x^2+z^2}^{2} zkz\, dy\, dz\, dx \\ &= \frac{15}{16\sqrt{2}} \int_{-\sqrt{2}}^{\sqrt{2}} \int_0^{\sqrt{2-x^2}} z^2(2 - x^2 - z^2)\, dz\, dx \\ &= \frac{15}{16\sqrt{2}} \int_{-\sqrt{2}}^{\sqrt{2}} \frac{2}{15} (2 - x^2)^{5/2}\, dx = \frac{5\pi}{16\sqrt{2}}. \end{aligned}$$

Thus, the center of mass is $(0, \frac{10}{7}, 5\pi/16\sqrt{2})$.

Finally, we find the centroid using formulas (4) with δ constant. Again by symmetry, $\bar{x} = 0$. We have

$$\bar{y} = \frac{\iiint_R y\, dV}{\iiint_R dV} = \frac{1}{\pi}\int_{-\sqrt{2}}^{\sqrt{2}}\int_0^{\sqrt{2-x^2}}\int_{x^2+z^2}^{2} y\, dy\, dz\, dx$$

$$= \frac{1}{\pi}\int_{-\sqrt{2}}^{\sqrt{2}}\int_0^{\sqrt{2-x^2}}\left(2 - \frac{1}{2}x^4 - x^2z^2 - \frac{1}{2}z^4\right) dz\, dx$$

$$= \frac{1}{\pi}\int_{-\sqrt{2}}^{\sqrt{2}}\left(\frac{8}{5} - \frac{4}{15}x^2 - \frac{4}{15}x^4\right)\sqrt{2-x^2}\, dx = \frac{4}{3},$$

$$\bar{z} = \frac{\iiint_R z\, dV}{\iiint_R dV} = \frac{1}{\pi}\int_{-\sqrt{2}}^{\sqrt{2}}\int_0^{\sqrt{2-x^2}}\int_{x^2+z^2}^{2} z\, dy\, dz\, dx$$

$$= \frac{16\sqrt{2}}{15\pi}$$

(see the integral for m). Thus, the centroid is $(0, \frac{4}{3}, 16\sqrt{2}/15\pi)$. ■

As with double integrals, the centroid concept can make the evaluation of certain triple integrals a simple task. Rewriting formulas (4) with δ constant, we get

$$\iiint_R x\, dV = \bar{x}V, \qquad \iiint_R y\, dV = \bar{y}V, \qquad \iiint_R z\, dV = \bar{z}V. \tag{5}$$

Thus, if the region R has known volume V and centroid $(\bar{x}, \bar{y}, \bar{z})$, and the integrand is linear, we have

$$\iiint_R (ax + by + cz + d)\, dV = (a\bar{x} + b\bar{y} + c\bar{z} + d)V. \tag{6}$$

Example 16.27 A solid occupying the region $R = [0, 3] \times [0, 2] \times [0, 4]$ has density $\delta(x, y, z) = 2x - 3y + z + 5$. Find its mass.

Solution The centroid of the region R is $(\bar{x}, \bar{y}, \bar{z}) = (\frac{3}{2}, 1, 2)$, and its volume is $3(2)(4) = 24$. Therefore, the mass is

$$m = \iiint_R \delta(x, y, z)\, dV$$

$$= \iiint_R (2x - 3y + z + 5)\, dV$$

$$= (2\bar{x} - 3\bar{y} + \bar{z} + 5)V$$

$$= \left[2\left(\frac{3}{2}\right) - 3(1) + 2 + 5\right](24) = 168.$$ ■

Exercises 16.4

Find the volume of the solid bounded by the given surfaces.

1. $x=0,\ \ y=0,\ \ z=0,\ \ x+y+z=a\ \ (a>0)$
2. $x=0,\ \ y=0,\ \ z=0,\ \ 20x+15y+30z=60$
3. $x=0,\ \ y=0,\ \ z=0,\ \ \dfrac{x}{2}+\dfrac{y}{3}+\dfrac{z}{5}=1$
4. $x=0,\ \ y=0,\ \ z=-1,\ \ \dfrac{x}{a}+\dfrac{y}{b}+\dfrac{z}{c}=1\ \ (a,\ b,\ c>0)$
5. $x=0,\ \ x=2,\ \ y=0,\ \ y=3,\ \ z=y+3,\ \ z=-x-1$
6. $x=1,\ \ x=2,\ \ z=0,\ \ z=4,\ \ y=x-1,\ \ y=z+9$
7. $x=0,\ \ x=r,\ \ y=0,\ \ y=\sqrt{r^2-x^2},\ \ z=0,\ \ z=\sqrt{r^2-x^2-y^2}$
8. $x=-a,\ \ x=a,\ \ y=-b,\ \ y=b,\ \ z=x+y+c,\ \ z=x+y-c$ $(a>0,\ b>0,\ c>0)$
9. $x^2+y^2=4,\ \ x^2+y^2+z^2=9$ (inside both)
10. $y=e^{x+z},\ \ y=0,\ \ x=0,\ \ x=\ln 2,\ \ z=0,\ \ z=1$

A solid with density δ occupies the region R. Find the mass and center of mass of the solid.

11. $R=[0,1]\times[0,1]\times[0,1];\ \ \delta(x,y,z)=xy+z$
12. $R=[0,a]\times[0,b]\times[0,c];\ \ \delta(x,y,z)=x^2+y^2+z^2$
13. $R=\{(x,y,z)|x^2+y^2+z^2\le a^2\};\ \ \delta(x,y,z)=k(a-\sqrt{x^2+y^2+z^2})$
14. $R=\{(x,y,z)|x^2+y^2+z^2\le 1\};$ δ varies as the distance from the plane $z=-1$
15. Find the volume of the tetrahedron bounded by the coordinate planes and the plane
$$\frac{x}{a}+\frac{y}{b}+\frac{z}{c}=1.$$
16. Find the volume of the ellipsoid
$$\frac{x^2}{a^2}+\frac{y^2}{b^2}+\frac{z^2}{c^2}=1.$$
17. Find the centroid of the tetrahedron in Problem 15.
18. Suppose the tetrahedron in Problem 15 has density $\delta(x,y,z)=2x+y$. Find its mass and center of mass.
19. A pyramid has vertices $(0,0,0)$, $(a,0,0)$, $(a,b,0)$, $(0,b,0)$, and $(0,0,c)$. Find its volume.
20. Find the centroid of the pyramid in Problem 19.
21. Find the volume bounded by $z=3x^2$, $z=4-x^2$, $y=0$, and $y+z=6$.
22. Find the volume bounded by $z=y^2$, $z=4$, $x=0$, and $x=7-y-z$.
23. If a tetrahedron has vertices (a_1,b_1,c_1), (a_2,b_2,c_2), (a_3,b_3,c_3), and (a_4,b_4,c_4), then its centroid is
$$\left(\frac{1}{4}(a_1+a_2+a_3+a_4),\ \frac{1}{4}(b_1+b_2+b_3+b_4),\ \frac{1}{4}(c_1+c_2+c_3+c_4)\right).$$
Verify this for the tetrahedron in Problem 15.

SECTION 16.5

Triple Integrals in Cylindrical and Spherical Coordinates

Triple integrals over regions in cylindrical or spherical coordinates are defined in much the same way as those in rectangular coordinates. Rather than give the definitions, we will show what changes are wrought in going to the iterated integrals in the other coordinate systems.

Recall that the triple integral $\iiint_R f(x, y, z)\, dV$ in rectangular coordinates becomes an iterated integral in which dV is replaced by the product of dx, dy, and dz (in one order or another). This is represented schematically in Figure 16.38.

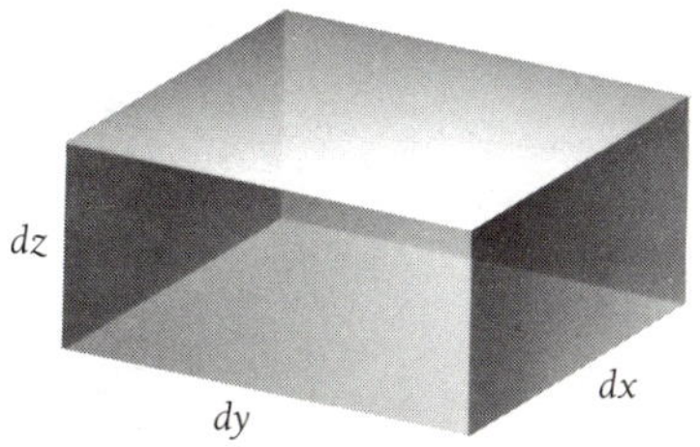

FIGURE 16.38 $dV = dx\, dy\, dz$

Cylindrical Coordinates

In cylindrical coordinates, the triple integral $\iiint_R f(x, y, z)\, dV$ becomes an iterated integral in which dV is replaced by the product of r, dr, $d\theta$, and dz (in one order or another). This is represented in Figure 16.39, using a "rectangular" region in cylindrical coordinates.

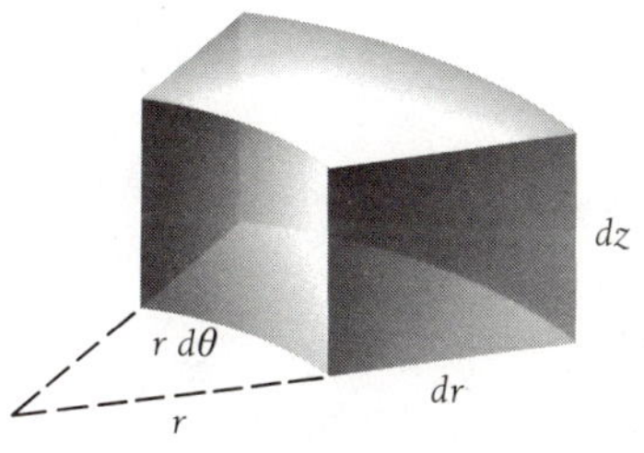

FIGURE 16.39 $dV = r\, dr\, d\theta\, dz$

We also can verify this fact by using a three-variable Jacobian. The rectangular coordinates are related to the cylindrical coordinates by the equations

$$x = r \cos \theta, \qquad y = r \sin \theta, \qquad z = z,$$

so that the Jacobian of x, y, z relative to r, θ, z is

$$\frac{\partial(x, y, z)}{\partial(r, \theta, z)} = \begin{vmatrix} \frac{\partial x}{\partial r} & \frac{\partial x}{\partial \theta} & \frac{\partial x}{\partial z} \\ \frac{\partial y}{\partial r} & \frac{\partial y}{\partial \theta} & \frac{\partial y}{\partial z} \\ \frac{\partial z}{\partial r} & \frac{\partial z}{\partial \theta} & \frac{\partial z}{\partial z} \end{vmatrix}$$

$$= \begin{vmatrix} \cos\theta & -r\sin\theta & 0 \\ \sin\theta & r\cos\theta & 0 \\ 0 & 0 & 1 \end{vmatrix}$$

$$= r\cos^2\theta + r\sin^2\theta = r.$$

■ **Example 16.28** Evaluate $\iiint_R rz\,dV$ for R the region inside the cylinder $r = 2$, above the xy-plane, and below the paraboloid $z = 9 - r^2$ (Figure 16.40).

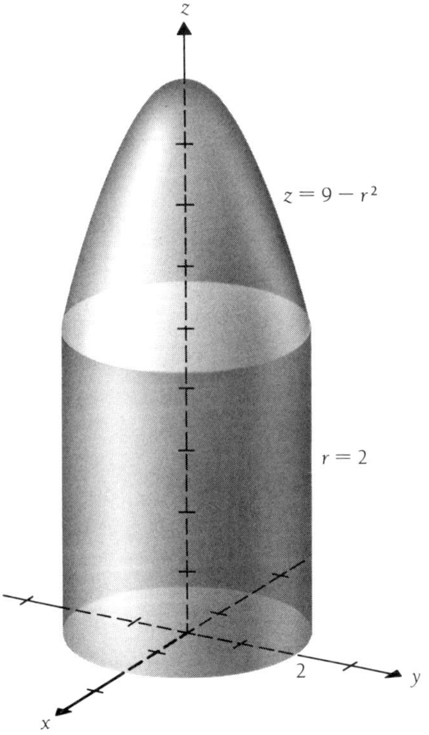

FIGURE 16.40

Solution We can describe R as

$$R = \{[r, \theta, z] | 0 \le r \le 2,\ 0 \le \theta \le 2\pi,\ 0 \le z \le 9 - r^2\}.$$

Hence, we get the iterated integral

$$\iiint_R rz\, dV = \int_0^{2\pi}\int_0^2\int_0^{9-r^2} (rz)r\, dz\, dr\, d\theta$$

$$= \int_0^{2\pi}\int_0^2 \left[\frac{1}{2}r^2z^2\right]_0^{9-r^2} dr\, d\theta$$

$$= \int_0^{2\pi}\int_0^2 \left(\frac{81}{2}r^2 - 9r^4 + \frac{1}{2}r^6\right) dr\, d\theta$$

$$= \int_0^{2\pi} \frac{2084}{35}\, d\theta = \frac{4168\pi}{35}.$$

■

■ **Example 16.29** Find the mass of the object occupying the region

$$R = \{[r, \theta, z] | 0 \le r \le a,\ 0 \le \theta \le 2\pi,\ 0 \le z \le L\}$$

(Figure 16.41) and having density $\delta(r, \theta, z) = 1/(2 + r)$.

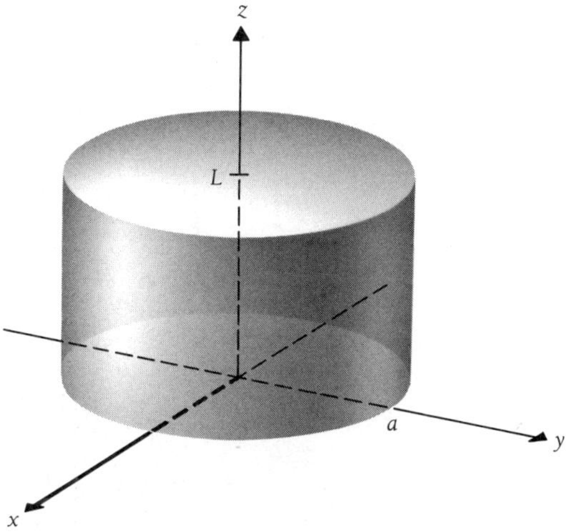

FIGURE 16.41

Solution The mass is

$$m = \iiint_R \delta\, dV = \int_0^{2\pi}\int_0^L\int_0^a \frac{1}{2+r} r\, dr\, dz\, d\theta$$

$$= \int_0^{2\pi}\int_0^L\int_0^a \left(1 - \frac{2}{2+r}\right) dr\, dz\, d\theta$$

$$m = \int_0^{2\pi} \int_0^L \Big[r - 2 \ln(2 + r) \Big]_0^a dz\, d\theta$$

$$= \left[a - 2 \ln \left(\frac{2 + a}{2} \right) \right] \int_0^{2\pi} \int_0^L dz\, d\theta$$

$$= L \left[a - 2 \ln \left(\frac{2 + a}{2} \right) \right] \int_0^{2\pi} d\theta$$

$$= 2\pi L \left[a - 2 \ln \left(\frac{2 + a}{2} \right) \right].$$ ■

■ **Example 16.30** Evaluate the iterated integral below by switching to cylindrical coordinates:

$$\int_{-2\sqrt{2}}^{2\sqrt{2}} \int_{-\sqrt{8-x^2}}^{\sqrt{8-x^2}} \int_{\sqrt{x^2+y^2}}^{\sqrt{16-x^2-y^2}} dz\, dy\, dx.$$

Solution By examining the limits of integration, we see that the given iterated integral is the triple integral $\iiint_R dV$ with R the region inside both the upper nappe of the cone $z^2 = x^2 + y^2$ and the sphere $x^2 + y^2 + z^2 = 16$. In cylindrical coordinates, we can describe R as

$$R = \{[r, \theta, z] | 0 \le r \le 2\sqrt{2},\ 0 \le \theta \le 2\pi,\ r \le z \le \sqrt{16 - r^2}\}.$$

See Figure 16.42. Hence, in cylindrical coordinates, the iterated integral becomes

$$\int_0^{2\pi} \int_0^{2\sqrt{2}} \int_r^{\sqrt{16-r^2}} r\, dz\, dr\, d\theta = \int_0^{2\pi} \int_0^{2\sqrt{2}} r(\sqrt{16 - r^2} - r)\, dr\, d\theta$$

$$= \int_0^{2\pi} \left[-\frac{1}{3} (16 - r^2)^{3/2} - \frac{1}{3} r^3 \right]_0^{2\sqrt{2}} d\theta$$

$$= \frac{32}{3} (2 - \sqrt{2}) \int_0^{2\pi} d\theta = \frac{64\pi}{3} (2 - \sqrt{2}).$$ ■

Spherical Coordinates

In spherical coordinates, a "rectangular" region is a curved "box," as shown in Figure 16.43. This region is approximately $d\rho$ in "length," $\rho\, d\phi$ in "height," and $r\, d\theta = \rho \sin \phi\, d\theta$ in "width," so that its volume is

$$dV = \rho^2 \sin \phi\, d\rho\, d\theta\, d\phi,$$

where the order of integration may be changed for convenience.

We also can derive this expression for dV by using the Jacobian. From Section 12.7, the equations relating rectangular coordinates to spherical coordinates are

$$x = \rho \sin \phi \cos \theta, \qquad y = \rho \sin \phi \sin \theta, \qquad z = \rho \cos \phi.$$

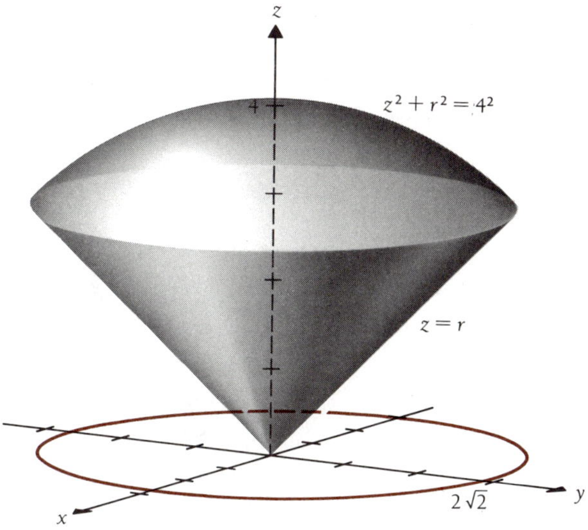

FIGURE 16.42

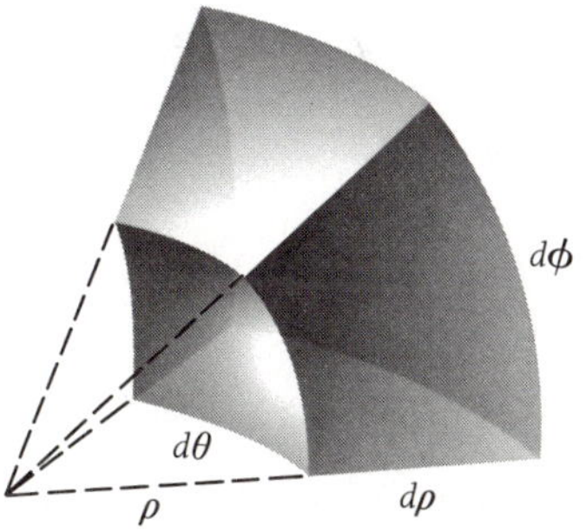

FIGURE 16.43 $dV = \rho^2 \sin\phi \, d\rho \, d\theta \, d\phi$

The Jacobian is therefore

$$\frac{\partial(x, y, z)}{\partial(\rho, \phi, \theta)} = \begin{vmatrix} \dfrac{\partial x}{\partial \rho} & \dfrac{\partial x}{\partial \phi} & \dfrac{\partial x}{\partial \theta} \\ \dfrac{\partial y}{\partial \rho} & \dfrac{\partial y}{\partial \phi} & \dfrac{\partial y}{\partial \theta} \\ \dfrac{\partial z}{\partial \rho} & \dfrac{\partial z}{\partial \phi} & \dfrac{\partial z}{\partial \theta} \end{vmatrix}$$

$$= \begin{vmatrix} \sin\phi\cos\theta & \rho\cos\phi\cos\theta & -\rho\sin\phi\sin\theta \\ \sin\phi\sin\theta & \rho\cos\phi\sin\theta & \rho\sin\phi\cos\theta \\ \cos\phi & -\rho\sin\phi & 0 \end{vmatrix}$$

$$= \rho^2 \sin\phi\cos^2\phi\cos^2\theta + \rho^2\sin^3\phi\sin^2\theta + \rho^2\sin\phi\cos^2\phi\sin^2\theta + \rho^2\sin^3\phi\cos^2\theta$$

$$= \rho^2\sin\phi(\cos^2\phi + \sin^2\phi)(\cos^2\theta + \sin^2\theta)$$

$$= \rho^2\sin\phi.$$

Example 16.31 Evaluate $\iiint_R 2\theta\, dV$ for R the region

$$R = \{[\rho, \theta, \phi] \mid 0 \le \rho \le a,\ 0 \le \theta \le \pi/2,\ 0 \le \phi \le \pi/2\}.$$

Solution The region R is the portion of the sphere of radius a contained in the first octant (Figure 16.44). We get the iterated integral

$$\begin{aligned}\iiint_R 2\theta\, dV &= \int_0^a \int_0^{\pi/2} \int_0^{\pi/2} 2\theta\rho^2 \sin\phi\, d\phi\, d\theta\, d\rho \\ &= \int_0^a \int_0^{\pi/2} 2\theta\rho^2(-\cos\phi)\Big|_0^{\pi/2} d\theta\, d\rho \\ &= \int_0^a \int_0^{\pi/2} 2\theta\rho^2\, d\theta\, d\rho \\ &= \int_0^a \theta^2\rho^2\Big|_0^{\pi/2} d\rho \\ &= \left(\frac{\pi}{2}\right)^2 \int_0^a \rho^2\, d\rho = \frac{\pi^2 a^3}{12}.\end{aligned}$$

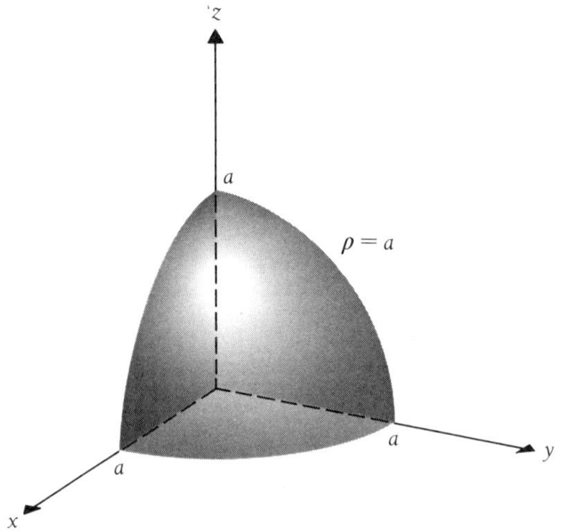

FIGURE 16.44 ■

Example 16.32 Find the mass of a sphere of radius a whose density at each point is proportional to the distance from the surface of the sphere.

Solution We can describe the sphere as the region

$$R = \{[\rho, \theta, \phi] \mid 0 \le \rho \le a,\ 0 \le \theta \le 2\pi,\ 0 \le \phi \le \pi\}.$$

The distance from point $[\rho, \theta, \phi]$ to the surface of the sphere is $a - \rho$, so the density is $\delta(\rho, \theta, \phi) = k(a - \rho)$. The mass is therefore

$$m = \iiint_R \delta\, dV = \int_0^{\pi} \int_0^{2\pi} \int_0^{a} k(a - \rho)\rho^2 \sin\phi \, d\rho\, d\theta\, d\phi$$

$$= \int_0^{\pi} \int_0^{2\pi} k\left[\frac{a}{3}\rho^3 - \frac{1}{4}\rho^4\right]_0^a \sin\phi\, d\theta\, d\phi$$

$$= \int_0^{\pi} \int_0^{2\pi} \frac{ka^4}{12} \sin\phi\, d\theta\, d\phi$$

$$= \int_0^{\pi} \frac{\pi k a^4}{6} \sin\phi\, d\phi$$

$$= \frac{\pi k a^4}{6}\left[-\cos\phi\right]_0^{\pi} = \frac{\pi k a^4}{3}.$$

■

Example 16.33 Evaluate the iterated integral of Example 16.30 by switching to spherical coordinates.

Solution The region R of Example 16.30 can be described in spherical coordinates as

$$R = \{[\rho, \theta, \phi] | 0 \le \rho \le 4,\ 0 \le \theta \le 2\pi,\ 0 \le \phi \le \pi/4\}$$

(Figure 16.45).

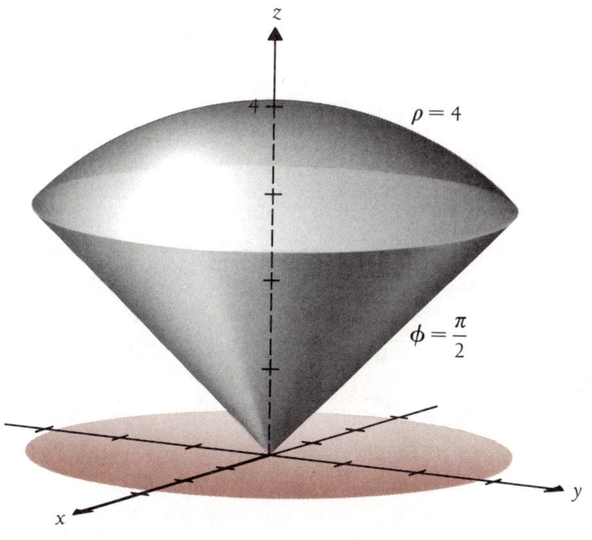

FIGURE 16.45

The integral therefore becomes

$$\int_0^{\pi/4}\int_0^{2\pi}\int_0^4 \rho^2 \sin\phi \, d\rho \, d\theta \, d\phi = \int_0^{\pi/4}\int_0^{2\pi} \frac{64}{3}\sin\phi \, d\theta \, d\phi$$

$$= \int_0^{\pi/4} \frac{128\pi}{3}\sin\phi \, d\phi$$

$$= \frac{128\pi}{3}\left(1 - \frac{\sqrt{2}}{2}\right) = \frac{64\pi}{3}(2 - \sqrt{2}).$$ ■

Exercises 16.5

Integrate.

1. $\iiint\limits_R r\cos\theta \, dV;\quad R = \{[r, \theta, z] | 0 \le r \le 4,\ 0 \le \theta \le \pi,\ 0 \le z \le r\}$
2. $\iiint\limits_R rz\sin\theta \, dV;\quad R = \{[r, \theta, z] | 0 \le r \le 2,\ 0 \le \theta \le 2\pi,\ r^2 \le z \le 4\}$
3. $\iiint\limits_R \frac{1}{\rho}\sin\theta \, dV;\quad R = \{[\rho, \theta, \phi] | 0 \le \rho \le 1,\ 0 \le \theta \le \pi/2,\ 0 \le \phi \le \pi/2\}$
4. $\iiint\limits_R 2\rho \, dV;\quad R = \{[\rho, \theta, \phi] | 0 \le \rho \le \cos\theta,\ -\pi/2 \le \theta \le \pi/2,\ 0 \le \phi \le \pi\}$

Find the volume:

5. Inside the cylinder $r = a\cos\theta$, above the plane $z = 0$, and below the sphere $r^2 + z^2 = a^2$
6. Inside the sphere $r^2 + z^2 = a^2$, below the cone $z = r$, and above the plane $z = 0$
7. Inside the cone $\phi = \alpha$ and below the sphere $\rho = a$
8. Inside the sphere $\rho = a$ and between the planes $\theta = \alpha$ and $\theta = \beta$ (in the first octant, with $0 \le \alpha < \beta \le \pi/2$)

In Problems 9–14, find the mass and center of mass of the given solid.

9. A sphere of radius a and density proportional to the distance from a given diameter
10. A sphere of radius a and density proportional to the distance from the center
11. A circular cylinder of radius a and length L, with density varying as the distance from one end
12. A circular cylinder of radius a and length $2a$, with density varying as the distance from the center of one end
13. A homogeneous hemispherical shell of inner radius a and outer radius b
14. A torus with cross-sectional diameter $2a$, outside diameter $2a + 2b$, and density proportional to the distance from the center of symmetry
15. Find the volume of the intersection of the two cylinders $x^2 + y^2 \le a^2$ and $y^2 + z^2 \le b^2$.

16. Find the volume of the intersection of the three cylinders $x^2+y^2 \le a^2$, $x^2+z^2 \le a^2$, and $y^2+z^2 \le a^2$.

17. A cylindrical hole is drilled through a solid sphere along a diameter. If the length of the resulting hole is 6 inches, find the volume of the solid so formed.

18. Find the volume enclosed by the surface $\rho = a(1-\cos\phi)$.

Evaluate by switching to cylindrical or spherical coordinates.

19. $\displaystyle\int_0^{\sqrt{2}}\int_0^{\sqrt{2-x^2}}\int_{\sqrt{x^2+y^2}}^{\sqrt{4-x^2-y^2}} dz\,dy\,dx$

20. $\displaystyle\int_0^{2}\int_0^{\sqrt{4-x^2}}\int_0^{8-x^2-y^2} dz\,dy\,dx$

21. $\displaystyle\int_{-1}^{1}\int_0^{\sqrt{1-x^2}}\int_0^{y} (x^2+y^2)\,dz\,dy\,dx$

22. $\displaystyle\int_{-1}^{1}\int_{1-\sqrt{1-y^2}}^{1+\sqrt{1-y^2}}\int_0^{x} dz\,dx\,dy$

23. Switch to **(a)** rectangular and **(b)** spherical coordinates:

$$\int_0^{2\pi}\int_0^{2}\int_0^{\sqrt{9-r^2}} r^2\,dz\,dr\,d\theta.$$

SECTION 16.6

Chapter Review

Important Concepts

Multiple integrals
Interior point
Interior
Region
Polygonal region
Circular region
Regions of Types I and II
Rectangular region
Double Riemann sum
Double integral
Iterated integrals
Changing order of integration
Solid below f
Volume
Area
Mass
Lamina
Centroid
Polar rectangle
Jacobian
Triple Riemann sum
Triple integral

Review Exercises

Let R be the region in the plane bounded by the curves $y=x^2$, $x=0$, and $y=4$. Find the following:

1. The area of R
2. The centroid of R
3. The mass of a plate covering R, with density $\delta(x, y) = y+1$
4. The center of mass of the plate in Problem 3
5. The volume over R under the plane $x+3y-z+2=0$
6. The volume over R under the surface $z=x^2+y^2$

Let R be the region in the first quadrant of the plane, inside the circle $x^2+y^2=9$. Find the following:

7. The area of R
8. The centroid of R
9. The mass of a plate covering R, with density $\delta(x, y) = \sqrt{x^2+y^2}$
10. The center of mass of the plate in Problem 9
11. The volume under $z=\sqrt{50-x^2-y^2}$ over R

Let R be the region in space bounded by the surfaces $x = 0$, $y = 0$, $z = 0$, and $x = 4 - y^2 - z^2$. Find the following:

12. The volume of R

13. The centroid of R

14. The mass of a solid filling R, with density $\delta(x, y, z) = z + 1$

15. The center of mass of the solid in Problem 14

Let R be the region in the first octant bounded by the surfaces $x = 0$, $y = 0$, $z = 12$, and $z = \sqrt{3x^2 + 3y^2}$. Find the following:

16. The volume of R

17. The centroid of R

18. The mass of a solid filling R, with density $\delta(x, y, z) = 16 - z$

19. The center of mass of the solid in Problem 18

Let R be the region in the first octant bounded by the surfaces $x = 0$, $y = 0$, $z = 2\sqrt{x^2 + y^2}$, and $z = \sqrt{16 - x^2 - y^2}$. Find the following:

20. The volume of R

21. The centroid of R

22. The mass of a solid filling R, with density $\delta(x, y, z) = 1 + \sqrt{x^2 + y^2 + z^2}$

23. The center of mass of the solid in Problem 22

Additional Exercises

1. Two cylinders of the same diameter intersect at an angle of 60°. Find the volume of the intersection.
2. Three cylinders of the same diameter intersect at right angles. Find the volume of the intersection.
3. A certain star is a sphere of radius a and has density inversely proportional to the distance from the center. Find its mass.
4. A spherical shell of inner radius a and outer radius b has density inversely proportional to the distance from the center. Find the mass of the shell.
5. If the weight density of the atmosphere is P pounds per cubic mile at sea level and it varies inversely as the fourth power of the distance from the center of the earth, find the total weight of the earth's atmosphere. Assume that the earth is a sphere of radius 4000 miles.
6. The force of gravitational attraction between two particles of masses m_1 and m_2 separated by a distance r has magnitude proportional to $m_1 m_2/r^2$ and is directed along the line joining the centers of the two particles. Show that the total gravitational force exerted by a homogeneous sphere of mass M and radius R on a particle of mass m at a distance d from the center of the sphere, $d > R$, is directed toward the center of the sphere and has magnitude proportional to mM/d^2.
7. Is the result of Problem 6 true if $d < R$?
8. Generalize Problem 6 to a nonhomogeneous sphere, replacing the center by the center of mass.
9. Does the result of Problem 8 hold for any object, whether a sphere or not?
10. Prove that the union of open sets is an open set.
11. Prove that the intersection of closed sets is a closed set.

12. Prove that the union of a finite number of closed sets is a closed set.
13. Prove that the intersection of a finite number of open sets is an open set.
14. Prove that a finite set is closed.
15. A point P is a **limit point** of set S if and only if S is an approach to P. Prove that a closed set contains all of its limit points.

CHAPTER 17

Vector Calculus

INTRODUCTION

The calculus of vector fields was developed for the most part during the 19th century, primarily by physicists. With vector calculus, they could describe gravitation, electricity and magnetism, fluid dynamics, and many other natural phenomena. Their work was regarded as the crowning touch in the application of calculus to the physical sciences and the description of natural laws. In fact, a century ago, scientists believed that they had just about found out all the secrets that nature held. However, as is ever the case in science, questions began to arise that had not been answered. About the beginning of the 20th century there occurred a revolution in physics just as profound as the one that Newton had started. Such revolutionary ideas as Heisenberg's uncertainty principle and Einstein's theory of relativity opened up entirely new ways of looking at the universe, and scientists agreed that they did not really understand nature all that well. The calculus and other mathematical disciplines continue to be applied to the problem of describing the universe, with the very small (quantum mechanics) and the very large (cosmology) being areas of intense research activity today.

In this chapter, we further develop the concept of a vector field, which was introduced in Chapter 15. We will discuss divergence and curl and introduce surface integrals. The major theorems of this chapter are Green's theorem, the divergence theorem, and Stokes's theorem. The major applications of these results to physical phenomena will be mentioned.

SECTION 17.1

Green's Theorem

In this section, we return to the evaluation of path integrals, begun in Chapter 15. The English mathematician George Green (1793–1841) introduced in 1828 results that have evolved into what is now known as Green's theorem. It relates certain kinds of path integrals to double integrals. Our purpose in this section is to present Green's theorem and its applications.

Simple Closed Curves

In Section 15.4 we discussed closed curves. A **simple** closed curve is a closed curve that does not cross itself (see Figure 17.1). That is, if C is a simple closed curve, parametrized by the vector function $\mathbf{r}(t)$, $t \in [a, b]$, then $\mathbf{r}(a) = \mathbf{r}(b)$, and if $t_1, t_2 \in (a, b)$ and $t_1 \neq t_2$, then $\mathbf{r}(t_1) \neq \mathbf{r}(t_2)$.

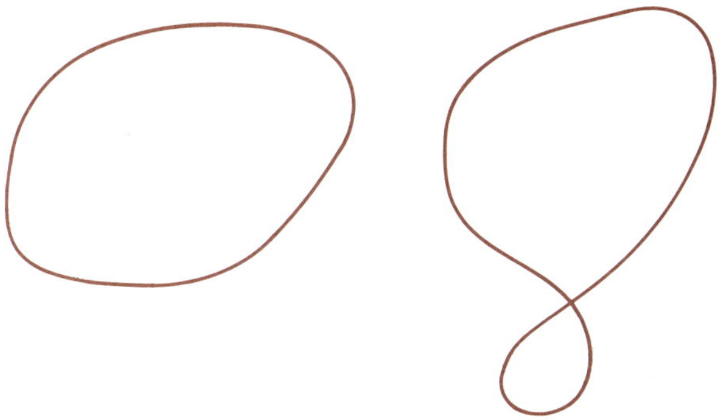

(a) Simple (b) Not simple

FIGURE 17.1 Closed curves

A simple closed curve in the plane bounds a portion of the plane, usually called the "inside" of the curve [see Figure 17.2(a)]. The curve, together with its "inside," constitute a region in the plane that we call the **region enclosed by the curve** [see Figure 17.2(b)].

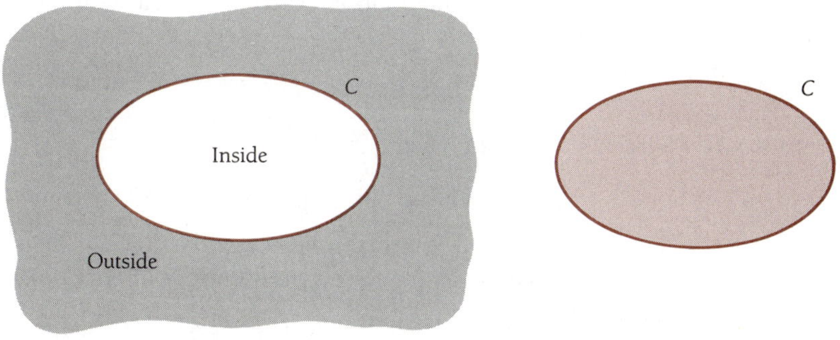

(a) A simple closed curve has an "inside" and an "outside"

(b) The region enclosed by a simple closed curve

FIGURE 17.2

We will call a region in the plane a **standard region** if it can be described both as a Type I region and as a Type II region (as defined in Section 16.1). A standard region is shown in Figure 17.3. The boundary of any standard region is a simple closed curve.

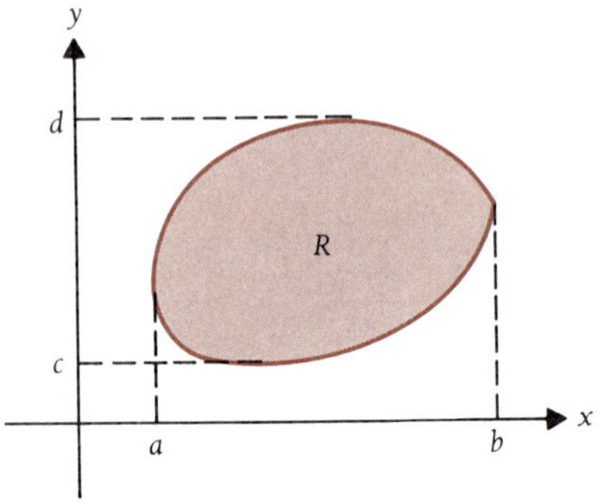

FIGURE 17.3 A standard region R

At any point of a smooth curve, there are two unit tangent vectors, pointing in opposite directions. If a unit tangent vector can be chosen at each point such that the unit tangent vector is a continuous function of position along the curve, then the curve is said to be **oriented**. A smooth simple closed curve can be oriented in two directions, clockwise and counterclockwise. The concept easily extends to a piecewise smooth simple closed curve.

Green's Theorem

Let C be a piecewise smooth simple closed curve, oriented in the counterclockwise direction. Green's theorem relates the path integral over C to a double integral over the region enclosed by C. We will use the symbol $\oint_C$ to denote a path integral over a piecewise smooth simple closed path C, oriented in the counterclockwise direction. We usually refer to $\oint_C$ simply as a **closed-path integral**.

THEOREM 17.1 **Green's Theorem** Let

$$\mathbf{F}(x, y) = \langle P(x, y), Q(x, y) \rangle$$

be a continuously differentiable vector field defined on an open set D, and let C be a piecewise smooth simple closed curve lying in D, enclosing a region R that lies in D. Then

$$\oint_C \mathbf{F} \cdot d\mathbf{r} = \iint_R (Q_x - P_y)\, dA.$$

Proof To avoid a lot of details that are beyond our scope, we will prove the theorem only for the case that R is a standard region, although the theorem holds for virtually any kind of region.

The region R can be described as a Type I region [see Figure 17.4(a)]:

$$R = \{(x, y) | a \leq x \leq b \text{ and } f(x) \leq y \leq g(x)\},$$

and as a Type II region [see Figure 17.4(b)]:

$$R = \{(x, y) | h(y) \leq x \leq k(y) \text{ and } c \leq y \leq d\}.$$

Now,

$$\iint_R (Q_x - P_y)\, dA = \iint_R Q_x\, dA - \iint_R P_y\, dA. \tag{1}$$

Using the Type II description of R, we have

$$\begin{aligned}\iint_R Q_x\, dA &= \int_c^d \int_{h(y)}^{k(y)} Q_x(x, y)\, dx\, dy \\ &= \int_c^d Q(x, y)\Big|_{h(y)}^{k(y)} dy \\ &= \int_c^d [Q(k(y), y) - Q(h(y), y)]\, dy. \end{aligned} \tag{2}$$

Now let $C = C_1 - C_2$, with parametrizations

$$C_1: \quad \mathbf{r}_1(t) = \langle k(t), t \rangle, \qquad t \in [c, d],$$
$$C_2: \quad \mathbf{r}_2(t) = \langle h(t), t \rangle, \qquad t \in [c, d]$$

[see Figure 17.4(b)].

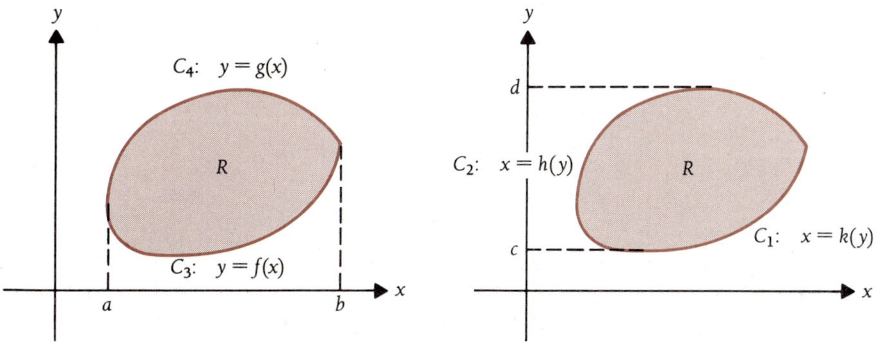

(a) The standard region R described as Type I

(b) The standard region R described as Type II

FIGURE 17.4

Then,

$$\begin{aligned}\oint_C Q(x, y)\, dy &= \int_{C_1 - C_2} Q(x, y)\, dy \\ &= \int_{C_1} Q(x, y)\, dy - \int_{C_2} Q(x, y)\, dy \\ &= \int_c^d Q(k(t), t)\, dt - \int_c^d Q(h(t), t)\, dt. \end{aligned} \tag{3}$$

Comparing equations (2) and (3), we see that

$$\oint_C Q\, dy = \iint_R Q_x\, dA. \tag{4}$$

Similarly, using the Type I description of R, we have

$$\begin{aligned}\iint_R P_y\, dA &= \int_a^b \int_{f(x)}^{g(x)} P_y(x, y)\, dy\, dx \\ &= \int_a^b P(x, y)\Big|_{f(x)}^{g(x)}\, dx \\ &= \int_a^b [P(x, g(x)) - P(x, f(x))]\, dx. \end{aligned} \tag{5}$$

Now let $C = C_3 - C_4$, with parametrizations

$$\begin{aligned} C_3\colon &\quad \mathbf{r}_3(t) = \langle t, f(t) \rangle, \qquad t \in [a, b], \\ C_4\colon &\quad \mathbf{r}_4(t) = \langle t, g(t) \rangle, \qquad t \in [a, b] \end{aligned}$$

[see Figure 17.4(a)]. Then,

$$\begin{aligned}\oint_C P(x, y)\, dx &= \int_{C_3 - C_4} P(x, y)\, dx \\ &= \int_{C_3} P(x, y)\, dx - \int_{C_4} P(x, y)\, dx \\ &= \int_a^b P(t, f(t))\, dt - \int_a^b P(t, g(t))\, dt. \end{aligned} \tag{6}$$

Comparing equations (5) and (6), we see that

$$\oint_C P\, dx = -\iint_R P_y\, dA. \tag{7}$$

Finally, we combine equations (4) and (7) with (1), and get

$$\begin{aligned}\oint_C \mathbf{F} \cdot d\mathbf{r} &= \oint_C (P\,dx + Q\,dy) \\ &= \oint_C P\,dx + \oint_C Q\,dy \\ &= -\iint_R P_y\,dA + \iint_R Q_x\,dA \\ &= \iint_R (Q_x - P_y)\,dA,\end{aligned}$$

and the theorem is proved.

An alternative form of Green's theorem is

$$\oint_C (P\,dx + Q\,dy) = \iint_R (Q_x - P_y)\,dA. \tag{8}$$

Let's consider some examples.

■ **Example 17.1** Evaluate the path integral

$$\oint_C [(x^2 + y^2)\,dx + (x^2 - y^2)\,dy]$$

for C the circle $x^2 + (y-3)^2 = 9$ in two ways:

(a) Directly by parametrizing C
(b) By using Green's theorem

Solution **(a)** We will parametrize the circle C as

$$\mathbf{r}(t) = \langle 3\cos t,\ 3 + 3\sin t\rangle, \qquad t \in [0, 2\pi].$$

Then

$$\begin{aligned}&\oint_C [(x^2 + y^2)\,dx + (x^2 - y^2)\,dy] \\ &= \int_0^{2\pi} \{[9\cos^2 t + (3 + 3\sin t)^2](-3\sin t) \\ &\qquad + [9\cos^2 t - (3 + 3\sin t)^2](3\cos t)\}\,dt \\ &= \int_0^{2\pi} [(18 + 18\sin t)(-3\sin t) \\ &\qquad + (9\cos^2 t - 9 - 18\sin t - 9\sin^2 t)(3\cos t)]\,dt\end{aligned}$$

$$= \int_0^{2\pi} (-54 \sin t - 54 \sin^2 t + 27 \cos^3 t - 27 \cos t$$
$$- 54 \sin t \cos t - 27 \sin^2 t \cos t)\, dt$$
$$= 54 \cos t\Big|_0^{2\pi} - 54 \int_0^{2\pi} \frac{1}{2}(1 - \cos 2t)\, dt + 27 \int_0^{2\pi} (1 - \sin^2 t) \cos t\, dt$$
$$- 27 \sin t\Big|_0^{2\pi} - 54\left(\frac{1}{2}\right) \sin^2 t\Big|_0^{2\pi} - 27\left(\frac{1}{3}\right) \sin^3 t\Big|_0^{2\pi}$$
$$= 0 - 27\left(t - \frac{1}{2}\sin 2t\right)\Big|_0^{2\pi} + 27\left(\sin t - \frac{1}{3}\sin^3 t\right)\Big|_0^{2\pi} - 0 - 0 - 0$$
$$= -54\pi.$$

(b) Let R be the region enclosed by the circle. Since $P(x, y) = x^2 + y^2$, $P_y = 2y$, and since $Q(x, y) = x^2 - y^2$, $Q_x = 2x$. Then Green's theorem says

$$\oint_C [(x^2 + y^2)\, dx + (x^2 - y^2)\, dy] = \oint_C (P\, dx + Q\, dy)$$
$$= \iint_R (Q_x - P_y)\, dA$$
$$= \iint_R (2x - 2y)\, dA.$$

Using the fact that the region R has centroid $(\bar{x}, \bar{y}) = (0, 3)$ and area $A = 9\pi$, we get

$$\oint_C [(x^2 + y^2)\, dx + (x^2 - y^2)\, dy] = (2\bar{x} - 2\bar{y})A$$
$$= (-6)(9\pi) = -54\pi.$$

It is easy to see why Green's theorem is preferred here! ■

■ **Example 17.2** Evaluate

$$\oint_C [(e^{x^3} + y)\, dx + (x^2 + \arcsin y^2)\, dy]$$

for C the rectangle with vertices (1, 2), (4, 2), (4, 3), and (1, 3).

Solution Using Green's theorem again, with $R = [1, 4] \times [2, 3]$, we have

$$\oint_C [(e^{x^3} + y)\, dx + (x^2 + \arcsin y^2)\, dy] = \iint_R (2x - 1)\, dA$$
$$= \int_1^4 \int_2^3 (2x - 1)\, dy\, dx$$
$$= \int_1^4 (2x - 1)\, dx = 12.$$
■

Example 17.3 Show that $\oint_C (-y)\,dx$, $\oint_C x\,dy$, and $\frac{1}{2}\oint_C [(-y)\,dx + x\,dy]$ all represent the area enclosed by the simple closed curve C.

Solution We use Green's theorem. Let C enclose the region R. For $\oint_C (-y)\,dx$, we have $P = -y$ and $Q = 0$. Hence,

$$\oint_C (-y)\,dx = \iint_R (Q_x - P_y)\,dA$$
$$= \iint_R dA = A.$$

For $\oint_C x\,dy$, we have $P = 0$ and $Q = x$, so

$$\oint_C x\,dy = \iint_R (Q_x - P_y)\,dA$$
$$= \iint_R dA = A.$$

The third integral is just the average of the other two. ■

Example 17.4 Use Example 17.3 to find the area enclosed by the ellipse

$$C: \quad \mathbf{r}(t) = \langle a \cos t, b \sin t\rangle, \qquad t \in [0, 2\pi].$$

Solution We will use $A = \oint_C x\,dy$ with $x = a \cos t$, $y = b \sin t$:

$$A = \oint_C x\,dy$$
$$= \int_0^{2\pi} (a \cos t)\, d(b \sin t)$$
$$= \int_0^{2\pi} ab \cos^2 t\, dt$$
$$= ab \int_0^{2\pi} \frac{1}{2}(1 + \cos 2t)\, dt$$
$$= \frac{1}{2} ab \left[t + \frac{1}{2}\sin 2t\right]_0^{2\pi} = \pi ab.$$ ■

If $\mathbf{F}(x, y) = \langle P(x, y), Q(x, y)\rangle$ is a gradient field, then Green's theorem automatically gives 0 as the value of $\oint_C \mathbf{F} \cdot d\mathbf{r}$, because $Q_x - P_y = 0$.

We will consider some other versions of Green's theorem later.

Exercises 17.1

Evaluate each path integral.

1. $\oint_C [(xy + 1)\,dx + (x^2 + y^2)\,dy]$;
   C is the square with vertices (0, 0), (1, 0), (1, 1), (0, 1)

2. $\oint_C [(xy + 1)\,dx + (x^2 + y^2)\,dy]$;
   C is the triangle with vertices (0, 0), (1, 0), $(\frac{1}{2}, 1)$

3. $\oint_C [(2xy + 3)\,dx + (2x^2 - xy)\,dy]$;
   C is the circle $(x - 2)^2 + (y - 1)^2 = 1$

4. $\oint_C [(2xy + 3)\,dx + (2x^2 - xy)\,dy]$;
   C is the circle $(x + 1)^2 + y^2 = 4$

5. $\oint_C [(\sec x + 2y)\,dx + (x^2 - \arctan y)\,dy]$;
   C is the circle $(x + 2)^2 + (y - 1)^2 = 9$

6. $\oint_C \left[x \ln y\,dx + \left(1 + \frac{1}{2y}\right) x^2\,dy\right]$;
   C is the ellipse $(x - 3)^2 + 2(y - 4)^2 = 4$

7. $\oint_C (x \ln y\,dx + y \ln x\,dy)$;
   C is the rectangle with vertices (1, 1), (3, 1), (3, 2), (1, 2)

8. $\oint_C [(1 + y^2)e^{xy}\,dx + (1 + xy)e^{xy}\,dy]$;
   C is the curve of Problem 7

9. $\oint_C (x^3\,dx + y^3\,dy)$;
   C is the path along $y = x^2$ from $(-2, 4)$ to $(2, 4)$, then along $y = 4$ to $(-2, 4)$

10. $\oint_C (x^4 y\,dx + xy^4\,dy)$;
    C is the path along $y = x$ from (0, 0) to (1, 1), then along $y = \sqrt{x}$ to (0, 0)

Find the area enclosed by each curve C. (See Example 17.3.)

11. C is the graph of $\mathbf{r}(t) = \langle 2 \cos t, \sin t\rangle$, $t \in [0, 2\pi]$.
12. C is the graph of $\mathbf{r}(t) = \langle 1 + \cos t, 2 + \sin t\rangle$, $t \in [0, 2\pi]$.
13. $C = C_1 - C_2$; C_1 is the graph of $\mathbf{r}_1(t) = \langle t, 0\rangle$, $t \in [0, 2\pi]$; C_2 is the graph of $\mathbf{r}_2(t) = \langle t - \sin t, 1 - \cos t\rangle$, $t \in [0, 2\pi]$.
14. $C = C_1 - C_2$; C_1: $\mathbf{r}_1(t) = \langle t, 0\rangle$, $t \in [0, 4]$; C_2: $\mathbf{r}_2(t) = \langle t, t^2 - 4t\rangle$, $t \in [0, 4]$

15. Suppose regions R_1 and R_2 are standard regions sharing a portion of the same boundary (see the accompanying figure). Let R_1 have boundary $C_1 + C_2$, and let R_2 have boundary $C_3 - C_1$. Use Green's theorem to show that

$$\oint_{C_2+C_3} (P\,dx + Q\,dy) = \iint_{R_1 \cup R_2} (Q_x - P_y)\,dA.$$

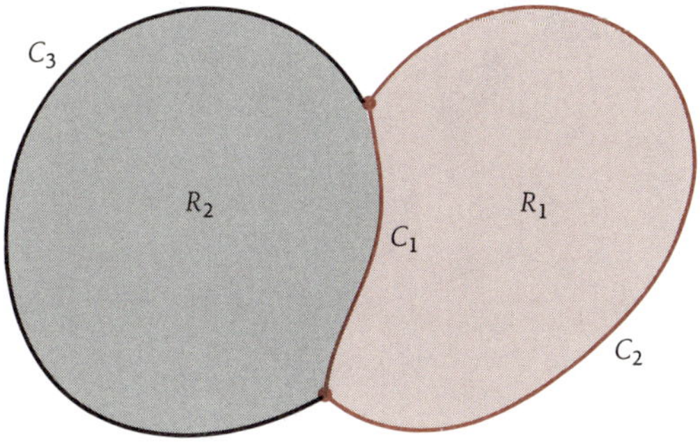

16. Using Problem 15, show how Green's theorem applies to any piecewise smooth simple closed curve enclosing a region that is a union of standard regions.

17. Show that if R is a standard region, then the intersection of R with any line parallel to a coordinate axis is either empty, a single point, or a single closed interval.

18. Show the converse of Problem 17. That is, show that if the intersection of a region R with any horizontal or vertical line is either empty, a single point, or a single closed interval, then R is a standard region.

SECTION 17.2

Surface Integrals

Starting with the double integral, we will show how to find the areas of certain surfaces in space. This will lead us to the concept of a surface integral.

Surface Area

In Section 8.4, we used definite integrals to find the areas of surfaces of revolution. Using double integrals, we can find the area of any surface that is the graph of a continuously differentiable function.

Let f be a continuously differentiable function of two variables, and let R be a region lying in the domain of f. Let S be the surface $z = f(x, y)$, with $(x, y) \in R$. We seek a double integral representing the area of S.

We begin by partitioning R into rectangular subregions, with the i,jth rectangular subregion R_{ij} having area ΔA_{ij}. Let S_{ij} be the portion of S lying over R_{ij} (see Figure

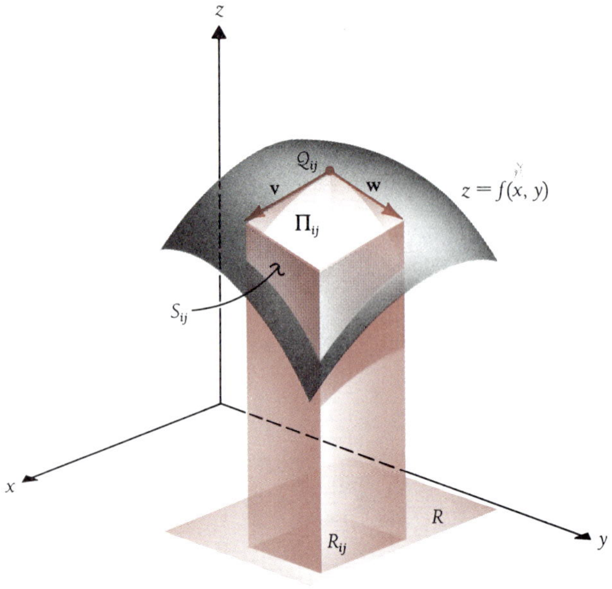

FIGURE 17.5

17.5). Let Q_{ij} be the point $(x_{i-1}, y_{j-1}, f(x_{i-1}, y_{j-1}))$. Let Π_{ij} be the portion of the plane tangent to S at Q_{ij} that lies over R_{ij}; Π_{ij} is a parallelogram (think of it as a plane section of a rectangular prism). Let ΔP_{ij} be the area of Π_{ij}; then ΔP_{ij} is an approximation to the area of S_{ij}. The area of S is therefore approximately

$$\sigma \approx \sum_{i=1}^{n} \sum_{j=1}^{m} \Delta P_{ij}.$$

To calculate the area ΔP_{ij}, we let $\mathbf{v}$ and $\mathbf{w}$ be vectors represented by arrows from Q_{ij} lying along the sides of Π_{ij} (see Figure 17.5). Then from the equation of the tangent plane (Section 14.6), we get

$$\mathbf{v} = \langle \Delta x_i, 0, f_x(x_{i-1}, y_{j-1})\Delta x_i \rangle,$$
$$\mathbf{w} = \langle 0, \Delta y_j, f_y(x_{i-1}, y_{j-1})\Delta y_j \rangle.$$

Then the vector $\mathbf{v} \times \mathbf{w}$ is normal to S, and its magnitude is the area

$$\begin{aligned}\Delta P_{ij} &= \|\mathbf{v} \times \mathbf{w}\| \\ &= \|\Delta x_i \Delta y_j \langle -f_x(x_{i-1}, y_{j-1}), -f_y(x_{i-1}, y_{j-1}), 1\rangle\| \\ &= \sqrt{[f_x(x_{i-1}, y_{j-1})]^2 + [f_y(x_{i-1}, y_{j-1})]^2 + 1}\,\Delta A_{ij}.\end{aligned}$$

It follows that

$$\sigma \approx \sum_{i=1}^{n} \sum_{j=1}^{m} \sqrt{[f_x(x_{i-1}, y_{j-1})]^2 + [f_y(x_{i-1}, y_{j-1})]^2 + 1}\,\Delta A_{ij},$$

which is a double Riemann sum whose limit is the double integral

$$\sigma = \iint_R \sqrt{[f_x(x, y)]^2 + [f_y(x, y)]^2 + 1}\, dA. \tag{1}$$

This is the surface area of S.

Example 17.5 Find the area of the portion of the plane $2x + 3y - 5z = 7$ lying over the region

$$R = \{(x, y) | 0 \le x \le 2 \text{ and } 0 \le y \le x^2\}$$

(Figure 17.6).

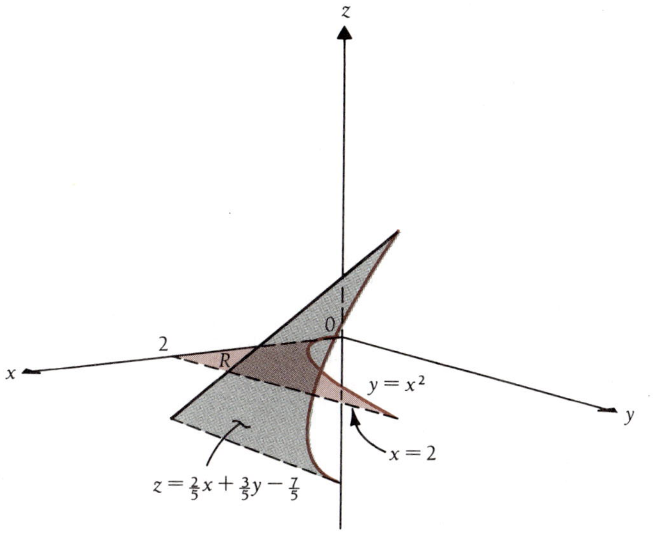

FIGURE 17.6

Solution Solving the equation of the plane for z, we get

$$z = f(x, y) = \frac{2}{5}x + \frac{3}{5}y - \frac{7}{5},$$

so that $f_x = \frac{2}{5}$ and $f_y = \frac{3}{5}$. Thus,

$$\sqrt{f_x^2 + f_y^2 + 1} = \sqrt{\frac{4}{25} + \frac{9}{25} + 1} = \frac{1}{5}\sqrt{38}.$$

Then by equation (1), the surface area is

$$\sigma = \iint_R \frac{1}{5}\sqrt{38}\, dA$$
$$= \frac{1}{5}\sqrt{38}\int_0^2 \int_0^{x^2} dy\, dx$$
$$= \frac{1}{5}\sqrt{38}\int_0^2 x^2\, dx = \frac{8}{15}\sqrt{38}.$$

■

■ **Example 17.6** Find the area of the part of the paraboloid $z = 16 - x^2 - y^2$ that lies above the xy-plane (Figure 17.7).

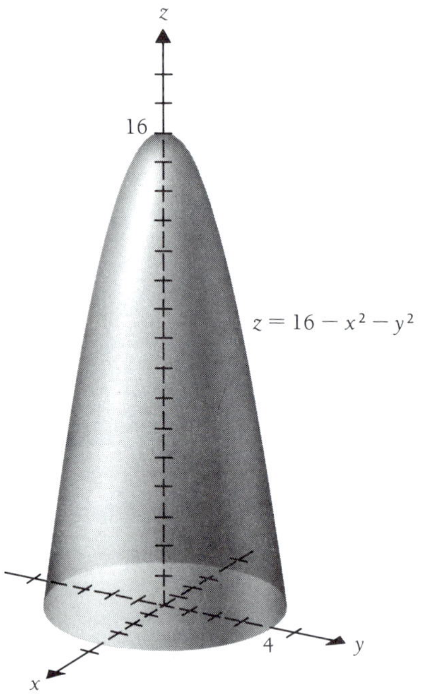

FIGURE 17.7

Solution The part of the paraboloid above the xy-plane is the portion over the region

$$R = \{(x, y) | x^2 + y^2 \le 16\}.$$

From $z = f(x, y) = 16 - x^2 - y^2$, we have $f_x = -2x$, $f_y = -2y$, and

$$\sqrt{f_x^2 + f_y^2 + 1} = \sqrt{4x^2 + 4y^2 + 1}.$$

The surface area is

$$\begin{aligned}\sigma &= \iint_R \sqrt{4x^2+4y^2+1}\,dA \\ &= \int_0^{2\pi}\int_0^4 \sqrt{4r^2+1}\,r\,dr\,d\theta \qquad \text{(Using polar coordinates)} \\ &= \int_0^{2\pi} \frac{1}{12}(4r^2+1)^{3/2}\Big|_0^4\,d\theta \\ &= \frac{\pi}{6}(65^{3/2}-1).\end{aligned}$$

■

Surface Integrals

Before we introduce the surface integral, we need to describe surfaces in general.

DEFINITION 17.1 If R is a region in the plane and f is a function of two variables defined on R, let S be the surface that is the graph of f,

$$S:\quad z=f(x,y), \qquad (x,y)\in R.$$

We say S is **smooth** if and only if f is continuously differentiable.

At each point of a smooth surface, there are two unit normal vectors that point in opposite directions. If a unit normal vector is chosen at each point so that the normal vector is a continuous function of position on the surface, the surface is said to be **oriented**.

We saw in Section 14.6 that the vector

$$\mathbf{n}(x,y)=\frac{\langle -f_x(x,y), -f_y(x,y), 1\rangle}{\sqrt{(f_x)^2+(f_y)^2+1}}$$

is a unit normal vector for the surface S: $z=f(x,y)$ when f is continuously differentiable. This unit normal is often called the **upper** unit normal, because its component in the **k** direction is positive. The vector $-\mathbf{n}$ is also a unit normal vector, in the opposite direction from $\mathbf{n}$. The choice of $\mathbf{n}$ or $-\mathbf{n}$ determines the orientation of the surface S. Unless otherwise specified, we will always use the upper unit normal vector to a surface.

A closed surface, such as a sphere or ellipsoid, has a normal vector that is sometimes called the **outer** unit normal vector. It points away from the "inside" of the surface.

These considerations of orientation also can be applied to surfaces that are piecewise smooth.

Now we can define a surface integral.

DEFINITION 17.2

Let G be a continuous function of three variables, defined at each point (x, y, z) of the smooth oriented surface S: $z = f(x, y)$, $(x, y) \in R$. The **surface integral of G over S** (also called the **integral of G over S with respect to surface area**) is defined by

$$\iint_S G(x, y, z)\, d\sigma = \iint_R G(x, y, f(x, y))\sqrt{(f_x)^2 + (f_y)^2 + 1}\, dA \tag{2}$$

Example 17.7 Evaluate $\iint_S yz\, d\sigma$ for the surface

$$S: \quad z = x, \qquad (x, y) \in R = \{(x, y) | 0 \le x \le 4 \text{ and } 0 \le y \le x\}.$$

Solution From $z = f(x, y) = x$, we have $f_x = 1$, $f_y = 0$, and $\sqrt{f_x^2 + f_y^2 + 1} = \sqrt{2}$. Hence, by equation (2),

$$\begin{aligned}\iint_S yz\, d\sigma &= \iint_R yx\sqrt{2}\, dA \\ &= \sqrt{2}\int_0^4 \int_0^x yx\, dy\, dx \\ &= \frac{\sqrt{2}}{2}\int_0^4 x^3\, dx = 32\sqrt{2}.\end{aligned}$$

■

Applications of Surface Integrals

If we compare the definition of surface integral with formula (1), we see that

$$\sigma = \iint_S d\sigma \tag{3}$$

is just the surface area of S.

If a sheet of material (such as metal) occupies the surface S and has density $\delta(x, y, z)$, then the mass of the sheet is

$$m = \iint_S \delta(x, y, z)\, d\sigma. \tag{4}$$

The center of mass of the sheet would have coordinates

$$\bar{x} = \frac{\iint_S x\delta \, d\sigma}{\iint_S \delta \, d\sigma}, \qquad \bar{y} = \frac{\iint_S y\delta \, d\sigma}{\iint_S \delta \, d\sigma}, \qquad \bar{z} = \frac{\iint_S z\delta \, d\sigma}{\iint_S \delta \, d\sigma}. \tag{5}$$

The centroid of the surface S would be given by the same formulas as (5), but with δ constant.

Example 17.8 A sheet of material occupying the surface

$$S: \quad z = 2 - x^2 - y^2, \qquad x^2 + y^2 \le 2,$$

has density $\delta(x, y, z) = x^2 + y^2 + z$ (Figure 17.8). Find its mass and center of mass.

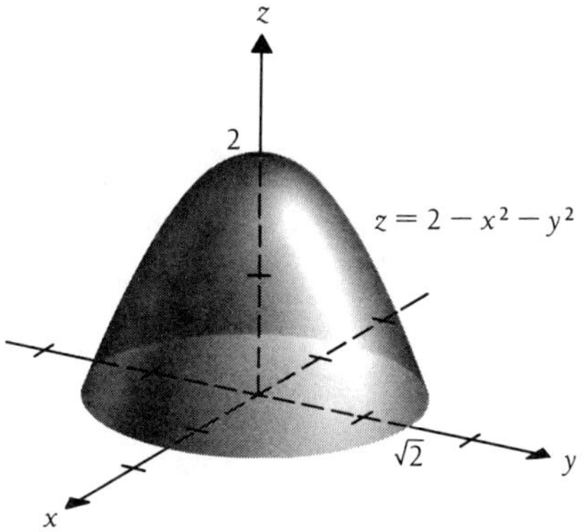

FIGURE 17.8

Solution With $z = f(x, y) = 2 - x^2 - y^2$, we have $f_x = -2x$, $f_y = -2y$, and therefore, $\sqrt{f_x^2 + f_y^2 + 1} = \sqrt{4x^2 + 4y^2 + 1}$. Then the mass of the surface is

$$\begin{aligned} m &= \iint_S (x^2 + y^2 + z) \, d\sigma \\ &= \iint_R (x^2 + y^2 + 2 - x^2 - y^2) \sqrt{4x^2 + 4y^2 + 1} \, dA \\ &= \int_0^{2\pi} \int_0^{\sqrt{2}} 2\sqrt{4r^2 + 1} \, r \, dr \, d\theta \qquad \text{(Switching to polar coordinates)} \\ &= \frac{1}{4} \int_0^{2\pi} \int_1^9 u^{1/2} \, du \, d\theta \qquad (u = 4r^2 + 1) \\ &= \frac{13}{3} \int_0^{2\pi} d\theta = \frac{26\pi}{3}. \end{aligned}$$

By symmetry, $\bar{x} = \bar{y} = 0$. From formulas (5),

$$\begin{aligned}
\bar{z} &= \frac{1}{m} \iint_S z\delta \, d\sigma \\
&= \frac{1}{m} \iint_S z(x^2 + y^2 + z) \, d\sigma \\
&= \frac{3}{26\pi} \iint_R (2 - x^2 - y^2) 2\sqrt{4x^2 + 4y^2 + 1} \, dA \\
&= \frac{3}{26\pi} \int_0^{2\pi} \int_0^{\sqrt{2}} (2 - r^2) 2\sqrt{4r^2 + 1} \, r \, dr \, d\theta \\
&= \frac{3}{13\pi} \int_0^{2\pi} \int_1^9 \left(2 - \frac{u-1}{4}\right) u^{1/2} \frac{1}{8} \, du \, d\theta \qquad (u = 4r^2 + 1) \\
&= \frac{3}{13\pi(32)} \int_0^{2\pi} \int_1^9 (9 - u) u^{1/2} \, du \, d\theta = \frac{111}{130}.
\end{aligned}$$

The center of mass is thus $(0, 0, \frac{111}{130})$. ■

Another important application of surface integrals arises in physical situations such as the flow of a fluid. For example, let $\mathbf{F}(x, y, z)$ be a vector field interpreted as the velocity of a fluid (such as air), and let S be a surface in the fluid (such as a screen). If $\mathbf{n}$ is a unit normal vector of S, then $\mathbf{F} \cdot \mathbf{n}$ is the component of the flow of the fluid in the direction of $\mathbf{n}$, that is, through the surface S in the direction of $\mathbf{n}$. The surface integral

$$\iint_S \mathbf{F} \cdot \mathbf{n} \, d\sigma \tag{6}$$

is called the **flux of F across S** and measures the amount of flow of the fluid through S.

For a surface that is the graph of a function of two variables, S: $z = f(x, y)$, $(x, y) \in R$, and a vector field $\mathbf{F} = \langle P, Q, M \rangle$, the flux integral (6) has a particularly simple form. Using the upper unit normal $\mathbf{n} = \langle -f_x, -f_y, 1 \rangle / \sqrt{f_x^2 + f_y^2 + 1}$, we have

$$\begin{aligned}
\iint_S \mathbf{F} \cdot \mathbf{n} \, d\sigma &= \iint_R \frac{\langle P, Q, M \rangle}{\sqrt{f_x^2 + f_y^2 + 1}} \cdot \langle -f_x, -f_y, 1 \rangle \sqrt{f_x^2 + f_y^2 + 1} \, dA \\
&= \iint_R \langle P, Q, M \rangle \cdot \langle -f_x, -f_y, 1 \rangle \, dA,
\end{aligned}$$

or

$$\iint_S \mathbf{F} \cdot \mathbf{n}\, d\sigma = \iint_R (-f_x P - f_y Q + M)\, dA. \tag{7}$$

Example 17.9 Find the flux of the vector field $\mathbf{F}(x, y, z) = \langle x, y^2, z \rangle$ through the triangular surface with vertices $(a, 0, 0)$, $(0, a, 0)$, and $(0, 0, a)$.

Solution The triangular surface is

$$S: \quad z = a - x - y, \qquad (x, y) \in R,$$

with

$$R = \{(x, y) | 0 \leq x \leq a \text{ and } 0 \leq y \leq a - x\}$$

(Figure 17.9). With $z = f(x, y) = a - x - y$, $f_x = f_y = -1$. Hence, by equation (7), the flux is

$$\begin{aligned}\iint_S \mathbf{F} \cdot \mathbf{n}\, d\sigma &= \iint_R (x + y^2 + a - x - y)\, dA \\ &= \iint_R (a - y + y^2)\, dA \\ &= \iint_R (a - y)\, dA + \iint_R y^2\, dA\end{aligned}$$

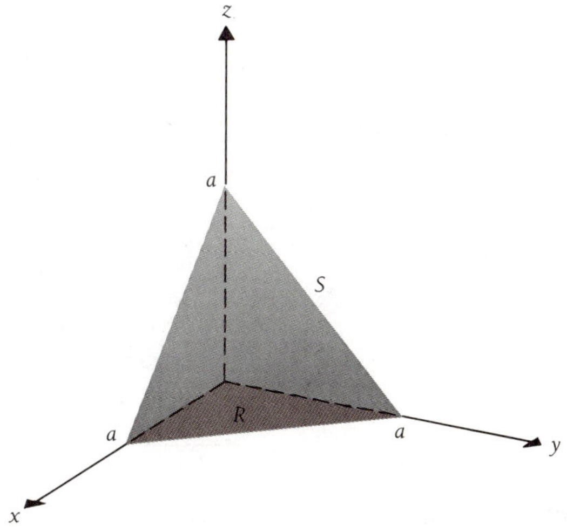

FIGURE 17.9

$$\iint_S \mathbf{F}\cdot\mathbf{n}\,d\sigma = (a-\bar{y})A + \int_0^a \int_0^{a-x} y^2\,dy\,dx$$
$$= \left(a - \frac{a}{3}\right)\left(\frac{1}{2}a^2\right) + \int_0^a \frac{1}{3}(a-x)^3\,dx$$
$$= \frac{1}{3}a^3 - \frac{1}{12}(a-x)^4\Big|_0^a$$
$$= \frac{1}{3}a^3 + \frac{1}{12}a^4.$$

■

■ Example 17.10 Find the flux of the vector field $\mathbf{F}(x, y, z) = \langle -x^2, -y^2, 2z\rangle$ out of the box with faces $x = -1$, $x = 1$, $y = -1$, $y = 2$, $z = -1$, $z = 1$ (Figure 17.10).

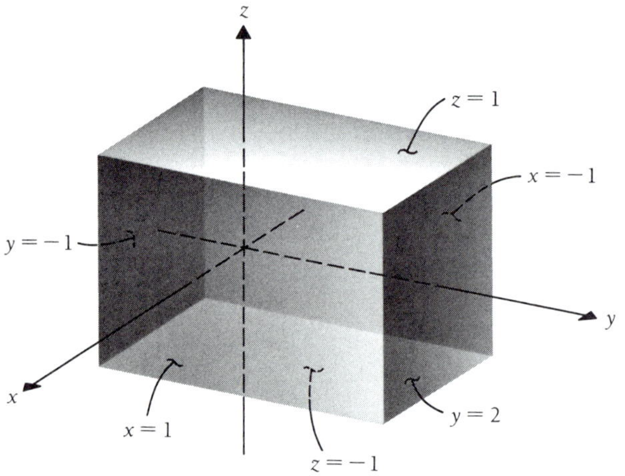

FIGURE 17.10

Solution We will compute the flux through each face separately, and then add the results together. We use the fact that the flux through each face is the value of $\mathbf{F}\cdot\mathbf{n}$ evaluated at that face times the area of the face. Note that $\mathbf{n}$ is the outer unit normal in each case.

FACE	$\mathbf{n}$	$\mathbf{F}\cdot\mathbf{n}$	FLUX	
$x = -1$	$-\mathbf{i}$	x^2	$(-1)^2(3)(2) =$	6
$x = 1$	$\mathbf{i}$	$-x^2$	$-1^2(3)(2) =$	-6
$y = -1$	$-\mathbf{j}$	y^2	$(-1)^2(2)(2) =$	4
$y = 2$	$\mathbf{j}$	$-y^2$	$-2^2(2)(2) =$	-16
$z = -1$	$-\mathbf{k}$	$-2z$	$-2(-1)(2)(3) =$	12
$z = 1$	$\mathbf{k}$	$2z$	$2(1)(2)(3) =$	12
			Total $=$	12

Thus, the total flux is 12. ■

In Example 17.10, we can interpret the results as follows: When the flux through a face is positive, then the net flow through that face is in the outward direction. When the flux is negative, then the net flow is into the box through that face. In the example, the total flux is positive, meaning that "more" is flowing out of the box than is flowing into it. In an actual physical situation, the fluid that is flowing in would be in a compressed state, and the fluid that is flowing out would be in a decompressed state. Thus, the flow outward is more rapid than the flow inward.

Exercises 17.2

For each given surface S, find: ***(a)*** *The area* ***(b)*** *The centroid*

1. S: $z = 1 - x - y,\quad 0 \le x \le 1,\quad 0 \le y \le 1 - x$

2. S: $z = 1 + x + y,\quad 0 \le x \le 2,\quad 0 \le y \le 1$

3. S: $z = a + bx + cy,\quad 0 \le x \le p,\quad 0 \le y \le q$

4. S: $z = a + bx + cy,\quad x^2 + y^2 \le d^2$

5. S: $x + 3y - 4z = 12,\quad 0 \le x \le 1,\quad 2 \le y \le 5$

6. S: $x + 3y - 4z = 12,\quad 0 \le x \le 2,\quad 0 \le y \le x^2$

7. S: the triangular surface with vertices $(a, 0, 0)$, $(0, b, 0)$, and $(0, 0, c)$

8. S: the triangular surface with vertices (a_1, b_1, c_1), (a_2, b_2, c_2), and (a_3, b_3, c_3)

9. S: $z = \sqrt{1 - x^2 - y^2},\quad x^2 + y^2 \le 1$

10. S: $z = \sqrt{1 - x^2 - y^2},\quad x^2 + y^2 \le \frac{1}{2}$

11. S: $z = 3 + \sqrt{2xy},\quad 1 \le x \le 2,\quad 1 \le y \le 3$ (Use numerical methods.)

12. S: $z = 3 + \sqrt{2xy},\quad 0 \le x \le a,\quad 0 \le y \le b$

13. S: $z = \sqrt{a^2 - x^2 - y^2},\quad x^2 + y^2 \le a^2,\quad x \ge 0$

14. S: $z = \sqrt{a^2 - x^2 - y^2},\quad x^2 + y^2 \le a^2,\quad 0 \le y \le mx$

15. S: $z = \dfrac{x^2 + y^2}{a^2},\quad x^2 + y^2 \le a^2$

16. S: $z = \dfrac{x^2}{a^2} + \dfrac{y^2}{b^2},\quad 0 \le x \le a,\quad 0 \le y \le b$

For each given surface S and density δ*, find:* ***(a)*** *The mass* ***(b)*** *The center of mass*

17. S as in Problem 1; $\delta(x, y, z) = kx$

18. S as in Problem 1; $\delta(x, y, z) = k(x + y)$

19. S as in Problem 5; $\delta(x, y, z) = x + y + z$

20. S as in Problem 2; $\delta(x, y, z) = x^2 + 1$

21. S as in Problem 9; $\delta(x, y, z) = z^2$

22. S as in Problem 9; $\delta(x, y, z) = kz$

Find the flux of **F** *across S.*

23. S as in Problem 1; $\mathbf{F}(x, y, z) = \langle x, y, z \rangle$

24. S as in Problem 1; $\mathbf{F}(x, y, z) = \langle x^2, -y^2, 0 \rangle$

25. S as in Problem 7; $\mathbf{F}(x, y, z) = \langle y^2, x^2, 0 \rangle$

26. S as in Problem 7; $\mathbf{F}(x, y, z) = \langle \sin x, \cos y, z \rangle$

27. S: the box with faces $x=0$, $x=1$, $y=0$, $y=1$, $z=0$, $z=3$; $\mathbf{F}(x, y, z) = \langle e^{-x}, y^2, z^3 \rangle$

28. S as in Problem 27; $\mathbf{F}(x, y, z) = \langle 2x, 0, 2y \rangle$

29. The **average value** of a function g on a surface S is the surface integral

$$\bar{g} = \frac{1}{\sigma} \iint_S g(x, y, z)\, d\sigma.$$

Find the average density of the surface in:
(a) Problem 19 **(b)** Problem 21

30. Let S: $z = f(x, y)$, $(x, y) \in R$ be a smooth surface, and let the outer unit normal $\mathbf{n}$ of S be represented by its direction cosines,

$$\mathbf{n} = \langle \cos\alpha, \cos\beta, \cos\gamma \rangle.$$

That is, α, β, and γ are the angles between $\mathbf{n}$ and the standard unit vectors $\mathbf{i}$, $\mathbf{j}$, and $\mathbf{k}$. Show that

$$\sec\gamma = \sqrt{f_x^2 + f_y^2 + 1}.$$

[Thus, we can write $\iint_S g(x, y, z)\, d\sigma = \iint_R g(x, y, f(x, y)) \sec\gamma\, dA$.]

SECTION 17.3

The Divergence Theorem

The divergence theorem is a higher-dimensional analog of Green's theorem. In this section, we discuss the divergence of a vector field, and then develop the divergence theorem and its applications.

The Divergence of a Vector Field

Let $\mathbf{F}(x, y, z) = \langle P(x, y, z), Q(x, y, z), R(x, y, z) \rangle$ be a continuously differentiable vector field. The **divergence of F** is the scalar function

$$\operatorname{div} \mathbf{F} = P_x + Q_y + R_z. \tag{1}$$

The divergence of $\mathbf{F}$ also can be denoted using the vector differential operator

$$\nabla = \left\langle \frac{\partial}{\partial x}, \frac{\partial}{\partial y}, \frac{\partial}{\partial z} \right\rangle$$

as

$$\operatorname{div} \mathbf{F} = \nabla \cdot \mathbf{F}, \tag{2}$$

for

$$\begin{aligned}\text{div } \mathbf{F} &= P_x + Q_y + R_z \\ &= \frac{\partial P}{\partial x} + \frac{\partial Q}{\partial y} + \frac{\partial R}{\partial z} \\ &= \left\langle \frac{\partial}{\partial x}, \frac{\partial}{\partial y}, \frac{\partial}{\partial z} \right\rangle \cdot \langle P, Q, R \rangle \\ &= \nabla \cdot \mathbf{F}.\end{aligned}$$

For a physical interpretation of the divergence of a vector field, suppose that **F** represents the velocity of a fluid. Then the divergence of **F** at the point (x, y, z) is a measure of the tendency of the fluid to flow away from the point (x, y, z). If div **F** is positive at a point P, we call P a **source** in the fluid; if div **F** is negative at P, we call P a **sink**. If div **F** is zero at every point, the fluid is called **incompressible**. Most of the fluids with which we are familiar, such as water under normal conditions, are incompressible, or very nearly so.

Example 17.11 Compute the divergence of:

(a) $\mathbf{F}(x, y, z) = \langle x^2y, xy - z^2, yz - 4 \rangle$
(b) $\mathbf{F}(x, y, z) = \langle e^x \sin yz, e^x \cos y \ln z, \ln(1 + e^x z) \rangle$

Solution

(a) $\text{div } \mathbf{F} = \nabla \cdot \mathbf{F}$
$\quad = 2xy + x + y.$

(b) $\text{div } \mathbf{F} = \nabla \cdot \mathbf{F}$
$\quad = e^x \sin yz - e^x \sin y \ln z + \dfrac{e^x}{1 + e^x z}.$ ■

The Divergence Theorem

Green's theorem in the plane relates a path integral over a simple closed curve to a double integral over the region enclosed by the curve. The divergence theorem does the same thing, but one dimension higher; it relates a surface integral over a simple closed surface in space to a triple integral over the region enclosed by the surface.

THEOREM 17.2

The Divergence Theorem Let S be a piecewise smooth simple closed surface, with outer unit normal $\mathbf{n}$, enclosing a region T in space, and let **F** be a vector field continuously differentiable on an open set containing T. Then

$$\iint_S \mathbf{F} \cdot \mathbf{n}\, d\sigma = \iiint_T \text{div } \mathbf{F}\, dV.$$

Proof The idea of our proof is similar to that of Green's theorem. We will prove it only for a standard region T—that is, a region that can be described as being enclosed by an upper surface and a lower surface, in the direction of any one of the coordinate axes. Using $\mathbf{F} = \langle P, Q, R\rangle$ and $\mathbf{n} = \langle \cos\alpha, \cos\beta, \cos\gamma\rangle$, the statement of the divergence theorem can be rewritten as

$$\iint_S (P\cos\alpha + Q\cos\beta + R\cos\gamma)\, d\sigma = \iiint_T (P_x + Q_y + R_z)\, dV.$$

The theorem will be proved if we can establish all three of the equations

$$\iint_S P\cos\alpha\, d\sigma = \iiint_T P_x\, dV,$$
$$\iint_S Q\cos\beta\, d\sigma = \iiint_T Q_y\, dV,$$
$$\iint_S R\cos\gamma\, d\sigma = \iiint_T R_z\, dV.$$

These three are similar, so we will establish only the last one.

Let T be described as

$$T = \{(x, y, z) \mid (x, y) \in U \text{ and } g(x, y) \le z \le h(x, y)\}$$

(Figure 17.11). That is, U is the projection of T into the xy-plane, and the surfaces $S_1: z = h(x, y)$ and $S_2: z = g(x, y)$ enclose T above and below, respectively. We can

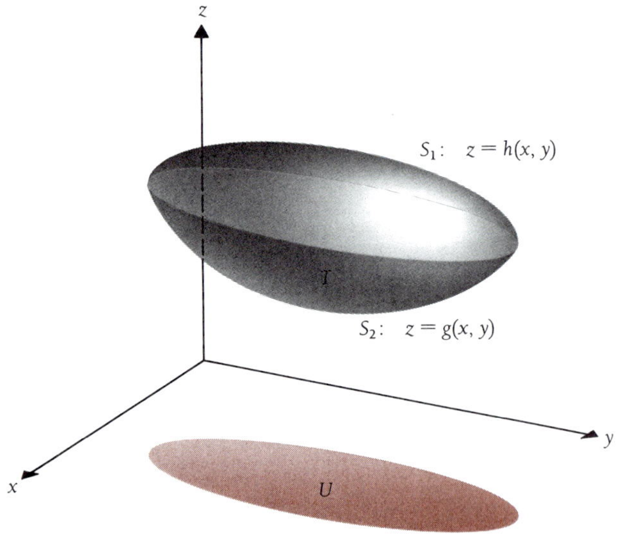

FIGURE 17.11

write $S = S_1 + S_2$. Then

$$\iint_S R \cos \gamma \, d\sigma = \iint_{S_1} R \cos \gamma \, d\sigma + \iint_{S_2} R \cos \gamma \, d\sigma.$$

For S_1, the outer normal is upward, and by Problem 30 of Section 17.2,

$$\iint_{S_1} R(x, y, z) \cos \gamma \, d\sigma = \iint_U R(x, y, h(x, y)) \cos \gamma \sec \gamma \, dA$$
$$= \iint_U R(x, y, h(x, y)) \, dA.$$

For S_2, the outer normal is downward, and $\cos \gamma$ is negative, so

$$\iint_{S_1} R(x, y, z) \cos \gamma \, d\sigma = \iint_U R(x, y, g(x, y)) \cos \gamma(-\sec \gamma) \, dA$$
$$= -\iint_U R(x, y, g(x, y)) \, dA.$$

Thus,

$$\iint_S R(x, y, z) \cos \gamma \, d\sigma = \iint_U [R(x, y, h(x, y)) - R(x, y, g(x, y))] \, dA$$
$$= \iint_U \Big[R(x, y, z) \Big]_{g(x, y)}^{h(x, y)} dA$$
$$= \iiint_T R_z(x, y, z) \, dV.$$

This completes our proof.

■ **Example 17.12** Find the flux of the vector field $\mathbf{F}(x, y, z) = \langle -x^2, -y^2, 2z \rangle$ out of the box with faces $x = -1$, $x = 1$, $y = -1$, $y = 2$, $z = -1$, $z = 1$ (see Example 17.10).

Solution By the divergence theorem, the flux is

$$\iint_S \mathbf{F} \cdot \mathbf{n} \, d\sigma = \iiint_T \operatorname{div} \mathbf{F} \, dV$$
$$= \iiint_T (-2x - 2y + 2) \, dV,$$

with T the region

$$T = \{(x, y, z) | -1 \leq x \leq 1, -1 \leq y \leq 2, -1 \leq z \leq 1\}.$$

Now, the volume of T is $V = 12$, and the centroid of T is $(\bar{x}, \bar{y}, \bar{z}) = (0, \frac{1}{2}, 0)$. Hence, the flux is

$$\begin{aligned}\iint_S \mathbf{F} \cdot \mathbf{n}\, d\sigma &= (-2\bar{x} - 2\bar{y} + 2)V \\ &= (0 - 1 + 2)(12) = 12.\end{aligned}$$

■

■ **Example 17.13** Find the flux of the vector field $\mathbf{F}(x, y, z) = \langle x, y, z \rangle$ through the unit sphere $S\colon x^2 + y^2 + z^2 = 1$.

Solution We will use the divergence theorem directly. Let T be the unit ball. Since the divergence of $\mathbf{F}$ is $\nabla \cdot \mathbf{F} = 1 + 1 + 1 = 3$, the flux is

$$\iint_S \mathbf{F} \cdot \mathbf{n}\, d\sigma = \iiint_T 3\, dV,$$

Carl Friedrich Gauss (1777–1855) was a child prodigy of the first magnitude. When he was 3 years old, he corrected his father's arithmetic; by the age of 5, he had mastered French, Latin, and Greek in addition to his native German. His precocity attracted the attention of the Duke of Brunswick, who paid for his education. At the age of 19, Gauss discovered a ruler-and-compass construction of a regular 17-sided polygon. This result convinced him that he should study mathematics instead of languages.

Gauss's early creative work was in number theory, where he recognized the importance of the law of quadratic reciprocity and gave its first complete proof. When the asteroid Ceres was discovered and then lost because of cloud cover, Gauss astonished the world by finding its position from the very inadequate observational data. He was rewarded by being appointed professor of astronomy at the University of Göttingen, where he remained for the rest of his life. The position required a certain amount of surveying work, in response to which Gauss invented a general theory of surfaces. (Further work in this direction by Riemann and others culminated in Einstein's theory of relativity.) In his work on the hypergeometric equation in 1812 is found the first systematic discussion of the convergence of infinite series. Gauss used the divergence theorem in his work with Wilhelm Weber on magnetism and electricity, and together, they built a functioning telegraph in 1833.

Early in his life, Gauss was recognized as the greatest living mathematician. Many feel that he was the greatest pure mathematician that the human race has yet produced.

which is just 3 times the volume of the unit ball,

$$3\left(\frac{4}{3}\right)\pi = 4\pi.$$

Notice that the flux of this vector field happens to be equal to the surface area of S. ■

The divergence theorem is also called **Gauss's theorem**, in honor of the great German mathematician C. F. Gauss.

Exercises 17.3

Find the divergence of **F**.

1. $\mathbf{F}(x, y, z) = \langle xy, x^2y^2, e^{xz}\rangle$
2. $\mathbf{F}(x, y, z) = \langle 1 + xz, y - z, x + y + z\rangle$
3. $\mathbf{F}(x, y, z) = \langle -x, -y, -z\rangle$
4. $\mathbf{F}(x, y, z) = (x^2 + y^2 + z^2)^{-1}\langle x, y, z\rangle$
5. $\mathbf{F}(x, y, z) = \langle \sin x, \cos y, \tan z\rangle$
6. $\mathbf{F}(x, y, z) = \langle yz, xz, xy\rangle$
7. $\mathbf{F}(x, y, z) = \nabla(x^4y^3z^2)$
8. $\mathbf{F}(x, y, z) = \nabla(e^{xy}\sqrt{z})$

Find the flux of each vector field **F** *out of the surface S.*

9. $\mathbf{F}(x, y, z) = \langle 2e^{-x}z, xy^2, z^3\rangle$; S is the box with faces $x = 0$, $x = 1$, $y = 0$, $y = 2$, $z = 0$, $z = 3$.
10. $\mathbf{F}(x, y, z) = \langle 2e^{-x}z, xy^2, z^3\rangle$; S is the unit box with faces $x = 1$, $x = -1$, $y = 1$, $y = -1$, $z = 1$, $z = -1$.
11. $\mathbf{F}(x, y, z) = \langle 2e^{-x}z, xy^2, z^3\rangle$; S is the unit sphere $x^2 + y^2 + z^2 = 1$.
12. $\mathbf{F}(x, y, z) = \langle 2e^{-x}z, xy^2, z^3\rangle$; S is the sphere of radius r, $x^2 + y^3 + z^2 = r^2$.
13. $\mathbf{F}(x, y, z) = \langle xy, \sin z, \cos y\rangle$; S is the unit box in Problem 10.
14. $\mathbf{F}(x, y, z) = \langle xy, \sin z, \cos y\rangle$; S is the unit sphere in Problem 11.
15. $\mathbf{F}(x, y, z) = \langle x, y, xyz\rangle$; S is the cylinder $x^2 + y^2 = 1$, $0 \le z \le 4$.
16. $\mathbf{F}(x, y, z) = \langle x, y, xyz\rangle$; S is the tetrahedron with faces $x = 0$, $y = 0$, $z = 0$, and $\frac{x}{a} + \frac{y}{b} + \frac{z}{c} = 1$ (assume a, b, and c are positive).
17. $\mathbf{F}(x, y, z) = \langle -2xy, -3z^2, xz\rangle$; S is the tetrahedron in Problem 16.
18. $\mathbf{F}(x, y, z) = \langle -2xy, -3z^z, xz\rangle$; S is the cylinder in Problem 15.
19. $\mathbf{F}(x, y, z) = \langle -x, -y, -z\rangle$; S is the unit sphere in Problem 11.
20. $\mathbf{F}(x, y, z) = \langle -x, -y, -z\rangle$; S is the unit box in Problem 10.
21. $\mathbf{F}(x, y, z) = \langle x^2, y^2, z^2\rangle$; S is the unit sphere in Problem 11.
22. $\mathbf{F}(x, y, z) = \langle x^2, y^2, z^2\rangle$; S is the unit box in Problem 10.
23. $\mathbf{F}(x, y, z) = (x^2 + y^2 + z^2)^{-1}\langle x, y, z\rangle$; S is the unit sphere in Problem 11.
24. $\mathbf{F}(x, y, z) = (x^2 + y^2 + z^2)^{-1}\langle x, y, z\rangle$; S is the unit box in Problem 10.
25. Find the divergence of the central force field

$$\mathbf{F}(x, y, z) = K(x^2 + y^2 + z^2)^{-n/2}\langle x, y, z\rangle,$$

with K a constant and n a positive integer (see Problem 4 in the Additional Exercises at the end of Chapter 15).

26. Find the flux of the central force field of Problem 25 out of the sphere $x^2 + y^2 + z^2 = r^2$ of radius r.

27. The divergence of the gradient of a function f, $\nabla \cdot \nabla f = \nabla^2 f$, is called the **Laplacian** of f. Show that

$$\nabla^2 f = f_{xx} + f_{yy} + f_{zz}.$$

28. Show that $\operatorname{div}(\mathbf{F} + \mathbf{G}) = \operatorname{div} \mathbf{F} + \operatorname{div} \mathbf{G}$.

29. Show that $\operatorname{div} c\mathbf{F} = c \operatorname{div} \mathbf{F}$.

30. Show that $\operatorname{div} f\mathbf{F} = f \operatorname{div} \mathbf{F} + \nabla f \cdot \mathbf{F}$.

31. Show that $\operatorname{div} f\nabla f = f\nabla^2 f + |\nabla f|^2$. (See Problem 27.)

32. Compute $\operatorname{div}(\nabla f \times \nabla g)$.

33. A vector field $\mathbf{F}$ is **solenoidal** if and only if $\iint_S \mathbf{F} \cdot \mathbf{n}\, d\sigma = 0$ for every closed surface S. Show that $\mathbf{F}$ is solenoidal if and only if $\operatorname{div} \mathbf{F} = 0$.

34. If the closed smooth surface S encloses a region of volume V, and $\mathbf{r} = \langle x, y, z \rangle$, show that

$$V = \frac{1}{3} \iint_S \mathbf{r} \cdot \mathbf{n}\, d\sigma.$$

35. Let $\mathbf{F}$ be a conservative force field with potential function f, and let the closed surface S enclose the region T. If $\mathbf{F}$ is solenoidal, show that

$$\iiint_T \mathbf{F} \cdot \mathbf{F}\, dV = \iint_S f\mathbf{F} \cdot \mathbf{n}\, d\sigma.$$

[*Hint*: Use Problems 30 and 33.]

36. Let C be a simple closed smooth curve in the plane enclosing a region R, and let $\mathbf{n}$ be the unit normal vector to C opposite to the curvature vector. If $\mathbf{F}$ is a two-dimensional vector field and $\mathbf{n}$ is continuous, show that

$$\oint_C \mathbf{F} \cdot \mathbf{n}\, ds = \iint_R \operatorname{div} \mathbf{F}\, dA.$$

(This is actually a version of Green's theorem that is sometimes called the **divergence theorem in the plane**.)

SECTION 17.4

Stokes's Theorem

Stokes's theorem—a generalization of Green's theorem—relates a path integral over a curve in space to a surface integral over a surface bounded by the curve. In fact, Green's theorem does just that, if one considers the region in the plane enclosed by the curve to be a flat, horizontal surface bounded by the curve. In order to state the generalization, named for the English mathematician and physicist George Stokes (1819–1903), we need to introduce the concept of the curl of a vector field.

The Curl of a Vector Field

Let $\mathbf{F} = \langle P, Q, R\rangle$ be a differentiable vector field. The **curl** of $\mathbf{F}$ is defined by

$$\textbf{curl } \mathbf{F} = \langle R_y - Q_z, P_z - R_x, Q_x - P_y\rangle. \tag{1}$$

Then we also have

$$\textbf{curl } \mathbf{F} = \nabla \times \mathbf{F},$$

since

$$\begin{aligned}\textbf{curl } \mathbf{F} &= \langle R_y - Q_z, P_z - R_x, Q_x - P_y\rangle \\ &= \left\langle \frac{\partial R}{\partial y} - \frac{\partial Q}{\partial z}, \frac{\partial P}{\partial z} - \frac{\partial R}{\partial x}, \frac{\partial Q}{\partial x} - \frac{\partial P}{\partial y}\right\rangle \\ &= \begin{vmatrix} \mathbf{i} & \mathbf{j} & \mathbf{k} \\ \frac{\partial}{\partial x} & \frac{\partial}{\partial y} & \frac{\partial}{\partial z} \\ P & Q & R \end{vmatrix} \\ &= \nabla \times \mathbf{F}.\end{aligned}$$

For a physical interpretation of the curl, we again represent $\mathbf{F}$ as the velocity of a fluid. Since the gradient ∇ is in the general direction away from the origin and $\mathbf{F}$ is in the direction of flow, $\textbf{curl } \mathbf{F} = \nabla \times \mathbf{F}$ is a vector at right angles to both these directions. The magnitude of the curl will be related to the sine of the angle between the directions of ∇ and $\mathbf{F}$, and hence is an indication of the tendency of the fluid to circulate or rotate (curl) about the point at which it is measured. If the curl of a vector field is zero at every point, the field is called **irrotational.**

■ **Example 17.14** Find the curl of the vector field:

(a) $\mathbf{F}(x, y, z) = \langle x, y, z\rangle$ **(b)** $\mathbf{F}(x, y, z) = \langle y, -x, 0\rangle$
(c) $\mathbf{F}(x, y, z) = \langle z \cos x, x \sin y, y\rangle$

Solution (a) This vector field is directed away from the origin [see Figure 17.12, page 1026, in which arrows representing some vectors $\mathbf{F}(x, y, 0)$ have been drawn with their tails at $(x, y, 0)$]. Its curl is

$$\begin{aligned}\nabla \times \mathbf{F} &= \begin{vmatrix} \mathbf{i} & \mathbf{j} & \mathbf{k} \\ \frac{\partial}{\partial x} & \frac{\partial}{\partial y} & \frac{\partial}{\partial z} \\ x & y & z \end{vmatrix} \\ &= 0\mathbf{i} + 0\mathbf{j} + 0\mathbf{k} = \mathbf{0}.\end{aligned}$$

Thus, this vector field is irrotational, having no tendency to rotate at all.

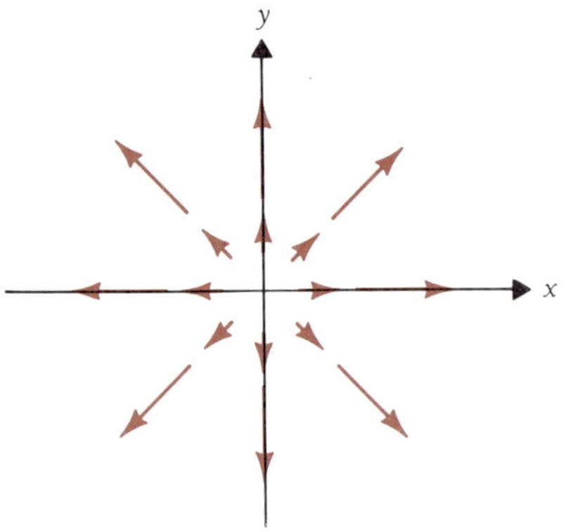

FIGURE 17.12 Diagram of the vector field $\langle x, y, z\rangle$

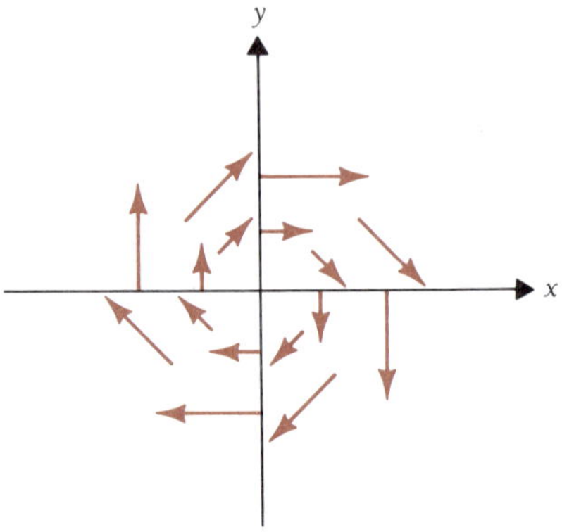

FIGURE 17.13 Diagram of the vector field $\langle y, -x, 0\rangle$

(b) This vector field appears to rotate markedly (see Figure 17.13). Its curl is

$$\begin{aligned}\nabla \times \mathbf{F} &= \begin{vmatrix} \mathbf{i} & \mathbf{j} & \mathbf{k} \\ \dfrac{\partial}{\partial x} & \dfrac{\partial}{\partial y} & \dfrac{\partial}{\partial z} \\ y & -x & 0 \end{vmatrix} \\ &= 0\mathbf{i} + 0\mathbf{j} + (-1 - 1)\mathbf{k} = -2\mathbf{k},\end{aligned}$$

with $\mathbf{k}$ being in the direction of the axis of rotation of the field.

(c) The curl here is

$$\begin{aligned}\nabla \times \mathbf{F} &= \begin{vmatrix} \mathbf{i} & \mathbf{j} & \mathbf{k} \\ \dfrac{\partial}{\partial x} & \dfrac{\partial}{\partial y} & \dfrac{\partial}{\partial z} \\ z\cos x & x\sin y & y \end{vmatrix} \\ &= (1 - 0)\mathbf{i} + (\cos x - 0)\mathbf{j} + (\sin y - 0)\mathbf{k} \\ &= \langle 1, \cos x, \sin y\rangle.\end{aligned}$$

■

■ **Example 17.15** Find the curl of a two-dimensional vector field.

Solution A two-dimensional vector field $\mathbf{F} = \langle P, Q\rangle$ may be regarded as a three-dimensional vector field by writing it as

$$\mathbf{F} = \langle P(x, y), Q(x, y), 0\rangle.$$

Its curl is then

$$\begin{aligned}\mathbf{curl}\ \mathbf{F} &= \langle 0 - Q_z, P_z - 0, Q_x - P_y\rangle \\ &= \langle 0, 0, Q_x - P_y\rangle.\end{aligned}$$

■

Using Example 17.15, we can state Green's theorem in terms of the curl as

$$\oint_C \mathbf{F} \cdot d\mathbf{r} = \iint_R \mathbf{curl\ F} \cdot \mathbf{k}\, dA.$$

It is this statement that we wish to generalize.

Stokes's Theorem

We will generalize Green's theorem by considering the smooth curve C to be in space, and we will let S be any smooth surface of which C is the boundary. We will choose the upper unit normal as the orientation of S, since that is compatible with the counterclockwise orientation of C. (To determine whether the two orientations are compatible, we use a right-hand rule: If the thumb of the right hand points in the direction of the unit normal of S, then the fingers will point in the direction of the orientation of the curve.) Then the closed-path integral over C is equal to the surface integral over S, as is stated by Stokes's theorem.

THEOREM 17.3

Stokes's Theorem Let C be a piecewise smooth simple closed curve in space, and let S be a smooth surface of which C is the boundary (see Figure 17.14). Let $\mathbf{n}$ be the upper unit normal of S, and let $\mathbf{F}$ be a vector field differentiable on an open set containing S. Then

$$\oint_C \mathbf{F} \cdot d\mathbf{r} = \iint_S \mathbf{curl\ F} \cdot \mathbf{n}\, d\sigma.$$

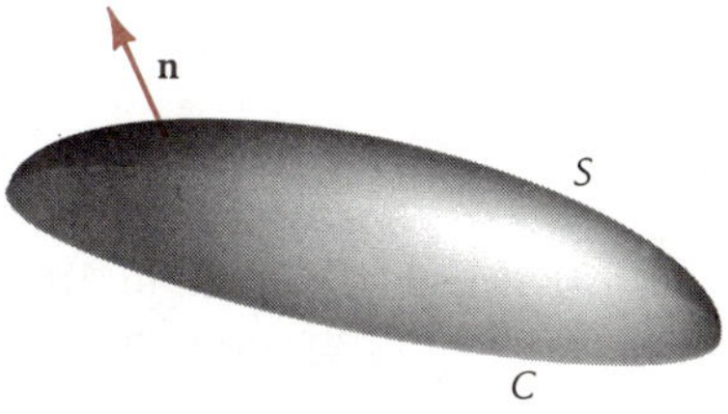

FIGURE 17.14

The proof of Stokes's theorem generally can be found in texts on vector analysis or advanced calculus. The proof is based on Green's theorem, but since it lends little insight into the workings of the theorem, we omit it.

■ **Example 17.16**

Evaluate the path integral $\oint_C \mathbf{F} \cdot d\mathbf{r}$, with $\mathbf{F}$ the vector field $\langle y, -x, z \rangle$ and C the circle $x^2 + y^2 = 1$, $z = 1$, in two ways:

(a) Directly by parametrizing the curve **(b)** By using Stokes's theorem

Solution **(a)** We can parametrize this curve by

$$\mathbf{r}(t) = \langle \cos t, \sin t, 1 \rangle, \qquad t \in [0, 2\pi].$$

Then,

$$\begin{aligned}\oint_C \mathbf{F} \cdot d\mathbf{r} &= \int_0^{2\pi} \langle \sin t, -\cos t, 1 \rangle \cdot \langle -\sin t, \cos t, 0 \rangle \, dt \\ &= \int_0^{2\pi} (-\sin^2 t - \cos^2 t) \, dt \\ &= \int_0^{2\pi} (-1) \, dt = -2\pi.\end{aligned}$$

(b) Using Stokes's theorem, we will need

$$\mathbf{curl\ F} = \langle 0, 0, -2 \rangle$$

[see Example 17.14(b)]. Also, S is the disc bounded by C, and the upper unit normal vector is $\mathbf{k}$. Then,

$$\begin{aligned}\oint_C \mathbf{F} \cdot d\mathbf{r} &= \iint_S \mathbf{curl\ F} \cdot \mathbf{n} \, d\sigma \\ &= \iint_S \langle 0, 0, -2 \rangle \cdot \langle 0, 0, 1 \rangle \, d\sigma \\ &= (-2) \iint_S d\sigma \\ &= (-2)(\text{Area of } S) = -2\pi.\end{aligned}$$

■

The path integral in Example 17.16 can be interpreted as work if $\mathbf{F}$ is regarded as a force field. If $\mathbf{F}$ is interpreted as the velocity of a fluid, then the path integral is called the **circulation** around the curve C. Since the result in the example is negative, then the circulation is in the direction opposite to the direction of orientation of the curve C.

■ **Example 17.17** Find the work done by the force field $\mathbf{F} = \langle x^2, xy, x + y + z \rangle$ in moving an object counterclockwise once around the curve C of intersection of the cone $z^2 = 4(x^2 + y^2)$ and the plane $z = y + 1$ (Figure 17.15).

Solution The work is

$$W = \oint_C \mathbf{F} \cdot d\mathbf{r},$$

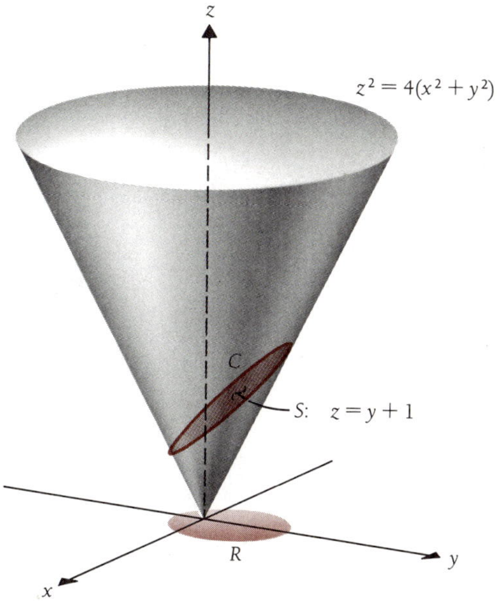

FIGURE 17.15

which by Stokes's theorem is

$$W = \iint_S \mathbf{curl\ F} \cdot \mathbf{n}\, d\sigma$$

for any smooth surface S of which C is the boundary. We will use the plane surface

$$S:\quad z = y + 1, \qquad (x, y) \in R,$$

with R the region enclosed by the projection of C onto the xy-plane. The projection is found by eliminating z from the two equations of the intersecting surfaces, or

$$(y + 1)^2 = 4(x^2 + y^2),$$

which is the ellipse

$$\frac{x^2}{\frac{1}{3}} + \frac{(y - \frac{1}{3})^2}{\frac{4}{9}} = 1.$$

Thus, R has centroid $(\bar{x}, \bar{y}) = (0, \frac{1}{3})$ and area $A = \pi(1/\sqrt{3})(\frac{2}{3}) = 2\pi/3\sqrt{3}$. Thus, by equation (7) of Section 17.2 with $f(x, y) = y + 1$, we have

$$W = \iint_S \langle 1, -1, y \rangle \cdot \mathbf{n}\, d\sigma$$

$$= \iint_R (1 + y)\, dA$$

$$W = (1 + \bar{y})A$$
$$= \left(1 + \frac{1}{3}\right)\left(\frac{2\pi}{3\sqrt{3}}\right) = \frac{8\pi}{9\sqrt{3}}.$$

■

Exercises 17.4

Find **curl F**.

1. $\mathbf{F} = \langle x + y - z, x - y + z, -x + y + z \rangle$
2. $\mathbf{F} = \langle xy, yz, zx \rangle$
3. $\mathbf{F} = \langle P(x), Q(y), R(z) \rangle$
4. $\mathbf{F} = \langle x^2 + y^2 + z^2, 3xyz, 6 \rangle$
5. $\mathbf{F} = \langle e^x, \sin(x + y) \rangle$
6. $\mathbf{F} = \langle x^2 - xy + y^2, 2xy \rangle$
7. $\mathbf{F} = \nabla(x^2 + yz)$
8. $\mathbf{F} = \langle x, y, xy \rangle + \nabla(z^2 + 2xy)$

Verify Stokes's theorem.

9. $\mathbf{F} = \langle x^2 + y^2, 2yz, z^2 - x^2 \rangle$; C: $x^2 + y^2 = 1, z = 0$;
   S: $z = \sqrt{1 - x^2 - y^2}, x^2 + y^2 \leq 1$
10. $\mathbf{F} = \langle x^2 + y^2 + z^2, x + y + z, 0 \rangle$; C: $\mathbf{r}(t) = \langle 2\cos t, 3\sin t, \sin t \rangle$, $t \in [0, 2\pi]$;
    S: $z = \sqrt{4 - x^2 - \frac{y^2}{3}}, \frac{x^2}{4} + \frac{y^2}{9} \leq 1$
11. $\mathbf{F}$ and C as in Problem 9; S: $z = 1 - x^2 - y^2, x^2 + y^2 \leq 1$
12. $\mathbf{F}$ and C as in Problem 10; S: $z = \frac{y}{3}, \frac{x^2}{4} + \frac{y^2}{9} \leq 1$

Find the work done by **F** *in moving an object around C in the counterclockwise direction.*

13. $\mathbf{F} = \langle y - x, z - y, x - z \rangle$; C as in Problem 9
14. $\mathbf{F} = \langle x^2, 0, y - z^2 \rangle$; C as in Problem 10
15. $\mathbf{F} = \langle xy, yz, zx \rangle$; C as in Example 17.17
16. $\mathbf{F} = \langle e^{yz}, \cos xz, \sin(x + y) \rangle$; C as in Example 17.17

Find the circulation of **F** *around the curve C in the counterclockwise direction.*

17. $\mathbf{F}(x, y, z) = \langle z, -x, y \rangle$; C: $x^2 + y^2 = 1, z = 4$
18. $\mathbf{F}(x, y, z) = \langle z, -x, y \rangle$; C: $x^2 + 2z^2 = 2, y = 2$
19. $\mathbf{F} = \langle y - x, z - y, x - z \rangle$; C: the intersection of $z^2 = x^2 + y^2$ and $x^2 + 4y^2 + 4z^2 = 4$
20. $\mathbf{F} = \langle x^2 + yz, x - y - z, x + y + z \rangle$; C as in Problem 19

Show the following:

21. $\operatorname{div}(\mathbf{curl\ F}) = 0$
22. $\mathbf{curl}(\mathbf{F} + \mathbf{G}) = \mathbf{curl\ F} + \mathbf{curl\ G}$
23. $\mathbf{curl}\ f\mathbf{F} = f\ \mathbf{curl\ F} + \nabla f \times \mathbf{F}$
24. $\mathbf{curl}(\mathbf{curl\ F}) = \nabla(\operatorname{div}\mathbf{F}) - \nabla^2\mathbf{F}$ (See Problem 27 in Section 17.3.)
25. $\operatorname{div}(\mathbf{F} \times \mathbf{G}) = \mathbf{G} \cdot \mathbf{curl\ F} - \mathbf{F} \cdot \mathbf{curl\ G}$
26. An object in uniform circular motion with position **r** and velocity **v** has angular velocity $\mathbf{w} = (1/\|\mathbf{r}\|^2)(\mathbf{r} \times \mathbf{v})$. Show that $\mathbf{curl\ v} = 2\mathbf{w}$.

27. Find the curl of the central force field

$$\mathbf{F}(x, y, z) = K(x^2 + y^2 + z^2)^{-n/2}\langle x, y, z\rangle.$$

28. Find the work done by the central force field **F** in Problem 27 in moving an object around any simple closed curve.

SECTION 17.5

Chapter Review

Important Concepts

Simple closed curve	Oriented surface	Divergence
Region enclosed by a curve	Upper unit normal	Divergence theorem
Standard region	Outer unit normal	Curl
Oriented curve	Surface integral	Irrotational vector field
Green's theorem	Flux	Stokes's theorem
Smooth surface		

Review Exercises

Find the work done by force **F** *in moving an object along curve C.*

1. $\mathbf{F} = \langle x \sin y, \sin x\rangle$; C: $\mathbf{r}(t) = \langle t, 3t^2\rangle$, $t \in [0, \pi/4]$
2. $\mathbf{F} = \langle 2x \ln y, x^2/y\rangle$; C: $y = x^2$ from $(1, 1)$ to $(2, 4)$
3. $\mathbf{F} = \langle xy, 3x^2\rangle$; C: $(x-1)^2 + y^2 = 1$, one circuit *clockwise*
4. $\mathbf{F} = \langle x^2, xy, x + y + z\rangle$; C: the intersection of $z^2 = 9(x^2 + y^2)$ and $z = y + 2$, one circuit counterclockwise
5. Show that $\oint_C [f(x)\, dx + g(y)\, dy] = 0$ if C is any circle.
6. Find $\oint_C (e^x \sin y\, dx + e^x \cos y\, dy)$, where C: $b^2x^2 + a^2y^2 = a^2b^2$.
7. For $\mathbf{r} = \langle x, y, z\rangle$, find $\nabla \cdot \mathbf{r}$, $\nabla \times \mathbf{r}$, and $\nabla\|\mathbf{r}\|$.

Let R be the region in the first quadrant bounded by $y = x^2$, $x = 0$, and $y = 4$. Find the surface area of the following:

8. The part of the plane $x + 3y - z + 2 = 0$ over R
9. The part of the paraboloid $z = x^2 + y^2$ over R
10. The part of $z = \sqrt{50 - x^2 - y^2}$ over R
11. Find the area of the part of the sphere $x^2 + y^2 + z^2 = 2a^2$ that is inside the cylinder $x^2 + y^2 = a^2$ (both pieces).
12. Find the lateral surface area of the portion of the cylinder $x^2 + (z-a)^2 = a^2$ that is inside the sphere $x^2 + y^2 + (z - 2a)^2 = 4a^2$.
13. Find the centroid of the surface $z = 2 - x^2 - y^2$, $z \geq 1$.

Find the flux of each vector field **F** *out of the given solid.*

14. $\mathbf{F} = \langle x^2, 3y^2, 3z\rangle$; $x^2 + y^2 \leq z \leq 4$
15. $\mathbf{F} = \langle x^2, y^2, z^2\rangle$; $x^2 + y^2 + z^2 \leq 1$
16. $\mathbf{F} = \langle x, y, z\rangle$; $x^2 + y^2 + z^2 \leq a^2$

17. Verify Stokes's theorem for $\mathbf{F} = \langle y, -x, -z \rangle$; S: $x^2 + y^2 \leq 9$, $z = 2$; and C: $x^2 + y^2 = 9$, $z = 2$.

18. Find $\iint_S y^2 z^2 \, d\sigma$, where S: $z = \sqrt{4 - x^2}$, $0 \leq y \leq 1$.

Additional Exercises

1. Set up the integral that gives the surface area of the ellipsoid

$$\frac{x^2}{a^2} + \frac{y^2}{b^2} + \frac{z^2}{c^2} = 1.$$

2. Find $\oint_C [(e^x - y)\, dx + (x - e^y)\, dy]$ for C any circle of radius 1.

3. Find

$$\oint_C \left(\frac{x}{x^2 + y^2}\, dy - \frac{y}{x^2 + y^2}\, dx \right)$$

for C any circle not enclosing the origin.

4. Find the centroid of the curve C: $\mathbf{r}(t) = \langle a \cos^3 t, a \sin^3 t \rangle$, $t \in [0, \pi/2]$.

5. If $\mathbf{F} = \langle x, y, z \rangle / \|\langle x, y, z \rangle\|$, show that div $\mathbf{F} = 2/\|\langle x, y, z \rangle\|$.

6. Find $\iiint_T (x^2 + y^2 + z^2)^{-1/2}\, dV$ for $T = \{(x, y, z) | x^2 + y^2 + z^2 \leq a^2\}$ by:
   (a) Direct computation (b) The divergence theorem and Problem 5

7. Show that if S is a sphere with outer normal $\mathbf{n}$, then for any continuously differentiable vector field $\mathbf{F}$, $\iint_S \mathbf{curl\ F} \cdot \mathbf{n}\, d\sigma = 0$.

8. Find $\int_C [(x - yz)\, dx + (y - zx)\, dy + (z - xy)\, dz]$ for C: $\mathbf{r}(t) = \langle \cos t, \sin t, t \rangle$, $t \in [0, 2\pi]$.

9. Compute $\nabla^2 f$ for:
   (a) $f(x, y, z) = x^2 + y^2 + z^2$ (b) $f(x, y, z) = x^2 - y^2 + z$
   (c) $f(x, y, z) = xyz$ (d) $f(x, y, z) = \sqrt{x^2 + y^2 + z^2}$

10. If $\nabla^2 f = 0$, the function f is called **harmonic**. Give some examples of harmonic functions.

11. Show that $\mathbf{F} = \langle e^x \sin x, y^{5/2} \rangle$ is irrotational.

12. Find the divergence of a two-dimensional vector field.

13. Show that no continuously differentiable vector field exists such that $\mathbf{curl\ F} = \langle x, y, z \rangle$.

14. Show that $\mathbf{curl}(\nabla f + \mathbf{curl\ F}) = \mathbf{curl}(\mathbf{curl\ F})$.

15. Show that a constant vector field is solenoidal.

16. Prove that $\mathbf{curl\ F} = \mathbf{0}$ for $\mathbf{F}$ any central force field.

17. Let f be a scalar function of three variables with continuous second-order partial derivatives, and let S be a smooth closed surface enclosing a region T of space. Prove that

$$\iiint_T \nabla^2 f\, dV = \iint_S f_{\mathbf{n}}\, d\sigma,$$

where $f_{\mathbf{n}}$ is the directional derivative of f in the direction of the outer unit normal $\mathbf{n}$ of S.

18. (a) Find $\operatorname{div}(f\nabla g)$.

(b) If f and g have continuous second-order partial derivatives, prove that

$$\iiint_T (f\nabla^2 g + \nabla f \cdot \nabla g)\, dV = \iint_S (f\nabla g) \cdot \mathbf{n}\, d\sigma.$$

(c) Prove that

$$\iiint_T (f\nabla^2 g - g\nabla^2 f)\, dV = \iint_S (f\nabla g - g\nabla f) \cdot \mathbf{n}\, d\sigma.$$

19. For $\mathbf{F} = \langle P, Q, R\rangle$, we define

$$\iint_S \mathbf{F}\, d\sigma = \iint_S P\, d\sigma + \iint_S Q\, d\sigma + \iint_S R\, d\sigma$$

and

$$\iiint_T \mathbf{F}\, dV = \iiint_T P\, dV + \iiint_T Q\, dV + \iiint_T R\, dV.$$

Prove that:

(a) $\displaystyle\iint_S \mathbf{F} \times \mathbf{n}\, d\sigma = -\iiint_T \nabla \times \mathbf{F}\, dV$ (b) $\displaystyle\iint_S f\mathbf{n}\, d\sigma = \iiint_T \nabla f\, dV$

20. Prove that if the smooth closed curve C bounds the smooth surface S, then

$$\oint_C f\nabla g \cdot d\mathbf{r} = \iint_S (\nabla f \times \nabla g) \cdot \mathbf{n}\, d\sigma.$$

21. An electric field $\mathbf{E}$ induces a magnetic field $\mathbf{B}$ such that **curl** $\mathbf{B} = \mathbf{E}$. Let S be a smooth surface with smooth boundary C. The **circulation** of the magnetic field around C is $\oint_C \mathbf{B} \cdot d\mathbf{r}$. Prove **Ampere's law**, that the flux of the electric field through S is equal to the circulation of the induced magnetic field around the boundary of S.

22. An electric field $\mathbf{E}$ induces a magnetic field $\mathbf{B}$ such that **curl** $\mathbf{E} = -d\mathbf{B}/dt$. Let S be a smooth surface with smooth boundary C. The **voltage drop** around C is $\oint_C \mathbf{E} \cdot d\mathbf{r}$. Prove **Faraday's law**, that the voltage drop around C is equal to the magnetic flux through S.

23. An electric field $\mathbf{E}$ satisfies div $\mathbf{E} = k\delta$, where k is a constant and δ is the charge density. Prove **Gauss's law**, that the flux of $\mathbf{E}$ through any smooth closed surface S is equal to the charge inside S.

24. If the flux of a vector field $\mathbf{F}$ through a closed surface S is negative, we can interpret it as a net flow inward; we say there is a **sink** inside S. If the flux is positive, we say there is a **source** inside S. The same language is used relative to a closed curve C in the plane for a two-dimensional vector field. Identify the locations of any sinks or sources in the following vector fields:

(a) $\mathbf{F} = \langle x, y\rangle$ (b) $\mathbf{F} = \langle -x, -y\rangle$ (c) $\mathbf{F} = \langle x, 0\rangle$
(d) $\mathbf{F} = \langle 0, y^2\rangle$ (e) $\mathbf{F} = \langle x, y, z\rangle$ (f) $\mathbf{F} = \langle x, -y, 0\rangle$

CHAPTER 18

Differential Equations

INTRODUCTION

The concept of differential equation was introduced in Section 4.6 in connection with the antiderivative. There and in Section 6.4, we studied some separable differential equations. In Section 4.7, we mentioned another type of differential equation as we discussed simple harmonic motion. In this chapter, we will discuss differential equations in more general terms and with more precision.

First, to describe what a differential equation is in general, we let y represent an n-fold differentiable function of x. An **ordinary differential equation** is an equation of the form

$$f(x, y, y', y'', \ldots, y^{(n)}) = 0, \tag{1}$$

with f a function of x, y, and the derivatives of y. The order of the highest derivative of y that appears in the equation is called the **order** of the equation.

A **solution** of equation (1) on an interval I is a function

$$y = F(x), \qquad x \in I,$$

such that

$$f(x, F(x), F'(x), F''(x), \ldots, F^{(n)}(x)) = 0$$

is true for all $x \in I$. The **general solution** of an equation of order n will normally involve n arbitrary constants called **constants of integration**. A **particular solution** is obtained by evaluating the constants of integration using **boundary conditions** of the form

$$F^{(i)}(x_0) = y_i.$$

We will restrict our attention to some particular first- and second-order differential equations and their applications. These will be certain equations of the forms

$$y' = f(x, y) \tag{2}$$

and

$$y'' = f(x, y, y'). \tag{3}$$

More complicated differential equations very quickly get beyond our scope, and even some of the above will defy solution.

It is not that solutions do not exist. If the function f is a function of two variables and is continuously differentiable in a region R containing a given interior point (x_0, y_0), then for sufficiently small h there is a unique solution $y = F(x)$, $x \in [x_0, x_0 + h]$, for equation (2) such that $F(x_0) = y_0$. Similarly, if the function f is a function of three variables and is continuously differentiable in a region R containing a given interior point (x_0, y_0, z_0), then for sufficiently small h there is a unique solution $y = F(x)$, $x \in [x_0, x_0 + h]$, for equation (3) such that $F(x_0) = y_0$ and $F'(x_0) = z_0$. The proofs of these existence and uniqueness theorems can be found in more advanced books on differential equations.

But finding the solutions may be difficult. We will consider several methods that apply to certain forms of equations, but other forms may involve methods we do not have space to consider—or as yet undiscovered methods!

Once we learn how to solve a differential equation, we will apply it to practical problems. The full power of mathematics in solving significant problems begins to appear as we study differential equations. You have already seen that even simple differential equations allow us to solve many interesting problems. Differential equations have long been commonly used in the physical and engineering sciences, and they are applied increasingly in the biological, social, and managerial sciences.

SECTION 18.1

Exact First-Order Equations

A first-order ordinary differential equation that can be written in the form

$$y' = \frac{g(x)}{f(y)}$$

is called **separable**, because it can be written in differential notation as

$$f(y)\,dy = g(x)\,dx.$$

Equations of this type and their applications were studied in Sections 4.6 and 6.4.

In this section, we study certain first-order equations that can be written in the form

$$y' = -\frac{P(x, y)}{Q(x, y)}, \tag{1}$$

or equivalently, in differential form,

$$P(x, y)\,dx + Q(x, y)\,dy = 0. \tag{2}$$

Exact Equations

If the differential form

$$P(x, y)\,dx + Q(x, y)\,dy \tag{3}$$

is exact, then the equation (1) or (2) is called an **exact equation.** Thus, equation (1) is an exact equation if and only if $P_y = Q_x$.

If equation (1) is exact, then expression (3) is the differential of some function f, where f is found as in Section 15.2. That is,

$$df = P(x, y)\,dx + Q(x, y)\,dy,$$

so that

$$df = 0. \tag{4}$$

It follows that

$$f(x, y) = C, \tag{5}$$

C a constant, is the solution of equation (1). More precisely, equation (5) defines y implicitly as a function F of x, and this function F is the general solution of equation (1).

■ **Example 18.1** Solve the equation

$$y' = \frac{-2xy}{x^2 + 2y}.$$

Solution This equation has the form (1) with $P(x, y) = 2xy$ and $Q(x, y) = x^2 + 2y$. Since $P_y = 2x = Q_x$, the equation is exact. Thus, we seek a function f such that

$$df = 2xy\,dx + (x^2 + 2y)\,dy.$$

By the methods of Section 15.2, we find that

$$f(x, y) = x^2y + y^2$$

is one such function. Thus, the general solution of our equation is

$$x^2y + y^2 = C, \qquad C \text{ a constant.}$$ ■

■ **Example 18.2** Solve the equation

$$xy'e^{xy} + ye^{xy} + 2 = 0,$$

subject to the boundary condition $y = 2$ when $x = 1$.

Solution This equation can be written in the form (2) as

$$(2 + ye^{xy})\,dx + xe^{xy}\,dy = 0.$$

This equation is exact (see Example 15.8), and is equivalent to $df = 0$ with

$$f(x, y) = e^{xy} + 2x.$$

The solution of the equation is therefore

$$e^{xy} + 2x = C.$$

If $y = 2$ when $x = 1$, then $e^{(1)(2)} + 2(1) = C$, or $C = 2 + e^2$. Thus, the specific solution we desire is

$$e^{xy} + 2x = 2 + e^2,$$

which also can be written as

$$y = \frac{1}{x}\ln(e^2 + 2 - 2x). \qquad \blacksquare$$

Integrating Factors

If equation (2) is not exact, it may be possible to find a function $I(x, y)$ such that multiplication through by $I(x, y)$ yields an exact equation. Such a function $I(x, y)$ is called an **integrating factor**. Under suitable conditions, it can be proved that an integrating factor exists, but finding the integrating factor is a process of trial and error. Consider an example or two.

■ **Example 18.3** Show that the differential equation

$$(x^2 + 2x + y^2)\,dx + 2y\,dy = 0$$

is not exact, but has integrating factor

$$I(x, y) = \frac{1}{x^2 + y^2}.$$

Solution The equation is not exact, since

$$\frac{\partial}{\partial y}(x^2 + 2x + y^2) = 2y \neq 0 = \frac{\partial}{\partial x}(2y).$$

Multiplying the equation through by $I(x, y) = 1/(x^2 + y^2)$, we get

$$\left(1 + \frac{2x}{x^2 + y^2}\right)dx + \frac{2y}{x^2 + y^2}\,dy = 0.$$

It is now evident that the equation is equivalent to $df(x, y) = 0$ with $f(x, y) = x + \ln(x^2 + y^2)$. The general solution is therefore

$$x + \ln(x^2 + y^2) = C. \qquad \blacksquare$$

Example 18.4 Show that the equation

$$(4xy + 2y^2)\,dx + (2x^2 + 2xy)\,dy = 0$$

is not exact. Then solve it by using the integrating factor $I(x, y) = x^2y$.

Solution Since

$$\frac{\partial}{\partial y}(4xy + 2y^2) = 4x + 4y \neq 4x + 2y = \frac{\partial}{\partial x}(2x^2 + 2xy),$$

the equation is not exact. Multiplying through the equation by x^2y, we get

$$(4x^3y^2 + 2x^2y^3)\,dx + (2x^4y + 2x^3y^2)\,dy = 0.$$

Since

$$\frac{\partial}{\partial y}(4x^3y^2 + 2x^2y^3) = 8x^3y + 6x^2y^2 = \frac{\partial}{\partial x}(2x^4y + 2x^3y^2),$$

this equation is exact. It is equivalent to the equation $df = 0$ with

$$f(x, y) = x^4y^2 + \frac{2}{3}x^3y^3,$$

found as in Section 15.2. The solution of the equation is therefore

$$x^4y^2 + \frac{2}{3}x^3y^3 = C.$$ ■

Further examples are found in the exercises.

Exercises 18.1

Determine which equations are exact, and solve those that are, subject to the boundary condition if given.

1. $(2xy - y^2)\,dx + (x^2 - 2xy)\,dy = 0$
2. $(12x^2 + 2x)\,dx + 3y^2\,dy = 0$
3. $y' = \dfrac{2y^2 - xy}{5xy - 2x^2}$
4. $y' = \dfrac{y^2 - xy}{4xy - 3y^2}$
5. $y' = -\dfrac{2x + ye^{xy}}{xe^{xy}}$; $y = 1$ when $x = 1$
6. $y' = \dfrac{\cos y}{x \sin y}$; $y = 0$ when $x = 1$
7. $\cos y\,dx + x \sin y\,dy = 0$; $y = 1$ when $x = 0$
8. $xe^{xy}\,dx + 2ye^{xy}\,dy = 0$; $y = 0$ when $x = 1$
9. $\left(\dfrac{1}{y} + \dfrac{y}{x^2}\right)dx - \left(\dfrac{x}{y^2} + \dfrac{1}{x}\right)dy = 0$; $y = 1$ when $x = 2$
10. $\left(y + \dfrac{1}{x + 2y}\right)dx + \left(x + \dfrac{2}{x + 2y}\right)dy = 0$; $y = 0$ when $x = 1$

Solve the following differential equations using the integrating factor given:

11. $(2xy + y^2)\,dx + (x^2 + xy)\,dy = 0$; $I(x, y) = x^2y$
12. $(4x^2 + 8xy + 3y^2)\,dx + (2x^2 + 2xy)\,dy = 0$; $I(x, y) = x^2$

13. $(xy - 2y^2)\,dx + (-2x^2 + 5xy)\,dy = 0;\quad I(x, y) = xy^{-7}$

14. $[1 + 2\sec(x + y)]\,dx + dy = 0;\quad I(x, y) = \cos(x + y)$

15. $-x\,dx + (2y\sqrt{x^2 + y^2} - y)\,dy = 0;\quad I(x, y) = \dfrac{1}{\sqrt{x^2 + y^2}}$

16. $(2xy^2 + y)\,dx + (2y^3 - x)\,dy = 0;\quad I(x, y) = \dfrac{1}{y^2}$

17. $\left(\dfrac{x^2 - y^2}{x} + \dfrac{2x}{x^2 - y^2}\right)dx - \dfrac{2y}{x^2 - y^2}\,dy = 0;\quad I(x, y) = \dfrac{1}{x^2 - y^2}$

18. $(5y^2 - 2xy)\,dx + (3x^2 - 8xy)\,dy = 0;\quad I(x, y) = x^4y^{-10}$

19. $\left(7x^2 + 8xy + \dfrac{5}{2}y^2\right)dx + (3x^2 + 3xy)\,dy = 0;\quad I(x, y) = x^{2/3}$

20. $(2xy + 5y^2)\,dx + (3x^2 + 8xy)\,dy = 0;\quad I(x, y) = x^{24}y^{38}$

Determine whether each equation has an integrating factor of the form x^my^n, and if so, give it.

21. $xy\,dx + (2x^2 + xy)\,dy = 0$

22. $(y^2 - xy)\,dx + (x^2 - 2xy)\,dy = 0$

SECTION 18.2

First-Order Linear Differential Equations

The first-order differential equation

$$y' = f(x, y)$$

is **linear** if and only if it can be written in the form

$$y' + P(x)y = Q(x). \tag{1}$$

Under certain conditions [for example, if $P(x) = ax$ and $Q(x) = bx$], equation (1) may turn out to be separable. Usually, we will need an integrating factor in order to solve the equation.

The Integrating Factor

For a first-order linear differential equation, the integrating factor is a function of x, constructed as follows: Beginning with equation (1), we let $\int P(x)\,dx$ represent any antiderivative of $P(x)$. The integrating factor is then

$$I(x) = e^{\int P(x)\,dx}. \tag{2}$$

To show that this integrating factor indeed works, we multiply equation (1) through by $I(x)$:

$$y'e^{\int P(x)\,dx} + P(x)ye^{\int P(x)\,dx} = Q(x)e^{\int P(x)\,dx}. \tag{3}$$

Now the left-hand side of equation (3) is seen to be the derivative of $ye^{\int P(x)\,dx}$; that is,

$$\frac{d}{dx}\left[ye^{\int P(x)\,dx}\right] = y'e^{\int P(x)\,dx} + ye^{\int P(x)\,dx}P(x). \tag{4}$$

Putting (4) into (3) gives us

$$\frac{d}{dx}\left[ye^{\int P(x)\,dx}\right] = Q(x)e^{\int P(x)\,dx}. \tag{5}$$

We may integrate equation (5), getting

$$ye^{\int P(x)\,dx} = \int Q(x)e^{\int P(x)\,dx}\,dx + C, \tag{6}$$

from which we solve for y:

$$y = e^{-\int P(x)\,dx}\int Q(x)e^{\int P(x)\,dx}\,dx + Ce^{-\int P(x)\,dx}. \tag{7}$$

This is the solution of equation (1).

■ **Example 18.5** Solve the equation $y' + 4xy = x^3$.

Solution This is a linear equation with $P(x) = 4x$ and $Q(x) = x^3$. We construct the integrating factor using equation (2):

$$\begin{aligned} I(x) &= e^{\int P(x)\,dx} \\ &= e^{\int 4x\,dx} \\ &= e^{2x^2}. \end{aligned}$$

Multiplying through by this factor, we get the equation

$$y'e^{2x^2} + 4xe^{2x^2}y = x^3e^{2x^2},$$

whose left-hand side is the derivative of ye^{2x^2}:

$$\frac{d}{dx}(ye^{2x^2}) = x^3e^{2x^2}.$$

Integrating this equation, we get

$$ye^{2x^2} = \int x^3e^{2x^2}\,dx + C.$$

Now, integrating by parts with $u = x^2$, $dv = xe^{2x^2}\,dx$, gives us

$$ye^{2x^2} = \left(\frac{1}{4}x^2 - \frac{1}{8}\right)e^{2x^2} + C.$$

Therefore, the general solution is

$$y = \frac{1}{4}x^2 - \frac{1}{8} + Ce^{-2x^2}.$$

■

Example 18.6 Solve the equation

$$xy' + 2y - x \sin x = 0$$

subject to the condition $y = 1$ when $x = \pi$.

Solution Putting the equation in the form (1), we write

$$y' + \left(\frac{2}{x}\right) y = \sin x.$$

Our integrating factor is

$$I(x) = e^{\int (2/x)\,dx} = e^{2 \ln x} = x^2.$$

Multiplying by $I(x)$, the equation becomes

$$x^2 y' + 2xy = x^2 \sin x,$$

which is the same as

$$\frac{d}{dx}(x^2 y) = x^2 \sin x.$$

Integrating, we get

$$\begin{aligned} x^2 y &= \int x^2 \sin x \, dx \\ &= (2 - x^2) \cos x + 2x \sin x + C. \end{aligned}$$

Dividing by x^2, we get

$$y = \frac{2 - x^2}{x^2} \cos x + \frac{2}{x} \sin x + \frac{C}{x^2}.$$

Now setting $x = \pi$, $y = 1$, we find

$$\begin{aligned} 1 &= \frac{2 - \pi^2}{\pi^2} \cos \pi + \frac{2}{\pi} \sin \pi + \frac{C}{\pi^2}, \\ 2 &= C. \end{aligned}$$

Thus, the solution we seek is

$$y = \frac{2 - x^2}{x^2} \cos x + \frac{2}{x} \sin x + \frac{2}{x^2}.$$ ■

Example 18.7 Show that the solution of a first-order linear differential equation does not depend upon the particular choice of antiderivative of $P(x)$ used in constructing the integrating factor.

Solution Using equation (2) for the integrating factor, we get equation (7) for the solution of equation (1). If we use another antiderivative to construct $I(x)$, say

$$I(x) = e^{\int P(x)\,dx + k},$$

then multiplying through equation (1) gives

$$\frac{d}{dx}\left[ye^{\int P(x)\,dx+k}\right] = Q(x)e^{\int P(x)\,dx+k}.$$

Integrating, we get

$$ye^{\int P(x)\,dx+k} = \int Q(x)e^{\int P(x)\,dx+k}\,dx + C,$$

so that

$$\begin{aligned} y &= e^{-\int P(x)\,dx-k}\left[\int Q(x)e^{\int P(x)\,dx+k}\,dx + C\right] \\ &= e^{-\int P(x)\,dx}e^{-k}\left[e^k \int Q(x)e^{\int P(x)\,dx}\,dx + C\right] \\ &= e^{-\int P(x)\,dx}\int Q(x)e^{\int P(x)\,dx}\,dx + C'e^{-\int P(x)\,dx} \end{aligned} \tag{8}$$

with $C' = Ce^{-k}$. Thus, solution (8) is the same as (7). ■

Exercises 18.2

Solve the following first-order linear differential equations, subject to the given boundary condition, if any.

1. $y' + y\sin x = \sin x$
2. $y' - \dfrac{y}{x} = 2x$
3. $y' - \dfrac{y}{x^2} = 2x;\quad y = 2$ when $x = \frac{1}{2}$
4. $y' - \dfrac{2y}{x^2} = 2x;\quad y = 2$ when $x = \frac{1}{2}$
5. $y' + \dfrac{y}{1-x} = x + 1;\quad y = 1$ when $x = 0$
6. $y' + \dfrac{y}{x-2} = e^x;\quad y = 0$ when $x = 1$
7. $x^2y' + 2y = 3x + 2;\quad y = 0$ when $x = 1$
8. $x^3y' - 4y = 3 - x;\quad y = 1$ when $x = 2$
9. $(2xy + 3x - 3)\,dx + (x - x^2)\,dy = 0$
10. $(7x^4 - y)\,dx - x^4\,dy = 0$
11. $xy' + (1 + x)y = 4$
12. $x^2y' + (1 + x)y = 3$
13. $y' + y\tan x = \cos x$
14. $y'\tan x + y = \cos x$
15. Show that the **Bernoulli equation**

$$y' + P(x)y = Q(x)y^n$$

becomes a first-order linear equation when we make the substitution $u = y^{1-n}$.

16. Refer to Problem 15. Solve the Bernoulli equation

$$y' + \frac{y}{x} = 2xy^3,$$

subject to the condition $y = 1$ when $x = 0$.

17. Show that the solution of $y' + P(x)y = Q(x)$, subject to $y = b$ when $x = a$, can be written as

$$y = be^{-R(x)} + \int_a^x e^{-[R(x)-R(t)]}\, Q(t)\, dt,$$

where $R(x) = \int_a^x P(t)\, dt$.

18. Show that if $y = f(x)$ is a solution of $y' + P(x)y = Q(x)$, then $y = f(x) + Ce^{-\int P(x)\, dx}$ is also a solution for each constant C.

SECTION 18.3

Homogeneous First-Order Differential Equations

Certain differential equations that can be written in the form

$$P(x, y)\, dx + Q(x, y)\, dy = 0$$

can be solved by a substitution method, provided the functions P and Q have a certain property called homogeneity.

Homogeneous Functions

A function f of two variables is said to be **homogeneous of degree n** if and only if

$$f(tx, ty) = t^n f(x, y)$$

for all x and y and any nonzero t.

Example 18.8 Determine whether the following functions are homogeneous, and if so, of what degree:

(a) $f(x, y) = x^2 + 3xy - 4y^2$ (b) $f(x, y) = x - 2y \sin \dfrac{x}{y}$

(c) $f(x, y) = \dfrac{x^2}{xy - y^2}$ (d) $f(x, y) = \dfrac{x+1}{y-1}$

Solution (a)
$$\begin{aligned} f(tx, ty) &= (tx)^2 + 3(tx)(ty) - 4(ty)^2 \\ &= t^2(x^2 + 3xy - 4y^2) \\ &= t^2 f(x, y), \end{aligned}$$

so f is homogeneous of degree 2.

(b) $$f(tx, ty) = tx - 2(ty)\sin\frac{tx}{ty} = t\left(x - 2y\sin\frac{x}{y}\right) = tf(x, y),$$

so f is homogeneous of degree 1.

(c) $$f(tx, ty) = \frac{(tx)^2}{(tx)(ty) - (ty)^2} = \frac{t^2x^2}{t^2(xy - y^2)} = f(x, y),$$

so f is homogeneous of degree 0.

(d) $$f(tx, ty) = \frac{tx + 1}{ty - 1},$$

which is not a power of t times f, so f is not homogeneous. ■

Homogeneous Equations

The first-order differential equation

$$P(x, y)\,dx + Q(x, y)\,dy = 0 \tag{1}$$

is said to be **homogeneous** or to **have homogeneous coefficients** if and only if P and Q are homogeneous functions of the same degree. Thus, equation (1) has homogeneous coefficients if and only if

$$\frac{P(tx, ty)}{Q(tx, ty)} = \frac{P(x, y)}{Q(x, y)}$$

for all x and y.

A homogeneous first-order differential equation may be solved by substituting

$$y = xu,$$

u a function of x, so that

$$dy = x\,du + u\,dx.$$

The substitution reduces the homogeneous equation to a separable equation.

The substitution

$$x = yv,$$

v a function of y, also can be used to reduce a homogeneous equation to a separable one.

Consider the following examples.

Example 18.9 Solve the equation

$$(x^2 - 3xy)\,dx + (3x^2 - y^2)\,dy = 0$$

subject to the condition $y = 0$ when $x = 1$.

Solution We note that $P(x, y) = x^2 - 3xy$ and $Q(x, y) = 3x^2 - y^2$ are both homogeneous of degree 2, and make the substitutions $y = xu$, $dy = x\,du + u\,dx$:

$$\begin{aligned}(x^2 - 3xy)\,dx + (3x^2 - y^2)\,dy &= 0,\\ (x^2 - 3x^2u)\,dx + (3x^2 - x^2u^2)(u\,dx + x\,du) &= 0,\\ (x^2 - x^2u^3)\,dx + (3x^3 - x^3u^2)\,du &= 0,\\ x^2(1 - u^3)\,dx + x^3(3 - u^2)\,du &= 0.\end{aligned}$$

Now we separate the variables and integrate:

$$\frac{1}{x}\,dx = \frac{3 - u^2}{u^3 - 1}\,du,$$

$$\frac{1}{x}\,dx = \left[\frac{3}{2}\frac{1}{u - 1} - \frac{3}{4}\frac{2u + 1}{u^2 + u + 1} - \frac{9}{4}\frac{1}{(x + \frac{1}{2})^2 + \frac{3}{4}} - \frac{1}{3}\frac{3u^2}{u^3 - 1}\right]du,$$

$$\ln|x| = \frac{3}{2}\ln|u - 1| - \frac{3}{4}\ln|u^2 + u + 1| - \frac{3\sqrt{3}}{2}\arctan\frac{2u + 1}{\sqrt{3}} - \frac{1}{3}\ln|u^3 - 1| + C.$$

Since $y = xu$, we have $u = y/x$, so

$$\ln|x| = \frac{3}{2}\ln\left|\frac{y - x}{x}\right| - \frac{3}{4}\ln\left|\frac{y^2 + xy + x^2}{x^2}\right| - \frac{3\sqrt{3}}{2}\arctan\frac{2y + x}{x\sqrt{3}} - \frac{1}{3}\ln\left|\frac{y^3 - x^3}{x^3}\right| + C.$$

Letting $x = 1$, $y = 0$, after some manipulation, we get $C = \pi\sqrt{3}/4$, so the final solution is

$$\ln|x| = \frac{3}{2}\ln\left|\frac{y - x}{x}\right| - \frac{3}{4}\ln\left|\frac{y^2 + xy + x^2}{x^2}\right| - \frac{3\sqrt{3}}{2}\arctan\frac{2y + x}{x\sqrt{3}} - \frac{1}{3}\ln\left|\frac{y^3 - x^3}{x^3}\right| + \frac{\pi\sqrt{3}}{4}.$$ ■

Example 18.10 Solve the equation

$$\frac{dy}{dx} = \frac{xy}{y^2 - x^2}.$$

Solution We can rewrite this equation in the form (1) as

$$\frac{x}{x^2 - y^2}\,dx + \frac{1}{y}\,dy = 0,$$

with the coefficients homogeneous of degree -1. This time, we use the substitution $x = yv$, $dx = y\,dv + v\,dy$, getting

$$\frac{yv}{y^2v^2 - y^2}(y\,dv + v\,dy) + \frac{1}{y}\,dy = 0,$$

$$\frac{v}{v^2 - 1}\,dv + \frac{1}{y}\left(\frac{v^2}{v^2 - 1} + 1\right)dy = 0.$$

Separating the variables and integrating,

$$\frac{v}{2v^2 - 1}\,dv = -\frac{1}{y}\,dy,$$

$$\frac{1}{4}\ln|2v^2 - 1| = -\ln|y| + C,$$

$$\ln|2v^2 - 1| = -4\ln|y| + 4C,$$

$$2v^2 - 1 = Ay^{-4} \qquad (A = e^{4C}).$$

Now $v = x/y$, so

$$\frac{2x^2}{y^2} - 1 = \frac{A}{y^4},$$

$$2x^2y^2 - y^4 = A.$$

This is the general solution. ■

Exercises 18.3

Determine whether the given equation is homogeneous, and if so, solve it, subject to the boundary condition, if any.

1. $(y^2 - 3xy)\,dx + x^2\,dy = 0$
2. $x^2\,dx + (x^2 + xy - y^2)\,dy = 0$
3. $(x^2 - 1)\,dx + (xy - y^2)\,dy = 0$
4. $\ln(xy)\,dx - \sin\dfrac{x}{y}\,dy = 0$
5. $y' = \dfrac{x + y}{x - y}$; $y = 1$ when $x = 1$
6. $y' = \left(\dfrac{xy}{x^2 + y^2}\right)^{1/2}$; $y = 0$ when $x = 1$
7. $y' = \dfrac{y}{x} + \tan\dfrac{y}{x}$; $y = \pi/4$ when $x = 1$
8. $2xyy'\cos\dfrac{y}{x} = 2y^2\cos\dfrac{y}{x} - x^2\sin\dfrac{y}{x}$; $y = 1$ when $x = 1/\pi$
9. Rework Example 18.9 using the substitution $x = yv$.
10. Rework Example 18.10 using the substitution $y = xu$.
11. Assuming that P and Q are both homogeneous of degree n, make the substitution $y = xu$ and rewrite equation (1) as a separable equation.

12. Differentiating the identity $P(tx, ty)/Q(tx, ty) = P(x, y)/Q(x, y)$ with respect to t, show that if P and Q are continuously differentiable and homogeneous of the same degree, then

$$\frac{xP_x + yP_y}{xQ_x + yQ_y} = \frac{P}{Q}$$

for all x and y. (This result is called **Euler's lemma.**)

13. Refer to Problem 12. Using Euler's lemma, show that if the first-order differential equation $P(x, y)\,dx + Q(x, y)\,dy = 0$ has homogeneous coefficients, then

$$I(x, y) = \frac{1}{xP(x, y) + yQ(x, y)}$$

is an integrating factor.

SECTION 18.4

Applications of First-Order Equations

Having developed the theory to solve a variety of first-order differential equations, we now turn to some applications. You have already seen many applications of separable differential equations in Sections 4.6 and 6.4. Some of the problems we will consider here may lead to separable equations, but most require the techniques developed thus far in this chapter for their solution.

Orthogonal Trajectories

Our first application is primarily geometric. Suppose we have a family of curves in the plane, given by equations of the form

$$f(x, y) = C.$$

That is, to each value of C there corresponds one curve of the family. Another curve that meets every curve of the family at right angles is called an **orthogonal trajectory** of the family. Using differential equations, it is possible to find the set of all orthogonal trajectories of a family of curves.

The general approach is as follows: Differentiating $f(x, y) = C$ implicitly, we find the slope of any curve in the family to be

$$y' = -\frac{f_x(x, y)}{f_y(x, y)}.$$

Thus, the slope of any orthogonal trajectory will be the negative reciprocal of this expression, namely

$$y' = \frac{f_y(x, y)}{f_x(x, y)}.$$

The set of solutions of this last differential equation is the family of orthogonal trajectories we seek.

■ **Example 18.11** Find the orthogonal trajectories of the curves

$$x^2 + 3xy - y^2 = C.$$

Solution Differentiating implicitly, we have

$$2x + 3y + 3xy' - 2yy' = 0,$$

so that

$$y' = -\frac{2x + 3y}{3x - 2y}.$$

Therefore, the orthogonal trajectories satisfy the differential equation

$$y' = \frac{3x - 2y}{2x + 3y},$$

which in differential notation is

$$(2y - 3x)\,dx + (2x + 3y)\,dy = 0.$$

Since

$$\frac{\partial}{\partial y}(2y - 3x) = 2 = \frac{\partial}{\partial x}(2x + 3y),$$

this equation is exact. If f is a function such that $df = (2y - 3x)\,dx + (2x + 3y)\,dy$, then

$$f_x = 2y - 3x;$$

therefore,

$$f(x, y) = 2xy - \frac{3}{2}x^2 + k(y).$$

Hence,

$$f_y = 2x + k'(y) = 2x + 3y,$$
$$k'(y) = 3y.$$

Therefore,

$$k(y) = \frac{3}{2}y^2 + C,$$

and

$$f(x, y) = 2xy - \frac{3}{2}x^2 + \frac{3}{2}y^2 + C.$$

Thus, the solution of our differential equation is

$$2xy - \frac{3}{2}x^2 + \frac{3}{2}y^2 + C = 0,$$

and each of the curves

$$4xy - 3x^2 + 3y^2 = K$$

$(K = -2C)$ is an orthogonal trajectory of the original family.

The curves of the original family are hyperbolas, as shown in color in Figure 18.1. The orthogonal trajectories are also hyperbolas, as shown in black in the figure.

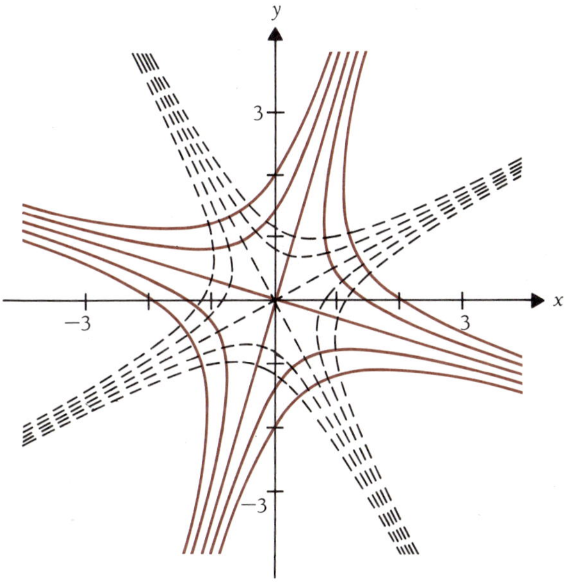

FIGURE 18.1 Curves $x^2 + 3xy - y^2 = C$ (color) and their orthogonal trajectories $4xy - 3x^2 + 3y^2 = K$ (black) ■

Mixing Problems

First-order differential equations arise in certain problems involving mixtures. Consider the next example.

■ **Example 18.12** A vat of brine consists of 10 pounds of salt dissolved in 25 gallons of water. A brine solution containing 0.2 pound of salt per gallon flows into the vat at the rate of 4 gallons per minute. The mixture, kept uniform by vigorous stirring, flows out of the vat at the rate of 3 gallons per minute. How much salt is in the vat at any time t? What is the concentration in the vat at time t?

Solution Let y be the amount of salt in the vat at time t. When $t = 0, y = 10$. At any time t, the volume of brine in the vat will be $25 + (4 - 3)t = 25 + t$ gallons, so the concentration will be $y/(25 + t)$. The rate at which salt is added to the vat is

$$(0.2 \text{ pound/gallon})(4 \text{ gallons/minute}) = 0.8 \text{ pound/minute},$$

and the rate at which salt leaves the vat is

$$\left(\frac{y}{25+t} \text{ pounds/gallon}\right)(3 \text{ gallons/minute}) = \frac{3y}{25+t} \text{ pounds/minute}.$$

Thus, the rate of change of y is

$$\begin{aligned}\frac{dy}{dt} &= \text{Rate in} - \text{Rate out}\\ &= 0.8 - \frac{3y}{25+t}.\end{aligned}$$

Written in the form

$$y' + \frac{3}{25+t}y = 0.8,$$

this equation is a linear first-order differential equation. To solve it, we find an integrating factor:

$$\begin{aligned}I(t) &= e^{\int [3/(25+t)]\,dt}\\ &= e^{3\ln(25+t)}\\ &= (25+t)^3.\end{aligned}$$

Multiplying our equation by this integrating factor, we get

$$y'(25+t)^3 + 3(25+t)^2y = 0.8(25+t)^3,$$

or

$$\frac{d}{dt}[y(25+t)^3] = 0.8(25+t)^3.$$

Integration with respect to t gives us

$$y(25+t)^3 = 0.2(25+t)^4 + C,$$

so that

$$y = 0.2(25+t) + C(25+t)^{-3}.$$

Using $y = 10$ when $t = 0$ gives us

$$\begin{aligned}10 &= 0.2(25) + C(25)^{-3},\\ C &= 5^7.\end{aligned}$$

Therefore,

$$y = 0.2(25+t) + 5^7(25+t)^{-3}$$

is the number of pounds of salt in the vat at time t. The concentration at time t is

$$\frac{y}{25+t} = 0.2 + 5^7(25+t)^{-4}.$$

Note that as t gets large, the concentration approaches 0.2, the concentration of the incoming brine. However, the vat gets pretty full, too! ∎

The method used in Example 18.12 also applies to certain motion problems, where resistance is met. Consider this example.

Example 18.13 A certain boat has an outboard motor that gives a maximum thrust of $100 - \frac{1}{2}v$ pounds at a speed of v miles per hour. The resistive force of the water is proportional to the boat's speed and is 20 pounds when the speed is 10 miles per hour. If the boat weighs 480 pounds, what is its maximum speed?

Solution Imagine the boat starting from rest at top acceleration. Then its maximum speed will be the limit of the speed as time t gets large. We will set up a differential equation relating acceleration to speed in order to find the speed as a function of time.

If the weight of the boat is 480 pounds, then its mass is $m = w/g = \frac{480}{32} = 15$ slugs. The resistive force on the boat at speed v is αv for some constant α; if the resistive force is 20 pounds at $v = 10$ miles per hour, then $\alpha = 2$. Thus, the total force on the boat at speed v is

$$F = 100 - \frac{1}{2}v - 2v = 100 - \frac{5}{2}v.$$

Now, from Newton's law $F = ma$, we find the acceleration of the boat to be

$$a = \frac{F}{m} = \frac{100 - \frac{5}{2}v}{15} = \frac{20}{3} - \frac{1}{6}v,$$

or

$$\frac{dv}{dt} + \frac{1}{6}v = \frac{20}{3}.$$

We can treat this equation either as separable or as first-order linear. Choosing the latter, we multiply by the integrating factor $e^{t/6}$:

$$\frac{dv}{dt}e^{t/6} + \frac{1}{6}e^{t/6}v = \frac{20}{3}e^{t/6},$$

$$\frac{d}{dt}(ve^{t/6}) = \frac{20}{3}e^{t/6},$$

$$ve^{t/6} = \frac{20}{3}(6e^{t/6}) + C,$$

$$v = 40 + Ce^{-t/6}.$$

Now as $t \to \infty$, $v \to 40$, so 40 miles per hour must be the maximum speed. ∎

Some miscellaneous applications appear in the exercises.

Exercises 18.4

Find the orthogonal trajectories of the following families of curves:

1. $y - x^2 = C$
2. $x^2 + 3y^2 = C$
3. $x^2 - 4y^2 = C$
4. $x^2 - xy - y^2 = C$
5. $2x^2 - 3xy - 2y^2 = C$
6. $x^2 - xy = C$
7. A vat contains 100 gallons of water in which 50 pounds of salt is dissolved. A solution containing 2 pounds of salt per gallon of water flows into the vat at the rate of 1 gallon per minute; the mixture, kept uniform by constant stirring, flows out at the same rate. Find the amount of salt in the vat at time t, and find the concentration. What is the "steady-state" concentration of the mixture?
8. Rework Problem 7 if the solution flows in at the rate of 3 gallons per minute and out at the same rate.
9. Rework Problem 7 if the solution flows in at 1 gallon per minute and out at the rate of 2 gallons per minute.
10. Rework Problem 7 if the incoming solution contains $\frac{1}{4}$ pound per gallon of salt, and the mixture flows out at the rate of 2 gallons per minute.
11. Glucose is injected into the bloodstream of a patient at the rate of k grams per minute. At the same time, metabolism removes glucose from the bloodstream at a rate proportional to the amount of glucose present. Find an expression for the amount of glucose in the bloodstream at any time t. What happens over a long period of time?
12. A chemical solution contains elemental silver. A titrate is steadily added to the solution at a rate of 0.5 gram per minute, and combines with the silver ions to form an insoluble precipitate. In this way, the titrate is removed from the solution at a rate proportional to the amount of titrate in the solution. Find an expression for the amount of titrate in the solution at time t.
13. Rework Example 18.13 if the maximum thrust is $P - kv$, the resistive force is $-\alpha v$, and the weight of the boat is W.
14. Rework Problem 13 if the resistive force is proportional to the product of the weight and the speed.
15. A freely falling object meets air resistance proportional to its speed. Find its speed as a function of time t. What is its terminal speed, $\lim_{t\to\infty} v$?
16. Rework Problem 15 if the air resistance is proportional to the square of the speed.
17. An electric circuit has resistance R, inductance L, and voltage E. When the switch is thrown closing the circuit, current i builds up at a rate satisfying the equation

$$L\frac{di}{dt} + Ri = E.$$

Find i as a function of time t. What is the maximum value of i? How long does it take for i to reach 90% of its maximum value?

18. When a tank of water drains through a hole under the force of gravity alone, **Torricelli's law** states that the velocity of the water issuing from the hole is $v = \sqrt{2gh}$, where g is the acceleration of gravity and h is the height of the water level above the hole. A hemispherical tank of radius R, initially full of water, drains through a hole in the bottom of radius r.

Find the depth of the water in the tank as a function of time. How long does it take the tank to drain?

19. Rework Problem 18 for an inverted conical tank of base radius R and height H.

20. A cylindrical tank standing on one end has radius R feet and is filled to depth H feet with water. A hole of radius r inches is opened in the bottom of the tank, and at the same time a hose starts pouring water into the tank at the constant rate of c cubic feet per minute. Use Torricelli's law (Problem 18) to express the depth of the water in the tank as a function of time. What happens in the long run?

SECTION 18.5

Second-Order Differential Equations

We are now ready to consider some second-order differential equations, the most general of which has the form

$$y'' = f(x, y, y'). \tag{1}$$

As we remarked in the introduction to this chapter, if the function f in equation (1) is continuously differentiable, then there is a unique solution for equation (1) satisfying the boundary conditions $y = y_0$ and $y' = z_0$ when $x = x_0$.

In this section, we will be concerned only with some special cases of equation (1), since the general case is beyond our scope. We will first consider two special forms, and then we will consider the theory of linear second-order differential equations.

Two Special Cases

The first special case is the form of equation (1) in which the dependent variable y is missing—that is,

$$y'' = f(x, y'). \tag{2}$$

When this happens, the substitution $u = y'$ yields the equation

$$u' = f(x, u), \tag{3}$$

which is a first-order differential equation. Then, if we are fortunate, we may solve equation (3) by methods we learned earlier, replace u by y', and finally solve another first-order equation.

Example 18.14 Solve the equation

$$y'' = \frac{(y')^2 + 1}{2xy'},$$

subject to the boundary conditions $y = 1$ and $y' = 2$ when $x = 1$.

Solution We note that y does not occur in the given equation, so we have an equation of the form (2). We let $u = y'$, and get

$$u' = \frac{u^2 + 1}{2xu}.$$

This is a separable equation and can be written as

$$\frac{2u}{u^2 + 1}\, du = \frac{1}{x}\, dx.$$

We find the solution to this equation as follows:

$$\begin{aligned} \ln(u^2 + 1) &= \ln x + C, \\ u^2 + 1 &= Ax, \qquad (A = e^C) \\ u^2 &= Ax - 1, \\ u &= \sqrt{Ax - 1}. \end{aligned}$$

Now, $u = y' = 2$ when $x = 1$, so

$$\begin{aligned} 2 &= \sqrt{A - 1}, \\ A &= 5. \end{aligned}$$

Thus,

$$u = \sqrt{5x - 1}.$$

That is,

$$y' = \sqrt{5x - 1},$$

so that

$$y = \frac{2}{3}\left(\frac{1}{5}\right)(5x - 1)^{3/2} + C.$$

Since $y = 1$ when $x = 1$, we get

$$\begin{aligned} 1 &= \frac{2}{15}(4)^{3/2} + C, \\ C &= -\frac{1}{15}. \end{aligned}$$

Thus,

$$y = \frac{2}{15}(5x - 1)^{3/2} - \frac{1}{15}.$$

This is the solution. ∎

The second special case is the form of equation (1) in which the independent variable x is missing—that is,

$$y'' = f(y, y'). \tag{4}$$

This time the substitution $u = y'$ and the identity

$$y'' = \frac{du}{dx} = \frac{du}{dy}\frac{dy}{dx} = u\frac{du}{dy} \tag{5}$$

turn equation (4) into

$$u\frac{du}{dy} = f(y, u), \tag{6}$$

which is a first-order equation for u as a function of y. If we can solve equation (6), then replacing u by y' and solving another first-order equation gives us a solution.

■ **Example 18.15** Solve the equation

$$y'' = -e^{-y}y',$$

subject to the boundary conditions $y = 0$ and $y' = 1$ when $x = 0$.

Solution This equation has the form (4), so we set $u = y'$ and $u\ (du/dy) = y''$:

$$\begin{aligned} u\frac{du}{dy} &= -e^{-y}u, \\ \frac{du}{dy} &= -e^{-y}, \\ u &= e^{-y} + C. \end{aligned}$$

Now, $u = y' = 1$ and $y = 0$ when $x = 0$, so $C = 0$. Thus,

$$\begin{aligned} u &= e^{-y}, \\ y' &= e^{-y}, \\ e^{y}\,dy &= dx, \\ e^{y} &= x + C, \end{aligned}$$

Setting $y = 0$ when $x = 0$, we find $C = 1$, so

$$e^{y} = x + 1,$$

or

$$y = \ln(x + 1)$$

is the final solution. ■

Second-Order Linear Equations

A second-order differential equation is **linear** if and only if it can be written in the form

$$y'' + p(x)y' + q(x)y = r(x). \tag{7}$$

Equation (7) is called **homogeneous** if and only if $r(x) = 0$. (Unfortunately, this use of the word "homogeneous" has nothing to do with the use of that word in the case of first-order equations.) For the remainder of this section, we will discuss the theory of the homogeneous linear second-order equation

$$y'' + p(x)y' + q(x)y = 0. \tag{8}$$

Our first result is easily seen to be true, so it will be stated without proof. (The proof is left as an exercise; see Problem 13.)

THEOREM 18.1

If $y = f(x)$ and $y = g(x)$ are two solutions of equation (8), then any linear combination

$$y = af(x) + bg(x)$$

is also a solution of equation (8).

Example 18.16

Verify that $y = \sin 2x$, $y = \cos 2x$, and $y = a \sin 2x + b \cos 2x$ are all solutions of the differential equation

$$y'' + 4y = 0.$$

Solution

For $y = \sin 2x$, we have $y' = 2 \cos 2x$ and $y'' = -4 \sin 2x$, so $y'' + 4y = 0$. If $y = \cos 2x$, then $y'' = -4 \cos 2x$, and again $y'' + 4y = 0$. For the last case, $y = a \sin 2x + b \cos 2x$, we have

$$y' = 2a \cos 2x - 2b \sin 2x,$$
$$y'' = -4a \sin 2x - 4b \cos 2x = -4y,$$

and again $y'' + 4y = 0$. ■

Two functions f and g are said to be **linearly dependent** if and only if there exist constants a and b, not both 0, such that the linear combination $af + bg$ is 0 for all x. If f and g are not linearly dependent, they are said to be **linearly independent**. Thus, functions f and g are linearly independent if and only if

$$af(x) + bg(x) = 0 \quad \text{for all } x \qquad \text{implies} \qquad a = b = 0.$$

A useful tool in determining whether two differentiable functions f and g are linearly independent is their **Wronskian**, defined by

$$W(x) = \begin{vmatrix} f(x) & g(x) \\ f'(x) & g'(x) \end{vmatrix}.$$

THEOREM 18.2 Two functions that are differentiable in interval I are linearly independent if their Wronskian is nonzero for some $x \in I$.

Proof Suppose the contrary: that the two functions f and g, differentiable in I, are linearly dependent. Then there exist constants a and b, not both 0, such that

$$af(x) + bg(x) = 0. \tag{9}$$

Differentiating, we also have

$$af'(x) + bg'(x) = 0. \tag{10}$$

Now equations (9) and (10) constitute a system of linear equations in the "unknowns" a and b. But a and b actually exist and are not both 0, so this system has a nontrivial solution, and the coefficient determinant

$$W(x) = \begin{vmatrix} f(x) & g(x) \\ f'(x) & g'(x) \end{vmatrix}$$

must be 0 for all $x \in I$. (Otherwise, we would be forced—by Cramer's rule, for example—to conclude that $a = b = 0$.) Thus, the assumption that f and g are linearly dependent forces the Wronskian to be 0 for all x. Therefore, if the Wronskian is nonzero for some $x \in I$, then f and g must be linearly independent.

■ **Example 18.17** Show that $y = \sin 2x$ and $y = \cos 2x$ are linearly independent.

Solution The Wronskian is

$$\begin{aligned} W(x) &= \begin{vmatrix} \sin 2x & \cos 2x \\ 2\cos 2x & -2\sin 2x \end{vmatrix} \\ &= -2\sin^2 2x - 2\cos^2 2x \\ &= -2 \neq 0, \end{aligned}$$

so they are independent by Theorem 18.2. Notice that in this case, the Wronskian is never 0. ■

Note that the converse of Theorem 18.2 is false. That is, two functions may be independent, even though the Wronskian is always 0.

For two linearly independent solutions of a linear second-order differential equation, we actually can say much more about the Wronskian.

THEOREM 18.3 If $y = f(x)$ and $y = g(x)$ are two linearly independent functions and each is a solution of equation (8), then their Wronskian is nowhere 0.

Proof Suppose the contrary: that for some x_0, the Wronskian of f and g is 0, or

$$W(x_0) = \begin{vmatrix} f(x_0) & g(x_0) \\ f'(x_0) & g'(x_0) \end{vmatrix} = 0.$$

Then the system of equations

$$\begin{cases} af(x_0) + bg(x_0) = 0 \\ af'(x_0) + bg'(x_0) = 0 \end{cases} \tag{11}$$

has a nontrivial solution; that is, there exist a and b, not both 0, satisfying equations (11). Let

$$h(x) = af(x) + bg(x).$$

Since f and g are both solutions of equation (8), so is h (by Theorem 18.1). By equations (11), we see that $h(x_0) = 0$ and $h'(x_0) = 0$. Now recall that only one solution of equation (8) satisfies a given set of boundary conditions; the solution $y = 0$ satisfies $y = 0$ and $y' = 0$ when $x = x_0$, so it must be that $h(x) = 0$. Thus, $af(x) + bg(x) = 0$, implying that f and g are linearly dependent. This contradiction forces us to conclude that the Wronskian is not 0 anywhere.

You may wonder where all of this is leading. We won't keep you in suspense: we can now show how to find all the solutions of equation (8).

THEOREM 18.4 If $y = f(x)$ and $y = g(x)$ are two linearly independent solutions of $y'' + p(x)y' + q(x)y = 0$, then every solution is a linear combination of f and g.

Proof Let $y = h(x)$ be any solution of equation (8). For any convenient point x_0, let $c = h(x_0)$ and $d = h'(x_0)$. Let a and b be the solution of the system

$$\begin{cases} af(x_0) + bg(x_0) = c \\ af'(x_0) + bg'(x_0) = d. \end{cases} \tag{12}$$

From Cramer's rule, we know that a and b exist, since the determinant of coefficients in system (12), which is also the Wronskian of f and g at x_0, is nonzero by Theorem 18.3. Now the function

$$H(x) = af(x) + bg(x)$$

is a solution of equation (8) satisfying

$$H(x_0) = af(x_0) + bg(x_0) = c = h(x_0),$$

and

$$H'(x_0) = af'(x_0) + bg'(x_0) = d = h'(x_0).$$

By the uniqueness of a solution satisfying given boundary conditions, $H(x) = h(x)$, and

$$h(x) = af(x) + bg(x).$$

This proves the theorem.

Theorem 18.4 tells us that once we have found two linearly independent solutions $y = f(x)$ and $y = g(x)$ of equation (8), then we have found every solution, because every solution has the form

$$y = af(x) + bg(x) \tag{13}$$

for some constants a and b. We call equation (13) the **general solution** of equation (8).

The only catch is that, in general, there is no way to go about finding two linearly independent solutions of equation (8). Only if the functions p and q are constants is there a procedure that always works; we will consider that case in the next section. Meanwhile, we must be content to know that linearly independent solutions of equation (8) exist, as our concluding theorem states.

THEOREM 18.5 The equation $y'' + p(x)y' + q(x)y = 0$ has two linearly independent solutions.

Proof Let x_0 be a convenient point. Then equation (8) has a solution $y = f(x)$ satisfying $f(x_0) = 1$ and $f'(x_0) = 0$. We know that equation (8) also has a solution $y = g(x)$ satisfying $g(x_0) = 0$ and $g'(x_0) = 1$. Now we claim that f and g are linearly independent, noting that the Wronskian at x_0 is not 0:

$$W(x_0) = \begin{vmatrix} f(x_0) & g(x_0) \\ f'(x_0) & g'(x_0) \end{vmatrix} = \begin{vmatrix} 1 & 0 \\ 0 & 1 \end{vmatrix} = 1.$$

Thus, by Theorem 18.2, our result is established.

■ **Example 18.18** Give the general solution of the equation $y'' + 4y = 0$.

Solution By Example 18.16, $y = \sin 2x$ and $y = \cos 2x$ are solutions, and they are linearly independent by Example 18.17. The general solution is therefore

$$y = a \sin 2x + b \cos 2x.$$

■

Exercises 18.5 *Solve the following equations, subject to the boundary conditions, if any:*

1. $y'' + y' = 0$
2. $y'' + 4y' = 0$
3. $y'' - 3y' = 0$; $y = 3$ and $y' = 1$ when $x = 0$
4. $y'' = 4 + (y')^2$; $y = 2$ and $y' = 0$ when $x = 1$

5. $y'' - y' = \cos x$
6. $xy'' + y' = x \cos x$
7. $y'' = 9y$; $y = 1$ and $y' = 1$ when $x = 0$
8. $yy'' = (y')^2$; $y = 1$ and $y' = 1$ when $x = 1$
9. $x^2y'' = y' + 1$
10. $x^3y'' = 4(y')^3$
11. Show that $y = e^x$ and $y = e^{-x}$ are solutions of $y'' - y = 0$. What is the general solution?
12. Show that $y = \cosh x$ and $y = \sinh x$ are solutions of $y'' - y = 0$.
13. Prove Theorem 18.1 using the fact that differentiation is a linear operator.
14. Show that the derivative of the Wronskian of f and g is

$$W'(x) = \begin{vmatrix} f(x) & g(x) \\ f''(x) & g''(x) \end{vmatrix}.$$

15. If f and g are solutions of equation (8), show that their Wronskian satisfies $W'(x) = -p(x)W(x)$.
16. With W as in Problem 15, let

$$P(x) = \int_{x_0}^{x} p(t)\, dt.$$

Show that

$$W(x) = W(x_0)e^{-P(x)}.$$

(This is known as **Abel's formula**, named after the Norwegian mathematician Niels Henrik Abel, 1802–1829, who discovered it.)

SECTION 18.6

Homogeneous Equations with Constant Coefficients

In Section 18.5, we introduced the homogeneous linear second-order equation

$$y'' + p(x)y' + q(x)y = 0.$$

In this section, we consider the case that the coefficients $p(x)$ and $q(x)$ are constant—that is, we consider the equation of the form

$$y'' + py' + qy = 0. \tag{1}$$

With the theory developed in Section 18.5, we can solve equations of this form.

The Auxiliary Equation

Our first aim is to find two linearly independent solutions to equation (1). Then the general solution will be the set of linear combinations of these two solutions. We begin by making a guess, and then we show how to correct our guess so that it works.

Let's guess that the function

$$y = e^{mx}$$

is going to be a solution to equation (1). Then it must satisfy the equation,

$$m^2e^{mx} + pme^{mx} + qe^{mx} = 0,$$
$$(m^2 + pm + q)e^{mx} = 0.$$

Since e^{mx} is never 0, we must have

$$m^2 + pm + q = 0. \tag{2}$$

That is, if $y = e^{mx}$ is going to be a solution of equation (1), then m must satisfy equation (2). This new equation is called the **auxiliary equation**, or **characteristic equation**, of the differential equation (1). The roots of the auxiliary equation are called **auxiliary roots**, or **characteristic roots**, of equation (1).

Thus, we know that if $y = e^{mx}$ is a solution of equation (1), then m is an auxiliary root of equation (1). The big question, however, is the other way around: If m is an auxiliary root of equation (1), then is $y = e^{mx}$ a solution? The answer is a qualified "yes," and the following theorems give the specific answers.

THEOREM 18.6 If the linear second-order equation $y'' + py' + qy = 0$ has two distinct real auxiliary roots m_1 and m_2, then the general solution of the equation is

$$y = ae^{m_1x} + be^{m_2x},$$

with a and b arbitrary constants.

Proof If m_1 is an auxiliary root of equation (1), then because equation (2) is satisfied, we see that $y = e^{m_1x}$ is a solution of equation (1). The same is true of $y = e^{m_2x}$.

The Wronskian of these two functions is

$$W(x) = \begin{vmatrix} e^{m_1x} & e^{m_2x} \\ m_1e^{m_1x} & m_2e^{m_2x} \end{vmatrix} = (m_2 - m_1)e^{m_1x}e^{m_2x},$$

which is not 0 if $m_1 \neq m_2$. Therefore, e^{m_1x} and e^{m_2x} are linearly independent, and the theorem is proved.

Example 18.19 Solve the equation $y'' + 3y' + 2y = 0$.

Solution The auxiliary equation is

$$m^2 + 3m + 2 = 0,$$

whose solutions are $m = -1$ and $m = -2$. Therefore, by Theorem 18.6, the general solution must be

$$y = ae^{-x} + be^{-2x}.$$

■

THEOREM 18.7 If the linear second-order equation $y'' + py' + qy = 0$ has a single auxiliary root m, then the general solution of the equation is

$$y = ae^{mx} + bxe^{mx}.$$

Proof If m is an auxiliary root, then $y = e^{mx}$ is a solution of equation (1). If m is the only auxiliary root, it must be that the discriminant of the auxiliary equation $m^2 + pm + q = 0$ is 0, or $p^2 - 4q = 0$. Hence, from the quadratic formula, we have

$$m = -\frac{p}{2}.$$

We can now verify by direct substitution that $y = xe^{mx}$ is also a solution of equation (1):

$$\begin{aligned} y'' + py' + qy &= (2me^{mx} + m^2xe^{mx}) + p(e^{mx} + mxe^{mx}) + qxe^{mx} \\ &= (m^2 + pm + q)xe^{mx} + (2m + p)e^{mx} \\ &= 0. \end{aligned}$$

Finally, the Wronskian of e^{mx} and xe^{mx} is

$$W(x) = \begin{vmatrix} e^{mx} & xe^{mx} \\ me^{mx} & e^{mx} + mxe^{mx} \end{vmatrix} = e^{2mx},$$

which is not 0, so e^{mx} and xe^{mx} are linearly independent, and the theorem is proved.

■ Example 18.20 Solve the equation

$$y'' + 4y' + 4y = 0,$$

subject to the conditions $y = 0$ and $y' = 1$ when $x = 0$.

Solution The auxiliary equation is

$$m^2 + 4m + 4 = 0,$$

which has only the root $m = -2$. Hence, by Theorem 18.7, the general solution of the equation is

$$y = ae^{-2x} + bxe^{-2x}.$$

Since $y = 0$ when $x = 0$, it follows that $a = 0$. Thus, our solution has the form

$$y = bxe^{-2x}.$$

Then,

$$y' = be^{-2x} - 2bxe^{-2x}.$$

Setting $y' = 1$ when $x = 0$, we get $b = 1$. Therefore, the particular solution we seek is

$$y = xe^{-2x}. \qquad \blacksquare$$

A Little Complex Analysis

Theorems 18.6 and 18.7 have taken care of the cases in which the auxiliary roots are real. If the auxiliary roots are complex, the exponential function will look somewhat strange. To make some sense of it, we introduce the formula

$$e^{ix} = \cos x + i \sin x, \tag{3}$$

due to the great Swiss mathematician Leonhard Euler.

Leonhard Euler (1707–1783) was the son of the pastor in the Swiss village of Riechen. His father sent him to the University of Basel to study theology, intending to make a preacher of him. At the university, Euler became acquainted with Daniel Bernoulli and thus with mathematics. Daniel's father Johann generously tutored Euler for 1 hour a week, and Euler worked hard so as not to annoy the irascible Johann with trivial questions. Efforts by Euler's father to win him back to theology were countered by the Bernoullis, and Euler became a mathematician, but remained true to his Calvinist faith all his life.

In 1727, Euler accepted a position with the Petrograd Academy in Russia. He wrote textbooks, reformed the system of weights and measures, and carried on other activities while maintaining vigorous research in mathematics. In 1736, he published a book on mechanics, which was the first of several superb books on advanced mathematics. In 1740, he accepted an offer from Frederick the Great and moved to the Berlin Academy. The pious Euler had a difficult time with the cynical Prussian king, and finally, in 1766, Euler returned to Russia, where Catherine the Great gave him a royal welcome and also one of her cooks. Shortly thereafter, Euler was afflicted with cataracts and lost his sight. But then, with fewer distractions, his output in mathematics actually increased, and he remained mathematically active until virtually the last second of his life.

Among Euler's many contributions to mathematics, his systematic organization of analysis ranks very high. This was done in three great works, *Infinite Processes* (1748), *Differential Calculus* (1755), and *Integral Calculus* (1766), in which the function concept was foremost for the first time. These works had a profound effect on all the mathematics that followed.

Euler was the most prolific mathematical researcher of all times. His bibliography contains 886 entries, which, averaged over his professional lifetime, works out to about 800 pages per year for 50 years. He popularized our current use of the symbols e, π, and i, and was first to state the famous relationship $e^{i\pi} + 1 = 0$.

To show a rationale for Euler's formula, we use the Maclaurin series for e^x, $\cos x$, and $\sin x$, namely (see Section 9.6)

$$e^x = 1 + x + \frac{x^2}{2!} + \frac{x^3}{3!} + \cdots,$$

$$\cos x = 1 - \frac{x^2}{2!} + \frac{x^4}{4!} - \frac{x^6}{6!} + \cdots,$$

$$\sin x = x - \frac{x^3}{3!} + \frac{x^5}{5!} - \frac{x^7}{7!} + \cdots.$$

Then, assuming that the Maclaurin series for e^x holds for complex numbers, we have

$$\begin{aligned} e^{ix} &= 1 + ix + \frac{(ix)^2}{2!} + \frac{(ix)^3}{3!} + \frac{(ix)^4}{4!} + \frac{(ix)^5}{5!} + \frac{(ix)^6}{6!} + \cdots \\ &= 1 + ix - \frac{x^2}{2!} - \frac{ix^3}{3!} + \frac{x^4}{4!} + \frac{ix^5}{5!} - \frac{x^6}{6!} - \cdots \\ &= \left(1 - \frac{x^2}{2!} + \frac{x^4}{4!} - \frac{x^6}{6!} + \cdots\right) + i\left(x - \frac{x^3}{3!} + \frac{x^5}{5!} - \cdots\right) \\ &= \cos x + i \sin x. \end{aligned}$$

Now let's return to our auxiliary equation. If it has complex roots, they must be conjugates of each other, of the form

$$m_1 = r + si, \qquad m_2 = r - si,$$

because the auxiliary equation has real coefficients. Then,

$$\begin{aligned} e^{m_1 x} &= e^{(r+si)x} \\ &= e^{rx}e^{isx} \\ &= e^{rx}(\cos sx + i \sin sx), \\ e^{m_2 x} &= e^{(r-si)x} \\ &= e^{rx}e^{-isx} \\ &= e^{rx}e^{i(-sx)} \\ &= e^{rx}[\cos(-sx) + i \sin(-sx)] \\ &= e^{rx}(\cos sx - i \sin sx). \end{aligned}$$

A linear combination of these two functions has the form

$$\begin{aligned} y &= ae^{m_1 x} + be^{m_2 x} \\ &= ae^{rx}(\cos sx + i \sin sx) + be^{rx}(\cos sx - i \sin sx) \\ &= (a + b)e^{rx} \cos sx + (a - b)ie^{rx} \sin sx \\ &= Ae^{rx} \cos sx + Be^{rx} \sin sx, \end{aligned}$$

with $A = a + b$ and $B = (a - b)i$. Note that A and B will be real if a and b are complex conjugates of each other.

All these considerations lead up to the next theorem.

THEOREM 18.8 If the linear second-order equation $y'' + py' + qy = 0$ has complex auxiliary roots $r + si$ and $r - si$, then the general solution of the equation is

$$y = Ae^{rx} \cos sx + Be^{rx} \sin sx,$$

with A and B arbitrary (real) constants.

Proof If $r + si$ is an auxiliary root, then

$$\begin{aligned}(r + si)^2 + p(r + si) + q &= 0,\\ (r^2 - s^2 + pr + q) + i(2rs + ps) &= 0.\end{aligned}$$

It follows that

$$r^2 - s^2 + pr + q = 0 \qquad \text{and} \qquad 2rs + ps = 0.$$

Now letting $y = e^{rx} \cos sx$, we have

$$\begin{aligned}y' &= re^{rx} \cos sx - se^{rx} \sin sx,\\ y'' &= (r^2 - s^2)e^{rx} \cos sx - 2rse^{rx} \sin sx.\end{aligned}$$

Hence,

$$\begin{aligned}y'' + py' + qy = (r^2 - s^2)e^{rx} \cos sx - 2rse^{rx} \sin sx +&\\ p(re^{rx} \cos sx - se^{rx} \sin sx) + q(e^{rx} \cos sx)&\\ = [(r^2 - s^2 + pr + q) \cos sx - (2rs + ps) \sin sx]e^{rx}&\\ = 0.&\end{aligned}$$

Thus, $y = e^{rx} \cos sx$ is a solution of equation (1). Similarly, $y = e^{rx} \sin sx$ is found to be a solution of equation (1). The Wronskian of these two functions is

$$\begin{aligned}W(x) &= \begin{vmatrix} e^{rx} \cos sx & e^{rx} \sin sx \\ (r \cos sx - s \sin sx)e^{rx} & (r \sin sx + s \cos sx)e^{rx} \end{vmatrix}\\ &= se^{2rx} \neq 0,\end{aligned}$$

so the theorem follows.

■ **Example 18.21** Solve the equation $y'' - 2y' + 5y = 0$.

Solution The auxiliary equation is

$$m^2 - 2m + 5 = 0,$$

which has roots

$$m = \frac{2 \pm \sqrt{4 - 20}}{2} = 1 \pm 2i.$$

Then, by Theorem 18.8, the general solution is

$$y = Ae^x \cos 2x + Be^x \sin 2x.$$

■

■ Example 18.22 Solve the equation

$$y'' + \omega^2 y = 0,$$

subject to the boundary conditions $y = 1$ and $y' = 3$ when $x = 0$.

Solution The auxiliary equation is

$$m^2 + \omega^2 = 0,$$

whose roots are $m = 0 \pm \omega i$. Thus, by Theorem 18.8, the general solution is

$$y = A \cos \omega x + B \sin \omega x.$$

Setting $y = 1$ when $x = 0$, we find $A = 1$. Then y has the form

$$y = \cos \omega x + B \sin \omega x.$$

Therefore,

$$y' = -\omega \sin \omega x + B\omega \cos \omega x.$$

Setting $y' = 3$ when $x = 0$, we get $3 = B\omega$, or $B = 3/\omega$. Thus, the solution is

$$y = \cos \omega x + \frac{3}{\omega} \sin \omega x.$$

This is an example of simple harmonic motion (see Section 4.7). ■

Exercises 18.6

Solve the following equations, subject to the boundary conditions, if any:

1. $y'' - 3y' + 2y = 0$
2. $y'' - 4y' + 3y = 0$
3. $y'' - 4y' + 4y = 0$
4. $y'' + 2y' + y = 0$
5. $y'' - 2y' + 10y = 0$
6. $y'' + 4y' + 5y = 0$
7. $y'' + y' - 2y = 0$; $y = 0$ and $y' = 4$ when $x = 0$
8. $y'' - 9y = 0$; $y = 1$ and $y' = 0$ when $x = 0$
9. $y'' + 6y' + 9y = 0$; $y = 0$ and $y' = 2$ when $x = 0$
10. $y'' - 8y' + 16y = 0$; $y = 4$ and $y' = 0$ when $x = 0$
11. $y'' + 6y' + 10y = 0$; $y = 4$ and $y' = 0$ when $x = 0$
12. $y'' - 10y' + 50y = 0$; $y = 0$ and $y' = 10$ when $x = 0$
13. $y'' - y' + y = 0$
14. $y'' - y' - y = 0$

15. (a) Show that $y = e^{mx}$ is a solution of $y'' - 2my' + m^2y = 0$.
(b) Find a second solution of $y'' - 2my' + m^2y = 0$ by substituting $y = ve^{mx}$, v a function of x, and finding v.

16. Prove that $e^{a+bi}e^{c+di} = e^{(a+bi)+(c+di)}$.

17. If f is a complex-valued function $f(x) = g(x) + ih(x)$, let $f'(x) = g'(x) + ih'(x)$. Show that:

(a) $\dfrac{d}{dx} e^{ix} = ie^{ix}$ (b) $\dfrac{d}{dx} e^{(a+bi)x} = (a+bi)e^{(a+bi)x}$

SECTION 18.7

Nonhomogeneous Linear Equations

A **nonhomogeneous** linear second-order differential equation with constant coefficients is of the form

$$y'' + py' + qy = f(x). \tag{1}$$

The homogeneous equation

$$y'' + py' + qy = 0 \tag{2}$$

is called the **complementary equation** of equation (1), and its general solution is called the **complementary solution** of equation (1). A function that is a solution of equation (1), by contrast, is called a **particular solution.**

If we know the complementary solution and some particular solution, then their sum is the general solution, as our main theorem shows.

THEOREM 18.9

If

$$y_c = ag(x) + bh(x)$$

is the complementary solution of equation (1) and

$$y_p = k(x)$$

is a particular solution, then the general solution of equation (1) is

$$\begin{aligned} y &= y_c + y_p \\ &= ag(x) + bh(x) + k(x). \end{aligned}$$

Proof Let $y = F(x)$ be any solution of equation (1). We can verify that the function

$$G(x) = F(x) - k(x)$$

is a solution of the complementary equation (2):

$$\begin{aligned} G''(x) + pG'(x) + qG(x) &= F''(x) - k''(x) + p(F'(x) - k'(x)) + q(F(x) - k(x)) \\ &= F''(x) + pF'(x) + qF(x) - [k''(x) + pk'(x) + qk(x)] \\ &= f(x) - f(x) \\ &= 0. \end{aligned}$$

Therefore, G is a linear combination of g and h, from the general solution of equation (2), or

$$G(x) = F(x) - k(x) = ag(x) + bh(x).$$

Thus,

$$F(x) = ag(x) + bh(x) + k(x),$$

and the theorem is proved.

■ **Example 18.23** Given that $y = x^2 + 1$ is a particular solution of $y'' + 3y' + 2y = 2x^2 + 6x + 4$, find the general solution.

Solution The complementary equation is

$$y'' + 3y' + 2y = 0,$$

and by Example 18.19, the solution to this equation is

$$y_c = ae^{-x} + be^{-2x}.$$

By Theorem 18.9, the general solution then must be

$$y = ae^{-x} + be^{-2x} + x^2 + 1.$$ ■

The Method of Undetermined Coefficients

We are now left with the problem of determining a particular solution for a given nonhomogeneous equation such as (1). The method that we will illustrate here is called the **method of undetermined coefficients.** It is a technique of guessing the form of a particular solution that depends on the form of $f(x)$ in equation (1). We will demonstrate with examples in the cases when $f(x)$ is a polynomial, an exponential, a sine or a cosine, or a product of these types of functions.

■ **Example 18.24** Solve the equation $y'' + 4y' + 3y = e^{2x}$.

Solution Since e^{2x} has only multiples of itself as its derivatives of any order, we guess that a function of the form $y = Ae^{2x}$ will be a particular solution of our equation. Substituting this expression for y, we get

$$\begin{aligned} 4Ae^{2x} + 8Ae^{2x} + 3Ae^{2x} &= e^{2x}, \\ 15Ae^{2x} &= e^{2x}. \end{aligned}$$

Thus, $15A = 1$, or $A = \frac{1}{15}$. Therefore, a particular solution is

$$y_p = \frac{1}{15} e^{2x}.$$

The complementary solution is found to be

$$y_c = ae^{-3x} + be^{-x},$$

so the general solution is

$$y = ae^{-3x} + be^{-x} + \frac{1}{15} e^{2x}.$$

■

■ Example 18.25 Solve the equation $y'' + 4y' + 3y = x^2$.

Solution We have the same complementary solution as in Example 18.24, namely

$$y_c = ae^{-3x} + be^{-x}.$$

To find a particular solution, we observe that the derivatives of x^2 involve x and constants. Consequently, we guess that a particular solution will be of the form

$$y_p = Ax^2 + Bx + C.$$

Puting this into our equation, we get

$$\begin{aligned} y_p'' + 4y_p' + 3y_p &= 2A + 4(2Ax + B) + 3(Ax^2 + Bx + C) \\ &= 3Ax^2 + (8A + 3B)x + (2A + 4B + 3C). \end{aligned}$$

In order for this last expression to equal x^2, we must have

$$\begin{aligned} 3A \qquad\qquad &= 1, \\ 8A + 3B \qquad &= 0, \\ 2A + 4B + 3C &= 0, \end{aligned}$$

so $A = \frac{1}{3}$, $B = -\frac{8}{9}$, and $C = \frac{26}{27}$. Thus, the particular solution is

$$y_p = \frac{1}{3} x^2 - \frac{8}{9} x + \frac{26}{27},$$

and the general solution is

$$y = ae^{-3x} + be^{-x} + \frac{1}{3} x^2 - \frac{8}{9} x + \frac{26}{27}.$$

■

■ Example 18.26 Solve the equation $y'' - 3y' + 2y = x \sin x$.

Solution The complementary solution is found to be

$$y_c = ae^{x} + be^{2x}.$$

For a particular solution, we note that $x \sin x$ will have as its derivatives combinations of such functions as $x \cos x$, $\sin x$, and $\cos x$. We guess that a particular solution will be of the form

$$y_p = (Ax + B)\sin x + (Cx + D)\cos x.$$

Then,

$$y_p' = (-Cx + A - D)\sin x + (Ax + B + C)\cos x$$

and

$$y_p'' = (-Ax - B - 2C)\sin x + (-Cx + 2A - D)\cos x,$$

so that

$$\begin{aligned} y_p'' - 3y_p' + 2y_p = {} & [(A + 3C)x + (-3A + B - 2C + 3D)]\sin x \\ & + [(-3A + C)x + (2A - 3B - 3C + D)]\cos x. \end{aligned}$$

This last expression will equal $x \sin x$ when

$$\begin{aligned} A \quad\quad + 3C \quad\quad &= 1, \\ -3A + B - 2C + 3D &= 0, \\ -3A \quad\quad + C \quad\quad &= 0, \\ 2A - 3B - 3C + D &= 0. \end{aligned}$$

Thus, $A = \frac{1}{10}$, $B = -\frac{3}{25}$, $C = \frac{3}{10}$, and $D = \frac{17}{50}$. The particular solution is therefore

$$y_p = \left(\frac{1}{10}x - \frac{3}{25}\right)\sin x + \left(\frac{3}{10}x + \frac{17}{50}\right)\cos x,$$

and the general solution is

$$\begin{aligned} y &= y_c + y_p \\ &= ae^x + be^{2x} + \left(\frac{1}{10}x - \frac{3}{25}\right)\sin x + \left(\frac{3}{10}x + \frac{17}{50}\right)\cos x. \end{aligned}$$ ■

These examples illustrate how to find a particular solution in the case that no term of $f(x)$ is a solution of the complementary equation. Our guess for y_p is a linear combination of $f(x)$ and its derivatives. Table 18.1 summarizes the common cases. In the table, $p_n(x)$ stands for a polynomial of degree n.

TABLE 18.1

$f(x)$	y_p
$p_n(x)$	$A_0 + A_1x + \cdots + A_nx^n$
$p_n(x)e^{mx}$	$(A_0 + A_1x + \cdots + A_nx^n)e^{mx}$
$p_n(x)e^{mx}\sin bx$ }	$(A_0 + A_1x + \cdots + A_nx^n)e^{mx}\sin bx$
$p_n(x)e^{mx}\cos bx$ }	$+ (B_0 + B_1x + \cdots + B_nx^n)e^{mx}\cos bx$

The question remains as to what to do when $f(x)$ is a solution of the complementary equation. Consider the next example.

Example 18.27 Solve the equation $y'' - 6y' + 9y = 2e^{3x}$.

Solution The complementary equation is $y'' - 6y' + 9y = 0$, whose auxiliary equation is $m^2 - 6m + 9 = 0$. Since there is only one auxiliary root, $m = 3$, the complementary solution is

$$y_c = ae^{3x} + bxe^{3x}$$

by Theorem 18.7. Now it is seen that $f(x) = 2e^{3x}$ is part of the complementary solution. If we ignore that fact and guess that a particular solution is of the form $y_p = Ae^{3x}$, then

$$y_p'' - 6y_p' + 9y_p = 9Ae^{3x} - 18Ae^{3x} + 9Ae^{3x} = 0,$$

rather than $2e^{3x}$. Thus, it is not appropriate to ignore the fact that $f(x)$ is part of the complementary solution. To overcome this difficulty, we note that when the auxiliary equation had only one root m, the two independent solutions were e^{mx} and xe^{mx}. That is, multiplying $f(x)$ by x might solve our problem. However, $xf(x) = 2xe^{3x}$ is still part of the complementary solution in this case, so we multiply by x again. That is, we make the guess

$$y_p = Ax^2e^{3x}.$$

Then,

$$\begin{aligned} y_p'' - 6y_p' + 9y_p &= A(9x^2 + 12x + 2)e^{3x} - 6A(3x^2 + 2x)e^{3x} + 9Ax^2e^{3x} \\ &= 2Ae^{3x}, \end{aligned}$$

and when we set

$$2Ae^{3x} = 2e^{3x},$$

we get $A = 1$. Thus, the general solution of the equation is

$$y = ae^{3x} + bxe^{3x} + x^2e^{3x}.$$ ■

Example 18.27 illustrates that when a term of $f(x)$ is part of the complementary solution, it should be multiplied by the least power of x that renders the product not part of the complementary solution.

Exercises 18.7 *Solve each equation, subject to the boundary conditions, if any.*

1. $y'' - 5y' + 4y = \sin 3x$
2. $y'' + y' - 6y = 4e^{3x}$
3. $y'' + y' - 6y = x^2e^{3x}$
4. $y'' - 5y' + 4y = (x^2 + 1)e^{-x}$
5. $y'' + 9y = x^3$; $y = 0$ and $y' = 1$ when $x = 0$
6. $y'' + y' - 2y = \cos x$; $y = 1$ and $y' = 4$ when $x = 0$

7. $y'' + 4y' + 5y = \sin x$

8. $y'' + 4y' + 5y = e^{-x} \cos x$

9. $y'' - y' + y = xe^x \sin x$

10. $y'' - y' - y = (x + 1)e^x$

11. Show that if the solution of $y'' + ay' + by = f_1(x)$ is $y = y_c + y_1$ and the solution of $y'' + ay' + by = f_2(x)$ is $y = y_c + y_2$, then the solution of $y'' + ay' + by = f_1(x) + f_2(x)$ is $y = y_c + y_1 + y_2$.

12. Solve the equations:
(a) $y'' + 9y = x^2 + e^x$ (b) $y'' + 9y = x + \cos 3x$

SECTION 18.8

Series Solutions of Differential Equations

It is often possible to find a solution of a differential equation expressed as a power series. We will illustrate in this section a method that works for equations of various orders—even in nonlinear cases.

Example 18.28 Find a series solution of

$$y' + ye^x = 4,$$

subject to the condition $y = 1$ when $x = 0$.

Solution Suppose $y = f(x)$ is a solution, with f given as a Maclaurin series,

$$f(x) = f(0) + f'(0)x + \frac{f''(0)}{2!}x^2 + \cdots.$$

Then we will have determined f if we can determine the constants $f(0), f'(0), f''(0)$, etc. From the given initial condition, we know $f(0) = 1$. From the equation, we have

$$f'(x) = 4 - f(x)e^x, \tag{1}$$

from which we find that

$$f'(0) = 4 - f(0)e^0 = 4 - 1 = 3.$$

Now differentiating equation (1), we get

$$f''(x) = -[f(x) + f'(x)]e^x, \tag{2}$$

from which we compute

$$f''(0) = -[f(0) + f'(0)]e^0 = -[1 + 3](1) = -4.$$

Now differentiating equation (2), we get

$$f'''(x) = -[f(x) + 2f'(x) + f''(x)]e^x, \tag{3}$$

from which we find

$$\begin{aligned} f'''(0) &= -[f(0) + 2f'(0) + f''(0)]e^0 \\ &= -[1 + 2(3) + (-4)](1) = -3. \end{aligned}$$

Continuing in this way, we can find as many of the constants $f^{(4)}(0), \ldots$ as we wish. Thus far, we have

$$f(x) = 1 + 3x + \frac{-4}{2!}x^2 + \frac{-3}{3!}x^3 + \cdots$$
$$= 1 + 3x - 2x^2 - \frac{1}{2}x^3 + \cdots.$$

This is the solution. ∎

Example 18.29 Solve the equation

$$y'' = 1 - y^2,$$

subject to the conditions $y = 1$ and $y' = 1$ when $x = 0$.

Solution Suppose $y = f(x)$ is the solution, with

$$f(x) = f(0) + f'(0)x + \frac{f''(0)}{2!}x^2 + \cdots.$$

From the given boundary conditions, we know that $f(0) = 1$ and $f'(0) = 1$. From the equation, we have

$$f''(x) = 1 - [f(x)]^2. \tag{4}$$

Therefore,

$$f''(0) = 1 - [f(0)]^2 = 1 - 1 = 0.$$

Differentiating equation (4), we get

$$f'''(x) = -2f(x)f'(x), \tag{5}$$

from which

$$f'''(0) = -2(1)(1) = -2.$$

Differentiating equation (5), we get

$$f^{(4)}(x) = -2f(x)f''(x) - 2[f'(x)]^2,$$

from which we find

$$f^{(4)}(0) = -2f(0)f''(0) - 2[f'(0)]^2$$
$$= -2(1)(0) - 2[1]^2 = -2.$$

Thus far, we have

$$f(x) = 1 + x + \frac{0}{2!}x^2 + \frac{-2}{3!}x^3 + \frac{-2}{4!}x^4 + \cdots$$
$$= 1 + x - \frac{1}{3}x^3 - \frac{1}{12}x^4 + \cdots.$$

This solution can be computed to as many terms as desired. ∎

Example 18.30 Find a series solution to

$$y'' = 1 - y,$$

subject to $y = 0$ and $y' = 4$ when $x = 0$.

Solution We let $y = f(x)$ be the solution, with

$$f(x) = f(0) + f'(0)x + \frac{f''(0)}{2!}x^2 + \cdots.$$

From the initial conditions, we know that $f(0) = 0$ and $f'(0) = 4$. From the equation, we have

$$f''(x) = 1 - f(x), \tag{6}$$

so that

$$f''(0) = 1 - f(0) = 1 - 0 = 1.$$

Differentiating equation (6), we get

$$f'''(x) = -f'(x), \tag{7}$$

so that

$$f'''(0) = -f'(0) = -4.$$

From equation (7), we find

$$f^{(4)}(x) = -f''(x), \tag{8}$$

so that

$$f^{(4)}(0) = -f''(0) = -1.$$

From equation (8),

$$f^{(5)}(x) = -f'''(x), \tag{9}$$

so that

$$f^{(5)}(0) = -f'''(0) = 4.$$

From equation (9),

$$f^{(6)}(x) = -f^{(4)}(x),$$

so that

$$f^{(6)}(0) = -f^{(4)}(0) = 1.$$

Thus far, we have

$$f(x) = 0 + 4x + \frac{1}{2!}x^2 + \frac{-4}{3!}x^3 + \frac{-1}{4!}x^4 + \frac{4}{5!}x^5 + \frac{1}{6!}x^6 + \cdots,$$

which we can write as

$$f(x) = 1 - \left(1 - \frac{1}{2!}x^2 + \frac{1}{4!}x^4 - \frac{1}{6!}x^6 + \cdots\right) + 4\left(x - \frac{1}{3!}x^3 + \frac{1}{5!}x^5 - \cdots\right)$$

$$= 1 - \cos x + 4 \sin x.$$

This can be verified as the solution to the original equation. ■

The advantage of this series solution method is that it gives us a way of constructing a solution when no other method we know applies. For further discussion of series solutions, we refer you to a textbook on differential equations.

Exercises 18.8

Find series solutions to each differential equation satisfying the given boundary conditions.

1. $y' + 3y = y^2$; $y = 1$ when $x = 0$
2. $y' = x^2 - y^2$; $y = 2$ when $x = 0$
3. $y'' = x^2y$; $y = 1$ and $y' = 2$ when $x = 0$
4. $y'' + y' = e^x$; $y = 0$ and $y' = 1$ when $x = 0$
5. $y' + y \sin x = 4$; $y = 1$ when $x = 0$
6. $y'' + y' = x^2 + y^2$; $y = 1$ and $y' = 0$ when $x = 0$
7. $y' - 4 = ye^x$; $y = 0$ when $x = 1$ [*Hint:* Use a Taylor series about $x = 1$.]
8. $y'' = 1 - y - y'$; $y = 1$ and $y' = 0$ when $x = 1$

SECTION 18.9

Applications of Second-Order Equations

Second-order differential equations have many applications in such topics as motion, electricity, and mechanics. In this section, we consider a few of the more common applications.

The Motion of a Pendulum

As our first application, we will consider a pendulum in motion.

■ **Example 18.31** A pendulum has length L, and its bob has mass m. Show that the motion of the pendulum does not depend on m.

Solution Imagine the pendulum in motion, being displaced slightly from its equilibrium position (see Figure 18.2). We will let the angle of displacement be y, and the arc of the displacement be s. The weight of the bob is a vertical force mg, whose projection along the tangent to the path is

$$\mathbf{F} = -mg(\sin y)\mathbf{T}, \tag{1}$$

where $\mathbf{T}$ is the unit tangent vector to the path. Now treating y and s as functions of t, the relation

$$s = Ly$$

yields

$$\frac{ds}{dt} = L\frac{dy}{dt}$$

and

$$\frac{d^2s}{dt^2} = L\frac{d^2y}{dt^2}. \tag{2}$$

By Newton's law, the force on the pendulum bob is the mass times the acceleration, or

$$\mathbf{F} = m\frac{d^2s}{dt^2}\mathbf{T}. \tag{3}$$

Combining equations (1), (2), and (3), we get

$$mL\frac{d^2y}{dt^2}\mathbf{T} = -mg(\sin y)\mathbf{T},$$

so that

$$\frac{d^2y}{dt^2} = -\frac{g}{L}\sin y. \tag{4}$$

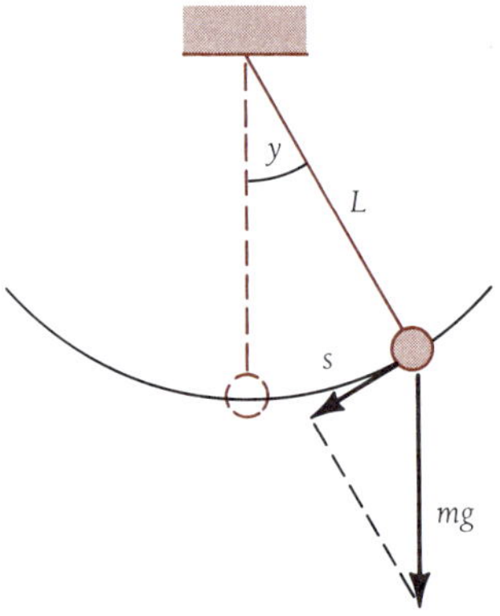

FIGURE 18.2 A pendulum

This equation governs the motion of the pendulum and does not involve the mass m of the pendulum bob. Thus, the motion of the pendulum does not depend on its mass, but only on its length.

If the displacement of the pendulum is so slight that $\sin y \approx y$ is a good approximation, then equation (4) becomes

$$y'' + \frac{g}{L}y = 0, \tag{5}$$

which is a case of simple harmonic motion. ■

Escape Velocity

A second application seeks to answer the following motion problem.

■ **Example 18.32** A projectile is fired vertically upward from the surface of the earth. What initial velocity must it have to escape from the earth's gravitational field? Assume that no force but gravity acts on the projectile after it is fired.

Solution To answer the question, we set up a differential equation. Let x be the distance from the projectile to the center of the earth and let m be the mass of the projectile. Then the force of gravity acting on the object is

$$F = -\frac{km}{x^2},$$

where k is a constant. The acceleration is therefore found from

$$-\frac{km}{x^2} = ma,$$

and is

$$a = \frac{d^2x}{dt^2} = -\frac{k}{x^2}.$$

This is the special form of a second-order equation in which the independent variable t (time) does not occur. Therefore, letting $v = dx/dt$ and $d^2x/dt^2 = dv/dt = (dv/dx)\,(dx/dt) = v\,(dv/dx)$, we get

$$v\frac{dv}{dx} = -\frac{k}{x^2}.$$

This equation is separable:

$$v\,dv = -\frac{k}{x^2}\,dx,$$

$$\frac{1}{2}v^2 = \frac{k}{x} + C.$$

Letting $v = v_0$ when $x = x_0$, we get

$$C = \frac{1}{2} v_0^2 - \frac{k}{x_0}.$$

Hence,

$$\frac{1}{2} v^2 = \frac{k}{x} + \frac{1}{2} v_0^2 - \frac{k}{x_0},$$
$$v^2 = \frac{2k}{x} + v_0^2 - \frac{2k}{x_0}.$$

For the projectile to escape from the earth's gravitational field means that v stays positive as $x \to \infty$; but as $x \to \infty$, $v^2 \to v_0^2 - (2k/x_0)$. Thus, we must have $v_0^2 > 2k/x_0$, or $v_0 > \sqrt{2k/x_0}$. To find k, we use the fact that at the surface of the earth, $x_0 \approx 3960$ miles and the acceleration of gravity is approximately 32 feet per second/second, or

$$\frac{k}{x_0^2} = 32,$$
$$k = 32x_0^2$$
$$= (32 \text{ ft/sec}^2)\left(\frac{1}{5280} \text{ mi/ft}\right)(3960 \text{ mi})^2$$
$$= \frac{(3960)^2}{165} \text{ mi/sec}^2.$$

Hence,

$$v_0 > \sqrt{\frac{2k}{x_0}} = \sqrt{\frac{2(3960)}{165}} \approx 6.93 \text{ mi/sec.}$$

Thus, the initial velocity must be greater than about 6.93 miles per second; that is, escape velocity is almost 7 miles per second. ■

Damped Vibrations

Our next application is a realistic look at vibratory motion, in which friction or other damping forces are at work.

■ **Example 18.33** An object, initially in simple harmonic motion, is subjected to a damping force proportional to the velocity of the object. Describe its subsequent motion.

Solution If the object is originally in simple harmonic motion, then its position x and acceleration a are related by an equation of the form

$$a = -\omega^2 x.$$

A damping force proportional to the velocity v produces another acceleration of the form

$$a = -\mu v,$$

where μ is the constant of proportionality. Thereafter, the total acceleration on the object is

$$a = -\omega^2 x - \mu v,$$

which is equivalent to the differential equation

$$\frac{d^2x}{dt^2} + \mu\frac{dx}{dt} + \omega^2 x = 0, \tag{6}$$

where t is time. This is a second-order linear equation with constant coefficients, with auxiliary equation

$$m^2 + \mu m + \omega^2 = 0.$$

The auxiliary roots are

(a) $m_1 = \dfrac{-\mu + \sqrt{\mu^2 - 4\omega^2}}{2}, \quad m_2 = \dfrac{-\mu - \sqrt{\mu^2 - 4\omega^2}}{2}, \quad \mu^2 > 4\omega^2;$

(b) $m = -\dfrac{\mu}{2}, \quad \mu^2 = 4\omega^2;$

(c) $m_1 = -\dfrac{\mu}{2} + \dfrac{1}{2}\sqrt{4\omega^2 - \mu^2}\, i, \quad m_2 = -\dfrac{\mu}{2} - \dfrac{1}{2}\sqrt{4\omega^2 - \mu^2}\, i, \quad \mu^2 < 4\omega^2.$

The solution of equation (6) is therefore of the form

(a) $x = Ae^{m_1 t} + Be^{m_2 t}$ if $\mu^2 > 4\omega^2$ (see Figure 18.3);

(b) $x = Ae^{mt} + Bte^{mt}$ if $\mu^2 = 4\omega^2$ (see Figure 18.4);

(c) $x = e^{-t\mu/2}(A \sin \frac{1}{2}\sqrt{4\omega^2 - \mu^2}\, t + B \cos \frac{1}{2}\sqrt{4\omega^2 - \mu^2}\, t)$ if $\mu^2 < 4\omega^2$ (see Figure 18.5).

Case (a) is called **overdamped**, case (b) is called **critically damped**, and case (c) is called **underdamped**.

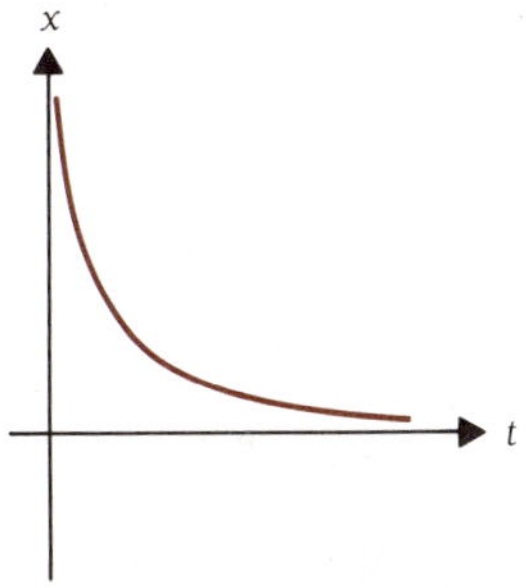

FIGURE 18.3
An overdamped vibration

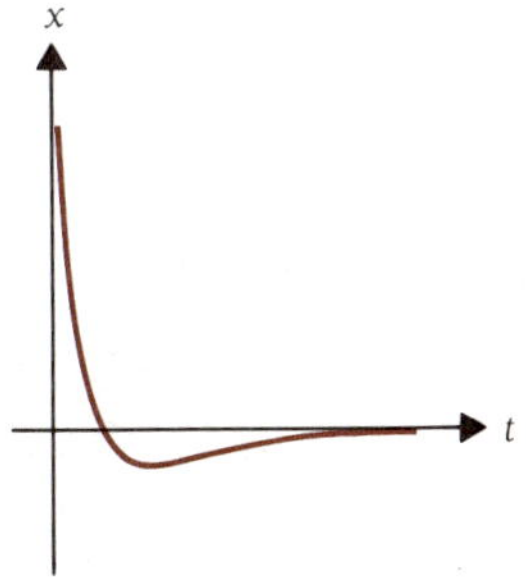

FIGURE 18.4
A critically damped vibration

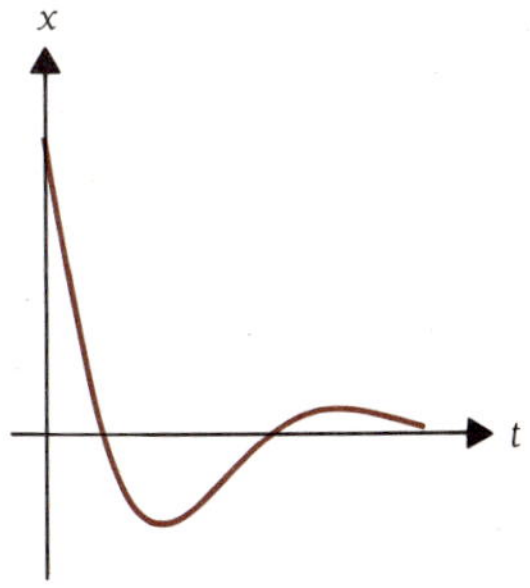

FIGURE 18.5
An underdamped vibration ■

Hanging Cables

Our last illustration describes the curve representing a hanging flexible cable.

Example 18.34 A flexible, uniform cable is fastened at two points. Describe the curve formed by the hanging cable.

Solution We will choose the lowest point on the cable as the origin O of our coordinate system (see Figure 18.6). Let the cable have mass m per unit length. At the lowest point on the cable, let H be the horizontal tension. At a point P on the cable, let T be the tension at an angle θ from the horizontal. Let s be the length of cable between P and O. Then the forces balance, so

$$H = T\cos\theta$$

and

$$smg = T\sin\theta.$$

Thinking of $y = f(x)$ as the curve along which the cable hangs, we have

$$y' = \tan\theta = \frac{T\sin\theta}{T\cos\theta} = \frac{smg}{H}.$$

Therefore,

$$\begin{aligned} y'' &= \frac{mg}{H}\frac{ds}{dx} \\ &= \frac{mg}{H}\sqrt{1 + (y')^2}. \end{aligned}$$

Letting $u = y'$, we then have

$$u' = \frac{mg}{H}\sqrt{1 + u^2},$$

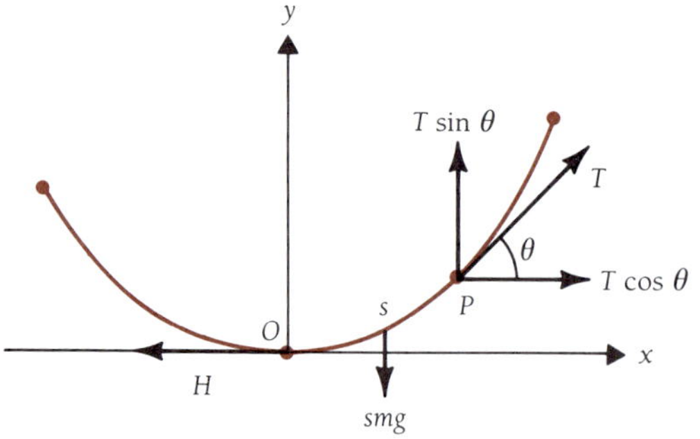

FIGURE 18.6 Forces on a hanging cable

or

$$\frac{du}{\sqrt{1+u^2}} = \frac{mg}{H}\,dx.$$

Integrating, we get

$$\ln(u + \sqrt{1+u^2}) = \frac{mgx}{H} + C,$$
$$u + \sqrt{1+u^2} = e^{mgx/H}e^C.$$

Now when $x = 0$, $y' = u = 0$, so $e^C = 1$. Thus,

$$u + \sqrt{1+u^2} = e^{mgx/H},$$
$$\sqrt{1+u^2} = e^{mgx/H} - u,$$
$$1 + u^2 = e^{2mgx/H} - 2ue^{mgx/H} + u^2,$$
$$2ue^{mgx/H} = e^{2mgx/H} - 1,$$
$$u = \frac{1}{2}(e^{mgx/H} - e^{-mgx/H}).$$

Thus,

$$y' = \sinh\frac{mgx}{H},$$

so

$$y = \frac{H}{mg}\cosh\frac{mgx}{H} + C.$$

Since $y = 0$ when $x = 0$, we get $C = -H/mg$. Thus,

$$y = \frac{H}{mg}\left(\cosh\frac{mgx}{H} - 1\right)$$

is the curve along which the cable hangs. This curve is called a **catenary**. ■

Exercises 18.9

1. Find the period of a pendulum whose motion is simple harmonic motion. Does the period depend on the amplitude?
2. Explain how the motion of a pendulum varies with its length under small vibrations.
3. A Foucault pendulum has length 32 feet. What will its period be if its motion is harmonic?
4. A pendulum under small vibrations has period 3 seconds. How long is it?
5. If the mass of the moon is $\frac{1}{80}$ that of the earth and its radius is 1080 miles, find the escape velocity from the surface of the moon.
6. One marble rolls down a frictionless inclined plane to the floor, and another marble falls from the top of the inclined plane to the floor. Show that both marbles have the same speed when they reach the floor.

7. Imagine a tunnel drilled through the earth along a diameter. Describe the motion of an object dropped into the tunnel from the earth's surface. (An object inside a hollow sphere has zero gravitational force acting on it.)

8. A cylindrical solid of weight W pounds and radius r feet floats in water with its axis vertical. It is depressed slightly from its equilibrium position and released. Ignoring friction forces, find the period of its vibration.

9. In Problem 8, suppose a small damping force proportional to the velocity is present. After how much time will the amplitude of the vibration be half of the initial displacement?

10. A cylindrical solid weighing 200 pounds floats in the water with its axis vertical. When depressed slightly and released, it vibrates with a period of 3 seconds. Assuming friction forces are negligibly small, what is the radius of the cylinder?

11. A flexible cable is fastened at two points of the same height and supports a horizontal roadbed of weight w per unit length. If the weight of the cable is negligible as compared to the weight of the roadbed, find an equation for the curve along which the cable lies.

12. A motion satisfying an equation of the form

$$\frac{d^2x}{dt^2} + \mu\frac{dx}{dt} + w^2x = f(t)$$

is called **forced vibratory motion**, with **forcing function** f. Describe the following forced vibratory motions, given that $x = 0$ and $x' = 4$ when $t = 0$:

(a) $\mu = 2, \quad w = 1, \quad f(t) = 3 \sin t$ (b) $\mu = 2, \quad w = 6, \quad f(t) = 4 \cos t$
(c) $\mu = 0, \quad w = 5, \quad f(t) = \sin 5t$ (d) $\mu = 0, \quad w = 1, \quad f(t) = 2t$

13. An electric circuit contains a capacitor of capacitance C farads, a resistor of resistance R ohms, a coil of inductance L henries, and a generator producing a voltage difference of $f(t)$ volts at time t, measured in seconds. If $q(t)$ is the charge in volts on the capacitor at time t, then

$$Lq''(t) + Rq'(t) + \frac{1}{C}q(t) = f(t).$$

If $i(t) = q'(t)$ is the current in amperes at time t, find q and i if $L = 3$, $R = 12$, $C = \frac{1}{12}$, $f(t) = \cos t$, $q(0) = 0$, and $i(0) = 1$.

14. Find q and i in Problem 13 if $L = 4$, $R = 5$, $C = \frac{1}{2}$, $f(t) = 1 - e^{-t}$, $q(0) = 20$, and $i(0) = 5$.

15. Show that for critically damped vibrations, x is 0 for at most one positive value of t.

SECTION 18.10

Chapter Review

Important Concepts

Ordinary differential equation
Order
Solution
General solution
Constants of integration
Particular solution
Boundary condition
Separable equation
Exact equation
Integrating factor
Linear differential equation
Homogeneous function
Homogeneous coefficients
Orthogonal trajectory

Mixing problems
Second-order differential equations
Homogeneous equation
Linearly dependent
Linearly independent
Wronskian
Auxiliary equation
Auxiliary root
Nonhomogeneous equation
Complementary equation
Complementary solution
Particular solution
Undetermined coefficients
Series solution
Motion of a pendulum
Escape velocity
Damped vibration
Catenary

Review Exercises

Solve each differential equation.

1. $[y + (2x + x^2y)e^{xy}]\, dx + (x + x^3e^{xy})\, dy = 0$
2. $[4(\cos x)(\ln y) + 2x]\, dx + \frac{4}{y} \sin x\, dy = 0$
3. $dy + (xy - x^3)\, dx = 0$
4. $dy + \left(\frac{y}{x} - x^3\right) dx = 0$
5. $(\sqrt{y^2 - x^2} - x)\, dx + (2\sqrt{y^2 - x^2} + y)\, dy = 0$
6. $[2x + y(x^2 + y^2)]\, dx + [2y + x(x^2 + y^2)]\, dy = 0$
7. $x^2y' + (1 + x)y = 2$
8. $x^2y' + 3y = 2x + 1$
9. $(x^2 - xy)\, dx + (x^2 - y^2)\, dy = 0$
10. $\sin \frac{y}{x}\, dx + \cos \frac{y}{x}\, dy = 0$
11. $xy'' + y' = x \sin x$
12. $x^2y'' = 1 - y'^2$
13. $y'' = 1 + y'^2$
14. $y'' = 16y + 1$
15. $y'' - 7y' + 12y = 0$
16. $4y'' - 4y' + y = 0$
17. $y'' + y' + 4y = 0$
18. $2y'' - 3y' + 2y = 0$
19. $y'' - 7y' + 12y = e^x$
20. $y'' - 7y' + 12y = e^{3x}$
21. $4y'' - 4y' + y = xe^{x/2}$
22. $4y'' - 4y' + y = x + e^{x/2}$
23. $y'' = x^2 + 4y^2$; $y = 1$ and $y' = 0$ when $x = 0$
24. $y'' = e^x - y'^2$; $y = 0$ and $y' = 2$ when $x = 0$
25. Find the orthogonal trajectories of the family of curves $x^2 + 4y^2 = C$.
26. A tank initially contains 20 gallons of a solution that is 30% alcohol. A solution of 50% alcohol runs into the tank at the rate of 2 gallons per minute, and the mixture, kept uniform by stirring, runs out at the same rate. Find the concentration of alcohol in the tank as a function of time.
27. A brick slides down an inclined plane, experiencing a friction force proportional to the speed of the brick. Find the terminal velocity of the brick as a function of the slope m of the plane and the coefficient of friction μ.
28. A conical tank of height h and radius r is situated with its vertex downward. The tank drains through a hole of radius a at the vertex. If the tank is full initially, how long does it take to drain?
29. What is the period, under small vibrations, of a pendulum of length 10 feet?
30. If the radius of Venus is 3800 miles and its mass is $\frac{4}{5}$ of the earth's mass, find the escape velocity from the surface of Venus.
31. A spring that is 1 foot long is stretched to a length of 13 inches when a 2 pound weight is hung on it. The weight is pulled downward, stretching the spring to 14 inches, and then

it is released. After 2 minutes, it is noted that the amplitude of the vibration has decreased to $\frac{1}{2}$ inch due to friction forces. Find the position of the weight as a function of time.

32. A certain block weighing 20 pounds displaces 10 pounds of water. When the block is released under water, it meets a resistive force equal in magnitude to twice the velocity. Describe its motion, and give its terminal velocity.

Additional Exercises

1. A cylindrical barrel standing upright contains 20 gallons of water in which 15 pounds of salt is dissolved. A bung hole of radius 1 inch is opened in the bottom of the barrel, and the brine begins to drain out at a rate governed by Torricelli's law (see Problem 18, Section 18.4). At the same time, a stream of pure water is directed into the barrel, flowing at the rate of 4 gallons per minute. If the radius of the barrel is 18 inches and the solution in the barrel is kept uniformly mixed, how much salt remains in the barrel after 5 minutes?
2. It takes 5 pounds of force to stretch a certain spring 1 inch beyond its natural length. The spring hangs from a block, and a 10 pound weight is hung on it. Then the block is displaced vertically, and its position as a function of time t is given by $f(t) = 2 \sin \pi t$. Describe the motion of the weight.
3. Rework Problem 2 in the case that the weight also encounters a resistive force equal in magnitude to half the velocity of the weight.
4. Find the orthogonal trajectories of the family of circles tangent to the x-axis at the origin.
5. Find the orthogonal trajectories of the family of circles with center at the origin.
6. Find the orthogonal trajectories of the level curves of the surface $z = x^2 - 4y^2$.

APPENDIX A

Determinants

A rectangular array of real numbers is called a **matrix**. A matrix with m rows and n columns is called an $\boldsymbol{m \times n}$ **matrix**. We use the symbol a_{ij} to represent the number occurring in the ith row and jth column of a matrix. We represent an $m \times n$ matrix by the notation

$$(a_{ij}) = \begin{bmatrix} a_{11} & a_{12} & \cdots & a_{1n} \\ a_{21} & a_{22} & \cdots & a_{2n} \\ \cdot & \cdot & & \cdot \\ \cdot & \cdot & & \cdot \\ \cdot & \cdot & & \cdot \\ a_{m1} & a_{m2} & \cdots & a_{mn} \end{bmatrix}.$$

A **square matrix** is a matrix with the same number of rows as columns. An $n \times n$ matrix is also called a **matrix of order** $\boldsymbol{n}$. We will consider only square matrices in this appendix.

Associated with a square matrix (a_{ij}) is a real number called the **determinant** of (a_{ij}), denoted by

$$\det(a_{ij}) = \begin{vmatrix} a_{11} & \cdots & a_{1n} \\ \cdot & & \cdot \\ \cdot & & \cdot \\ \cdot & & \cdot \\ a_{n1} & \cdots & a_{nn} \end{vmatrix}.$$

If (a_{ij}) has order n, we also say that $\det(a_{ij})$ has order n. When we refer to the elements, rows, or columns of $\det(a_{ij})$, we actually mean the elements, rows, or columns of the matrix (a_{ij}).

In order to describe how to find the real number that is the determinant of a given square matrix (a_{ij}), it is first necessary to introduce some apparently unrelated material.

Let S be the set of the first n positive integers, $S = \{1, 2, \ldots, n\}$. A **permutation** of S is a function that has domain and range both S and that has the property that distinct elements of S have distinct images. That is, a permutation of S is a one-to-one function from S onto itself. If σ is a permutation of S, then the list $\sigma(1), \sigma(2), \ldots, \sigma(n)$ is just a reordering of the list $1, 2, \ldots, n$. Thus, a permutation of S is just a reordering of the elements of S.

Each permutation σ of S has an **inverse** σ^{-1} that is also a permutation of S and has the property that $\sigma^{-1}(\sigma(k)) = k$ for all k. The **identity** permutation ι has the property that $\iota(k) = k$ for all k. The set of all permutations of S is denoted by S_n. When S has n elements, then S_n has $n!$ elements.

If $\sigma \in S_n$, let the images of $1, 2, \ldots, n$ be written in the order

$$\sigma(1), \sigma(2), \ldots, \sigma(n). \tag{1}$$

Whenever a number in this list (1) precedes a smaller number in the list, we say an **inversion** occurs. If the number of inversions in the list (1) is odd, we call σ an **odd permutation**; otherwise, we call σ an **even permutation**. The **signum** of a permutation is defined to be

$$\operatorname{sig}(\sigma) = \begin{cases} 1 & \text{if } \sigma \text{ is even} \\ -1 & \text{if } \sigma \text{ is odd.} \end{cases}$$

■ **Example A.1** Let $\sigma \in S_4$ with $\sigma(1) = 1, \sigma(2) = 4, \sigma(3) = 3$, and $\sigma(4) = 2$. Then the list (1) is the list

$$1, 4, 3, 2.$$

In this list, three inversions occur: 4 precedes 3, 4 precedes 2, and 3 precedes 2. Thus, σ is an odd permutation, and $\operatorname{sig}(\sigma) = -1$. ■

It can be shown that the signum of the composition of two permutations is the product of their signums. Since the identity permutation is even, it follows that a permutation and its inverse have the same signum.

Now we are ready to complete the definition of a determinant. Let (a_{ij}) be a square matrix of order n. For each $\sigma \in S_n$, we form the product

$$\operatorname{sig}(\sigma) a_{1\sigma(1)} a_{2\sigma(2)} \cdot \cdots \cdot a_{n\sigma(n)},$$

and then we find the sum of all such products. That is,

$$\det(a_{ij}) = \sum_{\sigma \in S_n} \operatorname{sig}(\sigma) a_{1\sigma(1)} a_{2\sigma(2)} \cdot \cdots \cdot a_{n\sigma(n)}. \tag{2}$$

Each product contains one element from each row and one element from each column.

The right-hand side of equation (2) is often called the **expansion** of $\det(a_{ij})$. Since S_n has $n!$ elements, there are $n!$ terms in the expansion of a determinant of order n.

Since the definition (2) is computationally useful only for determinants of small order, we will now consider some of the consequences of the definition. Many of these results will be useful in computing determinants.

THEOREM A.1 For a matrix (a_{11}) of order 1, the determinant is its single element,

$$\det(a_{11}) = a_{11}.$$

THEOREM A.2 For a matrix

$$(a_{ij}) = \begin{bmatrix} a_{11} & a_{12} \\ a_{21} & a_{22} \end{bmatrix}$$

of order 2, the determinant is

$$\det(a_{ij}) = \begin{vmatrix} a_{11} & a_{12} \\ a_{21} & a_{22} \end{vmatrix} = a_{11}a_{22} - a_{12}a_{21}.$$

Proof In S_2 there are two permutations, whose images in the order (1) are 1, 2 and 2, 1. The theorem now follows directly from equation (2).

Example A.2 If

$$(a_{ij}) = \begin{bmatrix} 3 & 7 \\ 5 & -2 \end{bmatrix},$$

then

$$\det(a_{ij}) = \begin{vmatrix} 3 & 7 \\ 5 & -2 \end{vmatrix} = 3(-2) - 7(5) = -41.$$

■

THEOREM A.3 For a matrix

$$(a_{ij}) = \begin{bmatrix} a_{11} & a_{12} & a_{13} \\ a_{21} & a_{22} & a_{23} \\ a_{31} & a_{32} & a_{33} \end{bmatrix}$$

of order 3, the determinant is

$$\det(a_{ij}) = \begin{vmatrix} a_{11} & a_{12} & a_{13} \\ a_{21} & a_{22} & a_{23} \\ a_{31} & a_{32} & a_{33} \end{vmatrix}$$
$$= a_{11}a_{22}a_{33} + a_{12}a_{23}a_{31} + a_{13}a_{21}a_{32} - a_{13}a_{22}a_{31} - a_{11}a_{23}a_{32} - a_{12}a_{21}a_{33}.$$

Proof In S_3 there are six permutations, whose images in the order (1) are

1, 2, 3	(Even);	3, 2, 1	(Odd);
2, 3, 1	(Even);	1, 3, 2	(Odd);
3, 1, 2	(Even);	2, 1, 3	(Odd).

The theorem now follows from equation (2).

An easy way to remember the pattern for a determinant of order 3 is the following diagram, in which the first two columns have been duplicated. Products of elements on the downward diagonals are added, and products of elements on upward diagonals are subtracted.

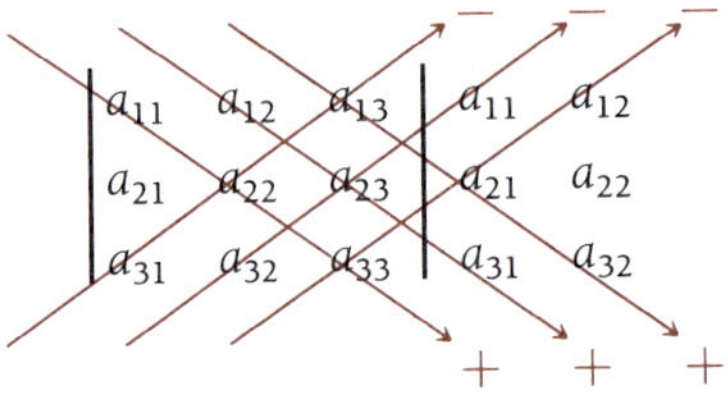

■ **Example A.3** If

$$(a_{ij}) = \begin{bmatrix} 4 & 1 & 0 \\ 3 & 2 & -1 \\ -5 & 7 & 6 \end{bmatrix},$$

then

$$\begin{aligned} \det(a_{ij}) &= \begin{vmatrix} 4 & 1 & 0 \\ 3 & 2 & -1 \\ -5 & 7 & 6 \end{vmatrix} \\ &= 4(2)(6) + 1(-1)(-5) + 0(3)(7) - 0(2)(-5) - 4(-1)(7) - 1(3)(6) \\ &= 48 + 5 + 0 - 0 - (-28) - 18 = 63. \end{aligned}$$

■

If (a_{ij}) is a matrix, its **transpose** is the matrix

$$(a_{ij})^{\mathrm{T}} = (a_{ji}).$$

That is, the transpose of a matrix is obtained by writing the rows as columns (and consequently, the columns as rows).

THEOREM A.4

$$\det(a_{ij})^{\mathrm{T}} = \det(a_{ij}).$$

Proof To help us keep track of the notational details, let $(a_{ij})^{\mathrm{T}} = (b_{ij})$ with $b_{ij} = a_{ji}$. Then

$$\begin{aligned}\det(a_{ij})^{\mathrm{T}} &= \det(b_{ij}) \\ &= \sum_{\sigma \in S_n} \operatorname{sig}(\sigma) b_{1\sigma(1)} \cdot \cdots \cdot b_{n\sigma(n)} \\ &= \sum_{\sigma \in S_n} \operatorname{sig}(\sigma) a_{\sigma(1)1} \cdot \cdots \cdot a_{\sigma(n)n}.\end{aligned}$$

Now we rearrange the factors in each term so that the first subscripts are in the natural order. This introduces σ^{-1} in the second subscript, and we get

$$\begin{aligned}\det(a_{ij})^{\mathrm{T}} &= \sum_{\sigma \in S_n} \operatorname{sig}(\sigma) a_{1\sigma^{-1}(1)} \cdot \cdots \cdot a_{n\sigma^{-1}(n)} \\ &= \sum_{\sigma^{-1} \in S_n} \operatorname{sig}(\sigma^{-1}) a_{1\sigma^{-1}(1)} \cdot \cdots \cdot a_{n\sigma^{-1}(n)}.\end{aligned}$$

This last equation holds because the set of inverses of all the permutations is simply the set of all permutations. Because σ^{-1} in the last sum is a dummy index, it can be replaced by σ, and the theorem follows.

The utility of Theorem A.4 is stated as a corollary.

COROLLARY Any property of determinants stated in terms of rows also holds when stated in terms of columns.

Thus, in the following theorems, the word "row" may be replaced throughout by the word "column," and another theorem results in each case.

THEOREM A.5 If two rows of a matrix are interchanged, the corresponding determinants differ in sign.

Proof Given matrix (a_{ij}), let (b_{ij}) be the matrix obtained by interchanging the kth and lth rows of (a_{ij}). In the expansion of $\det(b_{ij})$, a typical term is

$$\begin{aligned}&\operatorname{sig}(\sigma) b_{1\sigma(1)} \cdot \cdots \cdot b_{k\sigma(k)} \cdot \cdots \cdot b_{l\sigma(l)} \cdot \cdots \cdot b_{n\sigma(n)} \\ &\qquad = \operatorname{sig}(\sigma) a_{1\sigma(1)} \cdot \cdots \cdot a_{l\sigma(k)} \cdot \cdots \cdot a_{k\sigma(l)} \cdot \cdots \cdot a_{n\sigma(n)}.\end{aligned}$$

Now we rewrite the last expression so that the first subscripts are in order; that is, we interchange $a_{l\sigma(k)}$ and $a_{k\sigma(l)}$. But that will change the sign of the term, because the number of inversions will be changed by an odd number. To see that, note that an interchange of adjacent numbers creates or destroys exactly one inversion, and the interchange of a nonadjacent pair may be accomplished by an odd number of interchanges of adjacent pairs. Thus, each term is changed in sign, and $\det(b_{ij}) = -\det(a_{ij})$.

THEOREM A.6 If two rows of matrix (a_{ij}) are identical, then $\det(a_{ij}) = 0$.

Proof By Theorem A.5, interchanging the identical rows will change the sign of the determinant. But at the same time, the determinant is unchanged if the rows are identical. The only number that is unchanged when its sign is changed is 0.

THEOREM A.7 If each element of one row of matrix (a_{ij}) is multiplied by the constant c, then $\det(a_{ij})$ is multiplied by c.

Proof By the definition (2), multiplying each element of some row will cause c to appear once in each term of the expansion of the determinant, and the determinant is multiplied by c.

THEOREM A.8 If two rows of matrix (a_{ij}) are proportional, then $\det(a_{ij}) = 0$.

Proof If two rows are proportional, then each element of one of the rows is equal to a constant c times the corresponding element of the other row. Factoring out the constant c leaves us with a determinant in which two rows are identical, and the resulting determinant is 0 by Theorem A.6. The result then follows by Theorem A.7.

THEOREM A.9 If each element of the kth row of the matrix (a_{ij}) is a sum, say $a_{kj} = r_{kj} + s_{kj}$, let (b_{ij}) and (c_{ij}) be matrices identical to (a_{ij}) except that $b_{kj} = r_{kj}$ and $c_{kj} = s_{kj}$. Then $\det(a_{ij}) = \det(b_{ij}) + \det(c_{ij})$, or

$$\begin{vmatrix} a_{11} & \cdots & a_{1n} \\ \cdot & & \cdot \\ \cdot & & \cdot \\ \cdot & & \cdot \\ r_{k1} + s_{k1} & \cdots & r_{kn} + s_{kn} \\ \cdot & & \cdot \\ \cdot & & \cdot \\ \cdot & & \cdot \\ a_{n1} & \cdots & a_{nn} \end{vmatrix} = \begin{vmatrix} a_{11} & \cdots & a_{1n} \\ \cdot & & \cdot \\ \cdot & & \cdot \\ \cdot & & \cdot \\ r_{k1} & \cdots & r_{kn} \\ \cdot & & \cdot \\ \cdot & & \cdot \\ \cdot & & \cdot \\ a_{n1} & \cdots & a_{nn} \end{vmatrix} + \begin{vmatrix} a_{11} & \cdots & a_{1n} \\ \cdot & & \cdot \\ \cdot & & \cdot \\ \cdot & & \cdot \\ s_{k1} & \cdots & s_{kn} \\ \cdot & & \cdot \\ \cdot & & \cdot \\ \cdot & & \cdot \\ a_{n1} & \cdots & a_{nn} \end{vmatrix}.$$

Proof In the expansion of $\det(a_{ij})$, simply use the distributive law on each term.

THEOREM A.10 If each element of one row of matrix (a_{ij}) is increased by a constant multiple of the corresponding element of another row, then $\det(a_{ij})$ is unchanged.

Proof Let each element of the kth row of (a_{ij}) be increased by the constant c times the corresponding element of the lth row; then a_{kj} is replaced by $a_{kj} + ca_{lj}$. Then by Theorems A.9 and A.8, the determinant of the new matrix is the same as the original.

THEOREM A.11 If matrix (a_{ij}) has a row of 0's, then $\det(a_{ij}) = 0$.

Proof Using Theorem A.10, we can create a determinant with proportional rows. Then use Theorem A.8.

A **diagonal matrix** is a square matrix (a_{ij}) in which $a_{ij} = 0$ if $i \neq j$. That is, any nonzero elements must occur on the main diagonal:

$$\begin{bmatrix} a_{11} & 0 & \cdots & 0 \\ 0 & a_{22} & \cdots & 0 \\ \vdots & \vdots & \ddots & \vdots \\ 0 & 0 & \cdots & a_{nn} \end{bmatrix}$$

THEOREM A.12 If (a_{ij}) is a diagonal matrix, then $\det(a_{ij})$ is the product of the elements on the diagonal, or

$$\det(a_{ij}) = a_{11}a_{22} \cdot \cdots \cdot a_{nn}.$$

Proof In the expansion of $\det(a_{ij})$, every term except $a_{11}a_{22} \cdot \cdots \cdot a_{nn}$ has 0 as a factor.

The **elementary row operations** on a matrix are:

1. Interchanging two rows
2. Multiplying each element of a row by the same nonzero constant
3. Increasing each element of one row by a constant multiple of the corresponding element of another row

The **elementary column operations** are the same operations applied to columns. The effects of the elementary row and column operations on the determinant of a matrix are stated by Theorems A.5, A.7, and A.10.

Using elementary row and column operations, it is possible to transform any square matrix into a diagonal matrix. (A little experimentation will convince you that this is so; see the following example.) Hence, using elementary row and column operations and Theorems A.5, A.7, A.10, and A.12, we can compute the determinant of any square matrix.

■ **Example A.4** Compute the determinant of

$$\begin{bmatrix} 2 & 1 & 3 & 0 & 5 \\ 1 & 0 & 0 & -1 & 1 \\ 2 & -1 & 3 & 0 & 1 \\ 0 & 0 & 2 & 1 & 1 \\ 1 & 0 & 3 & 1 & -1 \end{bmatrix}.$$

Solution We will reduce the matrix to a diagonal matrix, keeping track of the effect of the row and column operations. Interchanging the first two rows gives us

$$\begin{vmatrix} 2 & 1 & 3 & 0 & 5 \\ 1 & 0 & 0 & -1 & 1 \\ 2 & -1 & 3 & 0 & 1 \\ 0 & 0 & 2 & 1 & 1 \\ 1 & 0 & 3 & 1 & -1 \end{vmatrix} = - \begin{vmatrix} 1 & 0 & 0 & -1 & 1 \\ 2 & 1 & 3 & 0 & 5 \\ 2 & -1 & 3 & 0 & 1 \\ 0 & 0 & 2 & 1 & 1 \\ 1 & 0 & 3 & 1 & -1 \end{vmatrix}.$$

Now we increase the elements of the second, third, and fifth rows by appropriate (negative) multiples of the elements of the first row, in order to produce 0's in the rest of the first column:

$$- \begin{vmatrix} 1 & 0 & 0 & -1 & 1 \\ 0 & 1 & 3 & 2 & 3 \\ 0 & -1 & 3 & 2 & -1 \\ 0 & 0 & 2 & 1 & 1 \\ 0 & 0 & 3 & 2 & -2 \end{vmatrix}.$$

Next, we increase the elements of the fourth and fifth columns by appropriate multiples of the first column to introduce 0's in the rest of the first row:

$$- \begin{vmatrix} 1 & 0 & 0 & 0 & 0 \\ 0 & 1 & 3 & 2 & 3 \\ 0 & -1 & 3 & 2 & -1 \\ 0 & 0 & 2 & 1 & 1 \\ 0 & 0 & 3 & 2 & -2 \end{vmatrix}.$$

Now we use the second row to get 0's in the rest of the second column:

$$- \begin{vmatrix} 1 & 0 & 0 & 0 & 0 \\ 0 & 1 & 3 & 2 & 3 \\ 0 & 0 & 6 & 4 & 2 \\ 0 & 0 & 2 & 1 & 1 \\ 0 & 0 & 3 & 2 & -2 \end{vmatrix},$$

and the second column to get 0's in the rest of the second row:

$$-\begin{vmatrix} 1 & 0 & 0 & 0 & 0 \\ 0 & 1 & 0 & 0 & 0 \\ 0 & 0 & 6 & 4 & 2 \\ 0 & 0 & 2 & 1 & 1 \\ 0 & 0 & 3 & 2 & -2 \end{vmatrix}.$$

Now we can apply Theorem A.7 to factor 6 out of the third row, getting

$$-6\begin{vmatrix} 1 & 0 & 0 & 0 & 0 \\ 0 & 1 & 0 & 0 & 0 \\ 0 & 0 & 1 & \frac{2}{3} & \frac{1}{3} \\ 0 & 0 & 2 & 1 & 1 \\ 0 & 0 & 3 & 2 & -2 \end{vmatrix}.$$

Next, we use the third row to get 0's in the rest of the third column:

$$-6\begin{vmatrix} 1 & 0 & 0 & 0 & 0 \\ 0 & 1 & 0 & 0 & 0 \\ 0 & 0 & 1 & \frac{2}{3} & \frac{1}{3} \\ 0 & 0 & 0 & -\frac{1}{3} & \frac{1}{3} \\ 0 & 0 & 0 & 0 & -3 \end{vmatrix},$$

and the third column to get 0's in the rest of the third row:

$$-6\begin{vmatrix} 1 & 0 & 0 & 0 & 0 \\ 0 & 1 & 0 & 0 & 0 \\ 0 & 0 & 1 & 0 & 0 \\ 0 & 0 & 0 & -\frac{1}{3} & \frac{1}{3} \\ 0 & 0 & 0 & 0 & -3 \end{vmatrix}.$$

Finally, we use the fourth column to produce 0's in the rest of the fourth row, getting

$$-6\begin{vmatrix} 1 & 0 & 0 & 0 & 0 \\ 0 & 1 & 0 & 0 & 0 \\ 0 & 0 & 1 & 0 & 0 \\ 0 & 0 & 0 & -\frac{1}{3} & 0 \\ 0 & 0 & 0 & 0 & -3 \end{vmatrix}.$$

Now, by Theorem A.12, the determinant is

$$-6(1)(1)(1)(-\tfrac{1}{3})(-3) = -6.$$

■

It should be pointed out that there are any number of ways to proceed in evaluating a determinant. It is usually best to use any or all of the theorems, adapting to the particular determinant at hand.

Exercises

Evaluate:

1. $\begin{vmatrix} 2 & -1 \\ 3 & 2 \end{vmatrix}$

2. $\begin{vmatrix} 5 & 4 \\ 3 & 2 \end{vmatrix}$

3. $\begin{vmatrix} \frac{1}{2} & \frac{4}{3} \\ \frac{2}{3} & \frac{2}{9} \end{vmatrix}$

4. $\begin{vmatrix} 1.63 & 2.85 \\ -2.06 & 1.72 \end{vmatrix}$

5. $\begin{vmatrix} 2 & -1 & 7 \\ 4 & 3 & -4 \\ 0 & 5 & 1 \end{vmatrix}$

6. $\begin{vmatrix} 7 & 2 & -1 \\ 6 & 4 & 3 \\ -3 & 0 & 7 \end{vmatrix}$

7. $\begin{vmatrix} 6 & 5 & 0.5 \\ 4 & 2 & 3 \\ -2 & -4 & 0 \end{vmatrix}$

8. $\begin{vmatrix} 1 & 2 & 4 \\ 1 & 3 & 9 \\ 1 & 4 & 16 \end{vmatrix}$

9. $\begin{vmatrix} 3 & 2 & 1 & -1 \\ 1 & 0 & -1 & 2 \\ 2 & -2 & 1 & 0 \\ 1 & 1 & -1 & 1 \end{vmatrix}$

10. $\begin{vmatrix} 1 & 0 & 0 & -1 \\ 0 & 1 & 1 & 1 \\ 1 & -1 & 0 & 0 \\ 1 & 1 & 1 & 1 \end{vmatrix}$

11. $\begin{vmatrix} 1 & 2 & 4 & 8 \\ 1 & 3 & 9 & 27 \\ 1 & 4 & 16 & 64 \\ 1 & 5 & 25 & 125 \end{vmatrix}$

12. $\begin{vmatrix} 3 & 1 & 1 & 1 \\ 1 & 3 & 1 & 1 \\ 1 & 1 & 3 & 1 \\ 1 & 1 & 1 & 3 \end{vmatrix}$

13. $\begin{vmatrix} 5 & 4 & 0 & 1 & 2 \\ 3 & 2 & 1 & 2 & 3 \\ 1 & 0 & -1 & -2 & -3 \\ -1 & -2 & 0 & 1 & 2 \\ -3 & -4 & 4 & 5 & 0 \end{vmatrix}$

14. $\begin{vmatrix} 3 & 1 & 1 & 1 & 1 \\ 1 & 3 & 1 & 1 & 1 \\ 1 & 1 & 3 & 1 & 1 \\ 1 & 1 & 1 & 3 & 1 \\ 1 & 1 & 1 & 1 & 3 \end{vmatrix}$

15. $\begin{vmatrix} r & s & s & \cdots & s \\ s & r & s & \cdots & s \\ s & s & r & \cdots & s \\ \cdot & \cdot & \cdot & & \cdot \\ \cdot & \cdot & \cdot & & \cdot \\ \cdot & \cdot & \cdot & & \cdot \\ s & s & s & \cdots & r \end{vmatrix}$

16. $\begin{vmatrix} 1 & 2 & 3 & \cdots & n \\ 1^2 & 2^2 & 3^2 & \cdots & n^2 \\ 1^3 & 2^3 & 3^3 & \cdots & n^3 \\ \cdot & \cdot & \cdot & & \cdot \\ \cdot & \cdot & \cdot & & \cdot \\ \cdot & \cdot & \cdot & & \cdot \\ 1^n & 2^n & 3^n & \cdots & n^n \end{vmatrix}$

A P P E N D I X B

Systems of Linear Equations

A **system of m linear equations in the n unknowns** $x_1, x_2, \ldots, x_n$ is a set of equations of the form

$$\begin{cases} a_{11}x_1 + a_{12}x_2 + \cdots + a_{1n}x_n = b_1 \\ a_{21}x_1 + a_{22}x_2 + \cdots + a_{2n}x_n = b_2 \\ \qquad \cdots \\ a_{m1}x_1 + a_{m2}x_2 + \cdots + a_{mn}x_n = b_m. \end{cases} \tag{1}$$

The numbers a_{ij} $(i = 1, \ldots, m; j = 1, \ldots, n)$ are called **coefficients**, and the numbers $b_1, \ldots, b_m$ are called **constant terms**. The $m \times n$ matrix

$$\begin{bmatrix} a_{11} & a_{12} & \cdots & a_{1n} \\ a_{21} & a_{22} & \cdots & a_{2n} \\ \cdot & \cdot & & \cdot \\ \cdot & \cdot & & \cdot \\ \cdot & \cdot & & \cdot \\ a_{m_1} & a_{m2} & \cdots & a_{mn} \end{bmatrix}$$

is called the **coefficient matrix** of the system, and the $m \times n + 1$ matrix

$$\begin{bmatrix} a_{11} & a_{12} & \cdots & a_{1n} & b_1 \\ a_{21} & a_{22} & \cdots & a_{2n} & b_2 \\ \cdot & \cdot & & \cdot & \cdot \\ \cdot & \cdot & & \cdot & \cdot \\ \cdot & \cdot & & \cdot & \cdot \\ a_{m_1} & a_{m2} & \cdots & a_{mn} & b_m \end{bmatrix}$$

is called the **augmented matrix** of the system.

A **solution** of the system (1) is a set of numbers $c_1, c_2, \ldots, c_n$ such that each equation in (1) is satisfied when $x_1 = c_1, x_2 = c_2, \ldots, x_n = c_n$.

Two systems of linear equations in the same number of unknowns are said to be **equivalent** in case every solution of either system is also a solution of the other system.

The most general method of solving a system of linear equations is to transform it into an equivalent system whose solutions are readily apparent. Certain operations on the equations of a system yield equivalent systems, including:

1. Rearranging the order of the equations
2. Multiplying both sides of an equation by the same nonzero constant
3. Adding to both sides of one equation the same multiple of the respective sides of another equation

It is not difficult to see that these operations on a system of equations are equivalent to the elementary row operations on the augmented matrix of the system (see Appendix A). Thus, rather than deal with the equations themselves, we deal with the augmented matrix of the system.

A matrix (a_{ij}) is said to be in **echelon form** whenever $a_{ij} = 0$ if $i > j$. That is, each element below the main diagonal is 0 in a matrix in echelon form. A system of equations is said to be in echelon form if its augmented matrix is in echelon form.

It is always possible to reduce a matrix to echelon form by using elementary row operations. Thus, it is always possible to transform a system of linear equations into an equivalent system in echelon form. The process of doing so is often called the **Gauss elimination process.** Once a system of linear equations is in echelon form, we can readily solve for the unknowns by working from the bottom up, as in the following example.

■ **Example B.1** Solve the system

$$\begin{cases} x + 3y + z + w = 2 \\ x - y + 4z - w = 6 \\ 2x + y + w = 4 \\ x - 2y - z + 2w = 8. \end{cases}$$

Solution The augmented matrix of this system is

$$\begin{bmatrix} 1 & 3 & 1 & 1 & 2 \\ 1 & -1 & 4 & -1 & 6 \\ 2 & 1 & 0 & 1 & 4 \\ 1 & -2 & -1 & 2 & 8 \end{bmatrix},$$

which we will reduce to echelon form using elementary row operations. By adding appropriate multiples of the elements of the first row to the respective elements of the other rows, we can get 0's in the rest of the first column:

$$\begin{bmatrix} 1 & 3 & 1 & 1 & 2 \\ 0 & -4 & 3 & -2 & 4 \\ 0 & -5 & -2 & -1 & 0 \\ 0 & -5 & -2 & 1 & 6 \end{bmatrix}.$$

Next, we multiply through the second row by -5, and through the third and fourth rows by 4, get 0's in the rest of the second column, and divide the second row through by 5:

$$\begin{bmatrix} 1 & 3 & 1 & 1 & 2 \\ 0 & 4 & -3 & 2 & -4 \\ 0 & 0 & -23 & 6 & -20 \\ 0 & 0 & -23 & 14 & 4 \end{bmatrix}.$$

Finally, we multiply through the third row by -1, get 0 as the last term in the third column, and divide through the last row by 8 to get

$$\begin{bmatrix} 1 & 3 & 1 & 1 & 2 \\ 0 & 4 & -3 & 2 & -4 \\ 0 & 0 & 23 & -6 & 20 \\ 0 & 0 & 0 & 1 & 3 \end{bmatrix}.$$

This is the augmented matrix of the system

$$\begin{cases} x + 3y + z + w = 2 \\ 4y - 3z + 2w = -4 \\ 23z - 6w = 20 \\ w = 3, \end{cases}$$

which is equivalent to the original system. Working from the bottom up, we find the solution:

$$\begin{aligned} w &= 3, \\ z &= \tfrac{1}{23}(6w + 20) = \tfrac{38}{23}, \\ y &= \tfrac{1}{4}(3z - 2w - 4) = -\tfrac{29}{23}, \\ x &= -3y - z - w + 2 = \tfrac{26}{23}. \end{aligned}$$

■

Some special cases may arise that require separate analysis. Suppose in reducing an augmented matrix, a row of the form

$$0 \quad 0 \quad \cdots \quad 0 \quad c$$

arises, with $c \neq 0$. Then the system of which that matrix is the augmented matrix contains the equation

$$0 = c,$$

which is an impossibility. The system therefore has no solution, and such a system is said to be **inconsistent**.

■ **Example B.2**

Solve the system

$$\begin{cases} x + 3y + z + w = 2 \\ x - y + 4z - w = 6 \\ 2x + 2y + 5z \phantom{{}-w} = 4 \\ x - 2y - z \phantom{{}-w} = 8. \end{cases}$$

Solution

We form the augmented matrix and reduce it as in Example B.1:

$$\begin{bmatrix} 1 & 3 & 1 & 1 & 2 \\ 1 & -1 & 4 & -1 & 6 \\ 2 & 2 & 5 & 0 & 4 \\ 1 & -2 & -1 & 0 & 8 \end{bmatrix} \rightarrow \begin{bmatrix} 1 & 3 & 1 & 1 & 2 \\ 0 & -4 & 3 & -2 & 4 \\ 0 & -4 & 3 & -2 & 0 \\ 0 & -5 & -2 & -1 & 6 \end{bmatrix}$$

$$\rightarrow \begin{bmatrix} 1 & 3 & 1 & 1 & 2 \\ 0 & -4 & 3 & -2 & 4 \\ 0 & 0 & 0 & 0 & -4 \\ 0 & -5 & -2 & -1 & 6 \end{bmatrix}.$$

At this point, we need proceed no further, since the third row gives the equation $0 = -4$. Thus, the system is inconsistent and has no solution. ■

Suppose that in reducing an augmented matrix, a row of 0's occurs:

$$0 \quad 0 \quad \cdots \quad 0 \quad 0.$$

Such a row corresponds to the equation

$$0 = 0,$$

which is perfectly true but of no help to us. What it tells us is that one of the original equations was unnecessary, and has in this way been dropped.

Example B.3 Solve the system

$$\begin{cases} 2x - y = 4 \\ x + 3y = -1 \\ 4x + 5y = 2. \end{cases}$$

Solution We reduce the augmented matrix:

$$\begin{bmatrix} 2 & -1 & 4 \\ 1 & 3 & -1 \\ 4 & 5 & 2 \end{bmatrix} \to \begin{bmatrix} 1 & 3 & -1 \\ 2 & -1 & 4 \\ 4 & 5 & 2 \end{bmatrix} \to \begin{bmatrix} 1 & 3 & -1 \\ 0 & -7 & 6 \\ 0 & -7 & 6 \end{bmatrix}$$

$$\to \begin{bmatrix} 1 & 3 & -1 \\ 0 & -7 & 6 \\ 0 & 0 & 0 \end{bmatrix}.$$

This last matrix is the augmented matrix of the system

$$\begin{cases} x + 3y = -1 \\ \quad -7y = 6 \end{cases}$$

(we dropped the equation $0 = 0$). The solution is easily found to be

$$y = -\tfrac{6}{7}, \qquad x = -3y - 1 = \tfrac{11}{7}.$$

Suppose that the last nonzero row of a matrix in echelon form has more than two nonzero terms, as in

$$0 \quad \cdots \quad 0 \quad a_k \quad a_{k+1} \quad \cdots \quad a_n \quad b.$$

Then the last equation in the corresponding system is

$$a_k x_k + a_{k+1} x_{k+1} + \cdots + a_n x_n = b.$$

We may solve for x_k in terms of $x_{k+1}, \ldots, x_n$, and hence solve for $x_1, x_2, \ldots, x_k$ in terms of the remaining unknowns. Since $x_{k+1}, \ldots, x_n$ may be arbitrary, the system has infinitely many solutions. Such a system is said to be **dependent.**

Example B.4 Solve the system

$$\begin{cases} x + 3y + z + w = 2 \\ x - y + 4z - w = 6 \\ 2x + y \qquad + w = 4 \\ x - 2y - z \qquad = 2. \end{cases}$$

Solution Working with the augmented matrix, we get

$$\begin{bmatrix} 1 & 3 & 1 & 1 & 2 \\ 1 & -1 & 4 & -1 & 6 \\ 2 & 1 & 0 & 1 & 4 \\ 1 & -2 & -1 & 0 & 2 \end{bmatrix} \rightarrow \begin{bmatrix} 1 & 3 & 1 & 1 & 2 \\ 0 & -4 & 3 & -2 & 4 \\ 0 & -5 & -2 & -1 & 0 \\ 0 & -5 & -2 & -1 & 0 \end{bmatrix}$$

$$\rightarrow \begin{bmatrix} 1 & 3 & 1 & 1 & 2 \\ 0 & 4 & -3 & 2 & -4 \\ 0 & 0 & -23 & 6 & -20 \\ 0 & 0 & -23 & 6 & -20 \end{bmatrix}$$

$$\rightarrow \begin{bmatrix} 1 & 3 & 1 & 1 & 2 \\ 0 & 4 & -3 & 2 & -4 \\ 0 & 0 & 23 & -6 & 20 \\ 0 & 0 & 0 & 0 & 0 \end{bmatrix}.$$

The corresponding system is

$$\begin{cases} x + 3y + z + w = 2 \\ \phantom{x + {}} 4y - 3z + 2w = -4 \\ \phantom{x + 4y - {}} 23z - 6w = 20. \end{cases}$$

Finally, we solve for x, y, and z in terms of w, getting

$$\begin{aligned} z &= \tfrac{1}{23}(6w + 20), \\ y &= \tfrac{1}{4}(3z - 2w - 4) = -\tfrac{1}{23}(7w + 8), \\ x &= -3y - z - w + 2 = \tfrac{1}{23}(50 - 8w). \end{aligned}$$

Now any value for w will yield a solution for the system. ■

The Gauss elimination process, or something like it, is the only universally applicable method for solving a system of linear equations. But for some special kinds of systems, special methods are available, as described below.

A **square** system of linear equations is a system with the same number of equations as unknowns. Thus, the coefficient matrix of a square system is a square matrix.

Let $D = \det(a_{ij})$ be the determinant of the coefficient matrix of a square system. The value of D is the key to the solution of the system. It can be shown that if $D \neq 0$, then the system has one and only one solution, but if $D = 0$, then the system is either inconsistent or dependent.

If $D \neq 0$, we can find the solution of the system in terms of determinants. For each j, let D_j be the determinant obtained from D by replacing the jth column of D by the column of constant terms. Then for each j, the solution of the system is given by $x_j = D_j/D$. This result is known as **Cramer's rule** (a proof can be found in most college algebra textbooks).

Finding a complete solution of a square system with n unknowns using Cramer's rule involves evaluating $n + 1$ determinants of order n. Thus, Cramer's rule is rather inefficient computationally. It does have the advantage of allowing one to solve for a particular unknown without having to solve for any of the others.

Example B.5 Solve for y in the system of Example B.1.

Solution The unknown y corresponds to the unknown x_2 in our more general notation, so we find D and D_2. We have

$$D = \begin{vmatrix} 1 & 3 & 1 & 1 \\ 1 & -1 & 4 & -1 \\ 2 & 1 & 0 & 1 \\ 1 & -2 & -1 & 2 \end{vmatrix} = 46$$

and

$$D_2 = \begin{vmatrix} 1 & 2 & 1 & 1 \\ 1 & 6 & 4 & -1 \\ 2 & 4 & 0 & 1 \\ 1 & 8 & -1 & 2 \end{vmatrix} = -58.$$

Hence,

$$y = \frac{D_2}{D} = -\frac{58}{46} = -\frac{29}{23},$$

as before. ■

Note that for the inconsistent system of Example B.2, $D = 0$, and for the dependent system of Example B.4, $D = 0$.

A **homogeneous** system of linear equations is a system in which all the constant terms are 0. A homogeneous system always has the solution

$$x_1 = x_2 = \cdots = x_n = 0,$$

which is called the **trivial** solution.

A square homogeneous system has nontrivial solutions if and only if the determinant of the coefficient matrix is 0.

Example B.6 Solve the system

$$\begin{cases} x + 2y - z = 0 \\ -x + y + 2z = 0 \\ -2x - y + 3z = 0. \end{cases}$$

Solution The determinant of coefficients is

$$D = \begin{vmatrix} 1 & 2 & -1 \\ -1 & 1 & 2 \\ -2 & -1 & 3 \end{vmatrix} = 0,$$

so nontrivial solutions exist. Working on the augmented matrix, we get

$$\begin{bmatrix} 1 & 2 & -1 & 0 \\ -1 & 1 & 2 & 0 \\ -2 & -1 & 3 & 0 \end{bmatrix} \to \begin{bmatrix} 1 & 2 & -1 & 0 \\ 0 & 3 & 1 & 0 \\ 0 & 3 & 1 & 0 \end{bmatrix}$$

$$\to \begin{bmatrix} 1 & 2 & -1 & 0 \\ 0 & 3 & 1 & 0 \\ 0 & 0 & 0 & 0 \end{bmatrix}.$$

Therefore, the original system is equivalent to the system

$$\begin{cases} x + 2y - z = 0 \\ \quad\;\; 3y + z = 0, \end{cases}$$

whose solution is

$$y = -\tfrac{1}{3}z, \qquad x = -2y + z = \tfrac{5}{3}z, \qquad z \text{ arbitrary.}$$ ■

Exercises

Solve:

1. $\begin{cases} x - 4y = 7 \\ 2x + y = 5 \end{cases}$

2. $\begin{cases} x - y + z = 7 \\ x + y + z = 9 \\ x + y - z = 6 \end{cases}$

3. $\begin{cases} x - 3y + 2z = 0 \\ 2x + y + z = 4 \\ x - 3y - z = 1 \end{cases}$

4. $\begin{cases} x + y - z - w = 1 \\ x + 2y + 3z - 6w = -1 \\ 2x + 3y + 2z - 7w = 0 \\ 2x - 3y - 2z + w = 4 \end{cases}$

5. $\begin{cases} x + 4y = 2 \\ 2x - y = -1 \\ -x + 5y = 3 \end{cases}$

6. $\begin{cases} x + 2y + 3z = 0 \\ x + 3y + 6z = 0 \\ y + 3z = 0 \end{cases}$

7. $\begin{cases} x + y + z + w = 5 \\ 2x + y + z + 2w = 8 \\ x + 3y + 3z + w = 9 \end{cases}$

8. $\begin{cases} x - y - 4z = 0 \\ 4x + 5y - z = 3 \\ 3x - 2y + z = -4 \\ x + y + 2z = 5 \end{cases}$

Mathematical Induction

The principle of mathematical induction is a property of the system of positive integers, $\mathbb{N} = \{1, 2, 3, \ldots\}$. It is fundamental to the logical development of the system of integers and is especially useful in proofs. It is based on an even more fundamental principle, called **well-ordering**.

Let S be a set of positive integers. By a **least element** of S we mean an integer $n \in S$ such that for any $x \in S$, $n \leq x$. That is, the least element of S is the element of S that comes first in the natural order of $\mathbb{N}$.

We state the well-ordering principle as an axiom.

AXIOM **The Well-Ordering Principle** Every nonempty subset of $\mathbb{N}$ has a least element.

We can now state and prove the principle of mathematical induction.

THEOREM C.1 **The Principle of Mathematical Induction** If S is a subset of $\mathbb{N}$ such that

1. $1 \in S$, and
2. $n + 1 \in S$ whenever $n \in S$,

then $S = \mathbb{N}$.

Proof Suppose, on the contrary, that $S \neq \mathbb{N}$. Then the complement of S is nonempty, so by the well-ordering principle it has a least element m. Thus, m is the least positive integer such that $m \notin S$. By item 1 in the theorem, $m \neq 1$. Hence, $m - 1 \in \mathbb{N}$, and since $m - 1 < m$, it must be that $m - 1 \in S$. But then $m = (m - 1) + 1 \in S$, by item 2. This contradiction forces us to conclude that $S = \mathbb{N}$.

A proof that employs the principle of mathematical induction is called a **proof by induction** or an **inductive proof**. Normally, an inductive proof is used to establish the validity of a sequence of statements $P_1, P_2, \ldots, P_n, \ldots$, in which each statement P_n involves the positive integer n.

Such an inductive proof normally proceeds by the following steps:

Step 1. Let S be the set of all $k \in \mathbb{N}$ such that P_k is true.

Step 2. Verify that $1 \in S$ by showing that P_1 is true. (This verifies item 1 of Theorem C.1.)

Step 3. Prove that $n+1 \in S$ whenever $n \in S$ by showing that if P_n is true, then P_{n+1} is also true. (This verifies item 2 of Theorem C.1, and is called the **inductive step**. Often, it is the only step of the inductive proof that is explicitly given.)

Step 4. Conclude that P_n is true for every $n \in \mathbb{N}$ by observing that $S = \mathbb{N}$, by Theorem C.1.

Consider some examples of inductive proofs.

Example C.1 Prove that

$$\sum_{i=1}^{n} i = \frac{n(n+1)}{2}.$$

Solution We will construct a proof by induction, using the above four steps. Let P_n be the statement

$$\sum_{i=1}^{n} i = 1 + 2 + \cdots + n = \frac{n(n+1)}{2}.$$

Let

$$\begin{aligned} S &= \{k \in \mathbb{N} \mid P_k \text{ is true}\} \\ &= \left\{k \in \mathbb{N} \mid 1 + 2 + \cdots + k = \frac{k(k+1)}{2}\right\}. \end{aligned}$$

This completes Step 1. Now, P_1 is the statement

$$\sum_{i=1}^{1} i = 1 = \frac{1(1+1)}{2},$$

which is clearly true. Thus, $1 \in S$. This completes Step 2. To carry out the inductive step, suppose $n \in S$. Then P_n is true, or

$$1 + 2 + \cdots + n = \frac{n(n+1)}{2}.$$

We wish to show that P_{n+1} is true, or

$$1 + 2 + \cdots + (n+1) = \frac{(n+1)(n+1+1)}{2}.$$

This is done as follows:

$$\begin{aligned} 1+2+\cdots+(n+1) &= 1+2+\cdots+n+(n+1) \\ &= \frac{n(n+1)}{2}+(n+1) \qquad \text{(By } P_n) \\ &= \frac{n(n+1)}{2}+\frac{2(n+1)}{2} \\ &= \frac{(n+1)(n+2)}{2}. \end{aligned}$$

Thus, P_{n+1} is true, and $n+1 \in S$. This completes Step 3. Thus, $S = \mathbb{N}$ by Theorem C.1, so P_n is true for all $n \in \mathbb{N}$. This completes the proof. ■

■ **Example C.2** Prove that for any $n \in \mathbb{N}$, $8^n - 3^n$ is divisible by 5.

Solution Let P_n be the statement

$$8^n - 3^n \text{ is divisible by } 5,$$

and let $S = \{k \in \mathbb{N} | P_k \text{ is true}\}$. Then P_1 is the statement

$$8 - 3 \text{ is divisible by } 5,$$

which is true, so $1 \in S$. If $n \in S$, then P_n is true, so $8^n - 3^n$ is divisible by 5. To show that P_{n+1} is true, we must show that $8^{n+1} - 3^{n+1}$ is divisible by 5. We can write

$$\begin{aligned} 8^{n+1} - 3^{n+1} &= 8^{n+1} - 8(3^n) + 8(3^n) - 3^{n+1} \\ &= 8(8^n - 3^n) + (8-3)3^n, \end{aligned}$$

each term of which is divisible by 5. Thus, P_{n+1} is true and $n+1 \in S$. The proof is therefore complete. ■

Our next and final example uses the well-ordering principle directly, and while it is a bit tongue-in-cheek, it illustrates the power of the principle.

■ **Example C.3** Prove that every positive integer has an interesting property.

Solution Suppose, on the contrary, that some positive integer has no interesting property. Then the set of such integers is nonempty, so it has a least element; call it n. Then n is the smallest positive integer that is utterly devoid of interesting properties, but that is interesting! ■

For some exercises, see Section 5.1.

APPENDIX D

Tables

TABLE I TRIGONOMETRIC FUNCTIONS (IN RADIANS FROM 0 TO $\pi/2 \approx 1.57$)

x	$\sin x$	$\cos x$	$\tan x$	$\cot x$	$\sec x$	$\csc x$	x	$\sin x$	$\cos x$	$\tan x$	$\cot x$	$\sec x$	$\csc x$
0.00	0.0000	1.0000	0.0000	—	1.0000	—	0.40	0.3894	0.9211	0.4228	2.3652	1.0857	2.5679
0.01	0.0100	1.0000	0.0100	100.00	1.0001	100.00	0.41	0.3986	0.9171	0.4346	2.3008	1.0904	2.5087
0.02	0.0200	0.9998	0.0200	49.993	1.0002	50.003	0.42	0.4078	0.9131	0.4466	2.2393	1.0952	2.4524
0.03	0.0300	0.9996	0.0300	33.323	1.0005	33.338	0.43	0.4169	0.9090	0.4586	2.1804	1.1002	2.3988
0.04	0.0400	0.9992	0.0400	24.987	1.0008	25.007	0.44	0.4259	0.9048	0.4708	2.1241	1.1053	2.3478
0.05	0.0500	0.9988	0.0500	19.983	1.0013	20.008	0.45	0.4350	0.9004	0.4831	2.0702	1.1106	2.2990
0.06	0.0600	0.9982	0.0601	16.647	1.0018	16.677	0.46	0.4439	0.8961	0.4954	2.0184	1.1160	2.2525
0.07	0.0699	0.9976	0.0701	14.262	1.0025	14.297	0.47	0.4529	0.8916	0.5080	1.9686	1.1216	2.2081
0.08	0.0799	0.9968	0.0802	12.473	1.0032	12.513	0.48	0.4618	0.8870	0.5206	1.9208	1.1274	2.1655
0.09	0.0899	0.9960	0.0902	11.081	1.0041	11.126	0.49	0.4706	0.8823	0.5334	1.8748	1.1334	2.1248
0.10	0.0998	0.9950	0.1003	9.9666	1.0050	10.017	0.50	0.4794	0.8776	0.5463	1.8305	1.1395	2.0858
0.11	0.1098	0.9940	0.1104	9.0542	1.0061	9.1093	0.51	0.4882	0.8727	0.5594	1.7878	1.1458	2.0484
0.12	0.1197	0.9928	0.1206	8.2933	1.0072	8.3534	0.52	0.4969	0.8678	0.5726	1.7465	1.1523	2.0126
0.13	0.1296	0.9916	0.1307	7.6489	1.0085	7.7140	0.53	0.5055	0.8628	0.5859	1.7067	1.1590	1.9781
0.14	0.1395	0.9902	0.1409	7.0961	1.0099	7.1662	0.54	0.5141	0.8577	0.5994	1.6683	1.1659	1.9450
0.15	0.1494	0.9888	0.1511	6.6166	1.0114	6.6917	0.55	0.5227	0.8525	0.6131	1.6310	1.1730	1.9132
0.16	0.1593	0.9872	0.1614	6.1966	1.0129	6.2767	0.56	0.5312	0.8473	0.6269	1.5950	1.1803	1.8826
0.17	0.1692	0.9856	0.1717	5.8256	1.0146	5.9108	0.57	0.5396	0.8419	0.6410	1.5601	1.1878	1.8531
0.18	0.1790	0.9838	0.1820	5.4954	1.0164	5.5857	0.58	0.5480	0.8365	0.6552	1.5263	1.1955	1.8247
0.19	0.1889	0.9820	0.1923	5.1997	1.0183	5.2950	0.59	0.5564	0.8309	0.6696	1.4935	1.2035	1.7974
0.20	0.1987	0.9801	0.2027	4.9332	1.0203	5.0335	0.60	0.5646	0.8253	0.6841	1.4617	1.2116	1.7710
0.21	0.2085	0.9780	0.2131	4.6917	1.0225	4.7971	0.61	0.5729	0.8196	0.6989	1.4308	1.2200	1.7456
0.22	0.2182	0.9759	0.2236	4.4719	1.0247	4.5823	0.62	0.5810	0.8139	0.7139	1.4007	1.2287	1.7211
0.23	0.2280	0.9737	0.2341	4.2709	1.0270	4.3864	0.63	0.5891	0.8080	0.7291	1.3715	1.2376	1.6974
0.24	0.2377	0.9713	0.2447	4.0864	1.0295	4.2069	0.64	0.5972	0.8021	0.7445	1.3431	1.2467	1.6745
0.25	0.2474	0.9689	0.2553	3.9163	1.0321	4.0420	0.65	0.6052	0.7961	0.7602	1.3154	1.2561	1.6524
0.26	0.2571	0.9664	0.2660	3.7591	1.0348	3.8898	0.66	0.6131	0.7900	0.7761	1.2885	1.2658	1.6310
0.27	0.2667	0.9638	0.2768	3.6133	1.0376	3.7491	0.67	0.6210	0.7838	0.7923	1.2622	1.2758	1.6103
0.28	0.2764	0.9611	0.2876	3.4776	1.0405	3.6185	0.68	0.6288	0.7776	0.8087	1.2366	1.2861	1.5903
0.29	0.2860	0.9582	0.2984	3.3511	1.0436	3.4971	0.69	0.6365	0.7712	0.8253	1.2116	1.2966	1.5710
0.30	0.2955	0.9553	0.3093	3.2327	1.0468	3.3839	0.70	0.6442	0.7648	0.8423	1.1872	1.3075	1.5523
0.31	0.3051	0.9523	0.3203	3.1218	1.0501	3.2781	0.71	0.6518	0.7584	0.8595	1.1634	1.3186	1.5341
0.32	0.3146	0.9492	0.3314	3.0176	1.0535	3.1790	0.72	0.6594	0.7518	0.8771	1.1402	1.3301	1.5166
0.33	0.3240	0.9460	0.3425	2.9195	1.0570	3.0860	0.73	0.6669	0.7452	0.8949	1.1174	1.3420	1.4995
0.34	0.3335	0.9428	0.3537	2.8270	1.0607	2.9986	0.74	0.6743	0.7385	0.9131	1.0952	1.3542	1.4830
0.35	0.3429	0.9394	0.3650	2.7395	1.0645	2.9163	0.75	0.6816	0.7317	0.9316	1.0734	1.3667	1.4671
0.36	0.3523	0.9359	0.3764	2.6567	1.0685	2.8387	0.76	0.6889	0.7248	0.9505	1.0521	1.3796	1.4515
0.37	0.3616	0.9323	0.3879	2.5782	1.0726	2.7654	0.77	0.6961	0.7179	0.9697	1.0313	1.3929	1.4365
0.38	0.3709	0.9287	0.3994	2.5037	1.0768	2.6960	0.78	0.7033	0.7109	0.9893	1.0109	1.4066	1.4219
0.39	0.3802	0.9249	0.4111	2.4328	1.0812	2.6303	0.79	0.7104	0.7038	1.0092	0.9908	1.4208	1.4078

TABLE I CONTINUED

x	$\sin x$	$\cos x$	$\tan x$	$\cot x$	$\sec x$	$\csc x$
0.80	0.7174	0.6967	1.0296	0.9712	1.4353	1.3940
0.81	0.7243	0.6895	1.0505	0.9520	1.4503	1.3807
0.82	0.7311	0.6822	1.0717	0.9331	1.4658	1.3677
0.83	0.7379	0.6749	1.0934	0.9146	1.4818	1.3551
0.84	0.7446	0.6675	1.1156	0.8964	1.4982	1.3429
0.85	0.7513	0.6600	1.1383	0.8785	1.5152	1.3311
0.86	0.7578	0.6524	1.1616	0.8609	1.5327	1.3195
0.87	0.7643	0.6448	1.1853	0.8437	1.5508	1.3083
0.88	0.7707	0.6372	1.2097	0.8267	1.5695	1.2975
0.89	0.7771	0.6294	1.2346	0.8100	1.5888	1.2869
0.90	0.7833	0.6216	1.2602	0.7936	1.6087	1.2766
0.91	0.7895	0.6137	1.2864	0.7774	1.6293	1.2666
0.92	0.7956	0.6058	1.3133	0.7615	1.6507	1.2569
0.93	0.8016	0.5978	1.3409	0.7458	1.6727	1.2475
0.94	0.8076	0.5898	1.3692	0.7303	1.6955	1.2383
0.95	0.8134	0.5817	1.3984	0.7151	1.7191	1.2294
0.96	0.8192	0.5735	1.4284	0.7001	1.7436	1.2207
0.97	0.8249	0.5653	1.4592	0.6853	1.7690	1.2123
0.98	0.8305	0.5570	1.4910	0.6707	1.7953	1.2041
0.99	0.8360	0.5487	1.5237	0.6563	1.8225	1.1961
1.00	0.8415	0.5403	1.5574	0.6421	1.8508	1.1884
1.01	0.8468	0.5319	1.5922	0.6281	1.8802	1.1809
1.02	0.8521	0.5234	1.6281	0.6142	1.9107	1.1736
1.03	0.8573	0.5148	1.6652	0.6005	1.9424	1.1665
1.04	0.8624	0.5062	1.7036	0.5870	1.9754	1.1595
1.05	0.8674	0.4976	1.7433	0.5736	2.0098	1.1528
1.06	0.8724	0.4889	1.7844	0.5604	2.0455	1.1463
1.07	0.8772	0.4801	1.8270	0.5473	2.0828	1.1400
1.08	0.8820	0.4713	1.8712	0.5344	2.1217	1.1338
1.09	0.8866	0.4625	1.9171	0.5216	2.1622	1.1279
1.10	0.8912	0.4536	1.9648	0.5090	2.2046	1.1221
1.11	0.8957	0.4447	2.0143	0.4964	2.2489	1.1164
1.12	0.9001	0.4357	2.0660	0.4840	2.2952	1.1110
1.13	0.9044	0.4267	2.1198	0.4718	2.3438	1.1057
1.14	0.9086	0.4176	2.1759	0.4596	2.3947	1.1006
1.15	0.9128	0.4085	2.2345	0.4475	2.4481	1.0956
1.16	0.9168	0.3993	2.2958	0.4356	2.5041	1.0907
1.17	0.9208	0.3902	2.3600	0.4237	2.5631	1.0861
1.18	0.9246	0.3809	2.4273	0.4120	2.6252	1.0815
1.19	0.9284	0.3717	2.4979	0.4003	2.6906	1.0772
1.20	0.9320	0.3624	2.5722	0.3888	2.7597	1.0729
1.21	0.9356	0.3530	2.6503	0.3773	2.8327	1.0688
1.22	0.9391	0.3436	2.7328	0.3659	2.9100	1.0649
1.23	0.9425	0.3342	2.8198	0.3546	2.9919	1.0610
1.24	0.9458	0.3248	2.9119	0.3434	3.0789	1.0573
1.25	0.9490	0.3153	3.0096	0.3323	3.1714	1.0538
1.26	0.9521	0.3058	3.1133	0.3212	3.2699	1.0503
1.27	0.9551	0.2963	3.2236	0.3102	3.3752	1.0470
1.28	0.9580	0.2867	3.3413	0.2993	3.4878	1.0438
1.29	0.9608	0.2771	3.4672	0.2884	3.6085	1.0408
1.30	0.9636	0.2675	3.6021	0.2776	3.7383	1.0378
1.31	0.9662	0.2579	3.7471	0.2669	3.8782	1.0350
1.32	0.9687	0.2482	3.9033	0.2562	4.0294	1.0323
1.33	0.9711	0.2385	4.0723	0.2456	4.1933	1.0297
1.34	0.9735	0.2288	4.2556	0.2350	4.3715	1.0272
1.35	0.9757	0.2190	4.4552	0.2245	4.5661	1.0249
1.36	0.9779	0.2092	4.6734	0.2140	4.7792	1.0226
1.37	0.9799	0.1994	4.9131	0.2035	5.0138	1.0205
1.38	0.9819	0.1896	5.1774	0.1931	5.2731	1.0185
1.39	0.9837	0.1798	5.4707	0.1828	5.5613	1.0166
1.40	0.9854	0.1700	5.7979	0.1725	5.8835	1.0148
1.41	0.9871	0.1601	6.1654	0.1622	6.2459	1.0131
1.42	0.9887	0.1502	6.5811	0.1519	6.6567	1.0115
1.43	0.9901	0.1403	7.0555	0.1417	7.1260	1.0100
1.44	0.9915	0.1304	7.6018	0.1315	7.6673	1.0086
1.45	0.9927	0.1205	8.2381	0.1214	8.2986	1.0073
1.46	0.9939	0.1106	8.9886	0.1113	9.0441	1.0062
1.47	0.9949	0.1006	9.8874	0.1011	9.9378	1.0051
1.48	0.9959	0.0907	10.983	0.0910	11.029	1.0041
1.49	0.9967	0.0807	12.350	0.0810	12.390	1.0033
1.50	0.9975	0.0707	14.101	0.0709	14.137	1.0025
1.51	0.9982	0.0608	16.428	0.0609	16.458	1.0019
1.52	0.9987	0.0508	19.670	0.0508	19.695	1.0013
1.53	0.9992	0.0408	24.498	0.0408	24.519	1.0008
1.54	0.9995	0.0308	32.461	0.0308	32.477	1.0005
1.55	0.9998	0.0208	48.078	0.0208	48.089	1.0002
1.56	0.9999	0.0108	92.620	0.0108	92.626	1.0001
1.57	1.0000	0.0008	1255.77	0.0008	1255.77	1.0000

TABLE II EXPONENTIAL AND LOGARITHMIC FUNCTIONS

n	e^n	e^{-n}	e^{10n}	e^{-10n}	$e^{n/10}$	$e^{-n/10}$	$\ln n$	$\ln 10n$
0.00	1.0000	1.0000	1.0000	1.0000	1.0000	1.0000	—	—
0.01	1.0101	0.9900	1.1052	0.9048	1.0010	0.9990	−4.6052	−2.3026
0.02	1.0202	0.9802	1.2214	0.8187	1.0020	0.9980	−3.9120	−1.6094
0.03	1.0305	0.9704	1.3499	0.7408	1.0030	0.9970	−3.5066	−1.2040
0.04	1.0408	0.9608	1.4918	0.6703	1.0040	0.9960	−3.2189	−0.9163
0.05	1.0513	0.9512	1.6487	0.6065	1.0050	0.9950	−2.9957	−0.6931
0.06	1.0618	0.9418	1.8221	0.5488	1.0060	0.9940	−2.8134	−0.5108
0.07	1.0725	0.9324	2.0138	0.4966	1.0070	0.9930	−2.6593	−0.3567
0.08	1.0833	0.9231	2.2255	0.4493	1.0080	0.9920	−2.5257	−0.2231
0.09	1.0942	0.9139	2.4596	0.4066	1.0090	0.9910	−2.4079	−0.1054
0.10	1.1052	0.9048	2.7183	0.3679	1.0101	0.9900	−2.3026	0.0000
0.11	1.1163	0.8958	3.0042	0.3329	1.0111	0.9891	−2.2073	0.0953
0.12	1.1275	0.8869	3.3201	0.3012	1.0121	0.9881	−2.1203	0.1823
0.13	1.1388	0.8781	3.6693	0.2725	1.0131	0.9871	−2.0402	0.2624
0.14	1.1503	0.8694	4.0552	0.2466	1.0141	0.9861	−1.9661	0.3365
0.15	1.1618	0.8607	4.4817	0.2231	1.0151	0.9851	−1.8971	0.4055
0.16	1.1735	0.8521	4.9530	0.2019	1.0161	0.9841	−1.8326	0.4700
0.17	1.1853	0.8437	5.4739	0.1827	1.0171	0.9831	−1.7720	0.5306
0.18	1.1972	0.8353	6.0496	0.1653	1.0182	0.9822	−1.7148	0.5878
0.19	1.2092	0.8270	6.6859	0.1496	1.0192	0.9812	−1.6607	0.6419
0.20	1.2214	0.8187	7.3891	0.1353	1.0202	0.9802	−1.6094	0.6931
0.21	1.2337	0.8106	8.1662	0.1225	1.0212	0.9792	−1.5606	0.7419
0.22	1.2461	0.8025	9.0250	0.1108	1.0222	0.9782	−1.5141	0.7885
0.23	1.2586	0.7945	9.9742	0.1003	1.0233	0.9773	−1.4697	0.8329
0.24	1.2712	0.7866	11.0232	0.0907	1.0243	0.9763	−1.4271	0.8755
0.25	1.2840	0.7788	12.1825	0.0821	1.0253	0.9753	−1.3863	0.9163
0.26	1.2969	0.7711	13.4637	0.0743	1.0263	0.9743	−1.3471	0.9555
0.27	1.3100	0.7634	14.8797	0.0672	1.0274	0.9734	−1.3093	0.9933
0.28	1.3231	0.7558	16.4446	0.0608	1.0284	0.9724	−1.2730	1.0296
0.29	1.3364	0.7483	18.1741	0.0550	1.0294	0.9714	−1.2379	1.0647
0.30	1.3499	0.7408	20.0855	0.0498	1.0305	0.9704	−1.2040	1.0986
0.31	1.3634	0.7334	22.1980	0.0450	1.0315	0.9695	−1.1712	1.1314
0.32	1.3771	0.7261	24.5325	0.0408	1.0325	0.9685	−1.1394	1.1632
0.33	1.3910	0.7189	27.1126	0.0369	1.0336	0.9675	−1.1087	1.1939
0.34	1.4049	0.7118	29.9641	0.0334	1.0346	0.9666	−1.0788	1.2238
0.35	1.4191	0.7047	33.1155	0.0302	1.0356	0.9656	−1.0498	1.2528
0.36	1.4333	0.6977	36.5982	0.0273	1.0367	0.9646	−1.0217	1.2809
0.37	1.4477	0.6907	40.4473	0.0247	1.0377	0.9637	−0.9943	1.3083
0.38	1.4623	0.6839	44.7012	0.0224	1.0387	0.9627	−0.9676	1.3350
0.39	1.4770	0.6771	49.4024	0.0202	1.0398	0.9618	−0.9416	1.3610
0.40	1.4918	0.6703	54.5982	0.0183	1.0408	0.9608	−0.9163	1.3863
0.41	1.5068	0.6637	60.3403	0.0166	1.0419	0.9598	−0.8916	1.4110
0.42	1.5220	0.6570	66.6863	0.0150	1.0429	0.9589	−0.8675	1.4351
0.43	1.5373	0.6505	73.6998	0.0136	1.0439	0.9579	−0.8440	1.4586
0.44	1.5527	0.6440	81.4509	0.0123	1.0450	0.9570	−0.8210	1.4816
0.45	1.5683	0.6376	90.0171	0.0111	1.0460	0.9560	−0.7985	1.5041
0.46	1.5841	0.6313	99.4843	0.0101	1.0471	0.9550	−0.7765	1.5261
0.47	1.6000	0.6250	109.9472	0.0091	1.0481	0.9541	−0.7550	1.5476
0.48	1.6161	0.6188	121.5104	0.0082	1.0492	0.9531	−0.7340	1.5686
0.49	1.6323	0.6126	134.2898	0.0074	1.0502	0.9522	−0.7133	1.5892

TABLE II CONTINUED

n	e^n	e^{-n}	e^{10n}	e^{-10n}	$e^{n/10}$	$e^{-n/10}$	$\ln n$	$\ln 10n$
0.50	1.6487	0.6065	148.4132	0.0067	1.0513	0.9512	−0.6931	1.6094
0.51	1.6653	0.6005	164.0219	0.0061	1.0523	0.9503	−0.6733	1.6292
0.52	1.6820	0.5945	181.2722	0.0055	1.0534	0.9493	−0.6539	1.6487
0.53	1.6989	0.5886	200.3368	0.0050	1.0544	0.9484	−0.6349	1.6677
0.54	1.7160	0.5827	221.4064	0.0045	1.0555	0.9474	−0.6162	1.6864
0.55	1.7333	0.5769	244.6919	0.0041	1.0565	0.9465	−0.5978	1.7047
0.56	1.7507	0.5712	270.4264	0.0037	1.0576	0.9455	−0.5798	1.7228
0.57	1.7683	0.5655	298.8674	0.0033	1.0587	0.9446	−0.5621	1.7405
0.58	1.7860	0.5599	330.2996	0.0030	1.0597	0.9436	−0.5447	1.7579
0.59	1.8040	0.5543	365.0375	0.0027	1.0608	0.9427	−0.5276	1.7750
0.60	1.8221	0.5488	403.4288	0.0025	1.0618	0.9418	−0.5108	1.7918
0.61	1.8404	0.5434	445.8578	0.0022	1.0629	0.9408	−0.4943	1.8083
0.62	1.8589	0.5379	492.7490	0.0020	1.0640	0.9399	−0.4780	1.8245
0.63	1.8776	0.5326	544.5719	0.0018	1.0650	0.9389	−0.4620	1.8405
0.64	1.8965	0.5273	601.8450	0.0017	1.0661	0.9380	−0.4463	1.8563
0.65	1.9155	0.5220	665.1416	0.0015	1.0672	0.9371	−0.4308	1.8718
0.66	1.9348	0.5169	735.0952	0.0014	1.0682	0.9361	−0.4155	1.8871
0.67	1.9542	0.5117	812.4058	0.0012	1.0693	0.9352	−0.4005	1.9021
0.68	1.9739	0.5066	897.8473	0.0011	1.0704	0.9343	−0.3857	1.9169
0.69	1.9937	0.5016	992.2747	0.0010	1.0714	0.9333	−0.3711	1.9315
0.70	2.0138	0.4966	1096.633	0.0009	1.0725	0.9324	−0.3567	1.9459
0.71	2.0340	0.4916	1211.967	0.0008	1.0736	0.9315	−0.3425	1.9601
0.72	2.0544	0.4868	1339.431	0.0007	1.0747	0.9305	−0.3285	1.9741
0.73	2.0751	0.4819	1480.300	0.0007	1.0757	0.9296	−0.3147	1.9879
0.74	2.0959	0.4771	1635.984	0.0006	1.0768	0.9287	−0.3011	2.0015
0.75	2.1170	0.4724	1808.042	0.0006	1.0779	0.9277	−0.2877	2.0149
0.76	2.1383	0.4677	1998.20	0.0005	1.0790	0.9268	−0.2744	2.0281
0.77	2.1598	0.4630	2208.35	0.0005	1.0800	0.9259	−0.2614	2.0412
0.78	2.1815	0.4584	2440.60	0.0004	1.0811	0.9250	−0.2485	2.0541
0.79	2.2034	0.4538	2697.28	0.0004	1.0822	0.9240	−0.2357	2.0669
0.80	2.2255	0.4493	2980.96	0.0003	1.0833	0.9231	−0.2231	2.0794
0.81	2.2479	0.4449	3294.47	0.0003	1.0844	0.9222	−0.2107	2.0919
0.82	2.2705	0.4404	3640.95	0.0003	1.0855	0.9213	−0.1985	2.1041
0.83	2.2933	0.4360	4023.87	0.0002	1.0865	0.9204	−0.1863	2.1163
0.84	2.3164	0.4317	4447.07	0.0002	1.0876	0.9194	−0.1744	2.1282
0.85	2.3396	0.4274	4914.77	0.0002	1.0887	0.9185	−0.1625	2.1401
0.86	2.3632	0.4232	5431.66	0.0002	1.0898	0.9176	−0.1508	2.1518
0.87	2.3869	0.4190	6002.91	0.0002	1.0909	0.9167	−0.1393	2.1633
0.88	2.4109	0.4148	6634.24	0.0002	1.0920	0.9158	−0.1278	2.1748
0.89	2.4351	0.4107	7331.97	0.0001	1.0931	0.9148	−0.1165	2.1861
0.90	2.4596	0.4066	8103.08	0.0001	1.0942	0.9139	−0.1054	2.1972
0.91	2.4843	0.4025	8955.29	0.0001	1.0953	0.9130	−0.0943	2.2083
0.92	2.5093	0.3985	9897.13	0.0001	1.0964	0.9121	−0.0834	2.2192
0.93	2.5345	0.3946	10938.0	0.0001	1.0975	0.9112	−0.0726	2.2300
0.94	2.5600	0.3906	12088.4	0.0001	1.0986	0.9103	−0.0619	2.2407
0.95	2.5857	0.3867	13359.7	0.0001	1.0997	0.9094	−0.0513	2.2513
0.96	2.6117	0.3829	14764.8	0.0001	1.1008	0.9085	−0.0408	2.2618
0.97	2.6379	0.3791	16317.6	0.0001	1.1019	0.9076	−0.0305	2.2721
0.98	2.6645	0.3753	18033.7	0.0001	1.1030	0.9066	−0.0202	2.2824
0.99	2.6912	0.3716	19930.4	0.0001	1.1041	0.9057	−0.0101	2.2925
1.00	2.7183	0.3679	22026.5	0.0000	1.1052	0.9048	0.0000	2.3026

TABLE III TABLE OF MANTISSAS (COMMON LOGARITHMS FROM 1.00 TO 9.99)

	0	1	2	3	4	5	6	7	8	9
1.0	0.0000	0.0043	0.0086	0.0128	0.0170	0.0212	0.0253	0.0294	0.0334	0.0374
1.1	0.0414	0.0453	0.0492	0.0531	0.0569	0.0607	0.0645	0.0682	0.0719	0.0755
1.2	0.0792	0.0828	0.0864	0.0899	0.0934	0.0969	0.1004	0.1038	0.1072	0.1106
1.3	0.1139	0.1173	0.1206	0.1239	0.1271	0.1303	0.1335	0.1367	0.1399	0.1430
1.4	0.1461	0.1492	0.1523	0.1553	0.1584	0.1614	0.1644	0.1673	0.1703	0.1732
1.5	0.1761	0.1790	0.1818	0.1847	0.1875	0.1903	0.1931	0.1959	0.1987	0.2014
1.6	0.2041	0.2068	0.2095	0.2122	0.2148	0.2175	0.2201	0.2227	0.2253	0.2279
1.7	0.2304	0.2330	0.2355	0.2380	0.2405	0.2430	0.2455	0.2480	0.2504	0.2529
1.8	0.2553	0.2577	0.2601	0.2625	0.2648	0.2672	0.2695	0.2718	0.2742	0.2765
1.9	0.2788	0.2810	0.2833	0.2856	0.2878	0.2900	0.2923	0.2945	0.2967	0.2989
2.0	0.3010	0.3032	0.3054	0.3075	0.3096	0.3118	0.3139	0.3160	0.3181	0.3201
2.1	0.3222	0.3243	0.3263	0.3284	0.3304	0.3324	0.3345	0.3365	0.3385	0.3404
2.2	0.3424	0.3444	0.3464	0.3483	0.3502	0.3522	0.3541	0.3560	0.3579	0.3598
2.3	0.3617	0.3636	0.3655	0.3674	0.3692	0.3711	0.3729	0.3747	0.3766	0.3784
2.4	0.3802	0.3820	0.3838	0.3856	0.3874	0.3892	0.3909	0.3927	0.3945	0.3962
2.5	0.3979	0.3997	0.4014	0.4031	0.4048	0.4065	0.4082	0.4099	0.4116	0.4133
2.6	0.4150	0.4166	0.4183	0.4200	0.4216	0.4232	0.4249	0.4265	0.4281	0.4298
2.7	0.4314	0.4330	0.4346	0.4362	0.4378	0.4393	0.4409	0.4425	0.4440	0.4456
2.8	0.4472	0.4487	0.4502	0.4518	0.4533	0.4548	0.4564	0.4579	0.4594	0.4609
2.9	0.4624	0.4639	0.4654	0.4669	0.4683	0.4698	0.4713	0.4728	0.4742	0.4757
3.0	0.4771	0.4786	0.4800	0.4814	0.4829	0.4843	0.4857	0.4871	0.4886	0.4900
3.1	0.4914	0.4928	0.4942	0.4955	0.4969	0.4983	0.4997	0.5011	0.5024	0.5038
3.2	0.5051	0.5065	0.5079	0.5092	0.5105	0.5119	0.5132	0.5145	0.5159	0.5172
3.3	0.5185	0.5198	0.5211	0.5224	0.5237	0.5250	0.5263	0.5276	0.5289	0.5302
3.4	0.5315	0.5328	0.5340	0.5353	0.5366	0.5378	0.5391	0.5403	0.5416	0.5428
3.5	0.5441	0.5453	0.5465	0.5478	0.5490	0.5502	0.5514	0.5527	0.5539	0.5551
3.6	0.5563	0.5575	0.5587	0.5599	0.5611	0.5623	0.5635	0.5647	0.5658	0.5670
3.7	0.5682	0.5694	0.5705	0.5717	0.5729	0.5740	0.5752	0.5763	0.5775	0.5786
3.8	0.5798	0.5809	0.5821	0.5832	0.5843	0.5855	0.5866	0.5877	0.5888	0.5899
3.9	0.5911	0.5922	0.5933	0.5944	0.5955	0.5966	0.5977	0.5988	0.5999	0.6010
4.0	0.6021	0.6031	0.6042	0.6053	0.6064	0.6075	0.6085	0.6096	0.6107	0.6117
4.1	0.6128	0.6138	0.6149	0.6160	0.6170	0.6180	0.6191	0.6201	0.6212	0.6222
4.2	0.6232	0.6243	0.6253	0.6263	0.6274	0.6284	0.6294	0.6304	0.6314	0.6325
4.3	0.6335	0.6345	0.6355	0.6365	0.6375	0.6385	0.6395	0.6405	0.6415	0.6425
4.4	0.6435	0.6444	0.6454	0.6464	0.6474	0.6484	0.6493	0.6503	0.6513	0.6522
4.5	0.6532	0.6542	0.6551	0.6561	0.6571	0.6580	0.6590	0.6599	0.6609	0.6618
4.6	0.6628	0.6637	0.6646	0.6656	0.6665	0.6675	0.6684	0.6693	0.6702	0.6712
4.7	0.6721	0.6730	0.6739	0.6749	0.6758	0.6767	0.6776	0.6785	0.6794	0.6803
4.8	0.6812	0.6821	0.6830	0.6839	0.6848	0.6857	0.6866	0.6875	0.6884	0.6893
4.9	0.6902	0.6911	0.6920	0.6928	0.6937	0.6946	0.6955	0.6964	0.6972	0.6981
5.0	0.6990	0.6998	0.7007	0.7016	0.7024	0.7033	0.7042	0.7050	0.7059	0.7067
5.1	0.7076	0.7084	0.7093	0.7101	0.7110	0.7118	0.7126	0.7135	0.7143	0.7152
5.2	0.7160	0.7168	0.7177	0.7185	0.7193	0.7202	0.7210	0.7218	0.7226	0.7235
5.3	0.7243	0.7251	0.7259	0.7267	0.7275	0.7284	0.7292	0.7300	0.7308	0.7316
5.4	0.7324	0.7332	0.7340	0.7348	0.7356	0.7364	0.7372	0.7380	0.7388	0.7396

TABLE III CONTINUED

	0	1	2	3	4	5	6	7	8	9
5.5	0.7404	0.7412	0.7419	0.7427	0.7435	0.7443	0.7451	0.7459	0.7466	0.7474
5.6	0.7482	0.7490	0.7497	0.7505	0.7513	0.7520	0.7528	0.7536	0.7543	0.7551
5.7	0.7559	0.7566	0.7574	0.7582	0.7589	0.7597	0.7604	0.7612	0.7619	0.7627
5.8	0.7634	0.7642	0.7649	0.7657	0.7664	0.7672	0.7679	0.7686	0.7694	0.7701
5.9	0.7709	0.7716	0.7723	0.7731	0.7738	0.7745	0.7752	0.7760	0.7767	0.7774
6.0	0.7782	0.7789	0.7796	0.7803	0.7810	0.7818	0.7825	0.7832	0.7839	0.7846
6.1	0.7853	0.7860	0.7868	0.7875	0.7882	0.7889	0.7896	0.7903	0.7910	0.7917
6.2	0.7924	0.7931	0.7938	0.7945	0.7952	0.7959	0.7966	0.7973	0.7980	0.7987
6.3	0.7993	0.8000	0.8007	0.8014	0.8021	0.8028	0.8035	0.8041	0.8048	0.8055
6.4	0.8062	0.8069	0.8075	0.8082	0.8089	0.8096	0.8102	0.8109	0.8116	0.8122
6.5	0.8129	0.8136	0.8142	0.8149	0.8156	0.8162	0.8169	0.8176	0.8182	0.8189
6.6	0.8195	0.8202	0.8209	0.8215	0.8222	0.8228	0.8235	0.8241	0.8248	0.8254
6.7	0.8261	0.8267	0.8274	0.8280	0.8287	0.8293	0.8299	0.8306	0.8312	0.8319
6.8	0.8325	0.8331	0.8338	0.8344	0.8351	0.8357	0.8363	0.8370	0.8376	0.8382
6.9	0.8388	0.8395	0.8401	0.8407	0.8414	0.8420	0.8426	0.8432	0.8439	0.8445
7.0	0.8451	0.8457	0.8463	0.8470	0.8476	0.8482	0.8488	0.8494	0.8500	0.8506
7.1	0.8513	0.8519	0.8525	0.8531	0.8537	0.8543	0.8549	0.8555	0.8561	0.8567
7.2	0.8573	0.8579	0.8585	0.8591	0.8597	0.8603	0.8609	0.8615	0.8621	0.8627
7.3	0.8633	0.8639	0.8645	0.8651	0.8657	0.8663	0.8669	0.8675	0.8681	0.8686
7.4	0.8692	0.8698	0.8704	0.8710	0.8716	0.8722	0.8727	0.8733	0.8739	0.8745
7.5	0.8751	0.8756	0.8762	0.8768	0.8774	0.8779	0.8785	0.8791	0.8797	0.8802
7.6	0.8808	0.8814	0.8820	0.8825	0.8831	0.8837	0.8842	0.8848	0.8854	0.8859
7.7	0.8865	0.8871	0.8876	0.8882	0.8887	0.8893	0.8899	0.8904	0.8910	0.8915
7.8	0.8921	0.8927	0.8932	0.8938	0.8943	0.8949	0.8954	0.8960	0.8965	0.8971
7.9	0.8976	0.8982	0.8987	0.8993	0.8998	0.9004	0.9009	0.9015	0.9020	0.9025
8.0	0.9031	0.9036	0.9042	0.9047	0.9053	0.9058	0.9063	0.9069	0.9074	0.9079
8.1	0.9085	0.9090	0.9096	0.9101	0.9106	0.9112	0.9117	0.9122	0.9128	0.9133
8.2	0.9138	0.9143	0.9149	0.9154	0.9159	0.9165	0.9170	0.9175	0.9180	0.9186
8.3	0.9191	0.9196	0.9201	0.9206	0.9212	0.9217	0.9222	0.9227	0.9232	0.9238
8.4	0.9243	0.9248	0.9253	0.9258	0.9263	0.9269	0.9274	0.9279	0.9284	0.9289
8.5	0.9294	0.9299	0.9304	0.9309	0.9315	0.9320	0.9325	0.9330	0.9335	0.9340
8.6	0.9345	0.9350	0.9355	0.9360	0.9365	0.9370	0.9375	0.9380	0.9385	0.9390
8.7	0.9395	0.9400	0.9405	0.9410	0.9415	0.9420	0.9425	0.9430	0.9435	0.9440
8.8	0.9445	0.9450	0.9455	0.9460	0.9465	0.9469	0.9474	0.9479	0.9484	0.9489
8.9	0.9494	0.9499	0.9504	0.9509	0.9513	0.9518	0.9523	0.9528	0.9533	0.9538
9.0	0.9542	0.9547	0.9552	0.9557	0.9562	0.9566	0.9571	0.9576	0.9581	0.9586
9.1	0.9590	0.9595	0.9600	0.9605	0.9609	0.9614	0.9619	0.9624	0.9628	0.9633
9.2	0.9638	0.9643	0.9647	0.9652	0.9657	0.9661	0.9666	0.9671	0.9675	0.9680
9.3	0.9685	0.9689	0.9694	0.9699	0.9703	0.9708	0.9713	0.9717	0.9722	0.9727
9.4	0.9731	0.9736	0.9741	0.9745	0.9750	0.9754	0.9759	0.9763	0.9768	0.9773
9.5	0.9777	0.9782	0.9786	0.9791	0.9795	0.9800	0.9805	0.9809	0.9814	0.9818
9.6	0.9823	0.9827	0.9832	0.9836	0.9841	0.9845	0.9850	0.9854	0.9859	0.9863
9.7	0.9868	0.9872	0.9877	0.9881	0.9886	0.9890	0.9894	0.9899	0.9903	0.9908
9.8	0.9912	0.9917	0.9921	0.9926	0.9930	0.9934	0.9939	0.9943	0.9948	0.9952
9.9	0.9956	0.9961	0.9965	0.9969	0.9974	0.9978	0.9983	0.9987	0.9991	0.9996

Solutions to Odd-Numbered Problems

Chapter 1 Section 1.1

1. Closed 3. 

5. $\{x|x>3\}$ 7. $(-\infty, 4]$ 9. Half-open

11. 

13. $[0, \pi)$ 15. Open 17. 

19. $[-4, 4]$ 21. $(-1, 3]$ 23. $(c-d, c+d)$ 25. $(0, \infty)$ 27. $(-\infty, -1)$ 29. $[-3, -2]$
31. $[-\frac{2}{3}, 1]$ 33. $\{x|2 \leq x \leq 6\}$ 35. $\{x|x \leq 0\}$ 37. $\{x||x-c|<d\}$ 39. $\{x|x \geq 0\}$
41. $\{x||x|<1\}$ 43. $\{x||x-1| \leq 1\}$ 45. $\{x||x-c|<d\}$ 47. $\{x||x-(a+\frac{1}{2})|<\frac{1}{2}\}$
49. $(-\infty, -5] \cup [1, \infty)$ 51. $(-\infty, 2) \cup (4, \infty)$ 53. $\varnothing$ (no solution) 55. $(-\infty, -1) \cup (1, \infty)$
57. $(-\infty, -\frac{8}{3}] \cup [-2, \infty)$ 59. $(-\infty, -1] \cup [4, \infty)$ 61. $[\frac{1}{2}, \infty)$ 63. $(-\infty, 1) \cup (\frac{5}{3}, \infty)$ 65. $(1, 2)$
67. $[-1, 1)$ 69. No 71. $|a| = |(a-b)+b| \leq |a-b| + |b|$ implies $|a| - |b| \leq |a-b|$.
73. L is a greatest lower bound of S if and only if *(1)* L is a lower bound of S, and *(2)* if b is any lower bound of S, then $b \leq L$.
75. lub = 5; glb = 3 77. lub = 0; no lower bound 79. lub = 1; glb = 0
81. If b_1 and b_2 are both least upper bounds, then $b_1 \leq b_2$ and $b_2 \leq b_1$, so $b_1 = b_2$.
83. Division by 0 occurs in going from line 4 to line 5.

Section 1.2

1. $\sqrt{17}$ 3. $\frac{1}{2}\sqrt{149}$ 5. $(\frac{3}{2}, 1)$ 7. $(-\frac{5}{4}, -\frac{5}{2})$ 9. $y = 3x + 7$ 11. $y = -x$ 13. $y = 2x + 4$
15. $y = -2x + 6$ 17. $y = -9x + 12$ 19. $y = -\frac{9}{34}x + \frac{2043}{340}$ 21. $y = 1$ 23. $y = b$ 25. $x = 1$
27. $x = 0$ 29. $y = 3x - 12$ 31. $y = -\frac{1}{3}x + (3\sqrt{3} + 5)/3$ 33. $y = -0.04x + 2.8472$
35. $y = 1.3x - 1.149$ 37. Slope $= \frac{1}{3}$, intercept $= -\frac{5}{3}$ 39. Slope $= \frac{3}{2}$, intercept $= \frac{1}{2}$

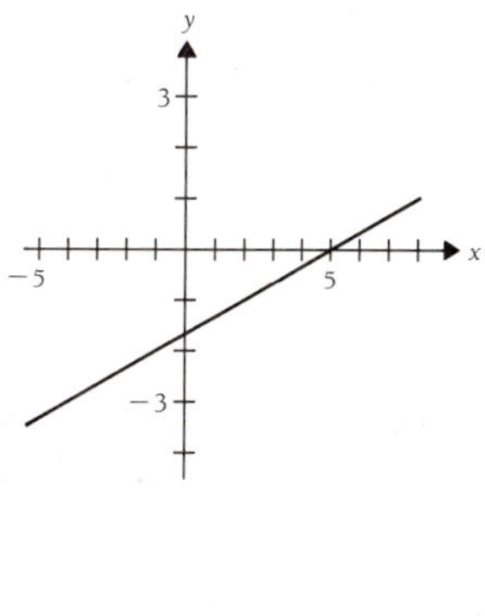

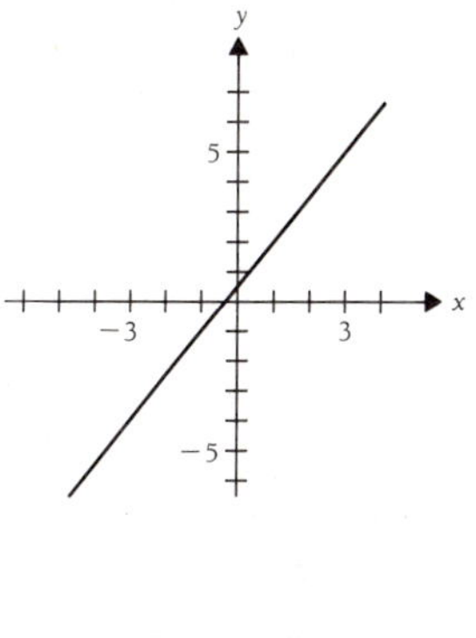

41. Slope $= \frac{3}{2}$, intercept $= 0$

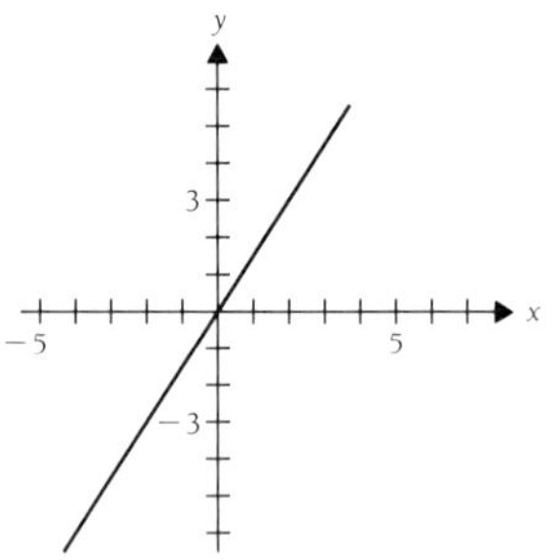

43. Slope $= -\frac{5}{7}$, intercept $= \frac{248}{35}$

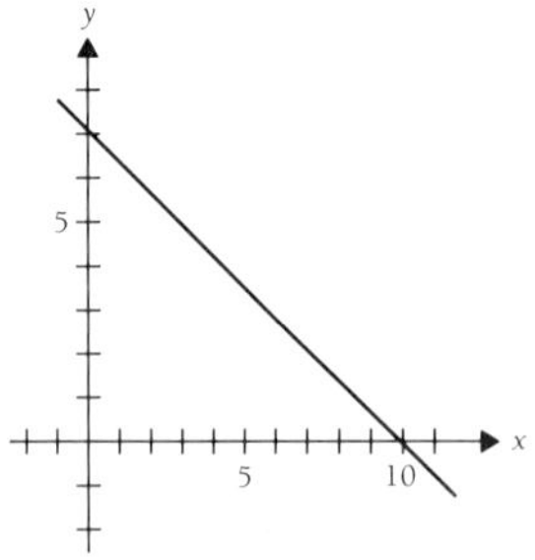

45. Since points $(a, 0)$ and $(0, b)$ satisfy the equation, and the equation is the equation of a line, it is the equation of the line with intercepts a and b.
47. x-intercept 4, y-intercept -3 49. $y = x$ 51. 10 square units
53. Lines AB and BC have different equations and different slopes, so the points are not collinear.
55. $y = \frac{4}{5}x$ 57. $|C|/\sqrt{A^2 + B^2}$ 59. $\left(\dfrac{2m + 4n}{m + n}, \dfrac{3m + n}{m + n}\right)$
61. The two-point form of a linear equation can be written as $(y_1 - y_2)x + (x_2 - x_1)y + (x_1 y_2 - x_2 y_1) = 0$, which is the equation given when the determinants are expanded.
63. The hypothesis implies $A_1 B_2 - A_2 B_1 = 0$, so that $A_1/B_1 = A_2/B_2$. That is, the two given lines have the same slope. Thus, they are the same line or they are parallel lines.
65. If the lines are concurrent, there is a point (a, b) such that

$$\begin{cases} A_1 a + B_1 b + C_1 = 0 \\ A_2 a + B_2 b + C_2 = 0 \\ A_3 a + B_3 b + C_3 = 0 \end{cases}$$

But this system in the "variables" a, b, and 1 has a solution if and only if the given determinant (the determinant of coefficients) is 0 (see Appendix B).

67. $\left(\dfrac{x_1 + x_2 + x_3}{3}, \dfrac{y_1 + y_2 + y_3}{3}\right)$

Section 1.3

1. $x^2 + y^2 = 9$ 3. $(x + 7)^2 + (y - 1)^2 = 4$ 5. $(x - 3)^2 + (y + \frac{2}{3})^2 = \frac{4}{9}$ 7. $(x - a)^2 + (y - 3)^2 = (1 + \sqrt{2})^2$
9. Circle; center $(0, 0)$, radius 2 11. Point circle; the point $(0, 0)$ 13. Circle; center $(-1, 0)$, radius 2
15. Circle; center $(3, 0)$; radius $\sqrt{17}$ 17. Imaginary circle; no graph 19. Point circle; the point $(2, -3)$
21. Circle; center $(a, 0)$, radius $|a|$ 23. Circle; center $(a/2, a/2)$, radius $|a|/\sqrt{2}$ 25. $x^2 + y^2 = 25$
27. $(x - 3)^2 + (y + 4)^2 = 25$ 29. $(x - 1)^2 + (y - 2)^2 = 9$ 31. $(x - 4)^2 + (y - 1)^2 = 16$
33. $(x - 1 \pm 2\sqrt{\frac{11}{5}})^2 + (y - 2 \pm \sqrt{\frac{11}{5}})^2 = 16$ 35. $(x - 2)^2 + (y - 2)^2 = 4$
37. $(x - 1)^2 + (y - 5)^2 = (\sqrt{26} - 2)^2$ or $(x - 1)^2 + (y - 5)^2 = (2 + \sqrt{26})^2$ 39. $(x - 1)^2 + (y \pm 1)^2 = 1$
41. $(x - 3)^2 + (y + 1)^2 = 1$ 43. $x^2 + y^2 - 3x - y = 0$ 45. $x = 2$ 47. $x = 5$ 49. $y = \pm(\sqrt{3}/3)(x - 4)$
51. Upper: $y = k + \sqrt{r^2 - (x - h)^2}$; lower: $y = k - \sqrt{r^2 - (x - h)^2}$

Section 1.4

1. Add 10. 3. Subtract 1 and then find the reciprocal. 5. $f(x) = 2x + 1$ 7. $f(x) = |3x - 1|$ 9. 3 11. $-\frac{7}{3}$ 13. $8b + 3$
15. 11 17. 8 19. $2x^2 + 4xh + 2h^2 - x - h + 8$ 21. $\mathbb{R}$ 23. $\mathbb{R}$ 25. $(-\infty, 3) \cup (3, \infty)$ 27. $[3, \infty)$ 29. $(3, \infty)$
31. $(-\infty, -2] \cup [2, \infty)$ 33. $F = kx$ 35. $r = k/t$ 37. $D = kv^2$ 39. $100(3950)^2/(4050)^2 \approx 95.12$ lb
41. $l \approx 1.23p^2$ or $p \approx 0.9\sqrt{l}$

43.

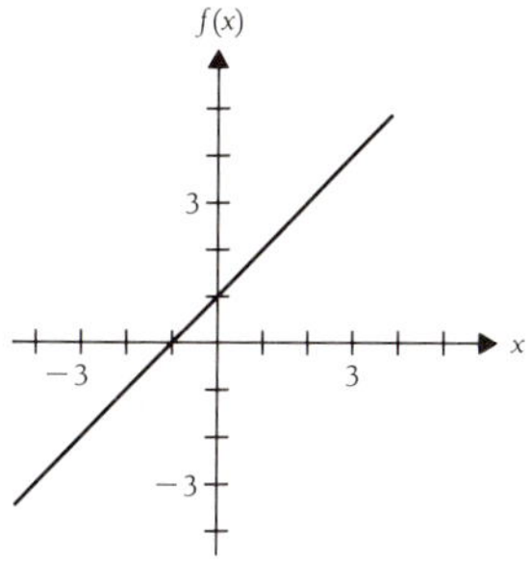

45.

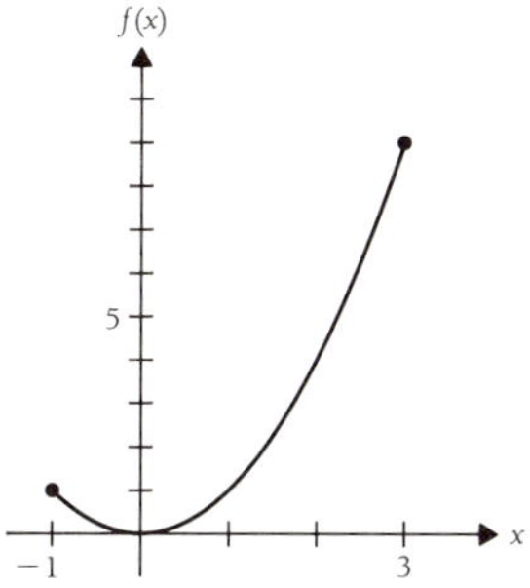

47.

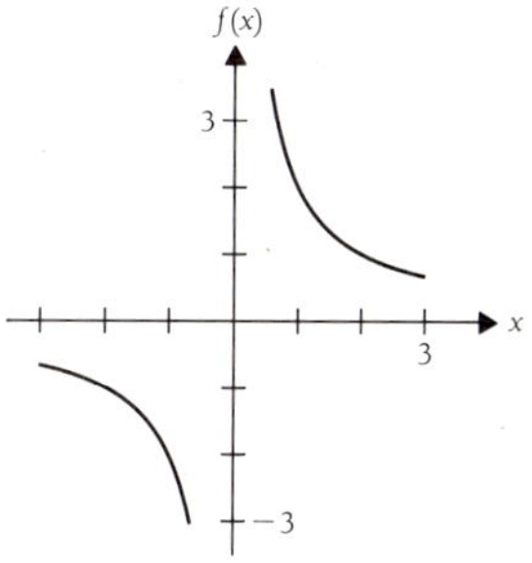

49.

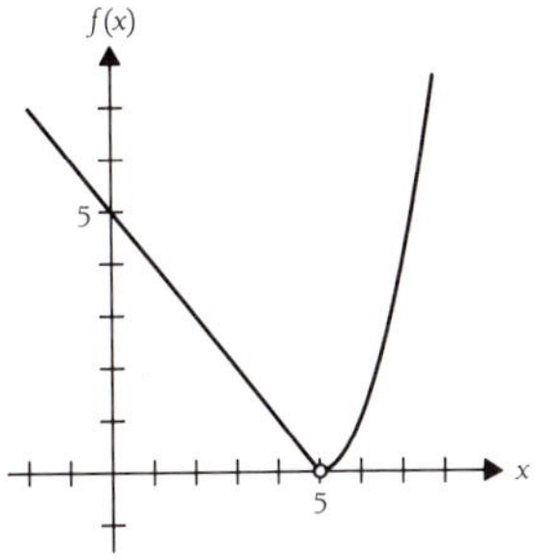

51.

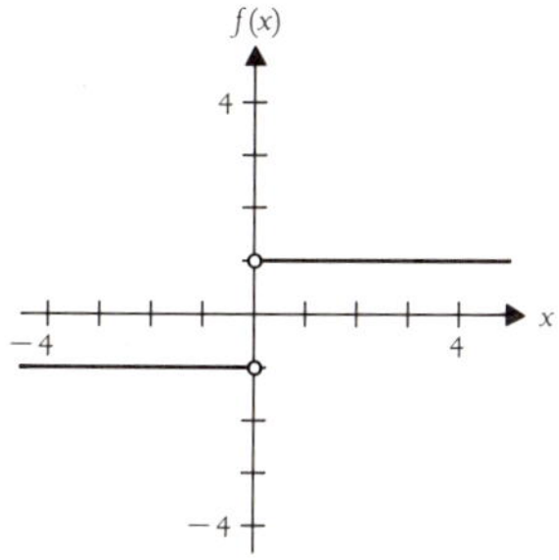

53.

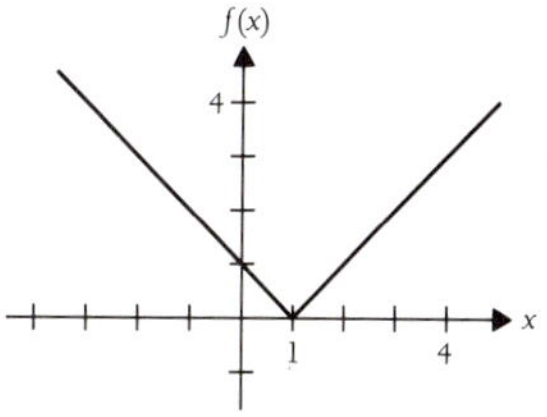

55.

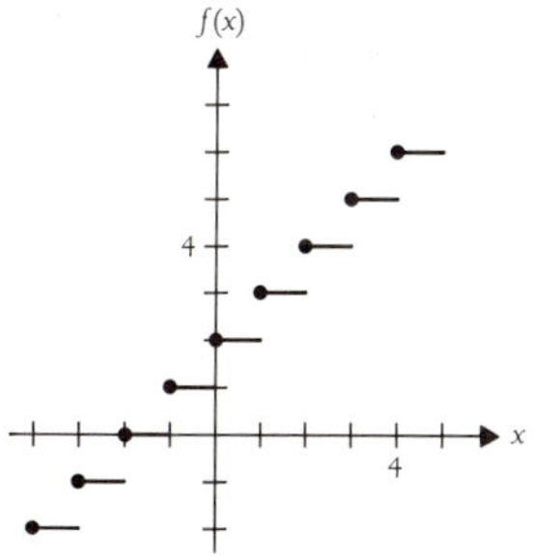

57.

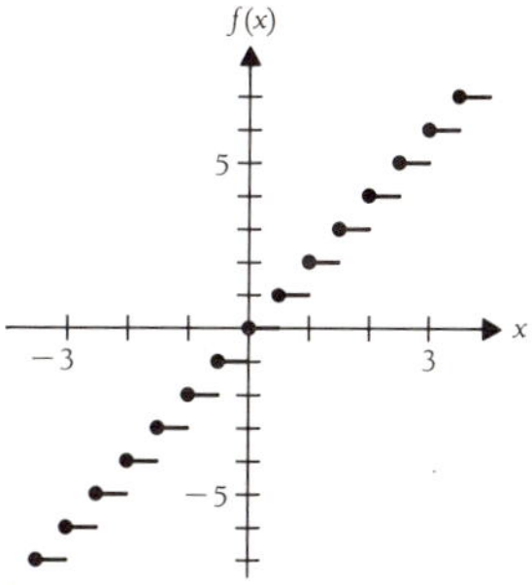

59.

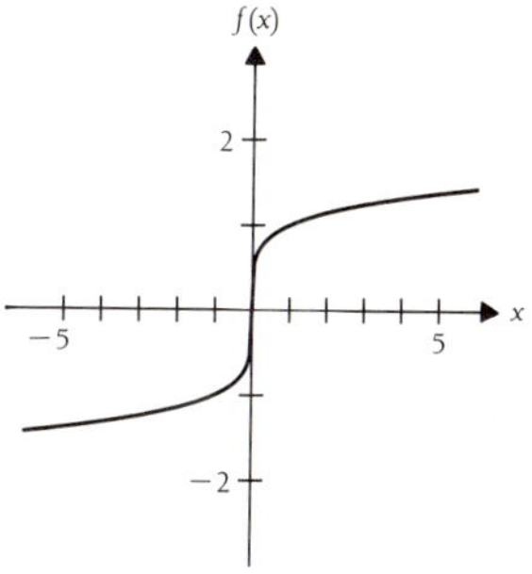

61.

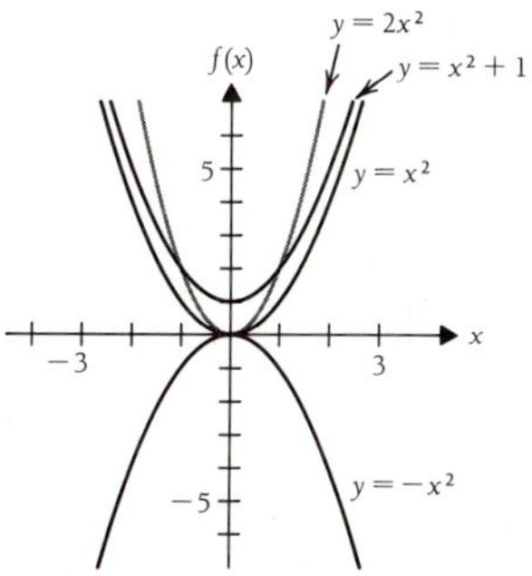

63.

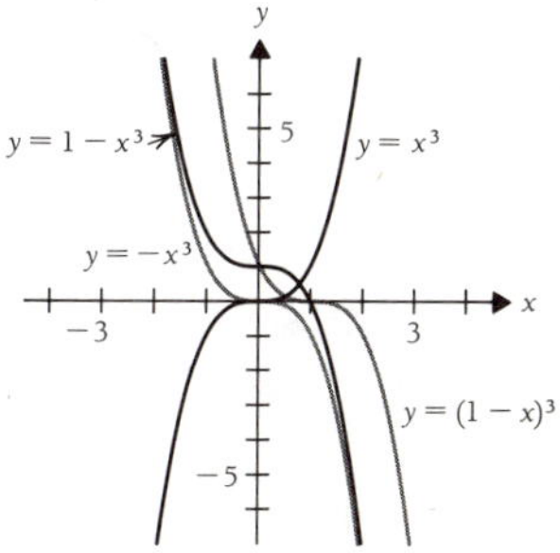

65.

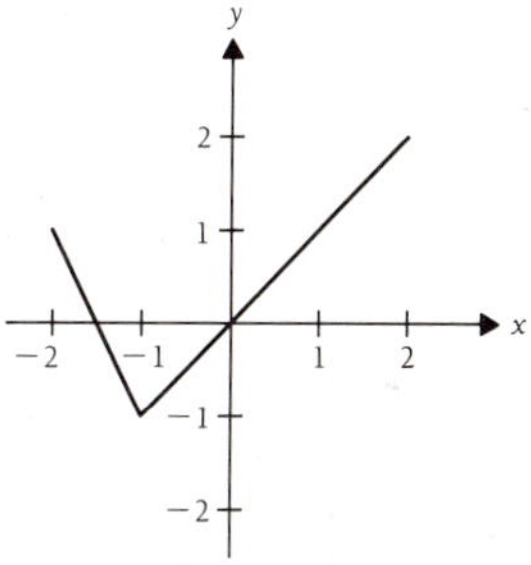

67.

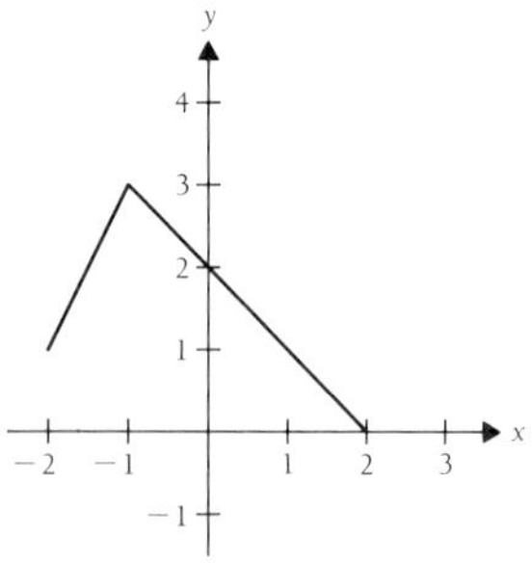

69.

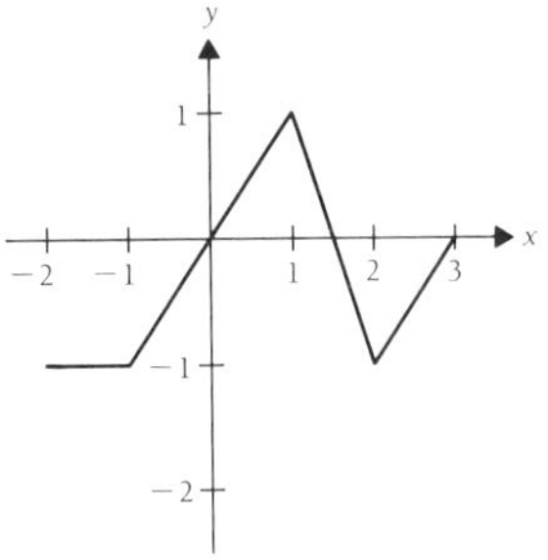

71.

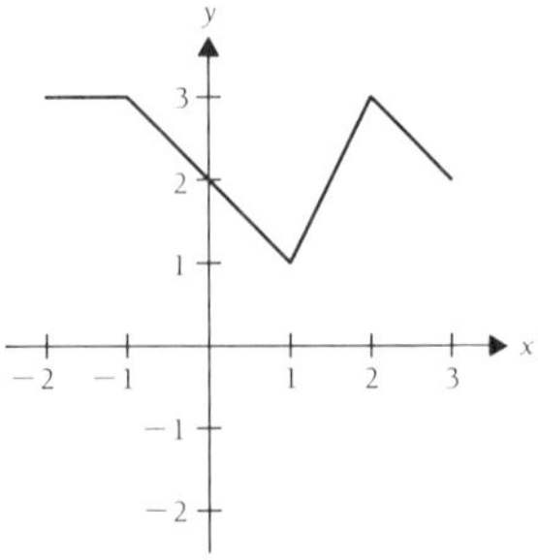

73. About the origin **75.** x-Axis **77.** Both axes and about the origin **79.** Both axes and about the origin **81.** Yes
83. $\lceil n \rceil$ is the least integer not less than n, which is n itself. If $n < x \leq n + 1$, then $n + 1$ is the least integer not less than x; that is, $\lceil x \rceil = n + 1$.

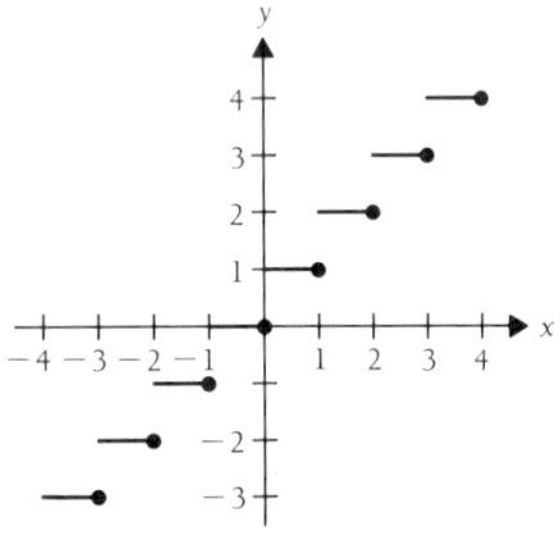

Section 1.5

1. $(f+g)(x) = 2x$, $(f-g)(x) = 6$, $(fg)(x) = x^2 - 9$, $(f/g)(x) = (x+3)/(x-3)$, $(g \circ f)(x) = x$, $(f \circ g)(x) = x$
3. $(f+g)(x) = x + (1/x)$, $(f-g)(x) = x - (1/x)$, $(fg)(x) = 1\ (x \neq 0)$, $(f/g)(x) = x^2\ (x \neq 0)$, $(g \circ f)(x) = 1/x$, $(f \circ g)(x) = 1/x$
5. $(f+g)(x) = |x| + |x-2|$, $(f-g)(x) = |x| - |x-2|$, $(fg)(x) = |x||x-2|$, $(f/g)(x) = |x|/|x-2|$, $(g \circ f)(x) = ||x| - 2|$, $(f \circ g)(x) = |x-2|$
7. $(f+g)(x) = |x| + x^2$, $(f-g)(x) = |x| - x^2$, $(fg)(x) = |x|x^2$, $(f/g)(x) = |x|/x^2$, $(g \circ f)(x) = |x|^2$, $(f \circ g)(x) = |x^2|$
9. $2x + 2$ **11.**

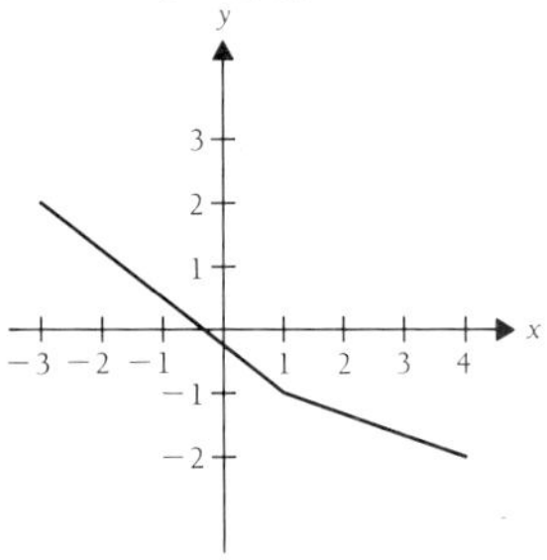

13.

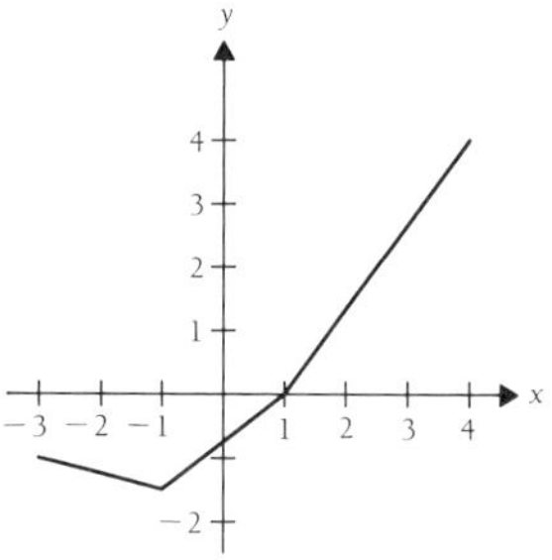

15.

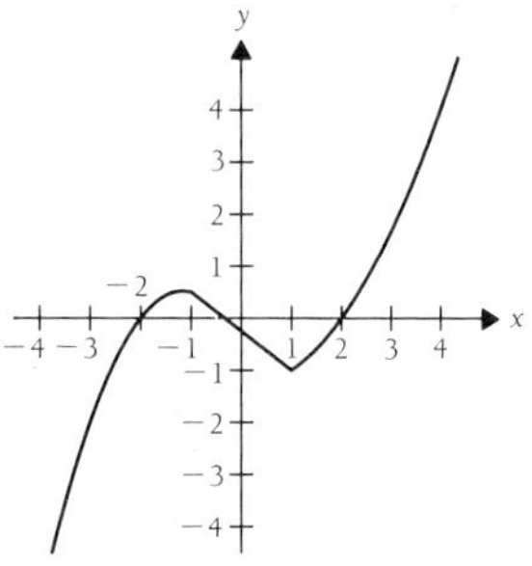

17.

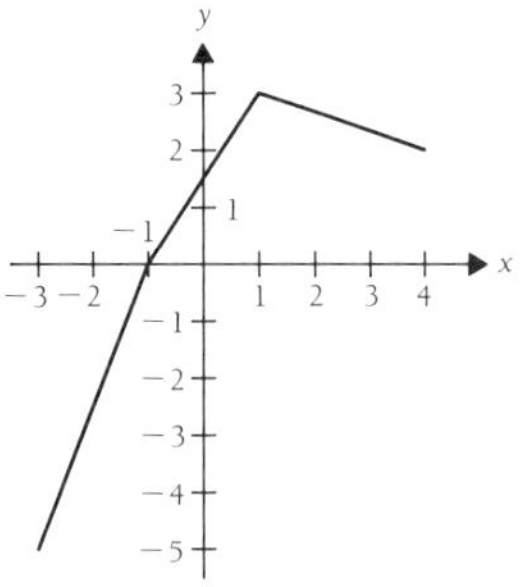

19. $f(x) = x^2$, $g(x) = x + 9$; and others **21.** $f(x) = 1/(x-3)$, $g(x) = \sqrt{x}$; and others
23. $f(x) = x$, $g(x) = x^3$, $h(x) = x - 2$; and others **25.** $f(x) = x + 1$, $g(x) = x^3$, $h(x) = \sqrt{x}$; and others **27.** $(-\infty, 1) \cup (1, \infty)$
29. $(-\infty, 0) \cup (0, 1) \cup (1, \infty)$ **31.** $(-\infty, 1]$ **33.** $[3, \infty)$ for each **35.** $[-2, 2]$ for $f + g$ and $g \circ f$; $[-2, -1) \cup (-1, 2]$ for f/g
37. Domain $(-\infty, 0) \cup (0, \infty)$; range $(-\infty, 3) \cup (3, \infty)$ **39.** Domain $(-\infty, -\frac{1}{2}) \cup (-\frac{1}{2}, \infty)$; range $(-\infty, \frac{1}{2}) \cup (\frac{1}{2}, \infty)$
41. Domain $[1, \infty)$; range $[0, \infty)$ **43.** Domain $\mathbb{R}$; range $[4^{1/3}, \infty)$ **45.** Yes **47.** Yes **49.** No **51.** Yes
53. $f(g(x)) = g(f(x)) = x$ **55.** $f(g(x)) = g(f(x)) = x$ **57.** $(x-1)/2$ **59.** $(x+1)^{1/3}$ **61.** $(2-x)/x$ **63.** $5x/(x-1)$

65.

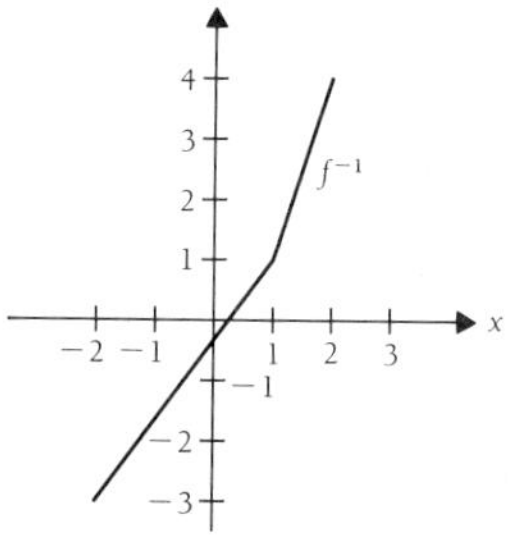

67.

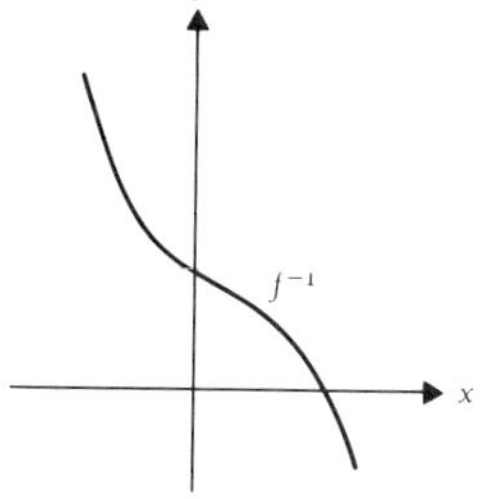

69. Restricted domain is $[2, \infty)$; $f^{-1}(x) = \sqrt{x} + 2$. (Other ways also.)
71. Restricted domain is $(0, \infty)$; $f^{-1}(x) = 1/\sqrt{x}$. (Other ways also.)
73. If $f(x_1) = f(x_2)$, then $x_1 = f^{-1}(f(x_1)) = f^{-1}(f(x_2)) = x_2$.
75. Because f^{-1} has inverse f; apply Problem 73.
77. A polynomial function of degree n has $n + 1$ coefficients. Requiring a point to lie on the graph of the function is a condition on the $n + 1$ coefficients. Repeating this requirement $n + 1$ times gives a system of $n + 1$ equations in $n + 1$ unknowns, which, in general, has at most one solution.
79. $-\frac{7}{4} < x < -\frac{3}{2}$ **81.** $(-\infty, 0) \cup [1, \infty)$ **83.** $(-\infty, 0) \cup [1, \infty)$ **85.** $(2, 3) \cup (3, \infty)$

Section 1.6

1. $-1/\sqrt{2}$ **3.** $\frac{1}{2}\sqrt{2 - \sqrt{2}}$ **5.** 2 **7.** $\frac{1}{4}(\sqrt{2} - \sqrt{6})$ **9.** 0.8415 **11.** 0.6131 **13.** 0.9878 **15.** 0.0945 **17.** $\sqrt{5}/3$ **19.** $\frac{3}{2}$
21. $-\frac{1}{9}$ **23.** $1/\sqrt{6}$ **25.** By definition, if $x = t$, then $P(x)$ is the point $(\cos x, \sin x)$.
27. $\cos(-x) = \cos(0 - x) = \cos 0 \cos x + \sin 0 \sin x = 1 \cdot \cos x = \cos x$

29. $\cos x = \cos\left[\frac{\pi}{2} - \left(\frac{\pi}{2} - x\right)\right] = \cos\frac{\pi}{2}\cos\left(\frac{\pi}{2} - x\right) + \sin\frac{\pi}{2}\sin\left(\frac{\pi}{2} - x\right) = 0 + 1 \cdot \sin\left(\frac{\pi}{2} - x\right) = \sin\left(\frac{\pi}{2} - x\right)$

31. $\sin x$ is the y-coordinate of $P(x)$; if x changes sign, then $\sin x$ changes sign. That is, $\sin(-x) = -\sin x$.

33.
$$\begin{aligned}\sin(x \pm y) &= \cos\left[\frac{\pi}{2} - (x \pm y)\right] \quad \text{[Formula (13)]}\\ &= \cos\left[\left(\frac{\pi}{2} - x\right) \mp y\right]\\ &= \cos\left(\frac{\pi}{2} - x\right)\cos y \pm \sin\left(\frac{\pi}{2} - x\right)\sin y\\ &= \sin x \cos y \pm \cos x \sin y \quad \text{(Problems 28 and 29)}\end{aligned}$$

35.
$$\begin{aligned}\sin 2x &= \sin(x + x) = \sin x \cos x + \cos x \sin x = 2 \sin x \cos x;\\ \cos 2x &= \cos(x + x) = \cos x \cos x - \sin x \sin x = \cos^2 x - \sin^2 x\\ &= (1 - \sin^2 x) - \sin^2 x = 1 - 2\sin^2 x\\ &= 1 - 2(1 - \cos^2 x) = 1 - 2 + 2\cos^2 x = 2\cos^2 x - 1;\\ \tan 2x &= \tan(x + x) = \frac{\tan x + \tan x}{1 - \tan x \tan x} = \frac{2 \tan x}{1 - \tan^2 x}\end{aligned}$$

37. $\cos x = \cos(2x - x) = \cos 2x \cos x + \sin 2x \sin x$, so that $\sin 2x \sin x = \cos x - \cos 2x \cos x$, whence $\sin 2x \dfrac{\sin x}{\cos x} = 1 - \cos 2x$, so that $\tan x = \dfrac{1 - \cos 2x}{\sin 2x}$.

39. $\frac{1}{2}[\cos(x - y) - \cos(x + y)] = \frac{1}{2}(\cos x \cos y + \sin x \sin y - \cos x \cos y + \sin x \sin y) = \frac{1}{2}(2 \sin x \sin y) = \sin x \sin y$;
$\frac{1}{2}[\sin(x + y) + \sin(x - y)] = \frac{1}{2}(\sin x \cos y + \cos x \sin y + \sin x \cos y - \cos x \sin y) = \frac{1}{2}(2 \sin x \cos y) = \sin x \cos y$;
$\frac{1}{2}[\cos(x + y) + \cos(x - y)] = \frac{1}{2}(\cos x \cos y - \sin x \sin y + \cos x \cos y + \sin x \sin y) = \frac{1}{2}(2 \cos x \cos y) = \cos x \cos y$

41.

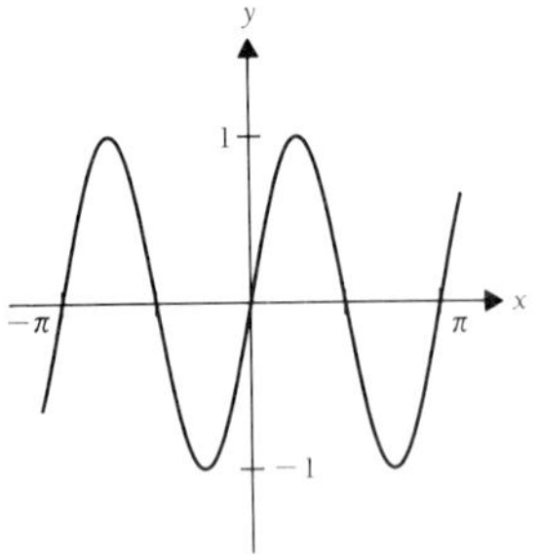

43.

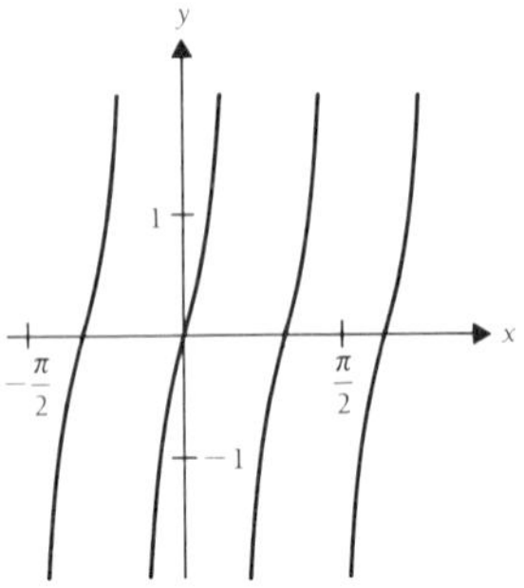

45.

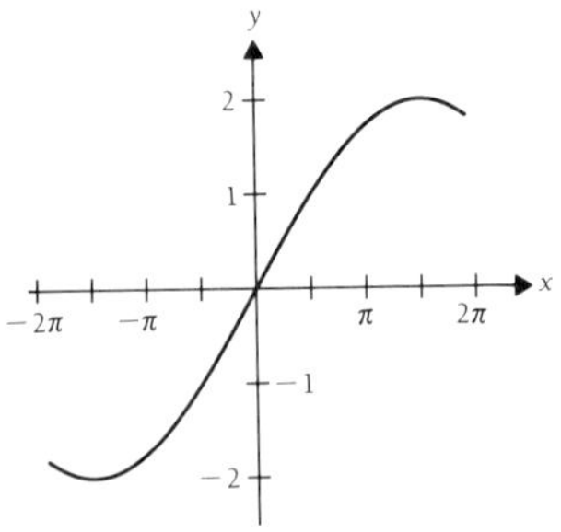

47.

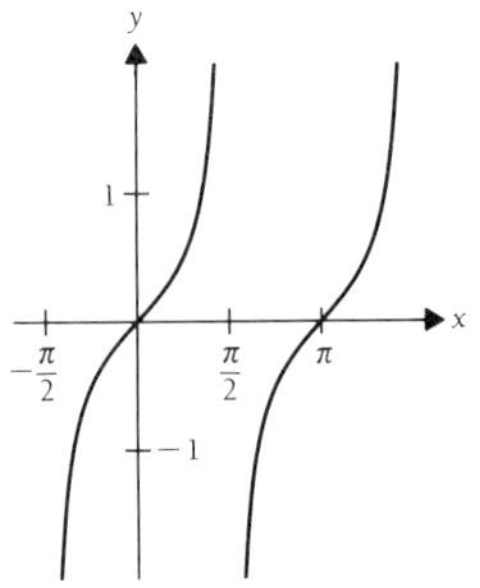

49.

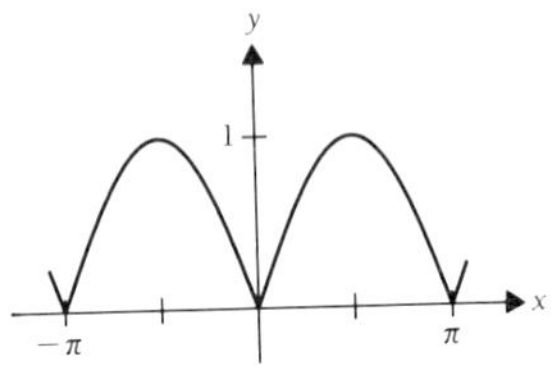

51.

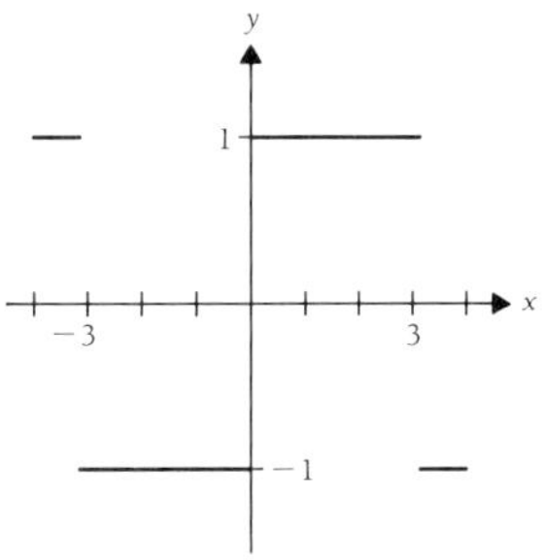

53.

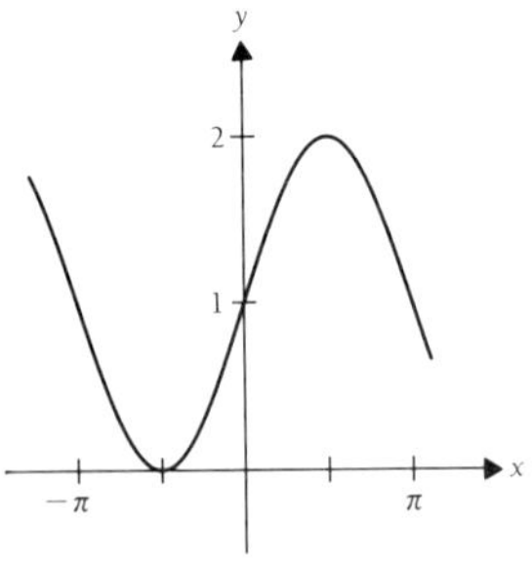

55.

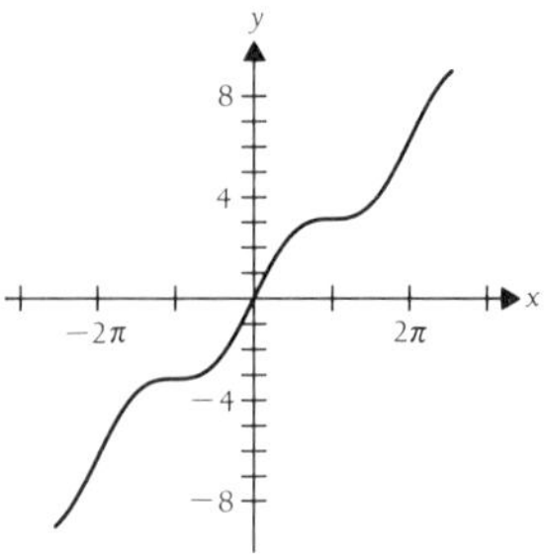

57. In the figure, let line l meet the x-axis at point P, and let Q be a point on the upper half of l. Let R be the foot of the perpendicular from Q to the x-axis. Then $\tan \phi = |QR|/|RP| = m$, the slope of l.

Review Exercises

1. 

3. 

5. 

7. 

9. 

11. 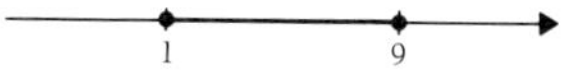

13. $\sqrt{5}$

15. (1, 2) 17. $y-1=7(x-3)$ 19. $x=3$ 21. $y=-x+6$ 23. $y-1=\frac{1}{4}(x-\frac{7}{2})$ 25. $(x-1)^2+(y+1)^2=2$
27. $(x-3)^2+(y-7)^2=64$ 29. Circle; center $(1,-2)$, radius $\sqrt{6}$ 31. Circle; center $(1,-2)$, radius 3
33. Domain $\mathbb{R}$; range $\mathbb{R}$ 35. Domain $\{x \mid x \neq 3\}$; range $\{x \mid x \neq 0\}$ 37. Domain $\{x \mid x \neq -1, x \neq 1\}$; range $(-\infty, -1] \cup (1, \infty)$
39. Domain $\{x \mid x \neq n\pi/2, n \in \mathbb{Z}\}$; range $\mathbb{R}$

41. 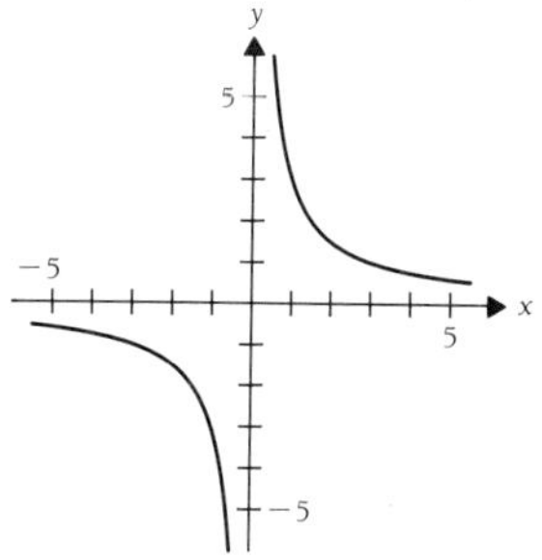

43. 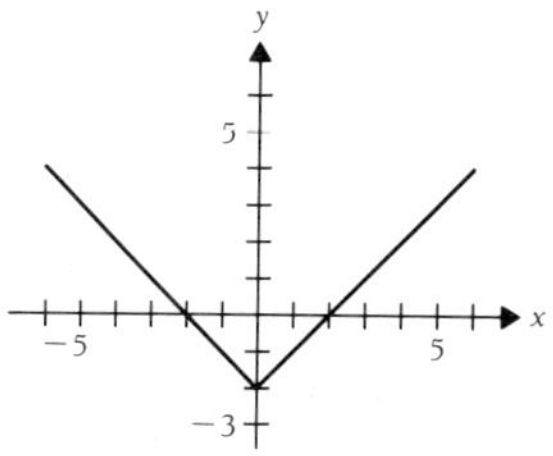

45. 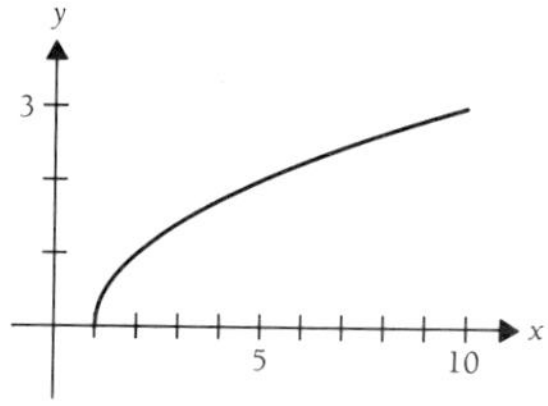

47. 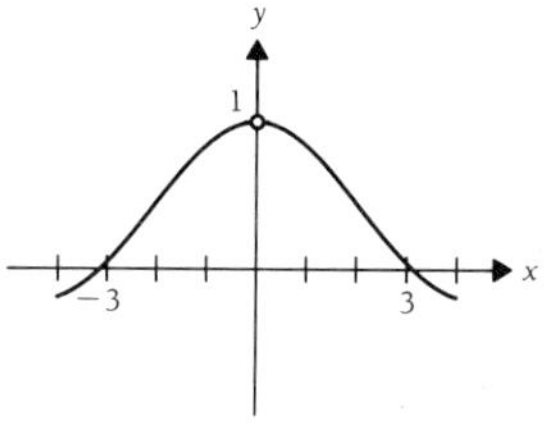

49. 16 51. $(x+2)^2$ 53. $-\sin 1 \approx -0.8415$ 55. $\frac{1}{4}x^2$ 57. $-(x+1)^{1/3}$ 59. $\cos^{-1}(x-1)$

Additional Exercises

1. The intersection of two closed intervals is a closed interval, a point, or empty. The union of two closed intervals is a closed interval if their intersection is not empty.
3. None 5. Only 0 7. $\{4, \frac{2}{3}\}$ 9. $\{-\frac{1}{6}, -\frac{7}{2}\}$ 11. $\varnothing$ 13. $\{\pm\pi/6 + n\pi \mid n \in \mathbb{Z}\}$
15. When both are positive, simply drop the absolute value signs. When both are negative, $|x+y| = -(x+y) = -x+(-y) = |x|+|y|$.
17. Use Problems 15 and 16. 19. 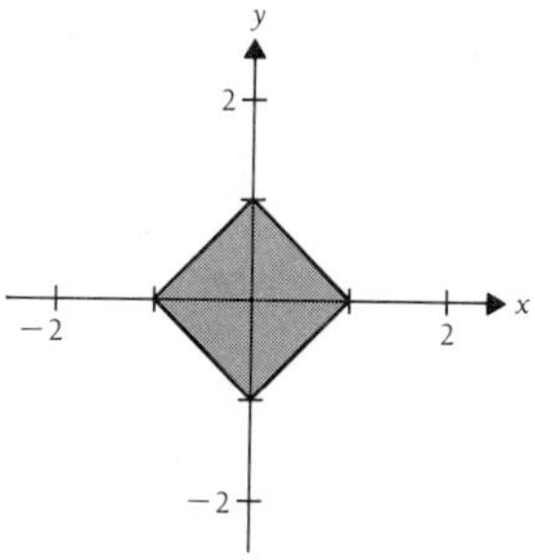

21. Outside 23. $0 \leq c \leq 2$

25. Substituting (x_i, y_i), $i = 1, 2, 3$, for (x, y) creates a determinant with two identical rows, so the equation is satisfied by each of the given points. Expansion gives the equation of a circle, so it is the circle through the three given points.

27. If $A_1B_2 + A_2B_1 = 0$, then the bisectors are the horizontal and vertical lines $x = (B_1C_2 - B_2C_1)/(A_1B_2 - A_2B_1)$ and $y = (C_1A_2 - C_2A_1)/(A_1B_2 - A_2B_1)$. Otherwise, let $m = [B_1B_2 - A_1A_2 \pm \sqrt{(A_1^2 + B_1^2)(A_2^2 + B_2^2)}]/(A_1B_2 + A_2B_1)$; then the bisector is $m(A_1B_2 - A_2B_1)x - (A_1B_2 - A_2B_1)y - m(B_1C_2 - B_2C_1) + C_1A_2 - C_2A_1 = 0$.

29. The inclination of l is $\alpha \pm \pi/2$, so the slope of l is $m = \tan(\alpha \pm \pi/2) = -\cot \alpha$. The coordinates of F are $(r \cos \alpha, r \sin \alpha)$. The equation of l is therefore

$$\begin{aligned} y - r \sin \alpha &= -\cot \alpha(x - r \cos \alpha), \\ y \sin \alpha - r \sin^2\alpha &= -x \cos \alpha + r \cos^2\alpha, \\ x \cos \alpha + y \sin \alpha &= r. \end{aligned}$$

31. $A = 3b/2$ 33. $I = 5t$ 35. \$200

37.

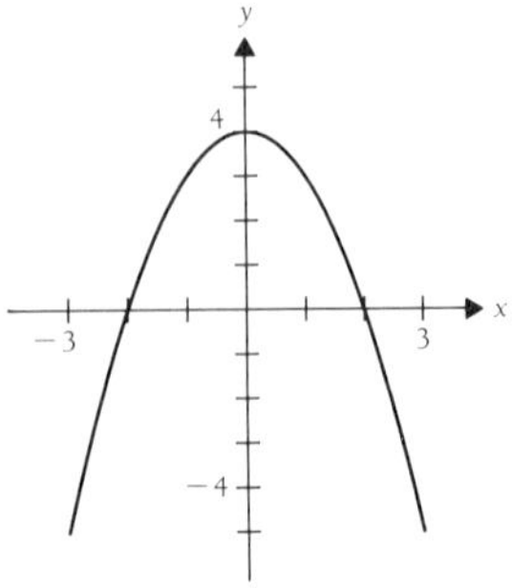

39.

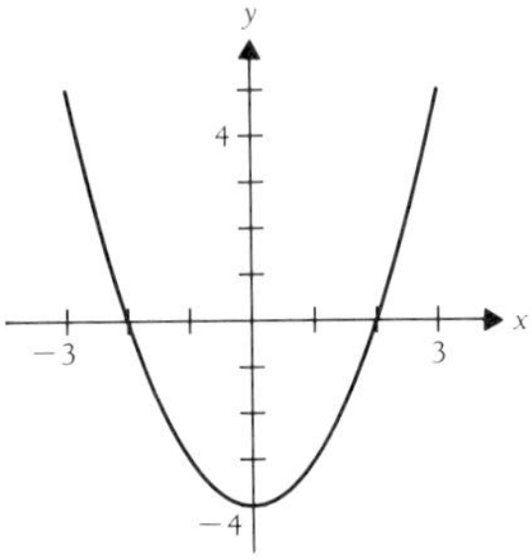

41.

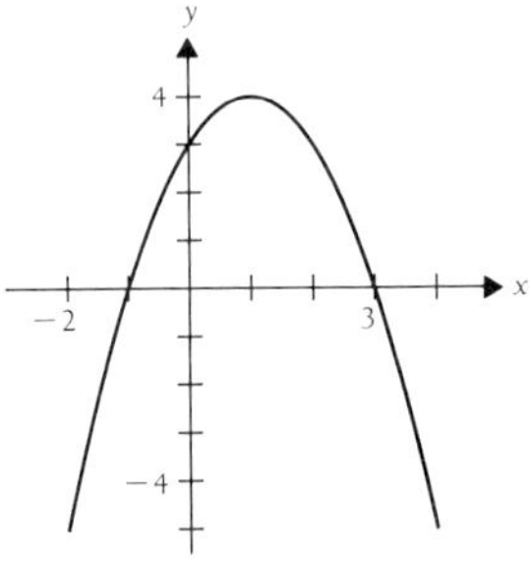

43.

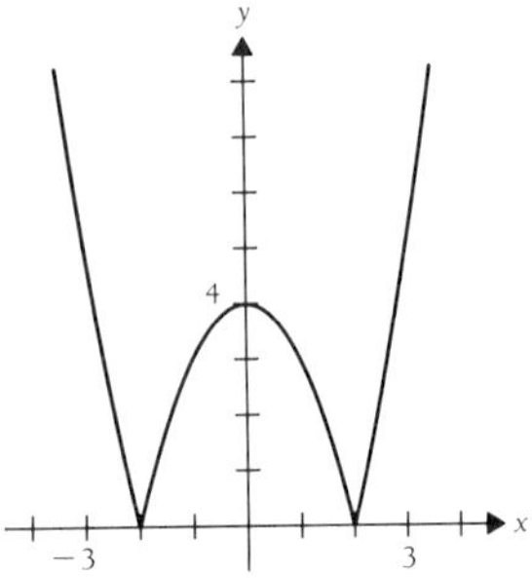

45.

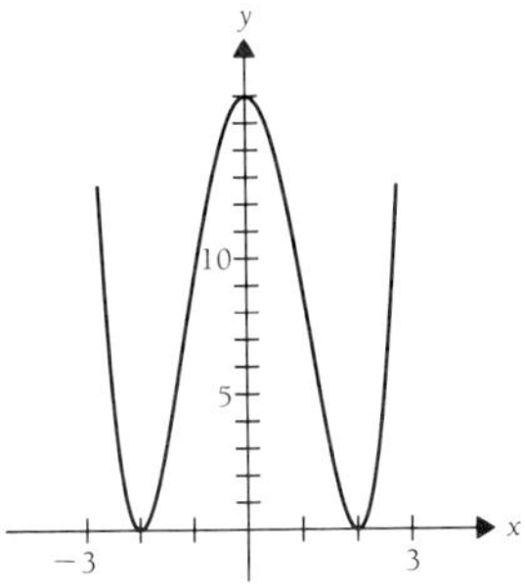

47. Its value is the nearest integer to x.

49.

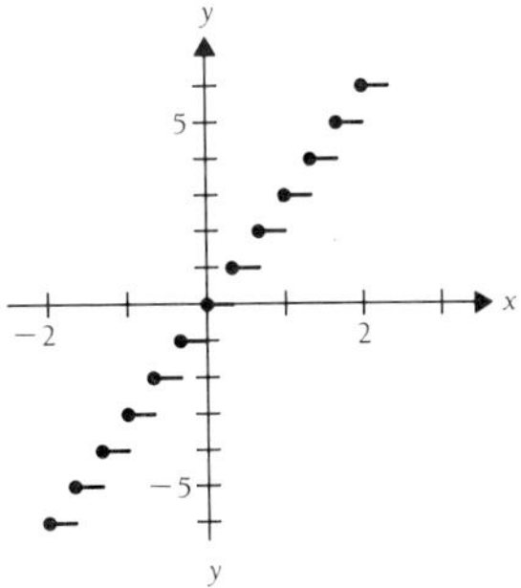

51.

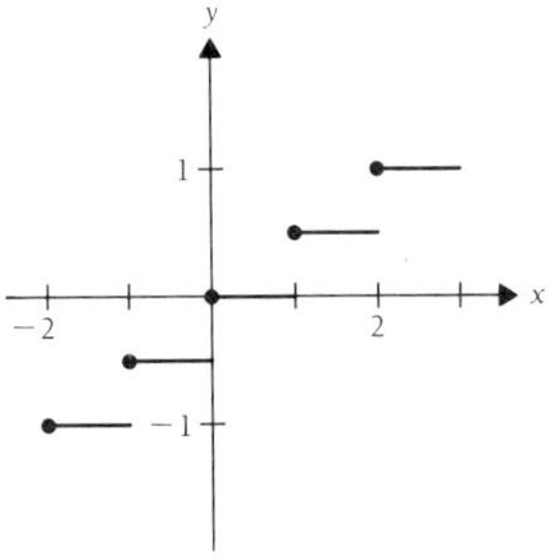

53.

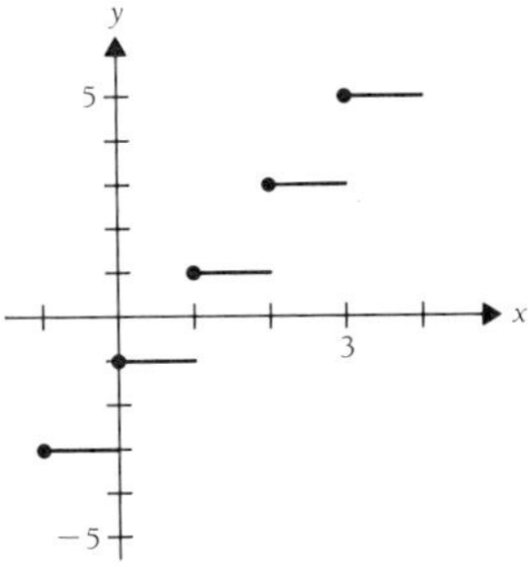

55.

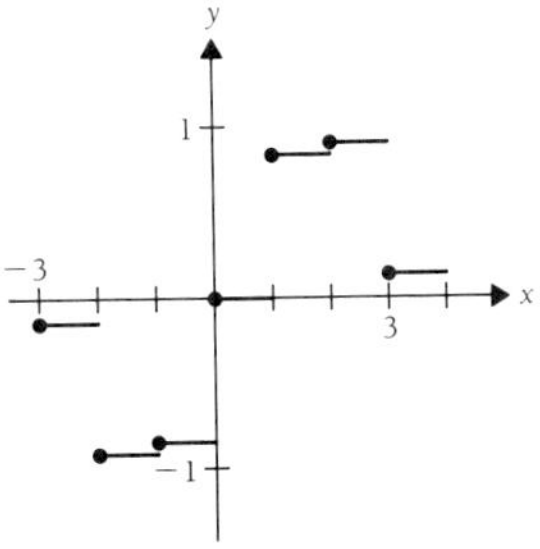

57.

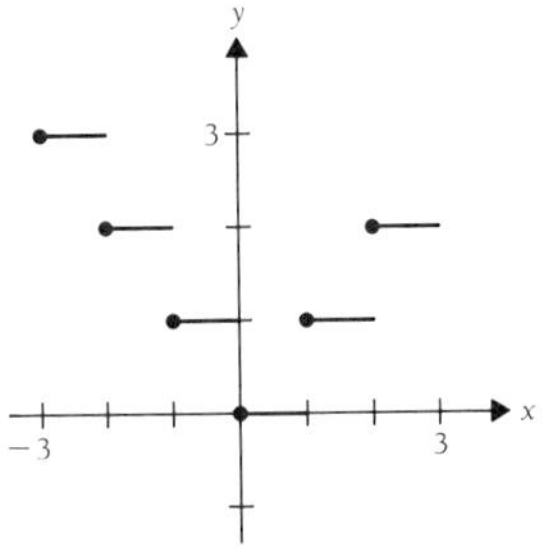

59. Domain $x \neq -3$; range $y \neq 1$ **61.** Domain $\mathbb{R}$; range $[-1, \infty)$ **63.** Domain $x \neq 5$; range $(0, \infty)$
65. Domain $\mathbb{R}$; range $\{0, 1, 2\}$ **67.** $2x - (1/x)$ **69.** $-2h/x(x+h) - h$ **71.** $1/x\ (x \neq 1)$ **73.** $f(x) = mx$
75. $F(x) = 1/x$ **77.** $f(x) = -\frac{5}{2}x^2 + \frac{13}{2}x$ **79.** $f^{-1}(x) = (x^{1/5} + 2)^{1/3}$

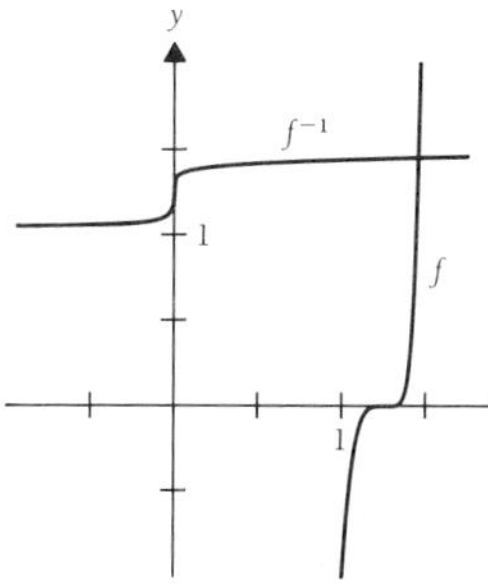

81. No **83.** Yes **85.** Yes **87.** 4 **89.** 3 **91.** 1 **93.** $f(x) = 1/(x-1)(x-2)(x-\sqrt{3})$
95. By the definition of the sine function as the y-coordinate of a point on the unit circle, it is evident that $\sin(x + 2n\pi) = \sin x$, $n \in \mathbb{Z}$. Since $p = 2\pi$ is the smallest positive number for which this is true for all x, the period is 2π.
97. Assume that a is positive. As ax increases from 0 to 2π, x increases from 0 to $2\pi/a$. Thus, the period of $\sin ax$ is $2\pi/a$.
99. If f is an even function, let $(x, f(x))$ be any point on the graph. Then the point $(-x, f(-x)) = (-x, f(x))$ is also on the graph, and this point is symmetrically opposite the first across the y-axis. If f is an odd function, let $(x, f(x))$ be any point on the graph. Then the point $(-x, f(-x)) = (-x, -f(x))$ is also on the graph, and this point is symmetric with the first through the origin.
101. Odd **103.** Neither **105.** Even **107.** Odd **109.** It is even. **111.** It is odd.

Chapter 2 Section 2.1

1. $2\pi - 3$ **3.** -2 **5.** -5 **7.** Does not exist **9.** 1 **11.** Does not exist **13.** 0 **15.** 1 **17.** 5
19. The x_i's approach $\sqrt{2} \approx 1.4142$ (if $x_1 > 0$) or $-\sqrt{2}$ (if $x_1 < 0$), which are the solutions of $x = g(x)$. **21.** $(9 \pm \sqrt{17})/2$
23. Within 0.0005 **25.** Within 0.00025 **27.** 0.0005 **29.** 0.0002 **31.** 0.007996 **33.** 0.00089973
35. Let $\varepsilon > 0$ be given. To make $|\sqrt{x} - 3| < \varepsilon$ by choosing $|x - 9| < \delta$, we note that if $|x - 9| < \delta$, then $9 - \delta < x < 9 + \delta$. Then $\sqrt{9-\delta} < \sqrt{x} < \sqrt{9+\delta}$ and $3 + \sqrt{9-\delta} < \sqrt{x} + 3 < 3 + \sqrt{9+\delta}$. In particular, $\sqrt{x} + 3 > 3 + \sqrt{9-\delta}$, so $1/(\sqrt{x}+3) < 1/(3 + \sqrt{9-\delta})$. Then $|\sqrt{x} - 3| = |(x-9)/(\sqrt{x}+3)| < \delta/(3+\sqrt{9-\delta})$. If we require $\delta/(3+\sqrt{9-\delta}) < \varepsilon$, then we eventually get $\delta(\delta - 6\varepsilon + \varepsilon^2) < 0$. Since $\delta > 0$, we must have $\delta - 6\varepsilon + \varepsilon^2 < 0$, or $\delta < 6\varepsilon - \varepsilon^2$. That is, if we choose $|x - 9| < 6\varepsilon - \varepsilon^2$, then $|\sqrt{x} - 3| < \varepsilon$.
37. -2 **39.** -15 **41.** (b), (c), and (d) **43.** About 1.5 **45.** About 1

Section 2.2

1. -8 **3.** Does not exist **5.** 1 **7.** $\frac{1}{2}$ **9.** $\frac{5}{3}$ **11.** Does not exist **13.** Does not exist **15.** $-\frac{1}{3}$ **17.** -6
19. Does not exist **21.** -1 **23.** $\frac{3}{2}$ **25.** 2 **27.** 0 **29.** $-\frac{1}{6}$ **31.** Does not exist **33.** $1/\sqrt{5}$ **35.** $-\frac{1}{4}$ **37.** Undefined
39. Does not exist **41.** Rule 2.2 **43.** Rule 2.4 **45.** Rule 2.1 **47.** Theorem 2.9
49. $\lim_{x\to c}[f(x) - g(x)] = \lim_{x\to c}[f(x) + (-1)g(x)] = \lim_{x\to c} f(x) + \lim_{x\to c}(-1)g(x) = \lim_{x\to c} f(x) + (-1)\lim_{x\to c} g(x) = \lim_{x\to c} f(x) - \lim_{x\to c} g(x)$.
51. First, consider the case that $M > 0$. Let $\varepsilon > 0$ be given. Then, since $\lim_{x\to c} g(x) = M$, there exists δ_1 such that if $0 < |x - c| < \delta_1$, then $0 < g(x) < M + \varepsilon$. Also, since $\lim_{x\to c} f(x) = 0$, there exists δ_2 such that if $0 < |x - c| < \delta_2$, then $|f(x)| < \varepsilon/(M + \varepsilon)$. Hence, for δ the lesser of δ_1 and δ_2 and $0 < |x - c| < \delta$, we have $0 < |f(x)|g(x) < \varepsilon$, or $|f(x)g(x)| < \varepsilon$, and the limit $\lim_{x\to c} f(x)g(x) = 0$ is proved. The case that $M < 0$ is similar, using $|g(x)|$ and $|M|$.
53. Since $M \neq 0$, choose ε_1 such that $0 < \varepsilon_1 < \frac{1}{2}|M|$. Then there exists δ_1 such that if $0 < |x - c| < \delta_1$, then $|g(x) - M| < \varepsilon_1$, so that $M - \varepsilon_1 < g(x) < M + \varepsilon_1$. Therefore, $|g(x)| > |M| - \varepsilon_1$, so that $1/|g(x)| < 1/(|M| - \varepsilon_1)$. Now given any $\varepsilon > 0$, there exists δ_2 such that if $0 < |x - c| < \delta_2$, then $|g(x) - M| < \varepsilon|M|(|M| - \varepsilon_1)$. Hence, if δ is the lesser of δ_1 and δ_2 and $0 < |x - c| < \delta$, then $|1/g(x) - 1/M| = |M - g(x)|/|Mg(x)| < \varepsilon|M|(|M| - \varepsilon_1)/|M|(|M| - \varepsilon_1) = \varepsilon$, and the limit is proved.

55. $\lim_{x\to c} f(x) = 0$ if and only if for every $\varepsilon < 0$ there exists $\delta < 0$ such that $|f(x) - 0| < \varepsilon$ when $0 < |x - c| < \delta$, if and only if for every $\varepsilon < 0$ there exists $\delta < 0$ such that $||f(x)| - 0| < \varepsilon$ when $0 < |x - c| < \delta$, if and only if $\lim_{x\to c}|f(x)| = 0$.
57. $f(x) = 1$ if x is rational, $f(x) = -1$ if x is irrational
59. Let $\varepsilon > 0$ be given. Since $\lim_{x\to a} f(x) = b$, for any $\varepsilon_1 > 0$ there exists $\delta_1 > 0$ such that $|f(x) - b| < \varepsilon_1$ when $0 < |x - a| < \delta_1$. Note that since f is increasing, if $x \neq a$, then $f(x) \neq b$. Thus, $0 < |f(x) - b| < \varepsilon_1$ when $0 < |x - a| < \delta_1$. From Figure 2.13, we also have the converse, so that $0 < |x - a| < \delta_1$ when $0 < |f(x) - b| < \varepsilon_1$. Letting $y = f(x)$, so that $x = f^{-1}(y)$, we have $|f^{-1}(y) - a| < \delta_1$ when $0 < |y - b| < \delta_1$, which says that $\lim_{y\to b} f^{-1}(y) = a$, or $\lim_{x\to b} f^{-1}(x) = a$.

Section 2.3

1. $\sqrt{2}$ 3. $1/\sqrt{3}$ 5. $\frac{2}{3}$ 7. $\frac{2}{5}$ 9. -4 11. 0 13. 1 15. Does not exist 17. 0 19. 0 21. sin 2 23. 1 25. 0 27. 8
29. Yes; $B = 1$ 31. Yes; $B = 1.56$ 33. Yes; $B = 1$ 35. Yes; $B = 1$
37. If $L > M$, let $\varepsilon = \frac{1}{2}(L - M)$. Then there exists δ such that both $|f(x) - L| < \varepsilon$ and $|g(x) - M| < \varepsilon$ when $0 < |x - c| < \delta$. But then $g(x) < M + \varepsilon = M + \frac{1}{2}(L - M) = \frac{1}{2}(L + M) = L - \frac{1}{2}(L - M) = L - \varepsilon < f(x)$, contradicting $f(x) \le g(x)$. Hence, $L \le M$.
39. Let $\varepsilon > 0$ be given. Then there exists δ such that both $|f(x) - L| < \varepsilon$ and $|g(x) - L| < \varepsilon$ when $0 < |x - c| < \delta$. That is, $L - \varepsilon < f(x) < L + \varepsilon$ and $L - \varepsilon < g(x) < L + \varepsilon$. But since $h(x)$ is between $f(x)$ and $g(x)$, we have $L - \varepsilon < h(x) < L + \varepsilon$. Thus, $|h(x) - L| < \varepsilon$, and the limit $\lim_{x\to c} h(x)$ exists, and equals L.

Section 2.4

1. 6 3. $\frac{1}{2}$ 5. 1 7. ∞ 9. Does not exist 11. ∞ 13. 1 15. 2 17. 1 19. 0 21. 1 23. 0 25. 0 27. 1 29. 1
31. 0 33. 3 35. n 37. 0 39. 0
41. $\lim_{x\to c^+} f(x) = \infty$ if and only if for every $B > 0$ there exists $\delta > 0$ such that if $0 < x - c < \delta$, then $f(x) > B$.
43. $\lim_{x\to c^+} f(x) = -\infty$ if and only if for every $B > 0$ there exists $\delta > 0$ such that if $0 < x - c < \delta$, then $f(x) < -B$.
45. $\lim_{x\to\infty} f(x) = \infty$ if and only if for every $B > 0$ there exists $M > 0$ such that if $x > M$, then $f(x) > B$.
47. $\lim_{x\to\infty} f(x) = -\infty$ if and only if for every $B > 0$ there exists $M > 0$ such that if $x > M$, then $f(x) < -B$.
49. Half of Theorem 2.16 was proved in Example 2.24. It remains to prove that if both the one-sided limits of f at c exist and equal L, then $\lim_{x\to c} f(x) = L$. Given $\varepsilon > 0$, there exists δ so that $|f(x) - L| < \varepsilon$ when $0 < x - c < \delta$ and when $0 < c - x < \delta$. But then if $0 < |x - c| < \delta$, one of these conditions must be satisfied, and $|f(x) - L| < \varepsilon$. Therefore, the two-sided limit exists and equals L.
51. Let $B > 0$ be given. Then there exists δ such that both $f(x) > B + 1$ and $|g(x) - L| < 1$ when $0 < |x - c| < \delta$. Hence, $f(x) + g(x) > B$, and the limit is proved.
53. Let $B > 0$ be given, and let $\varepsilon = \frac{1}{2}L$. Then there exists δ such that both $f(x) > 2B/L$ and $g(x) > \varepsilon$. Hence, $f(x)g(x) > (2B/L)(\frac{1}{2}L) = B$, and the limit is proved.
55. (a) $\lim_{x\to\infty}[\sqrt{x^2 + 2ax} - x] = a$
(b) $\lim_{x\to\infty}\sqrt{x^2 + 2ax} = \infty$ and $\lim_{x\to\infty} x = \infty$, but the limit of their difference is a, which is not necessarily 0.

Section 2.5

1. Removable discontinuity

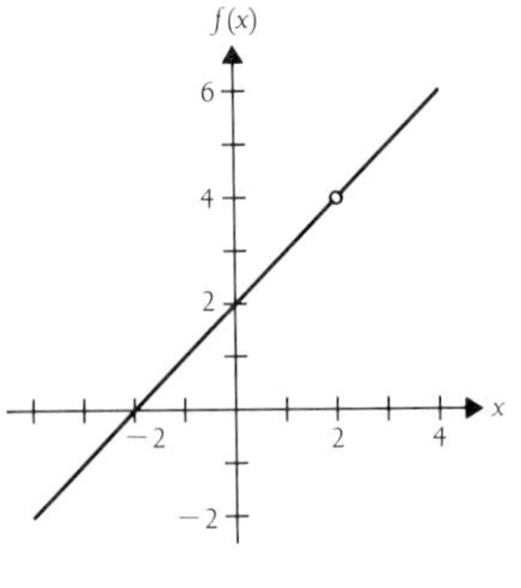

3. Continuous

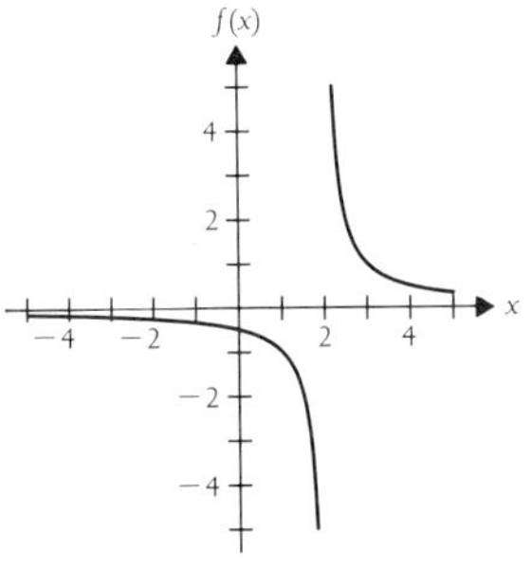

5. Removable discontinuity

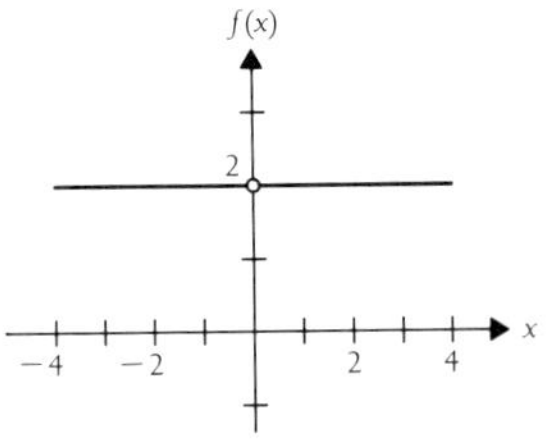

7. Continuous

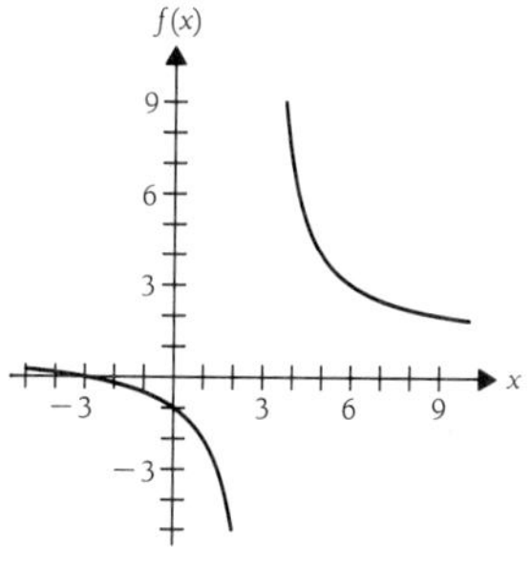

9. Infinite discontinuity

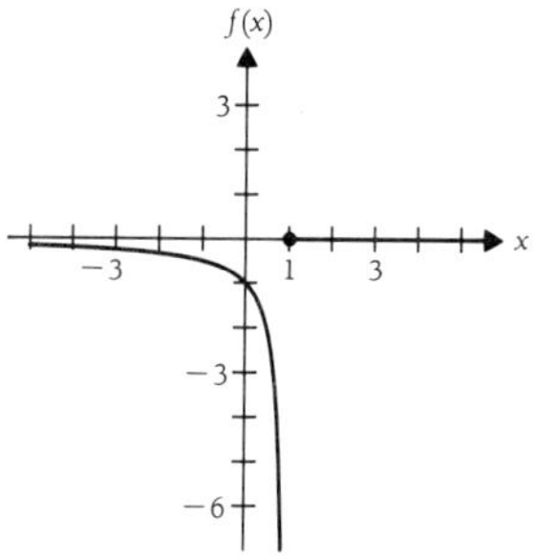

11. Infinite discontinuity

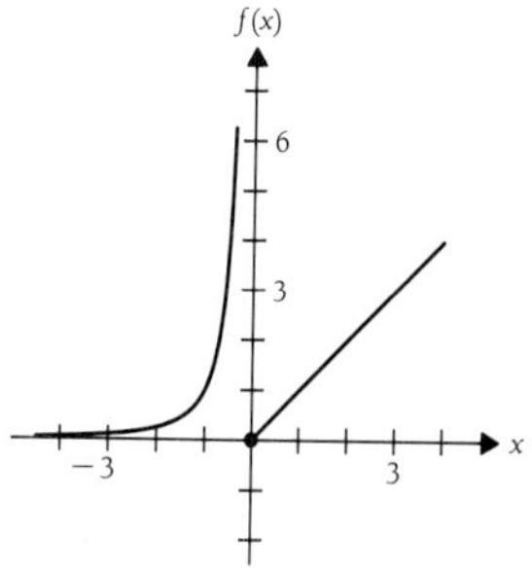

13. Infinite discontinuity

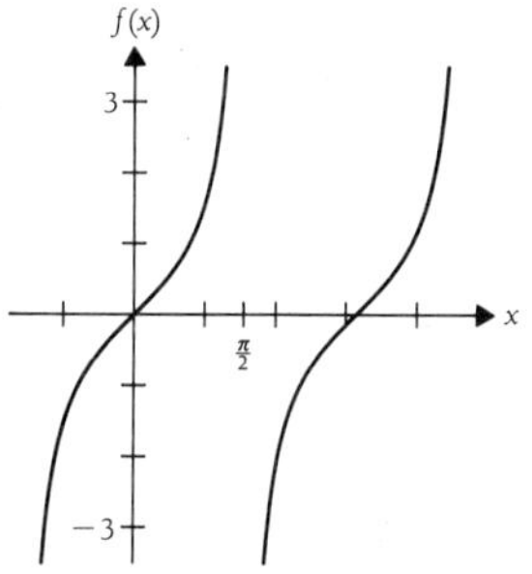

15. Infinite discontinuity

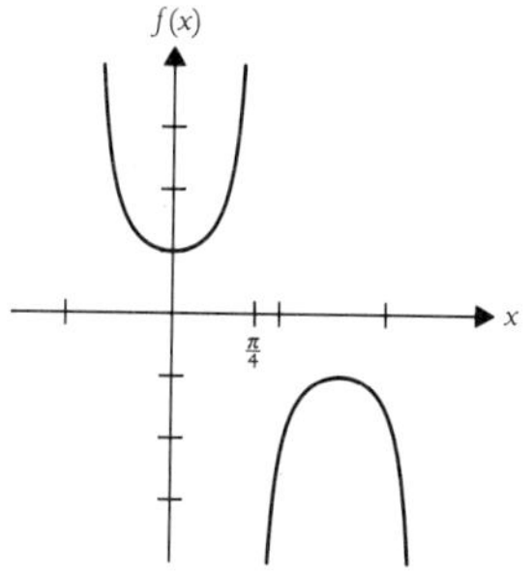

17. Continuous

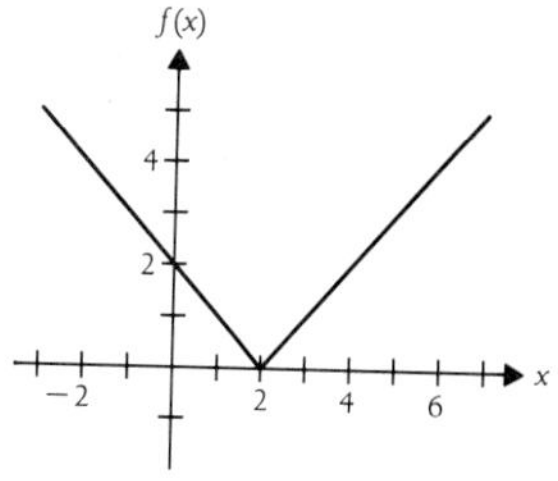

19. Removable discontinuity

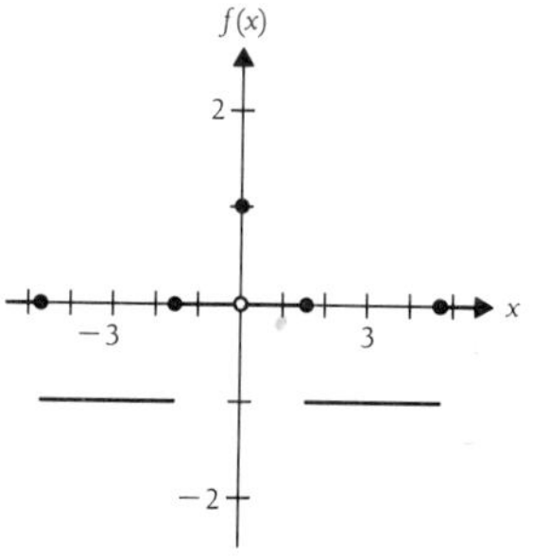

21. No discontinuities; continuous on $\mathbb{R}$ 23. Infinite discontinuity at -1 25. Infinite discontinuity at $(2n+1)\pi/4$
27. $[-2, 2]$ 29. Yes 31. Yes 33. Yes 35. Yes 37. Yes 39. Yes 41. $p(-1) = -1$, $p(0) = 4$ 43. Yes
45. $[1.4125, 1.41875]$ 47. $[-0.75, -0.6875]$ 49. $[0.6875, 0.75]$ 51. -1.21 53. 0.82 55. $L/2^n$
57. Let $K(x)$ be the number of kilometers traveled, and let $L(x)$ be the number of liters of fuel in the tank. Since K starts at 0 and ends positive, and L starts positive and ends at 0, the function $D = K - L$ is negative at the beginning and positive at the end. Therefore, at some point x_0, $D(x_0) = 0$, and $K(x_0) = L(x_0)$.
59. Since $\lim_{x\to c}[af(x) + bg(x)] = a \lim_{x\to c} f(x) + b \lim_{x\to c} g(x)$, $af + bg$ is continuous at c. Some $\lim_{x\to c} f(x)g(x) = \lim_{x\to c} f(x) \lim_{x\to c} g(x)$, fg is continuous at c. Since $g(c)$ is not 0, $\lim_{x\to c}[f(x)/g(x)] = [\lim_{x\to c} f(x)]/[\lim_{x\to c} g(x)]$, and f/g is continuous at c. Finally, $g \circ f$ is continuous at c by Theorem 2.10.
61. Polynomial and rational functions are continuous on their domains, so f and g are continuous at 0, but $g \circ f(x) = 1/x$ has an infinite discontinuity at 0.
63. Clearly, f is continuous on (c, d). Since $\lim_{x\to c} f(x) = f(c)$, we also have $\lim_{x\to c^+} f(x) = f(c)$, so f is continuous on $[c, d)$. Similarly, $\lim_{x\to d^-} f(x) = f(d)$, so f is continuous on $[c, d]$.

Review Exercises

1. 8 3. −1 5. 9 7. $\frac{1}{3}$ 9. Does not exist 11. Does not exist 13. ∞ 15. Does not exist 17. $\frac{1}{10}$ 19. 1 21. $-\frac{1}{2}$
23. 1 25. Does not exist 27. 3 29. ∞ 31. 0 33. 2 35. 0 37. $\frac{2}{3}$ 39. 0 41. $\mathbb{R}$
43. Infinite discontinuity at 0, −1, 1 45. Removable discontinuity at $\pi/2 + n\pi$; infinite discontinuity at $\pi/4 + n\pi$
47. Jump discontinuity at 0 49. Jump discontinuity at 0 51. Jump discontinuity at 0 53. Removable discontinuity at 0
55. Removable discontinuity at 0 57. Removable discontinuity at 0
59. Removable discontinuity at $\pi/2 + 2n\pi$; jump discontinuity at $n\pi$
61. A polynomial of odd degree has opposite signs as $x \to \infty$ and as $x \to -\infty$, so by continuity must cross the x-axis at least once.
63. The position increases continuously from 0° to 90° north latitude, and the temperature changes continuously from 100° to −20°. The difference must therefore be 0 at some point, at which point temperature equals position.

Additional Exercises

1. 0 3. 0 5. ∞ 7. ∞ 9. 6 11. 2 13. −1 15. Does not exist 17. 1 19. −1 21. 14 23. $B = 1/\varepsilon^2$ 25. $f(2) = 7$
27. $f(x) = |x|$
29. First, $f(x) = f(x+0) = f(x) + f(0)$, so $f(0) = 0$. Also, $f(a-b) + f(b) = f(a-b+b) = f(a)$, so $f(a-b) = f(a) - f(b)$. Then $\lim_{x\to c}[f(x) - f(c)] = \lim_{x\to c} f(x-c) = \lim_{z\to 0} f(z) = f(0) = 0$, so $\lim_{x\to c} f(x) = f(c)$, and f is continuous at c. This is true of any c, so f is continuous.
31. Noninteger values
33. No matter how small a δ is chosen, there are points within δ of any point c at which f takes on both 0 and 1 as values. Thus, if $\varepsilon < 1$, the definition of limit cannot be satisfied, and the limit $\lim_{x\to c} D(x)$ does not exist.
35. If x is rational, then $R(x)$ is positive, so $\sigma(R(x)) = 1$; if x is irrational, then $R(x) = 0$ and $\sigma(R(x)) = 0$. Thus, $\sigma(R(x)) = D(x)$.
37. $R(\sigma(0)) = 0$, but $R(\sigma(x)) = 1$ for $x \neq 0$. Thus, $R(\sigma(x))$ has a removable discontinuity at 0.
39. $\sin(R(x))$ is continuous at irrational x and discontinuous at rational x. 41. No
43. For $c \in (np, (n+1)p)$, $\lim_{x\to c} f(x) = \lim_{x\to c-np} f(x) = f(c-np) = f(c)$. Also, $\lim_{x\to 0^-} f(x) = \lim_{x\to p^-} f(x) = f(p) = f(0) = \lim_{x\to 0^+} f(x)$, so $\lim_{x\to 0} f(x) = f(0)$. Hence, $\lim_{x\to np} f(x) = \lim_{x\to 0} f(x) = f(0) = f(np)$. Thus, f is continuous on $\mathbb{R}$.
45. Yes. $R(p/q + 1) = R((p+q)/q) = 1/q = R(p/q)$ for all p/q. If x is irrational, so is $x + 1$, and $R(x+1) = 0 = R(x)$.
47. ∞ 49. 0 51. Does not exist 53. Length of $AB = 2r \sin(\pi/n)$
55. $C = \lim_{n\to\infty} 2rn \sin(\pi/n) = \lim_{n\to\infty} 2\pi r[\sin(\pi/n)/(\pi/n)] = 2\pi r$
57. The perimeter of an inscribed n-gon is $2rn \sin(\pi/n)$. The ratio of the circumference to the diameter is approximately the ratio of the perimeter of the n-gon to the diameter, which is $n \sin(\pi/n)$. This expression is independent of the diameter, and is therefore the same for all circles. Its limit as $n \to \infty$ is the ratio, which is π.
59. Both 61. Neither 63. Right 65. Neither
67. f is continuous on $[a, b]$ if and only if f is continuous in (a, b), left-continuous at a, and right-continuous at b.
69. [1.2079, 2] 71. [1.18, 2]
73. It involves a little more computation, but in most cases it gives better results.

Chapter 3 Section 3.1

1. 3 3. −1 5. m 7. 6 9. −1 11. −1 13. $\frac{2}{9}$ 15. $-\frac{1}{9}$ 17. Does not exist 19. 1 21. $y - 4 = 4(x-2)$
23. $y - 1 = -(x-1)$ 25. $y - 4 = \frac{1}{3}(x-8)$ 27. $y - 2 = \frac{1}{4}(x-3)$ 29. Does not exist 31. $y = x$ 33. Curve undefined
35. Corner 37. Corner 39. Corner (cusp) 41. $y - 4 = -\frac{1}{4}(x-2)$ 43. $y - 4 = -3(x-8)$ 45. $y = -x$
47. $x = 0$ (the y-axis) 49. $y - f(c) = -(1/m_{\tan})(x-c)$

Section 3.2

1. 144 fps 3. 112 fps 5. 144 fps 7. 129.6 fps 9. 96 fps 11. −32 fps² 13. −32 fps² 15. −32 fps² 17. −32 fps²
19. −32 fps² 21. −15,000 gal/hr 23. −10,000 gal/hr 25. −20,000 gal/hr 27. \$11.30/ton 29. \$120/ton
31. $P(x) = 240x - 0.101x^2 - 5000$ 33. \$118.80/ton 35. If $y = kx$, then $y'(x) = \lim_{h\to 0}[k(x+h) - kx]/h = \lim_{h\to 0} k = k$.
37. The average velocity over $[a, b]$ is the slope of line s; the instantaneous velocity at a is the slope of line t.

Section 3.3

1. 4; all x **3.** π; all x **5.** $2x$; all x **7.** $3 - 10x$; all x **9.** $3x^2$; all x **11.** $9x^2 + 4x$; all x **13.** $-1/x^2$; $x \neq 0$
15. $1/(x+1)^2$; $x \neq -1$ **17.** $\frac{1}{2}\sqrt{x-3}$; $x > 3$ **19.** $-\frac{1}{2}(x+2)^{3/2}$; $x > -2$ **21.** $\cos x$; all x **23.** $-2x/(x^2+1)^2$; all x
25. $v = 1 - 2t$; $a = -2$ **27.** $v = -32t$; $a = -32$ **29.** $v = 1 + 1/t^2$; $a = -2/t^3$ **31.** $v = 2\cos 2t$; $a = -4\sin 2t$ **33.** 2π
35. $dV/dr = 2\pi rh/3$; $dV/dh = \pi r^2/3$ **37.** $250 - 0.8x$; 10 **39.** 80; -48 **41.** 32 **43.** Does not exist **45.** $-\frac{3}{4}$ **47.** π
49. A constant function has slope 0. **51.**

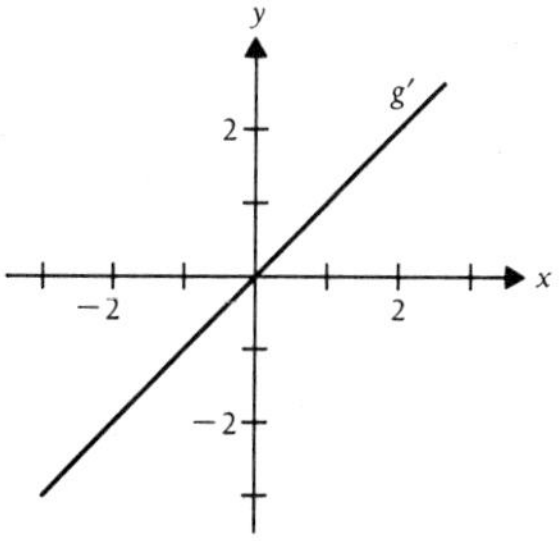

Section 3.4

1. $3x^2$ **3.** $8x^3$ **5.** 0 **7.** $4x + 3$ **9.** $2x - 2x^{-3}$ **11.** $4x - 6$ **13.** $(2x+3)\sin x + (x^2+3x+1)\cos x$
15. $2(2x+3)(x^2+3x+1)$ **17.** $1/(x+1)^2$ **19.** $-2/(2x-1)^2$ **21.** $-2(x^2+1)/(x^2-1)^2$ **23.** $(x\cos x - \sin x)/x^2$
25. $(x+1)(x^2+1) + x(x^2+1) + 2x^2(x+1)$ **27.** $[(1-x^2)\tan x + x(x^2+1)\sec^2 x]/(x^2+1)^2$ **29.** $2(1 + x\tan x)\sec x$
31. $(\tan^2 x + \sec^2 x)\sec x$ **33.** $-1/x^2 + \csc x \cot x$ **35.** $2x\tan x + x^2\sec^2 x$
37. $v = 80 - 32t$; $a = -32$; $v(0) = 80$; $a(0) = -32$ **39.** $v = -32t$; $a = -32$; $v(4) = -128$; $a(4) = -32$
41. $v = 4\cos t$; $a = -4\sin t$; $v(\pi) = -4$; $a(\pi) = 0$ **43.** $v = -4\cos t$; $a = 4\sin t$; $v(0) = -4$; $a(0) = 0$ **45.** $y = 2(x-1)$
47. $y - 1 = 2(x+1)$ **49.** $y = x$ **51.** $y = -(x - \pi)$ **53.** $y = -(x - \pi/2)$ **55.** $y + 1 = x$ **57.** $2s$ **59.** $3s^2$
61. $(\pm 1/\sqrt{3}, \pm\frac{1}{9}\sqrt{3})$ **63.** $D_x k = D_x kx^0 = k\, D_x x^0 = k(0x^{-1}) = 0$

65. $D_x[f(x) + g(x)] = \lim_{h\to 0} \dfrac{f(x+h) + g(x+h) - [f(x) + g(x)]}{h} = \lim_{h\to 0} \dfrac{f(x+h) - f(x)}{h} + \lim_{h\to 0} \dfrac{g(x+h) - g(x)}{h} = f'(x) + g'(x)$

67. $D_x[f(x) - g(x)] = \lim_{h\to 0} \dfrac{f(x+h) - g(x+h) - [f(x) - g(x)]}{h} = \lim_{h\to 0} \dfrac{f(x+h) - f(x)}{h} - \lim_{h\to 0} \dfrac{g(x+h) - g(x)}{h} = f'(x) - g'(x)$

69. $\dfrac{d}{dx}\left[\dfrac{f(x)}{g(x)}\right] = \dfrac{d}{dx}\left[f(x) \cdot \dfrac{1}{g(x)}\right] = f'(x) \cdot \dfrac{1}{g(x)} + f(x) \cdot \dfrac{-g'(x)}{[g(x)]^2} = \dfrac{f'(x)g(x) - f(x)g'(x)}{[g(x)]^2}$

71. $D_x \cot x = D_x \dfrac{1}{\tan x} = \dfrac{-D_x \tan x}{\tan^2 x} = \dfrac{-\sec^2}{\tan^2 x} = -\csc^2 x$

Section 3.5

1. $6(3x+5)$ **3.** $-7(1-x)^6$ **5.** $-64/x^5$ **7.** $3\sin^2 x\cos x$ **9.** $6x(x^2+4)^2$ **11.** $-5(2+4x^3)(2x+x^4)^{-6}$ **13.** $3\cos 3x$
15. $\frac{1}{2}\sec(x/2)\tan(x/2)$ **17.** $3\cos x - 3\sin 3x$ **19.** $-4\sin 4x + \sin x$ **21.** $\cos x\cos 4x - 4\sin x\sin 4x$
23. $8\sin^3 2x\cos 2x$ **25.** $2\cos(2x+3)$ **27.** $2x\cos x - x^2\sin x$ **29.** $\sec^2 x\cos(\tan x)$ **31.** $3x^2\sin(x^2) + 2x^4\cos(x^2)$

33. $\dfrac{8x^3 - 4x - 6(2x-1)(6x^2-1)}{(4x^3-2x)^4}$ **35.** $\dfrac{-90x^4 - 9x^8}{(x^4-6)^3}$ **37.** $\dfrac{4x(2x-3)(x^2+3) - 2(x^2+3)^2}{(2x-3)^2}$ **39.** $\dfrac{-140}{3x^5}\left(\dfrac{5}{3x^4}\right)^6$

41. $\dfrac{(x+2)^2}{(x+3)^3(x+4)^4} + \dfrac{2(x+1)(x+2)}{(x+3)^3(x+4)^4} - \dfrac{3(x+1)(x+2)^2}{(x+3)^4(x+4)^4} - \dfrac{4(x+1)(x+2)^2}{(x+3)^3(x+4)^5}$

43. $20(4x-9)^4(3x+2)^{-7} - 21(4x-9)^5(3x+2)^{-8}$ **45.** $40\sin 5x\cos 5x$ **47.** $-6\cos^2 2x\sin 2x$ **49.** $\sin 2x + 2x\cos 2x$

51. $2\cos^2 x - 4x\cos x\sin x$ **53.** $2\cos(2x-1)$ **55.** $\csc(x+1)[1 - (x+1)\cot(x+1)]$ **57.** $\dfrac{2 + \sin x - x\cos x}{(2+\sin x)^2}$

59. $\dfrac{2(x^2+1)\cos 2x - 2x\sin 2x}{(x^2+1)^2}$ **61.** $-[1 + \cos(x+1)\sin[x + \sin(x+1)]]$

63. $4[2x + (3x+1)^2 + 4x^3]^3[2 + 6(3x+1) + 12x^2]$ **65.** $18x^5$ **67.** $-2x \sin u$ **69.** $32 \sec^2 x$ **71.** $9u^2x^2$
73. $v = 3(2-t)^{-4}$; $a = 12(2-t)^{-5}$; $v(1) = 3$; $a(1) = 12$
75. $v = 12 \cos 3t$; $a = -36 \sin 3t$; $v(\pi/3) = -12$; $a(\pi/3) = 0$ **77.** -42 **79.** $y' = 2x\sqrt{x^2+4}$; $y'(0) = 0$
81. $y = \cos 3x = 4\cos^3 x - 3\cos x = (4\cos x \cos x - 3)\cos x$; $y' = [4\cos x(-\sin x) + 4(-\sin x)\cos x]\cos x + (4\cos x \cos x - 3)(-\sin x) = -12\cos^2 x \sin x + 3\sin x$
83. $(\pi/180)\cos x^\circ$

Section 3.6

1. $2x + 3$; 2 **3.** $(1-x)^{-2}$; $2(1-x)^{-3}$ **5.** $18(2x+7)^8$; $288(2x+7)^7$ **7.** $\sin 2x$; $2\cos 2x$ **9.** $a\cos ax$; $-a^2 \sin ax$
11. $\sec^2 x$; $2\sec^2 x \tan x$ **13.** $a\sec^2 ax$; $2a^2 \sec^2 ax \tan ax$ **15.** $n \sin^{n-1}x \cos x$; $n(n-1)\sin^{n-2}x \cos^2 x - n\sin^n x$
17. $2a \sin ax \cos ax = a \sin 2ax$; $2a^2 \cos 2ax$ **19.** $\sin x + x\cos x$; $2\cos x - x \sin x$
21. $\cot x - x\csc^2 x$; $-2\csc^2 x + 2x\csc^2 x \cot x$
23. $2\sin 2x/(1+\cos^2 x)^2$; $\dfrac{4\cos 2x}{(1+\cos^2 x)^2} + \dfrac{4\sin^2 2x}{(1+\cos^2 x)^3}$ **25.** $6(1-x)^{-4}$ **27.** $-a^3 \cos ax$ **29.** $360x - 48$ **31.** $7!(x+3)^{-8}$
33. $(-1)^n n!(x+1)^{-(n+1)}$
35. $a^n \sin ax$ if $n = 4m$; $a^n \cos ax$ if $n = 4m+1$; $-a^n \sin ax$ if $n = 4m+2$; $-a^n \cos ax$ if $n = 4m+3$
37. -32 **39.** $\frac{74}{1728}$ **41.** 0 **43.** $-\pi^2\sqrt{2}$
45. By induction on the degree of $p(x)$: If $p(x)$ has degree 1, say $p(x) = ax + b$, then $p'(x) = a$ and $p''(x) = 0$. Suppose that the $(n+1)$th derivative of a polynomial of degree n is 0. Let $p(x)$ be a polynomial of degree $n+1$. Then $p'(x)$ is a polynomial of degree n, so the $(n+1)$th derivative of p' is 0. That is, the $(n+2)$th derivative of p is 0. This completes the inductive step.

Section 3.7

1. 10.981; 10.8 **3.** 1.21π; 1.2π **5.** 1.05 **7.** 1.01 **9.** $7\frac{15}{16}$ **11.** $144\frac{1}{18}$ **13.** $3x^2\,dx$ **15.** $(2x+2)\,dx$
17. $8x(x^2+3)^3\,dx$ **19.** $-\dfrac{16x}{(x^2+4)^2}\,dx$ **21.** $4\cos 4x\,dx$ **23.** $-6\cos^2 2x \sin 2x\,dx$ **25.** $-2\tan x \sec^2 x\,dx$
27. $-3x^2 \csc(x^3)\,dx$ **29.** $A = s^2$; $dA = 2s\,ds = 2s(0.01s) = 0.02s^2 = 0.02A$ **31.** 3% **33.** 0.298 m^3 **35.** 103.8 in.^3
37. 173.2 ± 0.8 ft **39.** 19,040 to 19,360
41. As x changes in value by dx, the value along the tangent line changes by $f'(x)\,dx = dy$, and the value along a line with slope m changes by $m\,dx$. The relative error between Δy and dy is $(\Delta y - dy)/dx = [f(x+dx) - f(x)]/dx - f'(x)$, which approaches 0 as $dx \to 0$, while the relative error between Δy and $m\,dx$ is $(\Delta y - m\,dx)/dx = [f(x+dx) - f(x)]/dx - m$, which does not approach 0 as $dx \to 0$ unless $m = f'(x)$. Thus, for small dx, we can expect larger error if we use a linear approximation other than that along the tangent line.

Section 3.8

1. $-x/y$ **3.** -1 **5.** $2x/3y^2$ **7.** xy^3 **9.** $-\frac{1}{2}x^{-3/2}$ **11.** $x/\sqrt{x^2+1}$ **13.** $-\sqrt{y}/\sqrt{x}$ **15.** $(2y\sqrt{y}/\sqrt{x}) + (y/x)$ **17.** $\sec y$

19. $\dfrac{\sin y}{1 - x\cos y}$ **21.** $\dfrac{\cos x}{2y}$ **23.** $\dfrac{\cos(x+y) - y}{x - \cos(x+y)}$ **25.** $\dfrac{1 - y\cos xy}{x\cos xy - 1}$ **27.** $\dfrac{1}{1 + \cos y \sec^2(\sin y)}$ **29.** $\dfrac{4x}{|2x|}$ **31.** $\dfrac{3(3x-1)}{|3x-1|}$
33. $\dfrac{\sin x \cos x}{|\sin x|}$ **35.** $-\dfrac{x|y|}{y|x|}$ **37.** $-\dfrac{1}{y^3}$ **39.** $\frac{1}{2}x\sqrt{x}$ **41.** 0 **43.** $\dfrac{2(x+y)}{x^2}$ **45.** $-(1-x^2)^{-3/2}$ **47.** $\frac{1}{4}x^{-3/2}(x-1)^{-5/2}(4x-1)$
49. $x \in (0, \infty)$ **51.** $x \in (2n\pi, (2n+1)\pi)$ **53.** $y - 16 = \frac{8}{3}(x-8)$ **55.** $8y - 3\sqrt{3} = -\sqrt{3}(8x-1)$ **57.** $y - \frac{2}{3} = \frac{5}{4}(x - \frac{4}{3})$
59. $y = 4^{1/3}$ **61.** $y - 1/\sqrt{2} = -(x - 1/\sqrt{2})$ **63.** $y = 3^{1/4}/2^{3/2}$

65. $y = \sqrt{4 - x^2}$, $y = -\sqrt{4 - x^2}$

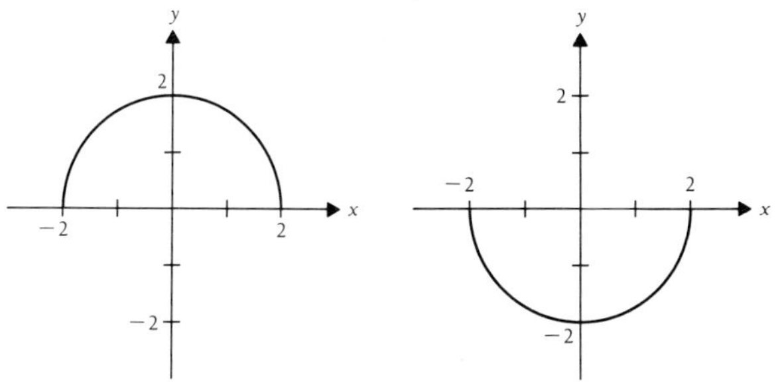

67. $y = 1 - |x|$, $x \in [-1, 1]$; $y = |x| - 1$, $x \in [-1, 1]$

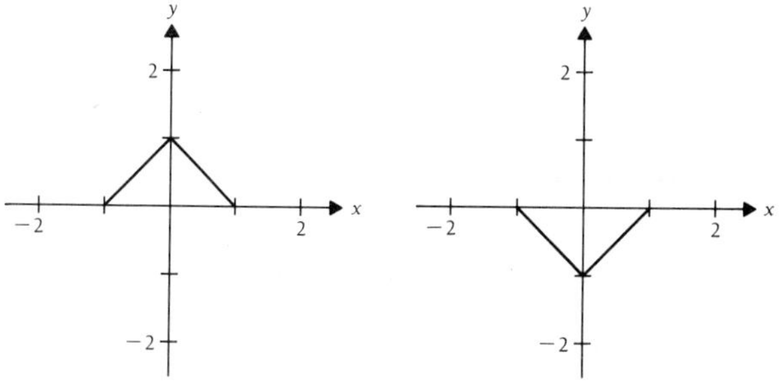

69. Let f be an even function. Then $f'(x) = D_x f(x) = D_x f(-x) = f'(-x)(-1)$, so that $f'(-x) = -f'(x)$. That is, f' is an odd function.

71.

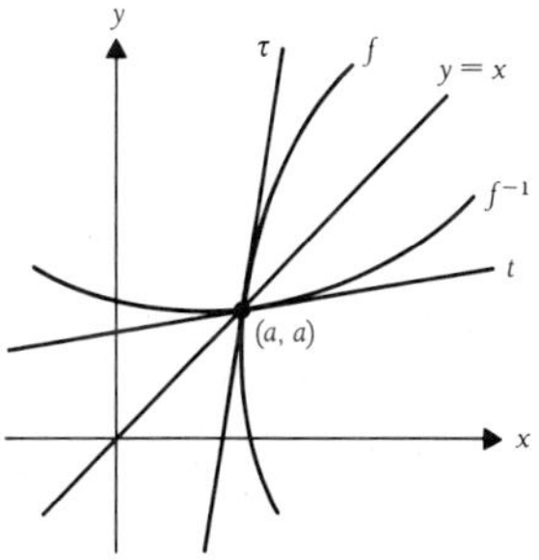

Review Exercises

1. 4 **3.** $-6x^{-3}$ **5.** $-2/x^2$ **7.** $\frac{3}{5}x^{-2/5}$ **9.** 2 **11.** $2x + 4$ **13.** $\frac{3}{2}\sqrt{x}$ **15.** $8(x+5)^7$ **17.** $3(1-x)^{-4}$
19. $(1/x^2 - 1)(x + 1/x)^{-2}$ **21.** $10x(x^2+4)^4 - 8x(x^2+5)^3$ **23.** $-(x+6)/x^3$ **25.** $-2(x^2 - 3x + 2)/(x+3)^5$
27. $5(x+4)^4(2x-1)^{-4} - 8(x+4)^5(2x-1)^{-5}$ **29.** $2 \cos x$ **31.** $2 \sin x \cos x$ **33.** $-2 \csc^2 2x$ **35.** $\csc 2x(1 - 2x \cot 2x)$
37. $3x/|x|$ **39.** $2(3+2x)/|3+2x|$ **41.** $\sin^2 x + 2x \sin x \cos x$ **43.** $-(\sin x \cos x)/|\cos x|$ **45.** $-2 - y/x$
47. $(2y - 4x\sqrt{y})/(4\sqrt{y} - x)$ **49.** $1/2\sqrt{x}(2 + 3y^2)$ **51.** $\sqrt{y}/\sqrt{x}$ **53.** $-y/x$ **55.** $\cos(x+y)/[2y - \cos(x+y)]$
57. $6(x^2+9)(5x^2+9)$ **59.** $\dfrac{-1}{4x\sqrt{x}}$ **61.** $9 \sec 3x\,(\tan^2 3x + \sec^2 3x)$ **63.** $2 \cos 2x - \sin x$ **65.** $2(x+y)/x^2$ **67.** $2y/x^2$
69. $(-1)^n n!(x+1)^{-(n+1)}$ **71.** $m(m-1)(m-2) \cdot \cdots \cdot (m-n+1)x^{m-n}$ **73.** T: $y - 5 = 2(x-1)$; N: $y - 5 = -\frac{1}{2}(x-1)$
75. T: $y - \sqrt{3} = (2/\sqrt{3})(x-2)$; N: $y - \sqrt{3} = -(\sqrt{3}/2)(x-2)$ **77.** T: $y - 2 = \frac{1}{4}(x-3)$; N: $y - 2 = -4(x-3)$
79. T: $y + \sqrt{2}/2 = (3/\sqrt{2})(x - \pi/4)$; N: $y + \sqrt{2}/2 = -(\sqrt{2}/3)(x - \pi/4)$
81. T: $y + \frac{3}{2} = (3\sqrt{3}/2)(x - \sqrt{3})$; N: $y + \frac{3}{2} = -(\frac{2}{9}\sqrt{3})(x - \sqrt{3})$ **83.** T: $y - \frac{3}{2} = (3\sqrt{5}/2)(x - \sqrt{5})$; N: $y - \frac{3}{2} = -(\frac{2}{15}\sqrt{5})(x - \sqrt{5})$

85. $v = 8t$; $a = 8$; $v(2) = 16$; $a(2) = 8$ **87.** $v = 5/t^2$; $a = -10/t^3$; $v(5) = \frac{1}{5}$; $a(5) = -\frac{2}{25}$
89. $v = \sin t + t\cos t$; $a = 2\cos t - t\sin t$; $v(\pi) = -\pi$; $a(\pi) = -2$
91. $v = -12\sin(3t-1)$; $a = -36\cos(3t-1)$; $v(0) = 12\sin 1$; $a(0) = -36\cos 1$ **93.** 2%; 4%
95. Increases by $(2B\cot B)$%

Additional Exercises

1. $x^3 - 3x^2 + 4 = (x-2)^2(x+1)$ **3.** $(x^2 - 4x + 4) - 0 = (x-2)^2$
5. Let $f(x) = p(x) - (mx+b) = p(x) - mx - b + 0 = p(x) - mx - b + [mc + b - p(c)] = p(x) - p(c) - m(x-c)$. Since $f(c) = 0$, $x - c$ is a factor of $f(x)$, or $f(x) = (x-c)q(x)$ for some polynomial $q(x)$. Hence, $p(x) - p(c) - m(x-c) = (x-c)q(x)$, so that

$$\frac{p(x)-p(c)}{x-c} - m = q(x).$$

Now,

$$q(c) = \lim_{x\to c} q(x) = \lim_{x\to c}\left[\frac{p(x)-p(c)}{x-c} - m\right] = p'(c) - m = 0,$$

so $x - c$ is a factor of $q(x)$. Thus, $q(x) = (x-c)r(x)$ for some $r(x)$, and $f(x) = (x-c)^2 r(x)$. Hence, $y = p(x)$ and $y = mx + b$ have a double intersection at $x = c$.
7. $D_x[c \cdot f(x)] = D_x c \cdot f(x) + c \cdot D_x[f(x)] = 0 \cdot f(x) + c \cdot f'(x) = c \cdot f'(x)$
9. $\dfrac{dy}{dx} = \dfrac{dy}{du}\dfrac{du}{dx}$, so $\dfrac{d^2y}{dx^2} = \dfrac{d}{dx}\left(\dfrac{dy}{du}\right)\dfrac{du}{dx} + \dfrac{dy}{du}\dfrac{d}{dx}\left(\dfrac{du}{dx}\right) = \left[\dfrac{d}{du}\left(\dfrac{dy}{du}\right)\dfrac{du}{dx}\right]\dfrac{du}{dx} + \dfrac{dy}{du}\dfrac{d}{dx}\left(\dfrac{du}{dx}\right) = \dfrac{d^2y}{du^2}\left(\dfrac{du}{dx}\right)^2 + \left(\dfrac{dy}{du}\right)\dfrac{d^2u}{dx^2}$
11. By induction, assume true for n. Then

$$\frac{d^{n+1}(u+v)}{dx^{n+1}} = \frac{d}{dx}\left(\frac{d^n u}{dx^n} + \frac{d^n v}{dx^n}\right) = \frac{d}{dx}\left(\frac{d^n u}{dx^n}\right) + \frac{d}{dx}\left(\frac{d^n y}{dx^n}\right) = \frac{d^{n+1}u}{dx^{n+1}} + \frac{d^{n+1}y}{dx^{n+1}}$$

13. If $y = f^{-1}(x)$, then $x = f(y)$, and differentiating implicitly, $1 = f'(y)(dy/dx)$. Thus, $f'(y) = 1/(dy/dx)$. Now differentiating $x = f(y)$ with respect to y, we get $dx/dy = f'(y)$. Hence, $dx/dy = 1/(dy/dx)$.
15. (0, 1) **17.** $(\sqrt{3}, 0)$, $(-\sqrt{3}, 0)$ **19.** $f'(2^+)$ does not exist; $f'(2^-) = -\infty$ (i.e., does not exist either)
21. $f'(2^+) = \infty$, $f'(2^-) = -\infty$ **23.** $f'(2^+) = \cos 2$, $f'(2^-) = \cos 2$ **25.** Use Theorem 2.16. **27.** $(c, 0)$ **29.** None
31. $(1, 1)$, $(-1, 1)$ **33.** $(n\pi, 0)$ **35.** $(0, 0)$ **37.** $(0, 0)$ **39.** $\sqrt{3}y - 3 = x + 1$ **41.** $y = 4$
43. If $f(c) = 0$, then $x - c$ is a factor of $f(x)$, or $f(x) = (x-c)q(x)$ for some polynomial $q(x)$. Then $f'(x) = q(x) + (x-c)q'(x)$, so that $f'(c) = q(c)$. But $f'(c) = 0$ means $q(c) = 0$, so $x - c$ is also a factor of $q(x)$. Thus, $q(x) = (x-c)r(x)$, and $f(x) = (x-c)^2 r(x)$. That is, c is a double zero of f.
45. If both

$$\lim_{x\to c^+}\frac{f(x)-f(c)}{x-c} = \infty \quad\text{and}\quad \lim_{x\to c^-}\frac{f(x)-f(c)}{x-c} = \infty, \quad\text{then}\quad \lim_{x\to c}\frac{f(x)-f(c)}{x-c} = \infty.$$

Hence, by Definition 3.1, there is a vertical tangent line at c. (We are assuming f is continuous.)

Chapter 4 Section 4.1

1. $\frac{4}{3}$ ft²/min **3.** $\frac{15}{4}$ m²/hr **5.** $\pi/4$ ft³/min **7.** $\frac{1}{2}$ ft/min **9.** $15/\sqrt{17}$ m/sec **11.** $\frac{1}{16}\sqrt{2}$ ft/sec **13.** $655/\sqrt{421}$ mph
15. $4(13 - 6\sqrt{2})^{1/2}$ knots **17.** Increasing at $4/15\sqrt{35}$ rad/sec **19.** $\frac{65}{6}$ ft/min **21.** 6 ft/min

23. Decreasing at $0.051k$ kg/cm^2/sec **25.** Increasing at \$200 per month **27.** Increasing at 5 per month **29.** 1 ft^3/sec
31. Increasing at 0.9% per month **33.** 747.9 mph

Section 4.2

1. None **3.** $0, \frac{3}{4}$ **5.** $2, -2$ **7.** $0, 3, -3$ **9.** 1 **11.** $2, -2$ **13.** $\pi/4 + n\pi$ **15.** $\pi/6 + 2n\pi, 5\pi/6 + 2n\pi, 3\pi/2 + 2n\pi$
17. $\pi + 2n\pi$ **19.** $n\pi/2$ **21.** 1 **23.** 2 **25.** $\cos x = 2/\pi$ **27.** $\sin 2x = 2/\pi$ **29.** Not differentiable at 0
31. Not continuous at 0 **33.** Not continuous at 2 **35.** $\sqrt{3}$
37. The equation of the secant line through $(a, f(a))$ and $(b, f(b))$ is

$$y = f(a) + \frac{f(b) - f(a)}{b - a}(x - a),$$

so the vertical distance between the line and the graph of f is

$$g(x) = f(x) - y = f(x) - f(a) - \frac{f(b) - f(a)}{b - a}(x - a).$$

39. By the mean value theorem, there exists c such that $f'(c) = [f(x) - f(a)]/(x - a)$. Now, simply solve for $f(x)$.
41. If $x_1 = 0$, then $x_2 = 1$, and $f'(1) = 0$, so computation must stop. **43.** After twelve iterations, the zero -2.1038 is found.
45. -0.6823 **47.** 3.8284 **49.** 4.6211 **51.** -1.5911 **53.** 0.7391 **55.** 0.8241 **57.** $-0.4552, 0.6452, 6.8100$
59. $-2.6458, -1.6180, 0.6180, 2.6458$ **61.** $-4.2361, -2.7417, 0.2361, 0.6667, 4.7417$
63. $-4.3594, -3.9711, -2.4475, -1.6854, -0.4908, 0.5611, 1.4712, 2.8175, 3.4146$
65. If $f(z) = 0$, then $g(z) = z - f(z)/f'(z) = z - 0 = z$.
67. Suppose $x_1 < z$ (the other case is similar). Then by the mean value theorem, there is some $c \in (x_1, z)$ such that $g(z) - g(x_1) = g'(c)(z - x_1)$. Now, $g(z) = z$, $g(x_1) = x_2$, and $|g'(c)| < 1$, so $|z - x_2| = |g'(c)||z - x_1| < |z - x_1|$. That is, x_2 is closer to z than x_1 is.
69. If f' is nonzero, then choose an interval $[c, d]$ containing z and lying in the original interval, and let m be the minimum absolute value of f' in $[c, d]$. Let $|f''(x)| < B$ for $x \in [c, d]$. Since f is continuous and has a zero z in $[c, d]$, we can choose a and b close enough to z so that $|f(x)| < m^2/10B$ for $x \in (a, b)$. Then $|g'(x)| = |f(x)f''(x)/[f'(x)]^2| < [m^2/10B][B/m^2] = 0.1$.
71. Given ε, choose n so that $(0.1)^n < \varepsilon/(b - a)$. Then apply Problem 70 n times.
73. For example, use $f(x) = \tan x$ on $(-\pi/2, \pi/2)$.

Section 4.3

1. No critical points;
increasing on $\mathbb{R}$;
no concavity;
no points of inflection

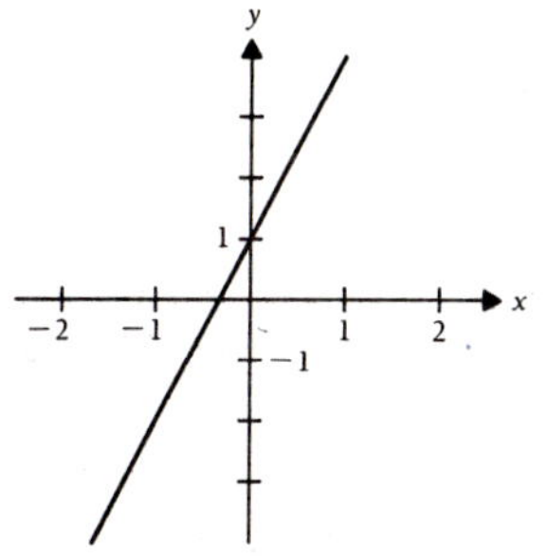

3. Critical points $-1, 1$;
increasing on $(-\infty, -1)$ and $(1, \infty)$,
decreasing on $(-1, 1)$;
concave down on $(-\infty, 0)$,
concave up on $(0, \infty)$;
point of inflection 0

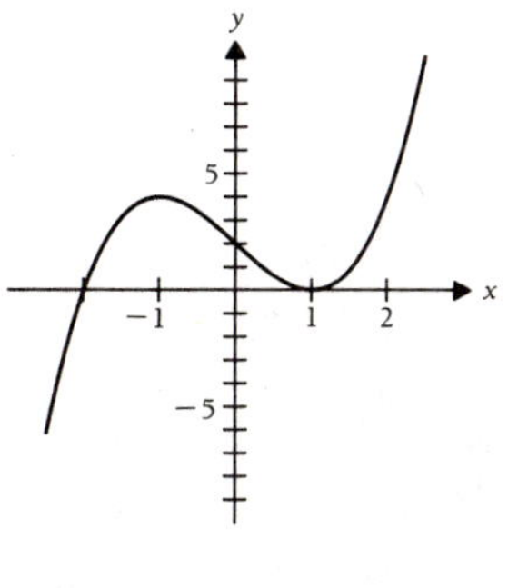

5. Critical points $4, -\frac{1}{2}$;
increasing on $(-\infty, -\frac{1}{2})$ and $(4, \infty)$,
decreasing on $(-\frac{1}{2}, 4)$;
concave down on $(-\infty, \frac{7}{4})$,
concave up on $(\frac{7}{4}, \infty)$;
point of inflection $\frac{7}{4}$

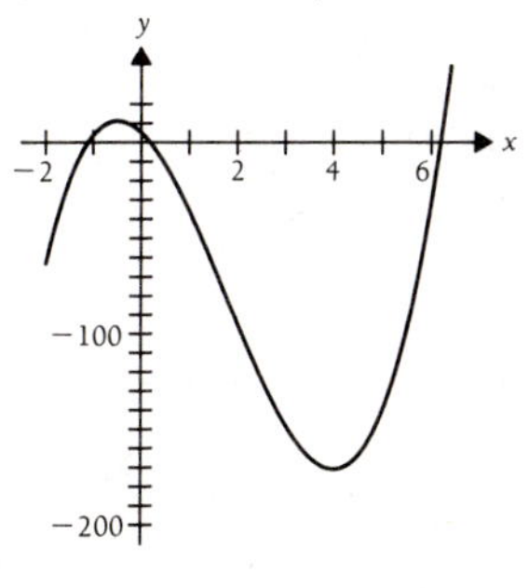

7. No critical points;
increasing on $(-\infty, 0)$ and $(0, \infty)$;
concave up on $(-\infty, 0)$,
concave down on $(0, \infty)$;
no point of inflection

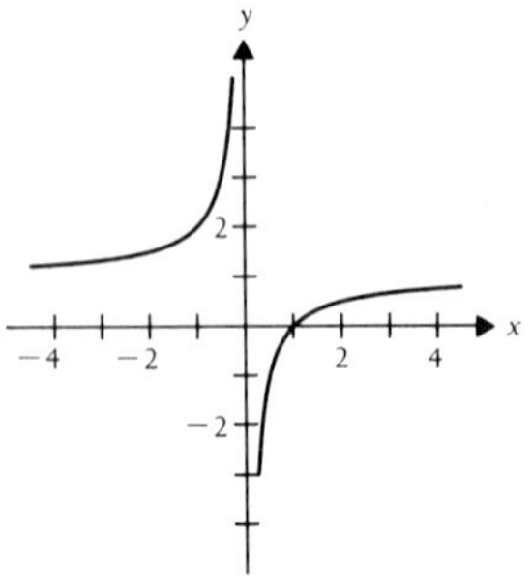

9. Critical points $-4, 0, 4$;
decreasing on $(-\infty, -4)$ and $(0, 4)$,
increasing on $(-4, 0)$ and $(4, \infty)$;
concave up on $(-\infty, -4)$ and $(4, \infty)$,
concave down on $(-4, 4)$;
inflection points $-4, 4$

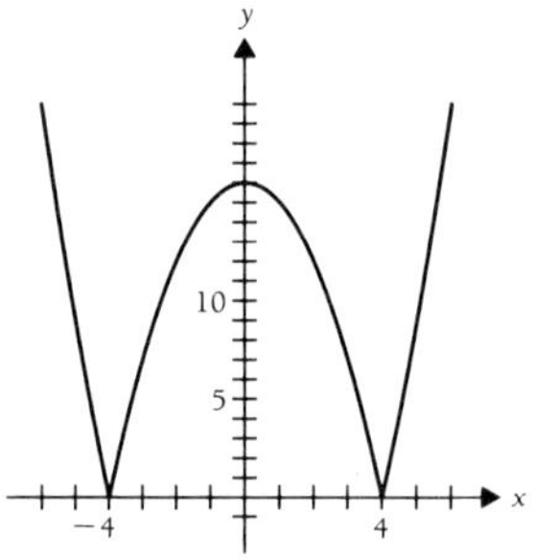

11. No critical points;
increasing on $(-\frac{1}{2}, \infty)$;
concave down on $(-\frac{1}{2}, \infty)$;
no inflection points

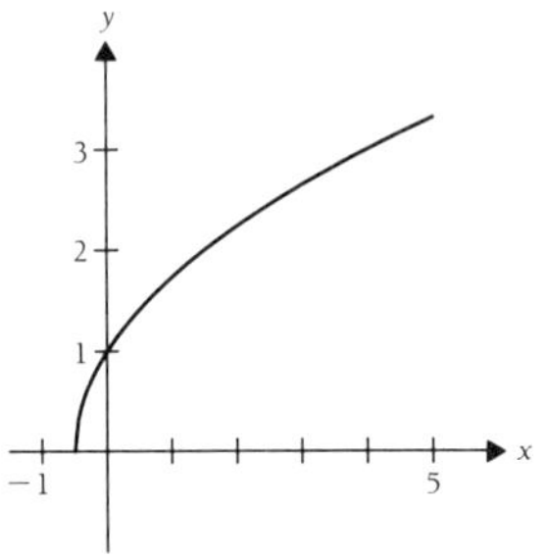

13. No critical points;
increasing on $\mathbb{R}$;
concave down on $(-\infty, 0)$,
concave up on $(0, \infty)$;
inflection point 0

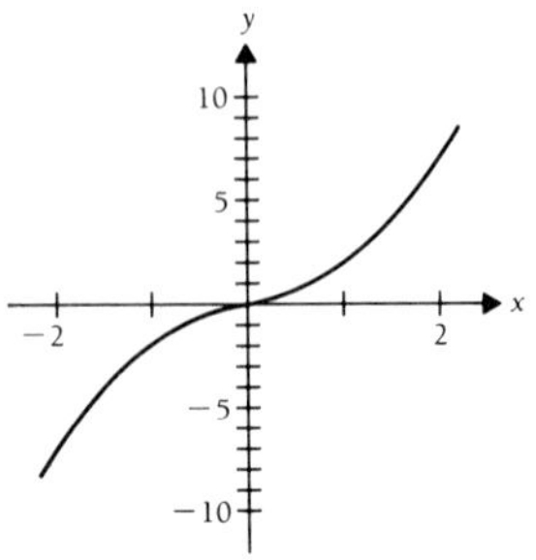

15. Critical points $\pi/4, 5\pi/4$;
increasing on $(0, \pi/4)$ and $(5\pi/4, 2\pi)$,
decreasing on $(\pi/4, 5\pi/4)$;
concave down on $(0, 3\pi/4)$ and $(7\pi/4, 2\pi)$,
concave up on $(3\pi/4, 7\pi/4)$;
inflection points $3\pi/4, 7\pi/4$

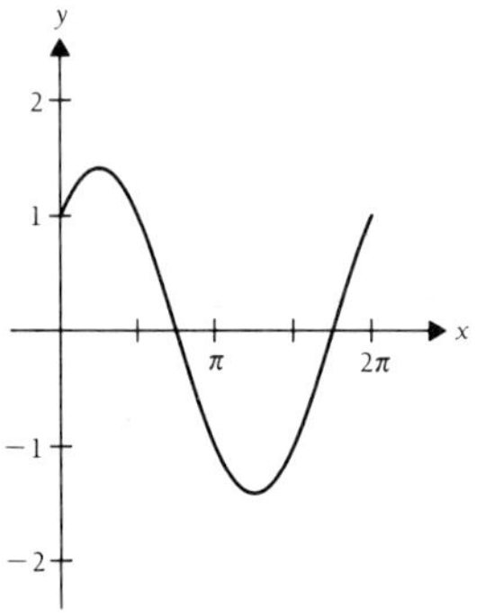

17. Critical points $\pm 2\pi/3 + 2n\pi$;
increasing on
$(-2\pi/3 + 2n\pi, 2\pi/3 + 2n\pi)$,
decreasing on
$(2\pi/3 + 2n\pi, 4\pi/3 + 2n\pi)$;
concave down on $(2n\pi, \pi + 2n\pi)$,
concave up on
$(\pi + 2n\pi, 2\pi + 2n\pi)$;
inflection points $n\pi$

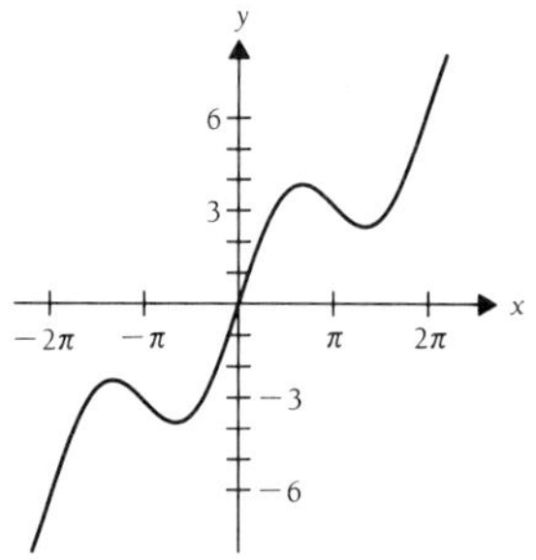

19. Critical points $n\pi/2$;
increasing on $(n\pi, \pi/2 + n\pi)$,
decreasing on $(\pi/2 + n\pi, \pi + n\pi)$;
concave down on
$(\pi/4 + n\pi, 3\pi/4 + n\pi)$,
concave up on
$(-\pi/4 + n\pi, \pi/4 + n\pi)$;
inflection points $\pi/4 + n\pi/2$

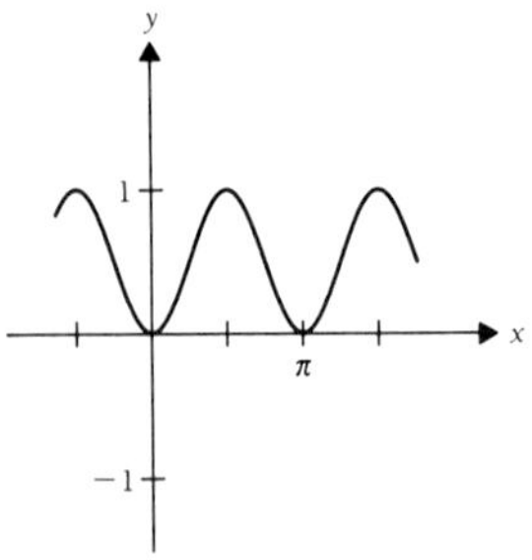

21. Critical points 1, 9;
increasing on (1, 3) and (3, 9),
decreasing on $(-\infty, -3)$, $(-3, 1)$, and $(9, \infty)$;
concave down on $(-\infty, -3)$ and (3, 13.214),
concave up on $(-3, 3)$ and $(13.214, \infty)$;
inflection point ≈ 13.214

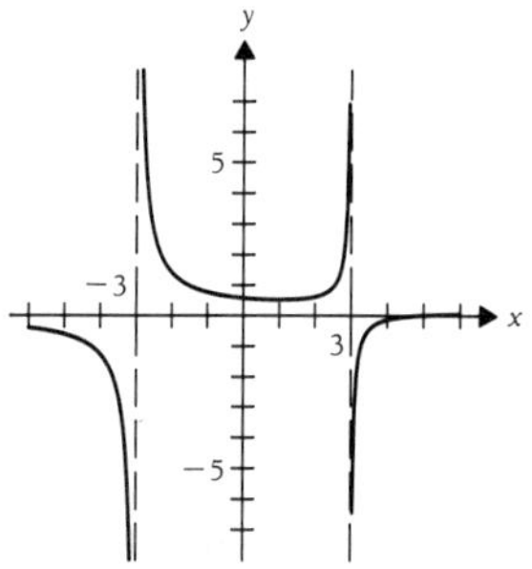

23. No critical points;
decreasing on $(-\infty, -3)$, $(-3, 3)$, and $(3, \infty)$;
concave down on $(-\infty, -3)$ and (0.786, 3),
concave up on $(-3, 0.786)$ and $(3, \infty)$;
inflection point ≈ 0.786

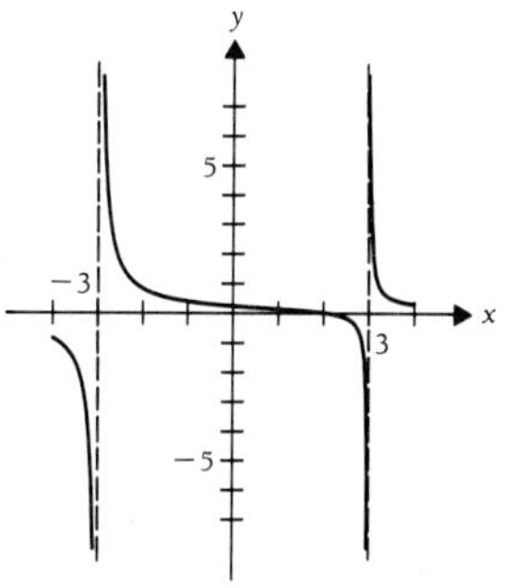

25. $y + 2 = -3(x - 1)$ 27. $y = 0$, $y + 7 = -15(x - 1)$, $y - 7 = -15(x + 1)$ 29. $x = 0$ 31. $y = -1(x - \pi)$
33. If $p(c_1) = p(c_2)$, then by Rolle's theorem there must be a point between c_1 and c_2 such that $p' = 0$ there. But c_1 and c_2 are consecutive critical points, so that cannot be.
35. p' can change sign only at a critical point. Thus, if $p'(a) < 0$, then p' must be negative throughout $(-\infty, c_0)$.
37. $f(x) = x^{5/3}$

Section 4.4

1. Absolute endpoint minimum of -8 at -3; absolute endpoint maximum of 16 at 5
3. Relative minimum of 0 at 1; relative maximum of 4 at -1 5. Absolute maximum of $3/4\sqrt[3]{4}$ at $1/\sqrt[3]{4}$
7. Endpoint maximum of $\frac{1}{2}$ at -2; endpoint minimum of $\frac{3}{2}$ at 2 9. Absolute maximum of 16 at 0
11. Relative endpoint maximum of 1 at 0; absolute minimum of 0 at 1; absolute endpoint maximum of $2^{2/3}$ at 3
13. Absolute minimum of $-\frac{1}{4}$ at -2; absolute maximum of $\frac{1}{4}$ at 2
15. Absolute endpoint minimum of 0 at 1 17. Absolute minimum of -1 at 0
19. Absolute endpoint minimum of 0 at -2; absolute endpoint minimum of 0 at 2
21. Relative endpoint maximum of 0 at -1; relative endpoint minimum of 0 at 1
23. Absolute maximum of 1 at $\pi/2 + n\pi$; absolute minimum of 0 at $n\pi$ 25. Relative maximum of 3 at 2
27. Absolute maximum of $\sqrt{2}$ at $\pi/4 + 2n\pi$; absolute minimum of $-\sqrt{2}$ at $5\pi/4 + 2n\pi$ 29. No extrema
31. Absolute maximum of 2 at $\pi/2 + 2n\pi$; relative maximum of 0 at $-\pi/2 + 2n\pi$; absolute minimum of $-\frac{9}{8}$ at $n\pi + (-1)^{n+1} \arcsin \frac{1}{4}$
33. If c is an inflection point of f, then f'' must change signs at c. That is, $(f')'$ changes sign at c, so f' changes from increasing to decreasing or vice versa. Therefore, f' has an extremum at c.
35. Yes

Section 4.5

1. $\frac{14}{3}$ in. by $\frac{35}{3}$ in. by $\frac{5}{3}$ in. 3. 20,000 yd^2 5. 100 ft wide by 80 ft deep
7. Given a rectangle with dimensions x and y, and perimeter P, $2x + 2y = P$, so that $y = \frac{1}{2}P - x$. Then the area is $A = xy = x(\frac{1}{2}P - x) = \frac{1}{2}Px - x^2$. The area is maximum when $dA/dx = \frac{1}{2}P - 2x = 0$, or $x = \frac{1}{4}P$, which makes the rectangle a square. Note that the second derivative of A is negative, so we do indeed have a maximum. It is absolute by Problem 36 of Section 4.4.
9. $\frac{32}{9}\sqrt{3}$ 11. $\frac{1}{2}b$ by $\frac{1}{2}h$ 13. $2\pi r^3/3\sqrt{3}$

15. Width: $816/(8 + 3\pi)$ in.; Height: $(408 + 102\pi)/(8 + 3\pi)$ in.
17. $S'(t) = \frac{3}{2}b^2 \csc t(\csc t - \sqrt{3} \cos t)$, which is 0 when $\csc t = \sqrt{3} \cot t$, or in other words, $\cos t = 1/\sqrt{3}$. This is true when $t = \cos^{-1}(1/\sqrt{3}) \approx 0.955$.
19. At $\pi L/(4 + \pi)$ 21. Base: 6.93 in. square; height: 10.4 in.
23. A can with radius r and height h has volume $V = \pi r^2 h$ and surface area $S = 2\pi rh + 2\pi r^2$. Substituting $h = V/\pi r^2$, we get $S = 2V/r + 2\pi r^2$. To minimize S, we set the derivative equal to 0: $dS/dr = -2V/r^2 + 4\pi r = 0$ when $r = (V/2\pi)^{1/3}$. But then $h = (V/\pi)(V/2\pi)^{-2/3} = 2(V/2\pi)^{1/3} = 2r$.
25. 44 trees 27. 100 29. 8 pianos, 20 times per year
31. $r(P) = f(P) - P$ is maximum when $r'(P) = f'(P) - 1 = 0$, or $f'(P) = 1$. 33. $\sqrt{32}$ ft 35. $(-0.724, -0.380)$
37. At $1/(1 + \sqrt[3]{6}) \approx 0.355$ of a block from the weaker light source
39. 10.811 ft 41. Either round to 16, or choose 16 four times out of five so that the average is 16.2.

Section 4.6

1. $x^2/2 + C$ 3. $2\sqrt{x} + C$ 5. $3x^2/2 + C$ 7. $x/2 + C$ 9. $x^2/2 - x^3/3 + C$ 11. $x^3/3 + 2x^2 + 3x + 1/x + C$
13. $-3 \cos x + C$ 15. $-\frac{1}{2} \cos 2x + C$ 17. $\frac{1}{2} \sec 2x + C$ 19. $-\frac{1}{4} \cot 4x + C$ 21. $x^3/3 - \cos x + C$
23. $2x\sqrt{x}/3 + \frac{1}{2} \cos 2x + C$ 25. $y^2 = 3x^2/2$ 27. $y^3 = 3x^2 + 1$ 29. $y^3 = 4 - 3 \cos x$ 31. $y^3 = 3x^4/4 - 3x + 8$
33. Dec. 1986 35. $\frac{1}{2}$ hr 37. 90 sec 39. $10(\sqrt[3]{4})$ ft

Section 4.7

1. $x = -4t$ 3. $x = 2t^2 + 24t$ 5. $x = t^2/2 - t^2/6 - 24t + 55$ 7. 484 ft 9. 56.25 ft 11. 400 ft/sec 13. 512 mm
15. 7.5 ft/sec 17. $x = 8 \sin \frac{1}{2}t$ 19. $\phi = \arcsin \frac{1}{3} \approx 0.3398$; $x = 6 \sin(2\pi t/3 + 0.3398)$
21. $x = A \sin(2\pi t/p + \phi)$, where $\phi = \arctan(2\pi x_0/pv_0)$ and $A = \sqrt{4\pi^2 x_0^2 + p^2 v_0^2}/2\pi$
23. $8\pi^2/3$ 25. $3(g/2)^{1/2}$, where g is the acceleration of gravity

Review Exercises

1. $\pm 1/\sqrt{3}$ 3. $\pi/2$ or $3\pi/2$
5. **(a)** $0, \pm\sqrt{5}/\sqrt{2}$ **(b)** Increasing on $(-\sqrt{5}/\sqrt{2}, 0)$ and $(\sqrt{5}/\sqrt{2}, \infty)$; decreasing on $(-\infty, -\sqrt{5}/\sqrt{2})$ and $(0, \sqrt{5}/\sqrt{2})$
(c) Concave up on $(-\infty, -\sqrt{5}/\sqrt{6})$ and $(\sqrt{5}/\sqrt{6}, \infty)$; concave down on $(-\sqrt{5}/\sqrt{6}, \sqrt{5}/\sqrt{6})$
(d) $\pm\sqrt{5}/\sqrt{6}$ **(e)** Relative maximum of 4 at 0; absolute minimum of $-\frac{9}{4}$ at $\pm\sqrt{5}/\sqrt{2}$
(f)

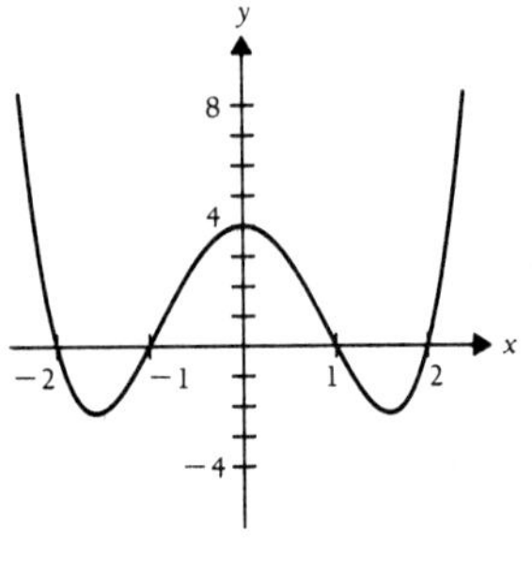

7. **(a)** $0, \pm\sqrt{3}$ **(b)** Increasing on $(-\sqrt{3}, \sqrt{3})$; decreasing on $(-\infty, -\sqrt{3})$ and $(\sqrt{3}, \infty)$
(c) Concave up on $(-\infty, -\sqrt{3}/\sqrt{2})$ and $(0, \sqrt{3}/\sqrt{2})$; concave down on $(-\sqrt{3}/\sqrt{2}, 0)$ and $(\sqrt{3}/\sqrt{2}, \infty)$
(d) $0, \pm\sqrt{3}/\sqrt{2}$ **(e)** Relative minimum of $-6\sqrt{3}$ at $-\sqrt{3}$; relative maximum of $6\sqrt{3}$ at $\sqrt{3}$
(f)

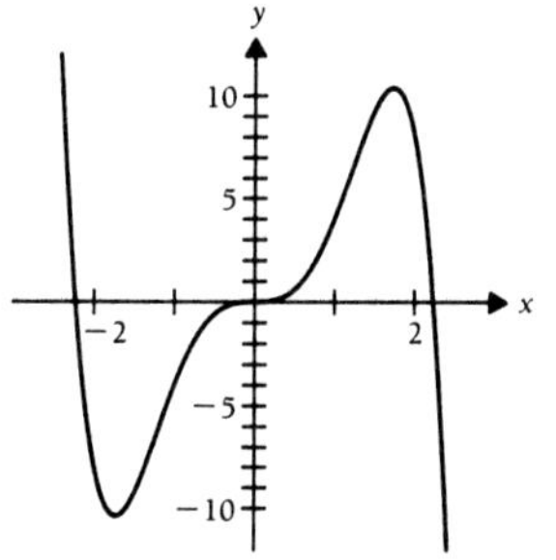

9. **(a)** $0, -2$ **(b)** Increasing on $(-\infty, -2)$ and $(0, \infty)$; decreasing on $(-2, -1)$ and $(-1, 0)$
(c) Concave down on $(-\infty, -1)$; concave up on $(-1, \infty)$ **(d)** No inflection points
(e) Relative maximum of -4 at -2; relative minimum of 0 at 0
(f)

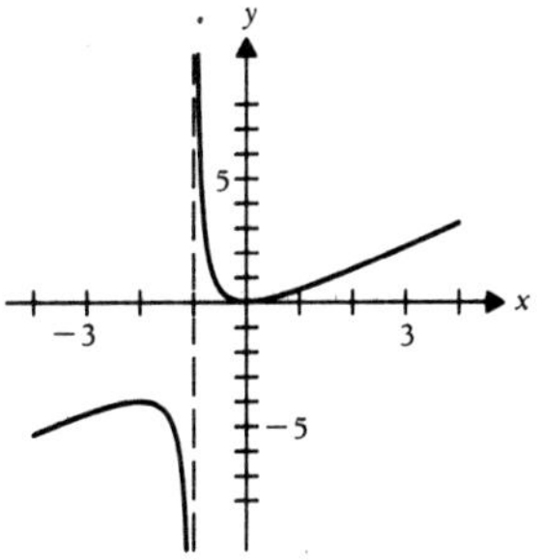

11. **(a)** $n\pi/2$ **(b)** Increasing on $(-\pi/2 + n\pi, n\pi)$; decreasing on $(n\pi, \pi/2 + n\pi)$
(c) Concave down on $(-\pi/4 + n\pi, \pi/4 + n\pi)$; concave up on $(\pi/4 + n\pi, 3\pi/4 + n\pi)$
(d) $\pi/4 + n\pi/2$ **(e)** Absolute maximum of 1 at $n\pi$; absolute minimum of -1 at $\pi/2 + n\pi$
(f)

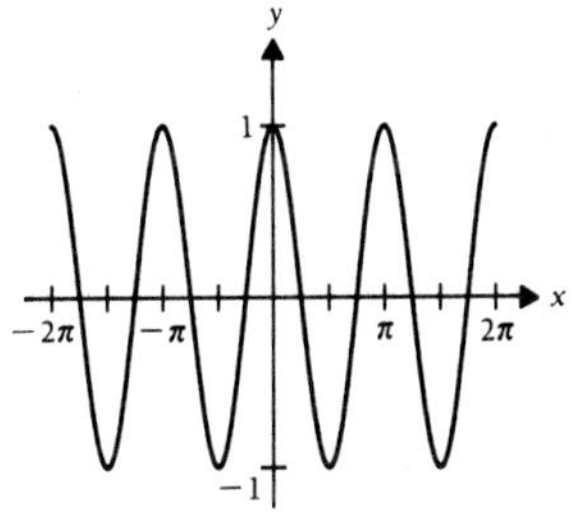

13. $s^2\sqrt{3}/8$ 15. 4500 items per week 17. $r = h = (V/\pi)^{1/3}$ 19. V is decreasing at the rate of $(k + 20)$ cm³/min 21. $\frac{25}{2}$ cm²
23. $\sqrt{70}$ ft from the point below the sign
25. Since $f'(x) = 1 + \cos x$ is never negative, the function is increasing, so has no extrema.
27. $y^2 = 2x^2 + 9$ 29. $y = -5/(3x^{5/3} + 5)$ 31. $\sin y = x^2/2$ 33. $2x^2 - x^3/3 - x^4/4 + C$ 35. $ax^2/2 + bx + C$
37. $2ax^{3/2}/3 + C$ 39. $3x + 3x^{2/3}/2 + C$ 41. $\frac{1}{4}\sin 4x + C$ 43. $x = 2 + 6t$ 45. $x = 9t - 3t^2$ 47. $x = 1/(1 - 3t)$
49. 272.25 ft 51. 96 ft/sec 53. At 8:51 PM on June 14; 21,367 km/hr 55. $x = (5/\pi)\sin 2\pi t$

Additional Exercises

1. $f(x) = x^2 + 1$ has $f'(0) = 0$, but nowhere does $f(a) = f(b) = 0$.
3. $f(x) = 6x^5 + 13x + 1$ has at least one real zero, because f is of odd degree. If f had two real zeros, then by Rolle's theorem, f' would be 0 somewhere between them. However, $f'(x) = 30x^4 + 13$ is never 0.

5. If there were two zeros of f, there would be another zero of f', by Rolle's theorem.
7. $f' = g'$ implies $f(x) = g(x) + C$. Setting $x = a$ gives $C = f(a) - g(a)$. Hence, $f(x) = g(x) + f(a) - g(a)$, and the conclusion follows.
9. If f is quadratic, then f'' is constant, so cannot be 0, as must be the case with an inflection point.
11. The three inflection points are $(1, 1)$, $(-2 + \sqrt{3}, \frac{1}{4}(1 + \sqrt{3}))$, and $(-2 - \sqrt{3}, \frac{1}{4}(1 - \sqrt{3}))$. The line on the last two is $y - \frac{1}{4}(1 + \sqrt{3}) = \frac{1}{4}(x + 2 - \sqrt{3})$, which is satisfied by the point $(1, 1)$.
13. Assuming $r > 0$, the maximum is $2/3\sqrt{3}\, r^2$ at $r/\sqrt{2}$.
15. If z is a double zero of f, then $f(x) = (x - z)^2 q(x)$ for some polynomial q. Then $f'(x) = 2(x - z)q(x) + (x - z)^2 q'(x) = (x - z)[2q(x) + (x - z)q'(x)]$, and z is also a zero of f'.
17. $A = 0$, $B = 2$ 19. $(2\sqrt{3} - 1)/(2\sqrt{3} + 1) \approx 55\%$ 21. $15/\sqrt{2}$ mph 23. (2.7335, 1.8633) 25. $r = 1.52$ in., $h = 1.85$ in.
27. \$48 29. $\frac{40}{13}$ mi from the spot on the river closest to the nearer ranch
31. A line joining the ranches divides the 8 mi stretch of river between them into two parts having the ratio 10 to 16.
33. So that $BD = \frac{1}{2}AC$ 35. Half the surface area 37. $(\frac{1}{4}, \frac{1}{8})$ 39. $C(x) = 3x - 0.05x^2 + 200$ 41. 256 ft 43. $\frac{625}{9}$ m
45. $(3\sqrt{3} - 4)/24$ ft/min 47. 13 49. 6.19 ft
51. If the denominator has greater degree, then $\lim_{x \to \infty} f(x) = 0$, and $y = 0$ (the x-axis) is a horizontal asymptote.
53. If the numerator of $f(x) = p(x)/q(x)$ has degree 1 greater than the degree of q, then long division produces $f(x) = mx + b + r(x)/q(x)$ with r having degree less than the degree of q. Since $r(x)/q(x)$ tends to 0 as $x \to \infty$, the line $y = mx + b$ is an oblique asymptote.

55.

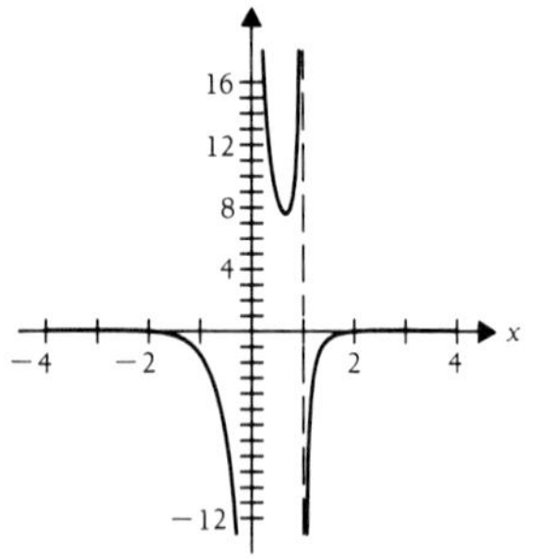

57.

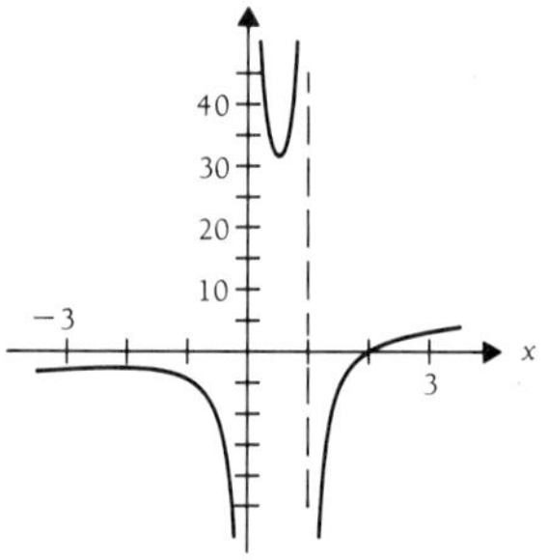

Chapter 5 Section 5.1

1. (a) $0, \frac{1}{4}, \frac{1}{2}, \frac{3}{4}, 1$ (b) m_i's: $0, \frac{1}{4}, \frac{1}{2}, \frac{3}{4}$; M_i's: $\frac{1}{4}, \frac{1}{2}, \frac{3}{4}, 1$ (c) $A_4 = \frac{3}{4}$; $B_4 = \frac{5}{4}$
(d) x_i's: $0, \frac{1}{8}, \frac{1}{4}, \frac{3}{8}, \frac{1}{2}, \frac{5}{8}, \frac{3}{4}, \frac{7}{8}, 1$; m_i's: $0, \frac{1}{8}, \frac{1}{4}, \frac{3}{8}, \frac{1}{2}, \frac{5}{8}, \frac{3}{4}, \frac{7}{8}$; M_i's: $\frac{1}{8}, \frac{1}{4}, \frac{3}{8}, \frac{1}{2}, \frac{5}{8}, \frac{3}{4}, \frac{7}{8}, 1$; $A_8 = \frac{7}{8}$; $B_8 = \frac{9}{8}$
3. (a) $0, \frac{1}{4}, \frac{1}{2}, \frac{3}{4}, 1$ (b) m_i's: $\frac{1}{4}, \frac{1}{2}, \frac{3}{4}, 1$; M_i's: $0, \frac{1}{4}, \frac{1}{2}, \frac{3}{4}$ (c) $A_4 = \frac{19}{8}$; $B_4 = \frac{21}{8}$
(d) x_i's: $0, \frac{1}{8}, \frac{1}{4}, \frac{3}{8}, \frac{1}{2}, \frac{5}{8}, \frac{3}{4}, \frac{7}{8}, 1$; m_i's: $\frac{1}{8}, \frac{1}{4}, \frac{3}{8}, \frac{1}{2}, \frac{5}{8}, \frac{3}{4}, \frac{7}{8}, 1$; M_i's: $0, \frac{1}{8}, \frac{1}{4}, \frac{3}{8}, \frac{1}{2}, \frac{5}{8}, \frac{3}{4}, \frac{7}{8}$; $A_8 = \frac{39}{16}$; $B_4 = \frac{41}{16}$
5. (a) $-2, -1, 0, 1, 2$ (b) m_i's: $-2, -1, 1, 2$; M_i's: $-1, 0, 0, 1$ (c) $A_4 = 6$; $B_4 = 14$
(d) x_i's: $-2, -\frac{3}{2}, -1, -\frac{1}{2}, 0, \frac{1}{2}, 1, \frac{3}{2}, 2$; m_i's: $-2, -\frac{3}{2}, -1, -\frac{1}{2}, \frac{1}{2}, 1, \frac{3}{2}, 2$; M_i's: $-\frac{3}{2}, -1, -\frac{1}{2}, 0, 0, \frac{1}{2}, 1, \frac{3}{2}$; $A_8 = \frac{17}{2}$; $B_8 = \frac{25}{2}$
7. (a) 1, 1.5, 2, 2.5, 3 (b) m_i's: 1.5, 2, 2.5, 3; M_i's: 1, 1.5, 2, 2.5 (c) $A_4 = 0.95$; $B_4 = 1.28$
(d) x_i's: 1, 1.25, 1.5, 1.75, 2, 2.25, 2.5, 2.75, 3; m_i's: 1.25, 1.5, 1.75, 2, 2.25, 2.5, 2.75, 3; M_i's: 1, 1.25, 1.5, 1.75, 2, 2.25, 2.5, 2.75; $A_8 = 1.02$; $B_8 = 1.19$
9. $2 + 4 + 6$ 11. $\frac{1}{2} + \frac{1}{3} + \frac{1}{4} + \frac{1}{5} + \frac{1}{6} + \frac{1}{7}$ 13. $1^2 + 2^2 + 3^2 + 4^2 + 5^2 + 6^2 + 7^2 + 8^2$ 15. $\frac{1}{2} + \frac{1}{3} + \frac{1}{4} + \frac{1}{5} + \frac{1}{6} + \frac{1}{7}$
17. $\sum_{i=0}^{4} (2i + 1)$ 19. $\sum_{i=1}^{7} \frac{1}{2^i}$ 21. $\sum_{i=1}^{30} 3i$ 23. $\sum_{n=1}^{49} \frac{n+1}{n}$ 25. $\sum_{i=1}^{n} f(x_i)$ 27. $\sum_{i=1}^{n} x_i^2$
29. If true for $n = k$, then $1 + 2 + 3 + \cdots + k = k(k + 1)/2$. Hence, $1 + 2 + 3 + \cdots + k + (k + 1) = k(k + 1)/2 + (k + 1) = [k(k + 1) + 2(k + 1)]/2 = (k + 1)(k + 2)/2$, and the statement is true for $n = k + 1$.
31. If true for $n = k$, then $1^3 + 2^3 + 3^3 + \cdots + k^3 = k^2(k + 1)^2/4$. Hence, $1^3 + 2^3 + 3^3 + \cdots + k^3 + (k + 1)^3 = k^2(k + 1)^2/4 + (k + 1)^3 = (k + 1)^2[k^2/4 + (k + 1)] = (k + 1)^2(k^2 + 4k + 4)/4 = (k + 1)^2(k + 2)^2/4$, and the statement is true for $n = k + 1$.

33. $\frac{9}{2}$ 35. 12 37. $\frac{8}{3}$ 39. $\frac{5}{6}$ 41. $\lim_{n\to\infty} \sum_{i=1}^{n} \frac{1}{n + i}$ 43. $\lim_{n\to\infty} \sum_{i=1}^{n} \frac{\pi}{n} \sin \frac{\pi i}{n}$

45. **(a)** $r^2 \sin \frac{\pi}{n} \cos \frac{\pi}{n}$ **(b)** $A_n = nr^2 \sin \frac{\pi}{n} \cos \frac{\pi}{n} = \pi r^2 \left(\frac{n}{\pi} \sin \frac{\pi}{n} \cos \frac{\pi}{n} \right)$

(c) $\lim_{n\to\infty} A_n = \lim_{n\to\infty} \pi r^2 \cdot \frac{\sin(\pi/n)}{\pi/n} \cdot \cos \frac{\pi}{n} = \pi r^2 \cdot 1 \cdot 1 = \pi r^2$

47. Inner area = 1; outer area = 2

Section 5.2

1. 1 **3.** $-\frac{3}{2}$ **5.** $\frac{7}{2}$ **7.** $-\frac{4}{3}$ **9.** $\frac{9}{2}$ **11.** 10 **13.** 10 **15.** $\pi/2$ **17.** $\int_0^2 x\,dx$ **19.** $\int_1^3 (1/x)\,dx$
21. Over $[n, n+1)$, $\lfloor x \rfloor = n$, so the area is just n. **23.** 0.6629 **25.** 0.3300 **27.** The area of the rectangle is $b - a$.

Section 5.3

1. $I - J$ **3.** $5K$ **5.** $-K$ **7.** $-2(I - J + K)$ **9.** 0 **11.** $\frac{1}{2}(I - J)$ **13.** $4P$ **15.** $P + R$ **17.** $3P - 5Q + 8R$ **19.** $-\frac{2}{3}P$
21. $-3P$ **23.** 0 **25.** 2 **27.** $\frac{1}{4}\sqrt{16 - \pi^2}$ **29.** $z \in [1, 2)$ **31.** $\frac{5}{6}$ **33.** 8, 0 **35.** 3, $\frac{3}{4}$
37. Interpreting $\int_0^n \lfloor x \rfloor\,dx$ as an area, the value is $1 + 2 + 3 + \cdots + (n-1) = (n-1)n/2$ by Problem 29 in Section 5.1.
39. If f is positive over $[a, b]$, then all absolute value signs can be dropped and equality holds. If f is negative, then all the signs are changed and again we have equality. Otherwise, $\int_a^b f(x)\,dx$ is a difference of areas, while $|f(x)|$ is positive, so that $\int_a^b |f(x)|\,dx$ is a sum of areas, and the inequality holds.

Section 5.4

1. $\frac{1}{7}$ **3.** $\frac{1}{5}$ **5.** $\frac{7}{8}$ **7.** $\frac{3093}{500{,}000}$ **9.** $-\frac{2}{3}$ **11.** 1 **13.** $-\frac{1}{42}$ **15.** $-\frac{2}{3}$ **17.** $\frac{2}{3}$ **19.** $\frac{62}{5}$ **21.** $\frac{381}{7}$ **23.** $\frac{4}{3}$ **25.** $1/\sqrt{2}$ **27.** $\frac{9}{2}$
29. 2 **31.** 0 **33.** $\sqrt{3}/4$ **35.** $3\sqrt{3}/8$ **37.** $\sqrt{2} - 1$ **39.** 2 **41.** $(\sqrt{2} - 1)/3$ **43.** $\sqrt{3}/2$ **45.** $x^3 + x$ **47.** $\cos 2x$
49. $f(g(x))g'(x)$ **51.** $1/2x$ **53.** $(\sin x)/2\sqrt{x} - 2x \sin x^4$ **55.** $F'(x) = f(x)$ is positive, so F is increasing.
57.

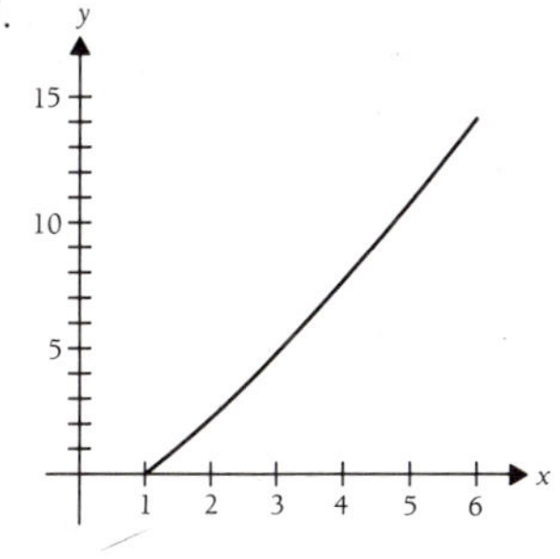

Section 5.5

1. $\frac{1}{8}(x+1)^8 + C$ **3.** $\frac{992}{5}$ **5.** $\frac{1}{8}(2x-1)^4 + C$ **7.** $\frac{1}{36}(4x-3)^9 + C$ **9.** 1752 **11.** $\frac{5}{27}(7^{9/5} + 2^{9/5})$ **13.** $\frac{3}{10}(x^2+1)^{5/3} + C$
15. $\frac{3}{5}(x^2-4)^{5/2} + C$ **17.** $\frac{3}{32}(65^{4/3} - 1)$ **19.** $\frac{4}{3}(7^{3/2})$ **21.** $\frac{2}{15}$ **23.** $\sqrt{2} - 1$ **25.** $(x^{2/3} + 4)^{3/2} + C$ **27.** $(x^{2/3} + 1)^{3/2} + C$
29. $\frac{2}{3}(x-4)^{3/2} + 8(x-4)^{1/2} + C$ **31.** $\frac{1023}{10} - \frac{511}{9}$ **33.** $\frac{2}{3}(x-1)^{3/2} + 6(x-1)^{1/2} + C$
35. $\frac{1}{28}(2x+3)^{7/2} - \frac{3}{10}(2x+3)^{5/2} + \frac{3}{4}(2x+3)^{3/2} + C$ **37.** $\frac{1}{3}$ **39.** $5\sqrt{3}/8$ **41.** $-\frac{1}{2}\cos x^2 + C$ **43.** $\frac{2}{3}\tan^{3/2} x + C$
45. There is no x outside the radical. **47.** We need x to the first power outside the radical.

Section 5.6

1. $\frac{9}{4}$ **3.** $\frac{1}{6}$ **5.** $\frac{1}{6}$ **7.** $\frac{1}{6}$ **9.** $2\sqrt{6}$ **11.** $10\sqrt{10}/3$ **13.** $\frac{40}{3}$ **15.** $\frac{16}{3}$ **17.** $\frac{1}{6}$ **19.** $\frac{32}{3}$ **21.** $8\sqrt{2}/3$ **23.** $\frac{27}{4}$ **25.** $\frac{34}{3}$ **27.** $\frac{19}{2}$
29. 1 **31.** 2 **33.** $\sqrt{2} - 1$ **35.** $2\sqrt{2}$

37. If f is positive, we just drop the absolute value signs. If f is negative, we change the sign of f and use the fact that $-f(x) = 0 - f(x)$. If f is partly positive and partly negative, we break up the integral and apply the above to each piece.
39. When f is entirely positive or entirely negative over $[a, b]$

Section 5.7

1. 3 **3.** $\frac{14}{9}$ **5.** $2/\pi$ **7.** $(4/5\pi)(\sqrt{2} - \frac{1}{2})$ **9.** $\frac{1}{3}(5\sqrt{5} - 2\sqrt{2})$ **11.** $(\frac{1}{90}\sqrt{3})(81\sqrt{3} - 7\sqrt{7})$ **13.** 6, 2, 6, 2 **15.** 24, 6, 24, 6
17. 3, 1, 3, 1 **19.** $-\frac{2000}{3}, -\frac{400}{3}, \frac{2000}{3}, \frac{400}{3}$ **21.** $-\frac{20}{3}, -\frac{5}{3}, \frac{34}{3}, \frac{17}{6}$ **23.** $0, 0, 6/\pi, 6/\pi$ **25.** $\frac{38}{3}$ **27.** $2 + 4 \sin 1$ **29.** \$37.02
31. \$20,365 **33.** 4 **35.** $\frac{8}{3}$ **37.** 47% **39.** 4 ft **41.** Change in velocity **43.** Change in cost **45.** Per unit change in earnings
47. Total number of births **49.** Stock B; the area under its graph is greater.

Review Exercises

1. $\frac{46}{3}$ **3.** $\int_1^2 x^{-2/3}\,dx = 3(2^{1/3} - 1)$ **5.** $\lim_{n\to\infty} \sum_{i=1}^{n} 5^{5/3} i^{2/3} n^{-5/3}$ **7.** 7 **9.** $25\pi/4$ **11.** 6 **13.** 0 **15.** $\frac{14}{3}$ **17.** $(\frac{6}{5})2^{2/3}$
19. $(2/\sqrt{3}) - 1$ **21.** $\frac{1}{20}(x + 4)^{20} + C$ **23.** $\frac{54}{289}$ **25.** $\sqrt{3}$ **27.** $\sqrt{3x^3 + 1}$ **29.** $1/(1 - x^2)$ **31.** $\frac{250}{3}$ **33.** $\frac{26}{3}$
35. $32 - (\frac{12}{5})4^{2/3}$ **37.** $4 - (\frac{3}{10})4^{2/3}$ **39.** 70 **41.** 2 **43.** Because $1/(2x - 1) > 1/2x$ over $[1, 2]$

Additional Exercises

1. $2(\sqrt{2} - 1)$ **3.** 0 **5.** $\frac{1}{10}(5^5 + 3^5)$ **7.** Substitute $u = x/c$.
9. $A'(x) = f(x)$, and the critical points of A are the zeros of f.
11. $\int_a^b |f(x)|\,dx$ **13.** $1/(n + 1)(n + 2)$
15. Let $\lim_{x\to a^+} f(x) = A$ and $\lim_{x\to b^-} f(x) = B$; and define g on $[a, b]$ by $g(a) = A$, $g(x) = f(x)$ if $x \in (a, b)$, and $g(b) = B$. Then g is continuous on $[a, b]$, so $\int_a^c g(x)\,dx$ and $\int_c^b g(x)\,dx$ exist for any $c \in (a, b)$. Also, $\lim_{t\to a^+} \int_a^t g(x)\,dx = 0$ by the continuity of g and the mean value theorem for integrals. Moreover, for $a < t < c < b$, $\int_t^c g(x)\,dx = \int_t^c f(x)\,dx$. Therefore, $\lim_{t\to a^+} \int_t^c f(x)\,dx = \lim_{t\to a^+} \int_t^c g(x)\,dx = \lim_{t\to a^+} [\int_a^c g(x)\,dx - \int_a^t g(x)\,dx] = \int_a^c g(x)\,dx$, and the limit exists. The left-hand limit at b is shown to exist similarly.
17. $\int_a^{a+p} f'(x)\,dx = f(a + p) - f(a) = f(a) - f(a) = 0$
19. $A(x) = \int_{g(x)}^{h(x)} f(t)\,dt = \int_0^{h(x)} f(t)\,dt - \int_0^{g(x)} f(t)\,dt$, so by the chain rule, $A'(x) = f(h(x))h'(x) - f(g(x))g'(x)$.
21. $2x^3(x^6 + 1)^5 - 3x^5(x^9 + 1)^5$

Chapter 6 Section 6.1

1. **(a)** $y = x - 1$ **(b)** $y - \ln 2 = \frac{1}{2}(x - 2)$ **(c)** $y - \ln 3 = \frac{1}{3}(x - 3)$ **(d)** $y - \ln 4 = \frac{1}{4}(x - 4)$
3. $\frac{1}{x}$ **5.** $\frac{a}{ax + b}$ **7.** $\frac{-1}{x^2 + x}$ **9.** $\frac{2x}{x^2 + 4}$ **11.** $\frac{12x^2}{4x^3 - 1}$ **13.** $\frac{3}{2(3x + 5)}$ **15.** $\frac{-1}{x - 1}$ **17.** $-\tan x$ **19.** $\csc x$
21. $\left[\frac{2}{x} + \frac{3x^2}{2(x^3 + 5)}\right] x^2\sqrt{x^3+5}$ **23.** $\left[\frac{x}{x^2+4} - \frac{15x^2}{2(x^3 - 1)}\right] \frac{\sqrt{x^2 + 4}}{\sqrt{(x^3 - 1)^5}}$ **25.** $\frac{x^2 + y^2 - 2x}{2(x^2 + y^2 + y)}$
27. Critical point: 0; absolute minimum of 0 at 0; increasing on $(0, \infty)$, decreasing on $(-\infty, 0)$; inflection points: $-1, 1$; concave down on $(-\infty, -1)$ and $(1, \infty)$, concave up on $(-1, 1)$
29. Let $f(x) = \sqrt{x} - \ln x$. Then $f'(x) = \frac{1}{2}\sqrt{x} - 1/x = (x - 2\sqrt{x})/2x\sqrt{x}$. The derivative f' is 0 at 4, negative in $(0, 4)$, and positive in $(4, \infty)$, so f has an absolute minimum at 4. Since $f(4) = \sqrt{4} - \ln 4 = 2 - 2\ln 2 > 0$, $f(x) > 0$ for all $x \in (0, \infty)$. That is, $\ln x < \sqrt{x}$.
31. $\ln|x - 3| + C$ **33.** $\ln 2 - \ln 3$ **35.** $\frac{1}{6}\ln(3x^2 + 5) + C$ **37.** $\frac{1}{3}(\ln x)^3 + C$ **39.** $\ln(1 + \sin x) + C$ **41.** $\frac{1}{2}\ln|\sec 2x| + C$
43. $11 - \sqrt{2} - \ln 3 + \ln(-1 + \sqrt{2})$ **45.** $\ln x$ is a one-to-one function, since it is increasing

47.

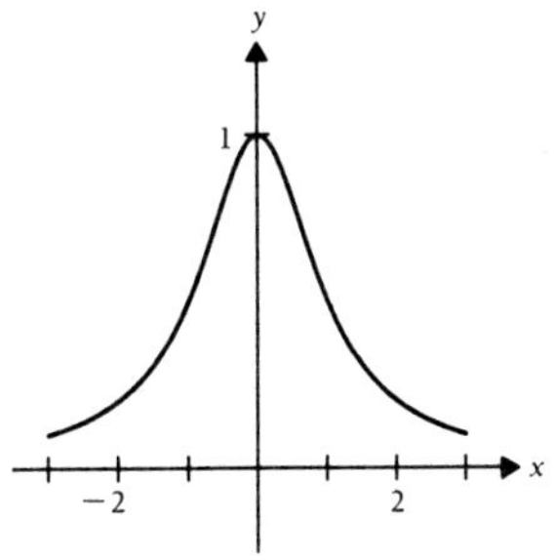

49. $1/(x^2+1)$
51. Let $y = T^{-1}(x)$. Then $T(y) = x$ and $T'(y)y' = 1$, so that $y' = 1/T'(y) = y^2 + 1 = [T^{-1}(x)]^2 + 1$.
53. 0.68060, 3.14154

Section 6.2

1. $2e^{2x}$ **3.** $-e^{1-x}$ **5.** $(1-x)e^{-x}$ **7.** $(3x^2+2x+3)e^{3x}$ **9.** $e^x \cos 2x - 2e^x \sin 2x$ **11.** $(3 \sin x + \cos x)e^{3x}$
13. $(2x - x^2 - 4)e^{-x}$ **15.** e **17.** $e^x \sec e^x \tan e^x$ **19.** $(2 \sec^2 x)e^{2 \tan x}$ **21.** $(e^{x+y} - y)/(x - e^{x+y})$ **23.** $(ye^{xy} - 1)/(1 - xe^{xy})$
25. No critical points; decreasing on $(-\infty, \infty)$; concave up on $(-\infty, \infty)$
27. Critical point: 0; increasing on $(-\infty, 0)$, decreasing on $(0, \infty)$; absolute maximum of 1 at 0; concave up on $(-\infty, -1/\sqrt{2})$ and $(1/\sqrt{2}, \infty)$, concave down on $(-1/\sqrt{2}, 1/\sqrt{2})$; inflection points: $-1/\sqrt{2}$, $1/\sqrt{2}$
29. $\frac{1}{2}e^{2x} + C$ **31.** $\frac{1}{4}(e^8 - 1)$ **33.** $1 - e^{-1}$ **35.** $2e^2 - 2e$ **37.** $\frac{1}{3}e^{x^3} + C$ **39.** $\ln(e^x + 4) + C$ **41.** $\sin e^x + C$
43. If $y = u^a$, then $\ln y = a \ln u$. Then $(1/y)\,dy = a(1/u)\,du$, so that $dy = ay(1/u)\,du = au^a(1/u)\,du = au^{a-1}\,du$.
45. $\sqrt{2}x^{\sqrt{2}-1}$ **47.** $2e^{2x} + 2ex^{2e-1}$ **49.** $\frac{1}{2} - 1/e$

Section 6.3

1. $-2^{-x} \ln 2$ **3.** $2x(\frac{1}{2})^{x^2} \ln \frac{1}{2}$ **5.** $2x/(\ln 2)(x^2+1)$ **7.** $(\cot x)/(\ln 10)$ **9.** $(1 + \ln x)x^x$ **11.** $(2/x)(\ln x)x^{\ln x}$
13. $2^x \ln 2 + 2x + (2x \ln 2x + x)(2x)^{x^2}$ **15.** $1/(x + 3^y - 3^y \ln 3)$ **17.** $y - 3 = (3 \ln 3)(x - 1)$ **19.** $y = x$
21. Critical point: 0; absolute maximum of 1 at 0 **23.** Critical point: $1/e$; absolute minimum of $e^{-2/e}$ at $1/e$
25. $-2^{-x}/\ln 2 + C$ **27.** $-2^{\cos x}/\ln 2 + C$ **29.** $e^x 2^{-x}/(1 - \ln 2) + C$ **31.** $\frac{1}{2} \ln x\, (\log_2 x)^2 + C$
33. Let $z = e^x$ and $w = e^y$. Then $x = \ln z$ and $y = \ln w$, so that $x - y = \ln z - \ln w = \ln(z/w)$. Hence, $e^{x-y} = z/w = e^x/e^y$.
35. $(b^x)^y = (e^{x \ln b})^y = e^{xy \ln b} = b^{xy}$
37. $L(x) = \log_b x$ is nonconstant and differentiable, and $L(xy) = \log_b xy = \log_b x + \log_b y = L(x) + L(y)$.
39. $x = \log_b b^x = \log_b b^y = y$

Section 6.4

1. 256,289 **3.** In year 2002 **5.** \$1822.12 **7.** 5.95 g **9.** 10.257 hr **11.** 3369 yr old **13.** \$1296.20 **15.** \$26.11
17. 99.98%
19. The growth rate is $f(P) = kP(L - P)$. The maximum growth rate must occur at a critical point of f; $f'(P) = k(L - 2P)$, which is 0 when $P = L/2$.
21. **(a)** $A_1 = A_0 + A_0(r)(1/n) = A_0(1 + r/n)$
(b) $A_2 = A_1 + A_1(r)(1/n) = A_1(1 + r/n) = A_0(1 + r/n)(1 + r/n) = A_0(1 + r/n)^2$

(c) Generalizing parts (a) and (b), the amount after k compounding periods must be $A_k = A_0(1 + r/n)^k$. Since in t years there are nt compounding periods, $A(t) = A_{nt} = A_0(1 + r/n)^{nt}$.
(d) Under continuous compounding, $A(t) = \lim_{n\to\infty} A_0(1 + r/n)^{nt} = A_0[\lim_{n\to\infty}(1 + r/n)^n]^t = A_0(e^r)^t = A_0e^{rt}$.

23. (a)

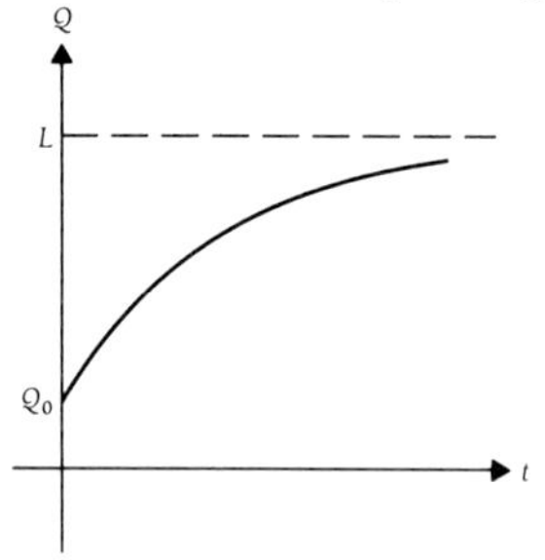

(b)

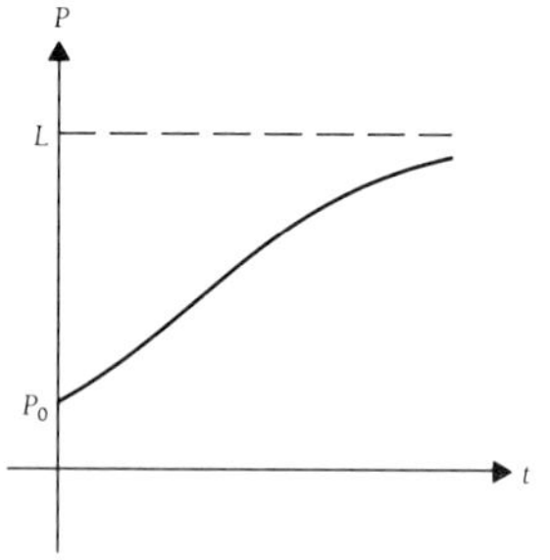

Section 6.5

1. 6 3. -1 5. $\frac{1}{2}$ 7. 3 9. 0 11. 0 13. 0 15. $-\infty$ 17. 1 19. -1 21. 1 23. $1/e$ 25. 1 27. 0 29. e 31. 1
33. sin 9 35. $\sqrt{3}$, by Theorem 2.10
37.

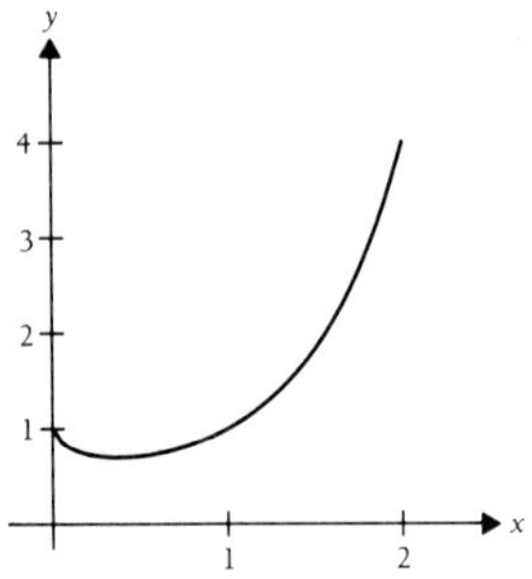

Section 6.6

1. $\pi/6$ 3. $-\pi/6$ 5. 0.7137 7. 0.8944
9. Let $y = \arcsin x$. Then $x = \sin y = \cos(\pi/2 - y)$, so $\arccos x = \pi/2 - y$. That is, $\pi/2 = y + \arccos x = \arcsin x + \arccos x$.
11. If $\sin x = a$, then $\arcsin a + 2n\pi$ and $(\pi - \arcsin a) + 2n\pi$ are all the possible solutions. The first of these forms is $\arcsin a$ plus an even multiple of π, and the second is $-\arcsin a$ plus an odd multiple of π. These numbers are precisely those given in the problem.
13. $2n\pi$ 15. $n\pi$ 17. $n\pi \pm \pi/4$ 19. $\pi/4 + n\pi$ 21. $-\pi/4 + n\pi = (-1)^{n+1}\pi/4 + n\pi/2$ 23. $\sqrt{1-x^2}$ 25. $x/\sqrt{1+x^2}$
27. $\sqrt{x^2-1}/x$ 29. $\sqrt{x^2-1}$
31. If $y = \arccos u$, then $u = \cos y$ and $du = -\sin y\, dy$. But $\sin(\arccos u) = \sqrt{1-u^2}$, so $dy = -du/\sqrt{1-u^2}$.
33. If $y = \operatorname{arccsc} u$, then $\csc y = u$ and $-\csc y \cot y\, dy = du$. Now, $\csc y = u$ and $\cot(\operatorname{arccsc} u) = \sqrt{u^2-1}$, so $dy = -du/u\sqrt{u^2-1}$.
35. $2/(x^2+4)$ 37. $2(\arcsin x)/\sqrt{1-x^2}$ 39. $(2e^{2\arctan x})/(1+x^2)$ 41. $(2\sqrt{1-x^2y^2} - y)/x$ 43. $-16x/(1+4x^2)^2$
45. $\arcsin(x/2) + C$ 47. $\sin^{-1}(x/4) + C$ 49. $\pi/4$ 51. $\frac{1}{3}\tan^{-1}3$ 53. $e^{\pi/6} - 1$ 55. $\sin(\ln|2x|)$
57. $\tan(2\arctan\frac{1}{5}) = [2\tan(\arctan\frac{1}{5})]/[1 - \tan^2(\arctan\frac{1}{5})] = 2(\frac{1}{5})/[1 - (\frac{1}{5})^2] = \frac{5}{12}$;
$\tan(4\arctan\frac{1}{5}) = \tan[2(2\arctan\frac{1}{5})] = 2\tan(2\arctan\frac{1}{5})/[1 - \tan^2(2\arctan\frac{1}{5})] = 2(\frac{5}{12})/[1 - (\frac{5}{12})^2] = \frac{120}{119}$;
$\tan(4\arctan\frac{1}{5} - \arctan\frac{1}{239}) = [\frac{120}{119} - \frac{1}{239}]/[1 + (\frac{120}{119})(\frac{1}{239})] = [120(239) - 119]/[119(239) + 120] = 1$.
Therefore, $\pi/4 = \arctan 1 = 4\arctan\frac{1}{5} - \arctan\frac{1}{239}$.

Section 6.7

1. $\cosh^2 x - \sinh^2 x = \frac{1}{4}(e^{2x} + 2 + e^{-2x}) - \frac{1}{4}(e^{2x} - 2 + e^{-2x}) = 1$
3. $\cosh^2 x - \sinh^2 x = 1$ implies $\cosh^2 x/\sinh^2 x - 1 = 1/\sinh^2 x$, so that $\coth^2 x - 1 = \text{csch}^2 x$.
5. $D_x \cosh x = D_x \frac{1}{2}(e^x + e^{-x}) = \frac{1}{2}(e^x - e^{-x}) = \sinh x$
7. $D_x \coth x = D_x[(\cosh x)/(\sinh x)] = (\sinh^2 x - \cosh^2 x)/\sinh^2 x = -\text{csch}^2 x$
9. $D_x \text{csch}\, x = D_x(1/\sinh x) = -\cosh x/\sinh^2 x = -\text{csch}\, x \coth x$
11. Integrate in Problem 4.
13. $\int \coth x\, dx = \int(\cosh x/\sinh x)\, dx = \int(1/u)\, du = \ln|u| + C = \ln|\sinh x| + C$
15. $3 \cosh 3x$ 17. $\sinh(x + y)/[\cos y - \sinh(x + y)]$ 19. $\frac{1}{2}\cosh 2x + C$ 21. $2 \sinh\sqrt{x + 1} + C$
23. Absolute minimum of 1 at 0 25. Concave up on $\mathbb{R}$
27.

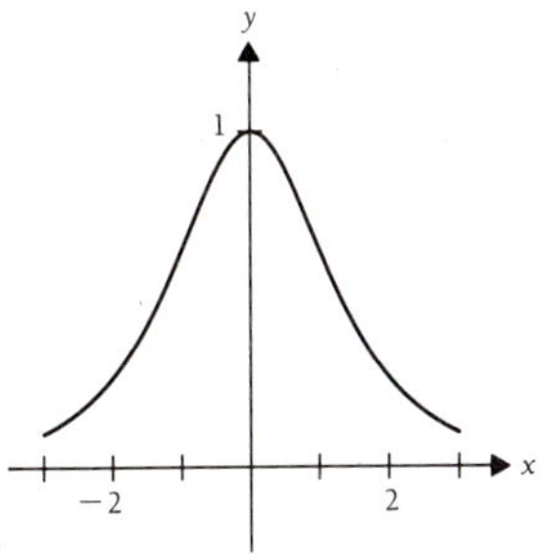

29. If $y = \sinh^{-1}x$, then $x = \sinh y = \frac{1}{2}(e^y - e^{-y})$, so $e^y - 2x - e^{-y} = 0$, and $e^{2y} - 2xe^y - 1 = 0$. Hence, solving the quadratic equation in e^y and discarding the extraneous negative root, $e^y = (2x + \sqrt{4x^2 + 4})/2 = x + \sqrt{x^2 + 1}$. Therefore, $y = \sinh^{-1}x = \ln(x + \sqrt{x^2 + 1})$.
31. $1/\sqrt{x^2 + 1}$ 33. $\sinh^{-1}x + C = \ln(x + \sqrt{x^2 + 1}) + C$
35. If $y = \tanh^{-1}x$, then $x = \tanh y = (e^y - e^{-y})/(e^y + e^{-y})$. Then $xe^y + xe^{-y} = e^y - e^{-y}$, so $(x - 1)e^y + (x + 1)e^{-y} = 0$. That is, $e^{2y} = (1 + x)/(1 - x)$, and $y = \frac{1}{2}\ln[(1 + x)/(1 - x)]$.

Review Exercises

1. $2 \ln 2$ 3. 0 5. 0 7. e 9. $(\ln \cos x - \tan x)e^x$ 11. $e^{\arctan(\ln x)}/x[1 + (\ln x)^2]$ 13. $(1 - xy - y^3)/(x^2 + xy^2 - 2y)$
15. $[4/(x + 3) - 3/(2 - x)](x + 3)^4(2 - x)^3$ 17. $4/\sqrt{1 - 16x^2}$ 19. $-\sin x \arcsin 2x + 2(\cos x)/\sqrt{1 - 4x^2}$ 21. $3/(x^2 + 9)$
23. $-(1/x^2)\arctan(1/x) - 1/x(x^2 + 1)$ 25. $-\frac{1}{4}e^{-4x} + C$ 27. $\cos 1 - \cos e$ 29. $2 \arctan x + C$ 31. $\frac{3}{4}\sin 1$
33. $-\sqrt{1 - x^2} + C$ 35. $y = Ce^{3x^2/2}$ 37. $r = Ce^{-2\cos\theta}$ 39. 16.98 days 41. 9582

Additional Exercises

1. $\pm 1/\sqrt{2}$ 3. $dx/dt = Abe^{bt} - Bbe^{-bt}$ and $d^2x/dt^2 = Ab^2e^{bt} + Bb^2e^{-bt} = b^2x$.
5. $i = E/R - Ae^{-Rt/L}$
7. If $a = dv/dt = g - kv$, then $[1/(g - kv)]\, dv = dt$ and $(-1/k)\ln(g - kv) = t + C$. Then $\ln(g - kv) = -kt - kC$, and $g - kv = e^{-kC}e^{-kt}$. Letting $v = 0$ when $t = 0$, we find $e^{-kC} = g$, so $g - kv = ge^{-kt}$. Now we solve for v to get the desired result.
9. $x = (g/k)t - (g/k^2)(1 - e^{-kt})$ 11. \$13,562 13. Yes, perhaps even more accurately.
15. Yes, using the decay rate of ln 2 divided by the half-life.

Chapter 7 Section 7.1

1. $\frac{1}{8}x^8 + C$ 3. $\frac{1}{8}(2x - 1)^4 + C$ 5. $-\ln|2 - x| + C$ 7. $\frac{1}{2}(e^6 - 1)$ 9. $10^x/(\ln 10) + C$ 11. $-\frac{1}{2}x^{-2} + e^x + C$
13. $16x + 8e^x + \frac{1}{2}e^{2x} + C$ 15. $-\frac{1}{3}\cos 3x + C$ 17. $-\frac{1}{2}\ln(\sqrt{3}/2)$ 19. $-\cos x + \ln(\cos x) + C$ 21. $\tan x + \sec x + C$
23. $\frac{1}{3}\tan 3x + C$ 25. $\frac{1}{6}\sec^2 3x + C$ 27. $\pi/6$ 29. $\pi/6$ 31. $-e^{\cos x} + C$ 33. $\frac{1}{3}\sin^3 x + C$ 35. $\frac{1}{6}\ln(3x^2 + 1) + C$

37. $\frac{1}{2}\ln(x^2+1)+\tan^{-1}x+C$ **39.** $-\frac{1}{3}\ln|9-3e^x|+C$ **41.** $\frac{1}{2}\arctan(\frac{1}{2}e^x)+C$
43. $\int \sin ax\,dx = (1/a)\int \sin u\,du = -(1/a)\cos u + C = -(1/a)\cos ax + C$
45. $\displaystyle\int \frac{dx}{x^2+a^2} = \int \frac{dx}{a^2[(x/a)^2+1]} = \frac{1}{a}\int \frac{du}{u^2+1} = \frac{1}{a}\arctan u + C = \frac{1}{a}\arctan\frac{x}{a} + C$
[We used the substitution $u = x/a$, $du = (1/a)\,dx$.]
47. $\int \tan ax\,dx = (1/a)\int \tan u\,du = (1/a)\ln|\sec u| + C = (1/a)\ln|\sec ax| + C$
49. Use the substitution $u = ax$. **51.** $\pi/3\sqrt{3}$ **53.** $\frac{1}{4}\ln(x^2+6)+(\sqrt{6}/4)\tan^{-1}(x/\sqrt{6})+C$ **55.** $\frac{1}{2}\arcsin 2x + C$
57. $\frac{1}{4}\arcsin(4x/3)+C$ **59.** Using formula (20), we get $\arcsin(x/a)|_0^{a/2} = \arcsin\frac{1}{2} = \pi/6$.

Section 7.2

1. $\sin x - x\cos x + C$ **3.** $(\frac{1}{2}x-\frac{1}{4})e^{2x}+C$ **5.** $\frac{1}{2}(e-1)$ **7.** $\frac{1}{4}\pi^2-2$ **9.** $\frac{3}{8}\sin x\sin 3x+\frac{1}{8}\cos x\cos 3x+C$
11. $\frac{1}{5}(2\sin x-\cos x)e^{2x}+C$ **13.** $x^2(\frac{1}{2}\ln x-\frac{1}{4})+C$ **15.** $\frac{2}{3}x(x+1)^{3/2}-\frac{4}{15}(x+1)^{5/2}+C$ **17.** $x\sin^{-1}x+\sqrt{1-x^2}+C$
19. $[1/(a^2+b^2)](a\sin bx - b\cos bx)e^{ax}+C$ **21.** $[b/(b^2-a^2)]\cos ax\sin bx - [a/(b^2-a^2)]\sin ax\cos bx + C$
23. $e^4x^5/5+C$ **25.** Let $u=(\ln x)^n$, $dv=dx$, and integrate by parts.
27. Let $u=x^n$, $dv=\sqrt{ax+b}\,dx$, and integrate by parts. In the resulting integral, let $(ax+b)^{3/2}=(ax+b)\sqrt{ax+b}$; multiply; separate into two integrals, one of which is the original; and solve for the original.
29. $e-2$ **31.** $\frac{1}{8}(4x^3-6x^2+6x-3)e^{2x}+C$ **33.** $(\frac{2}{21}x^2-\frac{8}{63}x+\frac{80}{567})(3x+5)^{3/2}+C$ **35.** $x\ln(x^2+1)-2x+2\tan^{-1}x+C$

Section 7.3

1. $\frac{1}{2}x-\frac{1}{4}\sin 2x+C$ **3.** $\frac{1}{3}\cos^3x-\cos x+C$ **5.** $\frac{1}{9}\sin^3 3x+C$ **7.** $-\frac{1}{4}\cos^4x+C$ **9.** $\frac{1}{8}x-\frac{1}{128}\sin 16x+C$ **11.** $3\pi/16$
13. $\cos x-\frac{2}{3}\cos^3x+C$ **15.** $\frac{1}{4}(-x+\sin 2x-\frac{1}{4}\sin 4x)+C$ **17.** $-\frac{1}{14}\cos 7x-\frac{1}{6}\cos 3x+C$
19. $\frac{1}{2}\ln|\sec 2x+\tan 2x|+C$ **21.** $\ln|\csc(x+1)-\cot(x+1)|+C$ **23.** $\frac{1}{3}\tan 3x+C$ **25.** $\frac{1}{4}\sec^2 2x+C$
27. $\frac{1}{5}\sec^5x-\frac{1}{3}\sec^3x+C$ **29.** $\frac{1}{4}\tan^4x+\frac{1}{6}\tan^6x+C$ **31.** $-\frac{1}{2}\csc 2x+C$ **33.** $\ln|\csc x-\cot x|+C$
35. $\frac{1}{2}x+\frac{1}{4}\sin 2x+C$ **37.** $\ln|\sec x+\tan x|-\sin x+C$ **39.** $(-2/\sqrt{\sec x})+C$ **41.** $\ln|1-\cos x|+C$ **43.** $x-\cos x+C$
45. $\displaystyle\int \csc x\,dx = \int \csc x\left(\frac{\csc x-\cot x}{\csc x-\cot x}\right)dx = \int\left(\frac{\csc^2x-\csc x\cot x}{\csc x-\cot x}\right)dx = \int \frac{1}{u}\,du = \ln|u|+C = \ln|\csc x-\cot x|+C$
47. Let $u=\sin^{n-1}x$ and $dv=\sin x\,dx$; integrate by parts; and in the resulting integral, replace $\cos^2x$ by $1-\sin^2x$, multiply out, and get two integrals, one of them the original, and solve for the original.
49. $\displaystyle\int \tan^n x\,dx = \int \tan^{n-2}x\tan^2x\,dx = \int \tan^{n-2}x(\sec^2x-1)\,dx = \int \tan^{n-2}x\sec^2x\,dx - \int \tan^{n-2}x\,dx = \int u^{n-2}\,du -$
$\displaystyle\int \tan^{n-2}x\,dx = \left(\frac{1}{n-1}\right)u^{n-1} - \int \tan^{n-2}x\,dx = \frac{1}{n-1}\tan^{n-1}x - \int \tan^{n-2}x\,dx$

Section 7.4

1. $\frac{1}{2}x\sqrt{1-x^2}+\frac{1}{2}\sin^{-1}x+C$ **3.** $\frac{1}{2}x\sqrt{a^2-x^2}+\frac{1}{2}a^2\sin^{-1}(x/a)+C$ **5.** $(1/\sqrt{3})[\ln|x+\sqrt{3}|-\ln\sqrt{3-x^2}]+C$
7. $\frac{1}{2}(a^2-x^2)+C$ **9.** $\frac{17}{36}+\frac{3}{16}\ln 3$ **11.** $\pi a^2/4$ **13.** $\frac{1}{2}x\sqrt{x^2+1}+\frac{1}{2}\ln|x+\sqrt{x^2+1}|+C$
15. $\frac{1}{7}(3+x^2)^{7/2}-\frac{3}{5}(3+x^2)^{5/2}+C$ **17.** $\ln|x+\sqrt{x^2+9}|+C$ **19.** $x/\sqrt{x^2+1}+C$ **21.** $x/8(4+x^2)+\frac{1}{16}\tan^{-1}(x/2)+C$
23. $1/\sqrt{2}+\frac{1}{2}\ln(1+\sqrt{2})$ **25.** $\frac{1}{2}x\sqrt{x^2-1}-\frac{1}{2}\ln|x+\sqrt{x^2-1}|+C$ **27.** $(3\sqrt{2}/2)-\frac{3}{2}\ln(1+\sqrt{2})$ **29.** $\frac{1}{3}(x^2-1)^{3/2}+C$
31. $3\sqrt{2}-\frac{1}{2}\ln(3+2\sqrt{2})$ **33.** $\frac{1}{2}\ln|x-2|-\frac{1}{2}\ln\sqrt{x^2-4}+C$ **35.** $(1/a)\ln|x-a|-(1/a)\ln\sqrt{x^2-a^2}+C$
37. $x+\frac{1}{2}\ln|(x-1)/(x+1)|+C$ **39.** $(\ln 3)/8$ **41.** $\ln|x+\sqrt{x^2-25}|+C$ **43.** $\ln|x+\sqrt{x^2-a^2}|-x/\sqrt{x^2-a^2}+C$
45. $\displaystyle-\frac{x}{2a(x^2-a^2)}-\frac{1}{2a^2}\ln\left|\frac{x-a}{\sqrt{x^2-a^2}}\right|+C$ **47.** $(1/a)\,\mathrm{arcsec}(x/a)+C$ **49.** $(1/a)\ln x-(1/a)\ln|a+\sqrt{x^2+a^2}|+C$
51. $(1/a)\tan^{-1}ax+C$ **53.** $\frac{1}{2}[a\sqrt{a^4+1}+(1/a)\ln(a^2+\sqrt{a^4+1})]$
55. $\sinh^{-1}(x/a)+C$ **57.** It works just like $x=a\sin\theta$, but with a sign change.

Section 7.5

1. $\sin^{-1}(x-1)+C$ **3.** $\frac{1}{4}(2x+3)\sqrt{x^2+3x}-\frac{9}{8}\ln|2x+3+2\sqrt{x^2+3x}|+C$ **5.** $\frac{1}{4}(3\sqrt{3}-1)+\frac{3}{8}\ln(1+2/\sqrt{3})$
7. $\ln|x+1+\sqrt{x^2+2x+2}|+C$ **9.** $\pi/8$ **11.** $\pi/\sqrt{3}$ **13.** $\sqrt{x^2+2x+2}+C$ **15.** $\dfrac{1}{\sqrt{5}}\arctan\dfrac{3x+2}{\sqrt{5}}+C$
17. $\dfrac{5\sqrt{10}-2\sqrt{3}}{6}+\dfrac{5\sqrt{3}}{18}\ln\left(1+\dfrac{\sqrt{6}}{\sqrt{5}}\right)$
19. If $a>0$ and $b^2-4ac>0$, then ax^2+bx+c becomes a difference of squares. A secant trigonometric substitution suffices.

Section 7.6

1. $\frac{1}{2}x^2+\frac{1}{2}\ln|x^2-1|+C$ **3.** $\frac{3}{2}\ln|x^2-1|+C$ **5.** $\ln|x-1|-\ln|x|+C$ **7.** $(x^2/2)-\frac{1}{2}\ln(x^2+1)+C$
9. $\frac{1}{2}\ln|x+1|-\frac{1}{4}\ln(x^2+1)+\frac{1}{2}\tan^{-1}x+C$ **11.** $\frac{3}{8}\tan^{-1}x+x/2(x^2+1)+x(1-x^2)/8(x^2+1)^2+C$
13. $\ln|(x+1)(x+2)(x+3)|+C$ **15.** $\frac{1}{3}\ln|x+1|+\frac{1}{6}\ln(x^2-x+1)+(1/\sqrt{3})\tan^{-1}[(2x-1)/\sqrt{3}]+C$
17. $1/(x+1)-2\ln|x+1|+2\ln|2x+1|+C$ **19.** $\frac{1}{3}\ln|x-2|+(1/\sqrt{3})\tan^{-1}[(x+1)/\sqrt{3}]-\frac{1}{6}\ln(x^2+2x+4)+C$
21. $(\frac{1}{2}a)\ln|(x-a)/(x+a)|+C$
23. $(\frac{1}{4}\sqrt{2})[\tan^{-1}(x\sqrt{2}+1)+\tan^{-1}(x\sqrt{2}-1)]+(\frac{1}{8}\sqrt{2})\ln(x^2+\sqrt{2}x+1)-(\frac{1}{8}\sqrt{2})\ln(x^2-\sqrt{2}x+1)+C$
25. $\frac{1}{2}\tan^{-1}(\frac{1}{2}e^x)+C$ **27.** $-1/(1+\ln x)+C$

Section 7.7

1. $2\sqrt{2x+9}+3\ln|\sqrt{2x+9}-3|-3\ln|\sqrt{2x+9}+3|+C$ **3.** $\sqrt{3}/2+\frac{1}{4}\ln(2-\sqrt{3})-\frac{1}{4}\ln(2+\sqrt{3})$
5. $\frac{2}{5}(x+1)^{5/2}-\frac{4}{3}(x+1)^{3/2}+2\sqrt{x+1}+C$ **7.** $2\sqrt{x+5}+2\ln|\sqrt{x+5}-2|-2\ln|\sqrt{x+5}+2|+C$
9. $\frac{1}{2}\ln(\sqrt{x+1}+\sqrt{x})-\frac{1}{2}\ln(\sqrt{x+1}-\sqrt{x})+\sqrt{x(x+1)}+C$ **11.** $2\sqrt{4+e^x}+2\ln|\sqrt{4+e^x}-2|-2\ln|\sqrt{4+e^x}+2|+C$
13. $\dfrac{2}{1-\tan(x/2)}-x+C$ **15.** $\dfrac{1}{\sqrt{3}}\left[\ln\left(2+\sqrt{3}-\tan\dfrac{\pi}{16}\right)-\ln\left(2-\sqrt{3}-\tan\dfrac{\pi}{16}\right)-\ln(\sqrt{3}+2)+\ln(\sqrt{3}-2)\right]$
17. $\dfrac{1}{2}\ln\left(\dfrac{2+2\sqrt{3}}{3}\right)-\dfrac{1}{2}\ln\dfrac{4}{3}+\dfrac{\pi}{6}$ **19.** $\dfrac{1}{\sqrt{5}}\left[\ln\left(2\tan\dfrac{x}{2}-1+\sqrt{5}\right)-\ln\left(2\tan\dfrac{x}{2}-1-\sqrt{5}\right)\right]+C$ **21.** $\dfrac{\pi}{4\sqrt{3}}$
23. $-\frac{1}{2}-\sqrt{2}-2\ln(1+\sqrt{2})+\ln 2$ **25.** $2\sqrt{x}-\frac{1}{2}x^{1/4}+2\ln(1+2x^{1/4})+C$ **27.** $\dfrac{1}{x}\sqrt{4x^2+2x-1}-\arcsin\dfrac{1-x}{x\sqrt{5}}+C$
29. $\frac{1}{2}\ln|\sec(2x-1)|+C$ **31.** $\frac{2}{3}-\frac{2}{9}\ln 4$ **33.** $-\frac{1}{12}(\sin^2 4x+2)\cos 4x+C$ **35.** $4\sin\frac{1}{2}x-2x\cos\frac{1}{2}x+C$
37. $\frac{1}{5}x^5\ln x-\frac{1}{25}x^5+C$
39. $\sin x=2\sin(x/2)\cos(x/2)=2(z/\sqrt{1+z^2})(1/\sqrt{1+z^2})=2z/(1+z^2)$, $\cos x=\cos^2(x/2)-\sin^2(x/2)=(1/\sqrt{1+z^2})^2(z/\sqrt{1+z^2})^2=(1-z^2)/(1+z^2)$

Section 7.8

1. Left: 0.7595, right: 0.6345 **3.** Left: 0.8259, right: 0.7426 **5.** 0.6970 **7.** 0.7842 **9.** 0.6933 **11.** 0.7854
13. Simpson's rule **15.** Simpson's rule **17.** 6.31656 **19.** 2.17363 **21.** 0.9037 **23.** $n\geq 260$ **25.** $n\geq 2$
27. Linear functions
29. Using $g(x)=ax^2+bx+c$ and $x_1=\frac{1}{2}(x_0+x_2)$, reduce both sides to $\frac{1}{3}a(x_2^3-x_0^3)+\frac{1}{2}b(x_2^2-x_0^2)+C(x_2-x_0)$. **31.** No
33. Left endpoint $=(1/n)(y_0+y_1+\cdots+y_{n-1})$ and Right endpoint $=(1/n)(y_1+y_2+\cdots+y_n)$. Their average is $\frac{1}{2}(\text{Left}+\text{Right})=(\frac{1}{2}n)(y_0+y_1+\cdots+y_{n-1}+y_1+\cdots+y_{n-1}+y_n)=(\frac{1}{2}n)(y_0+2y_1+2y_2+\cdots+2y_{n-1}+y_n)$, which is the trapezoidal rule.

Review Exercises

1. $\frac{1}{27}(3x+4)^9+C$ **3.** $\ln 2-\pi/4$ **5.** $-\frac{1}{2}e^{-x^2}+C$ **7.** $\frac{1}{2}(\arctan x)^2+C$ **9.** $\frac{2}{3}\sin(3x-5)+C$ **11.** $\pi^2/3-\frac{4}{27}$
13. $\frac{1}{2}(x^2+1)\arctan x-\frac{1}{2}x+C$ **15.** $x\,\mathrm{arcsec}\,2x-\frac{1}{2}\ln(2x+\sqrt{4x^2-1})+C$ **17.** $(2e^3+1)/9$ **19.** $x\log_3 x-x/(\ln 3)+C$
21. $\frac{1}{3}\sin^3x-\frac{1}{5}\sin^5x+C$ **23.** $\pi/4$ **25.** $\frac{1}{4}\tan 4x-x+C$ **27.** $\frac{1}{3}\sec^3x+C$ **29.** $\frac{1}{2}x\sqrt{x^2+b^2}+\frac{1}{2}b^2\ln(x+\sqrt{x^2+b^2})+C$
31. $\ln|x+\sqrt{x^2+8}|-(x/\sqrt{x^3+8})+C$ **33.** $\sqrt{8}-\sqrt{5}+2\ln(2+\sqrt{5})-2\ln(1+\sqrt{2})$ **35.** $x/\sqrt{1-x^2}-\arcsin x+C$
37. $\frac{1}{2}\ln|(1-\sqrt{1-x^2})/(1+\sqrt{1-x^2})|+C$ **39.** $\frac{1}{3}(x^2+2x-3)^{3/2}+C$ **41.** $\frac{1}{3}\sin^{-1}3x+C$
43. $\dfrac{1}{2b^2}\ln|x+b|-\dfrac{1}{6b^2}\ln|x^3+b^3|-\dfrac{1}{b^2\sqrt{3}}\arctan\dfrac{2x-b}{b\sqrt{3}}+C$ **45.** $\ln\frac{2}{5}-\frac{63}{200}$
47. $\frac{1}{2}x^2-5x+14\ln|x^2+5x-3|-\dfrac{80}{\sqrt{37}}\ln|2x+5-\sqrt{37}|+\dfrac{80}{\sqrt{37}}\ln|2x+5+\sqrt{37}|+C$
49. $\dfrac{1}{3}x^3\tan^{-1}x-\dfrac{x^2}{6}+\dfrac{1}{6}\ln(1+x^2)+C$ **51.** $\ln|x|-\dfrac{1}{x}\tan^{-1}x-\dfrac{1}{2}\ln(1+x^2)+C$
53. $-\frac{4}{3}(2-x)^{3/2}+\frac{2}{5}(2-x)^{5/2}+C$ **55.** $\frac{3}{56}(2x-1)^{7/3}+\frac{3}{16}(2x-1)^{4/3}+\frac{3}{8}(2x-1)^{1/3}+C$ **57.** $x+\cos x+C$
59. $\ln|1+\tan(x/2)|+C$ **61.** 1 **63.** 1.829, 1.429, 1.629, 1.611 **65.** 0.923

Additional Exercises

1. $10^{\ln x}/\ln 10+C$ **3.** $x^2/4+x(\sin 2x)/4+(\cos 2x)/8+C$
5. $\frac{4}{13}(3-x)^{13/4}-\frac{4}{3}(3-x)^{9/4}+\frac{12}{5}(3-x)^{5/4}-4(3-x)^{1/4}+C$ **7.** $\frac{4}{5}(1+\sqrt{x})^5-\frac{4}{3}(1+\sqrt{x})^3+C$
9. $(-x^4-2x^2-1)e^{-x^2}+C$ **11.** $\frac{1}{2}\ln|\sqrt{1+e^{2x}}-1|-\frac{1}{2}\ln|\sqrt{1+e^{2x}}+1|+C$
13. $(1/\sqrt{2})\ln|\tan(x/2+\pi/8)|+C$ **15.** $(\frac{3}{8}a^2)(ax+b)^{8/3}-(3b/5a^2)(ax+b)^{5/3}+C$ **17.** $e^x/(1+x)+C$ **19.** $\frac{2}{3}$
21. $\dfrac{1-\cos(a-b)\pi/2}{2(a-b)}+\dfrac{1-\cos(a+b)\pi/2}{2(a+b)}$ **23.** $\dfrac{\sin(a-b)\pi/2}{2(a-b)}+\dfrac{\sin(a+b)\pi/2}{2(a+b)}$ **25.** π **27.** $\frac{1}{2}\pi a^2$ **29.** $\pi/2$
31. $\ln(1-\cos\theta)+C$ **33.** $-\cot x+C$ **35.** $\frac{1}{2}\tan^2x+\ln|\cos x|+C$ **37.** $x\ln(x+\sqrt{x^2+1})-\sqrt{x^2+1}+C$
39. $-\frac{1}{2}\cos x^2+C$ **41.** $\frac{1}{2}\ln|u+1|-\frac{1}{2}\ln|u+3|+C$
43. $\frac{2}{11}(x+1)^{11/2}-\frac{10}{9}(x+1)^{9/2}+\frac{20}{7}(x+1)^{7/2}-4(x+1)^{5/2}+\frac{10}{3}(x+1)^{3/2}-2(x+1)^{1/2}+C$
45. $\ln(1+x+\sqrt{x})-(2/\sqrt{3})\tan^{-1}[(2\sqrt{x}+1)/\sqrt{3}]+C$ **47.** $x\ln x-2x+(x+1)\ln(x+1)+C$ **49.** 1.6094
51. 0.2527 **53.** $(x+\sqrt{6})(x-\sqrt{6})$ **55.** $(x^2+x+1)(x^2-x+1)$ **57.** $(x+2)(x+1)(x-2)$
59. Use partial fractions or trigonometric substitution. **61.** Integrate by parts. **63.** $\frac{32}{3}-\pi$
65. $3\sqrt{15}/8+13\pi/8-\frac{9}{4}\sin^{-1}(\frac{7}{8})-\sin^{-1}(\frac{11}{16})$ **67.** $2\pi-\sqrt{399}/16-4\sin^{-1}(\sqrt{7}/8)$ ft^2
69. **(a)** $(r_1^2+r_2^2)\dfrac{\pi}{2}-\dfrac{1}{2}\sqrt{(r_1+r_2+s)(-r_1+r_2+s)(r_1-r_2+s)(r_1+r_2-s)}+r_2^2\arcsin\dfrac{r_1^2-r_2^2-s^2}{2r_2s}-$
$r_1^2\arcsin\dfrac{r_1^2-r_2^2+s^2}{2r_1s}$
(b) πr_1^2 minus the answer in part (a)

Chapter 8 Section 8.1

1. $\frac{1}{4}\pi^2$ **3.** 8π **5.** $2304\pi/5$ **7.** 6π **9.** $2048\pi\sqrt{2}/3$ **11.** $\frac{1}{2}\pi(1-e^{-2b})$ **13.** $128\pi/5$ **15.** π **17.** $2\pi(1-e^{-b}-be^{-b})$
19. $\pi^2+4\pi/3$ **21.** 42π **23.** $128\pi/3$ **25.** $\pi/2$ **27.** $\frac{16}{3}$ **29.** $\frac{1}{3}\pi h(r_1^2+r_1r_2+r_2^2)$ **31.** $16\pi+6\sqrt{3}$ ft^3 **33.** $\frac{5}{3}\pi$ ft/min

Section 8.2

1. $\pi^2-2\pi$ **3.** $8\pi/3$ **5.** $2\pi e^e(e-1)-\frac{1}{2}\pi(e^2+1)$ **7.** $7\pi/6$ **9.** 8π **11.** $14\pi/15$ **13.** $56\pi/5$ **15.** $184\pi/7$ **17.** $4\pi^2$
19. $4\pi-\frac{1}{2}\pi^2$ **21.** $\pi/2$ **23.** $\dfrac{2\pi}{15}(1550-100\sqrt{5}+24\sqrt{3})$ **25.** $\dfrac{\pi}{15}(7546+100\sqrt{5}-60\sqrt{3})$ **27.** $\dfrac{\pi}{3}$ **29.** $\dfrac{\pi}{6}(12e-3e^2-7)$
31. $\dfrac{2\pi}{15}(41-24\sqrt{2})$ **33.** $\dfrac{\pi}{6}(16\sqrt{2}-21)$ **35.** $\dfrac{\pi}{\sqrt{2}}(4+\pi)$ **37.** $\pi(2\sqrt{2}-1)$ **39.** $320\pi/3$
41. **(a)** $\pi s^3/4$ **(b)** $\pi s^2(s+2r\sqrt{3})/4$ **(c)** $\pi s^3\sqrt{3}/4$ **(d)** $\pi s^2(s+2r)\sqrt{3}/4$ **(e)** $\pi s^3/2$ **(f)** $\pi s^2(2s+r\sqrt{3})/4$

Section 8.3

1. $\sqrt{17} + \frac{1}{4}\ln(4 + \sqrt{17})$ **3.** $\ln(1 + \sqrt{2})$ **5.** $\frac{3}{4} + \frac{1}{2}\ln 2$ **7.** $\frac{307}{8}$ **9.** $\sqrt{10} - \sqrt{2} + \ln(\sqrt{10} - 1) - \ln(\sqrt{2} - 1) - \ln 3$
11. $\frac{1}{54}(1625\sqrt{13} - 37\sqrt{37})$ **13.** $\frac{1}{2}(3\sqrt{13} - \sqrt{5}) + 2\ln(\sqrt{13} + 3) - 2\ln(\sqrt{5} + 1)$ **15.** 3π **17.** $\frac{1}{27}(13\sqrt{13} - \frac{125}{8})$ **19.** 2

Section 8.4

1. 2π **3.** $2\pi(\pi - 1)$ **5.** $4\pi\sqrt{5}$ **7.** $8\pi(4 + 500\sqrt{10})/1215$ **9.** $\pi(\pi a - 2r)r$ **11.** $\pi\left[\ln\left(\frac{9 + \sqrt{82}}{1 + \sqrt{2}}\right) - \frac{\sqrt{82}}{9} + \sqrt{2}\right]$

Section 8.5

1. At its midpoint **3.** At the center of the square **5.** $\frac{2}{3}$ ft from the heavy end **7.** The point $(\frac{7}{5}, \frac{16}{15})$
9. On the axis of symmetry, $\frac{1}{4}h$ from the base **11.** On the axis of symmetry, $\frac{2}{3}h$ from the base
13. On the axis of symmetry, up $(25\sqrt{5} + 1)/(25\sqrt{5} - 5)$ from the vertex **15.** $\frac{1745}{603}$

Section 8.6

1. $\pi^2 r^3$ **3.** $\frac{1}{2}\pi b^2 h$ **5.** $\pi h(2r + 3mh)\sqrt{m^2 + 1}$ **7.** $3\pi[4\sqrt{17} + \ln(4 + \sqrt{17})]$ **9.** $(\frac{4}{5}, \frac{2}{7})$
11. On the axis of symmetry, $\frac{1}{4}h$ from the base
13. If the triangle is placed in the first quadrant with its legs along the coordinate axes, the centroid is the point $(\frac{1}{3}a, \frac{1}{3}b)$.
15. On the altitude, $\frac{1}{3}h$ above the base
17. The centroid is found by integration in the usual way. Also, two of the medians of the triangle are $(a + b)y = hx$ and $(2a - b)y = 2h(x - \frac{1}{2}b)$, whose intersection is the same point.

Section 8.7

1. \$8689.30; \$26,187.43; \$61,424.31; \$132,382.69 **3.** \$21.55 **5.** \$558.94; \$289,154; \$79,354,000 **7.** 180,000 lb
9. 2159.845 lb **11.** 12.756 ft-lb **13.** 42,411.5 ft-lb **15.** 1 **17.** 22,125 **19.** No individual ever dies.
21. 319,452,805 metric tons **23.** 1992 **25.** \$28,700.48 **27.** 120 in.-lb **29.** 192 ft-lb

Section 8.8

1. $\frac{1}{4}$ **3.** Diverges **5.** Diverges **7.** Diverges **9.** Diverges **11.** Diverges **13.** $-\frac{1}{2}$ **15.** Diverges
17. If $p > 1$, then $\int_1^t x^{-p}\,dx = [1/(p - 1)](1 - 1/t^{p-1})$, which converges to $1/(p - 1)$ as $t \to \infty$. If $p = 1$, then $\int_1^t x^{-1}\,dx = \ln t$, which diverges as $t \to \infty$. If $p < 1$, then $\int_1^t x^{-p}\,dx = (t^{1-p} - 1)/(1 - p)$, which diverges as $t \to \infty$.
19. $1/k$ **21.** $\int_1^\infty (1/x)\,dx$ diverges by Problem 17.
23. The surface area is $S = \int_1^\infty 2\pi y\,ds = 2\pi\int_1^\infty(\sqrt{x^4 + 1}/x^3)\,dx$. Since $\sqrt{x^4 + 1}/x^3 > 1/x$, $S > 2\pi\int_1^\infty(1/x)\,dx$, which diverges by Problem 17. Thus, S is infinite.
25. Not necessarily **27.** By Problem 19, $\int_0^\infty ke^{-kx}\,dx = 1$.
29. $\int_0^\infty \cos x\,dx = \lim_{t\to\infty}(\sin t)$, which does not exist. Therefore, the improper integral diverges, but the limit is not infinite.

Review Exercises

1. $\frac{96}{5}$ **3.** $\frac{8}{27}(10\sqrt{10} - 1)$ **5.** **(a)** $384\pi/7$ **(b)** $576\pi/5$ **(c)** $3456\pi/35$ **(d)** 192π

7. **(a)** $\left(\frac{\int_0^4 kx(4 - x)(8 - x^{3/2})\,dx}{\int_0^4 k(4 - x)(8 - x^{3/2})\,dx}, \frac{\int_0^4 \frac{1}{2}k(4 - x)(8 - x^{3/2})^2\,dx}{\int_0^4 k(4 - x)(8 - x^{3/2})\,dx}\right)$

(b) $\left(\frac{\int_0^8 \frac{1}{2}ky(8 - y)^{4/3}\,dy}{\int_0^8 ky(8 - y)^{2/3}\,dy}, \frac{\int_0^8 ky^2(8 - y)^{2/3}\,dy}{\int_0^8 ky(8 - y)^{2/3}\,dy}\right)$ **9.** $2\pi n \int_0^1 \left(\frac{5x}{\sqrt{1 - x^2}} - x\right) dx$

11. No individuals enter the population. **13.** 3 **15.** $\frac{1}{3}$

Additional Exercises

1. **(a)** $(\frac{1}{2})^{1/5}$ **(b)** $(\frac{1}{2})^{1/4}$
3. Let solid S_i have plane cross-section area $A_i(y)$ at height y. The key is that $A_1(y) = A_2(y)$, and then $V_1 = \int_0^h A_1(y)\,dy = \int_0^h A_2(y)\,dy = V_2$.
5. $16r^3/3$ **7.** $S = \int_a^{a+h} 2\pi x\,ds$; use $x = \sqrt{r^2 - y^2}$ and $ds = (r/x)\,dx$ **9.** 589 lb **11.** 997.5 mi-lb **13.** 97,561 mi-lb
15. $-\frac{1}{2}mgR$
17. Not without modification. Just use exponential growth with the growth rate being the difference between the interest rate and the inflation rate.
19. If A is the area of the plate, the vertical coordinate of the centroid is $\bar{h} = \int_R hl\,dh/A$, so that $\int_R hl\,dh = \bar{h}A$. (Here, l is the length of the strip of the plate at level h that gives the area of the plate.) But the pressure on the plate is $F = \int_R whl\,dh = w\int_R hl\,dh = w\bar{h}A$, which is the pressure at the centroid times the area.
21. πa^2 **23.** $2r^3/3\sqrt{3}$ **25.** $\frac{1}{2}$ **27.** Diverges **29.** Diverges **31.** $\pi/2$ **33.** **(a)** 0.2835 **(b)** 0.6321 **(c)** 0.3679 **(d)** 0.1889

Chapter 9 Section 9.1

1. 0 **3.** 0 **5.** 1 **7.** 1 **9.** Does not exist **11.** Does not exist **13.** 0 **15.** 0 **17.** 0 **19.** 0 **21.** 0 **23.** 0 **25.** 1
27. 1 **29.** e **31.** 1 **33.** 0 **35.** 0 **37.** $10^{-1}, 10^{-2}, 10^{-3}, 10^{-4}, 10^{-5}$
39. $-5.44 \times 10^{-2}, -5.06 \times 10^{-3}, 8.27 \times 10^{-4}, -3.06 \times 10^{-5}, 3.57 \times 10^{-7}$
41. 2.59374, 2.70481, 2.71692, 2.71815, 2.71827 **43.** 0.43, 0.22, 0.14, 0.11, 0.09
45. If $|a_n| \to 0$, then we can make $||a_n| - 0| < \varepsilon$. But this is the same as $|a_n| < \varepsilon$, which is the same as $|a_n - 0| < \varepsilon$. This means that $a_n \to 0$.
47. From the proof, only the terms $a_1, a_2, \ldots, a_k$ do not necessarily lie in the interval $(L - \varepsilon, L + \varepsilon)$. An upper bound of the sequence is therefore the largest of these a_i's or $L + \varepsilon$, whichever is larger; a lower bound is the least of the a_i's or $L - \varepsilon$, whichever is smaller.
49. If $r > 1$, then $r = 1 + s$ with $s > 0$. Hence, for any n, $r^n = (1 + s)^n = 1 + ns + \cdots > 1 + ns$. Now, for any number B, however large, there exists an integer m such that $ms > B$. (This is called the **Archimedian property** of the real number system.) Therefore, $r^m > 1 + ms > B$, and $\{r^n\}$ diverges to ∞.
51. The bisection method yields a sequence of numbers $c_1, c_2, \ldots$ and a sequence of intervals $[a_1, b_1], [a_2, b_2], \ldots$ such that $c_1 = \frac{1}{2}(a_1 + b_1)$, $[a_2, b_2] = [a_1, c_1]$ or $[c_1, b_1]$ (whichever contains the root), $c_2 = \frac{1}{2}(a_2 + b_2)$, and so forth.
53. 1.00000, 1.54030, 1.57079, 1.57080, 1.57080, . . .

Section 9.2

1. $\frac{2}{9}$ **3.** $\frac{2}{3}$ **5.** $\frac{2}{3}$ **7.** 200 **9.** 10 **11.** $\frac{7}{2}$ **13.** 2 **15.** $\frac{1}{3}$ **17.** Diverges **19.** Diverges **21.** Diverges **23.** Diverges
25. Converges to $\frac{3}{2}$ **27.** Diverges **29.** Diverges **31.** $\frac{14}{99}$ **33.** $\frac{1175}{999}$ **35.** 180 cm **37.** At $16\frac{4}{11}$ min after 3
39. 2.929, 4.499, 5.187, 5.878
41. If Σa_n is a series such that for any $\varepsilon > 0$ there exists k such that if $m, n > k$ then $|s_m - s_n| < \varepsilon$, then Σa_n converges.
43. Let s_n be the partial sums of $\Sigma_{n=1}^{\infty} a_n$, and let t_n be the partial sums of $\Sigma_{n=1}^{\infty} b_n$. Let u_n be the partial sums of $\Sigma_{n=1}^{\infty} (a_n - b_n)$. Then $u_n = \Sigma_{i=1}^{n} (a_i - b_i) = \Sigma_{i=1}^{n} a_i - \Sigma_{i=1}^{n} b_i = s_n - t_n$. Hence, $\lim_{n\to\infty} u_n = \lim_{n\to\infty} (s_n - t_n) = \lim_{n\to\infty} s_n - \lim_{n\to\infty} t_n = s - t$.
45. The given series diverges, so it does not have any value.
47. Let $s = 1 + r + r^2 + r^3 + \cdots$. Then $rs = r + r^2 + r^3 + \cdots$, and subtracting gives $s - rs = 1$. That is, $s = 1/(1 - r)$. Here, we assumed $|r| < 1$.

Section 9.3

1. Convergent p-series **3.** Diverges; limit comparison with harmonic series
5. Diverges; comparison with harmonic series ($1/\sqrt{n}\ln n > 1/n$ if $n > 4$)
7. Converges; comparison with p-series $\Sigma_{n=1}^{\infty} n^{-2}$
9. Converges; integral test

11. Diverges; limit comparison with harmonic series **13.** Diverges; divergence theorem
15. Converges; comparison with p-series $\sum_{n=1}^{\infty} n^{-2}$
17. Diverges; divergence theorem **19.** Diverges; comparison with harmonic series
21. If $\int_1^\infty f(x)\,dx$ converges, say to V, then for every n, $\int_1^n f(x)\,dx < V$, and the sequence is bounded. If $\int_1^\infty f(x)\,dx$ diverges, then $\lim_{t\to\infty} \int_1^t f(x)\,dx = \infty$, and for any number B there exists K such that if $t > K$, then $\int_1^t f(x)\,dx > B$. Then there is an integer $n > K$ such that $\int_1^n f(x)\,dx > B$, and the sequence is unbounded.
23. No

Section 9.4

1. Absolutely convergent **3.** Absolutely convergent **5.** Divergent **7.** Absolutely convergent **9.** Absolutely convergent
11. Divergent **13.** Divergent **15.** Absolutely convergent **17.** Divergent **19.** Absolutely convergent
21. Absolutely convergent **23.** Absolutely convergent **25.** Absolutely convergent **27.** Absolutely convergent
29. Absolutely convergent **31.** Absolutely convergent **33.** 0.6456, 0.6688, 0.6768, 0.6808 **35.** 5
37. If the terms of a geometric series approach 0, then the ratio must be between -1 and 1, because the contrapositive is certainly true.
39. The divergence, ratio, and root tests fail; the integral test does not apply; and comparison tests do not work.
41. Rearranging a conditionally convergent series may yield a series that converges to a different sum.

Section 9.5

1. $[-\frac{1}{10}, \frac{1}{10})$ **3.** $(-1, 1)$ **5.** $(-1, 1)$ **7.** $[-1, 1]$ **9.** $(-1, 1)$ **11.** $(-1, 1)$ **13.** 0 only **15.** $(-\frac{2}{3}, \frac{2}{3})$ **17.** $[1, 3)$
19. $[0, 2)$ **21.** $(-1, 1)$ **23.** $(0, 2)$
25. By the root test (Theorem 9.21), if $\lim_{n\to\infty}|a_n x^n| = |x|/r$ is greater than 1, or $|x| > r$, then the series diverges; if the limit is less than 1, or $|x| < r$, then it converges. Thus, r is the radius of convergence.
27. $(-\infty, \infty)$ **29.** $(-1/e, 1/e)$ **31.** $(\frac{3}{2}, \frac{5}{2})$

Section 9.6

1. $1 - x + x^2 - x^3$ **3.** $1 - x^2$ **5.** $-1 + \frac{9}{2}(x - \pi/3)^2$ **7.** $x^3 + 3x - 5$
9. $1 + 2x + 4x^2/2! + 8x^3/3! + \cdots + 2^n x^n/n! + \cdots, (-\infty, \infty)$
11. $1 - x + x^2 - x^3 + \cdots + (-1)^n x^n + \cdots, (-1, 1)$
13. $1 + x^2/2! + x^4/4! + \cdots + x^{2n}/(2n)! + \cdots, (-\infty, \infty)$
15. $1 + x^2 + x^4/2! + x^6/3! + \cdots + x^{2n}/n! + \cdots, (-\infty, \infty)$ **17.** 0.540 **19.** 3.762 **21.** e^{-2} **23.** $e^2 - 1$ **25.** $\ln 2$
27. $2e^2$ **29.** $\cosh 1$ **31.** $\cosh 1 + \cos 1 - 2$
33. For any term $a_k x^k$ of the polynomial f, $f^{(k)}(0)/k! = a_k$.
35. If a polynomial function has degree n, then its $(n + 1)$th derivative is 0 and so are all higher derivatives.
37. Since f is continuous on $[a, b]$, it has a minimum m and a maximum M, so that $m < f(x) < M$ for all $x \in [a, b]$. Since g does not change sign, suppose $g(x)$ is positive (the negative case is similar). Then we have

$$mg(x) < f(x)g(x) < Mg(x)$$

for all $x \in [a, b]$, and it follows that

$$m\int_a^b g(x)\,dx = \int_a^b mg(x)\,dx < \int_a^b f(x)g(x)\,dx < \int_a^b Mg(x)\,dx = M\int_a^b g(x)\,dx$$

Therefore,

$$m < \frac{\int_a^b f(x)g(x)\,dx}{\int_a^b g(x)\,dx} < M.$$

Hence, by the intermediate value theorem, there exists $c \in [a, b]$ such that

$$f(c) = \frac{\int_a^b f(x)g(x)\,dx}{\int_a^b g(x)\,dx},$$

and the result follows.

39. Using the definition of the derivative, we get

$$f'(0) = \lim_{x\to 0} \frac{f(x) - f(0)}{x - 0} = \lim_{x\to 0} \frac{e^{-1/x^2}}{x} = \lim_{x\to 0} \frac{1/x}{e^{1/x^2}}$$
$$= \lim_{x\to 0} \frac{-1/x^2}{(2/x^3)e^{1/x^2}} = \lim_{x\to 0} \frac{x}{2e^{1/x^2}} = 0$$

by L'Hôpital's rule. Similarly, $f''(0) = 0$, $f'''(0) = 0$, and so on. Thus, the Maclaurin series for f is the zero series, converging to 0 for all x.

Section 9.7

1. $2(1-x)^{-3}$, $(-1, 1)$ **3.** $-(1+x)^{-2}$, $(-1, 1)$ **5.** $2(1+x)^{-3}$, $(-1, 1)$
7. $-(1/x)\ln(1-x)$, $[-1, 1)$ **9.** $(1/x^2)\ln(1+x^2)$, $[-1, 1]$ **11.** $x + (1-x)\ln(1-x)$, $[-1, 1)$
13. $-2x(1+x^2)^{-2}$, $(-1, 1)$ **15.** $(2-x)^{-2}$, $(0, 2)$ **17.** $2(2-x^2)^2$, $(-2, 2)$
19. Use the arctan x series (Example 9.53) at $x = 1$.
21. $\ln 2 = -\ln \frac{1}{2} = -\ln(1 - \frac{1}{2})$; use the $\ln(1+x)$ series with $x = -\frac{1}{2}$.
23. Substitute x^2 for x in the e^x series and integrate.
25. Multiply the sin x series by x^2 and integrate.
27. $1 + x^3 + x^6 + x^9 + \cdots$, $(-1, 1)$ **29.** $1 - 2x + 3x^2 - 4x^3 + \cdots$, $(-1, 1)$
31. $x^3 - \frac{1}{2}x^6 + \frac{1}{3}x^9 - \frac{1}{4}x^{12} + \cdots$, $(-1, 1]$ **33.** $\frac{4}{9}$ **35.** $\ln 4$
37. The second is much more rapid, in that the terms get smaller faster.

Section 9.8

1. $1 + \frac{2}{3}x - \frac{2\cdot 1}{2!3^2}x^2 + \frac{2\cdot 1\cdot 4}{3!3^3}x^3 - \frac{2\cdot 1\cdot 4\cdot 7}{4!3^4}x^4 + \cdots$, $(-1, 1)$

3. $1 - \frac{1}{3}x - \frac{2}{2!3^2}x^2 - \frac{2\cdot 5}{3!3^3}x^3 - \frac{2\cdot 5\cdot 8}{4!3^4}x^4 - \cdots$, $(-1, 1)$ **5.** $1 - x + x^2 - x^3 + x^4 - \cdots$, $(-1, 1)$

7. $x + \frac{1}{3}x^2 - \frac{2}{2!3^2}x^3 + \frac{2\cdot 5}{3!3^3}x^4 - \frac{2\cdot 5\cdot 8}{4!3^4}x^5 + \cdots$, $(-1, 1)$ **9.** $-x - \frac{1}{2}x^2 - \frac{1}{3}x^3 - \frac{1}{2}x^4 - \cdots$, $[-1, 1)$

11. $a + \frac{1}{2a}x^2 - \frac{1}{2!2^2a^3}x^4 + \frac{1\cdot 3}{3!2^3a^5}x^6 - \frac{1\cdot 3\cdot 5}{4!2^4a^7}x^8 + \cdots$, $(-a, a)$

13. Use the arcsine series from Example 9.58(c) with $x = 1$.
15. Use the series for $1/\sqrt{1+x}$ from Example 9.58(a) with $x = 1$.

17.
$$x^2y'' + xy' + y = x^2 \sum_{n=2}^{\infty} \frac{n(n-1)}{n^2+1}x^{n-2} + x\sum_{n=1}^{\infty} \frac{n}{n^2+1}x^{n-1} + \sum_{n=0}^{\infty} \frac{1}{n^2+1}x^n$$
$$= \sum_{n=2}^{\infty}\left[\frac{n(n-1)}{n^2+1} + \frac{n}{n^2+1} + \frac{1}{n^2+1}\right]x^n + \frac{1}{2}x + 1 + \frac{1}{2}x = 1 + x + \sum_{n=2}^{\infty} x^n$$
$$= 1 + x + x^2 + x^3 + x^4 + \cdots = \frac{1}{1-x}$$

Review Exercises

1. Converges to 0 **3.** Converges to 0 **5.** Converges to 0 **7.** Converges to 0 **9.** Diverges **11.** Converges to 2 **13.** Converges to $e/(e-1)$ **15.** Diverges **17.** Diverges **19.** Converges to $e^{10}-1$ **21.** Converges **23.** Diverges **25.** Converges to $\frac{14}{99}$ **27.** Converges to $\frac{4}{3}$ **29.** Converges to $1-\ln 2$ **31.** Converges to $-300/101^2$ **33.** Converges **35.** Converges **37.** Converges on $(-10, 10)$ to $10/(10-x)$ **39.** Converges on $[-1, 1]$ **41.** Converges on $(-e, e)$ **43.** Converges on $(-1, 1)$ to $(1/x)\ln(1-x)+1/(1-x)$ **45.** Converges on $(-\infty, \infty)$ to xe^x **47.** Converges on $(-\infty, \infty)$ to e^{5x}

49. $1+3x+\frac{3^2x^2}{2!}+\frac{3^3x^3}{3!}+\frac{3^4x^4}{4!}+\cdots, (-\infty, \infty)$ **51.** $5x-\frac{5^3x^3}{3!}+\frac{5^5x^5}{5!}-\frac{5^7x^7}{7!}+\cdots, (-\infty, \infty)$

53. $x-\frac{x^3}{2}+\frac{x^5}{3}-\frac{x^7}{4}+\cdots, (-1, 1]$ **55.** $1-\frac{1}{3}x+\frac{1\cdot 4}{2!3^2}x^2-\frac{1\cdot 4\cdot 7}{3!3^3}x^3+\frac{1\cdot 4\cdot 7\cdot 10}{4!3^4}x^4-\cdots, (-1, 1)$

57. $e^{-3}\left[1-3(x-1)+\frac{3^2}{2!}(x-1)^2-\frac{3^3}{3!}(x-1)^3+\cdots\right], (-\infty, \infty)$ **59.** $\frac{1}{4}-\frac{1}{3\cdot 6}+\frac{1}{5\cdot 8}-\frac{1}{7\cdot 10}+\cdots$

Additional Exercises

1. Converges to 0 **3.** Converges to 1 **5.** Converges to e^{-a} **7.** $x\in(-1, 1)$ **9.** $x\neq 0$ **11.** $x\in(-\frac{1}{2}, \frac{1}{2})$ **13.** Diverges **15.** Converges **17.** Diverges **19.** Converges **21.** $(-2, 2)$ **23.** $(-1, 1)$ **25.** $(-\infty, \infty)$ **27.** $(-\infty, \infty)$ **29.** $(-\frac{1}{2}, \frac{1}{2})$

31. Suppose that in every interval $(c-1/n, c+1/n)$ there is some z_n such that $f(z_n)\le 0$. Then $\{z_n\}\to c$, and $\{f(z_n)\}\to f(c)\le 0$, by continuity of f. This contradicts the fact that $f(c)>0$. Hence, some interval $(c-1/n, c+1/n)$ contains no z such that $f(z)\le 0$, or in other words, f is positive in $(c-1/n, c+1/n)$.

33. $a_n-a_{n+1}=\ln(n+1)-\ln n-1/(n+1)=\ln(1+1/n)-1/(n+1)$. Using the first two terms of the series for $\ln(1+x)$, $\ln(1+1/n)>1/n-1/2n^2$, so $a_n-a_{n+1}>(n-1)/2n^2(n+1)>0$ for $n>1$. Thus, $a_n>a_{n+1}$, and the sequence is decreasing. The upper bound is $a_1=1$. Geometrically, using outer rectangles, we see that $1+\frac{1}{2}+\frac{1}{3}+\cdots+1/n>\int_1^{n+1}(1/x)\,dx=\ln(n+1)>\ln n$, so $a_n>0$ and the sequence is bounded.

Chapter 10 Section 10.1

1. $\pi/4$ **3.** $\pi/2$ **5.** $\arctan\frac{3}{4}$ **7.** $\arctan(5/\sqrt{2})$ **9.** $1/\sqrt{17}$ **11.** $6/\sqrt{10}$ **13.** $2x'-y'+6=0$ **15.** $x'^2+8x'+23=3y'$

Section 10.2

1. $x^2=8y$ **3.** $x^2=-4(y-1)$ **5.** $x^2-2xy+y^2-6x-10y+9=0$

7. $V(0, 0)$; $F(0, 1)$; d: $y=-1$

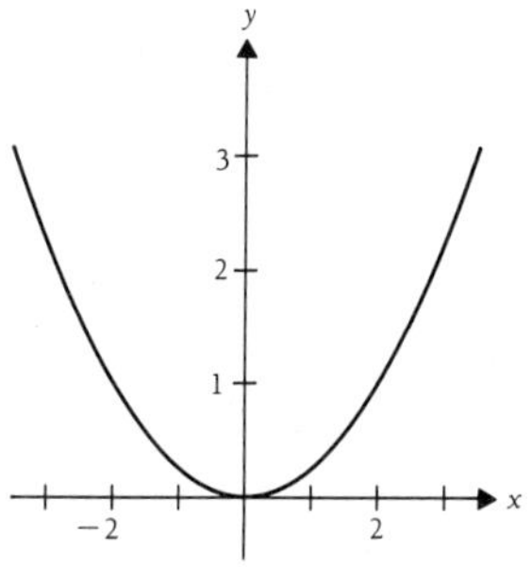

9. $V(-3, 3)$; $F(-3, \frac{9}{4})$; d: $y=\frac{15}{4}$

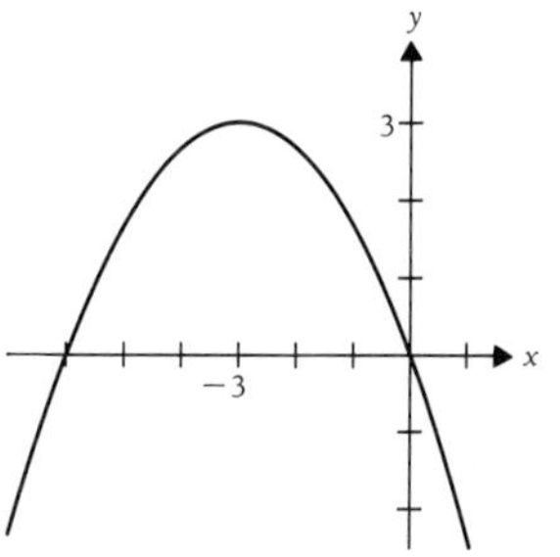

11. $V(1, \frac{7}{4})$; $F(1, \frac{3}{4})$; d: $y = \frac{11}{4}$

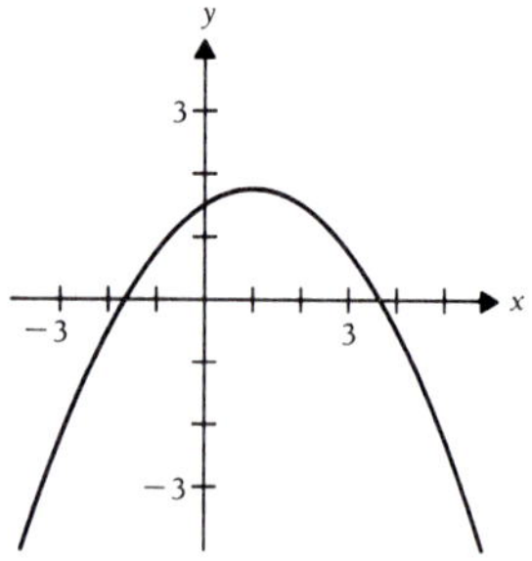

13. $x_0/2c$ **15.** $\arctan(2c/x_0)$ **17.** $|TF| = \sqrt{x_0^2 + (y_0 - c)^2} = \sqrt{4cy_0 + y_0^2 - 2cy_0 + c^2} = y_0 + c = |AF|$

Section 10.3

1. $y^2/16 + x^2/15 = 1$ **3.** $(x-2)^2/9 + y^2/5 = 1$ **5.** $84x^2 - 24xy + 91y^2 - 300x - 225y = 0$
7. $C(0, 0)$, $F(\pm\sqrt{5}, 0)$, $V(\pm 3, 0)$, $a = 3$, $b = 2$

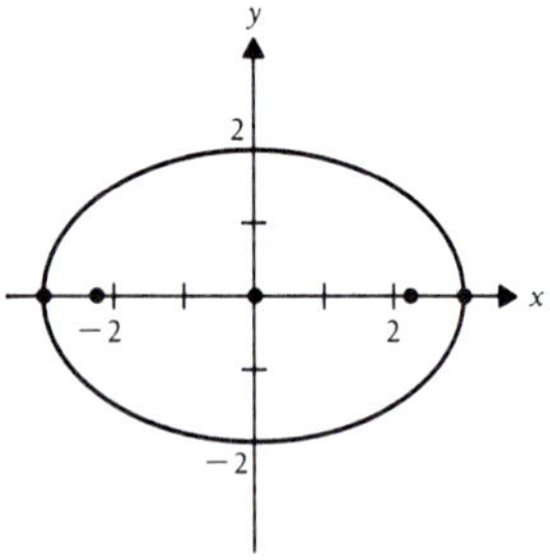

9. $C(0, -1)$, $F(\pm\sqrt{3}, -1)$, $V(\pm\sqrt{6}, -1)$, $a = \sqrt{6}$, $b = \sqrt{3}$

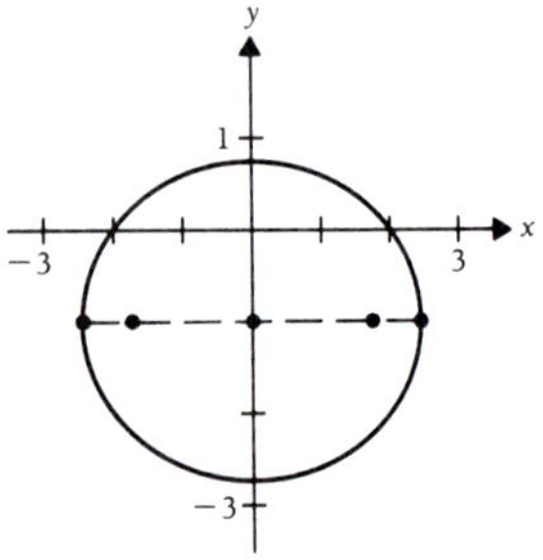

11. $C(-4, 0)$, $F(-4 \pm 2\sqrt{3}, 0)$, $V(-4 \pm 4, 0)$, $a = 4$, $b = 2$

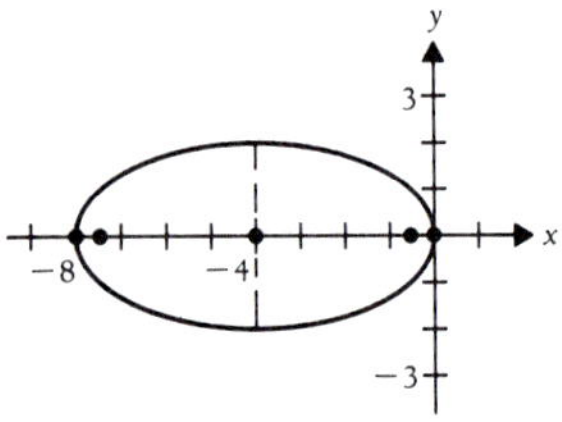

13. πab **15.** (a) $4\pi ab^2/3$ (b) $4\pi a^2b/3$ (c) $2\pi^2a^2b$ (d) $2\pi^2ab^2$ **17.** $-b^2x_0/a^2y_0$
19. The waves the diver makes will all be reflected toward the other focus and will all arrive there at the same time.

Section 10.4

1. $16x^2/9 - 16y^2/7 = 1$ **3.** $4(y - \frac{3}{2})^2 - x^2/2 = 1$ **5.** $xy = 1$ **7.** $C(0, 0)$, $F(\pm\sqrt{13}, 0)$, $V(\pm 3, 0)$, $y = \pm\frac{2}{3}x$
9. $C(0, 0)$, $F(\pm\sqrt{2}, 0)$, $V(\pm 1, 0)$, $y = \pm x$ **11.** $C(-4, 0)$, $F(-4 \pm 2\sqrt{5}, 0)$, $V(-4 \pm 4, 0)$, $y = \pm\frac{1}{4}(x + 4)$ **13.** b^2x_0/a^2y_0
15. Light striking the parabolic reflector from the direction parallel to its axis is reflected toward the focus of the parabola. There, it is reflected toward the other focus of the hyperbola, which is the eyepiece at the vertex of the parabola.
17. 677 m east and 180 m north of A

Section 10.5

1. $\pi/4$ **3.** $\pi/8$ **5.** $\pi/12$ **7.** $\frac{1}{2}$ arctan $\frac{3}{2}$

9.

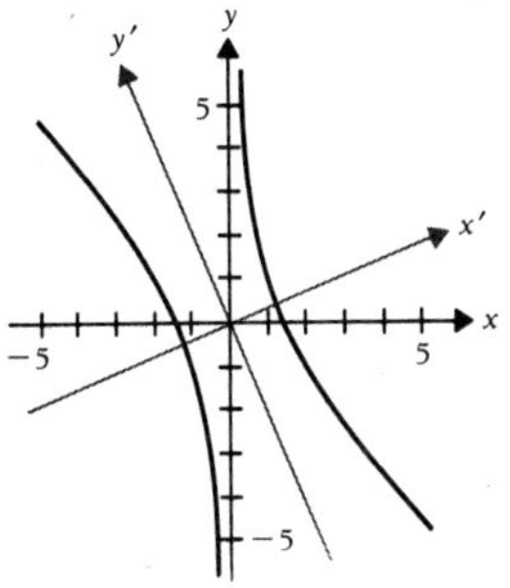

11.

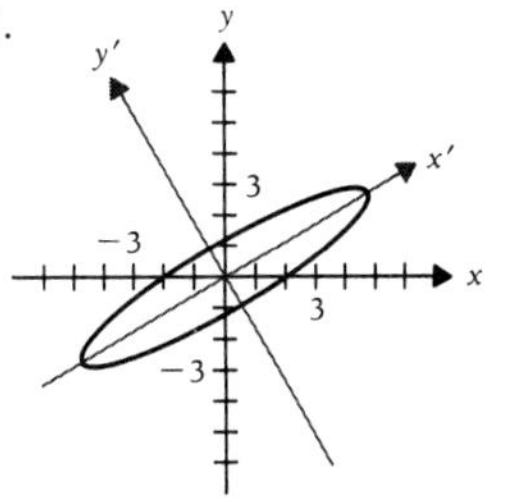

13.

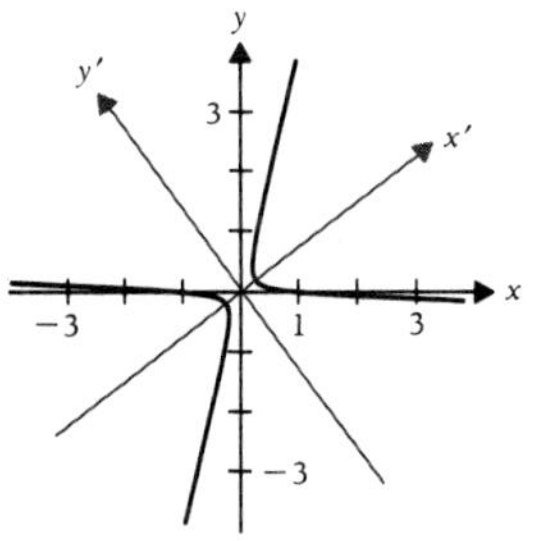

15.

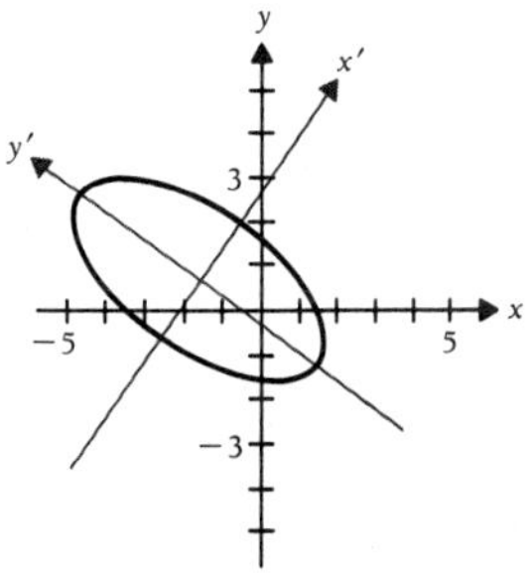

17. For $\alpha = \pi/2$, the transformation equations (3) become $x = -y'$, $y = x'$. The result follows immediately.
19. Multiply the first equation in (3) by cos α and the second by sin α and then add to get the first equation in (4). Multiplying the first by sin α and the second by cos α and subtracting, we get the second equation in (4).

Section 10.6

1. Parabola **3.** Hyperbola **5.** Two parallel lines **7.** Two parallel lines **9.** $\sqrt{5}/3$ **11.** 1
13. Under the translation $x = x' + h$, $y = y' + k$, the equation $ax^2 + bxy + cy^2 + dx + ey + f = 0$ becomes

$$ax'^2 + bx'y' + cy'^2 + (2ah + bk + d)x' + (2ck + bh + e)y' + (ah^2 + bhk + ck^2 + dh + ek + f) = 0,$$

so the discriminant $b^2 - 4ac$ is unchanged.

15. Let δ be the determinant of the original. If the first row of δ is multiplied by h and added to the last row, and the second row of δ is multiplied by k and added to the last row, then the same thing is done with columns, the determinant of the translated equation is obtained. Thus, the value δ is invariant under a translation.
17. Using $e = c/a$, $F_1 = (c, 0)$, and $d_1\colon x = a^2/c$, the equation $|PF_1| = e|Pd_1|$ is equivalent to the equation $(x^2/a^2) + (y^2/b^2) = 1$.

Review Exercises

1. $(x-1)^2 = 6y - 3$ 3. $(x-4)^2/4 + (y-3)^2/b^2 = 1,\ 0 < b < 2$ 5. $4y^2 - \frac{4}{3}(x-2)^2 = 1$
7. $3x^2 - y^2 - 30x + 6y + 54 = 0$

9.

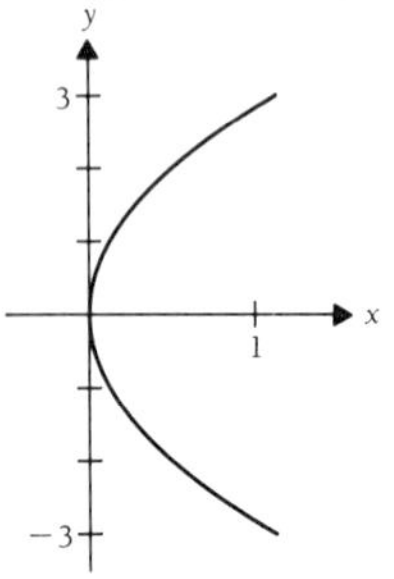

11.

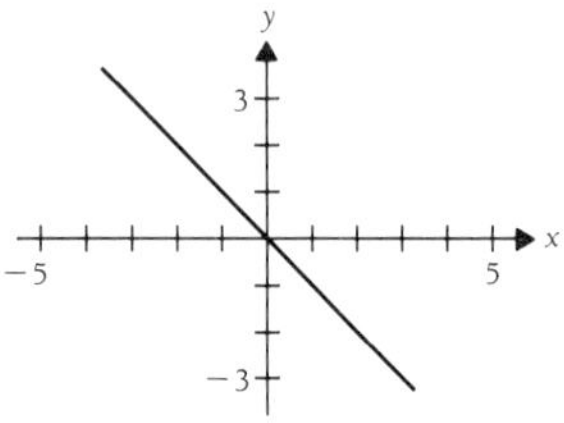

13.

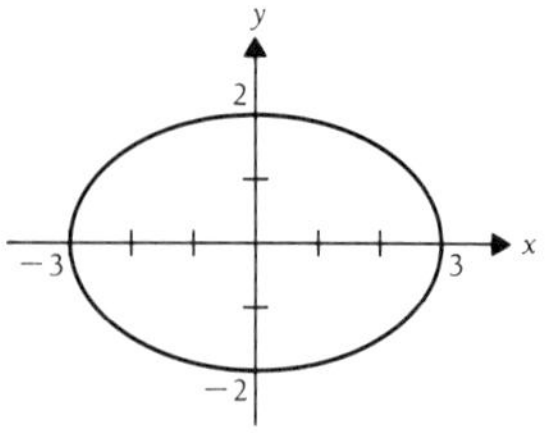

15.

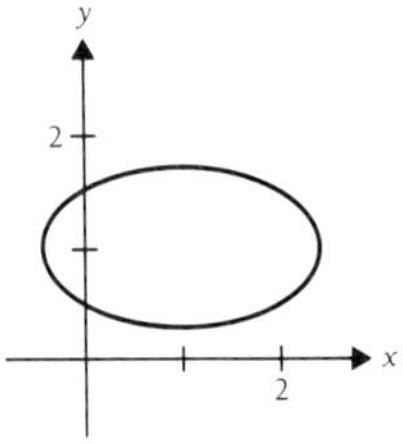

17.

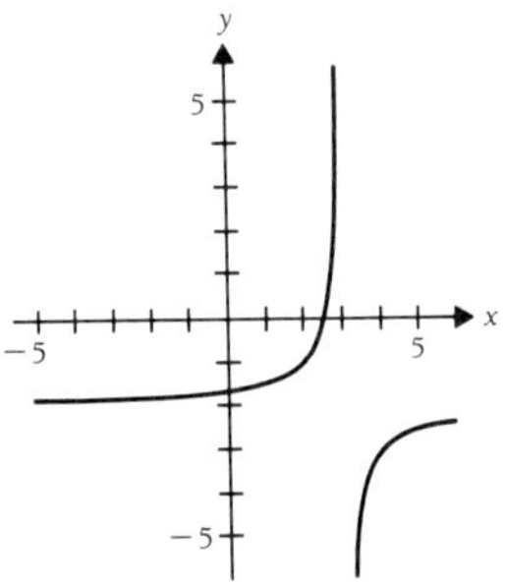

19.

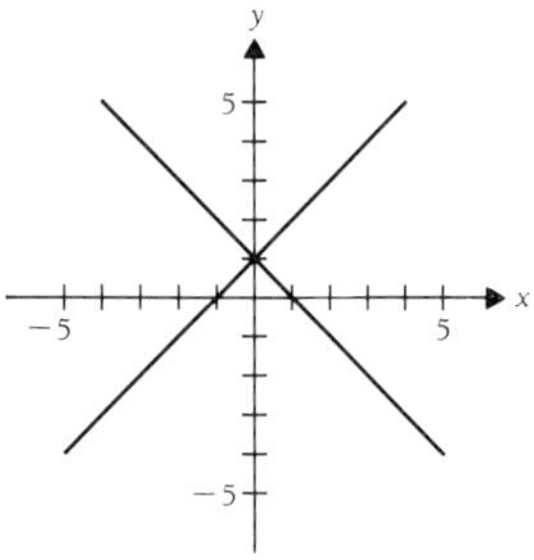

Additional Exercises

1. $\arctan(1/m)$ 3. (a) $4c$ (b) $2b^2/a$ (c) $2b^2/a$
5. Given the semiaxes a and b of the ellipse and the center C, construct circles of radii a and b at center C. For each radius of the circles, a point on the ellipse is constructed as in Problem 4.
7. The vertex 9. 4600 mi

Chapter 11 Section 11.1

1. $y = (11 - x)/3$

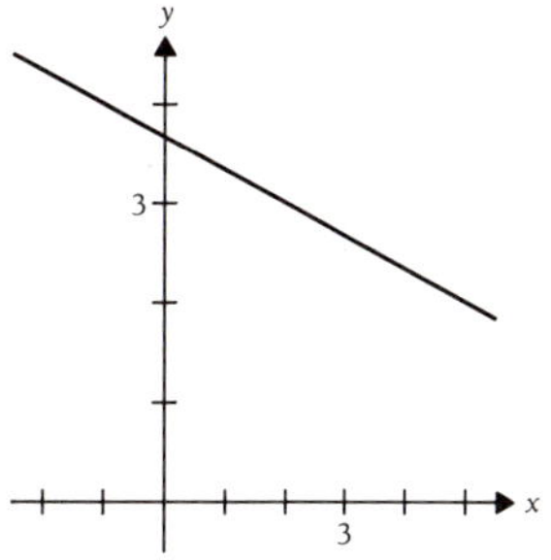

3. $x = (2 - y)^2$

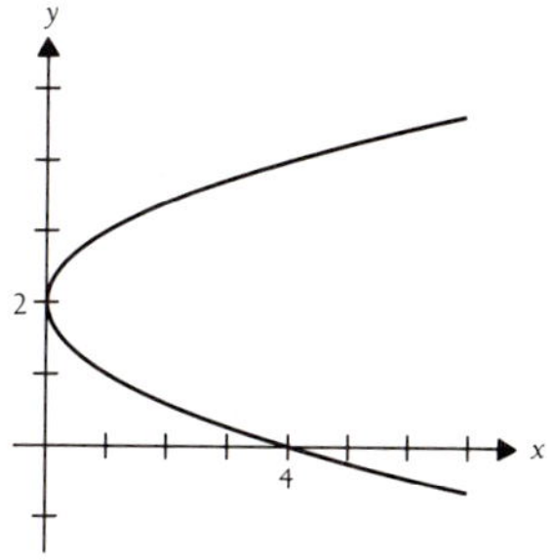

5. $x^2 + y^2 = 4$

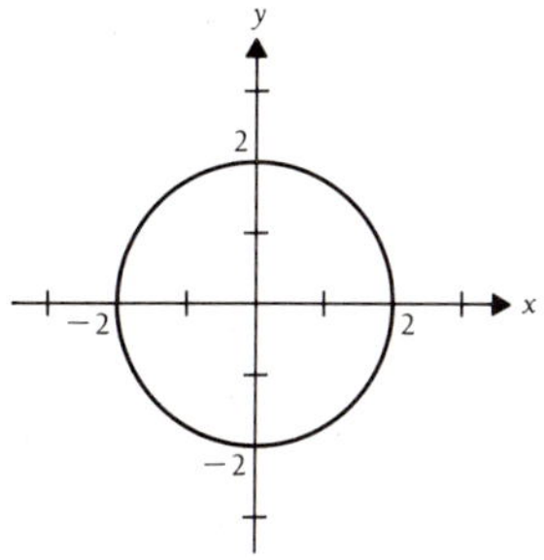

7. $y = 3 - \ln x$

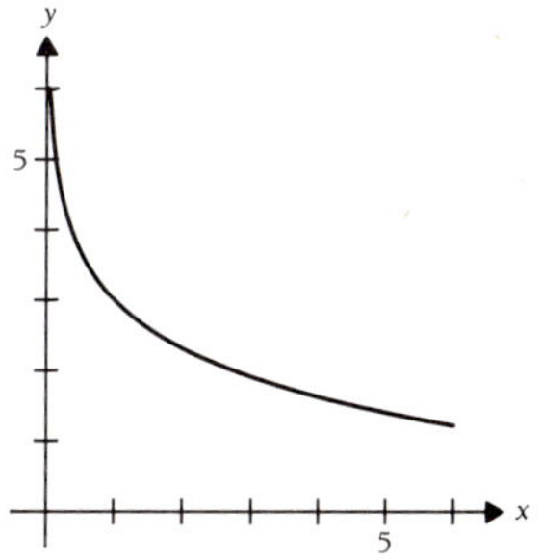

9. $-\frac{3}{4}$ **11.** $-\frac{1}{3}t^{-4}$ **13.** $-2 \tan t$ **15.** $(\sin t)/(1 - \cos t)$ **17.** $\frac{1}{4}t^{-3}$ **19.** $-\csc^3 t$ **21.** $-\cot^3 t$ **23.** $\frac{3}{2}\sqrt{37} + \frac{1}{4}\ln(6 + \sqrt{37})$
25. $\frac{2}{27}(13\sqrt{13} - 8)$ **27.** $8a$
29. Let $dx/dt = x'(t)$, $d^2x/dt^2 = x''(t)$, $dy/dt = y'(t)$, and $d^2y/dt^2 = y''(t)$; then use the definition of a 2×2 determinant.
31. $3\pi a^2$ **33.** $16\pi^2 a^2$ **35.** $6\pi^3 a^3$

Section 11.2

1. $[1, 5\pi/2]$, $[1, 9\pi/2]$, $[-1, 3\pi/2]$, $[-1, 7\pi/2]$
3. $[2, \pi]$, $[2, -\pi]$, $[-2, 2\pi]$, $[-2, 4\pi]$

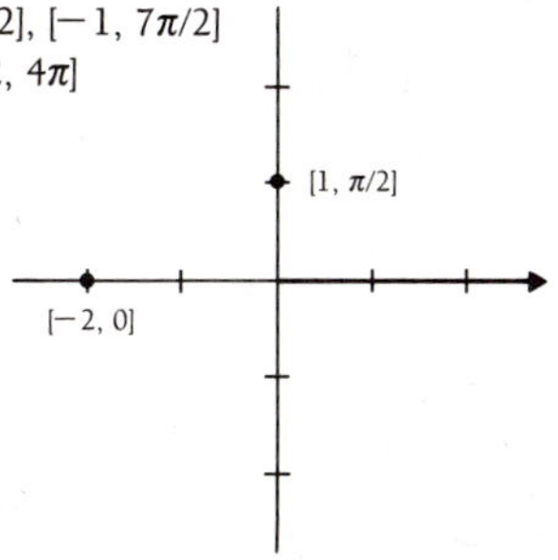

5. $(0, 0)$ **7.** $(2, 0)$ **9.** $(1, \sqrt{3})$ **11.** $(\frac{1}{2}(2 + \sqrt{2})^{1/2}, \frac{1}{2}(2 - \sqrt{2})^{1/2})$ **13.** $(2 \cos 2, -2 \sin 2)$ **15.** $(1, 1)$ **17.** $[1, \pi/2]$
19. $[2\sqrt{2}, \pi/4]$ **21.** $[\sqrt{3}, \pi/2]$ **23.** $[2, -\pi/6]$ **25.** $[\sqrt{85}, \tan^{-1}(\frac{6}{7})]$ **27.** $[\pi\sqrt{2}, -\pi/4]$ **29.** $\theta = \pi/4$ **31.** $r = 2$
33. $r^2 \sin 2\theta = 1$ **35.** $r = 1, 0 \le \theta \le \pi$ **37.** $y = 0$ **39.** $x^2 + y^2 = y$ **41.** $x = 1$ **43.** $(x^2 + y^2)^3 = y^2$

Section 11.3

1. $r(\cos\theta + \sin\theta) = 1$ **3.** $r = 3$ **5.** $r = 4\sin\theta$ **7.** $r = -2/(2+\cos\theta)$

9.
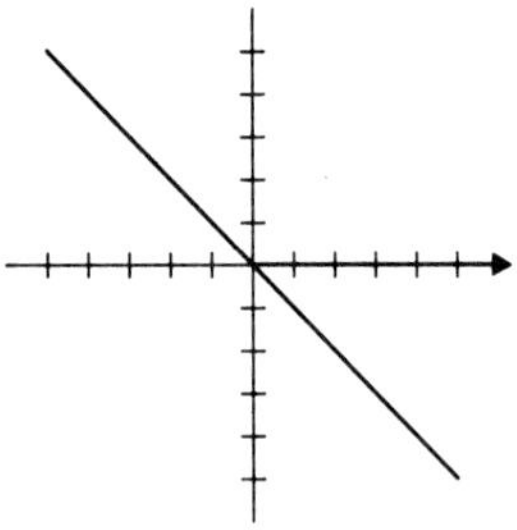

11.
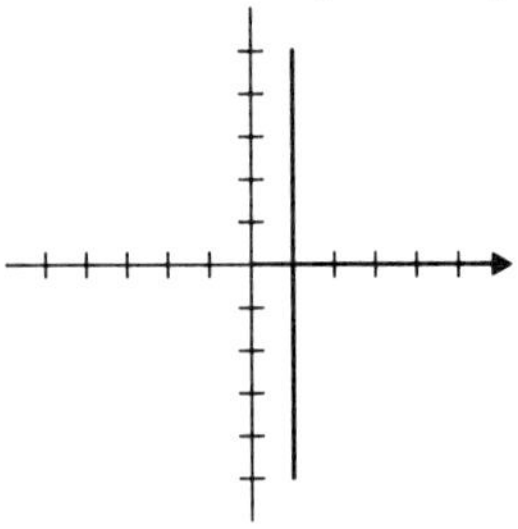

13.
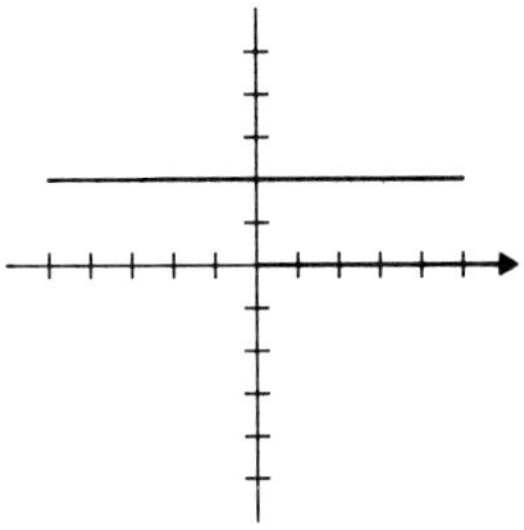

15.
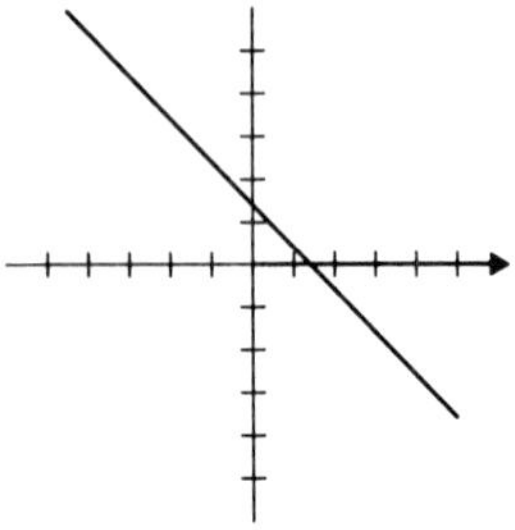

17.
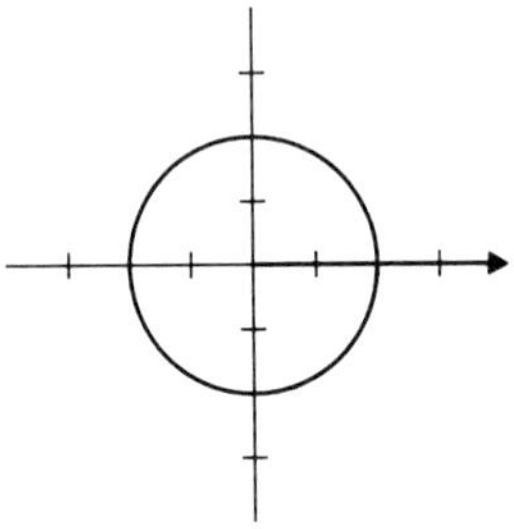

19.
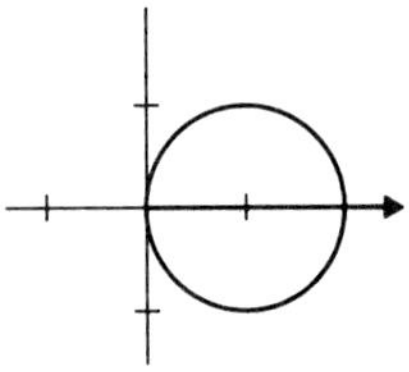

21.
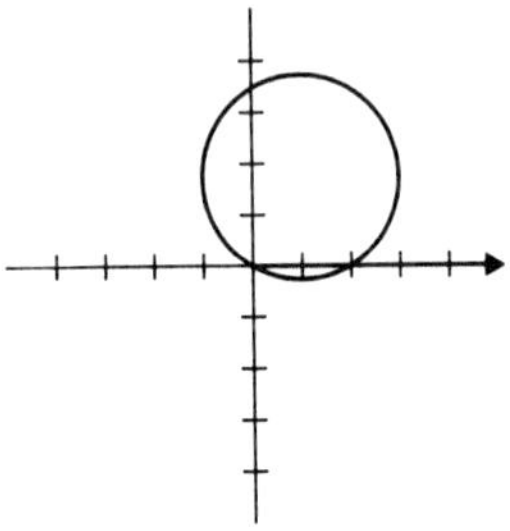

23.
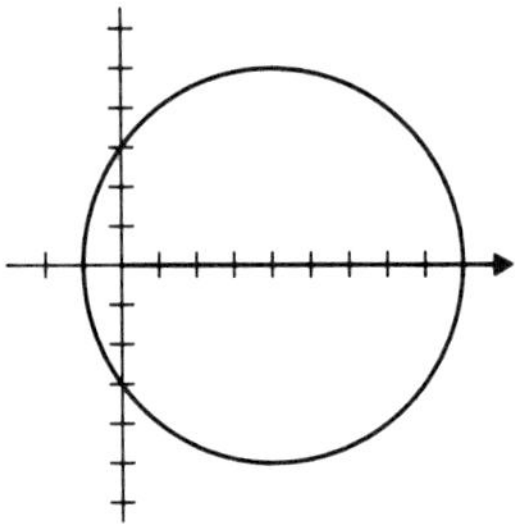

25.
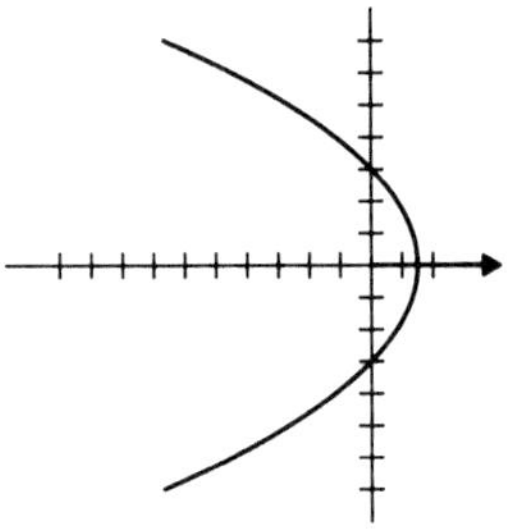

27.
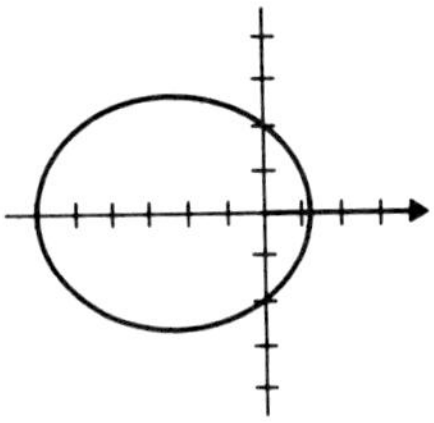

29.
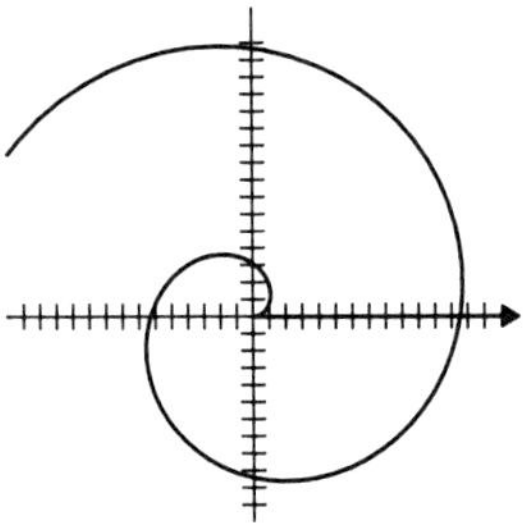

31.
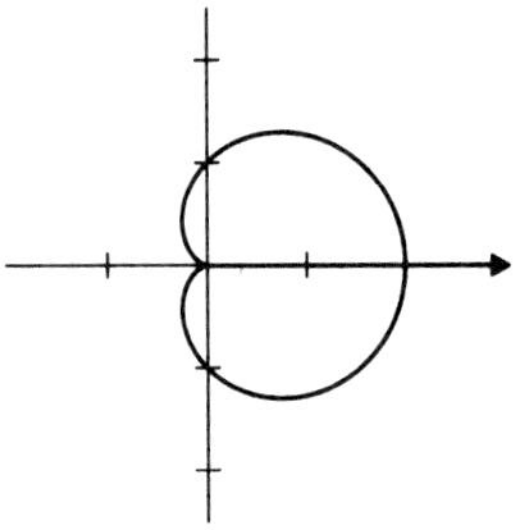

33.
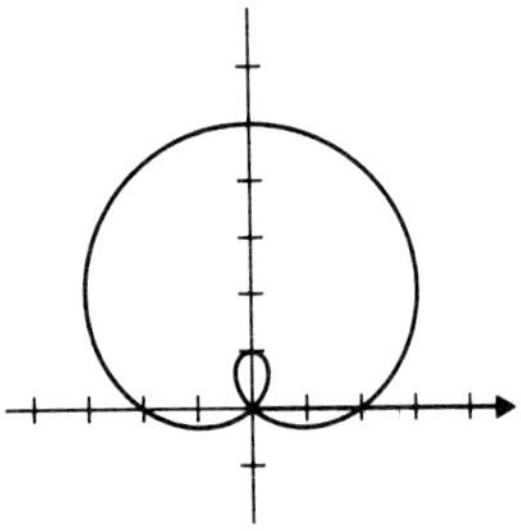

35.
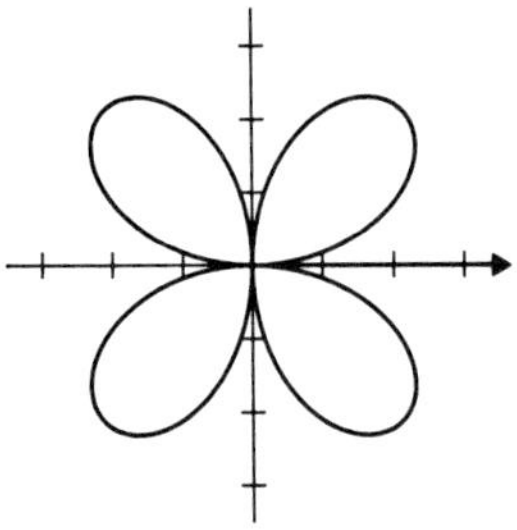

37.
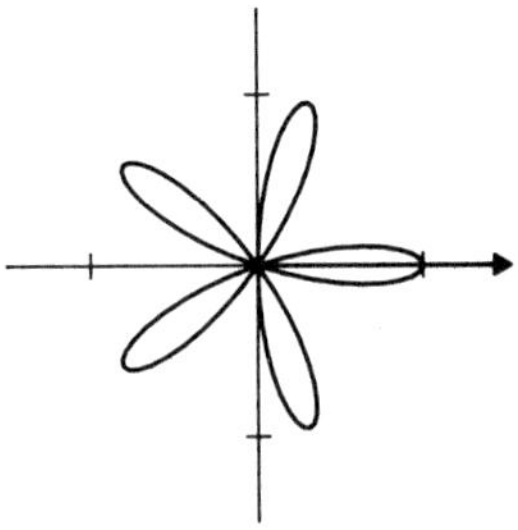

39.

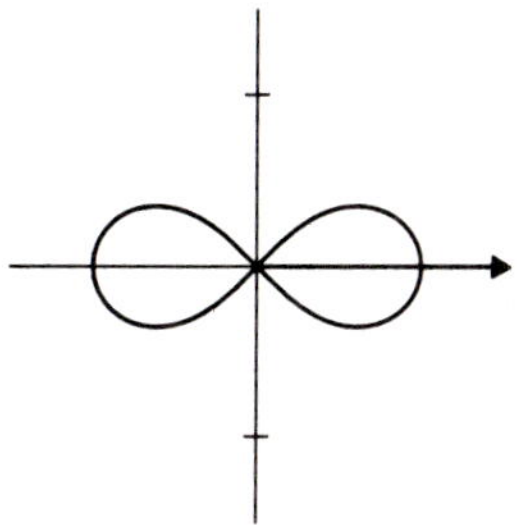

41.

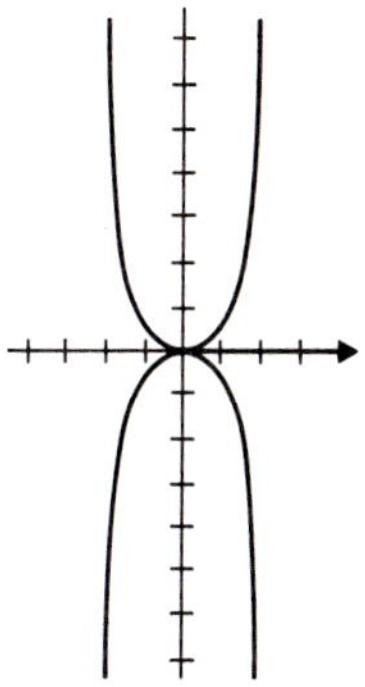

43.

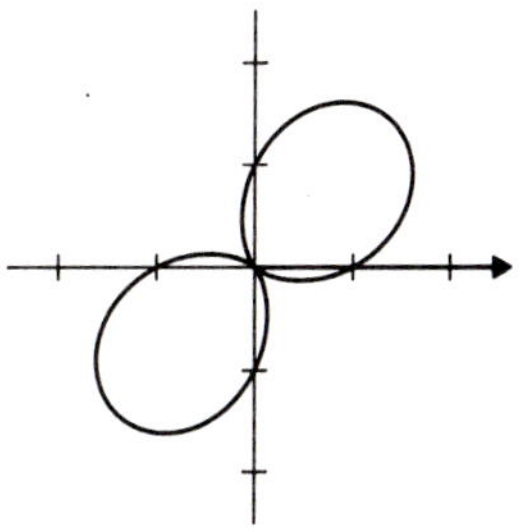

45. As $n\theta$ varies through an interval of length 2π, the curve creates two leaves, or rather four half-leaves. Thus, as θ varies from 0 to π, the curve creates $4(n/2) = 2n$ half-leaves. If n is odd, then the curve retraces the portion already sketched as θ varies from π to 2π, and the resulting curve has n leaves. If n is even, the curve creates the mirror image of the portion already sketched as θ varies from π to 2π, and the curve has $2n$ leaves.

47. $[4, 3\pi/2]$ **49.** $[\sqrt{2}/2, -\pi/4]$, $[0, 0]$ **51.** $[0, 0]$, $[2/\sqrt{7}, \pm\cos^{-1}(2/\sqrt{7})]$ **53.** $[0, 0]$, $[0.670, 0.335]$, $[0.670, 2.807]$

55. $[0, 0]$, $[1, 0]$, $[0.106, 2.034]$, $[1.894, 5.176]$, $[0.106, 1.107]$

Section 11.4

1. $m = -1/\sqrt{3}$; $\pi/2$ **3.** $m = \sqrt{3}$; $\pi/6$ **5.** $m = 1/\sqrt{3}$; $5\pi/6$ **7.** $m = \pi$; $\tan^{-1}\pi$ **9.** $m = 5/\sqrt{3}$; $\tan^{-1}(\sqrt{3}/2)$

11. Line is vertical; $\pi/2$ **13.** Horizontal: $[\frac{3}{2}, \pm\pi/3]$; vertical: $[2, 0]$, $[\frac{1}{2}, 2\pi/3]$, $[\frac{1}{2}, 4\pi/3]$

15. Horizontal: $[-1, \pm\pi/2]$, $[\frac{2}{3}, \pm\sin^{-1}(1/\sqrt{6})]$, $[\frac{2}{3}, \pm(\pi - \sin^{-1}(1/\sqrt{6}))]$; vertical: $[\pm 1, 0]$, $[\pm\frac{3}{4}, \pi/2 \pm \sin^{-1}(\frac{1}{2}\sqrt{2})]$

17. Horizontal: $[e^{2n\pi - 2\arctan(1/2)}, n\pi - \arctan\frac{1}{2}]$; vertical: $[e^{2n\pi + 2\arctan 2}, n\pi + \arctan 2]$

19. 7; $dr/d\theta = 0$ at the extrema of r.

Section 11.5

1. 9π **3.** 9 **5.** $9\pi/2$ **7.** $\pi/4$ **9.** $\pi/4$ **11.** $8\pi/3\sqrt{3}$ **13.** $\pi/3 + \sqrt{3}/2$ **15.** $\cos^{-1}\frac{1}{4} + 4 - \sqrt{15}$ **17.** 3π **19.** $8a$

21. $\sqrt{2}(e^{2\pi} - 1)$ **23.** $20\pi\sqrt{2}$ **25.** $8\pi a^3/3$ **27.** $(-4a/5, 4a/5)$

Review Exercises

1. $r = 3/(2 - \cos\theta)$ **3.**

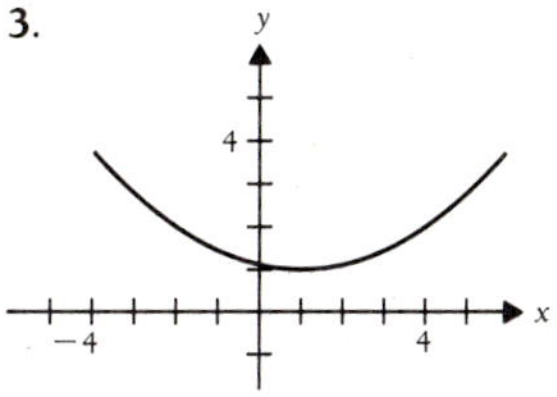

5.

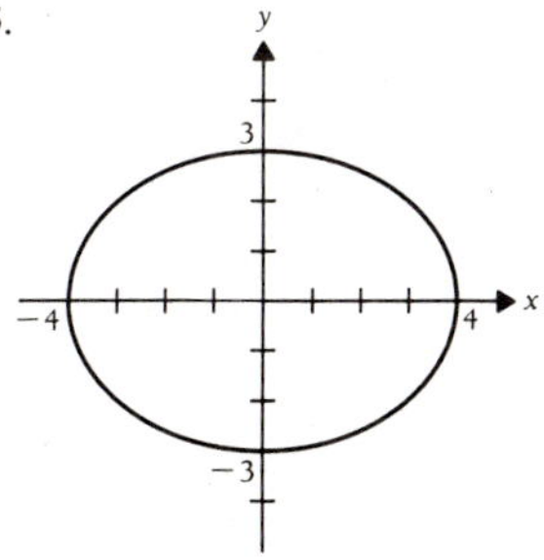

7.

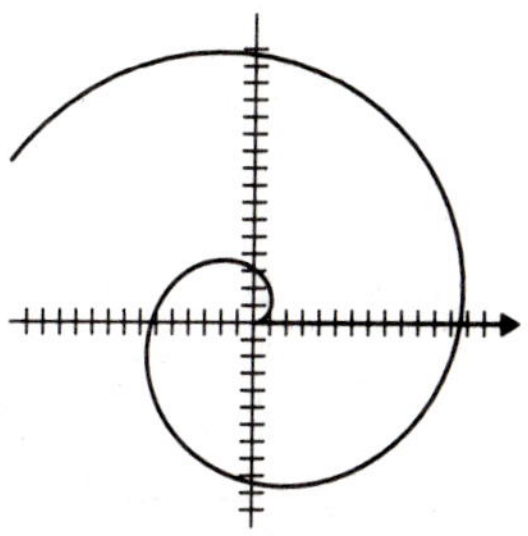

9.

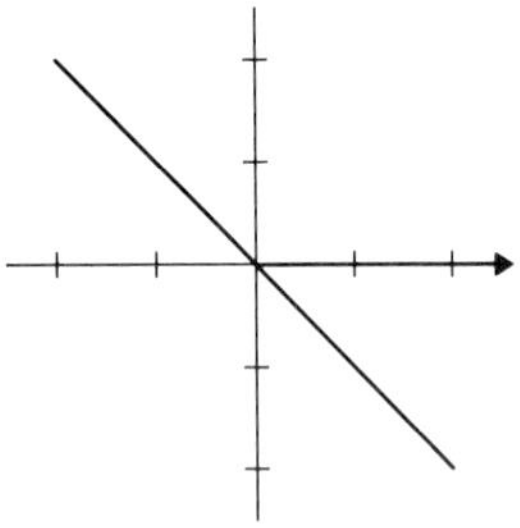

11.

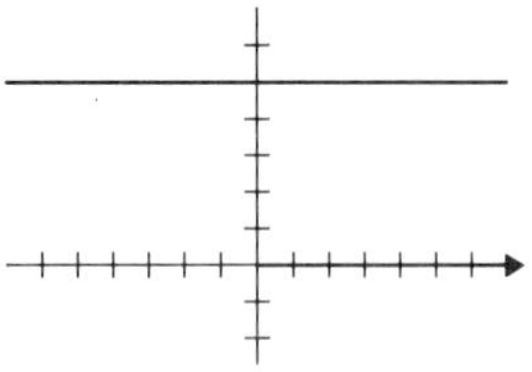

13.

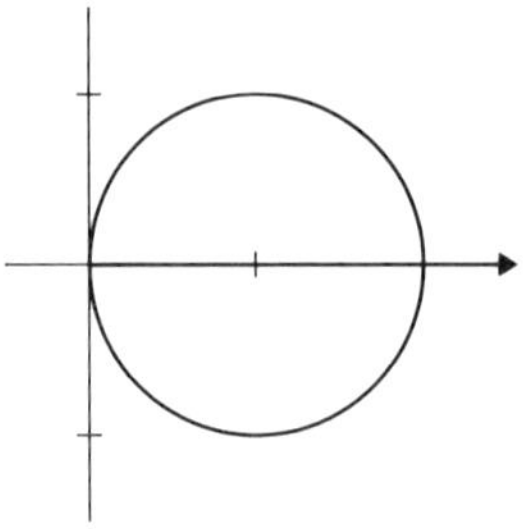

15.

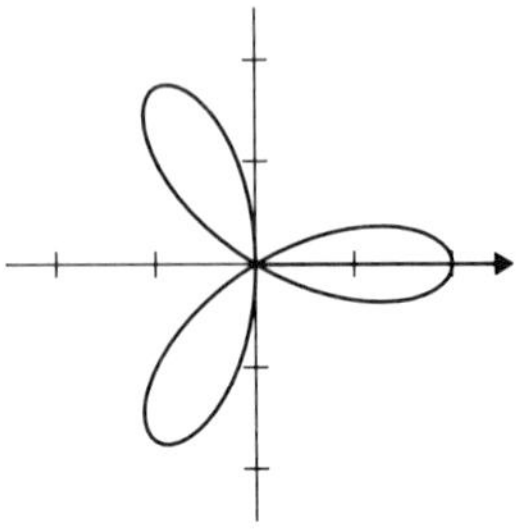

17.

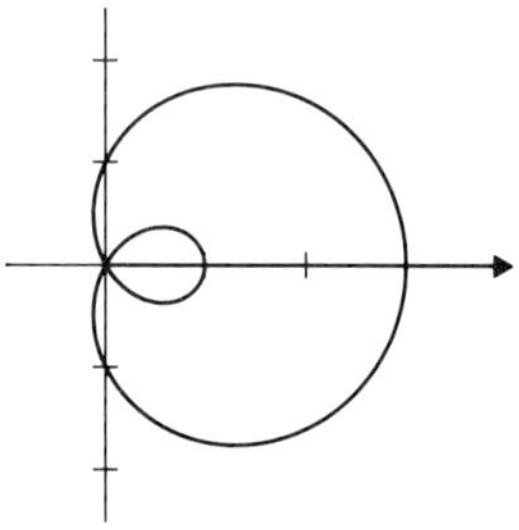

19.

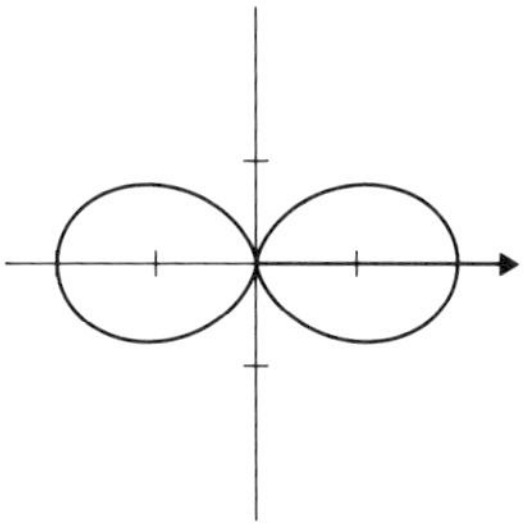

21.

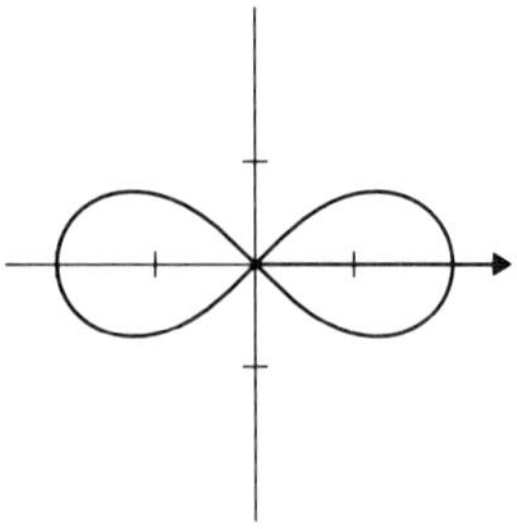

23.

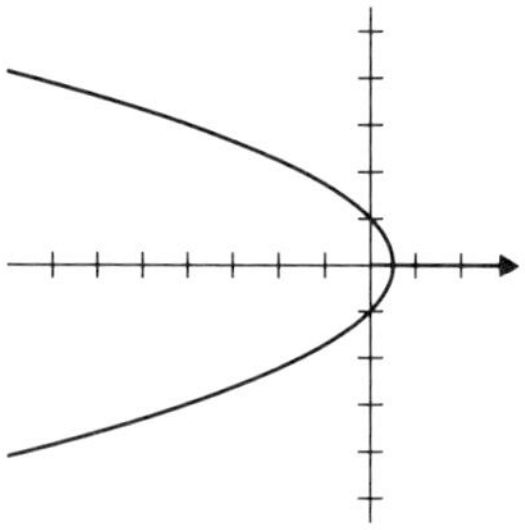

25.

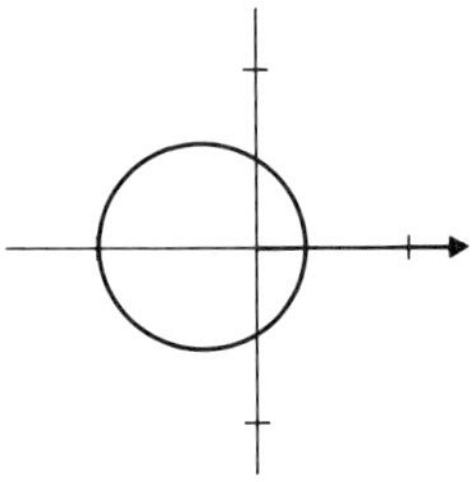

27.

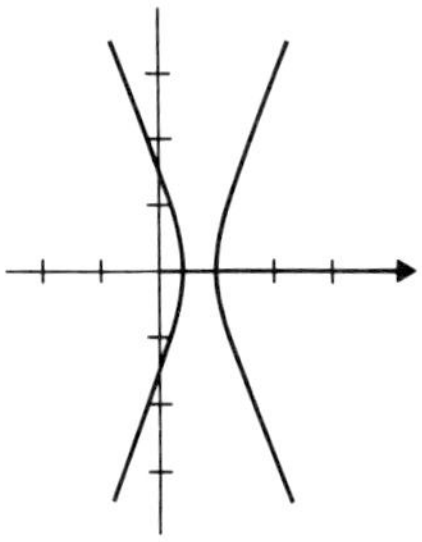

29.

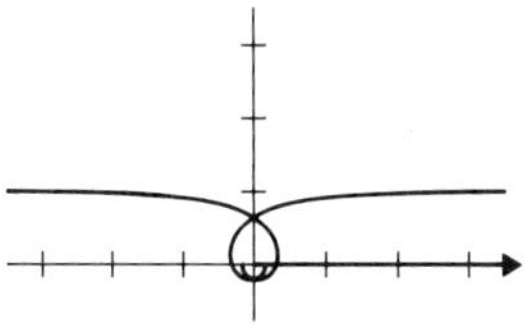

31. Horizontal: $[0, 0]$, $[4, \pi/2]$; vertical: $[2\sqrt{2}, \pi/4]$, $[2\sqrt{2}, 3\pi/4]$ **33.** None horizontal; vertical: $[\frac{1}{2}, 0]$
35. Horizontal: $[e^{3n\pi - 3\arctan(1/3)}, n\pi - \arctan\frac{1}{3}]$; vertical: $[e^{3n\pi + 3\arctan 3}, n\pi + \arctan 3]$
37. Horizontal: $[(9 + \sqrt{41})/4, \pm\cos^{-1}(3 - \sqrt{41})/8]$; vertical: $[1, 0]$, $[5, \pi]$, $[\frac{3}{2}, \pm\cos^{-1}\frac{3}{4}]$
39. Horizontal: $[0, 0]$, $[-1, \pi/2]$, $[3\sqrt{3}/2\sqrt{8}, \cos^{-1}(\sqrt{5}/\sqrt{8})]$, $[3\sqrt{3}/2\sqrt{8}, \pi - \cos^{-1}(\sqrt{5}/\sqrt{3})]$; vertical: $[0.987, 0.468]$, $[0.987, 2.674]$, $[-0.659, 1.287]$, $[-0.659, 1.855]$
41. $(a^2 + b^2/2)\pi$ **43.** $\pi a^2/2 - ab + (b^2 - a^2)\arctan(a/b)$ **45.** $\sqrt{a^4 - b^4} - b^2\cos^{-1}(b^2/a^2)$ **47.** $a^2/4$ **49.** $\sqrt{2} + \ln(1 + \sqrt{2})$
51. πa **53.** $(1/k)\sqrt{1 + k^2}(e^{k\beta} - e^{k\alpha})$

Additional Exercises

1. If $t = t_0$ at the point of intersection, then the angle θ of intersection satisfies $\tan\theta = \dfrac{(1+m^2)\sin t_0}{mt_0 - 1} - m$.

3. $\arctan \dfrac{a^2\sqrt{a^4 - b^4}}{2b^4 - a^4}$ 5.

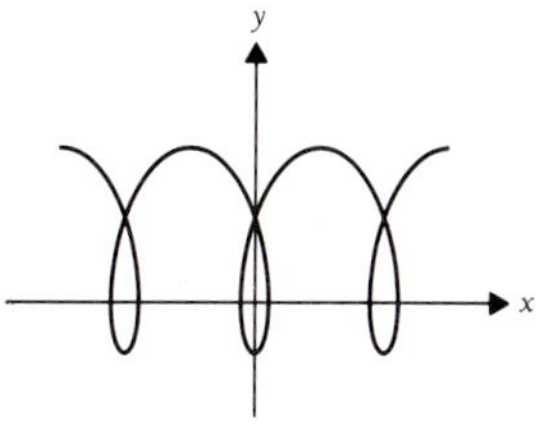

7. Let l be the vertical line from the center of the smaller circle. The distance from the y-axis to l is $(a-b)\cos t$. The arc of the smaller circle that has been in touch with the larger circle subtends the central angle at/b, so the tiny angle between l and the radius b shown in the figure is $(\pi/2) - (a-b)t/b$. Thus, x is as given. The y-coordinate is found similarly.
9. From the little triangle under the curve, we have $(t-x)^2 + y^2 = a^2$, and also $(dy/dt)/(dx/dt) = -y/(t-x)$. Differentiating the first equation with respect to t and then using the second equation to solve for dy/dt, we get $dy/dt = -y\sqrt{a^2 - y^2}/a^2$. Solving this equation by separating the variables, we get $y = 2ae^{-t/a}/(1 + e^{-2t/a}) = a\,\text{sech}(t/a)$. Then we get the value of x we seek by substituting in the original equation.
11. If the line segment QP is extended to meet the x-axis, then $\tan t = d/x$ and $x = d\cot t$. Solving the equations $y = x\tan t$ of the line segment OR and $x^2 + y^2 = dy$ of the circle simultaneously, we get $y = d\sin^2 t$.
13. $a^2/2$ if n is even; a^2 if n is odd
15. For a straight line, the parametric equations are both linear, so the second derivatives are 0 and the curvature is 0. For the circle $x = a\cos t$, $y = a\sin t$, the curvature is

$$\frac{(-a\sin t)(-a\sin t) - (a\cos t)(-a\cos t)}{[\sqrt{(-a\sin t)^2 + (a\cos t)^2}]^3} = \frac{1}{a}.$$

17. $d = \sqrt{r_1^2 + r_2^2 - 2r_1r_2\cos(\theta_2 - \theta_1)}$ 19. 320π ft^2

Chapter 12 Section 12.1

1. $\sqrt{26}$ 3. 1 5. Yes 7. $x^2 + y^2 + z^2 = 16$ 9. $(x-2)^2 + (y-1)^2 + z^2 = 1$ 11. $(x-1)^2 + (y-2)^2 + (z+2)^2 = 9$
13.

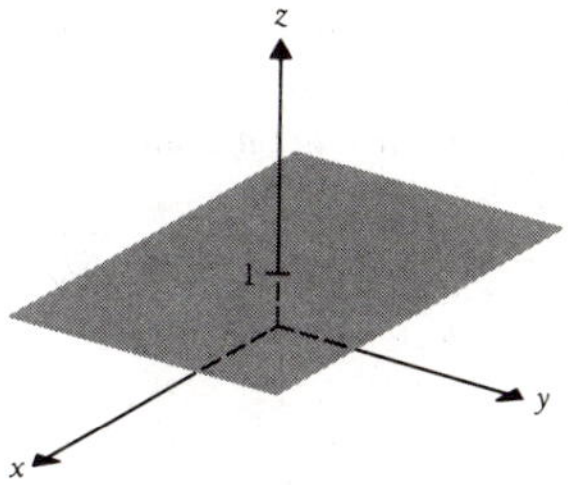

15.

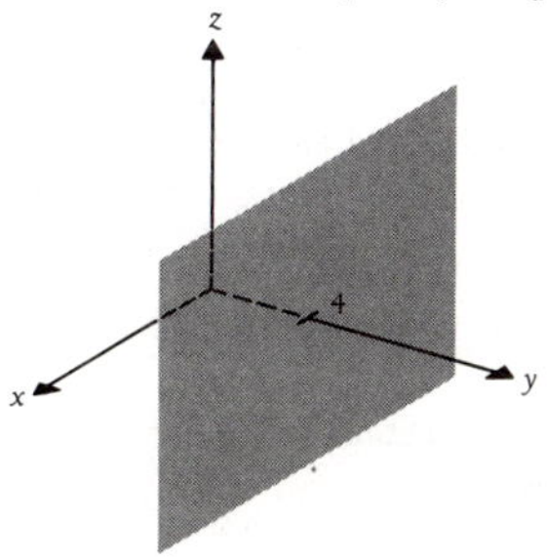

17. Extend the midpoint formula for the plane.

Section 12.2

1. $\sqrt{10}$ 3. $\langle 1, 4, 4\rangle$ 5. $\langle 2, 0, -6\rangle$ 7. $2\sqrt{10}$ 9. $\langle -1/\sqrt{10}, 0, 3/\sqrt{10}\rangle$ 11. $\langle -6, 0, 0\rangle$ 13. $\langle 12, -2, -10\rangle$
15. $-\frac{1}{6}, -\frac{1}{6}$ 17. $\langle 0, 1/\sqrt{26}, 5/\sqrt{26}\rangle$ 19. Omit the third coordinate. 21. Omit the third coordinate. 23. $\langle -2, 2\rangle$
25. $\langle 6, -3\rangle$ 27. $\langle 0, 2\rangle$ 29. $\sqrt{5}$ 31. $a = \frac{3}{2}, b = \frac{1}{2}$ 33. $3\mathbf{i} - \mathbf{j} + \mathbf{k}$ 35. $-\mathbf{i} - 5\mathbf{j}$ 37. $-\mathbf{i} + 13\mathbf{j}$ 39. $\sqrt{6}$ 41. $4\mathbf{i} - 3\mathbf{k}$

43. $\frac{11}{5}\langle 3, 0, 1\rangle - \frac{13}{5}\langle 1, 0, 2\rangle$ **45.** No
47. Let $P = (x_1, y_1, z_1)$ and $Q = (x_2, y_2, z_2)$. Then $\|\mathbf{v}(\overrightarrow{PQ})\| = \|\langle x_2 - x_1, y_2 - y_1, z_2 - z_1\rangle\| = \sqrt{(x_2 - x_1)^2 + (y_2 - y_1)^2 + (z_2 - z_1)^2} = |PQ|$.
49. $a(\mathbf{u} + \mathbf{v}) = a\langle u_1 + v_1, u_2 + v_2, u_3 + v_3\rangle = \langle au_1 + av_1, au_2 + av_2, au_3 + av_3\rangle = \langle au_1, au_2, au_3\rangle + \langle av_1, av_2, av_3\rangle = a\mathbf{u} + a\mathbf{v}$.
51. $\|a\mathbf{v}\| = \|\langle av_1, av_2, av_3\rangle\| = \sqrt{a^2v_1^2 + a^2v_2^2 + a^2v_3^2} = \sqrt{a^2}\sqrt{v_1^2 + v_2^2 + v_3^2} = |a|\,\|\mathbf{v}\|$.

Section 12.3

1. 11 **3.** -1 **5.** 11 **7.** 11 **9.** 22 **11.** -6 **13.** $\cos^{-1}(11/13\sqrt{2})$ **15.** 0 **17.** $11/\sqrt{13}$ **19.** $\langle -\frac{33}{26}, \frac{22}{13}, \frac{11}{26}\rangle$
21. $\langle 10, 9, -6\rangle$ **23.** $\langle -5, -2, -7\rangle$ **25.** $\langle -11, -11, 11\rangle$ **27.** $\langle -10, 1, -9\rangle$ **29.** $\sqrt{78}$ **31.** 25 **33.** $\sqrt{217}$ **35.** $\frac{1}{2}\sqrt{38}$
37. 1 **39.** $-3\mathbf{i} - 4\mathbf{j} - 7\mathbf{k}$ **41.** 0 **43.** 0 **45.** $\mathbf{k}$ **47.** $-\mathbf{j}$ **49.** 17
51. $(a\mathbf{u}) \cdot \mathbf{v} = \langle au_1, au_2, au_3\rangle \cdot \langle v_1, v_2, v_3\rangle = au_1v_1 + au_2v_2 + au_3v_3 = a(u_1v_1 + u_2v_2 + u_3v_3) = a(\mathbf{u} \cdot \mathbf{v})$
53. $$\mathbf{v} \times \mathbf{u} = \begin{vmatrix} v_1 & v_2 & v_3 \\ u_1 & u_2 & u_3 \\ \mathbf{i} & \mathbf{j} & \mathbf{k} \end{vmatrix} = -\begin{vmatrix} u_1 & u_2 & u_3 \\ v_1 & v_2 & v_3 \\ \mathbf{i} & \mathbf{j} & \mathbf{k} \end{vmatrix} = -(\mathbf{u} \times \mathbf{v})$$
55. $(\mathbf{u} + \mathbf{v}) \cdot (\mathbf{u} - \mathbf{v}) = \mathbf{u} \cdot \mathbf{u} - \mathbf{u} \cdot \mathbf{v} + \mathbf{v} \cdot \mathbf{u} - \mathbf{v} \cdot \mathbf{v} = \|\mathbf{u}\|^2 - \|\mathbf{v}\|^2$
57. If the vectors all lie in the same plane, then the parallelepiped they determine has 0 volume, so $\mathbf{u} \cdot \mathbf{v} \times \mathbf{w} = 0$.
59. If $\mathbf{u} \cdot \mathbf{v} \times \mathbf{w} = 0$, then the matrix of the system $a\mathbf{u} + b\mathbf{v} = \mathbf{w}$ has determinant 0. This means that the system has a nonzero solution for a and b, and $\mathbf{w}$ is a linear combination of $\mathbf{u}$ and $\mathbf{v}$ (see Appendix B).
61. $$W = \|\text{proj}(\mathbf{r}, \mathbf{F})\|\,\|\mathbf{r}\| = \left\|\left(\frac{\mathbf{F} \cdot \mathbf{r}}{\|\mathbf{r}\|^2}\right)\mathbf{r}\right\|\,\|\mathbf{r}\| = \frac{|\mathbf{F} \cdot \mathbf{r}|\,\|\mathbf{r}\|^2}{\|\mathbf{r}\|^2} = |\mathbf{F} \cdot \mathbf{r}|$$
63. -1.6×10^{-19} coulomb

Section 12.4

1. $\langle 1, 2, -3\rangle$ **3.** $\langle 2, 2, -2\rangle$ **5.** $\cos^{-1}(5/\sqrt{39})$ **7.** $\pi/2$ **9.** $x = 2, z = 3$ **11.** $x = t, y = 4t, z = -t$
13. $x = 4t, y = 2t, z = -t$ **15.** $x = 4 + t, y = -1 - t, z = 1 + t$ **17.** $x = 5t, y = t, z = 6t$
19. $x = 1 + 3t, y = -1 + 4t, z = t$ **21.** $x = 1 + 3t, y = 2, z = -5$ **23.** $x = 4, y = 3$ **25.** $\frac{4}{5}, \frac{3}{5}, 0$
27. $(3/\sqrt{38}, -5/\sqrt{38}, 2/\sqrt{38})$ and $(-3/\sqrt{38}, 5/\sqrt{38}, -2/\sqrt{38})$
29. **(a)** Point (0, 0, 0) **(b)** Point (2, 1, 6) **(c)** Point $(1, \frac{1}{2}, 3)$ **(d)** Point $(-2, -1, -6)$
(e) Segment with endpoints (0, 0, 0) and (2, 1, 6) **(f)** Ray from (0, 0, 0) through $(-2, -1, -6)$, not including (0, 0, 0)
31. **(a)** Point (0, 0, 0) **(b)** Point (6, 3, 18) **(c)** Point $(3, \frac{3}{2}, 9)$ **(d)** Point $(-6, -3, -18)$
(e) Segment with endpoints (0, 0, 0) and (6, 3, 18) **(f)** Ray from (0, 0, 0) through $(-6, -3, -18)$, not including (0, 0, 0)
33. $|OP_1| = |OP_2| = 1$, so the law of cosines says that $|P_1P_2|^2 = |OP_1|^2 + |OP_2|^2 - 2|OP_1||OP_2| \cos\theta = 1 + 1 - 2 \cdot 1 \cdot 1 \cos\theta = 2 - 2\cos\theta$. Also, $|P_1P_2|^2 = \|\mathbf{v}(\overrightarrow{P_1P_2})\|^2 = \|\mathbf{v}(\overrightarrow{OP_2}) - \mathbf{v}(\overrightarrow{OP_1})\|^2 = [\mathbf{v}(\overrightarrow{OP_2}) - \mathbf{v}(\overrightarrow{OP_1})] \cdot [\mathbf{v}(\overrightarrow{OP_2}) - \mathbf{v}(\overrightarrow{OP_1})] = \|\mathbf{v}(\overrightarrow{OP_2})\|^2 - 2\,\mathbf{v}(\overrightarrow{OP_2}) \cdot \mathbf{v}(\overrightarrow{OP_1}) + \|\mathbf{v}(\overrightarrow{OP_1})\|^2 = 2 - 2\langle\lambda_1, \mu_1, \nu_1\rangle \cdot \langle\lambda_2, \mu_2, \nu_2\rangle = 2 - 2(\lambda_1\lambda_2 + \mu_1\mu_2 + \nu_1\nu_2)$, so that $\cos\theta = \lambda_1\lambda_2 + \mu_1\mu_2 + \nu_1\nu_2$.
35. P lies in Π if and only if the vector $\langle x - x_1, y - y_1, z - z_1\rangle$ is perpendicular to $\langle a, b, c\rangle$, if and only if $a(x - x_1) + b(y - y_1) + c(z - z_1) = 0$.

Section 12.5

1. $-x + 5y + 8z - 22 = 0$ **3.** $7x + \frac{1}{2}y + 2z - 5 - \frac{1}{2}\sqrt{3} = 0$ **5.** $2x - y + z - 1 = 0$ **7.** $x + y - 3z - 1 = 0$
9. $8x - 26y + 3z = 0$ **11.** $x - 3y + 8z - 19 = 0$ **13.** $28x - 2y + 11z - 11 = 0$ **15.** $3x + 9y - 8z + 4 = 0$
17. **(a)** $2y + 3z - 16 = 0$ **(b)** $-2x + z + 4 = 0$ **(c)** $3x + y - 14 = 0$ **19.** $x = \frac{11}{3} + 2t, y = -\frac{5}{3} + 7t, z = 3t$
21. **(a)** $x = 0, y = t, z = -4 + 5t$ **(b)** $x = t, y = 0, z = -4 - t$ **(c)** $x = 1 - 5t, y = 1 - t, z = 0$

23. $x = -3t, y = 1 + 3t, z = 2t$ **25.** $x = 2 + t, y = 1 - 3t, z = -5 - t$ **27.** $x = t, y = -3t, z = 4t$
29. $x = 1 + 2t, y = 2t, z = -t$ **31.** $x = t, y = 2t, z = 3t$ **33.** $3x - 2y + z = 0$; $3x - 2y + z + 1 = 0$ **35.** $\pi/2$
37. $(\frac{17}{14}, -\frac{9}{14}, -\frac{5}{7})$
39. Each point satisfies the equation, and the equation is that of a plane, so it is the plane with the given intercepts.
41. Let $P = (x, y, z)$ and $P_i = (x_i, y_i, z_i)$, $i = 1, 2, 3$. If in the determinant we first subtract the second row from each other row and then expand by the last column, we get the determinant representing the vector triple product $\mathbf{v}(\overrightarrow{PP_1}) \cdot \mathbf{v}(\overrightarrow{P_1P_2}) \times \mathbf{v}(\overrightarrow{P_1P_3}) = 0$, which is the equation of the plane containing the points P_1, P_2, and P_3.
43. The line is parallel to the plane if and only if the point (x_1, y_1, z_1) on the line is not in the plane, or $Ax_1 + By_1 + Cz_1 \neq D$, and the direction vector $\langle a, b, c \rangle$ of the line is orthogonal to the normal vector $\langle A, B, C \rangle$ of the plane, or $aA + bB + cC = 0$.
45. The distance from the origin to the point

$$\left(-\frac{ad}{a^2 + b^2 + c^2}, \quad -\frac{bd}{a^2 + b^2 + c^2}, \quad -\frac{cd}{a^2 + b^2 + c^2}\right)$$

is

$$\sqrt{\frac{a^2d^2 + b^2d^2 + c^2d^2}{(a^2 + b^2 + c^2)^2}} = \frac{\sqrt{d^2}}{\sqrt{a^2 + b^2 + c^2}} = \frac{|d|}{\sqrt{a^2 + b^2 + c^2}}.$$

47. Let (x_1, y_1, z_1) be a point in $ax + by + cz + d_1 = 0$; then $ax_1 + by_1 + cz_1 = -d_1$. The distance between the planes is the distance from (x_1, y_1, z_1) to the plane $ax + by + cz + d_2 = 0$, which is

$$\frac{|ax_1 + by_1 + cz_1 + d_2|}{\sqrt{a^2 + b^2 + c^2}} = \frac{|d_2 - d_1|}{\sqrt{a^2 + b^2 + c^2}}.$$

49. The line has direction vector $\mathbf{d} = \langle a, b, c \rangle$ and the plane has normal vector $\mathbf{n} = \langle A, B, C \rangle$. If θ is the angle between the line and the plane, then the angle between $\mathbf{d}$ and $\mathbf{n}$ is $\pi/2 - \theta$, and satisfies $\cos(\pi/2 - \theta) = \mathbf{d} \cdot \mathbf{n}/\|\mathbf{d}\|\,\|\mathbf{n}\|$, which leads to the equation sought.

Section 12.6

1. Circular cylinder

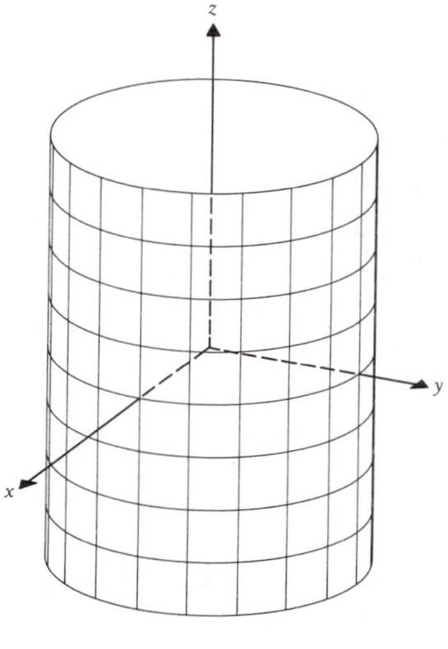

3. Cubic cylinder

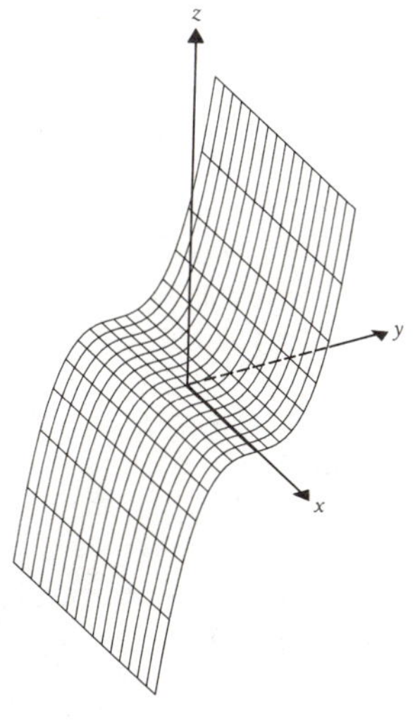

5. Surface of revolution

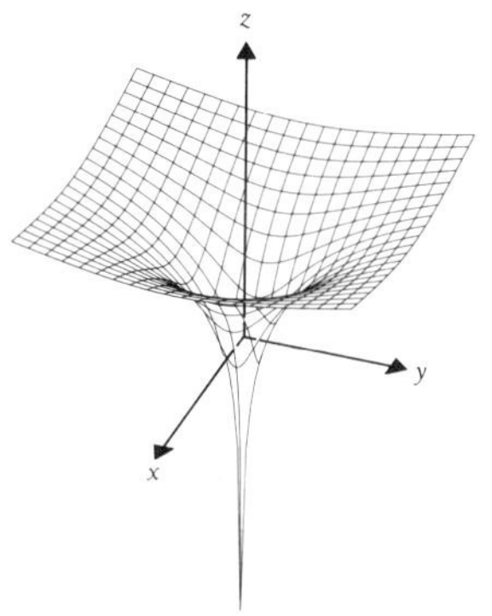

7. Sphere

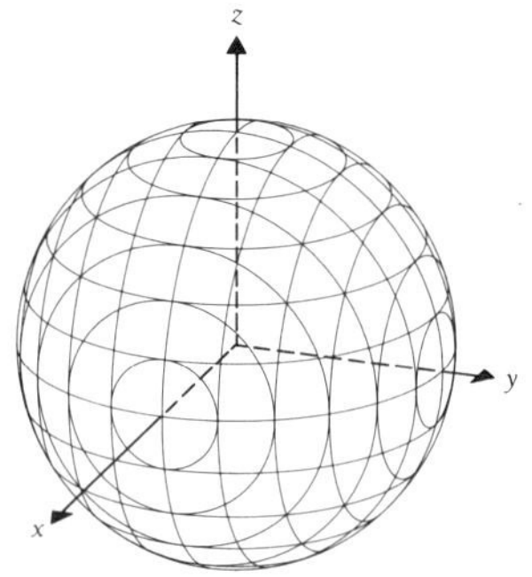

9. Circular hyperboloid of one sheet

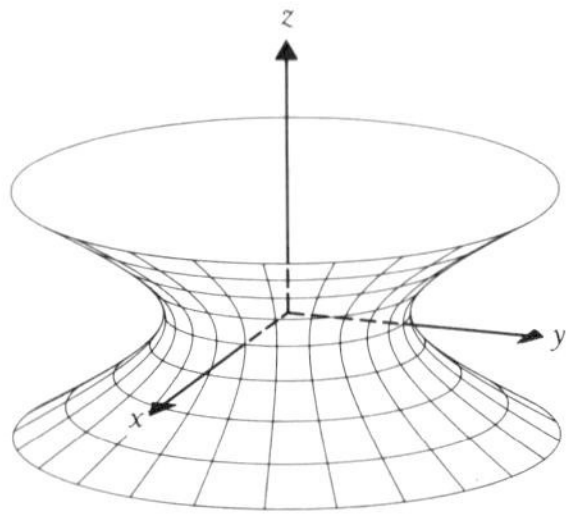

11. Circular paraboloid

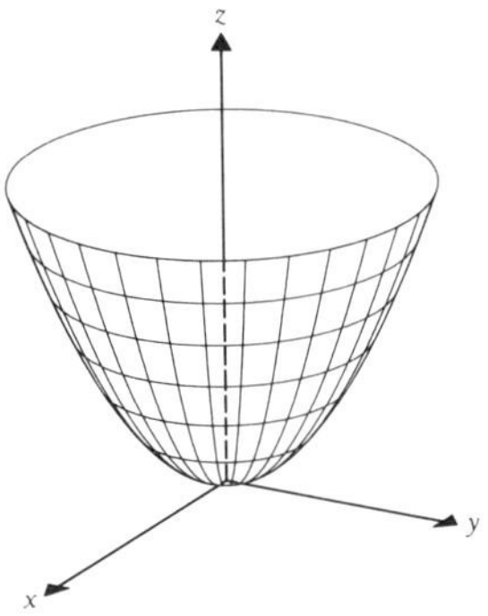

13. Circular cone

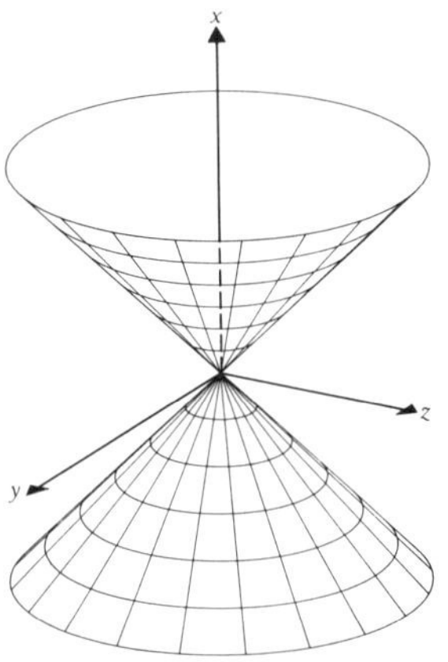

15. Two planes

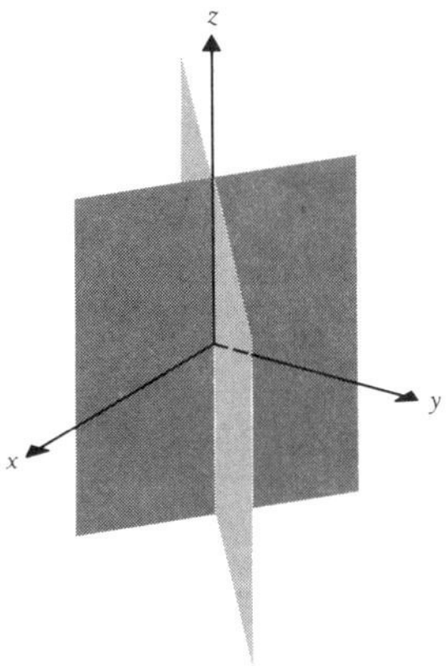

17. The yz-plane 19. The origin 21. No graph

Section 12.7

1. Cylindrical: $[2, \pi/3, 3]$; spherical: $[\sqrt{13}, \pi/3, \tan^{-1}\frac{2}{3}]$ 3. Cylindrical: $[2, \pi/2, -2]$; spherical: $[2\sqrt{2}, \pi/2, 3\pi/4]$
5. Rectangular: $(0, 2, 3)$; spherical: $[\sqrt{13}, \pi/2, \tan^{-1}\frac{2}{3}]$ 7. Rectangular: $(0, -1, 2)$; spherical: $[\sqrt{5}, -\pi/2, \tan^{-1}\frac{1}{2}]$

9. Rectangular: (0, 1, 0); cylindrical: $[1, \pi/2, 0]$ 11. Rectangular: (3, 0, 0); cylindrical: [3, 0, 0]
13. A circular cylinder; $r = 2$, $\rho = 2 \csc \phi$ 15. A circular cone; $r = 2z$, $\phi = \tan^{-1} 2$
17. Inverted circular paraboloid; $z = 4 - r^2$, $\rho^2 \sin^2\phi + \rho \cos \phi = 4$ 19. A circular cylinder; $x^2 + y^2 = 256$
21. A circular cone; $z^2 = 9(x^2 + y^2)$ 23. A sphere; $x^2 + y^2 + z^2 = 1$ 25. A sphere; $x^2 + y^2 + z^2 = 256$
27. The xy-plane; $z = 0$ 29. A sphere; $x^2 + y^2 + z^2 = z$
31. An "apple slice" whose section by the xy-plane is a circle and whose section by any plane containing the z-axis is a semicircle; $x = \sqrt{x^2 + y^2}\sqrt{x^2 + y^2 + z^2}$
33. $r = \sqrt{x^2 + y^2}$, $\theta = \tan^{-1}(y/x)$, $z = z$ 35. $\rho = \sqrt{x^2 + y^2 + z^2}$, $\theta = \tan^{-1}(y/x)$, $\phi = \tan^{-1}(\sqrt{x^2 + y^2}/z)$

Review Exercises

1. $\langle -1, 2, 3\rangle$ 3. $\langle 18, 19, -19\rangle$ 5. $\sqrt{38}$ 7. $\sqrt{14}$ 9. $(1/\sqrt{38})\langle 3, 5, -2\rangle$ 11. -37 13. $\cos^{-1}(-37/10\sqrt{19})$
15. $(-37/5\sqrt{2})$ 17. $\langle 19, -7, 11\rangle$ 19. 0 21. $4\mathbf{i} + \mathbf{j} + 2\mathbf{k}$ 23. -5 25. $\langle 2, 2, -3\rangle$ 27. $\langle 1, 0, 8\rangle$ 29. $\langle 16, -19, -2\rangle$
31. $16x - 19y - 2z + 22 = 0$ 33. $x = 4 + 16t$, $y = 4 - 19t$, $z = 5 - 2t$ 35. $\cos^{-1}(4/\sqrt{85})$
37. $x = 2 + 2t$, $y = 1 - 3t$, $z = -3 + 5t$ 39. $x = 2 + 2t$, $y = t$, $z = 3 - t$ 41. $(x - 1)^2 + y^2 + z^2 = 16$ 43. $x^2 + y^2 = 4$
45. Plane 47. Circular cone 49. Circular paraboloid 51. Sphere of radius 3
53. Hyperboloid of two sheets 55. Hyperboloid of two sheets 57. Parabolic cylinder
59. Cylindrical: $z = 4$; spherical: $\rho \cos \phi = 4$ 61. Cylindrical: $r^2 + z^2 = 25$; spherical: $\rho = 5$
63. Cylindrical: $r^2(36 \cos^2\theta - 9 \sin^2\theta) = 4z^2$; spherical: $(36 \cos^2\theta - 9 \sin^2\theta) \sin^2\phi = 4 \cos^2\phi$
65. $z = y$ 67. $x^2 + y^2 = y$ 69. $x^2 + y^2 = x$ 71. $-2/\sqrt{41}$, $6/\sqrt{41}$, $1/\sqrt{41}$ 73. $(1, 2, -1)$

Additional Exercises

1. $(\|\mathbf{v}\|\mathbf{u} + \|\mathbf{u}\|\mathbf{v}) \cdot (\|\mathbf{v}\|\mathbf{u} - \|\mathbf{u}\|\mathbf{v}) = \|\mathbf{v}\|^2(\mathbf{u} \cdot \mathbf{u}) - \|\mathbf{u}\|^2(\mathbf{v} \cdot \mathbf{v}) = \|\mathbf{v}\|^2\|\mathbf{u}\|^2 - \|\mathbf{u}\|^2\|\mathbf{v}\|^2 = 0$
3. $\mathbf{u} = \dfrac{1}{\|\mathbf{v} \times \mathbf{w}\|}(\mathbf{v} \times \mathbf{w})$
5. Reduce both sides to $\langle u_2v_1w_2 - u_2v_2w_1 - u_3v_3w_1 + u_3v_1w_3,\ u_3v_2w_3 - u_3v_3w_2 - u_1v_1w_2 + u_1v_2w_1,\ u_1v_3w_1 - u_1v_1w_3 - u_2v_2w_3 + u_2v_3w_2\rangle$
7. Use Problem 5. 9. $\dfrac{|abc|}{\sqrt{a^2b^2 + a^2c^2 + b^2c^2}}$ 11. Test $\mathbf{v}(\overrightarrow{P_1P_2})$ and $\mathbf{v}(\overrightarrow{P_1P_3})$ for parallelism.
13. Use Problem 12 to see if they intersect, and test their direction vectors for orthogonality.
15. If Problem 14 identifies all the generators, then $t = 0$ gives the origin on each one.
17. In each case, show that substituting the given expressions for x, y, and z satisfies the equation of the given quadric surface.
19. For example, let $P_0(x_0, y_0, z_0)$ be a point on $cz = x^2 - y^2$. We must show that there are unique values for s and t such that: *(1)* $\frac{1}{2}s + sct = x_0$, *(2)* $-\frac{1}{2}s + sct = y_0$, and *(3)* $2t = z_0$. From equation *(3)*, $t = z_0/2$. Putting this into equation *(1)* and solving for s, we eventually get $s = 1/(x_0 + y_0)$ or $s = 1/(x_0 - y_0)$. Now, the first value for s does not satisfy equation *(2)*, but the other does, so there is exactly one line of the regulus on P_0. The other cases are similar.
21. $\frac{1}{2}\pi a^4$
23. From point (x_0, y_0, z_0) to line $x = x_1 + at$, $y = y_1 + bt$, $z = z_1 + ct$, the distance is

$$\sqrt{\frac{[a(y_1 - y_0) - b(x_1 - x_0)]^2 + [a(z_1 - z_0) - c(x_1 - x_0)]^2 + [b(z_1 - z_0) - c(y_1 - y_0)]^2}{a^2 + b^2 + c^2}}.$$

Chapter 13 Section 13.1

1. $t \neq 0$, $t \neq 1$ 3. $t > 0$ 5. Line 7. Circular helix 9. Parabola
11. **(a)** $\langle 2t + t^2, t^2 + t^3, t + 1\rangle$ **(b)** $2t^3 + t^5 + 2t - 2$ **(c)** $\langle 2t^2 - t^4 + t^3, t^3 - t^2 - 4t, t^4\rangle$ **(d)** $\langle 2, 2t, 1\rangle$ **(e)** $\langle 2t, 3t^2, 0\rangle$ **(f)** $\langle 2 + 2t, 2t + 3t^2, 1\rangle$ **(g)** $6t^2 + 5t^4 + 2$ **(h)** $\langle 4t - 4t^3 + 3t^2, 3t^2 - 2t - 4, 4t^3\rangle$
13. **(a)** $\langle t, t^2 + \ln t, \tan t + \ln t\rangle$ **(b)** $t^2 \ln t + \tan t \ln t$ **(c)** $\langle (\ln t)^2 - t^2 \tan t, -t \ln t, t^3\rangle$ **(d)** $\langle 1, 1/t, \sec^2 t\rangle$

(e) $\langle 0, 2t, 1/t \rangle$ (f) $\langle 1, 2t + 1/t, \sec^2 t + 1/t \rangle$ (g) $2t \ln t + t + \sec^2 t \ln t + (1/t)\tan t$
(h) $\langle (2/t)\ln t - 2t \tan t - t^2 \sec^2 t, -\ln t - 1, 3t^2 \rangle$

15. (a) $\langle t^3 + 3t^4, t^3 - 3t^4, t^5 \rangle$ (b) $3t^2$ (c) $\langle 3, -3, 2t \rangle$ (d) $\langle 3t^2 + 12t^3, 3t^2 - 12t^3, 5t^4 \rangle$

17. (a) $\langle 2 \cos 2t \cos t - \sin 2t \sin t, 4 \cos 2t \sin t + 2 \sin 2t \cos t, \sin 2t + 2t \cos 2t \rangle$ (b) $2 \cos 2t$ (c) $\langle -\sin t, 2 \cos t, 1 \rangle$
(d) $\langle -2 \cos 2t \sin t, 4 \cos 2t \cos t, 2 \cos 2t \rangle$

19. $x = -1, y = -t, z = 2\pi + 2t$ 21. $y = 1$ 23. $\langle 2t + 1, t, 3t + 3 \rangle$ 25. $\langle e^t, \ln t + 1, -e^{-t} + 2/e \rangle$ 27. 3π

29. $\ln(1 + \sqrt{2})$

31. $(\mathbf{f} + \mathbf{g})'(t) = \langle (f_1 + g_1)'(t), (f_2 + g_2)'(t), (f_3 + g_3)'(t) \rangle = \langle f_1'(t) + g_1'(t), f_1'(t) + g_1'(t), f_1'(t) + g_1'(t) \rangle$
$= \langle f_1'(t), f_2'(t), f_3'(t) \rangle + \langle g_1'(t), g_2'(t), g_3'(t) \rangle = \mathbf{f}'(t) + \mathbf{g}'(t)$

33. $(\mathbf{f} \cdot \mathbf{g})'(t) = (f_1 g_1 + f_2 g_2 + f_3 g_3)'(t) = (f_1' g_1 + f_1 g_1' + f_2' g_2 + f_2 g_2' + f_3' g_3 + f_3 g_3')(t)$
$= (f_1' g_1 + f_2' g_2 + f_3' g_3)(t) + (f_1 g_1' + f_2 g_2' + f_3 g_3')(t) = \mathbf{f}'(t) \cdot \mathbf{g}(t) + \mathbf{f}(t) \cdot \mathbf{g}'(t)$

35. $\mathbf{f}(t) = \langle \sin t, \cos t, k \rangle$

37. (a) $\lim_{t\to c}[u(t)\mathbf{f}(t)] = \lim_{t\to c} \langle u(t)f_1(t), u(t)f_2(t), u(t)f_3(t) \rangle = \langle \lim_{t\to c} u(t)f_1(t), \lim_{t\to c} u(t)f_2(t), \lim_{t\to c} u(t)f_3(t) \rangle$
$= \langle \lim_{t\to c} u(t) \lim_{t\to c} f_1(t), \lim_{t\to c} u(t) \lim_{t\to c} f_2(t), \lim_{t\to c} u(t) \lim_{t\to c} f_3(t) \rangle = [\lim_{t\to c} u(t)][\lim_{t\to c} \mathbf{f}(t)]$
(b) $\mathbf{f} \times \mathbf{g}$ consists of products and differences of the components of **f** and **g**. The limits of products and differences are the products and differences of the limits. The formula follows.

39. By Problem 33, $(\mathbf{f}' \cdot \mathbf{f})' = \mathbf{f}'' \cdot \mathbf{f} + \mathbf{f}' \cdot \mathbf{f}' = \|\mathbf{f}'\|^2 + \mathbf{f} \cdot \mathbf{f}''$.

41. $\theta' = \dfrac{(\mathbf{f} \cdot \mathbf{g})[(\mathbf{f} \cdot \mathbf{f}')/(\mathbf{f} \cdot \mathbf{f}) + (\mathbf{g} \cdot \mathbf{g}')/(\mathbf{g} \cdot \mathbf{g})] - \mathbf{f}' \cdot \mathbf{g} - \mathbf{f} \cdot \mathbf{g}'}{\sqrt{(\mathbf{f} \cdot \mathbf{f})(\mathbf{g} \cdot \mathbf{g}) - (\mathbf{f} \cdot \mathbf{g})^2}}$

Section 13.2

1. $\mathbf{v} = \langle 1, -1, -3 \rangle$, $\|\mathbf{v}\| = \sqrt{11}$, $\mathbf{a} = \langle 0, 0, 0 \rangle$ 3. $\mathbf{v} = \langle 3, 2 \cos t, -2 \sin t \rangle$, $\|\mathbf{v}\| = \sqrt{13}$, $\mathbf{a} = \langle 0, -2 \sin t, -2 \cos t \rangle$

5. $\mathbf{v} = 3(\cos 3t)\mathbf{i} + 3(\sin 3t)\mathbf{j} + e^t\mathbf{k}$, $\|\mathbf{v}\| = \sqrt{9 + e^{2t}}$, $\mathbf{a} = -9(\sin 3t)\mathbf{i} + 9(\cos 3t)\mathbf{j} + e^t\mathbf{k}$

7. $\langle \frac{1}{4}(t^2 - 1), 2(t - 1), 1 + \ln t \rangle$ 9. $\langle 2t, \frac{1}{2}t^2 + t + 1, t^2 \rangle$ 11. $120\sqrt{6}/\pi$ rpm 13. $1250\pi^2/3$ cm/sec^2

15. If s is the speed and r is the radius of the tire, then $h_{max} = (s^2 + rg)^2/2s^2g \approx 16$ ft. 17. $\frac{1936}{9}$ ft

19. $\mathbf{a} = d\mathbf{v}/dt = (d\mathbf{v}/ds)(ds/dt) = \mathbf{v}(d\mathbf{v}/ds)$ 21. $D_t(\frac{1}{2}m\|\mathbf{v}\|^2) = D_t(\frac{1}{2}m\mathbf{v} \cdot \mathbf{v}) = \frac{1}{2}m(2\mathbf{v} \cdot D_t\mathbf{v}) = m\mathbf{a} \cdot \mathbf{v} = \mathbf{F} \cdot \mathbf{v}$

23. The force is $-\omega^2\mathbf{r}$; $\mathbf{r} \times (-\omega^2\mathbf{r}) = -\omega^2(\mathbf{r} \times \mathbf{r}) = \mathbf{0}$.

25. If $\|\mathbf{v}\|$ is constant, then $\mathbf{v} \cdot \mathbf{v}$ is constant, and its derivative is $2\mathbf{a} \cdot \mathbf{v} = 0$. Therefore, $\mathbf{a} \cdot \mathbf{v} = 0$, and **a** and **v** are orthogonal.

Section 13.3

1. $2/17\sqrt{17}$; $17\sqrt{17}/2$; $(-32, 0, \frac{25}{2})$; $y = 0$ 3. $\frac{4}{5}$; $\frac{5}{4}$; $(-\frac{1}{8}, -\sqrt{3}/2, -\pi/6)$; $37x + 15\sqrt{3}y - 8\sqrt{3}z = 78.7$

5. Tangential component: $\dfrac{81t^3 + 18t}{\sqrt{81t^4 + 36t^2 + 4}}$; normal component: 3

7. Tangential component: $\dfrac{8t}{\sqrt{1 + 8t^2}}$; normal component: $\dfrac{2\sqrt{2}}{\sqrt{1 + 8t^2}}$

9. $\dfrac{6}{|t|(\sqrt{4 + 9t^2})^3}$ 11. $\dfrac{|6x|}{(\sqrt{1 + 9x^4})^3}$

13. The area of the parallelogram of which two orthogonal unit vectors are adjacent sides is 1, so their cross product is a unit vector.

Review Exercises

1. $\langle 2t + t^2, \sin t + \cos t, e^t + e^{-t} \rangle$ 3. $2t^3 + \sin t \cos t + 1$ 5. $\langle 2 + 2t, \cos t - \sin t, e^t - e^{-t} \rangle$ 7. $6t^2 + \cos^2 t - \sin^2 t$

9. $\langle 2t^4, t^3 \sin t, t^3 e^t \rangle$ 11. $\sqrt{4 + \cos^2 t + e^{2t}}$ 13. $\left\langle 0, \dfrac{1}{2\sqrt{t}}, \dfrac{3\sqrt{t}}{2} \right\rangle$ 15. $\left\langle 0, \dfrac{-1}{4t\sqrt{t}}, \dfrac{3}{4\sqrt{t}} \right\rangle$ 17. $\dfrac{\langle 0, -9t, 3 \rangle}{(\sqrt{1 + 9t^2})^3}$

19. $\dfrac{3\sqrt{t}}{(\sqrt{1 + 9t^2})^3}$ 21. $\dfrac{\langle 0, -6t, 2 \rangle}{\sqrt{1 + 9t^2}}$ 23. $\dfrac{3\sqrt{t}}{\sqrt{1 + 9t^2}}$ 25. $\left\langle 1 + t + \dfrac{t^3}{6}, \dfrac{2t^3}{3}, \dfrac{t^2}{2} - \dfrac{t^3}{6} - t - 2 \right\rangle$

Additional Exercises

1. 2.1343
3. $dy/dt = -\operatorname{sech}(t/a)\tanh(t/a)$ and $dx/dt = 1 - \operatorname{sech}^2(t/a) = \tanh^2(t/a)$, so $dy/dx = -\operatorname{sech}(t/a)/\tanh(t/a)$. The equation of the tangent line is $y - a\operatorname{sech}(t/a) = (dy/dx)(x - t + a\tanh(t/a))$; setting $y = 0$ and solving for x gives $x = t$.
5. (a) $t = 0$ (b) $t = 1$ (c) $t = -1$ 7. $(\mathbf{f}\cdot\mathbf{g}\times\mathbf{h})' = (\mathbf{f}\cdot\mathbf{g})'\times\mathbf{h} + (\mathbf{f}\cdot\mathbf{g})\times\mathbf{h}' = \mathbf{f}'\cdot\mathbf{g}\times\mathbf{h} + \mathbf{f}\cdot\mathbf{g}'\times\mathbf{h} + \mathbf{f}\cdot\mathbf{g}\times\mathbf{h}$
9. $\pi/12$, $5\pi/12$ 11. 16,180 mph 13. 24.16 rpm
15. Generalize the triangle inequality and use the definition of integral as a limit of a sum.
17. Use $x = f(\theta)\cos\theta$, $y = f(\theta)\sin\theta$, and the formula in Example 13.21.
19. Using $x = \rho(t)\sin\phi(t)\cos\theta(t)$, $y = \rho(t)\sin\phi(t)\sin\theta(t)$, and $z = \rho(t)\cos\phi(t)$, simplify $(dx/dt)^2 + (dy/dt)^2 + (dz/dt)^2$.
21. Banking the curve at angle θ from the horizontal changes the component $\|\mathbf{n}\|$ of acceleration parallel to the road to $\|\mathbf{n}\|\cos\theta$, lessening the tendency to slide off the road.
23. $(2v^2/g)(\sin\theta - \cos\theta\tan\alpha)\cos\theta\sec\alpha$
25. The projectile lands downrange a distance $(v^2/g)\sin 2\theta$ and a distance $(2av^2/g^2)\sin^2\theta$ in the direction of the crosswind.

Chapter 14 Section 14.1

1. The entire plane 3. Disc, $x^2 + y^2 \le 9$ 5. $x^2 + y^2 \ge 9$ 7. $x^2 + y^2 \ge 9$ 9. $x^2 + y^2 \ge 9$ and $y \ne 0$ 11. $y > 0$
13. $y > -x$ 15. $x^2 + y^2 + z^2 \le 1$ 17. $z \ne 0$ 19. (a) f_4 (b) f_2 (c) f_1 (d) f_3 (e) f_5
21. (a) f_5 (b) f_2 (c) f_1 (d) f_4 (e) f_3 23. Circular paraboloid 25. Plane 27. Inverted circular paraboloid 29. Hemisphere

31.

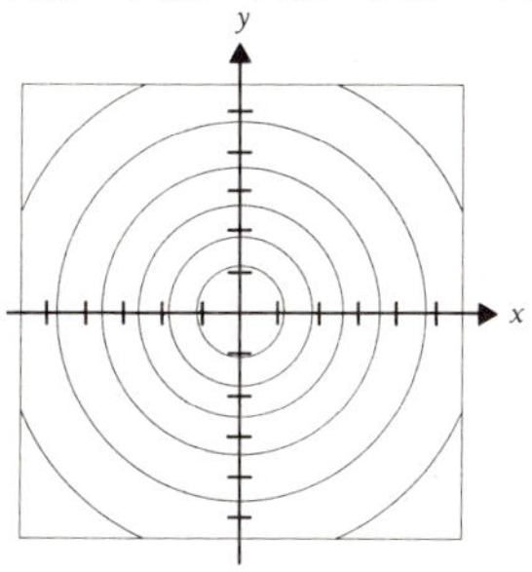

33.

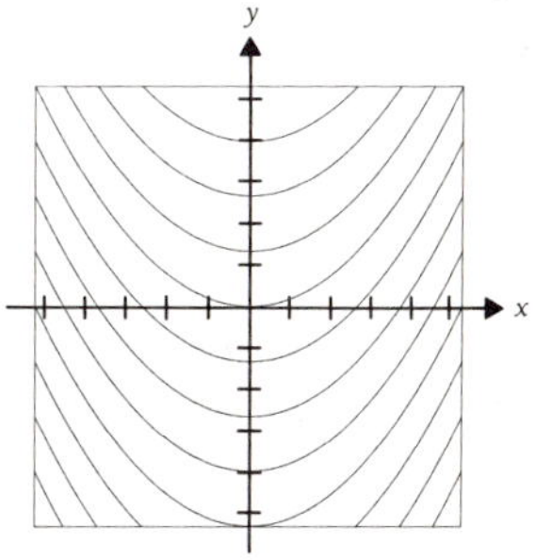

35.

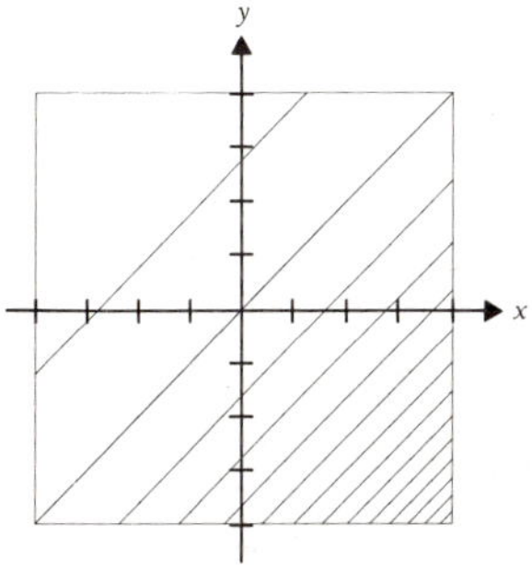

37.

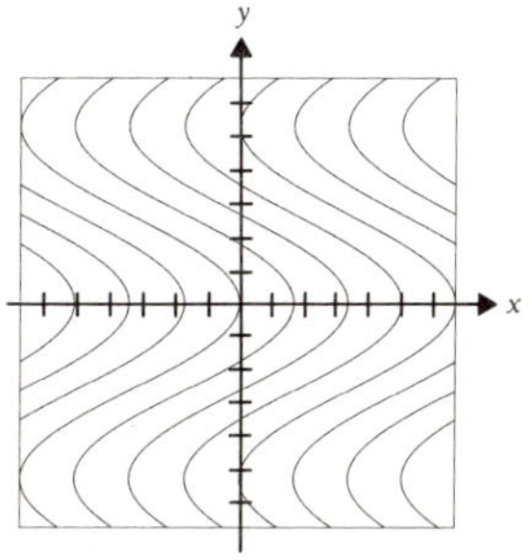

39.

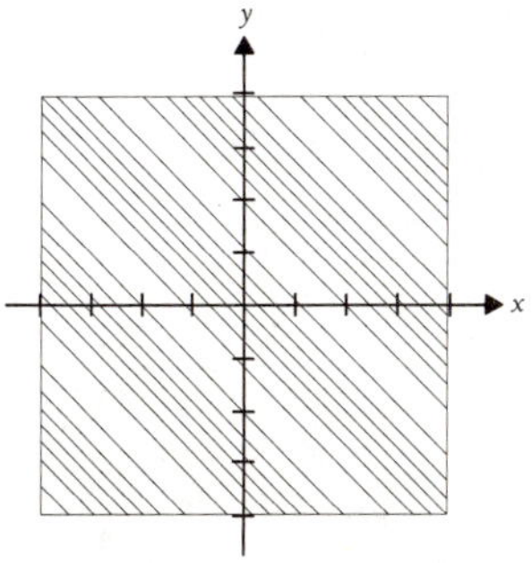

41. Circular hyperboloids, one cone 43. Circular paraboloids 45. $2x - y + x^2 - y^2$ 47. $(2x - y)(x^2 - y^2)$ 49. 2
51. -8 53. $xy - x^4$ 55. $(2x - y)(x^2 - y^2) - z^2$ 57. Yes; index 2 59. Yes; index 0
61. The polynomial is homogeneous of index n if and only if k^n is a factor of every term—that is, if and only if every term has total degree n.

Section 14.2

1. $\frac{2}{5}$ 3. 0 5. 2 7. $\frac{1}{5}$ 9. $\cos\frac{1}{2}$ 11. Does not exist 13. 0 15. $\frac{1}{2}$ 17. $(m - m^2)/(1 + m^2)$ 19. Does not exist
21. $\frac{1}{3}$ 23. $\lim_{x \to a} f(x, b) = \lim_{\substack{(x,y) \to (a,b) \\ y=b}} f(x, y) = L$; the other is similar. 25. Continuous 27. Not continuous
29. The circle and its interior form a closed disc, and it is bounded. 31. Yes 33. It is still bounded, and its complement is open.

Section 14.3

1. $f_x = 12x^2 - 3y^2$; $f_y = -6xy + 4y^3$; $f_{xx} = 24x$; $f_{xy} = f_{yx} = -6y$; $f_{yy} = -6y + 12y^2$
3. $f_x = y\ln(x^2 + y^2) + 2x^2y/(x^2 + y^2)$; $f_y = x\ln(x^2 + y^2) + 2xy^2/(x^2 + y^2)$; $f_{xx} = (2x^3y + 6xy^3)/(x^2 + y^2)^2$; $f_{xy} = f_{yx} = \ln(x^2 + y^2) + 2(x^4 + y^4)/(x^2 + y^2)^2$; $f_{yy} = (6x^3y + 2xy^3)/(x^2 + y^2)^2$
5. $f_x = 1/\sqrt{1 - (x + 2y)^2}$; $f_y = 2/\sqrt{1 - (x + 2y)^2}$; $f_{xx} = (x + 2y)/[\sqrt{1 - (x + 2y)^2}]^3$; $f_{xy} = f_{yx} = 2(x + 2y)/[\sqrt{1 - (x + 2y)^2}]^3$; $f_{yy} = 4(x + 2y)/[\sqrt{1 - (x + 2y)^2}]^3$
7. $f_x = 2xe^{x^2+y^2}$; $f_y = 2ye^{x^2+y^2}$; $f_{xx} = 2(1 + 2x^2)e^{x^2+y^2}$; $f_{xy} = f_{yx} = 4xye^{x^2+y^2}$; $f_{yy} = 2(1 + 2y^2)e^{x^2+y^2}$
9. $f_x = y\sin z$; $f_y = x\sin z$; $f_z = xy\cos z$; $f_{xx} = 0$; $f_{xy} = f_{yx} = \sin z$; $f_{yy} = 0$; $f_{xz} = f_{zx} = y\cos z$; $f_{yz} = f_{zy} = x\cos z$; $f_{zz} = -xy\sin z$
11. $f_x = 2x + 2y + 3z$; $f_y = 2x + 2y + 2z$; $f_z = 3x + 2y + 2z$; $f_{xx} = 2$; $f_{yy} = 2$; $f_{zz} = 2$; $f_{xy} = f_{yx} = 2$; $f_{xz} = f_{zx} = 3$; $f_{yz} = f_{zy} = 2$
13. $f_x = 1$; $f_y = 5$ 15. $f_x = 1$; $f_y = 0$ 17. (a) $\pi r^2/3$ (b) $2\pi rh/3$ 19. $\cos xyz - (3 + xyz)xyz\sin xyz$
21. $xf_x + yf_y = x(4Ax^3 + 4Bxy^2) + y(4Bx^2y + 4Cy^3) = 4f$
23. $xf_x + yf_y = x[xy^2(x - 2y)/(x + y)^2] + y[x^2y(y - 2x)/(x + y)^2] = x^2y^2(x - 2y + y - 2x)/(x + y)^2 = -f$
25. $x = 2 + t$, $y = 2$, $z = 1 - 4t$
27. (a) $f_x(0, y) = -y$; $f_x(0, 0) = 0$ (b) $f_y(x, 0) = x$; $f_y(0, 0) = 0$ (c) -1 (d) 1 (e) 1; $\lim_{x\to 0} f_{xy}(x, 0) = 1 \neq f_{xy}(0, 0)$
29. $-r\sin\theta$ 31. $r\cos\theta$ 33. $y/\sqrt{x^2 + y^2}$ 35. $x/(x^2 + y^2)$ 37. $z_{xx} = -a^2e^{-ay}\cos ax = az_y$
39. $\partial R/\partial r = -6c(3h^2 + 8rh + 8r^2)/(3rh + 4r^2)^2 < 0$; $\partial R/\partial h = -12cr^2/(3rh + 4r^2)^2 < 0$

Section 14.4

1. $f_x = 8x - 3y$; $f_y = 8y - 3x$ 3. $f_x = 2vx + 2uy + 4wx$; $f_y = 2ux - 2vy - 4wy$
5. $f_x = 2ry/\sqrt{1 - x^2y^2} + 2s + 2t + 2uy$; $f_y = 2rx/\sqrt{1 - x^2y^2} + 2s - 2t + 2ux$
7. $f_x = 2(u_1 + u_2 + u_3)(y + 2xy^2 + 3x^2y^3)$; $f_y = 2(u_1 + u_2 + u_3)(x + 2x^2y + 3x^3y^2)$
9. $f_x = \sin 2u - \sin 2v$; $f_y = \sin 2u + \sin 2v$; $f_z = \sin 2u - \sin 2v$ 11. $f_x = \frac{1}{2}(u + v)$; $f_y = \frac{1}{2}(v - u)$; $f_z = (\cos w)/z^2$
13. $\dfrac{4xt^3 - y}{t^2\sqrt{x^2 + y^2 - 9}}$ 15. $\dfrac{2u^2}{(1 - t)^2(u + v)^2} - \dfrac{v^2}{t^2(u + v)^2}$ 17. Increasing at 7 in.^{2}/min
19. Use the chain rule and the fact that $dx/dx = 1$. 21. $z_x = f_u + f_v$ and $z_y = f_u - f_v$; multiply 23. Use the chain rule.
25. $z_r^2 + z_\theta^2/r^2 = (f_x\cos\theta + f_y\sin\theta)^2 + (-f_x\sin\theta + f_y\cos\theta)^2 = f_x^2(\cos^2\theta + \sin^2\theta) + f_y^2(\sin^2\theta + \cos^2\theta) = f_x^2 + f_y^2$
27. f is not continuous at (0, 0), since the limit along $y = mx$ depends on m; so f is not differentiable at (0, 0). But $f_x(0, 0) = \lim_{h\to 0}[f(h, 0) - f(0, 0)]/h = 0$, and similarly for $f_y(0, 0)$, so the partial derivatives exist.
29. The partial derivatives are not defined at the origin. 31. No; f is not continuous at the origin.

Section 14.5

1. $\langle 3, -4\rangle$ 3. $\langle 8, 0\rangle$ 5. $\langle -3e^{-6}, 6e^{-6}\rangle$ 7. $\langle \frac{1}{2}, -\pi/8\rangle$ 9. $\langle 3, -1, -2\rangle$ 11. $\langle -1, 0, -1/e\rangle$ 13. $-1 + \frac{1}{2}\sqrt{3}$
15. $5\sqrt{2}$ 17. $-1/\sqrt{2}$ 19. 0 21. (a) (i) $-9\sqrt{2}$; (ii) $2\sqrt{65}$ (b) C: $y = -4x^2$

23. (a) (i) $-\sqrt{1+4e^4+9e^9}$; (ii) $(1-2e^2)/\sqrt{2}$ (b) C: $\langle -\ln(e^{-1}-kt), \frac{1}{2}\ln(e^{-4e^2}-4kt), \frac{1}{3}\ln(e^{-9e^3}-9kt)\rangle$
25. $\langle \frac{4}{5}, \frac{3}{5}\rangle$ 27. $\langle 1/\sqrt{2}, -1/\sqrt{2}, 0\rangle$, $\langle \frac{2}{3}, -\frac{1}{3}, -\frac{2}{3}\rangle$, or any linear combination of these
29. The derivative of a sum is the sum of the derivatives for each component.
31. Use the product rule for each component.
33. (a) ~1°19′ east of north; a 28°5′ slope (b) ~1°19′ west of south (c) 20°14′ slope
(d) ~1°19′ south of east or north of west

Section 14.6

1. N: $x=y=0$, $z=t$; T: $z=0$ 3. N: $x=1+6t$, $y=2+20t$, $z=23-t$; T: $6x+20y-z-23=0$
5. N: $x=y=0$, $z=t$; T: $z=1$ 7. N: $x=t$, $y=z=0$; T: $x=\sqrt{3}$ 9. (0, 0, 0) 11. None 13. None
15. Use the fact that the normal vector at (x_1, y_1, z_1) is $\langle -c^2x_1/a^2z_1, -c^2y_1/b^2z_1, -1\rangle$ and the fact that (x_1, y_1, z_1) is on the ellipsoid to simplify the equation of the tangent plane.
17. $\sin^{-1}(\sqrt{57}/38)$ 19. $\pm\sin^{-1}(\sqrt{6}/4)$ 21. $(\pm a, 0, 0)$, $(0, \pm b, 0)$, $(0, 0, \pm c)$ 23. $\cos^{-1}(1/\sqrt{10})$ 25. $\cos^{-1}(-1/\sqrt{17})$
27. $f_{\mathbf{u}} = \mathbf{u}\cdot\nabla f = 0$ means $\mathbf{u}$ is orthogonal to the normal vector, so it is tangent to the level curve.
29. Since ∇f is normal to the level surface, it must be normal to any curve lying in the surface.

Section 14.7

1. Absolute minimum of 0 at (0, 0) 3. None 5. Relative maximum of $\frac{2}{3}$ at (1, 0) 7. Relative maximum of 1 at (1, 1)
9. Relative minimum of -1 at $(\pm 1, 0)$ 11. Relative maximum of 1 at $(\pm 1, 0)$; absolute minimum of -1 at $(0, \pm 1)$
13. Absolute maximum of 1 at any point on $y=-x+2n\pi$, $n\in\mathbb{Z}$; absolute minimum of -1 at any point on $y=-x+(2n+1)\pi$, $n\in\mathbb{Z}$
15. Absolute maximum of 1 at $(0, 2n\pi)$, $n\in\mathbb{Z}$; absolute minimum of -1 at $(0, (2n+1)\pi)$, $n\in\mathbb{Z}$
17. Absolute maximum of 1 at (0, 0) 19. None 21. $4\times4\times2$ ft 23. $32/3\sqrt{3}$ 25. $a^3/27$ 27. $49\sqrt{66}/33$
29. \$63.68 per cassette player, \$63.78 per TV 31. Maximum $P=$ \$900 million, $A=$ \$10 million, $S=$ \$20
33. Using $c=p-a-b$, write $A=\frac{1}{4}\sqrt{p(p-2a)(p-2b)(2a+2b-p)}$. Setting the gradient of A equal to $\langle 0, 0\rangle$ gives $(p-2b)(p-2a-b)=0$ and $(p-2a)(p-2b-a)=0$. If $p=2b$ in the first equation, then $p=2a$ in the second, and $a=b$. However, that implies $c=0$, a minimum for A. Hence, $p=2a+b$ in the first equation, so $p=2b+a$ in the second, and again $a=b$. This time each equals $p/3$, and $c=p/3$. Thus, the triangle is equilateral.
35. Equilateral, with perimeter $3r\sqrt{3}$ 37. Buy 30 units of the second (\$10) commodity. 39. $y=6-3x/4$
41. $y=2.63-0.39x$ 43. (a) 1 (b) 0.29

Section 14.8

1. $(4, -3)$, $(-4, 3)$, $(3, 4)$, $(-3, -4)$ 3. $(\frac{1}{27}, \frac{2}{9}, \frac{4}{27})$ 5. $(\pm 2^{1/4}, \pm 2^{-1/4}, \pm 2^{-1/4})$, signs chosen independently
7. $(0, \pm 1, 0)$, $(\pm 1, 0, 0)$ 9. $(\frac{56}{83}, \frac{43}{83}, -\frac{59}{83})$ 11. $(\frac{38}{33}, -\frac{2}{3}, \frac{2}{3}, \frac{12}{11})$ 13. $\sqrt{3}/2$ 15. $(\pm\frac{1}{7}\sqrt{21}, \pm\frac{1}{7}\sqrt{21}, \mp\frac{2}{7}\sqrt{21})$
17. Maximum $20\frac{3}{4}$, minimum 14 19. 12
21. To maximize xyz subject to the constraint $x+y+z=S$, we get the equations $yz+\lambda=0$, $xz+\lambda=0$, $xy+\lambda=0$, and $x+y+z=S$. Eliminating λ from the first two, we get $(x-y)z=0$. Now if $z=0$, the product will not be very large, so $x=y$. Eliminating λ from the second and third equations produces $y=z$ in a similar fashion.
23. The coordinates of P are the averages of the coordinates of the four vertices.

Review Exercises

1. $A=\frac{1}{2}(b_1+b_2)h$ 3. All points in the plane; lines with slope -2 5. $y\neq 0$; parabolas with vertex at the origin
7. All points in space; planes with normal vector $3\mathbf{i}+4\mathbf{j}-\mathbf{k}$
9. The domains of $f+g$ and fg are the same, namely the intersection of the domains of f and g.
11. Does not exist 13. 3 15. Continuous on the entire plane 17. Continuous 19. $f_x=\sqrt{y}-y/2\sqrt{x}$; $f_y=x/2\sqrt{y}-\sqrt{x}$

21. $f_x = yx^{y-1}$; $f_y = x^y \ln x$ 23. $f_x = y^{x+z} \ln y$; $f_y = (x+z)y^{x+z-1}$; $f_z = y^{x+z} \ln y$ 25. $f_{xx} = 2$; $f_{xy} = f_{yx} = 0$; $f_{yy} = 2$
27. $f_{xx} = f_{yy} = e^{x+y} + e^{x-y}$; $f_{xy} = f_{yx} = e^{x+y} - e^{x-y}$
29. $f_{xx} = f_{yy} = f_{zz} = 0$; $f_{xy} = f_{yx} = z$; $f_{xz} = f_{zx} = y$; $f_{yz} = f_{zy} = x$ 31. $4u - 4v$ 33. $ve^{uvw}(wy \sec^2 xy + u)$
35. $-e^u \sin x \arcsin v + e^{u+x}/\sqrt{1-v^2}$ 37. $-2u \sin x + 2v \cos x + 6w \sec^2 x$ 39. $\langle \cos x \cos y, -\sin x \sin y \rangle$
41. $\langle yz \cos xyz, xz \cos xyz, xy \cos xyz \rangle$ 43. $2/\sqrt{5}$ 45. $4\sqrt{5}$ 47. $\frac{1}{2}\sqrt{6}$
49. N: $x = 1 + 10t$, $y = 1 + 8t$, $z = 9 - t$; T: $10x + 8y - z - 9 = 0$ 51. N: $x = t$, $y = t$, $z = 1 - t$; T: $x + y - z + 1 = 0$
53. Minimum at $(0, -1/\sqrt{3})$ 55. None 57. None 59. $\frac{1}{2}(11 - 6\sqrt{3})^{1/2}$ 61. 1
63. The angle between two spheres is the angle between their normal vectors at their point of intersection. Two intersecting spheres are generated by revolving two intersecting circles about their line of centers. The normal vectors at any point of intersection are the radial rays through that point, which are generated by the radial rays through a point of intersection of the two circles. Thus, the angle between them does not change.

Additional Exercises

1. Does not exist 3. $f_x = -(y \sin xy \ln y)y^{\cos xy}$; $f_y = [(1/y) \cos xy - x \sin xy \ln y]\, y^{\cos xy}$
5. $f_x = y \cos zw$; $f_y = x \cos zw$; $f_z = -xyw \sin zw$; $f_w = -xyz \sin zw$
7. Not continuous 9. (a) $\frac{1}{4}\sqrt{31}$ (b) $\frac{1}{2}\sqrt{23}$ 11. $3\sqrt{61} - 2$
13. Maximum $P = 225$; maximum $Q = 87.5$; maximum $P + Q = 616.67$ 15. The centroid 17. $\langle 6 \cos t, 6 \sin t, 6t \rangle$
19. The normal vector is $\mathbf{n} = \langle -f_x, -f_y, 1 \rangle$, which is continuous if and only if the partial derivatives f_x and f_y are continuous, if and only if f is continuously differentiable.

Chapter 15 Section 15.1

1. $x = 0$, $dx = 0.02$, $y = 0$, $dy = -0.03$; $f \approx 0$ 3. $x = 0$, $dx = -0.033$, $y = 2$, $dy = 0.025$; $f \approx 2.0085$
5. $x = 2$, $dx = 0.04$, $y = 0$, $dy = 0.13$, $z = 0$, $dz = -0.05$; $f \approx 2.04$
7. $x = 0$, $dx = -0.02$, $y = 1$, $dy = -0.03$, $z = 1$, $dz = -0.02$; $f \approx -0.06$ 9. 1.005 11. 0.96177
13. Increases about 1% 15. About 1.4π 17. Increased about 2.5% 19. Multiplied by c 21. Decreases about 5%

Section 15.2

1. Exact; $f = \frac{1}{2}(x^2 + y^2)$ 3. Exact; $f = x^2 y$ 5. Not exact 7. Exact; $f = x \ln y + y^2$ 9. Exact; $f = \frac{1}{2}(x^2 + y^2 + z^2)$
11. Exact; $f = x + e^y + \cos z$ 13. $x^2 - y^2$ 15. $x^3 + x^2 y - xy^2 - y^3$ 17. $x^2 - 3xy - 7xz + 5y^2 + 11yz - 13z^2$

Section 15.3

1. -34 3. -1 5. $\frac{8}{3}$ 7. 0 9. 0 11. 0 13. 1 15. 1 17. (a) $\frac{1}{2}$ (b) $-9\sqrt{6}$
19. $M_x = \lim_{n \to \infty} \sum_{i=1}^{\infty} x_i \Delta m_i = \lim_{n \to \infty} \sum_{i=1}^{\infty} x_i \delta \Delta s_i = \int_C \delta(x, y, z)\, ds$
21. Since $\mathbf{T} = \mathbf{r}'/\|\mathbf{r}'\|$, $\mathbf{F} \cdot d\mathbf{r} = \mathbf{F} \cdot \mathbf{r}'\, dt = \mathbf{F} \cdot (\mathbf{r}'/\|\mathbf{r}'\|)\|\mathbf{r}'\|\, dt = \mathbf{F} \cdot \mathbf{T}\, ds$. 23. -5440 ft-lb

Section 15.4

1. 0 3. 0 5. 0 7. -2 9. 8 11. 2 13. $\frac{1}{2}$ 15. $\frac{1}{2}$ 17. 0 19. 0 21. -13 23. 0 25. 0 27. $\frac{1}{2}$ 29. $-\frac{1}{2}$
31. Yes; $f = -xy$ 33. Yes; $f = z$ 35. Yes; $f = x + y - 2z^2$ 37. $\mathbf{F}(x, y, z) = k(x^2 + y^2 + z^2)^{-3/2}\langle -x, -y, -z \rangle$; yes

Review Exercises

1. (a) 52.167 (b) 28.809 3. (a) 3% (b) 2% 5. 11.5×11.5 ft 7. Exact; $f = \frac{1}{2}x^2 - yz$ 9. Exact; $f = xe^y + z$
11. Exact; $f = \frac{1}{2}x\sqrt{x^2 + y^2} + \frac{1}{2}y^2 \ln[(x + \sqrt{x^2 + y^2})/y]$ 13. Not exact 15. $f = (x + y)/(x - y)$ 17. $f = x^2 + x \tan^{-1} yz$
19. $-\frac{9}{140}$ 21. $\frac{1}{4} + \ln 2$ 23. $-\frac{3}{4}$ 25. -1 27. (a) 0 (b) 0 29. (a)–(d) $e - 1$ 31. $2k\pi$

Additional Exercises

1. **(a)** 0.2964 **(b)** $M_x = 0.2389$, $M_y = 0.2002$, $M_z = 0.7167$ **(c)** (0.8060, 0.6755, 2.4180) **(d)** $I_x = 0.2002$, $I_y = 0.1513$, $I_z = 1.8018$

3. $\left\langle -\frac{mv}{kq}\sin\frac{kqt}{m},\ b\cos\frac{kqt}{m},\ c\right\rangle$, an ellipse **5.** $-4000P\ln(1 + H/4000)$

Chapter 16 Section 16.1

1. 24 **3.** $\frac{4404}{35}$ **5.** $3\pi/2$ **7.** $\frac{1}{2}(e-1)$ **9.** $3(e^3-e)/2$ **11.** $5(e^2-e)/2$ **13.** $(3e^3+5e^2-8e)/2$ **15.** $3(e^3-e)/2$
17. $16\pi/3$ **19.** 2 **21.** 0 **23.** 0 **25.** It is the volume of a solid of height 1 over R.
27. Let B be a number such that $f(x, y) + B > 0$ over R, and let A be the area of R. Then $f(x, y) + B \le g(x, y) + B$, so $\iint_R [f(x, y) + B]\,dA \le \iint_R [g(x, y) + B]\,dA$ by the interpretation of the double integral as a volume. But $\iint_R [f(x, y) + B]\,dA = \iint_R f(x, y)\,dA + BA$ and $\iint_R [g(x, y) + B]\,dA = \iint_R g(x, y)\,dA + BA$, and the result follows.
29. $\iint_R f(x)g(y)\,dA = \int_c^d\int_a^b f(x)g(y)\,dx\,dy = \int_c^d g(y)[\int_a^b f(x)\,dx]\,dy = [\int_a^b f(x)\,dx][\int_c^d g(y)\,dy]$.
31. Let S be any set of points in the plane, and let B be the boundary of S. Let P be any point not in B. Suppose that every open disc containing P also contains a point of B. Then, by the definition of boundary, every open disc containing P contains a point in S and a point not in S, and that makes P a boundary point of S, contrary to fact. Hence, not every open disc containing P contains a point of B, so some open disc contains P and no point of B. That makes the complement of B open and hence B is closed.
33. (i) If S is an open set and P is any point of S, then there is an open disc that contains P and lies entirely in S. Hence, P cannot be a boundary point of S, so S contains none of its boundary points. (ii) If S contains none of its boundary points, let P be any point in S. If every open disc containing P contains a point not in S, then P would have to be a boundary point of S. But since S contains none of its boundary points, P is not a boundary point. Hence, there must be an open disc containing P that lies entirely in S, and therefore S is open.
35. **(a)** 1 **(b)** $e - 1/e$

Section 16.2

1. $\frac{14}{3}$ **3.** $\frac{5}{3}$ **5.** 2π **7.** $\pi a^4/2$ **9.** $\pi c^2/2ab$ **11.** **(a)** π **(b)** π **(c)** $(0, \frac{1}{4})$ **(d)** (0, 0)
13. **(a)** $\frac{32}{3}$ **(b)** 89.6 **(c)** $(\frac{2}{21}, \frac{124}{49})$ **(d)** $(0, \frac{12}{5})$
15. **(a)** ab **(b)** $\frac{ab(2a^2+3b)}{6}$ **(c)** $\left(\frac{3a^3+3ab}{4a^2+6b}, \frac{a^2b+2b^2}{2a^2+3b}\right)$ **(d)** $(\frac{1}{2}a, \frac{1}{2}b)$ **17.** **(a)** $\pi/2$ **(b)** $\frac{14}{15}$

Section 16.3

1. $\pi/4$ **3.** $\frac{2}{3}$ **5.** $\frac{2}{5}$ **7.** 1 **9.** $\frac{2}{3}$ **11.** $(5\pi/16, \frac{7}{12})$ **13.** $(\frac{1}{2}, 5/(8\pi - 6\sqrt{3}))$ **15.** $k\pi$; $(0, -\frac{2}{3})$
17. $\frac{1}{6}$; $(\frac{3}{4}\sin 2 - 3\cos 2 - \frac{3}{2}, \frac{11}{4} - \frac{3}{4}\cos 2 - 3\sin 2)$ **19.** $9\pi/8$ **21.** $\frac{32}{9}$ **23.** $\pi(e^2-1)/8$ **25.** $\pi/6$ **27.** $\pi/2$
29. $\sqrt{x^2-y^2}/(x^2+y^2)$

Section 16.4

1. $a^3/6$ **3.** 5 **5.** 39 **7.** $\pi r^3/6$ **9.** $4\pi(27 - 5\sqrt{5})/3$ **11.** $\frac{3}{4}$; $(\frac{5}{9}, \frac{5}{9}, \frac{11}{18})$ **13.** $k\pi a^4/3$; (0, 0, 0) **15.** $abc/6$
17. $(a/4, b/4, c/4)$ **19.** $abc/3$ **21.** $\frac{304}{15}$
23. If the vertices are (0, 0, 0), $(a, 0, 0)$, $(0, b, 0)$, and $(0, 0, c)$, then the centroid $(a/4, b/4, c/4)$ (see Problem 17) is clearly a case of this formula.

Section 16.5

1. 0 3. $\frac{1}{2}$ 5. $(3\pi - 4)a^3/9$ 7. $2\pi a^3(1 - \cos\alpha)/3$ 9. $k\pi^2 a^4/4$; $[0, 0, 0]$ 11. $\frac{1}{2}k\pi a^2 L^2$; $[0, 0, 2L/3]$
13. $\frac{2}{3}\pi(b^3 - a^3)$; $[0, 0, 3(b^4 - a^4)/8(b^3 - a^3)]$ 15. $8a^3/3$ 17. 36π 19. $8\pi(2 - \sqrt{2})/3$ 21. $\frac{2}{5}$
23. (a) $\int_{-2}^{2} \int_{-\sqrt{4-x^2}}^{\sqrt{4-x^2}} \int_0^{\sqrt{9-x^2-y^2}} \sqrt{x^2+y^2}\, dz\, dy\, dx$
(b) $\int_0^{2\pi}\int_0^{\tan^{-1}(2/\sqrt{5})} \int_0^3 \rho^4 \sin^3\phi \cos\phi\, d\rho\, d\phi\, d\theta + \int_0^{2\pi} \int_{\tan^{-1}(2/\sqrt{5})}^{\pi/2} \int_0^{2\csc\phi} \rho^4 \sin^3\phi\, d\rho\, d\phi\, d\theta$

Review Exercises

1. $\frac{16}{3}$ 3. $\frac{272}{15}$ 5. $\frac{796}{15}$ 7. $\dfrac{9\pi}{4}$ 9. $\dfrac{9\pi}{2}$ 11. $\dfrac{(250\sqrt{2} - 41\sqrt{41})\pi}{6}$ 13. $\left(\dfrac{4}{3}, \dfrac{32}{15\pi}, \dfrac{32}{15\pi}\right)$
15. $\left(\dfrac{140\pi + 256}{105\pi + 224}, \dfrac{52}{15\pi + 32}, \dfrac{10\pi + 32}{15\pi + 32}\right)$ 17. $\left(\dfrac{4\sqrt{3}}{\pi}, \dfrac{4\sqrt{3}}{\pi}, 9\right)$
19. $\left(\dfrac{128\sqrt{3}}{37\pi}, \dfrac{128\sqrt{3}}{35\pi}, \dfrac{288}{35}\right)$ 21. $\left(\dfrac{3(5 \arctan \frac{1}{2} - 2)}{\pi(5 - 2\sqrt{5})}, \dfrac{3(5 \arctan \frac{1}{2} - 2)}{\pi(5 - 2\sqrt{5})}, \dfrac{3}{10 - 4\sqrt{5}}\right)$
23. $\left(\dfrac{63(5 \arctan \frac{1}{2} - 2)}{5(5 - 2\sqrt{5})}, \dfrac{63(5 \arctan \frac{1}{2} - 2)}{5(5 - 2\sqrt{5})}, \dfrac{63}{40(5 - 2\sqrt{5})}\right)$

Additional Exercises

1. A cylinder of radius a making an angle of 60° with the vertical has equation $3z^2 - 2\sqrt{3}yz + 4x^2 + y^2 = 4a^2$. The volume of the intersection with a vertical cylinder is $32a^3/3\sqrt{3}$.
3. $2\pi ka^2$ 5. $128\pi P \times 10^9$ 7. No 9. Yes
11. Let S' denote the complement of the set S. If H and K are closed sets, then their complements H' and K' are open. Hence, by Problem 10, $H' \cup K'$ is an open set. But $H' \cup K' = (H \cap K)'$, so $H \cap K$ is a closed set.
13. If G_1 and G_2 are open sets, let P be any point in $G_1 \cap G_2$. Since each G_i is open, there is an open ball B_i containing P that lies entirely in G_i, $i = 1, 2$. But then there is an open ball B that contains P and is inside both B_1 and B_2, so that B lies inside $G_1 \cap G_2$. Hence, $G_1 \cap G_2$ is an open set.
15. Let S be a closed set and let P be a limit point of S. Then by definition of approach, every open set containing P must contain a point of S distinct from P. If P is in the complement of S, which is open, then there is an open set containing P that contains no point of S. Since this contradicts the fact that S is an approach to P, point P must lie in S. That is, S contains all its limit points.

Chapter 17 Section 17.1

1. $\frac{1}{2}$ 3. 3π 5. -54π 7. $3 \ln \sqrt{3} - 4 \ln 2$ 9. 0 11. 2π 13. 3π
15. $\oint_{C_2+C_3}(P\,dx + Q\,dy) = \oint_{C_1+C_2+C_3-C_1}(P\,dx + Q\,dy) = \oint_{C_1+C_2}(P\,dx + Q\,dy) + \oint_{C_3-C_1}(P\,dx + Q\,dy) =$
$\iint_{R_1}(Q_x - P_y)\,dA + \iint_{R_2}(Q_x - P_y)\,dA = \iint_{R_1 \cup R_2}(Q_x - P_y)\,dA$
17. If R is a standard region, then R can be described as $R = \{(x, y) | a \le x \le b \text{ and } f(x) \le y \le g(x)\}$. Now, any line parallel to the y-axis has the form $x = c$. If $c < a$ or $c > b$, then the intersection with R is empty. If $a \le c \le b$, the intersection is the interval $[f(c), g(c)]$, unless $f(c) = g(c)$, in which case the intersection is a point. The case of a horizontal line is treated analogously.

Section 17.2

1. (a) $\sqrt{3}/2$ (b) $(\frac{1}{3}, \frac{1}{3}, \frac{1}{3})$ 3. (a) $pq\sqrt{1 + b^2 + c^2}$ (b) $(p/2,, q/2, a + bp/2 + cq/2)$ 5. (a) $3\sqrt{26}/4$ (b) $(\frac{1}{2}, \frac{7}{2}, -\frac{1}{4})$
7. (a) $\frac{1}{2}\sqrt{a^2b^2 + a^2c^2 + b^2c^2}$ (b) $(a/3, b/3, c/3)$ 9. (a) 2π (b) $(0, 0, \frac{1}{2})$

11. **(a)** 2.9007 **(b)** (1.2819, 1.4030, 5.4133) **13.** **(a)** $2\pi a^2$ **(b)** $(0, 0, a/2)$
15. **(a)** $\left(\frac{\pi a}{6}\right)[(a^2+4)^{3/2}-a^3]$ **(b)** $\bar{x}=\bar{y}=0,\ \bar{z}=\dfrac{6(a^2+4)^{5/2}+10a^2(a^2+4)^{3/2}+2a^5}{40[(a^2+4)^{3/2}-a^3]}$
17. **(a)** $k\sqrt{3}/6$ **(b)** $(\frac{1}{2}, \frac{1}{4}, \frac{1}{4})$ **19.** **(a)** $45\sqrt{26}/16$ **(b)** $(-\frac{1}{180}, \frac{859}{540}, \frac{3813}{1080})$ **21.** **(a)** $2\pi/3$ **(b)** $(0, 0, \frac{3}{4})$ **23.** 2π
25. $(6a^2-3a^3+b^3)/12$ **27.** 12 **29.** **(a)** $\frac{15}{4}$ **(b)** 3

Section 17.3

1. $y+2x^2y+xe^{xz}$ **3.** -3 **5.** $\cos x-\sin y+\sec^2 z$ **7.** $12x^2y^3z^2+6x^4yz^2+2x^4y^3$ **9.** $60+18/e$ **11.** $\pi/2$ **13.** 0
15. 8π **17.** $abc(a-2b)/24$ **19.** -4π **21.** 0 **23.** 2π **25.** $K(3-n)/(x^2+y^2+z^2)^{n/2}$
27. $\nabla^2 f=\nabla\cdot\langle f_x, f_y, f_z\rangle=f_{xx}+f_{yy}+f_{zz}$ **29.** $\operatorname{div} c\mathbf{F}=\nabla\cdot c\mathbf{F}=c(\nabla\cdot\mathbf{F})=c\operatorname{div}\mathbf{F}$
31. $\operatorname{div} f\nabla f=\operatorname{div}\{ff_x, ff_y, ff_z\}=f_xf_x+ff_{xx}+f_yf_y+ff_{yy}+f_zf_z+ff_{zz}=f(f_{xx}+f_{yy}+f_{zz})+f_x^2+f_y^2+f_z^2=f(\nabla^2 f)+\|\nabla f\|^2$
33. $\mathbf{F}$ is solenoidal if and only if $\iiint_T \operatorname{div}\mathbf{F}\,dV=0$ for every solid T, if and only if $\operatorname{div}\mathbf{F}=0$.
35. If $\mathbf{F}$ is solenoidal, then $\operatorname{div}\mathbf{F}=0$. If $\mathbf{F}$ has potential f, then $\nabla f=\mathbf{F}$. By Problem 30, $\operatorname{div} f\mathbf{F}=f\operatorname{div}\mathbf{F}+\nabla f\cdot\mathbf{F}=\mathbf{F}\cdot\mathbf{F}$. Hence, $\iint_S f\mathbf{F}\cdot\mathbf{n}\,d\sigma=\iiint_T \operatorname{div} f\mathbf{F}\,dV=\iiint_T \mathbf{F}\cdot\mathbf{F}\,dV$.

Section 17.4

1. $\langle 0, 0, 0\rangle$ **3.** $\langle 0, 0, 0\rangle$ **5.** $\langle 0, 0, \cos(x+y)\rangle$ **7.** $\langle 0, 0, 0\rangle$ **9.** $\oint_C \mathbf{F}\cdot d\mathbf{r}=0$ **11.** $\oint_C \mathbf{F}\cdot d\mathbf{r}=0$ **13.** $-\pi$
15. $-8\pi/9\sqrt{3}$ **17.** $-\pi$ **19.** $\pi\sqrt{10}/5$
21. Assuming the second-order mixed partials of the components of $\mathbf{F}$ are equal, if $\mathbf{F}=\langle P, Q, R\rangle$, then $\operatorname{div}(\mathbf{curl\,F})=\operatorname{div}\langle R_y-Q_z, P_z-R_x, Q_x-P_y\rangle=(R_{yx}-Q_{zx})+(P_{zy}-R_{xy})+(Q_{xz}-P_{yz})=0$.
23. If $\mathbf{F}=\langle P, Q, R\rangle$, then $\mathbf{curl}\, f\mathbf{F}=\mathbf{curl}\langle fP, fQ, fR\rangle=\langle f_yR+fR_y-f_zP-fP_z, f_zP+fP_z-f_xR-fR_x, f_xQ+fQ_x-f_yP-fP_y\rangle=f\langle R_y-Q_z, P_z-R_x, Q_x-P_y\rangle+\langle f_yR-f_zQ, f_zP-f_xR, f_xQ-f_yP\rangle=f\,\mathbf{curl\,F}+\nabla f\times\mathbf{F}$.
25. If $\mathbf{F}=\langle P, Q, R\rangle$ and $\mathbf{G}=\langle S, T, U\rangle$, then $\operatorname{div}(\mathbf{F}\times\mathbf{G})=\operatorname{div}\langle QU-RT, RS-PU, PT-SQ\rangle=Q_xU+QU_x-R_xT-RT_x+R_yS+RS_y-P_yU-PU_y+P_zT+PT_z-S_zQ-SQ_z=S(R_y-Q_z)+T(P_z-R_x)+U(Q_x-P_y)-[P(U_y-T_z)+Q(S_z-U_x)+R(T_x-S_y)]=\mathbf{G}\cdot\mathbf{curl\,F}-\mathbf{F}\cdot\mathbf{curl\,G}$.
27. $\langle 0, 0, 0\rangle$

Review Exercises

1. $\frac{1}{6}-\frac{1}{6}\cos(3\pi^2/16)-3\pi/2\sqrt{2}+3\sqrt{2}$ **3.** -5π **5.** Use the fact that $f_y=g_x=0$ and Green's theorem.
7. $\nabla\cdot\mathbf{r}=3$; $\nabla\times\mathbf{r}=\langle 0, 0, 0\rangle$; $\nabla\|\mathbf{r}\|=(1/\|\mathbf{r}\|)\mathbf{r}$ **9.** 13.0046 **11.** $4\pi a^2(2-\sqrt{2})$ **13.** $\pi(5\sqrt{5}-1)/6$ **15.** 0 **17.** -18π

Additional Exercises

1. $\displaystyle\int_{-a}^{a}\int_{-(b/a)\sqrt{a^2-x^2}}^{(b/a)\sqrt{a^2-x^2}} \frac{\sqrt{a^4b^4+b^4(c^2-a^2)x^2+a^4(c^2-b^2)y^2}}{ab\sqrt{a^2b^2-b^2x^2-a^2y^2}}\,dy\,dx$ **3.** 0

5. $\operatorname{div}\dfrac{\langle x, y, z\rangle}{\|\langle x, y, z\rangle\|}=\operatorname{div}\dfrac{\langle x, y, z\rangle}{\sqrt{x^2+y^2+z^2}}=\dfrac{y^2+z^2+z^2+x^2+x^2+y^2}{(x^2+y^2+z^2)^{3/2}}=\dfrac{2(x^2+y^2+z^2)}{(x^2+y^2+z^2)^{3/2}}=\dfrac{2}{\sqrt{x^2+y^2+z^2}}=\dfrac{2}{\|\langle x, y, z\rangle\|}$
7. $\iint_S \mathbf{curl\,F}\cdot\mathbf{n}\,d\sigma=\iiint_T \operatorname{div}(\mathbf{curl\,F})\,dV=0$, since $\operatorname{div}(\mathbf{curl\,F})=0$. **9.** **(a)** 6 **(b)** 0 **(c)** 0 **(d)** $2/\sqrt{x^2+y^2+z^2}$
11. $\mathbf{curl\,F}=\langle 0, 0, 0\rangle$
13. If $\mathbf{F}=\langle P, Q, R\rangle$, then $\mathbf{curl\,F}=\langle R_y-Q_z, P_z-R_x, Q_x-P_y\rangle$. If $\mathbf{curl\,F}=\langle x, y, z\rangle$, then $R_y-Q_z=x$, and $R_{yx}=1+Q_{zx}=1+Q_{xz}$. Also, $P_z-R_x=y$, so $R_{xy}=P_{zy}-1=P_{yz}-1$. But $R_{xy}=R_{yx}$ implies $Q_{xz}-P_{yz}=-2$, so that $Q_x-P_y=-2z+C(x, y)\neq z$.
15. $\operatorname{div}\mathbf{F}=0$ **17.** $\iint_S f_{\mathbf{n}}\,d\sigma=\iint_S \nabla f\cdot\mathbf{n}\,d\sigma=\iiint_T \operatorname{div}\nabla f\,dV=\iiint_T \nabla^2 f\,dV$

19. (a) If S has equation $g(x, y, z) = 0$, then the unit normal vector is $\mathbf{n} = \langle g_x, g_y, g_z\rangle/(g_x^2 + g_y^2 + g_z^2)^{1/2}$, $\nabla \times \mathbf{F} = \langle R_y - Q_z, P_z - R_x, Q_x - R_y\rangle$, and $\mathbf{F} \times \mathbf{n} = \langle Qg_z - Rg_y, Rg_x - Pg_z, Pg_y - Qg_x\rangle/(g_x^2 + g_y^2 + g_z^2)^{1/2}$. Now, $\iint_S [(Qg_z - Rg_y)/(g_x^2 + g_y^2 + g_z^2)^{1/2}]\, d\sigma = \iint_S \langle 0, -R, Q\rangle \cdot \mathbf{n}\, d\sigma = \iiint_T \operatorname{div}\langle 0, -R, Q\rangle\, dV = -\iiint_T (R_y - Q_z)\, dV$, and similarly for the other two components.
(b) $\iint_S f\mathbf{n}\, d\sigma = \iint_S \langle f, f, f\rangle \cdot \mathbf{n}\, d\sigma = \iiint_T \operatorname{div}\langle f, f, f\rangle\, dV = \iiint_T \nabla f\, dV$
21. Circulation $= \oint_C \mathbf{B} \cdot d\mathbf{r} = \iint_S \textbf{curl}\, \mathbf{B} \cdot \mathbf{n}\, d\sigma = \iint_S \mathbf{E} \cdot \mathbf{n}\, d\sigma =$ Flux
23. Flux $= \iint_S \mathbf{E} \cdot \mathbf{n}\, d\sigma = \iiint_T \operatorname{div} \mathbf{E}\, dV = \iiint_T k\delta\, dV =$ Charge inside S

Chapter 18 Section 18.1

1. Exact; $x^2y - xy^2 = C$ 3. Not exact 5. Not exact 7. Not exact 9. Exact; $x^2 - y^2 = Cxy$ 11. $\frac{1}{2}x^4y^2 + 3x^3y^3 = C$
13. $\frac{1}{3}x^3y^{-6} - x^2y^{-5} = C$ 15. $y^2 - \sqrt{x^2 + y^2} = C$ 17. $(x^2 - y^2)(\ln x - C) = 1$ 19. $42x^{11/3} + 66x^{8/3}y + 33x^{5/3}y^2 = C$
21. Yes, $I(x, y) = y/x$

Section 18.2

1. $y = 1 + Ce^{\cos x}$ 3. $y = 2e^{2-(1/x)} + (e^{-1/x})\int_{1/2}^{x} 2te^{1/t}\, dt$ 5. $y = (1 - x)^2 - 2(1 - x)\ln(1 - x)$
7. $y = 2 - 2e^{(1-x)/x} + (e^{1/x})\int_1^x (3/t)e^{-1/t}\, dt$ 9. $y = 3(x - 1)\ln[(x - 1)/x] + C(x - 1)$ 11. $y = (Ce^{-x} + 4)/x$
13. $y = (x + C)\cos x$ 15. $u' + (1 - n)P(x)u = Q(x)$
17. Use $e^{R(x)}$ as the integrating factor, and let $f(x) = \int_c^x e^{R(t)}Q(t)\, dt$. Then $y = b$ when $x = a$ implies $c = a$.

Section 18.3

1. $y = 2x^3/(x^2 - C)$ 3. Not homogeneous 5. $\arctan(y/x) = \ln\sqrt{x^2 + y^2} - \ln\sqrt{2} + \pi/4$ 7. $\csc^2(y/x) = 2 - \ln x$
9. Same solution 11. $\dfrac{Q(1, u)}{P(1, u) + uQ(1, u)}\, du = -\dfrac{1}{x}\, dx$
13. Multiply through by the integrating factor, and then use Euler's lemma in the form $y(QP_y - PQ_y) = x(PQ_x - QP_x)$ to show that the resulting equation is exact.

Section 18.4

1. $2y + \ln x = C$ 3. $y = C/x^4$ 5. $(3x + 7y)^{49}(y - x)^3 = Cx^{42}$
7. $y = 200 - 150e^{-t/100}$ lb; concentration $= 2 - 1.5e^{-t/100}$ lb/gal; 2 lb/gal
9. $y = 2(100 - t) - 0.015(100 - t)^2$ lb; concentration $= 0.5 + 0.015t$ lb/gal, $t \le 100$
11. If c is the constant of proportionality at which glucose is removed, then the amount at any time is $y = k/c - Ae^{-ct}$, where A is a constant. Over a long time, the amount of glucose in the blood approaches the steady-state k/c.
13. $P/(\alpha + k)$
15. If the object has mass m and meets air resistance kv, then $v = (g/k)(1 - e^{-kt/m})$, where g is the acceleration of gravity. Terminal speed is g/k.
17. $i = (E/R)(1 - e^{-Rt/L})$; maximum $i = E/R$; time to reach 90% of maximum $i = (L \ln 10)/R$
19. $h = (H^4/2R^4)^{1/5}(R^2\sqrt{2H} - 5r^2\sqrt{g}\, t)^{2/5}$; time to drain tank $= R^2\sqrt{2H}/5r^2\sqrt{g}$

Section 18.5

1. $y = a + be^{-x}$ 3. $y = \frac{1}{3}(8 + e^{3x})$ 5. $y = ae^x + b - \frac{1}{2}(\sin x + \cos x)$ 7. $y = \frac{1}{3}(2e^{3x} + e^{-3x})$ 9. $y = a\int_b^x e^{-1/t}\, dt - x$
11. $y = ae^x + be^{-x}$
13. $[af(x) + bg(x)]'' + p(x)[af(x) + bg(x)]' + q(x)[af(x) + bg(x)] = a[f''(x) + p(x)f'(x) + q(x)f(x)] + b[g''(x) + p(x)g'(x) + q(x)g(x)] = 0$
15. $W' = fg'' - f''g = f(-pg' - qg) - (-pf' - qf)g = -p(fg' - f'g) = -pW$

Section 18.6

1. $y = ae^x + be^{2x}$ 3. $y = ae^{2x} + bxe^{2x}$ 5. $y = e^x(a\cos 3x + b\sin 3x)$ 7. $y = \frac{4}{3}(e^x - e^{-2x})$ 9. $y = \frac{1}{2}(x-1)e^{-3x}$
11. $y = e^{-3x}(4\cos x + 12\sin x)$ 13. $y = e^{x/2}[a\cos(x\sqrt{3}/2) + b\sin(x\sqrt{3}/2)]$
15. (a) $y'' - 2my' + m^2y = m^2e^{mx} - 2m(me^{mx}) + m^2e^{mx} = (m^2 - 2m^2 + m^2)e^{mx} = 0$
(b) If $y = ve^{mx}$, then $y' = (v' + mv)e^{mx}$ and $y'' = (v'' + 2mv' + m^2v)e^{mx}$. Hence, $y'' - 2my' + m^2y = v''e^{mx} = 0$ implies $v = ax + b$, and $y = (ax + b)e^{mx}$.
17. (a) $(e^{ix})' = (\cos x + i\sin x)' = -\sin x + i\cos x = i(\cos x + i\sin x) = ie^{ix}$
(b) $[e^{(a+bi)x}]' = [e^{ax}e^{ibx}]' = ae^{ax}e^{ibx} + e^{ax}ibe^{ibx} = (a+bi)e^{(a+bi)x}$

Section 18.7

1. $y = ae^x + be^{4x} - \frac{1}{50}(\sin 3x + 3\cos 3x)$ 3. $y = ae^{-3x} + be^{2x} + \frac{1}{54}(9x^2 - 21x + 11)e^{3x}$ 5. $y = \frac{1}{81}(29\sin 3x + 9x^3 - 6x)$
7. $y = e^{-2x}(a\cos x + b\sin x) - \frac{1}{12}(\sin x + 4\cos x)$ 9. $y = e^{x/2}[a\cos(x\sqrt{3}/2) + b\sin(x\sqrt{3}/2)] + e^x\sin x + (2-x)e^x\cos x$
11. The solutions as given imply that $y_c'' + ay_c' + by_c = 0$, $y_1'' + ay_1' + by_1 = f_1(x)$, and $y_2'' + ay_2' + by_2 = f_2(x)$. Hence, $(y_c + y_1 + y_2)'' + a(y_c + y_1 + y_2)' + b(y_c + y_1 + y_2) = 0 + f_1(x) + f_2(x)$.

Section 18.8

1. $y = 1 - 2x + x^2 + x^3 - \frac{5}{4}x^4 + \cdots$ 3. $y = 1 + 2x + \frac{1}{12}x^4 + \frac{1}{20}x^5 + \cdots$ 5. $y = 1 + 4x - \frac{1}{2}x^2 - \frac{4}{3}x^3 + \frac{1}{6}x^4 + \cdots$
7. $y = 4(x-1) + 2e(x-1)^2 + \frac{1}{3}(4e + 2e^2)(x-1)^3 + \frac{1}{6}(3e + 5e^2 + e^3)(x-1)^4 + \cdots$

Section 18.9

1. $2\pi\sqrt{L/g}$, independent of amplitude 3. 2π 5. 1.48 mi/sec
7. Using the fact that a body inside a hollow sphere has no gravitational force acting on it, an object inside the earth is attracted toward the center with a force proportional to the distance from the center. Therefore, the object dropped into the tunnel will experience harmonic motion and will oscillate from one end of the tunnel to the other.
9. If the damping force is small, we can assume that the vibration is underdamped [Example 18.33(c)]. The amplitude is half the original when $e^{-t\mu/2} = \frac{1}{2}$, or $t = (\ln 4)/\mu$.
11. $y = (w/2H)x^2$ 13. $q(t) = \frac{1}{25}(27te^{2t} - 3e^{2t} + 3\cos t + 4\sin t)$; $i(t) = \frac{1}{25}(21e^{2t} + 54te^{2t} - 3\sin t + 4\cos t)$
15. A critically damped vibration has displacement of the form $x = (A + Bt)e^{mt}$, which is 0 only if $A + Bt = 0$. This equation has at most one positive solution.

Review Exercises

1. $xy + x^2e^{xy} = C$ 3. $y = Ce^{x^2/2} - x^2 - 2$
5. $\ln y + 5\ln\dfrac{y - \sqrt{y^2 - x^2}}{x^2 + xy + y^2 - (x+y)\sqrt{y^2 - x^2}} - \arcsin\dfrac{x}{y} + 4\sqrt{2}\arctan\dfrac{x + y - \sqrt{y^2 - x^2}}{x\sqrt{2}} = C$
7. $y = \left(\dfrac{1}{x}\right)e^{1/x}\int_a^x \left(\dfrac{2}{t}\right)e^{-1/t}\,dt$ 9. $\dfrac{1}{\sqrt{3}}\arctan\dfrac{(2x+y)}{y\sqrt{3}} - \dfrac{1}{2}\ln(x^2 + xy + y^2) = C$ 11. $y = a\ln x - \sin x + \int_b^x \dfrac{1}{t}\sin t\,dt$
13. $y = \ln|\sec(x+a)| + b$ 15. $y = ae^{3x} + be^{4x}$ 17. $y = e^{-x/2}\left(a\cos\dfrac{x\sqrt{15}}{2} + b\sin\dfrac{x\sqrt{15}}{2}\right)$
19. $y = ae^{3x} + be^{4x} + \frac{1}{6}e^x$ 21. $y = ae^{x/2} + bxe^{x/2} + \frac{1}{24}x^3e^{x/2}$ 23. $y = 1 + 2x^2 + \frac{17}{12}x^4 + \frac{41}{45}x^6 + \cdots$ 25. $y = Cx^4$
27. $v = m/\mu$ 29. $\pi\sqrt{5}/2$ 31. $x = 2^{-t}\cos 4\sqrt{2}t$

Additional Exercises

1. About 11.5 lb **3.** $x = e^{-t/4}(69.75 \sin 3.99t - 5.81 \cos 3.99t) - 22.66 \sin \pi t + 5.81 \cos \pi t$
5. The family of lines through the origin

Appendix A

1. 7 **3.** $-\frac{7}{9}$ **5.** 190 **7.** 36 **9.** 9 **11.** 12 **13.** -72 **15.** $(r - s)^{n-1}[r + (n - 1)s]$

Appendix B

1. $x = 3, y = -1$ **3.** $x = \frac{41}{21}, y = \frac{3}{7}, z = -\frac{1}{3}$ **5.** $x = -\frac{2}{9}, y = \frac{5}{9}$ **7.** $x = 3 - w, y = 2 - z, z = z, w = w$

INDEX